2019 NORTH CAROLINA AGRICULTURAL CHEMICALS MANUAL

PESTICIDE USE AND SAFETY • APPLICATION EQUIPMENT • SPECIMEN IDENTIFICATION • FERTILIZER USE • INSECT CONTROL • WEED CONTROL • GROWTH REGULATORS • ANIMAL CONTROL • DISEASE CONTROL

This book is also available on the Extension Publications Catalog at content.ces.ncsu.edu/north-carolina-agricultural-chemicals-manual.

© College of Agriculture and Life Sciences, NC State University, Raleigh, NC

USEFUL PHONE NUMBERS

Carolinas Poison Center
1-800-222-1222
www.ncpoisoncenter.org
P.O. Box 32861, Charlotte, NC 28232-2861
Carolinas Poison Center can give advice on diagnosis and treatment of human illness resulting from toxic substances.

NCDA&CS Structural Pest Control and Pesticides Division
1-919-733-3556
The NCDA&CS Structural Pest Control and Pesticides Division provides information and help to aid in responding to emergencies involving pesticides.

National Pesticide Information Center
1-800-858-7378
npic.orst.edu
NPIC provides information by phone about pesticides Monday through Friday, 8 a.m. to 6 p.m., Central Time.

Pesticide Disposal Assistance Program
1-919-280-1061 (Raleigh, N.C.)
www.ncagr.gov/SPCAP/pesticides/PDAP/
PDAP gives information in disposal of unwanted pesticides.

MISUSE OF PESTICIDES

It is illegal to use any pesticide in a manner not permitted by its labeling. All recommendations for pesticide use included in this manual were legal at the time of publication, but the status of registration and use patterns are subject to change by actions of state and federal regulatory agencies.

Recommendations of specific chemicals are based on information on the manufacturer's label and performance in a limited number of trials. Because environmental conditions and methods of application by growers may vary widely, performance of the chemical will not always conform to the safety and pest control standards indicated by experimental data.

Recommendations for the use of agricultural chemicals are included in this publication as a convenience to the reader. The use of brand names and any mention or listing of commercial products or services in this publication does not imply endorsement by North Carolina State University nor discrimination against similar products or services not mentioned. Individuals who use agricultural chemicals are responsible for ensuring that the intended use complies with current regulations and conforms to the product label. Be sure to obtain current information about usage regulations and examine a current product label before applying any chemical. For assistance, contact your county North Carolina Cooperative Extension agent.

You can locate your local NC Cooperative Extension center's address and phone number at **ces.ncsu.edu**.

2019 North Carolina Agricultural Chemicals Manual
Table of Contents

I — PESTICIDE USE AND SAFETY INFORMATION
- Restricted Use Pesticides 2
- Local Need — 24(c) Registrations in N.C. 2
- The Safe Use of Pesticides 4

II — CHEMICAL APPLICATION EQUIPMENT
- Introduction 22
- Types of Equipment 22
- Cleaning Equipment 25
- Calibrating Chemical Application Equipment 25
- Useful Tables and Data 30

III — HOW TO SEND SPECIMENS FOR DISEASE, INSECT, AND WEED IDENTIFICATION
- How to Send Specimens 38
- Plant Disease and Insect Clinic 38
- What to Sample and How to Ship 38
- Disease Sample Submission and Instructions 39
- Insect Identification 39
- Plant and Weed Identification 40
- Soil Testing, Nematode Assay, Plant Tissue Nutrient Testing, Waste and Solution Analysis 40

IV — FERTILIZER USE
- Lime and Fertilizer Suggestions for Field, Pasture, and Hay Crops 42
- Fertilizer Suggestions for Tree Fruit 45
- Fertilizer Suggestions for Small Fruit 46
- Lime and Fertilizer Suggestions for Lawns 48
- Fertilizer Suggestions for Ornamental Plants in Landscapes 51
- Fertilizer Suggestions for Nursery Crops 52
- Lime and Fertilizer Suggestions for Vegetable Crops 53
- Fertilizer Rules and Regulations 57
- Nutrient Content of Fertilizer Materials 57
- Solubility of Selected Fertilizer Materials 61
- Mixing Herbicides with Nitrogen Solutions or Fluid Fertilizers 62
- How to Test Compatibility of Herbicides with Fluid Fertilizers 63
- Fertilizer Placement 64
- Livestock and Poultry Manure Production Rates and Nutrient Content 64
- Beneficial Use of Municipal Biosolids 68
- Certified Organic Farm Management Alternatives 69

V — INSECT CONTROL
- Relative Toxicity of Pesticides to Honey Bees 72
- Reducing the Risk of Pesticide Poisoning to Honey Bees 73
- Insect Control in Field Corn 74
- Insect Control in Grain Sorghum 77
- Insect Control in Small Grains 78
- Insect Control in Cotton 80
- Cotton Insect Resistance Management 83
- Insect Control in Peanuts 85
- Insect Control in Soybeans 87
- Insect Control on Flue-Cured and Burley Tobacco 90
- Insect Control for Commercial Vegetables 97
- Insect Control for Greenhouse Vegetables 132
- Insect Control for Livestock and Poultry 134
- Community Pest Control 140
- Industrial and Household Pests 144
- Arthropod Management for Ornamental Plants Grown in Greenhouses 147
- Arthropod Management for Ornamental Plants Grown in Nurseries and Landscapes 152
- Arthropod Control on Christmas Trees 159
- Commercial Turf Insect Control 164
- Insect Control for Wood and Wood Products 169
- Insect Control for the Home Vegetable Garden 173
- Control of Household Pests 178
- Insect Control for Home Lawns 187

VI — INSECT AND DISEASE CONTROL OF FRUITS
- Apple Spray Program 190
  - Relative Effectiveness of Various Fungicides for Apple Disease Control 192
  - Relative Effectiveness of Various Insecticides for Apple Insect and Mite Control 193
- Blueberry Management Program 195
- Caneberry Management Program 200
- Bunch Grape Management Program 206
- Muscadine Grape Management Program 216
  - Muscadine Disease Management Program 216
  - Relative Effectiveness of Various Fungicides for Muscadine Grape Disease Control 217
  - Muscadine Insect Management 217
- Peach and Nectarine Spray Guide 221
  - Relative Effectiveness and Safety of Various Insecticides for Peach Insects 223
  - Relative Effectiveness of Chemicals for Disease Control on Peaches and Nectarines 224
- Nematode Control on Peaches 224
- Commercial Pecan Insect Control 225
- Commercial Pecan Insect and Disease Spray Guide ... 230

Commercial Pecan Disease Control...............................233
Strawberry Disease Control...238
    Pre-Planting Disease Control....................................238
    Pre-Planting and Early Post-Planting: Nematode
    Management..238
    Planting and Early Post-Planting: Disease Control....239
    New Leaf Growth to Pre-Bloom: Disease
    Management..242
    Early Bloom and into Harvest: Disease Control.........243
    Relative Effectiveness of Various Chemicals for
    Strawberry Disease Control...247
Strawberry Insect Management.....................................248

VII — CHEMICAL WEED CONTROL

Chemical Weed Control in Field Corn 252
    Weed Response to Preemergence Herbicides..........262
    Weed Response to Postemergence Herbicides 263
Chemical Weed Control in Cotton 264
    Weed Response to Preplant, Preemergence, and
    Postemergence Overtop Herbicides in Cotton...........274
    Weed Response to Cotton Herbicides.........................275
    Weed Response to Burndown Herbicides for
    Conservation Tillage Cotton...276
Chemical Weed Control in Peanuts................................277
    Weed Response to Preplant Incorporated,
    Preemergence, and At-Cracking Herbicides..............282
    Weed Response to Postemergence Herbicides 283
    Brand Names and Formulations of Active
    Ingredients 284
Chemical Weed Control in Sorghum 285
Chemical Weed Control in Soybeans.............................288
    Weed Response to Preplant Incorporated and
    Preemergence Herbicides 298
    Weed Response to Postemergence Herbicides 299
Chemical Weed Control in Sunflowers 300
Chemical Weed Control in Tobacco 301
    Weed Response to Herbicides in Tobacco 303
Chemical Weed Control in Wheat, Barley, Oats,
    Rye, and Triticale 304
    Weed Response to Herbicides in Small Grains 309
Glyphosate Formulations..310
Herbicide Resistance Management................................311
Herbicide Modes of Action for Hay Crops, Pastures,
    Lawns and Turf 315
Chemical Weed Control in Clary Sage...........................317
Chemical Weed Control in Small Fruit Crops 319
Chemical Weed Control in Tree Fruits...........................326
Chemical Weed Control in Hay Crops and Pastures.....334
Chemical Weed Control in Lawns and Turf 340
Chemical Weed Control in Ornamentals 352
Chemical Weed Control in Vegetable Crops 355
Chemical Weed Control in Forest Stands 384
Forest Site Preparation, Stand Conversion,
    Timber Stand Improvement 386
Aquatic Weed Control..391

Biological Control of Aquatic Weeds with Triploid Grass
    Carp 393
Chemical Control of Aquatic Plants 393
Pond Dyes..397
Chemical Control of Specific Weeds 399
Chemical Control of Woody Plants................................402
Total Vegetation Control on Noncropland 404

VIII — PLANT GROWTH REGULATORS

Growth Regulators for Cotton..406
Guide for use of Defoliants on Cotton 406
Harvest Aids and Preharvest Desiccants 407
Growth-Regulating Chemicals for Apples.....................408
Growth Regulators for Floriculture Crops in
    Greenhouses 411
Growth Regulators for Woody Ornamental Crops.........438
Sucker Control for Flue-Cured Tobacco 438
Yellowing Agents for Flue-Cured Tobacco 440
Sucker Control in Burley Tobacco 440
Growth Regulators for Fruiting Vegetables...................441
Growth Regulators for Peanuts 441
Growth Regulators for Turfgrasses...............................442

IX — ANIMAL DAMAGE CONTROL

Animal Damage Control 448
Description of Potential Animal Pests...........................450
Animal Control Suggestions 453
Rodenticides..457
Fish Control 458

X — DISEASE CONTROL

Foliar Fungicides for Wheat Leaf Disease Control.........460
Seed Treatment for Wheat Foliar Disease Control.........462
Nematode Control in Corn..462
Fungicides for Control of Corn Foliar Diseases............462
Nematode Control on Cotton..464
Peanut Disease Control..464
Peanut Disease Management Calendar........................468
Soybean Disease Control...468
Tobacco Disease Control..473
Turfgrass Disease Control..475
Nematicides for Turf 498
Floral, Nursery, and Landscape Diseases.....................498
Disease Control for Forest, Christmas, and Ornamental
    Trees..512
Commercial Landscape and Nursery Crops Disease
    Control 519
Disease Control for Commercial Vegetables.................533

INDEX   **619**

ABBREVIATIONS   **625**

I — PESTICIDE USE AND SAFETY INFORMATION

Restricted Use Pesticides 2
    Restricted Use Pesticide 2
    Local Need — 24 (c) Registrations in North Carolina 2
Special Local Need Active 24 (c) Registrations in North Carolina 2
The Safe Use of Pesticides 4
    General Safety Instructions 4
    Hazard and Toxicity of Pesticides 5
    Pesticide Toxicity to People 5
    Pesticide Hazards to the Environment 5
    Hazardous Chemicals Right-to-Know Act 6
    North Carolina Worker Protection Standard Regulations 6
    Restricted Entry Intervals 7
    Preharvest Intervals 7
    Aerial Application Limitations 7
    Bee Protection 7
    Protecting Surface and Groundwater 8
    Chemigation 8
    Evaluating the Potential for Groundwater Contamination 11
    Proper Pesticide Storage 16
    Disposal of Pesticides 16
    Disposal of Empty Pesticide Containers 17
    Controlling Pesticide Drift 17
    Mixing and Loading Pesticides 18
    Cleaning Sprayer Systems 18
    Pesticide Record-Keeping Requirements 18

Restricted Use Pesticides

Wayne Buhler, Pesticide Education Specialist

Because of their potential to cause adverse effects on human health and the environment, many very hazardous pesticides are classified **RESTRICTED USE** by the U.S. Environmental Protection Agency (EPA). For an updated list of these pesticides, visit https://www.epa.gov/pesticide-worker-safety/restricted-use-products-rup-report and click on the link for "RUP Updated List." The label states **RESTRICTED USE PESTICIDE**, as indicated below.

Restricted Use Pesticide
For retail sale to and use only by certified applicators or by persons under their direct supervision and only for those uses covered by the certified applicator's certification.

Pesticide formulations labeled RESTRICTED USE PESTICIDE can only be sold in North Carolina by licensed dealers and purchased or used by licensed commercial applicators, public operators, certified or licensed structural pest control applicators, and certified private pesticide applicators, or by persons working under their direct supervision.

Local Need — 24 (c) Registrations in North Carolina

The North Carolina Department of Agriculture and Consumer Services (NCDA&CS) has registered a number of pesticides as Special Local Need products in North Carolina (Table 1-1). The state-approved labels have been submitted to EPA for review. These products can be used for the purposes listed in this manual and as detailed on the supplemental labels. It is important to remember, however, that the products listed here are for reference only, and any information given on the product labeling takes precedence over information given in this manual.

Application of a pesticide can begin as soon as the state has registered the Special Local Need. However, the applicator must have within his or her possession a copy of the supplemental labeling, including directions for use. Within 10 days, the state must notify EPA of what is registered, including a copy of the label. Within 90 days after the date of issuance of the Special Local Need registration by the state, EPA can disapprove the registration. If EPA does nothing within the 90 days, the registration automatically becomes a full federal registration for use only within the state.

If EPA should disapprove a 24 (c) label within the specified 90-day period, it is still legal for the grower to sell the agricultural commodity treated with the pesticide during the time the state label was effective because during that time, the application was legal. However, the commodity is subject to regulations governing tolerance, and residues must not exceed tolerance limits established for the pesticide.

Special Local Need - Active 24 (c) Registrations in North Carolina

The following table is provided for reference only and is subject to change at any time. If you have questions about a specific North Carolina 24 (c) registration, please contact Lee Davis, N.C. Department of Agriculture and Consumer Services Pesticide Registration Manager, (919) 733-3556 or send e-mail to: lee.davis@ncagr.gov.

Table 1-1. Special Local Need—24 (c) Registrations in North Carolina

Product Name	Use	SLN Number
Reflex	Nutsedge in pine seedling plantations	NC-950010
Witchaway Permviro Systems	Witchweed in various sites	NC-980005
Goal 2XL herbicide	Witchweed control in corn (allows planting of small grains within 10 months of last application)	NC-020001
Select 2EC herbicide	Annual and perennial weeds in kenaf	NC-020006
Actigard 50WG plant activator	Suppression of Tomato Spotted Wilt Virus (TSWV) in flue-cured tobacco	NC-020007
Orthene 97 insecticide	Thrips in peanuts (in furrow application)	NC-030003
Orthene 97 insecticide	Slash pine flower thrips, coneworms, coneborers, and seedbugs in southern pine seed orchards	NC-030004
Stinger herbicide	Broadleaf weeds in strawberries	NC-030005
Orthene 97 insecticide	Thrips on kenaf	NC-050003
Curfew fumigant	Parasitic nematodes and mole crickets in golf course turf	NC-050004
Intrepid 2F insecticide	Lepidoptera larvae on sweet potatoes	NC-060003
Brigade 2EC insecticide/miticide	Allows a different timing of application (pre-plant) at a higher rate	NC-070002
Select Max herbicide	Annual and perennial grasses in clary sage	NC-070003
Manzate Pro-Stick fungicide	Diseases on tobacco	NC-080002
Penncozeb 75DF dry flowable fungicide	Diseases on tobacco	NC-080003
Tenacity herbicide	Certain winter annual broadleaf weeds in DOT wildflower beds	NC-080005
Lorsban Advanced insecticide	Reduce PHI from 120 days to 60 days	NC-090001
Lorsban Advanced insecticide	Allows a higher application rate to be applied to tobacco	NC-090004
Dual Magnum herbicide	Reduce PHI to 60 days when used on tomatoes at 1.67 pints or fewer per acre	NC-100001
Admire Pro Systemic protectant	Reduces PHI to 60 days when used on sweet potatoes	NC-100003
Perm-Up 3.2 EC insecticide	Control regeneration weevils in conifer nurseries	NC-110001

Table 1-1. Special Local Need—24 (c) Registrations in North Carolina

Product Name	Use	SLN Number
Quadris Flowable fungicide	Control target spot on tobacco seedlings grown in greenhouses	NC-110003
Ridomil Gold SL	Allows for application in tobacco transplant water	NC-110004
Dual Magnum herbicide	Control pigweed and yellow nutsedge in various vegetable and berry crops	NC-110005
Select Max herbicide	Allows for use without an adjuvant when applied to certain vegetables	NC-110007
Corvus herbicide	Control glyphosate-resistant Palmer amaranth and other broadleaf weeds and grasses in field corn	NC-120001
Cotoran 4L herbicide	Control certain summer annual broadleaf weeds in DOT wildflower beds	NC-120002
Milestone herbicide	Control herbaceous broadleaf weeds and woody plants in forests and grazed areas in and around these sites	NC-120003
Gramoxone SL 2.0 herbicide	Suppression and/or control of Palmer amaranth in peanuts	NC-120004
Reflex herbicide	Control weeds in transplanted tomatoes and transplanted peppers	NC-120006
Linex 4L herbicide	Control cutleaf evening primrose and certain other weeds in clary sage	NC-120005
Milestone herbicide	Control or suppression of blackberry (briars), vines, and susceptible broadleaf weeds in newly planted longleaf pine	NC-120010
Prowl H2O herbicide	Control annual grass weeds and certain broadleaf weeds as they germinate in biofuel crops	NC-120011
Avipel Hopper Box (Dry) corn seed treatment	Protect field and sweet corn seed against consumption by black birds and crows	NC-130001
Avipel Liquid corn seed treatment	Protect field and sweet corn seed against consumption by black birds and crows	NC-130002
Vapam HL Soil fumigant	Establishes the use of a revised buffer zone table when the product is applied to peanuts using shank injection to raised beds	NC-130003
Sectagon 42 Agricultural fumigant	Establishes the use of a revised buffer zone table when the product is applied to peanuts using shank injection to raised beds	NC-130004
Metam CLR 42%	Establishes the use of a revised buffer zone table when the product is applied to peanuts using shank injection to raised beds	NC-130005
Malathion 8 Flowable Ag insecticide	Control of Spotted Wing Drosophila in blueberries	NC-130007
Malathion 8 Flowable Ag insecticide	Control of Spotted Wing Drosophila in caneberries	NC-130008
Bean Guard / Allegiance Fungicide	Seed treatment for control of seed and soilborne damping off diseases of clary sage	NC-140002
Sterilizing Gas 8	Control American foulbrood and other pests	NC-140003
Pyrimax 3.2L Herbicide	Control annual broadleaf and grass weeds in DOT wildflower beds	NC-150001
Reflex Herbicide	Control Palmer amaranth and other weeds in pumpkins and watermelons	NC-150002
Sonar Genesis Aquatic Herbicide	Control hydrilla on portions of the Eno River in Durham and Orange counties	NC-150003
Fierce Herbicide	Allows over-the-top applications to winter wheat	NC-150004
Armezon	Control weeds in miscanthus grown for nonfood and nonfeed biomass production	NC-150006
Valor SX Herbicide	Control weeds in clary sage	NC-150007
Sivanto Prime Insecticide	Control insects in Christmas trees	NC-160001
Profume Fumigant	Fumigation of non-edible commodities and for use under quarantine/regulatory treatment schedules	NC-160002
Profume	Fumigation of non-edible commodities	NC-160003
Anthem Flex Herbicide	Allows preemergence applications to winter wheat to control/suppress resistant and non-resistant Italian ryegrass	NC-160004
XtendiMax with VaporGrip Technology	Applicators must attend Auxin Herbicide-BMP training and observe 10 mph wind speed during application	NC-170001
Engenia Herbicide	Applicators must attend Auxin Herbicide-BMP training and observe 10 mph wind speed during application	NC-170002
Enlist Duo Herbicide	Applicators must attend Auxin Herbicide-BMP training and observe 10 mph wind speed during application	NC-170003
FeXapan Herbicide plus VaporGrip Technology	Applicators must attend Auxin Herbicide-BMP training and observe 10 mph wind speed during application	NC-170004
Beyond Clearfield Production System Herbicide	Control annual broadleaf and grass weeds in DOT wildflower beds	NC-170006
Treflan HFP Herbicide	Control weeds in turnips grown for their roots	NC-170007
Enlist One Herbicide	Applicators must attend Auxin Herbicide-BMP training and observe 10 mph wind speed during application	NC-180001
Parazone 3SL Herbicide	Control weeds, including Palmer amaranth, in clary sage	NC-180002
Axiom DF Herbicide	Control weeds, in NCDOT & NCDA&CS wildflower beds and pollinator research plots	NC-180003

The Safe Use of Pesticides

Wayne Buhler, Pesticide Education Specialist

General Safety Instructions

- Use pesticides only when needed.
- Always ask the advice of an expert on problems of pests and pesticides.
- Use the correct pesticide for the problem.
- Know any hazards that the pesticide might present.
- Read and follow the label.
- Commercial pest control operators, farmers, and other applicators of organophosphate and carbamate pesticides should contact their physician at the beginning of the season. At this time, you should inform the physician of the types of pesticides you will be using. The physician will determine the level of enzymes in your blood that may be affected by the use of certain pesticides. While discussing the pesticides to be used, review the signs and symptoms of pesticide poisoning.
- Know what to do in the event of an accident. *Plan ahead.* Call your physician or 911 immediately in the event of an accident.
- Have your physician's phone number programmed into your phone. In an emergency, time is extremely important.
- Take time to explain the safe pesticide use to employees.
- Check your application equipment for leaks or clogged lines, nozzles, and strainers.
- Calibrate equipment frequently for proper output, using water.
- Check respirator for cleanliness, clean filter, and proper fit.
- Check gloves and other protective clothing for holes and cleanliness before each use.
- Make sure plenty of clean water, detergent, towels, and a clean change of clothing are available.
- Do not permit delivery of pesticides unless a responsible representative is on hand to receive and properly store them.
- Make sure that people have been warned and livestock and pets that may be exposed have been removed from the area to be treated.
- Notify beekeepers who maintain beehives in the vicinity of a pesticide application.
- Cover food and water containers.
- Never eat, drink, or smoke when handling pesticides.
- Wash your hands before eating, smoking, or drinking.
- Make sure the time intervals between date of application and reentry, harvest, slaughter, or milking will comply with those given on the label.
- Rinse pesticide containers before recycling or disposal. (Put rinsate in sprayer tank.)

Select the Appropriate Product

The signal word indicates the acute (single, or short-term, exposure) toxicity and irritation potential of a formulated pesticide product. Each EPA-registered pesticide product is assigned one of the following signal words: **CAUTION, WARNING** or **DANGER (DANGER-POISON)**. The signal word is printed in bold letters on the front panel of a pesticide label.

CAUTION means the pesticide product is slightly toxic if eaten, absorbed through the skin, inhaled, or if it causes slight eye or skin irritation.

WARNING indicates the pesticide product is moderately toxic if eaten, absorbed through the skin, inhaled, or if it causes moderate eye or skin irritation.

DANGER means that the pesticide product is highly toxic. It is corrosive or causes severe burning to eyes or skin that can result in irreversible damage.

DANGER-POISON means the pesticide product is also highly toxic but only if it is eaten, absorbed through the skin, or inhaled. These products have a skull and crossbones symbol on the label.

The signal word may aid in choosing between equally effective pesticide products registered for a given pest. Contrary to common belief, the signal word **DOES NOT** tell you how well a pesticide will control a pest. For example, **"DANGER"** means the formulated pesticide product is more toxic to you, not that it is more toxic to pests than a product labeled **"CAUTION."**

Because of the risks involved in handling, many of the very hazardous pesticides have been **"RESTRICTED"** and can be bought and applied only by or under the direct supervision of certified or licensed individuals. These pesticides bear the words **"RESTRICTED USE PESTICIDE"** on the label.

Follow Label Directions

The label tells you how to use the pesticide properly and safely. Use a pesticide only on crops, animals, or other sites as the label directs. **Use the recommended rate, and apply the pesticide at the time and in the manner stated.**

Obey all precautions for using a pesticide safely, such as "Keep out of reach of children;" "Keep away from pets;" "Do not use near fire, sparks, or flame;" "Do not inhale, ingest, or allow to get on skin;" "Do not store near food, feed, seed, or animals;" "Do not contaminate water supplies;" and any other warnings on the label.

Wear Protective Clothing Called for on the Label

While people generally realize the danger in getting pesticides in the mouth or eyes or breathing gaseous fumes, they are frequently unaware of how harmful many pesticides are when absorbed through the skin. Anytime you handle or apply pesticides, wear at least a long-sleeved shirt and long-legged trousers made of a closely woven fabric, socks, and liquid-proof shoes. Wear personal protective equipment listed on the label.

With most **DANGER-** and **WARNING-**labeled pesticides, a respirator covering the nose and mouth and goggles or a face shield protecting the eyes are necessary. Again, the label will tell you the kind of protective equipment you need.

Use Proper Application Equipment

You cannot apply a pesticide properly or safely unless you have the correct equipment. Small jobs around the home with less hazardous pesticides may be done with simple equipment such as a pump-up sprayer, hose attachment sprayer, granular applicator, or hand duster. Larger jobs and those where more hazardous materials are used often require specially designed equipment. Contact your county North Carolina Cooperative Extension agent, pesticide dealer, or equipment dealer if you need further advice on proper equipment for applying pesticides.

Know Emergency First-Aid Procedures

In case of suspected poisoning — stomach cramps, dizziness, vomiting, heavy sweating — follow the label's first-aid advice and **IMMEDIATELY** call a doctor or take the person to a hospital. Take the pesticide label with you because the doctor needs it to prescribe the proper treatment.

If you spill a pesticide on yourself, remove any contaminated clothing immediately. Wash skin immediately with soap and water. **DO NOT USE AN ABRASIVE OR PETROLEUM-BASED CLEANER,** as this would allow the pesticide to penetrate your skin more easily. If an individual is exposed to pesticide vapors, get them to fresh air. Start artificial respiration if the person has stopped breathing. If a pesticide is splashed in the eyes or mouth, rinse it out with large quantities of clean water for at least 15 minutes. If swallowed, read the label to see if you should induce vomiting (this could be harmful, depending on the pesticide). Never give anything by mouth to an unconscious individual.

The doctor may want to call a poison control center for specific treatment of pesticide poisoning. (The number of the Carolinas Poison Center is **1-800-222-1222**.)

Hazard and Toxicity of Pesticides

Do not depend on toxicity values as the *only* factor to be considered regarding the hazards of a chemical to people or other animals. Pesticide users should be concerned with the *hazard(s)* associated with exposure to the chemical and not the *toxicity* of the material itself. Hazard and toxicity are not synonymous.

Toxicity is the inherent capacity of a substance to produce injury or death.

Hazard is a function of two primary variables, *toxicity* and *exposure*, and is the potential threat that injury will result from the use of a substance in a given formulation or quantity. Some hazards do not involve toxicity to people or other animals. For example, sulfur, oils, and numerous other chemicals are considered relatively non-hazardous to animals, but may pose considerable hazard to some plants (phytotoxicity).

A pesticide may be extremely toxic but present little hazard to the applicator or others when used:

- in a very dilute formulation;
- in a formulation that is not readily absorbed through the skin or readily inhaled;
- under conditions to which humans are not exposed;
- only by experienced applicators who are properly equipped to handle the chemical safely.

However, a chemical may exhibit a relatively low mammalian *toxicity* but present a *hazard* because it is normally used in a concentrated form, which may be readily absorbed or inhaled.

Pesticide Toxicity to People

Most pesticides are harmful to people if they are handled or applied in an unsafe way. A pesticide may harm a person if it is:

- swallowed (oral toxicity);
- breathed (inhalation toxicity);
- allowed to get on the skin or in the eye (dermal toxicity).

Children may be poisoned if a pesticide is left where they can eat it, play in it, or drink it. Pesticides should not be stored in unlabeled containers, such as soft drink bottles. A few pesticides give off harmful vapors that must not be breathed. Some applicators are poisoned when they allow pesticides to contact their skin. Oil-based pesticides (such as emulsifiable concentrates) penetrate the skin in greater quantity and more quickly than dusts, granules, or wettable powders. Sun-burnt or hot, sweaty skin with cuts or abrasions allows more rapid penetration.

Pesticide Hazards to the Environment

A pesticide may not affect people the same way it does the environment. Some pesticides may not harm the environment, even though they are moderately to highly hazardous to people. And some pesticides that are only slightly hazardous to people may cause greater environmental damage. A restricted use pesticide may be hazardous to people, the environment, or both. A given pesticide may be hazardous in the air (particle or vapor drift), soil, or water. It may leave harmful residues in food or injure non-target plants and animals, such as fish, bees, birds, other wildlife, and domestic animals.

Some pesticides are potentially more harmful to the environment because they last for a long time once they are applied. Others may accumulate in the body and cause poisoning. Most uses for persistent, accumulative pesticides (such as DDT) have been cancelled in the U.S. Some uses of other persistent pesticides are now restricted.

Wildlife Exposure: Managing the Risk

Wildlife may contact residues of pesticides applied to forests, aquatic habitats, farmland, rights-of-way, turf, and gardens. Pesticide poisonings to wildlife may be caused by runoff to surface water during rainfall, spray drift, foraging on pesticide-treated vegetation or insects, or consumption of pesticide-treated granules, baits, or seeds. Also, secondary poisoning occurs when an animal eats prey species that contain pesticide residues.

Fortunately, not all pesticides have detrimental effects on all wildlife, nor do pesticide residues necessarily have serious consequences for wildlife. Before using pesticides, get advice from wildlife, conservation, and pesticide professionals at universities and state and federal agencies on the choice and proper use of pesticides and alternative pest control strategies. Also, consider strategies to improve wildlife habitats.

Implementing the following suggestions will benefit wildlife while allowing you to control pests. As you look over the suggestions, keep in mind that you must also be in compliance with all pesticide product labels. North Carolina Pesticide Law of 1971 Article 52 G.S. 143, and 2 NCAC 9 L regulations.

Be Careful Around Natural Areas

- All wildlife need natural areas in which to feed, rest, reproduce, raise young, and take shelter. Create habitats by encouraging and promoting the growth of native vegetation. This also reduces the need for mowing.
- Plant disease- and insect-resistant trees and shrubs, thereby reducing the need for pesticides.
- Always store pesticides and wildlife feed separately, and do not feed wildlife near pesticide storage and mixing areas.

Wildlife Benefit When You Understand and Follow Pesticide Labels

- Keep wildlife habitats in mind when reading pesticide labels.
- Compare labels and select highly specific products that pose less risk to nontarget species. Read the label carefully, and use the lowest effective rate.
- Calibrate equipment carefully to assure that the pesticide is applied at labeled rates.
- Endangered Species Protection Bulletins are a part of EPA's Endangered Species Protection Program. Bulletins set forth geographically specific pesticide use limitations for the protection of threatened and endangered (listed) species and their designated critical habitat. Get the EPA *Endangered Species Protection Bulletins* from the Web at https://www.epa.gov/endangered-species (click on "Bulletins Live! Two"). If your pesticide label directs you to this Web site, you are required to follow the pesticide use limitation(s) found in the Bulletin for your intended application area, pesticide active ingredient or product and application month. Also, check regulation 2 NCAC 9L .2200 on pesticide use limitations to protect the Carolina heelsplitter freshwater mussel.
- Take heed of the label. The environmental wildlife precautions on labels are based on scientific and regulatory

actions. They must be followed. It's the law, it's good business, and it's the right thing to do.
- Consult the NCDA&CS Structural Pest Control and Pesticides Division or your county Cooperative Extension center for label clarification or to determine potential pesticidal impacts on wildlife. Consult natural resource agencies, natural heritage programs, and the Nature Conservancy for additional information about wildlife, native vegetation, and endangered species.

Be Alert for Wildlife Before and During Pesticide Application

- Avoid spraying near areas frequented by wildlife, especially flocks of birds, or, if possible, reduce the application rates.
- Homeowners should search for bird and mammal nests before spraying fruit or ornamental trees, shrubs, or lawns, and then avoid applications near those areas.
- Use mechanical, cultural, and biological pest control tactics when available and practical. For example, tillage, crop rotation, pest-resistant plants, natural predators, and trapping can help control pests.
- Scouting and pest identification are critical components of wise pesticide use. To save money and reduce impacts on wildlife, apply pesticides only when pests are present at unacceptable levels.
- Remember to guard against pesticide drift and runoff. Apply pesticides under low, directional wind conditions, and use adjuvants when appropriate. Buffer zones of unsprayed crops or grass strips adjacent to important habitats will help protect wildlife.
- Do not apply pesticides when heavy rain is imminent. Surface runoff may move some pesticides into ponds, streams, and wetlands inhabited by wildlife. In urban areas, such runoff may flow into storm drains leading directly to streams and rivers.
- Multiple pesticide applications may have cumulative effects, especially during breeding season. Reduce application frequency when possible, and target each application to the specific site of the pest instead of making applications over entire fields or lawns. Spot treating weeds and insects in lawns and gardens will reduce the amount of pesticide applied.
- Where practical, do not apply pesticides in and around field edges and corners, fencerows, set-aside acreage, nesting sites, vegetation near streams and wetlands, and areas dedicated to wildlife except to spot treat state-listed noxious weeds. Especially sensitive areas include endangered species habitats, native plant communities, and sinkholes.
- Check the label for instructions on incorporating or watering pesticide granules into the soil. These techniques allow the pesticide to reach the target pests more readily, and foraging birds are less likely to ingest them.
- Never spray leftover pesticides or wash equipment near wetlands, rivers, streams, creeks, potholes, ponds, marshes, sinkholes, wildlife habitats, or drains leading to these areas. Dispose of leftover pesticide as specified on the label.
- For rules and guidelines on protecting honey bees, see section on Bee Protection below and in Chapter 5, Reducing the Risk of Pesticides Poisoning to Honey Bees.
- For additional information, contact your county Cooperative Extension center or call the NCDA&CS Structural Pest Control and Pesticides Division at 919-733-3556.

This section was written by Henry Wade, Environmental Programs Manager, retired, NCDA&CS, Structural Pest Control and Pesticides Division.

Hazardous Chemicals Right-to-Know Act

The Hazardous Chemicals Right-To-Know Act (N.C.G.S. 95-173 et seq.) was adopted by the N.C. General Assembly in 1985. The purposes of this act are (1) to see that firefighters have all the information they need to respond to chemical emergencies and (2) to ensure that citizens have access to sufficient information to make informed judgments about hazards in their communities.

Public and private employers who normally use or store at least 55 gallons or 500 pounds of any hazardous chemical must comply with this law. Although the full requirements of this act do not apply to farms with 10 or fewer full-time employees, the employer must tell fire departments whom to contact in case of emergency.

If you receive a Safety Data Sheet (SDS) with a product you purchase, you will know the material has been classified as hazardous. The SDS gives health-related information, emergency and first-aid procedures, and other information needed to use, store, and dispose of the chemical properly.

North Carolina Worker Protection Standard Regulations

The Worker Protection Standard (WPS) requires the agricultural employer to reduce the risks of pesticide exposure for farmworkers by providing them with specific pesticide safety training, personal protective equipment, application notification, and a means to mitigate pesticide exposure through emergency assistance. The primary requirements for compliance are listed below.

- Provide annual pesticide safety training to both workers and handlers.
- Provide label-specified personal protective equipment to the appropriate employees, including respirators, fit tests, and the fit tests and training required for their proper use.
- Post warnings, oral notification, or both to inform employees of pesticide applications, restricted entry intervals (see below), and restricted areas.
- Provide the following safety information at a central location and at sites where decontamination supplies are located (if the decontamination supplies are at a permanent site).
    1. Emergency medical information (name, telephone numbers, and address of nearest medical facility);
    2. SDS sheets and pesticide-specific information (location of the area to be treated, product name, EPA registration number, active ingredient(s) of pesticide, time and date the pesticide is scheduled to be applied and time and date the application was completed, and the restricted-entry interval for the pesticide). Each day of application must be recorded as a separate application record. Recorded information must be retained for a period of 2 years and made available to any employee, treating medical professional, or designated representative who requests it.
    3. The WPS pesticide safety information poster. This no longer needs to be displayed as a poster as long as the following information is included: 7 concepts for preventing pesticides from entering your body; instructions for employees to seek medical help immediately if they have been poisoned, injured, or made ill by pesticides; name, address and phone of state or tribal pesticide regulatory authority; and name, telephone and address of nearby medical facility.
- Provide ample supplies at designated decontamination stations (single-use towels, soap, water, and clean change of clothing).

Restricted Entry Intervals

The restricted entry interval (REI) is the time after the end of a pesticide application when entry into the treated area is prohibited. The agricultural employer must not allow or direct any worker to enter or remain in the treated area before the REI has expired except under the conditions of *early entry*.

Early entry provisions of the Worker Protection Standard (WPS) allow trained and label-specified early entry PPE-equipped workers with a minimum age of 18 to enter a treated area during the REI to perform short-term activities with "limited contact," such as moving irrigation equipment or opening ventilation systems.

The following conditions must be met for "limited contact" early entry activities.

1. Worker's contact with the treated surfaces is minimal and limited to the feet, lower legs, hands, and forearms.
2. Pesticide product does not have a statement in the labeling requiring double notification.
3. PPE for early entry conforms with the label requirements or includes at least coveralls, chemical resistant gloves, socks, chemical resistant footwear, and eyewear, if eyewear is required by the label.
4. No hand labor, such as hoeing, picking, or pruning is performed.
5. The time in the area under REI for any worker does not exceed 8 hours in a 24-hour period.
6. Workers do not enter the area during the first 4 hours after application and not until applicable ventilation criteria and label specified inhalation exposure levels are reached.
7. Agricultural employer must give an oral or written notification of the specifics of the early entry. Notification must be given in a language that workers understand.

Preharvest Intervals

The preharvest interval is the time in days that must pass between the last application of a pesticide and harvesting a food or feed crop. The interval varies with the pesticide, depending largely on its persistence (how long the pesticide lasts) on or in the crop as well as on the pesticide's toxicity.

For example, one insecticide label for peach tree borer on peaches states: "Make only one application per season. Do not apply within 14 days before harvest." Fourteen days is the preharvest interval, and it means peaches cannot be harvested from an orchard where this pesticide has been applied until at least 14 days have passed. If harvested sooner, the peaches cannot be legally sold.

If you do not obey the preharvest interval, your crop can be seized and destroyed, and you can be fined. Most importantly, if crops are harvested and consumed before the preharvest interval has passed, people or animals may be poisoned.

Aerial Application Limitations

If you expect to have your crops sprayed by an aerial applicator, in order to obey N.C. Pesticide Regulations, you must carefully *plan the location of these crops*. Certain areas are restricted, and aerial pesticide applications cannot be made to these areas unless certain rules are followed.

DO NOT SPRAY (or otherwise allow pesticides to be deposited):

1. In any congested area unless permission is granted by appropriate authorities;
2. Within 300 feet of the premises of schools, hospitals, nursing homes, churches, or any building used for business or social activities if either the premises or the building is occupied by people;
3. Within 100 feet of a residence;
4. On a right-of-way of a public road or within 25 feet of the road, whichever is the greater distance;
5. In or near any body of water, if the pesticide is labeled as toxic or harmful to aquatic life, unless such aquatic life is the target of the pesticide.

Farmers can prevent potential problems by not planting crops near those areas just described and by always following directions on the pesticide label concerning the spraying of crops, especially near sensitive and restricted areas. The Bee Protection section below also reviews the limitations on aerial application.

Bee Protection

Anyone who hires an aerial applicator to apply a pesticide **labeled as toxic to bees shall notify all beekeepers with registered apiaries** located within 1 mile of the area to be treated. This notification must not take place less than 48 hours nor more than 10 days before the application.

A list of beekeepers with registered apiaries may be obtained in one of the following ways.

1. The NCDA&CS Plant Protection Section mails a list of registered apiaries to aerial applicators licensed in North Carolina quarterly. They will also provide the list on request (call 919-233-8214). This list will have the names of any beekeepers who have registered apiaries located within the required 1 mile from the target area.
2. The NCDA&CS Pesticide Section will mail a list of registered apiaries to farmers who have been identified by the beekeeper on the Apiary Registration Form as having farms within 1 mile of the applicable apiary. The failure of a beekeeper to list a farmer on the Apiary Registration Form does not relieve the farmer of responsibility for notifying the beekeeper of an aerial application of a bee-toxic pesticide within 1 mile of the registered apiary.

The list of registered apiaries is mailed on the first day of each quarter. Revised lists will be issued on the first day of each successive quarter. The lists of revised registered apiary locations will become effective on the fifth day of the first month of that quarter. The registration period will be effective for the calendar year and applies only to the listed apiary locations. Moving an apiary to a new site does not provide protection under the law, unless the new site is also registered.

The farmer can notify the beekeeper digitally, orally, or in writing of the approximate time the pesticide application will be made and the type of pesticide to be used. Digital communication may take place thru email or by texting. Oral notification can be by telephone or in person to the beekeeper or the alternate person designated on the apiary registration list. Acceptable written notification is by mail or by notice left at the beekeeper's residence or at an alternate location designated on the apiary registration list.

NCDA&CS has initiated a free voluntary beehive registration program in coordination with FieldWatch. Growers may register at http://www.fieldwatch.com as an applicator and view maps of the area surrounding their farms to help identify local beekeepers. Contact information is available for beekeepers, and this tool may be used to enhance communication with beekeepers to reduce the adverse effects of pesticide to bees. Growers may also see additional flags or signs to designate the location of apiaries.

Beekeepers who wish to voluntarily map and register their apiaries for free should first go to FieldWatch.com, then complete the registration process thru BeeCheck. Apiaries registered thru

Plant Industry will automatically be mapped and registered on BeeCheck.

Growers should be aware of label changes to 43 active ingredients that have been classified as acutely toxic to bees (products with an LD 50 of <11 micrograms per bee). New labels may place restrictions on pesticide applications when contract pollinators are on site. There are also application restrictions for neonicotinoid products being applied to crops and ornamentals that are blooming and attractive to bees.

Protecting Surface and Groundwater

Groundwater makes up 96 percent of the world's total water resources. Ninety percent of rural residents and 50 percent of the people in the United States depend on groundwater for drinking water. Although only very small amounts of pesticides have been detected in North Carolina's groundwater, we must reduce the likelihood of pesticides entering groundwater and surface water in order to avoid future water quality problems.

What can pesticide users do to prevent groundwater and surface water contamination?

Follow Label Directions Exactly

Pesticide labels provide valuable information concerning the pesticide's potential dangers of contaminating water and the environment. When applying a pesticide, the timing and placement instructions on the label must be followed correctly to ensure the pesticide is applied properly. Applying a pesticide when heavy rains are predicted could lead to water contamination. Likewise, placing a pesticide on top of the soil, when it should be incorporated, not only minimizes pest control but could lead to unnecessary runoff to surface water and perhaps groundwater.

Use Integrated Pest Management Practices

Cultural practices, such as crop rotation and cover crops, not only reduce pest populations, they also maintain and improve good soil and water conditions. Careful pest monitoring will also ensure that pesticides are used only when needed.

Prevent Spills and Back Siphoning

Pesticides spilled near wells, sinkholes, surface waters, or anywhere else can move into surface water and groundwater. Avoid mixing and loading pesticides near wells and other water sources. Use a long hose from the water source to the sprayer so if any spills occur, they will be farther away from the clean water supply. If a spill occurs, be sure to clean it up and move the contaminated soil to a place where it will not seep into the water or otherwise harm the environment. If contaminated soil cannot be applied to a labeled site that does not exceed the rate of application described on the product label, it must be disposed of at an approved waste disposal facility. When using water from a hose to dilute pesticides in a spray tank, do not allow the hose to be submerged in the spray tank, which can lead to a backflow situation where pesticides may be siphoned back into the water supply. The Chemigation section of this chapter also stresses regulations required to prevent the backflow of pesticides into water supplies.

Dispose of Wastes Properly

It's illegal and dangerous to dispose of pesticides improperly. If pesticides, their containers, or other hazardous materials are discarded where they can contaminate the water supply or environment, you (and your family) could drink pesticide-contaminated water. This contamination could move into your neighbors' or livestock's water supply as well as affect wildlife and conditions of soil and air. Don't take it for granted; many pesticides don't just disappear. Your responsibility for these hazardous materials includes proper disposal.

The guidelines for disposing of such materials can be found first on the pesticide label. Follow these instructions for disposal carefully. (See the sections in this chapter on Disposal of Pesticides and Disposal of Empty Pesticide Containers.)

Surface Water Protection

If there is more water in the soil than the soil can absorb, water (with pesticides in it) may flow into the groundwater or run off into streams, rivers, and lakes. Prolonged heavy rains and too much irrigation will also produce excess surface water. Pay attention to weather forecasts, maintain proper irrigation scheduling, and use strip crops to restrict potential surface water problems. The chemigation regulations in this manual are also designed to reduce both surface water and groundwater contamination.

Land Characteristics

Geology plays a key role in protecting groundwater and surface water. If groundwater is within a few feet of the soil surface, pesticides are much more likely to reach groundwater. If pesticides are applied to an area that drains into a sinkhole, irrigation or moderate rainfall may carry some of the pesticides directly into groundwater. You must select pesticides carefully when either groundwater is close to the soil surface or soil permeability is great (see Evaluating the Potential for Groundwater Contamination section).

N.C. Well Construction Standards — and in some cases more stringent local regulations relating to well location, casing, grouting, and other requirements — help ensure that groundwater is not contaminated. For example, N.C. regulations require that well casings be at least 12 inches above the soil surface and that casings be cemented at least 20 feet below the soil surface.

Soil Characteristics

Soil texture (sand, silt, clay), soil permeability, and soil organic matter all play a major part in pesticide movement. Soils containing large amounts of organic matter and clay, for example, will hold (absorb) some pesticides before they reach groundwater. But pesticides are more likely to move into groundwater through sandy soils — low in organic matter and clay — and loose, porous soils. Table 1-4 gives the relative leaching potential ratings for Southeastern U.S. soils.

Pesticide Characteristics

Some pesticides move into the soil more easily than others. Those with high water solubility are more likely to seep into the soil than those pesticides with extremely low water solubility.

Table 1-3 gives the relative mobility (movement) of certain pesticides, listed by common and brand names, in soils. Other sources of groundwater contamination include abandoned and uncontrolled waste sites, landfills, holding pits and ponds, leaking storage tanks, and septic tanks.

Chemigation

Applying pesticides to land, crops, or plants through an irrigation system is called chemigation. A limited number of pesticides are cleared for chemigation application. **Chemigation is only legal when the pesticide label has directions for such uses.**

The North Carolina Pesticide Board has adopted chemigation regulations to protect water resources from pesticide pollution by reducing the potential for backsiphoning or direct injection of pesticides into water sources. Farmers, greenhouse operators, nurserymen, golf course operators, turf growers, and others must comply with these regulations.

The types of irrigation equipment covered by these regulations include, but are not limited to, drip or trickle, center pivot, lateral move, traveler gun, and solid set systems. The regulations do not apply to hand-held hose-end sprayers that are constructed so that an interruption in water flow automatically prevents any backflow to the water supply. Protected water resources include, but are not limited to, private ponds, lakes, rivers, streams, canals, wells, and public water systems.

The following antipollution devices must be installed and maintained on an irrigation system used to apply any pesticide. Safety devices must meet the following qualifications.

Automatic low pressure drain—located on the bottom of the horizontal irrigation pipeline between the discharge side of the irrigation pump and the inlet side of the double check valves. This device shall be level and have an orifice size at least 3/16 the diameter of the irrigation pipe. The top of the drain shall not exceed beyond the inside surface of the bottom of the irrigation pipeline and shall be at least 2 inches above grade. The drain shall discharge at least 20 feet from any water supply. Furthermore, the discharge must be controlled to prevent it from reentering the water supply. In the event that the mainline check valves leak slowly, solution will drain away from rather than flow into the water supply.

Inspection port—located between the irrigation pump discharge side and the inlet side of the mainline check valves. The inspection port may be part of the vacuum relief valve. The purpose of the inspection port is to allow an individual to observe whether or not the mainline check valves are leaking.

Vacuum relief valve—located on the top of the horizontal irrigation pipeline between the discharge side of the irrigation pump and the inlet side of the double check valves. The orifice size of the valve shall be 3/16 of the diameter of the irrigation pipe. The purpose of the vacuum relief valve is to allow air into the pipeline when the water flow stops, preventing the creation of a vacuum that could lead to backsiphoning.

Double check valves—located between the irrigation pump discharge (or pressure side) and the point of pesticide injection into the irrigation pipeline. These valves must be within 10 degrees of horizontal. Double check valves prevent solution from draining or backsiphoning into the irrigation water source and polluting groundwater or surface water. Check valves must have positive closing action and a watertight seal. Note: For irrigation systems that contain media filters, refer to the section below entitled Chemigation Systems that Contain Media Filters.

Check valve—located on the pesticide line between the point of pesticide injection into the irrigation system and the pesticide injection unit. The check valve stops the flow of water from the irrigation system into the chemical supply tank. It should be constructed of chemically resistant materials. The check valve should always be flushed with clean water after injecting a chemical to prevent the deposition of chemical precipitates. Whenever a Dosatron, DosmaticPlus, or a similar metering device is used, the check valve needs to be positioned near the metering device on its outlet side. If this metering device is on a bypass line, the check valve must be located on the main irrigation line immediately downstream of the bypass line.

Flow interruption device, solenoid valve—located in the pesticide supply line between the pesticide injection unit and the pesticide supply tank or container. The solenoid valve provides a positive shut off on the chemical injection line. This prevents both chemical and water from flowing in either direction if the chemical pump is stopped. Because this valve will be subjected to different chemicals, it must be compatible with the chemicals being injected. The valve should be inspected often to assure that it is performing properly. A solenoid valve is not required with the Dosatron, DosmaticPlus, or a similar hydraulic injection device.

Functional systems interlock—The irrigation pump and the chemical injection pump must be interlocked or connected so that if the irrigation pump stops, the chemical injection pump will stop. The functional systems interlock ensures that a pesticide is applied with water through the irrigation system. The Dosatron, DosmaticPlus or a similar hydraulic injection device does not require the functional systems interlock.

Illegal Techniques

1. Some pesticide product labels prohibit application of the product by any irrigation system. Others prohibit applications through certain specific irrigation systems.
2. It is illegal to inject a pesticide into an irrigation system on the suction (or inlet) side of the irrigation pump.
3. It is illegal to connect an irrigation system directly to a public water system when applying any pesticide.

NOTE: When a public water system is used, the water must first be discharged into a reservoir tank. An air gap at least twice the inside diameter of the fill pipe must exist between the end of the fill pipe from the public water system and the top rim of the reservoir tank.

Chemigation Systems that Contain Media Filters

Some chemigators are using chemigation systems that have one or more sand-containing media filters. Surface water, and in some cases, groundwater, flows through these filters to remove debris that would clog the small orifices of the emitters on drip irrigation systems. **The injection of pesticides into the irrigation line must be on the outlet side of all media filters.** This prevents pesticide from passing through the media filters and contaminating the debris, which will be discharged into the environment whenever the media filters are backflushed.

Any water dumping or open dumping of pesticide dilutions is an illegal discharge of a pesticide in violation of NC Pesticide Board rule 2 NCAC 9L .0604. Further, pesticide product labels have enforceable language on the illegal disposal of pesticides.

Additionally, a check valve is required between the outlet side of all media filters and the point of pesticide injection into the irrigation line. If the injection system has bypass piping, a check valve would be positioned between the outlet side of all media filters and the inlet side of the bypass on the irrigation pipeline.

The purpose of the check valve mentioned above is to reduce the risk of media filter contamination if a backsiphonage occurs. Systems operating without this safeguard could dispose of pesticides unintentionally during a backflush cycle. This would be illegal. The chemigator can be fined for any illegal disposals.

Any chemigation system that is not in compliance with pesticide regulations will be issued a stop-use order. This order can only be released when a follow-up inspection indicates that the appropriate antipollution devices have been installed.

Hand-Held Hose-End Sprayers

Hand-held hose-end sprayers are allowed on the outlet side of a water hose. This device must contain a check valve that will prevent any backsiphoning from the pesticide reservoir into the water hose. The use of devices connected to a faucet or spigot that siphon pesticide from a reservoir or container is not permitted in North Carolina.

System Inspections

One of the requirements of the regulations is that the system operator must inspect the antisiphon devices and the functional systems interlock during periods of chemigation to ensure that they are functioning properly. If components of the system are defective, they must be repaired or replaced before any

Chapter I — 2019 N.C. Agricultural Chemicals Manual

chemigation is employed with a pesticide. Representatives of the NCDA&CS, Pesticide Section, may inspect an irrigation system used for chemigation at any time. If the system is not in compliance with the regulations, a stop-use order will be issued by the department, and the system must be inspected again by the departmental representative before the stop-use order can be removed.

For Additional Information

For a copy of regulation 2 NCAC 9L .2000 — Chemigation or the chemigation and fertigation brochure, visit the NCDA&CS Web site at http://www.ncagr.com/SPCAP/pesticides/sitemap.htm, or call the Structural Pest Control and Pesticides Division at 919-733-3556.

The chemigation section was written by Henry Wade, Environmental Programs Manager, retired, NCDA&CS, Structural Pest Control and Pesticides Division.

Evaluating the Potential for Groundwater Contamination

When estimating the groundwater contamination potential (GWCP) index for a pesticide at a given site, the characteristics of the soil at the site must also be evaluated. Soil properties are as important as a pesticide's chemical properties in determining mobility and risk to groundwater. See Table 1-4. Soil Leaching Potential (SLP) Indices. The GWCP index for a pesticide on a given soil is the mean of the PLP index and the SLP index, i.e.; GWCP = (PLP + SLP)/2. More detailed formulas are given at the end of the following table.

Table 1-2. Relative Pesticide Leaching Potential (PLP) Indices and Ratings for Commonly Used Pesticides

The PLP value will change with changes in application rate. Values in this table are calculated using average rates.*

KEY: Very High (VH) = 90 to 100, High (H) = 70 to 89, Moderate (M) = 50 to 69, Low (L) = 30 to 49, Very Low (VL) = 0 to 29

Common Name	Brand Name	Application Method	PLP	Rating
Herbicides				
2,4-D	Weedone	foliage	45	L
2,4-DB	Butyrac 200	foliage	41	L
acetochlor	Harness	soil	53	M
acifluorfen	Blazer	foliage	42	L
alachlor	Lasso	soil	55	M
ametryn	Evik	foliage	45	L
amitrole	Amizole	foliage	51	M
AMS	Ammate	foliage	74	H
arsenic acid	Hy-Yield	foliage	39	L
asulam	Asulox	foliage	47	L
atrazine	AAtrex	soil	56	M
benefin	Balan	soil	31	L
bensulfuron	Londax	foliage	20	VL
bensulide	Prefar	soil	57	M
bentazon	Basagran	foliage	48	M
bispyribac	Velocity	foliage	45	L
bromacil	Hyvar	soil	85	H
bromoxynil	Buctril	foliage	30	L
butachlor	Machete	soil	36	L
butylate	Sutan	soil	48	L
cacodylic	Rad-E-Cate	foliage	19	VL
carfentrazone	Aim	foliage	26	VL
carfentrazone	Aim	soil	23	VL
chlorimuron	Classic	soil	19	VL
chlorsulfuron	Glean	foliage	28	VL
clethodim	Envoy	foliage	40	L
clodinafop-prop	Discover	foliage	1	VL
clomazone	Command	soil	42	L
clopyralid	Stinger	foliage	46	L
cloransulam-me	First-Rate	soil	34	L
cloransulam-me	First-Rate	foliage	25	VL
cyanazine	Bladex	soil	41	L
cyanazine	Bladex	foliage	37	L
cycloate	Ro-Neet	soil	47	L
cyhalofop-bu	Clincher	foliage	1	VL
DCPA	Dacthal	soil	47	L
desmedipham	Betanex	foliage	27	VL
dicamba	Banvel	foliage	49	L
dichlobenil	Carson	soil	55	M
dichlorprop	DP-Amine	foliage	44	L
diclofop	Hoelon	foliage	50	M
diclosulam	Strongarm	soil	33	L
difenzoquat	Avenge	foliage	0	VL
diflufenzopyr	Distinct	foliage	21	VL
dimethenamid	Frontier	soil	46	L
diphenamid	Enide	soil	54	M
dimethipin	Harvaid	foliage	70	H
diquat	Diquat	foliage	15	VL
dithiopyr	Dimension	foliage	45	L
diuron	Karmex	soil	65	M
DSMA	Ansar	foliage	39	L
endothall	Aquathol	aquatic	64	M
EPTC	Double Play	soil	53	M
ethalfluralin	Sonalan	soil	25	VL
ethamesulfuron	Muster	foliage	28	VL
ethofumesate	Nortron	soil	52	M
fenarimol	Rubigan	foliage	44	L
fenoxaprop	Whip	foliage	24	VL
fluazifop	Fusilade	foliage	30	L
flufenacet	Define	soil	43	L
flumetsulam	Broadstrike	soil	50	M
flumiclorac	Resource	foliage	14	VL
flumioxazin	Valor	soil	45	L
flumioxazin	Valor	foliage	29	VL
fluometuron	Cotoran	soil	62	M
fluridone	Sonar	aquatic	34	L
fluroglycofen	Compete	foliage	18	VL
fluroxypyr	Starane u	foliage	45	L
fluthicet-me	Action	foliage	17	VL
fomesafen	Reflex	foliage	49	L
formasulfuron	Option	foliage	37	L
fosamine	Krenite	foliage	16	VL
glufosinate	Ignite	foliage	8	VL
glyphosate	Roundup	foliage	20	VL
halosulfuron-methyl	Permit	foliage	23	VL
haloxyfop	Galant	foliage	48	L
hexazinone	Velpar	soil	73	H
imazamethabenz	Assert	foliage	47	L
imazamox	Raptor	foliage	43	L
imazapic	Cadre	foliage	49	L
imazapyr	Arsenal	foliage	65	M
imazaquin	Scepter	foliage	44	L
imazethapyr	Pursuit	foliage	52	M

Table 1-2. Relative Pesticide Leaching Potential (PLP) Indices and Ratings for Commonly Used Pesticides

The PLP value will change with changes in application rate. Values in this table are calculated using average rates.*

KEY: Very High (VH) = 90 to 100, High (H) = 70 to 89, Moderate (M) = 50 to 69, Low (L) = 30 to 49, Very Low (VL) = 0 to 29

Common Name	Brand Name	Application Method	PLP	Rating
imazethapyr	Pursuit	soil	49	L
ioxynil	Totril	foliage	22	VL
isoxaflutole	Balance	soil	23	VL
isoxoben	Gallery	soil	47	L
lactofen	Cobra	foliage	25	VL
linuron	Lorox	foliage	42	L
linuron	Lorox	soil	48	L
MAA	MAA	foliage	37	L
MAMA	MAMA	foliage	37	L
MCPA	Chiptox	foliage	45	L
MCPB	Thistrol	foliage	51	M
mecoprop	Mecomec	foliage	58	M
mesotrione	Lexar	soil	39	L
metolachlor	Dual	soil	63	M
metribuzin	Sencor	soil	52	M
metsulfuron methyl	Ally	foliage	16	VL
molinate	Ordram	soil	52	M
MSMA	Daconate	foliage	35	L
napropamide	Devrinol	soil	52	M
naptalam	Alanap	foliage	63	M
nicosulfuron	Accent	foliage	30	L
norflurazon	Zorial	soil	51	M
oryzalin	Surflan	soil	41	L
oxadiazon	Ronstar	soil	39	L
oxyfluorfen	Goal	soil	24	VL
paraquat	Gramoxone	foliage	6	VL
pebulate	Tillam	soil	45	L
pendimethalin	Prowl	soil	20	VL
phenmedipham	Spin-Aid	foliage	21	VL
picloram	Tordon	soil	68	M
pinoxaden	Axial	foliage	8	VL
primisulfuron	Beacon	foliage	29	VL
prodiamine	Barricade	foliage	28	VL
prometon	Pramitol	soil	95	VH
prometryn	Caparol	soil	51	M
prometryn	Caparol	foliage	42	L
pronamide	Kerb	soil	48	L
propachlor	Ramrod	soil	50	M
propanil	Stam	foliage	27	VL
propham	Chem Hoe	soil	50	M
prosulfuron	Peak	foliage	27	VL
pyrazon	Pyramin	foliage	51	M
pyridate	Tough	foliage	28	VL
pyrithiobac	Staple	soil	41	L
quinclorac	Facet	foliage	47	L
quizalofop	Assure	foliage	31	L
rimsulfuron	Matrix	soil	41	L
rimsulfuron	Matrix	foliage	14	VL
sethoxydim	Poast	foliage	29	VL
siduron	Tupersan	soil	63	M
simazine	Princep	soil	62	M
sulfentrazone	Authority	soil	54	M
sulfometuron	Oust	foliage	37	L
sulfometuron	Oust	soil	42	L
sulfosulfuron	Certainty	foliage	23	VL
TCA	Varitox	soil	94	VH
tebuthiuron	Spike	soil	78	H
terbacil	Sinbar	foliage	51	M
terbutryn	Igran	foliage	34	L
thiazopyr	Visor	soil	57	M
thifensulfuron methyl	Pinnacle	foliage	20	VL
thiobencarb	Saturn	soil	45	L
topramezone	Impact	foliage	1	VL
tralkoxydim	Achieve	foliage	43	L
triallate	Avadex-2	soil	38	L
triasulfuron	Amber	foliage	21	VL
tribenuron	Express	foliage	18	VL
triclopyr	Garlon	foliage	54	M
trifluralin	Treflan	soil	25	VL
trifloxysulfuron	Envoke	foliage	21	VL
triflusulfuron	Debut	foliage	12	VL
vernolate	Vernam	soil	44	L
Growth Regulators, Defoliants, Desiccants				
chlormequat	Cycocel	foliage	4	VL
clofencet	Genesis	foliage	62	M
dimethipin	Harvade	foliage	72	H
ethephon	Super Boll	foliage	71	H
fumetralin	Prime+	foliage	22	VL
mefluidide	Embark	foliage	25	VL
mepiquat	PIX	foliage	4	VL
MH	Royal MH	foliage	49	L
pyraflufen	Ecopart	foliage	1	VL
sodium chlorate	Defol	foliage	80	H
thidiazuron	Dropp	foliage	30	L
tribufos	DEF	foliage	25	VL
Fungicides, Biocides				
azoxystrobin	Heritage	foliage	42	L
benomyl	Benlate	foliage	43	L
captan	Captan	foliage	54	M
carboxin	Evershield V	seed	13	VL
chlorothalonil	Bravo	foliage	33	L

Table 1-2. Relative Pesticide Leaching Potential (PLP) Indices and Ratings for Commonly Used Pesticides

The PLP value will change with changes in application rate. Values in this table are calculated using average rates.*
KEY: Very High (VH) = 90 to 100, High (H) = 70 to 89, Moderate (M) = 50 to 69, Low (L) = 30 to 49, Very Low (VL) = 0 to 29

Common Name	Brand Name	Application Method	PLP	Rating
copper hydroxide		foliage	41	L
DCNA	Botran	foliage	29	VL
dimethomorph	Acrobat	foliage	37	L
dodine	Syllit	foliage	1	VL
ethoprop	Mocap	soil	60	M
etridiazole	Terrazole	soil	17	VL
fenarimol	Rubigan	foliage	59	M
fenbutatin oxide	Vendex	foliage	38	L
fenhexamid	Elevate	foliage	14	VL
flutolanil	Moncut	foliage	39	L
fosethyl-Al	Alliette	foliage	9	VL
Iprodione	Rovral	foliage	27	VL
mancozeb	Dithane	foliage	38	L
maneb	Manzate	foliage	40	L
metalaxyl	Ridomil	soil	59	M
metiram	Polyram	foliage	4	VL
myclobutanil	Nova	foliage	32	L
oxamyl	Vydate	foliage	46	L
oxythioquinox	Morestan	foliage	24	VL
PCNB	Terrachlor	seed	28	VL
penconazole	Topaz	foliage	28	VL
propamocarb	Previcur	foliage	17	VL
propiconazole	Tilt	foliage	30	L
tebuconazole	Folicure	foliage	27	VL
terbufos	Counter	seed	31	L
thiabendazole	Mertect	foliage	37	L
thiophanate	Topsin	foliage	18	VL
thiophanate-methyl	Cercobin	foliage	41	L
trifloxystrobin	Stratego	foliage	8	VL
triflumizole	Procure	foliage	27	VL
triphenyltin acetate	Fentin Acetate	foliage	20	VL
triphenyltin chloride	Fentin Chloride	foliage	20	VL
triphenyltin hydroxide	Super Tin	foliage	20	VL
vinclozolin	Ronilan	foliage	18	VL
zineb	Dithane	foliage	35	L
ziram	Zirex	foliage	41	L
Insecticides, Acaricides, Miticides, Nematicides				
acephate	Orthene	foliage	52	M
acetamiprid	Assail	foliage	27	VL
aldicarb	Temik	soil	67	M
aldoxycarb	Standak	foliage	72	H
amitraz	Mitac	foliage	11	VL
azinphosmethyl	Guthion	foliage	22	VL
bendiocarb	Turcam	seed	26	VL
bifenthrin	Biflex	foliage	0	VL
carbaryl	Sevin	foliage	37	L
carbofuran	Furadan	foliage	54	M
carbofuran	Furadan	soil	73	H
carbosulfan	Advantage	foliage	11	VL
chlordimeform	Galecron	foliage	6	VL
chlorfenvinphos	Birlane	soil	40	L
chlorfenvinphos	Birlane	foliage	35	L
chlorobenzilate	Folbex	foliage	26	VL
chlorpyrifos	Lorsban	foliage	27	VL
chlorpyrifos	Lorsban	soil	30	L
clofentezine	Ovation	foliage	1	VL
cyfluthrin	Baythroid	foliage	0	VL
cyhalothrin	Karate	foliage	11	VL
cypermethrin	Ammo	foliage	0	VL
cyromazine	Trigard	foliage	39	L
deltamethrin	Decis	foliage	1	VL
diazinon	Diazinon	foliage	41	L
dicofol	Kelthane	foliage	33	L
dicrotophos	Bidrin	foliage	39	L
dietholate	Eradicane-Extra	soil	26	VL
diflubenzuron	Dimilin	foliage	0	VL
dimethoate	Dimethoate	foliage	47	L
disulfoton	Di-syston	foliage	33	L
endosulfan	Thiodan	foliage	18	VL
esfenvalerate	Asana	foliage	2	VL
ethion	Ethion	foliage	34	L
ethoprop	Mocap	soil	61	M
fenamiphos	Nemacur	soil	58	M
fenoxycarb	Logic	soil	0	VL
fenpropathrin	Danitol	foliage	0	VL
fenthion	Baytex	foliage	27	VL
fenvalerate	Asana XL	foliage	9	VL
flucythrinate	Pay-Off	foliage	0	VL
fluvalinate	Spur	foliage	0	VL
fonofos	Dyfonate	soil	29	VL
formetanate	Carzol	foliage	28	VL
hexythiazox	Hexygon	foliage	9	VL
hydramethylnon	Amdro	foliage	0	VL
imidacloprid	Timax	foliage	1	VL
indoxacarb	Steward	foliage	4	VL
lambda-cyhalothrin	Karate	foliage	0	VL
lindane	Lindane	foliage	55	M
malathion	Cythion	foliage	13	VL
methamidophos	Monitor	foliage	53	M
methidathion	Supracide	foliage	27	VL
methomyl	Lannate	foliage	43	L

Chapter I—2019 N.C. Agricultural Chemicals Manual

Table 1-2. Relative Pesticide Leaching Potential (PLP) Indices and Ratings for Commonly Used Pesticides

The PLP value will change with changes in application rate. Values in this table are calculated using average rates.*
KEY: Very High (VH) = 90 to 100, High (H) = 70 to 89, Moderate (M) = 50 to 69, Low (L) = 30 to 49, Very Low (VL) = 0 to 29

Common Name	Brand Name	Application Method	PLP	Rating
methoxychlor	Marlate	foliage	7	VL
methoxyfenozide	Intrepid	foliage	20	VL
methyl parathion	Penncap-M	foliage	11	VL
mevinphos	Phosdrin	foliage	30	L
monocrotophos	Azodrin	foliage	61	M
naled	Dibrom	foliage	25	VL
oxamyl	Vydate	foliage	46	L
oxydemeton-methyl	Metasystox-R	foliage	49	L
parathion	Ethyl-Parathion	foliage	19	VL
PCNB	Terraclor	soil	37	L
permethrin	Ambush	foliage	0	VL
phorate	Thimet	soil	48	L
phosmet	Imidan	foliage	37	L
phosphamidon	Dimecron	foliage	53	M
profenophos	Curacron	foliage	15	VL
propargite	Comite	foliage	39	L
propoxur	Baygon	foliage	49	L
pymetrozine	Fulfill	foliage	7	VL
spinosad	Tracer	foliage	11	VL
sulprophos	Bolstar	foliage	27	VL
temephos	Abate	aquatic	0	VL
terbufos	Counter	soil	31	L
thiamethoxam	Platinum	seed	54	M
thiamethoxam	Centric	foliage	12	VL
thiodicarb	Larvin	foliage	24	VL
tralomethrin	Scout	foliage	0	VL
trichlorfon	Dylox	foliage	38	L
trimethacarb	Landin	soil	38	L
zetamethrin	Fury	foliage	1	VL
Fumigants				
1,3-dichloropropene	Telone-2	soil	79	H
chloropicrin	Larvicide	soil	65	M
dazomet	Basamid	soil	92	VH
metam sodium	Vapam	soil	94	VH
methyl bromide	Brom-O-Gas	soil	100	VH
Molluscicides				
metaldehyde	Deadline Bullets	soil	41	L

*Formulae used to determine the values in Table 1-3 include the following:

$$PLP_{value} = \frac{(\text{Application rate of kg. ai/ha.})(\text{fraction hitting the soil})(T1/2)}{K_{oc}}$$

Where $T\frac{1}{2}$ = half-life of the parent compound under field conditions

K_{oc} = soil organic carbon binding value

$$PLP_{INDEX} = (\log PLP_{value})(14.3) + 57$$

Table 1-3. Relative Soil Leaching Potential (SLP) Indices and Ratings for Soils in the Southeastern U.S.

The SLP index will change slightly for the site where the soil series is located. Values in this table are calculated using the profile description of the original site where the soil was named.
KEY: Very High (VH) = 90 to 100, High (H) = 70 to 89, Moderate (M) = 50 to 69, Low (L) = 30 to 49, Very Low (VL) = 0 to 29

Soil	SLP	Rating	Soil	SLP	Rating	Soil	SLP	Rating	Soil	SLP	Rating
Alamance	69	M	Corolla	95	VH	Johnston	22	VL	Plummer	79	H
Alpin	84	H	Coxville	36	L	Kalmia	77	H	Ponzer	08	VL
Altavista	73	H	Craven	66	M	Kenansville	84	H	Portsmouth	26	VL
Appling	62	M	Creedmoor	64	M	Kureb	100	VH	Pungo	00	VL
Arapahoe	39	L	Croatan	14	VL	Lakeland	84	H	Rains	45	L
Argent	58	M	Cullowhee	77	H	Leaf	35	L	Rimini	84	H
Augusta	66	M	Dare	09	VL	Lenoir	32	L	Rion	64	M
Autryville	68	M	Davidson	55	M	Leon	84	H	Roanoke	57	M
Aycock	68	M	Deloss	29	VL	Liddell	47	L	Roper	16	VL

Table 1-3. Relative Soil Leaching Potential (SLP) Indices and Ratings for Soils in the Southeastern U.S.

The SLP index will change slightly for the site where the soil series is located. Values in this table are calculated using the profile description of the original site where the soil was named.
KEY: Very High (VH) = 90 to 100, High (H) = 70 to 89, Moderate (M) = 50 to 69, Low (L) = 30 to 49, Very Low (VL) = 0 to 29

Soil	SLP	Rating	Soil	SLP	Rating	Soil	SLP	Rating	Soil	SLP	Rating
Ballahack	20	VL	Dogue	67	M	Lignum	63	M	Rosman	55	M
Barclay	77	H	Dorovan	10	VL	Louisburg	81	H	Rumford	84	H
Bayboro	04	VL	Dothan	66	M	Lumbee	53	M	Seabrook	84	H
Baymeade	87	H	Dragston	80	H	Lynchburg	40	L	Stallings	58	M
Belhaven	10	VL	Duckston	98	VH	Lynn Haven	61	M	State	73	H
Bibb	75	H	Dunbar	35	L	Madison	68	M	Stockade	35	L
Blaney	73	H	Duplin	62	M	Mandarin	89	H	Tarboro	90	VH
Blanton	82	H	Durham	72	H	Marlboro	63	M	Tate	46	L
Bojac	82	H	Echaw	84	H	Marvyn	44	L	Tatum	64	M
Bonneau	70	H	Edneyville	54	M	Masada	37	L	Toisnot	68	M
Braddock	64	M	Emporia	67	M	Mayodan	63	M	Tomahawk	86	H
Bragg	64	M	Enon	77	H	Mccoll	24	VL	Tomotley	52	M
Brookman	07	VL	Evard	70	H	Mecklenburg	69	M	Torhunta	29	VL
Buncombe	87	H	Exum	69	M	Meggett	29	VL	Vance	66	M
Butters	83	H	Faceville	58	M	Munden	81	H	Varina	62	M
Byars	21	VL	Fanin	70	H	Nahunta	41	L	Vaucluse	76	H
Cainhoy	89	H	Foreston	81	H	Nankin	63	M	Wagram	71	H
Candor	79	H	Fork	72	H	Nason	41	L	Wahee	43	L
Cape Fear	13	VL	Fuquay	73	H	Nimmo	78	H	Wakulla	85	H
Caroline	64	M	Gaston	63	M	Nixonton	82	H	Wando	87	H
Cecil	60	M	Georgeville	59	M	Norfolk	67	M	Wasda	18	VL
Centenary	85	H	Gilead	62	M	Ocilla	53	M	Watauga	46	L
Chandler	47	L	Goldsboro	70	H	Onslow	70	H	Wedowee	64	M
Charleston	83	H	Goldston	78	H	Orangeburg	67	M	Weeksville	34	L
Chastain	40	L	Grantham	44	L	Ousley	84	H	Wehadkee	48	L
Chester	44	L	Grifton	61	M	Pacolet	63	M	White Store	61	M
Chewacla	47	L	Gritney	34	L	Pactolus	85	H	Wickham	75	H
Chipley	61	M	Hayesville	40	L	Pamlico	53	M	Wilbanks	43	L
Chowan	35	L	Helena	66	M	Pantego	12	VL	Wilkes	82	H
Clifton	40	L	Herndon	63	M	Pasquotank	52	M	Winnsboro	82	H
Colvard	88	H	Hullett	62	M	Paxville	40	L	Winton	72	H
Conaby	23	VL	Hyde	17	VL	Pender	69	M	Woodington	58	M
Conetoe	85	H	Invershiel	70	H	Perquimans	44	L	Worsham	43	L
Congaree	28	VL	Johns	78	H	Pinkston	73	H	Yaupon	72	H
									Yonges	57	M

SAMPLE CALCULATION FOR GWCP INDEX: Acetochlor (PLP index = 55 M) applied to an Alamance soil (SLP index = 69 M).
GWCP index = 55 + 69 / 2 = 62 M.

Table 1-4. Groundwater Contamination Potential (GWCP) Risk of Pesticide-Soil Combinations

Obtain numbers for PLP and SLP for your soil and pesticide from Tables 1-3 and 1-4, respectively.

Pesticide Leaching Potential (PLP) Rating	Soil Leaching Potential (SLP) Rating				
	0–29 Very Low	30–49 Low	50–69 Moderate	70–89 High	90–100 Very High
0–29 Very Low	Very Low Risk	Very Low Risk	Low Risk	Low Risk	Moderate Risk
30–49 Low	Very Low Risk	Low Risk	Low Risk	Moderate Risk	Moderate Risk
50–69 Moderate	Low Risk	Low Risk	Moderate Risk	Moderate Risk	High Risk
70–89 High	Low Risk	Moderate Risk	Moderate Risk	High Risk	High Risk
90–100 Very High	Moderate Risk	Moderate Risk	High Risk	High Risk	Very High Risk

Chapter I — 2019 N.C. Agricultural Chemicals Manual

Proper Pesticide Storage

Safe and proper storage can extend the shelf (storage) life of your pesticides, keep the containers in good condition, and keep the labels clean and legible.

Only rules for storing pesticides on the farm and in household situations are presented here. If you store restricted use pesticides in commercial storage facilities, you must meet additional requirements as outlined in Regulation 2 NCAC 9L.1903–.1913. Details on these requirements can be obtained at your county Cooperative Extension center or by calling the NCDA&CS Pesticide Section at 919-733-3556.

Household and Farm Situation Storage Rule (simplified)

1. This rule applies to all pesticides.
2. Store pesticides to prevent leaking and to aid inspection.
3. Do not store formulated pesticides in unlabeled containers. The following minimum information must be shown clearly and prominently on any containers of formulated pesticides:
   a. common chemical name (such as carbaryl for the product Sevin)
   b. percentage of each active ingredient
   c. EPA registration number
   d. signal word (DANGER, WARNING, CAUTION)
   e. use classification (restricted use or general use).
1. Do not store pesticides (formulated products or dilutions) in any food, feed, beverage, or medicine container that has previously been used for such purposes or that is specifically designed to contain only those products.
2. Do not store pesticides in a way that could contaminate foods, feeds, beverages, eating utensils, tobacco, or tobacco products, or otherwise result in accidental ingestion by people or domestic animals. In addition, pesticides should not be stored in such a way that could contaminate other pesticides, seeds, or fertilizers.
3. Store pesticides based on the following:
   a. storage recommendations, if any, on their labels, and
   b. labels on all other products, including nonpesticide products held in the same storage area.
4. When unattended, store pesticides to prevent unauthorized access.
5. Store pesticides in an area that is dry (does not accumulate water) and well-ventilated.
6. Pesticide storage areas should be free of combustible materials—such as gasoline, kerosene, or petroleum solvents other than those associated with pesticide application—and debris like waste paper, rags, or used cardboard boxes that may provide an ignition source. They must also be separated from other operations that present a fire hazard, such as welding or burning. Take appropriate care to reduce fire hazards when providing supplemental heating to storage areas during the winter.

Disposal of Pesticides

Reduce the Need for Disposal of Unwanted Pesticides

Because of the expenses, environmental hazards, and legal responsibilities associated with the disposal of pesticides and other hazardous waste, the best solution is to minimize or eliminate the need for disposal altogether. Careful planning of spray programs is essential in order to avoid purchasing more pesticides than will actually be needed for a particular application or season. However, in spite of careful planning, farmers (and homeowners) will occasionally need to dispose of unwanted pesticides. In addition, there are old, obsolete, or banned pesticide products in storage throughout the state that require disposal. Some of these materials are in containers that are structurally unsound and have incomplete or no labeling. The following information will aid in accomplishing disposal in the safest, most economical, and environmentally acceptable manner possible.

Donation of Excess Pesticides

If no longer needed, unopened containers of recently purchased pesticides may possibly be returned to the local dealer or manufacturer/formulator for a refund. If this cannot be arranged, the pesticide may be sold or donated to a neighbor or to someone who can and will use it properly. If donation is a possibility, there are two important factors that must be taken into consideration.

1. A pesticide designated **RESTRICTED USE** on the label must not be given to someone who is not a certified applicator.
2. A pesticide that has been banned or one for which all uses have been cancelled should not be donated to another person.

NCDA&CS Pesticide Disposal Assistance Program

If donation of unwanted pesticides is not a possibility, other options are available. One option is the NCDA&CS Pesticide Disposal Assistance Program (PDAP) for farmers and homeowners. This service is provided on an individual request basis with no fees charged. Specific requests for assistance should be made to Derrick Bell, NCDA&CS Pesticide Disposal Assistance Program Manager, (919) 280-1061.

Commercial Hazardous Waste Disposal

In some situations, the NCDA&CS may not be able to provide the disposal activity that is needed, but disposal assistance can still be provided. The NCDA&CS PDAP can provide other potential options or will furnish a listing of hazardous waste disposal companies upon request. Anyone contemplating this commercial option is advised to contact several firms, compare costs, and ask for and check references.

While all pesticides are hazardous substances, not all are legally classified by the U.S. Environmental Protection Agency (EPA) as Resource Conservation and Recovery Act (RCRA) hazardous wastes upon disposal. However, because pesticides cannot be legally disposed of at MSWLF/county landfills, an EPA-permitted hazardous waste disposal facility is an option that may be required.

Do not assume that disposal of a pesticide (once it is classified as a RCRA hazardous waste) at an EPA-permitted facility eliminates all further legal responsibility for that product. The person who generates a hazardous waste (farmer/homeowner) is legally and financially responsible for that material for as long as it remains in existence. Therefore, even though destructive disposal (by hazwaste incineration) may cost more than nondestructive disposal (by hazwaste landfilling), there are worthwhile, long-term benefits for using the incineration method. Regardless of the method, all paperwork (manifests) generated by the disposal should be retained permanently.

In summary, the disposal of some pesticides via commercial disposal companies involves RCRA regulations. There are stringent requirements for storage, transport, and disposal of hazardous waste and severe penalties for failure to comply with the regulations. For information regarding hazardous waste management and disposal contact:

Division of Waste Management, Hazardous Waste Section, N.C. Department of Environment Quality, 1646 Mail Service Center, Raleigh, NC 27699-1646.

Disposal of Excess Spray Solution from Tank and Equipment Rinses

As already indicated, proper planning and careful calculations should eliminate the need for disposal of large quantities of excess spray solutions. Small quantities should be sprayed out along field borders or on the row ends. Care must be taken not to exceed the labeled application rate.

Tank and equipment rinses should also be applied along field borders or on the row ends. If decontamination solutions, cleaners, detergents, ammonia, chlorine bleaches, etc., are used to remove residues, adequate dilution may be necessary to prevent soil and plant injury.

Disposal of Empty Pesticide Containers

Metal, Plastic, or Glass Containers of Liquid Formulations (5 Gallons or Fewer)

Before pesticide containers can be accepted for recycling or disposed of properly, they must be rinsed by one of the following methods.

Pressure rinsing

1. Drain the container into the spray tank for 30 to 60 seconds after the last amount starts to drip.
2. Insert tip of the pressure nozzle through the side of the pesticide container near its base.
3. While holding the container so the opening can drain into the spray tank, spray the inside of the container for at least 30 seconds.
4. Drain all rinse water into the spray tank.

Triple rinsing

1. Drain the container into the spray tank for 30 to 60 seconds after the last amount starts to drip.
2. Fill the containers 1/3 full with water, cap, and shake thoroughly. Empty this rinse water into the spray tank.
3. Repeat the above rinse procedure at least two more times, adding each amount of rinsate to the spray tank.
4. Punch holes in the bottom and sides of metal and plastic containers. The holes will prevent the containers from being reused and will indicate that they are indeed empty.

Properly-rinsed plastic pesticide containers can be delivered to a container recycling collection site. Nearly every North Carolina county has one or more of these sites. Call your county Extension center for directions to the nearest collection site. Properly rinsed metal or glass pesticide containers can be taken to a county solid waste collection system or to a county landfill.

Metal or Plastic Drums (30- to 55-Gallon)

1. Rinse and drain these containers into the spray tank.
2. Attempt to return drums to dealer or distributor (recycling). Plastic drums may be recycled at selected recycling centers. For more information, contact Environmental Programs Manager, NCDA&CS, Renee Woody, (919) 733-3556.
3. If you are not successful, attempt to donate drums to a drum reconditioner (contact Pesticide Disposal Assistance Program Manager, NCDA&CS, Derrick Bell, (919) 280-1061 for information on drum reconditioners operating in the state).
4. If steps 2 or 3 above fail, attempt to dispose of rinsed drums in the county sanitary landfill. This may not be possible in some cases because of the difficulty for landfill personnel to verify that the drums have been properly rinsed.
5. If the above procedures fail, contact the Solid and Hazardous Waste Management Branch, (919) 707-8200, for assistance with pesticide drum disposal.

Containers of Non-Liquid Pesticides

1. Shake container into the applicator tank until all the pesticide has been removed.
2. Tear open the container to make sure it is completely empty.
3. If the pesticide is a wettable powder and the container can be triple rinsed with water, do so. Add the rinsate to the tank.
4. Puncture, crush, or otherwise render the container incapable of being reused and then place in the solid waste collection system or carry to a county sanitary landfill facility.
5. If large containers can be returned for recycling, do so.

Controlling Pesticide Drift

Once discharged from the application equipment, pesticides may drift through the air and injure susceptible plants or sensitive animal life before the pesticides actually reach the target area. Movement through the air may be by spray drift, vapor, or dust.

If herbicides are allowed to drift, they can often cause extensive damage to susceptible crops. Drift of other pesticides—although not as likely to injure nearby crops—can damage livestock, bees, fish, and people, in addition to leaving illegal residues on crops.

Spray drift is the movement of airborne spray particles beyond the target. The amount of spray drift is influenced by the (1) droplet size, (2) amount of wind, and (3) height from which the spray is released.

When using ground equipment certain precautions can reduce spray drift.

1. Select nozzles that produce a minimum proportion of small droplets in relation to overall droplet size. The droplet size is influenced by pounds of pressure per square inch (psi) and the nozzle's design (opening size). Lower pressures and larger openings, for example, tend to produce larger droplets and less drift. Standard fan nozzles operate at 30 to 40 psi. Whirl chamber nozzles and low-pressure fan nozzles work at 15 to 20 psi. Select the largest nozzle opening that will provide enough gallons per acre for penetration, uniform coverage, mixing with the pesticide, and effective control.
2. Spray when wind velocity is at a minimum, when temperatures are moderate, and preferably before adjacent susceptible crops emerge. Where a susceptible crop has emerged, spray when the wind is moving away from the crop or consider leaving an untreated strip along the edge of the field.
3. Adjust the boom as close as possible to the target without losing uniform distribution of the pesticide. To bring the boom closer to the target, space the nozzles closer together or use wide-angle nozzles. The boom's height above the ground affects how long it takes a spray droplet to reach the ground. Wind velocities are usually lower closer to the ground.

Special nozzles that produce large droplets have been developed to control this drift from aerial application. Also, invert emulsions and drift control agents can be used.

Vapor drift occurs when vapor or fumes move from the area of application. Vapor drift may damage susceptible crops or may simply reduce the effectiveness of the herbicide or pesticide.

Ester formulations of 2,4-D may volatilize in hot weather and drift to susceptible crops. Therefore, select amine and sodium salts of 2,4-D that are less prone to volatilize and continue to take steps to prevent spray particle drift.

Fumigants like methyl bromide and chloropicrin drift after they turn into a gas. Make sure the covers or buildings where these chemicals are applied are airtight and application equipment lines and tanks do not leak.

Mixing and Loading Pesticides

People often get into trouble when mixing and loading pesticides. Some common problems are discussed briefly here.

Tank Mixes

Tank mixes (combinations of two or more pesticides in the spray tank at time of application) may fall into one of two categories.

- Instructions provided for such use on one or more labels of registered products.
- Tank mixes that are recommended by Cooperative Extension or are common agricultural practices.

Tank mixes recommended on labels are obviously consistent with the label.

Tank mixes will not be deemed "use inconsistent with the label" if:

- The products in the mix are applied at a dosage rate not to exceed the label instructions for use of any product in the mix used singly for the same set of pests on the same crop; and
- The label on one or more of the products does not explicitly instruct against such mixture.

The only mixtures proven effective and safe are those specified on product labels. The user applies all other mixtures at his or her own risk with respect to effects on crops and application equipment, applicator safety, environmental effects, and preharvest interval tolerances.

Compatibility

Some pesticides will not mix (are not compatible) with other pesticides or with liquid fertilizers in spray-tank mixtures. For example, wettable sulphur cannot be mixed with Lorsban or Morestan. Some herbicides are not compatible with liquid fertilizers and herbicide oils. Any time you plan to mix two or more pesticides, first make sure they are compatible. Follow the specific directions on the label to test for compatibility if you still have questions. Also remember that a pesticide may mix physically with another pesticide, but the activity of one or both may be altered (based on their chemical or biological incompatibility).

Adjuvants

Adjuvants are inert ingredients that are added to pesticide formulations or tank mixes to increase the effectiveness of the pesticide's active ingredients. Adjuvants may be wetting agents, emulsifiers, spreaders, stickers, penetrants, drift reduction agents, thickeners, buffers, and compatibility agents just to name a few. Adjuvants should not be used unless needed. Often, the pesticide formulation already contains the adjuvants needed for the application.

Many herbicide and other pesticide formulations require that a surfactant, penetrating agent, or other adjuvant be added to the spray tank to increase the pesticide's effectiveness. Read the label to find out what should or should not be added to the pesticide formulation to give you the best possible control.

Formulation Sequence

If you use more than one pesticide formulation (WP, WDG, DF, L, EC) in a spray tank, there is a proper order for adding them.

1. Add a small amount of water or other liquid carrier to the spray tank.
2. Dry materials go into the spray before liquid chemicals. If a wettable powder (WP) is used, put it in first as follows: Make a slurry with the wettable powder by adding a small amount of water to it until it forms a gravy-like consistency. Slowly add this slurry to the tank with the spray tank agitator (mixer) running.
3. Dry flowables (DF) or water-dispersible granules (WDG) go in second. Flowables should be premixed (1 part flowable to 1 part water) and poured slowly into the tank.
4. Liquid flowables (F or L) should be added third. Exception: When using Furadan 4F, this material should be put in last. Liquids should also be premixed (1 part liquid chemical to 2 parts water or liquid fertilizer) before blending in the tank. Many labels will give you the proper pesticide mixing sequence.
5. Emulsifiable concentrates (EC), should be combined last.

Safety Warning

When mixing and loading pesticides, you usually work with concentrated pesticides. Accident reports have shown that the danger of being poisoned may be greatest at this time. Wear the appropriate gloves and other protective clothing to avoid getting a pesticide on you or your clothing.

If a pesticide gets into your eyes, immediately rinse them with plenty of clean water; continue for at least 15 minutes. (The label on some pesticides calls for longer flushing times). If pesticides get on your skin, wash them off with water. Remove contaminated clothing and wash it separately from the family laundry before wearing any of it again. Clothing saturated with a highly hazardous pesticide (labeled DANGER) should be disposed of the same way you would discard the pesticide. Remember: Injuries from most pesticide accidents can be prevented if you know what to do and do it FAST. Take the person to a doctor if you suspect pesticide poisoning.

Cleaning Sprayer Systems

Most people are aware that spraying sensitive crops with a sprayer that has been used earlier to apply certain pesticides can lead to crop damage. Most pesticides can be washed out of sprayers. Dicamba (Banvel), 2,4-D, 2,4-DB, and MCPP are more difficult to wash out, however, and many crops are very sensitive to these herbicides. So it is best to have a separate sprayer to apply these herbicides. Certain crops are very sensitive to sulfonylurea herbicides (Classic, Canopy, Gemini, Glean, Harmony), but these can easily be washed out of sprayers if the proper procedure is used.

Before applying a pesticide with a sprayer that was previously used for some other pesticide, always wash out the sprayer thoroughly. It is best to wash out the sprayer immediately after use. Some pesticide labels give instructions on how to properly clean that pesticide out of the sprayer. If the label does not contain this information, that does not imply that residues of that particular pesticide in a sprayer will not harm other crops.

Pesticide Record-Keeping Requirements

The U.S. Department of Agriculture (USDA), through the 1990 Farm Bill, and the North Carolina Pesticide Board, through the N.C. Pesticide Law of 1971 (NCPL), require that applicators, dealers, and agricultural employers record certain pesticide information. The NCDA&CS, Structural Pest Control and Pesticides Division administers and enforces record keeping provisions. Records must be kept for certain lengths of time and made available to representatives of the NCDA&CS and USDA upon request.

Table 1-6 summarizes the pesticide application records required by federal and state regulations for certain pesticides and for compliance with the Worker Protection Standard (WPS). Readers are encouraged to read the actual laws and regulations for more detailed information. Dealers, certified applicators, licensed aerial and ground applicators, and agricultural employers all have some responsibilities for recording pesticide information.

Agricultural employers who hire pesticide handlers, workers, or both must display application information as required by the WPS. This information must be posted at a central location accessible by employees prior to application and kept for 30 days after the expiration of the restricted-entry interval (REI). Under state law, the time when the application was completed must be recorded. This and all other required record-keeping items must be maintained for a period of two years after the REI expires.

Pesticide dealers in North Carolina are required to keep sales records for all restricted-use pesticides. The 10 elements that are required for each restricted-use pesticide sale are:

1. Date of sale;
2. Initials of sales clerk;
3. Name of certified or licensed applicator;
4. Certification or license number from card;
5. Expiration date as shown on the card;
6. Product brand name;
7. EPA registration number;
8. Number of individual containers;
9. Size of individual containers;
10. Total quantity sold.

Table 1-5. Summary of Pesticide Record-Keeping Requirements for Growers and Applicators

Required Items	USDA Requirements for Private & Commercial Applicators (Restricted Use Pesticides)	NCPL[1] Requirements for Commercial Applicators & Public Operators (Restricted Use Pesticides)	NCPL[1] Requirements for Aerial Applicators (All Pesticides)	Federal & State Requirements for Agricultural Employers (WPS*) (Agricultural Use Pesticides)
Brand name/product name	✓	✓	✓	✓
EPA registration number	✓	✓	✓	✓
Total amount of pesticide used	✓	✓ amount/unit of measure (e.g. acre)	✓ amount of formulated product or active ingredient/acre PLUS amount of tank mix/acre	—
Date of application	✓	✓	✓	✓
Time of application		✓ Time completed	✓ Time completed	✓ Start time and completed
Description/location of treated area	✓	✓	✓	✓
Crop, commodity, or stored product	✓	✓	✓	✓
Size of area treated	✓	✓	✓	—
Name and address of property owner or operator	—	✓	✓	—
Name of applicator	✓	✓	✓	—
Name of licensee	✓ or name of supervisor	✓	✓ name of contractor and signature of record keeper	—
Certification number	✓	—	—	—
Active ingredients	—	—	—	✓
Restricted entry interval	—	—	—	✓
Record must be:	completed within 14 days of application and kept 2 years (commercial applicator, only, must furnish records to customer within 30 days)	kept 3 years	completed within 72 hours after application and kept 3 years	displayed, along with SDS, within 24 hours after end of application, posted for 30 days after the Restricted-Entry Interval expires and kept for 2 years after the REI expires.
		For these records, each day of application must be recorded as a separate application.		

*The federal Worker Protection Standard has been adopted by reference by the N.C. Pesticide Board. This standard requires that pesticide information be posted in a central location on an agricultural establishment.

[1] NCPL = North Carolina Pesticide Law

Source: North Carolina Department of Agriculture and Consumer Services

Record-keeping forms for the USDA restricted-use pesticide regulation and the Worker Protection Standard are available online at http:// pesticidesafety.ces.ncsu.edu/.

II — CHEMICAL APPLICATION EQUIPMENT

- Introduction 22
- Types of Equipment 22
  - Dusters 22
  - Granular Applicators 22
  - Broadcast Spreaders 22
  - Sprayers 23
  - Backpack Sprayers 24
  - Wick Applicators 24
  - GNSS and Variable Rate Technology 24
- Cleaning Equipment 25
- Calibrating Chemical Application Equipment 25
- Purpose 25
- Getting Started 25
- Calibrating a Sprayer 25
  - Preparing to Calibrate 25
- Calibration Methods 26
  - Basic Method 26
  - Nozzle Method 27
  - 128th Acre Method 27
  - Area Method 27
- Calibrating a Granular Applicator 27
  - Preparing to Calibrate 27
  - Calibration Methods 28
- Calibrating a Broadcast Spreader 28
  - Calibration Methods 28
- Calibration Variables 29
  - Speed 29
  - Pressure 29
  - Density 29
  - Band Application Versus Broadcast Application 29
  - Determining Upper and Lower Limits 29
- Useful Tables and Data 30

Chemical Application Equipment

G. T. Roberson, Extension Specialist, Biological and Agricultural Engineering

Introduction

The objective of a chemical application program can be summed up with a single sentence: *Put the right product on the right target at the right time.* To accomplish this objective, good management practices and proper equipment are necessary. In selecting agricultural chemical application equipment, follow these guiding principles:

1. Considering acreage, crops, and labor, select the equipment that will best fit your farming operation.
2. For medium to small farms, select smaller units. Large equipment is impractical on small farms with fields of irregular shape, leading to frequent breakdown and poor application.
3. If feasible, use separate spray equipment for certain types of weed-control chemicals to prevent accidental contamination and crop damage.
4. Remember that some chemical application equipment is designed for specific applications, and some components are not suitable for all jobs. For example, some spray nozzles recommended for insecticide application may not be suitable for herbicides.
5. Select a machine that is convenient to set up, hitch, and operate. Convenient features can help a busy operator complete an application on time.
6. Keep in mind that high-clearance, self-propelled sprayers often cannot be justified on small farms; consider hiring a custom operator if this type of machine is needed.
7. Use the proper pressure for spraying; avoid excessively high pressures. Check the chemical manufacturer's label for recommendations. You should also check the nozzle manufacturer's charts to ensure you have the correct droplet size for the pressure and application.
8. Check all adjustments for any type of application equipment carefully. Refer to your operator's manual for recommended settings. **Remember to CALIBRATE!**
9. Always wear the proper protective clothing as indicated on the chemical label and use all other recommended personal protective devices or practices when handling or using agricultural chemicals.

Types of Equipment

Dusters

Dusters are no longer used for field crops. Sprayers and granular applicators are generally preferred. However, small hand-held dusters and backpack dusters are still quite popular for use in gardens and around the home.

Small dusters for home or garden use are typically hand-held and hand-powered. Large dusters can be backpack units powered by small gasoline engines or battery-powered motors. They can hold from a few ounces to several pounds of material. The fan or air pump generates the necessary airflow to move the dust through the discharge tube to the nozzle. Aim the nozzle at the proper target and watch the application rate closely. It is easy to apply too much material in one small area, which can be hazardous.

Granular Applicators

Granular applicators are used for a wide range of jobs, from applying small volumes of pesticides at planting to applying fertilizer or lime using band or broadcast methods.

Granular applicators typically consist of a hopper, a metering system, drop tubes, and a diffuser. The metering system can work one of two ways: gravity flow or positive displacement. In the gravity flow type, the size of an orifice or opening is adjusted to control how the product flows out of the hopper. An agitator in the bottom of the hopper ensures smooth delivery. In the positive displacement type, the metering unit delivers a fixed volume of material for every revolution it makes. The drive mechanism is often connected to a ground wheel, so material is delivered whenever the wheel turns and the orifice or gate is open. Some units are driven by PTO, hydraulic drives, or small electric motors rather than ground drives. Positive displacement systems are generally more accurate than gravity flow units.

The metering device must function properly. Clean it regularly to remove caked material or other obstructions. Check the condition of components and replace them if they appear worn. Badly worn components may result in large application errors. Check the tubes and diffusers for leaks or blockage. Drop tubes or hoses should allow the material to fall freely without collecting in the tube. Keep the tubes as nearly vertical as possible to minimize blockage.

Broadcast Spreaders

Broadcast spreaders are widely used to apply fertilizer, lime, or amendments on lawns, gardens, or fields. These machines may be small hand-held or cart-mounted units for home or garden use, or they may be three-point hitch, trailer-mounted, or truck-mounted units for field use. Spreaders for home or garden use can be simple drop spreaders with a series of holes along the underside of the hopper to meter and spread the material, or they may be spinner spreaders, which use one or two rotating disks or spinners to spread the product across the swath. Larger field spreaders can be single-spinner, twin-spinner, or air-boom designs.

A large spinner spreader typically consists of a hopper, a drag chain or belt, a discharge gate, a chute, and one or two spinners. Some smaller units do not use the drag chain but rely on gravity flow through an orifice and an agitator to achieve the desired discharge rate. Larger broadcast spreaders can have a drag chain or belt driven by a ground wheel, hydraulic drive, or by PTO.

Spinner spreaders should produce a pattern that is heavy in the center and tapers to the edges. Desirable patterns for a spinner spreader are the triangle, oval, and flat top patterns shown in Figure 2-1. Proper spacing of the swaths in the field is critical to apply product correctly. Swath spacing should be measured as the width across the pattern to the points where each side achieves 50% of the intended application rate (Figure 2-1). This allows for overlapping the edges of the pattern so that a uniform rate is achieved across the field. If the swath spacing is too wide, some areas will not receive enough product between the passes of the spreader. If the swath spacing is too close, some areas will receive too much product. Be sure to check the spread pattern and maintain proper swath spacing in the field. The effects of improper swath spacing or a bad spread pattern for products such as fertilizer will often show up in the field as stripped or uneven crop development. It is very important to recognize the difference between the width of the spread pattern and the swath width. Figure 2-1 illustrates the spread pattern width is the maximum width the product is distributed, where swath width is the spacing of the applicator's passes through the field.

Air boom spreaders use a high volume air stream to suspend the product particles and convey them through tubes to diffusers spaced along the boom. The product is metered into an air chamber where the air stream catches the material and divides it into the tubes running to the diffusers. The product is uniformly distributed along the width of the boom with a very slight taper on the outside edges. As with the spinner spreader, proper swath spacing is critical. Because the air boom pattern has little taper, precise swath spacing must be maintained.

Smooth delivery of material is important. Check the discharge mechanism for blockage or wear. Check the drive mechanism to make sure it is functioning properly. Slipping wheels, worn belts, and worn chains can seriously affect performance and should be repaired. Check the spinners for holes in the bottom or in the vanes. Check for caked material on the vanes as well. Pay close attention to the speed of the spinner; excessively high or low speeds can cause improper application patterns. On air-boom spreaders, be sure to check the air chamber and tubes for blockages and leaks. Refer to your operator's manual for correct settings and adjustments on all machines.

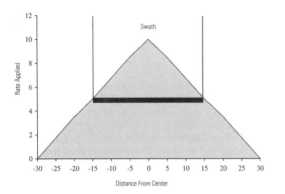

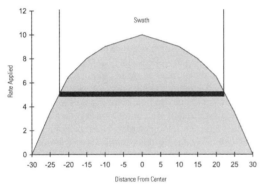

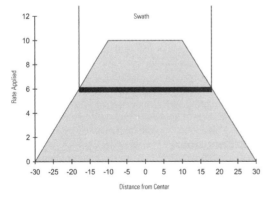

Figure 2-1. Desirable patterns for broadcast spinner spreader are the triangle, oval, and flat top, as shown above.

Sprayers

Most field sprayers in use today are hydraulic, which means the spray pressure is built up by the action of the pump on the spray mixture. Don't confuse the use of the term hydraulic here with the tractor hydraulic system. Sprayers range from low- to high-pressure and may be mounted, pull-type, or self-propelled models. There are also electrostatic and air-directed sprayers. Sprayer designs also include boom and boomless units to match a wide range of applications.

The basic components are the tank, pump, agitator, hoses, valves, fittings, and nozzles. Some modern units also incorporate electronic control systems to enhance accuracy and performance. When purchasing a new sprayer or when replacing components, choose carefully to ensure quality.

Two types of sprayer designs are currently available: tank mix and injection. The tank mix design consists of a large tank in which water and chemicals are mixed. Tank mix sprayers are less expensive and easy to operate. You should pay close attention to compatible tank mixes and the order of mixing for some products. Injection sprayers have a large tank that contains only water. An additional tank or tanks are used to hold the chemical. For products that are difficult to clean, you may want to have dedicated chemical tanks. While more costly to purchase, injection sprayers have the advantage of reduced risk since you never mix more than needed. The water tank is never contaminated with chemicals, so cleanup is simpler. Electronic controls are used to measure the flow of water and inject the proper quantity of chemical into the water stream before it is delivered to the nozzles. This design reduces contamination potential and eliminates the need for tank agitation.

Tanks are usually made of stainless steel, aluminum, fiberglass, or plastic and should include a large splash-proof filler. They should be resistant to rust and all chemicals placed in them. Check with your chemical supplier regarding compatibility.

A suction line connection and a drain should be connected at the bottom of the tank. Bottom suction makes the pump self-priming. When possible, use only one tank for the sprayer; multiple tanks are more difficult to agitate properly. The injection sprayer can use multiple tanks without problem since agitation is not required.

Pumps are available in brass, aluminum, cast iron, and corrosion-resistant alloys (Table 2-1, at end of chapter). Select a pump material that will not be corroded by the chemical you are using. The pump seal material (for example, viton or buna-N) should be compatible with the chemical.

Pump capacity is measured in gallons per minute (GPM) or gallons per hour (GPH). Select a pump with a capacity at least 50 percent larger than the sum of the nozzle outputs and the agitator. Oversizing the pump compensates for wear and keeps the system running longer. Pumps are usually attached directly to the tractor PTO. Secure the pump with an anchor bar and length of chain, leaving some slack rather than bolting the pump down rigidly. They should be allowed to "float" as they rotate, but should not be so loose that they twist the hoses. Secure the chain to a fixed position on the tractor, not one of the lift arms. Hydraulically driven pumps can be mounted on the sprayer frame. Characteristics of various types of pumps are listed in Table 2-1.

Jet agitators are usually used in sprayer tanks for tank mix sprayer designs. Place jet agitators in the bottom of the tank and connect to the high-pressure side of the control valve. This arrangement allows full flow to the agitator as well as full pressure control for the nozzles. Check with your sprayer supplier for proper size of agitator nozzles and flow rates. The return line from the pressure regulator is not adequate for agitation. You cannot ensure a consistent flow rate in the return line to get good agitation.

Every field sprayer should have a sprayer control unit. The control can be manual or electronic. The control should allow the operator to select the section of the boom or nozzles to use and regulate pressure. The control should also have a high accuracy pressure gauge. Measuring and controlling pressure accurately is essential for most sprayer designs. Never shortchange the quality of the pressure gauge.

Electronic control units for sprayers can enhance performance by improving application accuracy. These controls may include pressure sensors, flow rate sensors, or a combination of the two. These controls rely on solenoid or ball valves to control flow to the system and a throttling valve or bypass valve to regulate system pressure. These valves must be checked periodically to ensure proper performance. Some controls also include some type of ground speed sensor, such as GNSS (Global Navigation Satellite System), radar or wheel sensors. By maintaining accurate ground speed, the sprayer can compensate for changes in speed and maintain better accuracy.

Electronic or automatic controls can improve accuracy and performance of a sprayer; however, they are **no substitute for proper calibration.**

Hoses should be oil resistant and durable. They should have a test pressure twice the sprayer operating pressure. Suction hoses should be strong enough to prevent collapsing. The hoses should be large enough for proper flow while keeping pressure loss to a minimum (Table 2-2, at end of chapter). If a hose is too small, the pressure at the nozzles will be much less than that measured at the control valve, causing under-application. Suction hoses should be at least as large as the pump's intake port.

Sprayers should have a suction strainer to protect the pump. Surface area for a suction strainer should be 2 square inches per GPM. Mesh size is determined by the type of chemical used. Use strainers to protect valves and nozzles.

Strainers used on the pressure side should have an area of at least 1 square inch per GPM.

Nozzles are perhaps the most important part of the sprayer. Unfortunately, no one nozzle can cover every type of application. Some chemical labels will specify a particular nozzle for use in application. If it is on the label, be sure to use the nozzle specified. Nozzles vary according to capacity (gallons per minute or GPM), spray pattern angle, shape of spray pattern, and droplet size produced (Table 2-3, at end of chapter). Nozzles are usually equipped with strainers to protect them from abrasive particles. Even with protection, nozzles will wear. Check nozzles regularly and replace them if wear is observed.

Electrostatic sprayers offer better plant coverage and more efficient use of chemicals. Electrostatic sprayers use a high-voltage electrical charge to create an electrostatic field between the plant and the spray droplet. The plant and the spray droplet have opposite charges causing them to attract each other. The droplets cling to any exposed plant surface, top or bottom. More uniform coverage and potential for reduced spray volume make electrostatic spraying a cost-effective method of chemical application.

Air-directed sprayers fall into three groups: airblast, air boom, and air nozzle. Airblast sprayers have been widely used for years in vineyards and orchards. A nozzle or nozzle cluster sprays into the discharge of a high-volume fan. The airflow spreads the droplets onto the crop. Air boom sprayers are similar to conventional hydraulic sprayers. In addition to the sprayer boom and nozzles, an air tube is added with air ports or nozzles over the sprayer nozzles. The air blast from the air tube blows the spray down into the crop canopy, improving penetration and reducing drift. Air nozzles fall into two categories: pressure and induction. Air pressure nozzles use compressed air to atomize liquid and create the spray pattern rather than water pressure. Because droplet size and pattern are controlled by air pressure, there is no need to have different nozzles for each application rate. Air induction nozzles draw air into the spray stream to control drop formation. The drops formed are larger air-filled drops that are less likely to drift.

Drift control is an important consideration for any type of sprayer. Drift can be caused by several factors:

- Wind at or near the ground surface.
- High nozzle pressures, which in turn produce small droplet sizes.
- Ground speed of the sprayer can contribute to drift by creating added air turbulence around the boom.
- Evaporation of the liquid on hot days (vapor drift).
- Temperature inversions can contribute to drift by trapping spray particles which are then carried off when the inversion breaks down.

Often a combination of one or more of these factors is present. Drift contaminates other crops or the surrounding woods and streams. Electrostatic, air boom, and air nozzle sprayers all have drift-reducing potential. Conventional sprayers can also take advantage of drift reducing technology by incorporating spray shields, drift-reducing nozzles, or drift-reducing adjuvants (where allowed by the chemical label).

Backpack Sprayers

Backpack sprayers consist of a tank, a pump, and a spray wand with one or more nozzles. Some sprayers have a pressure-regulating valve or a pressure gauge to help the user maintain desired pressure. Small size, portability, and ease of use make the backpack sprayer a valuable tool for many users. Backpack sprayers are best suited for small acreage, spot spraying, hard-to-reach areas, and other areas where a larger sprayer is impractical.

Most backpack sprayers use hand pumps; however, some units have a small battery or engine-powered pumping system. Regardless, consistent and uniform pressure control is essential to ensure proper application.

Hand-operated sprayers should have a comfortably located, reversible handle (to allow for left- or right-hand use). The shoulder straps should distribute the load evenly across the shoulders. Consider a hip belt to help carry the weight of the larger units. The wand should be comfortable and allow for easy use of the trigger. The sprayer should also have removable screens to protect the pump and nozzles. These should be cleaned regularly. Finally, the sprayer should have a stable base to hold it upright for filling and mixing.

A portable containment and mixing system for backpack sprayers has been developed for use in pickup trucks and utility vehicles. This system secures the backpack to prevent tipping over and provides a means to shake the unit for mixing. Further, hand pumps and meters are provided to allow accurate mixing of doses for the sprayer.

Wick Applicators

A convenient method to apply contact herbicides is the wick applicator. In this design, a wick (rope, sponge, etc.) is soaked with a solution. Herbicide is applied to the plant target by brushing against the plant. The application rate is controlled by adjusting the chemical solution, the rate it is soaked by the wick, or the speed of travel over the target. Ropes are commonly used as wicks, which explains the name "rope wick applicator." Wick applicators offer the ability to apply chemical only to a selected target, e.g., weeds and grasses taller than the crop. Reduced chemical usage compared to broadcast spraying is possible.

GNSS and Variable Rate Technology

Improved accuracy, site-specific application rates, and higher profitability are among the advantages offered by Global Navigation Satellite Systems (GNSS) and variable rate applicators. GNSS refers to all satellite navigation systems world-wide, which also includes the Global Positioning System (GPS) maintained by the U.S. Department of Defense. GPS is a system of navigational satellites established by the military and made available for private use. GNSS signals alone are not accurate enough to guide or control field equipment. A correction signal or real-time correction source must be used to achieve the necessary accuracy. Depending on your location, you may wish to subscribe to a commercial correction signal service or use the free correction signal, Wide Area Augmentation System (WAAS) provided by the FAA or the free correction provided by some commercial services. Real Time Kinematic (RTK) correction is the most advanced correction service available and offers accuracy to within an inch. RTK correction can be achieved by operating your own base station, using a dealer base station network, or using the North Carolina Real Time Network (NCRTN).

To use a variable rate system, a machine such as a tractor, spreader truck, sprayer, or aircraft is equipped with a DGPS or RTK receiver. The operator uses the receiver and an on board computer to guide the equipment across the field. The ability to establish and maintain uniformly spaced swaths is an advantage with GNSS guidance systems. Once the swaths are established, a light bar, screen display, or automatic steering system is used to help the equipment operator stay on course. With the ability to determine position in a field accurately, the grower can take advantage of variable rate control systems and soil, weed, or yield maps prepared for the field. Information from these maps is used to determine the proper amount of material—fertilizer, lime, or pesticides—to apply to each area within the field. Therefore, only the areas that need or will benefit from higher application rates receive those rates, while other areas receive more moderate rates.

Control systems can be selected to manage single or multiple products. A single product (also called single channel) controller can be used for granular applicators, broadcast spreaders, or sprayers. Single channel controls for sprayers are used on tank mix sprayer systems. Multiple channel controllers can also be used on broadcast spreaders and sprayers. On broadcast spreaders, multiple channel controllers are used to control rates on multiple hopper spreaders to blend products on the go in the field as called for by prescription maps, crop sensors, or commands from the spreader operator. On sprayers, multiple channel controls can be used to inject pesticides into a constant stream of water before the spray is applied through the nozzles. This allows variable rates of pesticides to be applied as indicted by prescription maps, crop sensors, or operator commands. Another advantage of the computerized variable rate control systems is the ability to record a map of the product as it is applied to the field. Application record maps can be valuable assets in a complete management plan.

Cleaning Equipment

Sprayers, granular applicators, broadcast spreaders, and other application systems should be properly cleaned after use. Some chemicals are corrosive and may damage the components if left in the equipment. Other materials may contaminate the next application. Make sure to choose a cleaning agent recommended for the type of product you are using.

Rinsate in sprays can often be applied over the crop if it has been diluted properly. Check with your chemical supplier or local Cooperative Extension center for suggestions. Most sprayers can be equipped with an in-field rinse system for a few hundred dollars. The system consists of a rinse tank, control valves, and hoses to connect it to the sprayer circuit. A rinse nozzle must be installed inside the tank.

Wash the equipment on an approved wash pad or site. Protect components from rust and corrosion after washing with a light oil or other suitable material.

Calibrating Chemical Application Equipment

Purpose

To determine if the proper amount of chemical is being applied, the operator must measure the output of the application equipment, compare it to the intended output, and adjust the application accordingly to correct errors. This technique is known as *calibration*.

Calibration not only ensures accuracy, a critical factor with regard to many chemicals, but it can also save time and money and benefit the environment.

Getting Started

Careful and accurate control of ground speed is important for any type of chemical application procedure. From large self-propelled sprayers and spreaders to small, walk-behind or backpack units, precise ground speed is a key for success. Ground speed can be determined by one of two methods. The first method requires a test course and stopwatch. For this procedure, measure a suitable test course in the field and record the time it takes to cover the course with the equipment. The course should be between 100 and 300 feet long. Drive or walk the course at least twice (once in each direction) and average the times for greater accuracy. Calculate the speed with Equation 1 or refer to Table 2-6 at the end of the chapter.

$$\text{Equation 1. Ground Speed (MPH)} = \frac{\text{Distance} \cdot 60}{\text{Seconds} \times 88}$$

The second method is to use a true ground speed indicator such as a tractor-mounted radar or GNSS receiver. Do not rely on transmission speed charts and engine tachometers. They are not accurate enough for calibration.

Calibrating a Sprayer

Preparing to Calibrate

For calibration to be successful, several items need to be taken care of before going to the field. Calibration will not be worthwhile if the equipment is not properly prepared. Whenever possible, calibration should be performed using water only. If you must calibrate using spray mixture, do so on a site listed on the chemical label and with wind speeds less than 5 MPH. Follow the steps outlined below to prepare spraying equipment for calibration.

1. Inspect the sprayer. Be sure all components are in good working order and undamaged. On backpack sprayers, pay particular attention to the pump, control wand, strainers, and hoses. On boom sprayers, pay attention to the pump, control valves, strainers, and hoses. On airblast sprayers, be sure to inspect the fan and air tubes or deflectors as well. Be sure there are no obstructions or leaks in the sprayer.

2. Check the label of the product or products to be applied and record the following:
   - *Application Rate*, gallons per acre (GPA)
   - *Nozzle Type*, droplet size and shape of pattern
   - *Nozzle Pressure*, pounds per square inch (PSI)
   - *Type of Application*, broadcast, band, or directed

3. Next, determine some information about the sprayer and how it is to be operated. This includes:
   - *Type of Sprayer:* backpack, boom, or airblast. The type of sprayer may suggest the type of calibration procedure to use.
   - *Nozzle Spacing (inches):* For broadcast applications, nozzle spacing is the distance between nozzles.

- *Nozzle Spray Width (inches):* For broadcast applications, nozzle spray width is the same as nozzle spacing—the distance between nozzles. For band applications, use the width of the sprayed band if the treated area in the band is specified on the chemical label; use nozzle spacing if the total area is specified. For directed spray applications, use the row spacing divided by the number of nozzles per row. Some directed spray applications use more than one type or size of nozzle per row. In this case, the nozzles on each row are added together and treated as one. Spray width would be the row spacing.

- In most cases, a backpack sprayer uses a single nozzle. Some sprayers use mini-booms or multiple nozzles. The spray width is the effective width of the area sprayed, being sure to account for overlap. If you are using a sweeping motion from side to side, be sure to use the full width sprayed as you walk forward. If you are spraying on foliage in a row, use the row spacing. Dyes are available to blend with the spray to show what has been covered.

- *Spray Swath (feet):* The width covered by all the nozzles on the boom of a sprayer. For airblast or other boomless sprayers, it is the effective width covered in one pass through the field.

- *Ground Speed, miles per hour (MPH).* When using a backpack sprayer, walk a comfortable pace that is easy to maintain. Slow walking speeds will take longer to complete the task, while high speeds may be tiresome. Choose a safe, comfortable speed that will enable you to finish the job in a timely manner. On tractor-mounted sprayers, select a ground speed appropriate for the crop and type of sprayer used. Slow speeds will take longer to complete the task, while high speeds may be difficult to control and unsafe. Choose a safe, controllable speed that will enable you to finish the job in a timely manner. Ground speed can be determined from Equation 1 (page 23) or Table 2-6.

4. The *discharge rate, gallons per minute (GPM),* required for the nozzles must be calculated in order to choose the right nozzle size. Discharge rate depends on the application rate; ground speed; and nozzle spacing, spray width, or spray swath.

For applications using nozzle spacing or nozzle spray width (inches), use Equation 2 or Table 2-9 at the end of the chapter.

Equation 2. Discharge Rate =
$$\frac{Application\ Rate \times Ground\ Speed \times Nozzle\ Spray\ Width}{5{,}940}$$

For applications using the spray swath (feet):

Equation 3. Discharge Rate =
$$\frac{Application\ Rate \times Ground\ Speed \times Spray\ Swath}{495}$$

5. Choose an appropriate nozzle or nozzles from the manufacturer's charts and install on the sprayer. Check each nozzle to be sure it is clean and that the proper strainer is installed with it. Be sure to check the droplet size recommendation on the chemical label when you choose your nozzle. Spray nozzles are classified by the droplet size produced at the selected operating pressure as well as flow rate. *Do not confuse the flow rate color codes with the droplet size color codes.*

6. Fill the tank half full of water and adjust the nozzle pressure to the recommended setting. Measure the discharge rate for the nozzle. This can be done by using a flow meter or by using a collection cup and stopwatch. The flow meter should read in gallons per minute (GPM). If you are using the collection cup and stopwatch method, the following equation is helpful to convert ounces collected and collection time, in seconds, into gallons per minute.

Equation 4. Discharge Rate =
$$\frac{Ounces\ Collected \times 60}{Collection\ Time \times 128}$$

7. Whenever possible, calibrate with water instead of spray solution. Do not calibrate with spray solution unless required by the chemical label. Follow all recommendations on the label. If the spray solution has a density different than water, the rate can be corrected using the procedure shown in Calibration Variables.

8. On boom sprayers or sprayers with multiple nozzles, average the discharge rates of all the nozzles on the sprayer. Reject any nozzle that has a bad pattern or that has a discharge rate 10 percent more or less than the overall average. Install a new nozzle to replace the rejected one and measure its output. Calculate a new average and recheck the nozzles compared to the new average. Again, reject any nozzle that is 10 percent more or less than the average or has a bad pattern. When finished, select a nozzle that is closest to the average to use later as your quick check nozzle.

9. On backpack sprayers or sprayers with a single nozzle, compare the discharge rate of the nozzle on the sprayer to the manufacturer's tables for that nozzle size. Reject any nozzle that has a bad pattern or that has a discharge rate 10 percent more or less than the advertised rate. Install a new nozzle to replace the rejected one and measure its output.

Once the sprayer has been properly prepared for calibration, select a calibration method. When calibrating a sprayer, changes are often necessary to achieve the application rates needed. The sprayer operator needs to understand the changes that can be made to adjust the rate and the limits of each adjustment. The adjustments and the recommended approach are:

- *Pressure:* if the error in application rate is less than 10 percent, adjust the pressure.
- *Ground speed:* if the error is greater than 10 percent but less than 25 percent, change the ground speed of the sprayer.
- *Nozzle size:* if the error is greater than 25 percent, change nozzle size.

The goal is to have application rate errors less than 5 percent.

Calibration Methods

There are four methods commonly used to calibrate a sprayer:

- basic
- nozzle
- 128th acre
- area

The basic, nozzle, and 128th acre methods are time-based methods that require using a stopwatch or watch with a second hand to ensure accuracy. The area method is based on spraying a test course measured in the field. Each method offers certain advantages. Some are easier to use with certain types of sprayers. For example, the basic and area methods can be used with any type of sprayer. The 128th acre and nozzle methods work well for boom and backpack sprayers. Choose a method you are comfortable with and use it whenever calibration is needed.

Basic Method

1. Accurate ground speed is very important to good calibration with the basic method. For tractor-mounted sprayers, set the tractor for the desired ground speed and run the course at least twice. For backpack sprayers, walk the course and measure the time required. Walk across the course at least twice. Average the times required for the course distance

and determine ground speed from Equation 1 or Table 2-6 at the end of the chapter.

2. Calculate the application rate based on the average discharge rate measured for the nozzles, the ground speed over the test course, and the nozzle spacing, nozzle spray width, or spray swath on the sprayer.

When using nozzle spacing or nozzle spray width measured in inches, use the following equation:

Equation 5. Application Rate =
$$\frac{5{,}940 \times Discharge\ Rate}{Ground\ Speed \times Nozzle\ Spray\ Width}$$

For spray swath applications measured in feet:

Equation 6. Application Rate =
$$\frac{495 \times Discharge\ Rate}{Ground\ Speed \times Spray\ Swath}$$

3. Compare the application rate calculated to the rate required. If the rates are not the same, choose the appropriate adjustment and reset the sprayer.
4. Recheck the system if necessary. Once you have the accuracy you want, calibration is complete.

Nozzle Method

1. Accurate ground speed is very important to good calibration with the nozzle method. For tractor-mounted sprayers, set the tractor for the desired ground speed and run the course at least twice. For backpack sprayers, walk the course and measure the time required. Walk across the course at least twice. Average the times required for the course distance and determine ground speed from Equation 1 or Table 2-6 at the end of the chapter.
2. Calculate the nozzle discharge rate based on the application rate required, the ground speed over the test course, and the nozzle spacing, spray width, or spray swath of the sprayer. For nozzle spacing or spray width measured in inches use the following equations (or refer to Table 2-9 at end of chapter):

Equation 7. Discharge Rate =
$$\frac{Application\ Rate \times Speed \times Spray\ Width}{5{,}940}$$

For spray swath measured in feet:

Equation 8. Discharge Rate =
$$\frac{Application\ Rate \times Speed \times Spray\ Width}{495}$$

Set the sprayer and determine the average nozzle rate.

3. Compare the rate calculated to the average rate from the nozzles. If the two don't match, choose the appropriate adjustment and reset the system.
4. Recheck the system if necessary. Once you have the accuracy you want, calibration is complete.

128th Acre Method

1. The distance for one nozzle to cover 128th of an acre must be calculated. The nozzle spacing or spray width in inches is used to determine the spray distance. Spray distance is measured in feet. On backpack sprayers, be sure to measure the full width sprayed as you walk forward. Use Equation 9 or Table 2-14 at the end of the chapter.

Equation 9. Spray Distance =
$$\frac{4{,}084}{Spray\ Width}$$

2. Measure the spray distance on a test course in the field. Check the ground speed as you travel across the course. Be sure to maintain an accurate and consistent speed. Travel the course at least twice and average the time to cover the course.
3. For backpack sprayers, collect the output from the nozzle for the time measured in step 2. For tractor-mounted sprayers, park the sprayer, select the nozzle closest to the average, and collect the output for the time determined in step 4. Ounces collected will equal application rate in GPA.
4. Compare the application rate measured for the nozzle to the rate determined in step 3. If the rates are not the same, choose the appropriate adjustment and reset the system.
5. Recheck the system if necessary. Once you have the accuracy you want, calibration is complete.

Area Method

1. Determine the distance that can be sprayed by one tank using the full spray swath measured in feet.

Equation 10. Tank Spray Distance (ft) =
$$\frac{Tank\ Volume\ (gal) \times 43{,}560}{Application\ Rate\ (GPA) \times Swath\ (ft)}$$

2. Lay out a test course that is at least 10 percent of the tank spray distance from Step 1. Fill the sprayer tank with water only, mark the level in the tank, set the sprayer as recommended, and spray the water out on the course. Be sure to maintain an accurate and consistent speed.
3. After spraying the test course, carefully measure the volume of water required to refill the tank to the original level. Calculate the application rate as shown:

Equation 11. Application Rate (GPA) =
$$\frac{Volume\ Sprayed\ (gal) \times 43{,}560}{Test\ Course\ Distance\ (ft) \times Swath\ (ft)}$$

4. Compare the application rate measured to the rate required. If the rates are not the same, choose the appropriate adjustment method and reset the sprayer.
5. Recheck the system. Once you have the accuracy you want, calibration is complete.

Calibrating a Granular Applicator

Preparing to Calibrate

Granular application calibration is usually done with the chemical to be applied. It is difficult to find a blank material that matches the granular product. Extra care should be taken in handling this product. Minimize worker exposure and take precautions against spills during calibration.

To prepare for calibration, follow these steps:

1. Before calibrating, carefully inspect the equipment to ensure that all components are in proper working order. Check the hopper, the metering rotor, the orifice, and the drop tubes. Be sure there are no leaks or obstructions.
2. Determine the type of application required for the product:
   - Broadcast: treats the entire area (includes band applications based on broadcast rates).
   - Band: treats only the area under the band.
   - Row: treats along the length of the row.
3. Determine the application rate needed:
   - Broadcast: pounds per acre.
   - Band: pounds per acre of treated band width.
   - Row: pounds per acre or pounds per 1,000 feet of row length.
4. What type of drive system does the applicator use?
   - Independent: uses PTO, hydraulic, or electric motor drive.
   - Ground Drive: uses ground driven wheel.

5. Regardless of how the application rate is expressed or type of application, calibration is easier if the rate is expressed in terms of pounds per foot of row length. Use one of the following steps to determine the correct row rate in pounds per foot.

For broadcast and row applications (Application Rate = lb/ac):

Equation 12. Row Rate, lb/ft =
$$\frac{Application\ Rate \times Row\ Width\ (ft)}{43{,}560}$$

For banded applications (Application Rate = lb/ac of Treated Band):

Equation 13. Row Rate, lb/ft =
$$\frac{Application\ Rate \times Band\ Width\ (ft)}{43{,}560}$$

For directed (row) applications (Application Rate = lb per 1,000 ft):

Equation 14. Row Rate, lb/ft =
$$\frac{Application\ Rate}{1{,}000}$$

6. Choose a calibration distance to work with and measure a test course of this distance in the field you will be working in. Choose an area that is representative of field conditions. The calibration distance should be at least 50 feet but not more than 500 feet. Longer distances are generally more accurate.

7. Calculate the weight of material that should be collected for the calibration distance chosen.

Equation 15. Weight Collected =
Row Rate x Calibration Distance

8. Select a ground speed appropriate for the crop and type of equipment used. Slow speeds take longer to finish the task, while high speeds may be inefficient and unsafe. Consult your equipment manual for a recommended speed. Even ground-driven application equipment can be sensitive to changes in speed. Maintaining an accurate and consistent speed is very important. Choose a safe, controllable speed that will enable you to complete the job in a timely and efficient manner.

9. Set your equipment according to recommendations from the equipment or chemical manufacturer. Most equipment manufacturers and chemical manufacturers provide rate charts to determine the correct orifice setting or rotor speed for each applicator. Fill the hopper at least half full to represent average capacity for calibration.

10. Attach a suitable collection container to each outlet on the applicator. You should be able to collect all material discharged from the applicator. Locate a scale capable of weighing the samples collected in calibration. Some samples may be very small, so a low-capacity scale may be needed. An accurate scale for measurement is very important.

Calibration Methods

Two methods for calibrating granular applicators are commonly used. The first is the distance method. This method is preferred by many operators because it applies to any type of granular machine and is easy to perform. The second method is the time method. This method is similar to sprayer calibration and can be used for applicators driven by PTO, hydraulic, or electric motors.

Distance Method

1. On the test course selected in the field, collect the output from the applicator in a container as you travel the course and weigh the material collected. Record the time required to travel the course also. Run the course twice, once in each direction, and average the results for both weight and time.

2. Determine the weight of the product that should be collected for the calibration distance.

Equation 16. Weight Collected (lb) =
Row Rate (lb/ft) $\times$ Calibration Distance (ft)

Time Method

1. On the test course selected in the field, record the time required to travel the course. Run the course twice, once in each direction, and average the results. Accurate ground speed is very important to good calibration with the time method.

2. With the equipment parked, set the orifice control as recommended and run the applicator for the time measured to run the calibration distance. Collect and weigh the output of the applicator for this time measurement.

3. Determine the weight of the product that should be collected for the calibration distance.

Equation 17. Weight Collected (lb) =
Row Rate (lb/ft) $\times$ Calibration Distance (ft)

4. Compare the weight of the product actually collected during the time it took to cover the calibration distance to the weight expected for the calibration distance. If the rates differ by more than 10 percent, adjust the orifice, rotor speed, or ground speed and repeat. Bear in mind, speed adjustments are not effective for ground-driven equipment.

5. Repeat the procedure until the error is less than 10 percent.

Calibrating a Broadcast Spreader

1. Carefully inspect all machine components. Repair or replace any elements that are not in good working order.

2. Determine the type of drive system that is being used: ground drive or independent PTO. This may help determine the method of calibration.

3. Determine the application rate and the bulk density of the product to be applied.

4. Determine the spreader pattern and swath of the spreader. Check the pattern to ensure uniformity. To check the pattern, place collection pans across the path of the spreader. For drop spreaders, be sure to place a pan under each outlet. For centrifugal and pendulum spreaders, space the pans uniformly with one in the center and an equal number on each side. The pattern should be the same on each side of the center and should taper smoothly as you go to the outer edge. The swath would be set as the width from side to side where a pan holds 50 percent of the maximum amount collected in the center pan.

5. Fill the hopper half full to simulate average conditions.

6. Set the ground speed of the spreader.

7. Set the spreader according to the manufacturer's recommendations and begin calibration.

Calibration Methods

There are two common methods used to calibrate broadcast spreaders. The first method is the discharge method. To use this procedure, collect and measure the total discharge from the spreader as it runs across a test course. The second method, the pan method, is used on centrifugal and pendulum spreaders. The pattern test pans used to determine pattern shape and swath are used to determine the application rate.

Discharge Method

1. Determine the test distance to use. Longer distances may give better accuracy but may be difficult to manage. A distance of 300 to 400 feet is usually adequate. Use shorter distances if necessary to avoid collecting more material than you can reasonably handle or weigh.

2. Set the ground speed. Be sure to maintain a constant ground speed at all times.

3. If using a ground drive spreader, attach a collection bin to the discharge chute or under the outlets and collect all the material discharged from the spreader as it runs across the test distance. If using an independent drive spreader, record the time required to run the test course. Park the spreader at a convenient location and measure the discharge from the spreader for the time measured on the test distance. The course should be run twice and the times averaged for better accuracy.

4. Calculate the application rate (pounds per acre):

Equation 18. Application Rate, lb/A =
$$\frac{Weight\ Collected\ (lb) \times 43{,}560}{Distance\ (ft) \times Swath\ (ft)}$$

5. Compare the application rate measured to the rate required. Adjust and repeat as necessary.

Pan Method

1. Place pans in the field across the swath to be spread. Pans should be uniformly spaced to cover the full swath. One pan should be at the center of the swath with equal numbers of pans on each side. Use enough pans, 11 or more, to get a good measurement.

2. Make three passes with the spreader using the driving pattern to be used in the field. One pass should be directly over the center pan and the other passes at the recommended distance, lane spacing, to the left and right of the center pass.

3. Combine the material collected in the pans and determine the weight or volume collected. Divide by the number of pans used to determine the average weight or volume per pan.

4. Calculate the application rate.

If you are measuring the weight in the pans in grams:

Equation 19. Application Rate, lb/A =
$$\frac{13{,}829 \times Weight\ (grams)}{Pan\ Area\ (square\ inches) \times Collector\ Efficiency}$$

Collector Efficiency is a measure of how well the pans capture all the material. Bear in mind some particles will bounce out of the pan and some may bounce in. Collector efficiency is usually expressed as a decimal number from 0.0 to 1.0 (it can be higher than 1.- if there are a lot of particles bouncing in.) If you do not know the collector efficiency, use a value of 1.0.

If you are measuring the volume in the pans in cubic centimeters (cc):

Equation 20. Application Rate, lb/A =
$$\frac{13{,}829 \times Bulk\ Density\ (lb\ per\ cubic\ ft) \times Volume\ (cc)}{Pan\ Area\ (square\ inches) \times 62.4}$$

5. Compare the rate measured to the rate required.

Calibration Variables

Several factors can affect proper calibration. The ground speed of any type of PTO-powered machine can make a difference. On the other hand, ground-driven machines are usually only slightly affected by changes in ground speed. If using dry or granular material, product density will affect the discharge rate and may change the pattern for broadcast spreaders. For liquids, calibration can be affected by pressure, nozzle size, density, and viscosity of the liquid, and application type: band or broadcast. The following adjustments may help in adjusting these variables.

Speed

For PTO-powered equipment or other equipment in which the discharge rate is independent of ground speed, Equation 21 is useful.

Equation 21. New Application Rate =
$$Old\ Application \times \frac{Old\ Speed}{New\ Speed}$$

For ground-driven equipment, there should be little or no change in application rate when speed is changed.

Pressure

For liquids in sprayers, the discharge rate changes in proportion to the square root of the ratio of the pressures.

Equation 22. New Discharge Rate =
$$Old\ Discharge\ Rate = \sqrt{\frac{New\ pressure}{Old\ pressure}}$$

Density

For liquids in sprayers, the discharge rate changes if the specific gravity (S.G.) of the liquid changes. Use water for calibration and adjust as shown below. Calibrate with spray solution only if recommended by the supplier.

Equation 23. Water Discharge Rate =

Spray Discharge Rate $\times \sqrt{S.\ G.\ of\ Spray\ Solution}$

Band Application versus Broadcast Application

Some pesticide application recommendations are based on area of cropland covered. Other recommendations are based on area of land treated in the band covered. Check the label for the product you are using to see how it is listed.

Broadcast application is based on area of cropland covered. Nozzle spacing is the distance between nozzles. Band applications in which the area of covered cropland is used for calibration and those applications in which multiple nozzles per row are used are both treated like broadcast applications. Divide the row spacing by the number of nozzles used per row to get a nozzle spacing for calibration.

For band applications in which area of treated land—not cropland covered—is specified, use the width of the band at the ground as the spacing for calibration.

Determining Upper and Lower Limits

Upper and lower limits provide a range of acceptable error. To set these limits for a given sample size, use the equations below. First, however, you must decide upon the degree of accuracy you wish to achieve. Select a percent error: 2 percent, 5 percent, 10 percent, or any other level of accuracy.

Equation 24. Upper Limit =
$$Target\ Rate \times (1 + \frac{Percent\ Error}{100\%})$$

Equation 25. Lower Limit =
$$Target\ Rate \times (1 - \frac{Percent\ Error}{100\%})$$

Useful Tables and Data

Table 2-1. Pump Types and Characteristics

Pump Type	Available Construction Material	Typical Maximum Pressure (PSI)	Capacity (GPM)	Effect of Abrasives	Repairable
Gear	Bronze	80	3 to 7	Severe	No
Roller[1] (4 to 8)	Cast iron, Ni-resist, or bronze (nylon, Teflon, polypropylene, or rubber rollers)	75 to 150	6 to 68	Moderate	Yes
Centrifugal	Cast iron	50	40[2]	Moderate	Yes
Diaphragm	Cast iron or aluminum	75 to 850	5 to 62	Practically none	Yes
Piston	Cast iron	400 to 600	6 to 24	Moderately little	Yes

[1] Roller pumps are available with rubber rollers in lieu of nylon rollers. The rubber rollers have the longest life when used with abrasive materials, but the maximum pressure is limited to 75 psi. Pumps equipped with rubber rollers cost more.

[2] Capacity for a typical 1.25-inch suction size centrifugal pump. Flow is much greater than at low pressures.

Table 2-2. Normal Permissible Hose Flow Rates

Maximum Flow, GPM	2	4	8	12	20	40
Hose Size (inches)	0.375	0.5	0.625	0.75	1	1.25

Table 2-3. Nozzle Types and Materials

Nozzle Type	Nozzle Description
Flat fan	Typically used for broadcast applications.
Even	Typically used for band applications. Pattern is uniform across width. Do not overlap.
Whirl chamber	Wide-angle hollow cone used in place of flat fan. Minimum clogging and low drift.
Flooding	Wide spray patterns used for herbicides and fertilizers. Overlapping usually required.
Boomless	Wide swath of 30 feet or more. May be single or cluster nozzles. Coverage not as uniform as boom nozzles and more affected by wind.
Hollow cone	Very uniform distribution and better coverage of crop foliage.
Solid cone	Concentrates spray material on plants.
Off-center	Flat or cone pattern, used on end of boom to increase width of swath.
Double-outlet	Wider pattern angle and width.
Nozzle Material	**Material Description**
Brass	Most common material used, relatively inexpensive. High wear rate with abrasive materials.
Aluminum	Resistant to corrosives. High wear rate with abrasives.
Monel	Same as aluminum.
Plastic	Low cost, corrosion resistant, used for nonabrasive materials.
Stainless steel	Noncorrosive but relatively expensive. Low wear rate with abrasives.
Ceramic	Suitable for use with abrasives at high pressures. Long life.

Table 2-4. Weights and Measures

Weights

28.35 grams = 1 ounce
16 ounces = 1 pound = 453.6 grams
1 kilogram = 1,000 grams = 2.205 pounds
1 gallon water = 8.34 pounds = 3.8 kilograms
1 cubic foot water = 62.4 pounds = 28.3 kilograms
1 kilogram water = 33.81 ounces = 2.205 pounds
1 ton = 2,000 pounds = 907 kilograms
1 metric ton = 1,000 kilograms = 2,205 pounds

Length and Land Measure

1 inch = 2.54 centimeters
100 centimeters = 1 meter = 3.281 feet
16.5 feet = 5.5 yards = 1 rod
66 feet = 4 rods = 1 chain
1 mile = 5,280 feet = 1.609 kilometers
272.5 square feet = 30.25 square yards = 1 square rod
4,356 square feet = 16 square rods = 1 square chain
43,560 square feet = 160 square rods = 1 acre
43,560 square feet = 10 square chains = 1 acre
1 acre = 0.4047 hectare = 4,047 square meters
1 hectare = 2.471 acres = 10,000 square meters

Volume and Liquid Measure

3 teaspoons = 1 tablespoon = 14.8 cubic centimeters
2 tablespoons = 1 fluid ounce = 29.6 cubic centimeters
8 fluid ounces = 16 tablespoons = 1 cup = 236.6 cubic centimeters
2 cups = 32 tablespoons = 1 pint = 473.1 cubic centimeters
2 pints = 64 tablespoons = 1 quart = 946.2 cubic centimeters
1 liter = 1,000 cc = 0.2642 gallons = 33.81 ounces
4 quarts = 256 tablespoons = 1 gallon = 3,785 cubic centimeters
1 gallon = 128 fluid ounces = 231 cubic inches = 3,785 cubic centimeters
1 bushel = 1.244 cubic feet = 9.309 liquid gallons = 35.2 liters
1 cubic centimeter = 1 milliter

Table 2-5. Length of Row Required for 1 Acre

Row Spacing (inches)	Length or Distance
24	7,260 yards = 21,780 feet
30	5,808 yards = 17,424 feet
36	4,840 yards = 14,520 feet
42	4,149 yards = 12,447 feet
48	3,630 yards = 10,890 feet
54	3,227 yards = 9,860 feet
60	2,904 yards = 8,712 feet

Table 2-6. Travel Speed Chart

Miles per Hour	Time Required to Travel (in seconds)		
	100 feet	200 feet	300 feet
1	68	136	205
2	34	68	102
3	23	45	68
4	17	34	51
5	14	27	41
6	11	23	34
7	10	20	29
8	9	17	26
9	8	15	23
10	7	14	21

Table 2-7. Dilutions for Liquids and Dusts

Equivalent Quantities of Liquid Materials When Mixed by Parts

Water	1 to 100	1 to 200	1 to 400	1 to 800	1 to 1,600
100 gal	1 gal	2 qt	1 qt	1 pt	1 cup
50 gal	2 qt	1 qt	1 pt	1 cup	1/2 cup
5 gal	0.4 pt	0.2 pt	2 tbsp	5 tsp	2-1/2 tsp
1 gal	1.28 oz	0.64 oz	2 tsp	1 tsp	1/2 tsp

Example: If a recommendation calls for 1 part of the chemical to 800 parts of water, it would take 5 tsp in 5 gal of water to give 5 gal of a mixture of 1 to 800.

Equivalent Quantities of Dry Materials (Wettable Powders) for Various Quantities of Water

Water	Quantity of Material					
100 gal	1 lb	2 lb	3 lb	4 lb	5 lb	6 lb
50 gal	8 oz	1 lb	1.5 lb	2 lb	2.5 lb	3 lb
5 gal	3 tbsp	1.5 oz	2.5 oz	3.25 oz	4 oz	5 oz
1 gal	2 tsp	1 tbsp	1.5 tbsp	2 tbsp	5 tsp	1 oz

Example: If a recommendation calls for a mixture of 4 lb of a wettable powder to 100 gal of water, it would take 3.25 oz (approximately 6.5 tsp) to 5 gal of water to give 5 gal of the spray mixture of the same strength.

Note: Wettable pesticide materials vary considerably in density. Therefore, the teaspoonful (tsp) and tablespoonful (tbsp) measurements in this table are not exact dosage by weight but are within the bounds of safety and efficiency for mixing small amounts of spray.

Table 2-8. Chemical Formulations

Pounds of Active Ingredients Per Gallon, Pounds Per Pint of Liquid, and the Number of Pints for Various Per Acre Rates

Pounds of Active Ingredients In 1 Gallon of Commercial Products	Pounds of Active Ingredients per Pint	Pints of Commercial Product Needed for Each Acre to Give the Following Pounds of Active Ingredient per Acre					
		0.25	0.50	0.75	1	1.5	2
1.00	0.125	2.00	4.00	6.00	8.00	12.0	16.00
2.00	0.250	1.00	2.00	3.00	4.00	6.0	8.00
2.64	0.330	0.75	1.50	2.25	3.00	4.5	6.00
3.00	0.380	0.67	1.33	2.00	2.67	4.0	5.33
3.45	0.420	0.60	1.20	1.08	2.40	3.6	4.80
4.00	0.500	0.50	1.00	1.50	2.00	3.0	4.00
6.00	0.750	0.33	0.67	1.00	1.33	2.0	2.67

Available Commercial Materials in Pounds Active Ingredients Per Gallon Necessary to Make Various Percentage Concentration Solution

Pounds of Active Ingredients In 1 Gallon of Commercial Products	Pounds of Active Ingredients per Pint	Liquid Ounces of Commercial Product per 1 Gallon of Solution to Make:				
		0.5%	1%	2%	5%	10%
1.00	0.125	5.34	10.68	21.35	53.38	106.80
2.00	0.250	2.68	5.36	10.72	26.80	53.40
2.64	0.330	2.02	4.05	8.10	20.25	40.44
3.00	0.380	1.78	3.56	7.12	17.80	35.58
3.34	0.420	1.59	3.18	6.36	15.90	31.96
4.00	0.500	1.33	2.67	5.33	13.34	26.69
6.00	0.750	0.89	1.78	3.56	8.90	17.79

Based on 8.34 pounds per gallon (weight of water).

Chapter II—2019 N.C. Agricultural Chemicals Manual

Table 2-9. Required Nozzle Discharge Rates (Gallons per Minute) for Various Application Rates, Nozzle Spacings, and Travel Speeds

Travel Speed (MPH)	Nozzle Spacing (inches)	Application Rate (gallons per acre)							
		5	7.5	10	12.5	15	20	25	35
3	6	0.0152	0.0227	0.0303	0.0379	0.0455	0.0606	0.0758	0.106
	8	0.0202	0.0303	0.0404	0.0505	0.0606	0.0808	0.101	0.141
	10	0.0253	0.0379	0.0505	0.0631	0.0758	0.101	0.126	0.177
	12	0.0303	0.0455	0.0606	0.0758	0.0909	0.121	0.152	0.212
	14	0.0354	0.053	0.0707	0.0884	0.106	0.141	0.177	0.247
	16	0.0404	0.0606	0.0808	0.101	0.121	0.162	0.202	0.283
	18	0.0455	0.0682	0.0909	0.114	0.136	0.182	0.227	0.318
	20	0.0505	0.0758	0.101	0.126	0.152	0.202	0.253	0.354
	21	0.053	0.0795	0.106	0.133	0.159	0.212	0.265	0.371
	22	0.0556	0.0833	0.111	0.139	0.167	0.222	0.278	0.389
	24	0.0606	0.0909	0.121	0.152	0.182	0.242	0.303	0.424
	30	0.0758	0.114	0.152	0.189	0.227	0.303	0.379	0.53
	36	0.0909	0.136	0.182	0.227	0.273	0.364	0.455	0.636
	42	0.106	0.159	0.212	0.265	0.318	0.424	0.53	0.742
	48	0.121	0.182	0.242	0.303	0.364	0.485	0.606	0.848
	60	0.152	0.227	0.303	0.379	0.455	0.606	0.758	1.06
4	6	0.0202	0.0303	0.0404	0.0505	0.0606	0.0808	0.101	0.141
	8	0.0269	0.0404	0.0539	0.0673	0.0808	0.108	0.135	0.189
	10	0.0337	0.0505	0.0673	0.0842	0.101	0.135	0.168	0.236
	12	0.0404	0.0606	0.0808	0.101	0.121	0.162	0.202	0.283
	14	0.0471	0.0707	0.0943	0.118	0.141	0.189	0.236	0.330
	16	0.0539	0.0808	0.108	0.135	0.162	0.215	0.269	0.377
	18	0.0606	0.0909	0.121	0.152	0.182	0.242	0.303	0.424
	20	0.0673	0.101	0.135	0.168	0.202	0.269	0.337	0.471
	21	0.0707	0.106	0.141	0.177	0.212	0.283	0.354	0.495
	22	0.0741	0.111	0.148	0.185	0.222	0.296	0.370	0.519
	24	0.0808	0.121	0.162	0.202	0.242	0.323	0.404	0.566
	30	0.101	0.152	0.202	0.253	0.303	0.404	0.505	0.707
	36	0.121	0.182	0.242	0.303	0.364	0.485	0.606	0.848
	42	0.141	0.212	0.283	0.354	0.424	0.566	0.707	0.990
	48	0.162	0.242	0.323	0.404	0.485	0.646	0.808	1.130
	60	0.202	0.303	0.404	0.505	0.606	0.808	1.010	1.410
5	6	0.0253	0.0379	0.0505	0.0631	0.0758	0.101	0.126	0.177
	8	0.0337	0.0505	0.0673	0.0842	0.101	0.135	0.168	0.236
	10	0.0421	0.0631	0.0842	0.105	0.126	0.168	0.210	0.295
	12	0.0505	0.0758	0.101	0.126	0.152	0.202	0.253	0.354
	14	0.0589	0.0884	0.118	0.147	0.177	0.236	0.295	0.412
	16	0.0673	0.101	0.135	0.168	0.202	0.269	0.337	0.471
	18	0.0758	0.114	0.152	0.189	0.227	0.303	0.379	0.530
	20	0.0842	0.126	0.168	0.210	0.253	0.337	0.421	0.589
	21	0.0884	0.133	0.177	0.221	0.265	0.354	0.442	0.619
	22	0.0926	0.139	0.185	0.231	0.278	0.370	0.463	0.648
	24	0.101	0.152	0.202	0.253	0.303	0.404	0.505	0.707
	30	0.126	0.189	0.253	0.316	0.379	0.505	0.631	0.88
	36	0.152	0.227	0.303	0.379	0.455	0.606	0.758	1.060
	42	0.177	0.265	0.354	0.442	0.530	0.707	0.884	1.240
	48	0.202	0.303	0.404	0.505	0.606	0.808	1.010	1.410
	60	0.253	0.379	0.505	0.631	0.758	1.010	1.260	1.770

CHEMICAL APPLICATION EQUIPMENT

Table 2-10. Time (in Seconds) Needed To Collect 1 Pint (0.125 Gallon) of Liquid from a Single Nozzle at Various Application Rates, Nozzle Spacings, and Travel Speeds

Travel Speed (MPH)	Nozzle Spacing (inches)	Application Rate (gallons per acre)							
		5	7.5	10	12.5	15	20	25	35
3	6	495	330	247	198	165	124	99	70.7
	8	371	247	186	148	124	92.8	74.3	53
	10	297	198	148	119	99	74.3	59.4	42.4
	12	247	165	124	99	82.5	61.9	49.5	35.4
	14	212	141	106	84.9	70.7	53	42.4	30.3
	16	186	124	92.8	74.3	61.9	46.4	37.1	26.5
	18	165	110	82.5	66	55	41.3	33	23.6
	20	148	99	74.3	59.4	49.5	37.1	29.7	21.2
	21	141	94.3	70.7	56.6	47.1	35.4	28.3	20.2
	22	135	90	67.5	54	45	33.8	27	19.3
	24	124	82.5	61.9	49.5	41.3	30.9	24.8	17.7
	30	99	66	49.5	39.6	33	24.8	19.8	14.1
	36	82.5	55	41.3	33	27.5	20.6	16.5	11.8
	42	70.7	47.1	35.4	28.3	23.6	17.7	14.1	10.1
	48	61.9	41.3	30.9	24.8	20.6	15.5	12.4	8.84
	60	49.5	33	24.8	19.8	16.5	12.4	9.9	7.07
4	6	371	247	186	148	124	92.8	74.3	53
	8	278	186	139	111	92.8	69.6	55.7	39.8
	10	223	148	111	89.1	74.3	55.7	44.5	31.8
	12	186	124	92.8	74.3	61.9	46.4	37.1	26.5
	14	159	106	79.6	63.6	53	39.8	31.8	22.7
	16	139	92.8	69.6	55.7	46.4	34.8	27.8	19.9
	18	124	82.5	61.9	49.5	41.3	30.9	24.8	17.7
	20	111	74.3	55.7	44.5	37.1	27.8	22.3	15.9
	21	106	70.7	53	42.4	35.4	26.5	21.2	15.2
	22	101	67.5	50.6	40.5	33.8	25.3	20.3	14.5
	24	92.8	61.9	46.4	37.1	30.9	23.2	18.6	13.3
	30	74.3	49.5	37.1	29.7	24.8	18.6	14.8	10.6
	36	61.9	41.3	30.9	24.8	20.6	15.5	12.4	8.84
	42	53	35.4	26.5	21.2	17.7	13.3	10.6	7.58
	48	46.4	30.9	23.2	18.6	15.5	11.6	9.28	6.63
	60	37.1	24.8	18.6	14.8	12.4	9.28	7.72	5.3
5	6	297	198	148	119	99	74.3	59.4	42.4
	8	223	148	111	89.1	74.3	55.7	44.5	31.8
	10	178	119	89.1	71.3	59.4	44.5	35.6	25.5
	12	148	99	74.3	59.4	49.5	37.1	29.7	21.2
	14	127	84.9	63.6	50.9	42.4	31.8	25.5	18.2
	16	111	74	55.7	44.5	37.1	27.8	22.3	15.9
	18	99	66	49.5	39.6	33	24.8	19.8	14.1
	20	89.1	59.4	44.5	35.6	29.7	22.3	17.8	12.7
	21	84.9	56.6	42.4	33.9	28.3	21.2	17	12.1
	22	81	54	40.5	32.4	27	20.3	16.2	11.6
	24	74.3	49.5	37.1	29.7	24.8	18.6	14.8	10.6
	30	59.4	39.6	29.7	23.8	19.8	14.8	11.9	8.49
	36	49.5	33	24.8	19.8	16.5	12.4	9.9	7.07
	42	42.4	28.3	21.2	17	14.1	10.6	8.49	6.06
	48	37.1	24.8	18.6	14.8	12.4	9.28	7.42	5.3
	60	29.7	19.8	14.8	11.9	9.9	7.42	5.94	4.24

Table 2-11. Fumigant Application Rate Table

Application Rate (gal/acre)	Quantity per 100 Feet at Given Bandwidth													
	8-Inch Width		12-Inch Width		24-Inch Width		30-Inch Width		36-Inch Width		42-Inch Width		48-Inch Width	
	oz	cc	oz	cc	oz	cc	oz	cc	oz	cc	oz	cc	oz	cc
1	0.2	6	0.3	9	0.6	17	0.7	22	0.9	26	1.0	30	1.2	35
3	0.6	17	0.9	26	1.7	52	2.2	65	2.6	78	3.1	91	3.5	104
5	1.0	29	1.5	44	2.9	87	3.7	109	4.4	130	5.1	152	5.9	174
7	1.4	41	2.1	61	4.1	122	5.1	152	6.2	183	7.2	213	8.2	243
9	1.8	52	2.6	78	5.3	157	6.1	196	7.9	235	9.3	274	10.6	313
12	2.4	70	3.6	104	7.1	209	8.8	261	10.6	313	12.3	365	14.1	418
15	2.9	87	4.4	131	8.8	261	11.1	326	13.2	391	15.4	457	17.6	522
20	3.9	116	5.9	174	11.8	348	14.7	435	17.6	522	20.6	609	23.5	696

Table 2-12. Fertilizer Application Rate Table

Rate (lb/acre)	Pounds per 100 Feet of Row				
	24-inch Rows	30-inch Rows	36-inch Rows	42-inch Rows	48-Inch Rows
100	0.5	0.6	0.7	0.8	0.9
200	0.9	1.1	1.4	1.6	1.8
300	1.4	1.7	2.1	2.4	2.8
400	1.8	2.2	2.8	3.2	3.7
500	2.3	2.8	3.4	4.0	4.6
600	2.8	3.4	4.1	4.8	5.5
700	3.2	3.9	4.8	5.6	6.4
800	3.7	4.5	5.5	6.4	7.3
900	4.1	5.1	6.2	7.2	8.8
1,000	4.6	5.6	6.9	8.0	9.2
1,100	5.1	6.2	7.6	8.8	10.1
1,200	5.5	6.7	8.3	9.6	11.0
1,300	6.0	7.3	9.0	10.4	11.9
1,400	6.4	8.0	9.6	11.2	12.6
1,500	6.9	8.6	10.3	12.1	13.8

Table 2-13. Granular Application Rate Table

Rate (lb/acre)	Grams per 100 Feet of Row				
	24-in. rows	30-in. rows	36-in. rows	42-in. rows	48-in. rows
5	10	13	16	18	21
6	13	16	19	22	25
7	15	18	22	26	29
8	17	21	25	29	33
9	19	23	28	33	38
10	21	26	31	36	42
12	25	31	37	44	50
15	31	39	47	55	62
20	42	52	62	73	83
25	52	65	78	91	104
30	63	78	93	109	125
40	83	104	125	146	167
50	104	130	156	182	208
60	125	156	187	218	250
75	156	195	234	273	312

Table 2-14. Distance To Cover for Each Nozzle To Spray 1/128TH Acre

Average Nozzle Spacing (inches)	Distance to Cover (feet)
6	681
8	510
10	408
12	340
14	292
16	255
18	227
20	204
22	186
24	170
30	136
36	113
38	107
40	102
42	97
48	85

Note: Each ounce per nozzle represents 1 gallon per acre application rate.

III — HOW TO SEND SPECIMENS FOR DISEASE, INSECT, AND WEED IDENTIFICATION

How to Send Specimens for Disease, Insect, and Weed Identification 38
Plant Disease and Insect Clinic (PDIC) 38
What to Sample and How to Ship 38
Disease Sample Submission and Instructions 39
Insect Identification 39
Precautions and Limitations 40
Plant and Weed Identification 40
Soil Testing, Nematode Assay, Plant Tissue Nutrient Testing, Waste Analysis, and Solution Analysis 40

How to Send Specimens for Disease, Insect, and Weed Identification

B. Shew, Extension Plant Patholog

Plant Disease and Insect Clinic (PDIC)

Fees for Disease Diagnosis and Insect Identification Samples

Turf Sample Fees

- $120 out-of-state, all turf samples
- $60 in-state, golf courses
- $30 in-state, other turf, submitted by individual or business
- $20 in-state, other turf, submitted by Cooperative Extension agent with online submission form

Non-Turf Fees (all others)

- $75 out-of-state
- $30 in-state, submitted by individual or business
- $20 in-state, submitted by Cooperative Extension agent or NCDA&CS specialist with online submission form
- No charge for image-only samples submitted online. If a follow-up physical sample is necessary, the fees outlined above apply.

Please include a check made payable to NCSU with your sample.

Contact Information

Hours: 8 a.m. to 4:30 p.m., Monday through Friday

Plant Disease and Insect Clinic
State Courier #: 53-61-21
Campus Box 7211, Room 1227 Gardner Hall
100 Derieux Place
North Carolina State University
Raleigh, NC 27695-7211
http://www.ncsu.edu/pdic

General Information: 919-515-3619; plantclinic@ces.ncsu.edu

Insects: Matt Bertone, 919-515-9530; matt_bertone@ncsu.edu

Turf: Lee Butler, 919-513-3878; lee_butler@ncsu.edu

Sending Samples and Images Directly to the PDIC: Our database allows users to create an account, create a sample record, and upload sample information and images for diagnosis. For information and instructions on how to create an account, click the "new user" button at the top of the PDIC home page, http://www.ncsu.edu/pdic. Once you enter your sample online, you can track sample progress, review past reports and check status of your invoices. Log into the database by clicking the "login to database" button on the PDIC homepage.

We send reports as soon as the diagnosis is ready.

What to Sample and How to Ship

Visit our website at http://www.ncsu.edu/pdic and go to the "How to Submit a Sample" tab for instructions and an illustrated guide. Here are some things to remember:

- *Dead plants tell no tales* — Most plants that are totally dead, dry, or rotten are useless for diagnosis. Collect plants that show a range of the symptoms but are not yet dead. Likewise, collect live but diseased limbs and branches from trees or large shrubs.
- *More is better* — We may miss the main concern if you send only one plant. For bedding plants, young vegetables, and field crops, collect several plants for each problem to be diagnosed.
- *Getting to the root of the problem* — Many plant problems are related to the roots and soil. **Submit whole plants, including ROOTS AND SOIL whenever possible.** Dig (don't pull) plants up to keep roots intact. For large plants, submit a portion of the root ball and a pint of soil.
- *A GOOD picture is worth a thousand words* — Images help us to better understand the situation. Use our database to upload one or more images of the problem as it appears in the field, greenhouse, or landscape. Make sure images are in focus and clearly illustrate the problem, symptoms, or insect of concern from different angles and distances. Do not attempt to upload more than 10 MB of image files at one time.
- *A place for everything* — Keep soil *off* the foliage and *on* the roots. This keeps roots from drying out and the foliage free of contaminants. Put the roots and soil in a plastic bag and tie at the main stem. Loosely wrap foliage in newspaper, then pull the bag over the rest of the plant and tie again to keep foliage from drying out. Make sure foliage is dry for packaging.
- *Details, details* — The more you tell us about the situation on the PDIC form, the better. Please give complete information, including a client name and location, percent of planting affected, variety or cultivar, and insect collection information (if applicable).
- *History 101* — List all fertilizer, fungicide, herbicide, and insecticide applications made in the last 30 days. For field and vegetable crops, give the rotation history and planting date. For orchards and landscapes, indicate when the trees or other plants were transplanted or established.
- *Fresher is better* — Mail or deliver samples as soon as possible. On hot days or in very cold weather, store samples in an insulated cooler until you can send them. Avoid mailing on Fridays since many samples will degrade before delivery on Monday. Instead, hold samples over the weekend in a refrigerator or cooler. Succulent plants, fruits, vegetables, and mushrooms rot quickly and should be sent overnight express.
- *Fragile, handle with care* — Ship all samples in a crush-proof box. **Do not** send your sample in an envelope.
- *Be cool* — Never leave any sample or insect specimen in the sun or a closed car for even a few minutes; "cooked" samples may be impossible to diagnose.
- *We don't work for peanuts (or shredded paper)* — Please don't use foam peanuts to pad samples. Crumpled newspaper is cheaper, works well, and is easier for us to handle.
- *Noxious weeds* — Send no more than 1 quart of soil with samples from counties regulated under state and/or federal measures to prevent movement of noxious weeds. http://www.ncagr.gov/plantindustry/plant/weed/weedprog.htm
- *Out of state samples* — Out of state samples must be double bagged and sealed; samples that are not properly packaged will be destroyed upon receipt. **Do not send live insects or arthropods from out of state.**

Disease Sample Submission and Instructions

Turf Grasses

1. To diagnose turfgrass problems, we need **at least a 6-inch x 6-inch piece of the turf**, including the root system and soil. If using a golf course cup cutter, please send at least two plugs.
2. Collect samples from the border between healthy and diseased turf so that two-thirds of the sample is diseased and one-third is healthy.
3. Wrap the soil and roots in two to three layers of aluminum foil to **ensure that no soil leaks out during transit.**
4. **DO NOT store or transport the samples in plastic bags.** Instead, place the samples in a cardboard box and stuff it with newspaper or other packing material to hold the samples in place.
5. Download and print the turfgrass sample submission form from the clinic website: https://turfpathology.plantpath.ncsu.edu/diagnostics-lab/how-to-submit-a-sample/
6. Please fill out the sample submission form completely and describe the symptoms you are observing as accurately as possible. All of the information requested on the form is needed to make an accurate diagnosis.
7. **List all fertilizer, fungicide, herbicide, and insecticide applications made in the last 30 days.** Also, list any major cultural practices (aerification, topdressing, etc.) conducted in the last 30 days. These practices have a major impact on disease and insect development and provide valuable clues that will help us make an accurate diagnosis.

Field, Garden, or Landscape: Small Grains, Most Field Crops, Vegetable Plants, Small Shrubs, Perennials, and Annuals

1. Select plants with a range of symptoms but do not choose dead plants.
2. **Dig (do not pull) up several whole plants, leaving roots/soil intact.**
3. Wrap soil and roots in a plastic bag and tie the bag closed at the main stem.
4. Package and ship as described in general guidelines above.

Mushrooms, Fruits, and Vegetables

1. Wrap in dry newspaper, and then place in a plastic bag. Please ship overnight in a strong box with dry newspaper padding.

Small Plants in Greenhouse or Nursery

1. Send two to four entire plants. Leave in the pot if possible. For very large plants, remove most of the media and enclose root system with remaining media in a plastic bag.
2. For seedlings: send at least half the flat or up to a dozen plugs if the problem is scattered. Newspaper padding can help keep soil off leaves.
3. Label each bag with your code if the bag represents different types of samples (Good, Bad or A, B, C, etc.).
4. Package and ship as described in the general guidelines above.

Large Shrubs and Small Trees

1. Dig a generous shovel full of the small fine roots.
2. Seal these in a plastic bag along with 1 quart of soil.
3. In a separate plastic bag, include several branches showing symptoms. **Where cankers or diebacks are observed, we need stem or branch sections where live wood changes to dead wood.**
4. For trees and shrubs that are going to be removed, please also send the bottom 8 to 10 inches of the main trunk (cut below the soil line if possible, including some of the larger roots).

Large Trees

Decline of large trees often cannot be diagnosed by a clinic sample alone. However, with a good sample we can often identify leaf spots, vascular wilts, wood decays, root rots, and cankers. Images are invaluable aids to diagnosis and should be included with the sample if at all possible.

- *Cankers, swellings, diebacks* — Cut several affected branches to include several inches of healthy wood attached below the canker or dieback. Package in a plastic bag.
- *Vascular wilt diseases* — Find a wilting or dying but NOT dead branch, peel back the bark and look for dark tan, black, or greenish streaks in the sapwood. Send several branches with these symptoms.
- *Leaf spots* — Send several small branches with leaves attached. Seal loosely in a plastic bag. Don't add water to any sample.
- *Wood rots* — These can be identified ONLY by the presence of mushrooms, conks, or other fungal fruiting structures. Wrap mushrooms or conks loosely in several layers of newspaper; place the wrapped specimen in a plastic bag, add more newspaper padding, and ship in a crush-proof box. Fleshy mushrooms rot quickly. Please ship overnight and take photos as a back-up. Badly decayed samples are useless for identification.

Harvested Root Crops (sweet potato, potato, onion, etc.)

1. Collect six to eight roots/tubers showing the symptoms of concern.
2. Wrap each in newspaper separately or place in individual paper bags (do not use plastic bags).
3. **Package and ship in a crush-proof box with lots of newspaper padding.**

Houseplants or Interiorscape Plants

These plants typically have very few disease problems due to the dry air in most buildings. Most problems are related to improper moisture, temperature, nutrition, or light conditions. Please screen houseplants for these problems and avoid sending such samples to the clinic. When a disease or insect problem is suspected on a valuable specimen plant:

1. Send the whole plant if practical;
2. For plants too large to send, please send at least 2 cups of soil along with a generous handful of the smaller roots;
3. Cut several stems showing the symptoms of concern;
4. Package and ship as described in the general guidelines.

Insect Identification

- *Most insects* — Send cockroaches, termites, bugs, beetles, flies, wasps, ants, maggots, spiders, etc. in alcohol, 70% (or higher) concentration.
- *Mites, scales, aphids, and thrips* — Send alive on some of the affected foliage/stems, collected as you would a plant specimen. Place in a plastic bag when collected. If there are numerous individuals, save some in a vial of 70% (or higher) alcohol.

- *Butterflies and moths* — Freeze specimens or kill them with fumes of ethyl acetate (commonly found in nail polish remover) and package them lightly in tissue paper in a crush-proof box.
- *Caterpillars* — Send alive on some of the host plant in a plastic bag. If you cannot send alive on the host plant, please note the plant and send the caterpillar in alcohol. For larger caterpillars, it is advisable to drop the caterpillar into boiling water before putting in the alcohol (70% or higher concentration).
- *Grubs* — Send alive in a pint or two of soil enclosed in a plastic bag.

When in doubt, put specimens in alcohol, 70% or higher concentration. Collect several specimens if possible. Send insect specimens to the Plant Disease and Insect Clinic at the address given above.

Please *do not* wrap insects in tissue or cellophane, and then put them in an envelope. Also, don't stick specimens to paper with tape or place them in an empty or overcrowded vial; they may decay or get damaged in these situations. Insects that arrive badly damaged and/or without data (see below) will require more time to identify or may be too poor in quality to diagnose at all.

Important: Data submitted with insects should be complete and accurate. When filling out the clinic forms, please see that the following four pieces of information are provided with all specimens:

- *Date specimen collected* — Date actually found, not date received by agent or other second party.
- *Town and county where specimen was collected* — Location where the insect was actually found, which may not be the agent's or property owner's address. If not found within a town, give nearest town and distance and direction from it, or GPS coordinates.
- *Name of collector* — Name the specific person who actually captured or collected the insect. The agent or other second party should not be listed as the collector.
- *Where were the specimens, or on what were they feeding when collected?* — If a plant, specifically name the host plant, e.g., oak, tobacco, marigold, etc. If not on a plant, name spot they were found, e.g., windowsill, closet, in a box of dog food, etc.

We strongly advise sending in insect specimens with the information above. This information is very important for our diagnoses and records. Your sample identification and report may be delayed if we have to reach you to ask follow-up questions.

Human and animal pests

We **do not** accept nor search through blood, tissue, fecal, or other such samples. Potential organisms must be isolated and sent in 70% or higher concentration alcohol. For guidelines on submitting insects and other arthropods associated with human/animal health, please refer to this document:

http://go.ncsu.edu/pabkcu

Precautions and Limitations

Identifications and diagnoses are based on the material and information submitted. Time devoted to individual specimens must necessarily be limited, and the samples usually represent only a small percentage of the crop or problem. Reports reflect considered opinion and best judgment but may not always be statements of absolute fact or define the major problems affecting the crop or site.

Plant and Weed Identification

General plant identification services are provided by the NCSU Herbarium. Weed identification services (turfgrass weeds, cropland, and non-cropland), are provided by the Department of Crop Science. All forms can be found at: http://www.cals.ncsu.edu/plantbiology/ncsc/identification.htm

A complete plant or a specimen containing leaves, stem, roots, and flowers or fruit is absolutely necessary for a definite identification. It is impossible to identify a plant when only a leaf or stem is sent. Crushed or rotted specimens are especially hard to identify. Please prepare specimens properly and pack well, as described above, for clinic samples.

Soil Testing, Nematode Assay, Plant Tissue Nutrient Testing, Waste Analysis, and Solution Analysis

Information and forms are available in each North Carolina Cooperative Extension county center, by calling the NCDA&CS labs, and on the Web at

http://www.ncagr.gov/agronomi/index.htm

For a list of fees for various NCDA&CS services, see

http://www.ncagr.com/agronomi/fees.htm

NCDA&CS Agronomic Services Division,

Dr. Colleen M. Hudak-Wise, Director

Mailing Address:

1040 Mail Service Center, Raleigh NC 27699-1040

Physical Address:

4300 Reedy Creek Rd., Raleigh NC 27607-6465

Phone: 919-733-2655; FAX: 919-733-2837

IV — FERTILIZER USE

Lime and Fertilizer Suggestions for Field, Pasture, and Hay Crops	42
Fertilizer Suggestions for Tree Fruit	45
Fertilizer Suggestions for Small Fruit	46
Lime and Fertilizer Suggestions for Lawns	48
Fertilizer Suggestions for Ornamental Plants in Landscapes	51
Fertilizer Suggestions for Nursery Crops	52
Lime and Fertilizer Suggestions for Vegetable Crops	53
Fertilizer Rules and Regulations	57
Nutrient Content of Fertilizer Materials	57
Solubility of Selected Fertilizer Materials	61
Mixing Herbicides with Nitrogen Solutions or Fluid Fertilizers	62
How to Test Compatibility of Herbicides with Fluid Fertilizers	63
Fertilizer Placement	64
Livestock & Poultry Manure Production Rates and Nutrient Content	64
Beneficial Use of Municipal Biosolids	68
Certified Organic Farm Management Alternatives	69

Lime and Fertilizer Suggestions for Field, Pasture, and Hay Crops

M. Castillo, C. R. Crozier, K. Edmisten, L. Fisher, R. W. Heiniger,
D. L. Jordan, D. L. Osmond, A. Post, and R. Vann, Crop and Soil Sciences; D. H. Hardy, NCDA&CS

For best results, apply lime and nutrients (except nitrogen and boron) based on soil test information. If this information is unavailable, use these suggestions as general guidelines. Ranges in rates are given for P_2O_5 and K_2O. Sandy soils generally respond to higher rates of K_2O than clayey ones, but the opposite is true for P_2O_5. Also, attend to lime needs. For row crops (including corn, cotton, sorghum, soybean, and tobacco), experimental data across North Carolina demonstrate that starter P is not necessary on soils with very high P levels (P-index >100). On soils with P indexes between 50 and 100, starter band P may be useful with no-till management or when soils are cool and wet.

A liming program is essential to manage acidity in North Carolina soils. Base lime rates and frequency of applications on a routine soil sampling and testing program. In the absence of soil tests, most fields should receive 1 ton of lime per acre every 3 to 4 years. Liming more frequently at lesser rates may be necessary on sandy, light-colored soils where fertility changes more rapidly than in clay or organic soils, especially in years with excessive rainfall. Although liming is imperfect, over liming can cause micronutrient deficiencies, particularly manganese deficiency in small grain and soybeans grown on sandy soils. Dolomitic lime contains magnesium and is generally preferred over calcitic lime, especially on sandy soils. Lack of lime is the dominant soil fertility problem in North Carolina, even though liming is known to be a highly economical practice on either a short- or long-term basis.

Nitrogen rates are not based on soil testing. Use the RYE (Realistic Yield Expectation) for the soil and the appropriate nitrogen factor (NF) to establish the nitrogen rate. A range of NFs is found in Table 4-1A. The appropriate NF for each crop is determined by the soil mapping unit. Nitrogen rates based on RYE and NF are provided for each county by soil series at: http://nutrients.soil.ncsu.edu/yields/

Table 4-1A. Lime and Fertilizer Suggestions — Field Crops

Area of State or Soil Type	Optimum pH	Plant Nutrient Suggestions When Soil Test is Unavailable[a]			Remarks
		N factor	P_2O_5 lb/acre	K_2O lb/acre	
Corn Grain					
Mineral soils	6.0	0.80 to 1.0 lb/bu	10 to 20	80 to 100	Banded starter fertilizer with 20 to 30 pounds per acre of both N and P_2O_5 is recommended under no-till management or on cool, wet soils, especially when planting early. Starter P not likely to benefit on soils with very high initial P levels (P-index >100). Apply 1/4 to 1/3 of the N at planting. Sidedress remaining N when plants are 15 to 24 inches high. Under irrigation increase N rate by 10% to 15%. Starter band P if using no-till management or when soils are cool and wet. On deep sandy soils, apply K just before planting or use a split application of K at planting and at sidedress. Mineral soils that are sandy or greater than 18 inches to clay should receive 20 pounds S per acre. Test organic soils to determine copper needs.
Mineral-organic soils	5.5				
Organic soils	5.0				
Corn Silage					
Mineral soils	6.0	10 to 12 lb/ton	10 to 20	100 to 120	
Mineral-organic soils	5.5				
Organic soils	5.0				
Cotton					
Mineral soils	6.2	0.06 to 0.12 lb/lb lint	10 to 20	50 to 70	Apply 20 to 30 pounds N per acre at or before planting. Apply remaining N 2 to 3 weeks after first square. Starter P not likely to benefit on soils with very high initial P levels (P-index >100). After peanuts or soybeans, reduce total N by 25 to 30 pounds per acre. Apply 0.5 pound rate B at planting or as foliar spray at first bloom. On deep sands (more than 18 inches to clay), use higher end of pounds per RYE range and apply 1/2 the remaining N at early square and 1/2 just prior to bloom. Deep sands should also receive 20 to 25 pounds S per acre.
Mineral-organic soils	5.5	0.05 to 0.09			
Organic soils	5.0	0.03 to 0.06 lb/lb lint			
Peanut					
Coastal plain	6.0	0	10 to 20	0	To minimize Ca deficiency risk, reduce unnecessary application of other cations. Apply K based on soil test recommendations and incorporate. During the growing season, at early to mid-flowering (late June-early July), apply gypsum to all Virginia market types. Use soil testing to determine Ca need for small-seeded runner market types. For larger seeded "jumbo" runners, use half the rate as for Virginia market types. See *Peanut Information* series (http://www.peanuts.ncsu.edu/) for more information on gypsum products and rates. Apply Mg only as recommended based on soil test. Apply 0.5 pound B per acre (liquid or dry) at peak flower. Apply *Bradyrhizobia* inoculant to seed or in the seed furrow regardless of previous rotation history to ensure peanut is capable of fixing N. Inoculation is especially important on land where peanut has not been grown recently. Peanut taproot should have 10 or more nodules approximately 45 days after planting or supplemental N is needed. Apply 600 pounds/acre of ammonium sulfate (i.e., 125 pounds of N/acre) as soon as possible if roots are not nodulating effectively and foliage is N deficient. Zinc toxicity may occur in fields with soil test zinc indices greater than 250 or at lower indices with pH less than 6.0. Apply Mn at 0.5 to 1.0 pound Mn per acre when deficiency symptoms appear, often associated with high pH. See *Peanut Information* series for selection of appropriate B and Mn products if needed.
Small Grain, Grain					
Mineral soils	6.0	Wheat, Rye: 1.7 to 2.4 lb/bu	10 to 20	80 to 100	Apply 15 to 20 pounds N at planting (sometimes can be skipped if after soybean or peanut). If tiller density at spring green up (Feekes GS-3) is low, split N topdress (Feb., March). If tiller density is high, apply all N just before jointing. Use a tissue test at GS-5 to find the optimum nitrogen rate for wheat. Test organic soils to determine copper needs. Mineral soils that are sandy or greater than 18 inches to clay should receive 20 pounds S per acre. Sensitive to manganese deficiency when soil pH is greater than 6.2.
Mineral-organic soils	5.5	Barley, Triticale: 1.4 to 1.6 lb/bu			
Organic soils	5.0				
		Oats: 1.0 to 1.3 lb/bu	10 to 20	60 to 80	

Table 4-1A. Lime and Fertilizer Suggestions — Field Crops

Area of State or Soil Type	Optimum pH	Plant Nutrient Suggestions When Soil Test is Unavailable[a]			Remarks
		N factor	P_2O_5 lb/acre	K_2O lb/acre	
Sorghum, Grain					
Mineral soils	6.0	1.5 to 2.0 lb/bu	10 to 20	50 to 70	Apply 20% to 25% of the N before planting. Apply remainder as a topdressing. If used for silage, increase N and K_2O by 40 pounds per acre. Sensitive to manganese deficiency when soil pH is greater than 6.2.
Mineral-organic soils	5.5				
Organic soils	5.0				
Soybean					
Mineral soils	6.0	0	10 to 20	50 to 70	Fertilizer may be applied on preceding crop. Inoculate when planting in new land. If soil pH is low and lime is not applied, use 0.5 ounce of sodium molybdate per acre seed treatment to facilitate N fixation (not a recommended substitute for lime). Test organic soils to determine copper needs. Soybean is sensitive to manganese deficiency when soil pH is greater than 6.2.
Mineral-organic soils	5.5				
Organic soils	5.0				
Tobacco, Burley Greenhouse or Outdoor Float System - Float system: 5 to 7 days after seeding					
	5.5 to 6.0 (water)	See Remarks	See Remarks	See Remarks	75 to 100 ppm N from 20-10-20 or similar ratio fertilizer. Choose fertilizers with no more than 0.2% boron. Avoid fertilizers with 50% or more of N from urea. Test source water before using, and solutions during the season for nutrient levels, alkalinity, and conductivity.
Tobacco, Burley Greenhouse or Outdoor Float System - 4 weeks after seeding					
	5.5 to 6.0 (water)	See Remarks	See Remarks	See Remarks	75 to 100 ppm N from 20-10-20 or similar ratio fertilizer. Test solutions for nutrient concentrations.
Tobacco, Burley Field - Planting					
	6.0 (5.5 if history of Black Shank)	See Remarks	10 to 20	40 to 200	Apply 80 to 100 pounds N per acre under conditions where no manure or legume is involved. Avoid excess chloride by using only fertilizers formulated for tobacco.
Tobacco, Burley Field - Sidedressing					
	6.0 (5.5 if history of Black Shank)	See Remarks	0	0	Apply 100 to 150 pounds N per acre 2 to 3 weeks after transplanting. It may contain up to 75% ammonium-N. For most soils, total N application (planting + sidedressing) of 180 to 200 pounds per acre is adequate for optimum yield in the mountains and 250 pounds is adequate in the piedmont.
Tobacco, Flue-Cured Greenhouse - Float system: 5 to 7 days after seeding					
Coastal plain and piedmont	5.5 to 6.0 (water)	See Remarks	See Remarks	See Remarks	100 to 150 ppm N from 20-10-20 or similar ratio fertilizer. Choose fertilizer with no more than 0.2% boron. Avoid fertilizers with 50% or more of N from urea. Test source water before using, and solutions during the season for nutrient levels, alkalinity, and conductivity.
Tobacco, Flue-Cured Greenhouse - Float system: 4 weeks after seeding					
Coastal plain and piedmont	5.5 to 6.0 (water)	See Remarks	See Remarks	See Remarks	100 ppm N from 20-10-20 or similar ratio fertilizer or ammonium nitrate. Test solutions for nutrient concentrations.
Tobacco, Flue-Cured Field - Planting					
Coastal plain and piedmont	6.0	See Remarks and *Flue-Cured Tobacco Information*, AG-187	10 to 20	90 to 110	Apply 35 to 40 pounds N per acre from tobacco fertilizer containing up to 75% ammonium-N. If needed, apply P_2O_5 at or within 7 days after transplanting. Piedmont soils are more likely to require fertilizer P than those in the coastal plain.
Tobacco, Flue-Cured Field - Sidedressing					
0 to 10 in. to clay		See Remarks	0	See Remarks	Apply 20 pounds N per acre 2 to 3 weeks after transplanting. N source may contain up to 75% ammonium-N. N-P-K ratios of 1-0-0 or 1-0-1 are sufficient for most tobacco soils if 90 to 110 pounds of K_2O per acre were supplied by the base fertilizer at planting. Use plant tissue analysis to identify nutrient deficiencies that need to be corrected.
11 to 15 in. to clay		See Remarks	0	See Remarks	Apply 30 pounds N per acre under conditions described in remarks for 0 to 10 inches to clay. Use plant tissue analysis to identify nutrient deficiencies that need to be corrected.
Over 15 in. to clay		See Remarks and *Flue-Cured Tobacco Information*, AG-187	0	See Remarks	Apply 40 pounds N per acre under conditions described in remarks for 0 to 10 inches to clay. Use plant tissue analysis to identify nutrient deficiencies that need to be corrected.
Tobacco, Flue-Cured Field - Adjustment for leaching					
		See Remarks and *Flue-Cured Tobacco Information*, AG-187			Replace N and K that are lost by leaching as early as conditions permit. Adjustments are normally not needed on soils with clay less than 10 inches from the surface. Magnesium and sulfur leaching may also be a concern on deep sandy soils. Use plant tissue analysis as a guide for determining nutrient deficiencies that need to be corrected.

[a] Suggested rates may overestimate or underestimate actual needs. Take soil samples to assure accurate nutrient requirements.

Table 4-1B. Lime and Fertilizer Suggestions — Pasture and Hay Crops

Commodity	Purpose	Area of State	Optimum pH	Plant Nutrient Suggestions When Soil Test is Unavailable[1]			Remarks
				N Based on RYE	P_2O_5 lb/acre	K_2O lb/acre	
Alfalfa	Seeding	All	6.5	See Remarks	90	150	Apply 20 pounds N per acre before planting. Apply 3 pounds boron per acre. Use inoculant at seeding.
	Annual maintenance	All	6.5	0	45	130	Apply first topdressing in spring of second year, following seeding. Each annual topdressing should include 2 pounds boron per acre.
Ladino Clover/ Grass Mixture (>30% clover)	Seeding	All	6.5	See Remarks	90	150	Apply 20 pounds N per acre before planting.
	Annual maintenance	All	6.5	0	45	150	Apply first topdressing in spring of second year, following seeding. When legumes make up 30% of stands, yield will be similar to pure grass receiving 150 to 200 pounds of N per acre.
Bluegrass/ White Clover Mixture (>30% clover)	Annual maintenance	Mountain	6.0	0	40	90	
Tall Fescue Orchardgrass, Timothy, Prairiegrass	Seeding	All	6.5	See Remarks	45	70	Apply 40 to 60 pounds N per acre at planting.
	Annual maintenance	All	6.0	40 to 50 lb/ton[2]	40	90	Use higher N rate for maximum production. Apply 1/2 N in February to March and 1/2 N in August to September; or 1/3 N in February, 1/3 N in April, and 1/3 N in August.
Bermudagrass, hybrid and improved seed cultivars	Sprigging	All	6.0	See Remarks	45	90	Apply 60 to 80 pounds N per acre before planting. If sprigged in March, apply 1/2 N in May and 1/2 N in July first year. May apply additional 60 to 80 pounds N/acre if complete soil covered before Sept. 1. If planted with a companion legume, lime to soil pH 6.5.
	Annual maintenance	All	6.0	40 to 50 lb/ton[2]	40	180	Apply N in three or four applications, but not later than Sept. 1. If planted with a companion legume, lime to soil pH 6.5.
Sudan Hybrid Sorghum — Sudan hybrids Pearl Millet Ryegrass Small Grain	Seeding	All	6.0	See Remarks	40	90	Apply 50 to 60 pounds N per acre planting. Apply N in two to four applications following each harvest except the last one.
	Topdressing	All	6.0	40 to 60 lb/ton[2]	0	0	Apply N in 2 to 4 applications following each harvest except the last one.
Red Clover, Fescue, Orchardgrass Mixture (>30% clover)	Seeding	Piedmont	6.5	See Remarks	90	150	Apply 20 pounds N per acre before planting.
	Annual maintenance	Piedmont	6.5	0	25 to 35	60 to 80	Apply in second and third year if stand is adequate.
Switchgrass Flaccidgrass Caucasian Bluestem Eastern Gammagrass	Seeding	All	5.5 to 6.0	See Remarks	20	80	Apply 40 pounds N per acre after plants are 6 to 10 inches high to minimize grassy weed competition.
	Maintenance	All	5.5 to 6.0	Switchgrass: 30 to 40 lb/ton All others: 35 to 45 lb/ton	20	80	Apply 1/2 N in May, 1/2 N in July, or 1/3 each in May, June, and July.

[1] Suggested rates may overestimate or underestimate actual needs. Take soil samples to assure accurate nutrient requirements.

[2] Rate is for tons of dry matter produced. Recognize that while hay and silage crops remove most or all of the above ground vegetation and the nutrients contained therein, grazing animals excrete 90% or more of the nutrients they consume from feed. Therefore, consider how grazing animals are managed to uniformly distribute the excreta over the pasture. To minimize N loss to the environment, N requirements for pastures should be reduced by 25% to 50% compared to what would be used for hay-silage production; yields may or may not be reduced.

Fertilizer Suggestions for Tree Fruit

M. L. Parker, Horticultural Science; and D. H. Hardy, NCDA&CS

Table 4-2. Fertilizer Suggestions for Tree Fruit

Purpose	Material	Amount	Precautions and Remarks
Apples			
Preplant	Lime and P fertilizer	Depends on soil test.	Prepare soil as deep as possible before planting. Take soil samples at least 12 inches deep (preferably 0 to 8 inches and 8 to 16 inches) for lime and phosphorus recommendation. Apply one-half of total, adjusted for the depth of incorporation, and plow down; apply other half and work in well.
Improve and/or maintain growth of young trees	10-10-10 or its equivalent	1 lb for each year of age until tree begins bearing. Then as recommended from leaf analysis.	Apply before rainfall or irrigate after application before buds swell in the spring.
To raise boron level of tree	Solubor	1 lb/100 gal of spray at first cover	If leaf analysis shows a deficiency, use additional cover sprays as recommended from leaf analysis to reduce "cork spot." Dry years and large fruit may enhance the incidence of "cork spot."
Growth and fruit development	Nitrogen	1.25 lb of actual N for trees producing 10 to 15 bu of apples	Annual terminal growth should be about 12 inches. Use observations of growth, crop size, and fruit condition plus the leaf analysis to determine the yearly application.
Increase foliar level	Potassium	Apply according to leaf analysis. Rate dependent upon soil analysis.	Leaf analysis is a good indicator of the need for a soil application.
Increase calcium level of tree	Gypsum ($CaSO_4$)	15 to 50 lb/tree with a 6- to 10-ft radius.	Apply only as needed by low soil or tissue calcium. One application will usually last 3 to 5 years.
	Calcium nitrate	Apply in late fall or early spring at rate to supply recommended nitrogen.	Applied as soil applications to increase calcium supply and reduce "Bitter Pit."
Foliar application	Calcium nitrate	3 lb/100 gal for sprays two weeks apart and ending 2 to 3 weeks before harvest.	Apply to reduce the incidence of "Bitter Pit." Excessive use of $CaNO_3$ may result in excessive tree vigor, which may actually worsen Bitter Pit. Both soil and foliar applications may be needed on large-fruited varieties. Leaf analysis may be beneficial.
	Calcium chloride	Same time as calcium nitrate. 2 lb/100 gal water.	Apply to reduce incidence of "Bitter Pit." DO NOT apply when temperature is 85 degrees F or above.
Peaches			
Preplant	Lime and Phosphorus	Depends on soil test.	Apply dolomitic lime necessary to raise soil pH above 6.0. Apply phosphorus to raise levels to desired range as indicated by soil test.
Tree growth, first year	N, P_2O_5, & K_2O	5 lb/acre of each per application.	Broadcast 0.5 pound of 10-10-10 around trees after growth starts in spring (April). Repeat every 4 to 6 weeks until August on sandy soils. On heavier soils, apply 0.5 pound of 10-10-10 one month after planting and 0.5 pound of 10-10-10 in May.
Tree growth, second year	N, P_2O_5, and K_2O	10 lb/acre of each per application.	Double amounts used first year. Make first application before growth starts in March and repeat in May. In sandy soils or if leaching is severe, an additional application may be made in July.
Tree growth, third year	N, P_2O_5, and K_2O	30 lb/acre of each	Make first application of 15 pound of each before growth starts and repeat in 6 to 8 weeks. If leaching is severe, repeat in July.
Growth and fruit development of young, bearing trees	N, P_2O_5, & K_2O	70 lb/acre of each (Determine by soil and foliar analysis.)	Broadcast under trees 40 pounds/acre each of N, P_2O_5, and K_2O (for example 400 pounds/acre 10-10-10) before growth starts. Add 30 pounds/acre each of N, P_2O_5, and K_2O after fruit set. If soil phosphorus test is high, omit P in second application. If leaching is severe, apply 20 to 30 pounds N/acre after harvest.
Growth and fruit development of mature trees	N, P_2O_5, and K_2O	70 lb/acre of each (Determine by soil and foliar analysis.)	Broadcast under trees 40 pounds/acre each of N, P_2O_5, and K_2O (for example, 400 pounds/acre 10-10-10) before growth starts. Add 30 pounds/acre each of N, P_2O_5, and K_2O after fruit set. If soil test phosphorus is high, omit P in second application. If leaching is severe, apply 30 pounds N/acre after harvest.
Increase boron level of tree	Solubar		Apply boron to producing trees by spraying foliage with Solubor or equivalent product at a rate of 0.75 pounds Solubor/100 gallons of water once each year.
Soil pH maintenance	Dolomitic lime		Maintain soil pH above 6.0.

Fertilizer Suggestions for Small Fruit

B. Cline, G. E. Fernandez, and M. Hoffman, Horticultural Science; and K. Hicks, NCDA&CS

For best results, fertilize using soil and tissue test information. If unavailable, use the general suggestions below.

Table 4-3. Fertilizer Suggestions for Small Fruit

Purpose	Material	Amount	Precautions and Remarks
Blackberries			
Preplant	Lime, P_2O_5, and K_2O	Apply based on a recent soil test report.	Blackberries can be grown on a variety of soil types. Regardless of the soil type, however, organic matter additions, pH adjustments, and incorporation of phosphorus (P) and potassium (K) should be completed before planting to optimize productivity. Take a soil test 3 to 6 months prior to planting to ensure that soil amendments are added appropriately.
Growth first year	N	20 to 50 lb/acre	Apply N in a split application: 1 to 2 weeks after an early spring planting and again 30 days later. Optimally, N can be portioned out through the drip irrigation system on a weekly basis.
	P_2O_5 and K_2O	30 to 60 lb/acre	Assuming pre-plant P and K were applied according to soil test recommendations, additional P and K during the first year should not be needed. If this is not the case, apply P and K in 4-inch bands around but not closer than 6 inches from stems. Optimally, portion out P and K through the drip irrigation system on a weekly basis during the growing season.
Growth second year	N	50 to 80 lb/acre (use up to 80 lb/acre for sandy soil)	Apply N in a split application. Apply first application starting in March. Spread fertilizer uniformly in 4-inch bands around but not closer than 6 inches from stems. Optimally, N can be portioned out through the drip irrigation system on a weekly basis from March through May. Apply 10 to 30 lb/acre N after harvest. Use primocane leaf tissue analysis 10 to 14 days post-harvest to optimize N fertilization. Monitor growth and adjust N appropriately to achieve optimal growth rate; check excessive vegetative growth with a reduction in N rate.
	P_2O_5 and K_2O	30 to 60 lb/acre	P and K can be applied in the fall if this is most convenient. Optimally, apply P and K during the growing season as per N recommendations. Adjust P and K rates as needed according to tissue analysis.
Growth third year and mature planting	N	60 to 80 lb/acre	Apply N, P and K according to recommendations for second year. Use primocane leaf tissue analysis 10 to 14 days post-harvest to optimize and manage fertilization program.
	P_2O_5 and K_2O	30 to 60 lb/acre	
Blueberries			
Site modification one year to six months prior to planting	Raised beds, Sulfur, organic additions of peat moss or pine bark, P_2O_5, and K_2O	Apply based on a recent soil test report	Most NC soils require modification (raised beds, organic additions) to grow blueberries successfully, with the exception of certain high-organic, acid soils in the coastal plain. Blueberries require an acid soil with an organic matter content (HM%) above 3%, with good aeration and constant moisture. A soil pH of 5.0 or less is needed for highbush and southern highbush blueberries. Rabbiteye blueberries will tolerate slightly higher pH, up to 5.3. Sulfur at a rate of 1 pound per 100 square feet can be used to lower pH one point (for example, from 6.0 down to 5.0). Use 2 pounds per 100 square feet on heavier soils. Where needed, apply sulfur one year in advance and re-test prior to planting. Do not over-apply sulfur.
Growth first year	N, P_2O_5, and K_2O	40 to 80 lb 14-28-14/acre per application	Apply after first flush of growth and repeat every 4 to 6 weeks until mid-August. Extend application interval during dry periods until rainfall has totaled 4 inches Based on 1,360 plants per acre.
Growth second year	N, P_2O_5, and K_2O	Double first year amount at first application only	Use same schedule and amounts after first application as first year.
Growth and fruit development of bearing plants	N, P_2O_5, and K_2O	150 to 250 lb 14-28-14/acre	Apply 2/3 of this amount before bloom and 1/3 4 to 6 weeks later (early May). Apply 10 to 30 pounds additional N/acre immediately after harvest if more vigorous growth is desired. Apply 50 pounds per acre of diammonium phosphate (18-48-0) in mid-August to maintain plant vigor if P_2O_5 is low or leaching has been severe.
Grapes, Bunch			
Same as muscadine schedule. Petiole analysis can show which nutrients are limiting. Collect petioles opposite the first or second flower/fruit cluster at full bloom to veraiso. Contact your local Cooperative Extension agent for further information.			
Grapes, Muscadine			
Preplant	Lime, P_2O_5, and K_2O	Apply based on a recent soil test report.	Any pH adjustments and incorporation of phosphorus (P) and potassium (K) should be completed before planting to optimize productivity. Take a soil test 3 to 6 months prior to planting to ensure that soil amendments are added appropriately.
First year	N, P_2O_5, and K_2O	0.25 lb 10-10-10 per vine per application	Apply after growth starts (late April to early May) and repeat in June and July (but no later than mid-July as winter injury may occur). Broadcast in a circle at least 18 inches from the trunk.
Second year	N, P_2O_5, and K_2O	0.5 lb 10-10-10 per vine per application	Apply in March and again in May and early July. To minimize the potential for winter cold injury, Piedmont and foothills growers should omit the July fertilizer application. Do not put fertilizer closer than 21 inches from the trunk.
Third year	N, P_2O_5, and K_2O	0.75 lb 10-10-10 per vine per application	Apply in March, in May and again in late June. Piedmont and foothills growers should omit the late June fertilizer application. Do not put fertilizer closer than 21 inches from the trunk.
Mature vines	N, P_2O_5, and K_2O	200 lb 10-10-10/acre per application	Apply in March (near bud break), and again in late May. If more vigorous growth is desired, add an additional 20 pounds N per acre in late June (from 200 pounds 10-10-10). Omit this last application in the piedmont and foothills. In Eastern NC, an alternative fertilizer to 10-10-10 that shows promise involves the application of 6-6-18 in March and mid-to-late May at the rate of 333 pounds/acre per application (instead of 10-10-10 at 200 pounds/acre per application). A final application of calcium nitrate is applied in late June at 133 pounds /acre (provides 20 pounds N per acre). Use leaf tissue analysis to monitor nutrient uptake and fine-tune fertilization program. Tissue samples should be collected in early to mid-June. Collect a double fist full of mature leaves located opposite fruit clusters on fruiting shoots. Detach the petioles from the leaves before placing the leaf blades in a paper bag. Send samples to the Agronomic Division, NC Dept. of Agriculture and Consumer Services (see Chapter 3).
	B	1 lb Solubor/acre	For mature vineyards, a common recommendation has been to apply 1 pound of Solubor (20% boron) annually with 100 gallons of water per acre just before bloom. Boron deficiency is more likely on sandy soils with high pH. Excessive boron causes injury; do not exceed boron recommendations.

Table 4-3. Fertilizer Suggestions for Small Fruit

Purpose	Material	Amount	Precautions and Remarks
Raspberries			
Preplant	Lime, P_2O_5, and K_2O	Apply based on a recent soil test report.	Test before planting and apply P_2O_5, and K_2O and lime according to soil test.
Growth first year	N, P_2O_5, and K_2O	250 to 500 lb 10-10-10/acre	Fertilize 30 to 60 days after planting. Apply fertilizer in a band at the side of the row but not closer than 6 inches from stems. If using a drip system, the nutrients can be added via the drip system. Portion out the fertilizer at the recommended rates weekly or as needed.
Growth second year	N, P_2O_5, and K_2O	350 to 500 lb 10-10-10/acre	Apply fertilizer in a band at the side of the row but not closer than 6 inches from stems. If using a drip system, the nutrients can be added via the drip system. Portion out the fertilizer at the recommended rates weekly or as needed.
Growth third year and mature planting	N, P_2O_5, and K_2O	500 to 800 lb 10-10-10/acre	Apply fertilizer in a band at the side of the row but not closer than 6 inches from stems. If using a drip system, the nutrients can be added via the drip system. Portion out the fertilizer at the recommended rates weekly or as needed.
Strawberries, Matted-row			
Growth of new planting	N	30 to 40 lb N/acre	Apply in May and repeat in August or September on sandy soils. An additional 20 to 30 pounds N per acre may be applied in January.
	P_2O_5, K_2O, and lime	Depends on soil test	Test before planting and apply P_2O_5, K_2O, and lime based on soil test.
Growth and fruit development	N	30 to 40 lb N/acre	Apply in August or September and in sandy soils again in January.
	N, P_2O_5, and K_2O	300 to 400 lb 10-10-10	Apply after harvest. If soil test for P and K are high, 30 to 40 pounds of N may be used rather than 10-10-10.
Strawberries, Plasticulture			
Preplant (fall)	N	60 lb/acre	Broadcast and incorporate before bedding. Calcium ammonium nitrate (CAN) or a complete fertilizer (if P and K recommended) may be used.
	P_2O_5, K_2O, and lime	Refer to soil test. If not available, apply 60 lb P_2O_5 and 120 lb K_2O/acre.	Soil test 3 to 6 months prior to planting. Apply lime at least 3 months before planting. Broadcast and incorporate recommended nutrients before bedding.
Preharvest (spring)	N	1/2 to 1 lb/acre/day (3.5 to 7 lb/acre/week) based upon petiole nitrate test	Begin biweekly tissue testing when plants begin growing in the spring. Adjust rate or omit applications depending on tissue test interpretation. Weekly injection of fertilizer is preferred; however, biweekly applications are a common practice. N fertilizer needs to be greenhouse-grade to ensure solubility and avoid clogging the drip tape emitters [i.e., Calcium Nitrate, Potassium Nitrate, Urea Ammonium Nitrate (UAN)].
	Other nutrients	Depends on tissue test	Biweekly tissue tests will indicate need. If B is needed, apply at 1/8 pound B per acre. If tissue tests indicate the need for multiple nutrient supplements, do not assume that all fertilizer components will be compatible for a single injection. Perform a jar test with a small batch to check for mixture solubility and compatibility.

Lime and Fertilizer Suggestions for Lawns

Charles Peacock, Grady Miller, and Matt Martin, Crop and Soil Sciences

Suggested Establishment Fertilization

Collect a soil sample for NCDA&CS analysis and follow lime and fertilizer recommendations. If the soil has not been tested, incorporate 75 pounds of ground limestone, except for centipedegrass, and 15 pounds of 0-14-14 fertilizer (or equivalent) per 1,000 square feet into the soil to a depth of 4 to 8 inches before seeding. At seeding, apply 1 pound nitrogen per 1,000 square feet from a turf-grade fertilizer in which one-fourth to one-half of the nitrogen is slowly available (e.g., 12-4-8 or 16-4-8). Use half of this fertilization rate when establishing centipedegrass. For more information, see *Carolina Lawns* or the lawn maintenance calendar for your specific grass. These can be found at http://www.turffiles.ncsu.edu. You can also request copies at your local North Carolina Cooperative Extension center.

Table 4-4A. Suggested Maintenance Fertilization for Coastal Plain[1]

Lawn Grass Type	Monthly Nitrogen Application Rate per 1,000 Square Feet[2]												Total lb N/ 1,000 sq ft/yr
	Jan	Feb	March	April	May	June	July	Aug[3]	Sept	Oct	Nov	Dec	
Bermudagrass													
Basic				1		1		1					3
High				1	1	1	1	1	1				6
Centipedegrass[3]													
Basic					1								1
High					1			1					2
Fescue, Tall													
Basic		0.5							1		0.5		2
High		1	0.5						1	1	0.5		4
St. Augustinegrass													
Basic					1			1					2
High				0.5	1	0.5	1	0.5	0.5				4
Zoysiagrass (Emerald and Meyer cultivars)													
Basic					1			1					2
High				1		1		1					3
Zoysiagrass (other cultivars)													
Basic					1			1					2
High				1	0.5	1	0.5	1					4

[1] All rates are per 1,000 square feet. Multiply by 43.5 to convert to an acre basis. Follow table suggestions in the absence of soil test recommendations to the contrary. With the exception of centipedegrass, use a complete (N-P-K) turf-grade fertilizer in which 1/4 to 1/2 of the nitrogen is slowly available and that has a 3-1-2 or 4-1-2 analysis (e.g., 12-4-8, 16-4-8).

[2] In the absence of soil test recommendations, apply about 1 pound potassium per 1,000 square feet using 1.6 pounds of muriate of potash (0-0-60), 5 pounds of sul-po-mag (0-0-22), or 2 pounds of potassium sulfate (0-0-50) to bermudagrass, centipedegrass, and St. Augustinegrass.

[3] Centipedegrass should be fertilized very lightly after establishment. Fertilize established centipedegrass using a low-phosphorus, high-potassium fertilizer with an analysis approaching 1-1-2 or 1-1-3. Fertilizers absent of phosphorus are preferred if soils supporting centipedegrass exhibit moderate to high levels of phosphorus.

Table 4-4B. Suggested Maintenance Fertilization for Central Piedmont[1]

Lawn Grass Type	Monthly Nitrogen Application Rate per 1,000 Square Feet[2]												Total lb N/ 1,000 sq ft/yr
	Jan	Feb	March	April	May	June	July	Aug[3]	Sept	Oct	Nov	Dec	
Bermudagrass													
Basic					1		1		1				3
High				1	1	1	1	1	1				6
Centipedegrass[3]													
Basic					1								1
High					1			1					2
Fescue, Tall													
Basic		0.5							1		0.5		2
High		1	0.5						1	1	0.5		4
Kentucky Bluegrass													
Basic		0.5							1		0.5		2
High		1	0.5						1	1	0.5		4
Kentucky Bluegrass-Fine Fescue Mix													
Basic		0.5							1		0.5		2
High		1	0.5						1	1	0.5		4
Kentucky Bluegrass-Tall Fescue Mix													
Basic		0.5							1		0.5		2
High		1	0.5						1	1	0.5		4
Kentucky Bluegrass-Tall Fescue-Fine Fescue Mix													
Basic		0.5							1		0.5		2
High		1	0.5						1	1	0.5		4
Kentucky Bluegrass-Perennial Ryegrass Mix													
Basic		1							1		1		3
High		1	0.5						1	1	0.5		4
St. Augustinegrass													
Basic					1								2
High					1			1	1				3
Zoysiagrass (Emerald and Meyer cultivars)													
Basic					1								1
High				1				1					2
Zoysiagrass (other cultivars)													
Basic					1		1						2
High				1		1		1					3

[1] All rates are per 1,000 square feet. Multiply by 43.5 to convert to an acre basis. Follow table suggestions in the absence of soil test recommendations to the contrary. With the exception of centipedegrass, use a complete (N-P-K) turf-grade fertilizer in which 1/4 to 1/2 of the nitrogen is slowly available and that has a 3-1-2 or 4-1-2 analysis (e.g., 12-4-8, 16-4-8).

[2] In the absence of soil test recommendations, apply about 1 pound potassium per 1,000 square feet using 1.6 pounds of muriate of potash (0-0-60), 5 pounds of sul-po-mag (0-0-22), or 2 pounds of potassium sulfate (0-0-50) to bermudagrass, centipedegrass, St. Augustinegrass, and zoysiagrass.

[3] Centipedegrass should be fertilized very lightly after establishment. Fertilize established centipedegrass using a low-phosphorus, high-potassium fertilizer with an analysis approaching 1-1-2 or 1-1-3. Fertilizers absent of phosphorus are preferred if soils supporting centipedegrass exhibit moderate to high levels of phosphorus.

Table 4-4C. Suggested Maintenance Fertilization for the Mountains[1]

Lawn Grass Type	Monthly Nitrogen Application Rate per 1,000 Square Feet[2]												Total lb N/ 1,000 sq ft/yr
	Jan	Feb	March	April	May	June	July	Aug	Sept	Oct	Nov	Dec	
Bermudagrass													
Basic					1		1						2
High					1	1	1	1					4
Fescue, Tall													
Basic			0.5					1		0.5			2
High			1					1		1			3
Kentucky Bluegrass													
Basic			1					1					2
High			1					1					3
Kentucky Bluegrass-Fine Fescue Mix													
Basic			1					1					2
High			1					1		1			3
Kentucky Bluegrass-Tall Fescue Mix													
Basic			1					1					2
High			1					1		1			3
Kentucky Bluegrass-Tall Fescue- Fine Fescue Mix													
Basic			1					1					2
High			1					1		1			3
Kentucky Bluegrass- Perennial Ryegrass Mix													
Basic			1					1		0.5			2.5
High			1					1	1	0.5			3.5
Zoysiagrass (other cultivars)													
Basic					0.5			0.5					1
High					1		1						2

[1] All rates are per 1,000 square feet. Multiply by 43.5 to convert to an acre basis. Follow table suggestions in the absence of soil test recommendations to the contrary. With the exception of centipedegrass, use a complete (N-P-K) turf-grade fertilizer in which 1/4 to 1/2 of the nitrogen is slowly available and that has a 3-1-2 or 4-1-2 analysis (e.g., 12-4-8, 16-4-8).

[2] In the absence of soil test recommendations, apply about 1 pound potassium per 1,000 square feet using 1.6 pounds of muriate of potash (0-0-60), 5 pounds of sul-po-mag (0-0-22), or 2 pounds of potassium sulfate (0-0-50) to bermudagrass, St. Augustinegrass, and zoysiagrass.

Fertilizer Suggestions for Ornamental Plants in Landscapes

B. Fair, Horticultural Science

The fertilizer suggestions given in the table are intended as a general guide. They are not a replacement for soil analyses from samples taken on a regular basis to assure maintenance of good nutrient availability and soil pH. It is always a good idea to collect samples for a foliar analysis as well to determine if the plant itself is functioning properly. Prior to fertilizing any plants, make sure you have a clear objective for fertilizing. The most common reasons for fertilizing are - overcome a visible nutrient deficiency; eliminate a deficiency detected by a soil and/or foliar analysis; increase vegetative growth, flowering, or fruiting; and increase the vitality of the plant.

Fertilizer may not be needed or beneficial when: sufficient levels of all essential elements so that growth rate and condition of health are acceptable, high potential for certain pest problems, and/or herbicide damage with residual activity in plant.

Finally, be sure to apply fertilizer to only the target plants. Remove any fertilizer from adjacent hardscapes to prevent pollution of water resources.

Table 4-5. Fertilizer Suggestions for Ornamental Plants

Kind of Plant	Ratio	Type	Amount to Use	When to apply	Remarks
Trees	3-1-1 or 3-1-2	Slow Release	2 to 4 lb N/1,000 sq ft of root zone (at least 70% of dripzone) Do not exceed 6 lb N/1,000 sq ft per year	Apply slow-release split application between spring budbreak and leaf color in fall. Fall application does not necessarily predispose a plant to winter injury or promote additional growth.	Refer to ANSI A300 (Part 2) - 2011 Soil Management a. Modification, b. Fertilization, c. Drainage. Before applying fertilizer to any woody plant, you should outline your goals for fertilizing the plant(s). Acceptable reasons for fertilizing are to increase growth, flowering/fruiting, increase vigor/vitality, balance root and shoot growth, and to address a visible deficiency. Fertilizing with too much nitrogen can lead to increased herbivory or susceptibility to insects and some disease. Best application method is sub-surface liquid injection. Use broadcast dry granules only where no turf is present and only for nitrogen application. Be sure to irrigate or apply prior to a rain event. Phosphorous will bind to soil before plant roots are able to obtain sufficient quantities. Other methods include sub-surface dry drill-hole application, surface liquid application, foliar application, tree injection or implants. The last few methods should be used only when other options are impractical or under special circumstances. Always use products in accordance with manufacturer's recommendations.
		Quick Release	1 to 2 lb N/1,000 sq ft root zone	With sandy soils and areas with heavy rainfall or constant irrigation, multiple applications will be necessary. Split total dose evenly over these applications. Do not fertilize when plants are stressed. Do not fertilize during drought periods. **Do** fertilize when soil is moist. Only use quick release when a sufficient response cannot be accomplished with slow-release products. Slow-release products are always preferred for woody plants.	
Palms	3-1-3	15-5-15 including micronutrients, calcium and magnesium	1 to 1.5 lb N/1,000 sq ft in root zone (maybe 2 to 3 times out from the dripline on some palm species)		
Shrubs	3-1-1 or 3-1-2	Slow Release	2 to 4 lb N/1,000 sq ft of root zone (at least 70% of dripzone) Do not exceed 6 lb N/1,000 sq ft per year		
		Quick Release	1 to 2 lb N/1,000 sq ft root zone		
Flowers	1-2-2 or 1-2-1 or 1-1-1	Slow release Quick release	4 to 6 lb N/1,000 sq ft	Incorporate half slow release granular forms before planting; apply second half 6 weeks after planting If using liquid quick release form, apply every 1 to 4 weeks throughout growing season, using 1 qt. per sq ft	Refer to Horticultural Information Leaflet 551, *Bed Preparation and Fertilization Recommendations for Bedding Plants in the Landscape.*

Fertilizer Suggestions for Nursery Crops

A. V. LeBude, Horticultural Science, and K. Hicks, NCDA&CS

Table 4-6. Fertilizer Suggestions for Nursery Crops

Kind of Production	Amount to Use[1]	Remarks
Deciduous Tree Seedling Beds	Year 1: 50 lb nitrogen (N) per acre or 18 oz N per 1,000 sq ft of bed. Year 2: Beds receive 100lb N per acre.	Year 1: Surface apply after first true leaves appear. Base applications of other elements upon a soil test.
Field Production	Year 1: 50 lb of nitrogen per acre incorporated as preplant. Year 2: 0.5 to 1 oz N per plant. Year 3: 1 to 2 oz N per plant. Do not exceed 100 to 200 lb N per acre. For liquid fertilizer applications through drip irrigation, reduce N application by 1/2 rate of annual field grade fertilizer. Apply equal rates of N during several irrigation events.	Year 1: Incorporate nutrients into the soil before planting. For field grade fertilizers, apply 0.67 of total amount before bud swell, 0.33 in June. Base applications of other elements upon a soil test. For more information, see http://www.ces.ncsu.edu/depts/hort/nursery/cultural/cultural_docs/field-bmps/building_nursery_soils.pdf
Container Production	All essential elements must be provided when soilless mixes are used to produce nursery crops in containers. Controlled release fertilizers offer a consistent and reliable source of nutrients through the growing season. Formulations of nutrients and their release over time differ by fertilizer company and the optimal product chosen depends upon the species being grown as well as the location of the nursery in the state. Irrigation water quality can contribute significant amounts of some nutrients, such as calcium, magnesium, and iron. Test water quality of irrigation supplies at least once a year (e.g., mid-summer) to determine if nutrient adjustments are required. Diagnosis of whole production system nutritional problems requires analysis of foliar nutrient content (sample uppermost fully expanded leaves), collection of approximately 8 ounces of leachate solution (collect from containers approximately 30 minutes after irrigation), and an irrigation water sample (collect from the irrigation head). Current research suggests that dolomitic limestone rates depend upon the calcium and magnesium content in irrigation water. As a result, limestone may not be required if provided by irrigation, however if dolomitic limestone is needed add only 2 to 5 pounds per cubic yard of pine bark and sand potting mixes. Minor element supplements included in either NPK controlled release fertilizer products or from separate minor element packages are necessary in all pine bark mixes and should be incorporated if possible. Potting mixes containing composts, however, generally do not require dolomitic limestone or minor element supplements. Leachates, irrigation water, and foliar samples can be analyzed for $5 each by the Agronomic Division, NCDA&CS, 4300 Reedy Creek Road, Raleigh, NC 27607-6465.

[1] Rates may change with irrigation and soil type.

Lime and Fertilizer Suggestions for Vegetable Crops

J. R. Schultheis and J. M. Davis, Horticultural Science;
C. R. Crozier and D. L. Osmond, Crop and Soil Sciences

Important Notes:

1. Consult the Southeastern U.S. Vegetable Crop Handbook: http://www.thegrower.com/south-east-vegetable-guide/ for numerous crop management suggestions.
2. Since optimum fertilizer management practices vary widely due to the specific vegetable production system, soil type, weather, and previous management, plant tissue analysis may be needed to fine-tune decisions. Plant tissue samples can be analyzed for essential plant nutrients at a nominal fee by the N.C. Department of Agriculture and Consumer Services, Agronomic Services Lab, 1040 Mail Service Center, Raleigh, NC 27699-1040; (919) 733-2655. When using a private carrier to deliver the samples or dropping them off in person, use the physical address of 4300 Reedy Creek Road, Raleigh, NC 27607-6465. Consult the new plant tissue analysis guide for instructions for each specific crop http://www.ncagr.gov/agronomi/pdffiles/plantguide.pdf.
3. For most vegetables grown on light-textured soils, apply the total recommended P_2O_5 and K_2O together with 25 to 50 percent of the recommended nitrogen before planting. The remaining nitrogen can be sidedressed with a fertilizer containing nitrogen only. Sidedressing or topdressing potash (K_2O) is recommended only on extremely light sandy soils with very low cation exchange capacities.
4. It may be desirable to build up the phosphorus and potassium levels in infertile loam and silt loam soils more rapidly than provided by these recommendations. In such instances, add an additional 40 to 50 pounds of P_2O_5 and K_2O, respectively, to the recommendations listed in the table for soils testing low in phosphorus and potassium. Apply the additional amounts as a broadcast and plow down or broadcast and disk-in application.
5. In the absence of soil tests, use recommendations listed under medium phosphorus and medium potassium levels on light-textured soils that have been in intensive vegetable production.
6. For Piedmont growers producing vegetables on clay loam soils: Reduce the recommended nitrogen and potassium rates by 20 percent and increase the phosphorus rate by 25 percent of the rates indicated in this table.

Table 4-7. Lime and Fertilizer Suggestions for Vegetable Crops

Crop	Desirable pH	Nitrogen (N) lb/acre	Soil Phosphorus Level P_2O_5 lb/acre				Soil Potassium Level K_2O lb/acre				Total Amount of Nutrient Recommended and Suggested Methods of Application
			Low	Med	High	Very High	Low	Med	High	Very High	
Asparagus	6.5										
Growing crowns		100	200	100	50	0	200	150	50	0	Total recommended.
		50	200	100	50	0	100	75	50	0	Broadcast and disk in.
		50	0	0	0	0	100	75	0	0	Sidedress after cutting.
New Planting Crowns and direct seeding		50	200	100	50	0	200	100	100	0	Total recommended.
		0	200	100	50	0	100	75	50	0	Broadcast and plow down.
		50	0	0	0	0	100	25	50	0	Sidedress at first cultivation.
Cutting Bed or Nonhybrids		100	150	100	50	0	200	150	100	0	Total recommended.
		50	150	100	50	0	100	150	100	0	Broadcast and disk in.
		50	0	0	0	0	100	75	50	0	Sidedress at first cultivation.
New hybrids		100	200	150	100	0	300	225	150	0	Total recommended.
		50	200	150	100	0	150	100	75	0	Broadcast before cutting season.
		50	0	0	0	0	100	125	75	0	Sidedress after cutting.
		Apply 2 pounds boron (B) per acre every 3 years on most soils.									
Bean, Lima Single crop	6 to 6.5	70 to 110	120	80	40	20	160	120	80	20	Total recommended.
		25 to 50	80	40	20	0	120	80	60	0	Broadcast and disk-in.
		20	40	40	20	20	40	40	20	20	Band-place with planter.
		25 to 40	0	0	0	0	0	0	0	0	Sidedress 3 to 5 weeks after emergence.
Bean, Snap	6 to 6.5	40 to 80	80	60	40	20	80	60	40	20	Total recommended.
		20 to 40	40	40	0	0	40	40	0	0	Broadcast and disk-in.
		20 to 40	40	20	40	20	40	20	40	20	Band-place with planter.
Beet	6 to 6.5	75 to 100	150	100	50	0	150	100	50	0	Total recommended.
		50	150	100	50	0	150	100	50	0	Broadcast and disk-in.
		25 to 50	0	0	0	0	0	0	0	0	Sidedress 4 to 6 weeks after planting.
Broccoli	6 to 6.5	125 to 175	200	100	50	0	200	100	50	0	Total recommended.
		50 to 100	150	100	50	0	150	100	50	0	Broadcast and disk-in.
		50	50	0	0	0	50	0	0	0	Sidedress 2 to 3 weeks after planting.
		25	0	0	0	0	0	0	0	0	Sidedress every 2 to 3 weeks after first sidedressing.
		Apply 2 pounds boron (B) per acre with broadcast fertilizer.									

Table 4-7. Lime and Fertilizer Suggestions for Vegetable Crops

Crop	Desirable pH	Nitrogen (N) lb/acre	Recommended Nutrients Based on Soil Tests								Total Amount of Nutrient Recommended and Suggested Methods of Application
			Soil Phosphorus Level				Soil Potassium Level				
			Low	Med	High	Very High	Low	Med	High	Very High	
			P_2O_5 lb/acre				K_2O lb/acre				
Brussel Sprout, Cabbage, and Cauliflower	6 to 6.5	100 to 175	200	100	50	0	200	100	50	0	Total recommended.
		50 to 75	200	100	50	0	200	100	50	0	Broadcast and disk-in.
		25 to 50	0	0	0	0	0	0	0	0	Sidedress 2 to 3 weeks after planting.
		25 to 50	0	0	0	0	0	0	0	0	Sidedress if needed, according to weather.
		Apply 2 to 3 pounds boron (B) per acre and molybdenum (mo) per acre as 0.5 pound sodium molybdate per acre with broadcast fertilizer.									
Carrot	6 to 6.5	50 to 80	150	100	50	0	150	100	50	0	Total recommended.
		50	150	100	50	0	150	100	50	0	Broadcast and disk-in.
		25 to 30	0	0	0	0	0	0	0	0	Sidedress if needed.
		Apply 1 to 2 pounds boron (B) per acre with broadcast fertilizer.									
Celery	6 to 6.5	75 to 100	250	150	100	0	250	150	100	0	Total recommended.
		50	250	150	100	0	250	150	100	0	Broadcast and disk-in or drill deep.
		25 to 50	0	0	0	0	0	0	0	0	Sidedress 2 to 3 weeks after planting.
		Apply 2 to 3 pounds boron (B) per acre with broadcast fertilizer.									
Corn, Sweet	6 to 6.5	110 to 155	160	120	80	20	160	120	80	20	Total recommended.
		40 to 60	120	100	60	0	120	100	60	0	Broadcast before planting.
		20	40	20	20	20	40	20	20	20	Band-place with planter.
		50 to 75	0	0	0	0	0	0	0	0	Sidedress when corn is 12 to 18 in. tall.
		Apply 1 to 2 pounds boron (B) per acre with broadcast fertilizer. NOTE: On very light sandy soils, sidedress 40 pounds N per acre when corn is 6 in. tall and another 40 pounds N per acre when corn is 12 to 18 in. tall.									
Cucumber	6 to 6.5	80 to 160	150	100	50	25	200	150	100	25	Total recommended.
		40 to 100	125	75	25	0	175	125	75	0	Broadcast and disk-in.
		20 to 30	25	25	25	25	25	25	25	25	Band-place with planter 7 to 14 days after planting.
		20 to 30	0	0	0	0	0	0	0	0	Sidedress when vines begin to run, or apply in irrigation water.
		Drip fertilization: See "cucumber" in specific recommendations in the current Southeastern Vegetable Crop Handbook.									
Eggplant Bareground	6 to 6.5	100 to 200	250	150	100	0	250	150	100	0	Total recommended.
		50 to 100	250	150	100	0	250	150	100	0	Broadcast and disk-in.
		25 to 50	0	0	0	0	0	0	0	0	Sidedress 3 to 4 weeks after planting.
		25 to 50	0	0	0	0	0	0	0	0	Sidedress 6 to 8 weeks after planting.
		Apply 1 to 2 pounds boron (B) per acre with broadcast fertilizer.									
Plasticulture		145	250	150	100	0	240	170	100	0	Total recommended.
		50	250	150	100	0	100	100	100	0	Broadcast and disk in.
		95	0	0	0	0	140	70	0	0	Fertigate.
		Apply 1 to 2 pounds boron (B) per acre with broadcast fertilizer. Drip fertilization: See "eggplant" in specific recommendations in the current Southeastern Vegetable Crop Handbook.									
Endive, Escarole, Leaf Lettuce	6 to 6.5	75 to 125	200	150	100	0	200	150	100	0	Total recommended.
		50 to 75	200	150	100	0	200	150	100	0	Broadcast and disk-in.
		25 to 50	0	0	0	0	0	0	0	0	Sidedress 3 to 5 weeks after planting.
Iceberg Lettuce	6 to 6.5	85 to 175	200	150	100	0	200	150	100	0	Total recommended.
		60 to 80	200	150	100	0	200	150	100	0	Broadcast and disk-in.
		25 to 30	0	0	0	0	0	0	0	0	Sidedress 3 times beginning 2 weeks after planting.
Leafy Greens, Collard, Kale, Mustard	6 to 6.5	75 to 80	150	100	50	0	150	100	50	0	Total recommended.
		50	150	100	50	0	150	100	50	0	Broadcast and disk-in.
		25 to 30	0	0	0	0	0	0	0	0	Sidedress, if needed.
		Apply 1 to 2 pounds boron (B) per acre with broadcast fertilizer.									
Leek	6 to 6.5	75 to 125	200	150	100	0	200	150	100	0	Total recommended.
		50 to 75	200	150	100	0	200	150	100	0	Broadcast and disk-in.
		25 to 50	0	0	0	0	0	0	0	0	Sidedress 3 to 4 weeks after planting if needed.
		Apply 1 to 2 pounds boron (B) per acre with broadcast fertilizer.									

Table 4-7. Lime and Fertilizer Suggestions for Vegetable Crops

Crop	Desirable pH	Nitrogen (N) lb/acre	Soil Phosphorus Level — P₂O₅ lb/acre				Soil Potassium Level — K₂O lb/acre				Total Amount of Nutrient Recommended and Suggested Methods of Application
			Low	Med	High	Very High	Low	Med	High	Very High	
Cantaloupes and Mixed Melons Bareground	6 to 6.5	75 to 115	150	100	50	25	200	150	100	25	Total recommended.
		25 to 50	125	75	25	0	175	125	75	0	Broadcast and disk-in.
		25	25	25	25	25	25	25	25	25	Band-place with planter.
		25 to 40	0	0	0	0	0	0	0	0	Sidedress when vines start to run.
		Apply 1 to 2 pounds boron (B) per acre with broadcast fertilizer.									
Cantaloupes and Mixed Melons Plasticulture	6 to 6.5	75 to 150	150	100	50	25	200	150	100	25	Total recommended.
		25	150	100	50	25	100	75	50	25	Broadcast and disk in.
		50 to 100	0	0	0	0	100	75	50	0	Fertigate.
		Apply 1 to 2 pounds boron (B) per acre with broadcast fertilizer. Drip fertilization: See "muskmelon" in specific recommendations in the current Southeastern Vegetable Crop Handbook.									
Okra	6 to 6.5	100 to 200	250	150	100	0	250	150	100	0	Total recommended.
		50 to 100	250	150	100	0	250	150	100	0	Broadcast and disk-in.
		25 to 50	0	0	0	0	0	0	0	0	Sidedress 3 to 4 weeks after planting.
		25 to 50	0	0	0	0	0	0	0	0	Sidedress 6 to 8 weeks after planting.
		Apply 1 to 2 pounds boron (B) per acre with broadcast fertilizer. NOTE: Where plastic mulches are being used, broadcast 50 to 100 pounds nitrogen (N) per acre with recommended P₂O₅ and K₂O and disk incorporate prior to laying mulch. Drip fertilization: See "okra" in specific recommendations in the current Southeastern Vegetable Crop Handbook.									
Onion, Bulb	6 to 6.5	75 to 125	200	100	50	0	200	100	50	0	Total recommended.
		50 to 75	200	100	50	0	200	100	50	0	Broadcast and disk-in.
		25 to 50	0	0	0	0	0	0	0	0	Sidedress 4 to 5 weeks after planting.
Onion, Green		150 to 175	200	100	50	0	200	100	50	0	Total recommended.
		50 to 75	200	100	50	0	200	100	50	0	Broadcast and disk-in.
		50	0	0	0	0	0	0	0	0	Sidedress 4 to 5 weeks after planting.
		50	0	0	0	0	0	0	0	0	Sidedress 3 to 4 weeks before harvest.
		Apply 1 to 2 pounds boron (B) and 20 pounds sulfur (S) per acre with broadcast fertilizer.									
Parsley	6 to 6.5	100 to 175	200	150	100	0	200	150	100	0	Total recommended.
		50 to 75	200	150	100	0	200	150	100	0	Broadcast and disk-in.
		25 to 50	0	0	0	0	0	0	0	0	Sidedress after first cutting.
		25 to 50	0	0	0	0	0	0	0	0	Sidedress after each additional cutting.
Parsnip	6 to 6.5	50 to 100	150	100	50	0	150	100	50	0	Total recommended.
		25 to 50	150	100	50	0	150	100	50	0	Broadcast and disk-in.
		25 to 50	0	0	0	0	0	0	0	0	Sidedress 4 to 5 weeks after planting.
		Apply 1 to 2 pounds boron (B) per acre with broadcast fertilizer.									
Pea, English Spring plowed	5.8 to 6.5	40 to 60	120	80	40	0	120	80	40	0	Total recommended. Broadcast and disk-in before seeding.
Pea, Southern	5.8 to 6.5	16	96	48	0	0	96	48	0	0.	Broadcast and disk-in.
Pepper Bareground	6 to 6.5	100 to 130	200	150	100	0	200	150	100	0	Total recommended.
		50	200	150	100	0	200	150	100	0	Broadcast and disk-in.
		25 to 50	0	0	0	0	0	0	0	0	Sidedress after first fruit set.
		25 to 30	0	0	0	0	0	0	0	0	Sidedress later in season if needed.
Pepper Plasticulture		100 to 185	200	150	100	0	365	300	235	0	Total recommended
		50	200	150	100	0	100	100	100	0	Broadcast and disk in.
		50 to 135	0	0	0	0	265	200	135	0	Fertigage
		Drip fertilization: See "pepper" in specific commodity recommendations in *Plasticulture for Commercial Vegetables* (AG-489).									
Potato, Irish Loams and silt loams	5.8 to 6.2	100 to 150	110	90	70	50	200	150	50	50	Total recommended.
		85 to 135	60	40	20	0	200	150	50	50	Broadcast and disk in.
		15	50	50	50	50	0	0	0	0	Band-place with planter at planting.
Sandy loams and loamy sands		150	200	150	100	50	300	200	100	50	Total recommended.
		50	200	150	100	50	300	200	100	50	Broadcast and disk in.
		100	0	0	0	0	0	0	0	0	Sidedress 4 to 5 weeks after planting.
Pumpkin and Squash (Winter) Bareground	6 to 6.5	80 to 90	150	100	50	0	200	150	100	0	Total recommended.
		40 to 50	150	100	50	0	200	150	100	0	Broadcast and disk in.
		40 to 45	0	0	0	0	0	0	0	0	Sidedress when vines begin to run.

Table 4-7. Lime and Fertilizer Suggestions for Vegetable Crops

Crop	Desirable pH	Nitrogen (N) lb/acre	Soil Phosphorus Level P₂O₅ lb/acre				Soil Potassium Level K₂O lb/acre				Total Amount of Nutrient Recommended and Suggested Methods of Application
			Low	Med	High	Very High	Low	Med	High	Very High	
Pumpkin and Squash (Winter)		80 to 150	150	100	50	0	200	150	100	0	Total recommended.
		25 to 50	150	100	50	0	100	75	50	0	Disk in row
Plasticulture		55 to 100	0	0	0	0	100	75	50	0	Sidedress when vines begin to run.
Radish	6 to 6.5	50	150	100	50	0	150	100	50	0	Total recommended. Broadcast and disk-in or drill deep.
		Apply 1 to 2 pounds boron (B) per acre with broadcast fertilizer.									
Rutabaga and Turnip	6 to 6.5	50 to 75	150	100	50	0	150	100	50	0	Total recommended.
		25 to 50	150	100	50	0	150	100	50	0	Broadcast and disk in.
		25 to 50	0	0	0	0	0	0	0	0	Sidedress when plants are 4 to 6 in. tall.
		Apply 1 to 2 pounds boron (B) per acre with broadcast fertilizer.									
Spinach Fall	6 to 6.5	75 to 125	200	150	100	0	200	150	100	0	Total recommended.
		50 to 75	200	150	100	0	200	150	100	0	Broadcast and disk in.
		25 to 50	0	0	0	0	0	0	0	0	Sidedress or topdress.
Overwinter		80 to 120	0	0	0	0	0	0	0	0	Total recommended for spring application on overwintered crop.
		50 to 80	0	0	0	0	0	0	0	0	Apply in late February.
		30 to 40	0	0	0	0	0	0	0	0	Apply in late March.
Squash, Summer	6 to 6.5	100 to 130	150	100	50	0	200	150	100	0	Total recommended.
		25 to 50	150	100	50	0	150	100	50	0	Broadcast and disk in.
		50	0	0	0	0	0	0	0	0	Sidedress when vines start to run.
		25 to 30	0	0	0	0	0	0	0	0	Apply through irrigation system.
		Apply 1 to 2 pounds boron (B) per acre with broadcast fertilizer. Drip/trickle fertilization: See "summer squash" in specific recommendations in the current Southeastern Vegetable Crop Handbook.									
Sweet Potato	5.8 to 6.2	50 to 80	200	100	50	0	300	200	150	0	Total recommended.
		0	150	60	30	0	150	50	30	0	Broadcast and disk in.
		50 to 80	0	0	0	0	150	150	120	120	Sidedress 21 to 28 days after planting.
		Add 0.5 pounds of actual boron (B) per acre 40 to 80 days after planting.									
Tomato Bareground for sandy loams and loamy sands	6 to 6.5	80 to 90	200	150	100	0	300	200	100	0	Total recommended.
		40 to 45	200	150	100	0	300	200	100	0	Broadcast and disk in.
		40 to 45	0	0	0	0	0	0	0	0	Sidedress when first fruits are set and as needed.
TOMATO Bareground for loams and clay	6 to 6.5	75 to 80	200	150	100	0	250	150	100	0	Total recommended.
		50	200	150	100	0	250	150	100	0	Broadcast and plow down.
		25 to 30	0	0	0	0	0	0	0	0	Sidedress when first fruits are set and as needed.
		Apply 1 to 2 pounds boron (B) per acre with broadcast fertilizer.									
Tomato Plasticulture		130 to 210	200	150	100	0	420	325	275	0	Total recommended.
		50	200	150	100	0	295	220	125	0	Broadcast and disk in.
		80 to 160	0	0	0	0	295	220	125	0	Fertigate
		Apply 1 to 2 pounds boron (B) per acre with broadcast fertilizer. Drop fertilization: See "tomato" in specific recommendations in the current Southeastern Vegetable Crop Handbook.									
Watermelon Nonirrigated	6 to 6.5	75 to 90	150	100	50	0	200	150	100	0	Total recommended.
		50	150	100	50	0	200	150	100	0	Broadcast and disk in.
		25 to 40	0	0	0	0	0	0	0	0	Topdress when vines start to run.
Irrigated		100 to 150	150	100	50	0	200	150	100	0	Total recommended.
		50	150	100	50	0	150	150	100	0	Broadcast and disk in.
		25 to 50	0	0	0	0	0	0	0	0	Topdress when vines start to run.
		25 to 50	0	0	0	0	0	0	0	0	Topdress at first fruit set.
Plasticulture		125 to 150	150	100	50	0	200	150	100	0	Total recommended.
		25 to 50	150	100	50	0	100	75	50	0	Disk in row.
		100	0	0	0	0	100	75	50	0	Fertigation

NOTE: Excessive rates of N may increase hollow heart in seedless watermelons. Drip fertilization: See "watermelon" in specific recommendations in the current Southeastern Vegetable Crop Handbook.

Fertilizer Rules and Regulations

D. A. Crouse, C. R. Crozier, and D. L. Osmond, Crop and Soil Sciences; B. Bowers, NCDA&CS

Fertilizer, lime and landplaster are regulated by Plant Industry- NCDA&CS and rules and regulations in their entirety are found at http://www.ncagr.gov/plantindustry/pubs.htm.

Chlorine Guarantees for Tobacco Fertilizer

The maximum chlorine (Cl) guarantees permitted for tobacco plantbed fertilizer shall be:

1. For fertilizers with a nitrogen (N) guarantee up to and including 6 percent, 0.5 percent chlorine (Cl).
2. For fertilizers with a nitrogen (N) guarantee above 6 percent, 1 percent chlorine (Cl).

The maximum chlorine (Cl) guarantees permitted on field crop tobacco fertilizer shall be:

1. For fertilizer with a nitrogen (N) guarantee up to and including 4 percent, a maximum chlorine (Cl) guarantee of 2 percent.
2. For fertilizer with a nitrogen (N) guarantee greater than 4 percent, a maximum percent chlorine (Cl) guarantee not more than one-half of the respective total nitrogen (N) guarantee.

The maximum chlorine (Cl) permitted in tobacco top-dressers shall be 2 percent.

Size Standards for Agricultural Liming Material

Agricultural liming material shall conform to the following minimum screening standards:

1. Ninety percent must pass through U.S. Standard 20-mesh screen, with a tolerance of 5 percent.
2. For dolomitic limestone, 35 percent must pass through U.S. Standard 100-mesh screen; for calcitic limestone, 25 percent must pass through U.S. Standard 100-mesh screen, with a tolerance of 5 percent.

Additional Criteria for Agricultural Liming Material

1. A product must contain at least 6 percent magnesium (Mg) from magnesium carbonate to be classified as a dolomitic limestone.
2. There is no minimum calcium carbonate equivalent (CCE) requirement for limestone sold in North Carolina. However, the product must be labeled to show the amount necessary to equal that required from a liming material having a 90 percent CCE. Lime recommendations in North Carolina are based on 90 percent CCE. For example, a product having a CCE of 50 percent would be labeled "3,600 pounds of this material equals 1 ton of standard agricultural liming material."
3. Pelleted lime must slake down when it comes in contact with moisture, thereby meeting the size standards for agricultural liming materials given above.

Nutrient Content of Fertilizer Materials

C. R. Crozier and D. L. Osmond, Crop and Soil Sciences; R. Sherman, Biological and Agricultural Engineering; and A. V. LeBude, Horticultural Science

Although fertilizer materials may contain two or more nutrients, usually only one of the nutrients is commonly associated with the material. The content of that nutrient is generally well-known, but the contents of other associated nutrients are less well-known. Information on the contents of lesser-known nutrients in fertilizer materials exists but is often not readily available. Information presented below gives the nutrients and their contents that are normally found in fertilizer materials. In the case of micronutrient fertilizers, only the principle micronutrient is considered relevant. This information should be considered only as a guide, since actual nutrient contents may vary slightly, depending on the source of the fertilizer material. Additional information about the nutrient content of fertilizer and organic materials can be found in the NC State Extension publication AG-467 *Composting: A Guide to Managing Organic Yard Wastes*, or on the Internet at content.ces.ncsu.edu/composting-a-guide-to-managing-organic-yard-wastes

Table 4-8. Composition of Selected Micronutrient Fertilizers

Element	Fertilizer	Chemical Formula	Elemental Percentage (%)
Mn	Manganese sulfate	$MnSO_4 \cdot 3H_2O$	27
	Manganous oxide	MnO	41 to 68
	Manganese chloride	$MnCl_2$	7
	Manganese oxide	MnO_2	62 to 70
Mo	Sodium molybdate	$Na_2MoO_4 \cdot H_2O$	38
Cu	Copper sulfate	$CuSO_4 \cdot 5H_2O$	25
Zn	Zinc sulfate	$ZnSO_4 \cdot H_2O$	22 to 36
	Zinc oxide	ZnO	78
Fe	Ferrous sulfate	$FeSO_4 \cdot 7H_2O$	20
	Ferric sulfate	$Fe_2(SO_4)_3 \cdot 4H_2O$	23
B	Borax	$Na_2B_4O_7 \cdot 10H_2O$	11
	Solubor	$Na_2B_8O_{13} \cdot 4H_2O$	20

Table 4-9. Composition of Selected Organic Fertilizer Materials (see livestock & poultry manure section for additional information)

Nutrient Source	Analysis (%)[1]			Relative Rate of Nutrient Release
	N	P_2O_5	K_2O	
Animal tankage	7	9	0	Medium
Bone meal				
Raw	3	22	0	Very slow
Steamed	2	28	0	Slow
Castor pomace	5	1	1	Slow
Cotton seed meal	7	2	2	Slow
Dried blood	12	3	0	Fast
Feather meal	14	0.3	0.1	Fast
Hardwood ashes[2]	0.5	1	5	Slow
Linseed meal	5	2	2	Slow
Municipal yard and leaf compost[4]	0.5	0.4	0.8	Slow
Sheep wool pellets	5.8	0.9	2.4	Fast/Slow
Softwood ashes[2]	0.5	2	4	Slow
Tobacco stems	2	1	7	Slow
Turkey litter compost[3]	1.5	3.5	1	Slow
Food waste vermicompost	2.11	0.79	1.47	Slow
Animal manure vermicompost	1.67	3.04	0.55	Slow

[1] Percentages of nutrient elements may vary depending upon source. Values given are approximations.

[2] Plant-available N less than or equal to 0.25%.

[3] Compost products must be stabilized before use, or they may cause nutrient depletion.

[4] Municipal yard and leaf waste composts may contain measurable concentrations of metals, such as lead and zinc. A complete analysis of these soil amendments should be conducted before use.

Table 4-10. Composition of Typical N Solutions[1]

Contents (%)	No-Pressure Nonammonia Solutions[2]						Low Pressure Ammonia Solution	Aqua-Ammonia
Total N	16.0	19.0	21.0	28.0	30.0	32.0	37.0	20.0
NH_3	0	0	0	0	0	0	16.6	24.4
NH_4NO_3	45.8	54.3	60.0	40.0	42.2	44.3	66.8	0
Urea	0	0	0	30.0	32.7	35.4	0	0
Water	54.2	45.7	40.0	30.0	25.1	20.3	16.6	75.6
NO_3-N	8.0	9.5	10.5	7.0	7.4	7.7	11.7	0
NH_4-N	8.0	9.5	10.5	7.0	7.4	7.8	25.4[3]	20.0[3]
Urea-N	0	0	0	13.8	15.2	16.4	0	0
	Salt-out Temperatures (F)[4]							
	41	11	0	15	32	56		-58

[1] Values may vary depending on product.

[2] Several companies provide N solutions containing 3.5% to 5.0% sulfur.

[3] Includes N present as NH_3.

[4] Proprietary additives can alter salt-out temperatures; this should be verified with the dealer.

Table 4-11. Composition of Selected Fertilizer Materials[1]

Fertilizer Material	Chemical Formula	Nutrient Percentage (%)							
		NO_3-N	NH_4-N	Total N	P_2O_5	K_2O	Ca[4]	Mg	S
Ammonium nitrate	NH_4NO_3	17	17	34					
Monoammonium phosphate	$NH_4H_2PO_4$		11	11	48		1		2
Diammonium phosphate	$(NH_4)_2HPO_4$		16 to 18	16 to 18	46 to 48				
Ammonium sulfate	$(NH_4)_2SO_4$		21	21					24
Ammonium thiosulfate	$(NH_4)_2S_2O_3$			12					26
Anhydrous ammonia	NH_3		82	82					
Urea-Form[2]			38	38					
Calcium nitrate	$Ca(NO_3)_2$	15		15			19	1	
Nitrate of soda potash	$NaNO_3 \cdot KNO_3$	15		15		14			
Sodium nitrate	$NaNO_3$	16		16					
Urea	$CO(NH_2)_2 \cdot H_2O$		45 to 46	45 to 46					
Single superphosphate	$Ca(H_2PO_4)_2 + CaSO_4$				18 to 20		18 to 21		12
Triple superphosphate	$Ca(H_2PO_4)_2 \cdot H_2O$				42 to 50		12 to 14		1
Basic slag[3]	$5CaO \cdot P_2O_5 \cdot SiO$				2 to 17		3 to 33	3	
Potassium chloride	KCl					60 to 62			
Potassium nitrate	KNO_3	13		13		44			
Potassium sulfate	K_2SO_4					50 to 53		1	18
Potassium-magnesium sulfate	$K_2SO_4 \cdot 2MgSO_4$					22		11	23
Epsom salt	$MgSO_4 \cdot 7H_2O$						2	10	14
Gypsum	$CaSO_4 \cdot 2H_2O$						23		18.5

[1] Values may vary depending on product.

[2] Slow release N source.

[3] Lime value — about 0.67 agricultural limestone.

[4] Evaluate gypsum products marked as Ca sources based on guaranteed fertilizer analysis. Pure calcium sulfate is 29 percent calcium. Landplasters typically contain 70 to 85 percent calcium sulfate (21 to 25 percent calcium), while phosphogypsum and other products may contain 50 percent or less calcium sulfate (15 percent calcium). See Table 4-1 for rate recommendations for peanut.

Table 4-12. Composition of Selected Specialty, Alternative, or Fertilizer Efficiency Enhancer Materials

Not intended to be all inclusive, and inclusion does not imply endorsement. See product label for specific instructions. Some product evaluations at http://extension.agron.iastate.edu/compendium/index.aspx, "Compendium of Research Reports on Use of Non-Traditional Materials for Crop Production."

Product (Source)	Ingredient(s)	Primary Uses	Typical rate	Nutrient supply at typical rate (lb/A)	Comments
Accomplish LM (Loveland Products)	1% microorganisms, water-based fermentation mixture	With liquid fert., manure, or to surface residue	1 to 4 qt/ac		Non-rhizobial microbes plus mixture of compounds such as enzymes and organic acids resulting from fermentation.
AG-TEK™ BIO-D 4200 (Global Green Products)	Protein synthesized with aspartic acid, 4-0-0	In furrow and sidedress	1 to 2 qt/A	If 2 qt/A N-0.2	This is not a fertilizer replacement and must be used with a normal soil fertility program with full fertilizer.
Agrotain Plus (Koch)	NBPT+DCD (urease + nitrification inhibitors)	Mix with UAN solutions			Duration of urease and nitrification inhibition shorter at higher temperatures and perhaps organic matter-dependent. Affects urea and ammonium components of mixed N sources. Potential to reduce volatilization and leaching losses, especially in dry high pH situations or in sandy soils with heavy rainfall after application.
AVAIL (SFP)	Maleic itaconic copolymer 0-0-0	Broadcast or band with P fertilizer			Indirect mechanism, intended to increase soil CEC and influence soil cation interactions
Bio-forge (Stoller)	diformyl urea + potassium hydroxide (2-0-7)	In furrow	1 pt/A	N-0.025 K_2O-0.09	
Brandt Smart B-Mo (Brandt)	Boric acid + sodium molybdate (5% B + 0.5% Mo)	Foliar or soil	Foliar ½ to 1 pt/ac; soil 1 to 3 qt/ac	Foliar (1 pt/ac): 0.05 B + 0.005 Mo; Soil (3 qt/ac): 0.3 B + 0.03 Mo	For foliar, adjuvant is recommended & multiple applications may be needed if deficiency.
Environoc 401 Liquid (Biodyne)	Rhizobial inoculants plus fermentation residue				Unspecified on public website, but could include rhizobial and non-nodulating bacteria.
ESN (Agrium)	Urea (plastic-encapsulated) 44-0-0	broadcast		Each 1 lb 44-0-0: N-0.44	Temperature-dependent slow release, duration shorter at higher temperatures. Potential to reduce volatilization and leaching losses, especially in dry high pH situations or in sandy soils with heavy rainfall after application.
Excelis Maxx (Timac)	NBPT+DCD (urease + nitrification inhibitors)	Mix with UAN solutions or urea	UAN: 25 oz/ton, urea: 32 oz/ton		Duration of urease and nitrification inhibition shorter at higher temperatures and perhaps organic matter-dependent. Affects urea and ammonium components of mixed N sources. Potential to reduce volatilization and leaching losses, especially in dry high pH situations or in sandy soils with heavy rainfall after application.
Fertileader Elite (Timac Agro)	9-0-6 + Ca, B, patented ingredients	Foliar	3 pt/ac	N-0.3 K2O-0.2	Multiple products with highly soluble nutrients and patented ingredients.
Grasshopper (Grasshopper Fertilizer)	Multiple products (e.g. 30-8-10)	Foliar (also other products)	10 lb/A	10 lb of 30-8-10: N-3, P_2O_5-0.8, K_2O-1, B-0.005, Cu-0.005, Fe-0.05, Mn-0.005, Zn-0.01, Mo-0.0001	Multiple products with highly soluble ingredients and many with chelated micronutrients. Product list includes slow release N provided as methylene urea source.
GYP-Soil	Calcium sulfate (FGD source)	Broadcast			Considered equivalent sources of Ca and S as conventional gypsum when rate adjusted based on composition. Potential effects on soil physical properties and subsoil aluminum toxicity.
Instinct (Dow)	Nitrapyrin (nitrification inhibitor)	Mix with UAN solutions			Duration of nitrification inhibition lower at higher temperatures and lower in soils with high organic matter content. Affects urea and ammonium components of mixed N sources. Potential to reduce leaching losses, especially in sandy soils with heavy rainfall after application.
Invigor 8 (Genesis Ag)	0-37-37 derived from potassium acid phosphate	Seed treatment	1 lb per 20 units corn (approx. 1 lb/40 A)	0.01 lb P_2O_5, 0.01 lb K_2O	
K-Lime (Advanced Residuals Management LLC)	Paper mill by-product lime + K	Broadcast based on lime content	1 ton/ac ag lime	Verify label rate of lime and K with supplier	By product from paper mill processes. Need to verify ag lime equivalent and K concentration with supplier. Fine granular product.
Levitate (Loveland Products)	5-15-5 +Zn, fulvic acid	In furrow (IF), soil, foliar	IF:1 to 5 gpa, foliar 1 to 4 qt/ac	IF 5 gpa: 2.8N, 8.2 P2O5, 2.8 K2O, 0.8 Zn	
Limus (BASF)	NBPT + NPPT (two urease inhibitors)	Mix with UAN solutions			Duration of urease inhibition shorter at higher temperatures and perhaps organic matter-dependent. Affects urea component of mixed N sources. Potential to reduce volatilization losses, especially in dry high pH situations
MicroPack (Agriculture Solutions)	11-8-5 + B, Cu, Fe, Mn, Zn, Co, Mo	Foliar	2 qt/A, possibly 2 to 3 times	N-0.55, P_2O_5-0.4, K_2O-0.25, B-0.001, Cu-0.0025, Fe-0.005, Mn-0.0025, Zn-0.0025, Co-0.000025, Mo-0.000025	
Nachurs Solutions Multiple NPKS and micronutrient blends	Multiple products (e.g. HKW18 =3-18-18)	In furrow (IF), foliar	IF:2 to 6 gpa Foliar:1 to 3 gpa	2 gpa of 3-18-18: N-0.7, P_2O_5-4.2, K_2O-4.2	Multiple products with highly soluble ingredients formulated to minimize salt index.

Table 4-12. Composition of Selected Specialty, Alternative, or Fertilizer Efficiency Enhancer Materials

Not intended to be all inclusive, and inclusion does not imply endorsement. See product label for specific instructions. Some product evaluations at http://extension.agron.iastate.edu/compendium/index.aspx, "Compendium of Research Reports on Use of Non-Traditional Materials for Crop Production."

Product (Source)	Ingredient(s)	Primary Uses	Typical rate	Nutrient supply at typical rate (lb/A)	Comments
N-Boost 5 (Brandt)	Urea 5-0-0 & non-living yeast & bacteria	Foliar	2 to 3 qt/ac	0.2 to 0.3 lb N/ac	Foliar uptake of urea expected similar to urea solution.
Nutran	10-0-1 + 1% each S, Mn, Zn; 0.2% each B, Cu;	In furrow (IF), foliar	10 oz/ac	N-0.03; K,S,Mn,Zn-	Nutran
(Naturym LLC)	0.1% Fe			0.01 B,Cu-0.002; Fe-0.001	(Naturym LLC)
Nutrisphere (SFP)	Maleic itaconic copolymer 0-0-0	Band or broadcast with N fertilizers			Indirect mechanism, intended to increase soil CEC and influence soil cation interactions
N Zone (AgXplore International)	Adjuvant, glycols & polysaccharides, 0-0-0	Mix with urea, UAN, liquid manures			Marketed as penetrant aid for N sources. Most likely effect is with foliar application.
Optimize (Novozymes)	Rhizobial inoculants plus LCO promoter	Seed trt			Not intended to replace in-furrow inoculation on land not previously planted to soybean.
Pentilex (Biovante)	0-37-37	Seed trt	1 to 1.25gal /1000 lb	If 1 gal on 35k seed/A: P_2O_5-0.007, K_2O-0.007	
Quick-Sol	Sodium silicate	Soil, foliar	14 to 20 oz/A		Soluble silicon source, of potential benefit to some crops.
Rescue (Triangle)	7-4-9, & multiple other blends	Foliar	2 pt/ac	N-0.19, P2O5-0.11, K2O-0.25	Multiple products with highly soluble ingredients
Riser (Loveland Products)	7-17-3 + Cu, Fe, Mn, Zn	In furrow	2-5 gpa	If 3 gpa: N-2.2, P_2O_5-5.4, K_2O-1, Cu-0.02, Fe-0.06, Mn-0.02, Zn-0.3	
RSA IronMan + (Winfield Solutions)	6-0-0 from urea + 5% S, 6% Fe, 2% Mn, 1% Zn	Foliar	2 to 6 qt/ac	If 2 qt/ac: N-0.3, S-0.3, Fe-0.3, Mn-0.1, Zn-0.05	Max of 2 qt/10 gal water. Do not apply when temp > 90F, or when stress due to heat or lack of moisture.
Soygro [Soygro (Pty) Ltd]	Microbial inoculants				Unspecified on public website, but could include rhizobial and non-nodulating species.
Stay-N (Loveland Products)	Calcium polymer blend of glycols & polysaccharides, 0-0-0	Mix with urea, UAN, liquid manures			Marketed as aid for N sources. No specific urease, nitrification, or denitrification inhibitors. Ingredients similar to spray adjuvant/sticker-spreader products.
Sure-K (AgroLiquid)	2-1-6 with K as polyphos-phate & carbonate	Foliar, fertigation, or soil		For each 1 gpa: N-0.2, P2O5-0.1, K2O-0.9 lb/ac	Broadcast application to soil may reduce efficiency
TakeOff Sulfone (Verdesian)	5-20-15+14%S+0.5%Fe	Soil or foliar application	0.5 to 4 lb/A	If 2.5 lb/A: N-0.125, P_2O_5-0.5, K_2O-0.375, Fe-0.01	If foliar, do not exceed product concentration of 1.25% by volume in water.
Teprosyn Zn P (Helena)	9-15-0 + 18% Zn	Seed treatment			Derived from urea, calcium phosphate, and zinc phosphate. Teprosyn formulations also available with other nutrient compositions.

Solubility of Selected Fertilizer Materials

To be available to plants, at least some of a nutrient must be slightly soluble in the soil solution. The amount of substance that will dissolve at a given temperature in water is known as its solubility. Solubility of most chemicals is slightly higher at higher temperatures; that of others, especially ammonium and potassium nitrates, increases rapidly with temperature. The presence of other substances in the solution may either increase or decrease solubility. The solubility of selected pure fertilizer materials in water at 32 degrees F is shown below.

Table 4-13. Solubility of Selected Fertilizer Materials

Fertilizer Material	Chemical Formula	Solubility (lb/100 gal)	Salt Index (relative effect on the soil solution)
Ammonia	NH_3	750	47.1
Ammonium nitrate	NH_4NO_3	983	104.7
Ammonium sulfate	$(NH_4)_2SO_4$	592	69.0
Borax	$Na_2B_4O_7 \cdot 10H_2O$	25	
Calcium carbonate (limestone)	$CaCO_3$	0.050	4.7
Calcium metaphosphate	$Ca(PO_3)_2$	0.008	
Calcium nitrate	$Ca(NO_3)_2 \cdot 4H_2O$	1,117	52.5
Calcium sulfate	$CaSO_4 \cdot 2H_2O$	2	8.1
Copper sulfate	$CuSO_4 \cdot 5H_2O$	267	
Diammonium phosphate	$(NH_4)_2HPO_4$	209	29.9
Dicalcium phosphate	$CaHPO_4 \cdot 2H_2O$	0.168	8

Table 4-13. Solubility of Selected Fertilizer Materials

Fertilizer Material	Chemical Formula	Solubility (lb/100 gal)	Salt Index (relative effect on the soil solution)
Magnesia	MgO	0.005	1.4
Magnesium sulfate	$MgSO_4 \cdot 7H_2O$	580	44
Manganese sulfate	$MnSO_4 \cdot 4H_2O$	875	
Monoammonium phosphate	$NH_4H_2PO_4$	358	34.2
Monocalcium phosphate	$CaH_4(PO_4)_2 \cdot H_2O$	See below[1]	15.4
Potassium chloride	KCl	233	116.3
Potassium-magnesium sulfate	$K_2SO_4 \cdot 2MgSO_4$	200	
Potassium nitrate	KNO_3	108	73.6
Potassium sulfate	K_2SO_4	67	46.1
Sodium nitrate	$NaNO_3$	608	100.0
Urea	$CO(NH_2)_2$	559	75.4
Zinc sulfate	$ZnSO_4 \cdot 6H_2O$	584	

[1] Material decomposes with a small amount of water and is soluble in a large amount. The solubility varies with the conditions.

Mixing Herbicides with Nitrogen Solutions or Fluid Fertilizers

C. R. Crozier and W. Everman, Crop and Soil Sciences

Tank-mixing and applying herbicides with nonpressure nitrogen solutions or fluid fertilizers may offer savings in labor and time by eliminating at least one trip over the field. Effectiveness of some postemergence-directed herbicides for corn and sorghum is increased by applying them in nitrogen solution.

Herbicide labels will indicate if the product can be mixed with sprayable fertilizer. Herbicides may not always mix evenly throughout a sprayable fluid fertilizer, or the components may separate too quickly to make their combined use practical. Therefore, test every batch of fertilizer for compatibility before adding the herbicide. Batches can vary in pH, salt content, and salt concentration. Even these minor differences may affect compatibility. Sulfur-containing nitrogen solutions can be particularly troublesome.

The information presented below is intended as a guide only. The term "compatibility" as used here refers to chemical and physical compatibility and is not intended to supplant label directions.

A guide for making your own compatibility test follows. Also, herbicide labels give information on compatibility, tank-mix combinations, and procedures for testing compatibility with sprayable fluid fertilizer. Be sure to read the label!

Precautions

To help ensure successful results from application of herbicides and nonpressure nitrogen solutions or fluid fertilizers:

1. Always check compatibility by making a small scale test (directions given below) before mixing in field.
2. Use a compatibility agent if indicated on the herbicide label or if your small-scale test indicates the need.
3. Vigorously agitate tank contents while mixing and applying. The spray application equipment should use a high-capacity (eight-roller, internal gear, or centrifugal) pump. Spray tanks should be equipped with hydraulic jet agitators mounted in the bottom of the tank. Bypass lines usually do not provide adequate agitation. Jet agitators should be attached to a separate line from the pump.
4. Do not connect agitators to by-pass lines. For simple hydraulic jet agitators, a flow rate of 6 gallons per minute for every 100 gallons of tank capacity is sufficient. If volume-booster nozzles are used for agitation, the flow rate can be reduced to 2 to 3 gallons per minute for every 100 gallons of tank capacity. If a liquid-fertilizer type applicator is used with a metering pump, agitation must also be supplied. The best method is to use a separate power takeoff pump for extra circulation.
5. If a wettable powder or dry flowable is used, make a slurry with water and add slowly to the tank. Add wettable powder first, dry flowables second, flowables third, and liquids last.
6. A number of dry flowable herbicides now come packaged in water-soluble film packets. These packets usually will not dissolve in nitrogen solution or fluid fertilizer. When using nitrogen solution or fluid fertilizer as the carrier for a product packed in these film packets, slurry the herbicide in clean water before adding to the spray tank.
7. If a flowable product is used, premix one part flowable with one part water and add diluted mixture slowly to tank. The fluid fertilizer may be substituted for the water after compatibility has been checked.
8. Premix liquid products with two parts water or the fertilizer carrier before adding into the tank.
9. Reduce drift by choosing an appropriate nozzle size to put out the desired amount of solution without causing excessive pressure.
10. *Caution:* Take extra care in applying herbicide-fluid fertilizer mixtures to ensure that the correct herbicide rate is applied, that the herbicide is distributed uniformly, and that all directions concerning application of the herbicide are followed. Sprayer output will not be the same using fluid fertilizer as when using water as the carrier. Recalibrate sprayers for the fertilizer carrier. Do not apply herbicide-fluid fertilizer mixes overtop any crop except small grains, as injury will result.

How to Test Compatibility of Herbicides with Fluid Fertilizers

C. R. Crozier and W. Everman, Crop and Soil Sciences

Follow the compatibility test procedures on the herbicide label. If not on the label, follow the directions listed below.

1. Put 1 pint of fluid fertilizer in each of two 1-quart jars.
2. Following the adjacent table, add 0.25 to 0.375 teaspoon of a compatibility agent to one jar and shake for 5 or 10 seconds to mix. One-fourth teaspoon is equivalent to 2 pints/100 gallons of fluid fertilizer. Mark the jar "with" to indicate the compatibility agent has been added.
3. Next, add the proper amount of herbicide to each jar, according to the table. For herbicides used in small quantities per acre, one will need a greater volume of fluid fertilizer for the compatibility test. Adjust the herbicide/fertilizer ratio as follows:

    1 ounce dry flowable per acre = 0.7 teaspoon in 1 gallon
    1 fluid ounce liquid per acre = 0.25 teaspoon in 1 gallon

    If more than one herbicide is to be used in the mixture, add them separately with the wettable powders or dry flowables first, flowables second, and liquids last. Shake the jar gently for 5 to 10 seconds after each addition.
4. Let the jars stand for 5 minutes, then check for formation of large flakes, sludge, gels, or other precipitates, or to see if the herbicide remains as small, oily particles in the solution.
5. Allow both jars to stand for 30 minutes, checking them periodically. An emulsifiable concentrate normally will go to the top after standing, whereas wettable powders or dry flowables will either settle to the bottom of the jar or float to the top, depending on the density of the fertilizer and the herbicide. If the separate layers of fluid fertilizers and additives (herbicide and compatibility agent) can be resuspended by shaking, commercial application is possible.
6. If incompatibility of any form occurs in the jar with the compatibility agent added, do not mix the fluid fertilizer and herbicide in the same spray tank. If incompatibility occurs only in the jar without the compatibility agent, use of a compatibility agent is recommended.

Table 4-14. Amounts of Herbicide to Use to Test Compatibility[1]

Application Rate Product Per Acre	Add to 1 Pint of Fluid Fertilizer
Wettable powders or dry herbicides	
1 pound	1.4 teaspoon
2 pound	2.9 teaspoon
3 pound	4.3 teaspoon
4 pound	5.8 teaspoon
5 pound	7.2 teaspoon
Emulsifiable concentrates, flowables, liquids, or solutions	
1 pint	0.5 teaspoon
1 quart	0.9 teaspoon
2 quart	1.9 teaspoon
3 quart	2.9 teaspoon
1 gallon	3.8 teaspoon
5 quart	4.8 teaspoon

[1] This compatibility test is designed for 25 gallons of spray per acre with the maximum labeled rate of herbicide. For changes in spray volume or herbicide concentration, make the appropriate proportional change in the ingredients in the test. Regardless of spray volume, the amount of compatibility agents should be equal to 2 to 3 pints (0.25 teaspoon equals 2 pint; 0.375 teaspoon equals 3 pints) per 100 gallons of fertilizer.

Fertilizer Placement

C. R. Crozier and D. L. Osmond, Crop and Soil Sciences, and D. H. Hardy, NCDA&CS

Proper fertilizer placement can be as important as what kind of fertilizer you use. The first step, though, is a soil test to determine phosphorus and potassium needs and the use of realistic yield expectations to determine nitrogen. Once fertilizer needs are determined, fertilizer placement should be assessed.

Starter P (along with N) does not always lead to yield increases, but it can be an effective tool to enhance early season growth and reduce risks of losses associated with billbugs, competitive weeds, and summer droughts (see *SoilFacts: Starter Phosphorus Fertilizer and Additives in North Carolina Soils: Use, Placement, and Plant Response*, https://content.ces.ncsu.edu/starter-phosphorus-fertilizer-and-additives-in-nc-soils-use-placement-and-plant-response). Use of starter fertilizers is more important in no-till since soil warming is delayed, and the cooler temperatures can reduce the rate of crop growth. For fields that already have very high soil test P levels (P-index >100), numerous North Carolina tests have shown there is no advantage to applying additional P in a starter band. Even most mineral soils testing greater than a 50 P-Index generally do not need starter P.

Fertilizer placed in contact with or too close to seeds and young plants can cause salt injury, resulting in poor stands and slow starts. Nonuniform broadcasting of fertilizer with shallow mixing just before planting may give streaks through the field due to delayed germination or seedling injury. Salt injury is most severe in dry weather or following light rains that dissolve the fertilizer salts and leave highly concentrated salt solutions in the root zone. Nitrogen and potassium salts account for most of this injury.

To reduce salt injury risk, a side-band placement of starter fertilizer at planting is generally preferred over application directly in contact with the seed, commonly referred to as "pop-up" placement. Good results are often achieved for seeded crops like corn, cotton, sorghum, and soybeans using a 2 by 2-inch placement, placing the fertilizer 2 inches to the side and 2 inches below the planted seed. The risk of salt injury is also related to the amount of salt applied. Generally, a maximum rate of 80 pounds per acre of nitrogen alone, K_2O alone, or a combination of nitrogen plus K_2O is suggested for 2 x 2 band placement. When a greater rate is necessary, make a split application; broadcast part and apply the remainder in the row at seeding. If side placement is not possible or practical and starter fertilizer is placed directly in the seed furrow, the maximum rate should be much lower; 5 gpa of typical starter solution products contain less than 20 pounds N plus K2O per acre and is the maximum rate that should be used. An alternative method is to broadcast fertilizer prior to planting, with thorough mixing into the soil preferred, except for conservation tillage systems. The lack of soil mixing is an additional reason that banded starters are advantageous with no-till management.

For tobacco, place the band 3 to 5 inches from the transplants. This distance reduces the chance of placing plants in fertilizer bands. If side placement equipment is not available, place the fertilizer deep in the row so it will be 3 to 5 inches below the roots of the transplants.

Seedlings of small grain or plants such as clovers, grasses, and alfalfa respond very early to phosphate; consequently, it can be advantageous to place phosphate close to the seed as is done with the conventional grain drill.

Livestock & Poultry Manure Production Rates and Nutrient Content

D. A. Crouse, C. R. Crozier, and T. J. Smyth, Crop and Soil Sciences; and K. Hicks, NCDA&CS

The use of livestock and poultry manure as a fertilizer supplement in crop production has come full circle. Before the advent of inexpensive inorganic fertilizers, farmers routinely used manure to supply essential plant nutrients. Today, because of rising costs of commercial fertilizers and increasing emphasis on good manure management practices to protect water quality renewed interest has been focused on optimizing the nutrient benefits of manures.

Manure production and characteristics are influenced by on-farm management practices. For example, manure from open housing systems or from open storage ponds is diluted by rainfall. Drying and storing solid manures can increase ammonia loss and reduce the nitrogen content and fertilizer value. Microbial digestion in a waste treatment lagoon reduces total nitrogen by 50 to 85% and converts as much as 90% of the phosphorus to forms that settle into the sludge rather than remaining in lagoon liquid. All of the practices ultimately affect the total amount of recoverable nutrients that can be used in crop production.

Manure Production (Volume and Weights)

The total manure produced in volume or weight is related to the animal production system, the number of animals in that system, and the manure collection and treatment system. In liquid systems, the production of manure is often reported in gallons per animal per year or gallons per 1,000 birds capacity per year (Table 4-15). For dry or solid systems, the production is reported in tons per animal per year or tons per 1,000 birds capacity per year (Table 4-15).

Manure Nutrient Content

The manure nutrient content varies with animal age, diet, and waste (manure) management system. Nutrient data reported in Table 4-16 are statewide averages and may not reflect the actual nutrient content on any individual farm. Because of the variability of end products in the numerous different animal production and waste management systems manure must be sampled and analyzed within 60 days of application to land to determine its actual nutrient content. Manure samples can be analyzed for 11 essential plant nutrients for a nominal fee by the N.C. Department of Agriculture and Consumer Services. The Waste/Compost Analysis Laboratory (http://ncagr.gov/agronomi/uyrwaste.htm) provides details on collecting and submitting manure samples for waste analysis).

Plant Availability Coefficients

Two factors are important in the plant availability of nutrients in manure: mineralization potential and application method. Since the nutrients in manure are primarily in organic compounds, they must undergo microbial mineralization in order to become plant available. Once in the plant-available form, some nutrients are at risk of loss to the environment. For example, nitrogen is susceptible to gaseous losses due to volatilization of ammonia (a mineralized form of N). As a result, the availability of nitrogen for plant uptake is dependent on the mineralization potential of the manure. Further, the application method plays an important role in the availability of nutrients.

Chapter IV—2019 N.C. Agricultural Chemicals Manual

Subsurface placement such as soil incorporation tends to reduce volatile losses whereas surface placement such as irrigation tends to increase volatile losses. Most nutrients including phosphorus and potassium are not subject to volatile losses, meaning the application method has minimal impact on their availability. Plant availability coefficients (Table 4-17) are used to estimate the nutrients that would be available to the first crop, taking into consideration both the mineralization potential and the application method.

Calculating On-Farm Availability of Manure Nutrients

Land application of manures is an integral part of the overall soil-fertility management strategy on many farms. For farm nutrient budgeting, an assessment of the nutrients available for plant utilization must be completed. Using Tables 4-15 through 4-17, the calculation is simple:

$$\text{Total Plant Available Nutrients} = \frac{\text{No. of Animals}}{\text{Year}} \times \frac{\text{Weight or Volume}}{\text{Animal}} \times \frac{\text{Total Nutrients}}{\text{Weight or Volume}} \times \text{Availability Coefficient}$$

For example, consider a 900-sow farrow to finish operation where the lagoon effluent is soil incorporated following application but before planting of a grain crop.

$$\text{Total Plant Available N (PAN)} = \text{No. of Sows} \times \frac{\text{gallons}}{\text{sows/year}} \times \frac{\text{Total N}}{1{,}000 \text{ gal}} \times \frac{X \text{ lbs PAN}}{\text{lbs Total N}}$$

$$\text{Total Plant Available N (PAN)} = 900 \times \frac{10{,}478 \text{ gallons}}{\text{sows/year}} \times \frac{3.6 \text{ Total N}}{1{,}000 \text{ gal}} \times \frac{0.6 \text{ lbs PANs}}{\text{lbs Total N}}$$

$$\text{Total Plant Available N (PAN)} = 20{,}369 \text{ lbs of N available for the grain crop in the first year}$$

This same calculation can also be performed for other nutrients such as P, Zn, and Cu.

Management Considerations

Always apply manure as close to the period of maximum plant demand for nutrients as possible. Base manure application rates on the available portion of the nutrients and not the total concentration. Do not apply more than the receiver crop needs since excessive amounts not only waste valuable nutrients but may result in surface and/or groundwater pollution. Nutrient management planning guidelines (http://nutrients.soil.ncsu.edu) should be used to determine if application rates should be based on N or P content. Monitoring of Zn and Cu may be needed to avoid accumulation in soils to toxic levels. . Use soil testing (https://www.ncagr.gov/agronomi/sthome.htm) to predict nutrient and lime requirements and proper application rates of the manure. Use plant analysis (https://www.ncagr.gov/agronomi/uyrplant.htm) to monitor the crop nutritional status (actual uptake of soil nutrients) and effectiveness of the nutrient management program.

Additional manure management information is available in at go.ncsu.edu/crop-and-soil-publications

Table 4-15. Manure Volume and Weights in Typical North Carolina Animal Production Systems

Animal Production System	NCDA&CS Waste Code	Accumulated Manure
		gallons/animal/year[1]
Anaerobic Lagoon Liquid - Swine	ALS (except Farrow-to-Wean)	
Farrow-to-Wean (per sow)	ALF	3,203
Farrow-to-Feeder (per sow)		3,861
Farrow-to-Finish (per sow)		10,478
Wean-to-Feeder (per pig)		191
Wean-to-Finish (per pig)		776
Feeder-to-Finish (per pig)		927
Anaerobic Lagoon Sludge – Swine	ASS	
Farrow-to-Wean (per sow)		433
Farrow-to-Feeder (per sow)		522
Farrow-to-Finish (per sow)		1,417
Wean-to-Feeder (per pig)		30
Wean-to-Finish (per pig)		26.3
Feeder-to-Finish (per pig)		135
Dairy – Slurry	LSD	
Calf		1,876
Heifer		5,535
Milk Cow		7,749
		gallons/1,000 bird capacity/year[1,2]
Anaerobic Lagoon Liquid – Poultry	ALP	
Pullet (non-laying)		9,110
Pullet (laying)		22,201
Layer		25,373

Table 4-15. Manure Volume and Weights in Typical North Carolina Animal Production Systems

Animal Production System	NCDA&CS Waste Code	Accumulated Manure
		gallons/animal/year[1]
		gallons/1,000 bird capacity/year[1,2]
Anaerobic Lagoon Sludge – Poultry	ASP	
Pullet (non-laying)		1,659
Pullet (laying)		4,147
Layer		4,739
		tons/1,000 bird capacity/year[2]
Poultry Litter – Breeders	HBB	24
Poultry Litter – Broilers	HLB	
Whole House		7.2
Cake		4.0
Poultry Litter – Broiler Pullets	HBP	7.2
Poultry Litter – Layers	HLL	24
Poultry Litter – Layer Pullets	HLP	24
Poultry Litter – Turkeys	HLT	
Poult		5.3
Hen		17
Tom		25
Breeder		37
		tons/animal/year
Dairy – Scraped	SSD	
Calf		4.1
Heifer		12
Milk Cow		17
Beef – Scraped	SSB	
Stocker		1.5
Feeder		2.2
Brood Cow		3.0
Horse – Scraped	SSH	9.1

[1] To convert gallons to acre-inches, divide gallons by 27,154.

[2] Capacity is based on the maximum number of birds on the farm at any time, which is a measure of the size of the flock and not the total annual production on the farm.

Table 4-16. Total Nitrogen (N), Phosphorus (P AS P_2O_5), and Potassium (K AS K_2O) from Manure Sources

Production System	NCDA&CS Waste Code	N	P_2O_5	K_2O
		pounds of total nutrient per 1,000 gallons[1]		
Anaerobic Lagoon Liquid - Swine	ALS			
Boar		3.6	1.4	8.3
Farrow-to-Wean	ALF	2.4	0.9	4.1
Farrow-to-Feeder		3.6	1.4	8.3
Farrow-to-Finish		3.6	1.4	8.3
Wean-to-Feeder		3.6	1.4	8.3
Wean-to-Finish		3.6	1.4	8.3
Feeder-to-Finish		3.6	1.4	8.3
Anaerobic Lagoon Sludge – Swine	ASS	20.4	30.6	7.5
Anaerobic Lagoon Liquid - Poultry	ALP	3.1	1.0	13.8
Anaerobic Lagoon Sludge - Poultry	ASP	24.4	38.1	10.3
Dairy - Slurry	LSD	16.7	9.1	15.4

Chapter IV — 2019 N.C. Agricultural Chemicals Manual

Table 4-16. Total Nitrogen (N), Phosphorus (P AS P_2O_5), and Potassium (K AS K_2O) from Manure Sources

Production System	NCDA&CS Waste Code	N	P_2O_5	K_2O
		pounds of nutrient per ton		
Dairy - Scraped	SSD	11.2	7.0	9.8
Horse - Scraped	SSH	9.3	7.0	9.8
Beef - Scraped	SSB	13.0	8.3	13.6
		pounds of nutrient per ton		
Poultry Litter - Breeders	HBB	47.6	44.7	39.5
Poultry Litter - Broilers	HLB	57.8	40.0	48.6
Poultry Litter – Broiler Pullets	HBP	57.8	40.0	48.6
Poultry Litter – Layers	HLL	47.6	44.7	39.5
Poultry Litter – Layer Pullets	HLP	47.6	44.7	39.5
Poultry Litter - Turkeys	HLT	54.0	48.2	33.8

[1] To convert gallons to acre-inches, divide gallons by 27,154.

Table 4-17. First-Year Nutrient Availability Coefficients for Nitrogen (N), Phosphorus (P) and Potassium (K) from Manure Sources

Production System	NCDA&CS Waste Code	N		P		K	
		Broadcast or Irrigated	Incorporated or Injected	Broadcast or Irrigated	Incorporated or Injected	Broadcast or Irrigated	Incorporated or Injected
Anaerobic Lagoon Liquids							
Swine	ALS	0.5	0.6	1.0	1.0	1.0	1.0
Swine Farrow to Wean	ALF	0.5	0.6	1.0	1.0	1.0	1.0
Poultry	ALP	0.5	0.6	1.0	1.0	1.0	1.0
Other	ALO	0.5	0.6	1.0	1.0	1.0	1.0
Anaerobic Lagoon Sludges							
Swine	ASS	0.5	0.6	1.0	1.0	1.0	1.0
Poultry	ASP	0.5	0.6	1.0	1.0	1.0	1.0
Other	ASO	0.5	0.6	1.0	1.0	1.0	1.0
Slurries							
Beef	LSB	0.4	0.6	1.0	1.0	1.0	1.0
Dairy	LSD	0.4	0.6	1.0	1.0	1.0	1.0
Swine	LSS	0.4	0.6	1.0	1.0	1.0	1.0
Other	LSO	0.4	0.6	1.0	1.0	1.0	1.0
Scraped or Stockpiled Manure							
Beef	SSB	0.4	0.6	1.0	1.0	1.0	1.0
Dairy	SSD	0.4	0.6	1.0	1.0	1.0	1.0
Horse	SSH	0.4	0.6	1.0	1.0	1.0	1.0
Swine	SSS	0.4	0.6	1.0	1.0	1.0	1.0
Other	SSO	0.4	0.6	1.0	1.0	1.0	1.0
Poultry Litters							
Breeders	HBB	0.5	0.6	1.0	1.0	1.0	1.0
Broilers	HLB	0.5	0.6	1.0	1.0	1.0	1.0
Broiler Pullets	HBP	0.5	0.6	1.0	1.0	1.0	1.0
Layers	HLL	0.5	0.6	1.0	1.0	1.0	1.0
Layer Pullets	HLP	0.5	0.6	1.0	1.0	1.0	1.0
Turkeys	HLT	0.5	0.6	1.0	1.0	1.0	1.0
Other	HLO	0.5	0.6	1.0	1.0	1.0	1.0

FERTILIZER USE

Beneficial Use of Municipal Biosolids

J. G. White, Crop and Soil Sciences

Biosolids are the nutrient-rich, largely organic residuals of wastewater (sewage) treatment, i.e., sludge that has been generated at a water resource recovery facility (WRRF, a.k.a., wastewater treatment plant) and treated to meet US-EPA standards for land application. These standards include limitations on pathogens, potentially toxic metals, and disease-vector attractants. Biosolids can be recycled via land application to sustainably improve and maintain productive soils and foster plant growth. Most biosolids contain sufficient plant nutrients (e.g., nitrogen [N], phosphorus [P], potassium [K]) to have fertilizer value, and some have liming value (agricultural lime equivalent). Previous concerns regarding the levels of potentially toxic metals in biosolids have been alleviated by current US-EPA standards. However, most biosolids do contain small quantities of pharmaceuticals, personal care products, steroids, hormones, and flame retardants. These might pose risks to crops, the livestock or humans that consume them, and/or to the environment. Currently, these substances in biosolids are not regulated, and any potentially adverse or advantageous effects are not well characterized.

Biosolids are classified as either Class B, Class A, or Class A – Exceptional Quality (EQ) based on US-EPA regulatory reduction requirements for pathogens, vector attractants, and potentially toxic metals. Class B biosolids meet the least restrictive limits for concentrations of potentially toxic metals and have been treated to reduce pathogens to levels considered protective of human health. Land application of Class B biosolids requires site-use and access-restriction permits. Class A biosolids are further treated to reduce pathogens below detectable limits. Class A – Exceptional Quality biosolids meet stricter metal concentration requirements and may be applied by the general public in reasonable quantities without permit. Physically, North Carolina biosolids typically fall into one of four categories depending on their dry matter content and consistency. Class B slurries have very low solids content (2 – 5%). Class B cakes range from about 15 to 25% solids. Class A alkaline-treated biosolids range from ~35 to 80% solids, and Class A heat-treated biosolids have from 90 to 98% solids and are typically pelleted. Biosolids are also used in the production of commercial bulk compost.

Anyone who wants to distribute and land apply biosolids must comply with all relevant federal and state regulations. In some cases, a permit may be required and obtained from the North Carolina Department of Environmental Quality (NCDEQ). Typically, any WRRF wishing to apply biosolids to its own land or to private farmland must obtain a permit. To obtain a permit, sufficient data must be submitted to show that the biosolids meet the federal and state limits for pathogens, potentially toxic metals, and vector attraction, and that the proposed site is suitable to receive biosolids. Some of the factors considered in site selection include topography; soil type; distance from streams, wells, and property lines; flood hazard; and depth to bedrock and the water table. If the site is suitable, then application rates are calculated based on biosolids composition, crop nutrient requirements, previous history of biosolids application, and soil pH. The primary purposes of these criteria are to: 1) protect ground and surface water from contamination by biosolids nitrogen and phosphorus, and 2) foster crop nutrient-use efficiency. If a property owner wishes to apply certain biosolids to his or her land, he or she must obtain a permit. The preparation of permit applications is the responsibility of the municipality, industry, and land owner. The process is somewhat involved, and we do not attempt to outline it here. Information on permitting may be obtained from any NCDEQ regional office or at the NCDEQ Division of Water Resources - Water Quality Permitting Section - Non-Discharge Permitting Branch (http://deq.nc.gov/about/divisions/water-resources/water-resources-permits/wastewater-branch/non-discharge-permitting).

Many biosolids are treated with alkaline materials to raise sludge pH to reduce pathogens and vector attraction ("lime stabilized"). These biosolids typically have liming value that should be identified before use. If these biosolids are land-applied based on nutrient content (e.g., N, P) without regard to their agricultural lime equivalent, over-liming may occur and negatively affect plant growth.

Since most biosolids produced in North Carolina are suitable for use on agricultural land, we encourage interested farmers to contact their local Cooperative Extension center (https://www.ces.ncsu.edu/local-county-center/) and/or their NCDEQ Regional Office (http://deq.nc.gov/contact/regional-offices) about possible sources of biosolids in their area. Table 4-18 shows representative average characteristics of some biosolids available in North Carolina. The user should obtain actual, current analyses of any product that is to be land-applied in order to determine application rates that optimize plant nutrient use efficiency, moderate soil pH, and protect water quality.

Table 4-18. Plant Nutrient* and Metal Contents of Some Municipal Biosolids from or Available in North Carolina[1]

Biosolids Source	Solids	Total Nitrogen*	Plant-available Nitrogen[2]	P_2O_5*	K_2O*	Ca*	Mg*	Cu*	Zn*	Ni*	Cr	Pb	Cd	Mo*	As
	%	Constituents in pounds per dry ton of biosolids													
Cape Fear Public Utility Authority (Wilmington): Class B[3]	15	76	24	76	3	38	4	0.46	1.4	0.04	0.08	0.05	0.004	0.02	0.02
Cary: "Enviro Gems" pellets Class A Exceptional Quality[3]	90	135	40	149	13	23	12	0.46	1.2	0.03	0.05	0.03	0.004	0.02	0.01
Charlotte: Class B[3]															
Irwin Creek WWTP[4]	14	122	40	83	4	46	8	0.6	1.5	0.08	0.24	0.15	0.02	0.03	0.01
McAlpine Creek WWTP	17	110	36	133	4	36	14	0.42	1.5	0.06	0.18	0.08	0.004	0.02	0.01
McDowell Creek WWTP	14	131	47	163	7	48	12	0.32	1.3	0.03	0.12	0.04	0.004	0.01	0.01
Elizabeth City: Class B[3]	2	122	26	55	24	330	8	0.26	0.64	0.02	NA	0.03	0.002	0.01	0.01
Goldsboro: compost Class A Exceptional Quality[3]		45	7	8	22	31	10	0.18	0.28	0.04	NA	0.01	0.001	0.003	0.02
Milwaukee Metro Sewage District: "Milorganite" Class A Exceptional Quality[3]	95	100	32	92	8	42	14	0.5	1	0.06	0.46	0.09	0.002	0.02	0.01

Table 4-18. Plant Nutrient* and Metal Contents of Some Municipal Biosolids from or Available in North Carolina[1]

Biosolids Source	Solids	Total Nitrogen*	Plant-available Nitrogen[2]	P_2O_5*	K_2O*	Ca*	Mg*	Cu*	Zn*	Ni*	Cr	Pb	Cd	Mo*	As
Orange County: Class B[3]	3	222	47	206	21	42	11	0.54	1.6	0.03	0.04	0.01	0.001	0.01	0.004
Raleigh (Neuse River Resource Recovery Facility): "Raleigh Plus" Class A Exceptional Quality[3]	60	27	8	7	10	743	6	0.01	0.2	0.02	0.02	0.02	0.002	0.004	0.01
Raleigh (Neuse River Resource Recovery Facility): Class B[3]	3	144	56	275	180	20	6	0.5	1	0.05	0.05	0.04	0.01	0.03	0.03
Rocky Mount: Tar River Regional WWTP	4	15.4	3.5	4	0.5	4	0.4	0.03	0.04	0.06	NA	0.01	0.001	0.002	0.003
Winston-Salem Archie Elledge WWTP Pellets: Class A	93	79	23	73	4	39	8	0.56	3	0.05	0.1	0.08	0.004	0.02	0.01

* Essential plant nutrient

[1] Results vary between samples. This table represents average or median values. Data are from 2015 Annual Residual Sampling Summaries submitted to NC-DEQ Division of Water Quality except: Goldsboro Compost, which is from prior year data, and Milorganite, which is from www.milorganite.com.

[2] Plant-available nitrogen (PAN) estimated for the year of application. PAN depends on waste processing and land application methods and N forms in the biosolids. PAN is some fraction of total N = organic N + inorganic N (ammonium + nitrate + nitrite). Values here are for surface application; if injected or incorporated, PAN will be somewhat greater. For current methods used to estimate PAN, see the North Carolina Department of Agriculture and Consumer Services Waste and Compost Analysis Guide (http://www.ncagr.gov/agronomi/pdffiles/wasteguide.pdf).

[3] Class B biosolids meet minimum US-EPA land-application standards for pathogens, potentially toxic metals, and vector-attraction; Class A biosolids meet stricter pathogen standards; Class A Exceptional Quality biosolids are Class A biosolids with very low metal content.

[4] WWTP, Waste water treatment plant, a.k.a. water resource recovery facility.

Certified Organic Farm Management Alternatives

C. R. Crozier, S.C. Reberg-Horton, and A. Woodley, Crop and Soil Sciences

Although crop nutritional requirements are the same for organic and conventional farms, organic producers need to be more creative due to the limitations on allowable inputs. See the "North Carolina Organic Grain Production Guide" http://www.cefs.ncsu.edu/resources/organicgrainfinal.pdf for additional information. Since use of some soil amendments is limited to cases of nutrient deficiency, organic producers should maintain soil testing and plant tissue analysis records documenting specific nutrient deficiencies that need correction. With tissue testing, the appropriate plant part must be collected at the proper growth stage as specified by laboratory guidelines. See a detailed plant tissue analysis guide at www.ncagr.gov/agronomi/pdffiles/plantguide.pdf or contact your local Extension center for more details.

Certain inputs are allowable on organic production systems, if applied according to guidelines. These include many (not all) natural and certain synthetic materials. Current USDA regulations can be found at http://www.ams.usda.gov/AMSv1.0/nop. Find the section on "Organic Standards" and select "National List and Petitioned Substances," then find "National List and Petitions" and select "Current National List (eCFR)" to see a detailed listing of allowed and prohibited substances. See a list of commercially available materials that have been reviewed by the Organic Materials Review Institute (OMRI) at http://www.omri.org/omri-lists. In all cases, input use should be considered prohibited unless included in the farm plan and confirmed by the certifying authority prior to application.

Critical aspects of soil fertility management include pH, major nutrients (N, P, K, S, Ca, Mg), and micronutrients (especially B, Cu, Mn, Zn; but also, Fe, Mo, Cl, etc.). A summary of soil fertility parameters and organic management options is given in Table 4-19 below.

Specific guidelines must be followed when applying composts and manures in organic production systems. Materials must be applied at agronomic rates in compliance with any applicable nutrient management guidelines (http://nutrients.soil.ncsu.edu/), and which avoid excess nutrients. Raw animal manures must either be: 1) composted according to specific criteria specified in the USDA National List, 2) applied to land used for a crop not intended for human consumption, 3) incorporated into the soil at least 90 days prior to the harvest of an edible product not contacting soil or soil particles, or 4) incorporated into the soil at least 120 days prior to the harvest of an edible product that does contact soil or soil particles. The guidelines for compost production for organic agriculture state that the initial C:N ratio must be between 25:1 and 40:1; and a temperature between 131 and 170 degrees Fahrenheit must be achieved and maintained for at least 3 days for in-vessel or static aerated pile systems, or for at least 15 days during which there are at least 5 turnings for windrow systems. Composts not meeting these criteria must be applied based on other raw manure criteria, which also apply to animal waste lagoon liquids and solids, and stockpiled poultry litter. Human or industrial wastes such as lime stabilized sludges or composted biosolids are not allowed. Ashes of manures may not be used, but ashes from other untreated plant and animal materials may be applied if not combined with any prohibited substances. Avoid over-reliance on animal manures, since this could lead to accumulation of excess P, Cu, and Zn in soils. Sporadic use of manures in conjunction with more frequent use of legume cover crops, green manures or other N sources is an excellent way to supply several plant nutrients in appropriate amounts.

Table 4-19. Soil Fertility Parameters and Management Options

Parameter	Problem Documentation	Supply Options[1]	Not Allowed
pH	Soil test	Standard calcitic or dolomitic agricultural ground limestone; pH can be lowered by adding elemental sulfur.	Hydrated or burnt lime [$Ca(OH)_2$, CaO], industrial wastes, slags, lime-stabilized biosolids
Major Nutrients			
Nitrogen (N)	Tissue analysis	Legumes, manures[3], animal by-products (blood, fish), plant by-products (cotton, apple, fermentation wastes), mined sodium nitrate ($NaNO_3$)[3]	Synthetic fertilizers, sewage sludges, municipal waste composts
Phosphorus (P)	Soil test, tissue analysis	Manures[3], rock phosphate, animal by-products (bone meal; fish, shrimp, and oyster scraps; leather)	Processed rock phosphates
Potassium (K)	Soil test, tissue analysis	Manures[3], plant by-products (ash, dried seaweed), greensand, sulfate of potash (K_2SO_4)[4], possibly muriate of potash (KCl)[3,4]	KCl if excess chloride
Sulfur (S)	Tissue analysis	Manures[3], plant by-products (cotton motes, peanut meal), elemental sulfur[4], gypsum ($CaSO_4$), Epsom salt ($MgSO_4$)[4], sulfate of potash (K_2SO_4)[4]	Synthetic fertilizers
Calcium (Ca)	Soil test, tissue analysis	Standard calcitic or dolomitic agricultural ground limestone, gypsum ($CaSO_4$), bone meal, ash	$Ca(OH)_2$, CaO, calcium nitrate [$Ca(NO_3)_2$]
Magnesium (Mg)	Soil test, tissue analysis	Standard dolomitic agricultural ground limestone, Epsom salts ($MgSO_4$)[4], sulfate of potash magnesium, bone meal, plant by-products (cottonseed meal, wood ash)	Synthetic fertilizers
Micronutrients[2]			
Boron (B)	Tissue analysis	Manures, animal and plant by-products, soluble boron fertilizers[4]	
Copper (Cu)	Soil test, tissue analysis	Manures, animal and plant by-products, sulfates & oxides[4]	chlorides
Manganese (Mn)	Soil test, tissue analysis	Manures, animal and plant by-products, sulfates & oxides[4]	chlorides
Zinc (Zn)	Soil test, tissue analysis	Manures, animal and plant by-products, sulfates & oxides[4]	chlorides
Co, Fe, Mo, Se	Tissue analysis[5]	Manures, animal and plant by-products, sulfates, carbonates, oxides, or silicates[4]	Chlorides, nitrates

[1] Inputs must be on the National Organic Program or the OMRI-approved source list and approved by the certifying agents.

[2] Avoid over-application of micronutrients since toxicities can occur.

[3] Note restrictions.

[4] Documentation of nutrient deficiency required.

[5] Deficiencies of Co, Mo, and Se are not common in North Carolina, and these elements are not included in routine tissue analysis performed by the North Carolina Department of Agriculture and Consumer Services, although Mo can be included upon request and for an additional fee. Consult a Cooperative Extension center for information regarding private agricultural laboratories.

V — INSECT CONTROL

Relative Toxicity of Pesticides to Honey Bees	72
Reducing the Risk of Pesticide Poisoning to Honey Bees	73
Insect Control in Field Corn	74
Insect Control in Grain Sorghum	77
Insect Control in Small Grains	78
Insect Control on Cotton	80
Cotton Insect Resistance Management	83
Insect Control on Peanuts	85
Insect Control in Soybeans	87
Insect Control on Flue-Cured and Burley Tobacco	90
Insect Control for Commercial Vegetables	97
Relative Effectiveness of Insecticides and Miticides for Insect and Mite Control on Vegetables	129
Preharvest Intervals for Pyrethroid Insecticides in Vegetable Crops	131
Insect Control for Greenhouse Vegetables	132
Insect Control for Livestock and Poultry	134
Community Pest Control	140
Industrial and Household Pests	144
Arthropod Management for Ornamental Plants Grown in Greenhouses	147
Arthropod Management for Ornamental Plants Grown in Nurseries or Landscapes	152
Arthropod Control on Christmas Trees	159
Commercial Turf Insect Control	164
Insect Control for Wood and Wood Products	169
Insect Control for the Home Vegetable Garden	173
Control of Household Pests	178
Insect Control for Home Lawns	187

Relative Toxicity of Pesticides to Honey Bees

David R. Tarpy, Professor and Extension Apiculturist

Most pesticides are at least somewhat toxic to honey bees and other pollinators, although the degree of toxicity varies considerably from product to product. Insecticides are generally the most likely to cause a bee kill; herbicides, fungicides, and defoliants present relatively minor danger to bees if used according to label directions. **Check the pesticide label** for the relative toxicity of the active ingredient to bees and other pollinators (Table 5-1A) and apply with caution around beehives or when pollinators are actively foraging.

Table 5-1A. Relative Toxicity of Pesticides to Honey Bees

	Highly Toxic	Moderately Toxic	Relatively Non-toxic
LD50	Less than 2 micrograms per bee	Between 2 and 11 micrograms per bee	Above 11 micrograms per bee
Precautionary statement.	This product is highly toxic to bees exposed to direct treatment or residues on blooming crops or weeds. Do not apply this product or allow it to drift to blooming crops or weeds if bees are visiting the treatment area.	This product is toxic to bees exposed to direct treatment or residues on blooming crops or weeds. Do not apply this product if bees are visiting the treatment area.	No statement required.

Table 5-1B. Pesticide Use Inside and Around Honey Beehives

Pests	Chemical (Brand)	Formulation	Precautions and Remarks (Always follow product label directions for handling, product application, and disposal)
Tracheal Mite	menthol (Mite-A-Thol)	Crystalline granules	Both products generate vapors that kill tracheal mites. Apply onto inner cover/top super according to label directions. Best if used when ambient temperatures are above 70 degrees F for menthol and 50 degrees F for formic acid. Use gloves when handling crystals or gel packets.
	formic acid (Mite-Away Quick Strips)	Various delivery methods	
Varroa Mite	tau-fluvalinate (Apistan)	Plastic strip; pesticide-impregnated	Strips contain contact poison to kill mites. Use protective gloves when handling strips. Hang strips in brood-chamber according to label directions. Caution should be used, as mites have evolved a resistance to this particular chemical, and it may not be effective in many instances.
	formic acid (Mite-Away Quick Strips)	Various delivery methods	Product generates vapors to kill mites. Kills mites in sealed brood cells. Treat colonies according to label directions.
	coumaphos (Check-Mite+)	Plastic strip; pesticide-impregnated	For varroa mites, product should be used only when fluvalinate-resistance has been confirmed by NCDA Bee Inspectors. Caution should be exercised, as mites have evolved a resistance to this particular chemical and may not be effective in many instances.
	amitraz (Apivar)	Plastic strip; pesticide-impregnated	Strips contain active ingredient to kill mites upon contact. Use protective gloves when handling strips.
	thymol (ApiLife VAR or Apiguard)	Pesticide-impregnated vermiculite tablets or gel	Essential oils volatilize to kill mites outside of brood cells.
	sucrose octonoate (Sucrocide)	Liquid; mix with water	Spray all adult bees with fine mist; must be completely wetted to kill mites.
Small Hive Beetle (adults)	coumaphos (Check-Mite+)	Plastic strip; pesticide-impregnated	Use protective gloves when handling strips. Attach to cardboard or other material as specified on label direction and place strip-side down on bottom board to kill adult beetles. Application for varroa mites (see above) is not simultaneously effective for SHB.
(pupae)	permethrin (GardStar)	Liquid; mix with water	For ground treatment around hive(s) only. Kills larvae/pupae during soil-inhabiting phase of beetle life cycle. Mix and apply to soil according to label directions.
Wax Moth	paradichlorobenzene (Para-Moth)	Crystalline granules	Use to prevent infestation of stored hive equipment (drawn-comb) only. Do not use in hives containing honey bees. Use protective gloves when handling crystals. Store product in sealed container when not in use.

Always follow label directions, which require the removal of honey from beehives prior to most pesticide treatments.

Reducing the Risk of Pesticide Poisoning to Honey Bees

Precautions for the Pesticide Applicator

1. Always read and follow any warning statements regarding honey bees on the pesticide label.
2. If more than one product gives good control of the target pest, select a pesticide from the moderately toxic or relatively non-toxic groups instead of the highly toxic group from Table 5-1A.
3. Avoid applying any bee-toxic pesticides on blooming plants that attract bees. Keep pesticide drift from nearby blooming weeds that are attracting bees.
4. Time of pesticide application is very important. Apply pesticides that are toxic to bees in the late afternoon (after 3 p.m.) or in the evening if at all possible. Most honey bees have stopped foraging and have returned to their hives by 3 p.m. This allows maximum time for the active ingredient to break down before the bees come into contact with it the next day.
5. Select the safest formulation of the pesticide that is available for the intended use. "Drifting" of the pesticide from the target pest and/or crop to areas frequented by bees should be minimized and formulation selection is the key to this problem.
    a. "Dusts" almost always drift more than other pesticide formulations and are generally more dangerous to bees than are sprays or granular applications.
    b. Spray formulations are usually safer to bees than dusts, but there are differences among the spray formulation types. Generally, water-soluble formulations are safer than are emulsifiable-formulations, and fine sprays are less dangerous than are coarse sprays. Sprays of undiluted technical pesticide (ULV) may be more dangerous than diluted sprays.
    c. **Granular applications generally are the least likely to drift and accidentally kill bees**. Consider a granular formulation if it is suitable for controlling the target pest.
6. The mode of pesticide application is also important, particularly from a drifting standpoint. Aerial applications are generally more dangerous than applications by ground equipment. If a pesticide application is being made by air, it is the contractor's responsibility to notify any beekeepers that have *registered* apiaries (one or more hives of bees) within 1/2 mile of the area to be aerially sprayed. These regulations are defined in the N.C. Pesticide Laws, and the person responsible for the notification is the person who contracts for the aerial application.
7. Never apply any pesticide directly over a beehive. The NC Department of Agriculture & Consumer Services provides a voluntary program (DriftWatch) where you can check for apiaries near your location: http://ncagr.gov/pollinators/Driftwatch.htm
8. Notify beekeepers who have beehives near an area to be treated with a pesticide so that they may attempt to protect their bees.
9. Follow proper precautions in disposing of unused pesticides and pesticide containers. Be particularly careful not to contaminate water with pesticides, as the water may be collected by bees and result in bee kills.

Precautions for the Beekeeper

1. If your bees are located in any area where pesticides are commonly used, then identify yourself as a beekeeper to your neighbors who may use pesticides. The NC Department of Agriculture & Consumer Services provides a voluntary program (DriftWatch) where you can map your apiary location: http://ncagr.gov/pollinators/Driftwatch.htm
2. Identify your apiaries with your name and address or telephone number if the apiary is not associated with your residence so that you may be notified if pesticides are to be used by a neighboring individual.
3. Explain the importance of your bees in the pollination of crops being grown on nearby fields to those growers so that they may consider the value of the bees in pollination before applying any pesticides that may kill the pollinating insects.
4. Be aware of the precautions that apply to the pesticide applicator (above) so that you can serve as a resource in providing solutions to reducing bee kills.
5. Do not place apiaries in areas used to grow crops that require heavy and frequent usage of pesticides.
6. Register your apiary locations with the N.C. Department of Agriculture if aerial applications of pesticides are used in your apiary locations.
7. As a very last resort, move your beehives if possible when bee-toxic pesticides are being applied near your apiary. Covering the hives (e.g., with wet burlap) is usually not possible for large apiaries and can cause bees to overheat or suffocate.

Additional resources

NC State Extension – Pesticide Stewardship: https://pesticidestewardship.org/pollinator-protection/

NC Department of Agriculture – Protecting NC Pollinators: http://ncagr.gov/spcap/bee/

Reducing pesticide poisoning in bees (OSU): https://catalog.extension.oregonstate.edu/sites/catalog/files/project/pdf/pnw591.pdf

Mitigating pesticides (USDA): http://www.xerces.org/wp-content/uploads/2014/04/NRCS_Pesticide_Risk_Reduction_TechNote.pdf

Insect Control in Field Corn

D. D. Reisig and A.S. Huseth, Entomology and Plant Pathology

Table 5-2. Insect Control in Field Corn

Insecticide, Mode of Action Code, and Formulation	Per Acre Amount	Per Acre Active (lb)	Acres/gal (lb)	Preharvest Interval (PHI) (Days)	Precautions and Remarks
Annual White Grub — At Planting Seed Treatments/In Furrow					
bifenthrin, MOA 3 (Capture) LFR	3.4 to 13.6 oz	0.047 to 0.062	38 to 9.4	30	Provides control alone, without addition of seed treatment
clothianidin, MOA 4A (Poncho) 600 FS	—	0.025 mg per kernel	—	—	0.5 and 1.25 mg per kernel rate can provide improved control under high pest pressure or slow grow off conditions.
thiamethoxam, MOA 4A (Cruiser) 5 FS	—	0.025 mg per kernel	—	—	0.5 and 1.25 mg per kernel rate can provide improved control under high pest pressure or slow grow off conditions.
Billbug — At Planting Seed Treatments					
clothianidin, MOA 4A (Poncho) 600 FS	—	1.25 mg per kernel	—	—	Must be special-ordered from a seedsman. In most situations, these products will provide adequate control. Corn planted near previous year's corn, corn planted mid-April, and corn near good overwintering habitats are most at risk. In these situations, these products will not provide adequate control.
thiamethoxam, MOA 4A (Cruiser) 5 FS	—	1.25 mg per kernel	—	—	
Brown Stink Bug					
beta-cyfluthrin, MOA 3 (Baythroid XL) 1.0 EC	2.8 fl oz	0.022	45.7	21	Seedling injury mainly occurs in no-till. On larger plants, apply to stages just prior to tasseling. On tall corn, use ground application only at 15+ gallons volume per acre or if by air, work with applicator to ensure adequate coverage in the zone where the ear is forming. Results may be poor to mediocre depending on application. Applications can be effective up to, or less than, one week after treatment. Bifenthrin is the superior pyrethroid (MOA 3).
bifenthrin (Brigade, Discipline, Sniper, and others) 2 EC	6.4 fl oz	0.10	20	30	
bifenthrin, MOA 3 + zeta-cypermethrin, MOA 3 (Hero) 1.24 EC	10.3 fl oz	0.1	12.4	60 (forage) 30 (grain and stover)	
bifenthrin, MOA 3 + zeta-cypermethrin, MOA 3 (Steed) 1.5 EC	4.7 fl oz	0.055	27.2	60 (forage) 30 (grain and stover)	
cyfluthrin, MOA 3 (Tombstone) 1.0 EC	2.8 fl oz	0.044	45.7	21	
lambda-cyhalothrin, MOA 3 (Karate, Lambda-cyhalothrin, Silencer) 1.0 EC	3.84 fl oz	0.03	33.3	21	
(Warrior II and Karate Z) 2.08 CS	1.92 fl oz	0.03	66.7	21	
zeta-cypermethrin, MOA 3 (Mustang Max) 0.8 EC	4.0 fl oz	0.025	32	30	
zeta-cypermethrin, MOA 3 + bifenthrin, MOA 3 (Hero) 1.24 EC	10.3 fl oz	0.033 + 0.066	12.4	30	
Corn Leaf Aphid					
pyrethroids, MOA 3 and pyrethroid combinations	(see brown stink bug above for rates)	—	—	—	
Corn Earworm — In Whorl					
Bt transgenic corn, MOA 11 (Agrisure Viptera and Leptera)	—	—	—	—	This is transgenic corn seed. Observe the refuge specifications on the label. Corn earworm is not a yield-limiting pest in timely planted corn.
chlorantraniliprole, MOA 28 (Prevathon) 0.43 SC	14 to 20 fl oz	0.047 to 0.067	9.1 to 6.4	14	
Cutworm — Postemergence					
beta-cyfluthrin, MOA 3 (Baythroid XL) 1.0 EC	1.6 to 2.8 fl oz	0.017 to 0.022	80 to 45.7	21	Best to direct spray to the plant base and use at least 15 gallons volume per acre by ground. Pyrethroids are suggested for organic soils. Use higher rates for heavier infestations or aerial application.
bifenthrin (Brigade, Discipline, Sniper, and others) 2 EC	2.1 to 6.4 fl oz	0.033 to 0.10	61 to 20	30	
bifenthrin, MOA 3 + zeta-cypermethrin, MOA 3 (Hero) 1.24 EC	2.6 to 6.1 fl oz	0.25 to 0.06	49.2 to 21	60 (forage) 30 (grain and stover)	
bifenthrin, MOA 3 + zeta-cypermethrin, MOA 3 (Steed) 1.5 EC	2.5 to 3.5 fl oz	0.029 to 0.041	51.2 to 36.6	60 (forage) 30 (grain and stover)	
Bt transgenic corn, MOA 11 (Agrisure Viptera, Herculex, Leptra, PowerCore, Optimum Intrasect, SmartStax, Trecepta)	See remarks	—	—	NA	This is transgenic corn seed. Observe the refuge specifications on the label.
chlorpyrifos, MOA 1B (Lorsban) 4 E	2 pt	1	4	14 (silage) 35 (grain)	Do not feed Lorsban treated corn until 35 days post treatment.
cyfluthrin, MOA 3 (Tombstone) 1.0 EC	0.8 to 1.6 fl oz	0.013 to 0.025	160 to 80	21	
esfenvalerate, MOA 3 (Asana XL) 0.66 EC	5.8 to 9.6 fl oz	0.03 to 0.05	22.1 to 13.3	21	

Table 5-2. Insect Control in Field Corn

Insecticide, Mode of Action Code, and Formulation	Per Acre Amount	Per Acre Active (lb)	Acres/gal (lb)	Preharvest Interval (PHI) (Days)	Precautions and Remarks
Cutworm — Postemergence (continued)					
gamma-cyhalothrin, MOA 3 (Declare) 1.25 EC	0.77 to 1.28 fl oz	0.0075 to 0.0125	166.2 to 100	21	
lambda-cyhalothrin, MOA 3 (Karate, Lambda-cyhalothrin, Silencer) 1.0 EC	1.9 to 3.2 fl oz	0.015 to 0.025	67.4 to 40	21	
(Warrior II and Karate Z) 2.08 CS	1 to 1.6 fl oz	0.015 to 0.025	128 to 80	21	
methoxyfenozide, MOA 18A (Intrepid) 2F	4 to 8 fl oz	0.06 to 0.12	32 to 16	21	
zeta-cypermethrin, MOA 3 (Mustang Max) 0.8 EC	1.3 to 2.8 fl oz	0.008 to 0.0175	98.5 to 45.7	30	
European Corn Borer					
beta-cyfluthrin, MOA 3 (Baythroid XL) 1.0 EC	1.6 to 2.8 oz	0.017 to 0.022	80 to 45.7	21	Must be applied before borers enter stalk. Apply by ground only and into plant whorls with at least 25 gallons water per acre. Use 30 psi or less. A surfactant may improve whorl penetration.
bifenthrin, MOA 3 (Brigade, Discipline, Sniper, and others) 2 EC	2.1 to 6.4 oz	0.033 to 0.10	61 to 20	30	
bifenthrin, MOA 3 + zeta-cypermethrin, MOA 3 (Hero) 1.24 EC	4.0 to 10.3 oz	0.4 to 0.10	32 to 12.4	60 (forage) 30 (grain and stover)	
bifenthrin, MOA 3 + zeta-cypermethrin, MOA 3 (Steed) 1.5 EC	3.5 to 4.7 oz	0.041 to 0.055	36.6 to 27.2	60 (forage) 30 (grain and stover)	
Bt transgenic corn, MOA 11 (Agrisure Viptera, Genuity VT Double/Triple PRO, Herculex, Leptra, Optimum Intrasect, PowerCore, SmartStax, Trecepta)	See remarks	—	—	NA	This is transgenic corn seed. Plants will express Bt endotoxin. Observe the refuge specifications on the label.
chlorantraniliprole, MOA 28 (Prevathon) 0.43 SC	14 to 20 fl oz	0.047 to 0.067	9.1 to 6.4	14	
chlorpyrifos, MOA 1B (Lorsban) 15 G	6.5 lb	1	0.154	35	Apply by air or ground. Will handle whorl infestations, but effectiveness decreases with stalk boring. Rainfall soon after enhances control.
(Lorsban) 4 E	2 pt	1	4	35	
cyfluthrin, MOA 3 (Tombstone) 1.0 EC	1.6 to 2.8 fl oz	0.025 to 0.044	80 to 45.7	21	Must be applied before borers enter stalk. Apply by ground only and into plant whorls with at least 25 gallons water per acre. Use 30 psi or less. A surfactant may improve whorl penetration.
esfenvalerate, MOA 3 (Asana XL) 0.66 EC	9.6 fl oz	0.05	13.3	21	
gamma-cyhalothrin, MOA 3 (Declare) 1.25 EC	1.02 to 1.54 fl oz	0.01 to 0.015	125.5 to 83.1	21	
lambda-cyhalothrin, MOA 3 (Karate, Lambda-cyhalothrin, Silencer) 1.0 EC	2.6 to 3.8 fl oz	0.02 to 0.03	49.2 to 33.7	21	
(Warrior II and Karate Z) 2.08 CS	1.28 to 1.92 fl oz	0.02 to 0.03	100 to 66.7	21	
methoxyfenozide, MOA 18A (Intrepid) 2F	4 to 8 fl oz	0.06 to 0.12	32 to 16	21	
spinosad, MOA 5 (Blackhawk) 4 SC	1.67 to 3.3 fl oz	0.038 to 0.075	76.6 to 38.8	28	
zeta-cypermethrin, MOA3 (Mustang Max) 0.8 EC	2.7 to 4.0 fl oz	0.017 to 0.025	47.4 to 32	30	
Fall Armyworm — In Whorl					
Bt transgenic corn, MOA 11 (Agrisure Viptera, Genuity VT Double/Triple PRO, Leptra, PowerCore, SmartStax, Trecepta)	—	See remarks	—	—	This is transgenic corn seed. Observe the refuge specifications on the label.
chlorantraniliprole, MOA 28 (Prevathon) 0.43 SC	14 to 20 fl oz	0.047 to 0.067	9.1 to 6.4	14	Use a minimum of 15 gallons per acre by ground for whorl treatment (not by air). Low pressure spray and addition of surfactant may help liquid to penetrate into whorl. Application to large caterpillars may not give satisfactory results.
Grasshopper					
bifenthrin, MOA 3 (Brigade, Discipline, Sniper, and others) 2 EC	2.1 to 6.4 fl oz	0.033 to 0.10	61 to 20	30	Apply by air or ground uniformly over foliage as a broadcast treatment. Early morning treatment preferred. Use higher rates for heavy infestation. Grasshoppers are often confined to field margins.
chlorpyrifos, MOA 1B (Lorsban) 4 E	0.5 to 1 pt	0.25 to 0.5	16 to 8	21	
pyrethroids, MOA 3 and pyrethroid combinations	(see European corn borer above for rates)	—	—	—	

Table 5-2. Insect Control in Field Corn

Insecticide, Mode of Action Code, and Formulation	Per Acre Amount	Per Acre Active (lb)	Acres/gal (lb)	Preharvest Interval (PHI) (Days)	Precautions and Remarks
Sod Webworm, Chinch Bug					
bifenthrin, MOA 3 (Brigade, Discipline, Sniper, and others) 2 EC	2.1 to 6.4 fl oz	0.033 to 0.1	61 to 20	30	Apply to base of seedlings as a directed spray or over the row. Seldom an economic problem. Use higher rates for chinch bugs. Drop nozzles at 15 gallons per acre or above will give better results.
pyrethroids, MOA 3 and pyrethroid combinations	(see European corn borer above for rates)	—	—	—	
clothianidin, MOA 4A (Poncho) 600 FS	—	0.25 to 1.25 mg per kernel	—	—	1250 rate must be special-ordered from a seedsman.
thiamethoxam, MOA 4A (Cruiser) 5 FS	—	0.5 to 1.25 mg per kernel	—	—	
carbaryl, MOA 1A (Sevin XLR Plus) 4 EC	2 pt	1	4	14	Apply to base of seedlings as directed spray or over the row. Seldom an economic problem. Use higher rates for chinch bugs. Drop nozzles at 15 gallons/acre or above will give better results.
chlorpyrifos, MOA 1B (Lorsban) 4 E	1 pt	0.5	8	21	
Sugarcane Beetle — At Planting Treatments					
clothianidin, MOA 4A (Poncho) 600 FS	—	1.25 mg per kernel	—	—	This seed treatment, combined with an in-furrow insecticidal granular or liquid application will still provide only fair control. 1250 rate must be special-ordered from a seedsman.
clothianidin, MOA 4A + in-furrow insecticide, MOA 1B (Poncho 500) + (various, e.g., chlorpyrifos (Lorsban), phosphorothioic acid + bifenthrin (SmartChoice), tebupirimphos + cyfluthrin (Aztec), terbofos (Counter), etc.)	—	—	—	—	See recommendations for seed treatment above. Granular insecticide alone or 500 rate of seed treatment alone will not provide adequate control without granular insecticide. Expect only fair control.
True Armyworm — In Whorl and on Foliage					
bifenthrin, MOA 3 (Brigade, Discipline, Sniper, and others) 2 EC	2.1 to 6.4 oz	0.04 to 0.16	61 to 20	30	Apply into plant whorls where caterpillars are located and use a minimum of 15 gallons water per acre. Treat when caterpillars are small. Aerial application is satisfactory when caterpillars are not in whorl (post-tassel). Armyworm problems are usually confined to no-till planted corn seedlings. Consult county agent for scouting information.
chlorantraniliprole, MOA 28 (Prevathon) 0.43 SC	14 to 20 fl oz	0.047 to 0.067	9.1 to 6.4	14	
chlorpyrifos, MOA 1B (Lorsban) 4 EC	2 pt	1	4	35	
methomyl, MOA 1A (Lannate) 2.4 LV (Lannate) 90 SP	0.75 to 1.5 pt 0.25 to 0.5 lb	0.23 to 0.45 0.23 to 0.45	10.7 to 5.3 4 to 2	3 (forage) 21 (fodder)	
pyrethroids, MOA 3 and pyrethroid combinations	(see European corn borer above for rates)	—	—	—	
spinosad, MOA 6 (Blackhawk) 4 SC	1.67 to 3.3 fl oz	0.038 to 0.075	76.6 to 38.8	7 (forage or seed) 28 (grain)	
Western or Northern Corn Rootworm — At Planting, Seed Treatments					
clothianidin, MOA 4A (Poncho) 600 FS	—	1.25 mg/kernel	—	—	Must be special-ordered from a seedsman. Rootworms mainly a problem in Piedmont and mountain regions where corn is not rotated.
Bt transgenic corn, MOA 11 (Agrisure, Herculex XTRA, Genuity VT Triple PRO, Optimum Intrasect XTRA, SmartStax)	—	See remarks	—	—	This transgenic corn is designed to prevent root injury from rootworm larvae. Usually only needed in corn following corn. Observe the refuge specifications on the label. There is known resistance to the traits Cry3Bb1 and mCry3A (Agrisure, Genuity VT Triple PRO). No known resistance to products with Cry34AB1/Cry35Ab1 (Herculex XTRA, Optimum Intrasect XTRA, SmartStax).
chlorpyrifos, MOA 1B (Lorsban) 15 G	8 oz/1,000 ft	*	—	—	Apply granules 6- to 7-inch band over the open seed furrow and in front of the planter press wheel at planting time. Consult product label for incorporation instructions. Terbufos may be applied directly into the seed furrow. Do not apply phorate into seed furrow as seedling injury may occur. Terbufos may interact with Beacon herbicide and injure plants. Consult label.
phorate, MOA 1B (Thimet) 20 G	6 oz/1,000 ft				
tefluthrin, MOA 1A (Force) 3.0 G (Force) CS	4 to 5 oz/1,000 ft 0.46 to 0.57 oz/1,000 ft	*	—	—	
terbufos, MOA 1B (Counter) 20 G	6 oz/1,000 ft	*	—	—	
Wireworm — At Planting Treatments					
bifenthrin, MOA 3 (Capture) LFR	3.4 to 13.6 oz	0.047 to 0.062	—	—	Apply as an in-furrow spray, microstream, or t-band.
clothianidin, MOA 4A (Poncho) 600 FS	0.5 to 1.25 mg/kernel	—	—	—	1250 rate must be special-ordered from a seedsman.
thiamethoxam, MOA 4A (Cruiser) 5 FS	0.5 to 1.25 mg/kernel	—	—	—	
thiamethoxam, MOA 4A + chlorantraniliprole, MOA 28 (Lumivia)	—	0.25 mg + 0.25 mg per kernel	—	—	

Table 5-2. Insect Control in Field Corn

Insecticide, Mode of Action Code, and Formulation	Per Acre Amount	Per Acre Active (lb)	Acres/gal (lb)	Preharvest Interval (PHI) (Days)	Precautions and Remarks
Wireworm — At Planting Treatments (continued)					
phorate, MOA 1B (Thimet) 20G	6 oz/1,000 ft	—	—	—	Apply only in T-band over open furrows. Results may be poor if approximately 50% fails to fall with the seed (into seed furrows); however, in-furrow application may reduce stand.
tefluthrin, MOA 1A (Force) 3.0 G (Force) CS	4 to 5 oz/1,000 ft 0.46 to 0.57 oz/1,000 ft	*	—	—	T-band or in-furrow. If T-banded, some granules must fall with seed for wireworm control. Wireworm control is improved when used in-furrow. Terbufos may interact with Beacon herbicide when used in-furrow.
terbufos, MOA 1A (Counter) 20 G	6 oz/1,000 ft	*	—	—	

* For 30-inch or wider row spacings.

PRECAUTIONS: Always use pesticides according to label directions. Be mindful of reducing the impact of pesticides on wildlife and groundwater.

Insect Control in Grain Sorghum

D. D Reisig and A.S. Huseth, Entomology and Plant Pathology

Table 5-3. Insect Control in Grain Sorghum

Insecticide, Mode of Action Code, and Formulation	Per Acre Amount	Per Acre Active (lb)	Acres/gal (lb)	Preharvest Interval (PHI) (Days)	Precautions and Remarks
Aphid (including sugarcane aphid) — At Planting					
clothianidin, MOA 4A (Poncho) 600 FS	5.1 to 6.4 oz/cwt	See label	—	—	Follow label instructions for mixing.
clothianidin, MOA 4A + *Bacillus firmus* (for nematodes) (Poncho/VOTiVO)	6.13 fl oz/cwt	See label	—	—	
imidacloprid, MOA 4A (Gaucho) 480 FS (Gaucho) 600 FS	8 fl oz/cwt 6.4 fl oz/cwt	See label	—	45 (forage)	
thiamethoxam, MOA 4A (Cruiser) 5 FS	5.1 to 7.6 fl oz	See label	—	45 (forage)	
Aphid (excluding sugarcane aphid) — Foliar					
beta-cyfluthrin, MOA 3 (Baythroid XL) 1.0 EC	1.6 to 2.8 oz	0.017 to 0.022	80 to 45.7	21	Ground application with at least 15 gallons water per acre is preferred. Aerial application should use at least 5 gallons water per acre. At least 300 aphids per plant are necessary to justify treatment.
chlorpyrifos, MOA 1B (Lorsban) 75 WG	0.5 to 1 pt	0.25 to 0.5	16 to 8	28	
chlorpyrifos, MOA 1B + lambda-cyhalothrin, MOA 3 (Cobalt Advanced) 75 WG	11 to 38 fl oz	See label	11.6 to 3.4	30 to 60 (See label)	
cyfluthrin, MOA 3 (Tombstone) 1.0 EC	1.3 to 2.8 oz	0.2 to 0.044	98.5 to 45.7	14	
dimethoate, MOA 1B (Dimethoate) 4 EC	0.5 to 1 pt	0.25 to 0.5	16 to 8	28	
lambda-cyhalothrin, MOA 3 (Karate, Lambda-cyhalothrin, Silencer) 1.0 EC (Warrior II and Karate Z) 2.08 CS	2.56 to 3.84 fl oz 1.28 to 1.92 fl oz	0.02 to 0.03 0.02 to 0.3	50 to 33.3 100 to 66.7	30 30	
zeta-cypermethrin, MOA 3 (Mustang Max) 0.8 EC	3.2 to 4.0 fl oz	0.02 to 0.25	40 to 32	14 (grain) 45 (forage)	
Aphid (sugarcane aphid only) — Foliar					
flupyradifurone, MOA 4D (Sivanto) 200 SL	4 to 7 fl oz	0.052 to 0.091	32 to 18.3	21 (grain) 7 (forage)	A maximum of 28 ounces per acre can be used in a season.
sulfoxaflor, MOA 4C (Transform) 50 WG	0.75 to 1.5 oz	0.024 to 0.047	171 to 85	14 (grain) 7 (forage)	Note that this use is provided only under a Section 18 emergency exemption set to expire Nov. 30, 2018. This exemption is expected to be granted again in 2019, but check the label for confirmation. A maximum of 3 ounces per acre can be used in a season.
Chinch Bug — At Planting					
clothianidin, MOA 4A (Poncho) 600 FS	5.1 to 6.4 oz/100 lb seed	See label	—	—	Follow label instructions for mixing.
imidacloprid, MOA 4A (Gaucho) 480 FS (Gaucho) 600 FS	8 fl oz/cwt 6.4 fl oz/cwt	See label	—	45 (forage)	
imidacloprid, MOA 4A (Gaucho) 480 FS (Gaucho) 600 FS	8 fl oz/cwt 6.4 fl oz/cwt	See label	—	45 (forage)	

Table 5-3. Insect Control in Grain Sorghum

Insecticide, Mode of Action Code, and Formulation	Per Acre Amount	Per Acre Active (lb)	Acres/gal (lb)	Preharvest Interval (PHI) (Days)	Precautions and Remarks
thiamethoxam, MOA 4A (Cruiser) 5 FS	7.6 fl oz	See label	—	45 (forage)	
Chinch Bug — Foliar					
carbaryl, MOA 1A (Sevin XLR Plus) 4 EC	3 pt	1.5	2.7	21	Apply to base of plants where insects congregate. Begin applications when insects migrate from small grains or grass weeds to sorghum. Expect fair control from pyrethroids (MOA 3).
chlorpyrifos, MOA 1B (Lorsban) 75 WG	0.67 to 1.33 lbs	0.5 to 1.0	1.5 to 0.75	28	
pyrethroids, MOA 3 and pyrethroid combinations	(use highest labeled rates)	See label	—	—	
Corn Earworm/Webworm — In Heads					
Bacillus thuringiensis, MOA 11B2 (Various)	—	—	—	0	Best when larvae are small.
beta-cyfluthrin, MOA 3 (Baythroid XL) 1.0 EC	1.3 to 2.8 fl oz	0.01 to 0.022	98.5 to 45.7	14	Ground application with at least 15 gallons water per acre is preferred. Aerial application should use at least 5 gallons water per acre. Use higher rates by air for serious infestation. Threshold is one medium to large earworm or armyworm per head or three webworms per head. See label for Asana. Entrust is OMRI listed.
carbaryl, MOA 1A (Sevin XLR Plus) 4 EC	3 pt	1.5	2.7	21	
chlorantraniliprole, MOA 28 (Prevathon) 0.43 SC	14 to 20 oz	0.047 to 0.067	9.1 to 6.4	14	
cyfluthrin, MOA 3 (Tombstone) 1.0 EC	1.3 to 2.8 fl oz	0.02 to 0.044	98.5 to 45.7	14	
esfenvalerate, MOA 3 (Asana XL) 0.66 EC	5.8 to 9.6 fl oz	0.03 to 0.05	22 to 13.3	21	
lambda-cyhalothrin, MOA 3 (Karate, Lambda-cyhalothrin, Silencer) 1.0 EC (Warrior II and Karate Z) 2.08 C	2.6 to 3.8 fl oz 1.28 to 1.92 fl oz	0.02 to 0.03 0.02 to 0.03	49.2 to 33.7 100 to 66.7	30 30	
methomyl, MOA 1A (Lannate) 2.4 LV (Lannate) 90 SP	0.75 to 1.5 pt 0.25 to 0.5 lb	0.23 to 0.45 0.23 to 0.45	10.7 to 5.3 4 to 2	14 14	
spinosad, MOA 5 (Blackhawk) 4 SC (Entrust) 80 WP	1.7 to 3.0 oz 1 to 2 oz	0.039 to 0.068 0.05 to 0.01	75.3 to 42.7 16 to 8	21 (grain) 3 (forage)	
zeta-cypermethrin, MOA 3 (Mustang Max) 0.8 EC	1.8 to 4.0 oz	0.011 to 0.025	71.1 to 32	14 (grain) 45 (forage)	
Fall Armyworm					
chlorantraniliprole, MOA 28 (Prevathon) 0.43 SC	14 to 20 oz	0.047 to 0.067	9.1 to 6.4	14	Difficult to control—ground application only with high volume. Direct spray into whorls. Treat at 80% infestation (1 worm per plant) or 40% infestation (multiple worms per plant). Treat when worms are small. Addition of surfactant and application when dew is on plant may be helpful. Entrust is OMRI listed.
chlorpyrifos, MOA 1B (Lorsban) 75 WG	0.67 to 1.33 oz	0.5 to 1	191 to 96.2	See label	
methomyl, MOA 1A (Lannate) 2.4 LV (Lannate) 90 SP	0.75 to 1.5 pt 0.25 to 0.5 lb	0.23 to 0.45 0.23 to 0.45	10.7 to 5.3 4 to 2	14 14	
spinosad, MOA 5 (Blackhawk) 4 SC (Entrust) 80 WP	1.7 to 3.0 oz 1 to 2 oz	0.039 to 0.068 0.05 to 0.01	75.3 to 42.7 16 to 8	21 (grain) 3 (forage)	

Insect Control in Small Grains

D. D. Reisig and A.S. Huseth, Entomology and Plant Pathology

Table 5-4. Insect Control in Small Grains

Insecticide, Mode of Action Code, and Formulation	Per Acre Amount	Per Acre Active (lb)	Acres/gal (lb)	Preharvest Interval (PHI) (Days)	Precautions and Remarks
Aphid — At Planting					
imidacloprid, MOA 4A (Gaucho) 480 FS (Gaucho) 600 FS (Gaucho) XT	1 to 3 fl oz/cwt 0.8 to 2.4 fl oz/cwt 3.5 fl oz/cwt	See label	—	45 (forage)	Early season protection against aphids. Has shown barley yellow dwarf suppression. Most effective on early planted grains. Acknowledge plant-back restrictions. See Hessian fly section.
thiamethoxam, MOA 4A (Cruiser) 5 F)	0.75 to 1.33 fl oz/cwt	See label	—	45 (forage)	
Aphid — Foliar					
beta-cyfluthrin, MOA 3 (Baythroid XL) 1.0 EC	1.8 to 2.4 fl oz	0.014 to 0.019	71.1 to 53.3	7 (forage) 30 (harvest)	
cyfluthrin, MOA 3 (Tombstone) 1.0 EC	1.8 to 2.4 fl oz	0.028 to 0.038	71.1 to 53.3	30	
dimethoate, MOA 1B (Dimethoate) 4 EC	0.5 to 0.75 pt	0.25 to 0.37	16 to 10.7	35	Will not reduce barley yellow dwarf virus infection. Consult local Extension agent for scouting and threshold suggestions. Keep lambda-cyhalothrin away from waterways.
lambda-cyhalothrin, MOA 3 (Karate, Lambda-cyhalothrin, Silencer) 1.0 EC (Warrior II and Karate Z) 2.08 C	2.56 fl oz 1.28 fl oz	0.02 0.03	50 100	30 30	
zeta-cypermethrin, MOA 3 (Mustang Max) 0.8 EC	3.2 to 4.0 fl oz	0.02 to 0.025	40 to 32	14	

Table 5-4. Insect Control in Small Grains

Insecticide, Mode of Action Code, and Formulation	Per Acre Amount	Per Acre Active (lb)	Acres/gal (lb)	Preharvest Interval (PHI) (Days)	Precautions and Remarks
Cereal Leaf Beetle					
beta-cyfluthrin, MOA 3 (Baythroid XL) 1.0 EC	1.0 to 1.8 fl oz	0.008 to 0.014	128 to 71.1	7 (forage) 30 (harvest)	Use where beetle eggs/larvae are above threshold. Application of insecticide with topdress fertilizer for preventative control is not advised. Lower rates should only be used where population densities are above threshold, but moderate.
carbaryl*, MOA 1A (Sevin XLR Plus) 4 EC	1 pt	0.5	8	21	
chlorpyrifos, MOA 1B + lambda-cyhalothrin, MOA 3 (Cobalt Advanced) 2.63 EC	11 to 25 fl oz	See label	11.6 to 2.3	30	
cyfluthrin, MOA 3 (Tombstone) 1.0 EC	1.0 to 1.8 fl oz	0.016 to 0.028	128 to 71.1	30	
gamma-cyhalothrin, MOA 3 (Declare) 1.25 EC	1.02 to 1.54 oz	0.01 to 0.015	125.5 to 83.1	30	
lambda-cyhalothrin, MOA 3 (Karate, Lambda-cyhalothrin, Silencer) 1.0 EC (Warrior II and Karate Z) 2.08	2.56 fl oz 1.92 fl oz	0.02 0.03	50 66.7	30 30	
methomyl, MOA 1A (Lannate) 2.4 LV (Lannate) 90 SP	1 to 2 pt 0.25 to 0.5 lb	0.22 to 0.45 0.22 to 0.45	8 to 4	7 7	
zeta-cypermethrin, MOA 3 (Mustang Max) 0.8 EC	1.6 to 4.0 fl oz	0.011 to 0.025	80 to 32	14	
Hessian Fly— Fall Generation					
imidacloprid, MOA 4A (Gaucho) 600 FS (Gaucho) XT (Rancona Crest)	1.2 to 2.4 fl oz/cwt 3.5 fl oz/cwt 5.0 to 8.3 fl oz/cwt	See label	—	45 (forage)	Early season protection against Hessian fly. Seed usually treated by seedsman. Acknowledge plant-back restriction.
thiamethoxam, MOA 4A (Cruiser) 5 FS	0.75 to 1.33 oz/cwt	See label	—	45 (forage)	
Hessian Fly— Fall and Late Winter Generations					
beta-cyfluthrin, MOA 3 (Baythroid XL) 1.0 EC	2.4 fl oz	0.019	53.3	3 (forage) 30 (harvest)	Apply to fields with high egg count in fall; preferable at or before the 2 to 3 leaf stage. In spring, apply to infested fields as flies emerge. Use high rates for heavy infestations. Recent NCSU experiments suggest that a resistant variety or seed treatment are far superior to foliar sprays as rescue treatments.
cyfluthrin, MOA 3 (Tombstone) 1.0 EC	2.4 fl oz	0.038	53.3	30	
lambda-cyhalothrin, MOA 3 (Karate, Lambda-cyhalothrin, Silencer) 1.0 EC (Warrior II and Karate Z) 2.08 EC	3.8 fl oz 1.92 fl oz	0.03 0.03	33.7 66.7	30 30	
zeta-cypermethrin, MOA 3 (Mustang Max) 0.8 EC	4 fl oz	0.025	32	14	
True Armyworm — Spring					
beta-cyfluthrin, MOA 3 (Baythroid XL) 1.0 EC	1.8 to 2.4 fl oz	0.013 to 0.019	71.1 to 53.3	3 (forage) 30 (harvest)	Apply by air or ground when armyworms are at 2 per square foot or greater. Use higher rates when caterpillars are very numerous. High volume (3 to 5 gallons per acre) may be beneficial in thickly planted wheat. Poor performance may result when temperatures are cool or when rainfall washes residues from plants. Best to apply when conditions are warm (60 degrees F plus) and armyworms are active. Carbaryl may stimulate aphid populations. Entrust is OMRI listed.
carbaryl, MOA 1A (Sevin XLR Plus) 4 EC	1.5 pt	0.75	5.3	21	
chlorantrapiliprole, MOA 28 (Prevathon) 0.43 SC	14 to 20 oz	0.047 to 0.067	9.1 to 6.4	21	
cyfluthrin, MOA 3 (Tombstone) 1.0 EC	1.8 to 2.4 fl oz	0.028 to 0.038	71.1 to 53.3	30	
gamma-cyhalothrin, MOA 3 (Declare) 1.25 EC	1.02 to 1.54 oz	0.01 to 0.015	125.5 to 83.1	30	
lambda-cyhalothrin, MOA 3 (Karate, Lambda-cyhalothrin, Silencer) 1.0 EC (Warrior II and Karate Z) 2.08	2.6 to 3.8 fl oz 1.28 to 1.92 fl oz	0.02 to 0.03 0.02 to 0.03	49.2 to 33.7 100 to 66.7	30 30	
methomyl, MOA 1A (Lannate) 2.4 LV (Lannate) 90 SP	1.5 pt 0.5 lb	0.45 0.45	5.3 2	7 7	
spinosad, MOA 5 (Blackhawk) 4 SC (Entrust) 80 WP	1.1 to 3.0 oz 1 to 2 oz	0.026 to 0.068 0.05 to 0.01	116.4 to 42.7 16 to 8	3 (forage) 21 (harvest)	
zeta-cypermethrin, MOA 3 (Mustang Max) 0.8 EC	1.6 to 4.0 fl oz	0.011 to 0.025	80 to 32	14	
Wireworm — At Planting					
imidacloprid, MOA 4A (Gaucho) 480 FS (Gaucho) 600 FS (Gaucho) XT (Rancona Crest)	1 fl oz/cwt 0.8 fl oz/cwt 3.5 fl oz/cwt 8.3 fl oz/cwt	See label	—	45 (forage)	See remarks under Aphids. Seed treatments must be applied by seedsman.
thiamethoxam, MOA 4A (Cruiser) 5 FS	0.75 fl oz/cwt	See label	—	45 (forage)	

CAUTION: Always use pesticides according to label directions. Be mindful of reducing the impact of pesticides on wildlife and groundwater.

Insect Control on Cotton

D. D. Reisig and A.S. Huseth, Entomology and Plant Pathology

NOTE: Use the Mode of Action (MOA) codes following each insecticide to combat the development of insecticide resistance. Active ingredients sharing the same letter/number have the same mode of action.

Table 5-5A. Insect Control on Cotton

Insect Insecticide, Mode of Action (MOA), and Formulation	Per Acre		Acres/gal (lb)	Pre-harvest Interval (days)	Precautions and Remarks
	Amount	Active (lb)			
Beet Armyworm					
chlorantraniliprole, MOA 28 (Prevathon) 0.43 SC	14 to 27 oz	0.047 to 0.09	9.1 to 4.8	21	Bollgard II, Bollgard 3, TwinLink, WideStrike and WideStrike 3 varieties show high resistance to beet armyworm damage, unless larvae move to cotton from late burned-down weed hosts (see Bollworm/Budworm section for Bt cotton notes). Refer to labels for seasonal total active ingredient restrictions for all products.
emamectin benzoate, MOA 6 (Denim) 0.16 EC	6 to 8 oz	0.0075 to 0.01	21.3 to 16	21	
indoxacarb, MOA 22 (Steward) 1.25 SC	9.2 to 11.3 oz	0.09 to 0.11	14 to 11.5	14	
methoxyfenozide, MOA 18A (Intrepid) 2F	4.0 oz	0.06	33	14	
methoxyfenozide, MOA 18A + spinetoram, MOA 5 (Intrepid Edge) 3F	4.0 to 8.0 oz	0.094 to 0.188	32 to 16	28	
spinosad, MOA 5 (Blackhawk) 4 SC	2.4 to 3.2 oz	0.054 to 0.072	53.3 to 40	28	
chlorantraniliprole, MOA 28 (Prevathon) 0.43 SC	14 to 27 oz	0.047 to 0.09	9.1 to 4.8	21	
Bollworm[a] /Tobacco Budworm					
Bollgard II, MOA 11B2 (various varieties)	—	See remarks	—	—	Cry1Ac and Cry2Ab proteins in Bollgard II have low to moderate activity against bollworm and high activity against other pest caterpillar species on cotton except cutworms. No activity against insects other than caterpillars. Note that some populations of bollworm are resistant to both the Cry1Ac and Cry2Ab proteins.
Bollgard3, MOA-11B2 (various varieties)	—	See remarks	—	—	Cry1Ac and Cry2Ab proteins in Bollgard 3 have low to moderate activity against bollworm and high activity against other pest caterpillar species on cotton except cutworms. No activity against insects other than caterpillars. Note that some populations of bollworm are resistant to both the Cry1Ac and Cry2Ab proteins, but there is no known Vip3A resistance.
TwinLink, MOA 11B2 (various varieties)	—	See remarks	—	—	Cry1Ab and Cry2Ae proteins in TwinLink have low to moderate activity against bollworm and high activity against other pest caterpillar species on cotton except cutworms. No activity against insects other than caterpillars. Note that some populations of bollworm are resistant to both the Cry1Ab and Cry2Ae proteins.
WideStrike, MOA 11B2 (various varieties)	—	See remarks	—	—	Cry1Ac and Cry1F proteins in WideStrike have low to moderate activity against bollworm and high activity against other pest caterpillar species on cotton except cutworms. No activity against insects other than caterpillars. Note that some populations of bollworm are resistant to the Cry1Ac protein and that Cry1F is not lethal to bollworm.
WideStrike 3, MOA 11B2 (various varieties)	—	See remarks	—	—	Cry1Ac, Cry1F and Vip3A proteins in WideStrike 3 high activity in combination against all pest caterpillar species on cotton except cutworms. No activity against insects other than caterpillars. Although some populations of bollworm are resistant to the Cry1Ac protein and Cry1F is not lethal to bollworm, there is no known Vip3A resistance.
bifenthrin, MOA 3 (Brigade, Fanfare, Discipline, Sniper and others) 2 EC	6.4 oz	0.1	20	14	High rate of bifenthrin must be used for effective caterpillar control. Addition of acephate to bifenthrin does not always increase control and can add to the potential to flare other pests.
chlorantraniliprole, MOA 28 + lambda-cyhalothrin MOA 3 (Besiege) 1.25	6.5 to 12.5 oz	0.063 to 0.12	19.8 to 10.4	14	This insecticide is most effective when applied before larvae are present at the beginning of an egg-lay event.
chlorantraniliprole, MOA 28 (Prevathon) 0.43 SC	14 to 27 oz	0.047 to 0.09	9.1 to 4.8	21	This insecticide is most effective when applied before larvae are present at the beginning of an egg-lay event.
indoxacarb, MOA 22 (Steward) 1.25 SC	9.2 to 11.3 oz	0.09 to 0.11	13.9 to 11.4	14	Steward must be applied to early stage larvae for effective control.
methoxyfenozide, MOA 18A + spinetoram, MOA 5 (Intrepid Edge) 3F	6.0 to 8.0 oz	0.140 to 0.188	21.3 to 16	28	
spinosad, MOA 5 (Blackhawk) 4 SC	2.4 to 3.2 oz	0.054 to 0.073	74 to 55	28	
Cotton Aphid					
acetamiprid, MOA 4A (Assail, Strafer Max) 70 WP	0.6 to 1.1 oz	0.025 to 0.05	28 to 14	28	Due to a high potential for cotton aphid resistance to insecticides and because of the routine presence of significant levels of predators, parasites and pathogens that limit cotton aphid build-ups, treat for cotton aphids only as a last resort. In 2012, cotton aphid resistance to the neonicotinoid insecticide class was confirmed in North Carolina. Try to limit the use of this class of insecticides, especially for stink bugs. All insecticides in this section are nicotinoids except for Carbine. Belay cannot be applied foliar after pinhead square formation. Transform was available during 2018 as a Section 18 Emergency Exemption. There is no guarantee that this insecticide will be available during 2019.
clothianidin, MOA 4A (Belay) 2.13 WDG	3 to 4 oz	0.05 to 0.067	42.6 to 31.8	21	
flonicamid, MOA 9C (Carbine) 50 WG	1.4 to 2.8 oz	0.044 to 0.089	22.7 to 11.2	30	
imidacloprid, MOA 4A (Trimax Pro, Admire Pro, other generics) 4.0 F	1 to 1.5 oz	0.03 to 0.047	128 to 85	14	
sulfoxaflor MOA 4C (Transform)	0.75 to 1 oz	0.023 to 0.031	171 to 128	14	
thiamethoxam, MOA 4A (Centric) 40 WG	1.25 to 2.5 oz	0.03 to 0.06	13.3 to 8	21	

Table 5-5A. Insect Control on Cotton

Insect / Insecticide, Mode of Action (MOA), and Formulation	Per Acre Amount	Per Acre Active (lb)	Acres/gal (lb)	Pre-harvest Interval (days)	Precautions and Remarks
European Corn Borer					
Bollgard II, MOA 11B2 (various varieties)	—	See remarks	—	—	This is transgenic cotton seed.
Bollgard 3, MOA 11B2 (various varieties)	—	See remarks	—	—	
TwinLink, MOA 11B2 (various varieties)	—	See remarks	—	—	
TwinLink Plus, MOA 11B2 (various varieties)	—	See remarks	—	—	
WideStrike, MOA 11B2 (various varieties)	—	See remarks	—	—	
WideStrike 3, MOA 11B2 (various varieties)	—	See remarks	—	—	
beta-cyfluthrin, MOA 3 (Baythroid XL) 1.0 EC	1.6 to 2.6 oz	0.013 to 0.021	77 to 47.6	0	
bifenthrin, MOA 3 (Brigade, Fanfare, Declare, Discipline, Sniper and others) 2 EC	3.2 oz	0.05	40	14	European corn borers are generally more of a problem in rank, non-Bt cotton. Other materials listed for bollworm may provide some control.
lamda-cyhalothrin, MOA 3 (Warrior) 2.08 CS (Warrior II, Silencer) 1 EC	1.6 oz / 3.2 to 5.12 oz	0.025 / 0.025 to 0.04	80 / 40 to 25	21	
zeta-cypermethrin, MOA 3 (Mustang Max) 0.8 EC	2.9 to 3.55 oz	0.018 to 0.025 oz	44.4 to 32	14	
Fall Armyworm					
chlorpyrifos, MOA 1B (Lorsban) 4 E	1 to 2 pt	0.5 to 1	8 to 4	14	Various rates and combinations may be recommended, depending upon cotton phenology and the age distribution and population levels of larvae. Pyrethroids keep some fall armyworms from hatching. Bollgard II and WideStrike varieties show high resistance to fall armyworm damage.
emamectin benzoate, MOA 6 (Denim) 0.16 EC	8 to 12 oz	0.01 to 0.015	16 to 10.7	21	
indoxacarb, MOA 22 (Steward) 1.25 SC	9.2 to 11.3 oz	0.09 to 0.11	14 to 11.5	14	
lambda-cyhalothrin, MOA 3 +chlorantraniliprole, MOA 28 (Besiege) 1.25 ZC	6.5 to 12.5 oz	0.063 to 0.12	19.8 to 10.4	14	
methomyl, MOA 1A (Lannate) 2.4 LV (Lannate) 90 SP	1.5 pt / 0.5 lb	0.45 / 0.45	5.3 / 2	15 / 15	
chlorantraniliprole, MOA 28 (Prevathon) 0.43 SC	14 to 27 oz	0.047 to 0.09	4.8 to 27	21	Various rates and combinations may be recommended, depending upon cotton phenology and the age distribution and population levels of larvae. Pyrethroids keep some fall armyworms from hatching. Bollgard II, Bollgard 3, WideStrike, and WideStrike 3 varieties show high resistance to fall armyworm damage.
methoxyfenozide, MOA 1BA (Intrepid) 2F	4 to 10 oz	0.06 to 0.16	33 to 12.5	14	
methoxyfenozide, MOA 18A + spinetoram, MOA 5 (Intrepid Edge) 3F	6.0 to 8.0 oz	0.140 to 0.188	21.3 to 16	28	
novaluron, MOA 15 (Diamond) 0.83 EC	6 to 12 oz	0.04 to 0.08	21.3 to 10.7	30	
spinosad, MOA 5 (Blackhawk) 4 SC	2.4 to 3.2 oz	0.054 to 0.072	53.3 to 40	28	
Plant Bug					
acephate, MOA 1B (Orthene and other brands) 75 S / 90 S / 97 ST	0.3 to 1.3 lb / 0.25 to 1 lb / 0.25 to 1 lb	0.25 to 1 / 0.225 to 0.9 / 0.24 to 0.97	3.3 to 0.77 / 4 to 1 / 4 to 1	21 / 21 / 21	Prebloom treatment not recommended if square retention is in excess of 80%. If square retention is less than 80%, confirmation of threshold levels of plant bugs should be met prior to treatment. Note that Belay cannot be applied to foliar after pinhead square formation.
acetamiprid, MOA 4A (Assail) 70 WP	1.1 oz	0.5	14	28	Postbloom treatment more likely in low-spray environment, such as with Bt cottons. Neonicotinoids (MOA 4A) tend to be less effective mid- to late-season, but control can be erratic, as they will sometimes work season-long. In general, imidacloprid tends to be the least effective of the neonicotinoids. Some populations appear to be developing resistance to pyrethroids (MOA 3) and organophosphates (MOA 1D). **Rotating insecticide modes of action is critical for long-term management of this insect.** Nearly any insecticide can be improved by an immediate follow-up insecticide spray within 3 days of the initial spray.

Fields adjacent to corn, potatoes, weedy areas, ditch banks, and other sources of plant bugs may be at higher risk of plant bug injury.

Likelihood of damage levels of plant bugs on cotton generally higher in northeastern North Carolina counties.

Bidrin is toxic to humans. Be sure to follow label directions and observe 6-day reentry interval.

Transform was available during 2018 as a Section 18 Emergency Exemption. There is no guarantee that this insecticide will be available during 2019. |
chlorpyrifos, MOA 1B (Lorsban) 4 EC	6.1 oz	0.19	21	14	
clothianidin, MOA 4A (Belay) 2.13 SC	3 to 4 oz	0.05 to 0.067	42.6 to 31.8	21	
dicrotophos, MOA 1B (Bidrin) 8 EC	6 to 8 oz	0.375 to 0.5	21 to 16	10	
dicrotophos, MOA 1B + bifenthrin MOA 3 (Bidrin XP II) 5 EC	8 to 12 oz	0.313 to 0.54	16 to 9.3	30	
flonicamid, MOA 9C (Carbine) 50 WG	1.7 to 2.8 oz	0.054 to 0.089	75.3 to 45.7	30	
imidacloprid, MOA 4A (Admire Pro, Trimax Pro, other generics) 4 F	1.3 to 1.7 oz	0.047 to 0.061	98 to 75	14	
methomyl, MOA 1A (Lannate) 2.4 LV (Lannate) 90 SP	12 oz / 0.25 lb	0.225 / 0.225	10.7 / 4	15 / 15	
novaluron, MOA 15 (Diamond) 0.83 EC	9 to 12 oz	0.06 to 0.08	14 to 11	30	

Table 5-5A. Insect Control on Cotton

Insect Insecticide, Mode of Action (MOA), and Formulation	Per Acre Amount	Per Acre Active (lb)	Acres/gal (lb)	Pre-harvest Interval (days)	Precautions and Remarks
Plant Bug (continued)					
oxamyl, MOA 1A (Vydate)	8 to 32 oz	0.125 to 0.5	16 to 4	14	Transform was available during 2018 as a Section 18 Emergency Exemption. There is no guarantee that this insecticide will be available during 2019.
pyrethroids, MOA 3 and pyrethroid combinations	(see European corn borer above for rates)		—	—	
sulfoxaflor MOA 4C (Transform)	2 to 2.25 oz	0.063 to 0.071	64 to 57	14	
thiamethoxam, MOA 4A (Centric) 40 WG	2 to 2.5 oz	0.05 to 0.0625	64 to 51	21	
Soybean Looper					
chlorantraniliprole, MOA 28 (Prevathon) 0.43 SC	20 to 29 oz	0.067 to 0.097	6.4 to 4.4	21	Bollgard II, Bollgard 3, TwinLink, WideStrike, and WideStrike 3 varieties show high resistance to looper damage.
+ chlorantraniliprole, MOA 28 + lambda-cyhalothrin, MOA 3 (Besiege) 1.25 ZC	10.0 to 12.5 oz	0.098 to 0.12	12.8 to 10.4	14	
emamectin benzoate, MOA 6 (Denim) 0.16 EC	6 to 12 oz	0.01 to 0.015	10.6 to 16	21	
indoxacarb, MOA 22 (Steward) 1.25 SC	6.7 to 9.2 oz	0.065 to 0.09	19 to 14	14	
methoxyfenozide, MOA 18A (Intrepid) 2 F	4 to 10 oz	0.098 to 0.16	33 to 12.5	14	
methoxyfenozide, MOA 18A + spinetoram, MOA 5 (Intrepid Edge) 3F	4.0 to 8.0 oz	0.094 to 0.188	32 to 16	28	
spinosad, MOA 5 (Blackhawk) 4 SC	2.4 to 3.2	0.054 to 0.073	74 to 54	28	
Spider Mite					
abemectin, MOA 6 (Zephyr, Abemectin) 0.15 EC	8 to 16 oz	0.01 to 0.019	15 to 7.9	20	Control often unnecessary because of beneficial arthropods and fungi. Apply with 20-plus gallons of water (applies to all chemicals).
bifenthrin, MOA 3 (Brigade, Fanfare, Sniper, Declare, Discipline and others) 2 EC	3.8 oz	0.06	33	14	
dicofol, MOA UNC (Dicofol) 4 E	0.8 to 1.6 qt	0.8 to 1.6	5 to 2.5	14	
entoxazole, MOA 10B (Zeal) 72 WP	0.66 to 1 oz	0.03 to 0.045	45 to 30	28	
fenpropathrin, MOA 3 (Danitol) 2.4 EC	10.7 to 16 oz	0.2 to 0.3	12 to 8	21	
fenpyroximate, MOA 21A (Portal, Fujimite) 0.4 E	12 to 16 oz	0.037 to 0.05	10.8 to 8	14	Use 1.5 to 2X the amount of product if applied by aircraft.
propargite, MOA 12C (Comite) 6.55 L	1 qt	1.6	4	14	
spiromesifen, MOA 23 (Oberon) 2 SC	6 to 16 oz	0.094 to 0.25	21.3 to 8	30	Use 6 ounces only in early season to control low populations.
Stink Bug					
acephate, MOA 1B (Orthene) 75 S (Orthene and others) 97 S	1 lb 0.75 lb	0.75 0.75	1.3 1	21	Stink bugs may be more prevalent on unsprayed or less sprayed Bt cottons.
dicrotophos, MOA 1B (Bidrin) 8 EC	4 to 8 oz	0.25 to 0.5	32 to 16	10	Bidrin is extremely toxic to humans. **Be sure to observe the 6-day reentry interval.**
dicrotophos, MOA 1B + bifenthrin, MOA 3 (Bidrin XP II) 5EC	8.0 to 12.8 oz	0.313 to 0.54	16 to 9.3	30	Product contains 4.0 pounds dicrotophos and 1.0 pound bifenthrin per gallon. Toxic to humans; be sure to follow label directions and observe 6-day reentry interval.
oxamyl, MOA 1A (Vydate) 3.77 SL	17 oz	0.5	7.5	21	
pyrethroids, MOA 3 and pyrethroid combinations	(see European corn borer above for rates)		—	—	Pyrethroids provide good to excellent control green and brown marmorated stink bugs, but are **less effective against brown stink bugs**. Bifenthrin is more effective than other pyrethroids against brown stink bugs and provides a residual advantage over Bidrin.
Thrips (at planting treatment)					
imidacloprid, MOA 4A (Gaucho Grande 600 FS, Acceleron-I)	—	0.375 mg/seed	—	—	Seed treatments with, or without an in-furrow insecticide, may require a supplemental foliar treatment for thrips control. Supplemental sprays are less likely in late planted (after May 20) cotton. Note that resistance to neonicotinoids (imidacloprid and thiamethoxam) has been confirmed in tobacco thrips throughout the state. Variable control should be expected. During 2019, Deltapine is offering Gaucho as a base treatment. Aeris may be requested at the dealer level.
thiamethoxam, MOA 4A (Cruiser) 5 FS	—	0.34 mg/seed	—	—	
abamectin, MOA 6, + thiamethoxam MOA 4A (Avicta Duo 500FS, Avicta Complete, Acceleron-N)	—	0.15 abamectin + 0.375 thiamethoxam mg/seed	—	—	
imidacloprid, MOA 4A + thiodicarb, MOA 1A (Aeris) 48DS	—	0.375 imidacloprid + 0.375 thiodicarb mg/seed	—	—	

Table 5-5A. Insect Control on Cotton

Insect Insecticide, Mode of Action (MOA), and Formulation	Per Acre		Acres/gal (lb)	Pre-harvest Interval (days)	Precautions and Remarks
	Amount	Active (lb)			
Thrips (at planting treatment) (continued)					
imidacloprid (MOA 4A) + clothianidin (MOA 4A) + thiodicarb (MOA 1A) + *Bacillus firmus* (biological) (Aeris/Poncho/VOTiVO)	—	0.375 imidacloprid + 0.375 thiodicarb + 0.424 clothianidin mg/seed + 2 x 10^9 cfu/ml *B. fermis* units	—	—	
imidacloprid, MOA 4A (Admire Pro) 4.6F (Wrangler) 4.0F	7.4 to 9.2 8.5 to 10.5	0.27 to 0.33 0.27 to 0.33	17.3 to 13.9 15.1 to 12.2	—	Apply liquid into open furrow directly onto seed before furrow closure. Works best in combination with another at-planting treatment, such as a seed treatment. Note that resistance to imidacloprid has been confirmed in tobacco thrips throughout the state. Variable control should be expected.
Thrips (post-emergence)					
acephate, MOA 1B (Orthene) 75 S (Orthene) 90 S (Orthene) 97 S (Orthene) 97 ST[b]	3 to 4 oz 0.2 lb 2.5 to 3 oz 6 oz	0.14 to 0.19 0.18 0.15 to 0.18 0.375	5.3 to 4 5 6.4 to 5.3 2.67	21	Not suggested to replace at-plant insecticides in conventional cotton. With the high thrips populations often found in North Carolina, consider at least 0.25 pound a.i. per acre the standard rate for Orthene. Pyrethroids do not provide adequate thrips control on cotton.
cyantraniliprole MOA 28 (Exirel)	13.5 to 20.5 oz	0.088 to 0.133	9.5 to 6.2	7	
dicrotophos, MOA 1B (Bidrin) 8 EC	4 oz	0.25	32	10	
dimethoate, MOA 1B (Dimethoate) 4 EC	8 oz	0.25	16	10	
methamidophos, MOA 1B (Monitor) 4 EC	6.4 oz	0.2	20	50	
spinetoram, MOA 5 (Radiant) 1 SC	1.5 to 3 oz	0.01 to 0.02	85 to 43	28	Provides improved control of western flower thrips, as well as good control of tobacco thrips. Use higher rates for improved control.

[a] Lowest labeled rates for bollworms and budworms

[b] 2 (ee) state local need label for higher rates

NOTE: Upper or lower rate ranges do not indicate equivalent activity.

Cotton Insect Resistance Management

D. D. Reisig and A. S. Huseth, Entomology Department

Resistance occurs when some insects in a population survive a chemical treatment and are therefore able to pass on an inherited gene(s) for this survival to their offspring. Because these offspring are better able to survive the insecticide than those that are not resistant, the resistant individuals increase their numbers faster in the presence of the insecticide. After several generations, the resistant insects can outnumber the susceptible ones, and the insecticide becomes ineffective. Because the alleles that allow insects to survive an insecticide are often initially present in very few individuals out of a very large population of susceptible insects, resistance development may take years. Five to 20 years would be a common range for effectiveness of many insecticides.

Insects vary greatly in their ability to develop resistance to insecticides. For example, cotton aphids have been able to develop resistance to various classes of chemicals rapidly, while the boll weevil remains susceptible to several organophosphate insecticides after more than 50 years of exposure.

Insects develop resistance to insecticides in several ways. Some are able to break down (metabolize) insecticides, while others are able eliminate the toxins. Some can sequester insecticides (move them to a less harmful place in or on the body), and still others can avoid the toxin (behavioral resistance). The above are examples of different modes of action (MOA). Unfortunately, once an insect develops resistance to one insecticide, in most cases the insect is also resistant to others in the same class or group of insecticides sharing the same mode of action. For example, if tobacco budworms are resistant to the pyrethroid Baythroid, they are also resistant to the pyrethroid Warrior. To make matters worse, some insects may be resistant to several classes of insecticides, such as is presently the case with plant bugs in the Midsouth. In North Carolina some populations of cotton aphids (neonicotinoid class) and bollworms/corn earworms (pyrethroid class) have developed resistance to these chemical classes that were initially very effective.

As you can see from the table below, many different kinds of possible insecticide resistance have been identified. Most have complicated, hard-to-remember names. To make it easy to recognize different classes or modes of actions that can lead to resistance development, each chemical has been identified with a number, and occasionally subdivided with a letter. Products sharing the same number or letter and number combination have the same mode of action (for additional detail see: http://pested.okstate.edu/pdf/insecticide 20moa.pdf).

One major strategy in managing resistance is to avoid using products with the same mode of action (sharing the same number in the table) in the same year. Also, tank mixing insecticides with different modes of action may delay resistance development, but can also exacerbate development of resistance in the case of pre-mixed products, when additional insecticide may not be needed or is included at a low rate. Additionally, if only a single class of insecticides is listed for control of an insect (e.g., Assail, Centric, and Trimax Pro – all neonicotinoids – for cotton aphids), one should try to either limit insecticide use to a single spray or try to avoid treatment. One final strategy in minimizing insect resistance to insecticides is to avoid unneeded treatments by following recommended thresholds.

Listed below are the economically important cotton pests found in North Carolina, followed by the chemical and brand names and mode of action.

Table 5-5B. Cotton Insecticide Modes of Action (MOA); Insecticide Resistance Action Committee Designations

Insect	Chemical Name (Brand Name)	Mode of Action
Beet Armyworm	clorantraniliprole (Prevathon)	2B
	emabectin benzoate (Denim)	6
	indoxacarb (Steward)	22
	methoxyfenozide (Intrepid)	18A
	spinosad (Blackhawk)	5
Bollworm/Tobacco Budworm	*Bacillus thuringiensis* var. *kurstaki* (Bt toxin expressed by various varieties)	11B2
	chlorantraniliprole (Prevathon)	3 + 28
	chlorantraniliprole + lambda-cyhalothrin (Besiege)	22
	indoxacarb (Steward)	18A + 5
	methoxyfenozide + spinetoram (Intrepid Edge)	5
	spinosad (Blackhawk)	
Cotton Aphid	acetamiprid (Assail)	4A
	clothianidin	4A
	flonicamid (Carbine)	9C
	imidacloprid (Trimax Pro)	4A
	sulfoxaflor (Transform)	4C
	thiamethoxam (Centric)	4A
European Corn Borer	*Bacillus thuringiensis* var. *kurstaki* (Bt toxin expressed by various varieties)	11B2
	beta-cyfluthrin (Baythroid XL)	3
	bifenthrin (Brigade, Fanfare, Discipline, Sniper and others)	3
	lambda-cyhalothrin (Warrior, Warrior II)	22
	zeta-cypermethrin (Mustang Max)	3
		3 + 28
Fall Armyworm	chlorantraniliprole (Prevathon)	28
	chlorantraniliprole + lambda-cyhalothrin (Besiege)	18A + 5
	chlopyrifos (Lorsban)	1B
	emamectin benzoate (Denim)	6
	indoxacarb (Steward)	22
	methomyl (Lannate)	1A
	methoxyfenozide (Intrepid)	18A
	methoxyfenozide + spinetoram (Intrepid Edge)	18A + 5
	novaluron (Diamond)	15
	spinosad (Blackhawk)	5
Plant Bug	acephate (Orthene, and others)	1B
	acetamiprid (Assail)	4A
	chlopyrifos (Lorsban)	1B
	clothianidin (Belay)	4A
	dicrotophos (Bidrin)	1B
	flonicamid (Carbine)	9C
	imidacloprid (Trimax Pro, Admire Pro)	4A
	methomyl (Lannate)	1A
	novaluron (Diamond)	5
	oxamyl (Vydate)	1A
	pyrethroids (various)	3
	sulfoxflor (Transform)	4C
	thiamethoxam (Centric)	4A
Soybean & Cabbage Looper	chlorantraniliprole (Prevathon)	28
	clorantraniliprole + lambda-cyhalothrin (Besiege)	3 + 28
	emamectin benzoate (Denim)	3
	indoxacarb (Steward)	22
	methoxyfenozide (Intrepid)	18A
	methoxyfenozide + spinetoram (Intrepid Edge)	18A + 5
	spinosad (Blackhawk)	5
Spider Mite	abemectin (Zephyr, Abemectin)	6
	bifenthrin (Brigade, Capture, Discipline, Sniper and others)	3
	dicofol (Dicofol)	UNC*
	entoxazole (Zeal)	10B
	fenpropathrin (Danitol)	3
	fonpyroximate (Portal)	21A
	propargate (Comite)	12C
	spiromesfen (Oberon)	23
Stink Bug	acephate (Orthene, and others)	1B
	dicrotophos (Bidrin)	18
	dicrotophos + bifenthrin (Bidrin XP II)	18 + 3
	oxamyl (Vydate)	1A
	pyrethroids	3
Thrips (At-Planting)	imidacloprid	4A
	thiamethoxam	4A
	thiamethoxam + abamectin	4A + 6
	imidacloprid + thiodicarb	4A + 1A
	imidacloprid + clothianidin + thiodicarb (Aeris/Poncho/VOTiVO)	4A + 1A
Thrips (Postemergence)	acephate (Orthene, and others)	1B
	cyantraniliprole (Exirel)	28
	dicrotophos (Bidrin)	1B
	dimethoate (Dimethoate)	1B
		5

*UNC: Compound with unknown mode of action.

Insect Control on Peanuts

R. L. Brandenburg, Entomology and Plant Pathology

Table 5-6. Insect Control on Peanuts

Insecticide and Formulation	Amount of Formulation Per Acre	Precautions and Remarks
Seasonal Control of Thrips and Leafhoppers		
Thrips at Planting		
acephate (Orthene 97) (generics available)	0.75 to 1 lb	Apply as in-furrow spray in 3 to 5 gallons of water per acre. State (24c) label must be in possession at time of application.
phorate (Thimet) (generics available)	5.0 lb of 20% granules	
fluopyram + imidacloprid (Velum Total)	14 to 18 fl oz	Application rate appropriate for in-furrow spray during planting directed on or below seed, or chemigation into root-zone through low-pressure drip or trickle irrigation.
imidacloprid (Admire Pro)	7.0 to 10.5 fl oz	In furrow spray during planting, directed on or below seed.
thiamethoxam + mefenoxam + fludioxonil + azoxystrobin (Cruiser Maxx Peanuts)	treated peanut seed	Suppression only
aldicarb (AgLogic 15GG & AgLogic 15G)	7.0 lb	Apply granules in the seed furrow and cover with 1 inch or more of soil. May provide suppression of nematodes when applied according to specific label directions.
Thrips foliar postemergence		
acephate (Orthene) 97 (generics available)	0.375 to 0.75 lb	Do not feed or graze livestock on treated vines. Apply 10 to 50 gallons spray solution per acre to foliage. Do not apply more than 4.125 pounds per acre (4 pounds a.i. per acre) per season.
beta-cyfluthrin (Baythroid XL)	2.8 oz	
bifenthrin (Brigade)	2.1 to 6.4 fl oz	Pre-harvest interval of 14 days.
spinetoram (Radiant SC)	1.5 to 3.0 fl oz	Suppression only. See 2(ee) recommendation.
Control of Specific Pests		
Beet Armyworm		
Bacillus thuringiensis (Xentari)	0.5 to 2 lb	Apply to small caterpillars. Use highest rate for larger worms or high populations; 0 day harvest restriction.
methomyl (Lannate LV)	1.25 to 3 pt	Apply broadcast in sufficient water for good coverage when worms are small. Do not apply within 21 days of harvest. See fall armyworm for additional restrictions.
methoxyfenozide + spinetoram (Intrepid Edge)	4 to 8 fl oz	Application rate varies with timing. Lower rates appropriate for light infestations, smaller larvae and/or small plants.
indoxacarb (Steward)	9.2 to 11.3 oz	Do not apply more than 45 ounces per acre per crop. 14-day preharvest interval.
spinosad (Blackhawk)	1.7 to 3.3 fl oz	Do not apply more than 12.4 fluid ounces per season or make more than three applications. 3-day preharvest interval.
bifenthrin (Brigade)	2.1 to 6.4 fl oz	Pre-harvest interval of 14 days.
chlorantraniliprole (Prevathon)	14.0 to 20.0 fl oz/A	Make no more than 4 applications per crop per year.
Corn Earworm, Southern Armyworm, Green Cloverworm, Velvetbean Caterpillar		
acephate (Orthene) 97 (generics available)	0.75 to 1 lb	Do not feed or graze livestock on acephate-treated vines. Do not apply within 14 days of harvest (digging).
Bacillus thuringiensis (Dipel DF) (Dipel ES) (Xentari)	0.5 to 2 lb 1 to 2 pt 0.5 to 2 lb	For velvetbean caterpillar control only. Apply to small caterpillars and use highest rate for larger worms and/or high populations; 0-day harvest restriction. Xentari also controls southern armyworm.
esfenvalerate (Asana XL)	2.9 to 5.8 oz	Do not feed Asana-treated vines or graze livestock on treated plants.
fenpropathrin (Danitol) 2.4 EC	10.67 to 16 fl oz	Do not exceed 2.67 pints per acre per season. Use 10 to 50 gallons per acre by ground and 5 to 10 gallons per acre by air. Repeat no more often than every 7 days. Do not apply within 14 days of digging and do not feed or graze vines within 14 days of last application.
indoxacarb (Steward)	9.2 to 11.3 oz	Do not apply more than 45 ounces per acre per crop. 14-day preharvest interval. For corn earworm.
lambda-cyhalothrin (Karate Z)	1.28 to 1.92 oz	Do not feed or graze livestock on Karate-treated plants.
methomyl (Lannate LV)	0.75 to 3 pt	Apply to foliage when four or more worms are present per foot of row and preferably when worms are small. Do not apply methomyl within 21 days of harvest. Do not feed methomyl-treated vines to livestock. Use minimum of 3 gallons of water for aerial application.
methoxyfenozide + spinetoram (Intrepid Edge)	4 to 8 fl oz	Application rate varies with timing. Lower rates appropriate for light infestations, smaller larvae and/or small plants.
spinosad (Blackhawk)	2 to 3 fl oz	Do not apply more than 9 fluid ounces per season or make more than three applications. 3-day preharvest interval.
bifenthrin (Brigade)	2.1 to 6.4 fl oz	Pre-harvest interval of 14 days.
chlorantraniliprole+ lambda-cyhalothrin (Besiege)	6.0 to 10.0 fl oz/A	Pre-harvest interval 14 days. Do not exceed a total of 31 fluid ounces of Besiege per acre per year.
chlorantraniliprole (Prevathon)	14.0 to 20.0 fl oz/A	Make no more than 4 applications per crop per year.
cyantraniliprole (Exirel)	10.0 to 20.5 fl oz/A	Pre-harvest interval of 14 days.
Budworm, Tobacco		
cyantraniliprole (Exirel)	10.0 to 20.5 fl oz/A	Pre-harvest interval of 14 days.
Cutworm		
chlorpyrifos (Lorsban) 15 G	1.33 lb	Apply in 16- to 18-inch band over row when infestation is first seen. May be applied by air. Do not graze or feed immature crop to livestock.
esfenvalerate (Asana XL)	5.8 to 9.6 oz	Do not feed treated vines to livestock.
indoxacarb (Steward)	9.2 to 11.3 oz	Do not apply more than 45 ounces per acre per crop. 14 day preharvest interval.

Table 5-6. Insect Control on Peanuts

Insecticide and Formulation	Amount of Formulation Per Acre	Precautions and Remarks
Cutworm (continued)		
lambda-cyhalothrin (Karate Z)	0.96 to 1.6 oz	Do not use treated vines or hay for animal feed.
methomyl (Lannate LV)	1.5 to 3 pt	Do not apply within 21 days of harvest. Do not feed treated vines to livestock.
bifenthrin (Brigade)	2.1 to 6.4 fl oz	Pre-harvest interval of 14 days.
chlorantraniliprole + lambda-cyhalothrin (Besiege)	5.0 to 8.0 fl oz/A	Pre-harvest interval 14 days. Do not exceed a total of 31 fluid ounces of Besiege per acre per year.
cyantraniliprole (Exirel)	10.0 to 20.5 fl oz/A	Pre-harvest interval of 14 days.
Fall Armyworm		
acephate (Orthene) 97 (generics available)	0.75 to 1 lb	Do not apply within 14 days of harvest (digging). Do not feed or graze livestock on vines treated with acephate. Apply 10 to 50 gallons spray solution per acre. Do not apply more than 4.13 pounds per acre (4 pounds a.i. per acre per season).
fenpropathrin (Danitol) 2.4 EC	10 2/3 to 16 fl oz	Do not exceed 2.67 pints per acre per season. Repeat no more often than every 7 days. Do not apply within 14 days of digging and do not feed or graze vines within 14 days of last application.
indoxacarb (Steward)	9.2 to 11.3 oz	Do not apply more than 45 ounces per acre per crop. 14 day preharvest interval.
lambda-cyhalothrin (Karate Z)	1.28 to 1.92 oz	
methomyl (Lannate LV)	0.75 to 1.5 pt	Effective against all sizes of worms. Use minimum of 3 gallons of water for aerial application. Do not apply within 21 days of harvest. Do not feed methomyl-treated vines to livestock.
methoxyfenozide + spinetoram (Intrepid Edge)	4 to 8 fl oz	Application rate varies with timing. Lower rates appropriate for light infestations, smaller larvae and/or small plants.
spinosad (Blackhawk)	1.7 to 3.3 fl oz	Do not apply more than 12.4 fluid ounces per season or make more than three applications. 3-day preharvest interval.
bifenthrin (Brigade)	2.1 to 6.4 fl oz	Pre-harvest interval of 14 days.
chlorantraniliprole + lambda-cyhalothrin (Besiege)	6.0 to 10.0 fl oz/A	Pre-harvest interval of 14 days. Do not exceed a total of 31 fluid ounces of Besiege per acre per year.
chlorantraniliprole (Prevathon)	14.0 to 20.0 fl oz/A	Make no more than 4 applications per crop per year.
cyantraniliprole (Exirel)	10.0 to 20.5 fl oz/A	Pre-harvest interval of 14 days.
Leafhoppers		
acephate (Orthene) 97 (generics available)	0.75 to 1 lb	See remarks under Thrips.
esfenvalerate (Asana XL)	2.9 to 5.8 oz	Do not feed livestock Asana-treated vines or graze livestock on treated plants.
fenpropathrin (Danitol) 2.4 EC	6 to 10.67 fl oz	Do not exceed 2 2/3 pints per acre per season. Repeat no more often than every 7 days. Do not apply within 14 days of digging and do not feed or graze vines within 14 days of last application.
chlorantraniliprole + lambda-cyhalothrin (Besiege)	6.0 to 10.0 fl oz/A	
lambda-cyhalothrin (Karate Z)	0.96 to 1.6 oz	Do not use treated vines or hay for animal feed.
methomyl (Lannate LV)	0.75 to 3 pt	Do not apply within 21 days of harvest. Do not use treated vines as feed.
bifenthrin (Brigade)	2.1 to 6.4 fl oz	Pre-harvest interval of 14 days.
Lesser Cornstalk Borer		
chlorpyrifos (Lorsban, Pilot) 15 G (generics available)	7 to 14 lb	
chlorantraniliprole + lambda-cyhalothrin (Besiege)	10.0 fl oz/A	Pre-harvest interval 14 days. Do not exceed a total of 31 fluid ounces of Besiege per acre per year.
chlorantraniliprole (Prevathon)	14 to 20.0 fl oz/A	See 2 (ee) Label recommendation.
cyantraniliprole (Exirel)	13.5 to 20.5 fl oz/A	Pre-harvest interval of 14 days.
Southern Corn Rootworm		
chlorpyrifos (Lorsban, Pilot) 15 G (generics available)	13.3 lb	Apply in a 16- to 18-inch band over the row just before pegging.
phorate (Thimet) 20 G (generics available)	10 lb	
Spider Mite		
propargite (Comite) 73 L	2 pt	Apply in at least 26 gallons of water per acre. Spider mite outbreaks are less likely to develop if foliar insecticides are not used during July and August and copper fungicides are used for Cercospora leafspot. Do not apply propargite within 14 days of harvest.
fenpropathrin (Danitol) 2.4 EC	10.67 to 16 fl oz	Do not exceed 2.67 pints (42 2/3 fluid ounces) per acre per season. Use 10 to 50 gallons per acre by ground and 5 to 10 gallons per acre by air. Repeat no more often than every 7 days. Do not apply within 14 days of digging and do not feed or graze vines within 14 days of last application.
bifenthrin (Brigade)	5.1 to 6.4 fl oz.	Pre-harvest interval of 14 days.

Chapter V — 2019 N.C. Agricultural Chemicals Manual

Insect Control in Soybeans

D. D. Reisig and A. S. Huseth, Entomology and Plant Pathology

Table 5-7. Insect Control on Soybeans

Insect Insecticide and Formulation	per Acre Amount of Formulation	Active (lb)	Acres/gal. (lb)	Preharvest Interval (PHI) (Days)	Precautions and Remarks
Bean Leaf Beetle					
acephate, MOA 1B (Orthene) 97 S	0.75 to 1 lb	0.75 to 1	1.25 to 1	14	Treat when defoliation reaches threshold levels or buildup is obvious. Threshold is 30% prebloom defoliation or 15% defoliation 2 weeks prior to bloom through podfill. Pod skinning by this insect can be a concern in soybeans grown for seed. Selected pyrethroids will suppress bean leaf beetle. Tolerance can quickly develop if chemistries are not rotated. **In the premixed products listed, the effective chemistries are in MOA's 3 and 1B.**
acetamiprid, MOA 4A + bifenthrin, MOA 3 (Justice) 1.8 EC	5 fl oz	See label	25.6	30	
beta-cyfluthrin, MOA 3 (Baythroid XL) 1.0 EC	2.8 fl oz	0.022	45.7	45	
bifenthrin, MOA 3 (Brigade, Discipline, Sniper, and others) 2 EC	4 to 6.4 fl oz	0.062 to 0.10	32 to 20	30	
chlorantraniliprole, MOA 28 + lambda-cyhalothrin, MOA 3 (Besiege) 1.25 SC	5 to 8 fl oz	See label	25.6 to 16	21	
chlorpyrifos, MOA 1B + gamma-cyhalothrin, MOA 3 (Cobalt Advanced) 2.63 EC	19 to 24 fl oz	See label	6.7 to 5.3	30	
chlorpyrifos, MOA 1B (Lorsban) 4 E	1 to 2 pt	0.5 to 1	8 to 4	14	
cyfluthrin, MOA 3 (Tombstone) 2 E	1.6 to 2.8 fl oz	0.025 to 0.04	80 to 45.7	45	
diflubenzuron, MOA 15 + lambda-cyhalothrin, MOA 3 (DoubleTake) 3 SC	4 fl oz	See label	32	30	
imidacloprid, MOA 4A + cyfluthrin, MOA 3 (Leverage 360) 3.0 SE	2.8 fl oz	See label	45.7	45	
lambda-cyhalothrin, MOA 3 (Warrior, Lambda-cyhalothrin, Silencer) 1.0 EC (Karate Z and Warrior II) 2.08 CS	1.92 to 3.2 fl oz 0.96 to 1.6 fl oz	0.015 to 0.025 0.015 to 0.025	66.7 to 40 133.3 to 80	30 30	
lambda-cyhalothrin, MOA 3 + thiamethoxam, MOA 4A (Endigo ZC) 2.06 SE	4 to 4.5 fl oz	See label	32 to 28.4	30	
Beet Armyworm					
chlorantraniliprole, MOA 28 (Prevathon) 0.43 SC	14 to 20 fl oz	0.047 to 0.067	9.1 to 64	1	Ground application only for larger caterpillars. Control of large armyworms is difficult. Chlorantraniliprole, indoxacarb, methoxyfenozide, spinetoram, and spinosad are the superior products.
chlorantraniliprole, MOA 28 + lambda-cyhalothrin, MOA 3 (Besiege) 1.25 SC	9 fl oz	0.04 + 0.02	14.2	21	
indoxacarb, MOA 22 (Steward) 1.25 EC	5.6 to 11.3 fl oz	0.06 to 0.11	22.9 to 11.3	21	
methomyl, MOA 1A (Lannate) 2.4 LV (Lannate) 90 SP	1.5 pt 0.5 lb	0.45 0.45	5.3 2	14 14	
methoxyfenozide, MOA 18A (Intrepid) 2 F	4 to 8 fl oz	0.06 to 0.12	32 to 16	14 (grain) 7 (hay)	
methoxyfenozide, MOA 18A + spinetoram, MOA 5 (Intrepid Edge) 3F	4.0 to 6.4 oz	See label	32 to 20	28	
spinosad, MOA 5 (Blackhawk) 4 SC	1.7 to 2.2 fl oz	0.04 to 0.05	75.3 to 58.2	28	
Corn Earworm					
chlorantraniliprole, MOA 28 (Prevathon) 0.43 SC	14 to 20 fl oz	0.047 to 0.067	9.1 to 6.4	1	Treat when earworm numbers exceed threshold as determined by scouting. Be sure worms are present and 3/8 to 1/2 inch in size when treatment is applied. Use low rates for light infestations. Use higher rates by air. Go to Web page https://www.ces.ncsu.edu/wp-content/uploads/2017/08/CEW-calculator-v0.006.html for an online threshold calculator. At $10.00 per bushel, the plant compensates due to the low caterpillar levels needed to reach threshold at $10.00 and above.
chlorantraniliprole, MOA 28 + lambda-cyhalothrin, MOA 3 (Besiege) 1.25 SC	6 to 9 fl oz	See label	21.3 to 14.2	21	
chlorpyrifos, MOA 1B (Lorsban) 4 EC	1.5 to 2 pt	0.75 to 1	5.3 to 4	14	
chlorpyrifos, MOA 1B + gamma-cyhalothrin, MOA 3 (Cobalt Advanced) 2.63 EC	16 to 38 fl oz	See label	8 to 3.4	30	
indoxacarb, MOA 22 (Steward) 1.25 EC	5.6 to 11.3 fl oz	0.06 to 0.11	22.9 to 11.3	21	
methoxyfenozide, MOA 18A + spinetoram, MOA 5 (Intrepid Edge) 3F	4.0 to 6.4 oz	See label	32 to 20	28	
spinosad, MOA 5 (Blackhawk) 4 SC	1.7 to 2.2 fl oz	0.04 to 0.05	75.3 to 58.2	28	
Grasshopper					
acephate, MOA 1B (Orthene 97)	0.25 to 0.5 lb	0.25 to 0.5	4 to 2	14	Apply by air or ground uniformly over foliage as a broadcast treatment. Early morning treatment is preferred. Use higher rates for heavy infestations. Diflubenzuron is not effective to control adult grasshoppers. See label for additional instructions and suggestions.
chlorpyrifos, MOA 1B + gamma-cyhalothrin, MOA 3 (Cobalt Advanced) 2.63 EC	10 to 13 fl oz	See label	12.8 to 9.8	30	
chlorpyrifos, MOA 1B (Lorsban) 4 E	1 to 2 pt	0.5 to 1	8 59 4	14	
diflubenzuron, MOA 15 (Dimilin) 2L, 25W	2 fl oz 0.25 lb.	0.06 0.06	64 8	21	

Chapter V—2019 N.C. Agricultural Chemicals Manual

Table 5-7. Insect Control on Soybeans

Insect Insecticide and Formulation	per Acre		Acres/gal. (lb)	Preharvest Interval (PHI) (Days)	Precautions and Remarks
	Amount of Formulation	Active (lb)			
Green Cloverworm					
Bacillus thuringiensis, MOA 11B2 (Various)	—	—	—	0	Treat when defoliation reaches threshold. This insect is seldom an economic pest. See label of specific Bt products. Thresholds are listed under bean leaf beetle.
beta-cyfluthrin, MOA 3 (Baythroid XL) 1.0 EC	1.6 to 2.8 fl oz	0.0125 to 0.022	80 to 45.7	45	
chlorantraniliprole, MOA 28 (Prevathon) 0.43 SC	14 to 20 fl oz	0.047 to 0.067	9.1 to 6.4	1	
chlorantraniliprole, MOA 28 + lambda-cyhalothrin, MOA 3 (Besiege) 1.25 SC	5 to 8 fl oz	See label	25.6 to 16	21	
cyfluthrin, MOA 3 (Tombstone) 2E	1.6 to 2.8 fl oz	0.025 to 0.04	80 to 45.7	45	
esfenvalerate, MOA 3 (Asana XL) 0.66 EC	5.8 to 9.6 fl oz	0.03 to 0.05	22.1 to 13.3	21	
gamma-cyhalothrin, MOA 3 (Declare) 1.25 EC	1.54 fl oz	0.015	83.1	21	
indoxacarb, MOA 22 (Steward) 1.25 EC	8 to 11.3 fl oz	0.08 to 0.11	16 to 11.3	21	
lambda-cyhalothrin, MOA 3 (Warrior, Lambda-cyhalothrin, Silencer) 1.0 EC (Karate Z and Warrior II) 2.08 CS	1.92 to 3.2 fl oz 0.96 to 1.6 fl oz	0.015 to 0.025 0.015 to 0.025	66.7 to 40 133.3 to 80	30 30	
lambda-cyhalothrin, MOA 3 + thiamethoxam, MOA 4A (Endigo ZC) 2.06 SE	3.5 to 4 fl oz	See label	36.6 to 32	30	
methoxyfenozide, MOA 18A + spinetoram, MOA 5 (Intrepid Edge) 3F	4.0 to 6.4 oz	See label	32 to 20	28	
spinosad, MOA 5 (Blackhawk) 4 SC	1.1 to 2.2 fl oz	0.025 to 0.05	116.4 to 58.2	28	
zeta-cypermethrin, MOA 3 (Mustang Max) 0.8 EC	2.8 to 4 fl oz	0.0175 to 0.025	45.7 to 32	21	
zeta-cypermethrin, MOA 3 + bifenthrin, MOA 3 (Hero) 1.24 EC	10.3 fl oz	0.033 + 0.066	12.4	30	
Kudzu Bug					
acephate, MOA 1B (Orthene) 97 S	1 lb	1	1	14	Bifenthrin is the superior product (MOA 3).
bifenthrin, MOA 3 (Brigade, Discipline, Sniper, and others) 2 EC	4 to 6.4 fl oz	0.062 to 0.10	32 to 20	30	
bifenthrin, MOA 3 + acetamiprid, MOA 4a (Justice) 1.8 EC	5 fl oz	See label	25.6	30	
bifenthrin, MOA 3 + imidacloprid, MOA 4A (Brigadier) 2 E (Swagger) 1 F	6.1 fl oz 12.2 fl oz	See label See label	21 10.5	7 18	
gamma-cyhalothrin, MOA 3 (Declare) 1.25 EC	1.54 fl oz	0.015	83.1	21	
lambda-cyhalothrin, MOA 3 (Karate, Lambda-cyhalothrin, Silencer) 1.0 EC (Karate Z, Warrior II) 2.08 CS	3.84 fl oz 1.92 fl oz	0.03 0.03	33.3 66.7	30 30	
lambda-cyhalothrin, MOA 3 + thiamethoxam, MOA 4A (Endigo ZC) 2.06 SE	3.5 to 4.5 fl oz	See label	36.6 to 28.4	30	
zeta-cypermethrin, MOA 3 (Mustang Maxx) 0.8 EC	4 fl oz	0.025	32	21	
zeta-cypermethrin, MOA 3 + bifenthrin, MOA 3 (Hero) 1.24 EC	6.4 to 10.3 fl oz	See label	20 to 12.4	30	
Soybean Looper					
chlorantraniliprole, MOA 28 (Prevathon) 0.43 SC	14 to 20 fl oz	0.047 to 0.067	9.1 to 6.4	1	Treat when thresholds are reached or when buildup is obvious. Threshold is 15% defoliation in soybeans 2 weeks prior to flowering, but can be increased to 20% during R6 when growing conditions are ideal. Ground application is superior. **Resistance is occurring in this species and has been documented in the Blacklands for MOA 3 18A, and 28;** insecticides work best on small caterpillars. The most consistent insecticides in Blacklands are those containing MOA 5. (Intrepid Edge and Radiant).
chlorantraniliprole, MOA 28 + lambda-cyhalothrin, MOA 3 (Besiege) 1.25 SC	10 fl oz	See label	12.8	30	
chlorpyrifos, MOA 1B + gamma-cyhalothrin, MOA 3 (Cobalt Advanced) 2.63 EC	20 to 38 fl oz	See label	6.4 to 3.4	30	
indoxacarb, MOA 22 (Steward) 1.25 EC	5.6 to 11.3 fl oz	0.06 to 0.11	22.9 to 11.3	21	
methoxyfenozide, MOA 18A (Intrepid) 2F	4 to 8 fl oz	0.06 to 0.12	32 to 16	7 (hay) 14 (grain)	
methoxyfenozide, MOA 18A + spinetoram, MOA 5 (Intrepid Edge) 3F	4.0 to 6.4 oz	See label	32 to 20	28	
spinetoram, MOA 5 (Radiant) 1 SC	2 to 4 fl oz	0.016 to 0.12	64 to 32	7 (hay) 14 (grain)	
spinosad, MOA 5 (Blackhawk) 4 SC	1.1 to 2.2 fl oz	0.025 to 0.05	116.4 to 58.2	28	
spinosad, MOA 5 + gamma-cyhalothrin, MOA 3 (Consero)	2 to 3 fl oz	See label	64 to 42.7	See label	
Spider Mite					
bifenthrin, MOA 3 (Brigade, Discipline, Sniper, and others) 2 EC	5.12 to 6.4 fl oz	0.08 to 0.10	25 to 20	18	Miticides registered on soybean often provide erratic control. Two applications may be needed for high populations. The only true miticidal product listed is etoxazole, which has activity on the immatures.
chlorpyrifos, MOA 1B (Lorsban) 4E	1 to 2 pints	0.5 to 1	8 to 4	28	
etoxazole, MOA 11A1 (Zeal) SC	2 to 6 fl oz	0.045 to 0.135	64 to 21		

Table 5-7. Insect Control on Soybeans

Insect Insecticide and Formulation	per Acre		Acres/gal. (lb)	Preharvest Interval (PHI) (Days)	Precautions and Remarks
	Amount of Formulation	Active (lb)			
Stink Bug (Brown, Brown Marmorated, Green, and Southern Green)					
acephate, MOA 1B (Orthene) 97 S	0.5 to 1 lb	0.5 to 1	2 to 1	14	Treat when bug numbers exceed threshold. Go to https://soybeans.ces.ncsu.edu/stink-bug-economic-threshold-calculator/ for a threshold table. Acephate and the highest rates of pyrethroids are preferred for brown stink bug, with bifenthrin the preferred pyrethroid. Stink bugs are often late-season pests so be aware of the preharvest interval of insecticides. **In the premixed products listed, the effective chemistries are in MOA's 3 and 1B.**
bifenthrin, MOA 3 (Brigade, Discipline, Sniper, and others) 2 EC	2.1 to 6.4 fl oz	0.033 to 0.10	61 to 20	30	
chlorpyrifos, MOA 1B + gamma-cyhalothrin, MOA 3 (Cobalt Advanced) 2.63 EC	20 to 38 fl oz	See label	6.4 to 3.4	30	
cyfluthrin, MOA 3 (Tombstone) 2E	1.6 to 2.8 fl oz	0.025 to 0.04	80 to 45.7	45	
diflubenzuron, MOA 15 + lambda-cyhalothrin, MOA 3 (DoubleTake) 3 SC	4 fl oz	See label	32	30	
gamma-cyhalothrin, MOA 3 (Declare) 1.25 EC	1.54 fl oz	0.015	83.1	21	
imidacloprid, MOA 4A + cyfluthrin, MOA 3 (Leverage 360) 3.0 SE	2.8 fl oz	See label	45.7	45	
lambda-cyhalothrin, MOA 3 (Warrior, Lambda-cyhalothrin, Silencer) 1.0 EC (Karate Z and Warrior II) 2.08 CS	1.92 to 3.2 fl oz 0.96 to 1.6 fl oz	0.015 to 0.025 0.015 to 0.025	66.7 to 40 133.3 to 80	30 30	
lambda-cyhalothrin, MOA 3 + thiamethoxam, MOA 4A (Endigo ZC) 2.06 SE	4 to 4.5 fl oz	See label	32 to 28.4	30	
zeta-cypermethrin, MOA 3 (Mustang Max) 0.8 EC	4 fl oz	0.025	32	21	
zeta-cypermethrin, MOA 3 + bifenthrin, MOA 3 (Hero) 1.24 EC	10.3 fl oz	0.033 + 0.066	12.4	21	
Velvetbean Caterpillar					
Bacillus thuringiensis, MOA 11B2 (various)	—	—	—	0	See specific labels for use rates.
pyrethroids, MOA 3	—	—	—	—	
chlorantraniliprole, MOA 28 (Prevathon) 0.43 SC	14 to 20 fl oz	0.047 to 0.067	9.1 to 6.4	1	
chlorantraniliprole, MOA 28 + lambda-cyhalothrin, MOA 3 (Besiege) 1.25 SC	5 to 9 fl oz	See label	25.6 to 14.2	21	
diflubenzuron, MOA 15 (Dimilin) 2L	2 to 4 fl oz	0.06 to 0.125	64 to 32	21	
methoxyfenozide, MOA 18A (Intrepid) 2F	4 to 8 fl oz	0.06 to 0.12	32 to 16	7 (hay) 14 (grain)	
methoxyfenozide, MOA 18A + spinetoram, MOA 5 (Intrepid Edge) 3F	4.0 to 6.4 oz	See label	32 to 20	28	
spinetoram, MOA 5 (Radiant) 1 SC	2 to 4 fl oz	0.016 to 0.12	64 to 32	7 (hay) 14 (grain)	
spinosad, MOA 5 (Blackhawk) 4 SC	1.1 to 2.2 fl oz	0.025 to 0.05	116.4 to 58.2	28	
Grape Colaspis, Blister Beetle, Japanese Beetle, Mexican Bean Beetle, Spotted Cucumber Beetle, Three Cornered Alfalfa Hopper					
acephate, MOA 1B (Orthene) 97 S	0.75 to 1 lb	0.75 to 1	1.25 to 1	14	These insects are rarely pests; exercise care in determining if a problem exists. Do not spray Mexican bean beetle when many eggs and pupae are present; wait 4 to 5 days. Thrips have never been demonstrated to reduce soybean yields in North Carolina. Three cornered alfalfa hopper girdle mainstems when plants are below 10 inches tall and petioles when plants are larger. Treatments for three-cornered alfalfa hopper only impact yield when applied to seedling soybeans.
pyrethroids, MOA 3 combinations	(see corn earworm above for rates)	—	—	—	

CAUTION: Always use pesticides according to label directions. Be mindful of reducing the impact of pesticides on wildlife and groundwater.

Insect Control on Flue-Cured and Burley Tobacco

H. J. Burrack, Entomology and Plant Pathology

The Insect Resistance Action Committee (**IRAC**) has grouped insecticides sharing the same mode of action (**MOA**) into categories. The categories are listed following insecticide and formulation names. To minimize the likelihood of resistance development, avoid successive treatment with insecticides having the same MOA. The Organic Materials Registry Institute (**OMRI**) lists products acceptable for use in organic production. These products are identified in the Precautions and Remarks section.

Sanitation is important in controlling greenhouse pests. Keep all trash, equipment, etc., out of and away from the greenhouse. Growing plants other than tobacco can introduce difficult-to-control pests. Leaving the empty greenhouses open during cold periods and closed during the summer can help reduce insect pests.

In general, information is provided for the most commonly used formulations of active ingredients available in multiple formulations. Carefully check the label of the product you plan to use in the event that it differs from those listed. The label is the law! Residues of some pesticides are a concern for purchasers. Growers are encouraged to discuss insecticide options with their purchasers before treating to reduce potential residue concerns.

Flue-Cured and Burley Tobacco — Greenhouse

Table 5-8A. Insect Control on Flue-Cured and Burley Tobacco in Greenhouses

Insecticide, Formulation[1] and IRAC Group	Amount of Formulation	Restricted Entry Interval (REI) (hours)	Preharvest Interval (PHI) (days)	Precautions and Remarks[3]
Green peach aphid				
acephate, IRAC 1B (Orthene) 97 PE	Rate per 1,000 sq ft 3/4 tbsp (3/4 lb/acre)	24	3	There are many formulations of acephate. Apply in 3 gallons water per 1,000 sq ft. Even and thorough coverage is necessary for good control. Do not use more than 4 1/8 lb/acre Orthene (4 lb AI/acre). This includes greenhouse, transplant water, soil, and foliar applications.
imidacloprid, IRAC 4A (Admire Pro)	Rate per 1,000 plants 0.5 fl oz	12	14	**Only apply imidacloprid to control aphids in the greenhouse if tobacco will be transplanted within a week.** This application replaces tray drench applications for field control of aphids and flea beetles described below. There are many other formulations of imidacloprid.
thiamethoxam, IRAC 4A (Platinum) 75 SG (Platinum) SC	Rate per 1,000 plants 0.17 to 0.43 oz 0.5 to 1.3 fl oz	12	None given	**Only apply thiamethoxam to control aphids in the greenhouse if tobacco will be transplanted within a week.** This application replaces tray drench applications for field control of aphids and flea beetles described below.
Tobacco flea beetle				
acephate, IRAC 1B (Orthene) 97 PE	Rate per 1,000 sq ft 3/4 tbsp (3/4 lb/acre)	24	3	There are many formulations of acephate. Apply in 3 gallons water per 1,000 square feet. Even and thorough coverage is necessary for good control. Do not use more than 4 1/8 lb/acre Orthene (4 lb AI/acre). This includes greenhouse, transplant water, soil, and foliar applications.
cyantraniliprole, IRAC 28 (Verimark) SC	Rate per acre equivalent 10 to 13.5 fl oz	4	None given	**Verimark can be applied as a greenhouse tray drench prior to transplant. Applications earlier than one week before transplant have not been tested for efficacy. If Verimark is used to control insects in the greenhouse, it should not be reapplied prior to transplant.**
imidacloprid, IRAC 4A (Admire Pro)	Rate per 1,000 plants 0.5 fl oz	12	14	**Only apply imidacloprid to control aphids in the greenhouse if tobacco will be transplanted within a week.** This application replaces tray drench applications for field control of aphids and flea beetles described below. There are many other formulations of imidacloprid.
thiamethoxam, IRAC 4A (Platinum) 75 SG (Platinum) SC	Rate per 1,000 plants 0.27 to 0.43 oz 0.5 to 1.3 fl oz	12	None given	**Only apply thiamethoxam to control aphids in the greenhouse if tobacco will be transplanted within a week.** This application replaces tray drench applications for field control of aphids and flea beetles described below.
Slugs or snails				
hydrated or air-slaked lime		—	—	Apply lime in a band 3 to 4 inches wide around margins of beds.
metaldehyde bait (Deadline Bullets)	0.2 to 0.6 lb		12	At dusk scatter bait around margins of beds and in walkways and open spaces. TO AVOID PLANT INJURY, DO NOT PUT BAIT ON PLANTS.
iron phosphate bait (Sluggo)	0.5 to 1 lb		0	OMRI listed TO AVOID PLANT INJURY, DO NOT PUT BAIT ON PLANTS.

[1] Some insecticides are available in several formulations. Those listed are generally the most commonly used or available. Other formulations may or may not be suitable for use on tobacco or for a specific pest. Check labels carefully.

Chapter V — 2019 N.C. Agricultural Chemicals Manual

Flue-Cured and Burley Tobacco — Field

Table 5-8B. Insect Control on Flue-Cured and Burley Tobacco in the Field

Insecticide, Formulation[1] and IRAC Group	Amount of Formulation Per Acre	Restricted Entry Interval (REI) (hours)	Preharvest Interval (PHI) (days)	Precautions and Remarks
Green peach aphid				
Aphids are primarily pretopping pests. Greenhouse or transplant treatments may provide control through topping, and additional foliar treatments are not typically needed. Post topping, aphids are most common on suckers or regrowth. Sucker management via contact materials or hand removal is often sufficient to control post topping aphid populations. The threshold for green peach aphids in the field is 10% of plants scouted with 50 or more aphids on the upper leaves. Organically acceptable aphid control materials are generally less effective than conventional materials, soaphid control in organic production should be initiated upon first aphid appearance, and treatment should continue on 7 to 10 day intervals until topping. **Data on specific organic aphid controls are limited.** Organic tobacco with aphid populations should be topped as early as feasible. Post topping sucker control is very important for aphid control in organic tobacco.				
acephate, IRAC 1B (Orthene) 97	0.75 lb	24 If significant foliar contact will occur, gloves must be worn for 14 days after treatment.	3	**TRANSPLANT WATER APPLICATION.** Apply in a minimum of 100 gallons of transplant water/acre. To avoid plant injury, do not exceed 0.75 pound a.i. acephate per acre. **SUPPRESSION ONLY,** but may not provide suppression through topping. Continue to scout plants post transplant. Do not use more than 4 1/8 lb/acre Orthene (4 lb AI/acre). This includes greenhouse, transplant water, soil, and foliar applications.
acephate, IRAC 1B (Orthene) 97 PE	0.5 lb	24 If significant foliar contact will occur, gloves must be worn for 14 days after treatment.	3	**FIELD FOLIAR APPLICATIONS.** Use at least 25 gallons per acre at 60 PSI. Using hollow cone or small solid cone nozzles cover entire plant with spray. If control 4 days after treatment is not adequate, choose another MOA for subsequent applications. Do not use more than 4 1/8 lb/acre Orthene (4 lb AI/acre). This includes greenhouse, transplant water, soil, and foliar applications.
chlorantraniliprole + thiamethoxam, IRAC 28 + IRAC 4A (Durivo)	**Rate per 1,000 plants** 0.6 to 1.6 fl oz	12	None given	**TRANSPLANT WATER APPLICATION.** Apply no more than 0.2 pound chlorantraniliprole per acre per crop, which includes applications of Coragen, Beseige, and Durivo.
imidacloprid, IRAC 4A (Admire Pro)	**Rate per 1,000 plants** 0.6 fl oz	12	14	**TRANSPLANT WATER APPLICATION.** Rate is per 1,000 plants and should be converted for transplant water applications based on plant population. Proper calibration of application equipment is essential for effective transplant water applications. A metered or pressurized application system is recommended. Several concentrations of imidacloprid (1.6F, 2F, 4F, and 4.6F) are available. Carefully read the label to determine the correct rate for target pests.
imidacloprid, IRAC 4A (Admire Pro)	**Rate per 1,000 plants** 0.5 fl oz	12	14	**GREENHOUSE TRAY DRENCH APPLICATION.** Rate is per 1,000 plants. Apply no more than 5 days before transplanting. Immediately after application, wash the material off the plants onto the potting soil. The lowest label rate is sufficient for aphid and flea beetle management. See below for recommendations for areas with high incidence of Tomato Spotted Wilt Virus (TSWV). Several concentrations of imidacloprid (1.6F, 2F, 4F, and 4.6F) are available. Carefully read the label to determine the correct rate for target pests.
imidacloprid, IRAC 4A (Admire Pro) (several products) 2F	0.7-1.4 fl oz 1.6 to 3.2 fl oz	12	14	**FIELD FOLIAR APPLICATION.** Avoid using only Group 4A insecticides as foliar field applications for aphids on plants which were treated in the greenhouse with imidacloprid or thiamethoxam. Several concentrations of imidacloprid (1.6F, 2F, 4F, and 4.6F) are available. Carefully read the label to determine the correct rate for target pests.
thiamethoxam, IRAC 4A (Platinum) 75 SG (Platinum) SC	**Rate per 1,000 plants** 0.17 oz 0.5 fl oz	12	None given	**TRANSPLANT WATER APPLICATION.** Use lower label rate for aphids. Rate is per 1,000 plants and should be converted for transplant water applications based on plant population. Proper calibration of application equipment is essential for effective transplant water applications. A metered or pressurized application system is recommended. Make only one application of thiamethoxam per season. Thiamethoxam is also the active ingredient in Actara.
thiamethoxam, IRAC 4A (Platinum) 75 SG (Platinum) SC	**Rate per 1,000 plants** 0.17 oz 0.5 fl oz	12	None given	**GREENHOUSE TRAY DRENCH APPLICATION.** Use lower label rate for aphids. Rate is per 1,000 plants. Apply no more than 5 days before transplant. Immediately after application, wash the material off the plants onto the potting soil OR apply in transplant water.
thiamethoxam, IRAC 4A (Actara) 25 WDG	2 to 3 oz	12	14	**FIELD FOLIAR APPLICATION.** **Make only one application of thiamethoxam per season. Thiamethoxam is also the active ingredient in Platinum.**
acetamiprid, IRAC 4A (Assail) 30 SG	1.5 to 4 oz	12	7	**FIELD FOLIAR APPLICATION** Make no more than 4 applications of acetamiprid per season, and do not apply more than once every 7 days. Avoid using only Group 4A insecticides as foliar field applications for aphids on plants which were treated in the greenhouse with imidacloprid or thiamethoxam.
pymetrozine, IRAC 9B (Fulfill) 50 WG	2.75 oz	12	14	**FIELD FOLIAR APPLICATION.** Make no more than 2 applications of pymetrozine per year.
lambda-cyhalothrin, IRAC 3A (Warrior) (Karate Xeon)	2.5 to 3.0 oz 0.96 to 1.92 fl oz	24	40	**FIELD FOLIAR APPLICATION.** NOTE LONG PREHARVEST INTERVAL.
azadirachtin, IRAC UN (Aza Direct)	1 to 2 pt	4	0	**FIELD FOLIAR APPLICATION.** Optimal pH range 5.6 to 6.5. **OMRI** listed. Limited data.
pyrethrins IRAC 3 (Pyganic) 1.4 EC (Pyganic) 5.0 EC	16 to 64 fl oz 4.5 to 18 fl oz	12	0	**FIELD FOLIAR APPLICATION.** Pyganic should be buffered to pH 5.5 to 7. **OMRI** listed. Limited data.

Table 5-8B. Insect Control on Flue-Cured and Burley Tobacco in the Field

Insecticide, Formulation[1] and IRAC Group	Amount of Formulation Per Acre	Restricted Entry Interval (REI) (hours)	Preharvest Interval (PHI) (days)	Precautions and Remarks
Green peach aphid (continued)				
petroleum oil (Saf-T-Side)	1 to 2 gal	4	0	**FIELD FOLIAR APPLICATION.** **OMRI** listed. Limited data.
sorbitol octanoale (SucraShield)	0.8 to 1.0% v/v	48	0	**FIELD FOLIAR APPLICATION.** **OMRI** listed. Limited data.
rosemary and peppermint oil (Ecotec)	2 to 4 pt	0	0	**FIELD FOLIAR APPLICATION.** **OMRI** listed. Limited data.
Tobacco flea beetle				
Greenhouse or transplant treatments may provide control through topping, and additional foliar treatments are not typically needed. The threshold for foliar treatments on small, recently planted tobacco is 4 beetles per plant. Flea beetle populations may increase near harvest and require management if populations exceed 60 beetles per fully grown plant. Good coverage is required for effective flea beetle control in large plants. Use appropriate equipment and sufficient water volume to achieve coverage from the base to the top of the plant.				
acetamiprid, IRAC 4A (Assail) 30 SG	2.5 to 4 oz	12	7	**FIELD FOLIAR APPLICATION.** Make no more than 4 applications of acetamiprid per season, and do not apply more than once every 7 days. Avoid using only Group 4A materials for season long control of insects with more than 1 generation. Following treatments of Group 4A materials, rotate to a different MOA before making additional applications of a Group 4A material.
acephate, IRAC 1B (Orthene) 97	0.75 lb	24 If significant foliar contact will occur, gloves must be worn for 14 days after treatment.	3	**TRANSPLANT WATER APPLICATION.** Apply in a minimum of 100 gallons of transplant water/acre. To avoid plant injury, do not exceed 0.75 pound a.i. acephate per acre. Do not use more than 4 1/8 lb/acre Orthene (4 lb AI/acre). This includes greenhouse, transplant water, soil, and foliar applications.
acephate, IRAC 1B (Orthene) 97 PE	0.5 lb	24 If significant foliar contact will occur, gloves must be worn for 14 days after treatment.	3	**FIELD FOLIAR APPLICATION.** Use at least 25 gallons per acre at 60 PSI. Using hollow cone or small solid cone nozzles cover entire plant with spray. If control 4 days after treatment is not adequate, choose another MOA for subsequent applications. Do not use more than 4 1/8 lb/acre Orthene (4 lb AI/acre). This includes greenhouse, transplant water, soil, and foliar applications. Some purchasers may have concerns about acephate residues. Discuss acephate usage with purchaser prior to making applications.
chlorantraniliprole + thiamethoxam, IRAC 28 + IRAC 4A (Durivo)	**Rate per 1,000 plants** 1.0 to 1.6 fl oz	12	None given	**TRANSPLANT WATER APPLICATION.** Apply no more than 0.2 pound chlorantraniliprole per acre per crop, which includes applications of Coragen, Beseige, and Durivo.
cyantraniliprole, IRAC 28 (Verimark) SC	10 to 13.5 fl oz	4	None given	**GREENHOUSE TRAY DRENCH APPLICATION.** Rate is per acre.
imidacloprid, IRAC 4A (Admire Pro)	**Rate per 1,000 plants** 0.6 fl oz	12	14	**TRANSPLANT WATER APPLICATION.** Proper calibration of application equipment is essential for effective transplant water applications. A metered or pressurized application system is recommended. Several concentrations of imidacloprid (1.6F, 2F, 4F, and 4.6F) are available. Carefully read the label to determine the correct rate for target pests.
imidacloprid, IRAC 4A (Admire Pro)	**Rate per 1,000 plants** 0.5 fl oz	12	14	**GREENHOUSE TRAY DRENCH APPLICATION.** Rate is per 1,000 plants. Apply no more than 5 days before transplanting. Immediately after application, wash the material off the plants onto the potting soil. The lowest label rate is sufficient for aphid and flea beetle management. See below for recommendations for areas with high incidence of Tomato Spotted Wilt Virus (TSWV). Several concentrations of imidacloprid (1.6F, 2F, 4F, and 4.6F) are available. Carefully read the label to determine the correct rate for target pests.
imidacloprid, IRAC 4A (Admire Pro) (several products) 2F	0.7 to 1.4 fl oz 1.6 to 3.2 fl oz	12	14	**FIELD FOLIAR APPLICATION.** Avoid using only Group 4A insecticides as foliar field applications for aphids on plants which were treated in the greenhouse with imidacloprid or thiamothoxam. Several concentrations of imidacloprid (2F, 4F, and 4.6F) are available. Carefully read the label to determine the correct rate for target pests.
thiamethoxam, IRAC 4A (Platinum) 75 SG (Platinum) SC	**Rate per 1,000 plants** 0.27 oz 0.8 fl oz	12	None given	**TRANSPLANT WATER APPLICATION.** Use lower label rate for aphids. Rate is per 1,000 plants and should be converted for transplant water applications based on plant population. Proper calibration of application equipment is essential for effective transplant water applications. A metered or pressurized application system is recommended.
thiamethoxam, IRAC 4A (Platinum) 75 SG (Platinum) SC	**Rate per 1,000 plants** 0.27 oz 0.8 fl oz	12	None given	**GREENHOUSE TRAY DRENCH APPLICATION.** Use lower label rate for aphids. Rate is per 1,000 plants. Apply no more than 5 days before transplant. Immediately after application, wash the material off the plants onto the potting soil OR apply in transplant water.
lambda-cyhalothrin, IRAC 3A (Warrior) 1CS (Karate Xeon)	2.5 to 3.0 oz 0.96 to 1.92 fl oz	24	40	**FIELD FOLIAR APPLICATION.** NOTE LONG PREHARVEST INTERVAL.
thiamethoxam, IRAC 4A (Actara) 25 WDG	2 to 3 oz	12	14	**FIELD FOLIAR APPLICATION.** Make only 1 application of thiamethoxam per season. Thiamethoxam is also the active ingredient in Platinum.

Table 5-8B. Insect Control on Flue-Cured and Burley Tobacco in the Field

Insecticide, Formulation[1] and IRAC Group	Amount of Formulation Per Acre	Restricted Entry Interval (REI) (hours)	Preharvest Interval (PHI) (days)	Precautions and Remarks
Armyworm				
Armyworms are typically most common late in the growing season. Preventative treatment is not recommended.				
chlorantraniliprole, IRAC 28 (Coragen)	3.5 to 7 fl oz	4	1	**FIELD FOLIAR APPLICATION.** Make no more than 4 applications per season (with at least 3 days between applications), and apply no more than 15.4 fl oz per season.
lambda-cyhalothrin, IRAC 3A (Warrior) (Karate Xeon)	2.5 to 3.0 oz 0.96 to 1.92 fl oz	24	40	NOTE LONG PREHARVEST INTERVAL.
Budworm				
The threshold for tobacco budworm is 10% infested plants. This threshold is very conservative, and budworms should not be treated unless infestations exceed 10%. Coverage is important for budworm management. Use 1 to 3 full cone nozzles 6 to 12 inches above bud and a minimum of 25 gallons water per acre.				
acephate, IRAC 1B (Orthene) 97 PE	0.75 lb	24	3	There are many formulations of acephate. Do not use more than 4 1/8 lb/acre Orthene (4 lb AI/acre). This includes greenhouse, transplant water, soil, and foliar applications. Acephate has some activity against tobacco budworms, but other products are more effective. Some purchasers may have concerns about acephate residues. Discuss acephate usage with purchaser prior to making applications.
Bacillus thuringiensis, IRAC 11 DIPel DF	0.5 to 1 lb	4	0	There are many B.t. formulations, including Agree, Biobit, Condor, Crymax, Deliver, Dipel, Javelin, and Lepinox. Highest labeled rates are generally needed for budworm control. DiPel DF and many other B.t. formulations are **OMRI** listed, but not all Bt formulations are **OMRI** listed. Carefully read the label to determine if a material is acceptable for use on organically certified plants.
chlorantraniliprole, IRAC 28 (Coragen)	5.0 to 7.0 fl oz	4	1	**TRANSPLANT WATER APPLICATION.** Rate is per acre. Transplant applications of Coragen may suppress tobacco budworm populations for 4 to 7 weeks. Proper calibration of application equipment is essential for effective transplant water applications. A metered or pressurized application system is recommended. Apply no more than 15.4 fluid ounces of Coragen or more than 0.2 pound chlorantraniliprole per acre per crop, which includes applications of Coragen, Beseige, and Durivo.
chlorantraniliprole, IRAC 28 (Coragen)	3.5 to 5.0 fl oz	4	1	**FIELD FOLIAR APPLICATION.** Make no more than 4 applications per season (with at least 3 days between applications), and apply no more than 15.4 fluid ounces of Coragen or more than 0.2 pound chlorantraniliprole per acre per crop, which includes applications of Coragen, Beseige, and Durivo. Some purchasers may have concerns about chlorantraniliprole residues, particularly if used later in the growing season. Discuss chlorantraniliprole usage with purchaser prior to making applications.
chlorantraniliprole + thiamethoxam, IRAC 28 + IRAC 4A (Durivo)	Rate per 1,000 plants 1.6 fl oz	12	None given	**TRANSPLANT WATER APPLICATION.** Transplant applications of Durivo may suppress tobacco budworm populations for 4 to 7 weeks. Proper calibration of application equipment is essential for effective transplant water applications. A metered or pressurized application system is recommended. Apply no more than 0.2 pound chlorantraniliprole per acre per crop, which includes applications of Coragen, Beseige, and Durivo.
lambda-cyhalothrin, IRAC 3A (Warrior) 1CS (Karate Xeon)	2.5 to 3.0 oz 0.96 to 1.92 fl oz	24 24	40 40	To avoid build-up of resistance, rotate use of this product with other insecticides. NOTE THE LONG PREHARVEST USE RESTRICTION.
lambda-cyhalothrin + chloratraniliprole IRAC 3 + 28 (Besiege)	5.0 to 9.0 fl oz	24	40	NOTE THE LONG PREHARVEST USE RESTRICTION. Apply no more than 0.2 pound chlorantraniliprole per acre per crop, which includes applications of Coragen, Beseige, and Durivo.
lambda-cyhalothrin + thiamethoxam, IRAC 3 + 4A (Endigo) ZC	4.0 to 4.5 fl oz	24	40	NOTE THE LONG PREHARVEST USE RESTRICTION. Apply no more than 0.2 pound chlorantraniliprole per acre per crop, which includes applications of Coragen, Beseige, and Durivo.
spinosad, IRAC 5 (Blackhawk)	1.6 to 3.2 oz	4	3	While spinosad is a naturally derived active ingredient, Blackhawk is <u>not</u> **OMRI** listed.
Cutworm				
Preventative insecticide applications are not recommended for cutworms because they are infrequent pests and rescue materials are effective. Scout fields in the first 4 weeks following transplant for cutworm injury and treat if 10% of plants are clipped. Cutworm treatments should be applied in a directed spray over rows in the late afternoon or at dusk, when cutworms are most likely to be active.				
acephate, IRAC 1B (Orthene) 97 PE	0.75 lb	24	3	There are many formulations of acephate. Do not use more than 4 1/8 lb/acre Orthene (4 lb AI/acre). This includes greenhouse, transplant water, soil, and foliar applications. Some purchasers may have concerns about acephate residues. Discuss acephate usage with purchaser prior to making applications.
chlorantraniliprole, IRAC 28 (Coragen)	3.5 to 5 fl oz	4	1	Make no more than 4 applications per season (with at least 3 days between applications). Apply no more than 15.4 fluid ounces of Coragen or more than 0.2 pound chlorantraniliprole per acre per crop, which includes applications of Coragen, Beseige, and Durivo. Some purchasers may have concerns about chlorantraniliprole residues, particularly if used later in the growing season. Discuss chlorantraniliprole usage with purchaser prior to making applications.
lambda-cyhalothrin, IRAC 3A (Warrior) (Karate Xeon)	2.5 to 3 oz 0.96 to 1.92 fl oz	24	40	NOTE LONG PREHARVEST INTERVAL.

Table 5-8B. Insect Control on Flue-Cured and Burley Tobacco in the Field

Insecticide, Formulation[1] and IRAC Group	Amount of Formulation Per Acre	Restricted Entry Interval (REI) (hours)	Preharvest Interval (PHI) (days)	Precautions and Remarks
Cutworm (continued)				
lambda-cyhalothrin + chlorantraniliprole IRAC 3 + 28 (Besiege)	5.0 to 9.0 fl oz	24	40	NOTE LONG PREHARVEST USE RESTRICTION. Apply no more than 15.4 fluid ounces of Coragen or more than 0.2 pound chlorantraniliprole per acre per crop, which includes applications of Coragen, Besiege, and Durivo. Some purchasers may have concerns about chlorantraniliprole residues, particularly if used later in the growing season. Discuss chlorantraniliprole usage with purchaser prior to making applications.
Grasshopper				
acephate, IRAC 1B (Orthene) 97	0.25 to 0.5 lb	24	3	Nymphs (young) are more easily controlled than adults. There are many formulations of acephate. Do not use more than 4 1/8 lb/acre Orthene (4 lb AI/acre). This includes greenhouse, transplant water, soil, and foliar applications. Some purchasers may have concerns about acephate residues. Discuss acephate usage with purchaser prior to making applications.
Hornworm				
Treat for hornworms when 5 or more larvae longer than 1 inch and without cocoons are found per 50 plants. Hornworm larvae with cocoons should be considered 1/5 of a larva when counting. If treatment is necessary during harvesting, be certain to follow all labeled preharvest intervals.				
acephate, IRAC 1B (Orthene) 97 PE	0.5 lb	24	3	There are many formulations of acephate. Do not use more than 4 1/8 lb/acre Orthene (4 lb AI/acre). This includes greenhouse, transplant water, soil, and foliar applications. Some purchasers may have concerns about acephate residues, particularly if used later in the growing season. Discuss acephate usage with purchaser prior to making applications.
Bacillus thuringiensis, IRAC 11 DiPel DF	0.5 to 1 lb	4	0	There are many *B.t.* formulations, including Agree, Biobit, Condor, Crymax, Deliver, Dipel, Javelin, and Lepinox. Highest labeled rates are generally needed for budworm control. DiPel DF and many but not all Bt formulations are **OMRI** listed. Carefully read the label to determine if a material is acceptable for use on organically certified plants.
chlorantraniliprole, IRAC 28 (Coragen)	3.5 to 5 fl oz	4	1	**FIELD FOLIAR APPLICATION.** Because they are not frequent pests before topping, transplant water applications of Coragen for hornworms alone are not recommended. Make no more than 4 applications per season (with at least 3 days between applications). Apply no more than 15.4 fluid ounces of Coragen or more than 0.2 pound chlorantraniliprole per acre per crop, which includes applications of Coragen, Besiege, and Durivo. Lower label rates of Coragen are likely sufficient for hornworms. Some purchasers may have concerns about chlorantraniliprole residues, particularly if used later in the growing season. Discuss chlorantraniliprole usage with purchaser prior to making applications.
lambda-cyhalothrin + chloratraniliprole IRAC 3 + 28 (Besiege)	5.0 to 9.0 fl oz	24	40	NOTE THE LONG PREHARVEST USE RESTRICTION. Apply no more than 0.2 pound chlorantraniliprole per acre per crop, which includes applications of Coragen, Besiege, and Durivo.
lambda-cyhalothrin + thiamethoxam, IRAC 3 + 4A (Endigo) ZC	4.0 to 4.5 fl oz	24	40	NOTE THE LONG PREHARVEST USE RESTRICTION. Apply no more than 0.2 pound chlorantraniliprole per acre per crop, which includes applications of Coragen, Besiege, and Durivo.
spinosad, IRAC 5 (Blackhawk)	1.6 to 3.2 oz	4	3	While spinosad is a naturally derived active ingredient, Blackhawk is <u>not</u> OMRI listed.
Japanese beetle				
Infestations may be spotty within fields and do not typically require treatment.				
acephate, IRAC 1B (Orthene) 97	0.75 lb	24	3	There are many formulations of acephate. Do not use more than 4 1/8 lb/acre Orthene (4 lb AI/acre). This includes greenhouse, transplant water, soil, and foliar applications. Some purchasers may have concerns about acephate residues. Discuss acephate usage with purchaser prior to making applications.
lambda-cyhalothrin + chlorantraniliprole IRAC 3 + 28 (Besiege)	5.0 to 9.0 fl oz	24	40	NOTE THE LONG PREHARVEST USE RESTRICTION. Apply no more than 0.2 pound chlorantraniliprole per acre per crop, which includes applications of Coragen, Besiege, and Durivo.
lambda-cyhalothrin + thiamethoxam, IRAC 3 + 4A (Endigo) ZC	4.0 to 4.5 fl oz	24	40	NOTE THE LONG PREHARVEST USE RESTRICTION. Apply no more than 0.2 pound chlorantraniliprole per acre per crop, which includes applications of Coragen, Besiege, and Durivo.
imidacloprid, IRAC 4A (Admire Pro) (several products) 2F	1.4 fl oz 3.2 fl oz	12	14	**FIELD FOLIAR APPLICATION.** Avoid using only Group 4A materials for season long control of insects with more than 1 generation. Following treatments of Group 4A materials, rotate to a different MOA before making additional applications of a Group 4A material.
thiamethoxam, IRAC 4A (Actara) 25 WDG	2 to 3 oz	12	14	**Make only one application of thiamethoxam per season. Thiamethoxam is also the active ingredient in Platinum.**
Slug				
Slugs are only potential pests in the greenhouse and shortly following transplant. They do not present a risk to larger plants.				
iron phosphate bait (Sluggo)	20 to 44 lb	0	—	**OMRI** listed. TO AVOID PLANT INJURY, DO NOT PUT BAIT ON PLANTS.
metaldehyde bait (Deadline Bullets)	12 to 40 lb	12	—	Apply at dusk to soil surface between rows and around margins of field. DO NOT PUT BAIT ON PLANTS.

Table 5-8B. Insect Control on Flue-Cured and Burley Tobacco in the Field

Insecticide, Formulation[1] and IRAC Group	Amount of Formulation Per Acre	Restricted Entry Interval (REI) (hours)	Preharvest Interval (PHI) (days)	Precautions and Remarks
Stink bug				
Stink bugs rarely cause economic damage to tobacco and rarely require treatment.				
acephate, MOA 1B (Orthene) 97	0.75 lb	24	3	There are many formulations of acephate. Do not use more than 4 1/8 lb/acre Orthene (4 lb AI/acre). This includes greenhouse, transplant water, soil, and foliar applications. Some purchasers may have concerns about acephate residues. Discuss acephate usage with purchaser prior to making applications.
bifenthrin, IRAC 3 (Capture LFR)	3.4 to 6.8 fl oz	12	Do not apply after Layby	**FIELD FOLIAR APPLICATION.** NOTE THE LONG PREHARVEST USE RESTRICTION.
bifenthrin + imidacloprid, IRAC 3, 4A (Brigadier) 2SC	6.4 fl oz	12	Do not apply after Layby	**FIELD FOLIAR APPLICATION.** NOTE THE LONG PREHARVEST USE RESTRICTION.
lambda-cyhalothrin, IRAC 3A (Warrior) 1CS (Karate Xeon)	2.5 to 3 oz 0.96 to 1.92 fl oz	24 24	40 40	To avoid build-up of resistance, rotate use of this product with other modes of action. NOTE THE LONG PREHARVEST USE RESTRICTION.
lambda-cyhalothrin + chlorantraniliprole IRAC 3 + 28 (Besiege)	5.0 to 9.0 fl oz	24	40	NOTE THE LONG PREHARVEST USE RESTRICTION. Apply no more than 0.2 pound chlorantraniliprole per acre per crop, which includes applications of Coragen, Besiege, and Durivo.
Tomato spotted wilt virus (TSWV) suppression				
The materials below act on the thrips vector of TSWV. In addition to these materials, applications of acibenzolar-S-methyl (Actigard 50WG) timed to predicted thrips flights are also effective at suppressing TSWV. Consult the TSWV and Thrips Risk Forecasting Tool (http://climate.ncsu.edu/products/tobacco_tswv/index.php) for recommendation on timing Actigard applications.				
chlorantraniliprole + thiamethoxam, IRAC 28 + IRAC 4A (Durivo)	Rate per 1,000 plants 1.6 fl oz	12	None given	**TRANSPLANT WATER APPLICATION.** Transplant applications of Durivo may suppress tobacco budworm populations for 4 to 7 weeks. Proper calibration of application equipment is essential for effective transplant water applications. A metered or pressurized application system is recommended. Apply no more than 0.2 pound chlorantraniliprole per acre per crop, which includes applications of Coragen, Beseige, and Durivo. Thiamethoxam may be less effective at suppressing TSWV than imidacloprid.
imidacloprid, IRAC 4A (Admire Pro)	Rate per 1,000 plants 0.8 to 1.2 fl oz	12	14	**TRANSPLANT WATER APPLICATION.** Rate is per 1,000 plants and should be converted for transplant water applications based on plant population. Proper calibration of application equipment is essential for effective transplant water applications. A metered or pressurized application system is recommended. Several concentrations of imidacloprid (1.6F, 2F, 4F, and 4.6F) are available. Carefully read the label to determine the correct rate for target pests. Imidacloprid may be more effective at suppressing TSWV than thiamethoxam.
imidacloprid, IRAC 4A (Admire Pro)	Rate per 1,000 plants 0.8 fl oz	12	14	**GREENHOUSE TRAY DRENCH APPLICATION.** Rate is per 1,000 plants. Apply no more than 5 days before transplanting. Immediately after application, wash the material off the plants onto the potting soil. Several concentrations of imidacloprid (1.6F, 2F, 4F, and 4.6F) are available. Carefully read the label to determine the correct rate for target pests.
thiamethoxam, IRAC 4A (Platinum) 75 SG (Platinum) SC	Rate per 1,000 plants 0.27 to 0.43 oz 0.8 to 1.3 fl oz	12	None given	**TRANSPLANT WATER APPLICATION.** Rate is per 1,000 plants and should be converted for transplant water applications based on plant population. Proper calibration of application equipment is essential for effective transplant water applications. A metered or pressurized application system is recommended. Thiamethoxam may be less effective at suppressing TSWV than imidacloprid.
thiamethoxam, IRAC 4A (Platinum) 75 SG (Platinum) SC	Rate per 1,000 plants 0.27 to 0.43 oz 0.8 to 1.3 fl oz	12	None given	**GREENHOUSE TRAY DRENCH APPLICATION.** Use lower label rate for aphids. Rate is per 1,000 plants. Apply no more than 5 days before transplant. Immediately after application, wash the material off the plants onto the potting soil OR apply in transplant water. Thiamethoxam may be less effective at suppressing TSWV than imidacloprid.
Vegetable weevil				
acephate, IRAC 1B (Orthene) 97	0.5 to 0.75 lb	24	3	Treat plants in late afternoon for best control. Spray a band over center of row using a good volume of water. Do not use more than 4 1/8 lb/acre Orthene (4 lb AI/acre). This includes greenhouse, transplant water, soil, and foliar applications. Some purchasers may have concerns about acephate residues. Discuss acephate usage with purchaser prior to making applications.
lambda-cyhalothrin, IRAC 3A (Warrior) 1CS (Karate Xeon)	2.5 to 3 oz 0.96 to 1.92 fl oz	24 24	40 40	NOTE THE LONG PREHARVEST USE RESTRICTION.
lambda-cyhalothrin + chlorantraniliprole IRAC 3 + 28 (Besiege)	5.0 to 9.0 fl oz	24	40	NOTE THE LONG PREHARVEST USE RESTRICTION. Apply no more than 0.2 pound chlorantraniliprole per acre per crop, which includes applications of Coragen, Besiege, and Durivo.

Table 5-8B. Insect Control on Flue-Cured and Burley Tobacco in the Field

Insecticide, Formulation[1] and IRAC Group	Amount of Formulation Per Acre	Restricted Entry Interval (REI) (hours)	Preharvest Interval (PHI) (days)	Precautions and Remarks
Wireworm				
Wireworm treatments should be applied pretransplant in fields with a history of significant damage. If fields do not have a history of wireworm injury, greenhouse tray drench or transplant water treatments of imidacloprid or thiamethoxam will also suppress wireworm damage if they are present.				
bifenthrin + imidacloprid, IRAC 3, 4A (Brigadier 2SC)	6.4 fl oz	12	Do not apply after Layby	Use as described above for transplant water treatments for imidacloprid. Brigadier is not intended for greenhouse use. Data on wireworm control are limited.
bifenthrin, IRAC 3 (Capture LFR)	3.4 to 6.8 fl oz	12	Do not apply after Layby	Apply as a pretransplant soil treatment and incorporate into 4 inches of soil OR apply in transplant water at 3.4 to 6.8 fluid ounces per acre. Data on wireworm control are limited.
chlorantraniliprole + thiamethoxam, IRAC 28 + IRAC 4A (Durivo)	**Rate per 1,000 plants** 1.6 fl oz	12	None given	**TRANSPLANT WATER APPLICATION.** Transplant applications of Durivo may suppress tobacco budworm populations for 4 to 7 weeks. Proper calibration of application equipment is essential for effective transplant water applications. A metered or pressurized application system is recommended. Apply no more than 0.2 pound chlorantraniliprole per acre per crop, which includes applications of Coragen, Beseige, and Durivo. Thiamethoxam may be less effective at suppressing TSWV than imidacloprid.
ethoprop, IRAC 1B (Mocap) 15 G	13 to 40 lb (broadcast) 3.2 lb per 100 row ft (banded)	48	NA	**Preplant soil application only.** Rates depend on application timing and target pests. Mocap is highly toxic to humans. Use extreme caution when mixing and applying.
imidacloprid, IRAC 4A (Admire Pro)	**Rate per 1,000 plants** 1.2 fl oz	12	14	**GREENHOUSE TRAY DRENCH APPLICATION.** Rate is per 1,000 plants. Apply no more than 5 days before transplanting. Immediately after application, wash the material off the plants onto the potting soil. Several concentrations of imidacloprid (1.6F, 2F, 4F, and 4.6F) are available. Carefully read the label to determine the correct rate for target pests. Data on wireworm control are limited.
thiamethoxam, IRAC 4A (Platinum) 75 SG (Platinum) SC	**Rate per 1,000 plants** 0.43 oz 1.3 fl oz	12	None given	**GREENHOUSE TRAY DRENCH APPLICATION.** Use lower label rate for aphids. Rate is per 1,000 plants. Apply no more than 5 days before transplant. Immediately after application, wash the material off the plants onto the potting soil OR apply in transplant water. Data on wireworm control are limited.

[1] Some insecticides are available in several formulations. Those listed are generally the most commonly used or are readily available. Other formulations may or may not be suitable for use on tobacco or a specific pest. Check labels carefully.

[2] Many soil-applied insecticides can injure plants under certain conditions. Some soil-applied insecticides are very soluble and pose a threat to surface and groundwater; check labels carefully for warnings.

[3] Tobacco purchasers are concerned about pesticide residues in cured leaf. Use caution in making applications of methomyl, acephate, Group 3 (pyrethroid) insecticides. Select other materials when available.

More production information is available at http://www.tobacco.ces.ncsu.edu.

Chapter V — 2019 N.C. Agricultural Chemicals Manual

Insect Control for Commercial Vegetables
J. F. Walgenbach, G.G. Kennedy, and A. Huseth, Entomology and Plant Pathology

Read the pesticide label before application. High pressure (200 psi) and high volume (50 gallons per acre) aid in vegetable insect control. Ground sprays with airblast sprayers or sprayers with hollow cone drop nozzles are suggested. Incorporate several methods of control for best results. In recent years, the number of generic products has increased significantly. For brevity, these generic products typically are not listed within each section. The trade names listed are intended to aid in identification of products and are neither intended to promote use of specific trade names nor to discourage use of generic products. A list of active ingredients and generic brand names appears in a separate table at the end of this section.

The Insecticide Resistance Action Committee (IRAC) classifies insecticides based on their mode of action (MOA), with insecticides in the same MOA having the same mode of action. Effective insecticide resistance management involves the use of alternations, rotations, or sequences of different insecticide MOA classes. To prevent the development of resistance, it is important not to apply insecticides with the same MOA to successive generations of the same insect.

Table 5-9. Insect Control for Commercial Vegetables

CROP / Insect	Insecticide, Mode of Action Code, and Formulation	Amount of Formulation Per Acre	Restricted Entry Interval (REI)	Pre harvest Interval (PHI) (Days)	Precautions and Remarks
Asparagus					
Aphid	acetamiprid (Assail) 30 SG	2.5 oz	12	1	
	dimethoate 400, MOA 1B	1 pt	48 hrs	180	Do not exceed 5 pints per acre per year.
	malathion, MOA 1B (various) 57 EC	2 pt	12 hrs	1	Aphid colonies appear by early September.
	pymetrozine, MOA 9B (Fulfill) 50 WDG	2.75 oz	12 hrs	—	For aphid control on ferns after harvest.
Asparagus beetle, Japanese beetle, Grasshopper	carbaryl, MOA 1A (Sevin) 50 WP (Sevin) 80 S (Sevin) XLR Plus	2 to 4 lb 1.25 to 2.5 lb 1 to 2 qt	12 hrs	1	Low rate to be used on seedlings or spears. Do not apply more often than once every 3 days. With established beetle populations, 3 consecutive weekly sprays are required. Manage beetles and grasshoppers in the fall. The use of carbamates may result in aphid buildup.
	acetamiprid (Assail) 30 SG	2.5 oz	12	1	
	dimethoate 400, MOA 1B	1 pt	48 hrs	180	Do not exceed 5 pints per acre per year.
	malathion, MOA 1B (various) 57 EC	2 pt	12 hrs	1	Apply as needed.
	methomyl, MOA 1A (Lannate) 2.4 LV	1.5 pt	48 hrs	1	Leave a row on edge of field near overwintering sites of asparagus beetles fern out. This will attract and hold beetles for that directed insecticide spray (trap and destroy).
	pyrethroid, MOA 3				See table 5-9B for a list of registered pyrethroids and pre-harvest intervals.
	spinetoram, MOA 5 (Radiant) 1 SC	4 to 8 fl oz	4 hrs	60	For asparagus beetle only. This use is only for asparagus ferns; do not apply within 60 days of spear harvest.
Beet armyworm, Cutworm, Yellow-striped armyworm	*Bacillus thuringiensis*, MOA 11A (Dipel) DF	0.5 to1 lb	4 hrs	0	
	chlorantraniliprole, MOA 28 (Coragen) 1.67SC	3.5 to 5 fl oz	4 hrs	1	
	cyantraniliprole, MOA 28 (Exirel) 0.83EC	7 to 13.5 fl oz	12 hr	1	Do not make applications within 25 ft of water sources.
	methomyl, MOA 1A (Lannate) 2.4 LV (Lannate) 90 SP	1.5 to 3 pt 0.5 to 1 lb	48 hrs	1	
	spinetoram, MOA 5 (Radiant) 1 SC	4 to 8 fl oz	4 hrs	60	This use is only for asparagus ferns; do not apply within 60 days of spear harvest.
	spinosad MOA 5 (Entrust 2SC)	4 to 6 fl oz	4 hrs	60	This use is only for asparagus ferns; do not apply within 60 days of spear harvest. OMRI approved.
Beans (Snap, Lima, Pole)					
Aphid	acetamiprid MOA 4A (Assail) 30SG	2.5 to 5.3 oz	12 hrs	7	
	dimethoate 4 EC, MOA 1B	0.5 to 1 pt	48 hrs	0	On foliage as needed. Re-entry interval of 48 hours
	imidacloprid, Soil treatment (Admire Pro) 4.6 F (various) 2F	7 to 10.5 fl oz 16 to 24 fl oz	12 hrs	21	See label for soil application instructions. Also controls leafhoppers and thrips
	Foliar treatment Admire Pro 4.6 F (various) 1.6 F	1.2 fl oz to 3.5 fl oz	12 hrs	7	
	sulfoxaflor (Transform) 50 WG	0.75 to 1.0 oz	24 hrs	7	
	flupyradifurone (Sivanto Prime) 200 SL	7 to 14 fl oz	4 hrs	7	
	spirotetramat, MOA 23 (Movento) 2 SC	4 to 5 fl oz	24 hrs	1 (succulent) 7 (dried)	

Table 5-9. Insect Control for Commercial Vegetables

CROP Insect	Insecticide, Mode of Action Code, and Formulation	Amount of Formulation Per Acre	Restricted Entry Interval (REI)	Pre harvest Interval (PHI) (Days)	Precautions and Remarks
Beans (Snap, Lima, Pole) (continued)					
Thrips	acephate, MOA 1B (Orthene) 97 PE	0.5 to 1 lb	24 hrs	14	Lima beans may be treated and harvested the same day. Do not apply more than 2 pounds a.i. per acre per season.
	acetamiprid MOA 4A (Assail) 30SG	2.5 to 5.3 oz	12 hrs	7	
	pyrethroid, MOA 3		12 hrs		See table 5-9B for a list of registered pyrethroids and pre-harvest intervals.
	methomyl, MOA 1A (Lannate) 90 SP (Lannate) 2.4 LV	0.5 lb 1.5 pt	48 hrs	1	
	novaluron MOA 15 (Rimon) 0.83 EC	12 fl oz	12 hrs	1	Effective against immature thrips only.
	spinetoram, MOA 5 (Radiant) 1 SC	5 to 6 fl oz	4 hrs	3 (succulent); 28 (dried)	Do not apply more than 28 fluid ounces per acre per season on succulent beans or more than 12 fluid ounces on dried beans.
	spinosad, MOA 5 (Blackhawk)	2.5 to 3.3 oz	4 hrs	3 (succulent); 28 (dried)	Do not apply more than 20 ounces per acre per season on succulent beans or more than 8.3 ounces on dried beans.
Corn earworm, European corn borer, Lesser cornstalk borer, Looper	chlorantraniliprole, MOA 28 (Coragen) 1.67 SC	3.5 to 5 fl oz	4 hrs	1	
	novaluron MOA 15 (Rimon) 0.83 EC	6 to 12 fl oz	12 hrs	1	
	spinetoram, MOA 5 (Radiant) 1 SC	4.5 to 6 fl oz	4 hrs	3 (succulent); 28 (dried)	Do not apply more than 28 fluid ounces per acre per season on succulent beans or more than 12 fluid ounces on dried beans.
	spinosad, MOA 5 (Blackhawk)	1.7 to 3.3 oz	4 hrs	3 (succulent); 28 (dried	Do not apply more than 20 ounces per acre per season on succulent beans or more than 8.3 ounces on dried beans.
	pyrethroid, MOA 3		12 hrs		See table 5-9B for a list of registered pyrethroids and pre-harvest intervals.
Cowpea curculio	pyrethroid, MOA 3				See table 5-9B for a list of registered pyrethroids and pre-harvest intervals. Control may be poor in areas where resistant populations occur, primarily in the Gulf Coast areas. Addition of piperonyl-butoxide-synergist (Exponent) may improve control of pyrethroids.
Cucumber beetle, Bean leaf beetle, Japanese beetle, Cutworm	carbaryl, MOA 1A (Sevin) 50 WP 80 S XLR Plus	4 lb 2.5 lb 1 qt	12 hrs	3 (succulent) 21 (dried)	
	pyrethroid, MOA 3		12 hrs		See table 5-9B for a list of registered pyrethroids and pre-harvest intervals.
Grasshopper	pyrethroid, MOA 3		12 hrs		See table 5-9B for a list of registered pyrethroids and pre-harvest intervals.
Leafminer	cryomazine, MOA 17 (Trigard) 75 WP	2.66 oz	12 hrs	7	
	naled, MOA 1B (Dibrom) 8 EC	1 pt	48 hrs	1	Re-entry interval is 48 hours
	spinetoram, MOA 5 (Radiant) 1 SC	4 to 8 fl oz	4 hrs	3 (succulent); 28 (dried)	Do not apply more than 28 fluid ounces per acre per season on succulent beans or more than 12 fluid ounces on dried beans.
	spinosad, MOA 5 (Blackhawk)	2.5 to 3.3 oz	4 hrs	3 (succulent); 28 (dried)	Do not apply more than 20 ounces per acre per season on succulent beans or more than 8.3 ounces on dried beans.
Lygus bug	pyrethroid, MOA 3		12 hrs		See table 5-9B for a list of registered pyrethroids and pre-harvest intervals.
	carbaryl, MOA 1A (Sevin) 50 WP 80 S XLR Plus	3 lb 1.875 lb 1.5 qt	12 hrs	3 (succulent) 21 (dried)	On foliage when pods begin to form.
	dimethoate, MOA 1B (Dimethoate) 4 EC	1 pt	48 hrs	7	Do not apply if bees are visiting area to be treated when crops or weeds are in bloom.
Mexican bean beetle	acetamiprid MOA 4A (Assail) 30SG	2.5 to 5.3 oz	12 hrs	7	See table 5-9B for a list of registered pyrethroids and pre-harvest intervals.
	pyrethroid, MOA 3		12 hrs		See table 5-9B for a list of registered pyrethroids and pre-harvest intervals.
	carbaryl, MOA 1A (Sevin) 50 WP (Sevin) 80 S (Sevin) XLR Plus	1 to 2 lb 0.625 to 1.25 lb 1 qt	12 hrs	3 (succulent) 21 (dry)	On foliage as needed. Use low rate on young plants.
	novaluron MOA 15 (Rimon) 0.83 EC	9 to 12 oz	12 hrs	1	Controls immature stages only.
Potato leafhopper	acetamiprid MOA 4A (Assail) 30SG	2.5 to 5.3 oz	12 hrs	7	
	carbaryl, MOA 1A (Sevin) 50 WP (Sevin) 80 S (Sevin) XLR Plus	4 lb 2.5 lb 1 qt	12 hrs	3 (succulent) 21 (dry)	On foliage as needed.

Table 5-9. Insect Control for Commercial Vegetables

CROP / Insect	Insecticide, Mode of Action Code, and Formulation	Amount of Formulation Per Acre	Restricted Entry Interval (REI)	Pre harvest Interval (PHI) (Days)	Precautions and Remarks
Beans (Snap, Lima, Pole) (continued)					
Potato leafhopper (continued)	dimethoate 4 EC, MOA 1B	0.5 to 1 pt	48 hrs	7	
	methomyl, MOA 1A (Lannate) 90 SP (Lannate) 2.4 L	0.5 lb 1.5 to 3 pt	48 hrs	1 1 to 3	Do not graze before 3 days or use for hay before 7 days.
	pyrethroid, MOA 3		12 hrs		See table 5-9B for a list of registered pyrethroids and their reentry and pre-harvest intervals.
Seedcorn maggot, Wireworm	Use seed pretreated with insecticide for seedcorn maggot control.				Seed can be purchased pretreated. Pretreated seed will not control wireworms.
	bifenthrin MOA 3 (Empower) 1.15G	3.5 to 8.7 lb	9 days	9	Apply preplant broadcast incorporated in the top 1 to 3 inches of soil.
	chlorpyrifos MOA 1B (Lorsban) 4E	2 pts	24 hrs		Can be applied preplant broadcast incorporated in the top 1 to 3 inches of soil, or at planting as a T-band application. For at planting application, apply 1.8 fluid ounces per 1,000 feet of row at 30-inch row spacing. Apply the spray in a 3 to 5 inch wide band over the row behind the planting shoe and in front of the press wheel to achieve shallow incorporation. Do not make more than one application per year or apply more than 1 pound ai per acre.
Spider mites	bifenazate MOA 20D (Acramite) 4 SC	16 to 24 fl oz	12 hrs	3	
	acequinocyl MOA 20B (Kanemite) 15 SC	31 fl oz	12 hrs	7	
	fenpyroximate MOA 21A (Portal) 0.4 EC	2 pt	12 hrs	1	For use on snap bean only.
Stink bug, Kudzu bug	pyrethroid, MOA 3		12 hrs		See table 5-9B for a list of registered pyrethroids and pre-harvest intervals.
	naled, MOA 1B (Dibrom) 8 EC	1.5 pt/100 gal water	48 hrs	1	
Whiteflies	acetamiprid MOA 4A (Assail) 30 SG	4.0 to 5.3 oz	12 hrs	7	
	buprofezin, MOA 16 (Courier) 40 SC	9 to 13.6 fl oz	12 hrs	14	For use on snap beans only.
	flupyradifurone (Sivanto Prime)	10.5 to 14 fl oz	4 hrs	7	
	imidacloprid, MOA 4A Soil treatment (Admire Pro) 4.6 F (various) 2 F	7 to 10.5 fl oz 16 to 24 fl oz	12 hrs	21	See label for soil application instructions.
	Foliar treatment (Admire Pro) 4.6 F (various) 1.6 F	1.2 fl oz 3.5 fl oz	12 hrs	7	
	spirotetramat, MOA 23 (Movento)	4 to 5 oz	24 hrs	1 (succulent) 7 (dry)	PHI is 1 day for succulent beans and 7 days for dry beans.
Beet					
Aphid	flonicamid, MOA 9C (Beleaf) 50SG	2 to 2.8 oz	12 hrs	7	
	imidacloprid, MOA 4A Soil treatment (Admire Pro) 4.6 F (various) 2 F	4.4 to 10.5 fl oz 10 to 24 fl oz	12 hrs	21	See label for soil application instructions. Will also control flea beetle.
	flupyradifurone, MOA 4D (Sivanto) 200 SL	7.0 to 10.5 fl oz	4 hrs	1	
	Foliar treatment (Admire Pro) 4.6 F (various) 1.6 F	1.2 fl oz 3.5 fl oz	12 hrs	7	
	thiamethoxam, MOA 4A (Platinum) 75 SG	1.7 to 2.17 oz	12 hrs		Platinum may be applied to direct-seeded crops in-furrow at seed or transplant depth, post seeding or transplant as a drench, or through drip irrigation. Do not exceed 12 ounces per acre per season of Platinum. Check label for plant-back restrictions for a number of crops.
	(Actara) 25 WDG	1.5 to 3 oz	12 hrs	7	
Armyworm, Beet webworm	chlorantraniliprole MOA 28 (Coragen) 1.67 SC	3.5 to 5 fl oz	4 hrs	1	
	methoxyfenozide MOA 18 (Intrepid) 2F	6 to 16 fl oz	4 hrs	7	
	spinetoram, MOA 5 (Radiant) 1 SC	6 to 8 fl oz	4 hrs	7	Do not apply more than 32 fluid ounces per acre per season.
	spinosad, MOA 5 (Blackhawk)	1.7 to 3.3 oz	4 hrs	3	

Table 5-9. Insect Control for Commercial Vegetables

CROP / Insect	Insecticide, Mode of Action Code, and Formulation	Amount of Formulation Per Acre	Restricted Entry Interval (REI)	Pre harvest Interval (PHI) (Days)	Precautions and Remarks
Beet (continued)					
Blister beetle, Flea beetle	carbaryl, MOA 1A (Sevin) 50 WP 80 S XLR	3 lb 1.875 lb 1 qt	12 hrs	7	
	pyrethroid, MOA 3		12 hrs		See table 5-9B for a list of registered pyrethroids and pre-harvest intervals.
Leafminer	spinetoram, MOA 5 (Radiant) 1 SC	6 to 10 fl oz	4 hrs	7	Control will be improved with addition of a spray adjuvant.
Broccoli, Brussels Sprouts, Cabbage, Cauliflower, Kohlrabi					
Aphid	Where whitefly resistance is an issue (or any other insect with a high potential for resistance to Group 4A MOA insecticides), a foliar-applied Group 4A insecticide program and a soil-applied Group 4A program should not be used in the same season. Also, if using a foliar-applied program, avoid using a block of more than 3 consecutive applications of any products belonging to Group 4A insecticides.				
	acetamiprid, MOA 4A (Assail) 30 SG	2 to 3 oz	12 hrs	7	
	clothianidin, MOA 4A (Belay) 50WD	4.8 to 6.4 oz (soil) 1.6 to 2.1 oz (foliar)	12 hrs	21 (soil) 7 (foliar)	Soil application at planting only.
	dimethoate 4 EC, MOA 1B	0.5 to 1 pt	48 hrs	7	
	flonicamid, MOA 9C (Beleaf) 50SG	2 to 2.8 oz	12 hrs	0	
	flupyradifurone, MOA 4D (Sivanto) 200 SL	7.0 to 12.0 fl oz	4 hrs	1	
	imidacloprid, MOA 4A Soil treatment (Admire Pro) 4.6 F (various) 2 F	4.4 to 10.5 fl oz 10 to 24 fl oz	12 hrs	21	Do not follow soil applications of Admire with foliar applications of any neonicotinoid insecticide. Use only one application method. See label for soil application instructions. Imidacloprid also controls whiteflies.
	Foliar treatment (Admire Pro) 4.6 F (various) 1.6 F	1.3 fl oz 3.75 fl oz	12 hrs	7	Imidacloprid also controls whiteflies. Not effective against flea beetle.
	pymetrozine, MOA 9B (Fulfill) 50 WDG	2.75 oz	12 hrs	7	
	spirotetramat, MOA 23 (Movento) 2 SC	4 to 5 fl oz	24 hrs	1	Do not exceed 10 fluid ounces per season. Requires surfactant.
	thiamethoxam MOA 4A Soil treatment (Platinum) 75SG Foliar treatment (Actara) 25WDG	1.66 to 3.67 oz 1.5 to 3.0 oz	12 hrs	30 0	Platinum may be applied to direct-seeded crops in-furrow at seed or transplant depth, postseeding or transplant as a drench, or through drip irrigation. Do not exceed 3.67 ounces per acre per season. Thiamethoxam also controls whiteflies and certain thrips species.
Diamondback moth, Cabbage looper, Imported cabbageworm, Corn earworm, Cross-striped cabbageworm, Cabbage webworm, Armyworms	Insecticide-resistant populations, widespread in Southeastern U.S., may not be controlled with some registered insecticides. To manage resistance, avoid transplants from Georgia and Florida and avoid repeated use of the same materials for extended periods of time. Repeated use of pyrethroid insecticides destroys natural enemies and often aggravates diamondback moth problems. Do not allow populations to increase to large densities before initiating treatments.				
	Bacillus thuringiensis, MOA 11A (Dipel) 2X (Dipel) 4 L (Javelin) WG (Xentari) WDG	8 oz 1 to 2 qt 0.5 to 1 lb 0.5 to 1 lb	4 hrs	0	On foliage every 7 days. On summer or fall plantings, during periods when eggs and larvae are present. This usually occurs when true leaves appear; on other plantings, it may occur later. A spreader-sticker will be helpful. **Not effective against Cabbage Webworm**
	chlorantraniliprole, MOA 28 (Coragen) 1.67 SC	3.5 to 5 fl oz	4 hrs	3	Foliar or soil application. See label for soil application instructions.
	cyclaniliprole, MOA 28 (Harvanta) 50 SL	10.9 to 16.4 fl oz	4 hrs		
	cyantraniliprole, MOA 28 (Verimark) 1.67SC	5 to 10 fl oz	12 hrs	NA	Verimark is for soil application only. Apply at planting only. See label for application options.
	(Exirel) 0.83SE	7 to 17 fl oz	12 hrs	1	Exirel is for foliar application only. Use higher rates for cabbage looper.
	emamectin benzoate, MOA 6 (Proclaim) 5 WDG	3.2 to 4.8 oz	12 hrs	7	
	indoxacarb, MOA 22 (Avaunt eVo) 30 WDG	2.5 to 3.5 oz	12 hrs	3	Add a wetting agent to improve spray. Do not apply more than 14 ounces (0.26 pound a.i.) per acre per crop. The minimum interval between sprays is 3 days.
	novaluron, MOA 15 (Rimon) 0.83 EC	6 to 12 fl oz	12 hrs	7	Use lower rates when targeting eggs or small larvae, and use higher rates when larvae are large. Make no more than 3 applications, or 24 fluid ounces per acre per season.
	spinetoram, MOA 5 (Radiant) 1 SC	5 to 10 fl oz	4 hrs	1	
Flea beetle	acetamiprid, MOA 4A (Assail) 30 SG	2 to 3 oz	12 hrs	7	

Table 5-9. Insect Control for Commercial Vegetables

CROP / Insect	Insecticide, Mode of Action Code, and Formulation	Amount of Formulation Per Acre	Restricted Entry Interval (REI)	Pre harvest Interval (PHI) (Days)	Precautions and Remarks
Broccoli, Brussels Sprouts, Cabbage, Cauliflower, Kohlrabi (continued)					
Flea beetle (continued)	clothianidin, MOA 4A (Belay) 50WDG	4.8 to 6.4 oz (soil) 1.6 to 2.1 oz (foliar)	12 hrs	7 (foliar)	Soil applications may only be made at planting.
	cyantraniliprole, MOA 28 (Verimark) 1.67SC	6.75 to 13.5 fl oz	4 hrs	1	Verimark is for at planting soil application only. See label for application options.
	(Exirel) 0.83SE	13.5 to 20.5 fl oz	12 hrs	1	Exirel is for foliar application only.
	dinotefuran, MOA 4A Foliar treatment (Venom) 70 SG (Scorpion) 35SL	1 to 4 oz 2 to 7 fl oz	12 hrs	1	See label for soil application options.
	Soil treatment (Venom) 70 SG (Scorpion) 35SL	5 to 6 oz 9 to 10.5 fl oz		21	
	pyrethroid MOA 3		12 hrs		See table 5-9B for a list of registered pyrethroids and pre-harvest intervals.
Harlequin bug, Stink bug	clothianidin, MOA 4A (Belay) 50WDG	4.8 to 6.4 oz (soil) 1.6 to 2.1 oz (foliar)	12 hrs	NA 7 (foliar)	Soil application at planting only.
	dinotefuran, MOA 4A (Venom) 70 SG (Scorpion) 35 SL	3 to 4 oz 2 to 7 fl oz	12 hrs	1	Do not exceed 6 ounces of Venom per season.
	pyrethroid, MOA 3		12 hrs		See table 5-9B for a list of registered pyrethroids and pre-harvest intervals.
Yellowmargined leaf beetle	pyrethroid, MOA 3		12 hrs		Applications need to be made at the first sign of infestation. Problems are most common in spring and fall months along the gulf coast areas.
Root maggot	chlorpyrifos, MOA 1B (Lorsban) 4 EC (Lorsban) 75 WG	2 pt/100 gal 1.33 lb	24 hrs	—	Directed spray to transplants: Spray the base of the plant immediately after transplanting, using a minimum of 40 gallons per acre.
	chlorpyrifos, MOA 1B (Lorsban) 4 EC (Lorsban) Advanced (Lorsban) 75 WG	4 to 4.5 pts 4 to 4.5 pts 3 lb	24 hrs	—	Preplant incorporate: Apply as a broadcast spray to the soil surface in a minimum spray volume of 10 gal or more and incorporate into the top 2 to 4 inches of soil on the day of application.
	chlorpyrifos, MOA 1B (Lorsban) 4 EC	1.6 to 2.75 oz/ 1,000 ft row	24 hrs	—	Direct seeded: Apply in a 4-inch wide band behind planter shoe and in front of press wheel for shallow incorporation.
	(Lorsban) 15 G	4.6 to 9.2 oz/ 1,000 ft row	24 hrs		Direct seeded: Place across seed row in 4-inch band behind planter shoe and in front of press wheel.
	cyantraniliprole MOA 28 (Verimark) 1.67 SC	10 to 13.5 fl oz	4	—	Apply to soil at planting as an in-furrow spray, transplant tray drench, transplant water, hill drench, surface band, or soil shank.
	diazinon, MOA 1B (Diazinon 50 W) 50 WP	0.25 to 0.5 lb/ 50 gal	4 days	—	Transplant water: Apply in transplant water or drench water at 4 to 6 ounces per plant at transplanting.
Thrips	acetamiprid (Assail) 30 SG	4.0 oz	12 hrs	7	Efficacy will vary depending on thrips species.
	dimethoate 4 EC, MOA 1B	0.5 to 1 pt	48 hrs	7	
	imidacloprid, MOA 4A (Admire Pro) 4.6F (various) 2F (various) 1.6 F	1.3 fl oz 3.0 fl oz 3.75 fl oz	12 hrs	7	Check label for rates for other formulations. Foliar applications only.
	methomyl, MOA 1A (Lannate) 2.4 LV	1.5 fl oz	48 hrs	1	
	novaluron, MOA 15 (Rimon) 0.83 EC	6 to 12 fl oz	12 hrs	7	Make no more than 3 applications, or 24 fluid ounces, per acre per season.
	spinetoram, MOA 5 (Radiant) 1 SC	6 to 10 fl oz	4 hrs	1	
Whitefly	acetamiprid, MOA 4A (Assail) 30 SG	2.5 to 4.0 oz	12 hrs	7	Use a spreader stick to improve control.
	dinotefuran, MOA 4A Foliar treatment (Venom) 70 SG (Scorpion) 35SL	1 to 4 oz 2 to 7 fl oz	12 hrs	1	Do not follow soil applications with foliar applications of any neonicotinoid insecticide. Use only one application method. Do not apply more than 6 ounces per acre per season using foliar applications, or 12 ounces per acre per season using soil applications. Soil applications may be applied by: a narrow band below or above the seed line at planting; a post-seeding or transplant drench with sufficient water to ensure incorporation to the root zone; or through drip irrigation.
	Soil treatment (Venom) 70 SG (Scorpion) 35SL	5 to 6 oz 9 to 10.5 fl oz		21	
	flupyradifurone, MOA 4D (Sivanto) 200 SL	10.5 to 14.0 fl oz	4 hrs	1	

Table 5-9. Insect Control for Commercial Vegetables

CROP / Insect	Insecticide, Mode of Action Code, and Formulation	Amount of Formulation Per Acre	Restricted Entry Interval (REI)	Pre harvest Interval (PHI) (Days)	Precautions and Remarks
Broccoli, Brussels Sprouts, Cabbage, Cauliflower, Kohlrabi (continued)					
Whitefly (continued)	spiromesifen, MOA 23 (Oberon) 2 SC	7 to 8.5 fl oz	12 hrs	7	Do not exceed 25.5 fluid ounces per acre per season.
	spirotetramat, MOA 23 (Movento) 2 SC	4 to 5 fl oz	24 hrs	1	Do not exceed 10 fluid ounces per season. Requires surfactant.
	pyriproxyfen, MOA 7 (Knack) 0.86EC	8 to 10 fl oz	12 hrs	7	Only treat whole fields, and do not plant any crop other than those that Knack is registered within 30 days after the last application.
Carrot					
Aphid, Leafhopper	imidacloprid, MOA 4A Soil treatment (Admire Pro) 4.6 F (various) 2 F	4.4 to 10.5 fl oz 10 to 24 fl oz	12 hrs	21	Must be applied to the soil. May be applied via chemigation into the root zone through low-pressure drip, trickle, micro-sprinkler, or equivalent equipment; in-furrow spray or shanked-in 1 to 2 inches below seed depth during planting; or in a narrow band (2 inches or fewer) 1 to 2 inches directly below the eventual seed row in a bedding operation 14 or fewer days before planting. Higher rates provide longer lasting control. See label for information on approved application methods and rate per 100 row feet for different row spacings.
	Foliar treatment (Admire Pro) 4.6 F (various) 1.6 F	1.2 fl oz 3.5 fl oz	12 hrs	7	
	thiamethoxam, MOA 4A (Platinum) 75 SG	1.66 to 3.67 oz	12 hrs	30	Platinum may be applied to direct-seeded crops in-furrow at seeding, immediately after seeding with sufficient water to ensure incorporation into the root zone, or through trickle irrigation.
	(Actara) 25 WDG	1.5 to 3 oz	12 hrs	7	Actara is applied to foliage. Do not exceed 4 ounces Actara per acre per season.
	flonicamid, MOA 29 (Beleaf) 50SG	2 to 2.8 fl oz	12 hrs	3	
	flupyradifurone, MOA 4D (Sivanto) 200 SL	7.0 to 10.5 fl oz	4 hrs	7	
Armyworm, Parsleyworm	pyrethroid, MOA 3		12 hrs		See table 5-9B for a list of registered pyrethroids and pre-harvest intervals.
	carbaryl, MOA 1A (Sevin) 80 S (Sevin) XLR Plus	1.25 lb 1 qt	12 hrs	7	On foliage as needed.
	chlorantraniliprole, MOA 28 (Coragen) 1.67 SC	3.5 to 5 fl oz	4 hrs	1	Coragen may be used for foliar or drip chemigation.
	methomyl, MOA 1A (Lannate) 2.4 LV (Lannate) 90 SP	0.75 to 1.5 pt 0.25 to 0.5 lb	48 hrs	1	
	methoxyfenozide, MOA 18 (Intrepid) 2 F	4 to 10 fl oz	4 hrs	1	Use higher rates against large larvae.
	spinetoram, MOA 5 (Radiant) 1 SC	6 to 8 fl oz	4 hrs	3	Radiant will not control leafhoppers. Do not make more than 4 applications per year.
Leafminer	spinetoram, MOA 5 (Radiant) 1 SC	6 to 8 fl oz	4 hrs	3	
Wireworm	diazinon, MOA 1B (Diazinon) (AG 500)	4 qt	3 days	—	Broadcast and incorporate preplant.
Celery					
Aphid, Leafhopper, Flea beetle	imidacloprid, MOA 4A (Admire Pro) 4.6 F (various) 2 F	7 to 10.5 fl oz 16 to 24 fl oz	12 hrs	21	Apply via chemigation into the root zone, as an in-furrow spray at planting on/or below the seed, or as a post-seeding or transplant drench.
	flonicamid, MOA 9C (Beleaf) 50SG	2 to 2.8 oz	12 hrs	0	Will not control flea beetle
	flupyradifurone, MOA 4D (Sivanto) 200 SL	10.5 to 12.0 fl oz	4 hrs	1	Will not control flea beetle
	spirotetramat, MOA 23 (Movento) 2SC	4 to 5 fl oz	24 hrs	3	Do not exceed 10 fluid ounces per season. Not for flea beetle. Requires surfactant. Will not control flea beetle
	tolfenpyrad, MOA 21A (Torac) 1.29 EC	17 to 21 fl oz	12 hrs	1	
Armyworm, Corn earworm, Looper	chlorantraniliprole, MOA 28 (Coragen) 1.67 SC	3.5 to 5 fl oz	4 hrs	1	Foliar or drip chemigation. Drip chemigation must be applied uniformly to the root zone. See label for instructions.
	emamectin benzoate, MOA 6 (Proclaim) 5 WDG	2.4 to 4.8 oz	12 hrs	7	Do not make more than 2 sequential applications without rotating to another product with a different mode of action.
	methomyl, MOA 1A (Lannate) 2.4 LV	3 pt	48 hrs	7	Methomyl may induce leafminer infestations.
	methoxyfenozide, MOA 18 (Intrepid) 2 F	4 to 10 fl oz	4 hrs	7	For early season applications only to young crop and small plants. For mid- to late-season applications and to heavier infestations and under conditions in which thorough coverage is more difficult. Do not apply more than 16 fluid ounces per application, and do not exceed 64 fluid ounces per season. See Rotational Crop Restrictions on label.

Table 5-9. Insect Control for Commercial Vegetables

CROP Insect	Insecticide, Mode of Action Code, and Formulation	Amount of Formulation Per Acre	Restricted Entry Interval (REI)	Pre harvest Interval (PHI) (Days)	Precautions and Remarks
Celery (continued)					
Armyworm, Corn earworm, Looper (continued)	pyrethroid, MOA 3		12 hrs		See table 5-9B for registered pyrethroids and pre-harvest intervals.
	spinetoram, MOA 5 (Radiant) 1 SC	5 to 10 fl oz	4 hrs	1	Use higher rates for armyworms.
Leafminer	abamectin, MOA 6 (Agri-Mek) 0.15EC	1.75 to 3.5 fl oz	12 hrs	7	
	chlorantraniliprole, MOA 28 (Coragen) 1.67 SC	5 to 7.5 fl oz	4 hrs	1	Foliar or drip chemigation. Drip chemigation must be applied uniformly to the root zone. See label for instructions.
	cryomazine, MOA 17 (Trigard 75WP)	2.66 oz	12 hrs	7	
	spinetoram, MOA 5 (Radiant) 1 SC	6 to 10 fl oz	4 hrs	1	
Collard, Kale, Mustard Greens					
Aphid	acetamiprid, MOA 4A (Assail) 30 SG	2 to 3 oz	12 hrs	7	
	clothianidin, (Belay) 50 WDG	4.8 to 6.4 oz (soil) 1.6 to 2.1 oz (foliar)	12 hrs	 7 (foliar)	Soil application at planting only. Foliar applications.
	flonicamid, MOA 9C (Beleaf) 50SG	2 to 2.8 oz	12 hrs	0	
	flupyradifurone, MOA 4D (Sivanto) 200 SL	10.5 to 12.0 fl oz	4 hrs	1	
	imidacloprid, MOA 4A Soil treatment (Admire Pro) 4.6 F (various) 2 F	4.4 to 10.5 fl oz 10 to 24 fl oz	12 hrs	21	See label for soil application instructions. Admire Pro will also control flea beetle.
	Foliar treatment (Admire Pro) 4.6 F (various) 1.6 F	3.8 fl oz	12 hrs	7	
	pymetrozine, MOA 9B (Fulfill) 50 WDG	2.75 oz	12 hrs	7	
	spirotetramat, MOA 23 (Movento) 2SC	4 to 5 oz	24 hrs	1	Do not exceed 10 fluid ounces per season. Requires surfactant.
Diamondback moth, Caterpillars, including Cabbage looper, Imported cabbageworm, Cross-striped cabbageworm, Cabbage webworm, Armyworm	Insecticide-resistant populations may not be controlled with some registered insecticides. To manage resistance, avoid transplants from Georgia and Florida, and avoid the repeated use of the same materials for extended periods of time. Use of pyrethroid insecticides destroys natural enemies and aggravates diamondback moth problems. Do not allow populations to increase to large densities before treatments are initiated.				
	Bacillus thuringiensis, MOA 11A (Crymax) WDG (Dipel) 2 X, DF (Dipel) (Xentari) WDG	0.5 to 1.5 lb 8 oz 1 pt 0.5 to 1 lb	4 hrs	0	Use a spreader/sticker. Do not apply insecticides with the same mode of action more than twice to any generation of diamondback moth. After two applications, rotate to an insecticide with a different mode of action.
	chlorantraniliprole, MOA 28 (Coragen) 1.67 SC	3.5 to 4 fl oz	4 hrs	1	Foliar or drip chemigation. Drip chemigation must be applied uniformly to the root zone. See label for instructions.
	cyclaniliprole, MOA 28 (Harvanta) 50 SL	10.9 to 16.4 fl oz	4 hrs	1	
	emamectin benzoate, MOA 6 (Proclaim) 5 WDG	2.4 to 4.8 oz	12 hrs	14	
	flubendiamide, MOA 28 (Belt) 4SC	2 to 2.4 fl oz	12 hrs	1	
	indoxacarb, MOA 22 (Avaunt eVo) 30 WDG	3.5 oz	12 hrs	3	Do not apply Avaunt eVo more than twice to any generation of diamondback moth. After 2 applications, rotate to an insecticide with a different mode of action. Do not make more than 6 applications (4 in GA), or exceed 14 ounces per season per crop.
	spinetoram, MOA 5 (Radiant) 1 SC	5 to 10 fl oz	4 hrs	1	
Flea beetle	acetamiprid, MOA 4A (Assail) 30SG	2 to 4 oz	12 hrs	7	
	carbaryl, MOA 1A (Sevin) 50 WP (Sevin) 80 S (Sevin) XLR	3 lb 1.875 lb 1 qt	12 hrs	14	

Table 5-9. Insect Control for Commercial Vegetables

CROP Insect	Insecticide, Mode of Action Code, and Formulation	Amount of Formulation Per Acre	Restricted Entry Interval (REI)	Pre harvest Interval (PHI) (Days)	Precautions and Remarks
Collard, Kale, Mustard Greens (continued)					
Flea beetle (continued)	dinotefuran, MOA 4A Foliar treatment (Venom) 70 SG (Scorpion) 35SL Soil treatment (Venom) 70 SG (Scorpion) 35SL	 1 to 4 oz 2 to 7 fl oz 5 to 6 oz 9 to 10.5 fl oz	12 hrs	 7 21	Do not follow soil applications with foliar applications. Use only 1 application method. Do not apply more than 6 ounces per acre per season using foliar applications, or 12 ounces per acre per season using soil applications. Soil applications may be applied by: a narrow band below or above the seed line at planting; a post-seeding or transplant drench with sufficient water to ensure incorporation to the root zone; or through drip irrigation.
	pyrethroid, MOA 3		12 hrs		See table 5-9B for a list of registered pyrethroids and pre-harvest intervals.
Grasshopper	pyrethroid, MOA 3		12 hrs		See table 5-9B for a list of registered pyrethroids and pre-harvest intervals. May flare diamond back moth populations.
Harlequin bug, Stink bug, Yellowmargined leaf beetle	clothianidin, MOA 4A (Belay) 50 WDG	4.8 to 6.4 oz (soil); 1.6 to 2.1 oz (foliar)	12 hrs	7 (foliar)	Soil application at planting only.
	dinotefuran, MOA 4A Foliar treatment (Venom) 70 SG (Scorpion) 35SL Soil treatment (Venom) 70 SG (Scorpion) 35SL	 1 to 4 oz 2 to 7 fl oz 5 to 6 oz 9 to 10.5 fl oz	12 hrs	 7 21	Do not follow soil applications with foliar applications. Use only one application method. Do not apply more than 6 ounces per acre per season using foliar applications, or 12 ounces per acre per season using soil applications. Soil applications may be applied by: a narrow band below or above the seed line at planting; a post-seeding or transplant drench with sufficient water to ensure incorporation to the root zone; or through drip irrigation.
	pyrethroid, MOA 3		12 hrs		See table 5-9B for a list of registered pyrethroids and pre-harvest intervals.
	thiamethoxam, MOA 4A (Actara) 25WDG	3 to 5.5 oz	12 hrs	7	
	dinotefuran MOA 4A (Venom) 70SG (Scorpion) 35SL	1 to 4 oz 2 to 7 fl oz	12 hrs	7	Dinotefuran recommendations are for foliar applications.
Root maggot	chlorpyrifos, MOA 1B (Lorsban) 4 EC (Lorsban) 75WDG	1.6 to 2.75 fl oz 1.1 to 1.8/ 1,000 ft row	24 hrs	—	For directed-seeded crops, apply as a 4-inch band over the row after planting. For transplanted crops, apply as a directed spray immediately after transplanting.
Whitefly	acetamiprid, MOA 4A (Assail) 30 SG	2.5 to 4.0 oz	12 hrs	7	Apply against adults, before nymphs are present. Use a spreader stick to improve control.
	flupyradifurone, MOA 4D (Sivanto) 200 SL	10.5 to 14.0 fl oz	4 hrs	1	Do not make more than 3 applications or apply more than 28 fluid ounces per season.
	pyriproxyfen, MOA 7C (Knack) 0.86 EC	8 to 10 fl oz	12 hrs	7	Do not apply Knack more than twice per season or exceed 0.134 pound per acre per season.
	spiromesifen, MOA 23 (Oberon) 2 SC	7 to 8.5 fl oz	12 hrs	7	Do not make more than 3 applications or apply more than 25.5 fluid ounces per season.
	spirotetramat, MOA 23 (Movento) 2 SC	4 to 5 fl oz	24 hrs	1	Do not exceed 10 fluid ounces per season. Requires surfactant.
Corn, Sweet					
Corn earworm, Fall armyworm, European corn borer	transgenic sweet corn varieties expressing *Bt* protein				Highly effective against European corn borer. Effectiveness against corn earworm will vary among BT traits; additional insecticide applications may be required to prevent damage to the ear tips of some varieties.
	pyrethroid, MOA 3		12 hrs		Check label for variety limitations and grazing restrictions. Also, instances of corn earworm resistance to pyrethroids are becoming more prevalent in recent years. To protect ears, begin sprays when tassel shoots first appear. The frequency of sprays will vary depending on location and intensity of earworm populations, ranging from daily to twice weekly in higher elevations. Corn earworms and fall armyworms present in the late whorl stage must be controlled before tassel emergence to prevent migration to ears.
	chlorantraniliprole MOA 28 (Coragen) 1.67 SC	3.5 to 5 fl oz	4 hrs	1	
	methomyl, MOA 1A (Lannate) 90 SP (Lannate) 2.4 LV	4 to 6 oz 0.75 to 1.5 pt	48 hrs	0	Do not use methomyl for European corn borer control.
	indoxacarb, MOA 22 (Avaunt eVo) 30 WDG	2.5 to 3.5 oz	12 hrs	3	For control of fall armyworm and European corn borer in whorl stage only. Do not apply more than 14 ounces Avaunt eVo (0.26 lb a.i.) per acre per crop. Minimum interval between sprays is 3 days. Make no more than 4 applications per season.

Table 5-9. Insect Control for Commercial Vegetables

CROP / Insect	Insecticide, Mode of Action Code, and Formulation	Amount of Formulation Per Acre	Restricted Entry Interval (REI)	Pre harvest Interval (PHI) (Days)	Precautions and Remarks
Corn, Sweet (continued)					
Corn earworm, Fall armyworm, European corn borer (continued)	spinetoram, MOA 5 (Radiant) 1 SC	3 to 6 fl oz	4 hrs	1	Do not apply more than 36 ounces per acre per year.
	spinosad, MOA 5 (Blackhawk)	1.7 to 3.3 oz	4 hrs		
Cutworm	pyrethroid, MOA 3		12 hrs		See table 5-9B for a list of registered pyrethroids and pre-harvest intervals.
Flea beetle, Grasshopper, Japanese beetle, Rootworm beetle	pyrethroid, MOA 3		12 hrs		See table 5-9B for a list of registered pyrethroids and pre-harvest intervals.
Sap beetle	pyrethroid, MOA 3		12 hrs		See table 5-9B for a list of registered pyrethroids and pre-harvest intervals.
	carbaryl, MOA 1A (Sevin) 50 WP (Sevin) 80 S (Sevin) XLR Plus	2 lb 1.25 lb 1 qt	12 hrs	2	Infestations usually associated with prior ear damage. Populations build on overmature and damaged fruit and vegetables. Sanitation is important.
Southern corn billbug, Rootworm, Wireworm	*Seed treatments:* clothianidin, MOA 4A (Poncho 600) imidacloprid, MOA 4A (Gaucho 600)	1.13 fl oz per 80,000 seeds 4 to 8 oz per cwt seed	—		Seed treatments are applied by commercial seed treaters only. Not for use in hopper bins, slurry mixes, or any other type of on-farm treatment.
	pyrethroid, MOA 3		12 hrs		See table 5-9B for a list of registered pyrethroids and pre-harvest intervals.
	chlorpyrifos, MOA 1B (Lorsban) 4 E	4 pt	24 hrs	0	Preplant incorporation treatment. For postemergence treatment use 2 to 3 pints.
	terbufos, MOA 1B (Counter) 15 G	Banded: 6.5 to 13 lb (40 in. row spacing) OR 8 to 16 oz/1,000 ft row In-Furrow: 6.5 lb (40 in. row) OR 8 oz/10 ft row	—	—	Place granules in a 7-inch band over the row directly behind the planter shoe in front of press wheel. Place granules directly in the seed furrow behind the planter shoe. Rotation is advised.
Stink bug	pyrethroids, MOA 3				See table 5-9B for a list of registered pyrethroids and pre-harvest intervals.
	methomyl, MOA 1A (Lannate) 90SP	0.5 lb	48 hrs	0	Re-entry interval is 48 hours.
Cucurbit Crops (Cucumber, Cantaloupe, Pumpkin, Squash, Watermelon)					
Insecticide applications in cucurbits should be made in late evening to protect pollinating insects. Refer to the section of this chapter on Reducing the Risk of Pesticide Poisoning to Honey Bees for more information about protecting pollinators.					
Aphid	Where whitefly resistance is an issue (or any other insect with a high potential for resistance to Group 4A MOA insecticides), a foliar applied Group 4A insecticide program and a soil-applied Group 4A program should not be used in the same season. Also, if using a foliar-applied program, avoid using a block of more than 3 consecutive applications of any products belonging to Group 4A insecticides.				
	acetamiprid MOA 4A (Assail) 30SG	2.5 to 4.0 oz	12 hrs	0	Do not exceed 0.5 pound per acre per season.
	clothianidin, MOA 4A (Belay) 50 WDG	4.8 to 6.4 oz (soil) 1.6 to 2.1 oz (foliar)	12 hrs	7 (foliar)	Soil application at planting only. See label for application options. Do not use an adjuvant with foliar applications.
	cyantraniliprole MOA 28 (Verimark) 1.67 SC	10 to 13.5 fl oz	4 hrs	1	Applied to the soil at planting or later via drip irrigation system. See label for application options.
	flonicamid, MOA 29 (Beleaf) 50 SG	2 to 2.8 oz	12 hrs	0	
	flupyradifurone, MOA 4D (Sivanto Prime) 200 SL Soil application Foliar application	21 to 28 fl oz 7 to 14 fl oz	4 hrs	21 1	Soil applications through drip irrigation, injected below the seed level at planting, or drench at transplanting.
	pymetrozine, MOA 9B (Fulfill) 50 WDG	2.75 oz	12 hrs	0	Apply before aphids reach damaging levels. Do not exceed 5.5 ounces per acre per season.
	thiamethoxam, MOA 4A (Platinum) 75 SG (Actara) 25WDG	1.66 to 3.67 oz 1.5 to 3 oz	12 hrs	30 0	Platinum is for soil application and may be applied to direct-seeded crops in-furrow at seed or transplant depth, post seeding or transplant as a drench, or through drip irrigation. Do not exceed 8 ounces per acre per season of Platinum. Check label for plant-back restrictions for a number of crops. Actara is for foliar application only.

Table 5-9. Insect Control for Commercial Vegetables

CROP / Insect	Insecticide, Mode of Action Code, and Formulation	Amount of Formulation Per Acre	Restricted Entry Interval (REI)	Pre harvest Interval (PHI) (Days)	Precautions and Remarks
Cucurbit Crops (Cucumber, Cantaloupe, Pumpkin, Squash, Watermelon) (continued)					
Insecticide applications in cucurbits should be made in late evening to protect pollinating insects. Refer to the section of this chapter on Reducing the Risk of Pesticide Poisoning to Honey Bees for more information about protecting pollinators.					
Armyworm, Cabbage looper	*Bacillus thuringiensis*, MOA 11A (Crymax) WDG, (Dipel) 2X (Xentari) WDG	0.5 to 1.5 lb 8 oz 0.5 to 1 lb	4 hrs	0	On foliage as needed.
	chlorantraniliprole, MOA 28 (Coragen) 1.67 SC	3.5 to 5 fl oz	4 hrs	1	Coragen may be used for foliar or drip chemigation.
	cyclaniliprole, MOA 28 (Harvanta) 50SL	10.9 to 16.4 fl oz	4 hrs	1	
	cyantraniliprole, MOA 28 (Verimark) 1,67SC	5 to 13.5 fl oz	4 hrs	1	Verimark is for soil application only. It may be applied to the soil at planting at 6.75 to 13.5 ounces, or via drip chemigation at 5 to 10 fluid ounces. Do not make more than 2 soil or chemigation applications per season. See label for application options.
	(Exirel) 0.83SE	7 to 17 fl oz	12 hrs	1	Exirel is for foliar application only. Use higher rates for cabbage looper.
	indoxacarb, MOA 22 (Avaunt eVo) 30WDG		12 hrs		
	methoxyfenozide, MOA 18 (Intrepid) 2 F	4 to 10 fl oz	4 hrs	3	Use higher rates against large larvae.
	novaluron, MOA 15 (Rimon) 0.83EC	9 to 12 fl oz	12 hrs	1	
	spinetoram, MOA 5 (Radiant) 1 SC	5 to 10 fl oz	4 hrs	3	
Cucumber beetle	acetamiprid MOA 4A (Assail) 30SG	2.5 to 5.3 oz	12 hrs	0	Do not exceed 0.5 pound per acre per season.
	carbaryl MOA 1A (Sevin) 50 WP (Sevin) 80 S (Sevin) XLR Plus	2 lb 1.25 lb 1 qt	12 hrs	3	
	clothianidin, MOA 4A (Belay) 50 WDG	4.8 to 6.4 oz (soil) 1.6 to 2.1 oz (foliar)	12 hrs	21 (foliar)	Soil application at planting only. Do not use an adjuvant with foliar applications.
	dinotefuran, MOA 4A Foliar treatment (Venom) 70 SG (Scorpion) 35SL Soil treatment (Venom) 70 SG (Scorpion) 35SL	1 to 4 oz 2 to 7 fl oz 5 to 6 oz 9 to 10.5 fl oz	12 hrs	1 21	Do not make both a soil and foliar application, use one or the other. At planting applications are most effective against cucumber beetle. Will also control whiteflies and squash bug.
	imidacloprid, MOA 4A (Admire Pro) 4.6 F (various) 2F	7 to 10.5 fl oz 16 to 24 fl oz	12 hrs	21	Must be applied to the soil. See label for information on approved application methods. Will also control aphids and whiteflies.
	pyrethroid, MOA 3		12 hrs		See table 5.9B for a list of registered pyrethroids and pre-harvest intervals.
Leafminer	abamectin, MOA 6 (Agri-mek) 0.7 SC	1.75 to 3.5 fl oz	12 hrs	7	Do not use more than 6 applications per season.
	cyromazine, MOA 17 (Trigard) 75 WS	2.7 oz	12 hrs	0	
	chlorantraniliprole, MOA 28 (Coragen) 1.67 SC	2 to 3.5 fl oz	4 hrs	1	For foliar or drip chemigation. Drip chemigation must be applied uniformly to the root zone. See label for instructions.
	cyclaniliprole, MOA 28 (Harvanta) 50SL	10.9 to 16.4 fl oz	4 hrs	1	
	spinetoram, MOA 5 (Radiant) 1 SC	5 to 10 fl oz	4 hrs	3	
Pickleworm, Melonworm	pyrethroid, MOA 3		12 hrs		See table 5-9B for a list of registered pyrethroids and pre-harvest intervals.
	carbaryl, MOA 1A (Sevin) 50 WP (Sevin) 80 S (Sevin) XLR Plus	2 lb 1.25 lb 1 qt	12 hrs	3	On foliage when worms appear in blossoms. Repeat as needed. Protect pollinators. Rarely a problem before July.
	chlorantraniliprole, MOA 28 (Coragen) 1.67 SC	2 to 3.5 fl oz	4 hrs	1	For foliar or drip chemigation. Drip chemigation must be applied uniformly to the root zone. See label for instructions.

Table 5-9. Insect Control for Commercial Vegetables

CROP / Insect	Insecticide, Mode of Action Code, and Formulation	Amount of Formulation Per Acre	Restricted Entry Interval (REI)	Pre harvest Interval (PHI) (Days)	Precautions and Remarks
Cucurbit Crops (Cucumber, Cantaloupe, Pumpkin, Squash, Watermelon) (continued)					
Insecticide applications in cucurbits should be made in late evening to protect pollinating insects. Refer to the section of this chapter on Reducing the Risk of Pesticide Poisoning to Honey Bees for more information about protecting pollinators.					
	cyantraniliprole, MOA 28 (Verimark) 1,67SC	5 to 13.5 fl oz	4 hrs	1	Verimark is for soil application only. It may be applied to the soil at planting at 6.75 to 13.5 ounces, or via drip chemigation at 5 to 10 fluid ounces. Do not make more than two soil or chemigation applications per season. See label for application options. Exirel is for foliar application only.
	(Exirel) 0.83SE	7 to 13.5 fl oz	12 hrs	1	
	cyclaniliprole, MOA 28 (Harvanta) 50SL	10.9 to 16.4 fl oz	4 hrs	1	
	methoxyfenozide, MOA 18 (Intrepid) 2 F	4 to 10 fl oz	4 hrs	3	
	spinetoram, MOA 5 (Radiant) 1 SC	5 to 10 fl oz	4 hrs	3	
Spider mite	abamectin, MOA 6 (Agri-mek) 0.7 SC	1.75 to 3.4 fl oz	12 hrs	7	
	bifenazate, MOA 20D (Acramite) 50 WS	0.75 to 1.0 lb	12 hrs	3	Do not make more than 1 application per season.
	etoxazole, MOA 10B (Zeal) 72 WSP	2 to 3 oz	12 hrs	7	Does not kill adults
	fenpyroximate MOA 21 (Portal) 0.4EC	2 pt	12 hrs	3	Fenpyroximate is only registered on cucumber, not other cucurbits. Do not make more than 2 applications per season.
	spiromesifen, MOA 23 (Oberon) 2 SC	7 to 8.5 fl oz	12 hrs	7	
Squash bug	Squash bug is a common pest of cantaloupe, pumpkin and squash. Although cucumber and watermelon are occasionally reported as hosts of squash bug, rarely do infestations occur.				
	acetamiprid, MOA 4A (Assail) 30 SG	5.3 oz	12 hrs	0	Assail is most effective against newly laid eggs and nymphs.
	clothianidin, MOA 4A (Belay) 50SDG	4.8 to 6.8oz (soil); 1.6 to 2.1oz (foliar)	12 hrs	At planting / 7	See application instructions and precautionary bee statement above under aphid.
	dinotefuran, MOA 4A (Venom) 70 SG (Scorpion) 35 SL	3 to 4 oz / 2 to 7 fl oz	12 hrs	1	Do not exceed 6 ounces Venom per acre per season.
	pyrethroid, MOA 3		12 hrs		See table 5-9B for a list of registered pyrethroids and pre-harvest intervals.
Squash vine borer	Squash vine borer only attacks squash and pumpkin, and is more common in home gardens as opposed to commercial plantings.				
	acetamiprid, MOA 4A (Assail) 30 SG	5.3 oz	12 hrs	0	
	chlorantraniliprole, MOA 28 (Coragen) 1.67 SC	3.5 to 5 fl oz	4 hrs	1	Foliar or drip chemigation. Drip chemigation must be applied uniformly to the root zone. See label for instructions.
	pyrethroid, MOA 3		12 hrs		See table 5-9B for a list of registered pyrethroids and pre-harvest intervals.
Thrips	methomyl, MOA 1A (Lannate) 2.4 LV (Lannate) 90 SP	0.75 to 1.5 pt 0.25 to 0.5 lb	48 hrs	0	
	spinetoram, MOA 5 (Radiant) 1 SC	6 to 10 fl oz	4 hrs	3	
Whitefly	acetamiprid, MOA 4A (Assail)	1.1 to 2.3 oz	12 hrs	0	
	buprofezin, MOA 16 (Courier) 40 SC	9 to 13.6 oz	12 hrs	7	Use sufficient water to ensure good coverage. Do not apply more than twice per crop cycle.
	chlorantraniliprole MOA 28 (Coragen) 1.67 SC	5 to 7.5 fl oz	4 hrs	1	May be applied foliar or through drip irrigation. Drip chemigation must be applied uniformly to the root zone.
	cyantraniliprole, MOA 28 (Verimark) 1,67SC	10 fl oz	4 hrs	1	Verimark is for soil application only. It may be applied to the soil at planting at 6.75 to 13.5 ounces, or via drip chemigation at 5 to 10 fluid ounces. See label for application options. Exirel is for foliar application only. Use an adjuvant for best results.
	(Exirel) 0.83SE	13.5 to 20.5 fl oz	12 hrs	1	
	dinotefuran, MOA 4A Foliar treatment (Venom) 70 SG (Scorpion) 35SL Soil treatment (Venom) 70 SG (Scorpion) 35SL	1 to 4 oz 2 to 7 fl oz 5 to 6 oz 9 to 10.5 oz	12 hrs	1 21	Do not follow soil applications with foliar applications. Use only 1 application method. Do not apply more than 6 ounces per acre per season using foliar applications, or 12 ounces per acre per season using soil applications. Soil applications may be applied by: a narrow band below or above the seed line at planting; a post-seeding or transplant drench with sufficient water to ensure incorporation to the root zone; or through drip irrigation.
	flupyradifurone, MOA 4D (Sivanto Prime) 200 SL	10.5 to 14.0 fl oz	4 hrs	1	

Table 5-9. Insect Control for Commercial Vegetables

CROP Insect	Insecticide, Mode of Action Code, and Formulation	Amount of Formulation Per Acre	Restricted Entry Interval (REI)	Pre harvest Interval (PHI) (Days)	Precautions and Remarks
Cucurbit Crops (Cucumber, Cantaloupe, Pumpkin, Squash, Watermelon) (continued)					
Insecticide applications in cucurbits should be made in late evening to protect pollinating insects. Refer to the section of this chapter on Reducing the Risk of Pesticide Poisoning to Honey Bees for more information about protecting pollinators.					
Whitefly (continued)	imidacloprid, MOA 4A (Admire Pro) 4.6 F (various) 2 F	7 to 10.5 oz 16 to 24 fl oz	12 hrs	21	Must be applied to the soil. May be applied preplant; at planting; as a post-seeding drench or hill drench; subsurface sidedress; or by chemigation using low pressure drip or trickle irrigation. See label for information on approved application methods. Will also control aphids and cucumber beetles.
	pyriproxyfen, MOA 7C (Knack) 0.86 EC	8 to 10 oz	12 hrs	7	Do not make more than 2 applications per season, and do not make applications closer than 14 days apart.
	spiromesifen, MOA 23 (Oberon) 2 SG	7 to 8.5 fl oz	12 hrs	7	Apply against adults, before nymphs are present. Do not exceed 3 applications per season.
	thiamethoxam, MOA 4A (Platinum) 75 SG	1.66 to 3.67 fl oz	12 hrs	30	Platinum is for soil application and may be applied to direct-seeded crops in-furrow at seed or transplant depth, postseeding or transplant as a drench, or through drip irrigation. Do not exceed 11 ounces per acre per season of Platinum. Check label for plant-back restrictions for a number of crops.
	(Actara) 25WDG	3 to 5.5 oz		0	Actara is for foliar application.
Wireworm	diazinon, MOA 1B (Diazinon) AG 500	3 to 4 qt	3 days	—	Broadcast on soil before planting and thoroughly work into upper 6 inches.
Eggplant					
Aphid	Where whitefly resistance is an issue (or any other insect with a high potential for resistance to Group 4A MOA insecticides), avoid making foliar applications of Group 4A insecticides when a soil-applied Group 4A program is used; i.e., do not make both foliar and soil applications of Group 4A insecticides. Also, if using a foliar-applied program, avoid using a block of more than 3 consecutive applications of any products belonging to Group 4A insecticides.				
	acetamiprid, MOA 4A (Assail) 30 SG	2 to 4 oz	12 hrs	7	Thoroughly cover foliage to effectively control aphids. Do not apply more than once every 7 days, and do not exceed a total of 7 ounces per season.
	flonicamid, MOA 29 (Beleaf) 50 SG	2 to 4.8 oz	12 hrs	0	
	flupyradifurone, MOA 4D (Sivanto) 200 SL	7.0 to 12.0 fl oz	4 hrs	1	
	imidacloprid, MOA 4A Soil treatment (Admire Pro) 4.6 F (various) 2 F	7 to 10.5 oz 16 to 24 fl oz	12 hrs	21	See label for soil application instructions. For short-term protection of transplants at planting, apply Admire Pro (0.44 ounces per 10,000 plants) not more than 7 days before transplanting by 1) uniformly spraying on transplants, followed immediately by sufficient overhead irrigation to wash product into potting media; or 2) injection into overhead irrigation system with adequate volume to thoroughly saturate soil media.
	Foliar treatment (Admire Pro) 4.6 F (various) 1.6 F	1.3 to 2.2 fl oz 3.75 fl oz	12 hrs	0	
	pymetrozine, MOA 9B (Fulfill) 50 WDG	2.75 oz	12 hrs	14	Apply before aphids reach damaging levels. Do not exceed 5.5 ounces per acre per season.
	spirotetramat, MOA 23 (Movento) 2 SC	4 to 5 fl oz	24 hrs	1	Do not exceed 10 fluid ounces per season. Requires surfactant.
	thiamethoxam, MOA 4A Soil treatment (Platinum) 75 SG	1.66 to 3.67 oz	12 hrs	30	Platinum may be applied to direct-seeded crops in-furrow at seed or transplant depth, postseeding or transplant as a drench, or through drip irrigation. Do not exceed 8 ounces per acre per season. Check label for plant-back restrictions for a number of plants.
	Foliar treatment (Actara) 25 WDG	2 to 3 oz	12 hrs	0	Actara is for foliar application.
Blister beetle	pyrethroid, MOA 3		12 hrs		See table 5-9B for a list of registered pyrethroids and pre-harvest intervals.
Colorado potato beetle	Resistance to many insecticides is widespread in Colorado potato beetle. To reduce risk of resistance, scout fields and apply insecticides only when needed to prevent damage to the crop. Crop rotation will help prevent damaging Colorado potato beetle infestations. If control failures or reduced levels of control occur with a particular insecticide, do NOT make a second application of the same insecticide at the same or higher rate. If an additional insecticide application is necessary, a different insecticide representing a different MOA class should be used. Do NOT use insecticides belonging to the same class 2 years in a row for Colorado potato beetle control.				
	abamectin, MOA 6 (Agri-Mek) 0.7 SC	1.75 to 3.5 fl oz	12 hrs	7	Apply when adults and small larvae are present but before large larvae appear. For resistance management, use the higher rate.
	acetamiprid, MOA 4A (Assail) 30 SG	2 to 4 oz	12 hrs	7	Do not apply more than once every 7 days, and do not exceed 7 ounces of formulation per season.
	pyrethroid, MOA 3		12 hrs		See table 5-9B for a list of registered pyrethroids and pre-harvest intervals.
	chlorantraniliprole, MOA 28 (Coragen) 1.67 SC	3.5 to 5 fl oz	4 hrs	1	Foliar or drip chemigation. Drip chemigation must be applied uniformly to the root zone. See label for instructions.
	cyclaniliprole, MOA 28 (Harvanta) 50SL	10.9 to 16.4 fl oz	4 hrs	1	

Table 5-9. Insect Control for Commercial Vegetables

CROP / Insect	Insecticide, Mode of Action Code, and Formulation	Amount of Formulation Per Acre	Restricted Entry Interval (REI)	Pre harvest Interval (PHI) (Days)	Precautions and Remarks
Eggplant (continued)					
Colorado potato beetle (continued)	dinotefuran, MOA 4A Foliar treatment (Venom) 70 SG (Scorpion) 35SL Soil treatment (Venom) 70 SG (Scorpion) 35SL	 1 to 4 oz 2 to 7 fl oz 5 to 6 oz 9 to 10.5 fl oz	12 hrs	 1 21	Do not follow soil applications with foliar applications on any neonicotinoid insecticide. Use only 1 application method. Do not apply more than 6 ounces per acre per season using foliar applications, or 12 ounces per acre per season using soil applications. Soil application may be applied by: 1) a narrow band below or above the seed line at planting; 2) a post-seeding or transplant drench with sufficient water to ensure incorporation to the root zone; or 3) drip irrigation.
	imidacloprid, MOA 4A Soil treatment (Admire Pro) 4.6 F (various) 2 F	 7 to 10.5 fl oz 16 to 24 fl oz	12 hrs	 21	See application methods under Aphids, Thrips.
	Foliar treatment (Admire Pro) 4.6 F (various) 1.6 F	 1.3 fl oz 3.75 fl oz	12 hrs	 0	
	novaluron, MOA 15 (Rimon) 0.83 EC	9 to 12 fl oz	12 hrs	1	
	spinetoram, MOA 5 (Radiant) 1 SC	5 to 10 fl oz	4 hrs	1	
	sulfoxaflor (Closer) 2 SC	1.5 to 2.0 fl oz	12 hrs	1	
	thiamethoxam, MOA 4A (Platinum) 75 SG	1.66 to 3.67 oz	12 hrs	30	See application methods under Aphids.
	(Actara) 25 WDG	2 to 3 oz	12 hrs	0	
Eggplant lace bug	imidacloprid, MOA 4A Foliar treatment (Admire Pro) 4.6 F (various) 1.6 F	 1.3 to 2.2 fl oz 3.8 to 6.2 fl oz	12 hrs	 0	
	malathion, MOA 1B (various brands) 57 EC	3 pt	12 hrs	3	
Flea beetle	pyrethroid, MOA 3		12 hrs		See table 5-9B for a list of registered pyrethroids and pre-harvest intervals.
	carbaryl, MOA 1A (Sevin) 50 WP (Sevin) 80 S (Sevin) XLR Plus	 2 lb 1.25 lb 1 lb	12 hrs	3	
	clothianidin, MOA 4A (Belay) 50WDG	4.6 to 6.8 oz (soil); 1.6 to 2.1 fl oz (foliar)	12 hrs	7 (foliar)	Soil application at planting only.
	cyantraniliprole, MOA 28 (Verimark) 1.67SC	6.75 to 13.5 fl oz	4 hrs	1	Verimark for soil application only. Apply at planting or via drip chemigation. See label for application options.
	dinotefuran, MOA 4A Foliar treatment (Venom) 70 SG (Scorpion) 35SL Soil treatment (Venom) 70 SG (Scorpion) 35SL	 1 to 4 oz 2 to 7 fl oz 5 to 6 oz 9 to 10.5 fl oz	12 hrs	 1 21	Do not follow soil applications with foliar applications on any neonicotinoid insecticide. Use only 1 application method. Do not apply more than 6 ounces per acre per season using foliar applications, or 12 ounces per acre per season using soil applications. Soil application may be applied by: 1) a narrow band below or above the seed line at planting; 2) a post-seeding or transplant drench with sufficient water to ensure incorporation to the root zone; or 3) drip irrigation.
	thiamethoxam, MOA 4A (Platinum) 75 SG	1.66 to 3.67 oz	12 hrs	30	See application methods under Aphids.
	(Actara) 25 WDG	2 to 3 oz	12 hrs	0	
Hornworm, European corn borer, Beet army worm, Corn earworm	chlorantraniliprole, MOA 28 (Coragen) 1.67 SC	3.5 to 4 fl oz	4 hrs	1	Foliar or drip chemigation. Drip chemigation must be applied uniformly to the root zone. See label for instructions.
	cyantraniliprole, MOA 28 (Verimark) 1.67SC (Exirel) 0.83SE	 5 to 10 fl oz 7 to 13.5 fl oz	 4 hrs 12 hrs	 1 1	Verimark is for soil application only. Applications made at planting and/or via drip chemigation. See label for application options. Exirel is for foliar application only.
	cyclaniliprole, MOA 28 (Harvanta) 50SL	10.9 to 16.4 fl oz	4 hrs	1	
	indoxacarb, MOA 22 (Avaunt eVo) 30 WDG	2.5 to 3.5 oz	12 hrs	3	Do not apply more than 14 ounces per acre per season.
	methomyl, MOA 1A (Lannate) 2.4 LV	1.5 to 3 pt	48 hrs	5	

Table 5-9. Insect Control for Commercial Vegetables

CROP / Insect	Insecticide, Mode of Action Code, and Formulation	Amount of Formulation Per Acre	Restricted Entry Interval (REI)	Pre harvest Interval (PHI) (Days)	Precautions and Remarks
Eggplant (continued)					
Hornworm, European corn borer, Beet army worm, Corn earworm (continued)	methoxyfenozide, MOA 18 (Intrepid) 2 F	4 to 16 fl oz	4 hrs	1	Apply at rates of 4 to 8 fluid ounces early in season when plants are small. Apply at rates of 8 to 16 ounces to large plants or when infestations are heavy. During periods of continuous moth flights, retreatments at 7 to 14 days may be required. Do not apply more than 16 fluid ounces per application or 64 fluid ounces of Intrepid 2F per acre per season.
	spinetoram, MOA 5 (Radiant) 1 SC	5 to 10 fl oz	4 hrs	1	
	pyrethroid, MOA 3		12 hrs		See table 5-9B for a list of registered pyrethroids and pre-harvest intervals.
Leafminer	abamectin, MOA 6 (Agri-Mek) 0.15 EC	8 to 16 fl oz	12 hrs	7	Use low rates for low to moderate infestations, and high rates for severe infestations
	chlorantraniliprole, MOA 28 (Coragen) 1.67 SC	5 to 7.5 fl oz	4 hrs	1	Foliar, soil, or drip chemigation. Drip chemigation must be applied uniformly to the root zone. See label for application instructions.
	oxamyl, MOA 1A (Vydate) 2 L	1 to 2 qt	48 hrs	7	
	spinetoram, MOA 5 (Radiant) 1 SC	5 to 10 fl oz	4 hrs	1	
Stink bug, leaffooted bug	dinotefuran, MOA 4A Foliar treatment (Venom) 70 SG (Scorpion) 35SL Soil treatment (Venom) 70 SG (Scorpion) 35SL	1 to 4 oz 2 to 7 fl oz 5 to 6 oz 9 to 10.5 fl oz	12 hrs	1 21	
	pyrethroid MOA 3		12 hrs		See table 5-9B for a list of registered pyrethroids and preharvest intervals.
	thiamethoxam, MOA 4A (Actara) 25 WDG	3 to 5.5 oz	12 hrs	0	Do not exceed 11 ounces Actara per acre per season.
Spider mite	abamectin, MOA 6 (Agri-Mek) 0.7 SC	1.75 to 3.5 fl oz	12 hrs	7	Use low rates for low to moderate infestations, and high rates for severe infestations.
	acequinocyl, MOA 20B (Kanemite) 15SC	31 fl oz	12 hrs	1	
	bifenazate, MOA 20D (Acramite) 50 WS	0.75 to 1.0 lb	12 hrs	3	Do not make more than 1 application per season.
	etoxazole, MOA 10B (Zeal)	2 to 3 oz	12 hrs	7	Do not make more than 1 Zeal application per season.
	fenpyroximate MOA 21 (Portal) 0.4EC	2 pts	12 hrs	3	Do not make more than 2 applications per season.
	hexakis, MOA 12B (Vendex) 50 WP	2 to 3 lb	48 hrs	3	
	spiromesifen, MOA 23 (Oberon) 2 SG	7 to 8.5 fl oz	12 hrs	7	
Thrips	dinotefuran, MOA 4A Foliar treatment (Venom) 70 SG (Scorpion) 35SL Soil treatment (Venom) 70 SG (Scorpion) 35SL	1 to 4 oz 2 to 7 fl oz 5 to 6 oz 9 to 10.5 fl oz	12 hrs	1 21	See Whitefly for application instructions. Soil applications are more effective against thrips than foliar applications.
	cyantraniliprole, MOA 28 (Verimark) 1.67SC	5 to 10 fl oz	4 hrs	1	Soil applications of Verimark will suppress western flower thrips. Foliar applications of Exirel are less effective.
	cyclaniliprole, MOA 28 (Harvanta) 50SL	10.9 to 16.4 fl oz	4 hrs	1	Foliar applications will help suppress western flower thrips when used in a rotational program.
	imidacloprid, MOA 4A Admire Pro 4.6 F (various) 2 F	7 to 10.5 fl oz 16 to 24 fl oz	12 hrs	21	See Aphids for application instructions.
	methomyl, MOA 1A (Lannate) 2.4 LV	1.5 to 3 pt	48 hrs	3	
	spinetoram, MOA 5 (Radiant) 1 SC	6 to 10 fl oz	4 hrs	1	
Whitefly	acetamiprid, MOA 4A (Assail) 30 SG	2.5 to 4 oz	12 hrs	7	Begin applications when significant populations of adults appear. Do not wait until heavy populations have become established. Do not apply more than once every 7 days, and do not exceed 4 applications per season. Do not apply more than 7 ounces per season.

Table 5-9. Insect Control for Commercial Vegetables

CROP / Insect	Insecticide, Mode of Action Code, and Formulation	Amount of Formulation Per Acre	Restricted Entry Interval (REI)	Pre harvest Interval (PHI) (Days)	Precautions and Remarks
Eggplant (continued)					
Whitefly (continued)	chlorantraniliprole, MOA 28 (Coragen) 1.67 SC	5 to 7.5 fl oz	12 hrs	1	For foliar or drip chemigation. Drip chemigation must be applied uniformly to the root zone. See label for instructions.
	cyantraniliprole, MOA 28 (Verimark) 1.67SC	6.75 to 13.5 fl oz	4 hrs	1	Verimark for soil application only. Apply at planting or via drip chemigation. See label for application options.
	(Exirel) 0.83SE	13.5 to 20.5 fl oz	12 hrs	1	Exirel for foliar application only.
	dinotefuran, MOA 4A Foliar treatment (Venom) 70 SG (Scorpion) 35SL	1 to 4 oz 2 to 7 fl oz	12 hrs	1	Use only 1 application method (foliar or soil) of Group 4A insecticides. Soil applications may be applied in a narrow band on the plant row in bedding operations, as a post-seeding or transplant drench, as a side-dress after planting and incorporated 1 or more inches, or through a drip irrigation system.
	Soil treatment (Venom) 70 SG (Scorpion) 35SL	5 to 6 oz 9 to 10.5 fl oz		21	
	flupyradifurone, MOA 4D (Sivanto) 200 SL	10.5 to 14.0 fl oz	4 hrs	1	
	imidacloprid, MOA 4A (Admire Pro) 4.6 F (various) 2 F	7 to 10.5 fl oz 16 to 24 fl oz	12 hrs	21	Do not follow soil applications with applications of other neonicotinoid insecticides (Assail or Venom). See Aphids for application methods and restrictions.
	pyriproxyfen, MOA 7C (Knack) 0.86 EC	8 to 10 fl oz	12 hrs	14	Knack prevents eggs from hatching. It does not kill whitefly adults. Applications should begin when 3 to 5 adults per leaf are present. Do not make more than 2 applications per season, and do not apply a second application within 14 days of the first application. Do not exceed 20 fluid ounces of Knack per acre per season. Check label for plant-back restrictions.
	spirotetramat, MOA 23 (Movento) 2SC	4 to 5 fl oz	24 hrs	1	Do not exceed 10 fl oz per season. Requires surfactant.
	spiromesifen, MOA 23 (Oberon) 2 SC	7 to 8.5 fl oz	12 hrs	7	Do not exceed 3 applications or 25.5 fluid ounces per season.
	thiamethoxam, MOA 4A (Platinum) 75 SG	1.66 to 3.67 oz	12 hrs	30	Platinum is for soil applications and may be applied to direct-seeded crops in furrow at seed or transplant depth, at postseeding or transplant as a drench, or through drip irrigation. Do not exceed 11 ounces per acre per season. Check label for plant-back restrictions for a number of plants.
	(Actara) 25WDG	3 to 5.5		0	Actara is for foliar application.
Hops					
Aphids and leafhoppers	imidacloprid, MOA 4A (Admire) 4.6 F (generics) 2	2.8 fl oz 6.4 fl oz	12 hrs	12 hrs	
	pymetrozine, MOA 9B (Fulfill) 50 WDG	4 to 6 oz	12 hrs	14	For aphids only. Will not control leafhoppers.
	spirotetramat, MOA 23 (Movento) 2 F	5 to 6 fl oz	24 hrs	7	Do not exceed 12.5 fl oz per acre per season. Will also control twospotted spider mite.
	malathion, MOA 1B 5 EC 8 EC	1 pt 0.63 pt	12 hrs 12 hrs	10 10	May suppress twospotted spider mite.
	pyrethrins, MOA 3 (Pyganic) 1.4 EC (Pyganic) 5 EC	16 to 64 fl oz 4.5 to 17 fl oz	12 hrs 12 hrs	0 0	**OMRI approved.** Pyrethrins degrade very quickly in sunlight. Do not expect residual control.
Japanese beetle	bifenthrin, MOA 3 (Brigade) 2 EC (Brigade) WD	3.8 to 6.4 fl oz 9.6 to 16 of oz	12 hrs 12 hrs	14 14	See Table of Generic Insecticides for other bifenthrin products.
	imidacloprid, MOA 4A (Admire) 4.6 F (generics) 2	2.8 fl oz 6.4 fl oz	12 hrs 12 hrs	28 28	
Armyworms, cutworms, loopers, leafroller	*Bacillus thuringiensis*, MOA 11A (Dipel) DF, MOA (Crymax) WDG	0.5 to 1 lb 0.5 to 1.5 lb	4 hrs 4 hrs	0 0	
	bifenthrin, MOA 3 (Brigade) 2 EC (Brigade) WD	3.8 to 6.4 fl oz 9.6 to 16 of oz	12 hrs 12 hrs	14 14	See Table of Generic Insecticides for other bifenthrin products.
	chlorantraniliprole, MOA 28 (Coragen) 1.67 SC	3.5 to 5 fl oz	4 hrs	0	Foliar or drip chemigation. Drip chemigation must be applied uniformly to the root zone. See label for instructions.

Table 5-9. Insect Control for Commercial Vegetables

CROP / Insect	Insecticide, Mode of Action Code, and Formulation	Amount of Formulation Per Acre	Restricted Entry Interval (REI)	Pre harvest Interval (PHI) (Days)	Precautions and Remarks
Hops (continued)					
Armyworms, cutworms, loopers, leafroller (continued)	spinosad, MOA 5 (Entrust) SC	4 to 6 fl oz	4 hrs	1	OMRI approved.
	spinetoram, MOA 5 (Delegate) 25WG	2.5 to 4 oz	4 hrs	1	
Spider mites	abamectin, MOA 6 (Agri-Mek) 0.7 SC	1.75 to 3.5 fl oz	12 hrs	7	Do not exceed 48 fluid ounces per acre per season, or more than 2 sequential applications.
	acequinocyl, MOA 20B (Kanemite) 15 SC	31 fl oz	12 hrs	1	The use of a surfactant/adjuvant with Kanemite on tomatoes is prohibited.
	Bifenazate, MOA 20D (Acramite) 50 WS	0.75 to 1.0 lb	12 hrs	3	Do not make more than 1 application per season.
	etoxazole, MOA 10B (Zeal) 72 WSP	3 to 4 oz	12 hrs	7	Apply when mites are low, because Zeal is primarily an ovicide/larvicide.
	fenpyroximate MOA 21 (Portal) 0.4EC	2 pts	12 hrs	3	Do not make more than 2 applications per season.
	hexythiazox, MOA 10A (Savvy) 50 DF	4 to 6 oz	12 hrs	—	May be applied up to burr formation in hop vines. Apply when mites are low, because Savvy is primarily an ovicide, and also sterilizes females.
	Mineral Oil (TriTek) Various brands	1 to 2% soln	4 hrs		**OMRI approved.** TriTek is the only emulsified formulation of oil. All others do not contain an emulsifier
Lettuce					
Aphid	acetamiprid, MOA 4A (Assail) 30 SG	2 to 4 oz	12 hrs	7	Do not apply more than once every 7 days, and do not exceed 4 applications per season.
	clothianidin, MOA 4A (Belay) 2.13 SC	4.8 to 6.8 oz (soil); 1.6 to 2.1 oz (foliar)	12 hrs	7 (foliar)	Soil application at planting only.
	dimethoate 4 EC, MOA 1B	0.5 pt	48 hrs	14	
	flonicamid, MOA 29 (Beleaf) 50 SG	2 to 2.8 oz	12 hrs	0	
	flupyradifurone, MOA 4D (Sivanto) 200 SL	10.5 to 12.0 fl oz	4 hrs	1	
	imidacloprid, MOA 4A Soil treatment (Admire Pro) 4.6 F (various) 2 F	4.4 to 10.5 fl oz 10 to 24 fl oz	12 hrs	21	Do not follow soil applications with foliar applications of any neonicotinoid insecticide. See label for soil application instructions.
	Foliar treatment (Admire Pro) 4.6 F (various) 1.6 F	1.3 fl oz 3.8 fl oz	12 hrs	7	
	pymetrozine, MOA 9B (Fulfill) 50 WDG	2.75 oz	12 hrs	7	Apply before aphids reach damaging levels. Do not exceed 5.5 ounces per acre per season.
	spirotetramat, MOA 23 (Movento) 2SC	4 to 5 fl oz	24 hrs	3	Do not exceed 10 fluid ounces per season. Requires surfactant.
	thiamethoxam, MOA 4A (Platinum) 75 SG	1.66 to 3.67 oz	12 hrs	30	Do not follow applications of Platinum with foliar applications of any neonicotinoid insecticide. Platinum may be applied to direct-seeded crops in-furrow at the seeding or transplant depth, or as a narrow surface band above the seedling and followed by irrigation. Post seeding, it may be applied as a transplant or through drip irrigation. Actara is applied as a foliar spray.
	(Actara) 25 WDG	1.5 to 3 oz	12 hrs	7	
	tolfenpyrad, MOA 21A (Torac) 1.29 EC	17 to 21 fl oz	12 hrs	1	Do not apply until at least 14 days after plant emergence or after transplanting to allow time for root establishment.
Armyworm, Cabbage looper, Corn earworm	*Bacillus thuringiensis*, MOA 11A (Crymax) WDG (Dipel) DF	0.5 to 1.5 lb 8 oz	4 hrs	0	Only target small armyworms with Bts.
	chlorantraniliprole, MOA 28 (Coragen) 1.67 SC	3.5 to 5 fl oz	4 hrs	1	Foliar or drip chemigation.
	cyantraniliprole, MOA 28 (Verimark) 1.67SC	5 to 13.5 fl oz	4 hrs	1	Verimark is for soil application only. Applications made at planting and/or via drip chemigation. Use higher rates (>10 fluid ounces) where cabbage looper is a concern. See label for application options. Exirel is for foliar application only. Use higher rates (>13.5 fluid ounces for cabbage looper.
	(Exirel) 0.83SE	7 to 17 fl oz	12 hrs	1	
	emamectin benzoate, MOA 6 (Proclaim) 5 WDG	3.2 to 4.8 oz	12 hrs	7	Do not make more than 2 sequential applications without rotating to another product with a different mode of action.

Table 5-9. Insect Control for Commercial Vegetables

CROP / Insect	Insecticide, Mode of Action Code, and Formulation	Amount of Formulation Per Acre	Restricted Entry Interval (REI)	Pre harvest Interval (PHI) (Days)	Precautions and Remarks
Lettuce (continued)					
Armyworm, Cabbage looper, Corn earworm (continued)	indoxacarb, MOA 22 (Avaunt eVo) 30 WDG	2.5 to 3.5 oz	12 hrs	3	Do not apply more than 14 ounces of Avaunt eVo (0.26 lb a.i.) per acre per crop. The minimum interval between sprays is 3 days.
	methoxyfenozide, MOA 18 (Intrepid) 2 F	4 to 10 oz	4 hrs	1	Low rates for early-season applications to young or small plants. For mid- and late-season applications, use 6 to 10 ounces.
	Pyrethroid, MOA 3		12 hrs		See table 5-9B for registered pyrethroids and pre-harvest intervals.
	spinetoram, MOA 5 (Radiant) 1 SC	5 to 10 fl oz	4 hrs	1	
Leafhopper	dinotefuran, MOA 4A (Venom) 70 SG	1 to 3 oz (foliar) 5 to 6 oz (soil)	12 hrs	7 21	Do not follow soil applications with foliar applications of any neonicotinoid insecticide. Use only 1 application method. Do not apply more than 6 ounces per acre (foliar) or 12 ounces per acre (soil). Soil applications may be applied by: 1. Narrow band below or above the seed line at planting; 2. post seeding or transplant drench with sufficient water to ensure incorporation; or 3. drip irrigation.
	dimethoate 4 EC, MOA 1B	0.5 pt	48 hrs	14	14-day interval for leaf lettuce.
	flupyradifurone, MOA 4D (Sivanto) 200 SL	7.0 to 10.5 fl oz	4 hrs	1	
	imidacloprid, MOA 4A (various) 1.6 F	3.75 fl oz	12 hrs	7	There is a 12-month plant-back restriction for a number of crops. Check label for restrictions.
	pyrethroid, MOA 3		12 hrs		See table 5-9B for registered pyrethroids and pre-harvest intervals.
	thiamethoxam, MOA 4A (Actara) 25 WDG	1.5 to 3 oz	12 hrs	7	
	tolfenpyrad, MOA 21A (Torac) 1.29 EC	14 to 21 fl oz	12 hrs	1	Do not apply until at least 14 days after plant emergence or after transplanting to allow time for root establishment.
Slugs	iron phosphate (Sluggo)	20 to 44 lbs	0 hrs	0	OMRI approved. Sluggo should be scattered around the perimeter of the crop to provide a protective barrier for slugs and snails. If slugs are inside the rows, scatter the bait on the soil around the plants and between rows. For smaller plantings use at 0.5 to 1 lb 1,000 square feet.
Melon (See Cucurbit Crops)					
Mustard Greens (See Collard, Kale, Mustard Greens)					
Okra					
Aphid	imidacloprid, MOA 4A Soil treatment (Admire Pro) 4.6 F (various) 2 F	7 to 10.5 fl oz 16 to 24 fl oz	12 hrs	21	See label for soil treatment instructions.
	Foliar treatment (Admire Pro) 4.6 F (various) 1.6 F	1.3 to 2.2 fl oz 3.8 fl oz	12 hrs	0	
	flonicamid, MOA 29 (Beleaf) 50 SG	2 to 2.8 oz	12 hrs	0	
	flupyradifurone, MOA 4D (Sivanto) 200 SL	7.0 to 12 fl oz	4 hrs	1	
	malathion, MOA 1B (various brands) 8 F (various brands) 25 WP	1.5 pt 6 lb	12 hrs	1	
	spirotetramat, MOA 23 (Movento) 2SC	4 to 5 fl oz	24 hrs	3	Do not exceed 10 fluid ounces per season. Not for flea beetle. Requires surfactant.
	sulfoxaflor (Closer) 2 SC	1.5 to 2.0 fl oz	12 hrs	7	
Blister beetle, Flea beetle, Japanese beetle	carbaryl, MOA 1A (Sevin) 50 WP (Sevin) 80 S (Sevin) XLR Plus	4 lb 2.5 lb 2 qt	12 hrs	3	On foliage as needed.
	pyrethroid, MOA 3		12 hrs		See table 5-9B for a list of registered pyrethroids and pre-harvest intervals.

Table 5-9. Insect Control for Commercial Vegetables

CROP / Insect	Insecticide, Mode of Action Code, and Formulation	Amount of Formulation Per Acre	Restricted Entry Interval (REI)	Pre harvest Interval (PHI) (Days)	Precautions and Remarks
Okra (continued)					
Corn earworm, Tobacco budworm, European corn borer	carbaryl, MOA 1A (Sevin) 50 WP (Sevin) 80 S (Sevin) XLR Plus	4 lb 2.5 lb 2 qt	12 hrs	3	On foliage as needed.
	chlorantraniliprole, MOA 28 (Coragen) 1.67 SC	2 to 3.5 fl oz	4 hrs	1	Foliar or drip chemigation. Drip chemigation must be applied uniformly to the root zone. See label for instructions.
	cyantraniliprole, MOA 28 (Verimark) 1,67SC	5 to 10 fl oz	4 hrs	1	Verimark is for soil application only. Applications made at planting and/or via drip chemigation. See label for application options.
	(Exirel) 0.83SE	7 to 17 fl oz	12 hrs	1	Exirel is for foliar application only. Rates >13.5 for loopers only.
	cyclaniliprole, MOA 28 (Harvanta) 50SL	10.9 to 16.4 fl oz	4 hrs	1	Foliar applications will help suppress western flower thrips when used in a rotational program.
	methoxyfenozide, MOA 18 (Intrepid) 2 F	8 to 16 fl oz	4 hrs	1	
	novaluron, MOA 15 (Rimon) 0.83 EC	9 to 12 fl oz	12 hrs	1	
	spinetoram, MOA 5 (Radiant) 1 SC	5 to 10 fl oz	4 hrs	1	For corn earworm only.
	pyrethroid, MOA 3		12 hrs		See table 5-9B for a list of registered pyrethroids and pre-harvest intervals.
Spider mites	bifenazate, MOA 20D (Acramite) 50 WP	0.75 to 1 lb	12 hrs	3	Do not make more than 1 application per season.
	fenpyroximate MOA 21 (Portal) 0.4EC	2 pt	12 hrs	3	Do not make more than 2 applications per season.
Stink bug, leaffooted bug	pyrethroid, MOA 3		12 hrs		See table 5-9B for a list of registered pyrethroids and pre-harvest intervals.
Whitefly	buprofezin, MOA 16 (Courier) 40 SC	9 to 13.6 fl oz	12 hrs	1	
	chlorantraniliprole, MOA 28 (Coragen) 1.67 SC	2 to 3.5 fl oz	4 hrs	1	Foliar or drip chemigation. Drip chemigation must be applied uniformly to the root zone. See label for instructions.
	cyantraniliprole, MOA 28 (Verimark) 1,67SC (Exirel) 0.83SE	6.75 to 13.5 fl oz 13.5 to 20.5 fl oz	4 hrs 12 hrs	1 1	Apply Verimark to at planting and/or later via drip irrigation or soil injection. See label for application options. Exirel is for foliar application.
	flupyradifurone, MOA 4D (Sivanto) 200 SL	10.5 to 14.0 fl oz	4 hrs	1	
	imidacloprid, MOA 4A Soil treatment (Admire Pro) 4.6 F (various) 2 F	 7 to 14 fl oz 16 to 32 fl oz	12 hrs	 21	See label for soil application instructions.
	Foliar treatment (Admire Pro) 4.6 F (various) 1.6 F	 1.3 to 2.2 fl oz 3.8 oz	12 hrs	 0	
	pyriproxyfen, MOA 7C (Knack) 0.86 EC	8 to 10 fl oz	12 hrs	1	Do not make more than 2 applications per season.
	spirotetramat, MOA 23 (Movento) 2SC	4 to 5 fl oz	24 hrs	3	Do not exceed 10 fluid ounces per season. Not for flea beetle. Requires surfactant.
Onion					
Armyworm, Cutworm	chlorantraniliprole MOA 28 (Coragen)	3.5 to 7.5 fl oz	4 hrs	1	
	methoxyfenozide MOA 18 (Intrepid) 2F	4 to 8 fl oz 8 to 12 fl oz	4 hrs	1	Green onion only. Use lower rates in early season on small plants; use higher rates in late season and heavy infestations.
	pyrethroid, MOA 3		12 hrs		See table 5-9B for a list of registered pyrethroids and pre-harvest intervals.
	spinetoram, MOA 5 (Radiant) 1 SC	5 to 10 fl oz	4 hrs	1	
Leafminer	cryomazine, MOA 17 (Trigard) 75 WS	2.66 oz	12 hrs	7	
	spinetoram, MOA 5 (Radiant) 1 SC	6 to 8 fl oz	4 hrs	1	
Onion maggot, Seed corn maggot	Onion seed pre-treated with cyromazine (Trigard) can be used to control onion and seed corn maggot.				
	chlorpyrifos, MOA 1B (Lorsban) 4 E	32 fl oz	24 hrs		Apply as in-furrow drench at planting. Use a minimum of 40 gal per acre and incorporate to a depth of 1 to 2 inches. Do not make more than 1 application per year.

Table 5-9. Insect Control for Commercial Vegetables

CROP / Insect	Insecticide, Mode of Action Code, and Formulation	Amount of Formulation Per Acre	Restricted Entry Interval (REI)	Pre harvest Interval (PHI) (Days)	Precautions and Remarks
Onion (continued)					
Onion maggot, Seed corn maggot (continued)	diazinon, MOA 1B (Diazinon) (AG 500)	2 to 4 qt	3 days		Furrow application; drench the seed furrow at planting time. Apply as a furrow treatment at time of planting. Use separate hoppers for seed and chemical.
	pyrethroid, MOA 3		12 hrs		See table 5-9B for a list of registered pyrethroids and pre-harvest intervals.
Thrips	acetamiprid MOA 4A (Assail) 70 WP	2.1 to 3.4 oz	12 hrs	7	
	methomyl, MOA 1A (Lannate) 2.4 LV	1.5 pt	48 hrs	7	
	spinetoram, MOA 5 (Radiant) 1 SC	6 to 8 fl oz	4 hrs	1	
	pyrethroid, MOA 3		12 hrs		See table 5-9B for a list of registered pyrethroids and pre-harvest intervals.
Pea, English and Snow Pea (Succulent and dried)					
Aphid	acetamiprid MOA 4A (Assail) 70 WP	1 to 2.3 oz	12 hrs	7	Also controls leafhoppers. Succulent peas only.
	pyrethroid, MOA 3		12 hrs		See table 5-9B for a list of registered pyrethroids and pre-harvest intervals.
	dimethoate, MOA 1B (Dimethoate) 400 (4E)	0.33 pt	48 hrs	0	Do not make more than 1 application per season, and do not feed or graze if a mobile viner is used, or for 21 days if a stationary viner is used. Re-entry interval is 48 hours.
	flupyradifurone, MOA 4D (Sivanto) 200 SL	7.0 to 10.5 fl oz	4 hrs	7	Will also control leafhopper
	imidacloprid, MOA 4A Soil treatment (Admire Pro) 4.6 F (various) 2 F	7 to 10.5 fl oz 16 to 24 fl oz	12 hrs	21	See label for soil application instructions.
	Foliar treatment (Admire Pro) 4.6 F (various) 1.6 F	1.2 fl oz 3.5 fl oz	12 hrs	7	
Armyworm, Cloverworm, Cutworm, Looper	chlorantraniliprole MOA 28 (Coragen) 1.67 SC	3.5 to 5 fl oz	4 hrs	1	
	pyrethroid, MOA 3		12 hrs		See table 5-9B for a list of registered pyrethroids and pre-harvest intervals.
	spinetoram, MOA 5 (Radiant) 1 SC	4 to 8 fl oz	4 hrs	3 (succulent); 28 (dried)	Not for cutworm.
	spinosad, MOA 5 (Blackhawk)	2.2 to 3.3 oz	4 hrs	3 (succulent); 28 (dried)	
Leafhopper, Lygus bug, Stink bug	dimethoate, MOA 1B (Dimethoate) 400 (4E)	0.33 to 1 pt	48 hrs	See label	Do not make more than 1 application per season. Do not feed or graze if a mobile viner is used, or for 21 days if a stationary viner is used.
	methomyl, MOA 1A (Lannate) 2.4 LV	1.5 to 3 pt	48 hrs	3	Apply to foliage as needed.
	pyrethroid, MOA 3		12 hrs		See table 5-9B for registered pyrethroids and pre-harvest intervals.
Seedcorn maggot	See **Beans** for control				
Pea (Cowpea, southernpeas)					
Aphid, Thrips	acetamiprid MOA 4A (Assail) 70 WP	1 to 2.3 oz	12 hrs	7	Succulent peas only.
	pyrethroid, MOA 3		12 hrs		See table 5-9B for registered pyrethroids and pre-harvest intervals.
	flupyradifurone, MOA 4D (Sivanto) 200 SL	7.0 to 10.5 fl oz	4 hrs	7	Will not control thrips.
	imidacloprid, MOA 4A Soil treatment (Admire Pro) 4.6 F (various) 2 F	7 to 10.5 fl oz 16 to 24 fl oz	12 hrs	21	See label for soil application instructions
	Foliar treatment (Admire Pro) 4.6 F (various) 1.6 F	1.3 fl oz 3.5 fl oz	12 hrs	7	
	spinetoram, MOA 5 (Radiant) 1 SC	5 to 8 fl oz	4 hrs	3 (succulent) 28 (dried)	Radiant is not effective against aphids.
	sulfoxaflor (Transform) 50 WG	0.75 to 1.0 oz	12 hrs	7	
	spinosad, MOA 5 (Blackhawk)	2.2 to 3.3 oz	4 hrs	3 (succulent); 28 (dried)	Blackhawk is not effective against aphids.

Table 5-9. Insect Control for Commercial Vegetables

CROP Insect	Insecticide, Mode of Action Code, and Formulation	Amount of Formulation Per Acre	Restricted Entry Interval (REI)	Pre harvest Interval (PHI) (Days)	Precautions and Remarks
Pea (Cowpea, southernpeas) (continued)					
Bean leaf beetle	carbaryl, MOA 1A (Sevin) 4 L (Sevin) 80 S	0.5 to 1 qt 0.625 to 1.25 lb	12 hrs	3	Do not feed treated foliage to livestock.
	pyrethroid, MOA 3		12 hrs		See table 5-9B for a list of registered pyrethroids and pre-harvest intervals.
Corn earworm, Loopers, European corn borer, Armyworm	chlorantraniliprole MOA 28 (Coragen) 1.67 SC	3.5 to 5 fl oz	4 hrs	1	
	methoxyfenozide, MOA 18 (Intrepid) 2 F	4 to 16 fl oz	4 hrs	7	Use lower rates on smaller plants and higher rates for mid- to late-season applications, against corn earworm. Do not apply more than 16 fluid ounces (0.25 pound a.i.) per acre per season.
	methomyl, MOA 1A (Lannate) 90SP	0.5 to 1 lb	48 hrs	1	Re-entry interval is 48 hr.
	pyrethroid, MOA 3		12 hrs		See table 5-9B for a list of registered pyrethroids and pre-harvest intervals.
	spinetoram, MOA 5 (Radiant) 1 SC	3 to 6 fl oz	4 hrs	3 (succulent) 28 (dried)	Do not apply more than 12 fluid ounces (0.188 a.i.) per acre per season.
Cowpea curculio	pyrethroids, MOA 3		12 hrs		See table 5-9B for a list of registered pyrethroids and pre-harvest intervals. Control may be poor in areas where resistant populations occur, primarily in parts of Alabama and Georgia. In areas where resistance is a problem, pyrethroid insecticides should be used at the highest labeled rate and synergized by tank-mixing with 1 pint piperonyl butoxide synergist per acre. In fields where resistance is a problem, applications every 3 to 5 days may be necessary to maintain control of the cowpea curculio population.
	methomyl, MOA 1A (Lannate) 90 SP	0.5 to 1 lb	48 hrs	1	Re-entry interval is 48 hours. Not effective against resistant cowpea curculio populations.
Stink bug	methomyl, MOA 1A (Lannate) 90SP	0.5 to 1 lb	48 hrs	1	Re-entry interval is 48 hours.
	pyrethroid, MOA 3		12 hrs		See table 5-9B for a list of registered pyrethroids and pre-harvest intervals. Control may be poor in areas where resistant populations occur, primarily in the Gulf Coast areas.
Leafminer	spinetoram, MOA 5 (Radiant) 1 SC	5 to 8 fl oz	4 hrs	3 (succulent); 28 (dried)	
	spinosad, MOA 5 (Blackhawk)	2.5 to 3.3 oz	4 hrs	3 (succulent); 28 (dried)	
Pepper					
Aphid, Flea beetle	acetamiprid, MOA 4A (Assail) 70 WP	0.8 to 1.2 oz	12 hrs	7	Do not apply more than once every 7 days, and do not exceed 4 applications per season.
	clothianidin, MOA 4A (Belay) 50WDG	4.8 to 6.4 oz (soil) 1.6 to 2.1oz (foliar)	12 hrs	7	Soil application at planting only.
	cyantraniliprole, MOA 28 (Verimark)	6.75 to 13.5 fl oz	4 hr	1	Apply to soil at planting, as a transplant tray drench, in transplant water or hill drench. After planting may be applied via drip irrigation.
	dinotefuran, MOA 4A Foliar treatment (Venom) 70 SG (Scorpion) 35SL Soil treatment (Venom) 70 SG (Scorpion) 35SL	 1 to 4 oz 2 to 7 fl oz 5 to 6 oz 9 to 10.5 fl oz	12 hrs	1 21	Do not follow soil applications with foliar applications. Use only 1 application method. Do not apply more than 6 ounces per acre per season using foliar applications, or 12 ounces per acre per season using soil applications. Soil applications may be applied by 1) a narrow band below or above the seed line at planting; 2) a post-seeding or transplant drench with sufficient water to ensure incorporation to the root zone; or 3) drip irrigation. For flea beetle control only.
	flonicamid, MOA 9C (Beleaf) 50 SG	2 to 4.8 oz	12 hrs	0	Will not control flea beetle.
	flupyradifurone, MOA 4D (Sivanto) 200 SL	7.0 to 12.0 fl oz	4 hrs	1	
	imidacloprid, MOA 4A Soil treatment (Admire Pro) 4.6 F (various) 2 F Foliar treatment (Admire Pro) 4.6 F (various) 1.6 F	 7 to 14 fl oz 16 to 32 fl oz 1.3 fl oz 3.8 fl oz	12 hrs 12 hrs	21 0	Where whitefly resistance is a concern, do not follow soil applications with foliar applications of any neonicotinoid. See label for soil application instructions. For short-term protection of transplants at planting, apply Admire Pro (0.44 oz/10,000 plants) not more than 7 days before transplanting by 1) uniformly spraying on transplants, followed immediately by sufficient overhead irrigation to wash product into potting media; or 2) injection into overhead irrigation system using adequate volume to thoroughly saturate soil media.
	oxamyl, MOA 1A (Vydate) 2 L	1 to 2 qt	48 hrs	7	
	pymetrozine, MOA 9B (Fulfill) 50 WDG	2.75 oz	12 hrs	0	Apply before aphids reach damaging levels. Do not exceed 5.5 ounces per acre per season. Not for flea beetle.

Table 5-9. Insect Control for Commercial Vegetables

CROP / Insect	Insecticide, Mode of Action Code, and Formulation	Amount of Formulation Per Acre	Restricted Entry Interval (REI)	Pre harvest Interval (PHI) (Days)	Precautions and Remarks
Pepper (continued)					
Aphid, Flea beetle (continued)	spirotetramat, MOA 23 (Movento) 2SC	4 to 5 fl oz	24 hrs	1	Do not exceed 10 fluid ounces per season. Requires surfactant. Will not control flea beetle.
	sulfoxaflor (Closer) 2 SC	1.5 to 2.0 fl oz	12 hrs	1	
	thiamethoxam, MOA 4A Soil treatment (Platinum) 75 SG	1.66 to 3.67 oz	12 hrs	30	Platinum may be applied to direct-seeded crops in-furrow seeding or transplant depth, post seeding or transplant as a drench, or through drip irrigation. Actara is applied as a foliar spray. Do not exceed 11 ounces per acre per season of Platinum or Actara. Check label for plant-back restrictions for a number of crops.
	Foliar treatment (Actara) 25 WDG	2 to 4 oz	12 hrs	0	
Armyworm, Corn earworm, Looper, Hornworm, European corn borer	*Bacillus thuringiensis*, MOA 11A (Dipel) DF (Xentari) WDG	0.5 to 1.5 lb 0.5 to 1 lb	4 hrs	0	Not effective against European corn borer.
	chlorantraniliprole, MOA 28 (Coragen) 1.67 SC	2 to 3.5 fl oz	4 hrs	1	Foliar or drip chemigation. Drip chemigation must be applied uniformly to the root zone. See label for instructions.
	cyantraniliprole, MOA 28 (Verimark) 1,67SC	5 to 10 fl oz	4 hrs	1	Verimark is for soil application only. Applications made at planting and/or via drip chemigation. See label for application options. Exirel is for foliar application only.
	(Exirel) 0.83SE	7 to 13.5 fl oz	12 hrs	1	
	cyclaniliprole, MOA 28 (Harvanta) 50SL	10.9 to 16.4 fl oz	4 hrs	1	
	emamectin benzoate, MOA 6 (Proclaim) 5 WDG	2.4 to 4.8 oz	12 hrs	7	Apply when larvae are first observed. Additional applications may be necessary to maintain control.
	indoxacarb, MOA 22 (Avaunt eVo) 30 WDG	2.5 to 3.5 oz	12 hrs	3	Use only higher rate for control of armyworm and corn earworm. Do not apply more than 14 ounces of Avaunt eVo (0.26 pound a.i. per acre per crop). Minimum interval between sprays is 5 days.
	methoxyfenozide, MOA 18 (Intrepid) 2 F	4 to 16 fl oz	4 hrs	1	Apply at rates of 4 to 8 fluid ounces early in season when plants are small. Apply at rates of 8 to 16 ounces to large plants or when infestations are heavy. During periods of continuous moth flights re-treatments at 7 to 14 days may be required. Do not apply more than 16 fluid ounces per application or 64 fluid ounces of Intrepid per acre per season.
	novaluron, MOA 15 (Rimon) 0.83 EC	9 to 12 fl oz	12 hrs	1	The use of a surfactant/adjuvant with Rimon is prohibited on pepper.
	spinetoram, MOA 5 (Radiant) 1 SC	5 to 10 fl oz	4 hrs	1	
	pyrethroid, MOA 3		12 hrs		See table 5-9B for a list of registered pyrethroids and pre-harvest intervals.
Blister beetle, Stink bug, Leaffooted bug	dinotefuran, MOA 4A Foliar treatment (Venom) 70 SG (Scorpion) 35SL	1 to 4 oz 2 to 7 fl oz	12 hrs	1	Do not combine foliar applications with soil applications, or vice versa. Use only 1 application method.
	Soil treatment (Venom) 70 SG (Scorpion) 35SL	5 to 6 oz 9 to 10.5 fl oz		21	
	pyrethroid, MOA 3		12 hrs		See table 5-9B for a list of registered pyrethroids and pre-harvest intervals.
	thiamethoxam, MOA 4A (Actara) 25WDG	3 to 5.5 oz	12 hrs	0	
Leafminer	abamectin, MOA 6 (Agri-Mek) 0.7 SC	1.75 to 3.5 fl oz	12 hrs	7	
	cyromazine, MOA 17 (Trigard) 75 WP	2.66 oz	12 hrs	0	
	dimethoate 4 EC, MOA 1B	0.5 pt	48 hrs	0	Re-entry interval is 48 hr.
	spinetoram, MOA 5 (Radiant) 1 SC	6 to 10 fl oz	4 hrs	1	
Pepper maggot	acephate, MOA 1B (Orthene) 97 PE	0.75 to 1 lb	24 hrs	7	See comments under European corn borer.
	dimethoate 4 EC, MOA 1B	0.5 to 0.67 pt	48 hrs	0	
	pyrethroid, MOA 3		12 hrs		See table 5-9B for registered pyrethroids and pre-harvest intervals.
Pepper weevil	acetamiprid, MOA 4A (Assail) 30 SG	4 oz	12 hrs	7	
	cyclaniliprole, MOA 28 (Harvanta) 50SL	16.4 fl oz	4 hrs	1	

Table 5-9. Insect Control for Commercial Vegetables

CROP / Insect	Insecticide, Mode of Action Code, and Formulation	Amount of Formulation Per Acre	Restricted Entry Interval (REI)	Pre harvest Interval (PHI) (Days)	Precautions and Remarks
Pepper (continued)					
Pepper weevil (continued)	oxamyl, MOA 1A (Vydate) 2 L	2 to 4 pt	48 hrs	7	
	thiamethoxam, MOA 4A (Actara) 25 WP	3 to 4 oz	12 hrs	0	Do not exceed 8 oz of Actara per acre per season.
	pyrethroid, MOA 3		12 hrs		See table 5-9B for registered pyrethroids and pre-harvest intervals.
Broad mite	abamectin, MOA 6 (Agri-Mek) 0.7 SC	1.75 to 3.5 fl oz	12 hrs	7	On foliage as needed.
	fenpyroximate MOA 21 (Portal) 0.4EC	2 pt	12 hrs	3	Do not make more than 2 applications per season.
	spiromesifen, MOA 23 (Oberon) 2 SG	7 to 8.5 fl oz	12 hrs	7	Do not exceed 3 applications per season.
	spirotetramat MOA 23 (Movento) 2 SC	4 to 5 fl oz	12 hrs	1	
Thrips	dinotefuran, MOA 4A Soil treatment (Venom) 70 SG (Scorpion) 35SL	5 to 6 oz 9 to 10.5 fl oz	12 hrs	21	See label for application instructions and restrictions.
	cyclaniliprole, MOA 28 (Harvanta) 50SL	16.4 fl oz	4 hrs	1	
	flonicamid, MOA 20D (Beleaf) 50 SG	2 to 4.8 fl oz	12 hrs	0	Is an option for insecticide-resistant western flower thrips. Do not exceed 8.4 oz per acer per season.
	imidacloprid, MOA 4A (Admire Pro) 4.6 F (various) 2 F	7 to 14 fl oz 16 to 32 fl oz	12 hrs	21	See Aphids for application instructions. Treating transplants before setting in the field, followed by drip irrigation may suppress incidence of tomato spotted virus. Imidacloprid is ineffective against western flower thrips.
	methomyl, MOA 1A (Lannate) 2.4 LV	1.5 pt	48 hrs	3	
	spinetoram, MOA 5 (Radiant) 1 SC	6 to 10 fl oz	4 hrs	1	Do not exceed 29 fluid ounces per acre per season. Control of thrips may be improved by adding a spray adjuvant. See label for instructions.
Potato, Irish					
Aphid	acetamiprid, MOA 4A (Assail) 30 SG	1.5 to 4 oz	12 hrs	7	Do not make more than 4 applications per season. Thorough coverage is important. Assail belongs to the same class of insecticides (neonicotinoid) as Admire Pro, Provado, Actara, and Platinum and Colorado potato beetle populations have the potential to become resistant to this class.
	clothianadin MOA 4A Belay 50 WDG	1.0 to 1.5 oz	12 hrs	7	Apply Belay 50 WDG as foliar spray when populations reach a threshold level. Do not apply more than 3 applications. Belay belongs to the same class of insecticides (neonicotinoid) as Admire Pro, Provado, Actara, and Platinum and Colorado potato beetle populations have the potential to become resistant to this class.
	flonicamid, MOA 29 (Beleaf) 50 SG	2 to 2.8 oz	12 hrs	7	
	flupyradifurone, MOA 4D (Sivanto) 200 SL	7.0 to 12.0 fl oz	4 hrs	1	
	dimethoate 4 EC, MOA 1B	0.5 to 1 pt	48 hrs	0	Do not apply more than 2 pints total per year.
	imidacloprid, MOA 4A (Admire Pro) 4.6F (various) 1.6 F	1.2 fl oz 3.75 fl oz	12 hrs	7	To minimize selection for resistance in Colorado potato beetle, do not use acetamiprid, imidacloprid, or thiamethoxam for aphid control if either of these compounds was applied to the crop for control of Colorado potato beetle. See comments on insecticide rotation under Colorado potato beetle.
	pymetrozine, MOA 9B (Fulfill) 50 WDG	2.75 oz	12 hrs	14	Allow at least 7 days between applications. Do not exceed a total of 5.5 ounces (0.17 lb a.i.) per acre per season.
	thiamethoxam, MOA 4A (Actara) 25 WDG	3 oz	12 hrs	14	To minimize selection for resistance in Colorado potato beetle, do not use imidacloprid or thiamethoxam for aphid control if either of these compounds was applied to the crop for control of Colorado potato beetle.

Table 5-9. Insect Control for Commercial Vegetables

CROP / Insect	Insecticide, Mode of Action Code, and Formulation	Amount of Formulation Per Acre	Restricted Entry Interval (REI)	Pre harvest Interval (PHI) (Days)	Precautions and Remarks
Potato, Irish (continued)					
Colorado potato beetle	Colorado potato beetle populations in most commercial potato-growing areas have developed resistance to many insecticides. As a result, insecticides that are effective in some areas, or were effective in the past, may no longer provide control in particular areas. Colorado potato beetle readily develops resistance to insecticides. The following practices help to reduce the risk of resistance developing:				
	CROP ROTATION AND INSECTICIDE ROTATION (the use of insecticides representing different modes of action IRAC MOA class number in different years and against different generations of potato beetle within a year) are essential if insecticide resistance is to be managed and the risks of control failures due to resistance minimized. If control failures or reduced levels of control are observed with a particular insecticide, do NOT make a second application of the same insecticide at the same or higher rate. If an additional insecticide application is necessary, a different insecticide representing a different IRAC MOA class number should be used. Because potato beetle adults will move between adjacent and nearby fields from one year to the next, it is important to maintain the same rotation schedule of insecticide MOA classes in adjacent fields and groups of nearby fields.				
	SCOUT FIELDS: All insecticide applications to the potato crop, regardless of the target insect pest, have the potential to increase the resistance of the Colorado potato beetle to insecticides. Unnecessary insecticide applications should be avoided by scouting fields for insect pests and applying insecticides only when potentially damaging insect populations are present.				
	SPOT TREATMENTS: Because overwintered potato beetles invade rotated fields from sources outside the field, potato beetle infestations in rotated fields occur first along field edges early in the season. Limiting insecticide applications to infested portions of the field will provide effective control and reduce costs. Growers are advised to keep accurate records on which insecticides have been applied to their potato crop for control of Colorado potato beetle and on how effective those insecticides were at controlling infestations. This will make choosing an insecticide and maintaining insecticide rotations easier. Monitoring the insecticide resistance status of local populations will also make insecticide selection easier.				
	abamectin, MOA 6 (Agri-Mek) 0.7 SC	1.75 to 3.5 fl oz	12 hrs	14	Apply when adults and/or small larvae are present but before large larvae appear. Do not exceed 2 applications per season. Apply in at least 20 gallons water per acre.
	acetamiprid, MOA 4A (Assail) 30 SG	1.5 to 4.0 oz	12 hrs	7	Apply when most of the egg masses have hatched and many small but few large larvae are present. An additional application should be used only if defoliation increases. Allow at least 7 days between foliar applications. To minimize selection for resistance, do not use foliar applications of any IRAC MOA class 4A insecticides if any IRAC MOA class 4A insecticides were applied to the crop as soil or seed piece treatments. See comments on insect rotation under Colorado potato beetle.
	chlorantraniliprole, MOA 28 (Coragen) 1.67	3.5 to 5 oz	4 hrs	14	Do not apply more than 15.4 ounces Coragen per acre per crop season. Coragen treated insects may take several days to die but stop feeding almost immediately after treatment.
	clothianadin MOA 4A (Belay) 50 WDG	1.9 to 2.8 fl oz	12 hrs	7	Apply Belay 50 WDG as foliar spray Apply when adults and/or small larvae are present but before large larvae appear. Do not apply more than 3 applications. Belay belongs to the same class of insecticides (neonicotinoid) as Admire Pro, Provado, Actara, and Platinum and Colorado potato beetle populations have the potential to become resistant to this class.
	cyantraniliprole, MOA 28 (Verimark) 1.67SC	6.75 to 13.5 fl oz	4 hr	NA	Apply in-furrow at planting. Do not apply any other MOA Group 28 insecticide for Colorado potato beetle control following an at-plant application for cyantraniliprole. When applied at 10 to 13.5 fluid ounces per acre will provide control of European corn borer in most years, except possibly in very early planted potatoes.
	dinotefuran, MOA 4A (Venom) 70 SG	1 to 1.5 oz (foliar) 6.5 to 7.5 oz (soil)	12 hrs	7	Soil treatment for preplant, preemergence, or at ground crack only application only. To minimize selection for resistance, do not use foliar applications of any IRAC MOA class 4A insecticides if any IRAC MOA class 4A insecticides were applied to the crop as soil or seed piece treatments. See comments on insecticide rotation under Colorado potato beetle.
	imidacloprid seed piece treatment, MOA 4A (Genesis) 240 g/L	0.4 to 0.6 fl oz/100 lb of seed tubers			Resistance has been reported and may reduce efficacy or duration of control. To minimize selection for resistance, do not use foliar applications of any IRAC MOA class 4A insecticides if any of these compounds were applied to the crop as soil or seed piece treatments. See label for specific instructions. For early planted potatoes control may be marginal because of the prolonged time between application and Colorado potato beetle emergence. Limit use to locations where Colorado potato beetles were a problem in the same or adjacent fields during the previous year. Do not apply other IRAC MOA class 4A insecticides to a field if seed pieces were treated with Genesis. See product label for restrictions on rotational crops.
	imidacloprid, MOA 4A Soil treatment (Admire Pro) 4.6 F (various) 2.0 F	0.74 fl oz/ 1,000 ft row	12 hrs	—	Resistance has been reported and may reduce efficacy or duration of control. See comments on insecticide rotation under Colorado potato beetle. Admire Pro applied in-furrow at planting time may provide season-long control. However, for early planted potatoes control may be marginal due to the prolonged time between application and Colorado potato beetle emergence. Use only in potato fields that have a history of potato beetle infestations. If potatoes are rotated to a field adjacent to one planted in potato last year, a barrier treatment may be effective. (See Vegetable IPM Insect Note #45.) Admire Pro may also be applied as a seed treatment. Check label for instructions regarding this use. Check label for restrictions on planting crops following Admire Pro treated potatoes. There have been reports of low levels of resistance to imidacloprid. To minimize selection for resistance, do not use foliar applications of any IRAC MOA class 4A insecticides if any of these compounds were applied to the crop as soil or seed piece treatments. See comments on insecticide rotation under Colorado potato beetle.

Table 5-9. Insect Control for Commercial Vegetables

CROP Insect	Insecticide, Mode of Action Code, and Formulation	Amount of Formulation Per Acre	Restricted Entry Interval (REI)	Pre harvest Interval (PHI) (Days)	Precautions and Remarks
Potato, Irish (continued)					
Colorado potato beetle (continued)	Foliar treatment (Admire Pro) 4.6 (various) 1.6 F	1.3 fl oz 3.75 fl oz	12 hrs	7	Apply when most of the egg masses have hatched and most larvae are small (1/8 to 3/16 in.). An additional application should be made only if defoliation increases. Allow at least 7 days between foliar applications. Do not exceed 5.6 fluid ounces of Admire Pro per field per acre per season. Regardless of formulation, do NOT apply more than a total of 0.31 pound imidacloprid per season. Foliar applications of imidacloprid should not be applied if soil application was used. There have been reports of resistance to imidacloprid. To minimize selection for resistance, do not use foliar applications of any IRAC MOA class 4A insecticides if any of these compounds were applied to the crop as soil or seed piece treatments. See comments on insecticide rotation under Colorado potato beetle.
	imidacloprid cyfluthrin premix, MOA 4A and 3 (Leverage) 2.7 SE	3 to 3.75 fl oz		7	There have been reports of low levels of resistance to imidacloprid. To minimize selection for resistance, do not use foliar applications of any IRAC MOA class 4A insecticides if any of these compounds were applied to the crop as soil or seed piece treatments. See comments on insecticide rotation under Colorado potato beetle. Apply when most of the egg masses have hatched and most larvae are small (1/8 to 3/16 inch). An additional application should be made only if defoliation increases. Leverage will control European corn borer if application coincides with egg hatch and presence of small corn borer larvae. Leverage should not be used in fields treated with Admire Pro.
	novaluron, MOA 15 (Rimon) 0.83 EC	9 to 12 fl oz	12 hrs	14	Novaluron is an insect growth regulator with activity against eggs and larvae. Larvae are killed as they molt to the next stage. Eggs present at the time of application ware killed. Adults exposed produce few eggs. Novaluron is most effective if directed against overwintered adults when egg numbers are increasing, and small larvae are just beginning to appear. Do not apply to successive generations of Colorado potato beetle. Do not apply more than 24 fl oz per season.
	spinosad, MOA 5 (Blackhawk) 36WG	1.7 to 3.3 oz		3	Apply when most egg masses have hatched and both small and large larvae are present. Thorough coverage is important. Do not apply more than a total of 0.33 pound a.i. (14.4 ounces of Blackhawk or 21 ounces of Radiant) per crop. Do not apply in consecutive generations of Colorado potato beetle and do not make more than 2 applications per single generation of Colorado potato beetle. Do not make successive applications less than 7 days apart. To minimize the potential for resistance, do NOT use spinosad or spinetoram if it either product was applied to a potato crop in the field or an adjacent field within the last year.
	spinetoram, MOA 5 (Radiant) 1 SC	6 to 8 fl oz	4 hrs	7	
	thiamethoxam seed piece treatment, MOA 4A (Cruiser) 5 FS	0.11 to 0.16 fl oz/100 lb			See label for specific instructions. Resistance to neonicotinoid insecticides has been reported and may reduce efficacy or duration of control by thiamethoxam. To minimize selection for resistance, do not use foliar applications of any IRAC MOA class 4A insecticides if any of these compounds were applied to the crop as soil or seed piece treatments. See comments on insecticide rotation under Colorado potato beetle. For early planted potatoes control may be marginal because of the prolonged time between application and Colorado potato beetle emergence. Limit use to locations where Colorado potato beetles were a problem in the same or adjacent fields during the previous year.
	thiamethoxam, MOA 4A (Platinum) 75 SG	1.66 to 2.67 oz	12 hrs	7	Resistance to neonicotinoid insecticides has been reported and may reduce efficacy or duration of control by thiamethoxam. To minimize selection for resistance, do not use foliar applications of any IRAC MOA class 4A insecticides if any of these compounds were applied to the crop as soil or seed piece treatments. See comments on insecticide rotation under Colorado potato beetle. See product label for restrictions on rotational crops. Platinum applied in-furrow at planting time may provide season-long control. For early planted potatoes control may be marginal because of the prolonged time between application and Colorado potato beetle emergence. Limit use to locations where Colorado potato beetles were a problem in the same or adjacent fields in the previous year.
	(Actara) 25 WDG	3 oz	12 hrs	7	Resistance to neonicotinoid insecticides has been reported and may reduce efficacy or duration of control by thiamethoxam. To minimize selection for resistance, do not use foliar applications of any IRAC MOA class 4A insecticides if any of these compounds were applied to the crop as soil or seed piece treatments. See label for rotational restrictions. Actara is applied as foliar spray. Apply when most of the eggs have hatched and most of the larvae are small (1/8 to 3/16 inch). An additional application should be made only if defoliation increases. Allow at least 7 days between applications. Do not make more than 2 applications of Actara per crop per season.

Table 5-9. Insect Control for Commercial Vegetables

CROP / Insect	Insecticide, Mode of Action Code, and Formulation	Amount of Formulation Per Acre	Restricted Entry Interval (REI)	Pre harvest Interval (PHI) (Days)	Precautions and Remarks
Potato, Irish (continued)					
Colorado potato beetle (continued)	thiamethoxam, MOA 4A F chlorantraniliprole, MOA 28 Premix (Voliam Flexi)	4 oz		14	Resistance to neonicotinoid insecticides has been reported and may reduce efficacy or duration of control by thiamethoxam. To minimize selection for resistance, do not use foliar applications of any IRAC MOA class 4A insecticides if any of these compounds were applied to the crop as soil or seed piece treatments. See comments on insecticide rotation under Colorado potato beetle. Voliam Flexi is applied as a foliar spray. Apply when most of the eggs have hatched and most of the larvae are small (1/8 to 3/16 inch.). An additional application should be made only if defoliation increases. Allow at least 7 days between applications. Do not exceed 8 ounces of Voliam Flexi. See label for rotational restrictions. Voliam Flexi can be expected to provide control of European corn borer if application is timed correctly. See European corn borer for correct timing.
European corn borer	The Atlantic variety of potato is very tolerant of injury by European corn borer larvae. Consequently, control is not recommended on Atlantic unless more than 30 percent of the stems are infested. Control on all other varieties is recommended when infestations reach 20 percent infested stems. Application timing is critical. Scout for eggs and treat when eggs hatch or at the first sign of larvae entering petioles. Several days of cool wet weather will kill larvae and may eliminate the need for insecticide applications. If this occurs, flag additional egg masses and apply insecticide at hatch.				
	pyrethroid, MOA 3		12 hrs		Apply when threshold is reached (usually during the first half of May). A second application may be needed if the percentage of infested stems increases substantially 7 to 10 days after the first application. Ground applications are usually more effective than aerial applications. See table 5-9B for a list of registered pyrethroids and pre-harvest intervals.
	chlorantraniliprole, MOA 28 (Coragen) 1.67	3.5 to 5 oz	4 hrs	14	Do not apply more than 15.4 ounces Coragen per acre per crop season.
	thiamethoxam, MOA 4A chlorantraniliprole MOA 28 Premix (Voliam Flexi)	4 oz	12 hrs	14	Voliam Flexi is applied as a foliar spray. Apply when most of the eggs have hatched and most of the larvae are small (1/8 to 3/16 inch.). An additional application should be made only if defoliation increases. Allow at least 7 days between applications. To minimize selection for resistance, do not use foliar applications of any IRAC MOA class 4A insecticides if any of these compounds were applied to the crop as soil or seed piece treatments. Do not exceed 8 ounces of Voliam Flexi. See label for rotational restrictions Voliam Flexi can be expected to provide control of Colorado potato beetle if application is timed correctly (see Colorado potato beetle section for correct timing).
	indoxacarb, MOA 22 (Avaunt eVo) 30 WDG	3.5 to 6.0 oz	12 hrs	7	Apply when threshold is reached (usually during the first half of May). A second application may be needed if the percentage of infested stems increases substantially 7 to 10 days after the first application. Ground applications are usually more effective than aerial applications. Do not apply more than 24 ounces of Avaunt eVo per acre per crop.
	spinetoram, MOA 5 (Radiant) 1 SC	6 to 8 fl oz	4 hrs	7	Do not apply more than a total of 0.25 pound a.i. (32 fluid ounces product) per crop.
Flea beetle	imidacloprid, MOA 4A Soil treatment (Admire Pro) 4.6 F (various) 2.0 F	0.74 fl oz/ 1,000 ft row	12 hrs	—	If imidacloprid or thiamethoxam resistant Colorado potato beetles occur in the field, application of imidacloprid to control flea beetles has the potential to further increase resistance levels. Imidacloprid applied in-furrow at planting time may provide season-long control of flea beetles. However, for early planted potatoes control may be marginal due to the prolonged time between application and crop emergence. Check label for restrictions on planting crops following Admire Pro treated potatoes.
	Foliar treatment (Admire Pro) 4.6 (various) 1.6 F	1.3 fl oz 3.75 fl oz	12 hrs	7	See comments for imidacloprid resistance in Colorado potato beetle.
	thiamethoxam seed piece treatment, MOA 4A (Cruiser) 5 FS	0.11 to 0.16 fl oz/100 lb	12 hrs		See label for specific instructions. For early planted potatoes control may be marginal because of the prolonged time between application and flea beetle emergence. Limit use to locations where Colorado potato beetles were a problem in the same or adjacent fields during the previous year. To minimize selection for resistance, do not use foliar applications of any IRAC MOA class 4A insecticides if any of these compounds were applied to the crop as soil or seed piece treatments. See comments on insecticide rotation under Colorado potato beetle.
	thiamethoxam, MOA 4A (Platinum) 2 SC	5 to 8 fl oz	12 hrs	7	Platinum applied in-furrow at planting time may provide season-long control. However, for early planted potatoes control may be marginal due to the prolonged time between application and crop emergence. Limit use to locations where Colorado potato beetles were not a problem in the same or adjacent fields during the previous year. See product label for restrictions on rotational crops. See comments for imidacloprid resistance in Colorado potato beetle.
	(Actara) 25 WDG	3 oz	12 hrs	7	Actara is applied as foliar spray. See comments for imidacloprid resistance in Colorado potato beetle.
	thiamethoxam MOA 4A chlorantraniliprole moa 28 (Voliam Flexi)	4 fl oz		14	Do not exceed a total of 8.0 fluid ounces per acre Voliam Flexi or 0.094 lb ai/acre of thiamethoxam-containing products or 0.2 pound ai/acre of chlorantraniliprole-containing products per growing season. If imidacloprid or thiamethoxam resistant Colorado potato beetles occur in the field, application of Voliam Flexi to control flea beetles has the potential to further increase resistance levels. See comments for imidacloprid resistance in Colorado potato beetle.
	pyrethroid, MOA 3		12 hrs		See table 5-9B for a list of registered pyrethroids and pre-harvest intervals.

Table 5-9. Insect Control for Commercial Vegetables

CROP Insect	Insecticide, Mode of Action Code, and Formulation	Amount of Formulation Per Acre	Restricted Entry Interval (REI)	Pre harvest Interval (PHI) (Days)	Precautions and Remarks
Potato, Irish (continued)					
Leafhopper	carbaryl, MOA 1A (Sevin) 50 WP (Sevin) 80 S (Sevin) XLR Plus	1 to 2 lb 0.625 to 1.25 lb 1 pt	12 hrs	7	On foliage when leafhoppers first appear. Repeat every 10 days as needed. Often a problem in the mountains.
	dimethoate, MOA 1B various – check label for rate, PHI and REI				
	imidacloprid cyfluthrin premix, MOA 4A and 3 (Leverage) 2.7 SE (Leverage) 360	3 to 3.80 fl oz 2.8 fl oz	7	7	There have been reports of low levels of resistance to imidacloprid. To minimize selection for resistance, do not use foliar applications of any IRAC MOA class 4A insecticides if any of these compounds were applied to the crop as soil or seed piece treatments. See comments on insecticide rotation under Colorado potato beetle. Apply when most of the egg masses have hatched and most larvae are small (1/8 to 3/16 inch). An additional application should be made only if defoliation increases. Leverage should not be used in fields treated with Admire Pro.
	methomyl, MOA 1A (Lannate) 2.4 LV	1.5 pt	48 hrs	6	
	pyrethroid, MOA				See table 5-9B for a list of registered pyrethroids and pre-harvest intervals.
Leafminer	dimethoate 4 EC, MOA 1B various – check k label for rate, PHI and REI				
	chlorantraniliprole, MOA 28 (Coragen) 1.67 SC	3.5 to 5 fl oz	4 hrs	14	
Blister beetle, Leaffooted bug, Plant bug, Stink bug, Vegetable weevil	carbaryl, MOA 1A (Sevin) 50 WP (Sevin) XLR Plus	2 to 4 lb 1 to 2 qt	12 hrs	7	On foliage as needed.
	pyrethroid, MOA 3		12 hrs		See table 5-9B for a list of registered pyrethroids and pre-harvest intervals.
Potato tuberworm	chlorantraniliprole, MOA 28 (Coragen) 1.67 SC	3.5 to 5 fl oz	4 hrs	14	Do not exceed 4 applications per acre per crop. Do not apply more than 15.4 ounces Coragen per acre per crop season. Minimum interval between applications is 5 days.
	methomyl, MOA 1A (Lannate) 2.4 LV	1.5 to 3 pt	48 hrs	6	Prevent late-season injury by keeping potatoes covered with soil. To prevent damage in storage, practice sanitation.
	pyrethroid, MOA 3		12 hrs		See table 5-9B for a list of registered pyrethroids and pre-harvest intervals.
Thrips	dimethoate 4 EC, MOA 1B	0.5 pt	48 hrs	0	
	spinetoram, MOA 5 (Radiant) 1 SC	6 to 8 fl oz	4 hrs	7	
	spinosad, MOA 5 (Blackhawk) 36WG	2.25 to 3.5 oz	4 hrs	3	Control may be improved by addition of an adjuvant to the spray mixture.
Wireworm	Planting in fields previously in corn, soybean, or fallow may increase risk of wireworm.				
	bifenthrin, MOA 3 (Capture LFR)	25.5 fl oz			In furrow at planting.
	clothianidin (Belay) 50 WDG	6 fl oz	12 hrs		In-furrow at planting.
	ethoprop, MOA 1B (Mocap) 15 G	1.4 lb per 1,000 row ft	48 hrs	90	In-furrow at planting.
	fipronil, MOA 2B (Regent) 4 SC	3.2 fl oz	0 hrs	90	In-furrow at planting. Do NOT use T-banding over the top of a closed furrow.
	phorate, MOA 1B (Thimet) 20 G	Row Treatment: 10 to 20 oz (38 in. row spacing)	12 hrs	90	Can contribute to insecticide-resistance problems with Colorado potato beetle.
Pumpkin, Squash (see Cucurbit Crops)					
Radish					
Aphid, Flea beetle	pyrethroid, MOA 3		12 hrs		See table 5-9B for a list of registered pyrethroids and pre-harvest intervals.
	flupyradifurone, MOA 4D (Sivanto) 200 SL	7.0 to 10.5 fl oz	4 hrs	7	Will not control flea beetle or leafminer.
	Foliar treatment - imidacloprid (Admire Pro) 4.6 F (various) 1.6 F	1.2 fl oz 3.5 fl oz	12 hrs	7	Will not control leafminer.
	flonicamid, MOA 29 (Beleaf) 50 SG	2 to 2.8 oz	12 hrs	0	

Table 5-9. Insect Control for Commercial Vegetables

CROP / Insect	Insecticide, Mode of Action Code, and Formulation	Amount of Formulation Per Acre	Restricted Entry Interval (REI)	Pre harvest Interval (PHI) (Days)	Precautions and Remarks
Radish (continued)					
Aphid, Flea beetle (continued)	thiamethoxam, MOA 4A (Platinum) 75 SG	1.7 to 2.17 oz	12 hrs	30	See label for soil application instructions.
	(Actara) 25 WDG	1.5 to 3 oz		7	
Root maggot, Wireworm	chlorpyrifos, MOA 1B (Lorsban) 4E	1 fl oz/1,000 linear ft	24 hrs	—	Water-based drench in-furrow planting. Use a minimum of 40 gal of water per acre.
	diazinon, MOA 1B (AG 500) 50 WP	2 to 4 qt 4 to 8 lb	3 days		Broadcast just before planting and immediately incorporate into the upper 4 to 8 inches of soil.
Spinach					
Aphid	acetamiprid, MOA 4A (Assail) 30SG	2 to 4 oz	12 hrs	7	Do not apply more than once every 7 days, and do not exceed 5 applications per season.
	clothianidin, MOA 4A (Belay) 50 WDG	4.8 to 6.0 oz (soil) 1.6 to 2.1 fl oz (foliar)	12 hrs	7	Soil application at planting only.
	cyantraniliprole, MOA 28 (Verimark) 1,67SC	6.75 to 10 fl oz	4 hrs	1	Soil applications made at planting only. See label for application options.
	flonicamid, MOA 9C (Beleaf) 50 SG	2 to 2.8	12 hrs	0	
	flupyradifurone, MOA 4D (Sivanto) 200 SL	10.5 to 12.0 fl oz	4 hrs	1	
	imidacloprid, MOA 4A Soil treatment (Admire Pro) 4.6 F (various) 2 F	4.4 to 10.5 fl oz 10 to 24 fl oz	12 hrs	21	Do not follow soil applications with foliar applications of any neonicotinoid insecticides. See label for soil application instructions.
	Foliar treatment (Admire Pro) 4.6 F (various) 1.6 F	1.2 fl oz 3.8 fl oz	12 hrs	7	
	pymetrozine, MOA 9B (Fulfill) 50 WDG	2.75 oz	12 hrs	7	Apply before aphids reach damaging levels. Use sufficient water to ensure good coverage.
	spirotetramat, MOA 23 (Movento) 2 SC	4 to 5 fl oz	24 hrs	3	Do not exceed 10 fluid ounces per season. Requires surfactant.
	thiamethoxam, MOA 4A Soil treatment (Platinum) 75 SG	1.7 to 2.17 oz	12 hrs	30	See label for soil application instructions.
	Foliar treatment (Actara) 25 WDG	1.5 to 3 oz	12 hrs	7	
	tolfenpyrad, MOA 21A (Torac) 1.29 EC	17 to 21 fl oz	12 hrs	1	
Leafminer	chlorantraniliprole, MOA 28 (Coragen) 1.67 SC	5 to 7.5 fl oz	4 hrs	1	Foliar or drip chemigation. Drip chemigation must be applied uniformly to the root zone. See label for instructions.
	cyclaniliprole, MOA 28 (Harvanta) 50SL	16.4 fl oz	4 hrs	1	
	cryomazine, MOA 17 (Trigard) 75 WP	2.66 oz	12 hrs	7	
	spinetoram, MOA 5 (Radiant) 1 SC	6 to 10 fl oz	4 hrs	1	Spray adjuvants may enhance efficacy against leafminers. See label for information on adjuvants.
Armyworm, Beet webworm, Corn earworm, Cutworm, Looper	chlorantraniliprole, MOA 28 (Coragen) 1.67 SC	3.5 to 5 fl oz	4 hrs	3	
	cyclaniliprole, MOA 28 (Harvanta) 50SL	16.4 fl oz	4 hrs	1	
	emamectin benzoate, MOA 6 (Proclaim) 5 SG	2.4 to 4.8 oz	12 hrs	7	
	indoxacarb, MOA 22 (Avaunt eVo) 30 SG	2.5 to 3.5 oz	12 hrs	3	
	methomyl, MOA 1A (Lannate) 90 SP (Lannate) 2.4 LV	0.5 lb 1.5 pt	48 hrs	7	Air temperature should be well above 32 degrees F. Do not apply to seedlings less than 3 in. in diameter.
	methoxyfenozide, MOA 18 (Intrepid) 2 F	4 to 10 fl oz	4 hrs	1	Use low rates for early-season applications to young or small plants and 6 to 10 oz for mid- to late-season applications.
	spinetoram, MOA 5 (Radiant) 1 SC	5 to 10 fl oz	4 hrs	1	
	pyrethroid, MOA 3		12 hrs		See table 5-9B for a list of registered pyrethroids and pre-harvest intervals.
Squash (see Cucurbit Crops)					

Table 5-9. Insect Control for Commercial Vegetables

CROP / Insect	Insecticide, Mode of Action Code, and Formulation	Amount of Formulation Per Acre	Restricted Entry Interval (REI)	Pre harvest Interval (PHI) (Days)	Precautions and Remarks
Sweetpotato					
Aphids, Leafhopper, Whitefly	Aphids, leafhoppers, and whiteflies are rarely a problem.				
	acetamiprid, MOA 4A (Assail) 30SG	1.5 to 4 oz	12 hrs	7	Do not make more than 4 applications per season. Do not apply more frequently than once every 7 days. Use 2.5 to 4 ounces for aphids.
	clothianidin, MOA 4A (Belay) 2.13 SC	9 to 12 oz (soil)	12 hrs	21	Soil application as an in-furrow or sidedress application. For sidedress applications, immediately cover with soil.
	flonicamid, MOA 29 (Beleaf) 50 SG	2 to 2.8 oz	12 hrs	7	
	flupyradifurone, MOA 4D (Sivanto) 200 SL	7.0 to 14.0 fl oz	4 hrs	1	For aphids and leafhopper use 7.0 to 10.5 fluid ounces, for whitefly use 10.5 to 14.0 fluid ounces.
	imidacloprid, MOA 4A (Admire Pro) 4.6 F (various) 1.6 F	Foliar: 1.2 fl oz 3.5 fl oz Soil: 4.4 to 10.5 fl oz	12 hrs	7 / 60	2 foliar applications may be needed to control heavy populations. Allow 5 to 7 days between applications. The Admire Pro 24C label includes an infurrow or side dress application 45 days after planting at 4.4 to 10.5 fl oz/acre.
	pymetrozine, MOA 9B (Fulfill) 50 WDG	2.75 to 5.5 oz	12 hrs	14	
	spirotetramat MOA 23 (Movento) 2 SC	4 to 5 fl oz	24 hrs	7	Will not control leafhopper. Requires surfactant
	thiamethoxam, MOA 4A (Actara) 25 WDG	3 oz		14	Two applications of Actara may be needed to control heavy populations. Allow 7 to 10 days between applications. Do not exceed a total of 6 ounces of Actara per crop per season.
Armyworm, Looper, Corn earworm, Hornworm	chlorantraniliprole, MOA 28 (Coragen) 1.67 SC	3.5 to 5 fl oz	4 hrs	1	Foliar application only on sweetpotato.
	chlorantraniliprole and lambda-cyfluthrin premix, MOA 28 and 3 (Besiege)	6 to 9 fl oz	24 hrs	14	Treat when a combination of moth pests and cucumber beetles are above threshold.
	methoxyfenozide, MOA 18 (Intrepid) 2 F	6 to 10 fl oz	4 hrs	7	Damaging earworm infestations may occur in August or September. If significant infestations are present on foliage during harvest, larvae may feed on exposed root. Do not make more than 3 applications or apply more than 30 fl oz of Intrepid per acre per season.
	novaluron, MOA 15 (Rimon) 0.83 EC	9 to 12 fl oz	12 hrs	14	Do not make more than 2 applications per crop per season.
	spinosad MOA 5 (Blackhawk)	1.7 to 3.5 oz	4 hrs	3	
	spinetoram, MOA 5 (Radiant) 1 SC	6 to 8 fl oz	4 hrs	7	
Cucumber beetle (adults), Japanese beetle (adults), Tortoise beetle	Cucumber beetle larvae (Diabrotica) are a serious pest of sweetpotato in LA and MS. Controlling adult cucumber beetles in areas with a history of diabrotica damage can reduce damage to roots. Foliage feeding by beetles rarely causes economic loss, and control is not warranted unless defoliation is severe.				
	pyrethroid, MOA 3		12 hrs		See table 5-9B for registered pyrethroids and pre-harvest intervals.
	carbaryl, MOA 1A (Sevin) 50 WP (Sevin) 80 S, WSB (Sevin) XLR Plus	4 lb 2.5 lb 2 qt	12 hrs	7	Treat for tortoise beetles only if significant defoliation is observed. Tortoise beetles are frequently present but rarely reach levels requiring treatment.
	spinetoram, MOA 5 (Radiant) 1 SC	6 to 8 fl oz	4 hrs	7	
Flea beetle, Wireworm, White grub	bifenthrin, MOA 3 (various) 2 EC	9.6 to 19.2 fl oz		21	Apply as broadcast, preplant application to the soil and incorporate 4 to 6 inches prior to bed formation. This use has been demonstrated to control overwintered wireworm populations and reduce damage to roots at harvest. Chlorpyrifos will not control whitefringed beetle or other grubs that attack sweetpotato. Research has shown that best control is achieved when chlorpyrifos is applied as a preplant application incorporated 4 to 6 inches deep prior to bed formation, followed by 1 or more soil-directed, incorporations of bifenthrin during routine cultivation. Bifenthrin should be directed onto each side of the bed from the drill to the middle of the furrow and incorporated with cultivating equipment set to throw soil toward the drill. The objective is to provide a barrier of treated soil that covers the bed and furrows. Foliar sprays of various insecticides that target adults to prevent egg laying have not been shown to provide any reduction in damage to roots by wireworm larvae at harvest. **Note:** Check the registration status of chlorpyrifos before using this active ingredient in 2019.
	chlorpyrifos, MOA 1B (Lorsban) 15 G (Lorsban) 4 E (Lorsban Advanced)	13.5 lb 4 pt 4 pt	24 hrs	125 (60 in NC for Lorsban Advanced only)	
	chlothianidin MOA 4A (Belay) 2.13 SL	12 fl oz	12 hrs		
	imidacloprid MOA 4A (Admire Pro) 4.6SC	10.5 fl oz or 0.75 fl oz per 1,000 ft	3 days	60 days (NC Only) 125 days elsewhere	
Fruit fly (vinegar fly)	pyrethrins, MOA 3 (Pyrenone)	1 gal/100,000 cu ft	12 hrs	—	Postharvest application in storage. Apply as a space fog with a mechanical or thermal generator. Do not make more than 10 applications.

Table 5-9. Insect Control for Commercial Vegetables

CROP / Insect	Insecticide, Mode of Action Code, and Formulation	Amount of Formulation Per Acre	Restricted Entry Interval (REI)	Pre harvest Interval (PHI) (Days)	Precautions and Remarks
Sweetpotato (continued)					
Sweetpotato weevil	pyrethroid, MOA 3		12 hrs		See table 5-9B for registered pyrethroids and pre-harvest intervals.
	phosmet, MOA 1B (Imidan) 70 W	1.33 lb	5 days	7	
Thrips	spinetoram, MOA 5 (Radiant) 1 SC	6 to 8 fl oz	4 hrs	7	
Whitefringed beetle	phosmet, MOA 1B (Imidan) 70 W	1.33 lb	5 days	7	Do not make more than 5 applications per season. Whitefringed beetle adults are active in July and August. Do not plant in fields with a recent history of whitefringed beetles.
Tomato					
Aphid, Flea beetle	acetamiprid, MOA 4A (Assail) 30 SG	2 to 4 oz	12 hrs	7	Do not apply more than once every 7 days, and do not exceed 5 applications per season.
	clothianidin, MOA 4A (Belay) 50 WDG	4.8 to 6.4 oz (soil) 1.6 to 2.1 oz (foliar)	12 hrs	7	Soil applications at planting only.
	cyantraniliprole, MOA 28 (Verimark) 1.67 SC	6.75 to 13.5 fl oz	4	1	Soil applications at planting will control aphids and flea beetles. See label for application options.
	dimethoate 4 EC, MOA 1B	0.5 to 1 pt	48 hrs	7	Do not exceed rate with dimethoate as leaf injury may result.
	flonicamid, MOA 29 (Beleaf) 50 SG	2 to 4.8 oz	12 hrs	0	Will not control flea beetle.
	flupyradifurone, MOA 4D (Sivanto) 200 SL	7.0 to 10.5 fl oz	4 hrs	1	Will not control flea beetle.
	imidacloprid, MOA 4A Soil treatment (Admire Pro) 4.6 F (various) 2 F	7 to 10.5 fl oz 16 to 24 fl oz	12 hrs	21	For short-term protection at planting. Admire Pro may also be applied to transplants in the planthouse not more than 7 days before planting at the rate of 0.44 (4.6 F formulation) or 1 ounce (2 F formulation) per 10,000 plants. See label for soil application instructions.
	Foliar treatment (Admire Pro) 4.6 F (various) 1.6 F	1.2 fl oz 3.75 fl oz	12 hrs	0	
	pymetrozine, MOA 9B (Fulfill) 50 WDG	2.75 oz	12 hrs	0	For aphids only.
	spirotetramat, MOA 23 (Movento) 2SC	4 to 5 fl oz	24 hrs	1	Do not exceed 10 fl oz per season. Requires surfactant.
	thiamethoxam, MOA 4A (Platinum) 75 SG	1.66 to 3.67 oz	12 hrs	30	Platinum may be applied to direct-seeded crops in-furrow seeding or transplant depth, post seeding or transplant as a drench, or through drip irrigation. Do not exceed 11 ounces per acre per season. Check label for plant-back restrictions for a number of crops.
	(Actara) 25 WDG	2 to 3 oz	12 hrs	0	Actara is for foliar applications.
	spinetoram, MOA 5 (Radiant) 1 SC	5 to 10 fl oz	4 hrs	1	
Colorado potato beetle	acetamiprid, MOA 4A (Assail) 30 SG	1.5 to 2.5 oz	12 hrs	7	
	chlorantraniliprole, MOA 28 (Coragen) 1.67 SC	3.5 to 5 fl oz	4 hrs	1	Foliar or drip chemigation. Drip chemigation must be applied uniformly to the root zone. See label for instructions.
	cyantraniliprole, MOA 28 (Verimark) 1.67 SC (Exirel) 0.83 SE	5 to 10 fl oz 7 to 13.5 fl oz	4 hrs 12 hrs	1 1	Apply Verimark to soil via drip irrigation or soil injection. Exirel is for foliar application.
	imidacloprid, MOA 4A Soil treatment (Admire Pro) 4.6 F (various) 2 F	7 fl oz 16 fl oz	12 hrs	21	Use Admire Pro for soil or transplant drench treatment and 1.6 F formulation for foliar applications.
	Foliar treatment (Admire Pro) 4.6 F (various) 1.6 F	1.2 fl oz 3.75 fl oz	12 hrs	0	
	spinetoram, MOA 5 (Radiant) 1 SC	5 to 10 fl oz	4 hrs	1	
	thiamethoxam, MOA 4A (Platinum) 75 SG	1.66 to 3.67 oz	12 hrs	30	Platinum may be applied to direct-seeded crops in-furrow seeding or transplant depth, post seeding or transplant as a drench, or through drip irrigation. Do not exceed 11 oz per acre per season of Platinum. Check label for plant-back restrictions for a number of crops.
	(Actara) 25 WDG	2 to 3 oz	12 hrs	0	Actara is for foliar applications.
Armyworm, Cabbage looper, Hornworm, Tomato fruitworm, Pinworm	*Bacillus thuringiensis*, MOA 11A (Dipel) DF, MOA (Crymax) WDG	0.5 to 1 lb 0.5 to 1.5 lb	4 hrs	0	
	pyrethroid, MOA				See table 5-9B for a list of registered pyrethroids and pre-harvest intervals.

Table 5-9. Insect Control for Commercial Vegetables

CROP / Insect	Insecticide, Mode of Action Code, and Formulation	Amount of Formulation Per Acre	Restricted Entry Interval (REI)	Pre harvest Interval (PHI) (Days)	Precautions and Remarks
Tomato (continued)					
Armyworm, Cabbage looper, Hornworm, Tomato fruitworm, Pinworm (continued)	chlorantraniliprole, MOA 28 (Coragen) 1.67 SC	3.5 to 5 fl oz	4 hrs	1	Foliar or drip chemigation. Drip chemigation must be applied uniformly to the root zone. See label for instructions.
	cyantraniliprole, MOA 28 (Verimark) 1,67SC	5 to 10 fl oz	4 hrs	1	Verimark is for soil application only. Applications made at planting and/or via drip chemigation after planting. See label for application options.
	(Exirel) 0.83SE	7 to 13.5 fl oz	12 hrs	1	Exirel is for foliar application only.
	cyclaniliprole, MOA 28 (Harvanta) 50SL	16.4 fl oz	4 hrs	1	
	emamectin benzoate, MOA 6 (Proclaim) 5 WDG	2.4 to 4.8 oz	12 hrs	7	
	indoxacarb, MOA 22 (Avaunt eVo) 30 WDG	2.5 to 3.5 oz	12 hrs	3	Do not apply more than 14 ounces of Avaunt eVo (0.26 pound a.i.) per acre per crop. The minimum interval between sprays is 5 days.
	methomyl, MOA 1A (Lannate) 2.4 LV	1.5 to 3 pt	48 hrs	1	Methomyl may induce leafminer infestation.
	methoxyfenozide, MOA 18 (Intrepid) 2 F	4 to 10 fl oz	4 hrs	1	Use low rates for early-season applications to young or small plants and 6 to 10 ounces for mid- and late-season applications. Intrepid provides suppression of pinworm only.
	novaluron, MOA 15 (Rimon) 0.83 EC	9 to 12 fl oz	12 hrs	1	Do not make more than 3 applications per season.
	spinetoram, MOA 5 (Radiant) 1 SC	5 to 10 fl oz	4 hrs	1	
Cutworm	pyrethroid, MOA 3		12 hrs		See table 5-9B for a list of registered pyrethroids and pre-harvest intervals.
Leafminer	abamectin, MOA 6 (Agri-Mek) 0.7 SC	1.75 to 3.5 fl oz	12 hrs	7	Do not exceed 48 fluid ounces per acre per season, or more than 2 sequential applications.
	chlorantraniliprole, MOA 28 (Coragen) 1.67 SC	5 to 7.5 fl oz	4 hrs	1	Foliar or soil chemigation. Drip chemigation must be applied uniformly to the root zone. See label for soil application instructions.
	cyclaniliprole, MOA 28 (Harvanta) 50SL	16.4 fl oz	4 hrs	1	
	cryomazine, MOA 17 (Trigard) 75 WP	2.66 oz	12 hrs	0	See label for plant-back restrictions.
	spinetoram, MOA 5 (Radiant) 1 SC	6 to 8 fl oz	4 hrs	1	Do not exceed 29 fl oz per acre per season.
Spider mite	abamectin, MOA 6 (Agri-Mek) 0.7 SC	1.75 to 3.5 fl oz	12 hrs	7	Do not exceed 48 fluid ounces per acre per season, or more than two sequential applications.
	acequinocyl, MOA 29 (Kanemite) 15 SC	31 fl oz	12 hrs	1	The use of a surfactant/adjuvant with Kanemite on tomatoes is prohibited.
	bifenazate, MOA 20D (Acramite) 50 WS	0.75 to 1.0 lb	12 hrs	3	Do not make more than 1 application per season.
	cyflumetofen, MOA 25 (Nealta) 1.67 SC	13.7 fl oz	12 hrs	3	Do not make more than 1 application before using an effective miticide with a different mode of action.
	fenpyroximate MOA 21 (Portal) 0.4EC	2 pts	12 hrs	3	Do not make more than 2 applications per season.
Spider mite (continued)	spiromesifen, MOA 23 (Oberon) 2 SG	7 to 8.5 fl oz	12 hrs	7	Do not exceed 3 applications per season.
Stink bug	pyrethroid, MOA 3		12 hrs		See table 5-9B for a list of registered pyrethroids and pre-harvest intervals.
	dinotefuran, MOA 4A Foliar treatment (Venom) 70 SG (Scorpion) 35 SL	1 to 4 oz 2 to 7 fl oz	12 hrs	1	
	Soil treatment (Venom) 70 SG (Scorpion) 35 SL	5 to 6 oz 9 to 10.5 fl oz		21	
	thiamethoxam, MOA 4A (Actara) 25 WDG	3 to 5.5 oz	12 hrs	0	Do not exceed 11 ounces Actara per acre per season.
Thrips	dimethoate 4 EC, MOA 1B	0.5 to 1 pt	48 hrs	7	

Table 5-9. Insect Control for Commercial Vegetables

CROP / Insect	Insecticide, Mode of Action Code, and Formulation	Amount of Formulation Per Acre	Restricted Entry Interval (REI)	Pre harvest Interval (PHI) (Days)	Precautions and Remarks
Tomato (continued)					
Thrips (continued)	dinotefuran, MOA 4A Foliar treatment (Venom) 70 SG (Scorpion) 35 SL	 1 to 4 oz 2 to 7 fl oz	12 hrs	 1	See comments under Whitefly for application instructions and restrictions.
	Soil treatment (Venom) 70 SG (Scorpion) 35 SL	 5 to 6 oz 9 to 10.5 fl oz		 21	
	cyclaniliprole, MOA 28 (Harvanta) 50SL	10.9 to 16.4 fl oz	4 hrs	1	Harvanta will help suppress western flower thrips when used in a rotational program.
	flonicamid MOA 9c (Beleaf) 50 SG	2.4 to 4.8 fl oz	12 hrs	0	Beleaf has shown good activity against insecticide resistant western flower thrips.
	imidacloprid (Admire Pro) 4.6 SC	0.44 fl oz per 10,000 plants	12 hrs	—	For suppression of TSWV, treat transplants in the planthouse not more than 7 days before planting in the field. Transplants should be treated with overhead irrigation immediately after planting to ensure movement of imidacloprid into the soil media. See label for instructions.
	methomyl, MOA 1A (Lannate) 2.4 LV	1.5 to 3 pt	48 hrs	1	On foliage as needed.
	novaluron, MOA 15 (Rimon) 0.83 EC	9 to12 fl oz	12 hrs	1	Do not make more than 3 applications per season.
	spinetoram, MOA 5 (Radiant) 1 SC	6 to 10 fl oz	4 hrs	1	Will control thrips on foliage, not in flowers.
Whitefly	For resistance management of whiteflies, do not follow a soil application of a neonicotinoid (MOA group 4A) with a foliar application of any neonicotinoid. Locally resistant populations may affect the performance of specific insecticides.				
	acetamiprid, MOA 4A (Assail) 30 SG	2.5 to 4 oz	12 hrs	7	Do not apply more than once every 7 days, and do not exceed 5 applications per season.
	buprofezin, MOA 16 (Courier) 40 SC	9 to 13.6 fl oz	12 hrs	1	Use sufficient water to ensure good coverage. Do not apply more than twice per crop cycle, and allow 28 days between applications.
	chlorantraniliprole, MOA 28 (Coragen) 1.67 SC	5 to 7.5 fl oz	4 hrs	1	Foliar or soil application. Drip chemigation must be applied uniformly to the root zone. See label for soil application instructions.
	cyantraniliprole, MOA 28 (Verimark) 1,67 SC (Exirel) 0.83 SE	6.75 to 13.5 fl oz 13.5 to 20.5 fl oz	4 hrs 12 hrs	1 1	Apply Verimark to at planting and/or later via drip irrigation or soil injection. See label for application options. Exirel is for foliar application.
	dinotefuran MOA 4A Soil treatment (Venom) 70 SG (Scorpion) 35 SL	 5 to 6 oz 9 to 10.5 fl oz	12 hrs	 21	Soil applications of Venom or Scorpion may be made in a narrow band under the plant row as a post transplant drench, as a soil incorporated sidedress after plants are established, or in drip irrigation water. See label for instructions.
	Foliar treatment (Venom) 70 SG (Scorpion) 35 SL	 1 to 4 oz 2 to 7 fl oz		 1	
	imidacloprid, MOA 4A (Admire Pro) 4.6 F (various) 2 F	 16 to 24 fl oz 7 to 10.5 fl oz	12 hrs	21	Apply through a drip irrigation system or as a transplant drench with sufficient water to reach root zone. As a sidedress, apply 2 to 4 inches to the side of the row and incorporate 1 or more in. Residual activity will increase with increasing rates applied. Use higher rate for late-season or continuous infestations. Trickle irrigation applications will also control aphids and stinkbugs.
	pyriproxyfen, MOA 7C (Knack) 0.86 EC	8 to 10 fl oz	12 hrs	1	Do not apply more than 2 applications per growing season, and do not make applications closer than 14 days.
	spiromesifen, MOA 23 (Oberon) 2 SC	7 to 8.5 fl oz	12 hrs	7	Do not make more than 3 applications per season.
	spirotetramat, MOA 23 (Movento) 2SC	4 to 5 fl oz	24 hrs	1	Do not exceed 10 fluid ounces per season. Requires surfactant.
	thiamethoxam, MOA 4A (Platinum) 75 SG (Actara) 25 WDG	1.66 to 3.67 oz 3 to 5.5 oz	12 hrs 12 hrs	30 0	Platinum may be applied to direct-seeded crops in-furrow seeding or transplant depth, post seeding or transplant as a drench, or through drip irrigation. Do not exceed 11 ounces per acre per season of Platinum. Check label for plant-back restrictions for a number of crops. Actara is for foliar applications.
Wireworm	diazinon, MOA 1B (Diazinon) AG 500 or 50 WP	2 to 4 qt	48 hrs	—	Broadcast before planting and incorporate. Wireworms may be a problem in fields previously in pasture, corn, or soybean.
Turnip					
Aphid, Flea beetle	clothianidin, MOA 4A (Belay) 50 WDG	4.8 to 6.4 oz (soil) 1.6 to 2.1 oz (foliar)	12 hrs	7 (Foliar)	Soil application as in in-furrow, sidedress application, seed or transplant drench, or chemigation. See label for application instructions.

Table 5-9. Insect Control for Commercial Vegetables

CROP / Insect	Insecticide, Mode of Action Code, and Formulation	Amount of Formulation Per Acre	Restricted Entry Interval (REI)	Pre harvest Interval (PHI) (Days)	Precautions and Remarks
Turnip (continued)					
Aphid, Flea beetle (continued)	cyantraniliprole, MOA 28 (Verimark) 1,67 SC	6.75 to 13.5 fl oz	4 hrs	4	Soil applications made at planting only. See label for application options.
	dimethoate 4 EC, MOA 1B	0.5 pt	48 hrs	14	
	flonicamid, MOA 29 (Beleaf) 50 SG	2 to 2.8 oz	12 hrs	0	For aphids only.
	flupyradifurone, MOA 4D (Sivanto) 200 SL	7.0 to 10.5 fl oz	4 hrs	7	Will not control flea beetle
	imidacloprid, MOA 4A Soil treatment (Admire Pro) 4.6 F (various) 2 F	4.4 to 10.5 fl oz 10 to 24 fl oz	12 hrs	21	See label for soil application instructions.
	Foliar treatment (Admire Pro) 4.6 F (various) 1.6 F	1.2 fl oz 3.8 fl oz	12 hrs	7	
	pymetrozine, MOA 9B (Fulfill) 50 WDG	2.75 oz	12 hrs	7	Will not control flea beetle.
	thiamethoxam, MOA 4A (Platinum) 75 SG	1.7 to 4.01 oz	12 hrs	Apply at plant	Platinum is for soil application and Actara for foliar application.
	(Actara) 25 WDG	1.5 to 3 oz	12 hrs	7	
Harlequin bug, Vegetable weevil, Yellow margined leaf beetle	clothianidin, MOA 4A (Belay) 50 WDG	4.8 to 6.0 oz (soil)	12 hrs	21	Soil application as in in-furrow, side dress application, seed or transplant drench, or chemigation. See label for application instructions.
	imidacloprid, MOA 4A Soil treatment (Admire Pro) 4.6 F (Various) 2 F	4.4 to 10.5 fl oz 10 to 24 fl oz	12 hrs	21	Soil applications of imidacloprid will not control harlequin bug past 20 days after application.
	Foliar treatment (Admire Pro) 4.6 F (Various) 2 F	1.2 fl oz 2.8 fl oz		7	
	thiamethoxam, MOA 4A (Platinum) 75 SG (Actara) 25 WDG	1.7 to 4.01 oz 1.5 to 3 oz	12 hrs	Apply at plant 7	Platinum is for soil application and Actara for foliar application. .
	pyrethroid, MOA 3		12 hrs		See table 5-9B for a list of registered pyrethroids and pre-harvest intervals.
Cabbage looper, Diamondback moth	Insecticide-resistant populations, widespread in the Southeast, may not be controlled with some registered insecticides. To manage resistance, avoid transplants from Georgia and Florida, and avoid the repeated use of the same materials for extended periods of time. Repeated use of pyrethroid insecticides often aggravates diamondback moth problems. Do not allow populations to increase to large densities before treatments are initiated.				
	Bacillus thuringiensis, MOA 11A (Crymax) WDG (Dipel) 2 X (Dipel) 4 L (Xentari) WDG	0.5 to 1.5 lb 8 oz 1 to 2 pt 0.5 to 1 lb	4 hrs	0	On foliage, every 7 days as needed.
	chlorantraniliprole, MOA 28 (Coragen)	3.5 to 5.0 fl oz	4 hrs	1	For turnip greens or root turnips.
	cyantraniliprole, MOA 28 (Verimark) 1,67 SC (Exirel) 0.83 SE	5 to 10 fl oz 7 to 13.5 fl oz	4 hrs 12 hrs	1 1	Verimark and Exirel are for greens only, not root turnips. Verimark is for soil application only. Applications made at planting and/or later via drip chemigation. See label for application options. Exirel is for foliar application only.
	emamectin benzoate, MOA 6 (Proclaim) 5 WDG	2.4 to 4.8 oz	12 hrs	14	For turnip greens only.
	indoxacarb, MOA 22 (Avaunt eVo) 30 WDG	2.5 to 3.5 oz	12 hrs	3	Avaunt eVo may be applied only to turnip greens, not root turnips.
	spinetoram, MOA 5 (Radiant) 1 SC	3 to 6 fl oz	4 hrs	1	
Root maggot	chlorpyrifos, MOA 1B (Lorsban) 4 E (Lorsban) 75 WDG	1 to 2 pt 1.1 to 1.8 oz per 1,000 ft row	24 hrs	21	Irrigation or rainfall after application will enhance activity.
Watermelon (see Cucurbit Crops)					

Relative Effectiveness of Insecticides and Miticides for Insect and Mite Control on Vegetables

J. F. Walgenbach, Entomology and Plant Pathology, and G. G. Kennedy, Entomology Research

Table 5-9A. Relative Effectiveness of Insecticides and Miticides for Insect and Mite Control on Vegetables

Not all insecticides listed are registered on all vegetable crops. Refer to label before applying to a specific crop. Ratings are based on a consensus of vegetable entomologists in the southeastern United States. Table continued on following page.

(E = very effective; G = effective; F = somewhat effective; I = ineffective or insufficient data)

Chemical class (IRAC)	Common name	Example Product	Flea Beetle	Colorado potato beetle*	Cucumber beetles	Corn earworm*	European corn borer	Fall armyworm	Cabbage looper	Imported cabbageworm	Diamondback moth*	Squash vine borer
1A	carbaryl	Sevin	E	F	G	F	G	F	F	G	F	F
	methomyl	Lannate	F	I	I	G	G	G	G	G	G	I
	oxamyl	Vydate	F	F	F	I	I	I	I	I	I	I
1B	malathion	Malathion	G	F	G	F	F	F	F	G	F	F
	chlorpyrifos	Lorsban	I	I	I	F	F	F	F	G	F	I
	acephate	Orthene	I	I	I	F	E	G	F	G	I	I
	diazinon	Diazinon	I	I	I	I	I	I	I	I	I	I
	dimethoate	Dimethoate	G	I	F	I	I	I	I	I	I	I
3	permethrin	Pounce	G	F	G	G	G	F	G	E	F	E
	alpha cypermethrin	Fastac	E	F	E	G	E	G	G	E	F	E
	zeta cypermethrin	Mustang Max	E	F	E	G	E	G	G	E	F	E
	cyfluthrin	Baythroid/Renounce	G	F	G	G	G	F	G	E	F	E
	lambda cyhalothrin	Karate	E	F	E	G	E	G	G	E	F	E
	esfenvalerate	Asana XL	G	G	G	G	G	F	G	E	F	G
	gamma cyhalothrin	Proaxis	E	F	E	G	E	G	G	E	F	E
	fenpropathrin	Danitol	G	I	G	G	G	F	F	E	F	G
	bifenthrin	Brigade	E	F	E	G	G	F	F	E	F	E
4A	imidacloprid	Admire	F	G	E	I	I	I	I	I	I	I
	acetamiprid	Assail	G	E	G	I	I	I	I	I	I	F
	clothianidin	Belay	E	E	G	I	I	I	I	I	I	I
	thiamethoxam	Platinum/Actara	E	G	G	I	I	I	I	I	I	I
	dinotefuran	Venom/Scorpion	E	E	G	I	I	I	I	I	I	I
4C	sulfoxaflur	Closer/Transform	I	I	I	I	I	I	I	I	I	I
4D	flupyradifurone	Savinto	I	I	I	I	I	I	I	I	I	I
5	spinosad	Blackhawk/Entrust	I	E	I	G	G	G	G	E	G	G
	spinetoram	Radiant	I	E	I	E	E	G	G	E	G	G
6	emamectin benzoate	Proclaim	I	I	I	G	G	G	E	E	E	G
	abamectin	AgriMek	I	E	I	I	I	I	I	I	I	I
7C	pyriproxyfen	Knack/Distance	I	I	I	I	I	I	I	I	I	I
9B	pymetrozine	Fulfill	I	I	I	I	I	I	I	I	I	I
29	flonicamid	Beleaf	I	I	I	I	I	I	I	I	I	I
10B	etoxazole	Zeal	I	I	I	I	I	I	I	I	I	I
11A	Bt	Dipel, various	I	I	I	F	F	F	G	E	G	F
15	novaluron	Rimon	I	E	I	E	E	E	G	E	F	G
16	buprofezin	Courier	I	I	I	I	I	I	I	I	I	I
17	cyromazine	Trigard	I	G	I	I	I	I	I	I	I	I
18	methoxyfenozide	Intrepid	I	I	I	G	G	E	E	E	F	G
20B	acequinocyl	Kanemite	I	I	I	I	I	I	I	I	I	I
20D	bifenazate	Acramite	I	I	I	I	I	I	I	I	I	I
21A	fenpyroximate	Portal	I	I	I	I	I	I	I	I	I	I
	tolfenpyrad	Torac	G	I	I	F	F	F	F	G	I	I
22A	indoxacarb	Avaunt	F	G	F	E	G	G	E	E	G	G
23	spiromesifen	Oberon	I	I	I	I	I	I	I	I	I	I
	spirotetramat	Movento	I	I	I	I	I	I	I	I	I	I
25	cyflumetofen	Nealta	I	I	I	I	I	I	I	I	I	I
28	chlorantraniliprole	Coragen	I	E	I	E	E	E	E	E	E	G
	cyantraniliprole	Verimark/Exirel	G	E	I	E	E	E	E	E	E	G
	cyclaniliprole	Harvanta	F	E	I	E	E	G	G	E	E	G

Table 5-9A. Relative Effectiveness of Insecticides and Miticides for Insect and Mite Control on Vegetables (continued)

Chemical class (IRAC)	Common name	Example Product	Beet armyworm*	Stinkbugs/ Harlequin bug	Squash bug	Aphids*	Thrips	Western Flower Thrips*	Leafminer	Maggots	Whiteflies*	Cutworms	Wireworms	White grubs	Spider mites*
1A	carbaryl	Sevin	I	I	I	I	F	I	I	I	I	F	I	I	I
	methomyl	Lannate	F	G	G	F	E	G	F	I	F	I	I	I	I
	oxamyl	Vydate	I	F	F	G	G	F	I	I	F	I	I	I	==
1B	malathion	Malathion	I	F	F	F	F	I	I	F	I	F	I	I	I
	chlorpyrifos	Lorsban	I	I	I	I	F	I	I	E	I	G	G	G	I
	acephate	Orthene	I	I	I	G	G	I	F	I	F	G	I	I	I
	diazinon	Diazinon	I	I	I	I	I	I	I	G	I	F	G	F	I
	dimethoate	Dimethoate	I	G	F	E	E	F	G	I	I	I	I	I	I
3	permethrin	Pounce	I	F	G	F	F	I	F	I	I	G	I	I	I
	zeta cypermethrin	Mustang Max	I	G	E	F	F	I	F	I	I	E	I	I	I
	cyfluthrin	Baythroid/Renounce	I	G	E	F	F	I	F	I	I	E	I	I	I
	lambda cyhalothrin	Karate	I	G	E	F	F	I	F	I	I	E	I	I	I
	esfenvalerate	Asana XL	I	F	G	F	F	I	F	I	I	G	I	I	I
	gamma cyhalothrin	Proaxis	I	E	E	F	F	I	F	I	I	E	I	I	I
	fenpropathrin	Danitol	I	E	E	F	F	I	F	I	I	G	I	I	F
	bifenthrin	Brigade	I	E	E	F	G	I	F	F	I	E	G	F	F
4A	imidacloprid	Admire	I	F	G	E	G	I	I	G	G	I	F	G	I
	acetamiprid	Assail	I	F	F	E	G	I	I	I	G	I	I	I	I
	clothianidin	Belay	I	G	G	G	I	I	F	G	F	I	F	G	I
	thiamethoxam	Platinum/Actara	I	G	G	E	F	I	F	G	G	I	F	F	I
	dinotefuran	Venom/Scorpion	I	G	G	F	G	I	F	I	G	I	I	I	I
4C	Sulfoxaflor	Closer/Transform	I	F	F	E	F	I	I	I	E	I	I	I	I
4D	flupyradifurone	Savinto	I	I	I	E	I	I	I	I	G	I	I	E	I
5	spinosad	Blackhawk/Entrust	G	I	I	I	G	G	E	I	I	F	I	I	I
	spinetoram	Radiant	G	I	I	I	E	G	E	I	I	F	I	I	I
6	emamectin benzoate	Proclaim	E	I	I	I	I	I	F	I	I	F	I	I	I
	abamectin	AgriMek	I	I	I	I	G	F	E	I	I	I	I	I	E
7C	pyriproxyfen	Knack/Distance	I	I	I	I	I	I	I	I	E	I	I	I	I
9B	pymetrozine	Fulfill	I	I	I	E	I	I	I	I	F	I	I	I	I
9C	flonicamid	Beleaf	I	I	I	E	G	E	I	I	I	I	I	I	I
10B	etoxazole	Zeal	I	I	I	I	I	I	I	I	I	I	I	I	G
11A	Bt	Dipel, various	F	I	I	I	I	I	I	I	I	I	I	I	I
15	novaluron	Rimon	E	F	F	I	G	G	G	I	G	I	I	I	I
16	buprofezin	Courier	I	I	I	I	I	I	I	I	G	I	I	I	I
17	cyromazine	Trigard	I	I	I	I	I	I	E	I	I	I	I	I	I
18	methoxyfenozide	Intrepid	E	I	I	I	I	I	I	I	I	I	I	I	I
20B	acequinocyl	Kanemite	I	I	I	I	I	I	I	I	I	I	I	I	E
20D	bifenazate	Acramite	I	I	I	I	I	I	I	I	I	I	I	I	E
21A	fenpyroximate	Portal	I	I	I	I	I	I	I	F	I	I	I	I	G
	tolfenpyrad	Torac	F	I	G	G	F	I	I	F	I	I	I	I	I
22	indoxacarb	Avaunt eVo	E	I	I	I	I	F	I	I	F	I	I	I	I
23	spiromesifen	Oberon	I	I	I	I	I	I	I	I	G	I	I	I	G
	spirotetramat	Movento	I	I	I	E	I	I	I	I	G	I	I	I	I
25	cyflumetofen	Nealta	I	I	I	I	I	I	I	I	I	I	I	I	G
28	chlorantraniliprole	Coragen	E	I	I	F	I	E	I	G	I	I	I	I	I
	cyantraniliprole	Verimark/Exirel	E	I	I	G	F	F	E	G	G	I	I	I	I
	cyclaniliprole	Harvanta	E	I	I	I	F	F	E	I	F	I	I	I	I

* Denotes that insecticide-resistant populations may occur in some areas and can affect the performance of insecticides.

Preharvest Intervals for Pyrethroid Insecticides in Vegetable Crops

Table 5-9B. Preharvest Intervals (in Days) for Pyrethroid Insecticides in Vegetable Crops

See Table 5-9A to compare relative efficacy of these products against specific insect pests. Read the pesticide label for specific rates and application instructions.

		Common Name/Example Product (Restricted Entry Interval – REI)										
		alpha cypermethrin Fastac (12 hrs)	beta cyfluthrin Baythroid XL (12 hrs)	bifenthrin Brigade (12 hrs)	cypermethrin Various brands (12 hrs)	cyfluthrin Tombstone (12 hrs)	esfenvalerate Asana XL (12 hrs)	fenpropathrin Danitol (24 hrs)	gamma cyhalothrin Proaxis (24 hrs)	lambda cyhalothrin Karate/Warrior (24 hrs)	permethrin Pounce (12 hrs)	zeta cypermethrin Mustang Max (12 hrs)
	Asparagus	NR	NR	NR	NR	NR	NR	NR	NR	NR	1	NR
Bulb Vegetables	Onions, Green	NR	NR	NR	7	NR	NR	NR	NR	NR	NR	7
	Onions, Dry Bulb	NR	NR	NR	7	NR	NR	NR	14	14	1	7
Brassica Leafy Vegetables	Broccoli, Brussels Sprout, Cabbage, Cauliflower, Kohlrabi	1	0	7	1	0	3	7	1	1	1	1
	Collard, Mustard Green	1	0	7	1	0	7	NR	NR	NR	1	1
Cereal Corn	Sweet Corn	3	0	1	NR	0	1	NR	1	1	1	3
Cucurbits	Cantaloupe, Watermelon	1	0	3	NR	0	3	7	NR	1	0	1
	Cucumber, Pumpkin, Summer Squash, Winter Squash	1	0	3	NR	0	3	7	NR	1	0	1
Fruiting Vegetables	Eggplant, Pepper	1	7	7	NR	0	7	3	5	5	3	1
	Tomato	1	0	1	NR	7	1	3	5	5	0	1
	Okra	1	NR	7	NR	NR	NR	NR	NR	NR	NR	1
Legumes	Edible-podded	1	NR	3	NR	NR	3	NR	7	7	NR	1
	Succulent Shelled Pea and Bean	1		3	NR		3	7	7	7	NR	1
	Dried Shelled Pea and Bean	21	7	14	NR	7	21	NR	21	21	NR	21
Leafy Vegetables, Except Brassica	Head and Leaf Lettuce	1	0	7	5[C]	0	7[A]	NR	1	1	1	1
	Spinach	1	0	40	NR	0	NR	NR	NR	NR	1	1
	Celery	1	0	NR	NR	0	NR	NR	NR	NR	3	1
Root and Tuber Vegetables	Beet, Carrot, Radish. Turnip	1	0	21	NR	0	7	NR	NR	NR	1	1
	Potato	1	0	21	NR	0	NR	NR	NR	7	14	1
	Sweetpotato	1	0	21	NR	0	NR	NR	NR	7	NR	1

NR Not registered

[A] Head lettuce only

Insect Control for Greenhouse Vegetables
J. F. Walgenbach and G. G. Kennedy, Entomology and Plant Pathology

Sound cultural practices, such as sanitation and insect-free transplants, help prevent insect establishment and subsequent damage. Separate plant production houses, use of yellow sticky traps, and timely sprays will help prevent whitefly buildup. Use of *Encarsia* parasites for whitefly and other biological control agents in conjunction with use of pesticides is encouraged. Unless a pesticide label specifically states that a product cannot be used in a greenhouse vegetable crop, the product can be used on those crops for which it is registered. However, pesticides behave differently in the field and the greenhouse, and for many products, information is not available on greenhouse crop phytotoxicity and residue retention. If unsure of the safety of a product to a crop, apply to a small area before treating the entire crop.

Table 5-10. Insect Control for Greenhouse Vegetables

CROP / Insect	Insecticide and Formulation	Amount of Formulation	Re Entry Interval	Pre Harvest Interval (PHI) (Days)	Precautions and Remarks
Cucumber					
Aphid	flonicamid, MOA 29 (Beleaf) 50SG	0.065 to 0.1 oz per 1000 sq ft	12 hrs	0	May be applied either to the soil as a drench or drip irrigation for preventive control, or sprayed onto plants as a rescue treatment.
	imidacloprid, MOA 4A (Admire Pro) 4.6 F	0.6 fl oz/1,000 plants	12 hrs	0	Apply in a minimum of 21 gallons water using soil drenches, micro-irrigation, or drip irrigation. Do not apply to immature plants as phytotoxicity may occur. Make only 1 application per crop per season.
	insecticidal soap (M-Pede) 49 EC	2 bsp/gal water	12 hrs	0	
Cabbage looper	*Bacillus thuringiensis*, MOA 11 (various)	0.5 to 1 lb OR 3 pt/100 gal water	4 hrs	—	
	spinosad, MOA 5 (Entrust) SC	3 fl oz/100 gal	4 hrs	1	Do not make more than 2 consecutive applications. OMRI approved.
Spider mite	insecticidal soap (M-pede) 49 EC	2 bsp/gal water	12 hrs		Use predatory mites.
	mineral oil (TriTek)	1 to 2 gal/100 gal	4 hrs	0	Begin applications when mite populations are low, and repeat at weekly intervals.
	fenpyroximate, MOA 21A (Akari) 5SC	1 to 2 pts per 100 gal	12 hrs	7	
Whitefly, Leafminer	acetamiprid, MOA 4A (Assail) 30 SG	0.1 oz per 1000 sq ft	12 hrs	0	
	flonicamid, MOA 20 (Beleaf) 30 SG	0.065 to 0.1 oz per 1000 sq ft	12 hrs	0	
	imidacloprid, MOA 4A (Admire Pro) 4.6 F	0.6 fl oz/1,000 plants	12 hrs	0	Apply in a minimum of 21 gallons water using soil drenches, micro-irrigation, or drip irrigation. Do not apply to immature plants as phytotoxicity may occur. Make only 1 application per crop per season.
	insecticidal soap (M-Pede) 49 EC	2 bsp/gal water	12 hrs	0	May be used alone or in combination. Acts as an exciter.
	Beauveria bassiana (Bontanigard) 22 WP (Mycotrol) WP	1 lb/100 gal water 0.25 lb/20 gal water	4 hrs	0	Apply when whiteflies observed. Repeat in 4- to 5-day intervals.
Lettuce					
Aphid, Leafminer, Whitefly	pymetrozine, MOA 9B (Fulfill) 50 WG	0.063 oz per 1000 sq ft	12 hrs	0	Will not control leafminer.
	pyrethrins, MOA 3 (Pyganic) 5EC	0.25 to 0.5 fl oz per gal water	12 hrs	0	May be used alone, or tank mixed with a companion insecticide (see label for details).
	malathion, MOA 1B (various) 57 EC / 25 WP	1 qt/100 gal water / 4 lb/100 gal water	24 hrs	14 / 14	
	insecticidal soap (M-Pede) 49 EC	2 bsp/gal water	12 hrs	0	May be used alone or in combination. Acts as an exciter. Insecticidal soaps can cause phytotoxicity under high temperatures or slow drying conditions. If unsure, apply to a small area before treating the entire crop.
	Beauveria bassiana (Mycotrol WP)	0.25 lb/20 gal water	4 hrs	0	Under high aphid or whitefly pressure, apply at 2- to 5-day intervals.
Cabbage looper	*Bacillus thuringiensis*, MOA 11 (Javelin) WG	0.5 to 1.25/100 gal water	4 hrs	0	
	spinosad, MOA 5 Entrust SC	3 fl oz/100 gal	4 hrs	1	Do not make more than 2 consecutive applications.
Slugs	iron phosphate (Sluggo)	0.5 to 1 lb/1,000 sq ft	4 hr	1	Scatter the bait around the perimeter of the greenhouse to provide a protective barrier. If slugs are within the crop, then scatter the bait on the ground around the plants. Do not make more than 3 applications within 21 days. Sluggo will control slugs and snails, while Bug-N-Sluggo will also control earwigs, cutworms, sowbugs and pillbugs. Both are OMRI approved.
	iron phosphate + spinosad (Bug-N-Sluggo)	0.5 to 1 lb/1,000 sq ft	4 hr	1	
Spider mite	insecticidal soap (M-Pede) 49 EC	2 tbsp/gal water	12 hrs	0	
	mineral oil (TriTek)	1 to 2 gal/100 gal	4 hrs	0	Begin applications when mite populations are low, and repeat at weekly intervals.
Tomato, Pepper					
Aphid	flonicamid, MOA 20 (Beleaf) 50 SG	0.1 oz per 1000 sq ft	12 hrs	0	May be applied to the soil as a drench or drip irrigation for preventive control, or as a spray for rescue treatments. Will also control whiteflies.
	imidacloprid, MOA 4A (Admire Pro) 4.6 F	0.6 fl oz/1,000 plants	12 hrs	0	Apply in a minimum of 21 gallons water using soil drenches, micro-irrigation, or drip irrigation. Do not apply to immature plants as phytotoxicity may occur. Make only 1 application per crop per season. Also controls whiteflies.

Table 5-10. Insect Control for Greenhouse Vegetables

CROP Insect	Insecticide and Formulation	Amount of Formulation	Re Entry Interval	Pre Harvest Interval (PHI) (Days)	Precautions and Remarks
Tomato, Pepper (continued)					
Aphid (continued)	malathion, MOA 1B (various) 10 A 57 EC 25 WP	1 lb/50,000 cu ft 1 qt/100 gal water 4 lb/100 gal water	12 hrs	15 hr 1 1	
	insecticidal soap (M-Pede) 49 EC	2 tbsp/gal water	12 hrs	0	May be used alone or in combination. Acts as an exciter.
	Beauveria bassiana (Mycotrol WP)	0.25 lb/20 gal water		0	Apply when whiteflies are observed. Repeat in 4-to 5-day intervals.
Armyworm, Fruitworm, Cabbage looper, Cutworm, Pinworm	*Bacillus thuringiensis*, MOA 11 (Javelin) WG (Agree) WP (Dipel) DF Xentari DF	0.5 lb to 1.25 lb/100 gal water 1 to 2 lb 0.5 to 1.25 0.5 to 1.5	4 hrs	0	
	chlorfenapyr MOA 13 (Pylon) 2SC	6.5 to 13 fl oz/100 gal water or per acre area	12 hrs	0	For use on tomatoes more than 1 inch in diameter at maturity. Do not make more than 2 applications at 5 to 10 day intervals before rotating to an insecticide with a different mode of action.
	cyantraniliprole, MOA 28 (Exirel) SE	7 to 13.5 fl oz per acre, or per 100 gal	12 hrs	1	
	spinosad, MOA 5 Entrust SC	3 fl oz/100 gal	4 hrs	1	Do not make more than 2 consecutive applications. Do not apply to seedling tomatoes or peppers grown for transplants.
Leafminer	cyantraniliprole, MOA 28 (Exirel) SE	13.5 to 20.5 fl oz per acre, or per 100 gal	12 hrs	1	
	diazinon, MOA 1B (Diazinon, Spectracide) (AG 500) 50 WP	4 to 8 oz/100 gal water	48 hrs	3	Keep ventilators closed for 2 hours or overnight. Plant injury may result if labeling directions are not followed. For use by members of N.C. Greenhouse Vegetable Growers Association only.
	spinosad, MOA 5 (Entrust) SC	10 fl oz/100 gal	4 hrs	1	Do not apply to seedlings grown for transplants.
Slug	metaldehyde (various) bait	Follow label directions	12 hr		Apply to soil surface around plants. Do not contaminate fruit.
	iron phosphate (Sluggo)	½ teaspoon per 9-inch pot		0	
Spider mite, broad mite, rust mite	acequinocyl, MOA 20B (Kanemite) 15 SC	31 fl oz/100 gal	12 hr	1	
	bifenazate (Floramite) SC,	4 to 8 fl oz/100 gal water (1/4 to 1/2 tsp/gal)	12 hr	3	For use on tomatoes more than 1 inch in diameter at maturity. Not registered on pepper. Not for rust mite
	mineral oil (TriTek)	1 to 2 gal/100 gal		0	Begin applications when mite populations are low, and repeat at weekly intervals.
	chlorfenapyr, MOA 13 (Pylon) 2 SC	9.8 to 13 fl oz/100 gal water or per acre area		0	For use on tomatoes more than1 inch in diameter at maturity. Do not make more than 2 applications at 5 to 10 day intervals before rotating to an insecticide with a different mode of action.
	fenpyroximate, MOA 21A (Akari) 5 SC	1 to 2 pts per 100 gal	12 hrs	1	
	insecticidal soap (M-Pede) 49 EC	2 tbsp/gal water	12 hrs	0	
Thrips, including western flower	*Beauveria bassiana* (Mycotrol WP)	0.25 lb/20 gal water		0	Use screens on intake vents. Apply when whiteflies observed. Repeat in 4- to 5-day intervals.
	chlorfenapyr, MOA 13 (Pylon) 2SC	9.8 to 13 fl oz/100 gal water or per acre area		0	For use on tomatoes more than1 inch in diameter at maturity. Do not make more than 2 applications at 5 to 10 day intervals before rotating to an insecticide with a different mode of action.
	cyantraniliprole, MOA 28 (Exirel) SE	13.5 to 20.5 fl oz per acre, or per 100 gal	12 hrs	1	For foliage-feeding thrips only, not those in flowers.
	spinosad, MOA 5 (Entrust) SC	5.5 fl oz/100 gal	4 hrs	1	Do not make more than 2 consecutive applications, and do not apply more than 6 times in a 12 month period against thrips. Do not apply to seedlings grown for transplants.
Whitefly	*Beauveria bassiana* (BontaniGard) 22 WP (Mycotrol) WP	1 lb/100 gal water 0.25 lb/20 gal water	4 hrs	0	Apply when whiteflies are observed. Repeat in 4- to 5-day intervals.
	buprofezin, MOA 16 (Talus) 40 SC	9 to 13.6 oz/100 gal water or per acre area	12 hrs	1	Insect growth regulator that affects immature stages of whiteflies. Will not kill adults. For use on tomatoes only.
	cyantraniliprole, MOA 28 (Exirel) 0.83 SE	13.5 to 20.5 fl oz/100 gal water or per acre area	12 hrs	1	
	flonicamid, MOA 29 (Beleaf) 50 SG	0.1 oz per 1,000 sq ft	12 hrs	0	For use on tomato only.
	imidacloprid, MOA 4A (Admire Pro) 4.6 F	0.6 fl oz/1,000 plants	12 hrs	0	Apply in a minimum of 21 gallons water using soil drenches, micro-irrigation, or drip irrigation. Do not apply to immature plants as phytotoxicity may occur. Make only 1 application per crop per season. Also controls aphids.
	insecticidal soap (M-Pede) 49 EC	2 tbsp/gal water	12 hrs	0	
	pyrethrins and PBO, MOA 3 (Pyganic) 5 EC	0.25 to 0.5 fl oz per gal	12 hrs	0	May be used alone, or tank mixed with a companion insecticide. (See label for details.)
	pyriproxyfen, MOA 7C (Distance) 0.86 EC	6 fl oz/100 gal water	12 hrs	<1	Insect growth regulator that affects immature stages of whiteflies. Will not kill adults. Do not use on tomatoes more than 1 inch in diameter. Do not apply on non-bell pepper.

Insect Control for Livestock and Poultry

W. Watson, Entomology and Plant Pathology

Table 5-11A. Insect Control for Cattle

Insect Insecticide and Formulation	Amount of Formulation to Use in Water	Dosage per Animal	Minimum Interval (Days) Between Application and Harvest	Precautions and Remarks
Cattle Grub—(a) Beef and non-lactating dairy animals				
				Make all grub treatment after heel fly season ends but before Oct. 1.
doramectin (Dectomax) injectable	—	1 cc/110 lb	35	Not for female dairy cattle over 20 months of age.
ivermectin injectable pour-on bolus	— — —	1 cc/110 lb 1 ml/22 lb See label	49 48 —	Not for female dairy cattle of breeding age. For calves older than 12 weeks of age.
moxidectin (Cydectin) 0.5 PO	—	5 ml/110 lb	0	Not for use on lactating dairy cattle.
tetrachloryinphos (Rabon) 3.0 D	—	3 to 4 oz		Applied to backline and rubbed into warbles.
Cattle Grub—(b) Dairy animals (also beef and non-lactating dairy animals)				
eprinomectin (Eprinex) pour-on	—	1 ml/22 lb	0	
Horn Fly—(a) Dairy and beef animals				
coumaphos (CoRal) 1 D	—	3 to 6 tbsp	0	Repeat as necessary.
cyfluthrin (CyLence) 1 PO	—	—	0	Follow label instructions.
diflubenzuron oral larvicide (Clarify)	—	—	—	In feed according to label.
eprinomectin (Eprinex) pour-on	—	1 ml/22 lb	0	Effective control for 7 days only.
methoprene (Altocid) liquid	—	5 lb/ton of feed	—	Mixed into liquid feed
methoprene mineral mix	—	—	0	Daily in feed according to label.
moxidectin (Cydectin) 0.5 PO	—	5 ml/100 lb	0	Not for use on lactating dairy cattle.
permethrin EC or PO	—	—	0	See label for rate and application directions.
permethrin + diflubenzuron	—	3 ml/110 lb		See label for rate and application directions.
pyrethrins 0.1 OS + synergist	—	1 to 2 oz	0	Oil sprays will harm skin if not applied properly. Apply oil solutions daily as a mist.
tetrachlorvinphos (Rabon) 7.76 D oral larvicide	— —	0.032 oz/100 lb body wt.	—	Daily in feed according to label.
SELF-APPLICATING DEVICES coumaphos (Co-Ral) permethrin tetrachlorvinphos (Rabon) 3 D tetrachlorvinphos + dichlorvos (RaVap) 23 EC	— 5 oz/1 gal oil	4 qts/13 gal fuel oil — — —	0	For dairy and beef animals. These devices aid in face fly and louse control. Follow all label instructions. Inspect and charge oilers and dust bags weekly as needed.
EAR TAGS abamectin (XP820) beta-cyfluthrin (CyGuard) coumaphos + diazinon (CoRal Plus, Corathon) cyfluthrin (Cutter Gold, CyLence Ultra) cypermethrin (Python, Magnum) diazinon (40%) (Patriot) diazinon (20%) (Optimizer) diazinon + chlorpyrifos (Warrior) lambda-cyhalothrin (Saber) permethrin (GardStar) pirimiphos-methyl (Dominator) cypermethrin + abamectin + PBO (Tri-Zap)	—	2/head for optimal control		These devices give season-long fly control. Some tags are not for use on lactating dairy cattle. Some tags are restricted from use on calves under the age of 3 months. Use according to label. Other ear tags are available. Contact Entomology Department, N.C. State University, for current tag list.
Horn Fly—(b) Beef animals				
lambda-cyhalothrin Aim Capsule		1 capsule (600 lb)		Smart Vet applicator required
gamma cyhalothrin (StandGuard) pour-on		10 ml < 600 lb 15 ml > 600 lb		Do not apply more than once in 2 weeks or more than 4 times in 6 months.
ivermectin PO bolus	— —	1 ml/22 lb	48 —	Not for female dairy cattle of breeding age. Controls horn flies for up to 28 days. Bolus for calves older than 12 weeks of age.
tetrachlorvinphos (Rabon) 50 WP	5 oz/5 gal	2 to 4 qt	0	
SELF-APPLICATING DEVICES tetrachlorvinphos + dichlorvos (RaVap) 23 EC	5 oz/1 gal oil	—	0	For beef only. These devices aid in face fly and louse control.

Table 5-11A. Insect Control for Cattle

Insect Insecticide and Formulation	Amount of Formulation to Use in Water	Dosage per Animal	Minimum Interval (Days) Between Application and Harvest	Precautions and Remarks
Lice—(a) Dairy and beef animals				
coumaphos (CoRal) 1 D 6.15%	—	3 to 6 Tbsp		
	2.5 oz/4 gal	—	0	Spray thoroughly—wet to skin.
cyfluthrin (CyLence) 1 PO	—	—	—	Follow label instructions.
eprinomectin (Eprinex) pour-on	—	1 ml/22 lb	0	Follow label instructions.
permethrin EC PO permethrin plus diflubenzuron (Cleanup II)	See label	— —	0	Follow label instructions. Spray entire animal, second treatment at 14 to 21 days. Pyrethroid and IGR blend to control all louse life stages. Follow label instructions.
tetrachlorvinphos (Rabon) 3 D	—	2 oz	0	
Lice—(b) Beef animals				
gamma cyhalothrin (StandGuard) pour-on		10 ml < 600 lb 15 ml > 600 lb		Do not apply more than once in 2 weeks or more than 4 times in 6 months.
coumaphos 6.15%	5 oz/4 gal	—	0	Spray—wet to skin.
doramectin (Dectomax) injectable	—	1 cc/110 lb	35	Not for female dairy cattle over 20 months of age.
ivermectin injectable pour-on bolus	— — —	1 cc/110 lb 1 ml/22 lb —	49 48 —	Not for female dairy cattle of breeding age. Injection ineffective for control of biting lice. Pour-on controls both biting and sucking lice. Bolus for calves older than 12 weeks of age.
lambda-cyhalothrin (Saber) 1 PO	—	—	0	Follow label instructions.
lambda-cyhalothrin Aim Capsule		1 capsule (600 lb)		Smart Vet applicator required
moxidectin (Cydectin) 0.5 PO	—	5 ml/110 lb	0	Not for lactating dairy cattle.
tetrachlorvinphos (Rabon) 50 WP	5 oz/5 gal	2 to 4 oz	0	Spray thoroughly.
tetrachlorvinphos + dichlorvos (RaVap) 23 EC	See label	—	0	Do not treat more often than every 10 days. Spray entire animal.
Note: Self-applicating devices under horn fly aid in louse control.				
Face Fly				
lambda-cyhalothrin Aim Capsule		1 capsule (600 lb)		Smart Vet applicator required
cyfluthrin (CyLence) 1 PO	See label	—	—	Follow label instructions.
permethrin EC PO	See label See label	—	0	Follow label instructions.
diflubenzuron oral larvicide (Clarify)	—	—	—	In feed according to label.
EAR TAGS abamectin (XP820) beta cyfluthrin (CyGuard) cyfluthrin (Cutter Gold, CyLence Ultra) coumaphos + diazinon (Corathon) cypermethrin (Python, Magnum) diazinon + chlorpyrifos (Warrior) diazinon (40%) (Patriot) fenvalerate (Ectrin) lambda-cyhalothrin (Saber) permethrin (GardStar) pirimiphos-methyl (Dominator) cypermethrin + abamectin+PBO (Tri-Zap)		2/head for optimal control	0 0	These devices give season-long fly control or aid in the control of face flies. Some tags are not for use on lactating dairy cattle. Use according to label. Other ear tags are available. Contact Entomology Department, N.C. State University, for current tag list.
Note: Self-applicating devices under horn fly aid in face fly control.				
Mange				
doramectin (Dectomax) injectable	—	1 cc/110 lb	35	Not for female dairy cattle over 20 months of age.
eprinomectin (Eprinex) pour-on	—	1 ml/22 lb	0	Follow label instructions.
ivermectin injectable pour-on bolus	— — —	1 cc/110 lb 1 ml/22 lb —	49 48 —	Not for female dairy cattle of breeding age. Injection ineffective for control of biting lice. Pour-on controls both biting and sucking lice. Bolus for calves older than 12 weeks of age.
moxidectin (Cydectin) 0.5 PO	—	5 ml/110 lb	0	Not for lactating dairy cattle.
permethrin EC or PO	See label	—	0	Follow label instructions. Spray entire animal, second treatment at 14 to 21 days.
Maggots in Wounds				
coumaphos 6.15%	See label	—	—	—
permethrin 0.5% (Catron IV)	—	—	—	Spray wound directly and thoroughly. Repeat 5-7 days until healed.
pyrethrin + PBO + Dinpropyl isocinchomeronate	See label	—	—	

Table 5-11A. Insect Control for Cattle

Insect Insecticide and Formulation	Amount of Formulation to Use in Water	Dosage per Animal	Minimum Interval (Days) Between Application and Harvest	Precautions and Remarks
Stable Fly, Horse Fly, Deer Fly				
pyrethrins 0.1 OS plus synergist				May give protection for short periods.
Mosquitoes; Dairy and beef animals				
permethrin			0	
Ticks—Dairy and beef animals				
coumaphos 6.15%	5 oz/4 gal	—	10	Not for use on lactating dairy animals. Spray animals thoroughly.
permethrin	See label	—	0	
tetrachlorvinphos (Rabon) 50 WP	4 lb/50 gal	0.5 to 1 gal	—	Do not treat lactating dairy animals. Treat about every 3 weeks during periods of heavy tick activity. Spray animals thoroughly.
tetrachlorvinphos + dichlorvos (Rabon + Vapona, RaVap) EC	1 qt/50 gal	—	0	Spray animals completely.
House Fly, Lesser House Fly, Stable Fly, Other Filth Flies—Premises: beef and dairy				
bifenthrin (ActiShield) 7.9L	See label	0.33 to 1 fl oz/1,000 sq ft	—	May be applied as crack and crevice treatment while animals are present.
chlorpyrifos (Durashield), 20 CS	See label	—	—	Restricted use insecticide.
cyfluthrin (Tempo, Countdown) 20 WP or 2 L	See label	—	—	Do not apply when animals are present.
Cypermethrin Fendona CS	See label	2 to 5 oz/1,000 sq ft	—	Microencapsulated for controlled release.
Deltamethrin (Annihilator Polyzone)	0.25-1.5 oz/gal	1 pt/10,666-64,000 sq ft	—	Do not apply when animals are present
dichlorvos (Vapona) 2 EC or 4 EC	—	—	—	Fog, mist, or surface spray. Remove livestock before treatment.
gamma-cyhalothrin (StandGuard) 5.9 MC	See label	—	—	
lambda-cyhalothrin (Grenade, OxyFly) 9.7 ER	See label	—	—	
permethrin 25 WP or EC	See label	—	—	
pyrethrins 0.1 OS + synergist	—	—	—	Fog or mist.
spinosad (Elector) 44.2 PSP	2 oz/10 gal water	See label	Lactating and non-lactating cattle may be present when applied	Do not use more than once each week. Do not make more than 5 consecutive applications.
tetrachlorvinphos (Rabon) 50 WP	4 lb/25 gal	0.5 to 1 gal/500 sq ft	—	
tetrachlorvinphos + dichlorvos (RaVap) 23 EC	5 oz/1 gal	1 gal/500 to1,000 sq ft	—	Surface treatment only. DO NOT use as a space spray.
LARVICIDE cyromazine (Neporex) 2 SG	See label	Spray or dry application: 1 lb/200 sq ft	21	For larval control in manure or animal bedding only.
BAIT MIXTURES dichlorvos (Vapona) imidacloprid (QuickBayt) cyantraniliprole (Cyanarox) methomyl (Golden Malrin, Apache) nithiazine (QuikStrike) strip spinosad (Elector Bait) Beauveria bassiana (balEnce Bait)				Do not apply baits in areas accessible to animals. Labeled for organic farming.

Table 5-11B. Insect Control for Sheep and Goats

Insecticide and Formulation	Amount of Formulation to Use in Water	Dosage per Animal	Minimum Interval (Days) Between Application and Harvest	Precautions and Remarks
Lice and Sheep Ked				
pyrethrin + PBO permethrin (Gordons) 0.25	See label — —	0.5 to 2.0 oz/100 lb	—	
Blow Fly, other maggots in wounds				
permethrin 0.5% (Catron IV)	—	—	—	Spray wound directly and thoroughly. Repeat 5-7 days until healed.

Table 5-11C. Insect Control for Swine

Insecticide and Formulation	Amount of Formulation to Use in Water	Dosage per Animal	Minimum Interval (Days) Between Application and Harvest	Precautions and Remarks
Cockroaches, Spiders				
cyfluthrin (Tempo) 20 WP or 2 L	See label	—	—	
House Fly, Stable Fly—Premises				
bifenthrin (ActiShield) 7.9 L	See label	0.33 to 1 fl oz/1,000 sq ft	—	May be applied as crack and crevice treatment while animals are present.
cypermethrin Fendona CS	See label	2 to 5 oz/1,000 sq ft		Microencapsulated for controlled release.
cyromazine (Neporex) 2 G	See label	Spray or dry application: 1 lb/200 sq ft	21	For larval control only in manure or animal bedding.
deltamethrin (Annihilator Polyzone)	0.25 to 1.5 oz/gal	1 pt/10,666-64,000 sq ft	—	Do not apply when animals are present
gamma-cyhalothrin (StandGuard) 5.9 MC	See label	—	—	
lambda-cyhalothrin (OxyFly) 97 ER	—	—	—	
Beauveria bassiana (balEnce)	See label	See label	—	Labeled for organic farming.
Lice				
ivermectin injectable pre mix	—	1 cc/75 lb 300 g/ton	18 5	Continually feed for 7 days. For feeder pigs and finish hogs ONLY.
permethrin		—	5	Spray entire animal until thoroughly wet.
phosmet (Prolate/Lintox 11.75%)			1	Retreat in 14 days.
tetrachlorvinphos (Rabon) 50 WP	7 oz/5 gal	1 to 2 qt	0	
Mange Mite				
doramectin (Dectomax) injectable	—	1 cc/ 75 lb	24	
ivermectin injectable pre mix (Ivomec only)	— —	1 cc/75 lb 300 g/ton	18 5	Continually feed for 7 days. For feeder pigs and finishing hogs ONLY.
permethrin EC 10 PO (Swine Guard)		— 3 ml/100 lb	5	Spray entire animal until thoroughly wet. See label for correct rates and treatment intervals.
phosmet (Prolate/Lintox 11.75%)	2 qt in 50 gal		1 to harvest	Retreat in 14 days
Maggots in Wounds				
permethrin 0.5% (Catron IV	—	—	—	Spray wound directly and thoroughly. Repeat 5-7 days until healed.
House Fly				
tetrachlorvinphos (Rabon oral larvicide)				See label.
Also see CATTLE		—	—	Treat according to label.

Table 5-11D. Insect Control for Horses

Insecticide and Formulation	Amount of Formulation to Use in Water	Precautions and Remarks
Bot		
ivermectin (Zimecterin, Eqvalan)		Follow all instructions.
Horse Fly, Deer Fly, Mosquito		
For materials and control suggestions see CATTLE section.		
House Fly, Stable Fly—Premises		
bifenthrin (ActiShield) 7.9 L	See label	May be applied as crack and crevice treatment while animals are present.
cypermethrin Fendona CS	See label	2-5 oz/1,000 sq ft
cyromazine (Neporex) 2G	See label	Spray or dry application to stall bedding or muck pile.
(Solitude IGR) 2.1		In feed to control fly larvae in manure.
gamma-cyhalothrin (StandGuard) 5.9 MC	See label	
lambda-cyhalothrin		
Beauveria bassiana (balEnce)	See label	Organic labeling.
Horn Fly, Face Fly, House Fly, Stable Fly		
cypermethrin (Tri-Tec 14)		Follow label instructions.
dichlorvos (Vapona) + pyrethrin + piperonyl butoxide		Follow label instructions.
permethrin (Ectiban, Atroban, Tech-Trol, Tech-Trol 12, Permectrin II)		Follow label instructions.
permethrin + piperonyl butoxide (Poridon) (Flysect-7)		Pour on for fly control. Spray.
pyrethrin + piperonyl butoxide		Follow label instructions.
tetrachlorvinphos (Rabon oral larvicide)		In feed, mixed, or top-dressed for control of fly larvae in manure.
AUTOMATIC SPRAY SYSTEMS resmethrin; natural pyrethrins + piperonyl buxide		Follow label instructions.
BAIT MIXTURES dichlorvos (Vapona), imidacloprid (QuickBayt), methomyl (Golden Malrin, Apache), nithiazine (QuikStrike) Strip, spinosad (Elector Bait)		Do not apply baits in areas accessible to animals.

Table 5-11E. Insect Control for Poultry

Insecticide and Formulation	Amount of Formulation in Water	Dosage	Precautions and Remarks
Chicken Mite			
permethrin	See label	—	Provide easy-to-clean roosts and nests with few hiding places. Apply sprays thoroughly to roosts and cracks in surrounding areas. Repeat application as necessary. Follow labels carefully. Treatment of birds as for northern mite also helps.
Northern Fowl Mite, Lice			
permethrin	—	1 gal spray/100 birds	
permethrin (Poultry Mite Tags)	—	2 tags/bird	Follow label directions.
tetrachlorvinphos (Rabon) 50 WP	6.5 oz/5 gal	1 gal/100 birds or 1 to 2 gal/1,000 sq ft of litter	Direct on birds. Thorough coverage and feather penetration is essential. Follow labels carefully. Use 100 to 125 psi for good penetration. Apply premises spray as necessary to reduce NFM/lice dislodged from birds.
3 D	—	1 lb/300 birds or 1 lb/100 sq ft of litter	Direct on birds. Thorough coverage and feather penetration is essential. Follow labels carefully. Use 100 to 125 psi for good penetration. Apply premises spray as necessary to reduce NFM/lice dislodged from birds.
tetrachlorvinphos + dichlorvos (RaVap) 23 EC	5 oz/1 gal	1 gal/100 birds; 1 to 2 gal/1,000 sq ft of litter	Direct on birds. Thorough coverage and feather penetration is essential. Follow labels carefully. Use 100 to 125 psi for good penetration. Apply premises spray as necessary to reduce NFM/lice dislodged from birds.
House Fly, Lesser House Fly, Stable Fly, Other Filth Flies—Premises			
bifenthrin (ActiShield) 7.9 L	See label	0.33 to 1 fl oz/1,000 sq ft	May be applied as crack and crevice treatment while animals are present.
chlorpyrifos (Durashield) 20 CS	See label		Restricted use insecticide. Surface treatment only. DO NOT use as a space spray.
cyfluthrin (Tempo, Countdown, Optem) 20 WP or 2 L	See label	—	Remove birds from building prior to treatment of interior surfaces.
cypermethrin Fendona CS	See label	2 to 5 oz/1,000 sq ft	Microencapsulated for controlled release.
deltamethrin (Annihilator Polyzone)	0.25 to 1.5 oz/gal	1 pt/10,666-64,000 sq ft	Remove birds from building prior to treatment of interior surfaces.
dichlorvos (Vapona) 40 EC	—		Fog, mist, or surface spray. See label.

Table 5-11E. Insect Control for Poultry

Insecticide and Formulation	Amount of Formulation in Water	Dosage	Precautions and Remarks
House Fly, Lesser House Fly, Stable Fly, Other Filth Flies—Premises (continued)			
gamma-cyhalothrin (StandGuard) 5.9 MC	See label		
lambda-cyhalothrin (Grenade, OxyFly) 9.7 ER	See label	—	See cyfluthrin.
permethrin	See label	—	
pyrethrins 0.1 OS + synergist	See label	—	Fog or mist.
spinosad (Elector PSP) 44.2 spray	2oz/10 gal water	Spray thoroughly, prevent runoff 5,000 to 10,000 sq ft	
Beauveria bassiana (balEnce) spray	—	—	Apply as directed. Organic labeling.
tetrachlorvinphos (Rabon) 50 WP	4 lb/25 gal	0.5 to 1 gal/500 sq ft	
tetrachlorvinphos + dichlorvos (RaVap) 23EC	5 to 10 oz/1 gal	1 gal/500 to1,000 sq ft	
tetrachlorvinphos + dichlorvos (RaVap) 23 EC	5 oz/1 gal		Apply larvicide as spot treatment.
tetrachlorvinphos (Rabon) 50 WP	4 lb/25 gal		Apply larvicide as spot treatment.
LARVICIDES cyromazine (Neporex) 2 G			For use in all poultry.
(Flyzine, Larvadex) premix	See label	1 lb/ton of feed	Approved as a manure treatment for broiler breeders and caged layers only. Feed continuously for 4 to 6 weeks.
(Larvadex) 2 SL		Spray or dry application: 1 lb/200 sq ft	For use as manure spray for broiler breeders and caged layers.
BAIT MIXTURES cyantraniliprole (Zyrox) dichlorvos (Vapona) imidacloprid (QuickBayt) methomyl (Golden Malrin, Apache) nithiazine (QuikStrike) bait strip spinosad (Elector Bait) Beauveria bassiana (balEnce Bait)	—		Do not apply baits in areas accessible to poultry. Use as directed.
Northern Fowl Mite			
permethrin 2.5%	See label	2.5 oz/gal	No more than 1 gal spray per 100 birds, apply directly to the vent region for thorough coverage.
spinosad (Elector) PSP 44.1%		3 oz/10 gal water	No more than 1 gal spray per 100 birds, apply directly to the vent region for thorough coverage.
Scaly-Leg Mite			
crude petroleum oil	Undiluted	Dip shanks	
Chigger			
permethrin	—	See label	Apply day before poultry is put on range. Repeat in 2 to 3 weeks.
Stick-Tight Flea			
permethrin	—	See label	May be applied to birds.
pyriproxyfen (Pyri-Shield) 1.3 EC	—	—	Use in tank mix with permethrin as premise treatment.
Vaseline	—	Rub into areas of head where pest is attached	Keep dogs and other animals out of poultry areas. Yards, nesting, and roosting areas should be cleaned frequently.
Bed Bug, Fowl Tick			
bifenthrin (ActiShield) 7.9 L	See label	0.33 to 1 fl oz/1,000 sq ft	May be applied as crack and crevice treatment while birds are present.
cyfluthrin (Tempo, Countdown) 20 WP or 2 L	See label	—	Remove birds prior to treatment.
cypermethrin Fendona CS	See label	2 to 5 oz/1,000 sq ft	Microencapsulated for controlled release.
dichlorvos (Vapona) 40 EC	—	—	Use according to label.
lambda-cyhalothrin (Grenade) 9.7 ER	See label	—	
permethrin	—	—	
Darkling Beetle (Lesser Mealworm)			
bifenthrin (ActiShield) 7.9 L	See label	0.33 to 1 fl oz/1,000 sq ft	May be applied as crack and crevice treatment while poultry are present.
carbaryl (Sevin) 80 WSP 43 SL	—	—	Limited to building exteriors; see label.
cyfluthrin (Tempo, Countdown, Optem) 20 WP or 2 L	See label		Remove birds prior to treatment.
cypermethrin Fendona CS	See label	2-5 oz/1,000 sq ft	Microencapsulated for controlled release
spinosad (Elector) PSP 44.1%		2-4 oz/5,000 sq ft	

Table 5-11E. Insect Control for Poultry

Insecticide and Formulation	Amount of Formulation in Water	Dosage	Precautions and Remarks
Darkling Beetle (Lesser Mealworm) (continued)			
gamma-cyhalothrin (StandGuard) 5.9 MC	See label		
tetrachloryinphos (Beetle Shield) 6%		1.5-4 oz/100 sq ft	Apply with a duster.
spinosad (Elector) 44.2 PSP	2 oz/10 gal water	See label	Do not use more than once each week. Do not make more than 5 consecutive applications.
imidacloprid (Credo, Dominion, Exile DB) 428 CS	3 fl oz/0.5 to 2 gal water	1 gal/1,000 sq ft	
lambda-cyhalothrin (Grenade) 9.7 ER	See label		Remove birds prior to treatment.
(OxyFly) 9.7 R	—	—	
permethrin	—	—	
pyriproxyfen (Pyri-Shield) 1.3 EC	1 fl oz/gal	1 gal/1,000 to 1,500 sq ft	This slow-acting insect growth regulator is most effective when used in combination with other insecticides.
spinosad (Elector PSP) 44.2 spray	See label		
tetrachlorvinphos (Rabon) 50 WP 3 D	4 lb/50 gal —	1 to 2 gal/1,000 sq ft 1 lb/100 sq ft	Do not treat houses with birds 6 weeks old or less.
tetrachlorvinphos + dichlorvos (RaVap) 23 EC	5 to 10 oz/1 gal	1 gal/500 to 1,000 sq ft	
zetacypemethrin (ZetaGard LBT) Granular	50 lb/house	See Label	6 weeks withdrawal period before slaughter
Imported Fire Ants			
See COMMUNITY PEST CONTROL			
Rodents			
See ANIMAL DAMAGE CONTROL chapter— Rodenticides			

Community Pest Control

M. Waldvogel, Entomology and Plant Pathology and M. Reiskind, Entomology Research

NOTE: Insecticides recommended for use by Certified Applicators only. For rodents, see Animal Damage Control, Chapter 9.

Table 5-12A. Community Pest Control — Mosquito Adults[1]

Read pesticide labels carefully. Some pesticide products are not approved for application to edible plants. Avoid spraying flowering plants when bees are actively foraging.
KEY: Dv 0.9 = 90% of the spray volume droplets are smaller than value given VMD = Volume Median Diameter; um = micrometer

TYPE OF APPLICATION Insecticide and Formulation	Mixing Instructions and Application Equipment	Application Rate at 10 mph	Droplet Size Requirements on Label (um)	Precautions and Remarks
Ground Application				
bifenthrin 7.9L	0.33 to 1.0 fl oz/gal water in backpack or hydraulic sprayer			Apply at a rate of 1 gallons per 1000 square feet for thorough coverage of lawns and/or ornamentals.
Clove oil (Nature-Cide)	1:9 to 1:39 dilution in water		Outdoors – apply to wet surfaces but not to the point of run-off.	Treat with mist or spray around landscape plants, turf, ground cover, under decks, around building foundations where mosquitoes may rest.
deltamethrin (Suspend Polyzone)	0.33 to 1.0 fl oz/gal water in backpack or hydraulic sprayer			Treat with mist or spray around landscape plants, turf, ground cover, under decks, around building foundations where mosquitoes may rest.
etofenprox (Aqua Zenivex E20)	Apply undiluted or up to 1:4.5 dilution	Varies with dilution	VMD-7-30 um Dx 0.9 < 50 um	Do not apply more than 0.18 lbs per acre per site per year. Do not make more than 25 applications per site per year.
garlic oil% (ATSB concentrate)	38 fl oz/gal water in a backpack or hydraulic sprayer			Apply at a rate of 15 ounces per 100 linear feet to vegetation 1 to 5 feet above the ground wetting both surfaces of foliage to the point of runoff. Do not apply with handheld or truck-mounted cold ULV or thermal foggers or by aircraft.
lamda-cyhalothrin (Cyonara 9.7, Demand CS, Cysmic CS)	0.8 fl. oz/gal. water in backpack or hydraulic sprayer			Treat resting areas on structures as well as surrounding shrubs.
malathion 96.5% concentrate (Fyfanon ULV)	Use undiluted on aerosol ULV sprayer.	2 to 4.3 fl oz	VMD < 30 um Dv 0.9 < 50 um	Do not spray when wind speed is more than 5 mph.
	Dilute 3.9 to 5.2 gal to 100 gal with No. 2 fuel or diesel oil; use in thermal fog sprayer.			Avoid direct application to vehicles; these insecticides may damage paint. Apply when air temperatures are cool and wind speed is 3 mph or less. Toxic to fish, aquatic invertebrates, and wildlife.
naled (Dibrom) 87.4% concentrate	10 fl oz to 10 gal No. 2 fuel or diesel oil; use in thermal fog sprayer.	80 gal/hr	VMD < 40 um Dv 0.9 < 75 um	Toxic to fish, aquatic invertebrates, and wildlife. Restricted Use Pesticide.

Table 5-12A. Community Pest Control — Mosquito Adults[1]

Read pesticide labels carefully. Some pesticide products are not approved for application to edible plants. Avoid spraying flowering plants when bees are actively foraging.
KEY: Dv 0.9 = 90% of the spray volume droplets are smaller than value given VMD = Volume Median Diameter; *um* = micrometer

TYPE OF APPLICATION Insecticide and Formulation	Mixing Instructions and Application Equipment	Application Rate at 10 mph	Droplet Size Requirements on Label (*um*)	Precautions and Remarks
naled (Dibrom) 87.4% concentrate (continued)	Dilute 0.5 gal to 5 gal with soybean oil or HAN; use in ULV sprayer.	6 to 12 fl oz/min	VMD < 40 um Dv 0.9 < 75 um	Do not directly apply to water or to areas where runoff into water is likely to occur.
permethrin 10% to 57% concentrate	Apply undiluted or mix with refined mineral or soybean oil.	0.31 to 15 oz/min depending on dilution	VMD = 150 to 300 um	Permethrin 57% is not for use in residential misting systems. Do not allow drift onto cropland, poultry ranges or potable water supplies. Do not use on crops used for food or forage.
(Permanone) 10% EC	Dilute 1:20 with water (6.5 fl oz/ 1 gal of water).			Treat surfaces using course wet spray. Spray to runoff.
permethrin (20%) and piperonyl butoxide (20%) (Aqua-Reslin)	Dilute 1 gal with 2 to 12 gal water	2.1 to 9 oz/min depending on dilution	VMD < 30 um Dv 0.9 < 50 um	Dilute with water only. Toxic to fish and aquatic invertebrates. Can be used as barrier spray on building foundations (maximum height of 3') and vegetation around structure but not within 100 feet of lakes and streams. Structural applications to areas other than foundation limited to crack & crevice.
permethrin and piperonyl butoxide (Permanone 31-66, Biomist 4+12 ULV)	Dilute 1 gal to 2.4 gal with light weight oil; use in ULV sprayer.	0.5 to 3 fl oz/min	VMD < 30 um Dv 0.9 < 50 um	
prallithrin (1%) and sumithrin (5%) and piperonyl butoxide (5%) (Duet)	Apply undiluted in aerosol ULV sprayer	2.5 to 7.5 oz/min	VMD = 8 to 30 um Dv 0.9 < 50um	Do not allow drift onto pastureland, rangeland, or potable water supplies.
pyrethrins (5%) and piperonyl butoxide (25%) (Aquahalt)	Apply undiluted or diluted with water and applied as an ultra low volume (ULV)	0.27 to 0.76 oz/ac	VMD < 30 um Dv 0.9 < 50um	Do not apply more than 0.2 pounds of pyrethrins/acre/year or 1.0 pound PBO/acre/year in any treated area
resmethrin (18%) + piperonyl butoxide (54%) Scourge	Dilute 0.67 gal with 1 gal of light mineral oil; use in ULV sprayer.	4.5 to 9 fl oz/min	VMD < 30 um Dv 0.9 < 50um	Restricted-Use Pesticide. Can be applied ULV or diluted with refined soybean oil, light mineral oil of 54 second viscosity or other suitable solvent or diluent.
sumithrin and piperonyl butoxide (Anvil 10+10 ULV or 2+2 ULV)	Use undiluted or dilute 10+10 formulation with light mineral oil.	1.3 to 18.6 oz/min	VMD < 30 um Dv 0.9 < 50um	
Fixed Wing Aerial Application				
etofenprox (Aqua Zenivex E20)	0.00175 to 0.007 oz (undiluted) per acre	Varies with dilution	VMD <60 um Dx 0.9 < 100 um	Do not apply at altitudes below 100 feet. Do not apply more than 0.10 lbs per acre per site per year. Do not make more than 25 applications per site per year.
malathion 96.5% concentrate (Fyfanon ULV)	Use undiluted	2.6 to 3 fl oz/acre	VMD <60 um Dx 0.9 < 100 um	Toxic to fish, aquatic invertebrates, and wildlife. Do not directly apply to water or to areas where runoff into water is likely to occur. Do not retreat a site more than 3 times in any one week except in emergencies. Do not spray by fixed wing aircraft below 100 feet or by helicopter below 75 feet.
naled (Dibrom) 87.4% concentrate	Use undiluted.	0.5 to 1 fl oz/acre	VMD = 60 um Dv 0.9 < 115um	Toxic to fish, aquatic invertebrates, and wildlife. Do not directly apply to water, except when necessary to target areas where adult mosquitoes are present or to areas where runoff into water is likely to occur. Not for use in or around homes.
Fixed Wing Aerial Application (continued)				
naled (Dibrom) 87.4% concentrate (continued)	Dilute 50 to 100 fl oz to 100 gal with No. 2 fuel oil or diesel oil.	1 gal/acre	VMD = 60 um Dv 0.9 < 115um	Toxic to fish, aquatic invertebrates, and wildlife. Do not directly apply to water, except when necessary to target areas where adult mosquitoes are present or to areas where runoff into water is likely to occur. Not for use in or around homes.
permethrin (20%) and piperonyl butoxide (20%) (Aqua-Reslin)	Dilute 1 gal with 2 to 12 gal water	2.1 to 9 oz/min depending on dilution	VMD < 60 um Dv 0.9 < 100um	Dilute with water only. Toxic to fish and aquatic invertebrates.
resmethrin (18%) + piperonyl butoxide (54%) (Scourge)	Dilute 0.67 gal with 1 gal of light mineral oil; use in ULV sprayer.	4.5 to 9 fl oz/min	VMD < 60 um Dv 0.9 < 100 um	Restricted-Use Pesticide. Can be applied ULV or diluted with refined soybean oil, light mineral oil of 54 second viscosity or other suitable solvent or diluent.
prallithrin (1%) and sumithrin (5%) and piperonyl butoxide (5%) (Duet)	Apply undiluted on aerosol ULV sprayer	0.41 to 1.24 oz/ac	VMD = < 60 um	Do not allow drift onto pastureland, rangeland, or potable water supplies.
pyrethrins (5%) and piperonyl butoxide (25%) (Aquahalt)	Apply undiluted or diluted with water and applied as an ultra low volume (ULV)	0.27 to 0.76 oz/ac	VMD < 60 um Dv 0.9 < 80um	Do not apply more than 0.2 pounds of pyrethrins/acre/year or 1.0 pound PBO/acre/year in any treated area
sumithrin and piperonyl butoxide (Anvil 10+10)	Use undiluted.	3.8 to 5.7 fl oz/acre	VMD < 60 um Dv 0.9 < 80 um	

[1] Avoid direct applications to flowering plants when pollinators are active. Do not allow drift onto adjoining non-target areas. When treating residential properties, cover or remove pet food and water sources, grills, swimming pools and children's toys. Note: Treatment of structures (exterior or interior) requires a P-phase Structural Pest Control License in North Carolina.

Table 5-12B. Community Pest Control — Mosquito Immatures and Other Pests

PEST Insecticide and Formulation	Mixing Instructions and Application Equipment	Application Rate Per Acre	Precautions and Remarks
Mosquito—Immatures			
Bacillus thuringiensis, var. *israelensis* (Bactimos, Teknar, Vectobac) 50 WP 2 WP 14.3% aqueous conc. 15% aqueous conc. 1.2% aqueous conc. 0.8% aqueous conc.	Dilute with sufficient water to obtain uniform coverage.	6 to 12 oz 4 to 16 oz 0.5 to 3 pt 0.5 to 3 pt 0.25 to 2 pt 0.5 to 2 pt	Only effective against larvae. Can be applied to all breeding habitats, including potable water supplies.
Bacillus thuringiensis, var. *israelensis* (Bactimos) briquets 10%			Use one briquet per 100 square feet of surface area regardless of depth.
(Bactimos, Teknar, Vectobac) granules 0.2% pellets 0.4%	Ready to use	—	Apply 4 to 10 pounds per acre with aircraft or ground equipment.
methoprene (Altosid) 20% EC	3 to 4 fl oz/gal water	1 gal	Apply when larvae are in 3rd and 4th instar. Methoprene will not kill pupae or adults.
(Altosid) briquet 2.1%, 8.6% pellet 4.2%	Ready to use	—	Water less than 2 feet; 1 briquet per 100 square feet; deeper or flowing water; 1 briquet per 10 cubic feet.
(Altosid) granule 0.27%, 1.5%			2.5- to 10-pound pellet per acre; use high rate in breeding sites with high organic content.
spinosad (Natular XRG) 2.5% granule	Ready to use	5 to 20 lb	
monomolecular surface film (Agnique MMF)	—	0.2 to 0.5 gal	Use in conjunction with indicator oil to avoid over treatment.
proprietary mosquito control oils (GB-1111, etc.)	—	1 to 5 gal	Dosage depends on amount of floatage and vegetation in water.
temephos (Abate) 43% EC 1 G 2 G 5 G	0.5 to 1.5 fl oz/gal water — — —	1 gal 5 to 10 lb 2.5 to 5 lb 1 to 2 lb	
Midge ("fuzzy bills")			
temephos (Abate) 1 G 2 G 5 G	— — —	5 to 10 lb 2.5 to 5 lb 2 lb	Double recommended rates for water high in organic content.
methoprene 20% EC (Strike)	4 to 5 oz/ 1 million gal wastewater	—	For use in wastewater treatment facilities. Uniformly apply at the influent side over a 24-hr period.
4.25% pellet (Strike)	—	5 to 10 lb/acre	Apply to natural and manmade aquatic habitats. High rate recommended for wastewater.
spinosad (Natular XRG) 2.5% granule	Ready to use	5 to 20 lb	
Tick			
acetamiprid-permethrin (Transport)	—	Apply 0.11% concentration of active ingredient to cover 1,000 sq. ft.	Do not apply more than 0.11% finished dilution per 1,000 square feet.
bifenthrin (Talstar) 0.2% G 7.9% L	Ready to use 1 fl oz/100 gal water	100 to 200 lb/acre	Do not allow public use of area during treatment. 1 gallon per 1,000 square feet.
carbaryl (Sevin) 50 WP	0.1 lb/10 gal water	870 gal	Keep children and pets off treated areas until they have dried.
Clove oil (Nature-Cide)	1:9 to 1:39 dilution in water		Outdoors – apply to wet surfaces but not to the point of run-off.
cyfluthrin (Tempo) 24% EC 20% WP	5.9 fl oz/40 to 100 gal water 7.7 oz	40 to 100 gal	
deltatamethrin (Suspend Polyzone) 4.75T L	0.25 to 1.5 fl oz/gal water	1 to 3 gal/1,000 sq ft	Do not allow public use of area during treatment
imidacloprid-cyfluthrin (Temprid)	0.075% - 0.15% fl oz/gal water		Apply at rate not to cause drip/run-off from site
permethrin (Permethrin SFR)	1 2/3 fl oz/gal of water	0.4 to 0.8 fl. Oz/1000 sq ft	Do not allow public use of area during treatment. 1 gallon per 1,000 square feet
rosemary oil, Geraniol, Wintergreen (Essentria IC3)	1 to 8 oz of Essentria IC3 per gallon of water	43 gal	2 gallons per 1,000 square feet
Imported Fire Ants			
acetamiprid-bifenthrin (Transport Mikron)		Apply 0.11% concentration of active ingredient to cover 1,000 sq. ft.	Do not apply more than 0.11% finished dilution per 1,000 square feet.
avermectin (Ascend, Black Flag Fire Ant Ender) 0.011% B	—	1 lb	For use on turf, lawns, and other noncrop areas, such as parks and golf courses. Apply when soil temperature is greater than 60 degrees F. Apply after dew or rainfall has dried for maximum effectiveness.
cyfluthrin (Tempo) 24% EC 20% WP	5.9 fl oz/40 to 100 gal water 7.7 oz	40 to 100 gal	

Table 5-12B. Community Pest Control — Mosquito Immatures and Other Pests

PEST Insecticide and Formulation	Mixing Instructions and Application Equipment	Application Rate Per Acre	Precautions and Remarks
deltatamethrin (Suspend Polyzone) 4.75T L	0.25 to 1.5 fl oz/gal water	1 to 3 gal/1,000 sq ft	Do not allow public use of area during treatment
fenoxycarb (Award) 1.0% B	—	1 to 1.5 lb	Uniformly distribute 1 to 3 tablespoons around the edge of each mound. For broadcast applications, apply 1 to 1.5 pounds per acre. May be used on pastures and grazed areas on horse farms if horses are not intended for human consumption.
fipronil (Topchoice Granular) 0.0143%	—	87 lb	For use on home lawns, golf courses, commercial and recreational turf, and sod farms. One application of 87 pounds of product/acre per year. Restricted-Use Pesticide.
hydramethylnon (Amdro, Amdro Pro) B	—	1 to 1.5 lb	Broadcast uniformly on pasture and range grass, lawns, turf, and nonagricultural lands. Or distribute 5 level tbsp 3 to 4 feet around base of each mound (do not exceed 1.5 pounds per acre). Cutting/baling restrictions for pastures with dairy or beef cows.
hydramethylnon 0.365% + S-Methoprene 0.25% (Extinguish Plus) B	—	1.5 lb	Broadcast uniformly on pasture and range grass, lawns, turf, and nonagricultural lands. Or distribute 2 to 5 level tablespoons 3 to 4 feet around base of each mound (do not exceed 1.5 pounds per acre).
indoxacarb (Advion) 0.045%	—	1.5 lb	For use in outdoor areas on noncroplands.
metaflumizone (Siesta) (0.0653% B	-	1.5 lb	Broadcast uniformly on target area or use 2 to 4 level tablespoons 3 to 4 feet around base of each mound (do not exceed 1.5 pounds per acre).
methoprene (Extinquish) 0.5% B	—	1 to 1.5 lb	For use on crop and noncroplands, such as parks, zoos, sports fields, and school grounds.
pyriproxyfen (Distance) 0.5% B	—	1 to 1.5 lb	For use in outdoor areas on noncroplands.
spinosad (Conserve) 0.15% B	-	4 lbs	May require 2 applications per year. (OMRI certified)

For treatment of individual ant mounds with liquid insecticides, refer to the section on insect control for home lawns.

Industrial and Household Pests

M. Waldvogel and P. Alder, Entomology and Plant Pathology

For Use by Licensed Pest Management Professionals

Space limitations preclude listing all pesticide formulations and trade names. Other products or formulations may be used. Some products may contain a mixture of active ingredients. Read the product label for specific information about the active ingredients, application rates, and detailed instructions on use—particularly on permitted sites for application.

Mention of pesticides in this section does not imply that chemicals are or should be the first or only means of pest control. Nonchemical methods, including exclusion and sanitation, are important to long-term pest management.

Table 5-13. Industrial and Household Pests—For use by licensed pest management professionals only

Pesticide	Boric acid (Niban, Perma-Dust, Intice)		Diatomaceous earth (Mother Earth D)	Silica gel (Drione, Tri-Die, Cimex Dust)	Sodium Tetraborate (Gourmet Liquid Ant Bait, Cymex, Dominant Ant Bait)		Methomyl (Flytek, Sysco Fly Bait)	Propoxur (PT 2, Invader)	Acephate (Orthene)	Dichlorvos (Nuvan)	
Chemical Class[1]	Inorganic						Carbamate		Organophosphate		
Formulation[2]	Bait	Dust	Dust[3]	Dust[3]	Bait	Dust	Bait[4]	Sprayable	Sprayable[7]	Strip	Sprayable
Pests											
ANTS	X	X	X	X	X[6]	X			X		X
BED BUGS			X	X		X		X		X	X
BEES				X					X		X
BOOKLICE	X		X	X							
BUGS (TRUE)[4]	X			X				X	X	X	
CARPET BEETLES			X			X				X	X
CENTIPEDES	X		X	X				X	X		
CLOTHES MOTHS			X							X	X
CLOVER MITES	X		X	X				X	X		
COCKROACHES	X	X	X	X				X	X	X	X
CRICKETS	X	X	X	X					X		X
EARWIGS	X	X	X	X				X	X	X	
FLEAS		X	X	X	X				X		X
FLIES	X		X	X			X			X	X
HORNETS/WASPS				X					X		X
LADY BEETLES			X	X							
MILLIPEDES	X		X	X				X	X		X
MOSQUITOES (adults)											X
STORED PRODUCT PESTS	X		X	X							
SCORPIONS			X	X							
SILVERFISH	X	X	X	X				X	X	X	X
SPIDERS			X					X	X	X	X
SOWBUGS			X					X	X		X
SPRINGTAILS			X	X				X	X		X
TICKS				X				X			

[1] Alternating uses of insecticides in different chemical classes can help reduce the likelihood of the pests developing resistance to one group or class of compounds.

[2] **Formulations:**

Aerosol includes Crack & Crevice. Bait may be granular, gel, or station. Sprayable may be concentrate or powder, some RTU formulations.

[3] Some formulations of diatomaceous earth and silica gel contain pyrethrins as a flushing agent.

[4] Not to be used in or around residences or other buildings where children may be present.

[5] True bugs includes boxelder bugs, stink bugs, kudzu bugs, and similar occasional invaders.

[6] Baits may be formulated as solids, dusts or liquids.

[7] Some formulations of Orthene may be applied indoors as crack & crevice treatment only.

Table 5-13 (continued). Industrial and Household Pests—For use by licensed pest management professionals only

Pesticide	Bifenthrin (Bifen, Talstar)	Cyfluthrin (Tempo Ultra, Ultrashield CS)	Cypermethrin[2] (Demon, Cynoff, Fendona, Talstar Xtra)		Deltamethrin (DeltaDust, DeltaGuard, Suspend)		Esfenvalerate (Onslaught)	Etofenprox (Zenprox)	Fenvalerate (Pyrid)	Lambda-cyhalothrin (Demand, 228L)	Permethrin (Flee, Dragnet, Prelude)		Phenothrin (Bedlam, Nyguard Plus[6])	Prallethrin (ULD Spy-300, Altocirrus Fog)	Pyrethrins and pyrethrum (Kicker, Pyrenone)			Sumithrin (Bedlam) Nyguard Plus[6])	Tetramethrin (CB Stinger)
Chemical Class										Pyrethroids[1]									
Formulation[3]	S, G	S	D	G	S	D	G	S	S	S	S	S	G	S	G	S	S	A[4]	S[4]	D[4]	S	S
Pests																						
ANTS	X	X	X	X	X	X	X	X	X	X	X	X	X	X	X	X	X	X	X	X	X	
BED BUGS		X				X		X	X	X		X		X		X			X	X		
BEES	X	X	X			X		X	X		X		X						X	X		X
BOOKLICE		X				X		X			X			X					X			
BUGS (TRUE)[5]	X	X			X			X			X		X			X	X				X	
CARPET BEETLES		X				X		X	X	X	X	X		X			X	X	X			
CENTIPEDES	X	X	X	X	X	X	X	X	X	X		X			X	X					X	
CLOTHES MOTHS		X				X		X								X						
CLOVER MITES		X	X		X		X						X		X	X	X	X	X			
COCKROACHES	X	X	X		X	X	X	X	X	X	X		X		X		X	X	X	X		
CRICKETS	X	X	X		X	X			X	X	X		X	X	X	X		X	X			
EARWIGS	X	X	X	X	X				X	X		X	X	X	X	X	X	X	X			
FLEAS	X	X	X	X	X	X		X	X	X		X		X		X	X	X	X	X	X	
FLIES/GNATS	X	X	X		X			X		X		X		X		X	X	X	X			
HORNETS/WASPS	X	X	X		X			X	X	X		X				X	X	X	X			X
LADY BEETLES										X						X						
MILLIPEDES	X	X	X	X	X		X	X	X	X		X	X	X	X							
MOSQUITOES (adults)	X	X	X			X		X	X	X		X						X	X	X		
STORED PRODUCT																		X				
PESTS		X						X		X	X	X		X				X	X			
SCORPIONS	X	X	X	X	X			X			X	X	X			X			X	X		
SILVERFISH	X	X	X		X			X	X	X	X	X	X	X		X			X	X		
SPIDERS	X	X	X	X	X		X	X	X	X	X	X		X		X			X	X		
SOWBUGS	X	X	X	X	X	X	X	X		X	X		X	X		X			X	X		
SPRINGTAILS	X	X	X			X	X	X	X	X	X		X			X			X	X		
TICKS	X	X	X	X	X	X	X	X	X	X	X		X		X	X			X	X		

[1] Alternating uses of insecticides in different chemical classes can help reduce the likelihood of the pest developing resistance to one class or group of compounds. Many pyrethroids can be tank-mixed with piperonyl butoxide products to enhance insecticidal activity.

[2] Some products use alpha-cypermethrin or zeta-cypermethrin which contain chemical isomers or cypermethrin. Talstar Xtra is a mixture of zeta-cypermethrin and bifentrin.

[3] **KEY TO FORMULATION SYMBOLS:**

  A = aerosol

  B = bait (granular or station)

  D = dust

  G = granular

  S = sprayable (concentrate or powder, some RTU formulations)

[4] Some formulations of pyrethrins contain piperonyl butoxide as a synergist.

[5] True bugs includes boxelder bugs, stink bugs, kudzu bugs, and similar occasional invaders.

[6] Nyguard Plus contains the IGR pyriproxyfen

Table 5-13 (continued). Industrial and Household Pests — For use by licensed pest management professionals only

Pesticide	Hydroprene (Gencor)[3]	Fenoxycarb (Altosid, Pre-Strike)[3]	Methoprene (Altosid, Kabat, Pharorid, Precor, Vigren)[3]	Pyriproxyfen (Archer, Ultracide) V[3]	Acetamiprid+Bifenthrin (Transport)	Dinotefuran (Advance), Alpine)	Imidacloprid (FlyBait, Maxforce, Premise, Temprid)[8]	Thiamethoxam (Optiguard)[9]	Clothianidin (Maxforce Impact, PestXPert Bed Bug)	Abamectin (Ascend, Avert, Advance)	Aluminum phosphide (Phostoxin)[5]	Chlorfenapyr (Phantom)[6]	Cyanotraniliprole (Zyrox)	d-Limonene (ProCitra-DL)	Fipronil (Maxforce F, TopChoice, Termidor)[7]	2-Phenyl Proprionate (EcoPCO EC)	Hydramethylnon (Amdro, Siege, MaxForce)	Indoxacarb (Advion, Arilion)	Rosemary Oil (Essentria IC3)	Sulfuryl floride (Vikane, Profume, Zythor)[9]
Chemical class[1]	Insect Growth Regulators				Neonicotinoids					Other Classes										
Formulation[2]	A,S	S	B	A,S	A,S	B,S	B,D,S	B	B,S	B,S	B	F	S	B	S	B,G,S	A,S	B	B,S	A	F
Pests																					
ANTS	X		X		X	X	X	X	X[9]	X		X			X	X	X	X	X	X	
BED BUGS				X	X	X	X		X[11]			X			X				X	X	X
BEES					X	X										X				X	
BOOKLICE						X														X	
BUGS (TRUE)[4]					X	X	X					X			X					X	
CARPET BEETLES					X	X									X					X	X
CENTIPEDES					X		X					X			X					X	
CLOTHES MOTHS					X	X						X									X
CLOVER MITES					X							X								X	
COCKROACHES	X			X	X	X	X	X	X	X		X	X	X	X	X	X	X	X	X
CRICKETS					X	X	X		X			X			X	X	X		X	X	
EARWIGS	X					X	X		X			X			X	X	X			X	
FLEAS			B	X	X	X									X	X	X				
FLIES/GNATS	X				X	X	X	X					X	X	X	X				X	
HORNETS/WASPS					X	X						X[10]			X	X	X			X	
LADY BEETLES					X	X		X				X			X	X	X			X	
MILLIPEDES					X	X		X				X			X	X	X	X		X	
MOSQUITOES (adults)	X	X	X		X	X									X		X			X	
STORED PRODUCT PESTS	X			X	X	X	X		X		X	X					X				X
SCORPIONS					X	X						X				X	X			X	
SILVERFISH					X	X	X					X			X	X	X			X	
SPIDERS					X	X								X	X	X				X	
SOWBUGS					X	X						X					X			X	
SPRINGTAILS					X	X						X					X			X	
TICKS					X							X			X	X				X	

[1] Alternating uses of insecticides in different chemical classes can help reduce the likelihood of the pest developing resistance to one class or group of compounds.

[2] **KEY TO FORMULATION SYMBOLS:**

A = Aerosol (includes Crack & Crevice)   B = Bait (granular, gel or station)

D = Dust   F = Fumigant

G = Granular   S = sprayable (concentrate or powder, some RTU formulations)

[3] IGR products are not typically effective against adult stage of pests; use with an adulticide to provide quicker control of pest population.

[4] True bugs includes boxelder bugs, stink bugs, kudzu bugs, and similar occasional invaders

[5] Requires an F-Phase Structural Pest Control License.

[6] Chlorfenapyr labeled for indoor use only for these pests or limited spot treatment outdoors.

[7] Termidor liquid formulations are labeled for outdoor use only; use other insecticide products indoors.

[8] Temprid contains both imidacloprid and cyfluthrin.

[9] Optigard not for use against pharaoh ants or carpenter ants.

[10] Phantom is not a knockdown insecticide for pests such as wasps.

[11] Use spray formulation only for bed bugs. Also contains Metofluthrin and Piperonyl Butoxide

Chapter V — 2019 N.C. Agricultural Chemicals Manual

ORNAMENTALS

Arthropod Management for Ornamental Plants Grown in Greenhouses
S. D. Frank, Entomology and Plant Pathology

Successful pest management programs use a combination of appropriate pest control tactics. Always follow label precautions when handling or applying pesticides. Make chemical control part of an integrated pest management program that includes monitoring and pest identification along with appropriate cultural, physical, horticultural, and biological controls.

Responsible pesticide use includes resistance management. A system has been developed by the inter-company Insecticide Resistance Action Committee (IRAC; www.irac-online.org) to help you rotate chemicals correctly. Pesticides have been assigned an IRAC classification number based on their mode of action. To rotate properly, choose a product with a different IRAC number for each successive application directed against the same pest. Follow resistance management instructions on the label.

The information in this chart is not a substitute for the label. Pesticide labels and restrictions change frequently. Read and understand all label information before using any pesticide. Do not use pesticides for uses other than those on the label. Check county and state regulations for any local restrictions on the use of products listed here before using them.

Table 5-14. Arthropod Management for Ornamental Plants Grown in Greenhouses
Permitted application sites: G = greenhouse, L = landscape, N = Nursery. (Trade names listed are common examples of products that contain the active ingredient, not an endorsement of a particular product.)

Insect or Mite	Pesticide common name (Trade name)	Minimum Hours Between Application and Reentry	IRAC Mode of Action Group	Permitted application sites
Aphid	abamectin (Avid)	12 hr	6	G, L, N
	acephate (Orthene)	24 hr	1B	G, L, N
	acetamiprid (TriStar)	12 hr	4A	G, L, N
	azadirachtin (Azatin)	4 hr	18B	G, L, N
	Beauveria bassiana (Botanigard/Naturalis)	4 hr	M	G, L, N
	bifenthrin (Talstar)	12 hr	3	follow label
	cyantraniliprole (Mainspring)	4 hr	28	G
	cyfluthrin (Decathlon)	12 hr	3A	G, L, N
	dinotefuran (Safari)	12 hr	4A	G, L, N
	flonicamid (Aria)	12 hr	9B	G, L, N
	fluvalinate (Maverik)	12 hr	3A	G, L, N
	horticultural oil (various)	4 hr		G, L, N
	imidacloprid (Marathon II)	12 hr	4A	G, N
	insecticidal soaps	12 hr		G, N, L
	kinoprene (Enstar II)	4 hr	7A	G
	neem oil (Various)	4 hr	UN	G, L, N
	permethrin (Astro, others)	12 hr	3	follow label
	pymetrozine (Endeavor)	12 hr	9B	G, L, N
	pyrethrins (various)	12 hr	3A	G, L, N
	pyrifluquinazon (Rycar)	12 hr	UN	G
	spinetoram + sulfoxaflor (XXpire)	12 hr	4C + 5	G, L, N
	spirotetramat (Kontos)	24 hr foliar (see exception for drench application)	23	G, N
	thiamethoxam (Flagship)	12 hr	4A	G, L, N
	tolfenpyrad (Hachi-Hachi)	12 hr	21A	G
Broad Mite	abamectin (Avid)	12 hr	6	G, L, N
	chlorfenapyr (Pylon)	12 hr	13	G
	fenpyroximate (Akari)	12 hr	21A	G, N
	pyridaben (Sanmite)	12 hr	21A	G, L, N
	spiromesifen (Judo)	12 hr	23	G, N
Caterpillar	acephate (Orthene)	24 hr	1B	G, L, N
	acetamiprid (Tri-Star)	12 hr	4A	G, L, N
	azadirachtin (Azatin)	4 hr	18B	G, L, N
	Bacillus thuringiensis var. kurstaki	4 hr	11B2	follow label
	Beauveria bassiana	12 hr		follow label

Table 5-14. Arthropod Management for Ornamental Plants Grown in Greenhouses

Permitted application sites: G = greenhouse, L = landscape, N = Nursery. (Trade names listed are common examples of products that contain the active ingredient, not an endorsement of a particular product.)

Insect or Mite	Pesticide common name (Trade name)	Minimum Hours Between Application and Reentry	IRAC Mode of Action Group	Permitted application sites
Caterpillar (continued)	bifenthrin (Talstar)	12 hr	3	follow label
	chlorfenapyr (Pylon)	12 hr	13	G
	cyantraniliprole (Mainspring)	4 hr	28	G
	cyfluthrin (Decathlon)	12 hr	3A	G, L, N
	diflubenzuron (Adept)	12 hr	15	G
	fluvalinate (Mavrik)	12 hr	3A	G, L, N
	insecticidal soaps	12 hr		G, L, N
	novaluron (Pedestal)	12 hr	15	G, N
	permethrin (Astro, others)	12 hr	3	Follow label
	pyridalyl (Overture)	12 hr	UN	G
	spinetoram + sulfoxaflor (XXpire)	12 hr	4C + 5	G, L, N
	spinosad (Conserve)	4 hr	5	G, L, N
	tolfenpyrad (Hachi-Hachi)	12 hr	21A	G
Cyclamen Mite	abamectin (Avid)	12 hr	6	G, L, N
	chlorfenapyr (Pylon)	12 hr	13	G
	fenpyroximate (Akari)	12 hr	21A	G, N
	pyridaben (Sanmite)	12 hr	21A	G, L, N
	spiromesifen (Judo)	12 hr	23	G, N
Fungus Gnat Adults	bifenthrin (Talstar)	12 hr	3	Follow label
	cyfluthrin (Decathlon)	12 hr	3A	G, L, N
	fluvalinate (Mavrik)	12 hr	3A	G, L, N
	insecticidal soaps	12 hr		G, L, N
	permethrin (Astro, others)	12 hr	3	Follow label
Fungus Gnat Larvae	acetamiprid (Tri-Star)	12 hr	4A	G, L, N
	azadirachtin (Azatin)	4 hr	18B	G, L, N
	Bacillus thuringiensis var. israelensis	4 hr	11A1	Follow label
	chlorfenapyr (Pylon)	12 hr	13	G
	cyromazine (Citation)	12 hr	17	G, L, N
	diflubenzuron (Adept)	12 hr	15	G
	kinoprene (Enstar II)	4 hr	7A	G
	pyriproxyfen (Distance)	12 hr	7C	G, L, N
	Steinernema feltiae (various; beneficial nematode)	0 hr	Biological	G, L, N
Leafminer	abamectin (Avid)	12 hr	6	G, L, N
	acephate (Orthene)	24 hr	1B	G, L, N
	acetamiprid (Tri-Star)	12 hr	4A	G, L, N
	azadirachtin (Azatin)	4 hr	18B	G, L, N
	cyromazine (Citation)	12 hr	17	G, L, N
	dinotefuran (Safari)	12 hr	4A	G, L, N
	fenoxycarb (Preclude)	12 hr	7B	G
	imidacloprid (Marathon II, others)	12 hr	4A	Follow label
	spinosad (Conserve)	4 hr	5	G, L, N
	thiamethoxam (Flagship)	12 hr	4A	G, L, N
Mealybug	acephate (Orthene)	24 hr	1B	G, L, N
	acetamiprid (Tri-Star)	12 hr	4A	G, L, N
	azadirachtin (Azatin)	4 hr	18B	G, L, N

Table 5-14. Arthropod Management for Ornamental Plants Grown in Greenhouses

Permitted application sites: G = greenhouse, L = landscape, N = Nursery. (Trade names listed are common examples of products that contain the active ingredient, not an endorsement of a particular product.)

Insect or Mite	Pesticide common name (Trade name)	Minimum Hours Between Application and Reentry	IRAC Mode of Action Group	Permitted application sites
Mealybug (continued)	*Beuveria bassiana*	12 hr		Follow label
	bifenthrin (Talstar)	12 hr	3	Follow label
	buprofezin (Talus)	12 hr	16	G, N
	cyfluthrin (Decathlon)	12 hr	3A	G, L, N
	dinotefuran (Safari)	12 hr	4A	G, L, N
	flonicamid (Aria)	12 hr	9B	G, L, N
	horticultural oil (various)	4 hr		G, L, N
	imidacloprid (Marathon II, others)	12 hr	4A	Follow label
	insecticidal soaps	12 hr		G, L, N
	kinoprene (Enstar II)	4 hr	7A	G
	neem oil (Various)	4 hr	UN	G, L, N
	permethrin (Astro, others)	12 hr	3	Follow label
	pyrifluquinazon (Rycar)	12 hr	UN	G
	spinetoram + sulfoxaflor (XXpire)	12 hr	4C + 5	G, L, N
	spirotetramat (Kontos)	24 hr foliar (see exception for drench application)	23	G, N
	thiamethoxam (Flagship)	12 hr	4A	G, L, N
Scale (Armored) check label to be sure it lists scale to be treated	acephate (Orthene)	24 hr	1B	G, L, N
	acetamiprid (Tri-Star)	12 hr	4A	G, L, N
	bifenthrin (Talstar)	12 hr	3	Follow label
	buprofezin (Talus)	12 hr	16	G, N
	dinotefuran (Safari)	12 hr	4A	G, L, N
	horticultural oil (various)	4 hr		G, L, N
	kinoprene (Enstar II)	4 hr	7A	G
	thiamethoxam (Flagship)	12 hr	4A	G, L, N
Scale (Soft) check label to be sure it lists scale to be treated	acephate (Orthene)	24 hr	1B	G, L, N
	acetamiprid (Tri-Star)	12 hr	4A	G, L, N
	bifenthrin (Talstar)	12 hr	3	Follow label
	buprofezin (Talus)	12 hr	16	G, N
	cyantraniliprole (Mainspring)	4 hr	28	G
	dinotefuran (Safari)	12 hr	4A	G, L, N
	horticultural oil (various)	4 hr		G, L, N
	imidacloprid (Marathon II, others)	12 hr	4A	Follow label
	kinoprene (Enstar II)	4 hr	7A	G
	neem oil (Various)	4 hr	UN	G, L, N
	pyriproxyfen (Distance)	12 hr	7C	G, L, N
	thiamethoxam (Flagship)	12 hr	4A	G, N
Shorefly	acephate (Orthene)	24 hr	1B	G, L, N
	azadirachtin (Azatin)	4 hr	18B	G, L, N
	bifenthrin (Talstar)	12 hr	3	Follow label
	diflubenzuron (Adept)	12 hr	15	G
	imidacloprid (Marathon II, others)	12 hr	4A	Follow label
	kinoprene (Enstar II)	4 hr	7A	G
	pyriproxyfen (Distance)	12 hr	7C	G, L, N
	spinetoram + sulfoxaflor (XXpire)	12 hr	4C + 5	G, L, N
Slugs	iron phosphate (bait)	Follow label	UN	Follow label
	metaldehyde (bait)	Follow label	UN	Follow label
	methiocarb (bait)	Follow label	1A	Follow label

Table 5-14. Arthropod Management for Ornamental Plants Grown in Greenhouses

Permitted application sites: G = greenhouse, L = landscape, N = Nursery. (Trade names listed are common examples of products that contain the active ingredient, not an endorsement of a particular product.)

Insect or Mite	Pesticide common name (Trade name)	Minimum Hours Between Application and Reentry	IRAC Mode of Action Group	Permitted application sites
Spider Mites	abamectin (Avid)	12 hr	6	G, L, N
	acequinocyl (Shuttle)	12 hr	20B	G, N
	bifenazate (Floramite)	12 hr	UN	G, L, N
	chlorfenapyr (Pylon)	12 hr	13	G
	clofentezine (Ovation)	12 hr	10A	G, N
	cyflumetofen (Sultan)	12 hr	25	G, L, N
	etoxazole (TetraSan)	12 hr	10B	G, L, N
	fenazaquin (Magus)	12 hr	21A	G, L, N
	fenpyroximate (Akari)	12 hr	21A	G, N
	hexythiazox (Hexygon)	12 hr	10B	G, L, N
	horticultural oil (various)	4 hr		Follow label
	insecticidal soaps	12 hr		Follow label
	pyridaben (Sanmite)	12 hr	21A	G, L, N
	spiromesifen (Judo)	12 hr	23	G, N
Thrips	abamectin (Avid)	12 hr	6	G, L, N
	acephate (Orthene)	24 hr	1B	G, L, N
	acetamiprid (Tri-Star)	12 hr	4A	G, L, N
	azadirachtin (Azatin)	4 hr	18B	G, L, N
Thrips (continued)	*Beuveria bassiana*	12 hr		Follow label
	bifenthrin (Talstar)	12 hr	3	Follow label
	chlorfenapyr (Pylon)	12 hr	13	G
	cyantraniliprole (Mainspring)	4 hr	28	G
	cyfluthrin (Decathlon)	12 hr	3A	G, L, N
	flonicamid (Aria)	12 hr	9B	G, L, N
	fluvalinate (Mavrik)	12 hr	3A	G, L, N
	horticultural oil (various)	4 hr		Follow label
	kinoprene (Enstar II)	4 hr	7A	G
	novaluron (Pedestal)	12 hr	5	G, N
	pyrethrins (various)	12 hr	3A	G, L, N
	pyridalyl (Overture)	12 hr	UN	G
	spinetoram + sulfoxaflor (XXpire)	12 hr	4C + 5	G, L, N
	spinosad (Conserve)	4 hr	5	G, L, N
	tolfenpyrad (Hachi-Hachi)	12 hr	21A	G
Whitefly	abamectin (Avid)	12 hr	6	G, L, N
	acephate (Orthene)	24 hr	1B	G, L, N
	acetamiprid (Tri-Star)	12 hr	4A	G, L, N
	azadirachtin (Azatin)	4 hr	18B	G, L, N
	Beuveria bassiana	12 hr		Follow label
	bifenthrin (Talstar)	12 hr	3	Follow label
	buprofezin (Talus)	12 hr	16	G, N
	cyantraniliprole (Mainspring)	4 hr	28	G
	cyfluthrin (Decathlon)	12 hr	3A	G, L, N
	dinotefuran (Safari)	12 hr	4A	G, L, N
	fenazaquin (Magus)	12 hr	21A	G, L, N

Table 5-14. Arthropod Management for Ornamental Plants Grown in Greenhouses

Permitted application sites: G = greenhouse, L = landscape, N = Nursery. (Trade names listed are common examples of products that contain the active ingredient, not an endorsement of a particular product.)

Insect or Mite	Pesticide common name (Trade name)	Minimum Hours Between Application and Reentry	IRAC Mode of Action Group	Permitted application sites
Whitefly (continued)	fenoxycarb (Preclude)	12 hr	7B	G
	flonicamid (Aria)	12 hr	9B	G, L, N
	fluvalinate (Mavrik)	12 hr	3A	G, L, N
	horticultural oil (various)	4 hr		G, L, N
	imidacloprid (Marathon II, others)	12 hr	4A	Follow label
	insecticidal soaps	12 hr		G, L, N
	kinoprene (Enstar II)	4 hr	7A	G
	neem oil (Various)	4 hr	UN	G, L, N
	novaluron (Pedestal)	12 hr	5	G, N
	permethrin (Astro, others)	12 hr	3	Follow label
	pyridaben (Sanmite)	12 hr	21A	G, L, N
	pyriproxyfen (Distance)	12 hr	7C	G, L, N
	pyrifluquinazon (Rycar)	12 hr	UN	G
	spinetoram + sulfoxaflor (XXpire)	12 hr	4C + 5	G, L, N
	spirotetramat (Kontos)	24 hr foliar (see exception for drench application)	23	G, N
	thiamethoxam (Flagship)	12 hr	4A	G, N
	tolfenpyrad (Hachi-Hachi)	12 hr	21A	G

Arthropod Management for Ornamental Plants Grown in Nurseries or Landscapes

S. D. Frank, Entomology and Plant Pathology

Successful pest management programs use a combination of appropriate pest control tactics. Always follow label precautions when handling or applying pesticides. Make chemical control part of an integrated pest management program that includes monitoring and pest identification along with appropriate cultural, physical, horticultural, and biological controls.

Responsible pesticide use includes resistance management. A system has been developed by the Insecticide Resistance Action Committee (IRAC; www.irac-online.org) to help you rotate chemicals correctly. Pesticides have been assigned an IRAC classification number based on their mode of action. To rotate properly, choose a product with a different IRAC number for each successive application directed against the same pest. Follow resistance management instructions on the label.

The information in this chart is not a substitute for the label. Pesticide labels and restrictions change frequently. The label will provide the most updated information. Read and understand all label information before using any pesticide. Do not use pesticides for uses other than those on the label. Check county and state regulations for any local restrictions on the use of products listed here before using them.

Table 5-15. Arthropod Management for Ornamental Plants Grown in Nurseries or Landscapes

Permitted application sites: G = greenhouse, L = landscape, N = Nursery. (Trade names listed are common examples of products that contain the active ingredient, not an endorsement of a particular product.)

Insect or Mite	Pesticide common name (Trade name)	Minimum Hours Between Application and Reentry	IRAC Mode of Action Group	Permitted application sites
Adelgid	acetamiprid (TriStar)	12 hr	4A	G, L, N
	chlorantraniliprole (Acelepryn)	4 hr	28	L
	dinotefuran (Safari)	12 hr	4A	G, L, N
	horticultural oil (various)	4 hr		G, L, N
	imidacloprid (Merit, Marathon, others)	12 hr	4A	Follow label
	insecticidal soap (various)	12 hr		G, L, N
	spirotetramat (Kontos)	24 hr foliar (see exception for drench application)	23	G, N
	thiamethoxam (Flagship)	12 hr	4A	G, N
Aphid	abamectin (Avid)	12 hr	6	G, L, N
	acephate (Orthene)	24 hr	1B	G, L, N
	acetamiprid (TriStar)	12 hr	4A	G, L, N
	azadirachtin (Azatin)	4 hr	18B	G, L, N
	bifenthrin + imidacloprid (Allectus)	12 hr	3 + 4A	L
	bifenthrin + clothianidin (Aloft)	12 hr	4 + 4A	L
	Beauveria bassiana (BotaniGard)	4 hr		G, L, N
	carbaryl (Sevin)	12 hr	1A	L, N
	clothianidin (Celero, Arena)	12 hr	4A	Follow label
	cyfluthrin (Decathlon)	12 hr	3	G, N
	fluvalinate (Mavrik)	12 hr	3	G, L
	horticultural oil (various)	4 hr		G, L, N
	imidacloprid (Merit, Marathon)	12 hr	4A	Follow label
	neem oil (Triact) 70	4 hr	18B	G, L, N
	permethrin (Astro, Perm-up, others)	12 hr	3	Follow label
	pymetrozine (Endeavor)	12 hr	9B	G, L, N
	pyrethrins (various)	12 hr	3A	G, L, N
	insecticidal soap (various)	12 hr Follow label directions		G, L, N
	spinetoram + sulfoxaflor (XXpire)	12 hr	4C + 5	G, L, N
	spirotetramat (Kontos)	24 hr foliar (see exception for drench application)	23	G, N
	thiamethoxam (Flagship)	12 hr	4A	G, N
Armored Scale (such as Juniper scale, Oystershell scale, Pine needle scale, Tea scale, Euonymus scale, White peach scale)	acephate (Orthene)	24 hr	1B	G, L, N
	acetamiprid (TriStar)	12 hr	4A	G, L, N
	bifenthrin (Talstar)	12 hr	3	Follow label
	buprofezin (Talus)	12 hr	16	G, L, N
	carbaryl (Sevin)	Follow label directions	1A	L, N
	dinotefuran (Safari)	12 hr	4A	G, L, N

Table 5-15. Arthropod Management for Ornamental Plants Grown in Nurseries or Landscapes

Permitted application sites: G = greenhouse, L = landscape, N = Nursery. (Trade names listed are common examples of products that contain the active ingredient, not an endorsement of a particular product.)

Insect or Mite	Pesticide common name (Trade name)	Minimum Hours Between Application and Reentry	IRAC Mode of Action Group	Permitted application sites
	horticultural oil (various)	4 hr		G, L, N
	insecticidal soap (various)	Follow label directions 12 hr		G, L, N
	neem oil (Triact) 70	4 hr	18B	G, L, N
	pyriproxyfen (Distance)	12 hr	7C	G, L, N
	spinetoram + sulfoxaflor (XXpire)	12 hr	4C + 5	G, L, N
Asian Ambrosia Beetle	permethrin (Astro, Perm-up, Permethrin Pro)	12 hr	3	Follow label
Bagworm	acephate (Orthene)	24 hr	1B	G, L, N
	acetamiprid (Tri-Star)	12 hr	4A	G, L, N
	azadirachtin (Azatin)	4 hr	18B	G, L, N
	bifenthrin + imidacloprid (Allectus)	12 hr	3 + 4A	L
	bifenthrin + clothianidin (Aloft)	12 hr	4 + 4A	L
	Bacillus thuringiensis kurstaki (BiobitHP, DiPel, or Foray)	4 hr	11B2	G, L, N
	bifenthrin (Talstar)	Follow label directions	3	G, L, N
	carbaryl (Sevin)	Follow label directions	1A	L, N
	chlorantraniliprole (Acelepryn)	4 hr	28	L
	fluvalinate (Mavrik)	Follow label directions	3	G, L
	indoxacarb (Provaunt)	12 hr	22	L
	novaluron (Pedestal)	12 hr	15	G, N
	spinetoram + sulfoxaflor (XXpire)	12 hr	4C + 5	G, L, N
	spinosad (Conserve SC)	4 hr	5	G, N
Bark Beetles	permethrin (Astro, Perm-up, others)	12 hr	3	Follow label
	bifenthrin (Onyx, Talstar)	Follow label directions	3	Follow label
Black Vine Weevil	acephate (Orthene)	Follow label directions	1A	G, L, N
	Beauveria bassiana (BotaniGard)	4 hr		G, L, N
	bifenthrin (Onyx, Talstar)	Follow label directions	3	Follow label
	cyfluthrin + imidacloprid (Discus)	12 hr	3 + 4A	N
	dinotefuran (Safari)	12 hr	4A	G, L, N
	fluvalinate (Mavrik)	Follow label directions	3	G, L
	imidacloprid (Merit, Marathon, others)	12 hr	4A	Follow label
Borers (Clearwing, flatheaded, and roundheaded borers are included in this section. Make sure label specifically lists the type of borer you are trying to control.)	azadirachtin (Azatin)	4 hr	18B	G, L, N
	chlorantraniliprole (Acelepryn)	4 hr	28	L
	cyfluthrin + imidacloprid (Discus)	12 hr	3 + 4A	N
	dinotefuran (Safari)	12 hr	4A	G, L, N
	imidacloprid (Merit, Marathon II, others)	12 hr	4A	Follow label
	bifenthrin (Onyx, Talstar)	Follow local regulations for landscape reentry	3	Follow label
	permethrin (Astro, Perm-up, Permethrin Pro)	12 hr	3	Follow label
Caterpillars (such as armyworm, budworm, eastern tent caterpillar, fall webworm, orangestriped oakworm, leafrollers)	acephate (Orthene)	24 hr	1B	G, L, N
	acetamiprid (Tri-Star)	12 hr	4A	G, L, N
	azadirachtin (Azatin)	4 hr	18B	G, L, N
	Bacillus thuringiensis kurstaki (DiPel)	4 hr	11B2	G, L, N
	bifenthrin (Onyx, Talstar)	Follow label directions	3	Follow label
	bifenthrin + imidacloprid (Allectus)	12 hr	3 + 4A	L
	bifenthrin + clothianidin (Aloft)	12 hr	4 + 4A	L
	carbaryl (Sevin)	12 hr	1A	L, N
	chlorantraniliprole (Acelepryn)	4 hr	28	L
	indoxacarb (Provaunt)	12 hr	22	L
	insecticidal soap (various)	Follow label directions		G, L, N
	novaluron (Pedestal)	12 hr	15	G, N
	permethrin (Astro, Perm-up, Permethrin Pro)	12 hr	3	Follow label
	spinetoram + sulfoxaflor (XXpire)	12 hr	4C + 5	G, L, N
	spinosad (Conserve SC)	4 hr	5	G, N
	tebufenozide (Confirm)	4 hr	18A	L, N

Table 5-15. Arthropod Management for Ornamental Plants Grown in Nurseries or Landscapes

Permitted application sites: G = greenhouse, L = landscape, N = Nursery. (Trade names listed are common examples of products that contain the active ingredient, not an endorsement of a particular product.)

Insect or Mite	Pesticide common name (Trade name)	Minimum Hours Between Application and Reentry	IRAC Mode of Action Group	Permitted application sites
Cricket	bifenthrin (Onyx, Talstar)	12 hr	3	Follow label
	cyfluthrin (Decathlon)	Follow label directions	3	G, N
	pyrethrins (Pyrenone)	Follow label directions	3	Follow label
	insecticidal soap (various)	Follow label directions		G, L, N
Eriophyid Mite	abamectin (Avid)	12 hr	6	G, L, N
	horticultural oil (various)	4 hr		G, L, N
	spiromesifen (Judo, Forbid)	12 hr	23	G, N
False Spider Mites (such as privet mite)	acequinocyl (Shuttle)	12 hr	20B	G, N
	bifenazate (Floramite)	12 hr	Un	G, N, L
	etoxazole (TetraSan)	12 hr	10B	G, N, L
	horticultural oil (various)	4 hr		G, N, L
	insecticidal soaps	12 hr		G, N, L
	spiromesifen (Judo, Forbid)	12 hr	23	follow label
Fungus Gnat Adults	bifenthrin (Talstar)	12 hr	3	follow label
	cyfluthrin (Decathlon)	12 hr	3A	G, L, N
	fluvalinate (Mavrik)	12 hr	3A	G, L, N
	insecticidal soaps	12 hr		G, L, N
	permethrin (Astro, others)	12 hr	3	Follow label
Fungus Gnat Larvae	acetamiprid (Tri-Star)	12 hr	4A	G, L, N
	azadirachtin (Azatin)	4 hr	18B	G, L, N
	Bacillus thuringiensis var. israelensis	4 hr	11A1	follow label
	chlorfenapyr (Pylon)	12 hr	13	G
	cyromazine (Citation)	12 hr	17	G, L, N
	diflubenzuron (Adept)	12 hr	15	G
	kinoprene (Enstar II)	4 hr	7A	G
	pyriproxyfen (Distance)	12 hr	7C	G, L, N
	Steinernema feltiae (various; beneficial nematode)	0 hr	Biological	G, L, N
Grasshopper	bifenthrin (Onyx, Talstar)	12 hr	3	Follow label
	carbaryl (Sevin) 5 bait	Follow label directions	1A	Follow label
	cyfluthrin (Decathlon)	Follow label directions	3	G, N
	insecticidal soap (various)	12 hr		G, L, N
Japanese Beetle (Adult) and other leaf-feeding scarab beetles	acetamiprid (Tri-Star)	12 hr	4A	G, L, N
	acephate (Orthene)	Follow label directions	1A	G, L, N
	azadirachtin (Azatin XL)	4 hr	18B	G, L, N
	bifenthrin (Onyx, Talstar)	Follow label directions	3	Follow label
	bifenthrin (Talstar, Onyx)	12 hr	3	Follow label
	bifenthrin + imidacloprid (Allectus)	12 hr	3 + 4A	L
	bifenthrin + clothianidin (Aloft)	12 hr	4 + 4A	L
	carbaryl (Sevin)	Follow label directions	3	L, N
	chlorantraniliprole (Acelepryn)	4 hr	28	L
	clothianidin (Arena)		4A	L
	cyfluthrin + imidacloprid (Discus)	12 hr	3 + 4A	N
	cyfluthrin (Decathlon) 20 WP	Follow label directions	3	G, N
	dinotefuran (Safari)	12 hr	4A	G, L, N
	imidacloprid (Merit, Marathon II, others)	12 hr	4A	Follow label
	permethrin (Astro, Perm-up, Permethrin Pro)	12 hr	3	Follow label
	spinetoram + sulfoxaflor (XXpire)	12 hr	4C + 5	G, L, N
	thiamethoxam (Flagship)	12 hr	4A	G, N

Table 5-15. Arthropod Management for Ornamental Plants Grown in Nurseries or Landscapes

Permitted application sites: G = greenhouse, L = landscape, N = Nursery. (Trade names listed are common examples of products that contain the active ingredient, not an endorsement of a particular product.)

Insect or Mite	Pesticide common name (Trade name)	Minimum Hours Between Application and Reentry	IRAC Mode of Action Group	Permitted application sites
Lacebugs	acephate (Orthene)	Follow label directions	1A	G, L, N
	Beauveria bassina (BotaniGard)	4 hr		G, L, N
	bifenthrin (Talstar, Onyx)	12 hr	3	Follow label
	bifenthrin + imidacloprid (Allectus)	12 hr	3 + 4A	L
	bifenthrin + clothianidin (Aloft)	12 hr	4 + 4A	L
	carbaryl (Sevin)	12 hr	1A	L, N
	chlorantraniliprole (Acelepryn)	4 hr	28	L
	cyfluthrin + imidacloprid (Discus)	12 hr	3 + 4A	N
	dinotefuran (Safari)	12 hr	4A	G, L, N
	imidacloprid (Merit, Marathon, others)	12 hr	4A	Follow label
	permethrin (Astro, Perm-up, Permethrin Pro)	12 hr	3	Follow label
	spinetoram + sulfoxaflor (XXpire)	12 hr	4C + 5	G, L, N
	soap (Olympic Insecticidal)	Follow label directions 12 hr		Follow label
	thiamethoxam (Flagship)	12 hr	4A	G, N
Leaf Beetles (such as cucumber beetle, elm leaf beetle, willow leaf beetle, and flea beetles including Altica spp.)	acephate (Orthene)	12 hr	1A	G, L, N
	acetamiprid (TriStar)	12 hr	4A	G, L, N
	bifenthrin (Onyx, Talstar)	12 hr	3	Follow label
	bifenthrin + imidacloprid (Allectus)	12 hr	3 + 4A	L
	bifenthrin + clothianidin (Aloft)	12 hr	4 + 4A	L
	carbaryl (Sevin)	12 hr	3	L, N
	chlorantraniliprole (Acelepryn)	4 hr	28	L
	cyfluthrin + imidacloprid (Discus)	12 hr	3 + 4A	N
	dinotefuran (Safari)	12 hr	4A	G, L, N
	imidacloprid (Merit, Marathon II, others)	12 hr	4A	Follow label
	spinetoram + sulfoxaflor (XXpire)	12 hr	4C + 5	G, L, N
	spinosad (Conserve SC)	4 hr	5	G, N
	thiamethoxam (Flagship)	12 hr	4A	G, N
Leafhoppers (such as potato leafhopper and sharpshooters)	acephate (Orthene)	Follow label directions	1A	G, L, N
	acetamiprid (TriStar)	12 hr	4A	G, L, N
	bifenthrin (Onyx, Talstar)	Follow label directions	3	Follow label
	bifenthrin + imidacloprid (Allectus)	12 hr	3 + 4A	L
	bifenthrin + clothianidin (Aloft)	12 hr	4 + 4A	L
	carbaryl (Sevin)	Follow label directions	1A	L, N
	clothianidin (Arena)	12 hr	4A	L
	cyfluthrin (Decathlon)	Follow label directions	3	G, N
	cyfluthrin + imidacloprid (Discus)	12 hr	3 + 4A	N
	dinotefuran (Safari)	12 hr	4A	G, L, N
	fluvalinate (Mavrik)	Follow label directions	3	G, L
	imidacloprid (Merit, Marathon II, others)	12 hr	4A	Follow label
	neem oil (Triact) 90 EC	4 hr	18B	G, L, N
	permethrin (Astro, Perm-up, Permethrin Pro)	12 hr	3	Follow label
	thiamethoxam (Flagship)	12 hr	4A	G, N
	insecticidal soap	Follow label directions		G, L, N
	spirotetramat (Kontos)	24 hr foliar (see exception for drench application)	23	G, N

Table 5-15. Arthropod Management for Ornamental Plants Grown in Nurseries or Landscapes

Permitted application sites: G = greenhouse, L = landscape, N = Nursery. (Trade names listed are common examples of products that contain the active ingredient, not an endorsement of a particular product.)

Insect or Mite	Pesticide common name (Trade name)	Minimum Hours Between Application and Reentry	IRAC Mode of Action Group	Permitted application sites
Leafminers (such as boxwood leafminer, holly leafminer, birch leafminer) Note this includes dipterous, lepidopterous, and coleopterus leafminers. Make sure leafminer to be treated is listed on label.	abamectin (Avid)	Follow label directions	6	G, L, N
	acephate (Orthene)	Follow label directions	1A	G, L, N
	acetamiprid (TriStar)	24 hr	4A	G, L, N
	azadirachtin (Azatin XL)	12 hr	18B	G, L, N
	bifenthrin (Onyx, Talstar)	Follow label directions	3	Follow label
	chlorantraniliprole (Acelepryn SC)	4 hr	28	L
	clothianidin (Arena)	12 hr	4A	L
	cyfluthrin + imidacloprid (Discus)	12 hr	3 + 4A	N
	dinotefuran (Safari)	12 hr	4A	G, L, N
	imidacloprid (Merit, Marathon, others)	12 hr	4A	Follow label
	permethrin (Astro, Perm-up, Permethrin Pro)	12 hr	3	Follow label
	pyriproxyfen (Distance)	12 hr	7C	G, L, N
	spinosad (Conserve SC)	4 hr	5	G, N
Mealybugs	acephate (Orthene)	12 hr	1A	G, L, N
	acetamiprid (TriStar)	24 hr	4A	G, L, N
	Beauveria bassina (BotaniGard)	4 hr		G, L, N
	bifenthrin (Onyx, Talstar)	Follow label directions	3	Follow label
	buprofezin (Talus)	12 hr	16	G, N
	carbaryl (Sevin)	Follow label directions	1A	L, N
	cyfluthrin (Decathlon) 20 WP	Follow label directions	3	G, N
	clothianidin (Arena, Celero)		4A	L
	cyfluthrin + imidacloprid (Discus)	12 hr	3 + 4A	N
	dinotefuran (Safari)	12 hr	4A	G, L, N
	fluvalinate (Mavrik) 22.3 F	Follow label directions	3	G, L
	imidacloprid (Merit, Marathon, others)	12 hr	4A	Follow label
	neem oil (Triact)	4 hr	18B	G, L, N
	permethrin (Astro, Perm-up, Permethrin Pro)	12 hr	3	Follow label
	insecticidal soap (various)	Follow label directions 12 hr		G, L, N
	horticultural oil (various)	4 hr		G, L, N
	spinetoram + sulfoxaflor (XXpire)	12 hr	4C + 5	G, L, N
	spirotetramat (Kontos)	24 hr foliar (see exception for drench application)	23	G, N
	thiamethoxam (Flagship)	12 hr	4A	G, N
Pillbug	bifenthrin (Onyx, Talstar)	12 hr	3	Follow label
	cyfluthrin (Decathlon) 20 WP	Follow label directions	3	G, N
Plantbugs	bifenthrin (Onyx, Talstar)	Follow label directions	3	Follow label
	cyfluthrin (Decathlon)	Follow label directions	3	G, N
	permethrin (Astro, others)	12 hr	3	Follow label
	spinetoram + sulfoxaflor (XXpire)	12 hr	4C + 5	G, L, N
	thiamethoxam (Flagship)	12 hr	4A	G, N
	insecticidal soap (various)	Follow label directions 12 hr		G, L, N
Psyllid	acephate (Orthene)	Follow label directions	1A	G, L, N
	acetamiprid (TriStar)	24 hr	4A	G, L, N
	azadirachtin (Azatin XL)	12 hr	18B	G, L, N
	Beauveria bassiana (BotaniGard)	4 hr		G, L, N
	dinotefuran (Safari)	12 hr	4A	G, L, N
	imidacloprid (Merit, Marathon, others)	12 hr	4A	Follow label
	insecticidal soap (various)	12 hr		G, L, N
	neem oil (Triact)	4 hr	18B	G, L, N
	spinosad (Conserve SC)	4 hr	5	G, N
	thiamethoxam (Flagship)	12 hr	4A	G, N

Table 5-15. Arthropod Management for Ornamental Plants Grown in Nurseries or Landscapes

Permitted application sites: G = greenhouse, L = landscape, N = Nursery. (Trade names listed are common examples of products that contain the active ingredient, not an endorsement of a particular product.)

Insect or Mite	Pesticide common name (Trade name)	Minimum Hours Between Application and Reentry	IRAC Mode of Action Group	Permitted application sites
Sawfly	acephate (Orthene)	Follow label directions	1A	G, L, N
	acetamiprid (TriStar)	24 hr	4A	G, L, N
	carbaryl (Sevin)	Follow label directions	1A	L, N
	chlorantraniliprole (Acelepryn SC)	4 hr	28	L
	cyfluthrin (Decathlon) 20WP	Follow label directions	3	G, N
	cyfluthrin + imidacloprid (Discus)	12 hr	3 + 4A	N
	imidacloprid (Merit, Marathon, others)	12 hr	4A	Follow label
	indoxacarb (Provaunt)	12 hr	22	L
	insecticidal soap (various)	12 hr		G, L, N
	spinetoram + sulfoxaflor (XXpire)	12 hr	4C + 5	G, L, N
	spinosad (Conserve SC)	4 hr	5	G, N
Slug, Snail	iron phosphate (bait)	follow label	UN	Follow label
	metaldehyde + carbaryl (Sevin) bait	Follow label directions	Follow label	Follow Label
	methiocarb (Mesurol)	24 hr	1A	Follow label
Soft Scale (such as fletcher scale, cottony maple scale, wax scale)	acetamiprid (Tri-Star)	12 hr	4A	G, L, N
	buprofezin (Talus)	12 hr	16	G, N
	cyfluthrin + imidacloprid (Discus)	12 hr	3 + 4A	N
	dinotefuran (Safari)	12 hr	4A	G, L, N
	horticultural oil (various)	4 hr		G, L, N
	imidacloprid (Merit, Marathon, others)	Follow label directions	4A	Follow label
	pyriproxyfen (Distance)	12 hr	7C	G, L, N
	spinetoram + sulfoxaflor (XXpire)	12 hr	4C + 5	G, L, N
	thiamethoxam (Flagship)	12 hr	4A	G, N
Sowbug	cyfluthrin (Decathlon)	Follow label directions	3	G, N
Spider Mite (such as twospotted, southern red, and spruce spider mite)	abamectin (Avid)	12 hr	6	G, L, N
	acequinocyl (Shuttle)	12 hr	20B	G, N
	bifenazate (Floramite)	12 hr	Un	G, L, N
	clofentezine (Ovation)	12 hr	10A	G, N
	cyflumetofen (Sultan)	12 hr	25	G, L, N
	etoxazole (TetraSan)	12 hr	10B	G, N, L
	fenazaquin (Magus)	12 hr	21A	G, L, N
	fenpyroximate (Akari)	12 hr	21A	G, N
	hexythiazox (Hexygon)	12 hr	10B	G, L, N
	horticultural oil (various)	4 hr		follow label
	insecticidal soaps	12 hr		follow label
	pyridaben (Sanmite)	12 hr	21A	G, L, N
	spiromesifen (Judo, Forbid)	12 hr	23	Follow label
Spittlebug	acephate (Orthene)	12 hr	1A	G, L, N
	cyfluthrin (Decathlon)	Follow label directions	11B2	G, N
	horticultural oil (various)	4 hr		follow label
	insecticidal soaps	12 hr		follow label
Thrips	abamectin (Avid)	12 hr	6	G, L, N
	acephate (Orthene)	24 hr	1B	G, L, N
	acetamiprid (Tri-Star)	12 hr	4A	G, L, N
	azadirachtin (Azatin)	4 hr	18B	G, L, N
	Beauveria bassina (BotaniGard)	4 hr		G, L, N
	bifenthrin (Onyx, Talstar)	Follow label directions	3	Follow label

Table 5-15. Arthropod Management for Ornamental Plants Grown in Nurseries or Landscapes

Permitted application sites: G = greenhouse, L = landscape, N = Nursery. (Trade names listed are common examples of products that contain the active ingredient, not an endorsement of a particular product.)

Insect or Mite	Pesticide common name (Trade name)	Minimum Hours Between Application and Reentry	IRAC Mode of Action Group	Permitted application sites
Thrips (continued)	cyfluthrin (Decathlon)	12 hr	3A	G, L, N
	flonicamid (Aria)	12 hr	9B	G, L, N
	fluvalinate (Mavrik)	12 hr	3A	G, L, N
	horticultural oil (various)	4 hr		Follow label
	novaluron (Pedestal)	12 hr	5	G, N
	spinetoram + sulfoxaflor (XXpire)	12 hr	4C + 5	G, L, N
	spinosad (Conserve SC)	4 hr	4	G, N
Twig Borer	bifenthrin (Onyx, Talstar)	12 hr	3	Follow label
Whitefly	abamectin (Avid)	12 hr	6	G, L, N
	acephate (Orthene)	12 hr	1A	G, L, N
	acetamiprid (TriStar)	12 hr	4A	G, L, N
	azadirachtin (Azatin)	4 hr	18B	G, L, N
	Beauveria bassina (BotaniGard)	4 hr		G, L, N
	bifenthrin (Onyx, Talstar)	12 hr	3	Follow label
	buprofezin (Talus)	12 hr	16	G, N
	cyfluthrin (Decathlon)	Follow label directions	3	G, N
	dinotefuran (Safari)	Follow label directions	4A	G, L, N
	fenazaquin (Magus)	12 hr	21A	G, L, N
	fluvalinate (Mavrik)	Follow label directions	3	G, L
	flonicamid (Aria)	12 hr	9B	G, L, N
	horticultural oil (various)	4 hr		G, L, N
	imidacloprid (Merit, Marathon, others)	12 hr	4A	Follow label
	insecticidal soap (various)	Follow label directions 12 hr		G, L, N
	neem oil (Triact)	4 hr	18B	G, L, N
	novaluron (Pedestal)	12 hr	5	G, N
	permethrin (Astro, others)	12 hr	3	Follow label
	pyridaben (Sanmite)	12 hr	21A	G, L, N
	pyriproxyfen (Distance) 11.2 EC	12 hr	7C	G, L, N
	spinetoram + sulfoxaflor (XXpire)	12 hr	4C + 5	G, L, N
	spirotetramat (Kontos)	24 hr foliar (see exception for drench application)	23	G, N
	thiamethoxam (Flagship)	12 hr	4A	G, N
White Grubs (in containers or landscape plants (not turf) such as oriental and Japanese beetle)	*Beauveria bassina* (BotaniGard)	4 hr		G, L, N
	chlorantraniliprole (Acelepryn)	4 hr	28	L
	clothianidin (Arena)	12 hr	4A	L
	dinotefuran (Safari)	12 hr	4A	G, L, N
	imidacloprid (Merit, Marathon, others)	12 hr	4A	Follow label
	thiamethoxam (Flagship)	12 hr	4A	G, N

Arthropod Control on Christmas Trees

J. R. Sidebottom, Entomology Forestry

Table 5-16. Arthropod Control on Christmas Trees

** N.C. label

Insect or Mite Insecticide and Formulations	Amount of Formulation per Gallon of Spray	Amount per 100 Gallons of Water	Minimum Interval (Hours) Between Application and Reentry	Precautions and Remarks
Adelgids (Balsam Woolly Adelgid, Cooley, Eastern Spruce Gall)				
bifenthrin (Talstar Nursery Flowable)		20 to 40 oz/acre	12	Will also control twig aphids and spider mites but not rust mites.
bifenthrin 25% (Sniper)		3.9 to 12.8 oz/acre	12	Will also control twig aphids and spider mites but not rust mites.
bifenthrin (OnyxPro)		1.8 to 14.4 oz/100 gal	12	
chlorpyrifos (Lorsban 4E, Nufos 4E, Warhawk - Clearform)		1 qt/acre	24	Do not treat plants under extreme heat or drought stress. Control is achieved only when eggs and crawlers are not present.
dinotefuran (Safari)		4 to 8 oz/100 gal	12	Do not apply more than 2.7 pounds per acre.
esfenvalerate (Asana XL or Adjourn)		5.8 to 9.6 oz/100 gal	12	Use full rate to control balsam woolly adelgid.
imidacloprid (Couraze 1.6 For Pasada 1.6F)		4 to 8 oz/acre OR 2 oz/100 gal	12	Adding a spray adjuvant may improve coverage. Do not apply more than 40 ounces per acre per year.
imidacloprid (Admire Pro)		1.4 to 2.8 oz/acre	12	
insecticidal soap (M-Pede)		1 to 2 gal/100 gal	12	May cause foliage discoloration.
lambda-cyhalothrin (Lambda-T, Silencer or Warrior II)		1.28 to .56 oz/acre	24	Maximum use 0.96 pints per acre per year
petroleum oil (Damoil)		2 to 4 gal/100 gal dormant use 1 to 3 gal/100 gal summer use	4 hr	
spirotetramat (Movento)		5 to 10 oz/acre	24	Maximum use 20 ounces per acre per year. Use adjuvant to increase penetration.
Ants *(Also see "Imported Fire Ant" under Home Lawns table)*				
bifenthrin (Talstar Nursery Flowable)		5 to 10 oz/acre	12	
carbaryl (Sevin SL)		1 qt/acre	12	
chlorpyrifos (Lorsban 4E, Nufos 4E, Warhawk - Clearform)		1 qt/acre	24	Do not treat plants under extreme heat or drought stress.
insecticidal soap (M-Pede)		1 to 2 gal/100 gal	12	May cause foliage discoloration.
Aphid (including Balsam Twig Aphid and Cinara Aphid)				
abamectin (Ardent 0.15 EC, Avid 0.15 EC, Reaper 0.15 EC)		8 oz/100 gal	12	Do not apply more than 16 ounces or less than 8 ounces per acre. To suppress aphids, spray must contact young immatures.
azadirachtin (Aza-Direct)		1 to 2 pts/acre	4	Under extremely heavy pest pressure up to 3.5 pints may be used.
Beauveria bassiana (Naturalis T&O)	0.3 to 1 oz/gal	30 to 100 oz/100 gal	4	Spray immediately after mixing.
bifenthrin (Talstar Nursery Flowable)		5 to 40 oz/acre	12	
bifenthrin 25% (Sniper)		3.9 to 12.8 oz/acre	12	Will also control twig aphid and spider mites but not rust mites.
bifenthrin (OnyxPro)		1.8 to 14.4 oz/100 gal	12	
carbaryl (Chipco Sevin SL)		1 qt/acre	12	
chlorpyrifos (Lorsban 4E, Nufos 4E, Warhawk - Clearform)		1 qt/acre	24	Do not treat plants under extreme heat or drought stress.
cinnamaldehyde (Cinnamite)	0.85 oz/gal	85 oz/100 gal	4	
dimethoate (Dimethoate 400 or Clean Crop)		1 to 1 1/2 pt/acre	10 days	
disulfoton (Di-Syston 15 G)		1 tsp/tree OR 20 to 30 lb/acre	48 where rainfall exceeds 25 in./year	Spread the granules in the root zone of the trees at the dripline and work into the soil or water thoroughly within 48 hours of application. Not for use in bare-ground plantations.
esfenvalerate (Asana XL or Adjourn)		5.8 to 9.6 oz/100 gal	12	
flupyradifurone (Sivanto Prime)		7 to 14 oz/acre	4	Not for use in bare-ground plantations. May also control balsam woolly adelgid
imidacloprid (Couraze 1.6F or Pasada 1.6F)		4 to 8 oz/acre or 2 oz/100 gal	12	Adding a spray adjuvant may improve control. Do not apply more than 40 ounces per acre per year.
imidacloprid (Admire Pro)		1.4 to 2.8 oz/acre	12	
insecticidal soap (M-Pede)		1 to 2 gal/100 gal	12	May cause foliage discoloration.
lambda-cyhalothrin (Lambda-T, Silencer or Warrior II)		1.28 to 2.56 oz/acre	24	Maximum use 0.96 pints per acre per year

Table 5-16. Arthropod Control on Christmas Trees

** N.C. label

Insect or Mite Insecticide and Formulations	Amount of Formulation per Gallon of Spray	Amount per 100 Gallons of Water	Minimum Interval (Hours) Between Application and Reentry	Precautions and Remarks
Aphid (including Balsam Twig Aphid and Cinara Aphid) (continued)				
petroleum oil (Damoil)		2 to 4 gal/100 gal dormant use 1 to 3 gal/100 gal summer use	4 hr	
pymetrozine (Endeavor)		Up to 10 oz/acre	12	
spirotetramat (Movento)		5 to 10 oz/acre	24	Maximum use 20 ounces per acre per year. Use adjuvant to increase penetration.
thiamethoxam (Flagship 25WP)		2 to 4 oz/100 gal or 4 to 8 oz/acre	12	Maximum use 8 ounces per acre per year
Bagworm				
azadirachtin (Aza-Direct)		1 to 2 pt/acre	4	Under extremely heavy pest pressure up to 3.5 pints may be used.
bifenthrin (Talstar Nursery Flowable)		5 to 10 oz/acre	12	
carbaryl (Sevin SL)		1 qt/acre	12	
diflourobenzamide (Dimilin 4L)		1 to 8 oz/acre	12	Apply to early instars in mid- to late June.
dimethoate (Dimethoate 400 or Clean Crop)		1 to 1 1/2 pt/acre	10 days	
lambda-cyhalothrin (Lambda-T, Silencer or Warrior II)		1.28 to 2.56 oz/acre	24	Maximum use 0.96 pints per acre per year
spinosad (Conserve SC)		4 to 16 oz/acre	4	
tebufenozide (Confirm or Mimic 2LV)		4 to 8 oz/acre	4	Apply to early instar larvae; foliage development should be minimum of 20%. Do not apply more than 16 ounces per acre per year.
Elongate Hemlock Scale and Cryptomeria Scale				
bifenthrin 25% (Sniper)		3.9 to 12.8 oz/acre	12	Will also control twig aphid and spider mites but not rust mites. Best results when mixed with a systemic
buprofezin (Talus 70 DF)		14 oz/acre	12	
buprofezin (Talus 40 SC)		21.5 oz/acre	12	
dimethoate (Dimethoate 400 or Clean Crop)		1 to 1 1/2 pt/acre	10 days	Best results when mixed with other materials.
esfenvalerate (Asana XL)		5.8 to 9.6 oz/100 gal	12	Best results when mixed with a systemic.
dinotefuran (Safari)		4 to 8 oz/100 gal	12	Do not apply more than 2.7 pounds per acre.
spirotetramat (Movento)		5 to 10 oz/are	24	Maximum use 20 ounces per acre per year. Use adjuvant to increase penetration.
European Pine Shoot Moth				
azadirachtin (Aza-Direct)		1 to 2 pts/acre	4	Under extremely heavy pest pressure up to 3.5 pints may be used.
chlorpyrifos (Lorsban 4E or Nufos 4E)		1 qt/acre	24	Do not treat plants under extreme heat or drought stress.
dimethoate (Dimethoate 400 or Clean Crop)		1 to 1 1/2 pt/acre	10 days	
phosmet (Imidan 70-WSB)		1.3 to 1.5 lb/acre	13 days	
Gypsy Moth				
azadirachtin (Aza-Direct)		1 to 2 pts/acre	4	Under extremely heavy pest pressure up to 3.5 pints may be used.
bifenthrin (Talstar Nursery Flowable)		10 to 20 oz/acre	12	
carbaryl (Sevin SL)		.75 to 1 qt/acre	12	
chlorpyrifos (Lorsban 4E, Nufos 4E, Warhawk - Clearform)		1 qt/acre	24	Do not treat plants under extreme heat or drought stress.
diflourobenzamide (Dimilin 4L)		0.5 to 2 oz/acre	12	Apply to early instar and prior to full leaf expansion.
flubendiamide (Belt SC)		3 to 5 oz/acre	12	Do not use more than 10 oz per acre.
insecticidal soap (M-Pede)		1 to 2 gal/100 gal	12	May cause foliage discoloration.
lambda-cyhalothrin (Lambda-T, Silencer or Warrior II)		1.28 to 2.56 oz/acre	24	Maximum use 0.96 pints per acre per year
phosmet (Imidan 70-WSB)		1.3 to 1.5 lb/acre	13 days	
spinosad (Conserve SC)		4 to 16 oz/acre	4	
spinosad (Blackhawk Naturalyte)		1.1 to 4.4 oz/acre	4	
tebufenozide (Confirm or Mimic 2LV)		4 to 8 oz/acre	4	Apply to early instar larvae after each foliage flush at approximately 25% foliage expansion. Allow at least 6 hours between application and rainfall to assure thorough spray drying.

Table 5-16. Arthropod Control on Christmas Trees

** N.C. label

Insect or Mite Insecticide and Formulations	Amount of Formulation per Gallon of Spray	Amount per 100 Gallons of Water	Minimum Interval (Hours) Between Application and Reentry	Precautions and Remarks
Midge (Douglas fir needle midge, pine needle midge)				
chlorpyrifos (Lorsban 4E)		1 qt/acre	24	Do not treat plants under extreme heat or drought stress.
esfenvalerate (Asana XL)		5.8 to 9.6 oz/100 gal	12	
azadirachtin (Aza-Direct)		1 to 2 pts/acre	4	Under extremely heavy pest pressure up to 3.5 pints may be used.
carbaryl (Sevin SL)		1 qt/acre	12	
diflourobenzamide (Dimilin 4L)		1 to 2 oz/acre	12	Apply when second generation instars are present or 70% of first generation pupal cases are empty.
dimethoate (Dimethoate 400 or Clean Crop)		1 to 1 1/2 pt/acre	10 days	
esfenvalerate (Asana XL or Adjourn)		5.8 to 9.6 fl oz	12	Apply as needed for control. Spray sufficient gallonage to obtain good coverage of entire tree.
phosmet (Imidan 70-WSB)		1.3 to 1.5 lb/acre	13 days	
tebufenozide (Confirm or Mimic 2LV)		8 oz/acre	4	Apply to early instar larvae after each foliage flush at approximately 25% foliage expansion. Allow at least 6 hours between application and rainfall to assure thorough spray drying.
permethrin (Permethrin 3.2 EC)		4 to 8 oz/acre	12	
Pine Chafer				
esfenvalerate (Asana XL or Adjourn)		5.8 to 9.6 oz/100 gal	12	
lambda-cyhalothrin (Lambda-T, Silencer or Warrior II)		1.28 to 2.56 oz/acre	24	Maximum use 0.96 pints per acre per year
Rosette Bud Mite				
dimethoate (various brands)		1.3 pt/100 gal	10 days	
spirotetramat (Movento)		5 to 10 oz/acre	24	Maximum use 20 ounces per acre per year. Use adjuvant to increase penetration.
Rust Mites				
abamectin (Ardent 0.15EC, Avid 0.15 EC, Reaper 0.15 EC))		4 oz/100 gal	12	
carbaryl (Sevin SL)		1 qt/acre	12	
chloropyridan (Sanmite)		4 oz/100 gal or 10.7 oz/acre	12	
chlorpyrifos (Lorsban 4E, Nufos 4E, Warhawk - Clearform)		1 qt/acre	24	Do not treat plants under extreme heat or drought stress.
dimethoate (Dimethoate 400 or Clean Crop)		1 to 1.5 pt/acre	10 days	
fenpyroximate (Akari 5SC)		24 oz/100 gal	12	
insecticidal soap (M-Pede)		1 to 2 gal/100 gal	12	May cause foliage discoloration.
petroleum oil (Damoil)		2 to 4 gal/100 gal dormant use 1 to 3 gal/100 gal summer use	4 hr	
spirodiclofen (Envidor 2SC)		18 to 24.7 oz/acre	12	Make only one application per season.
Sawflies (Redheaded pine, red pine, European pine)				
carbaryl (Sevin SL)		1 qt/acre	12	
chlorpyrifos (Lorsban 4E, Nufos 4E, Warhawk - Clearform)		1 qt/acre	24	
dinotefuran (Safari)		4 to 8 oz/100 gal	12	Do not apply more than 2.7 pounds per acre.
esfenvalerate (Asana XL or Adjourn)		5.8 to 9.6 oz/100 gal	12	
imidacloprid (Couraze 1.6F or Pasada 1.6F)		4 to 8 oz/acre or 2 oz/100 gal	12	Adding a spray adjuvant may improve control. Do not apply more than 40 ounces per acre per year.
imidacloprid (Admire Pro)		1.4 to 2.8 oz/acre	12	
insecticidal soap (M-Pede)		1 to 2 gal/100 gal	12	May cause foliage discoloration.
lambda-cyhalothrin (Lambda-T, Silencer or Warrior)		1.28 to 2.56 oz/acre	24	Maximum use 0.96 pints per acre per year
malathion (Malathion 8)	2 tbsp/gal	0.4 gal/100 gal	12	
phosmet (Imidan 70-WSB)		1.3 to 1.5 lb/acre	13 days	
spinosad (Conserve SC)		4 to 16 oz/acre	4	
spinosad (Blackhawk Naturalyte)		1.1 to 4.4 oz/acre	4	
thiamethoxam (Flagship 25WP)		2 to 4 oz/100 gal or 4 to 8 oz/acre	12	

Table 5-16. Arthropod Control on Christmas Trees

** N.C. label

Insect or Mite Insecticide and Formulations	Amount of Formulation per Gallon of Spray	Amount per 100 Gallons of Water	Minimum Interval (Hours) Between Application and Reentry	Precautions and Remarks
Scale (Pine needle, pine tortoise, spruce bud, black pine, stripped pine; see also Elongate Hemlock and Cryptomeria Scale)				
azadirachtin (Aza-Direct)		1 to 2 pts/acre	4	Under extremely heavy pest pressure up to 3.5 pints may be used.
carbaryl (Sevin SL)		1 qt/acre	12	Controls crawlers only.
chlorpyrifos (Lorsban 4E, Nufos 4E, Warhawk - Clearform)		1 qt/acre	24	Do not treat plants under extreme heat and drought stress. Apply when scale crawlers are active.
dinotefuran (Safari)		4 to 8 oz/100 gal	12	Do not apply more than 2.7 pounds per acre.
insecticidal soap (M-Pede)		1 to 2 gal/100 gal	12	May cause foliage discoloration
lambda-cyhalothrin (Lambda-T or Silencer)		2.58 to 5.12 oz/acre	24	Maximum use 1.92 pints per acre per year
petroleum oil (Damoil)		2 to 4 gal/100 gal dormant use 1 to 3 gal/100 gal summer use	4	
spirotetramat (Movento)		5 to 10 oz/are	24	Maximum use 20 ounces/acre/year. Use adjuvant to increase penetration.
thiamethoxam (Flagship 25WP)		2 to 4 oz/100 gal or 4 to 8 oz/acre	12	For soft scales. Maximum use 8 ounces per acre per year
Seed Bugs/Seed Chalcid				
esfenvalerate (Asana XL or Adjourn)		9.6 oz/100 gal	12	
lambda-cyhalothrin (Lambda-T, Silencer or Warrior)		2.56 oz/100 gal	24	For high volume spray. See label for other application methods.
permethrin (Permethrin 3.2 EC)		30 oz/acre	12	
phosmet (Imidan 70-WSB)		1.3 to 1.5 lb/acre	13 days	
Spider Mite (Spruce spider mites)				
abamectin (Ardent 0.15 EC, Avid 0.15 EC, Reaper 0.15 EC)		4 to 8 oz/100 gal	12	Do not apply more than 16 ounces or less than 8 ounces per acre.
azadirachtin (Aza-Direct)		1 to 2 pts/acre	4	Under extremely heavy pest pressure up to 3.5 pints may be used.
Beauveria bassiana (Naturalis T&O)	0.3 to 1 oz/gal	30 to 100 oz/100 gal	4	Spray immediately after mixing
bifenazate (Floramite)		2 to 8 oz/100 gal	12	Add an adjuvant like Silwet L-77 or Sylgard 309 to the Floramite solution. Do not apply more than 32 oz per acre per year.
bifenazate (Aramite-4SC)		12 to 16 oz/acre		
bifenthrin (Talstar Nursery Flowable)		5 to 40 oz/acre	12	
bifenthrin 25% (Sniper)		3.9 to 12.8 oz/acre	12	Will also control twig aphid and spider mites but not rust mites.
bifenthrin (OnyxPro)		1.8 to 14.4 oz/100 gal	12	
chloropyridan (Sanmite)		4 oz/100 gal or 10.7 oz/acre	12	
chlorpyrifos (Lorsban 4E, Nufos 4E, Warhawk - Clearform)		1 qt/acre	24	Do not treat plants under extreme heat or drought stress. If eggs are present, reapply in 7 to 10 days to control newly hatched nymphs.
clofentezine (Apollo SC)	—	4 to 8 oz/acre	12	Most effective when applied at first sign of mite activity and mite eggs.
cinnamaldehyde (Cinnamite)	2 tbsp/gal	85 oz/100 gal	4	
cyflumetofen (Sultan)		13.7 oz/100 gal	12	Do not make more than 2 applications per year. Use at least 100 gallons of water per acre and get thorough coverage. Do not tank mix with insect or plant growth regulators or carbamate, organophosphate, or pyrethroid insecticides.
dimethoate (Dimethoate 400 or Clean Crop)		1 to 1 1/2 pt/acre	48	
disulfoton (Di-Syston 15G)		1 tsp/tree 20 to 30 lb/acre	48 where rainfall exceeds 25 in/year	Spread the granules in the root zone of the trees at the dripline and work into the soil or water thoroughly within 48 hours of application. Not for use in bare-ground plantations.
etoxazole (TetraSan 5 WDG)		28 to 24 oz/100 gal	12	TetraSan kills mite eggs and nymphs but not adult mite. Treated adults will not produce viable eggs.
fenazaquin (Magister, Magus)		12 to 24 oz/100 gal	12	Do not exceed more than 24 oz per acre per year
fenpyroximate (Akari 5SC)		16 to 24 oz/100 gal	12	
hexythiazox (Savvy) 50 WP	3 to 6 oz/acre	2 oz/100 gal	12	Do not make more than one application per year.
insecticidal soap (M-Pede)		1 to 2 gal/100 gal	12	May cause foliage discoloration.
propargite (Omite 30 WS)		3 to 7.5 lb/acre	7 days	Make no more than three applications per year. Compatibility restrictions.
spirodiclofen (Envidor 2SC)		18 to 24.7 oz/acre	12	Make only one application per season.

Table 5-16. Arthropod Control on Christmas Trees

** N.C. label

Insect or Mite Insecticide and Formulations	Amount of Formulation per Gallon of Spray	Amount per 100 Gallons of Water	Minimum Interval (Hours) Between Application and Reentry	Precautions and Remarks
Spittlebug				
chlorpyrifos (Lorsban 4F)		1 qt/acre	24	Do not treat plants under extreme heat or drought stress.
esfenvalerate (Asana XL)		5.8 to 9.6 oz/100 gal	12	
lambda-cyhalothrin (Lambda-T, Silencer or Warrior)		1.28 to 2.56 oz/acre	24	Maximum use 0.96 pints per acre per year
Spruce Needle Miner				
chlorpyrifos (Lorsban 4E)		1 qt/acre	24	Do not treat plants under extreme heat or drought stress.
Weevils (pales, northern pine, pitch eating, root collar, white pine)				
azadirachtin (Aza-Direct)		1 to 2 pts/acre	4	Under extremely heavy pest pressure up to 3.5 pints may be used.
chlorpyrifos (Lorsban 4E or Nufos 4E)	2 tbsp/gal	3 qt/100 gal	24	Apply as a cut stump drench.
diflourobenzamide (Dimilin 4L)		4 to 8 oz/acre	12	Treat prior to egg deposition.
esfenvalerate (Asana XL or Adjourn)		5.8 to 9.6 oz/100 gal	12	
phosmet (Imidan 70-WSP)		1.3 to 1.5 lb/acre	13 days	
White Grubs				
chlorpyrifos (Lorsban 4E, Nufos 4E, Warhawk - Clearform)		1 qt/acre	24	Incorporate into the soil if possible.
imidacloprid (Admire Pro)		7 to 14 oz/acre	12	Maximum per season: 14 ounces per acre
thiamethoxam (Flagship 25WG)		8 oz/acre	12	Apply from adult flight through peak hatch of targeted species.
Zimmerman Pine Moth				
azadirachtin (Aza-Direct)		1 to 2 pts/acre	4	Under extremely heavy pest pressure up to 3.5 pints may be used.
dimethoate (Dimethoate 400 or Clean Crop)		1 to 1 1/2 pt/acre	10 days	
tebufenozide (Confirm or Mimic 2LV)		4 to 8 oz/acre	4	Apply to early instar larvae; foliage development should be minimum of 20%. Do not apply more than 16 ounces per acre per year.

Commercial Turf Insect Control

R. L. Brandenburg, Entomology and Plant Pathology

Table 5-17. Insect Control in Commercial Turf

Pest Insecticide and Formulation	Amount per 1,000 sq ft	Precautions and Remarks
Annual Bluegrass Weevil		
bifenthrin (Menace, Talstar, others) F, GC	0.25 to 0.5 fl oz	Monitor for adults, apply at peak activity. Use GC formulation for golf courses.
chlorantraniliprole (Acelepryn)	.28 fl oz	Apply approximately 7 to 14 days after adulticide to target larvae.
cyantraniliprole (Ference)	0.28 fl oz	Monitor for adults, apply at peak activity. Apply approximately 7 to 14 days after adulticide to target larvae.
indoxacarb (Provaunt) SC	0.28 fl oz	Monitor for adults, apply at peak activity. Apply approximately 7 to 14 days after adulticide to target larvae
lambda-chyalothrin (Battle, Scimitar, Cayonara)	0.23 fl oz	Monitor for adults, apply at peak activity.
Ant (also see Imported Fire Ant)		
bifenthrin[1] (Menace, Talstar, others) F, GC; G form also available	0.5 to 1 fl oz	Use GC formulation for golf courses.
carbaryl[1] (Sevin) 80 WSP	1 to 1.5 oz	
zeta-Cypermethrin, bifenthrin, and imadacloprid (Triple Crown)	20-35 fl oz/acre	
clothianidin + bifenthrin (Aloft) GC SC LC SC GC G LC G	0.27 to 0.54 fl oz 0.27 to 0.54 fl oz 1.8 to 3.6 lb 1.8 to 3.6 lb	
cyfluthrin (Tempo SC)	0.143 fl oz	Home lawns only.
cypermethrin[1] (Demon) TC	See label	
deltamethrin (Deltagard) G	2 to 3 lb/1,000 ft	
fipronil 0.0143 G (Top Choice, Taurus G)	2 lb	
hydramethylnon[1] (Maxforce G, Amdro)	See label	
lambda-cyhalothrin[1] (Battle, Scimitar, Cyonara)	See label	Do not make applications within 20 feet of any body of water. No reentry until spray has dried.
Bee and Wasp (Burrowing)		
carbaryl[1] (Sevin) 80 WSP	1.5 oz	
pyrethroids[1] (Advanced Garden, Battle, Deltagard, Menace, Scimitar, Talstar, Tempo)	See label	
Bermudagrass Mite		
abamectin (Divanem)	3.125 to 6.25 fl oz/acre	Tank mix with wetting agent and irrigate 0.1 to 0.25 in water post application. Applicator must be in possession of the 2(ee) label recommendation for restricted uses.
Billbug		
bifenthrin[1] (Menace, Talstar, others) F, GC; G form also available	0.25 to 0.5 fl oz	Use GC formulation for golf courses.
chlorantraniliprole (Acelepryn)	0.184 to 0.46 fl oz	
chlorpyrifos[1] (Dursban) 50 WSP, Pro	See label	For use on golf courses; check new label.
clothianidin (Arena) .5G 50 WDG	 14 to 22 oz 0.15 to 0.22 oz	
clothianidin + bifenthrin (Aloft) GC SC LC SC GC G LC G	0.27 to 0.44 fl oz 0.27 to 0.54 fl oz 1.8 to 3.6 lb 1.8 to 3.6 lb	
deltamethrin (Deltagard) G	2 to 3 lb/1,000 ft	
dinotefuran (Zylam) 20 SG	1 oz	
imidacloprid[1] (Merit) 75 WSP	3 to 4 level tsp	Make application prior to egg hatch.
lambda-cyhalothrin[1] (Battle, Scimitar, Cyonara)	See label	Observe restrictions near water.
thiamethoxam (Meridian) 0.33 G 25 WG	 60 to 80 lb/acre 12.7 to 17 oz/acre	Optimum control when applied from peak flight of adults to peak of egg hatch. Also suppresses mole crickets and chinch bugs.
zeta-cypermethrin, bifenthrin, and imidacloprid (Triple Crown)	10 to 20 fl oz/acre	
Chinch Bug		
acephate[1] (Orthene T, T&O) 75 S	1.2 to 2.4 oz	
bifenthrin[1] (Menace, Talstar, others) F, GC; G form also available	0.25 to 0.5 fl oz	Use GC formulation for golf courses.
carbaryl[1] (Sevin) 80 WSP	2.5 to 3 oz	
chlorantraniliprole (Acelepryn)	0.184 to 0.46 fl oz	Suppression.

Table 5-17. Insect Control in Commercial Turf

Pest Insecticide and Formulation	Amount per 1,000 sq ft	Precautions and Remarks
Chinch Bug (continued)		
clothianidin (Arena) .5G 50 WDG	 1.4 to 1.8 lb 0.2 to 0.3 oz	
clothianidin + bifenthrin (Aloft) GC SC LC SC GC G LC G	 0.27 to 0.44 fl oz 0.27 to 0.54 fl oz 1.8 to 3.6 lb 1.8 to 3.6 lb	
chlorpyrifos[1] (Dursban), 2E, 4E, 50 WP, Pro	See label	For use on golf courses; check new label.
cyfluthrin (Tempo SC)	0.2 fl oz	Home lawns only.
cypermethrin (Demon) TC	0.33 to 0.65 fl oz	
deltamethrin (Deltagard) G	2 to 3 lb/1,000 ft	
dinotefuran (Zylam) 20 SG	1 oz	For suppression.
lambda-cyhalothrin[1] (Battle, Scimitar, Cyonara)	See label	Do not make applications within 20 feet of any body of water. No reentry until spray has dried.
permethrin[1] (Astro)	0.4 to 0.8 fl oz	
zeta-Cypermethrin, bifenthrin, and imadacloprid (Triple Crown)	20 to 35 fl oz/acre	
Cutworm, Armyworm		
acephate[1] (Orthene T, T&O)	1.2 to 2.4 oz	Commercial and residential turf only.
azadirachtin[1] (Neemix, Turplex)	See label	
bifenthrin[1] (Menace, Talstar, others) F, GC; G form also available	0.18 to 0.25 fl oz	Use GC formulation for golf courses.
Bt products, various labels	See label	.
carbaryl[1] (Sevin) 80 WSP and baits	0.75 to 1.5 oz	Treat in late afternoon. Apply in adequate water for good coverage but do not flood or water in. Do not cut grass for 1 to 3 days after treatment.
chlorantraniliprole (Acelepryn)	0.046 to 0.092 fl oz	
chlorpyrifos[1] (Dursban) 4 E, 2 ES, 50 WP, Pro	See label	For use on golf courses; check new label.
clothianidin (Arena) .5G 50 WDG	 1.4 to 1.8 lb 0.2 to 0.3 oz	Cutworms only.
clothianidin + bifenthrin (Aloft) GC SC LC SC GC G LC G	 0.27 to 0.54 fl oz 0.27 to 0.54 fl oz 1.8 to 3.6 lb 1.8 to 3.6 lb	
cyfluthrin[1] (Tempo SC)	0.143 fl oz	Home lawns only.
deltamethrin (Deltagard) G	2 to 3 lb/1,000 ft	
dinotefuran (Zylam) 20 SG	1 oz	
entomogenous nematodes[1]	See label	Read and follow special application instructions. Effective only against small cutworms.
indoxacarb (Provaunt) SC	0.0625 to 0.25 fl oz	Not labeled for use on sod farms.
lambda-cyhalothrin[1] (Battle, Scimitar, Cyonara)	See label	Do not make applications within 20 feet of any body of water. No reentry until spray has dried.
spinosad A + D (Conserve) SC	1.25 fl oz	Rate varies with size and species.
trichlorfon (Dylox, Proxol) 80 SP	1.5 to 3 oz	
Earthworm		
		Usually not a problem. No effective controls available.
Fall Armyworm		
acephate[1] (Orthene, T, T&O)	0.5 to 1.2 oz	Water in immediately after application.
chlorantraniliprole (Acelepryn)	0.046 to 0.092 fl oz	
chlorpyrifos[1] (Dursban) 4 E, 2 E, 50WP, Pro	See label	For use on golf courses; check new label.
indoxacarb (Provaunt) SC	0.0625 to 0.25 fl oz	Not labeled for use on sod farms.
pyrethroids[1] (Advanced Garden, Battle, Deltagard, Menace, Scimitar, Talstar, Tempo, Cyonara)	See label	
spinosad A + D (Conserve SC)	1.25 fl oz	Rate varies with size and species.
Grasshopper		
acephate[1] (Orthene T, T&O)	0.5 oz	Do not mow turfgrass for at least 24 hours after application.
deltamethrin (Deltagard) G	2 to 3 lb/1,000 ft	
lambda-cyhalothrin[1] (Battle, Scimitar, Cyonara)	See label	Do not make applications within 20 feet of any body of water. No reentry until spray has dried.
Ground Pearl		
		No effective control—practice good management.

Table 5-17. Insect Control in Commercial Turf

Pest Insecticide and Formulation	Amount per 1,000 sq ft	Precautions and Remarks
Imported Fire Ant (See http://www.ncagr.gov/plantindustry/plant/entomology/documents/ncifaquarantine.pdf for latest quarantine areas.)		
acephate[1] (Lesco-Fate Orthene, T, T&O) 75 S	See label 1 to 2 tsp/mound	Distribute uniformly over mound. For best results apply in early morning or late afternoon.
bifenthrin[1] (Menace, Talstar, others) F; G form also available	—	Follow label directions.
clothianidin + bifenthrin (Aloft) GC SC LC SC GC G LC G	See label 0.27 to 0.44 fl oz 0.27 to 0.54 fl oz 1.8 to 3.6 lb 1.8 to 3.6 lb	
deltamethrin (Deltagard) G	2 to 3 lb/	
fenoxycarb (Award) [1] B	1 to 3 level tbsp 1 to 1.5 lb/acre	Single mound treatment. Apply uniformly with ground equipment.
fipronil (Topchoice, Fipronil, others) 0.0143	2 lb	Apply as a broadcast.
fipronil + bifenthrin + lamda-cyhalothrin (Taurus Trio G)	2 lb	Apply as a broadcast. Irrigate prior to treatment.
hydramethylnon[1] (Amdro) 0.88% bait (Maxforce G)	— See label	Uniformly broadcast 1 to 1.5 pound of bait per acre with ground equipment on pastures, range grasses, lawns, and nonagricultural lands. Or distribute uniformly 5 level tablespoons of bait 3 to 4 feet around base of each mound. Do not exceed 1.5 pounds per acre.
imidacloprid + bifenthrin (Allectus, Atera)	See label	Rate varies with pest. Different formulations for different sites.
indoxacarb (Advion) bait	1.5 lb/acre	Bait formulation.
lambda-cyhalothrin[1] (Battle, Scimitar, Cyonara)	See label	
metaflumizone (Siesta) bait	1.0 to 1.5 lbs/acre 2 to 4 tbsp/mound	Do not exceed 4 applications in a one-year period.
methoprene (Extinguish) 0.5 % bait	1.5 lb/acre	Mound or broadcast.
methoprene + hydramethylnon (Extinguish Plus)	1.5 lb/acre	
pyriproxyfen (Distance, Esteem)	See label	Mound or broadcast.
spinosad (Justice bait)	See label	
spinosad A + D (Conserve SC)	0.1 fl oz/gal/mound	Dilute 0.1 fluid ounce in 1 gallon water. Use 1 to 2 gallons per mound.
Leafhopper, Spittlebug		
acephate[1] (Orthene, T, T&O) 75 S	1 oz	
bifenthrin[1] (Menace, Talstar, others) F, GC; G form also available	0.25 to 0.5 fl oz	Use GC formulation for golf courses.
carbaryl[1] (Sevin) 80 WSP	0.75 to 1.5 oz	
chlorpyrifos[1] (Dursban) 4 E, 50 WSP, Pro	See label	For use on golf courses; check new label.
deltamethrin (Deltagard) G	2 to 3 lb	
Millipede		
bifenthrin[1] (Menace, Talstar, others) F, GC; G form also available	0.25 to 0.5 fl oz	Use GC formulation for golf courses.
carbaryl[1] (Sevimol) (Sevin) 80 WSP	 1.5 to 3 oz 0.75 to 1.5 oz	
chlorpyrifos[1] (Dursban) 2 E, Pro	See label	For use on golf courses; check new label.
cypermethrin (Demon) TC	See label	
lambda-cyhalothrin[1] (Battle, Scimitar, Cyonara)	See label	Do not make applications within 20 ft of any body of water. No reentry until spray has dried.
Mole Cricket		
acephate[1] (Orthene T, T&O, Lesco-Fate)	1 to 1.9 oz	Water soil before application. Do not water in.
bifenthrin[1] (Menace, Talstar, others) F, GC; G form also available	0.5 to 1 fl oz	Use GC formulation for golf course.
carbaryl[1] (Sevin) baits	See label	
cyfluthrin[1] (Tempo SC, Tempo Ultra)	0.2 fl oz	Home lawn use only.
deltamethrin (Deltagard) G	2 to 3 lb	
dinotefuran (Zylam) 20 SG	See label	
entomogenous nematodes[1]	See label	Various formulations now available. Adequate soil moisture critical for good control.
fipronil (Chipco Choice, others) 0.1 G (Top Choice, Fipronil, others) 0.0143	 12.5-25 lb/A 2 lb	Use slit placement equipment. Apply as a broadcast.
imidacloprid (Merit) 75 WP 0.5G	 4 level tsp 1.8 lb	Apply while crickets are less than ½ inch long (June, early July).

Table 5-17. Insect Control in Commercial Turf

Pest Insecticide and Formulation	Amount per 1,000 sq ft	Precautions and Remarks
Mole Cricket (continued)		
indoxacarb (Advion) Insect G	50 to 200 lb/acre	Not for use on sod farms. DO NOT water in after application.
indoxacarb (Provaunt)	0.275 oz	Two applications 2 to 4 weeks apart work best, following egg hatch.
lambda-cyhalothrin[1] (Battle, Scimitar, Cyonara)	See label	Do not make applications within 20 feet of any body of water. No reentry until spray has dried.
zeta-Cypermethrin, bifenthrin, and imadacloprid	20 to 35 fl oz/acre	
Slug, Snail		
mesurol 2 B	1 lb	Apply late in afternoon.
metaldehyde	See label	
Sod Webworm		
acephate[1] (Lesco-Fate, Orthene T, T&O) (Precise 4G)	0.5 to 1 oz 2.8 lb	Home lawns only. Irrigate immediately.
azadirachtin[1] (Azatrol, Neemix, Turplex)	0.5 fl oz	
Bacillus thuringiensis, various brands	1 to 2 lb/acre	
bifenthrin[1] (Menace, Talstar, others) F, GC; G form also available	0.18 to 0.25 fl oz	Use GC formulation for golf courses.
carbaryl[1] (Sevin) 80 WSP	2.5 to 3 oz	
clorantraniliprole (Acelepryn)	0.046 to 0.092 fl oz	
chlorpyrifos[1] (Dursban) 4 E, 2 E, 5 G, Pro	See label	For use on golf courses; check new label.
clothianidin (Arena) .5G 50 WDG	 14 to 22 oz 0.15 to 0.22 oz	
chlothianidin + bifenthrin (Aloft) GC SC LC SC GC G LC G	 0.27 to 0.54 fl oz 0.27 to 0.54 fl oz 1.8 to 3.6 lb 1.8 to 3.6 lb	
cyfluthrin[1] (Tempo SC, Tempo Ultra)	0.143 fl oz	Irrigate immediately after application. Do not apply to newly seeded stands or bentgrass.
deltamethrin (Deltagard) G	2 to 3 lb	
dinotefuran (Zylam) 20 SG	1 oz	
indoxacarb (Provaunt) SC	0.0625 to 0.25 fl oz	Not labeled for use on sod farms.
lambda-cyhalothrin[1] (Cyonara, Scimitar, Battle)	See label	Do not make applications within 20 feet of any body of water. No reentry until spray has dried.
permethrin[1] (Astro)	0.4 to 0.8 fl oz	
spinosad A + D (Conserve) SC	1.25 fl oz	Rate varies with size and species.
trichlorfon[1] (Dylox, Proxol) 80 SP	1.5 to 3 oz	
Sowbug, Pillbug		
bifenthrin[1] (Talstar) F, GC G form also available	0.25 to 0.5 fl oz	Use GC formulation for golf courses.
carbaryl[1] (Sevin) 80 WSP	0.75 to 1.5 oz	
cypermethrin[1] (Demon) TC	See label	
deltamethrin (Deltagard) G	2 to 3 lb	
lambda-cyhalothrin[1] (Battle, Cyonara, Scimitar)	See label	Do not make applications within 20 feet of any body of water. No reentry until spray has dried.
Sugarcane Beetle		
bifenthrin[1] (Talstar) F, GC G form also available	0.5 to 1.0 fl oz	Target adults early (Apr-May). Insecticide efficacy significantly reduced for fall population.
White Grub (May beetle, chafers, green June beetle, and others)		
B.t. subspecies galleriae (grubGoneG)	100 to 150 lbs/acre	
chlorantraniliprole (Acelepryn)	0.184 to 0.367 fl oz	Optimal control when applied at egg hatch. Use higher rates later in summer.
clothianidin (Arena) .5G 50 WDG	 14 to 22 oz 0.15 to 0.22 oz	Mole cricket suppression.
clothianidin + bifenthrin (Aloft) GC SC LC SC GC G LC G	 0.27 to 0.54 fl oz 0.27 to 0.54 fl oz 1.8 to 3.6 lb 1.8 to 3.6 lb	
dinotefuran (Zylam) 20 SG	1 oz	
imidacloprid[1] (Merit) 75 WP	3 to 4 level tsp	Make application prior to egg hatch. (Offers some suppression of caterpillars.)

Table 5-17. Insect Control in Commercial Turf

Pest Insecticide and Formulation	Amount per 1,000 sq ft	Precautions and Remarks
White Grub (May beetle, chafers, green June beetle, and others) (continued)		
thiamethoxam (Meridian) 0.33 G 25 WG	60 to 80 lb/acre 12.7 to 17 oz/acre	Optimum control when applied from peak flight of adults to peak of egg hatch. Also suppresses mole crickets and chinch bugs.
trichlorfon (Dylox, Proxol) 80 SP	3.75 oz	Can be used with some success as a rescue treatment in August and September. Apply at egg hatch.
White Grub, Green June Beetle (only)		
B.t. subspecies galleriae (grubGoneG)	100-150 lbs/acre	
carbaryl[1] (Sevin) 80 WSP	1 to 1.5 oz	
chlorantraniliprole (Acelepryn)	0.184 to 0.367 fl oz	Optimal control when applied at egg hatch. Use higher rates later in summer.
chlorpyrifos[1] (Dursban) 50 WSP, Pro	See label	For use on golf courses; see new label.
clothianidin (Arena) .5G 50 WDG	14 to 22 oz 0.15 to 0.22 oz	Mole cricket suppression.
clothianidin + bifenthrin (Aloft) GC SC LC SC GC G LC G	0.27 to 0.54 fl oz 0.27 to 0.54 fl oz 1.8 to 3.6 lb 1.8 to 3.6 lb	
dinotefuran (Zylam) 20 SG	1 oz	Apply at egg hatch.
imidacloprid[1] (Merit) 75 WP	3 to 4 level tsp	Make application prior to egg hatch. Do not use on sod farms. Offers some suppression of caterpillars.
thiamethoxam (Meridian) 0.33 G 25 WG	60 to 80 lb/acre 12.7 to 17 oz/acre	Optimum control when applied from peak flight of adults to peak of egg hatch. Also suppresses mole crickets and chinch bugs.
White Grub (Japanese beetle)		
B.t. subspecies galleriae (grubGoneG)	100-150 lbs per acre	
carbaryl[1] (Sevin) 80 WSP	3 oz	
chlorantraniliprole (Acelepryn)	0.184 to 0.367 fl oz	Optimal control when applied at egg hatch. Use higher rates later in summer.
clothianidin + bifenthrin (Aloft) GC SC LC SC GC G LC G	0.27 to 0.54 fl oz 0.27 to 0.54 fl oz 1.8 to 3.6 lb 1.8 to 3.6 lb	
clothianidin (Arena) .5G 50 WDG	14 to 22 oz 0.15 to 0.22 oz	Mole cricket suppression.
dinotefuran (Zylam) 20SG	1 oz per 1000 sq ft	Can be used with some success as a rescue treatment in August and September. Apply at egg hatch
imidacloprid[1] (Merit) 75 WP	3 to 4 level tsp	Make application prior to egg hatch. (Offers some suppression of caterpillars.)
zeta-Cypermethrin, bifenthrin, and imidacloprid (Triple Crown)	20 to 35 fl oz/acre	
thiamethoxam (Meridian) 0.33 G 25 WG	60 to 80 lb/acre 12.7 to 17 oz/acre	Optimum control when applied from peak flight of adults to peak of egg hatch. Also suppresses mole crickets and chinch bugs.
trichlorfon[1] (Dylox, Proxol) 80 SP	3.75 oz	Can be used with some success as a rescue treatment in August and September. Apply at egg hatch.

[1] Several tradenames available. Check label for active ingredient. Always follow label instructions.

Chapter V—2019 N.C. Agricultural Chemicals Manual

Insect Control for Wood and Wood Products
M. G. Waldvogel and P. Alder, Entomology and Plant Pathology

Space limitations preclude listing all pesticide formulations and trade names. Other products or formulations may be used—but only those products labeled for the intended use. Products labeled for outdoor use only should never be applied indoors. Some insecticides listed here are designated for professional use only; others may have different formulations for professionals and the general public. Read the product label for specific information about the active ingredient, application rates, and detailed instructions on use.

Mention of pesticides in this section does not imply that chemicals are or should be the first or only means of pest control. Nonchemical methods, including exclusion, proper sanitation/maintenance, and moisture reduction, are critical to controlling wood-destroying pests.

Table 5-18. Insect Control for Wood and Wood Products

Insect Insecticide	Formulation[1]	Use[2]	Precautions and Remarks
Carpenter Ant—(a) Indoors			
2-Phenethyl Propionate 1% Pyrethrins (EcoPCO AR-X-M)	Aerosol	P	Apply as directed on label.
abamectin (Advance 375)	Bait	P	Apply as directed on label.
acetamiprid + bifenthrin (Transport Mikron)	Sprayable	P	Apply as directed on label.
allethrin (Ortho)	Aerosol	G	Apply as directed on label.
avermectin (Advance)	Bait	P	Apply as directed on label.
bifenthrin (Ortho) (Talstar)	Aerosol Sprayable	G G, P	Apply as directed on label.
boric acid (Niban, MotherEarth G, Intice)	Bait	P	May be formulated as granular, gel or liquid. Apply as directed on label
chlorfenapyr (Phantom)	Sprayable	P	Apply as directed on label.
cyfluthrin (Bayer Advanced) (Tempo)	Sprayable	G P	Apply as directed on label.
cypermethrin (Cynoff, Cyper TC)	Dust, Liquid	P	Apply as directed on label.
deltamethrin (Bayer Advanced) (Suspend, D-Fend Dust, D-Foam)	Sprayable, Dust, Foam	G, P G P	Apply as directed on label. D-Fend is applied dry to cracks and crevices and voids. D-Foam is applied to voids where nests may be located.
dinotefuran (Alpine)	Foam & Spray	P	Apply as directed on label
esfenvalerate (Onslaught)	Sprayable	P	Apply as directed on label.
fipronil (Combat) (Maxforce)	Bait	G P	Bait where you see ant activity. Apply as directed on label.
hydramethylnon (Combat)	Bait	P	Bait where you see ant activity. Apply as directed on label.
imidacloprid (Masterline MaxxPro 2F)	Sprayable	P	Apply as directed on label.
imidacloprid + cyfluthrin (Temprid SC)	Sprayable	P	Apply as directed on label.
indoxacarb (Advion, Arilon)	Bait (gel) Sprayable	P P	Bait where you see ant activity. Apply as directed on label. Apply as directed on label
lambda-cyhalothrin (Demand) (Spectracide)	Sprayable Spray and foam	P G	Apply as directed on label.
permethrin (Dragnet, Masterline, Permethrin SFR)	Sprayable	P	Apply as directed on label.
prallethrin-Lambda-cyhalothrin (Spectracide)	Foam	G	Apply to galleries as directed on label.
sodium borate (Boracare, Borathor) (Spectracide, Terminate)	Sprayable, Dust	P G	Apply as directed on label.
thiamethoxam (Optigard)	Sprayable	P	Apply as foam to wall voids or infested wood.
Carpenter Ant—(b) outdoors			
acetamiprid + bifenthrin (Transport Mikron)	Sprayable	P	Apply outdoors only as pinstream, spot, crack and crevice, or perimeter spray.
abamectin (Advance)	Bait	P	Place bait around perimeter.
bifenthrin (Ortho) (Bifen, Talstar)	Sprayable	G P	Spray or inject into wood.
boric acid (Perma-Dust Niban)	Aerosol, Bait	P	Place bait granules around perimeter.
chlorfenapyr (Phantom)	Sprayable	P	Exterior use limited to spot (2 square feet) and crack and crevice treatments at points of entry.
cyfluthrin (Bayer Advanced) (Tempo)	Sprayable	G P	Treat into and around the nest, then seal holes.
cypermethrin (Demon TC, Cyper TC)	Sprayable	P	Course spray or inject into wood for localized infestations.
zeta-cypermethrin (Cynoff)	Dust		
deltamethrin (Bayer Advanced) (Suspend, D-Fend Dust, D-Foam)	Sprayable, Dust, Foam	G, P G P	Apply as directed on label. D-Foam is applied to voids where nests may be located. Treat into and around the nest.
dinotefuran (Alpine)	Foam & Spray	P	Apply as directed on label (apply to damaged shrubs, tree stumps, fences, etc.)
esfenvalerate (Onslaught)	Sprayable	P	Apply as directed on label.

Table 5-18. Insect Control for Wood and Wood Products

Insect Insecticide	Formulation[1]	Use[2]	Precautions and Remarks
Carpenter Ant—(b) outdoors (continued)			
fipronil (Maxforce, Termidor, Taurus)	Bait, Granular, Powder	P	Apply bait granules in ant foraging areas. Water area after applying granules.
hydramethylnon (Maxforce)	Bait	P	Apply granules along perimeter of building or nest. (Maxforce is for professional use.)
imidacloprid (Masterline MaxxPro 2F, Bayer Advanced)	Sprayable Liquid Foam	G	Apply to galleries as directed on label.
imidacloprid+cyfluthrin (Temprid SC)	Sprayable	P	Apply as directed on label.
Indoxacarb (Arilon, Advion)	Bait (Granular/gel) Sprayable	P	Apply as directed on label
lamda-cyhalothrin (Demand) (Spectracide)	Sprayable	P G	Apply as directed on label.
permethrin (Dragnet, Masterline)	Sprayable	P G	Apply as crack and crevice or spot treatment or paint onto surface. Application by drilling and injecting is also permitted.
sodium borate (Boracare, Borathor) (Spectracide)	Sprayable	P G	Spray, brush on, or inject into wood. For long-term protection, apply a water repellent stain to exterior wood surfaces 2 to 3 weeks after treatment.
Carpenter Bee			
bifenthrin (Ortho) (Bifen, Talstar)	Sprayable	G P	Apply as a coarse surface spray and into entrance hole. Seal entrance hole. Spectracide is for the general public.
boric acid (Perma-Dust PT 240)	Aerosol	P	Inject into entrance hole or tunnels with wood injector nozzle. Seal entrance hole.
carbaryl (Sevin)	Dust, Sprayable	G	Apply liquid as a coarse surface spray and into gallery entrance. Puff into and around entrance holes, using dust applicator. Seal with wood plugs, putty, or stainless steel wool.
chlorfenapyr (Phantom)	Sprayable	P	Apply as directed on label.
cyfluthrin (Bayer Advanced) (Tempo)	Sprayable	G P	Apply liquid as a surface spray and into entrance hole. Seal entrance hole.
cypermethrin (Demon TC, Cyper TC)	Sprayable	P	Course spray or inject into wood for localized infestations.
zeta-cypermethrin (Cynoff)	Dust	P	Apply dust formulation directly to galleries
lambda-cyhalothrin (Demand) (Spectracide)	Sprayable	P G	Spray or inject into wood. Seal holes in wood before injecting. Avoid runoff.
deltamethrin (Bayer Advanced) (Suspend, D-Fend Dust, D-Foam)	Sprayable, Dust, Foam	G P	Apply liquid as a coarse surface spray and into gallery entrance. Inject foam or puff into and around entrance holes, using dust applicator. Seal with wood plugs, putty, or stainless steel or copper wool.
imidacloprid (Premise) (Bayer Advanced)	Foam	P G	Apply to galleries as directed on label.
imidacloprid+cyfluthrin (Temprid SC)	Sprayable	P	Apply as directed on label.
permethrin (Dragnet, Masterline) (Permethrin 3.2)	Sprayable	P G	Spray or inject into wood. Seal holes in wood before injecting. Avoid runoff.
prallethrin-lambda-cyhalothrin (Spectracide)	Foam	G	Apply to galleries as directed on label.
sodium borate (Boracare, Borathor) (Spectracide)	Sprayable Dust	P G	Apply dust formulation directly to galleries.
Old House Borer			
aluminum phosphide (Phostoxin)	Fumigant	P	For infested furniture, stacked lumber, other wood products. Apply under gas-tight tarpaulins or in sealed chamber. Requires an FPhase N.C. Structural Pest Control License.
bifenthrin (Ortho) (Bifen, Talstar)	Sprayable	G P	
cyfluthrin (Bayer Advanced) (Tempo)	Sprayable	G P	Coarse spray, brush on, or inject into wood. Avoid excessive runoff.
cypermethrin (Demon TC, Cyper TC)	Sprayable	P	
deltamethrin (Bayer Advanced) (Suspend, D-Fend Dust, D-Foam)	Sprayable, Dust, Foam	G P	
imidacloprid+cyfluthrin (Temprid SC)	Sprayable	P	Apply as directed on label.
permethrin (Dragnet, Masterline) (Permethrin 3.2)	Sprayable	P G	
sodium borate (Boracare, Timbor) (Spectracide)	Sprayable Dust	P G	Spray, brush on, or inject into wood. For permanent protection, a water repellent should be applied to exterior surfaces 2 to 3 weeks after treatment.
sulfuryl fluoride (Vikane, Zythor)	Fumigant	P	Apply under gas-tight tarpaulins only. Hold for 20-24 hours at temperature above 60 degrees F. Requires an FPhase N.C. Structural Pest Control License.

Table 5-18. Insect Control for Wood and Wood Products

Insect Insecticide	Formulation[1]	Use[2]	Precautions and Remarks
Powderpost Beetle			
aluminum phosphide (Phostoxin)	Fumigant	P	For infested furniture, stacked lumber, other wood products. Apply under gas-tight tarpaulin or in a sealed chamber. Requires an FPhase N.C. Structural Pest Control License.
bifenthrin (Ortho) (Bifen, Talstar)	Sprayable	G P	Coarse spray, brush on, or inject into wood. Avoid excessive runoff.
chlorfenapyr (Phantom)	Sprayable	P	
cyfluthrin (Bayer Advanced) (Tempo)	Sprayable	G P	
cypermethrin (Demon TC, Cyper TC) zeta-cypermethrin (Cynoff)	Sprayable Dust	P	Coarse spray or inject into wood for localized infestations.
deltamethrin (Bayer Advanced) (Suspend, D-Fend Dust, D-Foam)	Sprayable, Dust, Foam	G P	Surface spray or inject foam or dust into galleries.
imidacloprid (Bayer Advanced)	Foam	G	Apply to galleries as directed on label.
imidacloprid+cyfluthrin (Temprid SC)	Sprayable	P	Apply as directed on label.
lambda-cyhalothrin (Demand) (Spectracide)	Sprayable	P G	Apply as directed on label.
permethrin (Dragnet, Masterline)	Sprayable	P G	
sodium borate (Boracare, Timbor) (Spectracide)	Sprayable Dust	P G	For long-term protection, apply a water repellent to exterior surfaces 2 to 3 weeks after treatment.
sulfuryl fluoride (Vikane, Zythor)	Fumigant	P	For infested furniture, stacked lumber, other wood products. Apply under gas-tight tarpaulin. Hold for 20 to 24 hours at a temperature above 60 degrees F. Requires an F-Phase N.C. Structural Pest Control License.
Termite—Drywood Species (Wood Treatment)			
acetarmiprid + bifenthrin (Transport)[3]	Sprayable	P	Coarse spray or drill and inject wood.
aluminum phosphide (Phostoxin)	Fumigant	P	Apply under gas-tight tarpaulins or in sealed chamber.
bifenthrin (Ortho) (Bifen, Talstar)	Sprayable	G P	Coarse spray or inject into wood.
cyfluthrin (Bayer Advanced) (Tempo)	Sprayable	G P	Coarse surface spray or inject wood.
lambda-cyhalothrin (Demand) (Spectracide)	Sprayable	P G	Apply as directed on label. Localized treatments. Spectracide is not recommended as a sole protection against termites.
cypermethrin (Demon TC, Cyper TC) zeta-cypermethrin (Cynoff)	Sprayable Dust	P	Coarse spray or inject into wood for localized infestations.
fipronil (Termidor, Taurus)	Sprayable, Foam, Dry	P	Coarse surface spray or inject wood.
deltamethrin (Bayer Advanced) (Suspend, D-Fend Dust, D-Foam)	Sprayable, Dust, Foam	G P	Surface spray or inject foam or dust into galleries.
dinotefuran (Alpine)	Foam, Spray	P	Apply as directed on label (can be used on infested shrubs, fence posts, utility poles, etc.)
imidacloprid (Premise) (Bayer Advanced)	Foam	P G	Apply to galleries as directed on label.
imidacloprid (Dominion, Premise)	Sprayable, Foam	P	Drill and inject spray or foam into voids.
imidacloprid+cyfluthrin (Temprid SC)	Sprayable	P	Apply as directed on label.
methyl bromide (Meth-O-Gas Q)	Fumigant	P	Apply under gas-tight tarpaulins only. **Regulatory use only.**
permethrin (Dragnet, Masterline)	Sprayable	P G	Coarse spray on wood for localized infestation.
sodium borate (Boracare, Timbor) (Spectracide)	Sprayable	P G	Coarse surface spray or inject wood.
sulfuryl fluoride (Vikane, Zythor)	Fumigant	P	Apply under gas-tight tarpaulins only. Hold for 20 to 24 hours at temperature above 60 degrees F. Requires an F-Phase N.C. Structural Pest Control License.
thiamethoxam (Optiguard)	Sprayable	P	Coarse spray or drill and inject into wood.
Termite—Subterranean Species (a) (Wood treatment)			
acetarmiprid + bifenthrin (Transport)[3]	Sprayable	P	
bifenthrin (Bifen, Talstar) (Ortho)	Sprayable	P G	For use only in voids or channels in damaged wood or to cracks and spaces between wooden members of structures.
boric acid (Perma-Dust PT 240)	Aerosol	P	Coarse surface spray or inject wood.
chlorantraniliprole (Altriset)	Spraying	P	Coarse spray around or inject into infested poles, trees and stumps (Outdoors)
chlorfenapyr (Phantom)	Sprayable	P	Coarse spray or inject into wood.

Table 5-18. Insect Control for Wood and Wood Products

Insect Insecticide	Formulation[1]	Use[2]	Precautions and Remarks
Termite—Subterranean Species (a) (Wood treatment) (continued)			
cyfluthrin (Tempo) (Bayer Advanced)	Sprayable	P G	Coarse spray, brush on, or inject into wood. Avoid excessive runoff.
zeta-cypermethrin (Cynoff)	Dust	P	Inject into wood for localized infestations.
lambda-cyhalothrin (Demand) (Spectracide)	Sprayable	P G	Apply as directed on label. Localized treatments. Spectracide is not recommended as a sole protection against termites.
deltamethrin (Bayer Advanced) (Suspend, D-Fend Dust, D-Foam)	Sprayable, Dust, Foam	P G	Coarse surface spray or inject wood with spray, dust or foam.
diflubenzuron (Exterra, Advance)	Bait	P	Above-ground stations used in conjunction with in-ground baiting systems.
dinotefuran (Alpine)	Foam and Spray	P	Apply as directed on label (can be used on infested shrubs, fence posts, utility poles, etc.).
esfenvalerate (Onslaught)	Sprayable	P	Apply as directed on label. (For use against swarming termites only).
fipronil (Termidor, Taurus)	Sprayable, Foam	P	Coarse spray or inject into wood.
imidacloprid (Premise)	Sprayable, Gel, Foam	P	Gel and foam formulations may be injected into voids or damaged wood.
imidacloprid+cyfluthrin (Temprid SC)	Sprayable	P	Apply as directed on label.
noviflumuron (Recruit IV AG)	Bait	P	Available only as part of the Sentricon in-ground system (see below).
permethrin (Dragnet, Masterline)	Sprayable	P	Coarse spray, brush on, or inject into wood. Avoid excessive runoff.
sodium borate (Boracare, Penetreat) (Spectracide)	Sprayable Dust	P G	Spray, brush on, or inject into wood. For long-term protection, apply a water repellent to exterior wood surfaces 2 to 3 weeks after treatment. Not a replacement for a soil treatment.
sulfluramid (FirstLine)	Bait	P	Above-ground stations used in conjunction with in-ground system.
Termite—Subterranean Species (b) Soil treatment			
acetamiprid + bifenthrin (Transport)[3]	Sprayable	P	Dig trenches 6 inches wide and at least 4 inches deep along the foundation. Never trench below the top of the footing. Depending upon the depth of footer, rodding may be needed. Dilutions and rates of applications vary among specific products. Vertical barriers usually require about 4 gallons of spray per 10 linear feet for each foot of depth along a foundation. Follow label restrictions on treatment in crawlspaces containing wells or cisterns. Follow instructions if "excavation and backfill" is permitted. Exercise extreme caution when treating crawlspaces. Wear appropriate protective equipment as specified on product label. General (broadcast) treatments of crawlspace soil for termites are prohibited, except as noted on the label. **NOTE:** Most termite infestations require treatment by a W-phase licensed structural pest control operator. Requirements for termite treatments are outlined in 2NCAC 34:.0503, .0505. Apply Premise or Bayer Advanced granules to trenches as a spot treatment. Bayer Advanced for the general public is available only in granular formations.
bifenthrin (Bifen, Talstar) (Ortho)	Sprayable	P G	
chlorfenapyr (Phantom)	Sprayable	P	
chlorantraniliprole (Altriset)	Spraying	P	
cyfluthrin (Bayer Advanced) (Tempo)	Sprayable	G P	
cypermethrin (Demon TC, Cyper TC)	Sprayable	P	Apply as directed on label.
lambda-cyhalothrin (Demand) (Spectracide)	Sprayable	P G	Apply as directed on label. Localized treatments.
fipronil (Termidor, Taurus, Ultrathor)	Sprayable	P	
imidacloprid (Premise) (Bayer Advanced)	Sprayable, Granular	P G	
imidacloprid+cyfluthrin (Temprid SC)	Sprayable	P	
indoxacarb (Arilon)	Sprayable	P	Use for spot or local treatment only (Arilon is not intended as sole protection against termites)
permethrin (Dragnet FT, MasterLine)	Sprayable	P G	
diflubenzuron (Advance, Exterra)	Bait	P	Termite monitoring and baiting program. Available only through manufacturer-authorized pest control companies.
hexaflumuron (Shatter)	Bait	P	Termite monitoring and baiting program. Available only through manufacturer-authorized pest control companies
novaluron (Trelona CTB)	Bait	P	Termite monitoring and baiting program. Available only through manufacturer-authorized pest control companies.
noviflumuron (Recruit HD)	Bait	P	Termite monitoring and baiting program. Available only through manufacturer-authorized pest control companies.

[1] Formulation designations: Aerosol = injectable or spray; Dust = dry application; Fumigant = gas in pressurized cylinder or pellets; Foam = Injectable foam; Sprayable = liquid concentrate or wettable powder for mixing with water or in a ready-to-use form

[2] Use designations: P = Professional applicator (licensed in structural pest control); G = General public use

* Several trade names available. Check label for active ingredient. Always follow label instructions.

INSECT CONTROL FOR HOME USE
Insect Control for the Home Vegetable Garden
J. F. Walgenbach, Entomology and Plant Pathology

Homeowner products are numerous, and names change frequently. Insecticides listed below are identified by the active ingredient. Brand names for homeowner products identify the active ingredient; always check the "active ingredients" portion of the product label to determine if the product is appropriate for your needs. Refer to the product label for rates and pre harvest intervals.

Table 5-19. Insect Control for the Home Vegetable Garden

Commodity Insect	Insecticide Active ingredient	Minimum Interval (Days) Between Last Application and Harvest	Precautions and Remarks
Asparagus			
Asparagus beetle, Japanese beetle, grasshopper, and aphid	carbaryl	1	Carbaryl will not control aphids.
	permethrin	3	
Bean			
Aphid	malathion	1	
	bifenthrin	3	
	cyfluthrin	7	
	insecticidal soap	0	
Corn earworm, Mexican bean beetle, bean leaf beetle, flea beetle, Japanese beetle, and cucumber beetle, potato leafhopper, fleahopper, lygus, and stink bug	carbaryl	3	
	spinosad	3	Will not control Japanese beetle, cucumber beetle or stink bug.
	bifenthrin	3	
	cyfluthrin	7	
	lambda-cyhalothrin	7	21-day preharvest interval for dried beans.
Spider mite	bifenthrin	3	
	malathion	1	
	insecticidal soap	0	Apply treatment at first sign of mites and speckled plants.
Whitefly	*Beauveria bassiana*	0	
	insecticidal soap	0	
Beet			
Flea beetle, beet webworm, and blister beetle	carbaryl	3 (14)	On foliage as needed. Fourteen days if tops used; 3 days if tops not used.
Broccoli, Cabbage, Cauliflower, Brussels Sprouts, Rutabaga			
Aphid	bifenthrin	7	
	cyfluthrin	3	
	malathion	7	
	insecticidal soap	0	
Cabbage looper, imported cabbageworm, diamondback moth, and cutworm	*Bacillus thuringiensis*	0	Start control program when worms are small and treat foliage every 5 to 7 days.
	carbaryl	3	On foliage as needed. Will not control cabbage looper. Carbaryl is suggested for cutworm.
	bifenthrin	7	
	esfenvalerate	3	
	lambda-cyhalothrin	1	
	spinosad	1	
Flea beetle and thrips	carbaryl	3	
	malathion	7	
	spinosad	1	For thrips only.
Harlequin bug	bifenthrin	7	On foliage as needed.
	lambda-cyhalothrin	1	On foliage as needed.
	malathion	7	On foliage as needed.

Table 5-19. Insect Control for the Home Vegetable Garden

Commodity / Insect	Insecticide Active ingredient	Minimum Interval (Days) Between Last Application and Harvest	Precautions and Remarks
Cantaloupe			
Aphid and thrips	cyfluthrin	0	
	Esfenvalerate	3	
	malathion	1	
	insecticidal soap	0	On foliage as needed.
Cucumber beetle (spotted and striped), pickleworm, squash bug, and squash vine borer	bifenthrin	3	
	esfenvalerate	3	
	cyfluthrin	0	
Spider mite	insecticidal soap	0	On foliage as needed.
Carrot			
Armyworm, leafminer, and leafhopper	Bacillus thuringiensis	0	B.t. will not control leafhoppers.
	carbaryl	0	On foliage as needed.
	cyfluthrin	0	
Celery			
Aphid, flea beetle, leafminer, and flea hopper	malathion	7	On foliage as needed.
	permethrin	3	On foliage as needed.
Collard			
Aphid and flea beetle	bifenthrin	7	
	malathion	7	On foliage as needed.
	insecticidal soap	0	On foliage as needed.
Cabbage looper, diamondback moth, and imported cabbageworm	Bacillus thuringiensis	0	Begin foliage treatments early and repeat as necessary. Include a spreader/sticker.
	spinosad	1	
Harlequin bug	malathion	7	
	bifenthrin	7	
	lambda-cyhalothrin	1	
	cyfluthrin	0	
Corn (Sweet)			
Corn earworm, sap beetle, flea beetle, and Japanese beetle	bifenthrin	1	
	esfenvalerate	1	
	carbaryl	2	
Corn earworm, European corn borer, and fall armyworm	Bacillus thuringiensis	0	Consult specific label. B.t. is effective while worms are feeding on the foliage.
	cyfluthrin	0	
	esfenvalerate	1	
	lambda-cyhalothrin	1	
	permethrin	3	
	spinosad	1	
Cucumber			
Cucumber beetle (spotted and striped), and pickleworm	bifenthrin	3	
	esfenvalerate	3	
	cyfluthrin	0	
Spider mite	insecticidal soap	0	On foliage as needed.
Whitefly	insecticidal soap	0	On foliage as needed.
	Beauveria bassiana	0	
Eggplant			
Aphid, flea beetle, whitefly, lace bug	bifenthrin	7	
	lambda-dyhalothrin	5	
	malathion	3	On foliage as needed.
Colorado potato beetle, hornworm, and corn earworm	Bacillus thuringiensis var. tennebrionus	0	For Colorado potato beetle only. Treat when small larvae are present. Not effective against adults or large larvae.
	spinosad	1	
spider mite	insecticidal soap	0	On foliage as needed.

Table 5-19. Insect Control for the Home Vegetable Garden

Commodity Insect	Insecticide Active ingredient	Minimum Interval (Days) Between Last Application and Harvest	Precautions and Remarks
Lettuce			
Aphid, leafhopper	bifenthrin	7	
	lambda-cyhalothrin	1	
	malathion	14 leaf, 7 head	On foliage as needed.
	insecticidal soap	0	On foliage as needed.
Cabbage looper, corn earworm, and leafhopper	*Bacillus thuringiensis*	0	On foliage as needed.
	spinosad	1	On foliage as needed.
	lambda-cyhalothrin	1	
Mustard Greens			
Aphid, Flea beetle	bifenthrin	7	
	malathion	7	On foliage as needed.
	insecticidal soap	0	On foliage as needed.
Cabbage looper, diamondback moth, and imported cabbageworm	*Bacillus thuringiensis*	0	Begin foliage treatments early and repeat as necessary.
	spinosad	1	
Okra			
Aphid and leafminer	bifenthrin	7	
	malathion		
Corn earworm, European corn borer, flea beetle, and stink bug	spinosad	1	
	bifenthrin	7	
	cyfluthrin	1	
	esfenvalerate	1	
	permethrin	1	
Onion			
Onion thrips	lambda-cyhalothrin	14	
	malathion	3 (Green)	
	insecticidal soap	0	
Peas			
Aphid and leafminer	insecticidal soap	0	
Pepper			
Aphid and thrips	esfenvalerate	1	
	malathion	3	
	insecticidal soap	0	
European corn borer, flea beetle, tomato fruitworm, hornworm, and stink bug	carbaryl	3	Will not control stink bug
	Bifenthrin	7	Excellent control of stink bug
	cyfluthrin	7	
	esfenvalerate	1	
	permethrin	3	
	spinosad	1	Will not control stink bug
Potato, Irish			
Aphid	cyfluthrin	0	
	esfenvalerate	0	
European corn borer, potato tuberworm	*Bacillus thuringiensis*	0	
	carbaryl	0	Apply when eggs begin to hatch, and every 5 days as needed.
	esfenvalerate	1	
	permethrin	3	
Potato leafhopper, potato flea beetle, Colorado potato beetle, and blister beetle	imidacloprid	21	Apply to the soil immediately at planting for long-term control.
	Bacillus thuringiensis var. *san diego* var. *tennebrionus*	0	For Colorado potato beetle only. Treat when small larvae are present. Not effective against adults or large larvae.
	carbaryl	0	On foliage as needed. Treat when most Colorado potato beetle eggs have hatched.

Pumpkin—See SQUASH AND PUMPKIN

Table 5-19. Insect Control for the Home Vegetable Garden

Commodity Insect	Insecticide Active ingredient	Minimum Interval (Days) Between Last Application and Harvest	Precautions and Remarks
Radish			
Aphid	malathion	7	On foliage as needed.
Flea beetle and imported cabbageworm	cyfluthrin	0	
Spinach			
Aphid, thrips, and leafminer	permethrin	1	
	malation	7	
	insecticidal soap	0	On foliage as needed.
Corn earworm and loopers	*Bacillus thuringiensis*	0	
	permethrin	1	
	spinosad	1	
Squash and Pumpkin			
Aphid	bifenthrin	3	
	malathion	1	
	insecticidal soap	0	
Cucumber beetle (spotted and striped), flea beetle, and leafhopper	esfenvalerate bifenthrin	3 3	
Pickleworm	esfenvalerate	3	
	spinosad	3	
Squash bug	bifenthrin	3	
Tomato			
Aphid, fleabeetle	bifenthrin	1	
	malathion	1	
	insecticidal soap	0	
Cutworm (surface type)	esfenvalerate	1	
Colorado potato beetle	*Bacillus thuringiensis* var. *san diego* var. *tennebrionus*	0	For Colorado potato beetle only. Treat when small larvae are present. Not effective against adults or large larvae.
	spinosad	1	
Spider mite	insecticidal soap	0	On foliage as needed.
Stink bug	cyfluthrin	7	Do not make more than 6 applications per season.
	lambda-cyhalothrin	5	
	malathion	1	
	permethrin	7	Do not apply on cherry tomatoes or varieties less than 1 inch in diameter.
Thrips	Spinosad	1	
	insecticidal soap	0	
Tomato fruitworm, cabbage looper, tobacco hornworm	*Bacillus thuringiensis*	0	Treat weekly, if necessary. Begin when fruits are 0.5 inch in diameter. Fruitworms are most serious after August 1
	carbaryl	3	
	cyfluthrin	7	Do not make more than 6 applications per season.
	esfenvalerate	1	
	lambda-cyhalothrin	5	
	permethrin	7	Do not apply on cherry tomatoes or varieties less than 1 inch in diameter.
	spinosad	1	
Whitefly	*Beauveria bassiana*	0	Apply when whiteflies observed. Repeat in 4- to 5-day intervals.
	malathion	1	
	pyrethrum products	0	
	insecticidal soap	0	

Table 5-19. Insect Control for the Home Vegetable Garden

Commodity / Insect	Insecticide Active ingredient	Minimum Interval (Days) Between Last Application and Harvest	Precautions and Remarks
Turnip, Turnip Greens			
Aphid, flea beetle	bifenthrin	7	
	malathion	7	On foliage as needed.
	insecticidal soap	0	
Cabbage looper, diamondback moth, imported cabbageworm	*Bacillus thuringiensis*	0	On foliage as needed.
	spinosad	1	
Harlequin bug	Gamma-cyhalothrin	1	On foliage as needed.
Watermelon			
Aphid	bifenthrin	3	
	malathion	1	
	insecticidal soap	0	On foliage as needed.
Cucumber beetle (spotted and striped)	bifenthrin	3	
	esfenvalerate	3	
	malathion	1	
Spider mite	bifenthrin	3	
	malathion	1	
	insecticidal soap	0	
Thrips	Spinosad	3	
	malathion	1	
	insecticidal soap	0	

Control of Household Pests

(Products for Use by the General Public)

P. Alder and M. G. Waldvogel, Entomology and Plant Pathology

Mention of pesticides in this section does not imply that chemicals are or should be the first or only means of control. Nonchemical methods, including exclusion and sanitation, are important to long-term pest management.

Space limitations preclude listing all pesticide formulations and trade names. Other appropriate products or formulations may be used.

Never use products that are not labeled for the intended use. Products labeled for outdoor use only should never be applied indoors. Read the product label for specific pest information about the active ingredient, application rates, and detailed instructions on the product's use.

NOTE: The insecticides listed below are identified by the common name. The brand names of most consumer insecticide products do not identify the specific chemical used, and the formulation and/or its contents may be changed by the manufacturer. Always check the "Active ingredients" portion of the product label to determine if the product is appropriate for your needs.

Table 5-20. Control of Household Pests–Products for Use by the General Public

Insecticide	Formulation	Precautions and Remarks
Ant (a) Indoors (*For information on carpenter ants, see* Insect Control for Wood and Wood Products)		
avermectin (Raid, Enforcer)	Bait Station	Place bait stations in areas where ants are active. Keep of out of reach of children and pets. Use dust formulations only in inaccessible areas.
bifenthrin (Ortho)	Liquid, Aerosol Spray	
borax/boric acid (Terro, Amdro, Ortho)	Bait, Dust, Bait Station	Treat ant-traveled areas. Re-treat as effectiveness diminishes. Some products are not suitable for use in residential kitchens or commercial food/feed preparation sites. Read the product label carefully. Remove food from storage areas before treating.
cyfluthrin (Raid)	Aerosol Spray	
cypermethrin (Ortho, Raid, Hot Shot, Black Flag, Ace, Combat, Enforcer)	Aerosol Spray, Liquid	Apply products as directed on the label
deltamethrin (Black Flag, Raid, Terro, Spectracide, Ortho)	Aerosol, Liquid, Dust	
dinotefuran (Hot Shot, Black Flag)	Bait	
hydramethylnon (Amdro, Combat)	Granular, Bait Station	
diatomaceous earth (PermaGuard, Hot Shot)	Dust	
borax (Terro)	Bait	
fipronil (Combat)	Bait	
lambda-cyhalothrin (Spectracide, Hot Shot, Black Flag)	Liquid, Aerosol Spray	
imiprothrin (Raid, Black Flag)	Aerosol Spray	Imiprothrin is usually formulated with other pesticides in these products.
indoxacarb (Hot Shot, Spectracide)	Bait Station	
lemongrass oil (Hot Shot)	Liquid, Aerosol Spray	
mint oil (Victor)	Aerosol Spray	
permethrin (Bengal)	Aerosol Spray	
d-phenothrin (Raid, Ortho)	Aerosol Spray	
prallethrin (Raid, Hot Shot, Black Flag)	Aerosol Spray	
pyrethrins, pyrethrum (Hot Shot, Black Flag, Ortho)	Aerosol Spray	
thiamethoxam (Raid)	Bait	
Ant (b) Outdoors (*Also see "Ant" and "Imported Fire Ant" under Home Lawns table.*)		
bifenthrin (Ortho, Amdro)	Granular, Aerosol Spray, Liquid, Bait Granules	Apply granular bait around nest. Place bait stations in areas where ants are active. Treat nest and surrounding area. May be applied along building perimeter.
borax (Terro)	Bait	
cypermethrin (Black Flag, Hot Shot, Enforcer)	Liquid	Apply chemicals as directed on the label.
deltamethrin (Amdro, Black Flag, Raid, Spectracide, Terro)	Liquid	
diatomaceous earth (PermaGuard)	Dust	
dinotefuran (Hot Shot, Black Flag)	Bait	
fipronil (Combat)	Bait	
hydramethylnon (Amdro, Combat)	Bait	
indoxacarb (Spectracide, Hot Shot, Ortho, Black Flag)	Bait Station, Bait Granules	
lambda-cyhalothrin (Spectracide, Hot Shot)	Liquid, Granular, Aerosol Spray	
lemongrass oil (Hot Shot)	Liquid, Aerosol Spray	
permethrin (Black Flag)	Liquid	
pyrethrins (Black Flag)	Aerosol Spray	

Table 5-20. Control of Household Pests—Products for Use by the General Public

Insecticide	Formulation	Precautions and Remarks
Bed Bug		
bifenthrin (Ortho)	Aerosol Spray, Liquid	
cypermethrin (Hot Shot, Raid Enforcer)	Liquid, Aerosol Spray	
cyflutrhin	Liquid	
deltamethrin (Black Flag, Enforcer, Ortho, Spectracide, Terro)	Aerosol Spray, Dust, Liquid	
diatomaceous earth (PermaGuard, Hot Shot)	Dust	
dichlorvos (Hot Shot)	Pest Strip	
d-phenothrin (Raid, Ortho)	Aerosol Spray	
imiprothrin (Hot Shot)	Liquid	
lambda-cyhalothrin (Hot Shot)	Liquid	
N-octyl bicycloheptene dicarboximide (Raid)	Aerosol Spray	
permethrin (Hot Shot)	Liquid	
pyrethrins (Enforcer, Black Flag)	Fogger, Aerosol Spray	
phenoxybenzl (Enforcer)	Aerosol Spray	
pralletrhin (Hot Shot)	Aerosol Spray	
silicon dioxide (Hot Shot)	Dust	
Bee (a) Indoors		
bifenthrin (Ortho)	Aerosol Spray	
cypermethrin (Black Flag, Hot Shot, Enforcer)	Liquid	
deltamethrin (Ortho, Raid, Spectracide, Terro)	Liquid, Dust	Apply only for sporadic invaders. If bees are found frequently, locate and remove the nest.
diatomaceous earth (PermaGuard)	Dust	
(Raid)	Aerosol Spray	Apply products as directed on the label.
pyrethrins (Black Flag)	Aerosol Spray	
Bee (b) Outdoors For carpenter bees, see section *Insect control for Wood and Wood Products*		
bifenthrin (Ortho)	Liquid	Apply after dark when insects have returned to nest. Some materials available in pressurized cans that propel an insecticide stream up to 10 feet. Re-treatment may be necessary.
carbaryl (Sevin)	Dust, Liquid, Powder	
cypermethrin (Hot Shot)	Liquid	
deltamethrin (Amdro, Black Flag, Spectracide)	Liquid	Apply products as directed on the label.
lambda-cyhalothrin (Spectracide, Cutter)	Liquid	
d-phenothrin (Raid, Ortho)	Aerosol Spray	
Booklouse (psocid) (Indoors and outdoors)		
bifenthrin (Ortho)	Liquid	Apply as a barrier spray along foundation and entry points (doors and windows). Read labels to determine which products are suitable for indoor use. Clean up moisture problems, which may attract insects indoors. Excess moisture may impede product effectiveness.
diatomaceous earth (PermaGuard)	Dust	
deltamethrin (Black Flag)	Liquid	
mint oil (Victor)	Aerosol Spray	
pyrethrins, pyrethrum	Aerosol Spray	
Boxelder Bug (Outdoors)		
bifenthrin (Ortho)	Liquid	Harmless insects become nuisances when searching indoors for hibernation sites in the fall. Treat door thresholds, window ledges, and other areas where the insects congregate or may gain entry.
cypermethrin (Black Flag)	Liquid	
deltamethrin (Amdro, Black Flag, Raid, Spectracide)	Liquid	
lambda-cyhalothrin (Spectracide)	Liquid	
d-phenothrin (Raid)	Aerosol Spray	
Brown Dog Tick (a) Indoors		
bifenthrin (Ortho)	Liquid	
cypermethrin (Black Flag, Ortho, Ace, Enforcer)	Aerosol Spray, Liquid	
deltamethrin (Amdro, Raid, Spectracide Terro)	Aerosol Spray, Liquid	
diatomaceous earth (PermaGuard)	Dust	
d-phenothrin (Raid, Ortho)	Raid	
imiprothrin (Black Flag)	Aerosol Spray	
lambda-cyhalothrin (Spectracide, Black Flag)	Aerosol Spray, Liquid	
lemongrass oil (Hot Shot)	Aerosol Spray, Liquid	
permethrin (Hot Shot, Bengal)	Aerosol Spray,	
pralletrhin (Black Flag)	Aerosol Spray	

Table 5-20. Control of Household Pests—Products for Use by the General Public

Insecticide	Formulation	Precautions and Remarks
Brown Dog Tick (a) Indoors (continued)		
pyrethrins (Black Flag)	Aerosol Spray	
tetramethrin (Raid)	Aerosol Spray	
Brown Dog Tick (b) Outdoors and under buildings		
bifenthrin (Amdro, Ortho)	Granules	
cypermethrin (Black Flag, Enforcer)	Liquid	
deltamethrin (Black Flag, Raid, Spectracide, Terro)	Aerosol Spray, Liquid	
diatomaceous earth (PermaGuard)	Dust	
eugenol (Bioganic, Raid)	Aerosol Spray, Dust	
lambda-cyhalothrin (Spectracide, Cutter)	Aerosol Spray, Granule, Liquid	
lemongrass oil (Hot Shot)	Aerosol Spray, Liquid	
permethrin (Black Flag)	Liquid	
pyrethrins (Black Flag)	Aerosol Spray	
Carpet Beetle (a) Nonfabric areas and infested areas of carpets only		
cypermethrin (Black Flag, Hot Shot, Ortho, Raid)	Aerosol Spray, Liquid	
diatomaceous earth (PermaGuard, Hot Shot)	Dust	
deltamethrin (Black Flag, Ortho, Spectracide, Terro)	Aerosol Spray, Dust, Liquid	
d-phenothrin (Raid, Ortho))	Aerosol Spray	
lambda-cyhalothrin (Spectracide, Hot Shot, Black Flag)	Liquid	
pyrethrins, pyrethrum	Aerosol Spray	
bifenthrin (Ortho)	Aerosol Spray, Liquid	
deltamethrin (Raid)	Liquid	
pyrethrins (Black Flag)	Aerosol Spray	
Carpet Beetle (b) On fabric		
diatomaceous earth (PermaGuard)	Dust	
pyrethrins, pyrethrum	Aerosol Spray, Liquid	
Centipede (a) Indoors		
bifenthrin (Ortho)	Liquid	
cyfluthrin (Raid)	Aerosol Spray	
cypermethrin (Black Flag, Hot Shot, Enforcer)	Liquid	
deltamethrin (Black Flag, Ortho, Raid, Spectracide, Terro)	Aerosol Spray, Dust, Liquid	
diatomaceous earth (PermaGuard, Hot Shot)	Dust	
lambda–cyhalothrin (Spectracide)	Aerosol Spray, Liquid	
imiprothrin (Black Flag, Raid)	Aerosol Spray, Liquid	
lemongrass oil (Hot Shot)	Aerosol Spray, Liquid	
Permethrin (Bengal)	Aerosol Spray	
prallethrin (Black Flag)	Aerosol Spray	
pyrethrins (Black Flag)	Aerosol Spray	
Centipede (b) Outdoors		
bifenthrin (Amdro, Ortho)	Granule, Liquid	Treat infested areas around building foundations, vents, and similar access points. Barrier sprays of 12 to 18 inches along perimeter may be effective.
cypermethrin (Black Flag, Hot Shot, Enforcer)	Liquid	
deltamethrin (Amdro, Black Flag, Spectracide, Terro)	Aerosol Spray, Liquid	
diatomaceous earth (PermaGuard)	Dust	
lambda-cyhalothrin (Hot Shot, Spectracide)	Aerosol Spray, Granule, Liquid	
lemongrass oil (Hot Shot)	Aerosol Spray, Liquid	
pyrethrins (Black Flag)	Aerosol Spray	
Chigger (Red bug) Outdoors		
bifenthrin (Amdro, Ortho)	Granule, Liquid	Apply to grass, bushes, and weeds in the infested areas. Thoroughly saturate soil, but avoid runoff into ponds, lakes, or other bodies of water. Repeat as needed. Apply labeled repellent products to shoes, ankles, and legs before entering suspected chigger-infested areas.
gamma-cyhalothrin (Spectracide)	Granule, Liquid	
lamda-cyhalothrin (Spectracide, Cutter)	Granule, Liquid	
deltamethrin (Black Flag)	Liquid	
Clothes Moth (a) Nonfabric areas and infested areas of carpet only, See Carpet Beetle		
Clothes Moth (b) On fabric, See Carpet Beetle		

Table 5-20. Control of Household Pests—Products for Use by the General Public

Insecticide	Formulation	Precautions and Remarks
Clothes Moth (c) In storage areas		
dichlorvos (Pest Strip) Ortho No-Pest Strip	Strip	Hang on strip in clothes closets or storage chests up to 1,000 cubic feet in capacity. Not for use in occupied rooms or in closets in occupied rooms. Follow label instructions carefully.
paradichlorobenzene (PDB) naphthalene	Crystals or similar solid	Effective repellents on clean fabric in airtight enclosures. Avoid contact with plastic buttons and zippers.
Clover Mite (a) Indoors		
bifenthrin (Ortho)	Liquid	
cypermethrin (Black Flag)	Liquid	
deltamethrin (Black Flag, Ortho, Raid, Spectracide, Terro)	Aerosol Spray Liquid	
diatomaceous earth (PermaGuard)	Dust	
lambda-cyhalothrin (Spectracide)	Liquid	
pyrethrins (Black Flag)	Aerosol Spray	
Clover Mite (b) Outdoors		
bifenthrin (Ortho)	Granular	Treat around points of entry, such as foundations, vents, windows, and doors. Maintain a 12-inch wide vegetation-free zone along foundation. Spray 1 to 2 feet high along the foundation wall and a 3- to 5-feet barrier on the grass or landscaped areas around the foundation. Water immediately after applying granules.
cypermethrin (Black Flag, Raid)	Liquid, Aerosol Spray	
deltamethrin (Black Flag, Raid, Spectracide, Terro)	Liquid, Aerosol Spray	
diatomaceous earth (PermaGuard)	Dust	Apply products as directed on the label.
lambda-cyhalothrin (Spectracide, Cutter)	Liquid	
pyrethrins (Black Flag)	Aerosol Spray	
Cockroach (a) Indoors		
avermectin (Enforcer, Raid)	Bait Station	Apply sprays along baseboards, under sinks, in cabinets and other infested areas. Remove and cover food, cooking, and eating utensils before spraying cabinets. Do not restock shelves until surface dries completely. Some products are not suitable for use in residential kitchens or commercial food/feed preparation sites. Read the product label carefully
bifenthrin (Ortho)	Aerosol Spray, Liquid	
boric acid (Enforcer, Hot Shot, Terro, Ortho)	Bait Station, Dust	
cyfluthrin (Raid)	Aerosol Spray	
cypermethrin (Black Flag, Hot Shot, Ortho, Raid, Ace, Enforcer)	Aerosol, Liquid	
imiprothin (Black Flag, Raid)	Aerosol Spray	Imiprothrin is formulated with other pesticides in these products.
diatomaceous earth (PermaGuard, Hot Shot)	Dust	Use diatomaceous earth in the same manner as boric acid powders. Some formulations contain pyrethrins and pyrethrum.
deltamethrin (Black Flag, Ortho, Raid, Spectracide, Terro)	Aerosol Spray, Dust, Liquid	
dinotefuran (Hot Shot, Black Flag)	Bait	Place bait stations in infested areas; follow label instructions. Keep out of reach of children and pets. Sanitation is critical; before using baits, eliminate other food sources. Place bait stations in cabinets under sinks, behind stoves and refrigerators. Slow acting but gives long-lasting control. Force small amounts into all hidden nesting areas with dust applicator. Avoid overapplication and inhalation of dust. Some formulations may contain pyrethrins or pyrethrum. Do not contaminate food preparation or storage sites.
fipronil (Combat)	Bait, Bait Station	
hydramethylnon (Combat, Ortho)	Bait	
hydroprene (Egg Stopper)	Bait Station	Hydroprene is an insect growth regulator and should be used with an adulticide.
imiprothrin (Black Flag, Raid)		
lambda-cyhalothrin (Spectracide, Black Flag)	Liquid, Aerosol Spray	
lemongrass oil (Hot Shot)	Liquid, Aerosol Spray	
permethrin (Bengal, Hot Shot)	Aerosol Spray	
prallethrin (Hot Shot, Raid, Black Flag)	Aerosol Spray	
pyrethrins (Black Flag)	Aerosol Spray	
tetramethrin (Hot Shot)	Fogger	
chlorpyrifos (Hot Shot)	Bait	Apply products as directed on the label.
Cockroach (b) Outdoors		
bifenthrin (Ortho)	Liquid	Some species of cockroaches can live indoors and outdoors. Cockroaches that live outdoors tend to hide under mulch, ivy, and similar cover. Treat groundcover and along foundation walls, patios, and other areas where cockroaches are seen. Certain products cannot be used on or around edible plants. Read product labels for any limitations.
cypermethrin (Black Flag, Hot Shot)	Liquid	
deltamethrin (Amdro, Black Flag, Raid, Spectracide, Terro,)	Aerosol Spray, Liquid	
diatomaceous earth (PermaGuard)	Dust	Apply products as directed on the label.
dinotefuran (Hot Shot, Black Flag)	Bait	
hydromethylnon (Combat, Ortho)	Bait, Granule	
lambda-cyhalothrin (Spectracide, Cutter)	Granule, Liquid	
lemongrass oil (Hot Shot)	Aerosol Spray, Liquid	
pyrethrins (Black Flag)	Aerosol Spray	

Table 5-20. Control of Household Pests—Products for Use by the General Public

Insecticide	Formulation	Precautions and Remarks
Cricket (Indoors and in crawlspaces)		
boric acid	Bait	Crickets enter homes through basements and similar areas. Some formulations may be sprinkled along foundation. Read product label before using outdoors.
cyfluthrin (Raid)	Aerosol Spray	
cypermethrin (Black Flag, Hot Shot, Ortho, Raid, Combat, Enforcer)	Aerosol Spray, Liquid	Treat along foundation walls, patios, and other areas where crickets are seen.
deltamethrin (Black Flag, Ortho, Raid, Spectracide, Terro)	Dust, Liquid	Apply products as directed on the label.
diatomaceous earth (PermaGuard, Hot Shot)	Dust	Apply in a light 2- to 4-inch band around foundation. Do not use excessive amounts, and do not apply to foliage of ornamentals or to food crops.
hydramethylnon	Granule	Imiprothrin is formulated with other pesticides in these products.
imiprothrin (Raid, Black Flag)	Aerosol Spray	Apply products as directed on the label.
lambda-cyhalothrin (Spectracide, Black Flag, Cutter)	Aerosol Spray, Granule, Liquid	
lemongrass oil (Hot Shot)	Aerosol Spray, Liquid	
permethrin (Bengal, Black Flag)	Aerosol Spray, Liquid	
pyrethrins (Hot Shot, Black Flag)	Aerosol Spray	
bifenthrin (Ortho)	Aerosol Spray, Liquid	
prallethrin (Hot Shot, Raid, Black Flag)	Aerosol Spray	
Earwig (a) Indoors		
bifenthrin (Ortho)	Aerosol Spray, Liquid	
cyfluthrin (Raid)	Aerosol Spray	
cypermethrin (Black Flag, Hot Shot, Raid Enforcer)	Liquid, Aerosol Spray	
diatomaceous earth (PermaGuard, Hot Shot)	Dust	
deltamethrin (Raid, Black Flag)	Liquid, Aerosol Spray	
d-phenothrin (Ortho)	Aerosol Spray	
imiprothrin (Raid, Black Flag)	Aerosol Spray	
lambda-cyhalothrin (Spectracide)	Liquid	
lemongrass oil (Hot Shot)	Aerosol Spray, Liquid	
pyrethrins (Black Flag)	Aerosol Spray	
Earwig (a) Indoors (continued)		
prallethrin (Hot Shot, Raid)	Aerosol Spray	
tetramethrin (Hot Shot)	Fogger	
Earwig (b) Outdoors		
bifenthrin (Amdro, Ortho)	Granular, Liquid	Repeat treatments at 14-day intervals if necessary. Granular formulations are for outdoor use only and must be watered in or applied before rain.
cypermethrin (Black Flag, Hot Shot, Raid, Enforcer)	Aerosol Spray, Liquid	
diatomaceous earth (PermaGuard)	Dust	
gamma-cyhalothrin (Spectracide)	Liquid, Granule	
lambda-cyhalothrin (Spectracide)	Liquid	
lemongrass oil (Hot Shot)	Aerosol Spray, Liquid	
pyrethrins (Black Flag)	Aerosol Spray	
Flea (a) Indoors		
bifenthrin (Ortho)	Aerosol Spray, Liquid	
boric acid	Dust	Treat pet sleeping quarters and other localized areas, such as under cushions and furniture, as specified on label. Vacuum carpets and furniture before applying; dispose of contents properly. Sprays may be used for general area treatment. Also treat cracks, crevices, and similar areas only. Foggers are only effective when used in conjunction with sprays to other critical areas. Treat infested animals with properly labeled product for lasting control.
cypermethrin (Black Flag, Hot Shot, Ortho, Ace, Enforcer)	Aerosol Spray, Liquid	
deltamethrin (Black Flag, Ortho, Raid, Spectracide, Terro)	Aerosol Spray, Dust, Liquid	Apply as directed on the label.
diatomaceous earth (Hot Shot)		
lamda-cyhalothrin (Amdro, Spectracide)	Aerosol Spray, Granule, Liquid	
lemongrass oil (Hot Shot)	Aerosol Spray, Liquid	
d-phenothrin (Raid)	Aerosol	
pyrethrins (Hot Shot, Black Flag)	Liquid, Fogger, Aerosol Spray	
tetramethrin (Hot Shot)	Fogger	
permethrin (Enforcer, Bengal)	Liquid	
prallethrin (Black Flag)	Aerosol Spray	
sumithrin (Enforcer)	Dust	
methoprene (Precor) pyriproxyfen	Aerosol Spray, Fogger, Liquid	Insect growth regulators that control immature fleas only. Usually formulated with an adulticide.

Table 5-20. Control of Household Pests—Products for Use by the General Public

Insecticide	Formulation	Precautions and Remarks
Flea (a) Indoors (continued)		
imiprothrin (Black Flag, Raid)	Aerosol Spray	
tetramethrin (Enforcer)	Aerosol Spray	
phenoxybenzyl (Enforcer)	Aerosol Spray	
Flea (b) Outdoors		
bifenthrin (Amdro, Ortho)	Liquid	Concentrate on kennels and shaded areas where animals tend to rest of congregate. Apply liquid formulations with sufficient spray volume to saturate soil. Granular formulations must be watered in or applied before rain, Repeat as needed at 4- to 6-week intervals.
cypermethrin (Black Flag, Hot Shot, Enforcer)	Liquid	
deltamethrin (Black Flag, Raid, Spectracide, Terro)	Aerosol Spray, Liquid	
diatomaceous earth (PermaGuard)	Dust	
gamma-cyhalothrin (Spectracide)	Granule, Liquid	
lambda-cyhalothrin (Enforcer, Spectracide, Cutter)	Aerosol Spray, Liquid	Apply as directed on the label.
lemongrass oil (Hot Shot)	Aerosol Spray, Liquid	
permethrin (Black Flag)	Liquid	
pyrethrins (Black Flag)	Aerosol Spray	
Flies (a) Indoors		
cypermethrin (Hot Shot, Enforcer)	Liquid	
dichlorvos (Vapona, Pest Strip, Ortho No-Pest Strip, Hot Shot)	Strip	Strips can only be used in unoccupied areas. Apply as a surface spray to areas or objects (such as garbage cans) infested with flies. Repeat treatments as may be necessary. See label before treating areas of vegetation.
lambda-cyhalothrin (Spectracide)	Aerosol Spray, Liquid	
lemongrass oil (Hot Shot)	Aerosol Spray	Sanitation in the area is essential for satisfactory control of flies.
permethrin (Bengal)	Aerosol Spray	
prallethrin (Black Flag)	Aerosol Spray	
pyrethrins (Black Flag)	Aerosol Spray, Liquid	
d-phenothrin (Raid, Black Flag)	Aerosol	
tetramethrin (Hot Shot)	Fogger	
deltamethrin (Amdro, Black Flag, Raid, Spectracide, Terro)	Aerosol Spray, Liquid	
Flies (b) Outdoors		
cypermethrin (Hot Shot, Enforcer)	Liquid	Apply as a surface spray to areas or objects (such as garbage cans) infested with flies. Repeat treatments may be necessary. See label before treating areas of vegetation.
cyfluthrin (Raid)	Aerosol Spray	
deltamethrin (Amdro, Black Flag, Raid, Spectracide, Terro)	Aerosol Spray, Liquid	
imidacloprid (Maxforce)	Bait	Sanitation in the area is essential for satisfactory control using any of these chemicals but particularly important with baits.
lambda-cyhalothrin (Spectracide)	Aerosol Spray, Liquid	
d-phenothrin (Raid, Black Flag)	Aerosol Spray	Use as directed.
prallethrin (Ultrakill)	Aerosol Spray	
pyrethrins (Black Flag)	Aerosol Spray	
Hornets, Mud Daubers, Wasps, Yellow Jackets (a) Indoors		
bifenthrin (Ortho)	Liquid	
cypermethrin (Black Flag, Hot Shot, Enforcer)	Liquid	
deltamethrin (Raid, Spectracide, Terro)	Liquid	
eugenol (Bioganic)	Aerosol Spray, Dust	
prallethrin (Raid, Hot Shot, Ultrakill, Spectracide)	Aerosol Spray	
d-phenothrin (Raid)	Aerosol Spray	
pyrethrins (Black Flag)	Aerosol Spray	
tetramethrin (Hot Shot, Terro)	Fogger, Aerosol Spray	
Hornets, Mud Daubers, Wasps, Yellow Jackets (b) Nest and adjacent areas		
bifenthrin (Ortho)	Liquid	Apply to nest or opening after dark when insects have returned to nest. Re-treatment may be necessary. Most are packaged in pressurized containers that direct an insecticide stream of up to 10 feet. For yellowjackets and other soil-dwelling wasps, apply chemical to nests in soil.
carbaryl (Sevin)	Dust, Liquid	
cyfluthrin (Raid)	Aerosol Spray	
cypermethrin (Black Flag, Hot Shot, Enforcer)	Liquid	
deltamethrin (Amdro, Black Flag, Raid, Spectracide, Terro)	Aerosol Spray, Liquid	
diatomaceous earth (PermaGuard)	Dust	
eugenol (Bioganic)	Aerosol Spray, Dust	
lambda-cyhalothrin (Hot Shot)	Liquid	
phenothrin (Raid)	Aerosol Spray	
prallethrin (Ultrakill, Hot Shot, Raid, Spectracide)	Aerosol	

Table 5-20. Control of Household Pests—Products for Use by the General Public

Insecticide	Formulation	Precautions and Remarks
Lice: body, head, crab (on person)		
malathion (Ovide)	Liquid	Shampoo formulations. Thoroughly treat infested areas of body with lotion. Wash infested clothing with strong soap and very hot water. Dry clean woolens. Ovide requires a physician's prescription. **Insecticidal treatment of furniture, carpets, or other areas of the home is not needed.**
permethrin (Nix)	Liquid	
pyrethrins (Black Flag)	Liquid	
Millipede (a) Indoors		
bifenthrin (Ortho)	Liquid	
cypermethrin (Black Flag, Hot Shot, Raid, Enforcer)	Liquid, Aerosol Spray	
diatomaceous earth (PermaGuard, Hot Shot)	Dust	
imiprothrin (Raid, Black Flag)	Aerosol Spray, Liquid	
lambda-cyhalothrin (Spectracide)	Liquid	
lemongrass oil (Hot Shot)	Aerosol Spray, Liquid	
mint oil (Victor)	Aerosol Spray	
deltamethrin (Black Flag, Ortho, Raid)	Aerosol Spray, Dust	
d-phenothrin (Ortho)	Aerosol Spray	
prallethrin (Hot Shot, Raid)	Aerosol Spray	
pyrethrins (Black Flag)	Aerosol Spray	
Millipede (b) Outdoors		
bifenthrin (Amdro, Ortho)	Granule, Liquid	Use as barrier treatment along foundation wall, door threshold, window ledges. Some sprays may damage vegetation under hot, humid conditions. Read label precautions. For lawn treatment, apply an insecticide band 10 to 15 feet wide. Apply liquid formulations with sufficient spray volume to saturate soil. Use granular formulations outdoors only; water in or apply before rain. Repeat as needed at 4- to 6-week intervals.
cypermethrin (Black Flag, Hot Shot, Enforcer)	Liquid	
diatomaceous earth (PermaGuard)	Dust	
gamma-cyhalothrin (Spectracide)	Granule, Liquid	
lambda-cyhalothrin (Cutter)	Liquid	
lemongrass oil (Hot Shot)	Aerosol Spray, Liquid	
pyrethrins (Black Flag)	Aerosol Spray	
Mosquitoes (a) Indoors		
cypermethrin (Black Flag, Hot Shot, Enforcer)	Liquid	
deltamethrin (Black Flag, Raid, Spectracide, Terro)	Liquid	
lambda-cyhalothrin (Spectracide)	Aerosol Spray, Granule, Liquid	
lemongrass oil (Hot Shot)	Aerosol Spray	
tetramethrin (Hot Shot)	Fogger	
permethrin (Bengal)	Aerosol Spray	
phenothrin (Raid, Black Flag)	Aerosol Spray	
prallethrin (Black Flag)	Aerosol Spray	
pyrethrins (Black Flag)	Aerosol Spray	
Mosquitoes (b) Outdoors (See also *Community Pest Control* Section)		
allethrin (Coleman)	Repellent Coil	
Bacillus thuringiensis (*Bti*) (Mosquito Dunks)	Solid	A biopesticide containing bacteria that kill mosquitoes and some biting flies. Place in small ponds, birdbaths, and ornamental pools (not swimming pools). Follow instructions for specifics of application.
bifenthrin (Ortho)	Liquid	Long-term control requires eliminating or cleaning mosquito breeding areas, such as discarded containers, ditches, and other artificial sources of standing water. Spraying nearby vegetation may eliminate some mosquito resting sites, but some formulation as may damage vegetation. Aerosols or foggers may be used for temporary relief when winds are insignificant. Use repellents on exposed body areas.
deltamethrin (Black Flag, Raid, Terro)	Aerosol Spray, Liquid	
cyfluthrin (Raid)	Aerosol Spray, Fogger	
cypermethrin (Black Flag, Hot Shot, Enforcer)	Liquid	
gamma–cyhalothrin (Spectracide)	Liquid, Granule	
lambda-cyhalothrin (Amdro, Spectracide, Cutter)	Aerosol Spray, Granule, Liquid	
permethrin (Black Flag)	Liquid	
pyrethrins (Black Flag)	Aerosol Spray	
Pantry Pests (Pests in food storage areas)		
cypermethrin (Black Flag)	Liquid	
deltamethrin (Black Flag, Ortho, Raid, Spectracide, Terro)	Dust, Liquid	
diatomaceous earth (PermaGuard)	Dust	
imiprothrin (Black Flag, Raid)	Aerosol Spray	Imoprothin is formulated with other pesticides in these products.
lambda-cyhalothrin (Spectracide)	Liquid	
mint oil (Victor)	Aerosol Spray	
deltamethrin (Black Flag, Raid)	Aerosol Spray, Liquid	
pyrethrins, pyrethrum	Aerosol Spray	
bifenthrin (Ortho)	Aerosol Spray	

Table 5-20. Control of Household Pests—Products for Use by the General Public

Insecticide	Formulation	Precautions and Remarks
Silverfish		
bifenthrin (Ortho)	Aerosol Spray Liquid, Dust	Apply to cracks and crevices, behind and underneath appliances. Spray along baseboards and other areas where silverfish are found.
cyfluthrin (Raid)	Aerosol Spray	
cypermethrin (Black Flag, Hot Shot, Ortho, Raid, Ace, Enforcer)	Aerosol Spray, Liquid	
deltamethrin (Amdro, Black Flag, Ortho, Raid, Spectracide, Terro)	Dust, Liquid	
diatomaceous earth (PermaGuard, Hot Shot)	Dust	
d-phenothrin (Ortho)	Aerosol Spray	
hydramethylnon	Bait	
imiprothrin (Raid, Hot Shot)	Aerosol Spray	Imoprothin is formulated with other pesticides in these products
lambda-cyhalothrin (Spectracide)	Aerosol Spray, Granule, Liquid	Follow label directions.
lemongrass oil (Hot Shot)	Aerosol Spray, Liquid	
mint oil (Victor)	Aerosol Spray	
deltamethrin (Black Flag, Raid)	Aerosol, Liquid	
permethrin (Bengal)	Aerosol Spray	
prallethrin (Black Flag)	Aerosol Spray	
pyrethrins (Black Flag)	Aerosol Spray	
Sowbugs and Pillbugs (a) Indoors		
bifenthrin (Ortho)	Liquid	Clean up breeding and hiding places, and treat thoroughly. Outdoor barrier treatments along foundation and door thresholds are usually sufficient. Some products are not suitable for use in residential kitchens or commercial food/feed preparation sites. Read the product label carefully.
cypermethrin (Black Flag, Hot Shot, Ortho, Raid, Combat, Enforcer)	Aerosol Spray, Liquid	
deltamethrin (Black Flag, Ortho, Raid, Spectracide)	Dust, Liquid	
diatomaceous earth (PermaGuard)	Dust	Follow label directions.
lambda-cyhalothrin (Spectracide)	Aerosol Spray, Liquid	
mint oil (Victor)	Aerosol Spray	
permethrin (Bengal)	Aerosol Spray	
pyrethrins (Black Flag)	Aerosol Spray	
Sowbugs and Pillbugs (b) Outdoors		
bifenthrin (Ortho)	Granular, Liquid	
cypermethrin (Black Flag, Hot Shot, Ortho)	Aerosol Spray, Liquid	
deltamethrin (Black Flag, Ortho, Raid, Spectracide, Terro)	Dust, Liquid	
diatomaceous earth (PermaGuard)	Dust	
lambda-cyhalothrin (Spectracide, Cutter)	Aerosol Spray, Granule, Liquid	
pyrethrins (Black Flag)	Aerosol Spray	
Spiders (a) Indoors		
bifenthrin (Ortho)	Dust, Liquid	Treat infested areas, along baseboards. Use foggers if rooms have been undisturbed for some time and spider populations are extensive. Some products are not suitable for use in residential kitchens or commercial food/feed preparation sites. Read the product label carefully.
cyfluthrin (Raid)	Aerosol Spray	
cypermethrin (Black Flag, Hot Shot, Ortho, Raid, Ace, Combat, Enforcer)	Aerosol Spray, Liquid	
diatomaceous earth (PermaGuard)	Dust	
d-phenothrin (Orthol)	Aerosol Spray	Imoprothin is formulated with other pesticides in these products.
imiprothrin (Raid Max, Black Flag)	Aerosol Spray	
lambda-cyhalothrin (Spectracide, Black Flag)	Aerosol Spray, Liquid	Follow label directions.
lemongrass oil (Hot Shot)	Aerosol Spray, Liquid	
mint oil (Victor)	Aerosol Spray	
deltamethrin (Black Flag, Ortho, Raid, Spectracide, Terro)	Aerosol Spray, Dust, Liquid	
permethrin (Bengal)	Aerosol Spray	
pyrethrins (Black Flag)	Aerosol Spray	
prallethrin (Hot Shot, Raid, Black Flag)	Aerosol Spray	

Table 5-20. Control of Household Pests—Products for Use by the General Public

Insecticide	Formulation	Precautions and Remarks
Spiders (b) Outdoors		
bifenthrin (Amdro, Ortho)	Granule, Liquid	Apply as a barrier treatment along foundation. Spray corners of decks, eaves, porches and other areas where spiders tend to build webs. Webbing can be knocked down as an alternative. Exercise caution when spray in crawlspace. Avoid inhaling spray. Follow label directions.
cypermethrin (Black Flag, Hot Shot, Enforcer)	Liquid	
deltamethrin (Amdro, Spectracide, Terro)	Liquid	
diatomaceous earth (PermaGuard)	Dust	
lambda-cyhalothrin (Spectracide, Cutter)	Aerosol Spray, Granule, Liquid	
lemongrass oil (Hot Shot)	Aerosol Spray, Liquid	
pyrethrins (Black Flag)	Aerosol Spray	
Springtails (Indoors and outdoors)		
bifenthrin (Ortho)	Granular, Liquid	Apply as a barrier spray along foundation and entry points. Some products may be used indoors for temporary relief. Clean up moisture conditions that may attract insects indoors. Excess moisture may impede product effectiveness. Use indoors for temporary relief. Some products are not suitable for use in residential kitchens or commercial food/feed preparation sites. Read the product label carefully. Imoprothin is formulated with other pesticides in these products. Follow label directions.
deltamethrin (Black Flag, Raid)	Aerosol Spray	
diatomaceous earth (PermaGuard)	Dust	
imiprothrin (Raid, Black Flag)	Aerosol Spray	
lambda-cyhalothrin (Amdro, Spectracide)	Granule, Liquid	
mint oil (Victor)	Aerosol Spray	
pyrethrins, pyrethrum	Aerosol Spray	
gamma-cyhalothrin (Spectracide)	Liquid, Granules	
Stinging Caterpillars See *Trees and Woody Ornamentals* Section		
Stink Bugs (Indoors and outdoors)		
bifenthrin (Ortho)		
cypermethrin (Black Flag, Ortho, Raid)	Aerosol Spray, Liquid	
deltamethrin (Raid, Spectracide, Terro)	Liquid	
d-phenothrin (Ortho)	Aerosol Spray	
gamma-cyhalothrin (Spectracide)		
imiprothrin (Raid)	Aerosol Spray	
lambda-cyhalothrin (Cutter)	Liquid	
Stored Food Pests See Pantry Pests.		
Ticks(Outdoors) See Brown Dog Tick and *Control of Insects on Pets* section		
Wasps, Yellow Jackets See Hornets, etc.		

Formulation Designations: Bait may be gel or granular; fogger is a total release aerosol; liquid for mixing with water or ready-to-use; powder for mixing with water.

Insect Control for Home Lawns

R. L. Brandenburg, Entomology and Plant Pathology

NOTE: Some products are for use only by professionals. Homeowner products are numerous, and names change frequently, so it is not possible to list all homeowner products by brand names. When choosing a product to use at home, look at the label and use this table to compare the name of the active ingredients.

Table 5-21. Insect Control for Home Lawns

Pest Insecticide and Formulation	Amount per 1,000 Sq Ft	Precautions and Remarks
Ant (Also see Imported Fire Ant)		
carbaryl* (Sevin) 50 WP, 80 WSP and baits	See label	Treat mounds and surrounding area or apply broadcast.
clothianidin + bifenthrin (Aloft LC) G SC	 1.8 to 3.6 lb 0.27 to 0.54 fl oz	Toxic to fish and aquatic invertebrates. Do not apply near or allow runoff to surface waters or intertidal areas.
hydramethylnon* (Maxforce G) bait	See label	
pyrethroids* (Advanced Lawn, Bug-B-Gone, Deltaguard, Scimitar, Talstar, Tempo, Wisdom and others) Some ants are susceptible to fire ant products.	See label	Many pyrethroids are toxic to fish and aquatic invertebrates. Apply these products only as specified on the label.
Armyworm, Fall Armyworm, Cutworm		
azadirachtin* (Azatrol, Neemix, Turplex, etc.)	See label	
carbaryl* (Sevin) 50 WP, 80 WSP and baits	See label	Apply as a coarse spray in sufficient water for good coverage. Treat when first injury noted. Repeat as needed. Do not water into soil. Do not cut grass for 1 to 3 days after treatment.
chlorantraniliprole (Acelepryn) G SC	 1.15 to 2.3 lb 0.046 to 0.092 fl oz	Toxic to aquatic invertebrates, oysters and shrimp.
indoxacarb (Provaunt) WDG	0.046 to 0.092 oz	
pyrethroids* (Advanced Lawn, Bug-B-Gone, Deltaguard, Menace, Scimitar, Talstar, Tempo, Wisdom and others)	See label	Many pyrethroids are toxic to fish and aquatic invertebrates. Apply these products only as specified on the label.
spinosad A and D (Conserve) SC	0.25 to 1.25 fl oz	Rate varies with size and species.
thiamethoxam + lambda-cyhalothrin (Tandem)	See label	Highly toxic to fish and aquatic invertebrates.
trichlorfon* (Dylox, Proxol) 80 SP	1.5 to 3 oz	
various entomogenous nematode and *B.t.* products	See label	
Bee and Wasp		
carbaryl* (Sevin) 50 WP	6 to 8 oz	Most of these are parasitic on soil pests, especially grubs; therefore they are beneficial. Sometimes there are so many bees and wasps burrowing in the soil that chemical treatments are necessary to prevent damage or reduce danger from stings. Spot spray ground nest openings. Bee, wasp, and hornet sprays in pressurized cans are also effective.
pyrethroids* (Advanced Lawn, Bug-B-Gone, Deltaguard, Scimitar, Talstar, Tempo, Wisdom and others)	See label	
Chinch Bug		
*Beauveria bassiana** (Naturalis-T)	See label	
carbaryl* (Sevin) 80 WSP	2.7 to 3.6 oz	
chlorantraniliprole (Acelepryn) G SC	 1.15 to 2.3 lb 0.184 to 0.46 fl oz	Suppression only. Toxic to aquatic invertebrates, oysters and shrimp.
clothianidin + bifenthrin (Aloft LC) G SC	 1.8 to 3.6 lb 0.27 to 0.54 fl oz	Toxic to fish and aquatic invertebrates. Do not apply near or allow runoff to surface or intertidal areas.
dinotefuran (Zylam 20 SG)	1.0 fl oz	For suppression, make application prior to hatching of first instar nymphs.
pyrethroids* (Advanced Lawn, Bug-B-Gone, Deltaguard, Menace, Scimitar, Talstar, Tempo, Wisdom and others)	See label	Many pyrethroids are toxic to fish and aquatic invertebrates. Apply these products only as specified on the label.
thiamethoxam + lambda-cyhalothrin (Tandem)	See label	Apply when insects are first observed. Repeat applications may be necessary. Highly toxic to fish and aquatic invertebrates.
Grub, Green June Beetle (only) (Also see White Grub)		
carbaryl* (Sevin) 80 WSP	1.8 oz	Apply to the soil surface but do not water in.
Grub, White (Japanese beetle, Southern chafer, European chafer, billbug, green June beetle)		
carbaryl* (Sevin) 80 WSP	3.6 oz	
chlorantraniliprole (Acelepryn) G SC	 1.15 to 2.3 lb 0.184 to 0.46 fl oz	Toxic to aquatic invertebrates, oysters and shrimp.
clothianidin (Arena) 0.25 G 50 WDG	 1.84 to 3.67 lb 0.14 to 0.29 fl oz	Toxic to fish and aquatic invertebrates. Do not apply near or allow runoff to surface waters or intertidal areas.
clothianidin + bifenthrin (Aloft LC) G SC	 1.8 to 3.6 lb 0.27 to 0.54 fl oz	Toxic to fish and aquatic invertebrates. Do not apply near or allow runoff to surface waters or intertidal areas.
dinotefuran (Zylam 20 SG)	1.0 fl oz	Make application prior to or during peak egg hatch.
imidacloprid (Advanced Lawn Grub Control, Merit, many others)	See label	
thiamethoxam (Meridian)	See label	Highly toxic to aquatic invertebrates.

Table 5-21. Insect Control for Home Lawns

Pest Insecticide and Formulation	Amount per 1,000 Sq Ft	Precautions and Remarks
Grub, White (Japanese beetle, Southern chafer, European chafer, billbug, green June beetle) (continued)		
thiamethoxam + lambda-cyhalothrin (Tandem)	See label	Highly toxic to fish and aquatic invertebrates.
trichlorfon* (Proxol/Dylox) 80 SP	3.75 oz	
various entomogenous nematodes	See label	Must be Heterorhabditid species to be effective.
Imported Fire Ant		
acephate* (Ortho Fire Ant Killer and others)	1 to 2 tsp/ mound	Distribute uniformly over mound. For best results apply early in morning or late afternoon.
avermectin B1 (Ascend, Award II) 0.011% bait	See label	Apply as a mound treatment or broadcast bait.
carbaryl (Sevin)	See label	Use as a mound drench.
clothianidin + bifenthrin (Aloft LC SC)	2.3 to 3.6 lb	Toxic to fish and aquatic invertebrates. Do not apply near or allow runoff to surface waters or intertidal areas.
d-limolene (Orange Guard)	See label	Mound treatment. Acceptable to organic growers. May also be used around fruit and vegetable gardens.
fipronil 0.0143 G (Taurus G, Top Choice)	2 lb	Apply as a broadcast.
fipronil (Maxforce FC) bait	See label	Apply as a mound treatment or broadcast bait.
fipronil + bifenthrin + lambda-cyahothrin (Taurus Trio G)	2 lb	Apply as a broadcast. Irrigate prior to treatment.
hydramethylnon* (Amdro Fire Ant Bait, Amdro Pro, Maxforce G)	See label	Follow label directions precisely. Use fresh bait. Repeat treatment usually required.
indoxacarb (Spectracide Fire Ant Once and Done) (Over 'n Out Fire Ant Killer Mound Treatment) (Advion)	See label	
metaflumizone (Siesta) bait	See label	Mound or broadcast bait.
methoprene (Extinguish) bait	See label	Mound or broadcast. Follow label directions. Repeat treatments usually required.
methoprene + hydromethylnon (Extinguish Plus, Amdro Firestrike) bait	See label	Follow label directions precisely. Repeat treatments usually required. Use fresh bait. Found in broadcast or mound treatment packaging.
pyrethroids (Bayer Advanced, Menace, Ortho Fire Ant Killer, Mound Treatment, Talstar One, Tempo, Wisdom and others)	See label	Many pyrethroids are toxic to fish and aquatic invertebrates. Apply these products only as specified on the label.
pyriproxyfen (Distance) bait	See label	Mound or broadcast bait.
spinosad (Come and Get It Fire Ant Bait by Fertilome, Entrust, Payback, Green Light Fire Ant Control with Conserve, Green Light Fire Ant Killer with Spinosad Mound Drench) AUBURN UNIVERSITY HAS AN EXCELLENT PUBLICATION FOR HOMEOWNERS	See label	Acceptable to organic growers. Follow label directions precisely. Repeat treatments usually required. Use fresh bait. May also be used around fruit and vegetable gardens. http://www.aces.edu/pubs/docs/A/ANR-0175-A/ANR-0175-A.pdf
Mole Cricket		
carbaryl* baits	See label	
clothianidin + bifenthrin (Aloft LC) G SC	 1.8 to 3.6 lb 0.27 to 0.54 fl oz	Toxic to fish and aquatic invertebrates. Do not apply near or allow runoff to surface waters or intertidal areas. Application should be made during peak adult flight and egg lay.
dinotefuran (Zylam 20 SG)	1.0 fl oz	Make application prior to or during peak egg hatch.
fipronil (Top Choice, Taurus G)	2 lb	Apply as a broadcast.
imidacloprid (Advanced Lawn Grub Control, Merit)	See label	
indoxacarb (Advion Insect Granules) bait	See label	
indoxacarb (Provaunt) WDG	0.275 oz	
pyrethroids* (Advanced Lawn, Bug-B-Gone, Deltaguard, Menace, Scimitar, Talstar, Tempo, Wisdom and others)	See label	Many pyrethroids are toxic to fish and aquatic invertebrates. Apply these products only as specified on the label.
thiamethoxam + lambda-cyhalothrin (Tandem)	See label	Apply from first egg hatch to peak egg hatch. Highly toxic to fish and aquatic invertebrates.
Various entomogenous nematode products	See labels	Require irrigation.
Slug, Snail		
iron phosphate (Natria) bait		Apply in late afternoon.
measurol 2% B	1 lb	Apply in late afternoon.
metaldehyde	See label	Apply in late afternoon.
Sod Webworm (also Burrowing Sod Webworm)		
carbaryl* (Sevin) 80 WSP 50 WP	 3.6 oz 6.4 oz	Do not water in sprays. Use 6 gallons water plus the insecticide per 1,000 square feet. Treat in late afternoon. Do not cut grass for 1 to 3 days after treatment. Granules must be watered in.
dinotefuran (Zylam 20 SG)	1.0 fl oz	
pyrethroids* (Advanced Garden, Deltagard, Scimitar, Talstar, Tempo, Wisdom and others)	See label	Many pyrethroids are toxic to fish and aquatic invertebrates. Apply these products only as specified on the label.
spinosad A and D (Conserve) SC	0.25 to 1.25 fl oz	Rate varies with size and species.
thiamethoxam + lambda-cyhalothrin (Tandem)	See label	Highly toxic to fish and aquatic invertebrates.
Sod Webworm (also Burrowing Sod Webworm) (continued)		
trichlorfon* (Dylox, Proxol) 80 SP	1.5 to 3 oz	Use sufficient water for good coverage.
various entomogenous nematode and *B.t.* products	See label	

* Several trade names available. Check label for active ingredient. Always follow label instructions.

VI — INSECT AND DISEASE CONTROL OF FRUITS

Apple Spray Program 190
    Relative Effectiveness of Various Fungicides for Apple Disease Control................................... 192
    Relative Effectiveness of Various Insecticides for Apple Insect and Mite Control 193
Blueberry Management Program 195
Caneberry Management Program.. 200
Bunch Grape Insect Management.. 206
Bunch Grape Disease Management 211
    Muscadine Disease Management Program.. 216
    Relative Effectiveness of Various Fungicides for Muscadine Grape Disease Control 217
    Muscadine Insect Management.. 217
Peach and Nectarine Spray Guide... 221
    Relative Effectiveness and Safety of Various Insecticides for Peach Insects........................... 223
Relative Effectiveness of Chemicals for Disease Control on Peaches and Nectarines 224
Nematode Control on Peaches 224
Commercial Pecan Insect Control
Commercial Pecan Insect and Disease Spray Guide
Strawberry Disease Control.. 238
    Pre-Planting Disease and Weed Management.. 238
    Pre-Planting and Early Post-planting: Nematode Management 238
    Fumigants 238
    Relative Efficacy of Currently Registered Fumigants or Fumigant Combinations
    for Managing Soilborne Nematodes, Diseases, and Weeds in Plasticulture Strawberries...... 239
    Planting and Early Post-Planting: Disease Control.. 239
    Fungicide Resistance Management Recommendations... 240
    New Leaf Growth to Pre-Bloom: Disease Management 242
    Early Bloom (10%) and into Harvest: Disease Management.. 243
    Fungicide Selection for Botrytis and Anthracnose Fruit Rot Management............................... 243
    Relative Effectiveness of Various Chemicals for Strawberry Disease Control........................ 247
Strawberry Insect Management.. 248

Apple Spray Program

J. F. Walgenbach, Entomology and Plant Pathology, and

S. M. Villani, Entomology and Plant Pathology

See *Integrated Orchard Management Guide for Commercial Apples in the Southeast* (AG-572) for more detailed information on apple disease and insect control. For a copy, contact Jim Walgenbach, 455 Research Drive, Mills River, NC 28759; jim_walgenbach@ncsu.edu. The guide is also available online at https://apples.ces.ncsu.edu

Many pesticides have brand name and generic formulations. In general, information is provided for the most commonly used formulations of active ingredients available in multiple formulations. Carefully check the label of the product you plan to use in the event that it differs from those listed. The label is the law!

Table 6-1. Apple Spray Program

Number and Time of Application	Amount of Fungicide and Insecticide Per Acre
Green-Tip Spray When buds show 0.25-inch new growth	**Fungicide:** NOTE: Captan will cause injury if applied with or too close to oil applications. **Apple Scab:** Apply dodine (Syllit FL) 1.5 pt + mancozeb (Koverall) 3 lb *OR* dodine (Syllit FL) 1.5 pt + Captan 80 WDG 2.5 lb *OR* cyprodinil (Vangard 75WG) 5 oz *OR* cyprodinil (Vangard 75WG) 3 to 5 oz + mancozeb (Koverall) 3 lb *OR* cyprodinil (Vangard 75WG) 3 to 5 oz + metiram (Polyram 80DF) 3 lb *OR* pyrimethanil (Scala SC) 7 to 10 oz *OR* pyrimethanil (Scala SC) 5 oz + mancozeb (Koverall) 3 lb *OR* pyrimethanil (Scala SC) 5 oz + metiram (Polyram 80DF) 3 lb *OR* fluopyram/pyrimethanil (Luna Tranquility) 11.2 to 16 fl oz + mancozeb (Koverall) 3 lb *OR* penthiopyrad (Fontelis SC) 16 to 20 fl oz *OR* penthiopyrad (Fontelis SC) 16 to 20 fl oz + mancozeb (Koverall) 3 lb *OR* fluxapyroxad (Sercadis) 4.5 fl oz *OR* fluxapyroxad (Sercadis) 4.5 fl oz + mancozeb (Koverall) 3 lb *OR* captan (Captan 80 WDG) 5 lb *OR* metiram (Polyram 80DF) 3 to 6 lb *OR* mancozeb (Koverall) 3 to 6 lb. **Fire Blight:** Apply copper hydroxide/copper oxychloride (Badge SC) 3.5 to 7 pt (rate for silver to green tip only) *OR* copper hydroxide/copper oxychloride (Badge SC) 0.5 to 1.5 pt *OR* copper hydroxide/copper oxychloride (Badge X2, OMRI listed) 3.5 to 7 lb (rate for silver to green tip only) *OR* copper hydroxide/copper oxychloride (Badge X2, OMRI listed) 0.5 to 1.5 lb *OR* copper hydroxide (Kocide 3000) 3.5 to 7 lb (discontinue rate at ½" green tip) *OR* basic copper sulfate (Cuprofix Ultra 40 Disperss) 5 to 7.5 lb (apply rate between silver and green tip). Other formulated copper products are available, but are too numerous to list in this publication. **Phytophthora Rots:** Apply phosphorous acid (ProPhyt) 2 to 4 pt/100 gal. Ridomil Gold SL may be applied as a soil drench or spray prior to bud break in the spring (i.e. silver tip) or in the fall after harvest. **Insecticide:** Apply 2 to 3 gallons oil per 100 gallons water. For improved control of San Jose scale, add 1 quart chlorpyrifos (Lorsban 4EC) *OR* 4 ounces pyriproxyfen (Esteem 35 WP). This is an important spray for mite eggs and San Jose scale. See petal fall to first cover spray as an alternative timing for scale control. If using mating disruption for codling moth and oriental fruit moth, dispensers (200 Isomate CM-OFM TT dispensers per acre or 1 CheckMate CM-OFM Puffer per acre) should be in place before bloom.
Half-Inch Green Spray One week after GREEN-TIP SPRAY	**Fungicide:** Use same materials as GREEN-TIP SPRAY. See warning about high application rates of copper. **Insecticide:** If an insecticide was not applied at green tip, use one of the products listed above; otherwise no insecticide is needed.
Tight Cluster Spray One week after HALF-INCH GREEN SPRAY	**Fungicide:** **Apple Scab:** Apply trifloxystrobin (Flint WG) 2.5 oz *OR* kresoxim-methyl (Sovran 50 WG) 3.2 to 6.4 oz *OR* fenbuconzole (Indar 2F) 6 to 8 fl oz *OR* fluopyram/pyrimethanil (Luna Tranquility) 11.2 to 16 fl oz *OR* fluopyram/trifloxystrobin (Luna Sensation) 4 to 5.8 fl oz *OR* fluxapyroxad/pyraclostrobin (Merivon) 4 to 5.5 fl oz *OR* penthiopyrad (Fontelis) 16 to 20 fl oz *OR* cyprodinil/difenoconazole (Inspire Super) 12 fl oz *OR* benzovindiflupyr (Aprovia) 5.5 to 7 fl oz *OR* fluxapyroxad (Sercadis) 4.5 fl oz *OR* captan (Captan 80WDG) 5 lb *Or* mancozeb (Koverall) 3 to 6 lb *OR* metiram (Polyram 80 DF) 3 to 6 lb. For resistance management of single-site/"systemic" fungicides, it is suggested that a ½ rate of a multi-site protectant fungicide be added to the tank mixture. **Powdery Mildew:** Apply myclobutanil (Rally 40 WSP) 5 to 10 oz *OR* fenbuconazole (Indar 2F) 8 fl oz *OR* triflumizole (Procure 480SC) 8 to 16 fl oz *OR* flutriafol (Topguard) 8 to 12 fl oz *OR* fluopyram/pyrimethanil (Luna Tranquility) 11.2 to 16 fl oz *OR* fluopyram/trifloxystrobin Luna Sensation (4 to 5.8 fl oz) *OR* fluxapyroxad/pyraclostrobin (Merivon) 4 to 5.5 fl oz *OR* penthiopyrad (Fontelis) 16 to 20 fl oz *OR* trifloxystrobin (Flint WG) 2 to 2.5 oz *OR* kresoxim-methyl (Sovran 50 WG) 4 to 6.4 oz *OR* benzovindiflupyr (Aprovia) 5.5 to 7 fl oz *OR* parrafinic oil (JMS Stylet Oil) 1 to 2 gal/100 gal *OR* sulfur (Microthiol Disperss) 10 to 20 lb. **Cedar Apple/Quince Rust:** Apply a DMI (FRAC GROUP 3) used for powdery mildew *OR* mancozeb (Koverall) 3 lb. **Black Rot/Frogeye Leaf Spot:** Captan (Captan 80 WDG) 2.5 to 5 lb *OR* thiophanate –methyl (Topsin 4.5 FL) 15 to 20 fl oz **Insecticide:** For rosy apple aphid, apply 5 ounces acetamiprid (Assail 30 SG) *OR* 10 to 14 oz flupyradifurone (Sivanto Prime). For rosy apple aphid and plant bugs, apply 5.4 ounces thiamethoxam (Actara 25 WP) *OR* 16 ounces fenpropathrin (Danitol 2.4 EC).
Pink Spray When blossom buds are pink, stems extended	**Fungicide:** Use same fungicides as TIGHT CLUSTER SPRAY. Make sure to rotate between different FRAC Groups. **Insecticide:** If an insecticide effective against rosy apple aphid and/or tarnished plant bug was not applied at TIGHT CLUSTER, apply one of the above materials.
Bloom Spray	**Fungicide:** Use same fungicide as TIGHT CLUSTER SPRAY. **Fire Blight Control:** Apply streptomycin (Firewall, Agrimycin) 24 oz *OR* kasugamycin (Kasumin 2L) 64 fl oz/100 gal *OR* oxytetracyline (Fireline) 12 oz/100 gal *OR* acibenzolar-S-methyl (Actigard 50WG) 1 to 2 oz + streptomycin (Firewall) 24 oz *OR* acibenzolar-S-methyl (Actigard 50WG) 1 to 2 oz + oxytetracyline (Fireline) 12 oz *OR* Bacillus mycoides (LifeGard) 4.5 oz/100 gal + streptomycin (Firewall) 24 oz *OR* copper octanoate (Cueva) 2 qt *OR* Bacillus amyloliquefaciens (Double Nickel LC) 1 to 2 qt + copper octanoate (Cueva) 2 qt *OR* Bacillus subtilis (Serenade Optimum) 14 to 20 oz *OR* copper sulfate pentahydrate (MasterCop) 0.5 to 1.5 qt *OR* copper hydroxide/copper oxychloride (Badge SC) 0.5 to 1.5 pt *OR* copper hydroxide/copper oxychloride (Badge X2, OMRI listed). Other formulated copper products are available, but are too numerous to list in this publication. Be aware that phytotoxicity to fruit and leaves may occur if copper is used during this timing. **Insecticide:** DO NOT USE an insecticide at BLOOM SPRAY. The exception is 8 oz methoxyfenozide (Intrepid 2F) in late bloom where green fruitworm is a problem.

Table 6-1. Apple Spray Program

Number and Time of Application	Amount of Fungicide and Insecticide Per Acre
Petal-Fall Spray When most petals have fallen	**Fungicide:** **Apple Scab, Powdery Mildew, Rusts:** Use same fungicide as TIGHT CLUSTER SPRAY. Make sure to rotate between different FRAC Groups. **Black and White Rot (Bot Rots):** Apply captan (Captan 80WDG) 5 lb *OR* trifloxystrobin (Flint 50WG) 2.5 oz *OR* kresoxim-methyl (Sovran 50 WG) 4 to 6.4 oz *OR* thiophanate-methyl (Topsin 4.5 FL) 15 to 20 fl oz *OR* fluazinam (Omega 500F) 13.8 fl oz. **Glomerella Leaf Spot and Bitter Rot:** Apply captan (Captan 80WDG) 5 lb *OR* phosphorous acid (ProPhyt) 4 pt + captan (Captan 80WDG) 3.75 lb *OR* fluxapyroxad/pyraclostrobin (Merivon) *OR* fluxapyroxad/pyraclostrobin (Merivon) + mancozeb (Koverall) 3 lb + phosphorous acid (ProPhyt) 4 pt *OR* boscalid/pyraclostrobin (Pristine) 14.5 to 18 oz *OR* boscalid/pyraclostrobin (Pristine) 14.5 to 18 oz + captan (Captan 80WDG) 2.5 lb *OR* trifloxystrobin (Flint 50WG) 3 oz + captan (Captan 80WDG) 2.5 lb *OR* kresoxim-methyl (Sovran 50 WG) + captan (Captan 80WDG) 2.5 lb *OR* fluazinam (Omega 500F) 13.8 fl oz. **Insecticide:** For plum curculio and Oriental fruit moth, apply 5 ounces indoxacarb (Avaunt 35WD) *OR* 3 pounds phosmet (Imidan 70 WP) or 4.5 ounces thiamethoxam (Actara 25 WP) *OR* 4 ounces clothianidin (Clutch). For preventive control of European red mite, use 3 ounces abamectin (Agri-Mek 0.7SC) *PLUS* 0.25% horticultural spray oil (not a superior-type oil). If rosy apple aphid control is needed, use Actara or Clutch, or add 2.8 ounces imidacloprid (Admire 4.6SC).
First Cover Spray 8 to 10 days after PETAL-FALL SPRAY	**Fungicide:** **Powdery Mildew, Glomerella Leaf Spot, Black Rot, White Rot, Bitter Rot:** Refer to fungicides for PETAL FALL Application. **Flyspeck/Sooty Blotch:** Apply captan (Captan 80WDG) 2.5 to 5 lb *OR* trifloxystrobin (Flint 50WG) 1.5 to 2.5 oz *OR* kresoxim-methyl (Sovran 50 WG) 4 to 6.4 oz *OR* thiophanate methyl (Topsin 4.5FL) 15 to 20 fl oz *OR* ziram (Ziram 76DF) 6 lb *OR* cyprodinil/difenoconazole (Inspire Super) 12 fl oz *OR* fenbuconazole (Indar 2F) 6 to 8 fl oz *OR* benzovindiflupyr (Aprovia) 5.5 to 7 fl oz *OR* fluopyram/trifloxystrobin (Luna Sensation) 4 to 5.8 fl oz *OR* fluxapyroxad/pyraclostrobin (Merivon) 4 to 5.5 fl oz *OR* boscalid/pyraclostrobin (Pristine) 14.5 to 18.5 fl oz *OR* copper octanoate (Cueva) 2 qt *OR* copper octanoate (Cueva) 2 qt + *Bacillus amyloliquefaciens* (Double Nickel LC) 1 to 2 qt **Fire Blight (shoot blight/rat-tail bloom):** Apply prohexadione calcium (Apogee) 18 to 36 oz *OR* Bacillus mycoides (Lifegard) 4.5 oz/100 gal *OR* copper octanoate (Cueva) 2 qt *OR* copper octanoate (Cueva) 2 qt + *Bacillus amyloliquefaciens* (Double Nickel LC) 1 to 2 qt *OR* copper hydroxide/copper oxychloride (Badge X2, OMRI listed) 0.5 to 1.5 lb *OR* copper hydroxide/copper oxychloride (Badge SC) 0.5 to 1.5 pt. Other formulated copper products are available, but are too numerous to list in this publication. Be aware that phytotoxicity to fruit and leaves may occur if copper is used during this timing. **Insecticide:** For codling moth, apply 3 ounces chlorantraniliprole (Altacor 35 WDG) *OR* 5 ounces spinetoram (Delegate 25 WDG). If preventative control of European red mite is desired but was not applied at petal fall, apply 4 ounces clofentozine (Apollo SC) *OR* 4 ounces hexythiazox (Savey 50 DF), 3 ounces etoxazole (Zeal 72 WD), *OR* 18 fl oz spirodiclofen (Envidor 2SC). Where control of San Jose scale is needed, apply 4 oz pyriproxyfen (Esteem 35 WP) *OR* 2.1 lbs buprofezin (Centaur 70WDG) *OR* 4 lb diazinon (50WP) *OR* 6 to 9 fl ounces spirotetramat (Movento 2SC). Movento will also provide season-long protection against woolly apple aphid.
Second Cover Spray 10 to 14 days after FIRST COVER SPRAY	**Fungicide:** Refer to relative effectiveness table for appropriate fungicides and to FIRST COVER SPRAYS (above) for summer disease control. Substitute ziram (Ziram 76DF) for mancozeb products as mancozeb has 77-day PHI. **Insecticide:** Same as FIRST COVER SPRAY for control of codling moth.
Third Cover Spray 10 to 14 days after SECOND COVER SPRAY	**Fungicide:** Refer to relative effectiveness table for appropriate fungicides and to FIRST COVER SPRAYS (above) for summer disease control. **Insecticide:** Same as first cover for codling moth. For tufted apple bud moth, apply 12 ounces methoxyfenozide (Intrepid 2 F), 3 ounces chlorantraniliprole (Altacor 35 WDG) *OR* 5 ounces spinetoram (Delegate 25 WDG). On plantings susceptible to dogwood borer, apply 1 pound chlorpyrifos (Lorsban 4 EC) to trunk at base of tree using a handgun application anytime between third cover spray and mid-July. If aphid or potato leafhopper control is needed, apply 2.8 fl ounces imidacloprid (Admire Pro 4.6SC), *OR* 2 ounces thiamethoxam (Actara 25WD), *OR* 10 to 14 oz flupyradifurone (Sivanto Prime), *OR* 2 fl ounces sulfoxaflor (Closer 2SC), *OR* 2.5 to 4 ounces acetamiprid (Assail 30WD)
Summer Cover Sprays 7- to 14-day intervals or as pest density and weather conditions dictate	**Fungicide:** Refer to relative effectiveness table for appropriate fungicides and to FIRST COVER SPRAYS (above) for summer disease control. **Insecticide:** Refer to relative effectiveness tables and AG-572 for appropriate insecticides and miticides for summer insect control. For second generation codling moth sprays (mid to late July), do not use the same insecticide used for first generation control (first and second cover sprays). Important sprays include codling moth in mid to late July, apple maggot in late July. For apple maggot, apply 2.8 ounces imidacloprid (Admire Pro) or 3 pounds phosmet (Imidan). In orchards not using mating disruption up to this point, apply 1.2 ounces/acre of Check OFM-F (sprayable pheromone for mating disruption) in mid to late July and again one month later for late-season oriental fruit moth control. For control of brown marmorated stink bug, sprays should be based on insect population densities and timing of infestation of first generation adults in mid-July in the piedmont and early August in the mountains; apply 5 oz thiamethoxam (Actara 25 WDG), *OR* 18 oz fenpropathrin (Danitol 2.4EC) *OR* 2.8 oz lambda-cyhalothrin (Karate 2.04EC). *OR* other pyrethroids listed in the efficacy table. 2 to 3 sprays may be needed depending on stink bug pressure.

[1] Do not follow oil with captan or sulfur for 14 days.

Further Information

Southern Appalachian Apples Extension Portal: https://apples.ces.ncsu.edu/

Producing Tree Fruit for Home Use. NC State Extension, AG-28; https://content.ces.ncsu.edu/producing-tree-fruit-for-home-use.

A Grower's Guide to Apple Insects and Diseases in the Southeast. (http://ipm.ncsu.edu/apple/contents.html).

Chapter VI—2019 N.C. Agricultural Chemicals Manual

Relative Effectiveness of Various Fungicides for Apple Disease Control
S. M. Villani, Entomology and Plant Pathology

Many pesticides have brand name and generic formulations. In general, information is provided for the most commonly used formulations of active ingredients available in multiple formulations. Carefully check the label of the product you plan to use in the event that it differs from those listed. The label is the law!

(E = excellent; G=Good; F = Fair; P = Poor; NC = No control; ND = No data)

Table 6-2. Relative Effectiveness of Various Fungicides for Apple Disease Control

Fungicide and Rate of Usage Per Acre	FRAC Code	Days Between Last Spray and Harvest	Apple Scab	Rusts	Brooks Spot	Black Rot/ White Rot	Glomerella/ Bitter Rot	Sooty Blotch and Flyspeck	Powdery Mildew
benzovindiflupyr (Aprovia) 5.5 to 7 fl oz	7	30	E	F	ND	G-E	P-F	G-E	F-G
captan (Captan 80WDG) 5 lb	M4	0	G	P	G	G	G	F-G	NC
cyprodinil (Vangard 75 WG) 5 oz	9	0	F	NC	NC	NC	NC	NC	NC
difenoconazole + cyprodinil (Inspire Super) 12 fl oz	3 + 9	14	E	E	F	G-E	P	G-E	F
dodine (Syllit 3.4 FL) 1.5 to 3.0 pt + mancozeb (Koverall) 3 lb OR + captan (Captan 80WDG) 2.5 lb	M7 M3 M4	See label 77 (see label)	E E	P-F P	n/a n/a	n/a n/a	n/a n/a	n/a n/a	NC n/a
fenbuconazole (Indar 2F) 8 fl oz	3	14	E	E	F	F-G	P	F-G	G
fluazinam (Omega 500F) 13.8 fl oz	29	28	F	F	ND	F-G	F-G	F	P-F
fluopyram + trifloxystrobin (Luna Sensation) 4.0 to 5.8 fl oz	7 + 11	14	E	F	G-E	G	G-E	G-E	E
fluopyram + pyrimethanil (Luna Tranquility) 11.2 to 16 fl oz	7 + 9	72	E	F	ND	ND	ND	ND	G
Fluxapyroxad (Sercadis) 4.5 fl oz	7	0	E	F	ND	ND	P	P-F	F-G
fluxapyroxad + pyraclostrobin (Merivon) 4.4 to 5.5 fl oz	7 + 11	0	E	F	G-E	G-E	E	G-E	E
kresoxim-methyl (Sovran 50 WG) 4 to 6.4 oz [1]	11	30	E	P-F	G-E	G	F-G	G	G
mancozeb (Koverall) 3 to 6 lb	M3	77	G	F-G	F-G	F	G-E	P-F	P
metiram (Polyram 80 DF) 6 lb	M3	see label	G	F-G	F	P-F	F	P-F	P
myclobutanil (Rally 40WSP) 5 to 10 oz [1]	3	14	G	E	P-F	P-F	P	P	E
penthiopyrad (Fontelis 1.67 SC) 14 to 20 fl oz	7	28	G-E	F	F	F-G	F-G	F-G	G
phosphorous acid (ProPhyt) 3 to 4 pt + mancozeb (Koverall) 3 lb OR + captan (Captan 80WDG) 3.75 lb	33 M3 M4	77 0	G G	F-G P	F-G G	F G	G-E G-E	P-F F-G	P P
phosphorous acid (ProPhyt) 3 to 4 pt + ziram (Ziram 76DF) 3 lb + captan (Captan 80WDG) 3.75 lb	33 M3 M4	14	G-E	F-G	G	G	G-E	F-G	P
phosphorous acid (ProPhyt) 4 pt + mancozeb (Koverall) 3 lb + fluxapyroxad + pyraclostrobin (Merivon) 5.5 fl oz	33 M3 7 + 11	77	E	F-G	G-E	G-E	E	E	G-E
phosphorous acid (ProPhyt) 4 pt + captan (Captan 80WDG) 2.5 lb + fluxapyroxad + pyraclostrobin (Merivon) 5.5 fl oz	33 M4 7 + 11	77	E	F	G-E	G-E	E	E	E
pyraclostrobin + boscalid (Pristine 38 WG) 14.4 to 18.4 oz	11 + 7	0	G-E	F	G-E	G-E	G-E	G-E	G
pyrimethanil (Scala 5 SC) 7 to 10 fl oz	9	72	F-G	NC	NC	NC	NC	NC	NC
sulfur (Microthiol Disperss) 10 to 20 lb	M2	see label	P-F	P-F	NC	NC	NC	P	G-E
thiophanate methyl (Topsin 4.5FL) 15 to 20 fl oz	1	1	—[3]	NC	G	G	P	G-E	P-F
trifloxystrobin (Flint 50WG) 1.5 to 3 oz	11	14	G	P-F	G-E	G	E	G	G-E
triflumizole (Procure 480SC) 8 to 16 fl oz [1]	3	14	F-G	E	G	P-F	P	P	G-E
ziram (Ziram 76 DF) 3 to 6 lb [2]	M3	14	F	G	F	G	G-E	P

[1] Use higher rate when the likelihood of disease is high.
[2] Combine Ziram with Topsin-M 70W at 8 to 12 ounces per acre to improve white rot, black rot, sooty blotch, and flyspeck control.
[3] Thiophanate methyl is not recommended for apple scab control in North Carolina because of apple scab resistance.

Relative Effectiveness of Various Insecticides for Apple Insect and Mite Control

J. F. Walgenbach, Entomology and Plant Pathology

(E – excellent; G – good; F – fair; P – poor; NC – no control or insufficient data)

Table 6-3A. Relative Effectiveness of Various Insecticides for Apple Insect and Mite Control

Insecticide, Brand Name, and Amount per Acre	IRAC MOA Group	Days Between Last Spray and Harvest	San Jose Scale	European Red Mite	Twospotted Spider Mite	Rosy Apple Aphid	Green Apple/ Spirea Aphids	White Apple Leafhopper	Coddling Moth	Tufted Apple Bud Moth	Redbanded Leafroller	Oriental fruit moth
carbaryl (Sevin XLR) 4 pt	1A	1	P	NC	NC	NC	NC	E	G	P	G	G
oxamyl (Vydate 2L) 2 qt	1A	14	F	P	P	G	G	E	P	P	P	P
methyomyl (Lannate LV) 1 qt	1A	14	NC	NC	NC	P	G	E	P	G	G	G
chlorpyrifos (Lorsban 50W) 3 lb[1]	1B	>100	E	NC	NC	G	NC	NC	NC	NC	NC	NC
diazinon (Diazinon 50WP) 4 lb	1B	21	G	P	P	F	F	F	F	P	F	F
phosmet (Imidan 70W) 3 lb	1B	7	P	NC	NC	NC	NC	NC	G	P	G	G
cyfluthrin (Tombstone 2EC) 2.4 oz	3	7	NC	NC	NC	G	G	E	F	E	E	E
esfenvalerate (Asana XL) 8 oz	3	21	NC	NC	NC	G	G	E	F	E	E	G
fenpropathrin (Danitol 2.4 EC) 16 oz	3	14	NC	G	G	E	G	E	F	E	E	E
gamma-cyhalothrin (Proaxis 0.5EC) 3 oz	3	14	NC	NC	NC	E	E	E	F	E	E	E
lambda-cyhalothrin (Karate 2.08CS) 2 oz	3	21	NC	NC	NC	E	E	E	F	E	E	E
permethrin (Ambush 2E) 8 oz	3	>100	NC	NC	NC	G	G	E	F	E	E	G
zeta-cypermethrin (Mustang Maxx 0.8EC)	3	14	NC	NC	NC	E	E	E	F	E	E	E
acetamiprid (Assail 30 SG) 5.0 oz	4A	7	P	NC	NC	E	EE	E	G	P	P	G
clothianidin (Belay 2.13 SC) 6 oz	4A	7	NC	NC	NC	E	E	E	NC	NC	NC	NC
imidacloprid (AdmirePro4) 2.8 oz	4A	7	NC	NC	NC	E	E	E	NC	NC	NC	NC
thiomethoxam (Actara 25 WP) 4.5 oz	4A	35	NC	NC	NC	E	E	E	P	NC	NC	NC
sulfoxaflor (Closer 2 SC)	4C	7	F	NC	NC	E	E	E	NC	NC	NC	NC
flupyradifurone (Sivanto 200SL) 10.5 oz	4D	14	NC	NC	NC	E	E	E	NC	NC	NC	NC
spinetoram (Delegate 25WDG) 5 oz	5	7	NC	NC	NC	NC	NC	NC	E	E	E	E
abamectin (Agri-Mek 0.7SC) 3.0 oz	6	28	NC	E	E	NC	NC	G	NC	NC	NC	NC
pyriproxyfen (Esteem 35 WP) 5 oz	7C	35	E	NC	NC	P	P	NC	F	P	G	G
clofentezine (Apollo SC) 4 oz	10A	45	NC	E	E	NC	NC	NC	NC	NC	NC	NC
hexythiazox (Savey 50DF) 4 oz[1]	10A	28	NC	E	E	NC	NC	NC	NC	NC	NC	NC
etoxazole (Zeal 72 WDG) 3 oz	10B	28	NC	G	G	NC	NC	NC	NC	NC	NC	NC
B. thuringiensis (various brands) 1 lb	11	0	NC	NC	NC	NC	NC	NC	P	G	G	P
novaluron (Rimon 0.83 EC) 20 oz	15	14	NC	NC	NC	NC	NC	NC	G	E	E	E
buprofezin (Centaur 70WDG) 34.5 oz	16	14	E	NC	NC	NC	NC	G	NC	NC	NC	NC
methozyfenozide (Intrepid 2 F) 16 oz	18	14	NC	NC	NC	NC	NC	NC	G	E	E	G
fenpyroximate (Portal 0.4EC) 2 pt	21A	14	NC	E	E	NC	NC	G	NC	NC	NC	NC
pyridaben (Nexter 75 WP) 4.4 oz	21A	25	NC	G	F	NC	P	G	NC	NC	NC	NC
indoxacarb (Avaunt 30WDG) 5 oz	22A	28	NC	NC	NC	NC	NC	E	P	G	G	G
spirotetramat (Movento 2CS) 7.5 oz	23	7	G	NC	NC	E	E	G	NC	NC	NC	NC
cyflumetofen (Nealta 1.67SC) 13.7 oz	25	7	NC	E	E	NC	NC	NC	NC	NC	NC	NC
chlorantraniliprole (Altacor 35WDG) 3 oz	28	14	NC	NC	NC	NC	NC	NC	E	E	E	E
cyantraniliprole (Exirel 0.83SE) 12 oz	28	3	NC	NC	NC	NC	F	NC	E	E	E	E
bifenazate (Acramite 50WS) 1 lb	UN	7	NC	E	E	NC	NC	NC	NC	NC	NC	NC
codling moth virus (CYD-X) 3 g	—	0	NC	NC	NC	NC	NC	NC	G	NC	NC	F
oil, superior-type 3 gal/100 gal	—		NC	E	E	NC	P	NC	NC	P	P	P	P

[1] Use prebloom only.

Table 6-3B. Relative Effectiveness of Various Insecticides for Apple Insect and Mite Control (continued)

Insecticide, Brand Name, and Amount per Acre	IRAC MOA Group	Days Between Last Spray and Harvest	Spotted Tentiform Leafminer	Tarnished Plant Bug	Stink Bugs	Apple Maggot	Plum Curculio	Japanese Beetle	Woolly Apple Aphid	Safety of Beneficials[2] Coccinellids (lady beetles)	Predatory mites
carbaryl (Sevin XLR) 4 pt	1A	1	P	G	P	G	G	E	NC	P	P
oxamyl (Vydate 2E) 2 qt	1A	14	G	G	P	NC	P	P	NC	G	G
methymyl (Lannate LV) 1 at	1A	14	G	E	G	NC	P	P	NC	P	P
chlorpyrifos (Lorsban 4EC) 1 qt	1B	>100	NC	NC	NC	NC	NC	NC	G	G	G
diazinon (Diazinon 50WP) 4 lb	1B	21	P	G	G	G	G	G	E	P	G
phosmet (Imidan 70W) 3 lb	1B	7	P	G	P	E	G	E	NC	P	G
cyfluthrin (Tombstone 2EC) 2.4 oz	3	7	G	E	E	G	G	G	NC	P	P
esfenvalerate (Asana XL) 8 oz	3	21	G	G	F	G	F	G	NC	P	P
fenpropathin (Danitol 2.4 EC) 16 oz	3	14	G	E	E	G	G	G	NC	P	P
gamma-cyhalothrin (Proaxis 0.5EC) 3 oz	3	14	G	E	E	G	G	G	NC	P	P
lambda-cyhalothrin (Karate 2.08CS) 2 oz	3	21	G	E	E	G	G	G	NC	P	P
permethrin (Ambush 2EC) 8 oz	3	>100	G	G	G	G	F	G	NC	P	P
zeta-cypermethrin (Mustang Maxx 0.8EC)	3	14	G	E	E	G	G	G	NC	P	P
acetamiprid (Assail 30 SG) 5 oz	4A	7	G	F	F	G	F	E	P	P	G
clothianidin (Belay 2.13SC) 6 oz	4A	7	G	G	G	G	G	G	P	P	F
imidacloprid (Admire Pro) 2.8 oz	4A	7	G	F	F	G	F	G	P	P	G
thiomethoxam (Actara 25 WP) 4.5 oz	4A	35	G	E	E	F	E	G	P	P	F
sulfoxaflor (Closer 2 SC)	4C	7	G	E	F	NC	F	NC	P	F	G
flupyradifurone (Sivanto 200SL) 10.5 oz	4D	14	G	F	P	NC	NC	NC	NC	F	G
spinetoram (Delegate 25WDG) 5 oz	5	7	E	NC	NC	P	NC	NC	NC	F	G
abamectin (Agri-Mek 0.7SC) 3.0 oz	6	28	E	NC	NC	NC	NC	NC	NC	F	G
pyriproxyfen (Esteem 35 WP) 5 oz	7C	35	G	NC	NC	NC	NC	NC	NC	G	G
clofentezine (Apollo SC) 4 oz	10A	45	NC	NC	NC	NC	NC	NC	NC	E	G
hexythiazox (Savey 50DF) 4 oz[1]	10A	28	NC	NC	NC	NC	NC	NC	NC	E	G
etoxazole (Zeal 72 WDG) 3 oz	10B	28	NC	NC	NC	NC	NC	NC	NC	E	G
B. thuringiensis (various brands) 1 lb	11A	0	NC	NC	NC	NC	NC	NC	NC	E	E
hexakis (Vendex 50W) 4 lb	12B	14	NC	NC	NC	NC	NC	NC	NC	E	G
novaluron (Rimon 0.83 EC) 20 oz	15	14	NC	NC	NC	NC	NC	NC	NC	P	P
buprofezin (Centaur 70WDG) 34.5 oz	16	14	NC	NC	NC	NC	NC	NC	NC	G	E
methoxyfenozide (Intrepid 2 F) 16 oz	18	14	E	NC	NC	NC	NC	NC	NC	E	E
fenpyroximate (Portal 0.4 EC) 2 pt	21A	14	NC	NC	NC	NC	NC	NC	NC	E	G
pyridaben (Nexter 75 WP) 4.4 oz	21A	25	NC	NC	NC	NC	NC	NC	NC	E	G
indoxacarb (Avaunt 30WDG) 5 oz	22A	28	F	NC	NC	P	E	G	NC	F	E
spirotetramat (Movento 2CS) 7.5 oz	23	7	P	NC	NC	NC	NC	NC	E	E	E
cyflumetofen (Nealta 1.67SC) 13.7 oz	25	7	NC	NC	NC	NC	NC	NC	NC	E	G
chlorantraniliprole (Altacor 35WDG) 3 oz	28	14	E	NC	NC	P	P	NC	NC	E	E
cyantraniliprole (Exirel 0.83SE) 12 oz	28	3	E	NC	NC	F	F	NC	NC	E	E
bifenazate (Acramite 50WS) 1 lb	UN	7	NC	NC	NC	NC	NC	NC	NC	E	G
codling moth virus (CYD-X) 3 g	—	0	NC	NC	NC	NC	NC	NC	NC	E	E
oil, superior-type 3 gal/100 gal	—	NC	NC	NC	NC	NC	NC	NC	NC	E	G

[1] Use prebloom only.

[2] Ratings for beneficial arthropods are based on toxicity to the organism; i.e., E implies excellent safety (low toxicity) to the beneficial and will result in conservation of natural enemies, while P implies high toxicity and elimination of natural enemies.

Blueberry Management Program

H. J. Burrack, W. O. Cline, and S. M. Villani, Entomology and Plant Pathology

The Insecticide Resistance Action Committee (IRAC) groups insecticides and the Fungicide Resistance Action Committee (FRAC) groups fungicides into mode of action (MOA) categories. These categories are listed following the pesticide and formulation names. To reduce the risk of resistance development, avoid successive applications of products with the same MOA. Organically acceptable insecticides (**OMRI** listed) are indicated in Precautions and Remarks.

Insecticides should only be applied if the pest of concern is present in economically damaging levels. If insect injury does not result in greater loss than the cost of treatment, treatment is not justified. Therefore, some degree of insect presence should be tolerated, and insecticides should **not** be applied on a scheduled basis as may be appropriate for fungicides. Note that insecticides listed are acceptable for use on fruit to be marketed in the United States. If fruit is to be exported, check with purchasers to ensure that the materials you intend to use are acceptable for use on fruit in their target markets.

Fungicides are mainly protectants, and are usually applied prior to the appearance of disease symptoms, based on past history of the particular disease threat on a given cultivar, location and plant growth stage. Not all diseases are present on every farm. To avoid applying fungicides unnecessarily, learn to identify diseases by their symptoms, and keep records of those that occur on your farm.

Many pesticides have brand name and generic formulations. In general, information is provided for the most commonly used formulations of active ingredients available in multiple formulations. Carefully check the label of the product you plan to use in the event that it differs from those listed. The label is the law!

Table 6-4. Blueberry Management Program

Season and Pest	Product Name, Mode of Action Code, and Formulation	Amount of Formulation Per Acre	Restricted Entry Interval (REI) (hours)	Pre harvest Interval (PHI) (Days)	Precautions and Remarks
Dormant					
Scale insects	Oil superior-type, IRAC Unknown	1 to 3% vol/vol	4	0	Oil may be applied dormant or delayed dormant. Apply as needed for scale infestations. Reduce to 1% rate just before bloom. Do not apply oil when temperatures are expected to be higher than 65 degrees F or lower than 30 degrees F within 24 hours. Do not use within 14 days of lime-sulfur or Captan. Use 200 to 400 gallons water per acre. Some oils are **OMRI** listed; check labels.
Gall midge	**Blueberry gall midge** adults are tiny flies, and larvae are tiny white, carrot-shaped maggots that feed inside flower buds and leaf buds. Blueberry gall midge can be extremely injurious, especially to rabbiteye cultivars. Flies lay eggs in flower buds on warm winter days when bud scales initially begin to separate. **Gall midge sprays should be timed to protect the earliest flower buds which can realistically be expected to survive anticipated spring cold events.** Gall midge sprays also typically provide suppression of pre-bloom thrips population.				
	spinosad (IRAC 5) (Entrust SC)	4 to 6 fl oz	4	3	Entrust is **OMRI** listed. Do not apply more than 29 fl oz Entrust SC (0.45 lb active ingredient) per acre per year.
	Entrust (80W)	1.25 to 2 oz			Do not apply more than 9 oz Entrust 80W (0.45 lb active ingredient) per acre per year.
	spinetoram (IRAC 5) (Delegate)	3 to 6 oz	4	3	
	diazinon (IRAC 1) (Diazinon AG500)	1 pt per 100 gal water	5 days	7	Only one foliar application is allowed per year.
Delayed Dormant					
Exobasidium leaf and fruit spot	calcium polysulfide (FRAC M2) (Lime-Sulfur solution)	5 gal per acre in 50 to 70 gal of total spray volume	48	—	Apply at delayed dormant 1 to 2 weeks before leaf and/or flower buds begin to break. Exobasidium is not specifically on the label. However, when applied for Phomopsis, suppression of Exobasidium has been observed. **DANGER – calcium polysulfide products are caustic and can cause injury.** Calcium polysulfide products are also corrosive to metals and may permanently discolor or stain non-metal sprayer parts. Do not mix lime-sulfur solutions with acids or phosphate fertilizer products because deadly and potentially extremely flammable hydrogen sulfide gas may be emitted.
	calcium polysulfide (FRAC M2) (Sulforix)	1 gal per acre in sufficient water for coverage	48	—	Do not apply lime-sulfur or Sulforix within 14 days of an oil spray. Do not apply when air temperatures are above 85 degrees F. As a precaution, do not apply within 14 days of a Dormex spray.
Pre-Bloom Sprays - Green-tip on vegetative and flower buds					
Twig blight Mummy berry	fenbuconazole (FRAC 3) (Indar 75 WP) (Indar 2F)	2 oz 6 fl oz	12 12	30 30	
	pyraclostrobin + boscalid (FRAC 11+7) (Pristine 38 W)	18.5 to 23 oz	12	0	Do not make more than 2 sequential applications with any combination of strobilurin fungicides (Abound or Pristine) before alternation with a fungicide that has a different mode of action (Captan, Ziram, Switch). Do not make more than 4 applications of strobilurin fungicides per season. Do not tank mix Pristine with any other product except Pristine may be tank mixed with products that contain only Captan as the active ingredient.
	propiconazole (FRAC 3) (Orbit 3.6E, Tilt 3,6E, Bumper 41.8EC, Propimax EC)	6 fl oz	12	30	
	metconazole (FRAC 3) (Quash 50 WDG)	2.5 oz	12	7	May be applied by ground (min. 20 gpa) or air (min 10 gpa). Do not apply more than twice in a row, or more than 7.5 ounces per season, or more than 3 times per season.
	azoxystrobin + propiconazole (FRAC 3+11) (Quilt Xcel)	14 to 21 fl oz	12	30	Do not apply more than 82 fluid ounces per acre per season. Quilt Xcel may be applied by ground or air (minimum of 15 gpa).

Table 6-4. Blueberry Management Program

Season and Pest	Product Name, Mode of Action Code, and Formulation	Amount of Formulation Per Acre	Restricted Entry Interval (REI) (hours)	Pre harvest Interval (PHI) (Days)	Precautions and Remarks
Pre-Bloom Sprays - Green-tip on vegetative and flower buds (continued)					
Exobasidium leaf and fruit spot	The fungus Exobasidium causes green-to-pink spots on fruit that do not ripen normally, and spots on leaves that are light green above and white below. Affected berries are unsightly and not marketable. Fungicides applied for other diseases may provide some control. The disease is most severe in shaded locations with dense foliage and poor ventilation. For images of this disease, see: http://ncblueberryjournal.blogspot.com/2011/07/exobasidium-fruit-and-leaf-spot.html				
Flower thrips	Thrips rarely require treatment in southern high bush blueberries in North Carolina but can reach damaging levels in rabbiteye blueberries and late-blooming norther highbush like Duke. Thrips present in densities greater than 2 per flower in open rabbiteye blooms may justify treatment. Begin sampling bloom clusters for thrips at Stage 3. Sample 2 to 3 times a week from Stage 3 up to bloom. A minimum of 10 flower clusters per acre should be observed and either placed in a closed plastic bag at room temperature, soaked in alcohol, or shaken onto a white sheet of paper.				
	acetamiprid (IRAC 4A) (Assail 30SG)	4.5 to 5.3 oz	12	1	
	spinosad (IRAC 5) (Entrust SC)	4 to 6 fl oz	4	3	Entrust is **OMRI** listed. Do not apply more than 29 fl oz Entrust SC (0.45 lb active ingredient) per acre per year.
	Entrust (80W)	1.25 to 2 oz			Do not apply more than 9 oz Entrust 80W (0.45 lb active ingredient) per acre per year.
	spinetoram (IRAC 5) (Delegate)	3 to 6 oz	4	3	
Bloom Treatments - 10% to 20% bloom					
Pesticides can harm pollinating insects, so if pesticide applications are necessary during bloom, they should be made in the evening when bees are not foraging and to allow for the longest amount of dry time possible. See Table 5-1A. Relative Toxicity of Pesticides to Honey Bees for more information on specific active ingredients effects on bees.					
Twig blight	Same as Pre-Bloom Sprays				
Mummy berry	Same as Pre-Bloom Sprays				If mummy berry disease pressure is high this year or in previous years, apply fungicides every 7 to 10 days from budbreak through bloom. Foliar sprays using 25 to 50 gallons per acre are most effective.
Flower blight	Anticipate flower blight caused by the fungus *Botrytis cinerea* when excessive rain occurs during bloom, or following a freeze event that injures blossoms.				
	fenhexamid (FRAC 17) (Elevate 50 WDG)	1.5 lb	12	0	Elevate may not be applied by air.
	cyprodinil + fludioxonil (FRAC 9+12) (Switch 62.5 WG)	11 to 14 oz	12	0	Switch may not be applied by air.
	captan (FRAC M4) (Captan 50 WP)	4 lb	48	0	
	captan (FRAC M4) (Captec 4L)	2 qt	48	0	
	captan + fenhexamid (FRAC M4+ 17) (CaptEvate) 68 WG	3.5 to 4.7 lb	48	0	CaptEvate may not be applied by air. Do not use CaptEvate for more than 2 consecutive sprays.
Bloom Treatments - Full bloom					
Mummy berry	Same as PRE-BLOOM SPRAYS				Note that Indar should not be used alone at full bloom or between bloom and harvest. Tank mix with Captan, Captec, or Ziram.
Fruit rots	captan (FRAC M4) (Captan 50 WP)	4 lb	48	0	
	captan (FRAC M4) (Captec 4L)	2 qt	48	0	
	ziram (FRAC M3) (Ziram 76 DF)	3 lb	48	approx. 30	Ziram cannot be applied later than 3 weeks after full bloom.
	azoxystrobin (FRAC 11) (Abound 2.08 E)	6 to 15.5 fl oz	4	0	Do not make more than 2 sequential applications of any combination of strobilurin fungicides (Abound or Pristine) before alternating with a fungicide that has a different mode of action (Captan, Ziram, Switch). Do not make more than 4 applications per season.
	cyprodinil + fludioxonil (FRAC 9+12) (Switch 62.5 WG)	14 oz	12	0	
	pyraclostrobin + boscalid (FRAC 11+7) (Pristine38 W)	18.5 to 23 oz	12	0	Do not tank mix Pristine with any other product (fungicide, insecticide, adjuvant, fertilizer, etc.) except Pristine may be tank mixed with products containing Captan as the sole active ingredient.
	metconazole (FRAC 3) (Quash 50 WDG)	2.5 oz	12	7	May be applied by ground (min. 20 gpa) or air (min. 10 gpa). Do not apply more than twice in a row, or more than 7.5 ounces per season, or more than 3 times per season.
Flower blight	Same as Bloom Treatments (10% to 20% bloom)				
Petal Fall Treatments - Immediately after Bloom					
Fruit rots	Same as Bloom Treatments				Fruit rot treatments should be applied 7 to 10 days apart.
Flower thrips	Same as Pre-Bloom Treatments				
Plum curculio	Plum curculio is an infrequent pest of North Carolina blueberries. Petal fall treatments of the materials below will be effective against both plum curculio and fruitworms.				
	bifenthrin (IRAC 3) (Brigade) WSB	16 oz	12	1	Note that there are residue concerns for some IRAC Group 3A materials on fruit intended for export.

Table 6-4. Blueberry Management Program

Season and Pest	Product Name, Mode of Action Code, and Formulation	Amount of Formulation Per Acre	Restricted Entry Interval (REI) (hours)	Pre harvest Interval (PHI) (Days)	Precautions and Remarks
Petal Fall Treatments - Immediately after Bloom (continued)					
Plum curculio (continued)	chlorantraniliprole (IRAC 28) (Altacor)	3.0 to 4.5 oz	4	1	
	esfenvalerate (IRAC 3A) (Asana XL 0.66EC)	9.6 fl oz	12	14	Note that there are residue concerns for some IRAC Group 3A materials on fruit intended for export.
	Indoxacarb (IRAC 22) (Avaunt)	6 oz	12	7	Do not make more than 4 applications of Avaunt per season. Do not use adjuvants with Avaunt. There are established residue levels for indoxacarb for export to the European Union but not for export to Canada.
	zeta cypermethrin + bifenthrin (IRAC 3) (Hero)	4 to 10.3 fl oz	12	1	Hero is a premixed material and contains more than 1 active ingredient. Check labels carefully for the maximum amount of active ingredient than can be applied per acre per season. Note that there are residue concerns for some IRAC Group 3A materials on fruit intended for export.
	kaolin (IRAC unknown) (Surround) WP	25 to 50 lb	4	0	Surround acts like a barrier and masks fruit from pest recognition. Because of this barrier, fruit should be washed after harvest, and Surround may be most appropriate for processing fruit.
Cranberry fruitworm Cherry fruitworm	Fruitworm adults can be monitored with pheromone traps, and fruit should be observed for egg laying or evidence of tunneling. Treatments for fruitworms are most effective when timed to egg hatch, as larvae feed inside fruit. With the exception of Altacor, the materials listed below are not expected to have activity against plum curculio.				
	acetamiprid (IRAC 4A) (Assail) 30 SG	4.5 to 5.3 oz	12	1	
	carbaryl (IRAC 1A) (Sevin) XLR	1.5 to 2 qt	12	7	There are many carbaryl formulations.
	chlorantraniliprole (IRAC 28) (Altacor)	3.0 to 4.5 oz	4	1	Altacor is also effective against plum curculio in blueberries.
	esfenvalerate, (IRAC 3A) (Asana XL) 0.66 EC	4.8 to 9.6 oz	12	14	Note that there are residue concerns for some IRAC Group 3A materials on fruit intended for export.
	indoxacarb (IRAC 22) (Avaunt)	3.5 to 6.0 oz	12	7	
	methoxyfenozide (IRAC 18) (Intrepid) 2F	10 to 16 fl oz	4	7	
	novaluron (IRAC 15) (Rimon) 0.83 EC	20 to 30 fl oz	12	8	Rimon is not labeled for cherry fruitworm.
	pyriproxyfen (IRAC 7) (Knack)	16 fl oz	12	7	Knack is an insect growth regulator and application must be timed carefully to egg hatch.
	spinosad (IRAC 5) (Entrust SC)	4 to 6 fl oz	4	3	Entrust is **OMRI** listed. Do not apply more than 29 fl oz Entrust SC (0.45 lb active ingredient) per acre per year.
	Entrust (80W)	1.25 to 2 oz			Do not apply more than 9 oz Entrust 80W (0.45 lb active ingredient) per acre per year.
	spinetoram (IRAC 5) (Delegate) WG	3 to 5 oz	4	3	
	tebufenozide (IRAC 18) (Confirm) 2F	16 fl oz	4	14	
Leaf spots	fenbuconazole (FRAC 3) (Indar 75 WP) (Indar 2F)	2 oz 6 fl oz	12	30	Indar should not be used alone at full bloom or alone between bloom and harvest. Tank mix with Captan, Captec, or Ziram. Indar is usually limited to 5 applications per acre per year.
	pyraclostrobin + boscalid (FRAC 11+7) (Pristine 38 W)	18.5 to 23 oz	12	0	Do not make more than 2 sequential applications with any combination of strobilurin fungicides (Abound or Pristine) before alternation with a fungicide that has a different mode of action (Captan, Ziram, Switch). Do not make more than 4 applications of strobilurin fungicides per season. Do not tank mix Pristine with any other product (fungicide, insecticide, adjuvant, fertilizer, etc.) except Pristine can be mixed with Captan.
	propiconazole (FRAC 3) (Orbit 3.6E, Tilt 3.6E, Banner 41.8 EC, Propimax EC)	6 fl oz	12	30	
	metconazole (FRAC 3) (Quash 50 WDG)	2.5 oz	12	7	May be applied by ground (minimum of 20 gpa) or air (minimum of 10 gpa). Do not apply more than twice in a row, or more than 7.5 ounces per season, or more than 3 times per season.
	azoxystrobin + propiconazole (FRAC 3+11) Quilt Xcel	14 to 21 fl oz	12	30	Do not apply more than 82 fluid ounces per acre per season. Quilt Xcel may be applied by ground or air (minimum of 15 gpa).
Fruit Ripening through Harvest					
Spotted wing drosophila	Spotted wing drosophila (SWD) females lay eggs in ripening and ripe soft skinned fruits, and larvae develop internally. Materials listed are likely to be effective against SWD based on current data. SWD treatments should begin when flies are present, and fruit start to ripen and continue weekly through the end of harvest. Rotate IRAC groups between successive sprays. Some management tools used for blueberry maggot are effective against SWD, and management of blueberry maggot and SWD should be integrated as much as feasible.				
	bifenthrin (IRAC 3A) (Brigade WSB)	5.3 to 16.0 oz	12	1	No more than 5 applications of Brigade can be made per season. Note that there are residue concerns for some IRAC Group 3A materials on fruit intended for export.
	cyantraniliprole (IRAC 28) (Exirel)	13.5 to 20.5 fl oz	12	3	Minimum number of days between treatments is five. Do not apply a total of more than 0.4 pound ai/A of CYAZYPYR® or cyantraniliprole containing products per year.

Table 6-4. Blueberry Management Program

Season and Pest	Product Name, Mode of Action Code, and Formulation	Amount of Formulation Per Acre	Restricted Entry Interval (REI) (hours)	Pre harvest Interval (PHI) (Days)	Precautions and Remarks
Fruit Ripening through Harvest (continued)					
Spotted wing drosophila (continued)	fenpropathrin (IRAC 3A) (Danitol 2.4 EC)	10.33 to 16 fl oz	24	3	No more than 2 applications of Danitol can be made per season.
	malathion, (IRAC 1B) (Malathion) 8F	2.5 pt	12	1	There are several malathion formulations. No more than 2 applications of Malathion 8F can be made per year. No more than 5 pounds of malathion active ingredient from any source can be applied per acre per year. Use caution if this is the material of choice for multiple insect pests.
	(Malathion) ULV	10 fl oz	12	1	No more than 5 applications of Malathion ULV can be made per year. No more than 5 pounds of malathion active ingredient from any source can be applied per acre per year. Use caution if this is the material of choice for multiple insect pests.
	methomyl (IRAC 1A) (Lannate)	12 to 24 fl oz	48	3	No more than 4 applications of Lannate can be made per year.
	phosmet (IRAC 1B) (Imidan)	1.33 lb	24	3	No more than 5 applications of Imidan can be made per year.
	pyrthreins (IRAC 3) (Pyganic 1.4 EC)	16 to 64 fl oz	12	0	Pyganic is **OMRI** listed, but has limited residual activity and should not be used as the only SWD materials.
	spinosad (IRAC 5) (Entrust SC)	4 to 6 fl oz	4	3	Entrust is **OMRI** listed. Do not apply more than 29 fl oz Entrust SC (0.45 lb active ingredient) per acre per year.
	Entrust (80W)	1.25 to 2 oz			Do not apply more than 9 oz Entrust 80W (0.45 lb active ingredient) per acre per year.
	spinetoram (IRAC 5) (Delegate)	3 to 6 oz	4	3	No more than 19.5 ounces of Delegate can be applied per acre per year.
	zeta cypermethrin (IRAC 3) (Mustang Max)	4 fl oz	12	1	No more than 6 applications of Mustang Max can be made per year. Note that there are residue concerns for some IRAC Group 3A materials on fruit intended for export.
	zeta cypermethrin + bifenthrin (IRAC 3) (Hero)	4 to 10.3 fl oz	12	1	No more than 46.35 fluid ounces of product can be applied per acre per year. Note that there are residue concerns for some IRAC Group 3A materials on fruit intended for export.
Blueberry maggot	Blueberry maggot fly activity typically begins in late May. Adults should be monitored with yellow sticky traps baited with ammonia food lures (ammonium acetate, ammonium carbonate, or ammonium bicarbonate). Check traps and change lures at least once per week. Treatments for blueberry maggot are not necessary unless adults have been observed in traps. Materials effective for SWD are also effective against blueberry maggot; additional treatments are not needed for blueberry maggot.				
	acetamiprid (IRAC 4A) (Assail)	2.5 to 5.3 oz	12	1	Assail is effective against blueberry maggot, but should not be used alone for SWD management.
	esfenvalerate (IRAC 3A) (Asana XL) 0.66 EC	4.8 to 9.6 oz	12	14	Note that there are residue concerns for some IRAC Group 3A materials on fruit intended for export.
	fenpropathrin (IRAC 3A) (Danitol 2.4 EC)	10.33 to 16 fl oz	24	3	No more than 2 applications of Danitol can be made per season.
	imidacloprid (IRAC 4A) (Admire Pro)	2.1 to 2.8 fl oz (foliar)	12	3	Many formulations of imidacloprid are available. Admire Pro can be applied as either a soil or foliar treatment. Soil treatments are not recommended for blueberry maggot. Imidacloprid is not effective against SWD.
	malathion, (IRAC 1B) (Malathion) 8F	2.5 pt	12	1	There are several malathion formulations. No more than 2 applications of Malathion 8F can be made per year. No more than 5 pounds of malathion active ingredient from any source can be applied per acre per year. Use caution if this is the material of choice for multiple insect pests.
	(Malathion) ULV	10 fl oz	12	1	No more than 5 applications of Malathion ULV can be made per year. No more than 5 pounds of malathion active ingredient from any source can be applied per acre per year. Use caution if this is the material of choice for multiple insect pests.
	phosmet (IRAC 1B) (Imidan)	1.33 lb	24	3	No more than 5 applications of Imidan can be made per year.
	spinosad (IRAC 5) (Entrust SC)	4 to 6 fl oz	4	3	Entrust is **OMRI** listed. Do not apply more than 29 fl oz Entrust SC (0.45 lb active ingredient) per acre per year.
	Entrust (80W)	1.25 to 2 oz			Do not apply more than 9 oz Entrust 80W (0.45 lb active ingredient) per acre per year.
	spinetoram (IRAC 5) (Delegate)	3 to 6 oz	4	3	No more than 19.5 ounces of Delegate can be applied per acre per year.
	zeta cypermethrin (IRAC 3) (Mustang Max)	4 fl oz	12	1	No more than 6 applications of Mustang Max can be made per year. Note that there are residue concerns for some IRAC Group 3A materials on fruit intended for export.
	zeta cypermethrin + bifenthrin (IRAC 3) (Hero)	4 to 10.3 fl oz	12	1	No more than 46.35 fluid ounces of product can be applied per acre per year. Note that there are residue concerns for some IRAC Group 3A materials on fruit intended for export.

Table 6-4. Blueberry Management Program

Season and Pest	Product Name, Mode of Action Code, and Formulation	Amount of Formulation Per Acre	Restricted Entry Interval (REI) (hours)	Pre harvest Interval (PHI) (Days)	Precautions and Remarks
Post Harvest					
Leaf Spots	Same as Petal Fall Treatments.				Leaf spot treatments should be applied every 2 weeks post harvest. Later leaf spot treatments may by be omitted if leaf spot incidence is low.
Blueberry bud mite	Only treat for blueberry bud mite if damage was a problem in the previous year. Many varieties are resistant to blueberry bud mite and do not typically require treatment.				
	Post harvest hedging, cultural control	NA	NA	NA	Summer topping or hedging immediately after harvest controls bud mite by removing old, infested fruiting twigs and is the control method of choice for early ripening cultivars.
	Variety selection	NA	NA	NA	Most highly susceptible blueberry varieties are no longer grown. Bud mite can occur on O'Neal and Legacy. Bud mite is generally only a problem on high bush, not rabbiteye varieties.
	oil superior-type IRAC unknown (many formulations)	2 gal	4	0	Bud mite treatments should be applied after harvest and again four weeks later.
Sharpnosed leafhoppers	Sharpnosed leafhopper vectors blueberry stunt disease. To reduce disease transmitting populations of sharpnosed leafhopper, treatments should be timed to their flight activity. Sharpnosed leafhoppers can be monitored with yellow sticky traps. If present, blueberry stunt infected plants should be removed from fields.				
	acetamiprid (IRAC 4A) (Assail)	2.5 to 5.3 oz	12	1	Allow 7 days between Assail treatments.
	esfenvalerate (IRAC 3A) (Asana XL) 0.66 EC	4.8 to 9.6 oz	12	14	Note that there are residue concerns for some IRAC Group 3A materials on fruit intended for export.
	imidacloprid (IRAC 4A) (Admire Pro)	2.1 to 2.8 fl oz (foliar)	12	3	Several formulations of imidacloprid are available. Admire Pro can be applied as either a soil or foliar treatment. Soil treatments are not recommended for sharpnosed leafhopper.
	thiamethoxam (IRAC 4A) (Actara)	3 to 4 oz	12	3	Allow 7 days between Actara treatments. Maximum 12 ounces per acre per season.
Japanese beetles	Japanese beetle feeding seldom requires treatment in North Carolina blueberries, and some pesticides applied for leafhoppers and other pests will also control Japanese beetles. Do not make additional pesticide treatments for Japanese beetle unless severe defoliation occurs.				
	acetamiprid (IRAC 4A) (Assail)	2.5 to 5.3 oz	12	1	Allow 7 days between Assail treatments.
	carbaryl (IRAC 1A) (Sevin 80S)	2.5 lb	12	7	
	esfenvalerate (IRAC 3A) (Asana XL) 0.66 EC	4.8 to 9.6 oz	12	14	Note that there are residue concerns for some IRAC Group 3A materials on fruit intended for export.
	imidacloprid (IRAC 4A) (Admire Pro)	2.1 to 2.8 fl oz (foliar)	12	3	Several formulations of imidacloprid are available. Admire Pro can be applied as either a soil or foliar treatment. Soil treatments are not recommended for Japanese beetle.
	phosmet (IRAC 1B) (Imidan)	1.33 lb	24	3	
Red humped and yellow necked caterpillars	Several species of caterpillars can feed on blueberries from late summer to early fall. These caterpillars can potentially defoliate bushes, but are often not widespread throughout the planting.				
	Hand removal	NA	NA	NA	Hand removal is often sufficient to control populations because they are typically clustered on single or a few bushes.
	Bacillus thuringiensis sub. *kurstaki* (Bt) (IRAC 11A) Dipel DF	0.5 to 1.0 lb	4	0	There are many Bt formulations. Dipel DF is **OMRI** listed.
	chlorantraniliprole (IRAC 28) (Altacor)	3.0 to 4.5 oz	4	1	
	esfenvalerate (IRAC 3A) (Asana 0.66EC)	4.8 to 9.6 fl oz	12	14	Note that there are residue concerns for some IRAC Group 3A materials on fruit intended for export.
	tebufenozide (IRAC 18) (Confirm 2F)	16 fl oz	4	14	
Red imported fire ants	Fire ant baits should be applied when ants are actively foraging and take a few to several weeks to be fully effective. Fire ants can also be treated earlier in the season if present.				
	pyriproxyfen (IRAC 7C) (Esteem Ant Bait 0.5% B)	1.5 to 2 lb	12	1	Do not water for 24 hours after application.
	methoprene (IRAC 7C) (Extinguish Ant Bait 0.5% B)	1 to 1.5 lb	4	0	Extinguish can be applied as a mound treatment or broadcast. Extinguish is labeled for use on cropland, but Extinguish Plus is **NOT** labeled for use on cropland. Read labels carefully.
	spinosad (IRAC 5) (Entrust SC)	4 to 6 fl oz	4	3	Entrust is **OMRI** listed. Do not apply more than 29 fl oz Entrust SC (0.45 lb active ingredient) per acre per year.
	Entrust (80W)	1.25 to 2 oz			Do not apply more than 9 oz Entrust 80W (0.45 lb active ingredient) per acre per year.
Blueberry flea beetle	Blueberry flea beetles are an occasional pest in North Carolina blueberries and are active primarily post harvest. Damage is typically not economically significant, but when new shoots are eaten in the fall, yield for the following year will be impacted.				
	acetamiprid (IRAC 4A) (Assail)	2.5 to 5.3 oz	12	1	Allow 7 days between Assail treatments.
	esfenvalerate (IRAC 3A) (Asana XL) 0.66 EC	4.8 to 9.6 oz	12	14	Note that there are residue concerns for some IRAC Group 3A materials on fruit intended for export.
	carbaryl (IRAC 1A) (Sevin 80S)	2.5 lb	12	7	

Table 6-4. Blueberry Management Program

Season and Pest	Product Name, Mode of Action Code, and Formulation	Amount of Formulation Per Acre	Restricted Entry Interval (REI) (hours)	Pre harvest Interval (PHI) (Days)	Precautions and Remarks
Post Harvest (continued)					
Blueberry flea beetle (continued)	imidacloprid (IRAC 4A) (Admire Pro)	2.1 to 2.8 fl oz (foliar)	12	3	Admire Pro can be applied as either a soil or foliar treatment. Soil treatments are not recommended for blueberry flea beetles.
	phosmet (IRAC 1B) (Imidan)	1.33 lb	24	3	No more than 5 applications of Imidan can be made per year.
	spinosad (IRAC 5) (Entrust SC)	4 to 6 fl oz	4	3	Entrust is **OMRI** listed. Do not apply more than 29 fl oz Entrust SC (0.45 lb active ingredient) per acre per year.
	Entrust (80W)	1.25 to 2 oz			Do not apply more than 9 oz Entrust 80W (0.45 lb active ingredient) per acre per year.
	spinetoram (IRAC 5) (Delegate)	3 to 6 oz	4	3	No more than 19.5 ounces of Delegate can be applied per acre per year.
	thiamethoxam (IRAC 4A) (Actara)	3 to 4 oz	12	3	Allow 7 days between Actara treatments. Maximum 12 oz per acre per season.
	zeta cypermethrin (IRAC 3) (Mustang Max)	4 fl oz	12	1	No more than 6 applications of Mustang Max can be made per year. Note that there are residue concerns for some IRAC Group 3A materials on fruit intended for export.
	zeta cypermethrin + bifenthrin (IRAC 3) (Hero)	4 to 10.3 fl oz	12	1	No more than 46.35 fluid ounces of product can be applied per acre per year. Note that there are residue concerns for some IRAC Group 3A materials on fruit intended for export.

Further Information
Southeast Regional Blueberry Integrated Management Guide, www.smallfruits.org
NC Small Fruit and Specialty Crop IPM, www.ncsmallfruitsipm.blogspot.com

Caneberry Management Program
H. J. Burrack and W. O. Cline, Entomology and Plant Pathology

The Insecticide Resistance Action Committee (IRAC) and Fungicide Resistance Action Committee (FRAC) group insecticides into mode of action categories. These categories are listed following the pesticide and formulation names. To reduce the risk of resistance development, avoid successive applications of insecticides with the same IRAC or FRAC designation for the same pest. Organically acceptable insecticides (**OMRI** listed) are indicated in Comments and Precautions.

Insecticides should only be applied if the pest of concern is present in economically damaging levels. If insect injury does not result in greater loss than the cost of treatment, treatment is not justified. Therefore, a degree of insect presence should be tolerated, and insecticides should not necessarily be applied on a scheduled basis as may be appropriate for fungicides.

Pesticides should not be applied when bees are actively foraging. If necessary, apply insecticides and fungicides in the evening when bees are not active. Pay attention to pesticide label information regarding pollinator protection.

Many insecticide active ingredients are available in generic formulations. Generic products generally work similarly to their brand name counterparts, but formulation changes can impact efficacy and plant response. In general, information is provided for the most commonly used formulations of active ingredients available in multiple formulations. Carefully check the label of the product you plan to use in the event that it differs from those listed. The label is the law! Chemical names are subject to change; please check the active ingredient for all materials.

Table 6-5. Caneberry Management Program

SEASON and Pest	Product Name, Mode of Action Code, and Formulation	Rate of Formulation per Acre	Restricted Entry Interval (hours)	Preharvest Interval (PHI) (days)	Comments and Precautions
Late Winter or Early Spring When new growth is less than 0.5 inch long					
Anthracnose, Spur blight, Cane blight	liquid lime-sulfur, FRAC M2 (Sulforix)	3 gal/100 gal	See label	See label	This is an important spray for anthracnose control. Make sure canes are thoroughly covered. Use a dilute spray solution.
	Copper-based products FRAC M1	See label	See label	See label	Many copper-based fungicides are available. May be used as a component in Bordeaux mixture, see footnote.
Raspberry crown borer	Removing infested plants is an important cultural control. In blocks with a history of raspberry crown borer, apply an insecticide either in late October to early November or early April (1 application only) to provide a barrier for larvae boring into canes as they emerge from overwintering hibernacula. Follow label instructions for water volume and application methods.				
	bifenthrin, IRAC 3A (Brigade WSB)	16 oz	12	3	Do not exceed 0.2 lb bifenthrin (32 oz Brigade 10WSB) per acre per season. There are several other formulations of bifenthrin.
	esfenvalerate, IRAC 3A (Asana XL)	9.6 fl oz	12	7	
	chlorantraniliprole, IRAC 28 (Altacor)	3-4.5 oz	4	3	

Table 6-5. Caneberry Management Program

SEASON and Pest	Product Name, Mode of Action Code, and Formulation	Rate of Formulation per Acre	Restricted Entry Interval (hours)	Preharvest Interval (PHI) (days)	Comments and Precautions
Just Before Blooms Open					
Anthracnose, Cane blight, Cane canker, Leaf spots, Spur blight	boscalid + pyraclostrobin (Pristine) 38 WDG, FRAC 7 + 11	18.5 to 23 oz	12	0	Do not make more than 4 applications collectively of the strobilurin fungicides (Abound, Quilt Xcel, Cabrio, Heritage, and Pristine) per season.
	captan, FRAC M4 (Captan 50W) (Captan 80WDG) (Captec 4L)	4 lb 2.5 lb 2 qt	see label	3	Do not apply more than 20 pounds of Captan 50W or 12.5 pounds Captan 80WDG per acre per season. Different formulations of captan have different re-entry intervals. Check label.
	captan FRAC M4 + fenhexamid, FRAC M4 + 17 (CaptEvate 68WDG)	3.5 lb	48	3	Registered for use on raspberries, but not other brambles.
	liquid lime sulfur (Sulforix)	2 qt/100 gal	2 qt/100 gal	see label	This should be a follow-up application to the later winter/early spring lime sulfur application. Apply prior to bloom. See label regarding phytotoxicity warnings.
	pyraclostrobin (Cabrio 20EG), FRAC 11	14 oz	12	0	Do not make more than 4 applications collectively of the strobilurin fungicides (Abound, Quilt Xcel, Cabrio, Heritage, and Pristine) per season. Strobilurin (Group 11) fungicides will also control rusts.
	azoxystrobin, FRAC 11 (Abound 2SC)	6 to 15.5 fl oz	4	0	Do not make more than 4 applications collectively of the strobilurin fungicides (Abound, Quilt Xcel, Cabrio, and Pristine) per season. Strobilurin (Group 11) fungicides will also control rusts.
	azoxystrobin + propiconazole (Quilt Xcel), FRAC 11 + 3	14 to 21 fl oz	12	30	Do not make more than 4 applications collectively of the strobilurin fungicides (Abound, Quilt Xcel, Cabrio, and Pristine) per season. Strobilurin (Group 11) fungicides will also control rusts.
Powdery mildew (powdery mildew should be more of a concern on raspberries than on blackberries)	azoxystrobin, FRAC 11 (Abound 2SC)	6 to 15.5 fl oz	4	0	Do not make more than 4 applications collectively of the strobilurin fungicides (Abound, Quilt Xcel, Cabrio, and Pristine) per season. Strobilurin (Group 11) fungicides will also control rusts
	azoxystrobin + propiconazole (Quilt Xcel), FRAC 11 + 3	14 to 21 fl oz	12	30	Do not make more than 4 applications collectively of the strobilurin fungicides (Abound, Quilt Xcel, Cabrio, and Pristine) per season. Strobilurin (Group 11) fungicides will also control rusts.
	myclobutanil (Rally 40WSP; Sonoma 40WSP), FRAC 3	1.25 to 3 oz	24	1	
	propiconazole (PropiMax EC), FRAC 3	6 fl oz	12	30	
	pyraclostrobin (Cabrio 20EG), FRAC 11	14 oz	12	0	Do not make more than 4 applications collectively of the strobilurin fungicides (Abound, Quilt Xcel, Cabrio, Heritage, and Pristine) per season. Strobilurin (Group 11) fungicides will also control rusts.
	parrafinic oil (Organic JMS Stylet Oil)	3 to 6 qt/100 gal	4	see label	**OMRI listed.** There is also a non-organic formulated JMS Stylet Oil product. **DO NOT apply oil with captan. See label for other phytotoxicity warnings**
	potassium bicarbonate (Milstop)	2 to 5 lb	1	0	**OMRI listed.**
	potassium salts of fatty acids (M-Pede)	1 to 2% v/v solution	12	0	**OMRI listed.** DO NOT apply with sulfur or within 3 days of a sulfur application.
	sulfur (Kumulus DF)	6 to 15 oz	24	0	**OMRI** listed for some manufacturers. DO NOT apply within 2 weeks of an oil treatment.
Strawberry clipper weevil	Strawberry clipper weevil females lay their eggs in flower buds and clip the pedicle, causing the bud to wilt and drop off the plant. However, many blackberry and raspberry varieties can compensate for bud injury, and strawberry clipper rarely requires treatment. **Insecticides effective against strawberry clipper weevil are toxic to bees. Do not apply insecticides when bees are foraging.**				
	acetamiprid, IRAC 4A (Assail 30SG)	4.5 to 5.3 oz	12	1	
	bifenthrin, IRAC 3A (Brigade 10WSB)	8 to 16 oz	12	3	Do not exceed 0.2 lb bifenthrin (32 oz Brigade 10WSB) per acre per season.
	carbaryl, IRAC 1A (Sevin XLR)	1 to 2 qt	12	7	
	fenpropathrin, IRAC 3A (Danitol 2.4EC)	10.66 to 16 fl oz	24	3	Do not exceed 32 fluid ounces per acre per season.
	spinosad, IRAC 5 (Entrust SC)	4 to 6 fl oz	4	3	Entrust is **OMRI** listed. Do not apply more than 29 fl oz Entrust SC (0.45 lb active ingredient) per acre per year.
	Entrust (80W)	1.25 to 2 oz			Do not apply more than 9 oz Entrust 80W (0.45 lb active ingredient) per acre per year.
Gall midge	Gall midge larvae can feed on developing buds, and damage can appear similar to cold injury. Fields with a history of gall midge damage may require treatment, but this is rare. Confirm gall midge presence before considering a treatment targeting this pest alone.				
	bifenthrin, IRAC 3A (Brigade 2EC)	3.2 to 6.4 fl oz	12	3	Do not exceed 12.8 fluid ounces Brigade per acre per season.
	fenpropathrin, IRAC 3A (Danitol 2.4EC)	10.66 to 16 fl oz	24	3	Do not exceed 32 fluid ounces per acre per season.

Table 6-5. Caneberry Management Program

SEASON and Pest	Product Name, Mode of Action Code, and Formulation	Rate of Formulation per Acre	Restricted Entry Interval (hours)	Preharvest Interval (PHI) (days)	Comments and Precautions
Just Before Blooms Open (continued)					
Fire ants	Fire ants can be nuisance pests in caneberry plantings. **Optimal fire ant control programs for fruit make use of spring and fall broadcast bait applications.** Twice-a-year bait applications may be best in year one of a program to thoroughly suppress the ant population. In subsequent years, a single bait application 8 to 10 weeks before harvest may provide adequate ant control. Ant baits work best when soil is moist, but not wet. Active ant foraging is essential. Foraging activity can be gauged by placing a food item, such as a potato chip, near the mound for 30 minutes or disturbing the mound. If ants are feeding on the chip within 30 minutes, conditions are right to apply baits. Ideally, temperatures should be warm and sunny. Avoid application of ant baits when conditions are expected to be cold, overcast, rainy or very hot. **Treatment of individual mounds is often a necessary complement to broadcast bait use if the goal is to obtain short-term elimination of fire ants.**				
	pyriproxyfen, IRAC 7D (Esteem Ant Bait)	1.5 to 2 lb	12	1	Esteem Fire Ant Bait will take several weeks to reach full efficacy.
	s-methoprene, IRAC 7A (Extinguish Professional Fire Ant Bait)	1 to 1.5 lbs/acre	4	0	To treat smaller areas, apply 3 to 5 tbsp/1000 sq ft or 3 to 5 tbsp/mound. Extinguish Professional Fire Ant Bait (0.5% methoprene) is a slow-acting bait; it will take several weeks for Extinguish Professional Fire Ant Bait to reach full efficacy. Extinguish Professional Fire Ant Bait is legal for use on 'crop land.' Caution, Extinguish bait with methoprene plus hydramethylnon is not labeled for use on crop land.
Bloom and Petal Fall					
	Pesticides may be hazardous to pollinators. When making any pesticide application during bloom, apply material in the evening when bees are not foraging to allow for as long a dry time as possible. See Table 5.1. Relative Toxicity of Pesticides to Bees for more information.				
Double blossom	Sprays during bloom are most important for control of double blossom. Begin sprays when first infected blossoms open and continue every 10 to 14 days through bloom. Rotate strobilurin (Group 11) fungicides with Switch to avoid resistance. It is important to protect primocanes as long as infected flowers continue to open.				
	azoxystrobin, FRAC 4 (Abound 2SC)	6.2 to 15.4 fl oz	4	0	
	boscalid + pyraclostrobin (Pristine 38WDG), FRAC 7 + 11	18.5 to 23 oz	12	0	Pristine will also control botrytis.
	cyprodinil + fludioxonil (Switch 62.5 WG), FRAC 9 + 12	11 to 14 oz	12	0	
	azoxystrobin + propiconazole (Quilt Xcel), FRAC 11 + 3	14 to 21 fl oz	12	30	
	Bordeaux mixture FRAC M1	See note at end of table	24	1	Crop injury may occur with Bordeaux mixture under slow drying conditions or in hot weather. Some injury often accompanies the use of copper fungicides; if injury is excessive, discontinue use.
Botrytis fruit rot	Apply at early bloom and repeat at full bloom. Rotate products to reduce the likelihood of resistance.				
	captan, FRAC M4 (Captan 50W) (Captan 80 WDG) (Captec 4L)	4 lb 2.5 lb 2 qt	48	3	
	fenhexamid (Elevate 50 WDG), FRAC 17	1.5 lb	12	0	
	iprodione (Rovral 4L), FRAC 2	1 to 2 pt	24	0	
	cyprodinil + fludioxonil (Switch 62.5 WG), FRAC 9 + 12	11 to 14 oz	12	0	
	captan + fenhexamide (CaptEvate) 68WDG, FRAC M4 + 17	3.5 lb	48	30	CaptEvate is registered on raspberries only.
	Bacillus amyloliquefaciens strain D747 (Double Nickel LC), FRAC 44	0.5 to 6 qt	4	0	**OMRI listed.**
	Bacillus subtilis strain QST 713 (Serenade Optimum), FRAC 44	14 to 20 oz	4	0	**OMRI listed.**
	copper octanoate (Cueva), FRAC M1	0.5 to 2 gal	4	0	**OMRI listed.**
	Streptomyces lydicus WYEC100 (Actinovate AG), FRAC 48	3 to 12 oz	1	0	**OMRI listed.**
Cane canker, Cane blight, Spur blight	captan, FRAC M4 (Captan 50W) (Captan 80 WDG) (Captec 4L)	4 lb 2.5 lb 2 qt	see label	3	Re-entry interval depends on product/formulation
	captan FRAC M4 + fenhexamid, FRAC M4 + 17 (CaptEvate 68WDG)	3.5 lb	48	3	Registered for use on raspberries, but not other brambles.
	pyraclostrobin (Cabrio 20EG), FRAC 11	14 oz	12	0	The strobilurin fungicides (Group 11) will also control rusts and powdery mildew.
	boscalid + pyraclostrobin (Pristine 38WDG), FRAC 7 + 11	18.5 to 23 oz	12	0	The strobilurin fungicides (Group 11) will also control rusts and powdery mildew.
	azoxystrobin, FRAC 11 (Abound 2SC)	6.2 to 15.4 fl oz	4	0	The strobilurin fungicides (Group 11) will also control rusts and powdery mildew.
Powdery mildew	See **"Just Before Blooms Open"**				
Cane canker, Cane blight, Spur blight (continued)	azoxystrobin + propiconazole (Quilt Xcel), FRAC 11 + 3	14 to 21 fl oz	12	30	The strobilurin fungicides (Group 11) will also control rusts and powdery mildew.

Table 6-5. Caneberry Management Program

SEASON and Pest	Product Name, Mode of Action Code, and Formulation	Rate of Formulation per Acre	Restricted Entry Interval (hours)	Preharvest Interval (PHI) (days)	Comments and Precautions
Bloom and Petal Fall (continued)					
Rednecked cane borer	Scout canes during winter pruning. If 10% or greater of the primocanes per row, or more, of the primocanes than will be removed through pruning, have rednecked cane borer galls, control is justified. Treat after first bloom or when adults are observed.				
	bifenthrin, IRAC 3A (Brigade 10WSB)	8 to 16 oz	12	3	Do not exceed 0.2 lb bifenthrin (32 oz Brigade 10WSB) per acre per season.
Strawberry clipper weevil	See **JUST BEFORE BLOOMS OPEN**				
Post-Bloom					
Anthracnose, Leaf spots, Rusts, Powdery mildew	See **JUST BEFORE BLOOMS OPEN**				Applications for these diseases should be made every 14 days after petal fall until harvest
Double blossom	See **BLOOM AND PETAL FALL**				Additional treatments may be needed to protect primocanes if infected flowers continue to open.
Japanese beetles	Caneberries can tolerate some foliar feeding by Japanese beetles, but little work has been done to determine when foliar feeding impacts yield. Do not use Japanese beetle pheromone traps as they attract beetles from outside fields. Treatment for Japanese beetles is not recommended unless significant foliar loss occurs.				
	carbaryl, IRAC 1A (Sevin) 50 WP (Sevin) 4 XLR	2 lb 2 qt	12	7	
	fenpropathrin, IRAC 3A (Danitol) 2.4 EC	10.66 to 16 fl oz	24	3	Do not exceed 32 fluid ounces Danitol per acre per season.
	malathion, IRAC 1B (Malathion 8F)	2 pt	24	1	Make no more than 4 applications per year.
	zeta-cypermethrin (Mustang Max) IRAC 3	4 fl oz	12	1	Do not make more than 6 applications per season.
Leafrollers	Leafrolling caterpillars can feed on caneberry foliage. Foliage damage is typically not economically significant, but caterpillars can occasionally form webs on fruit. If caterpillars are impacting fruit, treatment may be justified.				
	Bacillus thuringiensis (Bt), IRAC 11A (Dipel DF)	0.5 to 1 lb	4	0	Dipel DF is **OMRI** listed.
	chlorantraniliprole, IRAC 28 (Altacor)	3-4.5 oz	4	3	
	spinosad (IRAC 5) (Entrust SC)	4 to 6 fl oz	4	3	Entrust is **OMRI** listed. Do not apply more than 29 fl oz Entrust SC (0.45 lb active ingredient) per acre per year.
	Entrust (80W)	1.25 to 2 oz			Do not apply more than 9 oz Entrust 80W (0.45 lb active ingredient) per acre per year.
	spinetoram, IRAC 5 (Delegate)	3 to 6 oz	4	1	Do not exceed 19.5 ounces Delegate per acre per season.
Stink bugs, Plant bugs	Stink bug feeding does not typically damage berries, but they may be contamination pests during harvest. Plant bug and stink bugs may also feed on developing buds or shoots. Treatment is justified if insects are contaminating fruit.				
	esfenvalerate, IRAC 3A (Asana XL)	9.6 fl oz	12	7	Avoid applications when bees are foraging. Apply during evenings or early morning.
	fenpropathrin, IRAC 3A (Danitol 2.4EC)	10.66 to 16 fl oz	24	3	Do not exceed 32 fluid ounces per acre per season.
	thiamethoxam, IRAC 4A (Actara 25WDG)	3 oz	12	3	
Spider mites	There is no research-based treatment threshold for spider mites in caneberries, but treatment is recommended when a random sample of leaflets from the planting has an average of 10 motile mites. Leaflets should be examined with a minimum 10x hand lens to determine mite counts. Spider mites are more significant pests of raspberries than blackberries. Insecticides used against other pests may flare spider mite populations, particularly IRAC 1 and 3 materials. Observe plants for spider mites following treatment with these materials.				
	bifenazate, IRAC Unknown (Acramite 50WS)	1 lb	12	1	
	hexythiazox, IRAC 10A (Savey 50DG)	6 oz	12	3	Savey is primarily active against eggs and immature mites. Apply when populations are low.
	horticultural oils, IRAC Unknown (Saf-T-Side)	1 to 2% by volume	4	0	Summer oils are effective in moderating low mite populations pre-harvest. Use on a trial basis only until certain oil will not result in fruit finish problems. **Do not** use oils within 14 days of using any sulfur-containing material. **Do not** apply oils when temperatures will exceed 90 degrees F or dip below 50 degrees F.
	horticultural oils, IRAC Unknown (JMS Stylet Oil)	0.75 to 1.5% by volume	4	0	
	horticultural oils, IRAC Unknown (Organic JMS Stylet Oil)	0.75 to 1.5% by volume	4	0	Organic JMS Stylet Oil is **OMRI** listed.
Broad mites	Broad mites are emerging pests in blackberries and have been most problematic in primocane fruiting varieties. Significant populations have been observed in AR and NC. Broad mites cause leaf stunting and upward or downward cupping. Rule out other causes of leaf stunting (such as herbicide injury) and confirm broad mite presence before treating.				
	abamectin, IRAC 6 (Agri-Mek SC)				Agri-Mek SC has a Section 2(ee) label for use in caneberries, effective from July 6, 2016 to July 6, 2020 in AR, FL, IL, IN, NC, PA, and SC. Two Agri-Mek SC applications should be made 7 apart to be most effective.

Table 6-5. Caneberry Management Program

SEASON and Pest	Product Name, Mode of Action Code, and Formulation	Rate of Formulation per Acre	Restricted Entry Interval (hours)	Preharvest Interval (PHI) (days)	Comments and Precautions
Harvest					
Botrytis fruit rot	boscalid + pyraclostrobin (Pristine 38 WDG), FRAC 7 + 11	18.5 to 23 oz	12	0	
	cyprodinil + fludioxonil (Switch 62.5 WG), FRAC 9 + 12	11 to 14 oz	12	0	
	fenhexamid (Elevate 50 WDG), FRAC 17	1.5 lb	12	0	
	iprodione (several brands) 50 WG, FRAC 2 4F	1 to 2 lb 1 to 2 pt	24	0	
	captan, FRAC M4 (Captan 50W) (Captan 80 WDG) (Captec 4L)	4 lb 2.5 lb 2 qt	see label	3	Include captan in this spray if ripe rot is a problem. Pristine will also control ripe rot.
	Captan + fenhexamide (CaptEvate) 68WDG, FRAC M4 + 17	3.5 lb	48	3	CaptEvate is registered on raspberries only. Include captan in this spray if ripe rot is a problem. Pristine will also control ripe rot.
Flower thrips	Flower thrips can be a contamination pest at harvest. Fruit can be placed in a clear plastic bag before harvest and observed for flower thrips. There is no evidence at this time to suggest that flower thrips damage fruit or flowers and reduce yield.				
	acetamiprid IRAC 4A (Assail 30SG)	4.5 to 5.3 oz	12	1	
	spinosad (IRAC 5) (Entrust SC)	4 to 6 fl oz	4	3	Entrust is **OMRI** listed. Do not apply more than 29 fl oz Entrust SC (0.45 lb active ingredient) per acre per year.
	Entrust (80W)	1.25 to 2 oz			Do not apply more than 9 oz Entrust 80W (0.45 lb active ingredient) per acre per year.
	spinetoram, IRAC 5 (Delegate)	3 to 6 oz	4	1	Do not exceed 19.5 ounces Delegate per acre per season.
	zeta cypermethrin IRAC 3 (Mustang Max)	4.0 fl oz	12	1	
Spotted wing drosophila	Spotted wing drosophila (SWD) is a pest of soft skinned fruit. Female SWD lay eggs in ripe and ripening fruit, which can appear externally undamaged. Spotted wing drosophila are present in damaging densities during typical blackberry and raspberry harvest periods in North Carolina, so preventative treatment, beginning when fruit begin to change color, is recommended. Treatments should be applied at least every 7 days, and mode of action (IRAC code) should be rotated between treatments. In addition to insecticide treatments, growers should also employ good cultural practices including: harvesting as frequently as possible, removing all ripe fruit from plants at each harvest, and storing all harvested from at temperatures below 41°F for as long as feasible before marketing.				
	bifenthrin, IRAC 3A (Brigade 10WSB)	8 to 16 oz	12	3	Do not exceed 0.2 lb bifenthrin (32 oz Brigade 10WSB) per acre per season.
	fenpropathrin, IRAC 3A (Danitol 2.4EC)	10.66 to 16 fl oz	24	3	Do not exceed 32 fluid ounces per acre per season.
	malathion, IRAC 1B (Malathion 8F)	2 pt	24	1	Make no more than 4 applications per year.
	spinosad (IRAC 5) (Entrust SC) Entrust (80W)	4 to 6 fl oz 1.25 to 2 oz	4	3	Entrust is **OMRI** listed. Do not apply more than 29 fl oz Entrust SC (0.45 lb active ingredient) per acre per year. Do not apply more than 9 oz Entrust 80W (0.45 lb active ingredient) per acre per year.
	spinetoram, IRAC 5 (Delegate)	3 to 6 oz	4	1	Do not exceed 19.5 ounces Delegate per acre per season.
	zeta cypermethrin IRAC 3 (Mustang)	4.3 fl oz	12	1	Do not exceed 25.8 fluid ounces Mustang per acre per season.
Japanese beetles and green June beetles	carbaryl, IRAC 1A (Sevin) 50 WP (Sevin) 4 XLR	2 lb 2 qt	12	7	
	fenpropathrin, IRAC 3A (Danitol) 2.4 EC	10.66 to 16 fl oz	24	3	Do not exceed 32 fluid ounces Danitol per acre per season.
	zeta-cypermethrin (Mustang Max) IRAC 3	4 fl oz	12	1	Do not make more than 6 applications per season.
Just After Harvest and 14 Days Later					
Leaf spots	captan (Captan 50W) (Captan 80 WDG)	4 lb 2.5 lb	see label	3	
Japanese beetle	See Post Bloom				Japanese beetle treatments are necessary post harvest only if feeding is removing greater than 10% of foliage on primocanes.
Late October or Early November					
Raspberry crown borer	Raspberry crown borer treatments should be applied once per year, either in late fall or early spring. Applications during both time periods are not necessary.				
	bifenthrin, IRAC 3A (Brigade WSB) IRAC 3A	16 oz	12	3	
	chlorantraniliprole, IRAC 28 (Altacor)	3 to 4.5 oz	4	3	
	esfenvalerate, IRAC 3A (Asana XL)	9.6 fl oz	12	7	

Chapter VI—2019 N.C. Agricultural Chemicals Manual

Table 6-5. Caneberry Management Program

SEASON and Pest	Product Name, Mode of Action Code, and Formulation	Rate of Formulation per Acre	Restricted Entry Interval (hours)	Preharvest Interval (PHI) (days)	Comments and Precautions
Dormant					
Scale insects	Scale insects may be present on caneberries but are typically kept below economically damaging levels by parasitoids and predators. An open canopy minimizes scale populations. Examine plants after harvest and during pruning for scale, and if present in high numbers or resulting in sooty mold growth, consider a dormant season oil treatment.				
	horticultural oils IRAC Unknown (Saf T Side)	1 to 2% by volume	4	0	
	horticultural oils IRAC Unknown (JMS Stylet Oil)	0.75 to 1.5% by volume	4	0	
	horticultural oils IRAC Unknown (Organic JMS Stylet Oil)	0.75 to 1.5% by volume	4	0	Organic JMS Stylet Oil is **OMRI** listed.
Raspberry cane borer, Red neck cane borer	During winter pruning, examine canes for raspberry cane borer injury. Prune canes girdled by raspberry cane borer 2 to 3 cm below the lower girdle or gall. If evidence of boring is present below this cut, successive cuts should be made until no further injury is observed. Destroy or remove cuttings to prevent reinfestation.				
Special Rust Sprays					
Cane and leaf rust, Orange rust	Begin applications in the spring just before orange rust pustules are formed on the lower leaf of brambles (use wild blackberries as indicators). Continue at 10- to 14-day intervals until the mean temperature remains above 77 degrees F. Infections can also occur in the late summer and fall. Chemicals are not very effective once systemic infection occurs, however, fungicide applications can be effective in preventing additional new infections.				
	boscalid + pyraclostrobin (Pristine 38 WG), FRAC 7 + 11	18.5 to 23 oz	12	0	Where orange rust has been a problem, alternate Rally and Cabrio or Pristine or azoxystrobin at 14-day intervals. For late leaf rust, begin when symptoms first appear, and continue on a 14-day interval.
	myclobutanil (Rally 40 WSP) 40 WSP, DF, WDG, FRAC 3	1.25 to 3 oz	24	1	
	pyraclostrobin (Cabrio 20EG), FRAC 11	14 oz	12	0	Where orange rust has been a problem, alternate Rally and Cabrio or Pristine or azoxystrobin at 14-day intervals.
	azoxystrobin, FRAC 11 (Abound 2SC)	6.2 to 15.4 fl oz	4	0	
	azoxystrobin + propiconazole (Quilt Xcel), FRAC 11 + 3	14 to 21 fl oz	12	30	
Special Treatments for Phytophthora Root Rot					
Phytophthora root rot	mefenoxam (Ridomil Gold SL), FRAC 4	—	48	45	Apply 0.25 pint per 1,000 linear feet of row in a 3 feet wide band in the spring and fall after harvest. Ridomil Gold is registered for raspberries only. 45-day phi.
	fosetyl Al (Aliette WSP), FRAC 33	5 lb	12	60	Begin when growth is 1 to 3 inches long and continue at 45- to 60-day intervals through the growing season. Registered for blackberries and raspberries. Maximum of 4 applications per year. 60-day phi.
	phosphite fungicides, FRAC 33	See label	See label	See label	Several phosphorus acid products are registered for control of Phytophthora root rot, including Prophyt and Agri-Fos. See label for recommendations.
Preplant Treatments for Nematodes					
Nematodes	1,3 dichloropene 37% + chloropicrin 57% (Pic-Clor 60 EC)	19.5 to 44.5 gal	5	NA	Preplant interval should be 4 to 8 weeks, or longer if dissipation is slow. See label for additional information.
	metam sodium (Vapam, Sectagon II, Busan 1020)	75 gal	See label	NA	Preplant interval is a minimum of 4 weeks.

More information available at http://rubus.ces.ncsu.edu.

Bordeaux mixture recipe is available at http://www.smallfruits.org/SmallFruitsRegGuide/Guides/2015/2015BrambleSpray%20Guide12_22_14.pdf

Bunch Grape Insect Management

H. J. Burrack, Entomology and Plant Pathology

With a few exceptions, which are noted, wine grapes should be treated for insects only when damaging insect populations are present. Where treatment thresholds are known, these are provided. For many insect pests of wine grapes in the southeast, thresholds do not exist. Consult cooperative Extension personnel for management recommendations if insects for which there are no thresholds are present.

The Insecticide Resistance Action Committee (IRAC) and the Fungicide Resistance Action Committee (FRAC) group insecticides and fungicides into mode of action categories. These categories are listed following the pesticide and formulation names. To reduce the risk of resistance development, avoid successive applications of insecticides or fungicides with the same IRAC or FRAC code for the same pest. Organically acceptable insecticides (OMRI listed) are indicated in Precautions and Remarks.

Some insecticide active ingredients are available in several formulations and under several trade names. For simplicity, the most common trade names and associated rates are listed. This is not intended to encourage the use of these products over generic versions.

PLEASE NOTE: This table does not contain information on wine grape disease management. For wine grape disease guidance, please refer to information provided online by Virginia Tech University at https://pubs.ext.vt.edu/456/456-017/Section-3_Grapes-1.pdf

Table 6-6A. Bunch Grape Insect Management

When to Spray and Pest	Pesticide, Formulation, and IRAC, FRAC Code	Amount of Formulation to Use per Acre	Reentry Interval (REI) (hours)	Preharvest Interval (PHI) (days)	Precautions and Remarks
Bud Swell					
Grape flea beetle	Apply only if damaging numbers of adult beetles are present. If 4% or more of buds have been damaged by grape flea beetles, treatment is justified. Grape flea beetle adults emerge in early spring and feed on newly swollen buds and lay eggs. Larvae and adults from subsequent generations feed on leaves, but foliar feeding typically does not result in economically significant damage nor justify treatment.				
	carbaryl, IRAC 1A (Sevin XLR Plus)	1 to 2 qt	12	7	
	cyfluthrin IRAC 3 (Baythroid 2 EC)	2.4 to 3.2 fl oz	12	3	
	fenpropathrin, IRAC 3 (Danitol 2.4 EC)	8 fl oz	24	21	
	phosmet, IRAC 1B (Imidan 70WP)	1.33 to 2.125 lb	14 days	7 (rates of 1.33 lb per acre or less) 14 (more than 1.33 lb per acre)	Do not apply more than 6.5 pounds Imidan per acre per year.
Climbing cutworms	Scout for cutworm if damaged buds are observed. Look for cutworms at night. Cutworm treatment may be justified if greater than 4% of the buds examined are damaged and the variety does not have fruitful secondary buds. Spray in the evening if possible as cutworms are active at night.				
	Bacillus thuringiensis (Bt), IRAC 11 DiPel DF	0.5 to 2 lb	4	0	There are many Bt formulations. DiPel DF is **OMRI** listed, but not all formulations are organically acceptable. Read label carefully.
	carbaryl, IRAC 1A (Sevin XLR Plus)	1 to 2 qt	12	7	
	chlorantraniliprole IRAC 28 (Altacor)	2 to 4.5 oz	4	14	
	fenpropathrin, IRAC 3 (Danitol 2.4 EC)	10.66 to 21.33 fl oz	24	21	Do not exceed 2.66 pints of Danitol per acre per season. Make no more than 2 applications of Danitol per season.
	methoxyfenozide, IRAC 18 (Intrepid 2F)	12 to 16 fl oz	4	30	Minimum application for airblast sprayers of 40 gpa.
	spinetoram IRAC 5 (Delegate 25 WG)	3 to 5 oz	4	7	
	spinosad, IRAC 5 (Entrust SC) (Entrust 80W)	4 to 8 fl oz 1.25 to 2.5 oz	4	7	Do not apply more than 29 fl oz Entrust SC or 9 oz of Entrust 80W (0.45 lb spinosad) per acre per season. Entrust is **OMRI** listed.
Mealybugs	Scout for mealybugs and European red mite under bark during dormant season. Use a minimum 10x hand lens to observe European red mite (ERM).				
	buprofezin, IRAC 16 (Applaud 70DF)	9 to 12 oz	12	7	Apply when crawlers are active, or at 493 and 990 degree-days (base 50 F), starting at April 1 (early and peak activity of first generation).
	clothianidin (IRAC 4A) Belay	6 fl oz (foliar) 6 to 12 fl oz (soil)	12	0 (foliar) 30 (soil)	Belay can be applied either to the soil or as a foliar spray. Soil applications are typically active for a longer period of time but must be made early in the year. Soil applications are more effective when made via drip irrigation.
	cyfluthrin, IRAC 3 (Baythroid 2EC)	2.4 to 3.2 fl oz	12	3	
	dinotefuran (IRAC 4A) (Venom 20SG)	0.44 to 0.66 lb (foliar) 1.13 to 1.32 lb (soil)	12	1 (foliar) 28 (soil)	Venom can be applied either to the soil or as a foliar spray. Soil applications are typically active for a longer period of time but must be made early in the year. Soil applications are more effective when made via drip irrigation.
	horticultural oils (Omni Supreme Spray Oil) (JMS Stylet Oil) (Organic JMS Stylet Oil)	0.5 to 1% by volume 1 to 2% by volume 1 to 2% by volume			If mealybugs or ERM were of economic concern during the previous season and present during dormant scouting, a dormant oil treatment may be justified during bud swell. **Do not** apply oil treatments in combination with sulfur or within 30 days of sulfur application. **Do not** apply oils when temperature will exceed 90 degrees F or dip below freezing. Organic JMS Stylet Oil is **OMRI** listed.

Table 6-6A. Bunch Grape Insect Management

When to Spray and Pest	Pesticide, Formulation, and IRAC, FRAC Code	Amount of Formulation to Use per Acre	Reentry Interval (REI) (hours)	Preharvest Interval (PHI) (days)	Precautions and Remarks
Bud Swell (continued)					
Mealybugs (continued)	imidacloprid (IRAC 4A) (Admire Pro)	1 to 1.4 fl oz (foliar) 7 to 14 fl oz (soil)	12	0 (foliar) 30 (soil)	Admire Pro can be applied either to the soil or as a foliar spray. Soil applications are typically active for a longer period of time but must be made early in the year. Soil applications are more effective when made via drip irrigation.
At or Just Before Budburst					
Leafhopper/ sharpshooters (Pierce's Disease suppression)	Consider an early season soil application of a neonicotinoid (4A) insecticide for leafhoppers if plants symptomatic for Pierce's Disease have been observed in the vineyard or in nearby vineyards. This strategy provides longer term control than foliar treatments.				
	acetamiprid, IRAC 4A (Assail 30SG)	2.5 oz	12	7	Assail can only be applied as a foliar treatment.
	clothianidin (IRAC 4A) Belay	6 fl oz (foliar) 6 to 12 fl oz (soil)	12	0 (foliar) 30 (soil)	Belay can be applied either to the soil or as a foliar spray. Soil applications are typically active for a longer period of time but must be made early in the year. Soil applications are more effective when made via drip irrigation.
	cyfluthrin, IRAC 3 (Baythroid 2EC)	2.4 to 3.2 fl oz	12	3	
	dinotefuran (IRAC 4A) (Venom 20SG)	0.44 to 0.66 lb (foliar) 1.13 to 1.32 lb (soil)	12	1 (foliar) 28 (soil)	Venom can be applied either to the soil or as a foliar spray. Soil applications are typically active for a longer period of time but must be made early in the year. Soil applications are more effective when made via drip irrigation.
	imidacloprid (IRAC 4A) (Admire Pro)	1 to 1.4 fl oz (foliar) 7 to 14 fl oz (soil)	12	0 (foliar) 30 (soil)	Admire Pro can be applied either to the soil or as a foliar spray. Soil applications are typically active for a longer period of time but must be made early in the year. Soil applications are more effective when made via drip irrigation. There are several formulations of imidacloprid.
Prebloom					
Flea beetle	See **Bud Swell** recommendations				
Grape berry moth	Grape berry moth is present in North Carolina, but it is not uniformly distributed in the state. If grape berry moth presence is suspected, observe flowers and fruit for injury and consider monitoring moth presence with pheromone baited traps.				
	chlorantraniliprole IRAC 28 (Altacor)	2 to 4.5 oz	4	14	chlorantraniliprole IRAC 28 (Altacor)
	fenpropathrin, IRAC 3 (Danitol 2.4 EC)	10.6 fl oz	24	21	
	methoxyfenozide, IRAC 18 (Intrepid 2F)	12 to 16 fl oz	4	30	Minimum application of Intrepid for airblast sprayers is 40 gallons per acre.
	indoxacarb (IRAC 22) (Avaunt 30DG)	5 to 6 oz	12	7	
	methomyl, IRAC 1A (Lannate SP) (Lannate LV)	0.5 to 1 lb 1.5 to 3 pt	7	14	
	pheromone (SPLAT-GBM)	1 kg	4	0	Apply SPLAT-GBM mating disruption when temperatures between 60-80 degrees F and no rain is expected within 1 to 2 hours. For high population densities, apply 1.0kg/A as 1,000 point sources of 1.0 g (1/4 tsp) throughout an acre. For low-moderate populations, apply 1.0 kg as 250 point sources of 2.5 g (1/2 tsp). See application information on label.
	spinosad, IRAC 5 (Entrust SC) (Entrust 80W)	4 to 8 fl oz 1.25 to 2.5 oz	4	7	Do not apply more than 29 fl oz Entrust SC or 9 oz of Entrust 80W (0.45 lb spinosad) per acre per season. Entrust is **OMRI** listed.
	spinetoram, IRAC 5 (Delegate)	3 to 5 oz	4	7	
Leafhoppers/ Sharpshooters (Pierce's Disease suppression) Initiation of foliar treatments should be based on trap captures.	If foliar and soil applications of group 4A pesticides are part of a management plan for Pierce's disease (i.e., Admire Pro applied via drip and Venom foliar), at least one application of a different IRAC insecticide should occur as a rotation between these treatments. Synthetic pyrethroid insecticides (Group 3) and organophosphates (Groups 1A and 1B) are broad spectrum insecticides and have the potential to flare spider mite populations. Observe spider mites before and after treatments to determine if these populations increase.				
	acetamiprid, IRAC 4A (Assail WSP)	2.5 oz	12	7	
	clothianidin (IRAC 4A) Belay	6 fl oz (foliar) 6 to 12 fl oz (soil)	12	0 (foliar) 30 (soil)	Belay can be applied either to the soil or as a foliar spray. Soil applications are typically active for a longer period of time but must be made early in the year. Soil applications are more effective when made via drip irrigation.
	cyfluthrin (IRAC 3) (Baythroid 2EC)	2.4 to 3.2 fl oz	12	3	
	dinotefuran (IRAC 4A) (Venom 20SG)	0.44 to 0.66 lb (foliar) 1.13 to 1.32 lb (soil)	12	1 (foliar) 28 (soil)	Venom can be applied either to the soil or as a foliar spray. Soil applications are typically active for a longer period of time but must be made early in the year. Soil applications are more effective when made via drip irrigation.
	fenpropathrin, IRAC 3 (Danitol 2.4 EC)	5.33 to 10.66 fl oz	24	21	
	imidacloprid (IRAC 4A) (Admire Pro)	1 to 1.4 fl oz (foliar) 7 to 14 fl oz (soil)	12	0 (foliar) 30 (soil)	Admire Pro can be applied either to the soil or as a foliar spray. Soil applications are typically active for a longer period of time but must be made early in the year. Soil applications are more effective when made via drip irrigation.

Table 6-6A. Bunch Grape Insect Management

When to Spray and Pest	Pesticide, Formulation, and IRAC, FRAC Code	Amount of Formulation to Use per Acre	Reentry Interval (REI) (hours)	Preharvest Interval (PHI) (days)	Precautions and Remarks
Prebloom					
Leafhoppers/ Sharpshooters (Pierce's Disease suppression) Initiation of foliar treatments should be based on trap captures. (continued)	carbaryl, IRAC 1A (Sevin 80S)	1.25 to 2.5 lb	12	7	Foliar treatment
	malathion, IRAC 1B (Malathion 57EC) (Malathion 5)	3 pt 3 pt	12	3	Foliar treatment
Grape phylloxera (foliar)	Grape phylloxera has root feeding and foliar feeding forms. Rootstocks used in grape propagation are resistant to root feeding forms and do not require treatment. Foliar phylloxera may be problematic in European-American hybrid varieties (i.e., Vidal, Seyval, etc.) and cause distinctive, wart-like galls on leaves. The mobile crawler stage of phylloxera is susceptible to insecticide treatment, but closed galls are not. Scouting for galls and crawlers should begin once leaves are expanded. If infested leaves are found in susceptible varieties, insecticide treatments should be timed to crawler emergence.				
	acetamiprid, IRAC 4A (Assail WSP)	2.5 oz	12	7	
	fenpropathrin, IRAC 3 (Danitol 2.4 EC)	5.33 to 10.66 fl oz	24	21	
	imidacloprid, IRAC 4A (Admire Pro)	1 to 1.4 fl oz (foliar) 7 to 14 fl oz (soil)	12	0 (foliar) 30 (soil)	Admire Pro can be applied either to the soil or as a foliar spray. Foliar treatments are likely to be more effective against foliar feeding phylloxera than soil treatments because they can be more closely timed to crawler emergence.
	kaolin (Surround WP Crop Protectant)	25 lb	4		Surround is a barrier that reduces insect feeding. Harvest parameters may be altered and maturity may be delayed, especially in white wine varieties. Closely monitor harvest parameters to determine optimal time to harvest. Changes in harvest parameters can affect final taste. Wine grapes sprayed up to veraison will have minimal adherence to berries. Applications after veraison will adhere more on grape berries. Surround is **OMRI** listed.
	spirotetramat, IRAC 23 (Movento 2SC)	6 to 8 fl oz	24	7	Movento is also effective against root feeding phylloxera.
Bloom					
Flower thrips	Thrips treatment may be justified if populations exceed an average of 10 thrips per cluster. To sample for thrips, beat blossom clusters over a white surface and count the number of thrips dislodged onto the surface. Count immediately after beating the blossom cluster. Sample at least 10 blossom clusters from different locations in the vineyard. During periods of heavy thrips pressure, a second application may be needed, but make it only if thrips numbers remain high. Wait at least 5 days before making a second application.				
	azadirachtin, IRAC Unknown (Aza-Direct)	1 to 2 pt	4		Aza-direct is **OMRI** listed. Data on thrips control are limited.
	dinotefuran, IRAC 4A (Venom)	3 to 5 oz	12	1	Foliar treatment only. Soil treatments are unlikely to be effective against flower feeding thrips.
	pyrethrins, IRAC 3 (Pyganic 1.4 EC) (Pyganic 5 EC)	16 to 64 fl oz 4.5 to 18 fl oz	12	0	Pyganic 1.4 EC and Pyganic 5 EC are **OMRI** listed. Data on thrips control are limited. Pyganic should be buffered to a pH between 5.5 and 7.
	spinosad, IRAC 5 (Entrust SC) (Entrust 80W)	4 to 8 fl oz	4	7	Do not apply more than 29 fl oz Entrust SC or 9 oz of Entrust 80W (0.45 lb spinosad) per acre per season. Entrust is **OMRI** listed.
	spinetoram, IRAC 5, (Delegate)	3 to 5 oz	4	7	
Postbloom (immediately after bloom)					
Grape berry moth, grape flea beetle, and leafhoppers/ sharpshooters (Pierce's Disease suppression)	See **PREBLOOM** recommendations.				
European red mite, twospotted spider mite	Sample for mites weekly using a minimum 10x hand lens. If greater than 50% of leaves observed have spider mites and no predatory mites are present, treatment is justified. Fast moving predatory mites can be distinguished from slower moving spider mites through direct observation. Rotate miticides between IRAC codes to minimize selection for resistance. Miticides should be applied in at least 50 gpa spray volume to ensure adequate coverage.				
	abamectin, IRAC 6 (Agri-Mek 0.15EC) (many other formulations)	16 fl oz			
	bifenazate, IRAC Unknown (Acramite 50 WS)	1 lb	12	14	The reentry interval is 5 days for cane turning, tying, and girdling of table grapes.
	cyflumetofen, IRAC 25 (Nealta)	13.7 fl oz	12	14	Do not make more than 2 Nealta applications per season and rotate to another mode of action (IRAC code) between treatments.
	dicofol, IRAC Unknown (Difocol 4EC) (Kelthane 50WSP)	2.5 pt 2.5 lb	12	7	Use Kelthane at 1 pound per acre on small vines. Do not make more than 2 applications per season.
	etoxazole, IRAC 10B (Zeal)	3 oz	12	28	Zeal is a growth regulator and kills eggs and young mites. It is most effective if applied when mite populations are low.
	fenbutatin-oxide, IRAC 12B (Vendex 50WP)	2.5 lb	48	28	Do not make more than 2 applications of Vendex per season.

Table 6-6A. Bunch Grape Insect Management

When to Spray and Pest	Pesticide, Formulation, and IRAC, FRAC Code	Amount of Formulation to Use per Acre	Reentry Interval (REI) (hours)	Preharvest Interval (PHI) (days)	Precautions and Remarks
Postbloom (immediately after bloom) (continued)					
European red mite, twospotted spider mite (continued)	fenpyroximate, IRAC 21A (FujiMite 5EC)	2 pt	12	14	Do not apply more than 2 pints of FujiMite per acre per season.
	horticultural oils, IRAC Unknown many materials, including (Saf T Side) (Glacial Spray Fluid)	1 to 2% by volume	4	0	Some oils are **OMRI** listed; check label. **Do not** use in combination with or immediately before or after spraying with fungicides such as Captan or any product containing sulfur. **Do not** use with carbaryl or dimethoate. **Do not** use with any product whose label recommends the use of no oils. Do not use in combination with NPK foliar fertilizer applications.
	pyridiben, IRAC 21 (Nexter 75WPSB) (Pyramite 60 WP)	5.2 oz 13.2 oz	12	7	The maximum amount of pyridiben allowed per acre per season is 26.4 ounces. Do not make more than 2 applications of pyridiben per season.
	spirodiclofen, IRAC 23 (Envidor 2SC)	18 fl oz	12	14	
Grape phylloxera (foliar form)	See **Prebloom** recommendations.				
1st Cover Spray (7 to 10 days after Postbloom Spray)					
Grape berry moth, leafhoppers/ sharpshooters (Pierce's Disease suppression)	See **Prebloom** recommendations				If foliar and soil applications of group 4A pesticides are part of a management plan for Pierce's Disease (i.e., Admire Pro applied via drip and Venom foliar), at least 1 application of a different IRAC insecticide should occur as a rotation between these treatments. Current information indicates that in areas where Pierce's Disease is a problem, controlling leafhoppers and sharpshooters through July reduces the risk of Pierce's Disease. See labels for preharvest intervals.
Japanese beetle, Green June beetles	Do not use Japanese beetle traps. Japanese beetle foliar feed only warrants treatment if it occurs on leaves below the top trellis wire. Green June beetles only require treatment if damaging otherwise sound fruit. Green June beetles are attracted to overripe, rotting fruit. Removal of damaged fruit will help reduce Green June beetle populations.				
	acetamiprid, IRAC 4A (Assail WSP)	1.1 oz	12	7	Foliar applications of Group 4A insecticides should NOT be used following a long-acting soil application of any group 4A insecticide (i.e., Admire Pro, Venom, or Clutch).
	azadirectin, IRAC Unknown (Aza-Direct)	1 to 2 pt oz	4	0	Aza-Direct is **OMRI** listed. Data on Japanese beetle control are limited.
	carbaryl, IRAC 1A (Sevin XLR Plus)	2 qt	12	7	Synthetic organophosphates (Groups 1A and 1B) are broad spectrum insecticides and have the potential to flare spider mite populations. Observe spider mites before and after treatments to determine if these populations increase.
	fenpropathrin, IRAC 3 (Danitol 2.4 EC)	10.6 to 21.3 fl oz	24	21	Do not apply Danitol within 21 days of harvest.
	indoxacarb, IRAC 22 (Avaunt 30DG)	3.5 to 6 oz	12	7	Avaunt is also very effective against caterpillar pests such as cutworms and grape berry moth.
	kaolin clay (Surround WP)	25 to 50 lb			Surround may delay fruit maturity, and therefore, anticipated harvest date. Fruit harvest characters should be carefully monitored if Surround is used to ensure timely harvest. Surround is **OMRI** listed.
	malathion, IRAC 1B (Malathion 8F)	1.88 pt	24	3	Synthetic organophosphates (Groups 1A and 1B) are broad spectrum insecticides and have the potential to flare spider mite populations. Observe spider mites before and after treatments to determine if these populations increase. REI is 72 hours for girling and tying. Injury may occur to grapes for applications made after bloom. Check for phytotoxicity in a small area before treating an entire field.
	phosmet, IRAC 1B (Imidan 70 WP)	1.33 to 2.125 lb	14 days	7 (rates of 1.33 lb per acre or less) 14 (more than 1.33 lb per acre)	Do not apply more than 6.5 pounds Imidan per acre per year.
Closing					
Japanese beetle, GreenJune beetle	Same as **1st Cover**				
Grape berry moth, Leafhopper/ sharpshooter (Pierce's Disease suppression)	Same as **Prebloom**				**Foliar applications of Group 4A insecticides should NOT be used following a long-acting soil application of any Group 4A insecticide (i.e., Admire Pro, Venom, or Clutch).** Current information indicates that in areas where Pierce's Disease is a problem, controlling leafhoppers and sharpshooters **through July** reduces the risk of Pierce's Disease. See labels for preharvest intervals.
2nd and Subsequent Cover Sprays (10- to 14-day intervals until the Preharvest Spray)					
Phylloxera, Japanese and June beetles	Same as **1st Cover**				Check labels for preharvest intervals.

Table 6-6A. Bunch Grape Insect Management

When to Spray and Pest	Pesticide, Formulation, and IRAC, FRAC Code	Amount of Formulation to Use per Acre	Reentry Interval (REI) (hours)	Preharvest Interval (PHI) (days)	Precautions and Remarks
2nd and Subsequent Cover Sprays (10- to 14-day intervals until the Preharvest Spray) (continued)					
Grape berry moth, Leafhopper/ sharpshooter (Pierce's Disease suppression)	Same as **Prebloom**				Current information indicates that in areas where Pierce's Disease is a problem, controlling leafhoppers and sharpshooters through July reduces the risk of Pierce's Disease. If Venom was applied as a soil treatment during prebloom, a second soil application is not permitted, but a foliar spray of Venom is permitted at this time. See label for further restrictions.
Mites	Same as **Postbloom**				
Grape rootworm, Southern grape rootworm	Grape rootworm larvae feed on roots. Adults are small, black weevils and make distinctive chain-like feeding markings on leaves. Foliar feeding does not result in yield reduction, but root feeding may reduce plant vigor over time. Treatments should be timed to adult activity, which typically peaks in June or July. Grape rootworms are sporadic pests in North Carolina and should not be treated preventatively.				
	carbaryl, IRAC 1A (Sevin XLR Plus)	2 qt	12 hours	7	
Preharvest (10 to 14 days before harvest)					
Spotted wing drosophila	Spotted wing drosophila (SWD) is a recently detected invasive pest of ripe and ripening fruit. SWD do not necessarily cause significant damage in wine grapes. Growers are encouraged to monitor adult flies in vineyards and larvae in fruit and treat if present. Weekly or twice weekly treatments may be necessary to manage damaging populations, but preventative treatments are not recommended at this time. Insecticides effective against SWD may also have activity against leafhoppers and beetle pests.				
	beta-cyfluthrin, IRAC 3A (Baythroid XL 1EC)	1.6 to 3.2 fl oz	12	3	
	fenpropathrin, IRAC 3A (Danitol 2.4EC)	5.33 to 21.33 fl oz	24	12	
	imidacloprid + cyfluthrin, IRAC 4A + 3 (Leverage 2.4)	3 to 8 fl oz	12	3	
	malathion IRAC 1B (Malathion 5EC)	1.5 pt	12	3	
	spinosad, IRAC 5 (Entrust SC)	4 to 8 fl oz	4	7	Do not apply more than 29 fl oz Entrust SC (0.45 lb spinosad) per acre per season. Entrust is **OMRI** listed.
	spinetoram, IRAC 5 (Delegate)	3 to 5 oz	4	7	
	zeta-cypermethrin, IRAC 3 (Mustang)	2.15 to 4.3 fl oz	12	1	
Harvest					
Yellowjackets and bees	Check to make sure wasps are not nesting in vines. Spot treat or manually remove nests if present. Cover sprays for wasps or bees is not recommended, because treatments with short PHI will not provide control, and only foraging worker wasps or bees will be killed, leaving the rest of the nest for reinfestation. Damaged fruit should be removed to reduce attraction for other bees and wasps.				
Spotted wing drosophila	See **Preharvest**				
Multicolored Asian lady beetle	Multicolored Asian lady beetle, MALB, can be a contaminant pest at harvest. Sample at least 10 clusters per acre within a few days of harvest, place in a plastic bag for approximately 30 minutes and count beetles. Treatment thresholds vary by variety.				
	imidacloprid, IRAC 4A (Admire Pro)	1 to 1.4 fl oz	12	0	Several concentrations of imidacloprid (1.6F, 2F, 4F, and 4.6F) are available. Carefully read the label to determine the correct rate for target pests. Data on control with imidacloprid are limited.
Postharvest (14- to 21-day intervals from harvest until first killing frost)					
Grape root borer	Grape root borer is potentially the most significant pest of grapes in North Carolina, but they are not necessarily present in all vineyards. Grape root borer moths should be monitored with pheromone baited traps. If moths are confirmed within a vineyard, mating disruption is the most effective control tool.				
	mating disruption (Isomate GRB)	100 dispensers	NA	NA	Dispensers should be placed prior to the beginning of grape root borer moth flight activity and be left in the vineyard until the end of flight activity. Moth flight timing varies between vineyards, but can be as early as July and last until October. Pheromone baited traps can help determine grape root borer populations and flight activity, but traps will not be effective if mating disruption is underway.
	Soil mounding, cultural control	NA			Use clean cultivation, mound soil (July 1 to Aug. 1) or at first moth emergence when using pheromone traps) or using tightly-sealed plastic mulch 3 feet from the base of vines. This practice will inhibit adult emergence from the soil when well timed. Mounded soil needs to be removed by Sept. 1.
	chlorpyrifos (Lorsban Advanced) IRAC 1B	4.5 pt/100 gal water	24	35	**Chlorpyrifos registration is under federal review at the time of this writing. Check registration status before using.** Apply 2 quarts of mixture to soil at base of each vine. A single application should be sufficient, either pre or post harvest, depending upon grape root borer flight timing. Spray should not contact fruit or foliage. Application can be made with flood nozzles and low pressure (40 to 60 psi). The preharvest interval for Lorsban is 35 days and in North Carolina this prohibits use during the most effective application timing.

For further information, see www.smallfruits.org.

Bunch Grape Disease Management

S. M. Villani and W. O. Cline, Entomology and Plant Pathology and

M. Nita, Plant Pathology, Physiology, and Weed Science, Virginia Tech

The Fungicide Resistance Action Committee (FRAC) groups fungicides into mode of action categories. These categories are listed following the pesticide and formulation names. To reduce the risk of resistance development, avoid successive applications of fungicides with the same FRAC code for the same pest. Some active ingredients are available in several formulations and under several trade names. For simplicity, the most common trade names and associated rates are listed. This is not intended to encourage the use of these products over generic versions.

Please Note: Pierce's Disease is caused by an endemic xylem-limited bacterial pathogen (*Xylella fastidiosa*). Fungicides are not effective against this disease. Suppression of Pierce's Disease relies on early- to mid-season use of insecticides to control leafhoppers and sharpshooters that transmit the pathogen from infected vines to nearby healthy vines. For suggestions on control of these insect vectors, see table 6-6A, Bunch Grape Insect Management.

Table 6-6B. Bunch Grape Disease Management Program

When to Spray and Disease/Pest	Pesticide, Formulation, and FRAC Code	Amount of Formulation to Use per Acre	Reentry Interval (REI) (hours)	Preharvest Interval (PHI) (days)	Precautions and Remarks
Dormant Anthracnose, Black rot, Phomopsis	liquid lime sulfur, FRAC M2	10 gal	48	0	Needed only where anthracnose is a problem. Removal of mummies, rachises, and cankered/dead wood during the dormant period is most efficacious for reduction of black rot. A dormant application of lime sulfur may help reduce the overwintering inoculum of fungi that cause black rot and Phomopsis; however, in-season protection of young shoots is the best option.
Budburst Pierce's Disease	Various insecticides (see table 6-6A Insect Management)	--	--	--	See Bunch Grape Insect Management Table 6-6A for control of insect vectors (leafhoppers/sharpshooters) that spread Pierce's Disease bacteria (*Xylella fastidiosa*).
New Shoots (7- to 10-day interval beginning at 1-inch shoot growth until Prebloom Spray) Phomopsis, Black rot, Powdery mildew, Downy mildew	mancozeb, FRAC M3 (various formulations) + sulfur (various formulations), FRAC M2	see label see label	24 24	66 0	The main target in the new shoot protection is Phomopsis. A powdery mildew fungicide is generally not needed in the first spray (1-inch shoot growth) unless the disease has been a problem in previous years. Sulfur will control powdery mildew and is a very economical option. Avoid sulfur on sulfur sensitive varieties. Some sulfur injury may occur on sulfur-tolerant varieties if the temperature exceeds 85 degrees F. The activity of sulfur is reduced at temperatures less than 65 degrees F. Do not mix sulfur and oil since it may result in injury.
Prebloom Phomopsis, Black rot, Powdery mildew, Downy mildew	mancozeb (various formulations) FRAC M3 — PLUS — sulfur (various brands), FRAC M2 — PLUS (if conditions are favorable for powdery mildew development) — quinoxyfen (Quintec 2SC) FRAC 43 or metrafenone (Vivando) 2.5SC), FRAC 48 or myclobutanil (Rally 40 WSP), FRAC 3 or tebuconazole (Elite 45DF), FRAC 3 or triflumizole (Procure 480SC), FRAC 3 or tetraconazole (Mettle 125ME), FRAC 3 flutriafol (Rhyme), FRAC 3	see label see label 4 to 6.6 fl oz 10.3 to 15.4 fl oz 3 to 5 oz 4 to 8 fl oz 3 to 5 fl oz 4 to 5 fl oz	24 24 12 12 24 12 12 12 (5 days for cane work)	66 0 21 14 14 7 14 7	Fungicide applications during the pre-bloom period are amongst the most important for powdery mildew, phomopsis, downy mildew, and black rot control. Where black rot is a problem, combine mancozeb with a sterol inhibiting fungicide (FRAC 3, aka DMI, SI fungicides). Captan is weak on black rot. Myclobutanil and tebuconazole are more active on black rot than triflumizole. To minimize the risk of resistance of the powdery mildew fungus to sterol inhibiting (SI fungicides, FRAC 3), limit use to 2 to 3 applications per season, use the maximum labeled rate and combine with sulfur. Note: quinoxyfen and metrafenone will work only against powdery mildew. Use lower rate of quinoxyfen with 14 day spray interval. FRAC 3 (DMI) fungicides do not have activity against downy mildew. Make sure to include a downy mildew fungicide in the tank mixture if applying a FRAC 3 during this application.

Table 6-6B. Bunch Grape Disease Management Program

When to Spray and Disease/Pest	Pesticide, Formulation, and FRAC Code	Amount of Formulation to Use per Acre	Reentry Interval (REI) (hours)	Preharvest Interval (PHI) (days)	Precautions and Remarks
Prebloom Phomopsis, Black rot, Powdery mildew, Downy mildew *(continued)*	or mancozeb (various formulations), FRAC M3 — PLUS —	see label	24	66	This section shows other options for management for Phomopsis, black rot, powdery mildew, and downy mildew. **Resistant isolates of the pathogens causing downy mildew and powdery mildew to the QoI fungicides (Abound, Flint, Sovran, or Pristine) are widespread in the mid-Atlantic grape growing region. Do not rely on them for downy mildew and powdery mildew control. To help minimize risk of resistance, tank mix QoI fungicides with sulfur (but not on sulfur-sensitive varieties).**
	sulfur (various brands), FRAC M2 — PLUS —	see label	24	0	
	azoxystrobin (Abound), FRAC 11	10 to 15.5 fl oz	4	14	Do not make more than 2 sequential applications of Flint, Sovran, Abound, Pristine, or Luna Experience before rotating to a non-QoI fungicide.
	or kresoxim-methyl (Sovran 50 WG), FRAC 11	3.2 to 6.4 oz	12	14	For both Pristine and Luna Experience, REI is 12 hours, but for cane work, REI is 5 days. See labels regarding phytotoxicity warnings for Pristine to Concord, Worden, Fredonia, or other *V. labrusca* or *V. labrusca* hybrids.
	or trifloxystrobin (Flint), FRAC 11	1.5 to 4 oz	12	14	
	or mandestrobin (Intuity), FRAC 11	6 fl oz	12	10	These combination materials (Revus Top and Luna Experience) can be used by themselves; however, it is better to tank mix with broad-spectrum materials such as mancozeb, captan, and sulfur in order to minimize the risk of fungicide resistance development in your field.
	or boscalid + pyraclostrobin (Pristine 38W), FRAC 7 + 11	8 to 12.5 oz	12 (5 days for cane work)	14	
	or azoxystrobin + Flutriafol FRAC 11 + 3 (Topguard EQ)	5 to 8 oz	12 (5 days for cane work)	14	**A fungicide with downy mildew activity must be added to Inspire Super in this spray.** Do not apply more than 80 fluid ounces/acre per season.
	or mandipropamid, + difenoconazole, FRAC 40 + 3 (Revus Top)	7.0 fl oz	12	14	Do not make more than 2 sequential applications of Luna Experience or any other Group 7 or Group 3 fungicide before rotating to a fungicide in another group.
	or difenoconazole + cyprodinil (Inspire Super) FRAC 3 + 9	16 to 20 fl oz	12	14	Although Aprovia and Rhyme contain the same FRAC group as Botrytis fungicides, it does not have a label for Botrytis due to fungicide resistance risks.
	or fluopyram + tebuconazole (Luna Experience) FRAC 7+3	6 to 8.6 fl oz	12 (5 days for cane work)	14	
	or benzovindiflupyr (Aprovia) FRAC 7	8.6 to 10.5 fl oz	12	21	
Prebloom Downy mildew specific materials (not listed above)	mefenoxam + mancozeb (Ridomil Gold MZ), FRAC 4 + M3	2.5 lb	48	66	Ridomil MZ contains mefenoxam + mancozeb. Gavel 75 DF contains zoxamide + mancozeb.
	or zoxamide + mancozeb(Gavel 75 DF), FRAC 22 + M3	2 to 2.5 lb	48	66	
	or mandipropamid (Revus), FRAC 40	8 fl oz	4	14	Revus products are very good protective materials for downy mildew, but they do not have any curative activity.
	or fenamidone (Reason 500SC), FRAC 11	2.7 fl oz	12	30	Do not add a crop oil to Revus if the Revus application is within 2 weeks of a sulfur or captan application.
	or cyazofamid (Ranman) FRAC 21	2.1 to 2.75 fl oz	12	30	
	or ametoctradin + dimethomorph (Zampro) FRAC 45+40 or	11 to 14 fl oz	12	14	Note: Use Ridomil product only when the environmental condition is strongly favoring downy mildew development. They have a very good kick-back activity; however, they are at high risk for resistance development. Tanos is *required* to be tank-mixed with a broad-spectrum fungicide such as captan or mancozeb. Make no more than 1 application of Tanos before rotating to a fungicide with a different mode of action (i.e. different FRAC group).
	Phosphorous acid (e.g., phosphonate) Phostrol, Agri-Fos, Prophyt, FRAC 33	Please see the label	4	0	These phosphorus acid-based products have pre- and post-symptom activity, providing approximately 7 days protectant activity. They all have a 0-day PHI. Do not exceed a 0.6% spray solution concentration of Prophyt. Use lower rate of Agri-Fos in 100 gallons water per acre early in season, and higher rates in 150 to 200 gallons of water per acre in late season and when the canopy is thick. Other phosphorous acid (aka phosphite) fungicides may be available. See label for correct rates.
Prebloom Pierce's Disease	Various insecticides (see table 6-6A Insect Management)	--	--	--	See Bunch Grape Insect Management Table 6-6A for control of insect vectors (leafhoppers/sharpshooters) that spread Pierce's Disease bacteria (*Xylella fastidiosa*).
Bloom Phomopsis, Black rot, Powdery mildew	See **Prebloom** recommendations				A bloom spray should be made if the time interval between the last prebloom spray and the postbloom spray is more than 10 days. Note: if you have historical issues with ripe rot or bitter rot, mix either captan, a QoI (FRAC=11), or mancozeb to your bloom application.

Table 6-6B. Bunch Grape Disease Management Program

When to Spray and Disease/Pest	Pesticide, Formulation, and FRAC Code	Amount of Formulation to Use per Acre	Reentry Interval (REI) (hours)	Preharvest Interval (PHI) (days)	Precautions and Remarks
Bloom Botrytis	iprodione (various formulations), FRAC 2 *or*	See label	48	7	A spray for botrytis during bloom may be beneficial in wet seasons and in vineyards with a botrytis problem. Elevate, Endura, iprodione, Inspire Super, Luna Experience, Vangard, etc. should be rotated through the season to avoid resistance development. (Note: make sure to rotate FRAC code. Botrytis fungus is known for developing fungicide resistance.) See product labels for complete information on resistance management and use restrictions. Rate of Vangard and Scala depends on whether it is applied alone or in tank mixture with another fungicide. See label for specifics. Endura, Kenia, Inspire Super, and Luna Experience will also control powdery mildew. **A fungicide with downy mildew activity must be added to Inspire Super in this spray.** Do not apply more than 80 fluid ounces/acre per season. See labels regarding phytotoxicity warnings for Concord, Worden, Fredonia, or other *V. labrusca* or *V. labrusca* hybrids.
	cyprodinil (Vangard 75 WG), FRAC 9 *or*	5 to 10 oz	12	7	
	fenhexamid (Elevate 50 WDG), FRAC 9 *or*	1 lb	12	0	
	pyrimethanil (Scala SC), FRAC 17 *or*	9 to 18 fl oz	12	7	
	boscalid (Endura 30W), FRAC 7 *or*	8 oz	12	14	
	isofetamid (Kenja 400SC), FRAC 7 *or*	20 to 22 fl oz	12	14	
	difenoconazole + cyprodinil (Inspire Super) FRAC 3 + 9 *or*	16 to 20 fl oz	12	14	
	fluopyram + tebuconazole (Luna Experience) FRAC 7+3 *or*	6 to 8.6 fl oz	12 (5 days for cane work)	14	
	boscalid + pyraclostrobin (Pristine WG), FRAC 7 + 11 *or*	18.5 to 23 oz	12 (5 days for cane work)	14	
	cyprodinil + fludioxonil (Switch 62.5 EG) FRAC 9+12	11 to 14 oz	12	7	
Postbloom (7 to 10 days after the Prebloom Spray) Phomopsis, Black rot, Powdery mildew, Downy mildew, Bitter rot, Ripe rot	mancozeb (various formulations) FRAC M3 + sulfur (various brands), FRAC M2	See label See label	24 24	66 0	During the postbloom period, berries are susceptible to black rot, powdery mildew, and downy mildew infection. Continue to apply mancozeb and sulfur as the backbone of your fungicide program and tank mix with an appropriate single-site chemistry. For specific downy mildew products refer back to the **prebloom section.** See labels regarding phytotoxicity warnings for Concord, Worden, Fredonia, or other *V. labrusca* or *V. labrusca* hybrids and also for sulfur-sensitive varieties.
	PLUS one of the following (FRAC 3)				
	myclobutanil (Rally 40WSP), FRAC 3 *or*	3 to 5 oz	12	14	
	tebuconazole (Elite 45DF), FRAC 3 *or*	4 oz	12	14	
	triflumizole (Procure 480SC), FRAC 3 *or*	4 to 8 fl oz	12	14	
	tetraconazole (Mettle 125ME), FRAC 3 *or*	3 to 5 fl oz	12	14	
	Flutriafol (Rhyme) FRAC 3 *or*	4 to 5 fl oz	12 (5 days for cane work)	14	
	mancozeb (various formulations) FRAC M3 + sulfur (various brands), FRAC M2	See label See label	24 24	66 0	
	PLUS one of the following (FRAC 11)				
	azoxystrobin (Abound), FRAC 11 *or*				
	kresoxim-methyl (Sovran 50WG) FRAC 11 *or*	10 to 15.5 fl oz	4	14	
	trifloxystrobin (Flint Fungicide), FRAC 11 *or*	3.2 to 6.4 oz	12	14	
	mandestrobin (Intuity), FRAC 11	1.5 to 4 oz	12	14	
	or	6 fl oz	12	10	
	PLUS one of the following (pre-mixed formulations)				
	azoxystrobin + flutriafol (Topguard EQ), FRAC 11 + 3 *or*	5 to 8 fl oz	12 (5 days for cane work)	14	
	boscalid + pyraclostrobin (Pristine WG), FRAC 7 + 11	8 to 12.5 oz	12 (5 days for cane work)	14	

Table 6-6B. Bunch Grape Disease Management Program

When to Spray and Disease/Pest	Pesticide, Formulation, and FRAC Code	Amount of Formulation to Use per Acre	Reentry Interval (REI) (hours)	Preharvest Interval (PHI) (days)	Precautions and Remarks
Postbloom Pierce's Disease	Various insecticides (see table 6-6A Insect Management)	--	--	--	See Bunch Grape Insect Management Table 6-6A for control of insect vectors (leafhoppers/sharpshooters) that spread Pierce's Disease bacteria (*Xylella fastidiosa*).
1st Cover Spray (7 to 10 days after Postbloom Spray) Phomopsis, Black rot, Powdery mildew, Downy mildew, Bitter rot, Ripe rot	See **Postbloom** recommendations				**Note:** There is a 66-day PHI for mancozeb. Use captan in place of mancozeb in the last half of the season.
1st Cover Spray (7 to 10 days after Postbloom Spray) Downy mildew	See **Postbloom** recommendations				
1st Cover Pierce's Disease	Various insecticides (see table 6-6A Insect Management)	--	--	--	See Bunch Grape Insect Management Table 6-6A for control of insect vectors (leafhoppers/sharpshooters) that spread Pierce's Disease bacteria (*Xylella fastidiosa*).
Closing Botrytis	See **Bloom recommendations**				At closing, add Elevate, Endura, Rovral, or Vangard to the appropriate cover spray for botrytis control. See Bloom Spray for information on resistance management when using Elevate, Endura, Rovral, and Vangard.
Closing Pierce's Disease	Various insecticides (see table 6-6A Insect Management)	--	--	--	See Bunch Grape Insect Management Table 6-6A for control of insect vectors (leafhoppers/sharpshooters) that spread Pierce's Disease bacteria (*Xylella fastidiosa*).
2nd and Subsequent Cover Sprays (10- to 14-day intervals until the Preharvest Spray) Ripe rot, Bitter rot	captan (various formulations), FRAC M4 + sulfur (various formulations), FRAC M2 + Additional fungicides for powdery mildew, downy mildew, or ripe rot control (see **post bloom** recommendations)	See label See label	See label 24	0 0	If additional sprays are needed for powdery mildew control, use sulfur. On sulfur-intolerant varieties and when temperatures exceed 85 degrees F, use an SI fungicide (Rally, Elite, Procure, Rubigan, or Vivando) in rotation with quinoxyfen (Quintec 2 SC) to keep resistance from developing.
2nd Cover Pierce's Disease	Various insecticides (see table 6-6A Insect Management)	--	--	--	See Bunch Grape Insect Management Table 6-6A for control of insect vectors (leafhoppers/sharpshooters) that spread Pierce's Disease bacteria (*Xylella fastidiosa*).
Veraison Botrytis	Same as **Bloom Spray**				Switch may reduce the severity of sour rot. Please make sure to rotate mode of action groups against Botrytis.
Preharvest (10 to 14 days before harvest) Ripe rot, Bitter rot, Botrytis	captan (various formulations), FRAC M4 PLUS one of the following iprodione (various formulations), FRAC 2 50 WP or 4 F or cyprodinil (Vangard 75 WG), FRAC 9 or fenhexamide (Elevate 50 WDG), FRAC 17 or pyrimethanil (Scala SC), FRAC 9 or boscalid (Endura 30W), FRAC 7 or cyprodinil + fludioxonil (Switch 62.5 WDG), FRAC 7 + 12 or azoxystrobin (Abound 2 SC), FRAC 11 or kresoxim-methyl (Sovran 50 WG), FRAC 11 or trifloxystrobin (Flint 50 WG), FRAC 11 or boscalid + pyraclostrobin (Pristine 38W), FRAC 7 + 11	See label See label 10 oz 1 lb 18 fl oz 8 oz 11 to 14 oz 11 to 15.4 fl oz 3.2 to 6.4 oz 3 oz 18.5 to 23 oz	See label 12 12 12 12 12 12 4 12 12 12	0 7 7 0 7 14 7 14 14 14 14	The REI for captan varies with trade name. See Pristine label for additional re-entry restrictions.

Table 6-6B. Bunch Grape Disease Management Program

When to Spray and Disease/Pest	Pesticide, Formulation, and FRAC Code	Amount of Formulation to Use per Acre	Reentry Interval (REI) (hours)	Preharvest Interval (PHI) (days)	Precautions and Remarks
Preharvest (10 to 14 days before harvest) Downy mildew	Prophyt, FRAC 33 or Phostrol, FRAC 33 or Agri-Fos, FRAC 33 or mandipropamid (Revus), FRAC 40	2.4 pt 2.5 to 5 pt 1.5 to 2.5 qt 8 fl oz	4 4 4 4	0 0 0 14	Phosphite fungicides are not very good protectants, but they have pre- and post-symptom activity. All have a 0-day PHI. Other phosphite fungicides may be available. Do not exceed a 0.6% spray solution concentration of Prophyt. Use higher rate of Agri-Fos in 150 to 200 gallons of water per acre late in the season when the canopy is thick. Other phosphorous acid (aka phosphite) fungicides may be available. Check label for correct rates.
Postharvest (14- to 21-day intervals from harvest until first killing frost) Downy mildew	copper compounds (various formulations), FRAC M1 or mancozeb, FRAC M3 75 DF, or 80 WP	See label 1.5 to 4 lb	See label 24		Premature defoliation may predispose vines to winter injury. Use shorter spray intervals when conditions are favorable for disease development. Copper may cause injury under cool, slow-drying conditions. Use mancozeb on copper sensitive varieties for downy mildew control. Use JMS Stylet Oil for powdery mildew control on sulfur sensitive varieties. Do not use captan, sulfur, or copper within 2 weeks of a JMS Stylet Oil application. Prophyt or Phostrol can also be used for downy mildew control.
Postharvest (14- to 21-day intervals from harvest until first killing frost) Powdery mildew	sulfur (various formulations), FRAC M2 or JMS Stylet Oil, FRAC NC	See label 1.5 to 2% 1 to 2 gallons	See label 4		

For further information, see www.smallfruits.org.

Muscadine Disease Management Program
W. O. Cline, Entomology and Plant Pathology

For effective disease control, commercial growers should apply fungicides every two weeks from mid-May through mid-July, beginning prior to the onset of disease symptoms. Fungicides are generally not necessary in home plantings.

The Fungicide Resistance Action Committee (FRAC) groups fungicides into mode of action (MOA) categories. These categories are listed following the pesticide and formulation names. To reduce the risk of resistance development, avoid successive applications of products with the same MOA for the same pest. Organically acceptable insecticides (OMRI listed) are indicated in Precautions and Remarks.

Table 6-7A. Muscadine Disease Management Program

TIMING / Pest(s)	Pesticide and Formulation	Amount of Formulation Per Acre	Restricted Entry Interval (REI)	Minimum Interval (Days) Between Application and Harvest; Preharvest Interval (PHI)	Precautions and Remarks
Shoots 6 to 10 inches					
Black rot, Bitter rot, Angular leaf spot, Powdery mildew	azoxystrobin, FRAC 11 (Abound 2.08 SC)	11 to 15.4 fl oz	4 hrs	14	Do not make more than 2 sequential applications of strobilurin fungicides (Abound, Flint, or Pristine) before alternating with nonstrobilurin fungicides (Captan, Nova, Rally, or Topsin M).
	myclobutanil, FRAC 3 (Nova, Rally 40 W)	3 to 5 oz	24 hrs	14	
	pyraclostrobin + boscalid, FRAC 11+7 (Pristine 38 W)	8 to 12.5 oz*	12 hrs/5 days	14	The REI for Pristine is 12 hours for all crop uses except cane tying, cane turning or cane girdling. These operations, not normally performed on muscadines, require a 5 day (5d) re-entry interval. *Recommended rates; higher rates up to 23 ounces per acre can be used when disease pressure is high.
	thiophanate-methyl, FRAC 1 (Topsin M70 WSB)	1 to 1.5 lb	48 hrs	7	
	trifloxystrobin, FRAC 11 (Flint 50 WG)	2 oz	12 hrs	14	Do not apply Flint fungicide to Concord grapes or injury may occur.
	EBDCs, FRAC M3 (Manzate Prostick, Penncozeb 75 DF, Dithane M45)	1.5 to 4 lb	24 hrs	66	Cannot be used within 66 days of harvest.
Powdery mildew only	wettable sulfur, FRAC M2 (Microthiol, other brands) 80 to 92% S	2 to 5 lb	24 hrs	1	Must be applied every 7 to 10 days. Dilute sulfur in 100 gallons of water per acre. Sulfur corrodes sprayers and trellis wires.
Bloom					
Black rot, Bitter rot, Angular leaf spot, Powdery mildew	Same as **Shoots 6 to 10 inches** recommendations				
Fruit rots, Sooty blotch	The bronze fresh-market cultivar 'Fry' is susceptible to sooty blotch.				
	azoxystrobin, FRAC 11 (Abound 2.08 SC)	11 to 15.4 fl oz	4 hrs	14	
	captan, FRAC M4 (Captan 50 WP) (Captec 4L)	2 to 4 lb	48 hrs	2 (re-entry)	Do not make more than 2 sequential applications of strobilurin fungicides (Abound, Flint, or Pristine) before alternating with nonstrobilurin fungicides (Captan, Nova, Rally, or Topsin M).
	captan, FRAC M4 (Captec 4L)	2 qt	48 hrs	2 (re-entry)	
	pyraclostrobin + boscalid, FRAC 11+7 (Pristine 38 W)	8 to 12.5 oz	12 hrs/5 days	14	The REI for Pristine is 12 hours for all crop uses except cane tying, cane turning or cane girdling. These operations, not normally performed on muscadines, require a 5 day (5d) re-entry interval.
	trifloxystrobin, FRAC 11 (Flint 50 WG)	2 oz	12 hrs	14	Do not apply Flint fungicide to Concord grapes or injury may occur.
Every 2 weeks until harvest					
Same as sprays for BLOOM					Tank mix Topsin M or Nova, Rally with Captan or Captec, OR alternate Topsin M or Nova, Rally with Abound, Flint, or Pristine.

Relative Effectiveness of Various Fungicides for Muscadine Grape Disease Control

Table 6-7B. Relative Effectiveness of Various Fungicides for Muscadine Grape Disease Control
(— = ineffective or injurious; +++++ = very effective or very safe)

Fungicide	Angular Leafspot	Bitter Rot	Powdery Mildew	Ripe Rot	Macrophoma Rot	Black Rot	Plant Safety
azoxystrobin (Abound)	++++	+++	+++	++++	++++	++++	+++++
captan (Captan, Captec)	+++	+	++	+++	+++	+++	+++++
myclobutanil (Nova, Rally) 40 W	++	++	++++	–	+	++++	+++++
pyraclostrobin + boscalid (Pristine)	++++	+++	+++	++++	++++	++++	+++++
EBDCs (Manzate, Penncozeb, Dithane, others)	+++	+++	++	+++	+++	+++	++++
sulfur (various)	–	–	++++	–	–	–	+++
thiophanate-methyl (Topsin M)	+++	++	+++	–	+	+++	+++++
trifloxystrobin (Flint)	+++	+++	++++	++++	++++	++++	+++++

Muscadine Insect Management

H. J. Burrack, Entomology and Plant Pathology

Insect management differs from disease management because some insect injury can be tolerated before economic damage occurs. Apply insecticides only if potentially damaging populations are present. Sampling techniques and tools are described when available. The Insecticide Resistance Action Committee (IRAC) groups insecticides into mode of action (MOA) categories. These categories are listed following the pesticide and formulation names. To reduce the risk of resistance development, avoid successive applications of products with the same MOA for the same pest. Organically acceptable insecticides (OMRI listed) are indicated in Precautions and Remarks.

Many insecticide active ingredients come in several formulations, or generic versions of the same formulation. In general, information is provided for the most commonly used formulations of active ingredients available in multiple formulations. Carefully check the label of the product you plan to use in the event that it differs from those listed. The label is the law!

Table 6-7C. Muscadine Insect Management

Pest	Pesticide and Formulation	Amount of Formulation Per Acre	Restricted Entry Interval (REI), hours, unless otherwise noted	Minimum Interval Between Application and Harvest; Preharvest interval (PHI), days	Precautions and Remarks
Aphids	Aphids are not common pests in North Carolina muscadines and are typically only problematic in spring on new growth. Aphid populations in late summer do not typically justify treatment. Treatment is only justified when sooty mold is present or new growth is deformed.				
	acetamiprid, IRAC 4A (Assail 30 SG)	2.5 oz	12		
	imidacloprid, IRAC 4A (Admire Pro)	1 to 1.4 fl oz	12	0	There are many formulations of imidacloprid. Read labels carefully for rate information.
	fenpropathrin, IRAC 3 (Danitol 2.4) EC	10.66 to 21.33 fl oz	24	21	Do not exceed 2.66 pints of Danitol per acre per season. Make no more than 2 applications of Danitol per season.
Climbing Cutworms	Scout for cutworm if damaged buds are observed. Look for cutworms at night. Cutworm treatment may be justified if greater than 4% of the buds examined are damaged and the variety does not have fruitful secondary buds. Spray in the evening if possible as cutworms are active at night. Only treat if cutworms are present.				
	Bacillus thuringiensis (Bt), IRAC 11 (many formulations)	rates vary	4	0	Many Bt formulations are OMRI listed.
	carbaryl, IRAC 1A (Sevin XLR Plus)	1 to 2 qt	12	7	
	chlorantraniliprole IRAC 28 (Altacor)	2 to 4.5 oz	4	14	Use between 100 to 200 gallons per acre total spray volume.
	cyfluthrin, IRAC 3A (Baythroid 2 EC)	2.4 to 3.2 fl oz	12	3	
	fenpropathrin, IRAC 3 (Danitol 2.4 EC)	10.66 to 21.33 fl oz	24	21	Do not exceed 2.66 pints of Danitol per acre per season. Make no more than 2 applications of Danitol per season.
	methoxyfenozide, IRAC 18 (Intrepid 2F)	12 to 16 fl oz	4	30	Minimum application for airblast sprayers of 40 gpa.
	rynaxypyr, IRAC 28 (Altacor)	3 to 4.5 oz	4	14	Use between 100-200 gallons per acre total spray volume.
	spinosad, IRAC 5 (Entrust 80 WP) (Entrust 2SC)	1.25 to 2.5 oz 4 to 8 fl oz	4	7	Do not apply more than 29 fl oz Entrust SC or 9 oz of Entrust 80W (0.45 lb spinosad) per acre per season. Entrust is **OMRI** listed.
	spinetoram, IRAC 5 (Delegate)	3 to 5 oz	4	7	Do not exceed 5 applications of Delegate per year or 19.5 ounces per acre per crop year.
Grape Berry Moth	Grape berry moth is present in NC and can occasionally damage muscadine grapes, but it is not uniformly distributed in the state. If grape berry moth presence is suspected, observe flowers and fruit for injury and consider monitoring moths with pheromone baited traps.				
	bifenthrin, IRAC 3 (Brigade 2EC)	3.2 to 6.4 fl oz	12	30	

Table 6-7C. Muscadine Insect Management

Pest	Pesticide and Formulation	Amount of Formulation Per Acre	Restricted Entry Interval (REI), hours, unless otherwise noted	Minimum Interval Between Application and Harvest; Preharvest interval (PHI), days	Precautions and Remarks
Grape Berry Moth (continued)	bifenthrin + imidacloprid, IRAC 3 + 4A (Brigadier)	3.8 to 6.4 fl oz	12	30	
	clothianidin, IRAC 4A (Clutch 50 WDG)	3 oz	12	0	
	indoxacarb IRAC 22 (Avaunt)	5 to 6 oz	12	7	
	methoxyfenozide, IRAC 18 (Intrepid 2F)	12 to 16 fl oz	4	30	
	spinosad, IRAC 5 (Entrust 80 WP) (Entrust 2SC)	1.25 to 2.5 oz 4 to 8 fl oz	4	7	Entrust is **OMRI** listed.
	phosmet, IRAC 1B (Imidan 70 WP)	1.33 to 2.125 lb	14 days	7 (rates of 1.33 lb per acre or less) 14 (more than 1.33 lb per acre)	Do not apply more than 6.5 pounds Imidan per acre per year.
	methomyl, IRAC 1A (Lannate SP)	0.5 to 1 lb	7 days	1, fresh market 14, wine	
	fenpropathrin, IRAC 3 (Danitol 2.4 EC)	10.66 to 21.33 fl oz	24	21	Do not exceed 2.66 pints of Danitol per acre per season. Make no more than 2 applications of Danitol per season.
	pyriproxyfen IRAC 7C (Esteem 0.83EC)	16 fl oz	12	21	
	rynaxypyr, IRAC 28 (Altacor)	3 to 4.5 fl oz	4	14	Use between 100 to 200 gallons per acre total spray volume.
Grape Flea Beetle	Grape flea beetle larvae feed on developing buds during bud swell. If greater than 4% of buds observed are damaged by grape flea beetles, treatment may be justified. Apply only if damaging numbers of adult beetles are present.				
	bifenthrin + imidacloprid, IRAC 3 + 4A (Brigadier)	3.8 to 6.4 fl oz	12	30	
	carbaryl, IRAC 1A (Sevin XLR Plus)	2 qt	12	7	
	fenpropathrin, IRAC 3 (Danitol 2.4 EC)	10.66 to 21.33 fl oz	24	21	Do not exceed 2.66 pints of Danitol per acre per season. Make no more than 2 applications of Danitol per season.
	phosmet, IRAC 1B (Imidan 70 WP)	1.33 to 2.125 lb	14 days	7 (rates of 1.33 lb per acre or less) 14 (more than 1.33 lb per acre)	Do not apply more than 6.5 pounds Imidan per acre per year.
Grape rootworm, Southern grape rootworm	Grape rootworm larvae feed on roots. Adults are small, black weevils that make distinctive chain-like feeding markings on leaves. Foliar feeding does not result in yield reduction, but root feeding may reduce plant vigor over time. Treatments should be timed to adult activity, which typically peaks in June or July. Grape rootworms are sporadic pests in North Carolina and should not be treated preventatively.				
	carbaryl, IRAC 1A (Sevin XLR Plus)	2 qt	12	7	
Leafhoppers, Sharpshooters	Leafhoppers are important vectors of Pierce's Disease in *Vinifera* grapes, but Pierce's Disease is not a common problem of muscadine grapes. Var. Carlos has been observed with Pierce's Disease symptoms, but the disease does not appear to persist in plants over-winter. Therefore, leafhoppers should not be preventatively treated in muscadines. Large leafhopper populations can cause leaf stippling and yellowing, and populations of this size may result in economic damage and justify treatment.				
	acetamiprid, IRAC 4A (Assail 30 SG)	2.5 oz	12		
	bifenthrin, IRAC 3 (Brigade 2EC)	3.2 to 6.4 fl oz	12	30	
	bifenthrin + imidacloprid, IRAC 3 + 4A (Brigadier)	3.8 to 6.4 fl oz	12	30	
	dinotefuran, IRAC 4A (Venom)	1 to 3 oz	12	28	Venom may be applied as a foliar spray at 1 to 3 ounces or to the soil at 5 to 6 ounces. See label for details.
	clothianidin IRAC 4A (Clutch 50 WDG)	6 oz	12	30	Clutch is applied to the soil either via drip or trickle irrigation.
	fenpropathrin, IRAC 3 (Danitol 2.4 EC)	5.3 to 10.6 fl oz	24	21	Do not exceed 2.66 pints of Danitol per acre per season. Make no more than 2 applications of Danitol per season.
	imidacloprid, IRAC 4A (Admire Pro)	1 to 1.4 fl oz (foliar) 7 to 14 oz (soil)	12	0 30	Admire Pro is applied to as a foliar spray and as a soil treatment See label for application details.
	malathion, IRAC 1B (57 EC or Malathion 5)	3 pt	12	3	Malathion may cause injury to berries if applied after bloom. Rates are based on 200 gpa spray volumes.
	phosmet, IRAC 1B (Imidan 70 W)	1.33 to 2.125 lb	14 days	14	Do not apply more than 6.5 pounds Imidan per acre per year.
	thiamethoxam IRAC 4A (Actara)	1.5 to 3.4 oz	12	5	

Chapter VI—2019 N.C. Agricultural Chemicals Manual

Table 6-7C. Muscadine Insect Management

Pest	Pesticide and Formulation	Amount of Formulation Per Acre	Restricted Entry Interval (REI), hours, unless otherwise noted	Minimum Interval Between Application and Harvest; Preharvest interval (PHI), days	Precautions and Remarks
Japanese Beetle, Green June Beetle	Foliar feeding by Japanese beetles in established vineyards does not justify treatment unless it occurs on leaves below the top trellis wire. Fruit feeding by Japanese beetles is rare. Green June beetles are attracted to damaged, decomposing fruit but may also feed on undamaged fruit once in the vineyard. Removal of damaged fruit that will not be harvested will minimize Green June beetle populations and should be conducted before considering pesticide application.				
	acetamiprid, IRAC 4A (Assail 30 SG)	2.5 oz			Southeastern data for Assail on Japanese beetles are limited.
	bifenthrin, IRAC 3 (Brigade 2EC)	3.2 to 6.4 fl oz	12	30	
	bifenthrin + imidacloprid, IRAC 3 + 4A (Brigadier)	3.8 to 6.4 fl oz	12	30	
	carbaryl, IRAC 1A (Sevin XLR Plus)	2 qt	12	7	
	clothianidin, IRAC 4A (Clutch 50 WDG)	3 oz	12	0	
	fenpropathrin, IRAC 3 (Danitol 2.4 EC)	10.66 to 21.33 fl oz	24	21	Do not exceed 2.66 pints of Danitol per acre per season. Make no more than 2 applications of Danitol per season.
	imidacloprid, IRAC 4A (Admire Pro)	1 to 1.4 fl oz (foliar)	12	0	Admire Pro is applied to as a foliar spray and as a soil treatment. Soil treatments are not recommended for Japanese and green June beetles.
	indoxacarb IRAC 22 (Avaunt)	5 to 6 oz	12	7	
	Kaolin IRAC NA (Surround)	25 to 50 lb	4	0	Surround creates a barrier on plants and reduces insect attraction and feeding. In order to be effective, leaf and/or fruit surfaces should be coated thoroughly. Material should be reapplied at least 7 days apart to maintain coverage. **Not recommended** for use in fresh market grapes after bloom as white residue can remain until harvest.
	phosmet, IRAC 1B (Imidan 70 W)	1.5 to 3.4 oz	14 days	7 (rates of 1.33 lb per acre or less) 14 (more than 1.33 lb per acre)	Do not apply more than 6.5 pounds Imidan per acre per year.
Scale insects	Scale insects are occasional pests of muscadine grapes				
	buprofezin, IRAC 16 (Applaud 70DF)	9 to 12 oz	12	7	Apply when crawlers are observed.
	spirotetramat, IRAC 23 (Movento 2SC)	6 to 8 fl oz	24	7	
Spider Mites	Sample for mites using a minimum 10x hand lens. There is no clearly defined threshold for mites in muscadine grapes. Treatment for *Vinifera* grapes is recommended when greater than 50% of leaves are infested. Fast moving predatory mites can be distinguished from slower moving spider mites through direct observation. Some insecticides, such as carbaryl, can flare mite populations, and care should be used with these materials when mites are present. Rotate acaricides between MOAs to minimize selection for resistance.				
	abamectin, IRAC 6 (Agri-Mek 0.15EC) (many other formulations)	16 fl oz	12	28	Abamectin is a restricted use product. Do not reapply within 21 days of initial application. Abamectin is an EC (emulsifiable concentrate), which can cause phytotoxicity in some crops. Check for possible plant injury before treating an entire field.
	bifenazate, IRAC Unknown (Acramite 50 WS)	1 lb	12	14	The reentry interval is 5 days for can turning, tying, and girdling. Apply in a minimum spray volume of 50 gallons per acre.
	cyflumetofen, IRAC 25 (Nealta)	13.7 fl oz	12	14	Do not make more than 2 Nealta applications per season and rotate to another mode of action (IRAC Code) between treatments.
	dicofol, IRAC Unknown (Difocol 4EC) (Kelthane 50WSP)	2.5 pt 2.5 lb	12	7	Use Kelthane at 1 pound per acre on small vines. Do not make more than 2 applications per season.
	etoxazole, IRAC 10B (Zeal)	3 oz	12	28	Zeal is a growth regulator and kills eggs and young mites. It is most effective if applied when mite populations are low.
	fenbutatin-oxide, IRAC 12B (Vendex 50WP)	2.5 lb	48	28	Do not make more than 2 applications of Vendex per season.
	fenpyroximate, IRAC 21A (FujiMite 5EC)	2 pt	12	14	Do not apply more than 2 pints of FujiMite per acre per season.
	horticultural oils, IRAC Unknown many materials, including (Saf T Side) (Glacial Spray Fluid)	1 to 2% by volume	4	0	Some oils are **OMRI** listed; check label. **Do not** use in combination with or immediately before or after spraying with fungicides such as Captan or any product containing sulfur. **Do not** use with carbaryl or dimethoate. **Do not** use with any product whose label recommends the use of no oils. Do not use in combination with NPK foliar fertilizer applications.
	pyridiben IRAC 21 (Nextor 75 WSB)	8.8 to 10.67 oz	12	7	The maximum amount of pyridiben allowed per acre per season is 26.4 ounces. Do not make more than 2 applications of pyridiben per season.
	spirodiclofen, IRAC 23 (Envidor 2SC)	18 fl oz	12	14	

Table 6-7C. Muscadine Insect Management

Pest	Pesticide and Formulation	Amount of Formulation Per Acre	Restricted Entry Interval (REI), hours, unless otherwise noted	Minimum Interval Between Application and Harvest; Preharvest interval (PHI), days	Precautions and Remarks
Stink bugs	Stink bugs are not a common pest of muscadine grapes, and there is no evidence that they directly damage fruit. If present at harvest, they may contaminate fruit, and under this scenario, they may justify treatment.				
	fenpropathrin, IRAC 3 (Danitol 2.4 EC)	10.66 to 21.33 fl oz	24	21	Do not exceed 2.66 pints of Danitol per acre per season. Make no more than 2 applications of Danitol per season. Fenpropathrin (and other Group 3 materials
	phosmet, IRAC 1B (Imidan 70 W)	1.33 to 2.125 lb	14 days	14	Do not apply more than 6.5 pounds Imidan per acre per year.
Grape Root Borer	Grape root borer is potentially the most significant pest of grapes in North Carolina, but they are not necessarily present in all vineyards. Grape root borer moths should be monitored with pheromone baited traps. If moths are confirmed within a vineyard, mating disruption is the most effective control tool.				
	mating disruption (Isomate GRB)	100 dispensers	NA	NA	Dispensers should be placed prior to the beginning of grape root borer moth flight activity and be left in the vineyard until the end of flight activity. Moth flight timing varies between vineyards, but can be as early as July and last until October. Pheromone baited traps can help determine grape root borer populations and flight activity, but traps will not be effective if mating disruption is underway.
	Cultivation or soil mounding	NA	NA	NA	Use clean cultivation, mound soil (July 1 or at first moth emergence when using pheromone traps) or using tightly-sealed plastic mulch 3 feet from the base of vines. This practice will inhibit adult emergence from the soil when well timed. Mounded soil needs to be removed by Sept. 1.
	chlorpyrifos, IRAC 1A (Lorsban Advanced)	4.5 pt per 100 gal	24	35	**Chlorpyrifos registration is under federal review at the time of this writing. Check registration status before using.** Apply 2 quarts of mixture to soil at the base of each vine. Make a single application 35 days before harvest. Spray should not contact fruit or foliage. Application can be made with flood nozzles and low pressure (40 to 60 psi).
Grape tumid gallmaker	Grape tumid gall maker adults are small flies. Their larvae, or maggots, infest clusters, leaves, or stems and cause the plant to create galls. Galls can be quite large and are often reddish. Infestations usually localized and do not typically require chemical treatment.				
	Hand removal	NA	NA	NA	Removal of affected plant parts is generally sufficient to prevent further infestation.
	spirotetramat, IRAC 23 (Movento 2SC)	6 to 8 fl oz	24	7	
Red Imported Fire Ant	Bait treatments can effectively manage fire ants, but they typically take 2 to 4 weeks to reach full efficacy. Baits must be applied when ants are actively foraging. Test for foraging by placing food near the nest. Check for ant activity after 30 minutes.				
	pyriproxyfen IRAC 7C (Esteem Ant Bait)	1.5 to 2 lb	12	1	Do not exceed 0.22 pound of active ingredient per season.
	methoprene MOA 7A (Extinguish Professional Fire Ant Bait)	1 to 1.5 lb	4	0	

Further Information

Muscadine Grape Diseases and Their Control, Plant Pathology Information Note 145, https://www.ces.ncsu.edu/depts/pp/notes/Fruit/fdin012/fdin012.htm
Southeast Regional Muscadine Integrated Management Guide, www.smallfruits.org

Chapter VI—2019 N.C. Agricultural Chemicals Manual

Peach and Nectarine Spray Guide

D. F. Ritchie and J. Walgenbach, Entomology and Plant Pathology

Although many pesticides are registered for disease and insect control on peaches, the following spray program lists the ones that have performed well under North Carolina conditions. The rates of pesticides recommended should give control when pest pressure is moderate to severe, assuming they are applied correctly. Where the rate is given as a range, the lower rate can be used when pest pressure is low; the higher rate should be used when pest pressure is great. Thus, the following spray program is intended to be only a guide since pest and orchard conditions can vary from orchard to orchard and year to year.

The rates given are based on the use of rate per acre; 75 to 125 gallons of water per acre provides optimal spray coverage for pest/disease control in most orchards.

Note: For imported fire ant, treat active mounds off season with directed bait formulations like Clinch, Esteem, Extinguish, and Logic. Insect growth regulators will give complete control after 30 days. Always follow label directions for best results.

Table 6-8. Peach and Nectarine Spray Guide

When to Spray	Pest	Pesticide	REI (hrs)	PHI (days)	Formulation Per Acre	Remarks
Dormant Before buds swell in late winter	Leaf curl	**Fungicide:** chlorothalonil (Bravo Weather Stik) 6 F OR ziram (Ziram) 76 DF	12 48	NA 14	4 pt 5 lb	Other chemicals registered for leaf curl include copper-containing compounds (consult labels). Copper provides adequate leaf-curl control when at least 4 pounds/acre is applied before bud-swell. To control white peach scale, two dormant oil sprays 2 weeks apart are necessary. Oil will NOT control leaf curl.
	Scale insect, mite	**Insecticide:** oil, superior-type + chlorpyrifos (Lorsban) 4E OR pyriproxyfen (Esteem) 35WP	4 96 12	NA NA 14	4 gal 2 pt 4 to 5 oz	Addition of an insecticide with oil will improve control of scales. Delaying applications when crawlers appear approximately 6 wks after bloom is also an option. NOTE: Heavy reliance on pyrethroid insecticides may flare scale populations, particularly San Jose scale.
Bloom	Brown rot, blossom blight	**Fungicide:** captan (Captan, Captec) 50 WP, 4L OR chlorothalonil (Bravo Weather Stik, Echo 720) 6 F OR thiophanate-methyl (Topsin M, T-Methyl) + captan (Captan, Captec) 50WP, 4L OR cyprodinil (Vangard) 75 WG	24-96 12 12 12 12	0 NA 1 0 0	5 lb, 2.5 qt 3.125 pt 1.0 lb + 4 lb, 2 qt 5 oz	Fungicide sprays at full pink to early bloom and again at full bloom may reduce blossom blight, but another spray may be needed if bloom extends beyond 2 weeks. Demethylation inhibiting (DMI) fungicides (Elite, Indar, Nova, Orbit, Quash) are effective against blossom blight but are prone to resistance problems if used regularly. Resistance to any one of the DMI fungicides results in cross-resistance to the others. **It is recommended that DMI fungicides be saved for preharvest sprays and that they not be used in bloom and cover sprays.** Do not use more than 1 application of thiophanate-methyl or if resistant strains are present. Vangard is another alternative to DMI fungicides during bloom.
		Insecticide: None				
Petal-Fall After petals are off but before fruit are showing	Scab, brown rot	**Fungicide:** captan (Captan, Captec) 50 WP, 4L OR sulfur	12 24	0 0	5 lb, 2.5 qt 9 lb actual sulfur	Including a fungicide at petal fall may enhance scab control.
	Plum curculio, catfacing insects, Oriental fruit moth	**Insecticide:** Pyrethroid (see list in Table 6-9) OR phosmet (Imidan) 50 WP OR thiamethoxam (Actara) 25 WDG OR indoxacarb (Avaunt) 30WG	12-24 24 12 12	3-14 14 14 14	— 3 lb 5 oz 5 oz	For the list of pyrethroids registered see IRAC MOA Group 3A in Table 6-9. Imidan has a 4-day re-entry interval, 14 days for general public (i.e., pick-your-own customers). Do not apply more than 11 ounces of Actara per acre per season. Avaunt is not recommended for catfacing insects (plant bugs and stink bugs), plum curculio and oriental fruit moth only.
Shuck Split to Shuck Fall After fruit are showing, but before 75% of the fruit have shucks off	Scab, brown rot	**Fungicide:** captan (Captan, Captec) 50 WP, 4 L OR chlorothalonil (Bravo Weather Stik, Echo 720) 6 F OR sulfur	12 12 24	0 NA 0	5 lb, 2.5 qt 4 pt 9 lb actual sulfur	Very critical period for start of scab control. Tank-mix of thiophanate-methyl 0.75 pound a.i./acre (Topsin M, T-Methyl) with captan or sulfur first 2 sprays enhances scab control. **Chlorothalonil cannot be used later than shuck split.**
	Plum curculio, catfacing insect	**Insecticide:** Same as in Petal-Fall				When cool weather delays shuckoff, a second application of insecticide in 7 to 10 days may be necessary to control catfacing insects.

Table 6-8. Peach and Nectarine Spray Guide

When to Spray	Pest	Pesticide	REI (hrs)	PHI (days)	Formulation Per Acre	Remarks
Cover Sprays Begin 7 to 10 days after shuck fall, continue 10 to 14 days, stopping at least 2 weeks before harvest	Scab, brown rot	**Fungicide:** Same as SHUCK SPLIT				First through third cover sprays are very important for scab control on peach. Chlorothalonil used at shuck split can give 3 weeks of scab control. NOTE: **Chlorothalonil cannot be used after the shuck split spray.**
	Plum curculio, Stink bugs, Oriental fruit moth	**Insecticide:** Pyrethroid (see list in Table 6-9) OR	12-24	3-14	—	Be sure to match the insecticide with the insect pest present in the orchard at the time of application.
		indoxacarb (Avaunt) 30 WG OR	12	14	6 fl oz	Do NOT apply esfenvalerate or permethrin within 2 weeks of harvest.
		phosmet (Imidan) 50 WP OR	24	14	3 lb	Do NOT make more than four applications of Avaunt per season.
		spinetoram (Delegate) 25 WDG OR	12	7	5 fl oz	Do NOT apply phosmet within 3 weeks of harvest.
		chlorantraniliprole (Altacor) 35 WDG OR	4	10	3 oz	Delegate and Altacor are primarily for oriental fruit moth control.
		cyantraniliprole (Exirel) 0.83SE	12	3	17 oz	
	San Jose Scale	buprofezin (Centaur) 70WDG OR	12	14	25 oz	In most orchards treated with oil and a scale-active insecticide before bloom, additional control measures are usually not necessary. However, in orchards with a history of San Jose scale, targeting first generation crawlers at the second cover spray will help to minimize the potential for damage.
		pyriproxyfen (Esteem) 35WP	12	14	4 to 5 oz	
	Brown marmorated stink bug	pyrethroid (see list in Table 6-9)	12-24	3-14	—	Peaches are highly attractive to BMSB. In the mountains and piedmont areas, where BMSB populations are highest, control may be required from May through harvest.
Preharvest Begin 2 to 3 weeks before harvest; apply fungicides at 7 to 10-day intervals. In periods of high disease pressure, closer spray intervals may be necessary.	Brown rot	**Fungicide:** azoxystrobin (Abound) 2.08 F OR	4	0	12 to 15 fl oz	NOTE: Check product label for any preharvest interval (PHI) or reentry interval (REI) times and other restrictions. azoxystrobin (PHI = 0 day, RE = 4 hr) azoxystrobin + difenoconazole (PHI = 0 day, REI = 12 hr) difenoconazole + cyrodinil (PHI = 2 days, REI = 12 hr) fenbuconazole (PHI = 0 day, REI = 12 hr) trifloxystrobin + fluropyram (PHI = 1 day, REI = 12 hr) metconazole (PHI = 14 days, REI = 12 hr) penthiopyrad (PHI = 0 day, REI = 12 hr) propiconazole (PHI = 0 day, REI = 24 hr) pyraclostrobin + boscalid (PHI = 0 day, REI = 12 hr) pyraclostrobin + fluxapyroxad (PHI = 0 day, REI = 12 hr) tebuconazole (PHI = 0 day, REI = 12 hr) Preharvest use of propiconazole, azoxystrobin + difenoconazole, and difenoconazole + cyrodinil, is limited to 2 applications. Do not make more than 2 sequential applications of these fungicides before alternating with a fungicide having a different mode of action.
		azoxystrobin + difenoconazole (Quadris Top) 2.71SC OR	12	0	12 to 14 fl oz	
		difenoconazole + cyrodinil (Inspire Super) 2.82EW OR	12	2	16 to 20 fl oz	
		fenbuconazole (Indar) 75 WSP OR	12	0	2 oz pouch	
		trifloxystrobin + fluropyram (Luna Sensation) 4.2SC OR	12	1	5.0 to 7.6 fl oz	
		metconazole (Quash) 50 WDG OR	12	14	3.5 to 4.0 oz	
		penthiopyrad (Fontelis) 1.67SC OR	12	0	14 to 20 fl oz	
		propiconazole (Bumper, Orbit, PropiMax) 3.6 EC OR	24	0	4 fl oz	
		pyraclostrobin + boscalid (Pristine) 38 WG OR	12	0	10.5 to 14.5 oz	
		pyraclostrobin + fluxapyroxad (Merivon) 500 SC OR	12	0	4 to 6.7 fl oz	
		tebuconazole (Elite, Orius, Tebuzol) 45 DF OR	12	0	4 to 8 oz	
		tebuconazole + trifloxystrobin (Adament) 50 WG	24	1	4 to 6 oz	
	June beetle, Japanese beetle	**Insecticide:** carbaryl (Sevin) 80 W3P OR	12	3	2.5 lb	Admire has a 0-day preharvest interval, but a 12-hour re-entry interval.
		imidacloprid (Admire Pro) 4.6SC OR	12	0	1.4 fl oz	
		acetamiprid (Assail) 30SG	12	7	5.3 to 7 oz	
Borer Spray	Peachtree borer	**Insecticide:** chlorpyrifos (Lorsban) 4 EC OR	96	NA	1 qt/100 gal	Tree trunks and limbs should be sprayed to drip, after harvest or after August 1, whichever comes last. Best control results when applied the week of Sept. 1.
		pyrethroid (see Table 6-9)	12-24	3-14		
Special Spray	Spider mite	**Miticide:**			When mite populations increase to large numbers they may cause severe injury. Examine outer leaves for mites and mottled appearance, especially during hot, dry periods.	
		Preventive Spray: hexythiazox (Savey) 50 DF OR	12	28	3 to 6 oz	Apollo and Savey are ovicides and should be applied early in the season. Apollo and Savey have the same mode of action, so if resistance develops to one compound, populations will be resistant to both products. Do not apply more than once per year, and preferably use once every other year.
		clofentezine (Apollo) SC	12	21	4 to 8 oz	
		Curative Mite Spray: bifenazate (Acramite) 50 WP	12	3	1 lb	Do not apply within 3 days of harvest.
		abamectin (Agri-Mek) 0.7SC	12	21	2.25 to 4.25 fl oz	Include 0.25% horticultural oil (not dormant oil) or a nonionic surfactant with Agri-Mek. 21 day PHI.

Relative Effectiveness and Safety of Various Insecticides for Peach Insects

J. F. Walgenbach, Entomology and Plant Pathology

(E – excellent; G – good; F – fair; NC – no control or no data)

Table 6-9. Relative Effectiveness and Safety of Various Insecticides for Peach Insects

IRAC† MOA Group	Insecticide Formulation and Rate per 100 Gallons Water	Days Between Last Spray and Harvest	Plum Curculio	Oriental Fruit Moth	Peachtree Borer	Catfacing insects (stink bugs)	Scales (White Peach, San Jose)	Beetles (June, Japanese)	Safety*
1A	methomyl (Lannate 2.4 L) 1 pt	4	F	G	NC	F	NC	F	Danger, Poison
	carbaryl (Sevin 80 SP) 1.25 lb	3	F	F	NC	NC	NC	E	Caution
1B	phosmet (Imidan 50 WP) 1.5 lb	14	E	G	NC	NC	NC	F	Warning
	chlorpyrifos (Lorsban 4.0 EC) 3 qt	Prebloom and postharvest only	NC	NC	E	NC	E	NC	Danger, Restricted
3A	beta-cyfluthrin (Baythroid 1EC) 2.4 oz	7	G	E	F	G	NC	F	Warning, Restricted
	cyfluthrin (Tombstone 2EC) 2.4 oz	7	G	E	F	E	NC		Danger, Restricted
	esfenvalerate (Asana 0.66 EC) 5.8 oz	14	G	E	F	NC	NC	G	Warning, Restricted
	fenpropathin (Danitol 2.4 EC) 16 oz	3	G	E	F	E	NC	E	Warning, Restricted
	gamma-cyhalothrin (Proaxis 0.5EC) 3.8 oz	14	G	E	F	E	NC	E	Caution, Restricted
	lambda-cyhalothrin (Karate 2.08CS) 1.9 oz	14	G	E	F	E	NC	E	Warning, Restricted
	permethrin (Pounce 2.0 EC, 25 WP) 6 oz	7	G	E	F	G	NC	G	Warning, Restricted
	zeta-cypermethrin (Mustang Maxx)	14	G	E	F	E	NC	E	Warning, Restricted
4A	acetamiprid (Assail 30 SG) 7 oz	7	F	G	NC	F	F	G	Caution
	chlothianidin (Belay SL) 6 oz	21	G	NC	NC	G	NC	G	Caution
	dinotefuran (Scorpion 35SL) 5.25 oz (Venom 70SG) 4 oz	3	G	NC	NC	E	NC	G	Caution
	imidacloprid (Provado 1.6F) 3 oz	0	NC	NC	NC	F	NC	G	Caution
	thiamethoxam (Actara 25WDG) 2.5 oz	14	E	F	NC	E	F	G	Caution
16	buprofezin (Centaur 70WSB) 17 oz	14	NC	NC	NC	NC	E	NC	Caution
28	chlorantraniliprole (Altacor 35WDG) 2.5 oz	10	NC	E	F	NC	NC	NC	Caution
	cyantraniliprole (Exirel 0.83SE) 17 oz	3	F	E	F	NC	NC	F	Caution
22	indoxacarb (Avaunt 30 DG) 5 oz	14	E	G	NC	NC	NC	F	Caution
7C	pyriproxyfen (Esteem 35 WP) 5 oz	14	NC	F	NC	NC	E	NC	Caution
5	spinetoram (Delegate 25WDG) 2.5 oz	7	NC	E	NC	NC	NC	NC	Caution
23	spirotetramat (Movento 2SC) 8 oz	7	NC	NC	NC	NC	E	NC	Caution
NC	oil superior NC 2 gal	NC	NC	NC	NC	NC	E	NC	Caution

† Insecticide Resistance Action Committee (IRAC) mode of action (MOA) group.

* Relative Toxicity (Safety):

Danger = most toxic to man

Caution = least toxic to man

Restricted = restricted use compound; may be applied only by licensed pesticide operators

Relative Effectiveness of Chemicals for Disease Control on Peaches and Nectarines

D. F. Ritchie, Entomology and Plant Pathology

(E = excellent; G = good; F = Fair; P = poor; NC = no control; NA = not applicable; ND = no data)

Table 6-10. Relative Effectiveness of Chemicals for Disease Control on Peaches and Nectarines

Fungicide/Bactericide and Rate per Acre per 100 gallons	Mode of Action Code	Days Between Last Spray and Harvest	Reentry Interval (REI)* (hours)	Leaf Curl	Blossom Blight	Brown Rot	Peach Scab	Bacterial Spot
azoxystrobin (Abound) 2.08 F — 12 fl oz	11	0	4	NA	F-G	G	G	NC
azoxystrobin + difenoconazole (Quadris Top)2.71SC — 14 fl oz	11, 3	0	12	NA	G	G-E	G-E	NC
captan (Captan, Captec) 50 WP, 4L — 5lb, 2.5 qt	M4	0	24 to 96	NA	F	F-G	G	NC
chlorothalonil (Bravo Weather Stik, Echo 720) 6 F — 4 pt	M5	NA	12**	G	F-G	NA	G	NC
copper 3000; Cuprofix ULTRA 40D) — 4 to 8 lb ***	M1	at least 21	12 to 24	F-G	NA	NA	NA	G
cyprodinil (Vangard) 75 WG — 5 oz	9	NA	12	NA	G	NA	NA	NC
difenoconazole + cyrodinil (Inspire Super) 2.82EW — 20 fl oz	3, 9	2	12	NA	G	G-E	G-E	NC
fenbuconazole (Indar) 75 WSP — 2 oz	3	0	12	NA	G	G	P-F	NC
fenhexamid (Elevate) 50 WDG — 1.5 lb	17	0	12	NA	ND	F	P-F	NC
trifloxystrobin + fluropyram (Luna Sensation) 4.2SC 6.0 fl oz	11, 7	1	12	NA	G	G-E	G	NC
iprodione (Rovral) — 1.5 lb, 1.5 pt ****	2	NA	12	NA	G	NA	NC	NC
metconazole (Quash) 50WDG	3	14	12	NA	G	G	F	NC
myclobutanil (Rally)40 WP — 4 oz	3	0	24	NA	G	F	NC-P	NC
oxytetracycline (FireLine, Mycoshield) 17 WP — 0.75 lb	41	21	12	NA	NA	NA	NA	F-G
penthiopyrad (Fontelis) 1.67SC 20 fl oz	7	0	12	NA	ND	F-G	P-F	NC
propiconazole (Bumper, Orbit, PropiMax) 3.6 EC — 4 fl oz	3	0	24	NA	G	G	NC-P	NC
pyraclostrobin + boscalid (Pristine) 38 WG — 10.5 to 14.5 oz	11, 7	0	12	NA	G	G-E	F	NC
pyraclostrobin + fluxapyroxad (Merivon) 500SC — 6.7 fl oz	11, 7	0	12	NA	G	E	G	NC
pyrimethanil (Scala SC) 60 SC — 1 pt	9	2	12	NA	ND	P-F	ND	NC
Sulfur — 9 lb actual sulfur	M2	0	24	NA	P-F	P	F-G	NC
tebuconazole (Elite, Orius, Tebuzol) 45 DF, WP — 4 oz	3	0	12	NA	G	G	P	NC
tebuconazole + trifloxystrobin (Adament) 50WG	3, 11	1	24	NA	G	G	P-F	NC
thiophanate-methyl (Topsin M) 70 WP, 4.5FL — 1 lb, 1.5 pt + captan (Captan) 50 WP — 4 lb	1, M4	1	12 24 to 96	NA	G	F-G	G	NC
ziram (Ziram) 76 DF — 5 lb	M3	14	48	G	P	P	P-F	P

* REI = reentry interval. Hours between last spray and reentry without using personal protective equipment. This time interval can vary depending on product formulation, always consult label of product being used.

** Consult chlorothalonil label for REI precautions related to risk of eye injury.

*** Rate of copper stated is for dormant spray. Rates must be sequentially greatly reduced to lessen foliar injury when used during the growing season.

**** Rovral is not registered for use after bloom.

Mode of Action Codes – fungicides having the same code have a similar mode of action and thus are not appropriate mixing or alternating partners for use in resistance management.

Nematode Control on Peaches

D. F. Ritchie, Entomology and Plant Pathology

Preplant Soil Fumigation — In light, sandy soil where root-knot and ring nematodes are present, preplant soil fumigation is imperative. If the nematode assay indicates the presence of root-knot or ring nematodes, it may be advantageous to fumigate the entire orchard site in **October to mid-November** before planting the trees in late winter to early spring. If the nematode assay does not indicate the presence of root-knot or ring nematodes, an 8- to 10-foot strip to be used for the tree row may be fumigated.

Table 6-11. Preplant Soil Fumigation

Materials	Rate/treated acre*
1,3 dichloropropene (Telone II) OR	27 to 36 gallons
metam-sodium (Vapam, Sectagon II, Busan 1020) tarped	75 to 100 gallons

* Rate will vary depending on soil type. Follow manufacturer's directions for rate and application procedures.

Postplant Treatment (Bearing and Nonbearing Trees) — NO MATERIALS REGISTERED for postplant use.

Further Information

Southeastern Peach Growers' Handbook (http://www.ent.uga.edu/peach/peachhbk/toc.htm)

2017 Southeastern Peach, Nectarine and Plum Pest Management and Culture Guide. University of Georgia Bulletin 1171 (http://www.ent.uga.edu/peach/PeachGuide.pdf) updated annually

> The information in this section for pecans is from the 2018 Commercial Pecan Spray Guide published by the University of Georgia.
> For more information, visit pecans.uga.edu

COMMERCIAL PECAN INSECT CONTROL (BEARING TREES)

Will Hudson, Extension Entomologist

ORCHARD SURVEY PROCEDURES

Insect and mite infestation levels should be estimated at least weekly based on thorough orchard sampling. Sample trees in all segments of each orchard. A good method is to sample every fourth tree in every fourth tree row (about 10% of the trees). Sample each major cultivar represented in the orchard. Sample a minimum of 10 terminals per tree. Check all compound leaves and the nut clusters on each terminal. Check as high in the tree as possible. Foliar pest counts should be made on compound leaves surrounding the nut clusters. Nut clusters should be inspected carefully for the presence of pests or damage. Hickory shuckworm damage should be monitored mid-season by examining fallen nuts for a whitish spot on the side. Pecan weevil populations should be monitored by survey traps.

PEST	PESTICIDE	MOA	AMOUNT PER ACRE	REI/PHI (Hours or Days)	TIMING AND REMARKS
Phylloxera	*chlorpyrifos* 4E Lorsban, Chlorphos	1B	2 pt	24 H/ –	Treat trees with a recent history of heavy infestation and surrounding trees. Apply at budbreak with the first prepollination spray. Note: Other imidacloprid formulations are available. Read labels carefully to find the proper rate.
	Centric 40WG	4A	2-2.5 oz	12 H/ –	
	Provado 1.6F	4A	3.5 oz	12 H/ –	
	Trimax Pro	4A	1.3-2.6 oz	12 H/ –	
Spittlebugs	*imidacloprid* Trimax, Provado, many generics	4A	See label *Several formulations are available.*	12 H/ –	Spittlebug infestations are easily recognized by the white, frothy masses on terminals or nut clusters. Definite thresholds have not been established and treatment is seldom needed. Many generic imidacloprid formulations are available.
Pecan Nut Casebearer	*chlorpyrifos* 4E Lorsban, Chlorphos	1B	1.5 pt	24 H/ –	Light infestations causing occasional damage do not require control in most crop years. The most serious damage usually occurs in mid May. Adult emergence should be monitored with pheromone traps. Place traps in orchards by mid-April. Begin sampling for nut casebearer in the first week of May. Pay particular attention to orchards not under a spray program the preceding year and orchards with a recent history of nut casebearer problems. Try to time sprays to stop injury before more than one nut per cluster is infested. It is recommended that broad-spectrum contact insecticides, such as chlorpyrifos and the pyrethroids, not be used in early- or mid-season to conserve beneficial insect populations. (See Special Considerations section.)
	Intrepid 2F	18	4-8 oz	4 H/ –	
	Spintor 2SC	5	4-10 oz	4 H/ –	
	Dimilin 2L	15	8-16 oz	12 H/ –	
	clothianadin Belay	4A	3-6 oz	12 H/ –	
	methoxyfenozide + spinetoram Intrepid Edge	5 + 18	4-6.4 oz	4 H –	
	tolfenpyrad Apta	21	17-27 oz	12 H –	**DO NOT** apply more than 1 application. No more than 27 oz/A/season.
	abamectin + cyantraniliprole Minecto Pro	6 + 28	8-12 oz	12 H/ 21 D	No more than 2 consecutive applications, no more than 24 oz/season.

UGA Extension Bulletin 841 • 2018 Commercial Pecan Spray Guide

Chapter VI — 2019 N.C. Agricultural Chemicals Manual

COMMERCIAL PECAN INSECT CONTROL

PEST	PESTICIDE	MOA	AMOUNT PER ACRE	REI/PHI (Hours or Days)	TIMING AND REMARKS
Mites	*abamectin* Agri-Mek SC and others	6	2.25-4.25 oz	12 H/ –	A non-ionic surfactant or horticultural oil MUST be added to the tank.
	Acramite 4SC	Unclassified	12-16 oz	12 H/ –	Mites, especially the pecan leaf scorch mite, are normally late season pests. Mite damage appears as bronzed, scorched areas on the undersides of leaflets. Scorched areas begin at the leaflet midribs then spread out toward leaflet margins. Mites often build up on low limbs in the shaded, interior portions of trees then spread rapidly up and out. For heavy infestations, repeat the application in 5-7 days.
	Envidor 2SC	23	14-18 oz	12 H/ –	
	Portal	21A	2 pt	12 H/ –	
	pyridaben Nexter	21	5.2-10.67 oz	24 H/ –	Savey is an ovicide and should be tank-mixed with an adulticide. Zeal is primarily an ovicide/larvicide.
	Savey 50DF	10A	3-6 oz	12 H/ –	
	Zeal	10B	2-3 oz	12 H/ –	
Yellow Aphids	**FOLIAR APPLICATIONS**				Yellow aphids may be present in orchards throughout the growing season. Populations are usually highest in April-May and again in August-September. In early season, DO NOT treat yellow aphids if they are the only insect problem. Rely on beneficial insects to suppress early season populations.
	Assail 30SG	4A	2.5-9.6 oz	12 H/ –	
	clothianidin Belay	4A	3-6 fl oz	12 H/ –	
	flonicamid Beleaf, Carbine	9C	2-2.8 oz	12 H/ –	In prolonged dry periods, lower, chronic aphid populations may require treatment to prevent the build-up of unacceptable levels of honeydew and sooty mold. WEEKLY SCOUTING IS VERY IMPORTANT IN TIMING APHID SPRAYS, ESPECIALLY IN LATE SEASON. Rotate among classes of insecticides between treatments to avoid resistance development.
	flupyradifurone Sivanto 200 SL	4D	7.0-10.5 oz	4 H/ 7 D	
	imidacloprid Provado, many generics	4A	See label	12 H/ –	
	pymetrozine Fulfill	9B	4 oz	12 H/ –	It is suggested that pyrethroid materials (cypermethrin, bifenthrin, etc.) not be used, alone or in combination, in early- or mid-season applications.
	pyridaben Nexter	21	5.2-10.67 oz	24 H/ –	Many generic formulations of imidacloprid are available. Read label carefully for recommended rate. Imidacloprid alone may not control yellow and black-margined aphids.
	sulfoxaflor Closer	4C	1.5-2.75 oz	12 H/ 7 D	
	thiamethoxam Centric	4A	2-2.5 oz	12 H/ –	Admire can be applied through a drip irrigation system, as an emitter spot application, or as a shanked-in emitter adjacent application. See label for complete details. Apply Admire only to orchards where drip irrigation has been established for at least 5 years.
	tolfenpyrad Apta	21A	17-27 oz	12 H/ –	
	SYSTEMIC APPLICATIONS				DO NOT apply more than 1 application of Apta, no more than 27 oz/A/season.
	Admire Pro	4A	7-14 fl oz	12 H/ –	Use the 14 oz rate for black pecan aphid control.

UGA Extension Bulletin 841 • 2018 Commercial Pecan Spray Guide

Chapter VI—2019 N.C. Agricultural Chemicals Manual

COMMERCIAL PECAN INSECT CONTROL

PEST	PESTICIDE	MOA	AMOUNT PER ACRE	REI/PHI (Hours or Days)	TIMING AND REMARKS
Black Pecan Aphid	SAME INSECTICIDES AS FOR YELLOW APHIDS or *chlorpyrifos* Lorsban, generics	1B	Check label	24 H/ –	Black pecan aphids may cause damage as early as May but are usually a serious problem only in late season. Damage appears as yellow spots on leaflets. Damaged spots later turn brown and 2-4 damaged spots per leaflet can cause leaflet drop. Carefully check all compound leaves on 10 terminals per tree, on at least 10 trees per orchard for the presence of black pecan aphids. Prior to July 1, treat if 25% of terminals have 2 or more black aphids. After July 1, treat if 15% of terminals have more than one black aphid and nymph clusters are found. Concentrate checks on susceptible cultivars such as Schley, Sumner and Gloria Grande. Be sure to check all compound leaves on each terminal examined.
	gibberellic acid ProGibb 4%	N/A	10 oz	N/A	Gibberellic acid is a plant growth regulator that prevents damage from black pecan aphid feeding and inhibits establishment in the orchard. It does not affect aphids directly and will not control any other pest, including yellow aphids. Three applications should be made at 2-week intervals, beginning in mid-July, applying 10 oz each time.
Hickory Shuckworm	*chlorpyrifos* 4E Lorsban, Chlorfos	1B	1-1¼ pt	24 H/ –	Shuckworms are active throughout the season, but do not cause significant damage until June or later. Prior to shell hardening, larval feeding causes nuts to drop. After shells harden, feeding causes shucks to stick to the shells, reducing quality. If orchards have a history of shuckworm infestation, a spray should be applied in early June. In early August, 2-3 additional sprays should be applied. Initiate August sprays at half-shell hardening and repeat at 2-week intervals until shuck split if shuckworm activity continues. Chlorpyrifos and pyrethroids (Asana, Ambush, Mustang, etc.) applied for other pests will also control shuckworm. It is not necessary to spray in August if pecan weevil controls are applied. Please note the Special Considerations section regarding the use of pyrethroid materials. **DO NOT** apply more than 1 application, no more than 27 oz/A/season.
	clothianadin Belay	4A	3-6 oz	12 H/ –	
	Dimilin 2L	15	8-16 oz	12 H/ –	
	Intrepid 2F	18	4-8 oz	4 H/ –	
	methoxyfenozide + spinetoram Intrepid Edge	5 + 18	4-6.4 oz	4 H/ –	
	tolfenpyrad Apta	21A	17-27 oz	12 H/ –	
	abamectin + cyantraniliprole Minecto Pro	6 + 28	8-12 oz	24 H/ 21 D	No more than 2 consecutive applications, no more than 24 oz/season.
	chlorantraniliprole + lambda-cyhalothrin Besiege	3 + 28	6-12.5 oz	24 H/ –	Besiege contains a pyrethroid, and may flare aphids and mites if used in early or mid-season. The best fit is for late season shuckworm.

UGA Extension Bulletin 841 • 2018 Commercial Pecan Spray Guide

Chapter VI—2019 N.C. Agricultural Chemicals Manual

COMMERCIAL PECAN INSECT CONTROL

PEST	PESTICIDE	MOA	AMOUNT PER ACRE	REI/PHI (Hours or Days)	TIMING AND REMARKS
Pecan Weevil	Carbaryl 80S Sevin	1A	3 lb	24 H/ –	Pecan weevil emergence may extend from July into October. Peak emergence is normally between August 10 and September 20. Emergence should be monitored in each infested grove with traps, knockdown sprays or a combination of these methods. Trees known to have a recent history of weevil problems should be selected for monitoring If excessive nut drop results from pecan weevil feeding punctures before pecan shells begin to harden, spray at once. After pecan shells harden and nuts reach the "dough" or "gel" stage, treat when weevils emerge (especially following rains) and continue at 7-10 day intervals until emergence stops. APHID OR MITE POPULATIONS MAY BUILD UP WHERE CARBARYL IS USED. If these pests become a problem, apply aphicides or miticides as previously directed. **NOTE:** Several pyrethroids, (Asana, Ammo, Baythroid, Brigade, Mustang Max) as well as Imidan are labeled for pecan weevil control. If these materials are used for weevils, they can be expected to be most effective where weevil populations are low. They may be adequate to prevent feeding injury from weevils emerging prior to shell hardening but their use could be risky under heavy weevil pressure after nuts reach the gel stage and are subject to weevil oviposition. (See Special Considerations section). Several products are available that combine a pyrethroid insecticide with an aphicide. These products may help suppress aphids while providing weevil control. Brand names include Endigo, Leverage, and others.
	Carbaryl 4F Sevin XLR Various pyrethroids	1A	4-5 qt	24 H/ –	

UGA Extension Bulletin 841 • 2018 Commercial Pecan Spray Guide

COMMERCIAL PECAN INSECT CONTROL

KERNAL FEEDING HEMIPTERANS
(Stink bugs and Plant bugs)

A complex of true bugs (stink bugs and plant bugs) attack pecan. They may be present in orchards all year but normally cause their most serious injury from late August through September. Prior to shell hardening, feeding injury causes nut drop. After shell hardening, their feeding causes black, bitter spots on kernels, reducing quality. They can continue to feed, through the hardened shells, until nuts are harvested. The presence and numbers of stink bugs and plant bugs should be noted in surveys throughout the season. Special attention should be paid to the true bugs in late-season orchard surveys. **Treat when 1 stink bug is found per 40 terminals OR when 5 or more are found** per knockdown spray on a sheet covering 20% of the area under a tree. Sprays for these insects are difficult to time properly because the bugs move in and out of orchards. Close checking is required to detect damaging populations. No materials have consistently given excellent stink bug control, possibly due to the difficulty in timing sprays. The pyrethroids are labeled for stink bug control. Please note the pre-harvest use restrictions of the products.

FIRE ANTS

Fire ants have been known to protect pecan aphids by destroying beneficial insects in pecan orchards. Fire ants should be controlled or at least kept out of pecan trees. Lorsban 4E at 2 pts/A as a ground spray is labeled for fire ant control. Best approach is probably applying an ant bait in late spring.

SCALE INSECTS

Scale populations build slowly, but can reach damaging levels before becoming obvious. Examine fallen limbs carefully during the season for scale presence. Preferred treatment is 1%-2% horticultural oil spray, applied in November-December and again in February. For severe problems an application of Esteem in June may be necessary.

OTHER INSECT PESTS

Pests such as pecan leaf casebearer, leaf miners, walnut caterpillar, fall webworm, pecan budmoth, nut curculio, shoot curculio, Prionus root borers and others may occasionally cause economic injury to pecan. Growers should be able to identify these pests and their damage. Color photographs of all pecan pests and their injury can be found in the Southern Pecan Growers Handbook and online from the UGA Extension pecan team (Google search "ugapecans"). The publication is available at $30 per copy. For ordering information, visit: http://extension.uga.edu/publications/for-sale.cfm.

Specific controls for occasional pests not covered in this spray guide can be obtained from your local county Extension agent.

SPECIAL CONSIDERATIONS

Alternative Formulations. Some pesticides listed in this publication are available in formulations other than the ones listed. If different formulations are used, apply an equivalent amount of actual toxicant per acre.

Pest Resistance and Chemical Use. The aphids and mites which attack pecan have demonstrated the ability to become resistant to insecticides applied for their control. The rate at which this resistance develops depends on the chemical used, the frequency of use, the duration of use, and the rates used. Aphid and mite exposure to effective materials should be minimized to prolong the effective life of the chemicals. It is suggested that no insecticide be applied until it is absolutely necessary (this can be determined by thorough sampling) and that chemicals be alternated as much as possible. Resistance to neonicotinyl insecticides has developed in some areas for both yellow- and black-margined pecan aphids. This class of insecticides includes imidacloprid, thiamethoxam, acetamiprid, and clothianidin. These materials no longer provide adequate control of resistant populations. Aphid and mite populations may flare following application of Sevin or pyrethroids. Growers should be alert for this response, and limit applications of these materials to the minimum necessary for weevil or stink bug control.

Supplemental Control Measures. Beneficial insects such as lady beetles and lacewings provide natural assistance in suppressing aphid and mite populations. Beneficials are of particular value in early season. Elimination of unneeded early-season insecticide sprays conserves existing populations of beneficial insects and reduces the potential for severe aphid problems later in the season. The planting of leguminous cover crops in tree-row middles promotes the build up and retention of lady beetle populations in orchards. Crimson clover and Hairy vetch appear to be two of the best ground covers. If leguminous ground covers are planted, a herbicide strip should be maintained down each tree row and special attention should be paid to the increased water requirements that are likely to exist. Extraneous plant material resulting from the heavy growth of legumes must be removed or broken down prior to harvest or implementation of a program of row middle vegetation suppression (see Weed Control section).

UGA Extension Bulletin 841 • 2018 Commercial Pecan Spray Guide

Chapter VI — 2019 N.C. Agricultural Chemicals Manual

COMMERCIAL PECAN INSECT AND DISEASE SPRAY GUIDE
(NON-BEARING TREES)

Will Hudson, Extension Entomology
Jason Brock and Tim Brenneman, Plant Pathology

FOLIAR SPRAYS

TIME OF APPLICATION	PEST	PESTICIDE	MOA	AMOUNT PER ACRE	REI/PHI (Hours or Days)	INSTRUCTIONS AND REMARKS
Bud Break When first buds open.	Foliar disease	Fungicide + *chlorpyrifos* Chlorphos, Lorsban	1B	+ half rate 1-2 pt 4-8 oz	24 H/ –	Spray sufficient gallonage for thorough coverage. For fungicide options, refer to the Prepollination section for Pecan Disease Control.
	Pecan bud moth	Intrepid 2F	18	3-4 oz	4 H/ –	
		methoxyfenozide + spinetoram Intrepid Edge	5 + 18	4-6.4 oz	4 H/ –	
		abamectin + cyantraniliprole Minecto Pro	6 + 28	8-12 oz	12 H/ –	No more than 24 oz/season.
	Hickory shoot curculio	*chlorpyrifos* Lorsban, Chlorphos, etc.	1B	1.5-2 pt	24 H/ –	Apply sprays for shoot curculio at bud-break on the earliest cultivars and repeat at 10-14 day intervals.
Cover Sprays Three weeks after bud-break spray and every 4-6 weeks as needed.	Foliar disease	Fungicide + *chlorpyrifos* Chlorphos, Lorsban	1B	See above + 1-2 pt	24 H/ –	Spray sufficient gallonage for thorough coverage.
	Pecan bud moth	*chlorpyrifos* Chlorphos, Lorsban, etc.	1B	1.5-2 pt	24 H/ –	
		Dimilin 2L		8-16 oz		
		Imidan 70WSP		1.5 lb		
		Intrepid 2F	18	4-8 oz	4 H/ –	
		abamectin + cyantraniliprole Minecto Pro	6 + 28	8-12 oz	12 H/ –	

UGA Extension Bulletin 841 • 2018 Commercial Pecan Spray Guide

PECAN CHEMICALS: PRE-HARVEST INTERVALS AND OTHER RESTRICTIONS

CHEMICAL	MOA	REI/PHI (Hours or Days)	TIMING AND REMARKS
Acramite 4 SC	Undetermined	12 H/ 14 D	Only 1 spray per year.
Admire	4A	12 H/ –	Apply to soil between May 15 and July 15. Apply only to orchards that have been established on trickle irrigation for at least 5 years. **DO NOT** apply more than 32 fl oz of Admire per acre per season as a soil application. **DO NOT** apply more than 0.5 lb ai of Admire or Provado/A/season.
Ammo		–/ 21 D	Up to 0.8 lb ai/A/season may be applied prior to shuck split. **DO NOT** graze or feed cover crops.
Asana		–/ 21 D	**DO NOT** feed or graze livestock on treated orchard floors. **DO NOT** exceed 0.3 lb ai/A/season. **DO NOT** mix with fungicides containing triphenyltin hydroxide.
Assail	4A	12 H/ 14 D	**DO NOT** apply more than 4 times per season, nor more often than every 7 days.
Baythroid		–/ 14 D	No more than 2.8 fl oz/A/season.
Belay	4A	12 H/ 21 D	No more than 12 oz/season. **DO NOT** graze.
Carbaryl	1A	24 H/ 14 D	**DO NOT** apply more than a total of 15 qt/season.
Centric	4A	12 H/ 14 D	**DO NOT** exceed 5 oz/A/season. Allow at least 7 days between applications.
Closer		–/ 7 D	No more than 4 applications per season, and no more than 2 consecutive applications.
Desperado		–/ 7 D	No more than 2.2 gal/season; no aerial application.
Dimethoate		–/ 21 D	**DO NOT** graze livestock in treated groves.
Elast F			**DO NOT** apply after shucks open. **DO NOT** graze treated areas.
Enable		–/ 28 D	**DO NOT** apply after shuck split. **DO NOT** apply more than 48 oz/A. **DO NOT** graze treated areas.
Endosulfan			**DO NOT** apply after shuck split. **DO NOT** graze livestock in treated groves. **DO NOT** exceed 2 applications per year or 4 qt/A/year.
Envidor	23	12 H/ 7 D	Maximum of 1 application per season.
Fury/Mustang		–/ 21 D	**DO NOT** apply more than 0.3 lb ai/A/season or after shuck split. **DO NOT** graze or cut treated cover crops for feed.
Headline		–/ 14 D	**DO NOT** apply more than 28 fl oz/A/season.

UGA Extension Bulletin 841 • *2018 Commercial Pecan Spray Guide*

Chapter VI—2019 N.C. Agricultural Chemicals Manual

PECAN CHEMICALS: PRE-HARVEST INTERVALS AND OTHER RESTRICTIONS

CHEMICAL	MOA	REI/PHI (Hours or Days)	TIMING AND REMARKS
Imidan		3 D/ 14 D	**DO NOT** graze livestock in treated groves.
Intrepid	18	18/ 14 D	**DO NOT** graze livestock in treated areas or feed cover crops grown in treated areas. **DO NOT** apply more than 10 fl oz/application or 64 oz/season.
Kelthane		–/ 7 D	Applicators must be in enclosed cabs or cockpits.
Lorsban, Chlorphos	1B	24 H/ 28 D	**DO NOT** allow livestock to graze in treated orchards. Make no more than 5 applications per season.
Nexter	21A	24 H/ 7 D	No more than 10.67 oz/application nor more than 2 applications per season. No aerial applications.
Portal	21A	12 H/ 14 D	No more than one application per season.
Propimax			**DO NOT** apply after shuck split. **DO NOT** graze livestock in treated areas or cut treated areas for feed. **DO NOT** apply more than 32 fl oz/A/season.
Provado	4A	12 H/ –	**DO NOT** apply after 28 fl oz of Provado/A/year. **DO NOT** apply more than a total of 0.5 lb ai of Provado or Admire/A/season.
Quilt		–/ 45 D	**DO NOT** apply after shuck split. **DO NOT** graze livestock in treated areas or cut treated areas for feed. **DO NOT** apply more than 122 fl oz/A/season.
Savey	10A	12 H/ –	**DO NOT** graze livestock in treated areas. Only one application per season may be made.
Sovran		–/ 45 D	**DO NOT** apply more than 25.6 fl oz/A/season.
Stratego		–/ 30 D	**DO NOT** apply after shuck split. **DO NOT** apply more than 30 fl oz/A/season.
Sulfur			No time limitations.
TPTH			**DO NOT** use more than 45 oz (36 oz ai) of product per season. **DO NOT** apply after shucks begin to open. **DO NOT** graze dairy or meat animals in treated groves.
Topsin M			**DO NOT** apply after shuck split. **DO NOT** graze livestock in treated areas or cut treated areas for feed. **DO NOT** apply more than 3 lb/A/season.
Trimax Pro	4A	12 H/ 7 D	Maximum of 10.1 oz/A allowed per crop season. Allow at least 10 days between applications.
Zeal	10B	12 H/ 28 D	Maximum of 1 application per season.

**DO NOT graze livestock in treated groves where prohibited or until grazing restrictions have been met.

UGA Extension Bulletin 841 • 2018 Commercial Pecan Spray Guide

Chapter VI—2019 N.C. Agricultural Chemicals Manual

The information in this section for pecans is from the 2018 Commercial Pecan Spray Guide published by the University of Georgia. For more information, visit pecans.uga.edu

PECAN DISEASE CONTROL

Jason Brock and Tim Brenneman, Department of Plant Pathology

DISEASE	CHEMICAL & FORMULATION	MOA	RATE/ACRE	REI/PHI (Hours or Days)	COMMENTS
PREPOLLINATION APPLICATIONS: Every 10-14 Days From Bud Break Through Nut Set					
Scab; Downy Spot	*azoxystrobin* Abound Azaka	11	6-12 fl oz	4 H/ 45 D	See info below: MOA Group 11.
	difenoconazole + azoxystrobin Quadris Top	3 + 11	10-14 fl oz	12 H/ 45 D	See info below: MOA Group 3. See info below: MOA Group 11.
	difenoconazole + azoxystrobin	3 + 11	8-14 fl oz	12 H/ 21 D	See info below: MOA Group 3. See info below: MOA Group 11.
	dodine Elast 400F + FRAC group 3 fungicide	U12 + 3	25 fl oz + half rate	48 H/ Do not apply after shuck split	See info below: MOA Group 3. For any tank mix combination of Elast, TPTH, or a group 3 fungicide, the rates provided are the lowest recommended and will provide excellent control of scab under most conditions. When disease pressure is elevated, the rate of either mixing partner can be increased.
	dodine Elast 400F + TPTH	U12 + 30	25 fl oz + half rate	48 H/ Do not apply after shuck split or within 30 D of harvest	For any tank mix combination of Elast, TPTH, or a group 3 fungicide, the rates provided are the lowest recommended and will provide excellent control of scab under most conditions. When disease pressure is elevated, the rate of either mixing partner can be increased. See info below: MOA Group 30. See info below: MOA Group U12.
	fenbuconazole Enable 2F	3	8 fl oz	12 H/ Do not apply after shuck split or within 28 D of harvest	See info below: MOA Group 3.
	kresoxim-methyl Sovran	11	2.4-3.2 fl oz	12 H/ 45 D	See info below: MOA Group 11.
	metconazole Quash	3	2.5-3.5 oz/A	12 H/ 25 D	See info below: MOA Group 3.
	phosphorous acid Phostrol ProPhyt FungiPhite Reliant	33	2-5 pt 2-3 pt 2-3 pt 4 pt	4 H/ —	See info below: MOA Group 33.

MOA Group 3: Resistance risk is moderate. For best results, tank mix tebuconazole with a surfactant. Do not add a surfactant if mixing with other fungicides. Increasing the rate of a Group 3 fungicide will be important if reduced sensitivity is known or suspected. Stand-alone use is not recommended where reduced sensitivity is known or suspected.

MOA Group 11: Resistance risk is moderate. Do not make more than 2 sequential applications. If only using solo products, group 11 fungicides should not be used in more than 1/3 of the total number of fungicide applications. If using group 3 tank-mixed with other modes of action, they should not be used in more than 1/2 of the total number of fungicide applications.

MOA Group 30: Resistance risk is low.

MOA Group 33: Resistance risk is low. For best control apply in 100 gpa by ground. Do not apply in consecutive applications. Three to five applications are generally recommended. There is currently an unresolved issue regarding potential residues of these products in tree nuts exported to the EU. Growers who know their crop is going to that market should avoid use until the issue is resolved. Check labels for potential limitations on maximum number of applications or amount of active ingredient allowed per season. Do not use when there is a phosphate deficiency.

MOA Group U12: Resistance risk is low. Do not use on Moore, Van Deman, Barton, or Shawnee. Do not use a surfactant. Do not use with foliar zinc treatments.

Extension Bulletin 841 • 2018 Commercial Pecan Spray Guide

Chapter VI—2019 N.C. Agricultural Chemicals Manual

PECAN DISEASE CONTROL

DISEASE	CHEMICAL & FORMULATION	MOA	RATE/ACRE	REI/PHI (Hours or Days)	COMMENTS
	**PREPOLLINATION APPLICATIONS:** Every 10-14 Days From Bud Break Through Nut Set				
Scab; Downy Spot (continued)	*phosphorous acid + tebuconazole* Viathon	33 + 3	2-2.5 pt	12 H/ 0 D	See info below: MOA Group 33. See info below: MOA Group 3.
	propiconazole Orbit Propimax EC Bumper 41.8EC	3	6-8 fl oz	12 H/ Do not apply after shuck split	See info below: MOA Group 3.
	propiconazole + azoxystrobin Quilt Quilt Xcel	3 + 11	14-27.5 fl oz 14-21 fl oz	12 H/ Do not apply after shuck split or within 45 D of harvest	See info below: MOA Group 3. See info below: MOA Group 11.
	pyraclostrobin Headline	11	6-7 fl oz	12 H/ 14 D	See info below: Group 11.
	tebuconazole Folicur 3.6F Tebuzole 3.6F Monsoon Orius 3.6F Toledo 3.6F	3	6-8 fl oz	12 H/ Do not apply after shuck split	See info below: MOA Group 3.
	tebuconazole + azoxystrobin Custodia	3 + 11	8.6-17.2	12 H/ 45 D	See info below: MOA Group 3. See info below: MOA Group 11.
	tebuconazole + trifloxystrobin Absolute	3 + 11	5-7.67 fl oz	12 H/ Do not apply after shuck split or within 30 D of harvest	See info below: MOA Group 3. See info below: MOA Group 11.
	tetraconazole + triphenyltin hydroxide Minerva Duo	3 + 30	16 oz	48 H/ 30 D	See info below: MOA Group 3. See info below: MOA Group 30.
	thiophanate methyl[3] (Topsin M) + TPTH or + Elast	1 + 30 or + U12	1 lb + half rate or + 25 fl oz	3 D/ Do not apply after shuck split	See info below: MOA Group 1. See info below: MOA Group 30. See info below: MOA Group U12.

MOA Group 1: Risk for resistance is high. Use should be limited. When conditions are very favorable for scab, Topsin can be used in combination with either a full rate of TPTH or Elast. Limit the use to 1 or 2 applications per season.

MOA Group 3: Resistance risk is moderate. For best results, tank mix tebuconazole with a surfactant. Do not add a surfactant if mixing with other fungicides. Increasing the rate of a Group 3 fungicide will be important if reduced sensitivity is known or suspected. Stand-alone use is not recommended where reduced sensitivity is known or suspected.

MOA Group 11: Resistance risk is moderate. Do not make more than 2 sequential applications. If only using solo products, group 11 fungicides should not be used in more than 1/3 of the total number of fungicide applications. If using group 3 tank-mixed with other modes of action, they should not be used in more than 1/2 of the total number of fungicide applications.

MOA Group 30: Resistance risk is low.

MOA Group 33: Resistance risk is low. For best control apply in 100 gpa by ground. Do not apply in consecutive applications. Three to five applications are generally recommended. There is currently an unresolved issue regarding potential residues of these products in tree nuts exported to the EU. Growers who know their crop is going to that market should avoid use until the issue is resolved. Check labels for potential limitations on maximum number of applications or amount of active ingredient allowed per season. Do not use when there is a phosphate deficiency.

MOA Group U12: Resistance risk is low. Do not use on Moore, Van Deman, Barton, or Shawnee. Do not use a surfactant. Do not use with foliar zinc treatments.

UGA Extension Bulletin 841 • 2018 Commercial Pecan Spray Guide

PECAN DISEASE CONTROL

DISEASE	CHEMICAL & FORMULATION	MOA	RATE/ACRE	REI/PHI (Hours or Days)	COMMENTS
PREPOLLINATION APPLICATIONS: Every 10-14 Days From Bud Break Through Nut Set					
Scab; Downy Spot (continued)	*triphenyltin hydroxide* (TPTH)[1] + FRAC group 3 fungicide	30 + 3	half rate[2] + 4 fl oz	48 H/ 30 D	See info below: MOA Group 30. See info below: MOA Group 3.
Anthracnose	Anthracnose is a disease with a long latent period; symptom expression occurs many weeks after infection. Fungicides used for control of scab have been effective in suppressing anthracnose.				
POSTPOLLINATION APPLICATIONS: Every 10-21 Days From Nut Set To Shell Hardening					
Scab	*difenoconazole + azoxystrobin* Quadris Top	3 + 11	10-14 fl oz	2 H/ 45 D	See info below: MOA Group 3. See info below: MOA Group 11.
	dodine Elast 400F	U12	50 fl oz	48 H/ Do not apply after shuck split	See info below: MOA Group U12.
	dodine Elast 400F + FRAC group 3 fungicide[3]	U12 + 3	25 fl oz + 4-6 fl oz	48 H/ Do not apply after shuck split	See info below: MOA Group U12. See info below: MOA Group 3.
	dodine Elast 400F + TPTH	U12 + 30	25 fl oz + half rate[2]	48 H/ Do not apply after shuck split	See info below: MOA Group U12. See info below: MOA Group 30.
	phosphorous acid Phostrol ProPhyt Viathon FungiPhite Reliant	33	2-5 pt 2-3 pt 2 pt 2-3 pt 4 pt	4 H/ —	See info below: MOA Group 33.
	phosphorous acid + tebuconazole Viathon	33 + 3	2-2.5 pt	12 H/ 0 D	See info below: MOA Group 33. See info below: MOA Group 3.
	propiconazole + azoxystrobin Quilt Quilt Xcel	3 + 11 3 + 11	20-28 fl oz 20-21 fl oz	12 H/ Do not apply after shuck split or within 45 D of harvest	See info below: MOA Group 3. See info below: MOA Group 11.

MOA Group 3: Resistance risk is moderate. For best results, tank mix tebuconazole with a surfactant. Do not add a surfactant if mixing with other fungicides. Increasing the rate of a Group 3 fungicide will be important if reduced sensitivity is known or suspected. Stand-alone use is not recommended where reduced sensitivity is known or suspected.

MOA Group 11: Resistance risk is moderate. Do not make more than 2 sequential applications. If only using solo products, group 11 fungicides should not be used in more than 1/3 of the total number of fungicide applications. If using group 3 tank-mixed with other modes of action, they should not be used in more than 1/2 of the total number of fungicide applications.

MOA Group 30: Resistance risk is low.

MOA Group 33: Resistance risk is low. For best control apply in 100 gpa by ground. Do not apply in consecutive applications. Three to five applications are generally recommended. There is currently an unresolved issue regarding potential residues of these products in tree nuts exported to the EU. Growers who know their crop is going to that market should avoid use until the issue is resolved. Check labels for potential limitations on maximum number of applications or amount of active ingredient allowed per season. Do not use when there is a phosphate deficiency.

MOA Group U12: Resistance risk is low. Do not use on Moore, Van Deman, Barton, or Shawnee. Do not use a surfactant. Do not use with foliar zinc treatments.

UGA Extension Bulletin 841 • 2018 Commercial Pecan Spray Guide

Chapter VI—2019 N.C. Agricultural Chemicals Manual

PECAN DISEASE CONTROL

DISEASE	CHEMICAL & FORMULATION	MOA	RATE/ACRE	REI/PHI (Hours or Days)	COMMENTS
POSTPOLLINATION APPLICATIONS: Every 10-21 Days From Nut Set To Shell Hardening *(continued)*					
Scab	*tebuconazole*[4] + *trifloxystrobin* Absolute	3 + 11	5-7.67 fl oz	12 H/ Do not apply after shuck split or within 30 D of harvest	See info below: MOA Group 3. See info below: MOA Group 11.
	tetraconazole + triphenyltin hydroxide Minerva Duo	3 + 30	16 oz	48 H/ 30 D	See info below: MOA Group 3. See info below: MOA Group 30.
	TPTH + FRAC group 3 fungicide	30 + 3	half rate + 4-6 fl oz	48 H/ 30 D	See info below: MOA Group 30. See info below: MOA Group 3.
	triphenyltin hydroxide (TPTH)[1] Agri Tin Agri Tin Flowable Super Tin 80WP Super Tin 4L	30	7.5 oz 12 fl oz 7.5 oz 12 fl oz	48 H/ 30 D	See info below: MOA Group 30.
	ziram Ziram		6-8 lb	48 H/ 55 D	Ziram as a multi-site alternative in cases where resistance to other protectants is an issue.

MOA Group 3: Resistance risk is moderate. For best results, tank mix tebuconazole with a surfactant. Do not add a surfactant if mixing with other fungicides. Increasing the rate of a Group 3 fungicide will be important if reduced sensitivity is known or suspected. Stand-alone use is not recommended where reduced sensitivity is known or suspected.

MOA Group 11: Resistance risk is moderate. Do not make more than 2 sequential applications. If only using solo products, group 11 fungicides should not be used in more than 1/3 of the total number of fungicide applications. If using group 3 tank-mixed with other modes of action, they should not be used in more than 1/2 of the total number of fungicide applications.

MOA Group 30: Resistance risk is low.

MOA Group 33: Resistance risk is low. For best control apply in 100 gpa by ground. Do not apply in consecutive applications. Three to five applications are generally recommended. There is currently an unresolved issue regarding potential residues of these products in tree nuts exported to the EU. Growers who know their crop is going to that market should avoid use until the issue is resolved. Check labels for potential limitations on maximum number of applications or amount of active ingredient allowed per season. Do not use when there is a phosphate deficiency.

Powdery Mildew	For powdery mildew, the scab fungicide program can be adjusted if needed. The FRAC group 3 fungicides or mixes containing FRAC 3 fungicides are the best options. Combining sulfur (4-6 lb/A) with fungicides used for scab control is also an option. <u>DO NOT</u> mix sulfur with Elast.
Zonate Leaf Spot	For zonate leaf spot, the scab fungicide program can be adjusted if needed. The FRAC group 3 fungicides or mixes containing FRAC 3 fungicides are the best options. Topsin M also provides suppression of Zonate leaf spot.
Anthracnose	Anthracnose is a disease with a long latent period; symptom expression occurs many weeks after infection. Fungicides used for control of scab have been effective in suppressing anthracnose.

[1] TPTH is available as Agri Tin, Agri Tin Flowable, Super Tin 80WP, and Super Tin 4L.
[2] Half rates are 3.75oz for Agri Tin and Super Tin 80WP; 6 fl oz for Agri Tin Flowable and Super Tin 4 L.
[3] Thiophanate methyl is available as Topsin M 70WDG, Topsin M 70 WP, and Topsin M WSB, and Topsin M 4.5 FL (20 fl oz rate is equivalent to 1 lb of wettable powder). Topsin XTR is a premix of thiophanate methyl and tebuconazole.
[4] For tebuconazole, use a minimum of 6 fl oz in tank mixes for nut scab control.

NOTE: In orchards where any nuts have any amount of scab by mid-June or in orchards where 10% or more of the nuts have any amount of scab by early July, the following measures should be taken:

1. The interval between fungicide sprays should not exceed 14 days until shell hardening.
2. On varieties with a summer growth flush, the spray interval should be closed so that no more than 10 days pass from the onset of the growth flush until a fungicide spray is made.
3. If the 5-day forecast shows the probability for several days of rain, close the interval to have as much acreage as possible treated within 7 days of the storm.

UGA Extension Bulletin 841 • 2018 Commercial Pecan Spray Guide

PECAN DISEASE CONTROL

After Shell Hardening: Fungicide coverage for crop protection is necessary to shell hardening. Beginning in early August, monitor for shell hardening and adjust fungicide needs accordingly.

Foliar diseases: Maintaining leaf health past shell hardening is important. If leaf scab, zonate leaf spot, or another foliar disease is of concern, refer to the previous sections for fungicide options and recommendations. Pay attention to use limitations and fungicide resistance management guidelines. DO NOT use Topsin in consecutive applications for leaf disease control.

DISEASE	CHEMICAL & FORMULATION	MOA	RATE/ACRE	REI/PHI (Hours or Days)	COMMENTS
Phytophthora Shuck and Kernel Rot	A treatment is advised in orchards with a history of this disease (primarily Houston, Peach, and Macon counties) when wet weather and warm temperatures <86 °F occur between shell hardening and shuck split.				
	TPTH	30	full rate		
	phosphorous acid Fosphite Fungi-Phite KPhite Phiticide Phostrol Rampart Topaz	33	1-2 qt	4 H/ –	The phosphite (phosphorous acid based) fungicides listed are EPA approved and considered to be very safe products. However, there is currently an unresolved issue regarding potential residues of these products in tree nuts exported to the EU. This affects only nuts exported to the EU, but growers who know their crop is going to that market may want to consider not using phosphite fungicides until this issue is resolved. Check labels for potential limitations on maximum number of applications or amount of active ingredient allowed per season.
	MOA Group 11 fungicides	11	full rate		
	copper hydroxide Kocide 3000 Kocide 2000	M1	0.75-1.75 lb 1.5-3 lb	48 H/ –	Use higher rates when disease pressure is high and large, mature trees.

Restrictions and Fungicide Resistance Management Recommendations

- Follow label instructions for proper use of all fungicide products, including safe handling, tank mixing, application method, and resistance management.
- DO NOT apply more than 32 fl oz of propiconazole/A/season.
- DO NOT apply more than 32 fl oz of tebuconazole/A/season.
- DO NOT apply more than 1.5 qt of fenbuconazole/A/season.
- DO NOT use more than 45 oz of Agri Tin or Super Tin 80 WP or 72 fl oz of Agri Tin Flowable or Super Tin 4 L/A/season.
- DO NOT apply more than 1.6 lb (25.6 oz) of kresoxim methyl/A/season.
- DO NOT use Elast full season.
- If using a group 3 fungicide alone prepollination, DO NOT use mixes containing a group 3 fungicide postpollination.
- DO NOT make more than 2 sequential and 3 total applications of group 11 fungicides.
- DO NOT apply more than 3 lb of thiophanate methyl (2.1 lb ai)/A/season.

UGA Extension Bulletin 841 • 2018 Commercial Pecan Spray Guide

Strawberry Disease Control

F. J. Louws, Entomology and Plant Pathology
(with major input from Rebecca Melanson, Mississippi State University, Guido Schnabel, Clemson University and Chuck Johnson, Virginia Tech)

For more information and details, see the *Southeast Regional Strawberry Integrated Pest Management Guide*, which is online at http://www.smallfruits.org/ assets/documents/ipm-guides/StrawberryIPMGuide.pdf.

Pre-Planting Disease and Weed Management

Table 6-12A. Pre-Planting Disease and Weed Management

Management Options	Amount of Formulation per Acre	Effectiveness	Comments
Anthracnose Angular leaf spot Phytophthora crown rot Fusarium wilt (not reported in Eastern U.S.) Viruses	Disease-free plants	Importance: E Efficacy: E	Use of certified plants or plants produced in a similarly stringent program is the most important method to prevent these diseases.
Nematodes	Sample soil	Importance: G	Sample soils for nematode analysis through local state services to determine which fumigant or IPM management plan may be required.
Nematodes and soilborne pathogens (Pythium, Phytophthora, Fusarium, Rhizoctonia)	Crop rotation and cover crop selection	Importance: G Efficacy: G	Selected summer cover crops and rotating fields to other crops for 2 to 3 years can suppress nematode populations and reduce black root rot and other disease problems.
Weeds Root and crown rot disorders Nematodes (Black root rot; Phytophthora crown rot)	Pre-plant fumigation and laying down plastic mulch	Efficacy: E	See fumigation table below. Consult with custom applicators and/or Extension agents for product and rate recommendations.

Pre-Planting and Early Post-planting: Nematode Management

Table 6-12B. Pre-planting and Early Post-planting: Nematode Management

Management Options	Amount of Formulation per Acre	Effectiveness	REI	PHI	Comments
Nimitz or Fluensulfone 480EC	3.5 to 7 pt per treated acre	See comments	0 hr	0 days	Nimitz is a "traditional contact nematicide." It has not been extensively tested on strawberry in the Southeast and Mid-Atlantic states, but research on other crops in these areas and on strawberry elsewhere suggests moderate to good activity - not quite as effective as soil fumigant standards - against most major plant-parasitic nematode species. Apply via drip or incorporated spray at least 7 days before planting; only 1 application per year. Soil temperature must be 60 degrees F or above. Soil incorporation in the top 6 to 8 inches is critical. Irrigating (0.5 to 1 inches) 2 to 5 days after application is recommended.
Majestene (heat-killed *Burkholderia* spp. strain A396)	4 to 8 qt	See comments	4 hr	0 days	Majestene is a biological nematicide approved for organic strawberry production. It has not been extensively field-tested on strawberry in the Southeast and Mid-Atlantic states, but research to date suggests useful activity against root-knot, lesion, sting, stunt, ring, and reniform nematodes. Can be applied as a pre-plant incorporated, in-furrow or banded spray as long as spray volume is sufficient to thoroughly soak the root zone. However, Majestene can also be drip-applied prior to planting, at planting or shortly thereafter, and again later in the season. Higher rates are likely more effective, and repeated applications also increase the extent and duration of nematode control. If nematode populations are high, another product may also be necessary for control.

Fumigants

New labels require extensive risk mitigation measures including fumigant management plans (FMPs), buffer restrictions, worker protection safety standards and other measures. Details are on the labels and see http://www2.epa.gov/soil-fumigants. Some fumigants are registered on multiple crops but with crop or soil-type specific rates; others are registered for specific crops and/or in certain states only. Follow all labels carefully.

Relative Efficacy of Currently Registered Fumigants or Fumigant Combinations for Managing Soilborne Nematodes, Diseases, and Weeds in Plasticulture Strawberries

Table 6-12C. Relative Efficacy of Currently Registered Fumigants or Fumigant Combinations for Managing Soilborne Nematodes, Diseases, and Weeds in Plasticulture Strawberries[1]

Product	Rate per Treated Acre[2]		Relative Efficacy[3]			
	Volume (gal)	Weight (lb)	Nematodes	Disease	Nutsedge	Weeds: Annual
Telone II (1,3-dichloropropene; 1,3-D)	15 to 27	153 to 275	E	P	P	P
Telone EC[3]	9 to 24[5]	91 to 242[5]	E	P	P	P
Telone C17 (1,3-D + chloropicrin)	32.4 to 42	343 to 445	E	G	P	P
Telone C35 (1,3-D + chloropicrin)	39 to 50	437 to 560	E	E	P	F
InLine (1,3-D + chloropicrin)[3]	29 to 57.6 (See Label)	325 to 645 (See Label)	E	E	P	G
Pic-Clor 60 (chloropicrin + 1,3-D)	48.6	588	E	E	P	G
Pic-Clor 60 EC[4]	42.6	503	E	E	P	G
Pic-Clor 80	34	440	G	E	P	F
Metam potassium[6]	30 to 62	318 to 657	F	G	P	VG
Metam sodium[6] (MS)	37.5 to 75	379 to 758	F	G	P	VG
Chloropicrin + MS[6]	19.5 to 31.5 + 37.5 to 75	275-444 + 379-758	F	E	F	VG
Chloropicrin	48.6	150 to 350	P	E	ND	ND
Tri-Pic 100EC[4]	8 to 24	100 to 300	P	E	ND	ND
Paladin (dimethyl disulphide)[7]	35.0 to 51.3	310 to 455	VG	VG	VG	G
Paladin PIC-21	41.2 to 60.1	392 to 572	VG	E	VG	G
Paladin EC[3,7]	37.0 to 54.2	326 to 479	VG	VG	VG	G
Dominus (allyl isothiocyanate)[8]	25 to 40[5]	212 to 340[5]	F	G	P	G

[1] Fumigants with lower efficacy against weeds may require a complementary herbicide or hand-weeding program, although use of virtually impermeable film (VIF) or totally impermeable film (TIF) may increase weed control, particularly with chloropicrin + 1,3-D products or Paladin. Refer to the Herbicide Recommendation section of this guide for directions pertaining to herbicide applications. Telone can persist more than 21 days under cool or wet soil conditions.

[2] Rates can sometimes be reduced if products are applied with VIF or TIF.

[3] Efficacy Ratings: The efficacy of a management option is indicated by E = excellent, VG = very good, G = good, F = fair, P = poor, and ND = no data. These ratings are benchmarks; actual performance will vary.

[4] Product is formulated for application through drip lines under a plastic mulch; efficacy is dependent on good distribution of the product in the bed profile.

[5] Labelled rates are per *broadcast-equivalent* acre, NOT per treated acre.

[6] Metam potassium can be Metam KLR, K-Pam, Sectagon K54 or other registered formulations, and should be used in soils with high sodium content. Metam sodium can be Vapam, Sectagon 42, Metam CLR or other registered formulations.

[7] Paladin should be applied with 21% chloropicrin and VIF or TIF to enhance disease control, and has low efficacy on certain small seeded broadleaf weeds and grasses. Paladin may not be registered in all States.

[8] Dominus is registered but there is limited experience with the product through University or independent trials in our region; growers may want to consider this on an experimental basis. Planting interval is 10 days. The active ingredient allyl isothiocyanate is similar to the active ingredient in metam sodium products (methyl isothiocyanate) and is likely to behave in a similar manner with a similar pest control profile.

Planting and Early Post-Planting: Disease Control

General Pesticide Information

FRAC/IRAC/HRAC codes — these acronyms refer to industry-sponsored committees addressing resistance to crop protection materials; Fungicide Resistance Action Committee (FRAC), Insecticides Resistance Action Committee (IRAC) and Herbicide Resistance Action Committee (HRAC). Pesticides affect their target pest in a variety of ways, and the way a pesticide kills the target organism is called the ***mode of action*** (**MOA**). Although pesticides have different names and may have different active ingredients, they may have the same MOA. Over time, pests can become resistant to a pesticide, and typically this resistance applies to all pesticides with the same MOA. When rotating pesticides, it is important to select pesticides with different MOAs. The FRAC/IRAC/HRAC have organized crop protection materials into groups with shared MOAs and given them specific codes, which appear on pesticide labels. The code **U** means the MOA is unknown. *When selecting pesticides, avoid successive applications of materials in the same MOA group to minimize potential resistance development.* MOA categories are listed in this guide to aid in the development of resistance management programs. More information about this topic can be found at www.frac.info, www.irac-online.org, and www.hracglobal.com.

Organic Materials Review Institute (OMRI; www.omri.org) listed materials are acceptable for production systems certified as organic. Organically acceptable materials (**OMRI** listed) are in the comments section.

Generics: Many pesticide active ingredients are available in generic formulations. For brevity, these formulations are not generally listed. Listed trade names are included to aid in identifying products and are not intended to promote the use of these products or to discourage the use of generic products. Generic products generally work similarly to their brand name counterparts, but formulation changes can impact efficacy and plant response. As with any new chemical, read and follow all label instructions. Chemical names are subject to change; please check the active ingredient for all materials.

Chapter VI—2019 N.C. Agricultural Chemicals Manual

The Pesticide Environmental Stewardship website is located at http://pesticidestewardship.org/Pages/default.aspx. Information on proper pesticide use and handling, calibration of equipment, reading pesticide labels, disposal, handling spills, and other topics is presented.

Pre-plant dips: Several products are registered as plant dips to manage pathogens or to protect plants just prior to field setting, but only a limited amount of research has been done with plant dips. In general, these treatments are not recommended except under specific circumstances; for example, if a disease has been diagnosed to be on the transplants. Products not labeled for dip treatments should not be used for dips, since poor plant performance has been observed in research trials.

Abound or Azaka (FRAC 11) — Mix 5 to 8 fl oz/100 gal of water. Dip plants for 2 to 5 minutes. Transplant treated plants as quickly as possible. This treatment has been developed for bare root transplants with a known problem of anthracnose. The dip is a whole plant dip, and some growers do not re-use the water for fear of spreading bacterial angular leaf spot and other diseases. It is reasonable to expect these fungicides to have some Rhizoctonia suppressive activity, but there are no research results to demonstrate a benefit. For managing Rhizoctonia, a root dip should suffice, rather than dipping whole plants. Rhizoctonia (and the black root rot problem) builds up over time, and it is doubtful that a root dip would offer much benefit for season long control. Growers must ensure root dip waste is properly disposed.

Switch 62.5WDG (FRAC 9 + 12) — Switch offers options for treating plants known to be infected with Colletotrichum species and has shown good efficacy in reducing losses due to the crown rot pathogen in bare root transplants (Colletotrichum gloeosporioides). Use 5 to 8 fl oz/100 gal water. Wash transplants to remove excess soil prior to dipping. Completely immerse planting stock in dip solution. Dip or expose plants for a minimum of 2 to 5 minutes. Do not reuse solution. Growers must ensure proper disposal of root dip waste. Plant treated plants as quickly as possible. Delayed planting could cause plant stunting.

Phosphites (FRAC P07) — Dip plants in 2.5 lb/100 gal (Aliette), 2 pints/100 gal (ProPhyt), or 2.5 pints/100 gal (Phostrol) for 15 to 30 minutes and then plant within 24 hours after treatment. This treatment should help to suppress Pythium and Phytophthora problems.

Little data are available for other plant dip products, including **Oxidate**, and it is doubtful that they offer management of root diseases. In most cases, root pathogens are internal to the tissue and are not controlled by surface disinfectants.

Fungicide Resistance Management Recommendations

Botrytis cinerea (gray mold) historically has a high potential to develop resistance, and recent data suggest a high percentage of strains are resistant to several important fungicides. Therefore, it is important to give these recommendations serious consideration:

1. Limit the number of times fungicides of the same group (same FRAC code) are applied in a single year.
2. Tank-mix a broad spectrum fungicide such as **captan or thiram with Topsin-M (a benzimidazole fungicide) since Topsin-M no longer has Botrytis activity due to resistance, but is helpful for several early season foliar diseases if present.**
3. Resistance profiles vary from farm-to-farm. Sample gray mold populations for their resistance profile through Clemson University (see details below).

It is currently suggested that the strobilurin (now called QoI; FRAC code 11) fungicides (Abound, Azaka, Cabrio, Intuity, Merivon, Pristine, and Quadris Top) be saved for use in controlling anthracnose diseases when there is a high potential for disease pressure. Captan or thiram should help suppress anthracnose when utilized in Botrytis or other disease control applications, but the QoI fungicides are currently the most efficacious materials for control of anthracnose. Some of these QoI materials may have activity against multiple pathogens other than the anthracnose pathogens, but unless anthracnose occurs in conjunction with these other diseases of concern, it is suggested that the QoIs not be used. With only 4 to 5 total applications of the QoI fungicides per crop, it is imperative that they be utilized effectively. Also, resistance management is extremely important with the QoIs; make sure to follow all resistance management guidelines. Recently, we have documented reduced activity with azoxystrobin (Abound, Azaka) with certain strains of the anthracnose fruit rot (AFR) pathogen. Other strains appear to be resistant to all QoI fungicides. Cabrio, Merivon, or Pristine have offered better control of AFR in recent research efforts AND if the strains are not resistant to QoI fungicides.

Powdery mildew — Monitor the field for the first signs of powdery mildew (leaf distortion and discoloration). Mildew in the fall does not appear to cause significant damage and may not reappear in the spring. *Therefore, most growers will not need to spray for powdery mildew.* However, fields have been observed in the fall with severe foliar disease incidence, and plant productivity may then be hampered, justifying control measures. Likewise, if powdery mildew pressure occurs in the spring and affects the fruit, the fruit will have a dull appearance and be unmarketable unless managed well. High tunnels favor powdery mildew development. Certain fungicides such as the QoI materials and Protocol are registered for powdery mildew, but are not recommended due to resistance selection.

Anthracnose (Colletotrichum spp.) — Most plantings are rarely at risk for anthracnose. Thus, anthracnose fungicides may not be needed. In most cases, contaminated plant sources are identified before or soon after planting. Know your plant source. If present, anthracnose on plants can cause petiole lesions (black sunken areas) stunting and plant death. Fall fungicide applications will be required for *Colletotrichum* only if plant source problems are identified, usually appearing as symptomatic plants or assayed for quiescent infections. Research results show that QoIs are more effective against the fruit rot pathogen ('*acutatum*') compared to the crown rot pathogen ('*gloeosporioides*'). Captan, Topsin M or Switch are as effective as the QoIs for controlling the crown rot pathogen. In general, it is most effective to save the QoI (FRAC 11) chemistry for spring applications and protect the fruit if anthracnose ('*acutatum*') is known to be present. Failure in management of some 'acutatum' populations has been observed with Abound or similar azoxystrobin products (see above).

Table 6-13A. Planting and Early Post-Planting: Disease Management

Management Options	Amount of Formulation per Acre	Effectiveness	REI	PHI	Comments (FRAC/IRAC Code)
Red stele; *Phytophthora* crown/root rots					
mefenoxam (Ridomil Gold SL)	1 pt/treated A	VG	see label	0 days	Apply in sufficient water in drip applications to move the fungicide into the root zone. Use proportionately less Ridomil Gold for band treatments. REI varies and is dependent upon method of application. Do not exceed 3 pts/year. **FRAC–4**
mefenoxam (Ultra Flourish)	2 pt/treated A	VG	see label	0 days	Apply in sufficient water to move the fungicide into the root zone. Use proportionately less mefenoxam for band treatments. **Do not exceed 6 pts per crop. FRAC–4**
metalaxyl (MetaStar 2E and generics)	2 qt/treated A	VG	see label	0 days	Apply in sufficient water to move the fungicide into the root zone. **Do not exceed 6 qt/treated A/year. FRAC–4**
phosphites, e.g. Aliette WDG ProPhyt, Phostrol	Various rates; see label	F	see label	0 days	Rates differ for foliar and drip applications. Phosphite-based chemicals are not as effective as Ridomil Gold. Consider phosphites if the pathogen is known to be resistant to mefenoxam or if root systems are poor AND foliage is healthy for chemical uptake. **FRAC–P07**
***Rhizoctonia* sp.(seedling root rot; basal stem rot)**					
Abound, Azaka	0.40 to 0.80 fl oz/1,000 row feet	F	4 hr	0 days	This is a drip irrigation application method. Can be considered especially for plug plants with poor root systems or plants placed into non-fumigated beds or beds with excess water in heavy soils. **FRAC–11**
Powdery mildew only					
Powdery mildew is not a common problem at this time of year; it may come in on transplants but usually does not persist or present an economic problem in open fields. There is a greater risk of powdery mildew in high tunnels. FRAC 11 products or product mixtures with FRAC 11 fungicides are labeled for use against powdery mildew but are not recommended for powdery mildew management in order to optimize FRAC 11 fungicide use for anthracnose fruit rot control.					
Procure 50WS Procure 480SC	4 to 8 oz 4 to 8 fl oz	E	12 hr	1 day	Check label for prohibited rotational crops. Do not plant leafy or fruiting vegetables within 30 days after application. Do not plant bulb or root vegetables within 60 days after application. Do not plant cotton, small cereal grains and all other crops not registered within one year of application. **FRAC–3**
Rally 40WSP	2.5 to 5 oz	E	24 hr	0 days	Rally is registered for control of leaf spot, leaf blight, and powdery mildew. **FRAC–3**
Rhyme	5 to 7 fl oz	E	12 hr	0 days	Rhyme is registered for control of powdery mildew and for drip application to manage charcoal rot. **FRAC–3**
Sulfur (multiple formulations)	See label	G	24 hr	1 day	Spray as needed. Avoid using in middle of a hot sunny day that may cause leaf burning. See label. **FRAC–M2**
Quintec	4 to 6 fl oz	E	24 hr	1 day	Do not use more than 4 times per crop and no more than 2 times in a row. Rotate with other mildewcides. See label. **FRAC–13**
Gatten	6 to 8 fl oz	Unknown	0 hr	0 hr	Crop can be harvested when spray has dried. Do not use more than 5 times per year. **FRAC–U13**
Protocol	1.33 pt	G	24 hr	1 day	Premix of 2 active ingredients, thiophanate-methyl (**FRAC–1**) and propiconazole (**FRAC–3**). No more than 2 sequential applications should be made before alternating with fungicides that have a different mode of action.
Anthracnose fruit rot ('acutatum')					
Pristine WG	18.5 to 23 oz	E	12 hr	0 days	Premix of two active ingredients, pyraclostrobin (**FRAC–11**) and boscalid (**FRAC–7**). See resistance management notes above.
Merivon	5.5 to 8 fl oz	E	12 hr	0 days	Premix of two active ingredients, pyraclostrobin (**FRAC–11**) and fluxapyroxad (**FRAC–7**). See resistance management notes above.
Luna Sensation	4.0 to 7.6 fl oz	E	12 hr	0 days	Premix of 2 active ingredients, trifloxystrobin (**FRAC–11**) and fluopyram (**FRAC–7**). See resistance management notes above.
Cabrio 20EG	12 to 14 oz	E	24 hr	0 days	Active ingredient, Pyraclostrobin (**FRAC–11**)
Abound, Azaka	6.2 to 15.5 fl oz	E	4 hr	0 days	Failure in management of some 'acutatum' populations has been observed with Abound and similar products. **FRAC–11**
Intuity	6 fl oz	TBA	0 hr	0 days	See notes above to manage risk of developing fungicide resistance. **FRAC-11**. No more than 2 applications should be made per season for resistance management.
Tilt and multiple generics	4 fl oz	G	12 hr	0 days	No more than 2 sequential applications should be made before alternating with fungicides that have a different mode of action. **FRAC–3**
Quadris Top	12 to 14 fl oz	G	12 hr	0 days	Premix of 2 active ingredients, azoxystrobin (**FRAC–11**) and difenoconazole (**FRAC–3**). No more than 2 sequential applications should be made before alternating with fungicides that have a different mode of action.
Protocol	1.33 pt	G	24 hr	1 day	Premix of 2 active ingredients, thiophanate-methyl (**FRAC–1**) and propiconazole (**FRAC–3**). No more than 2 sequential applications should be made before alternating with fungicides that have a different mode of action.
Anthracnose crown rot ('gloeosporioides' crown rot)					
Captan 50W Captan 80WDG	3 to 6 lb (50W) 1.87 to 3.75 lb (80W)	F	24 hr	1 day	In plantings known to be infected with the anthracnose crown rot pathogen, consider applying captan plus Topsin-M at 10- to 14-day intervals, for a total of 2 to 3 applications in the fall. **FRAC–M4**
Captec 4L	1.5 to 3.0 qt/100 gal	F	24 hr	1 day	**FRAC–M4**
Topsin-M 70WP	1 lb	F	12 hr	1 day	For suppression only. See notes above on resistance management. **FRAC–1**
Quadris Top	12 to 14 fl oz	G	12 hr	0 days	Same as above. **FRAC–3 + 11**

Note: A treated acre is the amount of area under the plastic i.e. in most strawberry fields there is about one acre under plastic on two acres of land.

New Leaf Growth to Pre-Bloom: Disease Management

Table 6-13B. New Leaf Growth to Pre-bloom: Disease Management

Management Options	Amount of Formulation per Acre	Effectiveness	REI	PHI	Comments (FRAC/IRAC Code)
Botrytis crown rot					
Botrytis crown rot may occur during warm winter periods after early bloom is killed by frost and colonized by *Botrytis*. The pathogen typically grows down the flower stem (peduncle) and colonizes the upper crown tissue, causing death of the leaf petioles, particularly if plants are large or planted densely.					
Rovral 4F and generics (iprodione)	1.5 to 2 pt	VG	24 hr	see comments	Do not apply after first fruiting flower, and do not make more than 1 application of Rovral per season. Crown rot control during the early winter and prior to bloom may be the most effective use of the one Rovral application allowed in strawberries. **FRAC–2**
Switch 62.5 WG	11 to 14 oz	VG	12 hr	0 days	See resistance management information above. **FRAC–9 + 12**
Captan 50W	3 to 6 lb (50W)	F	24 hr	1 day	See notes below. **FRAC–M4**
Captan 80WDG	1.9 to 3.8 lb (80WDG)				
Botrytis					
Remove dead and dying leaves just before bloom		Importance: F Efficacy: G			Symptomatic leaf removal is effective but may not be economical if fungicides are heavily used for Botrytis management. If anthracnose fruit rot is present, hand-pruning plants may create more anthracnose disease problems. Do not use QoI fungicides - these should be saved for use as fruit develop and to avoid selection of resistant populations.
Leaf spots, Leaf blights and Powdery Mildew generally do not become economically important diseases in the fall or early spring. Thus, fungicides are generally not required for these problems. Thresholds have not been established, so the need for fungicides should be determined on a farm-by-farm basis depending on the disease pressure present. Phomopsis and leaf spot may be associated with plant sources; therefore, disease incidence can vary from year to year. Warm wet weather favors disease progress. See previous notes on powdery mildew under "Planting and Early Post-planting: Disease Management." In the spring, monitor fields closely observing the underside of strawberry leaves to determine if powdery mildew is present. FRAC 11 products or mixtures with FRAC 11 fungicides are labeled but not listed to manage powdery mildew and leaf spots in order to optimize FRAC 11 fungicide use for anthracnose fruit rot control.					
Phomopsis leaf blight					
Captan 50W	3 to 6 lb	F	24 hr	1 day	When foliar symptoms appear, make 1 or 2 captan applications plus Topsin-M at a 10- to 14-day interval for better control than captan products alone would provide. Do not apply more than 24 lb captan active ingredient per acre per year. **FRAC–M4**
Captan 80WDG	1.87 to 3.75 lb	F	24 hr	1 day	
Captec 4L	1.5 to 3.0 qt/100 gal	F	24 hr		
Topsin-M 70WP	1 lb	++	12 hr	1 day	See note above on resistance management. **FRAC–1**
Rally 40WSP	2.5 to 5 oz	++++	24 hr	0 days	Rally is registered for control of leaf spot, leaf blight, and powdery mildew. Do not apply more than 30 oz per acre. **FRAC 3**
Common leaf spot, leaf scorch, leaf blight (e.g. Mycosphaerella, Phomopsis, Gnomonia)					
Captan 50W or Captan 80 WDG plus Topsin-M 70WP	1 lb (50W); 1.6 lb (80WDG) 1 lb	G	24 hr 24 hr	1 day 1 day	When foliar symptoms appear, make 1 or 2 captan applications plus Topsin-M at a 10- to 14-day interval for better control than captan products alone would provide. Do not apply more than 24 lb captan active ingredient per acre per year. Do not tank mix captan products with highly alkaline pesticides, such as Bordeaux mixture. See resistance management notes above. **FRAC–M4, FRAC–1**
Captan 50W Captan 80 WDG	3 to 6 lb 1.87 to 3.75 lb	F	24 hr	1 day	**FRAC–M4**
Thiram	2.6 qt	F	24 hr 24 hr	1 day 3 days	**FRAC–M3**
Rally 40WSP	2.5 to 5 oz	VG	24 hr	0 days	Rally is registered for control of leaf spot, leaf blight, and powdery mildew. Do not apply more than 30 oz per year. **FRAC–3**
Powdery mildew only					
Procure 480SC	4 to 8 fl oz	E	12 hr	1 day	Check label for prohibited rotational crops. Do not plant leafy or fruiting vegetables within 30 days after application. Do not plant bulb or root vegetables within 60 days after application. Do not plant cotton, small cereal grains and all other crops not registered within 1 year of application. **FRAC–3**
Rally 40WSP	2.5 to 5 oz	E	24 hr	0 days	Rally is registered for control of leaf spot, leaf blight, and powdery mildew. Do not apply more than 30 oz per year. **FRAC–3**
Rhyme	5 to 7 fl oz	?	12 hr	0 days	Rhyme is registered for control of powdery mildew and for drip application to manage charcoal rot. **FRAC-3**
Quintec	4 to 6 fl oz	E	24 hr	1 day	Do not use more than 4 times per crop and no more than 2 times in a row. Rotate with other mildewcides. Rotation to non-registered crops less than 30 days after application is prohibited. **FRAC–13**
Torino	3.4 oz	?	4 hr	0 days	Do not make more than 2 applications per year. Do not apply more than once every 14 days. **FRAC–U06**
Gatten	6 to 8 fl oz	Unknown	0 hr	0 hr	Crop can be harvested when spray has dried. Do not use more than 5 times per year. **FRAC-U13**
Tilt and other generics	4 fl oz	G	12 hr	0 days	No more than 2 sequential applications should be made before alternating with fungicides that have a different mode of action. **FRAC–3**
Angular (bacterial) leaf spot (*Xanthomonas fragariae*)					
Basic copper sulfate (various formulations)	See labels	P	48 hr	0 hr	Angular (bacterial) leaf spot can be a serious problem during cool, wet conditions. These compounds provide some control unless conditions highly favor disease. Repeat applications at 7 to 10 day intervals. Discontinue when phytotoxicity appears, usually after 4 to 5 applications. NOTE: All copper sulfate, copper hydroxide and other copper products labeled for strawberry can be used, but check label for the proper rate because different products will contain different percentages of active ingredient. **FRAC–M1.**
copper hydroxide (various formulations)	See labels	P	24 hr	0 days	
copper salts of fatty and rosin acids (various formulations)	See labels	P	12 hr	0 days	
cuprous oxide (various formulations)	1.05 to 4.2 lbs a.i. (various formulations)	P	12 hr	0 days	

Table 6-13B. New Leaf Growth to Pre-bloom: Disease Management

Management Options	Amount of Formulation per Acre	Effectiveness	REI	PHI	Comments (FRAC/IRAC Code)
Angular (bacterial) leaf spot (*Xanthomonas fragariae*) (continued)					
Actigard 50WG	0.5 to 0.75 oz./a	P	12 hr	0 days	Labeled for suppression; Do not apply to stressed plants. **DO NOT EXCEED MAXIMUM RATE.** Actigard is a plant activator and has no direct activity on the bacteria. See supplemental label for details. **FRAC-21**
Red stele; Phytophthora crown/root rots					
mefenoxam (Ridomil Gold SL and other formulations)	1 pt	++++	12 hr	0 days	Strawberry plants initiate considerable root growth in the early spring. Time control applications in problem fields when new growth begins in the spring. Apply in sufficient water to move the fungicide into the root zone. Use proportionately less fungicide for band treatments (e.g., for drip applications). **FRAC-4**
Ultra Flourish	2 pt				
metalaxyl (MetaStar and generics)	2 qt/treated A	++++	48 hr	0 days	
phosphites (e.g., Aliette, ProPhyt, Phostrol)	Various rates; see label	++	12 hr	0 days	The phosphite-based chemicals are not as effective as Ridomil Gold. Consider phosphites if the pathogen is known to be resistant to mefenoxam or if strawberry plants have poor root systems but sufficient foliage for chemical uptake. **FRAC-33**

Early Bloom (10%) and into Harvest: Disease Management
F. J. Louws, Entomology and Plant Pathology

The primary diseases of concern at early bloom and into harvest are **Botrytis fruit rot (BFR)** and **anthracnose fruit rot (AFR)**. Most growers rarely experience anthracnose problems and may not need an anthracnose management program. Several **key principles** should be kept in mind:

1. Abound, Azaka, Cabrio, Intuity, Merivon, Pristine, and Luna Sensation belong to the same family of chemicals (QoI; FRAC 11 chemistry). Pyraclostrobin (Cabrio, Merivon, and Pristine) has offered better control of AFR in recent research efforts. No more than 2 applications of a FRAC 11 fungicide should be made per season for resistance management. Strategic timing is necessary. Pristine, Luna Sensation and Merivon also have a second chemical that has good broad spectrum activity against a number of diseases, especially those caused by Botrytis. QoI resistance has been found in 'acutatum' populations in the south. The problem tends to be plant-source-associated.

2. Captan, thiram, and Switch offer a broad spectrum of disease control. Switch is modest against AFR in NC research.

3. Polyoxin D zinc salt (PhD; OSO 5%SC) is as effective as captan for Botrytis at high label rates and can help reduce reliance on fungicides that have resistance concerns.

4. Elevate should not be used more than twice per season due to resistance concerns. It is effective against Botrytis but no other fungal pathogens.

5. High-risk fungicides of the same chemical class (FRAC group) should not be applied in consecutive applications.

6. CaptEvate is a premix of captan and Elevate which has good broad-spectrum activity.

7. Bloom sprays are the most important for managing Botrytis, because 90% of fruit infection occurs through the flower at bloom. Recent research suggests bloom sprays are also critical for AFR.

8. Fruit rot diseases develop rapidly during wet periods or in poorly ventilated locations. Control is easier when initiated before the problem develops. Spray coverage is important and dependent on nozzle condition, tractor speed, pressure, and plant density. Spray coverage can be checked with water sensitive cards.

Fungicide Selection for Botrytis and Anthracnose Fruit Rot Management

Management of Botrytis fruit rot (gray mold; **BFR**) and anthracnose fruit rot (**AFR**) caused by "*Colletotrichum acutatum*" has become more complex. Growers need to use products that work against resistant strains of BFR and manage AFR. We developed a new table to help with the decision process (see below).

Table 6-14A (below) shows our current understanding of the efficacy of fungicides for the Southeastern US (north of Florida). Efficacy in the table is indicated as follows: E = excellent, VG = very good, G = good, F = fair, P = poor. A large number of farms are experiencing problems with Botrytis strains that are resistant to one or more fungicide. (Color codes are similar to the codes in the MyIPM app).

BOTRYTIS CONTROL: *Botrytis cinerea* historically has a high potential to develop resistance. Therefore, it is important to give these recommendations serious consideration:

1. If a Botrytis spray is needed before bloom (e.g. to control Botrytis crown rot) use Rovral (FRAC 2).

2. Use members of any FRAC group (except M3 or M4) no more than twice per season (For example, if you used Fontelis once and Merivon once you maxed out the 2 applications for FRAC 7 fungicides).

3. Resistance profiles vary from farm-to-farm. Sample gray mold populations for their resistance profile through Clemson University (http://www.clemson.edu/extension/peach/commercial/diseases/index.html). Based on samples submitted to Clemson, the **Fungicide Decision Management Table** below shows a decision guide to manage Botrytis fruit rot. If you do not know your profile, it is best to avoid over-reliance on products where resistance is prevalent. **If in doubt, follow Decision Code E-1 since this will address the most common resistance issues for BFR control.** If you also have FRAC 11 resistance for AFR, follow Decision Code E-2.

4. Specific plant sources may be identified as having AFR infestations. In that case growers need to manage both BFR and AFR.

Chapter VI—2019 N.C. Agricultural Chemicals Manual

AFR CONTROL: Resistance to FRAC 11 fungicides (Pristine, Cabrio, Inuity, Merivon, Abound, Azaka, Luna Sensation) has been found in Florida, North Carolina and California; problems tend to be plant-source associated. Therefore, it is a good idea to use the FRAC 11 fungicides only in mixture at the lower labeled rate with the higher labeled rate of captan products (Captan or Captec) alternated with captan alone. If you know the resistance profile, see the **Fungicide Decision Management Table** below. Also, recently, we have documented reduced activity with azoxystrobin (Abound, Azaka) with certain strains of the AFR pathogen. Cabrio and FRAC 7+11 products have offered better control of AFR in recent research efforts and if the strains are not resistant to FRAC 11 fungicides.

FRAC 7+11 products can be used if your resistance profile shows the FRAC 7 component is still effective against BFR. If FRAC 7 resistance is diagnosed or you don't know, we recommend using Cabrio (plus captan). Like BFR, our data shows early bloom sprays are also critically important for AFR management.

For cases when there is no anthracnose and growers need to focus on gray mold control (most fields) follow Decision Code A below.

Options: For a reduced fungicide program, initiate applications at FIRST bloom as above but apply subsequent sprays before predicted wet weather that favors Botrytis; end applications about 26 to 30 days before expected final harvests. Increase the time between spray applications when dry weather persists. Research trials have documented that 4 sprays during bloom often are sufficient to offer season-long BFR control. Also, consult available forecasting models linked through the Strawberry IPM guide.

For cases when anthracnose is present and there is no known resistance within the Botrytis population follow Decision Code B-1.

Before predicted periods of cool and wet weather during bloom, use Switch (FRAC 12+9) for better Botrytis control. Use Switch with captan if Botrytis pressure is expected to be heavy. Switch also has decent anthracnose control. FRAC 7+11 products or Cabrio show the best efficacy against AFR under high anthracnose pressure in research studies and either can be used if there is no resistance to FRAC 7 fungicides (an active ingredient in FRAC 7+11 products). Also, if weather conditions (warm & wet) favor AFR, or you start to approach the upper limit of FRAC 11 fungicides allowed (4 to 5 applications), consider rotating to a tank-mix of captan + Tilt.

See the Strawberry IPM Guide (http://www.smallfruits.org/ipm-guides.html) for more detailed Information on total IPM Programs and download the MyIPM-SED app to learn more about disease/pest management and FRAC codes. Also see: Diagnosis tool: https://diagnosis.ces.ncsu.edu/strawberry/

Strawberry Disease Factsheets: https://strawberries.ces.ncsu.edu/strawberries-diseases/ for additional information.

Table 6-14A. Fungicide Selection for Botrytis and Anthracnose Fruit Rot Management

	FRAC	BFR	Botrytis Resistance	AFR
Captan or Captec	M4	G	None	G
Captevate	M4 + 17	VG	Prevalent for 'Elevate'	G
Thiram	M3	G	None	F
Fracture	M12	P	No data	No Data
Topsin M	1	Not effective	Widespread	Not effective
Rovral	2	G	Prevalent	Not effective
Tilt and generics	3	Not effective	Not applicable	F
Fontelis	7	VG	Prevalent	F
Kenja	7	VG	Not prevalent	Not effective
Scala	9	G	Prevalent	Not effective
Pristine	7 + 11	G	Prevalent	VG*
Merivon	7 + 11	VG	Prevalent	VG*
Luna Sensation	7 + 11	VG	Not Prevalent	VG*
Cabrio	11	Not effective	Widespread	VG*
Abound or Azaka	11	Not effective	Widespread	VG*
Switch	12 + 9	E	Not Prevalent	G
Elevate	17	E	Prevalent	Not effective
Ph-D, OSO, Tavano	19	G	Not Prevalent	No data

* Resistance issues to FRAC 11 Fungicides have been reported in FL, CA and NC in the last 3 years. Problems tend to be plant source related.

Table 6-14B. Fungicide Decision Management Table

Decision Code	Fungicide Resistance Issue		Sprays during bloom and fruit ripening					
	Botrytis	Anthracnose	1	2	3	4	5	6
A	No resistance	No Disease	12 + 9	7	thiram + 17	thiram + 19	captan	Go to 1
B-1	No resistance	No resistance	captan + 17	11 + 7	12 + 9	captan + 19	11 + 7	Go to 1
C-1	FRAC 7	No resistance	captan + 17	captan + 11	12 + 9	captan + 11	captan + 19	Go to 1
D-1	FRAC 17	No resistance	thiram + 11	captan	12 + 9	11 + 7	captan + 19	Go to 1
E-1	FRAC 7 + 17	No resistance	thiram + 11	12 + 9	captan	captan + 11	12 + 9	Go to 1
F-1	FRAC 12 + 9	No resistance	captan + 17	11 + 7	thiram	11 + 7	captan + 19	Go to 1
G-1	FRAC 12 + 9 + 17	No resistance	thiram + 11	captan	thiram	captan + 11	captan + 19	Go to 1
H-1	FRAC 12 + 9 + 7	No resistance	captan + 17	captan + 11	thiram	captan + 11	captan + 19	Go to 1
I-1	FRAC 12 + 9 + 7 + 17	No resistance	thiram + 11	captan	thiram	captan + 11	captan + 19	Go to 1
B-2	No resistance	FRAC 11	captan + 17	captan + 7	12 + 9	captan + 19	captan + 7	Go to 1
C-2	FRAC 7	FRAC 11	captan + 17	captan	12 + 9	captan + 17	12 + 9	Go to 1
D-2	FRAC 17	FRAC 11	captan + 7	12 + 9	captan + 7	12 + 9	captan + 19	Go to 1
E-2	FRAC 7 + 17	FRAC 11	12 + 9	captan	captan + 19	12 + 9	captan	Go to 1
F-2	FRAC 12 + 9	FRAC 11	captan + 17	captan + 7	thiram	captan + 19	captan	Go to 1
G-2	FRAC 12 + 9 + 17	FRAC 11	thiram + 7	captan	captan + 7	captan	captan + 19	Go to 1
H-2	FRAC 12 + 9 + 7	FRAC 11	captan + 17	captan	thiram	captan + 17	captan + 19	Go to 1
I-2	FRAC 12 + 9 + 7 + 17	FRAC 11	thiram	captan	captan + 19	captan	captan + 19	Go to 1

Decision Management Code Guidelines:

A: Botrytis is expected with no resistance and plants are verified to be anthracnose free.

B-1 to I-1: The anthracnose pathogen is known to be <u>sensitive</u> to FRAC 11 products

B-2 to I-2: The anthracnose pathogen is known to be <u>resistant</u> to FRAC 11 products

NOTE: For B-1 to I-1: If anthracnose is known to be absent, then the FRAC 11 products are **NOT** needed.

B-1: Botrytis is expected; no resistance is documented, and plants are verified to harbor the anthracnose pathogen

C-1: Botrytis is resistant to FRAC 7 products and plants are verified to harbor the anthracnose pathogen

Etc.,

Table 6-15. Early Bloom (10%) and into Harvest: Disease Management

Management Options	Amount of Formulation per Acre	Effectiveness	REI	PHI	Comments (FRAC/IRAC Code)
Switch also has decent anthracnose control. Pristine, Merivon, or Cabrio show the best efficacy against AFR under high anthracnose pressure in research studies and either can be used if there is no resistance to FRAC 7 fungicides. Also, if weather conditions (warm & wet) favor AFR, or you start to approach the upper limit of FRAC 11 fungicides allowed (4 to 5 applications), consider rotating to a tank mix of captan + Tilt.					
Botrytis gray mold					
Captan 50W Captan 80WDG	3 to 6 lb (50W) or 1.9-3.8 lb (80W)	G	24 hr	1 day	See suggested schedule above. Do not apply more than 24 lb of captan active ingredient per acre per year. **FRAC-M4**
Captec 4L	2.5 qt	G	24 hr	1 day	
Switch 62.5WG	11 to 14 oz	E	12 hr	0 days	Do not apply more than twice per season due to resistance management. See resistance management notes. **FRAC-12, FRAC-9**
Ph-D WDG OSO 5% SC	6.2 oz	G	4 hr	0 days	Do not apply more than twice per season due to resistance management. **FRAC-19**
Thiram	2.6 qt	G	24 hr	3 days	Make 3 to 5 applications at 10-day intervals. Thiram is a broad spectrum fungicide similar to captan. **FRAC-M3**
Elevate 50WDG	1.5 lb	E	4 hr	0 days	Do not apply more than twice per season due to resistance management. Under light pressure, 1.0 lb Elevate plus captan may be used (see label). **FRAC-17**
Fontelis	16 to 24 fl oz	E	12 hr	0 days	Do not apply FRAC-7 products more than twice per season due to resistance management. Some matted row cultivars may show phytotoxicity (see label). **FRAC-7**
Kenja 400SC	13.5 to 15.5 fl oz	E	12 hr	0 days	Do not apply FRAC-7 products more than twice per season due to resistance management. Some matted row cultivars may show phytotoxicity (see label). **FRAC-7**
CaptEvate 68 WDG	3.5 to 5.25 lb	E	24 hr	0 days	CaptEvate is a combination product of captan plus Elevate. Do not make more than 2 consecutive applications before switching to a fungicide with a different mode of action. Do not apply more than 21.0 lb/acre/season. With plastic mulch, do not apply within 16 feet of naturally vegetated or aquatic areas. **FRAC-M4, FRAC-17**
Scala	18 fl oz 9 fl oz	G	12 hr	1 day	Use lower rate only in a tank mix with another fungicide active against gray mold (e.g. captan or Thiram). **FRAC-9**

Table 6-15. Early Bloom (10%) and into Harvest: Disease Management

Management Options	Amount of Formulation per Acre	Effectiveness	REI	PHI	Comments (FRAC/IRAC Code)
Botrytis gray mold (contined)					
Luna Tranquility	16 to 27 fl oz	E	12 hr	0 day	Do not use any FRAC 9 or 7 products more than twice per season for resistance management. **FRAC-9, FRAC-7**
Luna Sensation	6 to 7.6 fl oz	E	12 hr	0 day	Do not use any FRAC 11 or 7 products more than twice per season for resistance management. **FRAC-11, FRAC-7**
Fracture	24.4 to 36.6 fl oz	P	4 hr	1 day	Active ingredient is a protein extract of sweet white Lupin seeds. Some efficacy can be expected at the highest rate. **FRAC-BM01**
Botrytis gray mold and Anthracnose fruit rot (acutatum)					
Products in this section are labeled for both Botrytis and anthracnose.					
Pristine WG	18.5 to 23 oz	E	12 hr	0 days	Do not apply more than 2 applications per acre per crop year. See page 20. **FRAC-11, FRAC-7**
Luna Sensation	6 to 7.6 fl oz	E	12 hr	0 day	Do not use any FRAC 11 or 7 products more than twice per season for resistance management. **FRAC-11, FRAC-7**
Merivon	8 to 11 fl oz	E	12 hr	0 days	Do not apply more than 2 applications per acre per crop year. See page 20. **FRAC-11, FRAC-7**
Captan 50W	3 to 6 lb (50W)	G	24 hr	1 day	For better control and resistance management, use captan applications plus Topsin-M (see label). See suggested schedule above. Do not apply more than 24 lb of captan active ingredient per acre per year. **FRAC-M4**
Captan 80 WDG	1.87 to 3.75 lb (80WDG)				
Anthracnose fruit rot (acutatum)					
Abound	6.2 to 15.5 fl oz	VG (failure found in some fields)	4 hr	4 hr	See notes to manage risk of developing fungicide resistance. In recent research, Abound and similar products have performed less well than Cabrio/Pristine. **FRAC-11**
Azaka	6.0 to 15.5 fl oz				
Intuity	6 fl oz	?	12 hr	1 day	See notes to manage risk of developing fungicide resistance. **FRAC-11** No more than 2 applications should be made per season for resistance management.
Luna Sensation	4 to 7.6 fl oz	E	12 hr	0 days	Do not use any FRAC 11 or 7 products more than twice per season for resistance management. **FRAC-11, FRAC-7**
Merivon	5.5 to 8 fl oz	VG	12 hr	0 days	See notes to manage risk of developing fungicide resistance. **FRAC-11, FRAC-7**
Pristine WG	18.5 to 23 oz	VG	12 hr	0 days	See notes to manage risk of developing fungicide resistance. **FRAC-11, FRAC-7**
Cabrio EG	12 to 14 oz	VG	12 hr	0 days	See notes to manage risk of developing fungicide resistance. **FRAC-11**
Tilt and multiple generics	4 fl oz	G?	12 hr	0 days	Registered for Anthracnose Fruit Rot only. No more than 2 sequential applications should be made before alternating with fungicides that have a different mode of action. Not registered for Anthracnose crown rot control. **FRAC-3**
Quadris Top	12 to 14 fl oz	G	12 hr	0 days	Premix of 2 active ingredients, azoxystrobin (**FRAC-11**) and difenoconazole (**FRAC-3**). No more than 2 applications should be made per season for resistance management.
Protocol	1.33 pt	G	24 hr	1 day	Premix of 2 active ingredients, thiophanate-methyl (**FRAC-1**) and propiconazole (**FRAC-3**). No more than 2 applications should be made per season for resistance management.
Anthracnose crown rot ('gloeosporioides')					
Captan 50W	3 to 6 lb (50W)	F	24 hr	1 day	In plantings known to be infected with the anthracnose crown rot pathogen, consider applying captan plus Topsin-M at 10- to 14-day intervals, for a total of 2 to 3 applications in the fall. **FRAC-M4**
Captan 80WDG	1.87 to 3.75 lb (80W)				
Captec 4L	2.5 qt	F	24 hr	1 day	**FRAC-M4**
Topsin-M 70WP	1 lb	F	12 hr	1 day	See note on resistance management. **FRAC-1**
Quadris Top	12 to 14 fl oz	G	12 hr	0 days	Same as above. **FRAC-3, FRAC-11**
Powdery mildew (only)					
Procure 50WS	4 to 8 oz	E	12 hr	1 day	Check label for prohibited rotational crops. Do not plant leafy or fruiting vegetables within 30 days after application. Do not plant bulb or root vegetables within 60 days after application. Do not plant cotton, small cereal grains and all other crops not registered within one year application. **FRAC-3**
Procure 480SC	4 to 8 fl oz				
Rally 40WSP	2.5 to 5 oz	E	24 hr	1 day	Rally is registered for control of leaf spot, leaf blight, and powdery mildew. Do not apply more than 30 oz per year. **FRAC-3**
Rhyme	5 to 7 fl oz	?	12 hr	0 days	Rhyme is registered for control of powdery mildew and for drip application to manage charcoal rot. **FRAC-3**
Gatten	6 to 8 fl oz	Unknown	0 hr	0 days	Crop can be harvested when spray has dried. Do not use more than 5 times per year. **FRAC-U13**
Quintec	4 to 6 fl oz	E	24 hr	1 day	Do not use more than 4 times per crop and no more than 2 times in a row. Rotate with other mildewcides. Rotation to all other crops within 1 year after application, unless Quintec is registered for use on those crops, is prohibited. **FRAC-13**
Torino	3.4 oz	VG	4 hr	0 days	Do not make more than 2 applications per year. Do not apply more than once every 14 days. **FRAC-U06**
Powdery mildew and Anthracnose (acutatum)					
Abound	6.2 to 15.5 fl oz	VG	4 hr	4 hr	See notes on page 20 to manage risk of developing fungicide resistance. **FRAC-11**
Azaka	6.0 to 15.5 fl oz				
Pristine WG	18.5 to 23 oz	VG	12 hr	0 days	See notes on page 20 to manage risk of developing fungicide resistance. **FRAC-11, FRAC-7**

Table 6-15. Early Bloom (10%) and into Harvest: Disease Management

Management Options	Amount of Formulation per Acre	Effectiveness	REI	PHI	Comments (FRAC/IRAC Code)
Powdery mildew and Anthracnose (acutatum) (continued)					
Luna Sensation	6-7.6 fl oz	E	12 hr	0 day	Do not use any FRAC 11 or 7 products more than twice per season for resistance management. **FRAC–11, FRAC–7**
Cabrio EG	12 to 14 oz	VG	12 hr	0 days	See notes on page 20 to manage risk of developing fungicide resistance. DO NOT EXCEED 1.5 QT/YEAR. **FRAC–11**
Tilt and multiple generics	4 fl oz	G	12 hr	0 days	Registered for Anthracnose Fruit Rot only. No more than 2 sequential applications should be made before alternating with fungicides that have a different mode of action. Not registered for Anthracnose crown rot control. **FRAC–3**
Intuity	4 fl oz	?	12 hr	1 days	See notes to manage risk of developing fungicide resistance. **FRAC–11**. No more than 2 applications should be made per season for resistance management.
Quadris Top	12 to 14 fl oz	G	12 hr	0 days	Premix of two active ingredients, azoxystrobin (**FRAC–11**) and difenoconazole (**FRAC–3**). No more than 2 applications should be made per season for resistance management.

Relative Effectiveness of Various Chemicals for Strawberry Disease Control

F. J. Louws, Entomology and Plant Pathology

Table 6-16. Effectiveness of Various Chemicals for Strawberry Disease Control

Fungicide	FRAC Code	Angular Leaf Spot	Anthracnose (crown root)	Anthracnose (fruit rot)	Botrytis crown	Botrytis fruit rot	Common leaf	Leaf blight	Leather rot	Mucor fruit rot	Phytophthora	Powdery Mildew	Red stele root rot	Rhizopus rot
copper (various)	M01	P^P	NC	NC	P	NC	P^P	NC	P^P	NC	NC	NC	NC	NC
sulfur (various)	M02	NC	NC	NC	NC	NC	NC	NC	NC	NC	NC	G^R	NC	NC
thiram (Thiram SC)	M03	NC	G	G	F	G	F	F	F	F	NC	NC	NC	F
captan (Captan 50W, others)	M04	NC	F	G	F	G	F	F	F	F	NC	NC	NC	F
thiophanate-methyl (Topsin M 70WP)	1	NC	G^R	NC	G^R	G^R	G	G	NC	XX	NC	F^R	NC	NC
iprodione (Rovral 4F)	2	NC	NC	NC	VG^R	VG^R	G	NC	NC	XX	NC	NC	NC	NC
flutriafol (Rhyme)	3	ND	ND	ND	ND	ND	ND	ND	ND	ND	ND	E	ND	ND
myclobutanil (Rally 40WSP)	3	NC	NC	NC	NC	NC	VG	VG	NC	NC	NC	E	NC	NC
propiconazole (Tilt, others)	3	NC	F	F	NC	NC	F	ND	NC	NC	NC	G^R	NC	NC
tetraconazole (Mettle 125ME)	3	ND	ND	ND	ND	ND	ND	ND	ND	ND	ND	E	ND	ND
triflumizole (Procure 50WS, Procure 480SC)	3	NC	NC	NC	NC	ND	ND	ND	NC	NC	NC	E^R	NC	NC
thiophanate-methyl + propiconazole (Protocol)	1 + 3	NC	G^R	G^R	G^R	G^R	G	G	NC	XX	NC	G^R	NC	NC
isofetamid (Kenja 400SC)	7	NC	NC	NC	E^R	E^R	NC	NC	NC	NC	NC	G^R	NC	NC
penthiopyrad (Fontelis)	7	NC	NC	NC	E^R	E^R	NC	NC	NC	NC	NC	G^R	NC	NC
fluopyram + pyrimethanil (Luna Tranquility)	7 + 9	NC	NC	NC	E^R	E^R	NC	NC	NC	NC	NC	G^R	NC	NC
pyrimethanil (Scala)	9	NC	NC	NC	G^R	G^R	NC	NC	NC	NC	NC	NC	NC	NC
Strobilurins:														
azoxystrobin (Abound; Azaka, others)	11	NC	G	G/E	F	F	F	NC	VG	NC	NC	F	NC	NC
mandestrobin (Intuity)	11	ND	ND	ND	ND	ND	ND	ND	ND	ND	ND	E	ND	ND
pyraclostrobin (Cabrio EG)	11	NC	G	VG/E	F	F	F	NC	VG	NC	NC	F	NC	NC
azoxystrobin + difenoconazole (Quadris Top)	11 + 3	NC	G	G	F	F	G	ND	F	NC	NC	G	NC	NC
azoxystrobin + propiconazole (QuiltXcel)	11 + 3	NC	VG	G	NC	NC	ND	ND	NC	NC	NC	G	NC	NC
pyraclostrobin + boscalid (Pristine)	11 + 7	NC	G	E	VG^R	VG^R	VG	VG	NC	ND	NC	F	NC	ND
pyraclostrobin + fluxapyroxad (Merivon)	11 + 7	NC	G	E	E^R	E^R	VG	VG	NC	ND	NC	F	NC	ND
trifloxystrobin + fluopyram (Luna Sensation)	11 + 7	NC	G	E	E^R	E^R	VG	VG	NC	ND	NC	F	NC	ND
cyprodinil + fludioxonil (Switch)	12 + 9	ND	G	F	VG	E	P	P	NC	ND	NC	NC	ND	ND
quinoxyfen (Quintec)	13	NC	NC	NC	NC	NC	NC	NC	NC	NC	NC	E	NC	NC
fenhexamide (Elevate 50 WDG)	17	NC	NC	NC	E^R	E^R	NC	NC	NC	NC	NC	NC	NC	NC
fenhexamide + captan (CaptEvate 68 WDG)	M04 + 17	NC	F	G	E^R	E^R	G	F	F	F	NC	NC	NC	F
polyoxin D (Ph-D; OSO; 5%SC)	19	ND	ND	ND	G	G	ND	ND	ND	ND	ND	ND	ND	ND

Table 6-16. Effectiveness of Various Chemicals for Strawberry Disease Control

Fungicide	FRAC Code	Angular Leaf Spot	Anthracnose (crown root)	Anthracnose (fruit rot)	Botrytis crown	Botrytis fruit rot	Common leaf	Leaf blight	Leather rot	Mucor fruit rot	Phytophthora	Powdery Mildew	Red stele root rot	Rhizopus rot
cyflufenamid (Torino)	U06	NC	NC	NC	NC	NC	NC	NC	NC	NC	NC	VG	NC	
mefenoxam (Ridomil Gold SL, Ultra Flourish)	4	NC	NC	NC	NC	NC	NC	NC	VG[R]	NC	VG	NC	VG	NC
metalaxyl (MetaStar 2E, others)	4	NC	NC	NC	NC	NC	NC	NC	VG[R]	NC	VG	NC	VG	NC
fosetyl-Al (Aliette, others)	P07	NC	NC	NC	NC	NC	NC	NC	F	NC	F	NC	F	NC
phosphites (ProPhyt, Phostrol, others)	P07	NC	NC	NC	NC	NC	NC	NC	F	NC	F	NC	F	NC
acibenzolar-S-methyl (Actigard)	21	P	NC	NC	NC	NC	NC	NC	NC	NC	NC	NC	NC	NC
BLAD (Fracture)	BM01	NC	ND	ND	P	P	ND	ND	NC	ND	NC	F	NC	ND
flutianil (Gatten)	U13	ND	ND	ND	ND	ND	ND	ND	ND	ND	ND	ND	ND	ND

[1] These ratings are benchmarks; actual performance will vary. Efficacy ratings do not necessarily indicate a labeled use for every disease.

[2] Efficacy Ratings: The efficacy or importance of a management option is indicated by E = excellent, VG = very good, G = good, F = fair, P = poor, NC = no control, and ND = no data. XX indicates that use of this chemical can increase the disease.

[P] Phytotoxicity could occur.

[R] Not effective if pathogen is resistant to the fungicide.

Strawberry Insect Management

H. J. Burrack, Entomology and Plant Pathology

Examine strawberry plants for insects and mites prior to and following transplant. Consider treating if damaging populations of early season pests, such as cutworms and spider mites, are present. Initiate a weekly insect and mite sampling program in early spring, prior to flowering. Base treatments on comparison of field counts to treatment thresholds, when available.

Insecticide Resistance Action Committee (IRAC) mode of action (MOA) groupings are listed following insecticide names. Materials in the same IRAC grouping have the same mode of action. When selecting insecticides, avoid successive applications of materials in the same IRAC group to minimize potential resistance development. Organically acceptable materials (**OMRI** listed) are noted under Precautions and Remarks.

Many insecticide active ingredients are available in generic formulations. Generic products generally work similarly to their brand name counterparts, but formulation changes can impact efficacy and plant response. In the following table, information is provided for the most commonly used formulations of active ingredients available in multiple formulations. Carefully check the label of the product you plan to use in the event that it differs from those listed. The label is the law!

Table 6-16. Strawberry Insect Control

Season Pest	Insecticide, Formulation, and IRAC Group	Amount of Formulation per Acre	Reentry Interval (hours)	Pre harvest interval (days)	Precautions and Remarks
Post Transplant					
Cyclamen mite	Cyclamen mites are rare in North Carolina strawberries and are typically introduced on infested plants. Inspect plants closely upon receipt and post transplant.				
	fenpyroximate IRAC 21 (Portal)	2 pt	12	1	
	imidacloprid, IRAC 4A (Admire Pro)	10.3 to 14 fl oz (soil)	12	14	Admire Pro can be applied to the soil in transplant water or through drip irrigation or as a foliar treatment. Foliar treatments are not recommended for use against cyclamen mites. Apply in transplant water or through irrigation. Do not apply when bees are foraging or within 10 days of bloom.
Crickets	Cricket feeding on foliage rarely requires treatment, but crickets may occasionally damage fruit when grown in high tunnels during the winter and early spring.				
	carbaryl, IRAC 1A (Sevin 4 XLR)	1 to 2 qt	12	7	Many formulations of carbaryl are available.
	malathion, IRAC 1B (Malathion) 57 EC	1.5 to 3 pt	12	3	
Cutworm	Small cutworms feed on leaves before damaging crowns. If cutworms are suspected but caterpillars are not observed, check plants in evening because larvae are nocturnal.				
	carbaryl, IRAC 1A (Sevin 4 XLR)	1 to 2 qt	12	7	Many formulations of carbaryl are available. Foliar applications for carbaryl can flare spider mites. Apply late in afternoon when plants clipped at the base are first noticed.
	chlorantraniliprole, IRAC 28 (Coragen)	3.5 to 7.5 fl oz	4	1	
	fenpropathrin, IRAC 3A (Danitol)	10.67 to 21.33 fl oz	24	2	Do not make more than 2 total applications.
	methoxyfenozide, IRAC 18 (Intrepid)	6 to 12 fl oz	4	3	

Table 6-16. Strawberry Insect Control

Season Pest	Insecticide, Formulation, and IRAC Group	Amount of Formulation per Acre	Reentry Interval (hours)	Pre harvest interval (days)	Precautions and Remarks
Post Transplant (continued)					
Cutworm (continued)	spinosad, IRAC 5 (Entrust SC) (Entrust 80W)	4 to 8 fl oz 1.25 to 2.5 oz	4	1	Rotate to a different class of insect control products after 2 successive applications of spinosad. Do not apply more than 18 fl oz Entrust SC or Entrust 80W (0.28 lb spinosad) per acre per crop. Both formulations of Entrust are **OMRI** listed.
	malathion, IRAC 1B (Malathion 8 Flowable)	1.5 to 2 pt	12	3	Malathion 8 Flowable can be applied via drip lines, allowing treatment under plastic. Other malathion formulations are labeled in strawberries may not be applied in the same way.
	Bacillus thuringiensis (Bt), IRAC 11B2 (Dipel DF)	0.5 to 1.0 lb	4	0	Dipel DF is **OMRI** listed.
Slugs and snails	metaldehyde (Deadline Bullets)	0.4 to 1.6 lb	NA	See label	Repeated applications may be necessary.
	iron phosphate (Sluggo)	10 to 44 lb	NA	1	Repeated applications may be necessary. Soil should be moist with no standing water when product is applied. Sluggo is **OMRI** listed.
Preharvest					
Red imported fire ants	Baits treatments will control entire mounts, but treatments take between 2 and 4 weeks to be fully effective. Treat active mounds off season or before picking begins with directed bait formulations. Ensure that ants are actively foraging before applying baits. If mounds develop during harvest, drench treatments may reduce activity temporarily. Consult your Cooperative Extension agent for mound drench recommendations.				
	pyriproxyfen, IRAC 7C (Esteem Ant Bait 0.5% B)	1.5 to 2 lb	12	1	Do not water for 24 hours after application.
	methoprene, IRAC 7C (Extinguish Ant Bait 0.5% B)	1 to 1.5 lb	4	0	Extinguish can be applied as a mound treatment or broadcast. Extinguish is labeled for use on cropland, but Extinguish Plus is **not** labeled for use on cropland. Read labels carefully.
Aphids	Aphids are typically infrequent pests in strawberries. If aphids are present preflowing in numbers greater than 10 per newly expanded leaf, they should be managed before bloom. Harvest period populations are often controlled by natural enemies. Aphids typically only warrant preventative treatment, via soil applied insecticides, in nursery production to prevent virus transmission.				
	flupyradifurone, IRAC 4D (Sivanto 200L)	7 to 10.5 fl oz	4	0	Do not make applications fewer than 10 days apart, apply in at least 10 gal per acre, and apply no more than 28 total fl oz per acre.
	imidacloprid, IRAC 4A (Admire Pro, soil) (Admire Pro, foliar)	10.5 to 14 fl oz 1.3 fl oz	12 12	14 7	Admire Pro can be used as a soil or foliar treatment. Soil applications should be made through the irrigation system. **Do not** make imidacloprid applications when bees are foraging or within 10 days of bloom.
	insecticidal soap (M-pede)	2.5 fl oz	12	0	Rate is per 100 gallons of water. Test for phytotoxicity effects on a limited area before widespread use. **OMRI** listed.
	malathion, IRAC 1B (Malathion) 57 EC	1.5 pt	12	3	Only 4 applications can be made per year. Several malathion formulations are labeled in strawberries.
	thiamethoxam, IRAC 4A (Actara)	1.5 to 3 oz	12	3	Do not apply material immediately prior to bud opening, during bloom, or when bees are foraging.
Strawberry weevil (clipper)	Preventative treatments for strawberry clipper are not recommended. **Materials effective against strawberry clipper are also toxic to bees.** Follow pollinator protection language on pesticide labels carefully.				
	acetamiprid, IRAC 4A (Assail 30 SG)	4 to 6.9 oz	12	1	Do not apply when bees are foraging
	bifenthrin, IRAC 3A (Brigade WSB)	6.4 to 32 oz	12	0	Do not apply when bees are foraging. Do not apply more than 80 ounces of product per acre per year.
	carbaryl, IRAC 1A (Sevin 4 XLR)	1 to 2 qt	12	1	Do not apply when bees are foraging, but Sevin XLR is relatively less bee toxic compared to other carbaryl formulations when dry.
	fenpropathrin, IRAC 3A (Danitol 2.4EC)	16 to 21.66 fl oz	24	2	Do not apply when bees are foraging.
	spinosad, IRAC 5 (Entrust SC) (Entrust 80W)	4 to 8 fl oz 1.25 to 2.5 oz	4	1	Rotate to a different class of insect control products after 2 successive applications of spinosad. Do not apply more than 18 fl oz Entrust SC or Entrust 80W (0.28 lb spinosad) per acre per crop. Both formulations of Entrust are **OMRI** listed.
Two-spotted spider mite	Coverage is important for spider mite management. Materials should generally be used at the high label rate, in high volumes of water (200 gallons per acre recommended), and applied using high pressure or electrostatic equipment. Mites should be treated if they exceed 5 per leaflet prior to harvest.				
	abamectin, IRAC 6 (Agri-Mek 0.15 EC)	16 oz	12	3	Make 2 applications 7 to 10 days apart when mites first appear. Do not exceed 64 fluid ounces per acre in a growing season. Do not apply in less than 100 gallons of water per acre. Do not repeat treatment within 21days of second application. Do not use in strawberry nurseries.
	acequinocyl, IRAC 20B (Kanemite 15 SC)	31 oz	12	1	Allow 21 days between treatments. Do not make more than 2 applications per season.
	bifenazate, IRAC Unknown (Acarmite 50WP) (Vigilant 4SC)	1 lb 12 to 16 fl oz	12	1	Make only 2 applications of bifenazate per year. Use in a minimum of 100 gallons per acre.
	cyflumetofen, IRAC 25 (Nealta)	13.7 fl oz	12	1	Use only 2 applications per year. Do not apply successive Nealta applications closer than 14 days apart.
	etoxazole, IRAC 10B (Zeal)	3 oz	12	1	Zeal is an ovicide/larvicide and should be applied early in the mite life cycle.
	fenpyroximate IRAC 21 (Portal)	2 pt	12	1	
	hexythiazox, IRAC 10A (Savey 50 WP)	6 oz	12	3	One application per season. Will control eggs and suppress small mites. Do not use in nurseries.
	mineral/petroleum oils (Organic JMS Stylet Oil)	0.75% by volume	4	4 hours	There are numerous oils registered in strawberries. Oils are effective only if very good coverage is achieved. Oils should not be applied 48 hours or less before freezing temperature, at temperatures over 90 degrees F, or to water-stressed plants. Because oils lack the residual activity of conventional acaricides, they may need to be applied repeatedly to control mites. Organic JMS Stylet Oil is **OMRI** listed.
	predatory mites (*Phytoseiulus persimilis* and others)	30,000 to 60,000	NA	NA	Release 2 to 3 mites per plant when mite populations are low. Predatory mite releases must be initiated at or before Twospotted spider mites reach threshold levels (2 to 5 mites per leaflet), and spider mite populations must be followed closely after predatory mite releases. Consult commercial insectaries for predatory mite release rate and species recommendations. Other predatory mite species may also provide good control of twospotted spider mites in NC strawberries.

Table 6-16. Strawberry Insect Control

Season Pest	Insecticide, Formulation, and IRAC Group	Amount of Formulation per Acre	Reentry Interval (hours)	Pre harvest interval (days)	Precautions and Remarks
Preharvest (continued)					
Two-spotted spider mite (continued)	rosemary & peppermint oils (Ecotec)	32 to 64% by volume	0	0	Because oils lack the residual activity of conventional acaricides, they may need to be applied repeatedly for control. Ecotec is **OMRI** listed.
	spiromesifen, IRAC 23 (Oberon 2SC)	16 fl oz	12	3	Do not apply more than 48 fluid ounces or make more than 3 applications per season.
Whitefly	spiromesifen, IRAC 23 (Oberon 2SC)	12 to 16 fl oz	12	3	Do not apply more than 48 fluid ounces or make more than 3 applications per season.
	thiamethoxam, IRAC 4A (Actara)	1.5 to 3 oz	12	3	**Do not** apply material immediately prior to bud opening, during bloom, or when bees are foraging.
Harvest					
Sap beetle	**Materials effective against sap beetles are toxic to bees.** Follow pollinator protection language on labels carefully.				
	Bait buckets and fruit removal	NA	NA	NA	Cultural control is the most effective form of sap beetle management. Sap beetles are attracted to the odor of overripe fruit. Thorough picking will reduce sap beetle populations and can eliminate the need for treatment. Culls should be disposed of offsite or buried. Bucket traps baited with rotting fruit or bread dough placed outside the field will attract sap beetles and can be used to determine when populations are present or to lure insects from field. Buckets should be checked and emptied at least weekly. Baits should be disposed of offsite or buried.
	novaluron IRAC 15 (Rimon 0.83 EC)	12 fl oz	12	1	Rimon is an insect growth regulator and is effective at reducing populations of immature sap beetles.
Spotted wing drosophila	Female spotted wing drosophila (SWD) lay eggs in ripening and ripe soft skinned fruits. SWD injury in spring bearing strawberries has been inconsistent in previous years, but summer and fall fruiting strawberries are at high risk of infestation. If SWD are active during strawberry harvest, treatments should be applied weekly and reapplied in the event of rain. **Many materials effective against SWD are toxic to bees.** Follow pollinator protection language on labels carefully. Apply SWD treatments in the evening or night, when bees are not actively foraging.				
	bifenthrin, IRAC 3A (Brigade WSB)	6.4 to 32 oz	12	0	There are many bifenthrin formulations. Do not apply when bees are foraging. Do not apply more than 80 ounces of product per acre per year. Brigade is effective against adult sap beetles.
	fenpropathrin, IRAC 3A (Danitol 2.4 EC)	16 to 21.33 fl oz	24	2	
	malathion, IRAC 1B (Malathion 57 EC)	1.5 to 3 pt	12	3	
	spinosad, IRAC 5 (Entrust SC) (Entrust 80W)	4 to 8 fl oz 1.25 to 2.5 oz	4	1	Rotate to a different class of insect control products after 2 successive applications of spinosad. Do not apply more than 18 fl oz Entrust SC or Entrust 80W (0.28 lb spinosad) per acre per crop. Both formulations of Entrust are **OMRI** listed.
	spinetoram, IRAC 5 (Radiant SC)	6 to 10 fl oz	4	1	
Corn earworm, European corn borer	Corn earworm and European corn borer larvae can feed on strawberry fruit. This damage is most common in warm years. Watch for eggs on strawberry fruit near the stem end. Adult moths can be monitored using pheromone traps conditions are appropriate for infestation. **Many materials effective against caterpillar pests are toxic to bees.** Follow pollinator protection language on labels carefully.				
	chlorantraniliprole IRAC 28 (Coragen)	3.5 to 5 fl oz	4	1	
	novaluron IRAC 15 (Rimon 0.83 EC)	9 to 12 fl oz	12	1	Rimon treatments must be timed to egg hatch.
	Bacillus thuringiensis (Bt), IRAC 11B2 (Dipel DF)	0.5 to 1.0 lb	4	0	Dipel DF is **OMRI** listed.
Tarnished plant bugs or Lygus bugs	Lygus bugs are typically only present in North Carolina strawberries at the end of the spring season, although they may be more problematic in day neutral, ever-bearing, or other strawberry season extension systems. Lygus bug injury results in malformed fruit and can resemble poor pollination. Lygus injury can be distinguished from poor pollination based on seed size. The seeds of Lygus damaged fruit are all the same size, while poor pollination results in varied seed sizes. **Many materials effective against lygus bugs are toxic to bees.** Follow pollinator protection language on labels carefully.				
	novaluron IRAC 15 (Rimon 0.83 EC)	9 to 12 fl oz	12	1	
	bifenthrin, IRAC 3A (Brigade WSB)	6.4 to 32 oz	12	0	There are many bifenthrin formulations. Do not apply when bees are foraging. Do not apply more than 80 ounces of product per acre per year. Brigade is effective against adult sap beetles.
	fenpropathrin, IRAC 3A (Danitol 2.4 EC)	16 to 21.33 fl oz	24	2	**Do not** apply when bees are foraging.
Flower thrips	Treatment is only necessary when thrips injury is present on berries. Thrips injury, which resembles bronzing on the stem end of berries, will typically not be present until the end of the season, if at all. **Materials effective against thrips are toxic to bees.** Follow pollinator protection language on labels carefully.				
	spinosad, IRAC 5 (Entrust SC) (Entrust 80W)	4 to 8 fl oz 1.25 to 2.5 oz	4	1	Rotate to a different class of insect control products after 2 successive applications of spinosad. Do not apply more than 18 fl oz Entrust SC or Entrust 80W (0.28 lb spinosad) per acre per crop. Both formulations of Entrust are **OMRI** listed.
	spinetoram, IRAC 5 (Radiant SC)	6 to 10 fl oz	4	1	
Spittlebug	Spittlebugs are occasional pests in strawberries and should only be treated if directly damaging fruit.				
	fenpropathrin, IRAC (Danitol 2.4 EC)	10.67 oz	24	2	Do not apply when bees are foraging. Do not make more than 2 applications.
	malathion (several products) 57 EC	1.5 pt	12	3	Do not apply when bees are foraging.

Further Information

Southeast Regional Strawberry Integrated Management Guide, http://www.smallfruits.org
Strawberry Growers Information Portal, http://strawberries.ces.ncsu.edu

VII — CHEMICAL WEED CONTROL

Chemical Weed Control in Field Corn	252
Weed Response to Preemergence Herbicides — Corn	262
Weed Response to Postemergence Herbicides — Corn	263
Chemical Weed Control in Cotton	264
Weed Response to Cotton Herbicides	274
Weed Response to Cotton Herbicides	275
Weed Response to Cotton Herbicides	276
Chemical Weed Control in Peanuts	277
Weed Response to Preplant Incorporated, Preemergence, and At-Cracking Herbicides in Peanuts	282
Weed Response to Postemergence Herbicides in Peanuts	283
Brand Names and Formulations of Active Ingredients	284
Chemical Weed Control in Sorghum	285
Chemical Weed Control in Soybeans	288
Weed Response to Preplant Incorporated and Preemergence Herbicides in Soybeans	298
Weed Response to Postemergence Herbicides in Soybeans	299
Chemical Weed Control in Sunflowers	300
Chemical Weed Control in Tobacco	301
Weed Response to Herbicides in Tobacco	303
Chemical Weed Control in Wheat, Barley, Oats, Rye, and Triticale	304
Weed Response to Herbicides in Small Grains	309
Glyphosate Formulations	310
Herbicide Resistance Management	311
Herbicide Modes of Action for Hay Crops, Pastures, Lawns and Turf	315
Chemical Weed Control in Clary Sage	317
Chemical Weed Control in Small Fruit Crops	319
Chemical Weed Control in Tree Fruit Crops	326
Chemical Weed Control in Hay Crops and Pastures	334
Chemical Weed Control in Lawns and Turf	340
Chemical Weed Control in Ornamentals	352
Chemical Weed Control in Vegetable Crops	355
Chemical Weed Control in Forest Stands	384
Forest Site Preparation, Stand Conversion, Timber Stand Improvement	386
Aquatic Weed Control	391
Biological Control of Aquatic Weeds with Triploid Grass Carp	393
Chemical Control of Aquatic Plants	393
Pond Dyes	397
Chemical Control of Specific Weeds	399
Chemical Control of Woody Plants	402
Total Vegetation Control in Noncropland	404

Chemical Weed Control in Field Corn

C. W. Cahoon, Crop and Soil Sciences Department

NOTES: A mode of action code has been added to the Herbicide and Formulation column of this table. Use MOA codes for herbicide resistance management. See Table 7-10, Herbicide Resistance Management, for details.

Control of witchweed is part of the State/Federal Quarantine Program. Contact the N.C. Department of Agriculture, Plant Industry Division, at 1-800-206-9333.

Table 7-1A. Chemical Weed Control in Field Corn

Herbicide, Mode of Action Code[1] and Formulation	Amount of Formulation Per Acre	Pounds Active Ingredient Per Acre	Precautions and Remarks
No-Till Burndown, Emerged annual weeds, top-kill and suppression of perennials			
glyphosate, MOA 9 (numerous brands and formulations)	See label	0.38 to 1.13 (lb a.e.)	Glyphosate is available as an isopropylamine salt and a potassium salt. Glyphosate formulations and application rates should be compared on the basis of pounds of glyphosate acid equivalent (a.e.) per gallon and per acre, respectively. The rate in the preceding column is expressed as a.e. See TABLE 7-10 for glyphosate rate conversions. Apply before crop emerges. Glyphosate rate depends on weed species and weed size; see labels for suggested rates. See comments on labels concerning nitrogen as the carrier. Weed control may be decreased when nitrogen or other liquid fertilizers are used as carriers. Apply in 10 to 20 gallons of water per acre using flat fan nozzles. For residual grass and broadleaf weed control, glyphosate can be tank mixed with most preemergence corn herbicides and herbicide combinations. See the section on Corn—Preemergence. Refer to specific product labels for application rates, weeds controlled, application directions, and precautions. Adjuvant recommendations vary according to the glyphosate product used. See label of brand used for specific recommendations. May tank mix glyphosate with Harmony SG at 0.45 to 0.9 ounce per acre to improve control of curly dock, Carolina geranium, henbit, and wild garlic. Tank mix can be applied any time prior to corn emergence. See Harmony SG label for details. May tank mix Resolve with glyphosate for improved control of Italian ryegrass and henbit. Glyphosate and the above glyphosate tank mixes will not control field pansy. A tank mix of Gramoxone plus atrazine should be used where field pansy is present. Glyphosate-resistant horseweed (marestail) is now common in eastern North Carolina counties. A tank mix of glyphosate or Gramoxone plus either 1.5 to 2 pints of 2,4-D or 0.5 pint of Clarity is suggested. Apply these tank mixes 7 to 14 days ahead of planting. If horseweed is present at planting time, a tank mix of Gramoxone plus atrazine is suggested.
paraquat, MOA 22 (Gramoxone Inteon) 2 SL (Firestorm) 3 SL (Parazone) 3 SL	2 to 4 pt 1.33 to 2.67 pt 1.33 to 2.67 pt	0.5 to 1 (lb a.e.)	Apply before, during, or after planting but before crop emerges using clean water or clear fertilizer solution as the carrier. Apply in a minimum of 10 GPA (20 to 40 preferred) using flat fan nozzles. Add either a nonionic surfactant at 1 pint per 100 gallons or a crop oil concentrate at 1 gallon per 100 gallons. Use 0.5 to 0.64 pound a.e. on weeds 1 to 3 inches, 0.75 pound a.e. on weeds 3 to 6 inches, and 1 pound a.e. on weeds 6 inches or taller. Use 0.5 pound a.e. for rye cover crop or 0.75 pound a.e. for wheat cover crop. Rainfast within 30 minutes. For residual grass and broadleaf weed control, paraquat can be tank mixed with most preemergence corn herbicides and herbicide combinations. See the section on Corn—Preemergence. Refer to specific product labels for application rates, weeds controlled, application directions, and precautions. Better and more consistent burndown will be achieved with mixtures of paraquat plus atrazine than with paraquat alone.
glufosinate-ammonium, MOA 10 (Liberty 280 SL)	32 to 43 fl oz	0.59 to 0.79	Liberty can be applied prior to emergence of transgenic or non-transgenic hybrids to control emerged weeds. See label for adjuvant use. If applied preplant, one application of Liberty at 29 to 43 fl oz per acre may be applied in-season to LibertyLink varieties only. Total for the year is not to exceed 87 fl oz per acre. Day time temperatures should be above 75 degrees F for control by Liberty. Thorough spray coverage is necessary: a minimum of 20 GPA with flat fan nozzles is suggested.
carfentrazone-ethyl, MOA 14 (Aim EC)	1.0 to 2.0 oz	0.016 to 0.031	Apply alone or with other herbicides or fertilizers as a burndown to control or suppress weeds. For optimum performance apply to actively growing weeds. Coverage is essential for good control. Broad-spectrum control of annual and perennial weeds requires a tank mix with herbicides such as glyphosate, Liberty, Gramoxone, 2,4-D or dicamba. Add either a nonionic surfactant at 2 pints per 100 gallons, or crop oil concentrate at 1 to 2 gallons per 100 gallons, or methylated seed oil.
pyraflufen-ethyl, MOA 14 (ET) 1 SL	0.5 to 2.0 fl oz	0.003 to 0.015 (lb a.i.)	ET can be used for suppression of small emerged summer annual and winter weeds. See label for adjuvant and spray volume recommendations. Research with ET is limited in North Carolina.
No-Till Burndown, Glyphosate-resistant horseweed (marestail)			
flumioxazin, MOA 14 (Valor SX) S1 WDG + paraquat, MOA 22 (Gramoxone Inteon) 2.5 SL or glyphosate, MOA 9 (numerous brands and formulations)	2.0 oz 2 to 4 pt See label	0.063 0.5 to 1.0 (lb a.e.) 0.38 to 1.13 (lb a.e.)	Corn may be planted 7 days after application if a minimum of 25% of the soil surface is covered with residue and a minimum of ¼ inch of rainfall has occurred between application and planting, otherwise corn must be planted 14 to 30 days after application. Can be applied with other herbicides, including Clarity, 2,4-D and atrazine. Apply with non-ionic surfactant at 1 quart per 100 gallons. Carefully follow label directions for sprayer cleaning after each day's use.
flumioxazin, MOA 14 + pyroxasulfone, MOA 15 (Fierce) 76 WDG + paraquat, MOA 22 (Gramoxone Inteon) 2.5 SL or glyphosate, MOA 9 (numerous brands and formulations)	3.0 oz 2 to 4 pt See label	0.063 + 0.08 0.5 to 1.0 (lb a.e.) 0.38 to 1.13 (lb a.e.)	Corn may be planted 7 days after application if a minimum of 25% of the soil surface is covered with residue and a minimum of ¼ inch of rainfall has occurred between application and planting, otherwise corn must be planted 14 to 30 days after application. Can be applied with other herbicides, including Clarity, 2,4-D and atrazine. Apply with non-ionic surfactant at 1 quart per 100 gallons. Carefully follow label directions for sprayer cleaning after each day's use.
glyphosate, MOA 9 (numerous brands and formulations) + atrazine, MOA 5 (numerous brands) + 2,4-D, MOA 4 (numerous brands)	See labels	0.75 to 1.13 (lb a.e.) + 1 to 2 + 0.75 to 1	See comments for glyphosate alone. Apply mixtures containing 2,4-D at least 7 to 14 days ahead of corn planting.

Table 7-1A. Chemical Weed Control in Field Corn

Herbicide, Mode of Action Code[1] and Formulation	Amount of Formulation Per Acre	Pounds Active Ingredient Per Acre	Precautions and Remarks
No-Till Burndown, Glyphosate-resistant horseweed (marestail) (continued)			
glyphosate, MOA 9 (numerous brands and formulations) + atrazine, MOA 5 (numerous brands) + dicamba, MOA 4 (numerous brands)	See labels	0.75 to 1.13 (lb a.e.) + 1 to 2 + 0.25	See comments for glyphosate alone. Mixtures containing dicamba may be applied to medium- to fine-textured soils before or during planting. Do not apply to coarse-textured soils with less than 2.5% organic matter. Avoid contact of the herbicide with the seed by planting corn at least 1.5 inches deep and ensuring the furrow is closed.
glufosinate-ammonium, MOA 10 (Liberty 280 SL) + atrazine, MOA 5 (numerous brands)	29 fl oz + 1 to 2 qt	0.53 + 1 to 2	Daytime temperatures should be above 75 degrees F for control by Liberty. Thorough spray coverage is necessary; a minimum of 20 GPA with flat-fan nozzles is suggested.
No-Till Burndown or Preemergence			
saflufenacil, MOA 14 (Sharpen)	2.0 to 3.5 fl oz	0.045 to 0.078	Sharpen can be used burndown for control of glyphosate-resistant horseweed. See label for adjuvant selection (burndown) and application rate based on soil texture and organic matter content. A tank mix of glyphosate, paraquat, or glufosinate should be used in burndown applications to provide control of grass and additional broadleaf weed species. Do not apply where an organophosphate or carbamate insecticide has been applied.
dimethenamid-p, MOA 15 + saflufenacil, MOA 14 (Verdict)	10 to 16 oz	0.44 to 0.65 (lb a.i.)	Verdict can be used to control a range of grass and broadleaf weeds, including glyphosate-resistant horseweed. See label for adjuvant selection (burndown) and application rate based on soil texture and organic matter content. Corn injury from Verdict can occur when organophosphate or carbamate insecticides are applied to corn.
imazethapyr, MOA 2 + saflufenacil, MOA 14 (Optill)	2.0 oz	0.085 (lb a.i.)	**Use only on clearfield hybrids.** See Optill label for application with other herbicides. Apply prior to corn emergence to prevent injury. Do not apply if an organophosphate or carbamate insecticide has been used.
rimsulfuron, MOA 2 + thifensulfuron-methyl, MOA 2 (Leadoff) 33.4 WDG	1.5 to 2.7 oz	0.031 to 0.056	Leadoff can be applied from fall up to planting to control a range of grass and broadleaf weeds. Including a non-selective herbicide to provide additional control of emerged weeds is recommended. See label for recommended adjuvants. The addition of atrazine will provide added residual and burndown activity.
thifensulfuron, MOA 2 + tribenuron, MOA 2 (FirstShot SG with TotalSol) 50 WDG	0.5 to 0.8 oz	0.008 to 0.013 + 0.008 to 0.013	FirstShot can be applied from fall up to 14 days before corn planting to control a range of broadleaf weeds. Including a non-selective herbicide to provide additional control of emerged weeds is recommended. See label for recommended adjuvants. The addition of atrazine will provide added residual and burndown activity.
oxyfluorfen, MOA 14 (Goal 2XL) 2L (GoalTender) 4L	1 to 2 pt 0.5 to 1 pt	0.25 to 0.5	Goal 2XL can be applied at 30 days or more prior to corn planting to control common winter and summer annual weed species. After application but prior to corn planting, at least three significant rainfalls (0.25 inch or greater) must occur, otherwise Goal 2XL must be incorporated into the soil to a depth of 2.5 inches to avoid crop injury. A tank mix combination with glyphosate, paraquat, glufosinate, 2,4-D, or dicamba is recommended. See label for tank mix recommendations. Do not use corn plants from a treated field for green chop, ensilage, forage, or fodder.
Preemergence, Annual grasses; control of suppression of yellow nutsedge			
acetochlor, MOA 15 (Harness) 7 EC (Surpass EC) 6.4 EC (Surpass NXT) 7 EC (TopNotch) 3.2 FME (Warrant) 3 ME	2.25 to 5 pt 1.25 to 2.75 pt 1.5 to 3 pt 4 to 6 pt	1.1 to 2.4 1.1 to 2.4 1.2 to 2.4 1.6 to 2.4	Controls most annual grasses, pigweed, and nightshade. Does not adequately control Texas panicum, seedling johnsongrass, and shattercane. Controls yellow nutsedge when incorporated; suppresses yellow nutsedge if applied preemergence. Do not apply to sands with less than 3% organic matter, loamy sands with less than 2% organic matter, or sandy loams with less than 1% organic matter if groundwater depth is 30 feet or less. Read label and adjust rates for soil texture, organic matter, and tillage system. Slightly higher rates can be used for no-till or minimum-till systems. On conventionally tilled soils. Harness and Surpass NXT can be used at rates of 2.5 to 3.4 pints on soils with 6% to 10% organic matter and 3.4 pints on soils with greater than 10% organic matter. Surpass can be used at rates up to 3.75 pints on soils with greater than 7% organic matter. These herbicides can be shallowly incorporated; see labels for details. See labels for rotational crops. May be tank mixed with atrazine or simazine for broadleaf weed control. Surpass and TopNotch contain the safener dichlormid. May be applied to emerged corn up to 11 inches tall. Does not control emerged weeds. See labels for tank mix options to control emerged weeds.
alachlor, MOA 15 (Micro-Tech) 4 FME	2 to 4 qt	2 to 4	Controls most annual grasses and pigweed. At higher rates, controls nightshade. Does not adequately control Texas panicum, seedling johnsongrass, and shattercane. Generally less effective on yellow nutsedge than dimethenamid or metolachlor. Read labels and adjust rates for soil texture and organic matter. May be shallowly incorporated; see labels for details. May be mixed with atrazine or simazine for broadleaf weed control. May be applied to emerged corn up to 5 inches tall. Does not control emerged weeds. See labels for tank mix options to control emerged weeds.
dimethenamid-P, MOA 15 (Outlook) 6.0 EC	12 to 21 fl oz	0.56 to 1.0	Use 12 to 18 fluid ounces on soils with less than 3% organic matter and 14 to 21 fluid ounces on soils with greater than 3% organic matter. Controls most annual grasses and pigweed. At higher rates, controls nightshade and yellow nutsedge. Better yellow nutsedge control if incorporated, see label for incorporation details. Does not adequately control Texas panicum, seedling johnsongrass, and shattercane. Read label and adjust rates for soil texture and organic matter. May be mixed with atrazine or simazine for broadleaf weed control. Do not apply to sandy soil with less than 3% organic matter where depth to groundwater is 30 ft or less. May be applied to emerged corn up to 12 inches tall. Does not control emerged weeds. See labels for tank mix options to control emerged weeds.
metolachlor, MOA 15 (Me-Too-Lachlor II) 7.8 EC (Parallel) 7.8 EC	1 to 2 pt	0.98 to 1.95	See comments for s-metolachlor products. Products containing s-metolachlor are more active on weeds per unit of formulated product than those containing metolachlor. In general, it takes 1.5 pints of a metolachlor product to get the activity one would get from 1 pint of an s-metolachlor product.
S-metolachlor, MOA 15 (Brawl) 7.64 EC (Cinch) 7.64 EC (Dual II Magnum) 7.64 EC (Medal II) 7.64 EC	1 to 2 pt	0.96 to 1.91	Controls most annual grasses and pigweed. At higher rates, controls nightshade and yellow nutsedge. Better yellow nutsedge control if incorporated; see label for incorporation details. Does not adequately control Texas panicum, seedling johnsongrass, and shattercane. Read labels and adjust rates for soil texture and organic matter. May be mixed with atrazine or simazine for broadleaf weed control. May be applied to emerged corn up to 40 inches tall. Direct if corn is taller than 5 inches Does not control emerged weeds. See labels for tank mix options to control emerged weeds.

Table 7-1A. Chemical Weed Control in Field Corn

Herbicide, Mode of Action Code[1] and Formulation	Amount of Formulation Per Acre	Pounds Active Ingredient Per Acre	Precautions and Remarks
Preemergence, Annual grasses; control of suppression of yellow nutsedge (continued)			
pyroxasulfone, MOA 15 (Zidua SC) 4.17 SC	2.5 to 6.5 fl oz	0.08 to 0.212	Use 2.5 to 4.5 fluid ounces of Zidua SC per acre on coarse soils, 3.25 to 5 fluid ounces of Zidua SC on medium soils, and up to 6.5 fluid ounces of Zidua SC per acre on fine textured soils. Controls most annual grasses, pigweed, and nightshade. Provides suppression of Texas panicum, seedling johnsongrass, and shattercane. May be mixed with atrazine or Sharpen for broadleaf weed control. May be applied to emerged corn up to the V4 stage. Does not control emerged weeds. See label for tank mix options to control emerged weeds.
pyroxasulfone, MOA 15 + fluthiacet-methyl, MOA 14 (Anthem) 2.15	5 to 13 oz	0.097 to 0.212 + 0.002 to 0.006	For Anthem, use 5 to 8 ounces per acre on coarse soils, 6.5 to 10 ounces on medium soils, and 9 to 13 ounces on fine textured soils. For Anthem Maxx, use 2.5 to 4 ounces per acre on coarse soils, 2.5 to 5.5 ounces for medium soils, and 3 to 6.5 ounces on fine-textured soils. For Anthem Flex, use 2.75 to 5 ounces per acre for coarse soils, 3 to 6 ounces for medium soils, and 3.5 to 7.28 for fine soils. See labels for rate adjustments according to organic matter content. Control most annual grasses, pigweed, and nightshade. Provides suppression of Texas panicum, seedling johnsongrass, and shattercane. May be applied to emerged corn through V4 stage. Will provide control of some small emerged broadleaf weeds (<2"). See label for tank mix options.
pyroxasulfone, MOA 15 + fluthiacet-methyl, MOA 14 (Anthem Maxx) 4.3	2.5 to 6.5 fl oz	0.082 to 0.212 + 0.002 to 0.006	
pyroxasulfone, MOA 15 + carfentrazone, MOA 14 (Anthem Flex) 4	2.75 to 7.28 fl oz	0.08 to 0.212 + 0.006 to 0.015	For Anthem Flex, use 2.75 to 5 ounces per acre for coarse soils, 3 to 6 ounces for medium soils, and 3.5 to 7.28 for fine soils. See label for rate adjustments according to organic matter content. Control most annual grasses, pigweed, and nightshade. Provides suppression of Texas panicum, seedling johnsongrass, and shattercane. Will provide control of some small emerged broadleaf weeds (<2"). See label for tank mix options.
Preemergence, Annual broadleaf weeds			
atrazine, MOA 5 (AAtrex 4L) 4 F (AAtrex Nine-O) 90 WDG	1 to 2 qt 1.1 to 2.2 lb	1 to 2	Do not exceed 1.6 pounds a.i. on highly erodible soils (as defined by the NRCS) with less than 30% plant residue cover. Do not exceed 2 pounds a.i. on any soil. See labels for comments on rotational crops. May be applied preplant incorporated; see labels for details. May be tank mixed with preemergence grass control herbicides. Generic brands of atrazine are available, including products containing 5 pounds per gallon. See label for details on set-back requirements from streams and lakes.
isoxaflutole, MOA 27 (Balance FLEXX) 2 L	3 to 6 fl oz	0.047 to 0.09	Controls most broadleaf weeds, some annual grasses. Do not exceed 3 fluid ounces on coarse soils with 1.5% organic matter or less. May be tank mixed with preemergence grass control herbicides. Addition of atrazine will help extend residual control of broadleaf weeds such as morningglory species. May be applied postemergence up to 2-leaf collar corn. See labels for comments on rotational crops. May be applied preplant incorporated; see labels for details. See label for pH, groundwater, and soil texture recommendations.
mesotrione, MOA 27 (Callisto) 4 F	6 to 7.7 oz	0.19 to 0.24	Controls pigweed, lambsquarters, jimsonweed, common ragweed, smartweed, velvetleaf, and nightshade. Does not control sicklepod or prickly sida. Not adequately effective on cocklebur or morningglory. Callisto is generally more effective when applied postemergence. No rotational restrictions for small grains or for other crops planted the following spring. Can mix with various preemergence grass control herbicides or with atrazine or atrazine-containing products. See precautions on label concerning use of Counter and Lorsban.
Preemergence, Most annual grasses and broadleaf weeds			
acetochlor, MOA 15 + atrazine, MOA 5 (Degree Xtra) 4.04 FME (FulTime NXT) 4.04 FME	2.9 to 3.7 qt	2 to 2.5 + 1 to 1.25	Controls most broadleaf weeds and annual grasses. Does not adequately control Texas panicum, seedling johnsongrass, or shattercane. Do not apply to sands with less than 3% organic matter, loamy sands with less than 2% organic matter, or sandy loams with less than 1% organic matter if groundwater depth is 30 ft or less. Read labels and adjust rates for soil texture and organic matter. See labels for comments on rotational crops. May be incorporated; see labels for details. See labels for details on set-back requirements from streams and lakes. Do not exceed 1.6 pounds a.i. atrazine on highly erodible soils (as defined by NRCS) with less than 30% plant residue cover. Degree Xtra and FulTime NXT contains 2.7 pounds acetochlor and 1.34 pounds atrazine per gallon. Harness Xtra and Keystone NXT contain 3.1 pounds acetochlor and 2.5 pounds atrazine per gallon. FulTime contains 2.4 pounds acetochlor and 1.6 pounds atrazine per gallon. These products and certain tank mixes may also be applied early postemergence; see labels for details.
acetochlor, MOA 15 + atrazine, MOA 5 (FulTime) 4 F	2.5 to 5 qt	1.5 to 3 + 1 to 2	
acetochlor, MOA 15 + atrazine, MOA 5 (Harness Xtra) 5.6 F (Keystone NXT) 5.6 F	1.4 to 3 qt	1.1 to 2.3 + 0.9 to 1.9	
acetochlor, MOA 15 + mesotrione, MOA 27 (Harness MAX) 3.85 L	55 to 88 fl oz	1.51 to 2.42 + 0.142 to 0.227	Controls pigweed, lambsquarters, jimsonweed, common ragweed, smartweed, velvetleaf, nightshade, and most annual grasses. Does not control Texas panicum, sicklepod or prickly sida. Not adequately effective on cocklebur or morningglory. Use lower rates on coarse textured soils with less than 3% organic matter and higher rates on finer textured soils with more than 3% organic matter. See label for specific details on use rates according to soil texture and organic matter content. On medium and fine textured soils, up to 95 fl oz per acre may be used in areas of heavy grass infestations. On soils with 6 to 10% organic matter use 81 to 95 fl oz per acre and on soils with more than 10% organic matter use 95 fl oz per acre.
atrazine, MOA 5 (AAtrex 4L) 4 F (AAtrex Nine-O) 90 WDG + simazine, MOA 5 (Princep 4L) 4 F (Princep Caliber 90) 90 WDG	2 to 3 pt 1.1 to 1.6 lb + 1 to 1.44 qt 1.1 to 1.6 lb	1 to 1.5 1 to 1.44 + 1 to 1.44	Controls most annual broadleaf weeds plus crabgrass, goosegrass, fall panicum, and foxtails. Does not control Texas panicum, broadleaf signalgrass, seedling johnsongrass, or shattercane. Can be incorporated; see labels for details. Can use a 1:2 ratio of atrazine to simazine on more severe annual grass problems. If using 1:2 ratio, atrazine rates are 0.66 to 0.96 pound a.i. and simazine rates are 1.34 to 1.92 pounds a.i. Read label and adjust rates to soil texture. See labels for rotational restrictions and other precautions. Generic brands of simazine and atrazine are available, including atrazine products containing 5 pounds per gallon. See atrazine label for details on set-back requirements from streams and lakes.
metolachlor, MOA 15 + atrazine, MOA 5 (Parallel Plus) 5.5 F	1.4 to 2.83 qt	0.95 to 1.9 + 1 to 2	See below comments for s-metolachlor plus atrazine. Products containing s-metolachlor are more active on weeds per unit of formulated product than those containing metolachlor.
S-metolachlor, MOA 15 + atrazine, MOA 5 (Bicep II Magnum) 5.5 F (Brawl II ATZ) 5.5 F (Cinch ATZ) 5.5 F (Medal II AT) 5.5 F	1.3 to 2.6 qt	0.78 to 1.56 + 1 to 2	Controls most broadleaf weeds and annual grasses. Does not adequately control Texas panicum, seedling johnsongrass, or shattercane. Rates for coarse-textured soils may not give sufficient control of heavy fall panicum, broadleaf signalgrass, and other grassy weeds. Cultivation and/or an additional herbicide application may be needed. Read label and adjust rates for soil texture and organic matter. Do not exceed 2.1 quarts on highly erodible soils (as defined by the NRCS) with less than 30% plant residue cover. See labels for comments on rotational crops. May be applied preplant incorporated; see labels for details. See label for details on set-back requirements from streams and lakes. These products contain 2.4 pounds of S-metolachlor and 3.1 pounds atrazine per gallon. May be applied early postemergence; see label for details.

Table 7-1A. Chemical Weed Control in Field Corn

Herbicide, Mode of Action Code[1] and Formulation	Amount of Formulation Per Acre	Pounds Active Ingredient Per Acre	Precautions and Remarks
Preemergence, most annual grasses and broadleaf weeds (continued)			
S-metolachlor, MOA 15 + atrazine, MOA 5 + mesotrione, MOA 27 (Lexar) 3.7 L (Lexar EZ) 3.7 L	3 to 3.5 qt	1.3 to 1.5 + 1.3 to 1.5 + 0.17 to 0.20	Controls most broadleaf weeds and annual grasses. Does not adequately control Texas panicum, seedling johnsongrass, or shattercane. May not adequately control cocklebur, morningglory, or sicklepod. Use 3 quarts on soils with less than 3% organic matter; use 3.5 quarts on soils with greater than 3% organic matter. Not recommended on soils with greater than 10% organic matter. See label for setback requirements from streams and lakes. May be applied postemergence to corn up to 12 inches tall; see label for tank mixes to control grasses when applying postemergence. Lexar and Lexar EZ contain 1.74 pounds S-metolachlor, 1.74 pounds atrazine, and 0.224 pound mesotrione per gallon. Lumax contains 2.68 pounds S-metolachlor, 1 pound atrazine, and 0.268 pound mesotrione per gallon. Lumax EZ contains 2.49 pounds S-metolachlor, 0.935 pound atrazine, and 0.249 pound mesotrione per gallon. Lexar EZ and Lumax EZ include an enhanced capsule-suspension formulation for better handling.
S-metolachlor, MOA 15 + atrazine, MOA 5 + mesotrione, MOA 27 (Lumax) 3.95 L	2.5 to 3 qt	1.68 to 2.01 + 0.63 to 0.75 + 0.168 to 0.201	
S-metolachlor, MOA 15 + atrazine, MOA 5 + mesotrione, MOA 27 (Lumax EZ) 3.67 L	2.7 to 3.25 qt	1.68 to 2.02 + 0.63 + 0.76 + 0.168 + 0.202	
isoxaflutole, MOA 27 + thiencarbazone-methyl, MOA 2 (Corvus)	3.33 to 5.6 fl oz	0.106 to 0.243 + 0.003 to 0.007 + 0.876 to 2.003	Controls most broadleaf weeds and annual grasses. Do not exceed 3.33 fluid ounces on coarse soils with 2.0% organic matter or less. The addition of atrazine will help extend residual control of broadleaf weeds such as morningglory species. May be applied postemergence up to 2-leaf collar corn. See label for pH, groundwater, and soil texture specific recommendations.
pyroxasulfone, MOA 15 + fluthiacet-methyl, MOA 14 + atrazine, MOA 5 (Anthem ATZ)	1.75 to 4 pt	0.985 to 2.253	Controls most broadleaf weeds and annual grasses. Does not adequately control Texas panicum, seedling johnsongrass, or shattercane. Do not exceed 4.0 pt on highly erodible soils (as defined by the NRCS) with less than 30% plant residue cover. See labels for comments on rotational crops. May be applied preplant on medium and fine textured soils; see label for details. See label for details on set-back requirements from streams and lakes. Use 1.75 to 2.0 pints on coarse soils, 2 to 2.75 pints on medium soils, and up to 4 pints on fine soils preemergence. May be applied postemergence up to 4-leaf corn; see label for details.
S-metolachlor, MOA 15 + atrazine, MOA 5 + mesotrione, MOA 27 + Bicyclopyrone, MOA 27 (Acuron) 3.44 L	2.5 to 3.0 qt	1.34 to 1.61 + 0.625 to 0.75 + 0.15 to 0.18 + 0.038 to 0.045	Controls most broadleaf weeds and annual grasses. Does not adequately control Texas panicum, seedling johnsongrass, or shattercane. May need a tank-mix for control of heavy infestations of broadleaf signalgrass. Use 2.5 quarts on soils with less than 3% organic matter; use 3.0 quarts on soils with greater than 3% organic matter. Not recommended on soils with greater than 10% organic matter. See label for setback requirements from streams and lakes. May be applied postemergence to corn up to 12 inches tall; see label for tank mixes to control grasses when applying postemergence. Acuron contains 2.14 pounds S-metolachlor, 1.0 pound atrazine, 0.24 pound mesotrione, and 0.06 pounds bicyclopyrone per gallon.
S-metolachlor, MOA 15 + mesotrione, MOA 27 + Bicyclopyrone, MOA 27 (Acuron Flexi) 3.26 L	2 to 2.25 qt	1.43 to 1.61 + 0.16 to 0.18 + 0.04 to 0.045	Acuron Flexi does not contain atrazine. Controls most broadleaf weeds and annual grasses. Does not adequately control Texas panicum, seedling johnsongrass, or shattercane. May need a tank-mix for control of heavy infestations of broadleaf signalgrass. Use 2 quarts on soils with less than 3% organic matter; use 2.25 quarts on soils with greater than 3% organic matter. Not recommended on soils with greater than 10% organic matter. See label for setback requirements from streams and lakes. May be applied postemergence to corn up to 30 inches tall or 8-leaf growth stage; see label for tank mixes to control grasses when applying postemergence. Acuron Flexi contains 2.86 pounds S-metolachlor, 0.32 pound mesotrione, and 0.08 pounds bicyclopyrone per gallon.
acetochlor, MOA 15 + flumetsulam, MOA 2 + clopyralid, MOA 4 (SureStart II) 4.25 L (TripleFLEX II) 4.25 L	1.5 to 3.0 pt	0.7 to 1.41 + 0.023 to 0.045 + 0.07 to 0.143	Controls most broadleaf weeds and annual grasses. Does not adequately control Texas panicum, seedling johnsongrass, or shattercane. Do not apply to sands with less than 3% organic matter, loamy sands with less than 2% organic matter, or sandy loams with less than 1% organic matter if groundwater depth is 30 feet or less. Read label and adjust rates for soil texture and organic matter. Tobacco, cotton, and peanuts cannot be planted for 18 to 26 months after application of SureStart II or TripleFLEX II, see label for comments on rotational crops. May be incorporated; see labels for details. This product and certain tank mixes may also be applied early postemergence; see label for details.
acetochlor, MOA 15 + mesotrione, MOA 27 + clopyralid, MOA 4 (Resicore) 3.29 L	2.25 to 3.0 qt	1.58 to 2.1 + 0.169 to 0.225 + 0.107 to 0.143	Controls most broadleaf weeds and annual grasses. Does not adequately control Texas panicum, seedling johnsongrass, or shattercane. Do not apply to sands with less than 3% organic matter, loamy sands with less than 2% organic matter, or sandy loams with less than 1% organic matter if groundwater depth is 30 ft or less. Read label and adjust rates for soil texture and organic matter. The addition of atrazine will improve weed control. Tobacco, cotton, and peanuts cannot be planted for 18 months after application of Resicore, see label for comments on rotational crops. May be incorporated; see labels for details. This product and certain tank mixes may also be applied early postemergence; see label for details.
pendimethalin, MOA 3 (Prowl) 3.3 EC (Prowl H$_2$O) 3.8 L + atrazine, MOA 5 (AAtrex 4L) 4 F (AAtrex Nine-O) 90 WDG	1.8 to 3.6 pt 2 to 4 pt 1 to 2 qt 1.1 to 2.2 lb	0.75 to 1.5 0.95 to 1.9 + 1 to 2	Annual grass control is more variable and sometimes less acceptable with this combination than with alternatives. Suggested for use only on fields with light annual grass pressure. Do not apply to excessively wet soils. Read label and adjust rates to soil texture and organic matter. Do not exceed 1.6 pounds a.i. atrazine on highly erodible soils (as defined by the NRCS) with less than 30% plant residue cover. See labels for comments on rotational crops. Do NOT incorporate. See atrazine label for details on set-back requirements from streams and lakes. Generic brands of atrazine (including products containing 5 pounds/gallon) and pendimethalin (such as Acumen, Helena Pendimethalin, Pendant, Pendimax, and Stealth, all containing 3.3 pounds/gallon) are available.
glufosinate-ammonium, MOA 10 (Liberty 280 SL)	32 to 43 fl oz	0.59 to 0.79 (lb a.i.)	Liberty can be applied prior to emergence of transgenic or nontransgenic hybrids. See comments under **No-Till Burndown** for more details.
Early Postemergence, Small annual broadleaf and grass weeds			
acetochlor, MOA 15 + atrazine, MOA 5 (Degree Xtra) 4.04 FME (FulTime NXT) 4.04 FME	2.9 to 3.7 qt	2 to 2.5 + 1 to 1.25	Apply as a very early postemergence application to weeds no larger than 2 leaves and before corn exceeds 11 inches. Adjust rates for soil types as specified on labels. See remarks concerning soil type limitations, set-back requirements from streams and lakes, and rotational crops under Field Corn —Preemergence. May be tank mixed with several other herbicides to control emerged weeds. If an atrazine-containing herbicide was applied earlier, the total amount of atrazine per acre per season should not exceed 2.5 pounds a.i.
acetochlor, MOA 15 + atrazine, MOA 5 (FulTime) 4 F	2.5 to 5 qt	1.5 to 3 + 1 to 2	

Table 7-1A. Chemical Weed Control in Field Corn

Herbicide, Mode of Action Code[1] and Formulation	Amount of Formulation Per Acre	Pounds Active Ingredient Per Acre	Precautions and Remarks
Early Postemergence, Small annual broadleaf and grass weeds (continued)			
acetochlor, MOA 15 + atrazine, MOA 5 (Harness Xtra) 5.6 L (Keystone NXT) 5.6 L	1.4 to 3 qt	1.1 to 2.3 + 0.9 to 1.9	
acetochlor, MOA 15 + mesotrione, MOA 27 (Harness MAX) 3.85 L	40 to 75 fl oz	1.1 to 2.06 + 0.103 to 0.193	May be applied to corn up to 11 inches in height. Apply to actively growing weeds before weeds exceed 3 inches in height. Add either 1 quart nonionic surfactant per 100 gallons spray solution or 1 gallon crop oil concentrate per 100 gallons spray solution. Controls most broadleaf weeds. Partial control of common ragweed and morningglory. Does not control sicklepod or prickly sida. Use lower rates on coarse textured soils with less than 3% organic matter and higher rates on finer textured soils with more than 3% organic matter. See label for specific details on use rates according to soil texture and organic matter content. On medium and fine textured soils, up to 75 fl oz per acre may be used in areas of heavy weed infestations. On soils with more than 6% organic matter use 75 fl oz per acre. Do not make a second application of Harness MAX within 14 days of the first application. See label for specified tank mixtures.
atrazine, MOA 5 (AAtrex 4L) 4 F (AAtrex Nine-O) 90 WDG	2 qt 2.2 lb	2	Atrazine can be sprayed overtop of corn as an early postemergence treatment. Must be applied before weeds are over 1.5 inches tall to be effective and before corn exceeds 12 inches tall. Not effective during drought. Add 1 quart per acre of crop oil concentrate. If an earlier application was made, the total atrazine applied may not exceed 2.5 pounds a.i. per acre per year. See label for details on set-back requirements from streams and lakes. May be tank mixed with preemergence grass control herbicides. When tank mixing, check respective labels for application rates, weeds controlled, specific application directions, and precautions. Generic brands are available, including products containing 5 pounds per gallon.
metolachlor, MOA 15 + atrazine, MOA 5 (Parallel Plus) 5.5 F	1.4 to 2.83 qt	0.95 to 1.9 + 1 to 2	See below comments for s-metolachlor plus atrazine products. Products containing s-metolachlor are more active on weeds per unit of formulated product than those containing metolachlor.
S-metolachlor, MOA 15 + atrazine, MOA 5 (Bicep II Magnum) 5.5 L (Brawl II ATZ) 5.5 F (Cinch ATZ) 5.5 L (Medal II AT) 5.5 F	1.6 to 2.6 qt	0.96 to 1.56 + 1.24 to 2	Apply as a very early postemergence application to weeds no larger than two leaves and before corn exceeds 5 inches See remarks for these products under Field Corn—Preemergence. If an atrazine-containing herbicide was applied earlier, the total amount of atrazine per acre per season should not exceed 2.5 pounds a.i. See label for set-back requirements from streams and lakes.
mesotrione, MOA 27 + atrazine, MOA 5 (Callisto Xtra) 3.7 L	20 to 24 fl oz	0.078 to 0.093 + 0.5 to 0.6	May be applied to corn up to 12 inches tall. Add nonionic surfactant at 1 quart per 100 gallons spray solution or crop oil concentrate at 1 gallon per 100 gallons spray solution. In addition, a spray grade urea ammonium nitrate or ammonium sulfate is also recommended. Do not use methylated seed oil or adjuvant blends containing methylated seed oil or severe crop injury may occur. Can be mixed with Liberty on Liberty Link corn or glyphosate on Roundup Ready corn.
S-metolachlor, MOA 15 + atrazine, MOA 5 + mesotrione, MOA 27 (Lexar) 3.7 L (Lexar EZ) 3.7 L	3 to 3.5 qt	1.3 to 1.5 + 1.3 to 1.5 + 0.17 to 0.20	May be applied to corn up to 12 inches tall. Add nonionic surfactant according to label directions. Do not apply to Counter-treated corn. Application to corn treated with other organophosphate insecticides may cause injury. Do not apply in liquid fertilizer or severe crop injury may occur. Lumax and Lumax EZ have less atrazine per gallon of product than Lexar and Lexar EZ. For this reason, postemergence control may not be as consistent.
S-metolachlor, MOA 15 + atrazine, MOA 5 + mesotrione, MOA 27 (Lumax) 3.95 L	2.5 to 3 qt	1.68 to 2.01 + 0.63 to 0.75 + 0.168 to 0.201	
S-metolachlor, MOA 15 + atrazine, MOA 5 + mesotrione, MOA 27 (Lumax EZ) 3.67 L	2.7 to 3.25 qt	1.68 to 2.02 + 0.63 + 0.76 + 0.168 + 0.202	
topramezone, MOA 27 + dimethenamid-P, MOA 15 (Armezon PRO) 5.35 EC	14 to 20 fl oz	0.01 to 0.016 + 0.57 to 0.82	May be applied to corn from emergence up to the 8-leaf stage or 30-inches-tall corn. Armezon PRO rates vary by soil texture and organic matter, refer to label for recommendations. Add methylated seed oil or crop oil concentrate at 0.5 to 1.0 gallon per 100 gallons of water or nonionic surfactant at 0.25 to 0.5 gallon per 100 gallons of water. Oil-type adjuvants are not recommended when tank-mixing with atrazine. Add nitrogen fertilizer at 1.25 to 2.5 gallons per 100 gallons of water or 8.5 to 17 pounds of ammonium sulfate per 100 gallons.
S-metolachlor, MOA 15 + atrazine, MOA 5 + mesotrione, MOA 27 + Bicyclopyrone, MOA 27 (Acuron) 3.44 L	2.5 to 3.0 qt	1.34 to 1.61 + 0.625 to 0.75 + 0.15 to 0.18 + 0.038 to 0.045	May be applied to corn up to 12 inches tall. Add nonionic surfactant according to label directions. Do not apply to Counter-treated corn. Application to corn treated with other organophosphate insecticides may cause injury. Occasional corn leaf burn may result, but this will not affect later growth or corn yield.
S-metolachlor, MOA 15 + mesotrione, MOA 27 + Bicyclopyrone, MOA 27 (Acuron Flexi) 3.26 L	2 to 2.25 qt	1.43 to 1.61 + 0.16 to 0.18 + 0.04 to 0.045	Acuron Flexi does not contain atrazine. May be applied postemergence to corn up to 30 inches tall or 8-leaf growth stage; see label for tank mixes to control grasses. Add nonionic surfactant according to label directions. Do not apply to Counter-treated corn. Application to corn treated with other organophosphate insecticides may cause injury. Occasional corn leaf burn may result, but this will not affect later growth or corn yield.

Table 7-1A. Chemical Weed Control in Field Corn

Herbicide, Mode of Action Code[1] and Formulation	Amount of Formulation Per Acre	Pounds Active Ingredient Per Acre	Precautions and Remarks
Early Postemergence, Small annual broadleaf and grass weeds (continued)			
acetochlor, MOA 15 + flumetsulam, MOA 2 + clopyralid, MOA 4 (SureStart II) 4.25 L (TripleFLEX II) 4.25 L	1.5 to 3.0 pt	0.7 to 1.41 + 0.023 to 0.045 + 0.07 to 0.143	Apply as a very early postemergence application to broadleaf weeds no larger 2 inches and before corn exceeds 11 inches. Established or germinated grass weeds present at application will not be controlled. May be tank mixed with several other herbicides to control emerged weeds. Adjust rates for soil types as specified on label. See remarks concerning soil type limitations and rotational crops under Field Corn — Preemergence. Tobacco, cotton, and peanuts cannot be planted for 18 to 26 months after application of SureStart II or TripleFLEX II, see label for comments on rotational crops.
acetochlor, MOA 15 + mesotrione, MOA 27 + clopyralid, MOA 4 (Resicore) 3.29 L	2.25 to 3.0 qt	1.58 to 2.1 + 0.169 to 0.225 + 0.107 to 0.143	Apply as a very early postemergence application to broadleaf weeds no larger 3 inches and before corn exceeds 11 inches. Control of emerged grass weeds will not be consistent. May be tank mixed with several other herbicides to control emerged weeds. Adjust rates for soil types as specified on label. See remarks concerning soil type limitations and rotational crops under Field Corn — Preemergence. Tobacco, cotton, and peanuts cannot be planted for 18 months after application of Resicore, see label for comments on rotational crops.
Postemergence, Annual broadleaf weeds			
bentazon, MOA 6 (Basagran) 4 SL	1.5 to 2 pt	0.75 to 1	Apply overtop or directed. Drop nozzles suggested after corn is 8 inches tall to ensure better weed coverage. Controls many broadleaf weeds such as cocklebur, jimsonweed, smartweed, velvetleaf, prickly sida, spurred anoda, and spreading dayflower. See label for weeds controlled and recommended weed size for treatment. Add crop oil concentrate at 2 pints per acre. May be tank mixed with atrazine.
acetochlor, MOA 15 + flumetsulam, MOA 2 + clopyralid, MOA 4 (SureStart) 4.25 L	1.5 to 3.0 qt		Apply as a very early postemergence application to broadleaf weeds no larger 2 inches and before corn exceeds 11 inches. Established or germinated grass weeds present at application will not be controlled. May be tank mixed with several other herbicides to control emerged weeds. Adjust rates for soil types as specified on label. See remarks concerning soil type limitations and rotational crops under Field Corn — Preemergence. Tobacco, cotton, and peanuts cannot be planted for 18 to 26 months after application of SureStart, see label for comments on rotational crops.
acetochlor, MOA 15 + mesotrione, MOA 27 + clopyralid, MOA 4 (Resicore) 3.29 L	2.25 to 3.0 qt	1.58 to 2.1 + 0.169 to 0.225 + 0.107 to 0.143	Apply as a very early postemergence application to broadleaf weeds no larger 3 inches and before corn exceeds 11 inches. Control of emerged grass weeds will not be consistent. May be tank mixed with several other herbicides to control emerged weeds. Adjust rates for soil types as specified on label. See remarks concerning soil type limitations and rotational crops under Field Corn — Preemergence. Tobacco, cotton, and peanuts cannot be planted for 18 months after application of Resicore, see label for comments on rotational crops.
bromoxynil, MOA 6 (Buctril) 2 EC (Buctril) 4 EC	1 to 1.5 pt 0.5 to 0.75 pt	0.25 to 0.38	Can be applied overtop of corn from four-leaf stage up to tasseling. Drop nozzles suggested after corn is 8 inches tall for better coverage on weeds. Controls most broadleaf weeds if treated when small. See label for weeds controlled and recommended weed size for treatment. Does not control sicklepod, prickly sida, spurred anoda, or croton. Marginally effective on pigweed and morningglory unless treated very timely. Crop oil or surfactant not necessary when applying Buctril alone. Can tank mix with Accent, atrazine, Banvel, Clarity, or 2,4-D for broader spectrum control. Primary advantage over 2,4-D or Clarity is safety when sensitive crops are nearby. Will cause some burn on corn foliage.
carfentrazone, MOA 14 (Aim EC) 2 L	0.5 fl oz	0.008	Controls velvetleaf, morningglory, redroot pigweed, lambsquarters, and nightshade. See label for weed size to treat. Apply before corn exceeds V8 stage (8 leaves with collars). Add nonionic surfactant at 1 quart per 100 gallons. May be mixed with 2,4-D amine, Accent, atrazine, Banvel, Callisto, Clarity or Distinct. May be mixed with Lightning for Clearfield corn, Liberty for Liberty Link corn, or glyphosate for Roundup Ready corn.
dicamba, dimethylamine salt, MOA 4 (Banvel) 4 SL (Diablo) 4 SL (Dicamba DMA Salt) 4 SL (Rifle) 4 SL (Sterling) 4 SL	0.5 pt	0.25	Apply overtop of corn from spike stage until 8 inches tall. On corn 8 to 36 inches tall, Banvel, Clarity, Engenia, FeXapan, and XtendiMax can be applied using drop nozzles. Carefully follow all precautions on label concerning drift to sensitive crops. Dicamba is more effective than 2,4-D on smartweed, sicklepod, nightshade, burcucumber, and pokeweed.
dicamba, diglycolamine salt, MOA 4 (Clarity) 4 SL (XtendiMax) 2.9 SL (FeXapan) 2.9 SL	0.5 pt 11 fl oz 11 fl oz	0.25	
dicamba, BAMPA salt, MOA 4 (Engenia) 5 SL	6.4 fl oz	0.25	
dicamba, sodium salt, MOA 4 + diflufenzopyr, sodium salt (Distinct) 76.4 WDG	4 oz	0.125 + 0.05	Apply to corn 4 to 36 inches tall. Drop nozzles suggested on corn taller than 10 inches Drop nozzles must be used on corn taller than 24 inches Rate can be increased to 6 ounces on corn shorter than 10 inches Add nonionic surfactant at 1 quart per 100 gallons spray solution plus either 5 quarts 30% UAN or 5 pounds ammonium sulfate per 100 gallons. Do not add crop oil. Control of annual weeds similar to that by Banvel or Clarity. Distinct may be somewhat more effective on perennial broadleaf weeds. Carefully follow all precautions on label concerning drift to sensitive crops.
dicamba, sodium salt, MOA 4 + diflufenzopyr, sodium salt + safener (Status) 61.1 WDG	5 to 10 oz	0.14 to 0.28 + 0.053 to 0.106	Apply to corn 4 to 36 inches tall. Drop nozzles suggested on corn taller than 24 inches Add nonionic surfactant at 1 quart per 100 gallons spray solution plus either 5 quarts 30% UAN or 5 pounds ammonium sulfate per 100 gallons. Do not add crop oil. Potential for crop injury from Status is much less than from dicamba products without safener.
dicamba, diglycolamine salt, MOA 4 + safener (DiFlexx) 4 SC	6 to 16 fl oz	0.19 to 0.5	Apply to corn from spike through V10 (10 leaf collar) stage or corn 36 inches tall, whichever comes first. Nonionic surfactant at 1 quart per 100 gallons spray solution or crop oil concentrate at 1 gallon per 100 gallons spray solution or methylated seed oil at 1 gallon per 100 gallons spray solution may be used to improve efficacy of DiFlexx, especially in dry growing conditions. If using one of the approved adjuvants, also add 2 to 4 quarts per acre urea ammonium nitrate or 8.5 to 17 pounds ammonium sulfate per 100 gallons spray solution. Do not use sprayable fluid fertilizer as the carrier. DiFlexx contains the new corn safener cyprosulfamide. Research on DiFlexx is limited in North Carolina.

Table 7-1A. Chemical Weed Control in Field Corn

Herbicide, Mode of Action Code[1] and Formulation	Amount of Formulation Per Acre	Pounds Active Ingredient Per Acre	Precautions and Remarks
Postemergence, Annual broadleaf weeds (continued)			
flumiclorac pentyl ester, MOA 14 (Resource) 0.86 EC	4 to 8 fl oz	0.027 to 0.054	Can be applied overtop of corn from the 2-leaf through the 10-leaf stage at 4 to 6 fluid ounces per acre. At 4 to 6 fluid ounces, Resource controls velvetleaf and small lambsquarters, ragweed, smooth pigweed, and Palmer amaranth. When applying overtop, add nonionic surfactant at 1 quart/100 gallons spray solution. Resource can be directed at 4 to 8 fluid ounces per acre. At 8 fluid ounces, Resource controls velvetleaf and small cocklebur, lambsquarters, ragweed, jimsonweed, Palmer amaranth, redroot and smooth pigweed, and prickly sida. See label for recommended weed sizes for treatment. When directing, add 2 pints per acre of crop oil concentrate. For broader spectrum control, Resource may be tank mixed with atrazine, Accent, Banvel, Buctril, Clarity, or 2,4-D. May be mixed with glyphosate on Roundup Ready corn, with Liberty on Liberty Link corn, and with Lightning on Clearfield corn.
mesotrione, MOA 27 (Callisto) 4 F	3 fl oz	0.094	Can be applied overtop or with drop nozzles until corn is 30 inches tall or has eight leaves. Add crop oil concentrate at 1 gallon per 100 gallons spray solution. Do not use methylated seed oil or adjuvant blends containing methylated seed oil. Controls most broadleaf weeds. Partial control of common ragweed and morningglory. Does not control sicklepod or prickly sida. Can tank mix with atrazine, Accent Q, or Steadfast Q. See precautions on labels of these products. Can be mixed with Liberty on Liberty Link corn or glyphosate on Roundup Ready corn. No rotational restrictions for small grains or other crops planted the following spring. Rainfast in 1 hour. See precautions on label concerning use of Counter and Lorsban.
thifensulfuron methyl, MOA 2 (Harmony SG) 50 WDG	0.125 oz	0.0039	Apply only to corn with 2 to 6 leaves (1 to 4 collars) but not larger than 4 collars or 12 in tall. Add either nonionic surfactant at 1 quart per 100 gallons or crop oil concentrate at 1 gallon per 100 gallons. Also, add a nitrogen-containing fertilizer according to label directions. Controls lambsquarters, pigweeds, smartweed, and velvetleaf; see label for recommended weed size for treatment. Harmony SG will also control curly dock and burcucumber. See label for comments concerning injury when used in conjunction with insecticides. May be mixed with atrazine.
2,4-D amine, MOA 4 (various brands) 3.8 SL	0.5 to 1 pt	0.24 to 0.48	Use 0.5 pint overtop when corn is 4 to 5 inches tall and weeds are small. Increase rate to 1 pint as corn reaches 8 inches Use drop nozzles and direct spray toward base of corn if over 8 inches tall. Do not cultivate for about 10 days after spraying as corn may be brittle. Reduce rate of 2,4-D if extremely hot or soil is wet. For better control of sicklepod and horsenettle, add a nonionic surfactant to 1 pint of 2,4-D and direct spray. Not adequately effective on smartweed, nightshade, burcucumber, or pokeweed. Use extreme caution to avoid drift to sensitive crops such as cotton and tobacco. **Use of ester formulations or acid + ester formulations (such as Weedone 638) or 2,4-D is not suggested if sensitive crops, especially cotton or tobacco, are located within 1 mile of the corn.**
rimsulfuron, MOA 2 + mesotrione, MOA 27 (Realm Q) 38.75% WDG	4 oz	0.0188 + 0.0781	Apply overtop or with drop nozzles to corn up to 20 inches or 7 leaf collars. Controls redroot and smooth cocklebur, common ragweed, pigweed, and velvetleaf. Partial control of morningglory. Also controls 1- to 2-inches fall panicum. Provides short-term residual control of lambsquarters, nightshade, redroot pigweed, and pigweed. Does not control sicklepod or prickly sida. Add 1 quart nonionic surfactant per 100 gallons and 2 quarts/acre UAN. May tank mix with other postemergence corn herbicides (except Basagran), including glyphosate on Roundup Ready corn and Liberty on Liberty Link corn. See label statement concerning sensitive hybrids. See rotational restrictions on label.
rimsulfuron, MOA 2 + thifensulfuron, MOA 2 (Resolve Q) 22.4% WDG	1.25 oz	0.014 + 0.0031	Apply overtop or with drop nozzles to corn up to 20 inches or 7 leaf collars. Controls redroot and smooth pigweed and velvetleaf. Suppresses cocklebur, smartweed, lambsquarters, common ragweed, and morningglory. Also controls 1- to 2-inches tall panicum. Provides short-term residual control of lambsquarters, nightshade, redroot pigweed, and smooth pigweed. Add 1 quart nonionic surfactant per 100 gallons and 2 quart/acre UAN. May tank mix with other postemergence corn herbicides (except Basagran), including glyphosate on Roundup Ready corn and Liberty on Liberty Link corn. See label statement concerning sensitive hybrids. See rotational restrictions on label.
tembotrione, MOA 27 (Laudis) 3.5 L	3 fl oz	0.082	Can be applied overtop or with drop nozzles to corn from emergence up to V8 stage. Add methylated seed oil at 1 gallon per 100 gallons of spray solution. Also add 1.5 quarts/A UAN. Controls most broadleaf weeds. Does not control sicklepod or prickly sida and only suppresses morningglory. Controls or suppresses some grasses. See label for weeds controlled and recommended size for treatment. Can tank mix with atrazine, Accent Q, Buctril, or Steadfast Q. Can tank mix with Liberty on Liberty Link corn or glyphosate on Roundup Ready corn. Rain-free in 1 hour. See label for rotational restrictions.
tembotrione, MOA 27 + dicamba, diglycolamine salt, MOA 4 + safener (DiFlexx Duo) 2.13 SC	24 to 40 fl oz	0.05 to 0.084 + 0.35 to 0.58	May be applied to corn from emergence up to, but not including, the V7 (seventh leaf collar) or 36 inches tall, whichever occurs first. May be applied directed with drop nozzles to corn in the V7 to V10 (7-10 collars), up to 36 inches tall, or up to 15 days prior to tassel, whichever comes first. Add methylated seed oil or crop oil concentrate at 1 gallon per 100 gallons of spray solution. Also add a high-quality urea ammonium nitrate at 1.5 quarts or spray-grade ammonium sulfate at 8.5 to 17 pounds per 100 gallons spray solution. Do not apply DiFlexx Duo with liquid fertilizers as the primary spray carrier. Do not apply DiFlexx Duo to corn that exhibits injury from previous herbicides applications. DiFlexx Duo contains the new corn safener, cyprosulfamide, to improve corn tolerance to dicamba. Research on DiFlexx Duo is limited in North Carolina.
tembotrione, MOA 27 + thiencarbazone-methyl, MOA 2 (Capreno) 3.45 SC	3 fl oz	0.068 + 0.013	Apply postemergence over the top prior to V6 stage. Use drop nozzles for better coverage of later stages of corn. Apply with crop oil concentrate at 1 gallon/100 gallons and UAN at 1.5 quarts/A. See comments under Laudis for spectrum of control and possible tank mixture. See label for precautions when organophosphate insecticide is used in corn.
topramezone, MOA 27 (Impact) 2.8 L (Armezon) 2.8 L	0.75 fl oz	0.016	Can be applied overtop or with drop nozzles to corn from emergence until 45 days prior to harvest. Add crop oil concentrate or methylated seed oil at 1 gallon per 100 gallons spray solution. Also add 1.25 to 2.5 gallons UAN per 100 gallons spray solution. See label for adjuvant recommendations in tank mixes. Controls most broadleaf weeds. Does not control sicklepod and only suppresses morningglory. Controls or suppresses some grasses. See label for weeds controlled and recommended size for treatment. Can tank mix with glyphosate on Roundup Ready corn, Liberty on Liberty Link corn, or Lightning on Clearfield corn. Rain-free in 1 hour. See label for rotational restrictions.
pyraflufen-ethyl, MOA 14 (ET)	0.5 to 2.0 fl oz	0.0008 to 0.0003 (lb a.i.)	ET can be used for limited suppression of small broadleaf weeds up to V4 stage. Do not apply with crop oil concentrate. Some leaf speckling can occur but is transient. See label for adjuvant and spray volume recommendations. Research with ET is limited in North Carolina.
Postemergence, Annual grasses, broadleaf weeds, and johnsongrass			
nicosulfuron, MOA 2 + rimsulfuron, MOA 2 (Steadfast Q) 37.7 WDG	1.5 oz	0.024 + 0.011	Apply to corn up to 20 inches tall with 6 or fewer leaf collars. Add 1 gallon crop oil concentrate per 100 gallons or 1 quart nonionic surfactant per 100 gallons. Also add nitrogen fertilizer according to label directions. Controls johnsongrass and most annual grasses and broadleaf weeds. May not adequately control crabgrass, goosegrass, Palmer amaranth, and sicklepod. See label for weeds controlled and weed size to treat. Do not apply to corn treated with Counter 15G; see label for precautions concerning other organophosphate insecticides. See label for comments on susceptible hybrids. Can tank mix with atrazine, Callisto, Clarity, or Distinct. See label precautions for tank mixes. Steadfast Q contains 25% nicosulfuron and 12.5% rimsulfuron.

Table 7-1A. Chemical Weed Control in Field Corn

Herbicide, Mode of Action Code[1] and Formulation	Amount of Formulation Per Acre	Pounds Active Ingredient Per Acre	Precautions and Remarks
Postemergence, Annual grasses, broadleaf weeds, and johnsongrass (continued)			
nicosulfuron, MOA 2 + mesotrione, MOA 27 (Revulin Q) 51.2 WDG	3.4 to 4.0 oz	0.031 to 0.036 + 0.078 to 0.092	Apply to corn up to 20 inches tall with 6 or fewer leaf collars. Add 1 gallon crop oil concentrate per 100 gallons or 1 quart nonionic surfactant per 100 gallons. Also add nitrogen fertilizer according to label directions. Controls johnsongrass and most annual grass and broadleaf weeds. May not adequately control goosegrass, sicklepod, and Palmer amaranth greater than 3 inches tall. See label for weeds controlled and weed size to treat. Do not apply to corn treated with Counter 15G; see label for precautions concerning tankmixing with foliar applied organophosphate insecticides. Can tank mix with atrazine, glyphosate, or glufosinate. See label precautions for tank mixes. Revulin Q contains 14.4% nicosulfuron and 36.8% mesotrione.
Postemergence, Annual broadleaf weeds and some annual grasses: **Clearfield Hybrids Only**			
imazethapyr, MOA 2 + imazapyr, MOA 2 (Lightning) 70 WDG	1.28 oz	0.042 + 0.014	USE ONLY ON CLEARFIELD HYBRIDS. Make only one application per year. Can be applied anytime up to 45 days prior to harvest. Use of drop nozzles will give better coverage in larger corn. Controls most annual broadleaf weeds and certain annual grasses; see label for weeds controlled and recommended growth stage for application. Not adequately effective on ragweed. Season-long sicklepod control may require a pre-emergence application of atrazine or a lay-by application of an appropriate herbicide. Suppresses yellow and purple nutsedge. Add either 1 quart per 100 gallons nonionic surfactant or 1 gallon per 100 gallons of crop oil concentrate or methylated seed oil. Also add 1 to 2 quarts per acre of UAN or 2.5 pounds per acre ammonium sulfate. May be tank mixed with most other corn herbicides; see labels for details. See Lightning label for rotational restrictions.
Postemergence, Annual grasses and annual broadleaf weeds: **Liberty Link Hybrids Only**			
glufosinate-ammonium, MOA 10 (Liberty 280) 2.34 SL	32 fl oz	0.59	Use only on Liberty Link hybrids. May be applied overtop from corn emergence up to V6 growth. Apply with drop nozzles to corn up to 36 inches tall. Rate depends on weed species and weed size; see label for details. Controls most annual grass and broadleaf weeds, but only marginally effective on goosegrass. Not effective on dayflower. Timing of application is critical for pigweed control. Use of drop nozzles in corn over 8 inches tall may improve spray coverage. If Liberty was applied preplant, one application of Liberty at 29 to 43 fl oz per acre may be applied in-season to LibertyLink varieties only. If Liberty was not used preplant, may make 2 applications in season. Total for the year is not to exceed 87 fl oz per acre. Add 3 pounds per acre of ammonium sulfate. Do not add surfactant or crop oil. May be tank mixed with most postemergence corn herbicides; see respective labels for details. Tank mixes of Liberty plus atrazine have been most effective.
Postemergence, Annual grasses and broadleaf weeds, johnsongrass, and suppression of perennial broadleaf weeds: **Glyphosate-tolerant Hybrids Only**			
glyphosate, MOA 9 (numerous brands and formulations)	See label	0.75 to 1.125 (lb a.e.)	Apply only to glyphosate-tolerant hybrids. Glyphosate is available as an isopropylamine salt and a potassium salt. Glyphosate formulations and application rates should be compared on the basis of pounds of glyphosate acid equivalent (a.e.) per gallon and per acre, respectively. See Table 7-10 for glyphosate rate conversions. Glyphosate controls most annual weeds; exceptions include dayflower, Florida pusley, and hemp sesbania. Timely application is critical for morningglory control. Glyphosate also controls johnsongrass and suppresses other perennial weeds. See label of brand you use for recommended rates and sizes of weeds to treat. Adjuvant recommendations vary according to glyphosate product. See label of brand used for specific recommendations. Apply overtop from corn emergence through the V8 stage (8 leaves with collars) or until corn reaches 30 inches, whichever comes first. Drop nozzles suggested on corn 24 to 30 inches Apply only with drop nozzles when corn is 30 to 48 inches Make multiple applications at least 10 days apart. Do not exceed a total of 2.25 pounds a.e. per acre in crop. For resistance management, do not rely entirely on glyphosate. Herbicides with other modes of action should be included in the program. Such herbicides can be preemergence, mixed with glyphosate, or lay-by. See comments on resistance management in Table 7-10. Any registered soil-applied herbicide or lay-by herbicide can be used on Roundup Ready corn. Aim, atrazine, Callisto, Clarity, Degree Xtra, Distinct, Harness, Harness Xtra, Impact, Laudis, Resolve, Resolve Q, Resource, Status or 2,4-D can be mixed with glyphosate applied postemergence. When using a tank mix, follow all directions and precautions on the respective labels, especially corn stage for application.
glyphosate, MOA 9 + mesotrione, MOA 27 (Callisto GT) 4.18 L	2 pt	0.95 + 0.095	Apply only to glyphosate-tolerant hybrids. Apply from corn emergence up to 30 inches or 8 leaf collars. Add nonionic surfactant at 1 quart per 100 gallons spray solution. Also add AMS according to label directions. Crop oil concentrate may be at 1 gallon per 100 gallons spray solution, but increases the risk for crop injury. Do not use methylated seed oil or adjuvant blends containing methylated seed oil or severe crop injury may occur.
glyphosate, MOA 9 + s-metolachlor, MOA 15 + mesotrione, MOA 27 (Halex GT) 4.39 L	3.6 to 4 pt	0.94 to 1.05 + 0.94 to 1.05 + 0.094 to 0.105	Apply only to glyphosate-tolerant hybrids. Apply from corn emergence up to 30 inches or 8 leaf collars. Add nonionic surfactant at 1 quart/100 gallons spray solution. Also add AMS according to label directions. Do not substitute UAN for AMS. May tank mix with atrazine. See precautions on label when using Halex GT in conjunction with insecticides.
Postemergence, most broadleaf weeds: **Enlist Hybrids Only**			
2,4-D choline, MOA 4 (Enlist One) 3.8 SL	1.5 to 2 pt	0.71 to 0.95	Apply only to hybrids designated as Enlist. Apply when weeds are small, and corn is no larger than V8 growth stage or 30 inches tall, whichever occurs first. For corn 30 to 48 inches tall, apply only using ground application equipment using drop nozzles and avoid spraying into whorl of corn plants. Controls most broadleaf weeds. Make 1 to 2 applications at least 12 days apart. The high rate may result in temporary, cosmetic injury in the form of spotting or temporary plant leaning but will not affect long-term crop development or yield.
2,4-D choline, MOA 4 + glyphosate, MOA 9 (Enlist Duo) 3.3 SL	3.5 to 4.75 pt	0.7 to 0.95 + 0.74 to 1.01	Apply only to hybrids designated as Enlist. Apply when weeds are small, and corn is no larger than V8 growth stage or 30 inches tall, whichever occurs first. For corn 30 to 48 inches tall, apply only using ground application equipment using drop nozzles and avoid spraying into whorl of corn plants. Controls most broadleaf and grass weeds. Make 1 to 2 applications at least 12 days apart. The high rate may result in temporary, cosmetic injury in the form of spotting or temporary plant leaning but will not affect long-term crop development or yield.
Postemergence, Yellow nutsedge			
bentazon, MOA 6 (Basagran) 4 SL	1.5 to 2 pt	0.75 to 1	Apply overtop of corn or directed when yellow nutsedge is 6 to 8 inches tall. If needed, make a second application 7 to 10 days later. Add 2 pints per acre of crop oil concentrate.

Table 7-1A. Chemical Weed Control in Field Corn

Herbicide, Mode of Action Code[1] and Formulation	Amount of Formulation Per Acre	Pounds Active Ingredient Per Acre	Precautions and Remarks
Postemergence, Yellow and purple nutsedge			
halosulfuron (Sandea) 75 WDG	0.67 to 1.33 oz	0.03 to 0.06	Apply overtop or with drop nozzles to corn from spike stage until layby. Add nonionic surfactant at 1 quart/100 gallons spray solution.
Postemergence, Yellow and purple nutsedge: Clearfield hybrids Only			
imazethapyr, MOA 2 + imazapyr, MOA 2 (Lightning) 70 WDG	1.28 oz	0.042 + 0.014	**Use only on Clearfield hybrids.** Apply when nutsedge is 1 to 3 inches tall. Add surfactant and nitrogen-containing fertilizer as specified on the label. See comments on Lightning in section on broadleaf weed control. Label claims suppression only.
Postemergence, Yellow and purple nutsedge; Glyphosate-tolerant Hybrids Only			
glyphosate, MOA 9 (numerous brands and formulations)	See label	0.75 to 1.125 (lb a.e.)	**Apply only to Roundup Ready hybrids.** 2 applications of glyphosate may be necessary for nutsedge control. See previous comments under Roundup Ready corn.
Postemergence, Annual grasses			
foramsulfuron, MOA 2 (Option) 35 WDG	1.5 oz	0.033	Apply overtop to corn in V1 to V6 stage. Can be applied with drop nozzles to corn through V8 stage. Add a methylated seed oil at 1.5 pints per acre and either 1.5 to 2 quarts per acre of 30% UAN or 1.5 to 3 pounds per acre of ammonium sulfate. Do not cultivate for 7 days before or after application. Controls most annual grasses; see comments on label for crabgrass and broadleaf signalgrass. Also controls small broadleaf weeds, such as burcucumber, cocklebur, pigweed, lambsquarters, common ragweed, velvetleaf, and nightshade. May be applied twice per season. May be tank mixed with atrazine, Callisto, or Distinct. Do not apply to corn treated with Counter, Dyfonate, or Thimet. See comments on label concerning sensitive hybrids.
nicosulfuron, MOA 2 (Accent) 75 WDG (Accent Q) 54.5 WDG	0.67 oz 0.9 oz	0.031	Can be applied overtop or with drop nozzles to corn up to 20 inches tall. If corn is 20 to 36 inches tall, apply only with drop nozzles and avoid spraying into the corn whorl. Do not apply if corn is greater than 36 inches Add either a crop oil concentrate at 1 gallon per 100 gallons or a nonionic surfactant at 1 quart per 100 gallons spray solution. See label concerning additional adjuvants. Do not cultivate for 10 days before application. Controls ryegrass, small broadleaf signalgrass, foxtails, fall panicum, Texas panicum, barnyardgrass, shattercane, and seedling johnsongrass. May not adequately control crabgrass and goosegrass. Also controls small burcucumber, jimsonweed, morningglory, pigweed, and smartweed. Can be applied twice, but do not exceed 1.33 ounces Accent or 1.8 ounces Accent Q per acre per year. Reduced rates may be applied under certain conditions; see label for details. May be tank mixed with atrazine, Callisto, Clarity, or Distinct for improved broadleaf control. See label for comments concerning injury when used in conjunction with insecticides.
Postemergence, Johnsongrass			
formasulfuron, MOA 2 (Option) 35WDG	1.5 oz	0.033	Apply before seedling and rhizome johnsongrass exceed 16 inches tall. See comments for Option under Annual Grasses.
nicosulfuron, MOA 2 (Accent) 75 WDG (Accent Q) 54.5 WDG	0.67 oz 0.9 oz	0.031	Apply when seedling johnsongrass is 4 to 12 inches tall, rhizome johnsongrass is 8 to 18 inches tall, or shattercane is 4 to 12 inches tall. See other comments for Accent under Annual Grasses.
Postemergence, Johnsongrass: Glyphosate-tolerant Hybrids Only			
glyphosate, MOA 9 (numerous brands and formulations)	See label	0.75 (lb a.e.)	**Apply only to glyphosate-tolerant hybrids.** See previous comments under glyphosate-tolerant hybrids.
Postemergence, Bermudagrass: Glyphosate-tolerant Hybrids Only			
glyphosate, MOA 9 (numerous brands and formulations)	See label	0.75 to 1.125 (lb a.e.)	**Apply only to glyphosate-tolerant hybrids.** See previous comments under glyphosate-tolerant hybrids. 2 applications are usually required for adequate control.
Lay-by, Annual broadleaf weeds; control or suppression of perennial broadleaf weeds			
2,4-D amine, MOA 4 (various brands) 3.8 SL	0.5 to 1 pt	0.24 to 0.48	Apply with drop nozzles. Do not apply to corn in the tassel to dough stage. May add 1 quart of nonionic surfactant per 100 gallons spray solution. Surfactant may increase control of sicklepod and perennial weeds. Corn hybrids vary in sensitivity; check with seed dealer for sensitivity of hybrid used. Use extreme caution to avoid drift to sensitive crops, such as cotton and tobacco. **Use of ester formulations or acid + ester formulations (such as Weedone 638) or 2,4-D is not suggested if sensitive crops, especially cotton or tobacco, are located within 1 mile of the corn.** Liquid nitrogen may be used as the carrier. When using 2,4-D amine, mix 1 pint of herbicide in at least 2 quarts of water, and add this mixture to the spray tank with considerable agitation until thoroughly mixed. Do not allow nitrogen-herbicide mixture to stand in the sprayer.
dicamba, dimethylamine salt, MOA 4 (Banvel) 4 SL	0.5 pt	0.25	Apply as directed spray using water as the carrier to corn up to 36 inches tall. Do not apply within 15 days of tassel emergence. Add nonionic surfactant at 1 pint per 100 gallons for Clarity or 2 pints per 100 gallons for Banvel, Distinct, or Status. See comments on labels concerning addition of UAN or AMS. Follow precautions on labels concerning drift to sensitive crops.
dicamba, diglycolamine salt, MOA 4 (Clarity) 4 SL	0.5 pt	0.25	
dicamba, sodium salt, MOA 4 + diflufenzopyr, sodium salt (Distinct) 61.1 WDG	4 oz	0.125 + 0.053	
dicamba, sodium salt MOA 4 + diflufenzopyr, sodium salt + safener (Status) 61.1 WDG	5 to 10 oz	0.14 to 0.28 + 0.053 to 0.107	
ametryn, MOA 5 (Evik) 80 WDG	2 lb	1.6	Apply as a directed spray after corn is at least 15 inches tall. Do not apply Evik within 3 weeks of tasseling. Add nonionic surfactant at 2 quarts per 100 gallons spray solution. Evik and Linex may be applied using liquid nitrogen as the carrier. Add surfactant when using nitrogen as the carrier.
linuron, MOA 7 (Linex) 4 L	1.25 to 1.5 pt	0.63 to 0.75	Apply as a directed spray after corn is at least 15 inches tall. Note that current labeled rates of Linex have been reduced from previous years. Use 2 quarts nonionic surfactant per 100 gallons of spray solution. Linex may be applied using liquid nitrogen as the carrier. Do not apply within 57 days of harvest.

Table 7-1A. Chemical Weed Control in Field Corn

Herbicide, Mode of Action Code[1] and Formulation	Amount of Formulation Per Acre	Pounds Active Ingredient Per Acre	Precautions and Remarks
Preharvest, Annual grasses and johnsongrass			
sodium chlorate (Defol 750) 7.5 L	3.2 qt	6	Apply on warm, sunny day at least 14 days before anticipated harvest. Apply by ground or air after corn reaches hard dough or dent stage. Add surfactant or crop oil according to label directions. Thorough spray coverage essential.
Preharvest, Broadleaf weeds			
2, 4-D, amine, MOA 4 (various brands)	1 to 2 pt	0.48 to 0.95	Suppresses perennial broadleaf weeds and controls many annual broadleaf weeds. Apply after hard dough or dent stage by ground or air. Avoid drift to sensitive crops.
carfentrazone, MOA 14 (Aim) 2 EC	1.9 fl oz	0.03	Desiccates morningglory, cocklebur, and pigweed. Apply 3 or more days ahead of harvest. Add 1 gallon crop oil concentrate per 100 gallons spray solution. Thorough coverage is critical; use minimum of 20 GPA for ground application. May be applied by air. For dense morningglory infestations, two applications at 1 ounce/acre may be more effective.
Preharvest, Annual grasses and broadleaf weeds			
paraquat, MOA 22 (Gramoxone Inteon) 2 SL	1.2 to 2 pt	0.3 to 0.5	Apply after black layer has formed and at least 7 days prior to harvest. Add nonionic surfactant at 1 quart per 100 gallons spray solution. Generic brands of paraquat containing 3 pounds active ingredient per gallon are available These products would be applied at 0.8 to 1.3 pints per acre.
Preharvest, Annual grasses, johnsongrass, and broadleaf weeds			
glyphosate, MOA 9 (numerous brands and formulations)	See label	0.75 (lb a.e.)	Apply after kernel fill is complete (black layer formed) and grain moisture is 35% or less. Apply at least 7 days prior to harvest. Maximum rate for aerial application varies by product; see label of brand used. Avoid drift to other crops and desirable vegetation.
Postharvest, Horsenettle and other perennial and annual broadleaf weeds			
2,4-D amine, MOA 4 (various brands) 3.8 SL + dicamba, MOA 4 (Banvel) 4 SL (Clarity) 4 SL	2 pt + 1 to 2 pt 1 to 2 pt	0.95 + 0.5 to 1	This is an effective way to reduce perennial broadleaf weeds in succeeding crops. Follow label precautions on dicamba label concerning drift to sensitive crops. Delay small grain seeding at least 20 days.
Postharvest, Bermudagrass, other annual and perennial weeds			
glyphosate, MOA 9 (numerous brands and formulations)	See label	1.5 (lb a.e.)	This is an effective way to reduce perennial weeds in succeeding crops. Apply at least 10 to 14 days before killing frost. Rate can be increased up to 3.75 pounds a.e. Include adjuvant according to the label for the brand used. Dicamba may be mixed with glyphosate.

Weed Response to Preemergence Herbicides — Corn

C. W. Cahoon, Crop and Soil Sciences Department

Ratings based upon average to good soil and weather conditions for herbicide performance and upon proper application rate, technique, and timing.

Table 7-1B. Weed Response to Preemergence Herbicides in Corn

Species	Acuron	Atrazine	Atrazine + Simazine	Bicep II Magnum, Brawl II, Cinch ATZ, or Medal II AT	Balance Flexx	Callisto	Corvus	Dual II Magnum, Brawl II, Cinch, or Medal II	Degree Xtra, Fultime, Fultime NXT, Harness Xtra, or Keystone NXT	Harness, Surpass, Surpass NXT, TopNotch	Lexar, Lexar EZ, Lumax, or Lumax EZ	Outlook	Prowl H$_2$O	Resicore	SureStart II or TripleFLEX II	Zidua SC, Anthem, Anthem Maxx, or Anthem Flex
Grasses:																
Bermudagrass	N	N	N	N	F	N	N	N	N	N	N	N	N	N	N	N
Broadleaf signalgrass	G	P	P	G	N	P	—	G	G	G	G	FG	P	G	G	GE
Crabgrass	E	G[1]	G[2]	E	F	F	E	E	E	E	E	E	F	E	E	E
Fall panicum	E	N	FG	E	FG	PN	GE	E	E	E	E	E	PF	E	E	E
Foxtails	E	F	FG	E	FG	PN	E	E	E	E	E	E	F	E	E	E
Goosegrass	E	F	FG	E	N	PN	GE	E	E	E	E	E	PF	E	E	E
Johnsongrass																
Seedling	F	N	N	PF	FG	N	—	PF	PF	PF	F	PF	PF	PF	PF	PF
Rhizome	N	N	N	N	F	N	—	N	N	N	N	N	N	N	N	N
Shattercane	PF	N	N	P	FG	N	—	P	P	P	PF	P	PF	P	P	P
Texas panicum	PF	N	N	PF	N	N	G	PF	PF	PF	PF	PF	PF	PF	PF	F
Sedges:																
Nutsedge																
Yellow	FG	N	N	F	N	PF	F	FG[3]	PF	PF	FG	F	N	PF	PF	F
Purple	N	N	N	N	N	—	—	N	N	N	N	N	N	N	N	N
Broadleaf Weeds:																
Balloon vine	—	G	GE	G	—	—	—	N	G	N	G	N	N	—	—	N
Burcucumber[4]	—	F	FG	F	F	—	—	N	F	N	F	N	N	N	N	N
Cocklebur	E	G	GE	G	N	PF	G	N	G	N	G	N	N	F	F	N
Eastern black nightshade	E	E	E	E	GE	E	E	F	E	F	E	F	N	GE	FG	F
Florida beggarweed	G	G	GE	G	—	—	GE	F	G	F	G	F	N	F	F	F
Florida pusley	E	E	E	E	—	—	GE	GE	E	GE	E	GE	G	F	F	G
Hemp sesbania	G	F	F	F	—	—	—	N	F	N	F	N	N	F	—	N
Jimsonweed	E	E	E	E	FG	G	G	N	E	N	E	N	N	FG	F	F
Lambsquarters	E	E	E	E	GE	E	E	F	E	F	E	FG	G	GE	FG	FG
Morningglory	G	G	G	G	N	FG	G	N	G	N	G	N	N	F	F	N
Pigweed	E	E	E	E	G	E	E	G	E	GE	E	GE	FG	GE	GE	E
Prickly sida	E	E	E	E	F	—	G	P	E	P	E	P	N	N	F	P
Ragweed																
Common	E	E	E	E	F	F	G	PF	E	PF	E	F	N	F	FG	F
Giant	G	FG	G	G	F	—	—	N	G	N	FG	N	N	F	FG	N
Sicklepod	G	G	GE	G	N	P	G	N	G	N	G	N	N	—	FG	N
Smartweed	G	G	GE	G	FG	GE	—	N	G	N	G	N	FG	FG	FG	F
Tropic croton	G	G	GE	G	N	PN	G	N	G	N	G	N	N	—	—	—
Velvetleaf	G	G	G	G	GE	E	G	N	G	N	G	N	N	GE	FG	F

[1] No control of smooth crabgrass.
[2] Poor to fair on smooth crabgrass.
[3] Dual is normally good on yellow nutsedge when incorporated
[4] Multiple flushes of germination; one application of any herbicide will seldom be adequate.

Key:

E = excellent control, 90% or better
G = good control, 80% to 90%
F = fair control, 50% to 80%
P = poor control, 25% to 50%
N = no control, less than 25%

Weed Response to Postemergence Herbicides — Corn

C. W. Cahoon, Crop and Soil Sciences Department

Ratings based upon average to good soil and weather conditions for herbicide performance and upon proper application rate, technique, and timing.

Table 7-1C. Weed Response to Postemergence Herbicides — Corn

Species	Accent or Accent Q	Aim	Armezon or Impact	Atrazine[1]	Banvel, Clarity, DiFlexx, Distinct, or Status	Basagran	Buctril	Callisto	Capreno	Enlist One or 2,4-D[2]	Evik[3]	Glyphosate[4]	Halex GT	Harmony SG	Laudis	Liberty[4]	Lightning[5]	Linex[3]	Resource	Revulin Q	Steadfast Q
Grasses:																					
Bermudagrass	N	N	N	N	N	N	N	N	N	N	N	F[8]	F[8]	N	N	NP	N	N	N	N	N
Broadleaf signalgrass	GE	N	N	F	N	N	N	P	GE	N	E	E	E	N	G	GE	G	GE	N	G	G
Crabgrass	PF	N	FG	FG[7]	N	N	N	F	GE	N	GE	E	E	N	G	G	PF	GE	N	PF	PF
Fall panicum	G	N	FG	P	N	N	N	P	E	N	GE	E	E	N	N	E	PF	GE	N	G	G
Foxtails	G	N	GE	G	N	N	N	P	E	N	E	E	E	N	GE	E	G	E	N	G	G
Goosegrass	P	N	F	G	N	N	N	P	E	N	GE	E	E	N	F	PF	P	G	N	P	P
Johnsongrass Seedling	E	N	N	P	N	N	N	P	E	N	GE	E	E	N	F	GE	GE	G	N	E	E
Rhizome	GE	N	N	N	N	N	N	N	N	N	P	E	E	N	N	F[12]	G[9]	NP	N	GE	G
Shattercane	E	N	F	P	N	N	N	—	E	N	G	E	E	N	F	—	G	FG	N	E	E
Texas panicum	G	N	FG	NP	N	N	N	—	GE	N	G	E	E	N	G	G	PF	GE	N	G	G
Sedges:																					
Nutsedge Yellow	P	N	—	PF	N	G[13]	N	F	F	N	F	F[8]	F[8]	N	F	P	F	F	N	P	P
Purple	N	N	—	N	N	N	N	F	F	N	PF	FG[8]	FG[8]	N	F	P	FG	P	N	N	N
Broadleaf Weeds:																					
Balloon vine	—	—	FG	G	G	P	—	—	G	—	—	E	E	—	—	—	—	—	P	—	—
Burcucumber[10]	F	N	FG	FG	F	P	F	—	F	P	F	E	E	PF	F	P	F	FG	F	F	F
Cocklebur	F	N	GE	E	E	E	E	E	GE	E	E	E	E	FG	GE	E	E	E	G	E	F
Eastern black nightshade	N	G	GE	GE	E	P	G	G	E	F	G	FG	GE	P	E	G	GE	PF	FG	G	P
Florida beggarweed	—	—	—	G	G	N	E	—	FG	E	G	G	—	—	E	—	E	P	—	—
Florida pusley	N	FG	F	G	G	PN	E	—	GE	G	E	P	—	GE	FG	FG	G	—	—	—	—
Hemp sesbania	F	—	—	FG	E	P	G	—	F	E	F	P	—	—	F	—	—	—	P	—	—
Jimsonweed	FG	N	GE	E	E	E	E	E	E	E	E	E	E	F	E	G	E	E	G	E	E
Lambsquarters	P	G	GE	E	E	FG	E	E	GE	E	E	E	E	E	E	G	E	G	E	E	F
Morningglory	F	G	FG	G	E	P	G	GE	GE	E	E	FG[11]	FG[11]	FG	F	E	G	E	FG	GE	G
Pigweed	G	G	GE	E	E	E	N	F	E	E	E	E	E	E	E	FG	E	E	—	E	G
Prickly sida	P	—	—	F	GE	G	F	E	P	G	G	GE	G	P	N	GE	G	GE	—	P	P
Ragweed Common	P	P	FG	GE	E	G	E	FG	GE	E	E	E	E	F	GE	E	PF	E	G	FG	P
Giant	P	N	FG	F	GE	GE	E	—	G	E	G	G	G	P	G	G	P	G	P	—	P
Sicklepod	F	—	P	G	GE	N	N	P	G	E	E	G	G	P	PF	E	F[14]	GE	N	P	F
Smartweed	G	—	GE	G	E	E	GE	G	E	F	G	G	G	E	.E	E	GE	GE	N	P	G
Tropic croton	—	—	—	G	GE	F	FG	—	G	G	G	G	G	P	—	—	P	G	P	—	—
Velvetleaf	F	E	GE	G	G	G	G	G	E	G	G	E	E	G	GE	G	GE	G	E	G	F

[1] Assumes addition of crop oil concentrate.
[2] Apply Enlist One to Enlist hybrids only.
[3] Apply directed only.
[4] Apply only to Liberty Link hybrids.
[5] Apply only to Clearfield corn hybrids.
[6] Apply only to glyphosate-resistant hybrids. See comments on resistance management in TABLE 7-10.
[7] No control of smooth crabgrass.
[8] Control is good with two applications of glyphosate.
[9] Follow-up treatment with Accent may be needed for acceptable control.
[10] Multiple flushes of germination; one application of any herbicide will seldom be adequate.
[11] With good application timing and a follow-up application as needed, morningglory control can be good.
[12] Liberty applied twice is usually good on johnsongrass.
[13] Two applications may be needed for good control.
[14] Sicklepod control by Lightning can be erratic. For more consistent control, mix atrazine, Banvel, Clarity, Distinct, Marksman, or 2,4-D with Lightning.

Key:
E = excellent control, 90% or better
G = good control, 80% to 90%
F = fair control, 50% to 80%
P = poor control, 25% to 50%
N = no control, less than 25%

Chemical Weed Control in Cotton

C. W. Cahoon, Assistant Professor and Extension Weed Specialist, Crop and Soil Sciences Department

Table 7-2A. Brand Names and Formulations of Active Ingredients Listed in Table 7-2B.

Active ingredient(s)	Brand name(s)[1]	Formulation[2]	Mode of Action
acetochlor	Warrant	3 CS	15
acetochlor + fomesafen, premix	Warrant Ultra	3.45 CS (2.82 + 0.63)	15 + 14
carfentrazone	Aim EC	2 EC	14
clethodim	Intensity One, Select Max, TAPOUT	0.97 EC	1
	Arrow 2EC, Avatar S2, Cleanse, Clethodim 2E; Clethodim 2EC, Dakota, Intensity, Select 2EC, Shadow, Tide USA Clethodim 2EC, Volunteer, Willowood Clethodim 2EC	2 EC	1
	Shadow Ultra	1 EC	1
	Shadow 3 EC, Section Three	3 EC	1
2,4-D, choline salt	Enlist One	3.8 S	4
2,4-D choline salt + glyphosate	Enlist Duo	3.3 S	4 + 9
2,4-D, dimethylamine salt	Numerous brand names	S	4
2,4-D, ethylhexyl ester	Numerous brand names	EC	4
dicamba, BAPMA salt	Engenia	5 S[3]	4
dicamba, diglycolamine salt	Clarifier, Clarity, Clash, Detonate, Dicamba DGA, Dicamba HD, Dicash DGA-4, Sterling Blue, Strut, Veritas	4[3] S	4
dicamba, diglycolamine salt with additive VaporGrip	XtendiMax with VaporGrip Technology	2.9 S[3]	4
dicamba acid + 2,4-D ester	Burnmaster	4.07 EC (1 + 3.077)	4 + 4
	Spitfire	3.57 EC (0.5 + 3.07)	4 + 4
dimethenamid-P	Outlook	6 EC	15
diuron	Direx 4L, Diuron 4L, Parrot 4L	4 F	7
	Diuron 80, Diuron 80 DF, Diuron 80 WDG, Karmex DF	80 WDG	7
fluazifop p-butyl	Fusilade DX	2 EC	1
flumiclorac pentyl ester	Resource	0.86 EC	14
flumioxazin	Outflank, Panther, RedEagle Flumioxazin, Rowel, Valor SX, Warfox	51 WDG	14
	Panther SC, Valor EZ	4 F	14
flumioxazin + pyroxasulfone, premix	Fierce	76 WDG (33.5 + 42.5)	14 + 15
fluometuron	Cotoran 4L, Sharda Fluometuron	4 F	7
fluridone	Brake	1.2 F	12
fluridone + fluometuron, premix	Brake FX	3.6 F	12 + 7
fluridone + fomesafen, premix	Brake F16	2.7 F	12 + 14
fomesafen	Agent 1.88, Battle Star, Foma 1.88, fomesafen 1.88, Fomesafen Sodium SC, Rumble, Shafen Star, Top Gun Flex, Vamos, Willowood Fomesafen 1.88	1.88 S	14
	Andros 2.0, Foma 2.0, Fomesafen 2 SL, Reflex, Ringside, Shafen, Top Gun, Willowood Fomesafen 2 SL	2 S	14
	Sinister	2.87 S	14
fomesafen + glyphosate, premix	Flexstar GT3.5	2.82 S (0.56 + 2.26[3])	14 + 9
glufosinate-ammonium	Agri Star Surmise, Cheetah, Forfeit 280, Glufosinate 280 SL, Kong Glufosinate 280, Liberty 280 SL, Reckon 280 SL, Refer 280 SL, Tide glufosinate, Total 2.3, Willowood Glufosinate 280SL	2.34 S	10
glyphosate	See Table 7-9.	S	9
glyphosate + 2.4-D choline salt	Enlist Duo	3.3 S[3]	9 + 4
glyphosate + S-metolachlor, premix	Sequence	5.25 EC (2.25[3] + 3)	9 + 15
lactofen	Boa, Cobra, Lactofen 2.0, Mongoose	2 EC	14
MSMA	MSMA 6 Plus, Target 6 Plus	6 S	17
	MSMA 6.6, Target 6.6	6.6 S	17
paraquat	Cyclone SL 2.0, Gramoxone SL 2.0	2 S	22
	Bonedry, Devour, Helmquat 3SL, Paraquat Concentrate, Para-Shot 3.0, Parazone 3SL, Quik-Quat, Willowood Paraquat 3SL	3 S	22
pendimethalin	Acumen, Framework, Helena Pendimethalin, Pavilion 3.3 EC, PendiPro 3.3 EC, Prowl 3.3 EC, Satellite 3.3, Stealth	3.3 EC	3
	Prowl H2O, Satellite Hydrocap	3.8 AS	3
prometryn	Caparol 4L	4 F	5
prometryn + trifloxysulfuron, premix	Suprend	80 WDG (79.3 + 0.7)	5 + 2
pyraflufen ethyl	ET Herbicide/Defoliant	0.208 EC	14
pyrithiobac sodium	Dupont Staple LX, Pysonex	3.2 S	2
pyroxasulfone	Zidua SC	4.17 SC	15
quizalofop p-ethyl	Assure II, Se-Cure EC, Targa	0.88 EC	1
rimsulfuron + thifensulfuron, premix	Leadoff	33.4 WDG (16.7 + 16.7)	2 + 2
	Crusher	50 WDG (25 + 25)	2 + 2

Table 7-2A. Brand Names and Formulations of Active Ingredients Listed in Table 7-2B.

Active ingredient(s)	Brand name(s)[1]	Formulation[2]	Mode of Action
S-metolachlor	Brawl, Charger Basic, Dual Magnum, EverpreX, Medal	7.62 EC	15
	Brawl II, Dual II Magnum, Medal II, Moccasin II Plus	7.64 EC	15
	Moccasin	8.0 EC	15
S-metolachlor + fomesafen, premix	Prefix	5.29 EC	14 + 15
saflufenacil	Sharpen	2.85 SC	14
sethoxydim	Poast Plus	1 EC	1
	Poast	1.5 EC	1
thifensulfuron + tribenuron, premix	Audit 1:1, Edition Broadspec, Firstshot, Rapport Broadspec	50 WDG (25 + 25)	2 + 2
	Harmony Extra SG	50 WDG (33.33 + 16.67)	2 + 2
	Nimble, Treaty Extra, T-Square, Volta Extra	75 WDG (50 + 25)	2 + 2
trifloxysulfuron	Envoke	75 WDG	2
trifluralin	Treflan HFP, Trifluralin 4 EC, Trifluralin 4 E.C., Trifluralin HF, Triflurex HFP, Trust	4 EC	3

[1] Brands registered for sale and use in cotton in North Carolina in 2018 (http://www.kellysolutions.com/nc/searchbychem.asp); labels for most products available at www.cdms.net.
[2] AS, aqueous suspension; CS, capsule suspension; EC, emulsifiable concentrate; F, flowable; S, solution; SC, suspension concentrate; WDG, water dispersible granule.
[3] Rate expressed as acid equivalent.

NOTE: A mode of action (MOA) code has been added to the Herbicide and Formulation column of the following table. Use MOA codes for herbicide resistance management. See Table 7-10, Herbicide Resistance Management, for details.

Table 7-2B. Chemical Weed Control in Cotton

Herbicide, Mode of Action Code and Formulation	Amount of Formulation Per Acre	Pounds Active Ingredient Per Acre	Precautions and Remarks
Early Preplant Burndown, Burndown of emerged annual weeds in no-till, strip-till, or stale seedbed systems, any variety			
glyphosate; MOA 9	See label	0.56 to 1.13 (lb a.e.)	Apply any time prior to planting to control emerged weeds. See labels for weeds controlled, application rates for specific weeds, and application directions and precautions. Does not adequately control cutleaf eveningprimrose, field pansy, or Carolina geranium, or wild radish. Glyphosate is available in several formulations. Glyphosate formulations and application rates should be compared on the basis of pounds of glyphosate acid equivalent (a.e.) per gallon and per acre, respectively. Rates in the preceding column are expressed as a.e. See Table 7-9 for glyphosate rate conversions. Adjuvant recommendations vary according to the glyphosate product used. See label of brand used for specific recommendations. Cover crops: Wheat < 12 inches: 0.56 lb a.e. Wheat > 12 inches: 0.75 lb a.e. Rye < 18 inches: 0.56 lb a.e. Rye > 18 inches: 0.75 lb a.e. **See comments under EARLY PREPLANT BURNDOWN—Glyphosate-resistant horseweed.**
glyphosate; MOA 9 + Aim (2 EC); MOA 14	See label + 0.5 to 1 fl oz	0.56 to 1.13 (lb a.e.) + 0.008 to 0.016	See comments for glyphosate alone. Aim contains carfentrazone; See Table 7-2A. Aim added to glyphosate will increase speed of control and may improve control of some species, but overall long-term control is generally not improved. This tank mix will not control cutleaf eveningprimrose, wild radish, or glyphosate-resistant horseweed. There is no waiting period between application and cotton planting.
glyphosate; MOA 9 + Burnmaster (4.07 EC) or Spitfire (3.57 EC); MOA 4	See label + 1.5 to 2 pt or 1.5 to 2 pt	0.56 to 1.13 (lb a.e.) + 0.76 to 1.02 (lb a.e.) or 0.67 to 0.89 (lb a.e.)	See comments for glyphosate alone. Burnmaster contains 1.0 lb a.e./gal of dicamba acid plus 3.07 lb a.e./gal of 2,4-D ester. Spitfire contains 0.5 lb a.e./gal of dicamba acid plus 3.07 lb a.e./gal of 2,4-D ester. Following application of either Burnmaster or Spitfire and a minimum of 1 inch rainfall, a waiting period of at least 30 days is required before planting. At the 2 pt rate, Burnmaster or Spitfire will control glyphosate-resistant horseweed.
glyphosate; MOA 9 + Clarity (4 S); MOA 4 or Engenia (5S); MOA 4 or XtendiMax (2.9S); MOA 4	See label + 8 fl oz or 6.4 fl oz or 11 fl oz	0.56 to 1.13 (lb a.e.) + 0.25 (lb a.e.) or 0.25 (lb a.e.) or 0.25 (lb a.e.)	See comments for glyphosate alone. Clarity and XtendiMax are formulated as the diglycolamine salt of dicamba; Engenia is formulated as the BAPMA salt. These salts are preferred rather than the dimethylamine salt. Other brands containing dicamba diglycolamine salt are listed in Table 7-2A. Following application of dicamba and a minimum of 1 inch rainfall, a waiting period of at least 21 days is required before planting any cotton not containing the XtendFlex trait. There is no waiting period for cotton with the XtendFlex trait if using Engenia or XtendiMax. Dicamba controls or suppresses several annual broadleaf weeds, and it suppresses Carolina geranium and curly dock. See Table 7-2E for weed response. Dicamba is somewhat less effective on cutleaf eveningprimrose than 2,4-D. This tank mixture will control glyphosate-resistant horseweed.
glyphosate; MOA 9 + 2,4-D; MOA 4 or Enlist One (3.8 S); MOA 4	See label + See label or 1 to pt	0.56 to 1.13 (lb a.e.) + 0.24 to 0.95 (lb a.e.) or 0.48 to 0.95 (lb a.e.)	See comments for glyphosate alone. Most, but not all, brands of 2,4-D may be applied at least 30 days ahead of cotton planting. Cotton containing the Enlist trait can be planted anytime following Enlist One application. 2,4-D is typically applied at 0.48 lb a.e. (1 pt/acre of 3.8 lb/gal formulation). See Table 7-2E for weed response. Excellent control of cutleaf eveningprimrose can be obtained with 2,4-D at 0.18 to 0.24 lb a.e. Glyphosate plus 2,4-D is not effective on Carolina geranium. At higher rates (0.95 lb a.e.; 2 pt/acre of 3.8 lb/gal formulation), this tank mix will control small glyphosate-resistant horseweed. Amine and ester formulations of 2,4-D mixed with glyphosate are similarly effective. An amine formulation is preferred if sensitive vegetation is nearby. **See comments under EARLY PREPLANT BURNDOWN—Glyphosate-resistant horseweed.**
Enlist Duo (3.3 S); MOA 9 + 4	3.5 to 4.75 pt	0.74 to 1.0 (lb a.e.) Glyphosate ± 0.7 to 0.95 (lb a.e.) 2,4-D	Enlist Duo contains 1.7 lb a.e./gal glyphosate and 1.6 lb a.e./gal 2,4-D as the choline salt. Apply at least 30 days ahead of planting any variety not containing the Enlist trait. Cotton containing the Enlist trait can be planted anytime following Enlist Duo application. Controls most weeds but not effective on Carolina geranium. **See comments under EARLY PREPLANT BURNDOWN—Glyphosate-resistant horseweed.**

Table 7-2B. Chemical Weed Control in Cotton

Herbicide, Mode of Action Code and Formulation	Amount of Formulation Per Acre	Pounds Active Ingredient Per Acre	Precautions and Remarks
Early Preplant Burndown, Burndown of emerged annual weeds in no-till, strip-till, or stale seedbed systems, any variety (continued)			
glyphosate; MOA 9 + ET (0.208 EC); MOA 14	See label + 0.5 to 2 fl oz	0.56 to 1.13 (lb a.e.) + 0.0008 to 0.0032	See comments for glyphosate alone. ET contains pyraflufen ethyl. ET added to glyphosate will increase speed of control and may improve control of some species, but overall long-term control is generally not improved. This tank mix will not control cutleaf eveningprimrose, wild radish, or glyphosate-resistant horseweed. There is no waiting period between application and cotton planting.
glyphosate; MOA 9 + Firstshot (50 WDG); MOA 2 + 2	See label + 0.8 oz	0.56 to 1.13 (lb a.e.) + 0.025	Firstshot is a 1:1 ratio premix of thifensulfuron plus tribenuron. Other brands of this premix are listed in Table 7-2A. This treatment should be applied at least 14 days prior to planting. Compared to glyphosate alone, the tank mix is more effective on Carolina geranium, curly dock, henbit, swinecress, Virginia pepperweed, wild mustard, and wild radish. Add nonionic surfactant according to the Firstshot label. This tank mix is not effective on cutleaf eveningprimrose or glyphosate-resistant horseweed.
glyphosate; MOA 9 + Harmony Extra (50 WDG); MOA 2 + 2	See label + 0.75 oz	0.56 to 1.13 (lb a.e.) + 0.023	Harmony Extra is a 2:1 ratio premix of thifensulfuron plus tribenuron formulated as 50 WDG. This treatment should be applied at least 14 days prior to planting. Compared to glyphosate alone, the tank mix is more effective on Carolina geranium, curly dock, henbit, swinecress, Virginia pepperweed, wild mustard, and wild radish. See Table 7-2E for weed response. Add nonionic surfactant according to the Harmony Extra label. This tank mix is not effective on cutleaf eveningprimrose or glyphosate-resistant horseweed. Other brands of thifensulfuron plus tribenuron 2:1 ratio, formulated as 75 WDG, include Nimble, Treaty Extra, T-Square, and Volta Extra. See Table 7-2A. The equivalent rate of these brands is 0.5 oz/acre.
glyphosate; MOA 9 + Leadoff (33.4 WDG); MOA 2 + 2	See label + 1.5 oz	0.56 to 1.13 (lb a.e.) + 0.031	Leadoff is a 1:1 ratio premix of rimsulfuron plus thifensulfuron. Can be applied from late fall to 30 days prior to planting. Controls emerged winter annual weeds plus provides residual control of later emerging winter weeds. See Leadoff label for adjuvant recommendations. 2,4-D can also be included in the mixture. Leadoff does not substitute for a flumioxazin (Valor, others) application. The best use of Leadoff is a late fall or winter application (December to early March) followed by another burndown application containing flumioxazin 2 to 4 weeks ahead of planting. Crusher 50 WDG also contains a 1:1 ratio of rimsulfuron plus thifensulfuron. The equivalent rate of Crusher is 1 oz/acre.
glyphosate; MOA 9 + Resource (0.86 EC); MOA 14	See label + 2 to 4 fl oz	0.56 to 1.13 (lb a.e.) + 0.013 to 0.027	See comments for glyphosate alone. Resource contains flumiclorac pentyl ester. Resource added to glyphosate will increase speed of control and may improve control of some species, but overall long-term control is generally not improved. This tank mix will not control cutleaf eveningprimrose, wild radish, or glyphosate-resistant horseweed. There is no waiting period between application and cotton planting.
glyphosate; MOA 9 + Valor SX (51 WDG); MOA 14	See label + 1 to 2 oz	0.56 to 1.13 (lb a.e.) + 0.031 to 0.063	See comments for glyphosate alone. Valor SX contains flumioxazin. Other brands of flumioxazin are listed in Table 7-2A. In no-till or stale seedbed system, a minimum of 14 days must pass, and a 1-inch rainfall must occur between flumioxazin application and cotton planting when flumioxazin is applied at 1 oz/acre; 21 days must pass when applied at 1.5 to 2 oz/acre. If a strip-till operation occurs between flumioxazin application and cotton planting, the waiting interval can be reduced to 14 days for 2 oz flumioxazin. However, strip-tilling after flumioxazin application will reduce or eliminate weed control in the tilled strip. Compared to glyphosate alone, the tank mix will improve control of cutleaf eveningprimrose and wild radish. However, this tank mix is less effective than glyphosate plus 2,4-D on primrose and wild radish. Dicamba or 2,4-D may be added to this mixture. Regardless of glyphosate product used, a nonionic surfactant at 1 qt/100 gal. is recommended on flumioxazin labels. Applied at 1 oz/acre, flumioxazin will give 2 to 4 weeks residual control of lambsquarters, pigweed, prickly sida, spurge, and Florida pusley. At 2 oz/acre, flumioxazin will give 6 to 8 weeks residual control of these species. Application to cover crop or dense stand of winter weeds may reduce residual control. This tank mixture will not control glyphosate-resistant horseweed. **See comments under EARLY PREPLANT BURNDOWN—Glyphosate-resistant horseweed.** **Carefully follow label directions for cleaning out the sprayer after each day's use.**
paraquat; MOA 22 2 lb/gal formulations 3 lb/gal formulations	2.6 to 4 pt 1.7 to 2.7 pt	0.65 to 1	Apply any time prior to planting to control emerged weeds. Add nonionic surfactant at 1 pt per 100 gal or crop oil concentrate at 1 gal per 100 gal. Follow directions and precautions on label. Usually not adequately effective on cutleaf eveningprimrose, horseweed, or larger wild mustard or wild radish. Apply 0.65 lb a.i. for wheat and 0.5 lb a.i. for rye cover crops. Best control of small grain cover crops will be achieved if paraquat is applied at the boot stage or later. See Table 7-2A for brands.
paraquat; MOA 22 2 lb/gal formulations 3 lb/gal formulations + diuron, MOA 7 4 F formulations 80 WDG formulations	2.6 to 4 pt 1.7 to 2.7 pt + 1 to 2 pt 0.63 to 1.25 lb	0.65 to 1 + 0.5 to 1	See comments for paraquat alone. See diuron label for use rates on various soils. Apply 15 to 45 days ahead of planting. If Cotoran is applied preemergence, reduce Cotoran rate to account for residual activity of diuron. When mixed with crop oil concentrate and applied in April, this combination has given good control of common weeds, including cutleaf eveningprimrose. See Table 7-2A for brands of paraquat and diuron.
Early Preplant Burndown, Glyphosate-resistant horseweed, any variety			
Enlist Duo; MOA 9 + 4 + Valor SX (51 WDG); MOA 14	4.75 pt + 2 oz	1.0 (lb a.e.) glyphosate + 0.95 (lb a.e.) 2,4-D + 0.063	Glyphosate-resistant horseweed is very common in eastern North Carolina and is beginning to become a problem in the Piedmont. See previous comments concerning waiting intervals between application of 2,4-D, dicamba, and flumioxazin and planting. The 2,4-D or dicamba is needed in the mixture to control emerged resistant horseweed, and the flumioxazin will control horseweed germinating after this application. Dicamba may be somewhat more effective than 2,4-D on horseweed. Better results will be obtained if treatment is applied prior to first of April. Enlist Duo contains 1.7 lb a.e./gal glyphosate and 1.6 lb a.e./gal 2,4-D as the choline salt. Current labeling does not allow tank mixing with other herbicides, such as Valor. Engenia contains dicamba BAPMA salt; Clarity and XtendiMax contain the dicamba diglycolamine salt. Enlist One contains the choline salt of 2,4-D.
Enlist One; MOA 4 + glyphosate; MOA 9 + Valor SX (51 WDG); MOA 14	2 pt + See label + 2 oz		

Table 7-2B. Chemical Weed Control in Cotton

Herbicide, Mode of Action Code and Formulation	Amount of Formulation Per Acre	Pounds Active Ingredient Per Acre	Precautions and Remarks
Early Preplant Burndown, Glyphosate-resistant horseweed, any variety (continued)			
glyphosate; MOA 9 + 2,4-D; MOA 4 + Valor SX (51 WDG); MOA 14	See label + See label + 2 oz	0.56 to 1.13 (lb a.e.) + 0.95 (lb a.e.) + 0.063	
glyphosate; MOA 9 + Engenia (5S); MOA 4 or XtendiMax (2.9S); MOA 4 + Valor SX (51 WDG); MOA 14	See label + 6.4 fl oz or 11 fl oz ± 1 to 2 oz	0.56 to 1.13 (lb a.e.) ± 0.25 (lb a.e.) or 0.25 (lb a.e.) ± 0.063	Valor SX contains flumioxazin; see Table 7-2A for other brands of flumioxazin. In no-till or stale seedbed systems, a minimum of 14 days must pass, and a 1-inch rainfall must occur between flumioxazin application and cotton planting when flumioxazin is applied at 1 oz/acre; 21 days must pass when applied at 1.5 to 2 oz/acre. If a strip-till operation occurs between flumioxazin application and cotton planting, the waiting interval can be reduced to 14 days for 2 oz flumioxazin. However, strip-tilling after flumioxazin application will reduce or eliminate weed control in the tilled strip.
glyphosate; MOA 9 + Clarity (4 SL); MOA 4 + Valor SX (51 WDG); MOA 14	See label + 8 fl oz + 1 to 2 oz	0.56 to 1.13 (lb a.e.) + 0.25 + 0.063	
glyphosate; MOA 9 + Sharpen (2.85 F); MOA 14	See label + 1.0 fl oz	0.56 to 1.13 (lb a.e.) + 0.022	Sharpen contains saflufenacil. After applying Sharpen, wait to plant cotton until at least 42 days and an accumulation of 1 inch of rainfall has occurred. Do not apply Sharpen to soils classified as sand with less than 1.5% organic matter. Do not mix flumioxazin with Sharpen. See Sharpen label for specifics on adjuvant selection.
Liberty (2.34 S); MOA 10	29 to 43 fl oz	0.53 to 0.79	Liberty (see Table 7-2A for other brands containing glufosinate) is recommended only for fields where growers have failed to control glyphosate-resistant horseweed and it is too late to use 2,4-D or dicamba. Best results with glufosinate will be obtained if sprayed when daytime temperatures exceed 75 degrees. If greater than 29 oz applied preplant, the seasonal total applied cannot exceed 72 fl oz.
At Planting Burndown, Burndown of cover crops and weeds at planting, any variety			
glyphosate; MOA 9	See label	0.56 to 1.13 (lb a.e.)	See Table 7-9 for glyphosate brands and rate conversions. See Table 7-2A for brands of paraquat. If an early burndown treatment was applied (see COTTON—Early Preplant Burndown), apply glyphosate or paraquat in combination with desired residual herbicides at planting. Glyphosate or paraquat may be tank mixed with registered preemergence herbicides and applied after planting but before cotton emergence. See suggested rates and precautions on labels of tank-mix partners. If an early burndown treatment was not used, apply glyphosate or paraquat 7 to 21 days ahead of planting. If weeds are emerged at planting, make a second application in combination with desired residual herbicides. See comments on residual herbicides under COTTON—Preemergence. Glyphosate and paraquat rates depend upon weed species and size; see labels for recommended rates. Add nonionic surfactant at 1 pt per 100 gal or crop oil concentrate at 1 gal per 100 gal spray solution to paraquat. Need for adjuvants with glyphosate depends upon brand used; see specific labels for details. Cover crops: Wheat < 12 inches: glyphosate, 0.56 lb a.e. or paraquat, 0.65 lb a.i. Wheat > 12 inches: glyphosate 0.75 lb a.e. or paraquat, 0.65 lb a.i. Rye < 18 inches: glyphosate, 0.56 lb a.e. or paraquat, 0.5 lb a.i. Rye > 18 inches: glyphosate, 0.75 lb a.e. or paraquat, 0.5 lb a.i.
paraquat; MOA 22 2 lb/gal formulations 3 lb/gal formulations	2.6 to 4 pt 1.7 to 2.7 pt	0.65 to 1	
Liberty (2.34 S); MOA 10	29 to 43 fl oz	0.53 to 0.79	See comments under EARLY PREPLANT BURNDOWN for more details. Suggested only if glyphosate-resistant horseweed is a problem. Liberty contains glufosinate. Other brands of glufosinate are listed in Table 7-2A.
At Planting Burndown, Burndown of cover crops and weeds at planting, Enlist varieties only			
glyphosate; MOA 9 + Enlist One (3.8 S); MOA 4	See label + 2 pt	1.0 (lb a.e.) glyphosate + 0.95 (lb a.e.) 2,4-D	Enlist varieties only. Enlist One is the only brand of 2,4-D approved for this application timing. Can be applied any time prior to planting or behind the planter. Controls most weeds but not effective on Carolina geranium. See website Enlisttankmix.com for approved adjuvants, drift reduction agents, and tank mixes; can be mixed with all commonly used preemergence herbicides. See Enlist One federal label and the North Carolina 24(c) Special Local Need label for details on drift management, including recommended nozzles and pressures, wind speed, boom height, temperature inversions, buffers, and susceptible plants. Applicators must be certified by NC Dept. Ag. for training on *Auxin Herbicides – Best Management Practices*. It is best for this application to follow an earlier burndown application. **See comments under EARLY PREPLANT BURNDOWN—Glyphosate-resistant horseweed.**
At Planting Burndown, Burndown of cover crops and weeds at planting, XtendFlex varieties only			
glyphosate; MOA 9 + Engenia (5S); MOA 4 or XtendiMax (2.9S); MOA 4	See label + 12.8 fl oz or 22 fl oz	0.56 to 1.13 (lb a.e.) ± 0.5 (lb a.e.) or 0.5 (lb a.e.)	XtendFlex varieties only. Engenia and XtendiMax contain dicamba and are the only brands of dicamba approved for this application timing. Can be applied any time prior to planting or behind the planter. It is best for this application to follow an earlier burndown application. See websites *Engeniatankmix.com* and *xtendimaxapplicationrequirements.com* for approved adjuvants, drift reduction agents, and tank mixes; can be mixed with all commonly used preemergence herbicides. See federal labels, supplemental labels for use in dicamba-tolerant cotton, and the North Carolina 24(c) Special Local Need labels for details on drift management, including recommended nozzles and pressures, wind speed, boom height, temperature inversions, buffers, and susceptible plants. Applicators must be certified by NC Dept. Ag. For training on *Auxin Herbicides – Best Management Practices*. **See comments under EARLY PREPLANT BURNDOWN—Glyphosate-resistant horseweed.**
Preplant Incorporated, Annual grasses and certain small-seeded broadleaf weeds, any variety			
Prowl 3.3 EC (3.3 EC); MOA 3	1.2 to 3.6 pt	0.5 to 1.5	Prowl 3.3 EC and Prowl H2O contain pendimethalin; Treflan contains trifluralin; see Table 7-2A for other brands of pendimethalin and trifluralin. Consult labels for application rates and for time, method, and depth of incorporation. Deep incorporation, especially on sandy soils, may cause stunting and delayed crop development. Incorporation of trifluralin can be delayed 24 hrs; pendimethalin incorporation can be delayed 7 days. Immediate incorporation is suggested.
Prowl H2O (3.8 AS); MOA 3	2 to 4 pt	0.95 to 1.9	
Treflan (4EC); MOA 3	1 to 2 pt	0.5 to 1	
Preemergence, Annual grasses and pigweed, any variety			
Warrant (3 CS); MOA 15	3 pt	1.125	Warrant contains acetochlor in an encapsulated formulation. It can be applied in combination with another preemergence herbicide such as diuron, fluometuron (Cotoran, others), or fomesafen (Reflex, others).

Table 7-2B. Chemical Weed Control in Cotton

Herbicide, Mode of Action Code and Formulation	Amount of Formulation Per Acre	Pounds Active Ingredient Per Acre	Precautions and Remarks
Preemergence, Annual grasses, pigweed, and lambsquarters, any variety			
Prowl 3.3 EC; MOA 3	2.4 to 3.6 pt	1.0 to 1.5	Prowl 3.3 EC and Prowl H2O Contain pendimethalin. See Table 7-2A for other brands of pendimethalin. See labels for rates on specific soils. May be mixed with diuron, fluometuron (Cotoran, others), fomesafen (Reflex, others), or pyrithiobac (Staple LX, others).
Prowl H2O (3.8AS); MOA3	2.1 to 4 pt	1.0 to 1.9	
Preemergence, Annual grasses and various broadleaf weeds, any variety			
Warrant Ultra (3.45 CS); MOA 15 + 14	3 pt	1.29	Warrant Ultra is a premix formulation containing 2.82 lb/gal of acetochlor plus 0.63 lb/gal fomesafen. Use preemergence only on coarse-textured soils.
Preemergence, Annual broadleaf weeds, any variety			
Brake (1.2F); MOA 12	16-32 fl oz	0.15 to 0.3	Label specifies to tank mix Brake with another residual herbicide when Brake is applied at less than 21 oz/acre. Suggested tank mixes include Cotoran, Direx, Reflex, or Warrant. If applied alone, Brake will be most effective at 32 oz. See label for rotational restrictions. Brake suggested primarily for fields with problem Palmer amaranth populations.
Brake F 16 (2.7F); MOA 12 + 14	16 fl oz	0.15 (fluridone) ± 0.1875 (fomesafen)	Brake F16 is a premix of fluridone plus fomesafen. See label for rotational restrictions. Suggested primarily for fields with problem Palmer amaranth populations.
Brake FX (3.6F); MOA 12 + 7	32 fl oz	0.15 (fluridone) ± 0.75 (fluometuron)	Brake FX is a premix of fluridone plus fluometuron. See label for rotational restrictions. Suggested primarily for fields with problem Palmer amaranth populations.
Cotoran (4 F); MOA 7	1 to 2 qt	1 to 2	Cotoran contains fluometuron; see Table 7-2A for other brands. Use lower end of rate range on lighter soils. May be tank mixed with pendimethalin (Prowl, others), fomesafen (Reflex, others), pyrithiobac (Staple LX, others), or Warrant.
diuron, MOA 7 4 F formulations 80 WDG formulations	1 to 2 pt 0.63 to 1.25 lb	0.5 to 1	See labels for rates on specific soils. May be mixed with pendimethalin (Prowl, others), fomesafen (Reflex, others), pyrithiobac (Staple LX, others), or Warrant. See rotational restrictions and maximum seasonal use rates on label. See Table 7-2A for brands of diuron.
Reflex (2 S); MOA 14	1 pt	0.25	Reflex contains fomesafen. Other brands containing 1.88 or 2 lb/gal of fomesafen are listed in Table 7-2A. Suggested primarily for control of Palmer amaranth and common ragweed. Label restricts preemergence application only to coarse-textured soils. May be tank mixed with diuron, fluometuron Cotoran, others), pendimethalin (Prowl, others), pyrithiobac (Staple LX, others), or Warrant. See labels for specific comments on tank mixing.
Sinister (2.87 S); MOA 14	0.7 pt	0.25	Sinister contains fomesafen. Suggested primarily for control of Palmer amaranth. Label restricts preemergence application only to coarse-textured soils. May be tank mixed with diuron, Cotoran, pendimethalin (Prowl, others), pyrithiobac (Staple LX, others), or Warrant. See labels for specific comments on tank mixing.
Staple LX (3.2 SL); MOA 2	1.7 to 2.1 fl oz	0.0425 to 0.053	Staple contains pyrithiobac. Other brands of pyrithiobac are listed in Table 7-2A. Do not apply pyrithiobac preemergence on soils with less than 0.5% organic matter. May tank mix with diuron, Cotoran, pendimethalin (Prowl, others), or fomesafen (Reflex, others). Palmer amaranth biotypes resistant to pyrithiobac are very common in North Carolina.
Post-emergence Overtop, Annual broadleaf weeds, any variety			
Envoke (75 WDG); MOA 2	0.1 oz	0.0047	Envoke contains trifloxysulfuron. May be applied overtop cotton after it has a minimum of 5 true leaves up to 60 days prior to harvest. On larger cotton, directed application is preferred for better coverage of weeds. Add nonionic surfactant at 0.25% by volume (1 qt per 100 gal). May make two applications, but do not exceed 0.0188 lb a.i./acre per year of trifloxysulfuron from the combined use of all trifloxysulfuron-containing products (Envoke and Suprend). Do not mix with other pesticides when applying overtop of cotton. See label for rotational restrictions and weeds controlled. Controls most broadleaf weeds with timely application; common exceptions include prickly sida, jimsonweed, copperleaf, and spurred anoda. Reduced growth of cotton, due to shortened internodes, is sometimes observed. Shortened internodes are more likely on smaller cotton. Envoke may also be applied overtop at 0.15 oz/acre if needed for larger weeds. Pyrithiobac (Staple, others) and Envoke are ALS inhibitors. Biotypes of Palmer amaranth, common ragweed, and cocklebur resistant to ALS inhibitors have been found in North Carolina; ALS-resistant Palmer amaranth is very common. To aid in resistance management, it is suggested that an ALS inhibitor be applied only once per year.
Staple LX (3.2 S); MOA 2	2.6 to 3.8 fl oz	0.065 to 0.095	Staple contains pyrithiobac. Other brands of pyrithiobac are listed in Table 7-2A. May be applied overtop of cotton from cotyledonary stage up to 60 days prior to harvest. Avoid application during or shortly after cool weather. Add nonionic surfactant at 0.25% by volume (1 qt per 100 gal). Do not add crop oil. May make 2 applications per year, not exceeding a total of 5.1 fl oz. May be tank mixed with most insecticides, but do not tank mix with any product containing malathion. Tank mixing with post-emergence grass control herbicides is discouraged. See label for rotational restrictions and weeds controlled. Timing of application is very important for most weeds. Apply before susceptible broadleaf weeds exceed 4 inches tall. Does not control tall morningglory, lambsquarters, or common ragweed. Only suppresses sicklepod. Pyrithiobac and Envoke are ALS inhibitors. Biotypes of Palmer amaranth, common ragweed, and cocklebur resistant to ALS inhibitors have been found in North Carolina; ALS-resistant Palmer amaranth is very common. To aid in resistance management, it is suggested that an ALS inhibitor be applied only once a year.
Envoke (75 WDG); MOA 2 + Staple LX (3.2 S); MOA 2	0.1 oz + 1.3 to 1.9 fl oz	0.0047 + 0.033 to 0.048	See comments for Envoke and Staple applied alone. Compared with Envoke alone, tank mix is more effective on eclipta, jimsonweed, and spurred anoda. Compared with Staple alone, tank mix is more effective on ragweed, lambsquarters, tall morningglory, and sicklepod. Add nonionic surfactant at 0.25% by volume.
Post-emergence Overtop, Annual grasses, any variety			
Assure II (0.88 EC); MOA 1	7 to 8 fl oz	0.05 to 0.06	Assure II contains quizalofop p-ethyl. Other brands are listed in Table 7-2A. Apply to actively growing grass not under drought stress. See label for maximum weed size to treat and suggested rate. Apply in 10 to 40 gpa. Add either crop oil concentrate at 1% by volume (1 gal per 100 gal) or nonionic surfactant at 0.25% by volume (1 qt per 100 gal). A second application may be made if needed. May use 5 oz per acre for seedling johnsongrass or shattercane.
Fusilade DX (2 EC); MOA 1	8 to 12 fl oz	0.125 to 0.188	Apply to actively growing grass not under drought stress. Suggested application rate varies by species and weed size; see label. Apply in 5 to 40 gpa at 40 to 60 psi. Add either crop oil concentrate at 1% by volume (1 gal per 100 gal) or nonionic surfactant at 0.25% by volume (1 qt per 100 gal). Second application may be made if necessary. May use 6 oz per acre for seedling johnsongrass or shattercane.

Table 7-2B. Chemical Weed Control in Cotton

Herbicide, Mode of Action Code and Formulation	Amount of Formulation Per Acre	Pounds Active Ingredient Per Acre	Precautions and Remarks
Post-emergence Overtop, Annual grasses, any variety (continued)			
Poast (1.5 EC) or Poast Plus (1.0 EC) MOA 1	16 fl oz or 24 fl oz	0.19	Apply to actively growing grass not under drought stress. Consult label for maximum grass size to treat. Apply in 5 to 20 gpa at 40 to 60 psi. Add 2 pt per acre of crop oil concentrate. A second application may be made if necessary. Consult label for special rates for early treatment or rescue treatment. Poast and Poast Plus contain sethoxydim. See Table 7-2A for other brands of sethoxydim.
Select (2 EC) or Select Max (0.97 EC) MOA 1	6 to 8 fl oz or 9 to 16 fl oz	0.094 to 0.125 or 0.068 to 0.121	Contains clethodim; see Table 7-2A for other brands. Apply to actively growing grass not under drought stress. See labels for maximum weed size to treat and suggested rate. Apply in 10 to 40 GPA. Add adjuvant according to label. A second application may be made if needed.
Post-emergence Overtop, Bermudagrass and Johnsongrass, any variety			
Assure II (0.88 EC); MOA 1	10 fl oz	0.07	Assure II contains quizalofop p-ethyl. Other brands are listed in Table 7-2A. Apply to actively growing bermudagrass when runners are up to 6 inches or johnsongrass 10 to 24 inches tall. A second application of 7 fl oz per acre may be applied if needed when bermudagrass regrowth is up to 6 inches or johnsongrass regrowth is 6 to 10 inches. Add either a crop oil concentrate at 1% by volume (1 gal per 100 gal) or a nonionic surfactant at 0.25% by volume (1 qt per 100 gal).
Fusilade DX (2 EC); MOA 1	12 fl oz	0.19	Apply to actively growing bermudagrass when runners are 4 to 8 inches long or johnsongrass 8 to 18 inches tall and before boot stage. If regrowth occurs, make a second application of 8 fl oz when bermudagrass runners are 4 to 8 inches or johnsongrass regrowth is 6 to 12 inches. Add either crop oil concentrate at 1 qt per acre or nonionic surfactant at 0.25% by volume (1 qt per 100 gal).
Poast (1.5 EC) or Poast Plus (1.0 EC) MOA 1	24 fl oz or 36 fl oz	0.28	Apply to actively growing bermudagrass before runners exceed 6 inches or to johnsongrass up to 25 inches tall. If regrowth occurs or new plants emerge, make a second application before bermudagrass runners exceed 4 inches or johnsongrass regrowth exceeds 12 inches. See labels for second application rates. Add 2 pt of crop oil concentrate per acre. Poast and Poast Plus contain sethoxydim. See Table 7-2A for other brands of sethoxydim.
Select (2 EC) or Select Max (0.97 EC) MOA 1	8 to 16 fl oz or 12 to 32 fl oz	0.125 to 0.25 or 0.091 to 0.24	Apply to actively growing bermudagrass when runners are up to 6 inches or johnsongrass is 12 to 24 inches tall. A second application may be applied if needed when bermudagrass regrowth is up to 6 inches or johnsongrass regrowth is 6 to 18 inches. See labels for second application rates. Add adjuvant according to label. Use the higher rate under heavy grass pressure or on larger grass. Contains clethodim; see Table 7-2A for other brands.
Post-emergence Overtop, Annual broadleaf weeds and most annual grasses, Glytol LibertyLink, Enlist, or XtendFlex Cultivars only			
Liberty (2.34 S); MOA 10	29 to 43 fl oz	0.53 to 0.79	Contains glufosinate; see Table 7-2A for other brands. Apply overtop or directed from cotton emergence until the early bloom stage. Good spray coverage is critical. Use flat-fan nozzles and a minimum of 15 GPA. Better coverage may be obtained on larger cotton with a semi-directed application. An adjuvant is not necessary. Application time of day is important. Two hours of sunshine before a morning application is suggested. Do not apply later than 1 hour before sunset. Multiple applications are allowed. Liberty at 29 fl oz can be applied three times, with a seasonal maximum of 87 fl oz. If applied at rates greater than 29 oz, only two applications are allowed and the total rate per season cannot exceed 72 fl oz. Liberty controls most annual grass and broadleaf weeds, although timing of application on pigweed (including Palmer amaranth) and grasses (especially goosegrass) is critical. Preemergence herbicides are encouraged to help in control of pigweed and grasses. Liberty is generally more effective on broadleaf weeds than grasses. Broadleaf weeds should be 2 to 3 inches tall and grasses 1 to 2 inches tall. Postemergence grass control herbicides, such as clethodim, fluazifop, quizalofop, and sethoxydim, should not be mixed with glufosinate. Applications of postemergence grass herbicides and glufosinate should be separated by at least 5 days.
Liberty (2.34 S); MOA 10 + Dual Magnum, 7.62 EC; MOA 15	29 to 43 fl oz + 1 to 1.33 pt	0.53 to 0.79 + 0.95 to 1.27	See comments for Liberty applied alone. Dual Magnum contains S-metolachlor; see Table 7-2A for other brands. S-metolachlor will not control emerged weeds, but it can provide residual control of susceptible species such as annual grasses and pigweed species. This treatment may cause foliar burn on the crop. Burn may be enhanced if applied to cotton with dew, under extremely high temperatures, or when mixed with insecticides or adjuvants. Several products containing metolachlor (not S-metolachlor) are available. Metolachlor products are less effective per unit of formulated product than those with S-metolachlor. In general, it takes 1.5 pt of a metolachlor product to give the activity one gets from 1 pt of S-metolachlor.
Liberty (2.34 L); MOA 10 + Outlook (6 EC); MOA 15	29 to 43 fl oz + 12 to 16 fl oz	0.53 to 0.79 + 0.56 to 0.75	See comments for Liberty applied alone. Outlook will not control emerged weeds, but it can provide residual control of susceptible species such as annual grasses and pigweed species. This treatment may cause foliar burn on the crop. Burn may be enhanced if applied to cotton with dew, under extremely high temperatures, or when mixed with insecticides or adjuvants.
Liberty (2.34 L); MOA 10 + Staple LX (3.2 S); MOA 2	29 to 43 fl oz + 1.3 to 3.8 fl oz	0.53 to 0.79 + 0.033 to 0.095	See comments for Liberty applied alone. Staple contains pyrithiobac. See Table 7-2A for other brands of glufosinate and pyrithiobac. See directions on pyrithiobac label concerning adjuvant usage. Staple will improve control of non-ALS-resistant Palmer amaranth plus provide residual control.
Liberty (2.34 L); MOA 10 + Warrant (3 CS); MOA 15	29 to 43 fl oz + 3 pt	0.53 to 0.79 + 1.125	See comments for Liberty applied alone. Warrant will not control emerged weeds, but it can provide residual control of susceptible species such as annual grasses and pigweed species. This treatment may cause foliar burn on the crop. Burn may be enhanced if applied to cotton with dew, under extremely high temperatures, or when mixed with insecticides or adjuvants.
Post-emergence Overtop, Annual broadleaf weeds and most annual grasses, Phytogen Widestrike varieties			
Liberty (2.34 S); MOA 10	29 fl oz	0.53	Liberty contains glufosinate; see Table 7-2A for other brands of glufosinate. Phytogen cultivars with the Widestrike trait can be treated with glufosinate. Tolerance to glufosinate in these cultivars is not complete, and varying levels of crop injury may be observed. Greater injury can be expected when glufosinate is mixed with insecticides or other herbicides. **Growers assume the liability** of crop injury when cotton with the Widestrike trait is treated with glufosinate. It is suggested that the rate not exceed 29 oz per application with a maximum of two applications per year. It is also suggested that glufosinate not be applied beyond the 8-leaf stage of cotton and that AMS not be included in the application. See above comments for use of glufosinate on LibertyLink cultivars, including the statement on application time of day.

Chapter VII—2019 N.C. Agricultural Chemicals Manual

Table 7-2B. Chemical Weed Control in Cotton

Herbicide, Mode of Action Code and Formulation	Amount of Formulation Per Acre	Pounds Active Ingredient Per Acre	Precautions and Remarks
Post-emergence Overtop, Annual grasses, broadleaf weeds, perennial grasses, and nutsedge; suppression of perennial broadleaf weeds, Enlist, Glytol LibertyLink, Roundup Ready Flex, or XtendFlex varieties only			
glyphosate; MOA 9	See labels	0.56 to 1.13 (lb a.e.)	Glyphosate formulations and application rates should be compared on the basis of pounds of glyphosate acid equivalent (a.e.) per gallon and per acre, respectively. Rates in the preceding column are expressed as a.e. See TABLE 7-9 for glyphosate rate conversions. Glyphosate controls most annual weeds; exceptions include dayflower, dove weed, Florida pusley, and hemp sesbania. Timely application is critical for morningglory control. Multiple applications are needed for nutsedge and bermudagrass. See label of brand you use for recommended rates and sizes of weeds to treat. Adjuvant recommendations vary according to glyphosate product. See label of brand used for specific recommendations. Glyphosate-resistant Palmer amaranth is common in North Carolina, and glyphosate-resistant common ragweed is present in several counties. Continued heavy reliance on herbicide programs based predominantly on glyphosate will enhance selection for resistant biotypes. Other chemistry, including preemergence herbicides, tank mixes with glyphosate, and layby herbicides in addition to glyphosate, is recommended as part of a resistance-management strategy. See section on Herbicide Resistance and TABLE 7-10A.
glyphosate; MOA 9 + Dual Magnum (7.62 EC); MOA 15	See labels + 1 to 1.33 pt	0.56 to 1.13 (lb a.e.) + 0.95 to 1.27	See comments for glyphosate applied alone. Dual Magnum contains S-metolachlor; see Table 7-2A for other brands. S-metolachlor will not control emerged weeds, but it can provide residual control of susceptible species such as annual grasses and pigweed species. This treatment may cause foliar burn on the crop. Burn may be enhanced if applied to cotton with dew, under extremely high temperatures, or when mixed with insecticide or adjuvant. Several products containing metolachlor (not S-metolachlor) are available. Metolachlor products are less effective per unit of formulated product than those with S-metolachlor. In general, it takes 1.5 pt of a metolachlor product to give the activity one gets from 1 pt of S-metolachlor.
glyphosate; MOA 9 + Envoke (75 WDG); MOA 2	See labels + 0.1 oz	0.56 to 1.13 (lb a.e.) + 0.0047	See comments for glyphosate and Envoke applied alone. See Envoke label and glyphosate label for suggestions on adjuvant usage. Tank mix can be applied from 5-leaf cotton stage until 60 days prior to harvest. For better crop safety, however, cotton should have at least 7 to 8 leaves at time of treatment.
glyphosate; MOA 9 + Outlook (6 EC); MOA 15	See labels + 12 to 16 fl oz	0.56 to 1.13 (lb a.e.) + 0.56 to 0.75	See comments for glyphosate applied alone. Can apply from first true leaf to mid-bloom stage. Outlook will not control emerged weeds, but it will provide residual control of susceptible species such as annual grasses and pigweed species. Optimum timing is 2- to 3-leaf cotton, and before weeds emerge. Make only one application per year. Outlook plus glyphosate may cause some foliar burn on cotton; mixing with insecticides or adjuvants may increase burn. Suggested rates are 12 oz on coarse soils, 14 oz on medium soils, and 16 oz on fine soils.
glyphosate; MOA 9 + Staple LX (3.2 S;) MOA 2	See labels + 1.3 to 3.8 fl oz	0.56 to 1.13 (lb a.e.) + 0.033 to 0.095	See comments for glyphosate applied alone. Staple contains pyrithiobac; see Table 7-2A for other brands. Can apply overtop from cotyledonary stage cotton until 60 days prior to harvest. See directions on pyrithiobac labels concerning adjuvant usage. Palmer amaranth resistant to both pyrithiobac and glyphosate is widespread in North Carolina.
glyphosate; MOA 9 + Warrant (3 CS); MOA 15	See labels + 3 pt	0.56 to 1.13 (lb a.e.) + 1.125	See comments for glyphosate applied alone. Apply after cotton is completely emerged but before first bloom. Warrant will not control emerged weeds, but it will provide residual control of susceptible species such as annual grasses and pigweed species. Optimum timing is 2- to 3-leaf cotton, and before weeds emerge. A second application can be made if directed to the soil surface. Warrant plus glyphosate may cause some foliar burn on cotton; mixing with insecticides or adjuvants may increase burn. Warrant rate can be increased to 4 pt on some soils; see label for details.
Sequence (5.25 L); MOA 9 + 15	2.5 to 3.5 pt	0.70 to 1.0 (lb a.e.) Glyphosate + 0.94 to 1.3 S-metolachlor	Sequence is a premix containing 2.25 lb a.e./gal glyphosate and 3 lb/gal S-metolachlor. See comments for glyphosate alone and glyphosate + S-metolachlor. Apply to cotton in cotyledonary stage up to 10-leaf stage, but not to cotton taller than 12 inches. Apply 2.5 pt to cotton less than 5 leaves. Can increase rate to 3.5 pt on cotton with 5 to 10 leaves.
Post-emergence Overtop, Annual grasses and broadleaf weeds. Enlist varieties only			
Liberty (2.34 S); MOA 10 + Enlist One (3.8 S); MOA 4	29 to 43 oz + 2 pt	0.53 to 0.79 + 0.95	**Enlist varieties only**. Enlist One contains the choline salt of 2,4-D. It is the only brand of 2,4-D registered for this use. Can be applied any time from cotton emergence to mid-bloom stage. Can be applied twice postemergence; allow minimum of 12 days between applications. See website *Enlisttankmix.com* for approved adjuvants, drift reduction agents, and tank mixes. Can be mixed with Dual, EverpreX, Moccasin, Staple LX, or Warrant for residual control. See Enlist One federal label and the North Carolina 24(c) Special Local Need label for details on drift management, including recommended nozzles and pressures, wind speed, boom height, temperature inversions, buffers, and susceptible plants. Applicators must be certified by NC Dept. Ag. for training on *Auxin Herbicides - Best Management Practices*.
Post-emergence Overtop, Annual and perennial grasses, broadleaf weeds, nutsedge. Enlist varieties only			
glyphosate; MOA 9 + Enlist One (3.8 S); MOA 4	See labels + 2 pt	1.0 (lb a.e.) + 0.95	**Enlist varieties only**. Enlist One contains the choline salt of 2,4-D. It is the only brand of 2,4-D registered for this use. Can be applied any time from cotton emergence to mid-bloom stage. Can be applied twice postemergence; allow minimum of 12 days between applications. See website *Enlisttankmix.com* for approved adjuvants, drift reduction agents, and tank mixes. Can be mixed with several brands of glyphosate; see website for approved mixes. Can be mixed with Dual, EverpreX, Moccasin, Staple LX, or Warrant for residual control. See Enlist One federal label and the North Carolina 24(c) Special Local Need label for details on drift management, including recommended nozzles and pressures, wind speed, boom height, temperature inversions, buffers, and susceptible plants. Applicators must be certified by NC Dept. Ag. for training on *Auxin Herbicides - Best Management Practices*.
Enlist Duo (3.3 S); MOA 9 + 4	4.75 pt	1.0 (lb a.e.) Glyphosate ± 0.95 (lb a.e.) 2,4-D	**Enlist varieties only**. Enlist Duo contains 1.7 lb a.e./gal glyphosate and 1.6 lb a.e./gal 2,4-D as the choline salt. Can be applied any time from cotton emergence to mid-bloom stage. Can be applied twice postemergence; allow minimum of 12 days between applications. See website *Enlisttankmix.com* for approved adjuvants, drift reduction agents, and tank mixes. As of August 2018, Moccasin (s-metolachlor) was the only residual herbicide approved for mixing with Enlist Duo. See Enlist Duo federal label and the North Carolina 24(c) Special Local Need label for details on drift management, including recommended nozzles and pressures, wind speed, boom height, temperature inversions, buffers, and susceptible plants. Applicators must be certified by NC Dept. Ag. for training on *Auxin Herbicides - Best Management Practices*.

Table 7-2B. Chemical Weed Control in Cotton

Herbicide, Mode of Action Code and Formulation	Amount of Formulation Per Acre	Pounds Active Ingredient Per Acre	Precautions and Remarks
Post-emergence Overtop, Annual and perennial grass, broadleaf weeds, nutsedge. XtendFlex varieties only			
Glyphosate; MOA 9 + Engenia (5 S); MOA 4 or XtendiMax (2.9 S); MOA 4	See label + 12.8 fl oz or 22 fl oz	0.56 to 1.13 (lb a.e.) ± 0.5 (lb a.e.) or 0.5 (lb a.e.)	**XtendFlex varieties only.** Engenia and XtendiMax contain dicamba. These are the only brands of dicamba registered for this use. Can be applied any time from cotton emergence to 7 days prior to harvest. Can be applied multiple times postemergence, not to exceed a total of 51.2 oz Engenia or 88 oz of XtendiMax. Only two postemergence applications suggested, preferably before first bloom. Separate applications by at least 7 days. See websites *Engeniatankmix.com* and *xtendimaxapplicationrequirements.com* for approved adjuvants, drift reduction agents, and tank mixes. As of August 2017, the only approved tank mix partners were Staple, Warrant, and selected brands of clethodim. See federal labels, supplemental labels for use in dicamba-tolerant cotton, and the North Carolina 24(c) Special Local Need labels for details on drift management, including recommended nozzles and pressures, wind speed, boom height, temperature inversions, buffers, and susceptible plants. Applicators must be certified by NC Dept. Ag. for training on *Auxin Herbicides - Best Management Practices.*
Post-emergence Overtop, Volunteer Roundup Ready corn, any variety			
Assure II (0.88 EC); MOA 1	5 to 8 fl oz	0.034 to 0.055	Contains quizalofop; see Table 7-2A for other brands. See above comments for quizalofop. See quizalofop label for application rates on various sizes of corn and for adjuvant recommendations. Can mix with glyphosate.
Fusilade DX (2 EC); MOA 1	4 to 6 fl oz	0.063 to 0.094	See above comments for Fusilade. See label for application rates on various sizes of corn and for adjuvant recommendations. Can mix with glyphosate.
Poast (1.5 EC) or Poast Plus (1.0 EC) MOA 1	16 fl oz or 24 fl oz	0.189	Contains sethoxydim; see Table 7-2A for other brands. See above comments for sethoxydim. See sethoxydim label for application rates on various sizes of corn and for adjuvant recommendations. Can mix with glyphosate.
Select (2 EC) or Select Max (0.97 EC) MOA 1	4 to 8 fl oz or 6 to 12 fl oz	0.063 to 0.125 or 0.045 to 0.106	Contains clethodim; see Table 7-2A for other brands. See clethodim labels for application rates on various sizes of corn and for adjuvant recommendations. Can mix with glyphosate.
Post-emergence Overtop, Volunteer Roundup Ready soybean, any variety			
Envoke (75 WDG); MOA 2	0.1 oz	0.0047	See above comments for Envoke. Cotton should have at least five leaves, and the soybean should have no more than four to five trifoliate leaves. Not adequately effective on soybean with the STS trait.
Post-emergence Directed, Cocklebur, small annual grasses, and nutsedge, any variety			
MSMA; MOA 17 6 lb/gal formulation 6.6 lb/gal formulation	 2.67 pt 2.5 pt	2	Do not apply overtop at these rates. MSMA can be directed alone or mixed with other postemergence broadleaf herbicides on cotton at least 3 inches tall up to first bloom. Do not apply MSMA after first bloom. Adequate control of nutsedge usually requires two applications. Follow label directions for use of adjuvants.
Post-emergence Directed, Annual broadleaf weeds, small annual grasses, and nutsedge, any variety			
Caparol (4 F); MOA 5 + MSMA; MOA 17 6 lb/gal formulation 6.6 lb/gal formulation	1.3 to 2.8 pt + 2.67 pt 2.5 pt	0.65 to 1.4 + 2	Apply 1.3 pt Caparol as directed spray only to cotton at least 6 inches tall. Increase to higher rate for the soil type after cotton is at least 12 inches tall. See label for rates on various soil types. Add 2 qt nonionic surfactant per 100 gal spray solution. Do not apply after first bloom. Aim at 1 fl oz or Cobra at 6 to 12.5 fl oz may be added to improve control of larger morningglory. Cotton should be at least 16 inches tall when applying Aim. Do not allow Aim to contact green stem tissue.
Cobra (2 EC); MOA 14 + MSMA; MOA 17 6 lb/gal formulation 6.6 lb/gal formulation	12.5 oz + 2.67 pt 2.5 pt	0.2 + 2	Apply as directed spray or with hooded sprayer. Cotton should be at least 6 to 8 inches tall, preferably larger. See Cobra label for weeds controlled, directions on weed size and application rates, and use of surfactant or crop oil. Do not apply MSMA after first bloom. Cobra contains lactofen; see Table 7-2A for other brands.
Cobra (2 EC); MOA 14 + diuron, 4 F; MOA 7 + MSMA; MOA 17 6 lb/gal formulation 6.6 lb/gal formulation	6 to 12.5 fl oz + 0.8 to 1.2 pt + 2.67 pt 2.5 pt	0.094 to 0.2 + 0.4 to 0.6 + 2	See Table 7-2A for brands of diuron. Apply as directed spray or with hooded sprayer. Cotton should be at least 12 inches tall. See Cobra label for weeds controlled and directions on weed size and application rates. Add 1 qt per acre of crop oil concentrate. See rotational restrictions on diuron label. Do not apply MSMA after first bloom.
Cotoran (4 L); MOA 7 + MSMA; MOA 17 6 lb/gal formulation 6.6 lb/gal formulation	1 to 2 qt + 2.67 pt 2.5 pt	1 to 2 + 2	Apply as a directed spray only to cotton at least 3 inches tall up to first bloom. Follow label directions for weed size and addition of surfactant. See Cotoran label for maximum application rates per season and rotational restrictions. S-metolachlor (Dual Magnum, others) may be added. Cotoran contains fluometuron; see Table 72-A for other brands.
diuron (4 F) or diuron (80 WDG); MOA 7 + MSMA; MOA 17 6 lb/gal formulation 6.6 lb/gal formulation	1.6 to 2.4 pt or 1 to 1.5 lb + 2.67 pt 2.5 pt	0.8 to 1.2 + 2	See Table 7-2A for brands of diuron. Apply as directed spray only to cotton at least 12 inches tall. Adjust rate according to soil type. See application precautions on label. Add nonionic surfactant at 1 to 2 quarts per 100 gal spray solution or crop oil concentrate at 1 gal per 100 gal spray solution. See label for rotational restrictions. Do not apply MSMA after first bloom. Aim at 1 fl oz or Cobra at 6 to 8 fl oz per acre may be added to improve control of larger morningglory. Cotton should be at least 16 inches tall when applying Aim. Do not allow Aim to contact green stem tissue.
Fierce (76 WDG); MOA 14 + 15 + MSMA; MOA17 6.0 lb/gal formulation 6.6 lb/gal formulation	3 oz + 2.67 pt 2.5 pt	0.143 + 2	Fierce is a premix containing 33.5% flumioxazin plus 42.5% pyroxasulfone. Can be applied with hooded sprayer after cotton is at least 6 inches tall. Do not allow spray solution to contact cotton. Can be applied with layby applicator after cotton is at least 16 inches tall, but do not contact more than the lower 2 inches of cotton stalk. Add non-ionic surfactant according to label. Do not use crop oil, methylated seed oil, or organo-silicant adjuvants. Controls emerged broadleaf weeds normally controlled by flumioxazin plus MSMA and gives residual control of annual grasses, Palmer amaranth and other pigweed species, nightshade, eclipta, Florida pusley, and most annual grass species.
Layby Pro (4 L); MOA 7 + MSMA; MOA 17 6 lb/gal formulation 6.6 lb/gal formulation	2 pt + 2.67 pt 2.5 pt	1.0 + 2	Layby Pro contains 2 lb/gal diuron plus 2 lb per gal linuron. Apply as directed spray only to cotton at least 15 inches tall. See application precautions on label. Add crop oil concentrate at 1 gal per 100 gal spray solution. See label for rotational restrictions. Do not apply MSMA after first bloom. Aim at 1 fl oz per acre may be added to improve control of larger morningglory. Do not allow Aim to contact green stem tissue.
Suprend (80 WDG); MOA 5 + 2 + MSMA; MOA 17 6 lb/gal formulation 6.6 lb/gal formulation	1 to 1.25 lb + 2.67 pt 2.5 pt	0.8 to 1 + 2	Suprend is a premix product containing 79.3% prometryn plus 0.7% trifloxysulfuron. Apply as directed spray to cotton at least 6 inches tall, preferably taller. Add nonionic surfactant at 1 qt per 100 gal spray solution. See rotational restrictions on label. Do not apply MSMA after first bloom. Do not exceed 0.0188 lb a.i./acre per year of trifloxysulfuron from the combined use of all trifloxysulfuron-containing products (Envoke and Suprend).

Table 7-2B. Chemical Weed Control in Cotton

Herbicide, Mode of Action Code and Formulation	Amount of Formulation Per Acre	Pounds Active Ingredient Per Acre	Precautions and Remarks
Post-emergence Directed, Annual broadleaf weeds, small annual grasses, and nutsedge, any variety (continued)			
Valor SX (51 WDG); MOA 14 + MSMA; MOA17 6.0 lb/gal formulation 6.6 lb/gal formulation	2 oz + 2.67 pt 2.5 pt	0.064 + 2	Valor contains flumioxazin; see Table 7-2A for other brands. Apply as directed spray only to cotton at least 16 inches tall. Direct the spray to the lower 2 inches of the cotton stem. Do not allow spray solution to contact green portion of stem. Add nonionic surfactant at 1 qt per 100 gal spray solution. Do not use crop oil concentrate, methylated seed oil, organo-silicone adjuvants, or any adjuvant product containing any of these. Do not apply MSMA after first bloom. No rotational restrictions of concern in North Carolina. May be applied under a hood on cotton at least 6 inches tall. Do not allow spray solution to contact cotton.
Post-emergence Directed, Annual grasses and broadleaf weeds, nutsedge, and suppression of perennial weeds; Enlist, Glytol LibertyLink, Roundup Ready Flex or XtendFlex varieties only			
Glyphosate; MOA 9	See labels	0.75 to 1.13 (lb a.e.)	Glyphosate alone can be directed up to 7 days prior to harvest. When using glyphosate alone, contact with the cotton is of no concern; the primary reason to direct is to obtain better coverage of weeds under the crop canopy. Use of other herbicides, in addition to glyphosate, is recommended to aid in resistance management. See the section on Herbicide Resistance Management and TABLE 7-10A. When tank mixing, follow directions on label of tank mix partner concerning cotton size for application, application directions (including allowable contact with cotton plant), and rotational restrictions. Glyphosate-resistant Palmer amaranth is widespread in North Carolina, and glyphosate-resistant common ragweed is present in several counties.
Glyphosate; MOA 9 + Aim (2 EC); MOA 14	See labels + 1 to 1.5 fl oz	0.75 to 1.13 (lb a.e.) + 0.016 to 0.024	Cotton should be at least 16 inches tall. Extreme care should be exercised in application; see directions and precautions on Aim label. Contact on green stem tissue will lead to severe injury. Add crop oil concentrate according to the Aim label. See comments on Aim label concerning sprayer clean-out. Compared to glyphosate alone, this combination controls larger morningglories. See above comments for glyphosate alone.
glyphosate; MOA 9 + Caparol (4 F); MOA 5	See labels + 0.5 to 1	0.75 to 1.13 (lb a.e.) + 1 to 2 pt	Caparol contains prometryn; see Table 7-2A for other brands. Direct to cotton at least 6 to 8 inches tall. Use 1 to 1.3 pt prometryn on cotton 6 to 12 inches tall; rate can be increased to 2 pt on cotton at least 12 inches tall. Add surfactant according to the label of the brand of glyphosate used. See precautions and rotational restrictions on prometryn label. Compared to glyphosate alone, this combination will improve control of larger morningglory and may provide residual control of small-seeded broadleaf weeds, such as pigweed. This mixture may give less control of larger grasses than glyphosate alone under drier conditions.
glyphosate MOA 9 + diuron (4 F) or diuron (80 WDG); MOA 7	See labels + 1 to 1.5 pt or 0.63 to 0.94 lb	0.75 to 1.13 (lb a.e.) + 0.5 to 0.75	See Table 7-2A for brands of diuron. Use 1 pt of diuron 4F or 0.63 lb diuron 80WDG on cotton 8 to 12 inches tall. Increase rate to 1.5 pt or 0.94 lb on cotton greater than 12 inches. See comments for glyphosate applied alone. Add surfactant according to the label of the glyphosate brand used. Compared to glyphosate alone, this combination controls larger morningglories and provides residual control of small-seeded broadleaf weeds, such as pigweed. This tank mix may give less control of larger grasses than glyphosate alone under dry conditions. See diuron label for rotational restrictions.
glyphosate MOA 9 + Dual Magnum (7.62 EC); MOA 9	See labels + 1 to 1.33 pt	0.75 to 1.13 (lb a.e.) + 0.95 to 1.27	Dual Magnum contains s-metolachlor; see Table 7-2A for other brands of s-metolachlor. Can be applied to cotton 3 inches tall through layby. See comments for glyphosate applied alone. S-metolachlor does not improve control of emerged weeds, but it can give residual control of annual grasses, pigweed species, and spreading dayflower plus suppression of yellow nutsedge. Do not apply to sand or loamy sand soils.
glyphosate MOA 9 + Envoke (75 DF); MOA 2	See labels + 0.1 to 0.2 oz	0.75 to 1.13 (lb a.e.) + 0.0047 to 0.0094	Direct to cotton from 6 inches tall through layby. Add nonionic surfactant according to the Envoke label. Compared to glyphosate alone, the combination is more effective on nutsedge and morningglory and provides residual control of susceptible broadleaf weeds. See comments above for glyphosate alone.
glyphosate; MOA 9 + Fierce (76 WDG); MOA 14 + 15	See label + 3 oz	0.75 to 1.13 (lb a.e.) + 0.143	Fierce is a premix containing 33.5% flumioxazin plus 42.5% pyroxasulfone. See comments for glyphosate applied alone and Fierce applied alone. Can be applied with hooded sprayer after cotton is at least 6 inches tall. Do not allow spray solution to contact cotton. Can be applied with layby applicator after cotton is at least 16 inches tall, but do not contact more than the lower 2 inches of cotton stalk. Add non-ionic surfactant according to premix label. Do not use crop oil, methylated seed oil, or organo-silicanic adjuvants.
glyphosate; MOA 9 + Outlook (6 EC); MOA 15	See labels + 12 to 16 fl oz	0.75 to 1.13 (lb a.e.) + 0.56 to 0.75	Can be directed to cotton up to mid-bloom. See comments for glyphosate applied alone. Outlook does not improve control of emerged weeds, but it can give residual control of annual grasses and pigweed species. Suggested rates are 12 oz on coarse soils, 14 oz on medium soils, and 16 oz on fine soils.
glyphosate; MOA 9 + Prefix (5.29 EC); MOA 14 + 15	See labels + 2 to 2.33 pt	0.75 to 1.13 (lb a.e.) + 1.32 to 1.54	Prefix contains fomesafen plus S-metolachlor. Use as a layby application to cotton with a minimum of 4 inches of bark on the stem. Do not use Prefix at layby if fomesafen (Reflex, others) was used preemergence. See Prefix label for suggestions on adjuvant.
glyphosate; MOA 9 + Reflex (2 L); MOA 14	See label + 1 to 1.5 pt	0.75 to 1.13 (lb a.e.) + 0.25 to 0.375	Reflex contains fomesafen; see Table 7-2A for other brands. Use as a layby application to cotton with a minimum of 4 inches of bark on the stem. Add surfactant or crop oil according to the fomesafen label. May include prometryn, diuron, S-metolachlor, Envoke, Layby Pro, or Suprend in the mixture. Do not use fomesafen at layby if fomesafen was used preemergence.
glyphosate MOA 9 + Suprend (80 WDG); MOA 5 + 2	See labels + 1 to 1.25 lb	0.75 to 1.13 (lb a.e.) + 0.8 to 1.0	Suprend contains prometryn plus trifloxysulfuron. Direct to cotton at least 6 to 8 inches tall. Add surfactant according to label of glyphosate brand used. See precautions and rotational restrictions on Suprend label. Compared to glyphosate alone, this combination will improve control of larger morningglory and nutsedge, and may provide residual control of small-seeded broadleaf weeds, such as pigweed. This mixture may give less control of larger grasses than glyphosate alone under drier conditions.
glyphosate; MOA 9 + Valor SX (51 WDG); MOA 14	See label + 1 to 2 oz	0.75 to 1.13 (lb a.e.) + 0.031 to 0.063	Valor contains flumioxazin; see Table 7-2A for other brands. Cotton should be at least 16 inches tall. Direct the spray to the lower 1 to 2 inches of the cotton stem; minimize cotton contact as much as possible. Do not allow spray solution to contact green portion of stem. See comments above for glyphosate alone. Add nonionic surfactant at 1 qt per 100 gal spray solution. DO NOT use crop oil concentrate, methylated seed oil, organo-silicone adjuvants, or any adjuvant product containing any of these. No rotational restrictions of concern in North Carolina. Compared with glyphosate alone, the combination will give better control of larger morningglories plus residual control of susceptible broadleaf weeds. May be applied under a hood on cotton at least 6 inches tall. Do not allow spray solution to contact cotton.
glyphosate; MOA 9 + Warrant (3.0 CS); MOA 9	See labels + 3 pt	0.75 to 1.13 (lb a.e.) + 1.125	Can be directed to cotton up to first bloom. See comments for glyphosate applied alone. Warrant does not improve control of emerged weeds, but it can give residual control of annual grasses and pigweed species.

Table 7-2B. Chemical Weed Control in Cotton

Herbicide, Mode of Action Code and Formulation	Amount of Formulation Per Acre	Pounds Active Ingredient Per Acre	Precautions and Remarks
Post-emergence Directed, Annual grasses and broadleaf weeds, nutsedge, and suppression of perennial weeds; Enlist, Glytol LibertyLink, Roundup Ready Flex or XtendFlex varieties only (continued)			
glyphosate; MOA 9 + Warrant Ultra (3.45 CS); MOA 9 + 14	See labels +	0.75 to 1.13 (lb a.e.) +	Warrant Ultra contains acetochlor plus fomesafen. Use as a layby application to cotton with a minimum of 4 inches of bark on the stem. Avoid contact with cotton foliage. Do not use Warrant Ultra at layby if Warrant Ultra or fomesafen (Reflex, others) was used preemergence. Do not exceed a maximum of 3 lb active acetochlor per year from all applications.
glyphosate; MOA 9 + Zidua (85 WDG); MOA 15	See labels + 0.75 to 2.1 oz	0.75 to 1.13 (lb a.e.) + 0.040 to 0.112	Labeled rate of Zidua is 0.75 to 1.5 oz on coarse- or medium-textured soils and 1.5 to 2.1 oz on fine-textured soils. Do not use on soils with greater than 10% organic matter. Direct to minimize contact with cotton foliage when cotton is from 5-leaf stage to beginning of bloom stage. Do not apply overtop. Zidua does not control emerged weeds but gives residual control of annual grasses and pigweed species.
Sequence (5.25 L); MOA 9 + 15	2.5 to 3.5 pt	0.70 to 1 (lb a.e.) glyphosate + 0.94 to 1.3 S-metolachlor	Sequence contains 2.25 lb a.e./gal glyphosate plus 3 lb/gal S-metolachlor. Direct to cotton up to 12 inches tall. Do not add adjuvants or tank mix with other products. Compared with glyphosate alone, Sequence will give residual control of annual grasses, pigweed species, and spreading dayflower plus suppression of nutsedge. See comments above for glyphosate alone.
Post-Emergence (hooded sprayers), Annual grasses, broadleaf weeds, and sedges; any variety			
glyphosate; MOA 9	See labels	0.75 (lb a.e.)	On varieties not resistant to glyphosate, hoods must be kept close to the ground so that no spray solution contacts the crop. Speed should not exceed 5 mph. Use 5 to 10 gpa and maximum pressure of 25 psi. Do not use liquid nitrogen as the carrier. Other herbicides as discussed in the section on directed application may be mixed with glyphosate to improve burndown and to provide residual control.
paraquat; MOA 22 2.0 lb/gal formulations 3.0 lb/gal formulations	 1.2 to 2.4 pt 0.8 to 1.6 pt	0.3 to 0.6	See Table 7-2A for brands of paraquat. Hoods should be kept as close to the ground as possible. Do NOT allow the spray solution to contact cotton plants. Apply in a minimum of 10 gpa at maximum pressure of 25 psi. Do not exceed 5 mph. It is suggested that cotton be at least 6 inches tall. Add nonionic surfactant or crop oil concentrate according to the paraquat label. Control will generally be much better if diuron, Cotoran, or prometryn is mixed with paraquat. Diuron, Cotoran, or prometryn may also provide residual control.
Harvest Aid, Annual grasses and broadleaf weeds			
glyphosate; MOA 9	See labels	0.75 to 1.5 (lb a.e.)	Apply to any cultivar after at least 60% of the bolls are open. May be tank mixed with some defoliants; see labels for details. Include nonionic surfactant according to the label of the glyphosate brand used. Can be applied to Roundup Ready Flex, Xtend Flex, Enlist, or GlyTol LibertyLink cotton up to 7 days before harvest.
paraquat; MOA 22 2.0 lb/gal formulations 3.0 lb/gal formulations	 1 to 2 pt 0.67 to 1.33 pt	0.25 to 0.5	See Table 7-2 for paraquat brands. Defoliate cotton as normal. After at least 75 to 80% of the bolls are open, the remaining bolls expected to be harvested are mature, and most of the cotton leaves have dropped, apply paraquat in a minimum of 20 gal per acre and add 1 pt nonionic surfactant per 100 gal. Wait 5 days before picking, then pick as soon as possible.

Weed Response to Cotton Herbicides

C. W. Cahoon, Assistant Professor and Extension Weed Specialist, Crop and Soil Sciences Department

Ratings based upon average to good soil and weather conditions for herbicide performance and upon proper application rate, technique, and timing.

Table 7-2C. Weed Response to Preplant Incorporated and Preemergence Herbicides in Cotton

Species	PPI Prowl or Treflan	PRE Brake	PRE Cotoran	PRE diuron	PRE Prowl	PRE Reflex	PRE Staple	PRE Warrant	Reflex + Cotoran	Reflex + diuron	Reflex + Warrant	diuron + Warrant	Cotoran + Warrant	Brake + Cotoran	Brake + diuron	Brake + Reflex
Bermudagrass	N	N	N	N	N	N	N	N	N	N	N	N	N	N	N	N
Broadleaf signalgrass	G	FG	P	P	F	FG	P	G	FG	FG	G	G	G	FG	FG	FG
Crabgrass	E	G	FG	FG	G	FG	P	E	FG	FG	E	E	E	G	G	G
Crowfootgrass	E	ND	FG	FG	G	ND	P	E	FG	FG	E	E	E	FG	FG	ND
Fall panicum	G	ND	F	P	F	ND	PF	E	F	ND	E	E	E	F	ND	ND
Foxtails	E	ND	FG	ND	G	ND	P	E	FG	ND	E	E	E	FG	ND	ND
Goosegrass	E	G	F	F	G	ND	PF	E	F	F	E	E	E	G	G	G
Johnsongrass																
Seedling	E	ND	P	P	G	ND	FG	F	ND	ND	F	F	F	ND	ND	ND
Rhizome	P	N	N	N	N	N	N	N	N	N	N	N	N	N	N	N
Texas panicum	G	PF	P	P	F	PF	N	P	PF	PF	P	P	P	PF	PF	PF
Nutsedge																
Purple	N	PF	N	N	N	ND	F	N	ND	ND	ND	N	N	PF	PF	PF
Yellow	N	FG	N	N	N	GE	F	PF	GE	GE	GE	PF	PF	FG	FG	GE
Cocklebur	N	G	FG	F	N	G	NP	N	G	G	G	F	FG	G	G	G
Common purslane	E	G	E	E	G	G	G	G	E	G	G	E	E	E	E	G
Common ragweed																
glyphosate-susceptible	N	FG	E	G	N	G	NP	P	E	G	G	G	E	E	G	GE
glyphosate-resistant	N	FG	E	G	N	G	NP	P	E	G	G	G	E	E	G	GE
Cowpea	N	ND	P	P	N	ND	FG	N	ND	ND	ND	P	P	ND	ND	ND
Crotalaria	N	ND	G	G	N	ND	ND	N	G	G	ND	G	G	G	G	ND
Eclipta	P	ND	G	G	P	GE	ND	FG	GE	GE	GE	G	G	G	G	GE
Florida beggarweed	P	ND	GE	G	N	P	G	P	GE	G	P	G	GE	GE	G	ND
Florida pusley	E	ND	FG	PF	FG	P	F	E	FG	PF	E	E	E	FG	PF	ND
Hemp sesbania	N	ND	P	P	N	P	P	N	P	P	P	P	P	ND	ND	ND
Jimsonweed	N	ND	G	G	N	G	FG	N	G	G	G	G	G	G	G	G
Lambsquarters	GE	ND	E	E	G	E	G	P	E	E	E	E	E	E	E	E
Morningglory																
tall	P	F	G	F	P	PF	P	P	G	F	PF	F	G	G	F	F
other species	P	F	G	F	P	PF	F	P	G	F	PF	F	G	G	F	F
Palmer amaranth																
glyphosate-susceptible	FG	E	F	G	PF	E	G	G	E	E	E	GE	G	E	E	E
glyphosate-resistant	FG	E	F	G	PF	E	G	G	E	E	E	GE	G	E	E	E
ALS-resistant	FG	E	F	G	PF	E	N	G	E	E	E	GE	G	E	E	E
glyphosate- & ALS-resistant	FG	E	F	G	PF	E	N	G	E	E	E	GE	G	E	E	E
Pigweed, redroot	E	E	GE	GE	FG	E	E	GE	E	E	E	GE	GE	E	E	E
Prickly sida	N	ND	G	F	N	ND	G	P	G	F	ND	F	G	F	F	ND
Sicklepod	N	P	G	F	N	P	PF	N	G	F	P	F	G	G	P	P
Smartweed	N	ND	G	G	N	ND	G	N	G	G	ND	G	G	G	G	ND
Spurge	N	ND	PF	F	N	ND	G	F	PF	F	F	F	F	F	F	ND
Spurred anoda	N	G	F	F	N	ND	E	N	F	F	ND	F	F	G	G	G
Tropic croton	N	G	FG	FG	N	FG	FG	N	FG	FG	FG	FG	FG	G	G	G
Velvetleaf	N	ND	F	PF	N	ND	E	N	F	PF	ND	PF	F	F	PF	ND
Volunteer peanuts	N	P	PF	P	N	P	P	N	PF	P	P	P	PF	PF	P	P

[1] Two applications may be required for acceptable control.
[2] Acceptable johnsongrass control can be obtained with two applications of Liberty.

Key:
- E = excellent control, 90% or better
- G = good control, 80% to 90%
- F = fair control, 50% to 80%
- P = poor control, 25% to 50%
- N = no control, less than 25%
- ND = data not available

Weed Response to Cotton Herbicides

C. W. Cahoon, Assistant Professor and Extension Weed Specialist, Crop and Soil Sciences Department

Ratings based upon average to good soil and weather conditions for herbicide performance and upon proper application rate, technique, and timing.

Table 7-2D. Weed Response to Postemergence Overtop Herbicides in Cotton

Species	Assure II	Fusilade	Poast, Poast Plus	Select	Enlist Duo or glyphosate + Enlist One	Envoke	Staple	glyphosate	glyphosate + Engenia or XtendiMax	glyphosate + Envoke	glyphosate + Staple	Liberty	Liberty + Enlist One	Liberty + Staple
Bermudagrass	G[1]	G[1]	F[1]	G[1]	F[2]	N	N	F[2]	F[2]	F[2]	F[2]	N	N	N
Broadleaf signalgrass	G	GE	E	E	E	N	N	E	E	E	E	G	G	G
Crabgrass	G	G	GE	GE	E	P	N	E	E	E	E	G	G	G
Crowfootgrass	G	F	FG	G	E	N	N	E	E	E	E	G	G	G
Fall panicum	GE	GE	E	E	E	NP	N	E	E	E	E	G	G	G
Foxtails	E	E	E	E	E	NP	N	E	E	E	E	G	G	G
Goosegrass	G	G	GE	GE	E	NP	N	E	E	E	E	P	P	P
Johnsongrass														
Seedling	E	E	E	E	E	P	P	E	E	E	E	G	G	G
Rhizome	E	GE	G	GE	GE	P	NP	GE	GE	GE	GE	F[2]	F[2]	F[2]
Texas panicum	G	G	E	E	E	NP	N	E	E	E	E	G	G	G
Nutsedge														
Purple	N	N	N	N	FG[2]	FG	PF	FG[2]	FG[2]	GE	FG[2]	PF	PF	F
Yellow	N	N	N	N	FG[2]	G	PF	F[2]	F[2]	E	FG[2]	PF	PF	F
Cocklebur	N	N	N	N	E	GE	G	E	E	E	E	E	E	E
Common purslane	N	N	N	N	FG	ND	F	FG	FG	ND	G	FG	FG	FG
Common ragweed														
glyphosate-susceptible	N	N	N	N	E	G	P	E	E	E	E	E	E	E
glyphosate-resistant	N	N	N	N	E	G	P	N	E	G	P	E	E	E
Cowpea	N	N	N	N	E	G	G	GE	E	E	E	G	GE	E
Crotalaria	N	N	N	N	G	ND	G	G	ND	G	G	ND	ND	G
Doveweed	N	N	N	N	P	ND	N	P	P	ND	P	P	P	P
Eclipta	N	N	N	N	E	PF	G	E	E	E	E	G	GE	E
Florida beggarweed	N	N	N	N	E	GE	G	E	E	E	E	G	GE	E
Florida pusley	N	N	N	N	G	P	NP	PF	G	PF	PF	F	G	F
Hemp sesbania	N	N	N	N	E	ND	GE	PF	E	ND	GE	ND	GE	ND
Jimsonweed	N	N	N	N	E	N	E	E	E	E	E	E	E	E
Lambsquarters	N	N	N	N	E	G	N	G	E	E	G	E	E	E
Morningglory														
tall	N	N	N	N	E	G	P	FG	E	E	FG	E	E	E
other species	N	N	N	N	E	G	G	FG	E	E	GE	E	E	E
Palmer amaranth														
glyphosate-susceptible	N	N	N	N	E	PF	F	E	E	E	E	G	E	G
glyphosate-resistant	N	N	N	N	G	PF	F	N	G	PF	F	G	E	G
ALS-resistant	N	N	N	N	E	N	N	E	E	E	E	G	E	G
glyphosate- & ALS-resistant	N	N	N	N	G	N	N	N	G	N	N	G	E	G
Pigweed, redroot	N	N	N	N	E	FG	G	E	E	E	E	G	E	GE
Prickly sida	N	N	N	N	G	N	F	FG	G	FG	FG	FG	G	FG
Sicklepod	N	N	N	N	E	E	PF	E	E	E	E	E	E	E
Smartweed	N	N	N	N	G	G	G	G	E	E	E	GE	GE	E
Spreading dayflower	N	N	N	N	ND	N	FG	P	P	P	FG	PF	PF	FG
Spurge	N	N	N	N	G	ND	FG	G	ND	G	G	FG	FG	G
Spurred anoda	N	N	N	N	E	P	G	E	E	E	E	P	GE	G
Tropic croton	N	N	N	N	E	PF	P	E	E	E	E	G	GE	G
Velvetleaf	N	N	N	N	E	G	G	E	E	E	E	F	GE	G
Volunteer peanuts	N	N	N	N	G	PF	P	F	E	FG	F	GE	GE	GE

[1] Two applications may be needed for adequate control.

[2] Acceptable control with two applications or a follow-up application of glyphosate.

Key:

E = excellent control, 90% or better

G = good control, 80% to 90%

F = fair control, 50% to 80%

P = poor control, 25% to 50%

N = no control, less than 25%

ND = data not available

Weed Response to Cotton Herbicides

C. W. Cahoon, Assistant Professor and Extension Weed Specialist, Crop and Soil Sciences Department

Ratings based upon average to good soil and weather conditions for herbicide performance and upon proper application rate, technique, and timing.

Table 7-2E. Weed Response to Postemergence-Directed Herbicides in Cotton

Species	Caparol + MSMA	Cobra + MSMA	Cotoran + MSMA	diuron+ MSMA	Suprend+ MSMA	Valor or Fierce + MSMA	glyphosate + Aim	glyphosate + Caparol	glyphosate + diuron	glyphosate + Suprend	glyphosate + Valor or Fierce	paraquat + diuron[1]
Bermudagrass	N	N	N	N	N	N	F[2]	F[2]	F[2]	F[2]	F[2]	P
Broadleaf signalgrass	F	F	F	F	F	F	E	GE	GE	GE	E	GE
Crabgrass	FG	PF	F	F	FG	F	E	GE	GE	GE	E	G
Crowfootgrass	FG	PF	F	F	FG	F	E	GE	GE	GE	E	G
Fall panicum	FG	PF	F	F	FG	F	E	GE	GE	GE	E	G
Foxtails	FG	PF	F	F	FG	F	E	GE	GE	GE	E	G
Goosegrass	FG	PF	F	F	FG	F	E	GE	GE	GE	E	G
Johnsongrass												
Seedling	FG	PF	F	F	FG	F	E	GE	GE	GE	E	G
Rhizome	P	P	P	P	P	P	GE	G	G	G	GE	P
Texas panicum	F	P	P	F	F	P	E	GE	GE	GE	E	G
Nutsedge												
Purple	F[3]	F[3]	F[3]	F[3]	E	F[3]	FG[2]	FG[2]	FG[2]	GE	FG[2]	PF
Yellow	FG[3]	FG[3]	FG[3]	G	E	G	F[2]	F[2]	F[2]	E	F[2]	PF
Cocklebur	E	E	E	E	E	E	E	E	E	E	E	E
Common purslane	FG	G	FG	G	ND	G	FG	GE	GE	E	GE	E
Common ragweed												
glyphosate-susceptible	E	E	GE	E	E	GE	E	E	E	E	E	G
glyphosate-resistant	E	E	GE	E	E	GE	PF	P	PF	G	P	G
Cowpea	G	FG	G	G	E	G	GE	GE	GE	E	E	E
Crotalaria	G	G	G	G	E	ND	G	G	G	G	ND	ND
Doveweed	N	N	N	N	ND	ND	P	P	ND	ND	G	E
Eclipta	G	E	G	E	E	E	E	E	E	E	E	G
Florida beggarweed	E	E	E	E	E	E	E	E	E	E	E	E
Florida pusley	F	F	F	F	F	FG	G	G	G	G	GE	P
Hemp sesbania	PF	F	PF	PF	ND	ND	GE	ND	ND	ND	ND	FG
Jimsonweed	G	GE	GE	G	G	E	E	E	E	E	E	E
Lambsquarters	G	F	G	G	GE	FG	GE	GE	GE	E	GE	G
Morningglory	G	E	G	GE	E	E	E	GE	GE	E	E	G
Palmer amaranth												
glyphosate-susceptible	F	G	FG	G	GE	G	E	E	E	E	E	GE
glyphosate-resistant	F	G	FG	G	GE	G	PF	PF	F	PF	PF	GE
ALS-resistant	F	G	FG	G	F	G	E	E	E	E	E	GE
glyphosate- & ALS-resistant	F	G	FG	G	F	G	PF	PF	F	P	PF	GE
Pigweed, redroot	G	G	G	GE	GE	GE	E	E	E	E	E	GE
Prickly sida	GE	GE	FG	GE	GE	GE	FG	G	G	G	GE	FG
Sicklepod	GE	PF	G	GE	E	GE	E	E	E	E	E	E
Smartweed	F	F	G	F	ND	G	GE	G	G	E	G	GE
Spreading dayflower	G	G	G	G	G	G	P	P	P	P	P	ND
Spurge	G	G	PF	G	ND	G	GE	GE	GE	E	G	ND
Spurred anoda	F	F	FG	F	ND	G	E	E	E	E	E	G
Tropic croton	G	E	G	G	GE	E	E	E	E	E	E	G
Velvetleaf	F	F	F	F	F	G	E	E	E	E	E	ND
Volunteer peanuts	FG	PF	FG	G	G	FG	FG	FG	G	FG	FG	F

[1] Apply only with a hooded sprayer and avoid all contact with cotton foliage or stems.
[2] Acceptable control with a follow-up application of glyphosate.
[3] Acceptable control with second application of MSMA or a follow-up application of glyphosate.

Key:
E = excellent control, 90% or better
G = good control, 80% to 90%
F = fair control, 50% to 80%
P = poor control, 25% to 50%
N = no control, less than 25%
ND = data not available

Chemical Weed Control in Peanuts

D. L. Jordan, Crop and Soil Sciences Department

Control of witchweed is part of the State/Federal Quarantine Program. Contact the N.C. Department of Agriculture, Plant Industry Division, at 1-800-206-9333. Note that the active ingredient and examples of tradenames are listed for each herbicide. This approach is used for the simplicity of presentation but does not imply that other formulations of the same active ingredient are not equally effective. See Table 7-3D for products other than those listed in 7-3A, 7-3B, and 7-3C that are equally effective.

Table 7-3A. Chemical Weed Control in Peanuts

Herbicide and Formulation	Pounds Active Ingredient Per Acre	Precautions and Remarks
Preplant Incorporated, Annual grasses and small-seeded broadleaf weeds		
alachlor, MOA 15 (Intrro 4 EC)	2 to 3 (2 to 3 qt)	Incorporate no deeper than 2 inches; see label for specific instructions. Unless shallowly incorporated, Intrro is more consistently effective when applied preemergence. Weak on Texas panicum. Do not apply more than 3 qt of Intrro per acre per season. **Before using Intrro, check with buyers to determine if there are marketing restrictions on Intrro-treated peanuts.**
acetochlor, MOA 15 (Warrant 3 ME)	0.94 to 1.5 (1.25 to 2 qt)	Apply and incorporate in top 2 inches of soil. Do not apply more than 4 qt of Warrant per acre per year.
ethalfluralin, MOA 3 (Sonalan 3 EC)	0.56 to 0.75 (1.5 to 2 pt)	Controls common annual grasses including Texas panicum. Use 3 pt Prowl or 2 pt ethalfluralin for control of broadleaf signalgrass, Texas panicum, and fall panicum. Incorporate 3 inches deep for Texas panicum; otherwise, incorporate 2 to 3 inches deep. See labels for maximum waiting period between application and incorporation. Immediate incorporation is best. Dual Magnum, Outlook, or Warrant may be tank mixed with Prowl or Sonalanto suppress yellow nutsedge.
pendimethalin, MOA 3 (Prowl H2O 3.8 EC) (Prowl 3.3 EC)	0.71 to 1.43 (1.5 to 3 pt) (1.7 to 3.5 pt)	
Preplant Incorporated, Annual grasses, small-seeded broadleaf weeds, and nutsedge		
dimethenamid, MOA 15 (Outlook 6.0 L)	0.75 to 1 (16 to 21 fl oz)	Apply and incorporate in top 2 inches of soil within 14 days of planting. Use high rate of Dual Magnum, Dual, or Outlook for yellow nutsedge and broadleaf signalgrass. Not effective on purple nutsedge. Weak on Texas panicum. May be tank mixed with Prowl or Sonalan.
metolachlor, MOA 15 (Dual Magnum 7.62 EC) (Dual 8 EC)	0.95 to 1.27 (1 to 1.33 pt) (1.5 to 2 pt)	
Preplant Incorporated, Broadleaf weeds and suppression of nutsedge		
diclosulam, MOA 2 (Strongarm 84 WDG)	0.024 (0.45 oz)	Effective on common cocklebur, morningglory, common ragweed, eclipta, and common lambsquarters. Suppresses yellow and purple nutsedge. Does not control sicklepod. More effective when applied in combination with Dual, Outlook, Warrant, Prowl, or Sonalan. See label for rotation restrictions, especially corn and grain sorghum. Growers are cautioned that Strongarm can occasionally injure cotton the following year on soils with a shallow hardpan (less than 10 inches) and/or loam soils. Cotton grown under early season stress resulting from conditions such as excessively cool, wet, dry, or crusted soils may be particularly susceptible to carryover of Strongarm. The rotation interval between applying Strongarm to peanut and then planting cotton is 18 months in Camden, Currituck, Pasquotank, and Perquimans counties. Some weed species have developed resistance to Strongarm including common ragweed and Palmer amaranth.
Preplant Incorporated, Annual grasses, broadleaf weeds, and suppression of nutsedge		
diclosulam, MOA 2 Strongarm + pendimethalin, MOA 3 (Prowl H2O 3.8 EC) (Prowl 3.3 EC) or ethalfluralin, MOA 3 (Sonalan 3 EC) or metolachlor, MOA 15 (Dual Magnum 7.62 EC) (Dual 8 EC) or dimethenamid (Outlook 6.0 L) or acetochlor (Warrant 3 ME)	0.024 (0.45 oz) + 0.71 to 1.43 (1.5 to 3 pt) (1.7 to 3.5 pt) or 0.56 to 0.75 (1.5 to 2 pt) or 0.95 to 1.27 (1 to 1.33 pt) (1.5 to 2 pt) or 0.75 to 1 (16 to 21 fl oz) or 0.95 to 1.5 (1.24 to 2 qt)	Effective on annual grasses, common cocklebur, common ragweed, eclipta, morningglory, and common lambsquarters. Suppresses purple and yellow nutsedge. Does not control sicklepod. See Strongarm label for rotation restrictions.
PPI followed by PRE, Annual grasses, broadleaf weeds, and suppression of nutsedge		
pendimethalin, MOA 3 (Prowl H2O 3.8 EC) (Prowl 3.3 EC) or ethalfluralin, MOA 3 (Sonalan 3 EC) or metolachlor, MOA 15 (Dual Magnum 7.62 EC) (Dual 8 EC) or dimethenamid, MOA 15 (Outlook 6.0L) or acetochlor, MOA 15 (Warrant 3 ME) followed by diclosulam, MOA 2 (Strongarm 84 WDG) or flumioxazin, MOA 14 (Valor SX 51 WDG)	0.71 to 1.43 (1.5 to 3 pt) (1.7 to 3.5 pt) or 0.56 to 0.75 (1.5 to 2 pt) or 0.95 to 1.27 (1 to 1.33 pt) (1.5 to 2 pt) or 0.75 to 1 (16 to 21 oz) or 0.95 to 1.5 (1.24 to 2 qt) 0.024 0.45 oz or 0.063 (2 oz)	Controls most broadleaf weeds. Will not control sicklepod and is marginal on certain large-seeded broadleaf weeds. **Do not incorporate Valor SX.** Valor SX should be applied to the soil surface immediately after planting. Significant injury can occur if flumioxazin is incorporated or applied 3 or more days after planting. Significant injury from Valor SX has been noted in some years even when applied according to label recommendations. However, injury is generally transient and does not affect yield. See previous comments about cotton response to Strongarm applied the previous year on some soils. Up to 3 oz per acre of Valor SX can be applied to peanut but injury potential increases. **See product label for sprayer cleanup before other uses.**

Table 7-3A. Chemical Weed Control in Peanuts

Herbicide and Formulation	Pounds Active Ingredient Per Acre	Precautions and Remarks
Split application (PPI + POST), Most broadleaf weeds and nutsedge		
imazethapyr, MOA 2 (Pursuit 2 AS)	0.031 + 0.031 (2 + 2 oz)	Effective on most common broadleaf weeds and yellow and purple nutsedge. Does not control eclipta, lambsquarters, ragweed, or croton. Pursuit will usually control seedling johnsongrass and foxtails. For control of other annual grasses, Pursuit may be tank mixed with Dual Magnum, Dual, Outlook, Prowl H2O, Prowl, or Sonalan and incorporated. See label for incorporation directions and rotational restrictions. Some weed species have developed resistance to Pursuit. Research in N.C. has generally shown more effective control of a broader spectrum of weeds with split applications of half of the Pursuit applied preplant incorporated followed by the other half applied early postemergence.
Preemergence, Annual grasses and small-seeded broadleaf weeds		
alachlor, MOA 15 (Intrro 4 EC)	2 to 3 (2 to 3 qt)	Apply as soon after planting as possible. All 4 herbicides are weak on Texas panicum. **Before using Intrro, check with buyers to determine if there are marketing restrictions on Intrro-treated peanuts.**
dimethenamid, MOA 15 (Outlook 6.0 L)	0.75 to 1 (16 to 21 fl oz)	
metolachlor, MOA 15 (Dual Magnum 7.62 EC) (Dual 8 EC)	0.95 to 1.27 (1 to 1.33 pt) (1.5 to 2 pt)	
acetochlor (Warrant 3 ME)	0.95 to 1.5 (1.25 to 2 qt)	
Preemergence, Broadleaf weeds		
flumioxazin, MOA 14 (Valor SX 51 WDG)	0.063 2 oz	Apply within 2 days after planting. Significant injury can occur if Valor SX is incorporated or applied 3 or more days after seeding. Controls carpetweed, common lambsquarters, Florida pusley, nightshade, pigweeds, prickly sida, and spotted spurge. Does not control sicklepod, yellow and purple nutsedge, or annual grasses. Morningglory control is marginal where Valor SX is applied at 2 oz per acre. Significant injury from Valor SX has been noted in some years even when applied according to label. However, injury is generally transient and does not affect yield. Injury may occur if excessive and forceful rainfall occurs when peanut is emerging. Peanut recovers from injury by midseason in most instances. Up to 3 oz per acre of Valor SX can be applied to peanut, but injury potential increases. **See product label for comments on sprayer cleanup before other uses.**
Preemergence, Annual grasses, broadleaf weeds, and suppression of nutsedge		
flumioxazin, MOA 14 (Valor SX 51 WDG) + metolachlor, MOA 15 (Dual Magnum 7.62 EC) (Dual 8 EC) or dimethenamid, MOA 15 (Outlook 6.0L) or acetochlor, MOA 15 (Warrant 3 ME)	0.063 (2 oz) + 0.95 to 1.27 (1 to 1.33 pt) 1.5 to 2 pt) or 0.75 to 1 (16 to 21 fl oz) or 0.94 to 1.5 (1.25 to 2 qt)	Apply within 2 days after planting. Significant injury can occur if applied 3 or more days after planting. The combination of Valor SX and Dual, Dual Magnum, Warrant, or Outlook does not control sicklepod but will control annual grasses (except Texas panicum) and will suppress yellow nutsedge. Valor SX and Warrant will not suppress yellow nutsedge. Significant injury from Valor SX has been noted in some years even when applied according to label recommendations. However, injury is generally transient and does not affect yield. Injury may occur if excessive and forceful rainfall occurs when peanut is emerging. Peanut recovers from injury by midseason in most instances. Up to 3 oz per acre of Valor SX can be applied to peanut but injury potential increases. **See product label for comments on sprayer cleanup before other uses.**
diclosulam, MOA 2 (Strongarm 84 WDG)	0.024 (0.45 oz)	Effective on common cocklebur, morningglory, common ragweed, eclipta, and common lambsquarters. Suppresses yellow and purple nutsedge. Does not control sicklepod. More effective when applied in combination with Dual, Dual Magnum, Outlook, Prowl, Sonalan, or Warrant. See label for rotation restrictions, especially corn and grain sorghum. See previous comments on possible cotton injury from Strongarm applied the previous year on some soils.
sulfentrazone, MOA 14 + carfentrazone, MOA 14 (Spartan Charge (0.35 + 3.15 F)	0.07 to 0.12 (3 to 5 fl oz)	Do not apply Spartan Charge after peanuts crack soil. Application immediately after planting is advised. See label for specific rates based on soil texture and organic matter content. See product label for comments on application with other herbicides. Rotation restriction for planting cotton following Spartan Charge at recommended rates for peanut is 12 months.
diclosulam, MOA 2 (Strongarm 84 WDG) + metolachlor, MOA 15 (Dual Magnum 7.62 EC) (Dual 8 EC) or dimethenamid, MOA 15 (Outlook 6.0 L) or acetochlor, MOA 15 (Warrant 3 ME)	0.024 (0.45 oz) + 0.95 to 1.27 (1 to 1.33 pt) 1.5 to 2 pt) or 0.75 to 1 (16 to 21 oz) or 0.94 to 1.5 (1.25 to 2 qt)	Effective on annual grasses, common cocklebur, common ragweed, eclipta, morningglory, and common lambsquarters. Suppresses purple and yellow nutsedge. Does not control sicklepod. See label for rotation restrictions. Some weed species have developed resistance to Strongarm. See previous comments on carryover potential to cotton on some soils and restrictions on planting corn or grain sorghum after use in peanut.
Preemergence, Most annual broadleaf weeds and nutsedge		
imazethapyr, MOA 2 (Pursuit 2 AS)	0.063 (4 fl oz)	Effective on most common broadleaf weeds and yellow and purple nutsedge. Does not control ragweed, eclipta, lambsquarters, or croton. Pursuit may be tank mixed with Dual, Dual Magnum, Warrant, or Outlook for annual grass control. See label for **rotational restrictions**. Some weed species have developed resistance to Pursuit. Research in N.C. has generally shown more effective control of a broader spectrum of weeds with split applications of half of the Pursuit applied preplant incorporated followed by the other half applied early postemergence.
Cracking stage, Emerged annual grasses and broadleaf weeds		
paraquat, MOA 22 (Gramoxone 2.5 SL) (Parazone 3 SL)	0.13 (8 oz) (5.4 oz)	Apply at ground cracking to control small emerged annual grasses and broadleaf weeds. May be tank mixed with Dual, Dual Magnum, Outlook, or Warrant for residual control. Tank mix may increase injury to emerged peanuts. Add 1 pint nonionic surfactant per 100 gallons spray solution. Follow safety precautions on label. Applying Basagran at 0.5 pt per acre will reduce injury.
Cracking stage and Postemergence, Additional residual control of annual grasses and certain small-seeded broadleaf weeds		
alachlor, MOA 15 (Intrro 4 EC)	2 to 3 (2 to 3 qt)	Use as a supplement to preplant or preemergence herbicides to provide additional residual control of annual grasses and certain small-seeded broadleaf weeds such as pigweed and eclipta. This treatment will not control emerged grasses or broadleaf weeds. See product labels for recommended tank mixtures with contact and systemic herbicides with foliar activity on weeds.
dimethenamid, MOA 15 (Outlook 6.0L)	0.75 to 1 (16 to 21 oz)	
metolachlor, MOA 15 (Dual Magnum 7.62 EC) (Dual 8 EC)	0.95 1 pt 1.5 pt	

Table 7-3A. Chemical Weed Control in Peanuts

Herbicide and Formulation	Pounds Active Ingredient Per Acre	Precautions and Remarks
Cracking stage and Postemergence, Additional residual control of annual grasses and certain small-seeded broadleaf weeds (continued)		
acetochlor, MOA 15 (Warrant 3 ME)	0.95 to 1.5 (1.25 to 2 qt)	
pyroxasulfone, MOA 15 (Zidua) 85 WG (Zidua) 4.25 SC	0.08 to 11 (1.5 to 2.1 oz) (2.4 to 3.3 fl oz)	
Cracking stage, Most annual broadleaf weeds and nutsedge		
imazethapyr, MOA 2 (Pursuit 2 AS)	0.063 (4 oz)	Effective on most common broadleaf weeds and yellow and purple nutsedge. Does not control ragweed, eclipta, lambsquarters, or croton. If weeds are emerged, add surfactant or crop oil according to label directions. See label for rotational restrictions. Pursuit may be tank mixed with paraquat. Some weed species have developed resistance to Pursuit.
Cracking stage, Some emerged broadleaf weeds and suppression of eclipta and yellow nutsedge		
diclosulam, MOA 2 (Strongarm 84 WDG)	0.024 (0.45 oz)	Strongarm can be applied through the cracking stage. Add 1 quart nonionic surfactant per 100 gallons. The spectrum of weeds controlled is much narrower when applied to emerged weeds. Strongarm will not control emerged common lambsquarters or pigweeds but will control common ragweed and morningglories and will suppress yellow nutsedge and eclipta. See product labels for information on mixing Strongarm with other herbicides. Some weed species have developed resistance to Strongarm. See product label for carryover potential to cotton, corn, and grain sorghum. Strongarm suppresses emerged marestail and dogfennel more effectively than other postemergence broadleaf herbicides when applied to small weeds.
Postemergence, Annual broadleaf weeds		
acifluorfen, MOA 14 (Ultra Blazer 2 L)	0.25 to 0.38 (1 to 1.5 pt)	Apply when weeds are small and actively growing. Use minimum of 20 GPA and high pressure (40 to 60 psi). See label for species controlled, maximum weed size to treat, and addition of surfactant. Do not apply more than 2 pints per acre per season. May make sequential applications of 0.25 pound followed by 0.25 pound per acre. Allow at least 15 days between sequential applications.
acifluorfen, MOA 14 (Ultra Blazer 2 L) + 2,4-DB, MOA 4 (Butyrac 200 2 L)	0.25 to 0.38 (1 to 1.5 pt) + 0.25 (16 fl oz)	Addition of 2,4-DB to Ultra Blazer improves control of certain weeds when weed size exceeds that specified on the Ultra Blazer label. See label suggestions on use of surfactant or crop oil. Apply when peanuts are at least 2 weeks old and before pod filling begins.
bentazon, MOA 6 (Basagran 4 L)	0.75 to 1 (1.5 to 2 pt)	Apply when weeds are small and actively growing. Use minimum of 20 GPA and high pressure (40 to 60 psi). See label for addition of oil concentrate, species controlled, and maximum weed size to treat. Basagran may also be applied at 1 pint per acre for control of cocklebur, jimsonweed, and smartweed 4 inches or less. Do not apply more than 4 pints of bentazon per acre per season.
bentazon, MOA 6 (Basagran 4 L) + acifluorfen, MOA 14 (Ultra Blazer 2 L)	0.5 to 1 (1 to 2 pt) + 0.25 to 0.38 (1 to 1.5 pt)	See above comments for Ultra Blazer and Basagran. See labels for weeds controlled, maximum weed size to treat, and use of adjuvants. Can be applied as a tank mixture or as Storm 4L.
bentazon, MOA 6 + acifluorfen, MOA 14 (Storm 4L)	0.5 + 0.25 (1.5 pt)	These rates of bentazon and acifluorfen (Ultra Blazer and Basagran) may not provide consistent control of lambsquarters, prickly sida, spurred anoda, and morningglory.
bentazon, MOA 6 (Basagran 4 L) + acifluorfen, MOA 14 (Ultra Blazer 2 L) + 2,4-DB, MOA 4 (Butyrac 200 2 L)	0.5 (1 pt) + 0.25 (1 pt) + 0.125 to 0.25 (8 to 16 fl oz)	Adding 2,4-DB will improve control of larger morningglory, cocklebur, common ragweed, pigweed, jimsonweed, and citron. Add surfactant or crop oil according to label directions. Apply when peanuts are at least 2 weeks old. Do not apply after pod filling begins. See comments for Ultra Blazer and Basagran alone.
bentazon, MOA 6 (Basagran 4 L) + 2,4-DB, MOA 4 (Butyrac 200 2 L)	0.75 to 1 1.5 to 2 pt) + 0.125 (8 fl oz)	Addition of 2,4-DB to Basagran improves control of morningglories. See above comments for Basagran. Add surfactant or crop oil according to label directions. Do not make more than 2 applications per year. Apply when peanuts are at least 2 weeks old and not within 45 days of harvest.
imazapic, MOA 2 (Cadre 2 AS) (Impose 2 AS)	0.063 (4 fl oz)	Controls most broadleaf weeds except ragweed, croton, lambsquarters, and eclipta. Apply before weeds exceed 2 to 4 inches; see label for specific weed sizes to treat. Add nonionic surfactant at 1 quart per 100 gallons or crop oil concentrate at 1 quart per acre. A soil-applied grass control herbicide should be used. However, Cadre will usually control escaped broadleaf signalgrass, large crabgrass, fall panicum, and Texas panicum but not goosegrass. Cadre can be mixed with Cobra, Ultra Blazer, and 2,4-DB. See label for rotational restrictions. Some weed species have developed resistance to Cadre.
imazethapyr, MOA 2 (Pursuit 2 L)	0.063 (4 fl oz)	Effective on most common broadleaf weeds and yellow and purple nutsedge. Does not control eclipta, lambsquarters, ragweed, or croton. Apply when weeds are 3 inches tall or less. Add surfactant or crop oil according to label directions. See label for rotational restrictions. Pursuit may be tank mixed with Basagran, Ultra Blazer, Gramoxone, and 2,4-DB. Some weed species have developed resistance to Pursuit.
2,4-DB, MOA 4 (Butyrac 200 2 L)	0.2 to 0.25 (12 to 16 fl oz)	Effective on cocklebur and morningglory; pitted morningglory may be only partially controlled. Best results achieved when applied to small weeds. May use two applications per year. Do not apply within 45 days before harvest.
lactofen, MOA 14 (Cobra 2 EC)	0.2 (12.5 fl oz)	Apply after peanuts have at least six true leaves. Apply to actively growing peanut. Controls most annual broadleaf weeds. See label for species controlled and maximum weed size to treat. Add nonionic surfactant at 1 quart per 100 gallons or crop oil concentrate or methylated seed oil at 1 to 2 pints per acre. See label on when to use various adjuvants. Allow at least 14 days between applications. Can be tank mixed with Basagran, Pursuit, Cadre, 2,4-DB, and/or Select.
lactofen, MOA 14 (Cobra 2 EC) + bentazon, MOA 6 (Basagran 4 L)	0.2 (12.5 fl oz) + 0.75 to 1 (1.5 to 2 pt)	See above comments for Basagran and Lactofen alone. See labels for weeds controlled, maximum weed size to treat, and use of adjuvants.

Table 7-3A. Chemical Weed Control in Peanuts

Herbicide and Formulation	Pounds Active Ingredient Per Acre	Precautions and Remarks
Postemergence, Annual broadleaf weeds (continued)		
lactofen, MOA 14 (Cobra 2 EC) + bentazon, MOA 6 (Basagran 4 L) + 2,4-DB, MOA 4 (Butyrac 200 2 L)	0.2 (12.5 fl oz) + 0.75 to 1 (1.5 to 2 pt) + 0.125 to 0.25 (8-16 oz)	Adding 2,4-DB will improve control of larger morningglory, cocklebur, common ragweed, jimsonweed, and citron. See above comments for bentazon, lactofen, and 2,4-DB. See labels for weeds controlled, maximum weed size to treat, and use of adjuvants.
lactofen, MOA 14 (Cobra 2 EC) + imazapic, MOA 2 (Cadre 2 AS) (Impose 2 AS)	0.2 (12.5 fl oz) + 0.063 (4 fl oz)	See above comments for imazapic and lactofen. See labels for weeds controlled, maximum weed size to treat, and use of adjuvants. Some weed species have developed resistance to Cadre.
lactofen, MOA 14 (Cobra 2 EC) + imazethapyr, MOA 2 (Pursuit 2 AS)	0.2 (12.5 fl oz) + 0.063 (4 fl oz)	See above comments for imazethapyr and lactofen. See labels for weeds controlled, maximum weed size to treat, and use of adjuvants. Some weed species have developed resistance to Pursuit.
Postemergence, Annual grasses and broadleaf weeds		
paraquat, MOA 22 (Gramoxone 2 SL) (Parazone 3 SL)	0.13 (8 fl oz) (5.4 fl oz)	See label for weeds controlled and maximum weed size to treat; best results if weeds 1 inches or less. A postemergence application may be made following an at-crack application. Do not make more than 2 applications per season, do not apply later than 28 days after ground cracking, and do not apply if peanuts are under stress or have significant injury from thrips feeding. Gramoxone is more effective when applied within 2 weeks after peanut emergence. Add 1 pint of nonionic surfactant per 100 gallons of spray solution. Will cause foliar burn on peanuts, but peanuts recover, and yield is not affected. Follow all safety precautions on label.
paraquat, MOA 22 (Gramoxone 2 SL) (Parazone 3 SL) + bentazon, MOA 6 (Basagran 4 L)	0.13 (8 oz) (5.4 oz) + 0.25 to 0.75 (0.5 to 1.5 pt)	See previous comments for paraquat alone. Adding Basagran improves control of common ragweed, prickly sida, smartweed, lambsquarters, and cocklebur and reduces injury to peanuts from paraquat. May be applied any time from ground cracking up to 28 days after ground cracking. Add 1 pint of nonionic surfactant per 100 gallons of spray solution.
paraquat, MOA 22 (Gramoxone 2 SL) (Parazone 3 SL) + bentazon, MOA 6 + acifluorfen, MOA 14 (Storm 4 L)	0.13 (8 fl oz) (5.4 fl oz) + 0.5 + 0.25 1 pt	See previous comments for paraquat alone. Storm improves control of common ragweed, smartweed, lambsquarters, common cocklebur, tropic croton, and spurred anoda. May be applied anytime from ground cracking up to 28 days after ground cracking. Add 0.5 pint of nonionic surfactant per 100 gallons of spray solution. The mixture of Gramoxone SL and Storm is more injurious than these herbicides applied alone.
Postemergence, Florida beggarweed		
chlorimuron, MOA 2 (Classic 0.25 DF)	0.008 (0.5 oz)	Use only for control of Florida beggarweed. Apply from 60 days after crop emergence to within 45 days of harvest. Application to peanuts less than 60 days old will result in crop injury and yield reduction. Apply before Florida beggarweed has begun to bloom and before it has reached 10 inches tall. Larger beggarweed may only be suppressed. Add 1 quart of nonionic surfactant per 100 gallons spray solution; do not add crop oil. May be tank mixed with 2,4-DB; see label for rates and precautions. **Recommended as a salvage treatment only.**
Postemergence, Yellow nutsedge		
bentazon, MOA 6 (Basagran 4 L)	0.75 to 1 (1.5 to 2 pt)	Apply when nutsedge is 6 to 8 inches tall. A repeat application 7 to 10 days later may be needed. Adding crop oil concentrate at 1 quart per acre will increase control. Do not apply more than 2 pints of Basagran per season. Not effective on purple nutsedge.
Postemergence, Yellow and purple nutsedge		
imazapic, MOA 2 (Cadre 2 AS) (Impose 2 AS)	0.063 (4 fl oz)	Apply postemergence when nutsedge is 4 inches or less. Add nonionic surfactant at 1 quart per 100 gallons or crop oil concentrate at 1 quart per acre. See label for rotational restrictions.
imazethapyr, MOA 2 (Pursuit 2 AS)	0.063 (4 fl oz)	Apply before nutsedge is larger than 3 inches tall. Add surfactant at 1 quart per 100 gallons or crop oil concentrate at 1 quart per acre. Do not mix with Basagran for nutsedge control. See label for rotational restrictions. A split application with half of the Pursuit applied preplant incorporated and half applied early postemergence may be more effective than applying all of the Pursuit at one time.
Postemergence, Annual grasses		
clethodim, MOA 1 (Select Max 0.97 EC) (Select 2 EC) sethoxydim, MOA 1 (Poast 1 EC) (Poast Plus 1.5 EC)	0.094 to 0.125 (9 to 16 fl oz) (6 to 8 fl oz) 0.19 (1.5 pt) (1 pt)	Apply Select and Poast to actively growing grass not under drought stress. Consult labels for maximum grass size to treat. Apply in 5 to 20 GPA at 40 to 60 psi. Do not cultivate within 7 days before or after application. Add 2 pints crop oil to Poast. See label for adjuvant use with Select or Select Max. Some broadleaf/sedge herbicides and fungicides can reduce the efficacy of Select and Poast when applied in tank mixtures. See product labels for specific instructions concerning compatibility with other chemicals. See *2017 Peanut Information* AG-331 for specific pesticides that reduce control by these herbicides.
Postemergence, Bermudagrass		
clethodim, MOA 1 (Select Max 0.97 EC) (Select 2 EC)	0.125 to 0.25 (12 to 32 fl oz) (8 to 16 fl oz)	Apply to actively growing bermudagrass before runners exceed 6 inches In most cases, a second application will be needed. Make second application if regrowth occurs. See comments under annual grasses for adjuvant selection and tank mixing for these herbicides.
sethoxydim, MOA 1 (Poast 1 EC) (Poast Plus 1.5 EC)	0.28 (2.25 pt) (1.5 pt)	

Table 7-3A. Chemical Weed Control in Peanuts

Herbicide and Formulation	Pounds Active Ingredient Per Acre	Precautions and Remarks
Postemergence, Rhizome johnsongrass		
clethodim, MOA 1 (Select Max 0.97 EC) (Select 2 EC)	0.125 to 0.25 (12 to 32 fl oz) (8 to 16 fl oz)	Apply to actively growing johnsongrass before it exceeds 25 inches tall. Add 2 pints per acre of crop oil concentrate. A second application of the same rates can be made if needed before new plants or regrowth exceeds 12 inches.
sethoxydim, MOA 1 (Poast 1 EC) (Poast Plus 1.5 EC)	0.28 (2.25 pt) (1.5 pt)	
Postemergence, Suppression of large Palmer amaranth and other pigweed species that are resistant to the ALS inhibiting herbicides imazapic, chlorimuron, imazethapyr, and diclosulam		
2,4-DB, MOA 4 (Butyrac 200 2 SL) + lactofen, MOA 14 (Cobra 2 EC) or acifluorfen, MOA 14 (Ultra Blazer 2 L)	0.25 (16 fl oz) + 0.20 (12.5 fl oz) or 0.38 (1.5 pt)	Suppresses and does not completely control Palmer amaranth and other pigweed species that exceed 8 inches. Suppression of weeds exceeding 12 inches will be less than suppression of smaller weeds. Do not expect suppression to exceed 60%. Applying 2,4-DB 3 to 4 days prior to Ultra Blazer or Cobra may be more effective than tank mixtures of 2,4-DB with Ultra Blazer or Cobra. Cobra is generally more effective on larger Palmer amaranth and other pigweed species than Ultra Blazer. Apply crop oil concentrate at 1 gallon per 100 gallons water with acifluorfen or lactofen. Do not apply adjuvant with 2,4-DB alone. See product labels for comments on spray volume and effects on peanut especially during pod set and pod fill. Higher spray volumes are more effective by increasing spray coverage of the contact herbicides Ultra Blazer and Cobra.
2,4-DB, MOA 4 (Butyrac 200 2 SL) then lactofen, MOA 14 (Cobra 2 EC) or acifluorfen, MOA 14 (Ultra Blazer 2 L)	0.25 (16 fl oz) then 0.20 (12.5 fl oz) or 0.38 (1.5 pt)	Two applications of 2,4-DB spaced 10 to 14 days apart will suppress Palmer amaranth and other pigweed species. Although suppression by 2,4-DB is lower than sequential or tank mix application of 2,4-DB and acifluorfen or lactofen within 2 weeks after application, suppression by sequential applications of 2,4-DB 4 to 5 weeks after initial application is only slightly lower than suppression by sequential or tank mix application of 2,4-DB and Ultra Blazer or Cobra. For more information on managing herbicide-resistant weeds in peanut, see AG-331, *2017 Peanut Information*.
2,4-DB, MOA 4 (Butyrac 200 2 L) then 2,4-DB, MOA 4 (Butyrac 200 2 L)	0.25 (16 oz) then 0.25 (16 oz)	
paraquat, MOA 22 (Gramoxone SL)	See comments	Apply in a roller/wiper implement. Best control achieved when at least 60% coverage of weed foliage occurs. Do not allow paraquat to contact peanut foliage. Mix 1 part Gramoxone SL (other formulations may not be labeled) with 1 to 1.5 parts water to prepare 40 to 50% solution. Add nonionic surfactant at 1 quart per 100 gallons. Adjust equipment to apply up to 2 pints per acre of the herbicide-water mixture.
Postemergence, Late-season residual control of annual grasses and certain small-seeded weeds		
dimethenamid, MOA 15 (Outlook 6.0 L)	0.75 to 1 (16 to 21 fl oz)	Will not control emerged grasses or weeds; apply following a cultivation or appropriate postemergence herbicide if emerged grasses or broadleaf weeds are present. Benefit likely only on very sandy fields heavily infested with annual grasses that receive above normal rainfall during the first 4 to 5 weeks of the growing season. Lay-by of Dual Magnum, Outlook, or Warrant may also be of value in fields with a history of eclipta problems; the application must be made before eclipta emerges. Rates are on a broadcast basis; apply in an 18-inch band to row middles. See labels for preharvest intervals.
metolachlor, MOA 15 (Dual Magnum 7.62 EC)	0.64 to 0.84 (0.67 to 0.88 pt)	
acetochlor, MOA 15 (Warrant 3 ME)	0.95 to 1.5 (1.25 to 2 qt)	
Postemergence, Harvest aid for morningglory control		
Carfentrazone, MOA 14 (Aim) 2 EC	0.015 to 0.031 (1.0 to 2.0 oz)	Aim desiccates annual morningglory. Apply with 1 quart nonionic surfactant per 100 gallons water or 1 gallon crop oil concentrate per 100 gallons water. Apply within 7 days of digging and vine inversion.

Weed Response to Preplant Incorporated, Preemergence, and At-Cracking Herbicides in Peanuts

D. L. Jordan, Crop and Soil Sciences Department

Ratings based upon average to good soil and weather conditions for herbicide performance and upon proper application rate, technique, and timing.

Table 7-3B. Weed Response to Preplant Incorporated, Preemergence, At-Cracking and Postemergence Herbicides in Peanuts

Herbicides Key: PPI = Preplant Incorporated; PRE = Preemergence; AC= At-Cracking; POST = Postemergence

	Prowl or Sonalan PPI	Prowl or Sonalan + Dual Magnum or Dual PPI	Prowl or Sonalan + Outlook PPI	Dual Magnum or Dual PPI	Warrant PPI	Outlook PPI	Strongarm PPI or PRE	Prowl or Sonalan + Strongarm PPI	Dual Magnum, Dual or Outlook PPI or PRE	Pursuit PPI + POST	Dual Magnum or Dual PRE	Intrro PRE	Warrant PRE	Outlook PRE	Valor SX PRE	Prowl or Sonalan PPI + Valor SX PRE	Dual Magnum, Dual, Outlook or Warrant + Valor SX PRE	Dual Magnum or Dual AC[1] or POST[1]	Intrro AC[1] or POST[1]	Outlook AC[1] or POST[1]	Zidua AC[1] or POST[1]	Gramoxone SL AC or POST[1]	Strongarm AC[2]	Gramoxone SL + Strongarm AC[2]
Bermudagrass	N	N	N	N	N	N	N	N	N	N	N	N	N	N	N	N	N	N	N	N	N	P	N	P
Black nightshade	N	F	F	F	F	F	N	N	F	G	F	FG	FG	F	E	E	E	F	FG	F	F	PF	N	G
Broadleaf signalgrass	G	E	E	G	FG	FG	P	G	G	G	G	FG	FG	FG	P	G	FG	G	FG	FG	FG	E	N	GE
Carpetweed	G	G	G	FG	FG	FG	G	G	G	FG	FG	FG	FG	G	—	G	G	FG	FG	G	G	FG	—	G
Cocklebur	N	N	N	N	N	N	G	G	G	GE	N	N	N	N	PF	PF	PF	N	N	N	N	E	E	E
Common ragweed	N	P	PF	PF	PF	F	G	G	GE	P	PF	PF	F	FG	G	GE	PF	PF	F	F	F	F	E	E
Crabgrass	E	E	E	E	E	E	P	E	E	F	E	E	E	E	PF	E	E	E	E	E	E	G	N	G
Crowfootgrass	E	E	E	E	E	E	—	—	—	—	E	E	E	E	PF	G	G	E	E	E	E	E	N	GE
Dayflower	P	GE	—	GE	—	—	G	G	GE	—	GE	—	—	—	F	F	GE	GE	—	—	—	—	—	G
Eclipta	N	G	G	G	FG	G	GE	GE	GE	P	FG	FG	FG	FG	G	G	GE	FG	FG	FG	FG	FG	NP	FG
Fall panicum	G	E	E	E	E	E	P	E	E	PF	E	E	E	E	PF	FG	GE	E	E	E	E	E	N	GE
Florida beggarweed	N	PF	PF	F	F	F	F	F	F	P	F	F	F	F	G	GE	E	F	F	F	F	E	FG	G
Foxtails	E	E	E	E	E	E	P	E	E	G	E	E	E	E	PF	E	E	E	E	E	E	E	N	GE
Goosegrass	E	E	E	E	E	E	P	E	E	PF	E	E	E	E	PF	GE	E	E	E	E	E	E	N	GE
Jimsonweed	N	N	N	N	N	N	GE	GE	GE	G	N	N	N	N	G	G	GE	N	N	N	N	E	—	E
Johnsongrass, Seedling	G	G	G	PF	PF	PF	N	G	PF	GE	PF	PF	PF	PF	N	FG	PF	PF	PF	PF	PF	E	N	GE
Johnsongrass, Rhizome	P	PF	PF	N	N	N	N	P	N	FG	N	N	N	N	N	N	N	N	N	N	N	P	N	P
Lambsquarters	G	NG	G	F	F	FG	FG	GE	GE	FG	F	F	F	FG	GE	GE	GE	F	F	FG	FG	F	N	G
Morningglory	P	P	P	N	N	N	G	G	G	G	N	N	N	N	FG	G	G	N-P	N-P	N-P	P	F	GE	E
Nutsedge, Yellow	N	G	FG	G	N	FG	FG	FG	G	FG	FG	P	N	F	P	PF	FG	FG	P	F	F	PF	PF	G
Nutsedge, Purple	N	N	N	N	N	N	FG	FG	FG	N	N	—	N	P	P	P	N	N	N	N	N	PF	NP	PF
Pigweed	G	E	E	G	G	G	G	E	E	E	G	GE	G	GE	E	E	E	G	GE	GE	GE	E	NP	E
Prickly sida	N	P	P	P	P	P	FG	FG	FG	G	P	P	P	P	FG	G	G	P	P	P	P	F	—	G
Purslane	G	GE	GE	G	FG	G	—	G	G	—	G	G	FG	G	G	GE	GE	GE	P	P	P	—	—	—
Sicklepod	N	NP	NP	NP	NP	NP	P	P	P	P	NP	PF	NP	NP	P	PF	PF	NP	PF	NP	NP	G	N	G
Smartweed	N	N	N	N	N	N	G	G	G	G	N	N	N	N	—	—	—	N	N	N	N	G	—	E
Spurge spp.	P	F	F	PF	P	PF	—	—	—	P	F	P	P	F	G	G	G	N	N	N	N	F[2]	—	F[2]
Spurred anoda	N	N	N	N	N	N	FG	FG	FG	G	N	N	N	N	F	FG	FG	N	N	N	N	P	—	G
Texas millet	G	G	G	PF	PF	PF	P	G	PF	PF	PF	PF	PF	PF	G	F	PF	PF	PF	PF	PF	F	E	N	GE
Tropic croton	N	N	N	N	N	N	PF	PF	PF	P	N	N	N	N	—	—	—	N	N	N	N	F	—	F
Velvetleaf	N	N	N	N	N	N	GE	GE	GE	FG	N	N	N	N	F	FG	FG	N	N	N	N	F	—	FG

[1] Residual control only.

[2] Assumes weeds are 1- to 2-inches tall or smaller.

Key:
E = excellent control, 90% or better   G = good control, 80% to 90%
F = fair control, 50% to 80%   P = poor control, 25% to 50%
N = no control, less than 25%

Weed Response to Postemergence Herbicides in Peanuts

D. L. Jordan, Crop and Sciences Department

Ratings based upon average to good soil and weather conditions for herbicide performance and upon proper application rate, technique, and timing.

Table 7-3C. Weed Response to Postemergence Herbicides — Peanuts

Herbicides Key: PPI = Preplant Incorporated; PRE = Preemergence; AC = At-Cracking; POST = Postemergence

Species	Butyrac 200	Gramoxone SL[1]	Gramoxone SL + Basagran	Gramoxone SL + Storm	Basagran	Basagran + Butyrac 200	Ultra Blazer	Ultra Blazer + Butyrac 200	Ultra Blazer + Basagran[2]	Storm	Storm + Butyrac 200	Pursuit + Butyrac 200	Cadre or Impose	Cobra	Cobra + Basagran	Cobra + Basagran + Butyrac 200	Cobra + Cadre or Impose	Cobra + Pursuit	Poast or Poast Plus	Select or Select Max
Bermudagrass	N	P	P	P	N	N	N	N	P	N	N	N	N	N	N	N	N	N	FG	G
Black nightshade	N	PF	PF	G	P	P	G[1]	G[1]	G[1]	G[1]	G[1]	G	G	G[1]	G[1]	G[1]	G	G	N	N
Broadleaf signalgrass	N	GE	E	GE	N	N	NP	NP	P	NP	NP	G	G	N	N	N	G	G	E	E
Carpetweed	P	FG	FG	G	P	P	GE	E	E	G	G	FG	FG	G	G	G	G	G	N	N
Cocklebur	E	G	E	E	E	E	E	E	E	E	E	E	E	G	G	G	E	E	N	N
Common ragweed	PF	F	G	E	G[4]	G[4]	E[1]	E[1]	E[1]	E[1]	E[1]	P	PF	E	E	E	E	E	N	N
Crabgrass	N	G	G	G	N	N	N	N	N	N	N	FG	FG	N	N	N	FG	FG	GE	GE
Crowfootgrass	N	GE	G	GE	N	N	P	P	P	P	P	P	G	N	N	N	G	P	F	G
Dayflower	—	G	G	FG	G	G	—	—	G	FG	FG	—	G	—	G	G	G	—	N	N
Eclipta	P	F	F	FG	FG	FG	G	G	G	FG	FG	P	F	G	G	G	G	G	N	N
Fall panicum	N	GE	G	GE	N	N	PF	PF	P	PF	P	PF	G	N	N	N	G	PF	E	E
Florida beggarweed	P	G	GE	G	N	P	PF	PF	F	P	P	P	F	F	F	F	F	F	N	N
Foxtails	N	GE	G	GE	N	N	PF	PF	P	PF	PF	G	G	N	N	N	G	G	E	E
Goosegrass	N	GE	G	GE	N	N	N	N	N	N	N	F	N	N	N	N	F	N	GE	GE
Jimsonweed	P	G	E	E	E	E	E	E	E	E	E	G	E	E	E	E	E	E	N	N
Johnsongrass, Seedling	N	GE	GE	GE	N	N	P	P	P	P	P	GE	E	N	N	N	E	GE	E	E
Johnsongrass, Rhizome	N	P	P	P	N	N	N	N	N	N	N	F	FG	N	N	N	FG	F	G	GE
Lambsquarters	PF	F	G	G	FG	G[4]	G	G	GE	G	G	P	PF	P	FG	G	PF	P	N	N
Morningglory, Pitted	FG	F	FG	E	P	G	E	E	E	E	E	G	GE	G	G	G	GE	G	N	N
Morningglory, Others	E	F	FG	E	P	E	GE	E	E	GE	E	E	G	G	G	E	G	E	N	N
Nutsedge, Yellow	N	PF	FG	G	G[3]	G	N	N	G	F	F	F	G	N	G[3]	G[3]	G	F	N	N
Nutsedge, Purple	N	PF	PF	PF	NP	P	N	N	P	N	N	FG	G	N	P	P	G	FG	N	N
Pigweed	PF	F	G	E	N	P	E	E	E	E	E	E	E	E	E	E	E	E	N	N
Prickly sida	F	F	G	G	G	G	E	F	G	FG	G	P	G	G	G	G	G	G	N	N
Purslane	FG	—	G	G	G	G	E	E	E	GE	GE	FG	—	E	E	E	E	E	N	N
Sicklepod	G[3]	G	G	G	N	G[6]	NP	G[6]	NP	NP	G[6]	G[6]	E	P	P	G[6]	E	F	N	N
Smartweed	PF	G	E	E	E	E	GE	E	E	E	E	G	F	F	E	E	F	G	N	N
Spurge spp.	P	F[1]	F[1]	F[1]	P	E	F[1]	P	F[1]	PF[1]	PF[1]	PF[1]	—	F[1]	F[1]	F[1]	—	N	N
Spurred anoda	P	F	FG	G	P	G	GE	P	P	G	F	F	F	G	F	G	GE	G	F	N	N
Texas millet	N	GE	G	GE	N	N	NP	NP	NP	NP	NP	NP	G	P	N	N	N	G	NP	E	E
Tropic croton	PF	F	F	G	F	F	G	G	G	G	G	P	P	G	G	G	G	G	N	N
Velvetleaf	P	F	G	FG	G	G	PF	PF	FG	FG	FG	FG	G	G	G	G	G	G	N	N

[1] Assumes weeds are 1 to 2 inches tall or smaller.
[2] Assumes optimum rates and ratios of Basagran and Blazer; see labels.
[3] Two applications, 10 to 14 days apart.
[4] Assumes optimum conditions and addition of crop oil concentrate.
[5] Ratings assume weeds in one- to two-leaf stage.
[6] Assumes follow-up treatment with 2,4-DB.

Key:

E = excellent control, 90% or better
G = good control, 80% to 90%
F = fair control, 50% to 80%
P = poor control, 25% to 50%
N = no control, less than 25%

Brand Names and Formulations of Active Ingredients

Table 7-3D. Brand names and formulations of active ingredients listed in Table 7-3A.

Active ingredient(s)	Brand name(s)[a]	Formulation	Mode of Action
Acetochlor	Warrant	3 ME	15
Acifluorfen	Ultra Blazer		22
Bentazon	Basagran		
Carfentrazone	Aim EC	2 EC	14
Clethodim	Intensity One, Select Max, Shadow Ultra, TAPOUT	0.97 EC	1
	Arrow 2EC, Agri Star Clethodim 2EC, AmTide Clethodim 2 EC, Avatar, Avatar S2, Clethodim, Dakota, Grassout Max, Intensity, PS Clethodim, Section 2 EC, Select, Shadow, Tide Clethodim, Volunteer, Willowood Clethodim	2 EC	1
	Shadow 3 EC, Section Three	3 EC	1
Dimethenamid-P	Outlook	6 EC	15
Flumioxazin	Outflank, Panther, Rowel, Valor SX, Warfox	51 WDG	14
Lactofen	Cobra	2 EC	14
Paraquat	Gramoxone SL, Gramoxone Inteon	2 S	22
	Bonedry, Devour, Firestorm, Helmquat, Paraquat Concentrate, Para-Shot, Parazone, Willowood Paraquat	3 S	22
Pendimethalin	Acumen, Framework, Helena Pendimethalin, Pendipro 3.3 EC, Prowl 3.3 EC, Stealth	3.3 EC	3
	Prowl H_2O, Satellite Hydrocap	3.8 AS	3
Pyroxasulfone	Zidua	85 WG, 4.25 SC	15
S-metolachlor	Brawl, Dual Magnum, Medal	7.62 EC	15
	Brawl II, Cinch, Dual II Magnum, Medal II	7.64 EC	15
Metolachlor	Parallel PCS, Stalwart	8 EC	15
Sulfentrazone + carfentrazone	Spartan Charge		14
Sethoxydim	Nufarm Sethoxydim SPC, Poast Plus	1 EC	1
	Poast	1.5 EC	1
2,4-DB	Albaugh Butyrac 175 Broadleaf Herbicide, 2,4-DB 175 Herbicide	1.75 SC	4
	Albaugh Butyrac 200 Broadleaf Herbicide, 2,4-DB 200 Herbicide, 2,4-DB 200	2 SC	4

[a] Brands listed were registered for sale and use in peanut in North Carolina in 2015 according to http://www.kellysolutions.com/nc/searchbychem.asp

[b] AS, aqueous suspension; CS, capsule suspension; EC, emulsifiable concentrate; F, flowable; ME, microencapsulated; S, solution; SC, suspension concentrate; WDG, water dispersible granule.

Chapter VII—2019 N.C. Agricultural Chemicals Manual

Chemical Weed Control in Sorghum

W. J. Everman, Crop and Soil Sciences Department

NOTE: A mode of action code has been added to the Herbicide and Formulation column of this table. Use MOA codes for herbicide resistance management. See Table 7-10, Herbicide Resistance Management, for details.

Table 7-4. Chemical Weed Control in Sorghum

Herbicide, Mode of Action Code[1], and Formulation	Amount of Formulation Per Acre	Pounds Active Ingredient Per Acre	Precautions and Remarks
Grain Sorghum No-Till Burndown, Emerged annual broadleaf and grass weeds, suppression or control of perennials			
glyphosate, MOA 9 (numerous brands and formulations)	See label	0.56 to 1.13 (lb a.e.)	Glyphosate is available as an isopropylamine salt and a potassium salt. Glyphosate formulations and application rates should be compared on the basis of pounds of glyphosate acid equivalent (a.e.) per gallon and per acre, respectively. Rate in the preceding column is expressed as a.e. See Table 7-10 for glyphosate rate conversions. Apply before crop emerges. Glyphosate rate depends upon weed species and weed size; see labels for suggested rates. Higher rates can be applied for perennial weeds; see labels for details. See comments on labels concerning nitrogen as the carrier. Apply in 10 to 20 gallons of water per acre using flat fan nozzles. For residual grass and broadleaf weed control, glyphosate products may be tank mixed with most preemergence herbicides. See the section on Grain Sorghum—Preemergence. Refer to specific product labels for application rates, weeds controlled, application directions, and precautions. Adjuvant recommendations vary according to the glyphosate product used. See label of brand used for specific recommendations.
Grain Sorghum No-Till Burndown, Emerged annual broadleaf and grass weeds, top-kill of perennials			
paraquat, MOA 22 (Gramoxone Inteon) 2 SL	2 to 4 pt	0.5 to 1	Apply before, during, or after planting but before crop emerges using clean water or clear fertilizer solution as the carrier. Apply in a minimum of 10 GPA (20 to 40 preferred) using flat fan nozzles. Add either a nonionic surfactant at 1 pint per 100 gallons or crop oil concentrate at 1 gallon per 100 gallons. Use 0.5 to 0.64 pound a.i. on weeds 1 to 3 inches, 0.75 pound a.i. on weeds 3 to 6 inches, and 1 pound a.i. on weeds 6 inches or larger. Use 0.5 pound a.i. for rye cover crop or 0.75 pound a.i. for wheat cover crop. Rainfast within 30 minutes. For residual grass and broadleaf weed control, paraquat can be tank mixed with most preemergence sorghum herbicides and herbicide combinations. See the section on Grain Sorghum—Preemergence, Conventionally Planted. Refer to specific product labels for application rates, weeds controlled, application directions, and precautions. Better control of emerged weeds will be obtained with tank mixtures of Gramoxone plus an atrazine-containing product. Generic brands of paraquat containing 3 pounds active per gallon may be applied at 1.3 to 2.7 pints.
Grain Sorghum No-Till Burndown or Preemergence			
saflufenacil, MOA 14 (Sharpen)	1.0 to 2.0 fl oz	0.027 to 0.054 (lb a.i.)	Sharpen can be applied to control glyphosate-resistant marestail prior to grain sorghum emergence. See label for application with other herbicides and specifics on adjuvant selection. To avoid injury potential with burndown or preemergence applications, consult local seed company for possible injury of grain sorghum hybrids or varieties.
thifensulfuron, MOA 2 + tribenuron, MOA 2 (FirstShot SG with TotalSol) 50 WDG	0.5 to 0.8 oz	0.008 to 0.013 + 0.008 to 0.013	FirstShot should be applied 14 days prior to sorghum planting. Tank mix with glyphosate or paraquat for broad spectrum burndown. See label for specific rates and soil type restrictions. See label for adjuvant recommendations.
Grain Sorghum Preemergence, Annual broadleaf weeds and certain annual grasses			
atrazine, MOA 5 (AAtrex) 4 F (AAtrex Nine-O) 90 WDG	1 to 2 qt 1.1 to 2.2 lb	1 to 2	Controls most broadleaf weeds and large crabgrass, crowfootgrass, foxtails, goosegrass, and sandbur. Does not control broadleaf signalgrass, fall panicum, Texas panicum, seedling johnsongrass, or shatter-cane. Do not use on sand, loamy sand, or sandy loam soils. Do not use on medium- or fine-textured soils with less than 1% organic matter. On highly erodible soils (defined by NRCS) with less than 30% plant residue cover, do not exceed 1.6 pounds active ingredient. See labels for details on set-back requirements from streams and lakes. See labels for comments on rotational crops. For improved grass control, atrazine may be tank mixed with s-metolachlor, alachlor, or dimethenamid if the seed have been properly treated with a safener. See comments for s-metolachlor, alachlor, or dimethenamid applied preemergence. Generic brands of atrazine are available.
Grain Sorghum Preemergence, Annual grasses and small-seeded broadleaf weeds			
alachlor, MOA 15 (Intrro) 4 EC (Micro-Tech) 4 FME	1.5 to 2.5 qt	1.5 to 2.5	Controls most annual grasses and pigweed. Does not control seedling johnsongrass, shattercane, or Texas panicum. Use only on grain sorghum planted with seed properly treated with a safener containing the active ingredient flurazole. Rate depends upon soil texture; see label for details. May be tank mixed with atrazine for broadleaf weed control. See comments for atrazine applied preemergence.
dimethenamid-P, MOA 15 (Outlook) 6.0 EC	12 to 21 fl oz	0.56 to 0.98	Use 12 to 18 fluid ounces on soils with less than 3% organic matter or 14 to 21 fluid ounces on soils with greater than 3% organic matter. Controls most annual grasses and pigweed. Does not control seedling johnsongrass, shattercane, or Texas panicum. Use only on grain sorghum planted with seed properly treated with Concep or Screen protectant. Rate depends upon soil texture and organic matter; see label for details. May be tank mixed with atrazine for broadleaf weed control. See comments for atrazine applied preemergence.
metolachlor, MOA 15 (Me-Too-Lachlor) 8 EC (Parallel) 7.8 EC (Parallel PCS) 8 EC (Stalwart) 8 EC	1 to 1.67 pt	1 to 1.67	See comments for s-metolachlor products. Products containing s-metolachlor are more active on weeds per unit of formulated product than those containing metolachlor. In general, it takes 1.5 pints of a metolachlor product to get the activity one would get from 1 pint of an s-metolachlor product.
S-metolachlor, MOA 15 (Brawl) 7.62 EC (Brawl II) 7.64 EC (Cinch) 7.64 EC (Dual Magnum) 7.62 EC (Dual II Magnum) 7.64 EC (Medal) 7.62 EC (Medal II) 7.64 EC	1 to 1.67 pt	0.95 to 1.6	Controls most annual grasses and pigweed. Does not control seedling johnsongrass, shattercane, or Texas panicum. Use only on grain sorghum planted with seed properly treated with Concep or Screen protectant. Rate depends upon soil texture and organic matter; see label for details. May be tank mixed with atrazine for broadleaf weed control. See comments for atrazine applied preemergence.

Chapter VII—2019 N.C. Agricultural Chemicals Manual

Table 7-4. Chemical Weed Control in Sorghum

Herbicide, Mode of Action Code[1], and Formulation	Amount of Formulation Per Acre	Pounds Active Ingredient Per Acre	Precautions and Remarks
Grain Sorghum Preemergence, Annual grasses and broadleaf weeds			
alachlor, MOA 15 + atrazine, MOA 5 (Bullet) 4 FME (Lariat) 4 F	2.5 to 4 qt	1.56 to 2.5 + 0.94 to 1.5	Controls most annual grasses and broadleaf weeds. Does not control seedling johnsongrass, shattercane, or Texas panicum. Use only on grain sorghum planted with seed properly treated with Concep or Screen protectant. Rate depends upon soil texture and organic matter; see label for details. See label for comments on rotational crops and details on set-back requirements from streams and lakes.
dimethenamid-P, MOA 15 + atrazine, MOA 5 (Guardsman Max) 5F	2.5 to 4.6 pt	0.5 to 1 + 1 to 1.9	Controls most annual grasses and broadleaf weeds. Does not control seedling johnsongrass, shattercane, or Texas panicum. Use only with Concep-treated seed. Apply only to medium- or fine-textured soils. Rate depends on soil texture and organic matter; see label for details. See label for comments on rotational crops and set-back requirements from streams and lakes. May be applied postemergence to sorghum up to 12 inches tall.
S-metolachlor, MOA 15 + atrazine, MOA 5 (Bicep II Magnum) 5.5 F (Brawl II ATZ) 5.5 F (Cinch ATZ) 5.5 F (Medal II AT) 5.5 F	1.6 to 2.1 qt	0.96 to 1.26 + 1.24 to 1.63	Controls most annual grasses and broadleaf weeds. Does not control seedling johnsongrass, shattercane, or Texas panicum. Use only with Concep-treated seed. Apply only to medium- and fine-textured soils with at least 1% organic matter. See label for comments on rotational crops and details on set-back requirements from streams and lakes. May be applied postemergence to sorghum up to 12 inches tall.
dimethenamid-P, MOA 15 + saflufenacil, MOA 14 (Verdict) + dimethenamid-P, MOA 15 (Outlook) 6.0 EC	10 to 18 fl oz + + 4 to 12 fl oz	0.045 to 0.08 + 0.4 to 0.7 + 0.19 to 0.56	Tank mix with Outlook at 4 to 10 fluid ounces on coarse soils or 6 to 12 fluid ounces on medium to fine soils to ensure grass control. The tank mix controls most annual grasses and broadleaf weeds. Does not control seedling johnsongrass, shattercane, or Texas panicum. Use only with Concep-treated seed. Apply only where organic matter is greater than 1.5%. Rate depends on soil texture; see label for details. See label for comments on rotational crops. DO NOT apply to emerged sorghum, or severe injury may occur.
Grain Sorghum Postemergence, Annual grass and small broadleaf weeds			
quinclorac, MOA 4, 26 (Facet L) 1.5 L	22 to 32 fl oz	0.26 to 0.375	Apply from preemergence up to 12 inches tall. Grass and broadleaf weeds must be under 2 inches tall. Controls small barnyardgrass, broadleaf signalgrass, large crabgrass, and foxtail species. See label for list of broadleaf weeds controlled. Add 1 quart per acre of crop oil concentrate or 1 to 2 pints per acre of methylated seed oil. May be tankmixed with atrazine, 2,4-D, dicamba, Peak, or Buctril. See label for rotation restrictions.
Grain Sorghum Postemergence, Annual broadleaf weeds			
atrazine, MOA 5 (AAtrex) 4 F (AAtrex Nine-O) 90 WDG	1.2 qt 1.3 lb	1.2	Apply after sorghum reaches the three-leaf stage but before it exceeds 12 inches tall. Do not use on sand or loamy sand soil. Broadleaf weeds must be 4 inches tall or less. See label for list of weeds controlled. Add 1 quart per acre of crop oil concentrate. If a postemergence application is made following an at-planting application, do not exceed a total of 2.5 pounds active ingredient per acre per season. See label for details on set-back requirements from streams and lakes. Generic brands of atrazine are available.
bentazon, MOA 6 (Basagran) 4 SL	1.5 to 2 pt	0.75 to 1	Apply overtop or directed any time prior to heading. See label for weeds controlled and recommended weed size for treatment. Adding crop oil concentrate at 1 to 2 pt per acre will improve control. Do not apply more than 2 pints Basagran per acre per season. Basagran also controls or suppresses yellow nutsedge. May be tank mixed with atrazine. When tank mixing, see respective labels for application rates, directions, and precautions.
bromoxynil, MOA 6 (Buctril) 2 EC (Buctril 4 EC) 4 EC	1.5 pt 0.75 pt	0.375	Can apply overtop of sorghum from the four-leaf stage until the preboot stage. Use of drop nozzles is suggested after sorghum is 6 to 8 inches tall to ensure better weed coverage. An adjuvant is not needed. Controls cocklebur, morningglory, lambsquarters, ragweed, jimsonweed, smartweed, velvetleaf, and very small pigweed. See label for recommended weed size for treatment. Do not apply when sorghum foliage is wet. May be tank mixed with atrazine, Banvel, Clarity, or 2,4D. When tank mixing, see respective labels for application rates, directions, and precautions.
bromoxynil, MOA 6 + pyrasulfotole MOA 27 (Huskie) 2.06 EC	13 to 16 oz	0.21 to 0.26	Apply overtop between 3 leaf stage up to 30 inches and/or flag leaf emergence, whichever comes first. Best control occurs when weeds are 4 inches tall or less. Controls cocklebur, morningglory, lambsquarters, ragweed, jimsonweed, smartweed, velvetleaf, and pigweed. Transitory leaf burn will occur after Huskie application to grain sorghum. Stunting and yellowing can also occur; however these symptoms generally dissipate within 21 days and do not affect yield. May be tank mixed with atrazine, 2,4-D or dicamba as needed for additional broadleaf control. May also be tankmixed with Bicep II Magnum, Dual II Magnum, Guardsman Max, Outlook, Starane, and Warrant for additional control. When tank mixing, see respective labels for application rates, directions, and precautions.
carfentrazone, MOA 14 (Aim) 2 EC	0.5 fl oz	0.008	Apply from sorghum emergence through six-leaf stage. Add nonionic surfactant according to label directions. Controls small lambsquarters, morningglory, pigweed. Aim at rates up to 1 fluid ounce can be applied with drop nozzles.
dicamba, MOA 4 (Banvel) 4 S L (Clarity) 4 SL	0.5 pt 0.5 pt	0.25	Apply from spike stage until sorghum is 8 inches tall. May be tank mixed with atrazine or Buctril. When tank mixing, see respective labels for application rates, directions, and precautions. **Carefully follow all precautions on labels to avoid drift to sensitive crops.**
dicamba, MOA 4 + atrazine, MOA 5 (Marksman) 3.2 F	2 pt	0.20 + 0.53	Controls most broadleaf weeds. Apply when sorghum has two to five leaves (about 2 to 8 inches tall). Do not add surfactant or crop oil. Do not apply in vicinity of dicamba-sensitive crops. See label for details on set-back requirements from streams and lakes.
prosulfuron, MOA 2 (Peak) 57 WDG	0.75 to 1 oz	0.027 to 0.036	Controls pigweed, lambsquarters, cocklebur, morningglory, jimsonweed, ragweed, smartweed, sicklepod, and velvetleaf. Apply to sorghum 5 to 30 inches tall. Use drop nozzles if sorghum is over 20 inches Add nonionic surfactant at 1 quart per 100 gallons or crop oil concentrate at 1 quart per acre. See label for rotational restrictions. May tank mix with atrazine, Banvel, Buctril, Marksman, or 2,4-D. See labels for details. See Peak label for rotational restrictions.
2,4-D amine formulation, MOA 4 (various brands) 3.8 SL	0.5 pt	0.24	Can apply overtop of sorghum 6 to 15 inches tall. Wait until secondary roots are well established. Sorghum is less tolerant of 2,4-D than is corn. Use drop nozzles as soon as possible and certainly after sorghum is 8 inches tall. Note that 2,4-D rates listed here are less than rates on most labels. Less than label-recommended rates are suggested to avoid injury to the crop. Do not apply during boot, flowering, or early dough stages. May be applied in nitrogen solution at lay-by. When mixing 2,4-D amine in nitrogen solution, add 1 pint of 2,4-D amine to 4 pints of water and mix. Then add this mixture to the nitrogen solution in the spray tank with considerable agitation until thoroughly mixed. Do not allow mixture to stand in sprayer. **Use extreme caution to avoid drift to sensitive crops such as cotton and tobacco. Ester formulations of 2,4D may be applied to sorghum. However, use of ester formulations of 2,4-D or acid/ester mixtures, such as Weedone 638, is not suggested if sensitive crops are located within 1 mile of the sorghum.**

Table 7-4. Chemical Weed Control in Sorghum

Herbicide, Mode of Action Code[1], and Formulation	Amount of Formulation Per Acre	Pounds Active Ingredient Per Acre	Precautions and Remarks
Grain Sorghum Postemergence-Directed, Annual grass and broadleaf weeds			
linuron, MOA 7 (Linex) 4 L	1 to 2 pt	0.5 to 1	Apply as directed spray in 25 to 40 gallons per acre of water. Add 1 pint of nonionic surfactant per 25 gallons of spray mixture. For application with precision directed equipment, apply 0.5 pound active per acre when sorghum is 12 inches tall and weeds are up to 2 inches tall. Apply 0.5 to 1 pound active per acre when sorghum is 15 inches tall and weeds are 2 to 4 inches tall.
Forage Sorghum Preemergence, Annual broadleaf weeds and certain annual grasses			
atrazine, MOA 5 (AAtrex) 4 F (AAtrex Nine-O) 90 WDG	1 to 2 qt 1.1 to 2.2 lb	1 to 2	Controls most broadleaf weeds and large crabgrass, crowfootgrass, foxtails, goosegrass, and sandbur. Does not control broadleaf signalgrass, fall panicum, Texas panicum, seedling johnsongrass, or shatter-cane. Do not use on sand, loamy sand, or sandy loam soils. Do not use on medium- or fine-textured soils with less than 1% organic matter. Do not exceed 1.6 pounds active ingredient per acre on highly erodible soils (as defined by the NRCS) with less than 30% plant residue cover. See labels for comments on rotational crops. See label for details on set-back requirements from streams and lakes. For improved grass control, atrazine may be tank mixed with Cinch, Dual Magnum, or Dual II Magnum; see comments for Cinch, Dual Magnum, or Dual II Magnum applied preemergence. Generic brands of atrazine are available.
metolachlor, MOA 15 (Me-Too-Lachlor) 8 EC (Parallel) 7.8 EC (Parallel PCS) 8 EC (Stalwart) 8 EC	1 to 1.67 pt	1 to 1.67	See comments for s-metolachlor products. Products containing s-metolachlor are more active on weeds per unit of formulated product than those containing metolachlor. In general, it takes 1.5 pints of a metolachlor product to get the activity one would get from 1 pint of an s-metolachlor product.
S-metolachlor, MOA 15 (Brawl) 7.62 EC (Brawl II) 7.64 EC (Cinch) 7.64 EC (Dual Magnum) 7.62 EC (Dual II Magnum) 7.64 EC (Medal) 7.62 EC (Medal II) 7.62 EC	1 to 1.67 pt	0.95 to 1.6	Controls most annual grasses and pigweed. Does not control seedling johnsongrass, shattercane, or Texas panicum. Use only on sorghum planted with seed properly treated with Concep or Screen protectant. Rate depends upon soil texture and organic matter; see label for details. May be tank mixed with atrazine for broadleaf weed control. See comments for atrazine applied preemergence.
S-metolachlor, MOA 15 + atrazine, MOA 5 (Bicep II Magnum) 5.5 F (Brawl II ATZ) 5.5 F (Cinch ATZ) 5.5 F (Medal II AT) 5.5 F	1.6 to 2.1 qt	0.96 to 1.26 + 1.24 to 1.63	Controls most annual grasses and broadleaf weeds. Does not control seedling johnsongrass, shattercane, or Texas panicum. Use only with Concep- or Screen-treated seed. Apply only to medium- and fine-textured soils with at least 1% organic matter. See label for comments on rotational crops and details on set-back requirements from streams and lakes.
Forage Sorghum Postemergence, Annual broadleaf weeds			
atrazine, MOA 5 (AAtrex) 4 F (AAtrex Nine-O) 90 WDG	1.2 qt 1.3 lb	1.2	Apply after sorghum reaches the three-leaf stage but before it exceeds 12 inches tall. Do not use on sand or loamy sand soil. Broadleaf weeds must be 4 inches tall or less. See label for list of weeds controlled. Add 1 quart per acre of crop oil concentrate. If a postemergence application is made following an at-planting application, do not exceed a total of 2.5 pounds active ingredient per acre per season. Do not graze or feed forage from treated areas for 21 days following application. See label for details on set-back requirements from streams and lakes. Generic brands of atrazine are available.
bentazon, MOA 6 (Basagran) 4 SL	1.5 to 2 pt	0.75 to 1	Apply overtop or directed any time prior to heading. See label for weeds controlled and recommended weed size for treatment. Adding crop oil concentrate at 1 to 2 pints per acre will improve control. Do not apply more than 2 pints per acre per season. Basagran also controls or suppresses yellow nutsedge. May be tank mixed with atrazine. When tank mixing, see respective labels for application rates and directions and precautions. Do not graze treated fields for at least 12 days following Basagran application.
bromoxynil, MOA 6 (Buctril) 2 EC (Buctril 4 EC) 4 EC	1.5 pt 0.75 pt	0.375	Can apply overtop of sorghum from the four-leaf stage until the preboot stage. See label for weeds controlled and recommended weed size for treatment. Do not apply when sorghum foliage is wet. May be tank mixed with atrazine, Banvel, Clarity, or 2,4-D. When tank mixing, see respective labels for application directions, precautions, and weeds controlled. Do not cut for feed or fodder or graze within 30 days of application.
carfentrazone, MOA 14 (Aim) 2 EC	0.5 fl oz	0.008	Apply from sorghum emergence through six-leaf stage. Add nonionic surfactant according to label directions. Controls small lambsquarters, morningglory, and pigweed.
dicamba, MOA 4 (Banvel) 4 SL (Clarity) 4 SL	0.5 pt 0.5 pt	0.25	Apply from spike stage until sorghum is 8 inches tall. May be tank mixed with atrazine or Buctril. When tank mixing, see respective labels for application rates and directions and precautions. Do not cut for silage prior to mature grain stage. Do not remove animals from treated areas for slaughter prior to 30 days after application. For lactating dairy animals, wait 7 days before grazing or 37 days before harvest for hay. There is no waiting period between treatment and grazing for non-lactating animals. **Carefully follow all precautions on labels to avoid drift to sensitive crops.**
dicamba, MOA 4 + atrazine, MOA 5 (Marksman) 3.2 F	2 pt	0.28 + 0.53	Controls most broadleaf weeds. Apply when sorghum has two to five leaves (is about 2 to 8 inches tall). Do not add surfactant or crop oil. Do not apply in the vicinity of dicamba-sensitive crops. See label for set-back requirements from streams and lakes.
2,4-D amine formulation, MOA 4 (various brands) 3.8 SL	0.5 pt	0.24	Can apply overtop of sorghum 6 to 15 inches tall. Wait until secondary roots are well established. Sorghum is less tolerant of 2,4-D than is corn. Note that 2,4-D rates listed here are less than rates on most labels. Less than label-recommended rates are suggested to avoid injury to the crop. Do not apply during boot, flowering, or early dough stages. Do not forage or feed sorghum fodder for 7 days following application. **Use extreme caution to avoid drift to sensitive crops such as cotton and tobacco. Ester formulations of 2,4D may be applied to sorghum. However, use of ester formulations of 2,4-D or acid/ester mixes, such as Weedone 638, is not suggested if sensitive crops, especially cotton and tobacco, are located within 1 mile of the sorghum.**

[1] Mode of Action (MOA) code developed by the Weed Science Society of America. See Table 7-10, Herbicide Resistance Management, for details.

Chemical Weed Control in Soybeans
W. J. Everman, Crop and Soil Sciences Department

NOTES: A mode of action code has been added to the Herbicide and Formulation column of this table. Use MOA codes for herbicide resistance management. See Table 7-10, Herbicide Resistance Management, for details.

Control of witchweed is part of the State/Federal Quarantine Program. Contact the N.C. Department of Agriculture, Plant Industry Division, at 1-800-206-9333.

Table 7-5A. Chemical Weed Control in Soybeans

Herbicide, Mode of Action Code[1], and Formulation	Amount of Formulation Per Acre	Pounds Active Ingredient Per Acre	Precautions and Remarks
Preplant (foliar application), Conventional or Reduced Tillage, Control or suppression of emerged weeds to reduce tillage operations			
glyphosate, MOA 9 (numerous brands and formulations)	See label	0.38 to 1.13 (lb a.e.)	Glyphosate is available as an isopropylamine salt and a potassium salt. Glyphosate formulations and application rates should be compared on the basis of pounds of glyphosate acid equivalent (a.e.) per gallon and per acre, respectively. Rate in the preceding column is expressed as a.e. See Table 7-10 for glyphosate rate conversions. Recommended rates depend upon weed species and size; see labels for details. Higher rates can be used for specific situations. Delay tillage at least 3 days after application. Adjuvant recommendations vary by glyphosate brand; follow directions on label of brand used. May add 0.75 to 1 pint of 2,4-D for improved control of specific broadleaf weeds. Delay planting at least 7 days after application of ester formulations of 2,4-D or 15 days after application of amine formulations. Use only a brand of 2,4-D with the preplant application included on the label. Follow all precautions on the 2,4-D label. Use of an ester formulation of 2,4-D is discouraged within 1 mile of cotton.
Preplant Incorporated, Annual grasses			
ethalfluralin, MOA 3 (Sonalan) 3 EC	1.5 to 3 pt	0.56 to 1.12	Controls common annual grasses plus pigweed and lambsquarters. Incorporate in top 2 to 3 inches of seedbed within 2 days of application; immediate incorporation suggested. For broadleaf weed control, Sonalan may be tank mixed with most broadleaf herbicides. When tank mixing, see respective labels for application rates, weeds controlled, specific application directions, and precautions.
metolachlor, MOA 15 (Me-Too-Lachlor) 8 EC (Parallel PCS) 8 EC (Parrlay) 8 EC (Stalwart) 8 EC	1 to 2 pt	1 to 2	See comments for s-metolachlor products. Products containing s-metolachlor are more active on weeds per unit of formulated product than those containing metolachlor. In general, it takes 1.5 pints of metolachlor product to get the activity one would get from 1 pint of s-metolachlor product.
pendimethalin, MOA 3 (Prowl) 3.3 EC (Prowl H$_2$O) 3.8 L	1.2 to 3.6 pt 1.5 to 3 pt	0.5 to 1.5 0.7 to 1.4	Controls common annual grasses plus pigweed and lambsquarters. Incorporate in top 2 to 3 inches of seedbed within 7 days of application; immediate incorporation suggested. For broadleaf weed control, pendimethalin may be tank mixed with most broadleaf herbicides. When tank mixing, see respective labels for application rates, weeds controlled, specific application directions, and precautions. Generic brands are available.
S-metolachlor, MOA 15 (Brawl) 7.62 EC (Cinch) 7.64 EC (Dual Magnum) 7.62 EC (Dual II Magnum) 7.64 EC (Medal) 7.62 EC (Medal II) 7.64 EC	1 to 2 pt	0.95 to 1.91	Controls annual grasses and pigweed. At higher rates, controls nightshade and yellow nutsedge. Better yellow nutsedge control if incorporated; see labels for incorporation details. Except for yellow nutsedge, preemergence application preferred. Does not adequately control Texas panicum, seedling johnsongrass, and shattercane. Read labels and adjust rates for soil texture and organic matter. These herbicides may be applied at rates up to 2.5 pints on soils with 6% to 20% organic matter. For broadleaf weed control, Smetolachlor may be tank mixed with most broadleaf herbicides; do not mix with Valor. When tank mixing, see respective labels for application rates, weeds controlled, specific application directions, and precautions.
trifluralin, MOA 3 (Treflan) 4 EC (Treflan HFP) 4 EC	1 to 2 pt	0.5 to 1	Controls common annual grasses plus pigweed and lambsquarters. Incorporate in top 2 to 3 inches of seedbed within 8 hr of application; immediate incorporation suggested. For broadleaf weed control, Treflan may be tank mixed with most broadleaf herbicides. When tank mixing, see respective labels for application rates, weeds controlled, specific application directions, and precautions. Generic brands are available.
Preplant Incorporated, Annual broadleaf weeds			
imazaquin, MOA 2 (Scepter) 70 WDG	2.8 oz	0.123	Controls most broadleaf weeds; a follow-up post-emergence herbicide application often needed for adequate sicklepod control. Follow all precautions on the label, including rotational restrictions. For annual grass control, Scepter may be tank mixed with alachlor, pendimethalin, S-metolachlor, or trifluralin. When tank mixing, see respective labels for application rates, weeds controlled, specific application directions, and precautions.
metribuzin, MOA 5 (Sencor) 75 WDG	0.33 to 0.67 lb	0.25 to 0.5	Controls many broadleaf weeds. Will not adequately control cocklebur or morningglory. Acceptable control of sicklepod may require a follow-up postemergence herbicide application. Activity of metribuzin is highly dependent upon soil texture and organic matter. Follow label directions for application rates, soil type restrictions, etc. Do not use on sand with less than 1% organic matter. Do not use on loamy sand or sandy loam soils with less than 0.5% organic matter. Some varieties are particularly sensitive to metribuzin; see labels for details. Soybeans may be injured when metribuzin is applied to soil treated with organophosphate insecticides and/or nematicides; see precautions on label. For annual grass control, Sencor may be tank mixed with alachlor, pendimethalin, S-metolachlor, or trifluralin. When tank mixing see respective labels for application rates, weeds controlled, specific application directions, and precautions.
dimethenamid-P, MOA 15 (Outlook) 6.0 EC	14 to 21 fl oz	0.66 to 0.98	Incorporate 2 inches deep. Not effective on purple nutsedge. Follow label carefully for use rates on various soil types. Do not apply to sandy soils if organic matter is less than 3% and depth to groundwater is 30 feet or less.
metolachlor, MOA 15 (Me-Too-Lachlor) 8 EC (Parallel PCS) 8 EC (Parrlay) 8 EC (Stalwart) 8 EC	1 to 2 pt	1 to 2	See comments for s-metolachlor products. Products containing s-metolachlor are more active on weeds per unit of formulated product than those containing metolachlor. In general, it takes 1.5 pints of a metolachlor product to get the activity one would get from 1 pint of an s-metolachlor product.
S-metolachlor, MOA 15 (Brawl) 7.62 EC (Cinch) 7.64 EC (Dual Magnum) 7.62 EC (Dual II Magnum) 7.64 EC (Medal) 7.62 EC (Medal II) 7.64 EC	1.33 to 2 pt	1.27 to 1.91	Incorporate 2 inches deep. Not effective on purple nutsedge.

Table 7-5A. Chemical Weed Control in Soybeans

Herbicide, Mode of Action Code[1], and Formulation	Amount of Formulation Per Acre	Pounds Active Ingredient Per Acre	Precautions and Remarks
Burndown, No-Till Planting, Emerged grass and broadleaf weeds			
glyphosate, MOA 9 (numerous brands and formulations)	See label	0.56 to 1.13 (lb a.e.)	Glyphosate is available as an isopropylamine salt and a potassium salt. Glyphosate formulations and application rates should be compared on the basis of pounds of glyphosate acid equivalent (a.e.) per gallon and per acre, respectively. The rate in the preceding column is expressed as a.e. See TABLE 7-10 for glyphosate rate conversions. Apply before crop emergence. Rate depends upon weed species and size; see labels for details. Higher rates (up to 3.75 pounds acid equivalent) may be used for perennial weeds. Adjuvant recommendations vary by glyphosate brand. See label of brand used for specific recommendations. For residual grass and broadleaf weed control, glyphosate may be tank mixed with most preemergence soybean herbicides. Refer to the label of the tank mix partner for application rates, directions, limitations, weeds controlled, and precautions.
glyphosate, MOA 9 + fomesafen, MOA 14 (Flexstar GT) 3.29 L	3 to 4.5 pt	1 to 1.55 (lb a.e.) 0.25 to 0.37	Apply before crop emergence. See label for adjuvant suggestions. May mix with 2,4-D or dicamba for improved burndown of specific weeds. See waiting intervals between application and planting on labels for 2,4-D or dicamba. Do not exceed 4.5 pints per acre of Flexstar GT per year. Also, do not exceed 0.375 pound a.i. of fomesafen per year from all sources.
paraquat, MOA 22 (Gramoxone Inteon) 2 SL	2 to 4 pt	0.5 to 1	Apply before crop emergence. Use 2 pints on weeds 1 to 3 inches, 3 pints on weeds 3 to 6 inches, and 4 pints on weeds 6 inches or taller. Use 2 pints for rye cover crop and 2.5 to 3 pints on wheat cover crops. Add crop oil concentrate or nonionic surfactant according to label directions. Generic brands of paraquat containing 3 pounds active per gallon are available. Apply these products at two-thirds of the rates mentioned here. Residual herbicides for grass and broadleaf weed control may be tank mixed with Gramoxone. Control of cutleaf eveningprimrose, wild radish, and most broadleaf weeds will be increased by adding 2,4-D at 0.75 to 1 pint. Delay planting at least 7 days after application of ester formulations of 2,4-D or 15 days after application of amine formulations of 2,4D. Use of ester formulations is discouraged if sensitive crops, especially cotton and tobacco, are located within 1 mile.
Burndown, No-Till Planting, Cutleaf eveningprimrose, wild radish, and vetch, plus other weeds controlled by glyphosate			
glyphosate, MOA 9 (numerous brands and formulations) + 2,4-D, MOA 4 (numerous brands and formulations) 3.8 SL	See label + 0.5 to 1 pt	0.56 to 1.13 (lb a.e.) + 0.24 to 0.48	See comments for glyphosate alone. Apply ester formulations of 2,4-D at least 7 days ahead of planting. Apply amine formulations of 2,4-D at least 15 days ahead of planting. Plant soybeans at least 1 inch deep. See comments on 2,4-D labels concerning use on coarse-textured soils with less than 1% organic matter. For 2,4-D formulations other than 3.8 pounds per gallon, adjust rate accordingly. Use 0.5 pint 2,4-D for primrose; use 1 pint for other weeds. Use of ester formulations is discouraged if sensitive crops, particularly cotton or tobacco, are located within 1 mile.
pyraflufen-ethyl, MOA 14 (ET) 1 SL	0.5 to 2.0 fl oz	0.003 to 0.015 (lb a.i.)	ET can be used for limited suppression of small emerged summer annual and winter weeds. See label for adjuvant and spray volume recommendations Research with ET is limited in North Carolina.
Burndown, No-Till Planting, Glyphosate-resistant horseweed plus other weeds			
glyphosate, MOA 9 (numerous brands and formulations) + 2,4-D, MOA 4 (numerous brands and formulations) 3.8 SL + flumioxazin, MOA 14 (Valor SX) 51 WDG	See label + 1.5 to 2 pt + 2 to 3 oz	0.56 to 1.13 (lb a.e.) + 0.71 to 0.95 + 0.064 to 0.096	Glyphosate-resistant horseweed (marestail) is relatively common in eastern North Carolina, and continued spread is anticipated. See comments for glyphosate alone. Acceptable control of glyphosate-resistant horseweed requires both a residual herbicide (Valor SX, Valor XLT, Envive, Leadoff, or Trivence) and either 2,4-D or Clarity. An alternative approach would be application of glyphosate plus either 2,4-D or Clarity preplant followed by Gramoxone plus a residual herbicide at planting. Do NOT till or otherwise disturb the soil surface following application of Valor SX, Valor XLT, Envive, Leadoff, or Trivence. 2,4-D rates suggested for horseweed should be applied at least 30 days ahead of planting. For 2,4-D formulations other than 3.8 pounds per gallon, adjust rate accordingly. Use of ester formulations of 2,4-D is discouraged if sensitive crops, especially cotton and tobacco, are located within 1 mile. Following application of Clarity and accumulation of at least 1 inch rainfall, delay soybean planting at least 14 days. Follow precautions on Clarity label concerning drift to sensitive crops. Leadoff must be applied at least 30 days ahead of planting. Trivence may be applied any time from fall through spring, up to 3 days after planting. See label for recommended adjuvants. Rates are dependent upon soil texture and organic matter, see label for specific rate recommendations. See label for crop rotation restrictions. Afforia may be applied at 2.5 to 3.75 oz 7 days prior to planting soybean, the rate must be reduced to 2.5 oz if applying less than 7 days prior to planting. Rates are dependent upon soil texture and organic matter, see label for specific rate recommendations. See label for crop rotation restrictions. Horseweed cannot be controlled with a burndown prior to planting double-crop soybeans because the combine cuts off the horseweed, leaving little to no foliage to spray. If horseweed is present in wheat, apply 0.75 to 0.9 ounce of either Harmony SG or Harmony Extra with TotalSol plus 3 ounces of Clarity in February or early March.
glyphosate, MOA 9 (numerous brands and formulations) + 2,4-D, MOA 4 (numerous brands and formulations) 3.8 SL + flumioxazin, MOA 14, + chlorimuron, MOA 2 (Valor XLT) 40.3 WDG	See label + 1.5 to 2 pt + 3 to 5 oz	0.56 to 1.13 (lb a.e.) + 0.71 to 0.95 + 0.056 to 0.094 + 0.019 to 0.032	
glyphosate, MOA 9 (numerous brands and formulations) + 2,4-D, MOA 4 (numerous brands and formulations) 3.8 SL + flumioxazin, MOA 14, + pyroxasulfone, MOA 15 (Fierce) 76 WDG	See label + 1.5 to 2 pt + 3 to 3.75 oz	0.56 to 1.13 (lb a.e.) + 0.71 to 0.95 + 0.06 to 0.079 + 0.0796 to 0.0996	
glyphosate, MOA 9 (numerous brands and formulations) + 2,4-D, MOA 4 (numerous brands and formulations) 3.8 SL + flumioxazin, MOA 14 + chlorimuron, MOA 2 + thifensulfuron, MOA 2 (Envive) 41.3 WDG	See label + 1.5 to 2 pt + 2.5 to 4 oz	0.56 to 1.13 (lb a.e.) + 0.71 to 0.95 + 0.046 to 0.074 + 0.0214 to 0.023 + 0.0045 to 0.007	

Table 7-5A. Chemical Weed Control in Soybeans

Herbicide, Mode of Action Code[1], and Formulation	Amount of Formulation Per Acre	Pounds Active Ingredient Per Acre	Precautions and Remarks
Burndown, No-Till Planting, Glyphosate-resistant horseweed plus other weeds (continued)			
glyphosate, MOA 9 (numerous brands and formulations)	See label	0.56 to 1.13 (lb a.e.)	
+	+	+	
2,4-D, MOA 4 (numerous brands and formulations) 3.8 SL	1.5 to 2 pt	0.71 to 0.95	
+		+	
rimsulfuron, MOA 2	1.5	0.0155	
+		+	
thifensulfuron, MOA 2 (Leadoff) 33.4 WDG		0.0155	
glyphosate, MOA 9 (numerous brands and formulations)	See label	0.56 to 1.13 (lb a.e.)	
+	+	+	
2,4-D, MOA 4 (numerous brands and formulations) 3.8 SL	1.5 to 2 pt	0.71 to 0.95	
+		+	
flumioxazin, MOA 14		0.048 to 0.070	
+		+	
chlorimuron, MOA 2		0.0146 to 0.021	
+		+	
metribuzin, MOA 5 (Trivence) 61.3 DG	6 to 8.7 oz	0.167 to 0.243	
glyphosate, MOA 9 (numerous brands and formulations)	See label	0.56 to 1.13 (lb a.e.)	
+	+	+	
2,4-D, MOA 4 (numerous brands and formulations) 3.8 SL	1.5 to 2 pt	0.71 to 0.95	
+	+		
flumioxazin, MOA 14		0.071 to 0.096	
+		+	
thifensulfuron, MOA 2	2.5 to 3.75 oz	0.008 to 0.012	
+		+	
tribenuron, MOA 2 (Afforia) 50.8 DG		0.008 to 0.012	
saflufenacil, MOA 14 (Sharpen) 3.42 SL	1.0 fl oz	0.027 (lb a.i.)	Sharpen can be applied to control glyphosate-resistant marestail. Applying Sharpen with other herbicides will broaden the spectrum of control. See label for specific information on adjuvant selection. **Interval between application and soybean planting for Sharpen varies by soil texture and organic matter content. See Sharpen label for specific information.**
glufosinate, MOA 10 (Liberty) 2.34 SL	29 to 36 fl oz	0.53 to 0.66	Liberty 280 SL can be applied prior to emergence of any transgenic or conventional soybean variety to control emerged weeds. See label for adjuvant use. In crop applications to **Liberty-Link soybeans** can be made at 22-29 fluid ounces following a burndown application with a maximum seasonal use of 65 ounces per acre. Thorough spray coverage is essential. Apply in minimum of 15 GPA; dense weed canopies require 20 to 40 GPA. Poor performance is likely if daytime temperatures are less than 75 degrees F or if weeds are drought stressed.
oxyfluorfen, MOA 14 (Goal 2XL) 2L (GoalTender) 4L	1 to 2 pt 0.5 to 1 pt	0.25 to 0.5	Goal 2XL can be applied at 7 days or more prior to soybean planting to control common winter and summer annual weed species. A tank mix combination with glyphosate, paraquat, glufosinate, 2,4-D or dicamba is recommended. See label for tankmix recommendations.
Burndown, No-Till Planting, Curly dock, vetch, and Carolina geranium plus other weeds controlled by glyphosate			
glyphosate, MOA 9 (numerous brands and formulations)	See label	0.56 to 1.13 (lb a.e.)	See comments for glyphosate alone. Harmony SG can be applied any time prior to soybean planting. Soybean planting should be delayed at least 14 days after application of Harmony Extra. Soybean may be planted 1 day after 0.5 ounce of FirstShot is applied, however planting should be delayed at least 7 days if higher rates are applied.
+	+	+	
thifensulfuron, MOA 2 (Harmony SG) 50 WDG	0.75 oz	0.023	
glyphosate, MOA 9 (numerous brands and formulations)	See label	0.56 to 1.13 (lb a.e.)	
+	+	+	
thifensulfuron, MOA 2		0.016	
+		+	
tribenuron, MOA 2 (Harmony Extra SG with TotalSol) 50 WDG	0.75 oz	0.008	
glyphosate, MOA 9 (numerous brands and formulations)	See label	0.56 to 1.13 (lb a.e.)	
+	+	+	
thifensulfuron, MOA 2		0.008 to 0.013	
+		+	
tribenuron, MOA 2 (FirstShot SG with TotalSol) 50 WDG	0.5 to 0.8 oz	0.008 to 0.013	

Table 7-5A. Chemical Weed Control in Soybeans

Herbicide, Mode of Action Code[1], and Formulation	Amount of Formulation Per Acre	Pounds Active Ingredient Per Acre	Precautions and Remarks
Burndown, No-Till Planting, Italian ryegrass, wheat, barley, and rye			
glyphosate, MOA 9 (numerous brands and formulations) + clethodim, MOA 1 (Select MAX) 0.97 EC	See label + 9 to 16 fl oz	0.56 to 1.13 (lb a.e.) + 0.067 to 0.12 (lb a.i.)	Apply to weeds 2 to 6 inches tall. See label for instructions on adjuvant used depending on glyphosate formulation.
Preemergence, No-Till or Conventional, Any Cultivar, Annual grasses			
alachlor, MOA 15 (Intrro) 4 EC (Micro-Tech) 4 FME	2 to 3 qt	2 to 3	Controls annual grasses except Texas panicum, shattercane, and seedling johnsongrass. Also controls pigweed and nightshade. For broadleaf weed control, alachlor may be tank mixed with most broadleaf herbicides; do not mix with Valor SX, Valor XLT, or Envive unless applied 14 or more days ahead of planting. When tank mixing, see respective labels for application rates, weeds controlled, specific application directions, and precautions. May also be shallowly incorporated; see labels for details. Generic brands of alachlor are available.
clomazone, MOA 13 (Command 3 ME) 3 FME	1.3 to 3.3 pt	0.5 to 1.25	Controls most annual grasses; shattercane and Texas panicum are only suppressed. Also controls a number of broadleaf weeds. See label for weeds controlled. Read the label carefully and follow all precautions on label pertaining to off-site movement, buffer zones, drift control agents, and rotational restrictions. For broader spectrum control, Command 3 ME may be tank mixed with most broadleaf herbicides. When tank mixing, see respective labels for application rates, weeds controlled, specific application directions, and precautions.
dimethenamid-P, MOA 15 (Outlook) 6.0 EC	12 to 21 fl oz	0.56 to 0.98	Use 12 to 18 fluid ounces on soils with less than 3% organic matter or 14 to 21 fluid ounces on soils with greater than 3% organic matter. Controls annual grasses except seedling johnsongrass, Texas panicum, and shattercane. Also controls pigweed and nightshade. See label for application directions and rates for various soils. May also be shallowly incorporated; see label for details. For broadleaf weed control, Outlook may be tank mixed with most broadleaf herbicides; do not mix with Valor SX, Valor XLT, or Envive unless applied 14 or more days ahead of planting. Do not apply to sandy soils if organic matter is less than 3% and depth to groundwater is 30 feet or less.
metolachlor, MOA 15 (Me-Too-Lachlor) 8 EC (Parallel PCS) 8 EC (Parrlay) 8 EC (Stalwart) 8 EC	1 to 2 pt	1 to 2	See comments for s-metolachlor products. Products containing s-metolachlor are more active on weeds per unit of formulated product than those containing metolachlor. In general, it takes 1.5 pints of a metolachlor product to get the activity one would get from 1 pint of s-metolachlor product.
S-metolachlor, MOA 15 (Brawl) 7.62 EC (Cinch) 7.64 EC (Dual Magnum) 7.62 EC (Dual II Magnum) 7.64 EC (Medal) 7.62 EC (Medal II) 7.64 EC	1 to 2 pt	0.95 to 1.91	Controls annual grasses except Texas panicum, shattercane, and seedling johnsongrass. Also controls pigweed and nightshade on mineral soils. May also be shallowly incorporated; see label for details. For broadleaf weed control, Smetolachlor may be tank mixed with most broadleaf herbicides; do not mix with Valor SX or Valor XLT unless applied 14 or more days ahead of planting. When tank mixing, see respective labels for application rates, weeds controlled, specific application directions, and precautions.
pendimethalin, MOA 3 (Prowl) 3.3 EC (Prowl H$_2$O) 3.8 L	1.2 to 3 pt 1.5 to 2.5 pt	0.5 to 1.2 0.7 to 1.2	Preemergence application of pendimethalin suggested only where annual grass pressure is expected to be light. Pendimethalin generally performs better when incorporated. For broadleaf weed control, pendimethalin may be tank mixed with most broadleaf herbicides. When tank mixing, see respective labels for application rates, weeds controlled, specific application directions, and precautions. Generic brands are available.
pyroxasulfone, MOA 15 (Zidua) 85 WG (Zidua SC) 4.17 SC	1.5 to 3.5 oz 2.5 to 5.75 fl oz	0.0796 to 0.186	Use 1.5 to 2.1 ounces of Zidua (2.5 to 3.5 fluid ounces of Zidua SC) per acre on coarse soils, 2 to 3 ounces on medium soils, and up to 3.5 ounces (5.75 fluid ounces of Zidua SC) per acre on fine textured soils. Controls most annual grasses, pigweed, and nightshade. Provides suppression of Texas panicum, seedling johnsongrass, and shattercane. May be tank mixed with most broadleaf herbicides; when tank mixing, see respective labels for application rates, weeds controlled, specific application directions, and precautions. May be applied to emerged soybean at the first trifoliate leaf stage to third trifoliate leaf stage. Do not apply from emergence through unifoliate stage or unacceptable injury may occur. Does not control emerged weeds. See label for tank mix options to control emerged weeds.
pyroxasulfone, MOA 15 + fluthiacet-methyl, MOA 14 (Anthem) 2.15 L	6 to 11 oz	0.101 to 0.185	Use 6 to 6.5 ounces per acre on coarse soils, 6.5 to 9.5 ounces on medium soils, and up to 11 ounces per acre on fine textured soils. Controls most annual grasses, pigweed, and nightshade. Provides suppression of Texas panicum, seedling johnsongrass, and shattercane. May be tank mixed with most broadleaf herbicides; when tank mixing, see respective labels for application rates, weeds controlled, specific application directions, and precautions. May be applied to emerged soybean up to third trifoliate leaf stage. May provide some control of emerged broadleaf weeds (less than 2 inches). See label for tank mix options to control emerged weeds.
Preemergence, No-Till or Conventional, Any Cultivar, Annual broadleaf weeds			
clomazone, MOA 13 (Command 3 ME) 3 FME	1.3 to 3.3 pt	0.5 to 1.25	Command controls selected broadleaf weeds such as balloonvine, velvetleaf, spurred anoda, prickly sida, croton, Pennsylvania smartweed, common ragweed, lambsquarters, and jimsonweed. It also controls most annual grasses. Command does not control pigweed, morningglory, sicklepod, nightshade, and ladysthumb. See label for specific weeds controlled and rates for specific weeds. Read label carefully and follow all precautions on label pertaining to off-site movement, buffer zones, drift control agents, and rotational restrictions. For broader spectrum control, Command may be tank mixed with a number of soil-applied herbicides; see label for details. When tank mixing, see respective labels for application rates, weeds controlled, specific application directions, and precautions.
flumetsulam, MOA 2 (Python) 80 WDG	0.8 to 1.33 oz	0.04 to 0.067	Controls most broadleaf weeds; control of ragweed, cocklebur, and morningglory can be variable. Rates of 1.25 to 1.33 ounces suggested for sicklepod. Acceptable control of sicklepod may require a follow-up postemergence herbicide application. May be mixed with registered soil-applied grass control herbicides. See label for weeds controlled, **rotational restrictions,** and restrictions on soil type and organic matter.

Table 7-5A. Chemical Weed Control in Soybeans

Herbicide, Mode of Action Code[1], and Formulation	Amount of Formulation Per Acre	Pounds Active Ingredient Per Acre	Precautions and Remarks
Preemergence, No-Till or Conventional, Any Cultivar, Annual broadleaf weeds (continued)			
flumioxazin, MOA 14 (Valor SX) 51 WDG (Rowel) 51 WDG	2 to 3 oz	0.063 to 0.094	Rate depends on weed species and soil texture; follow label directions when selecting rate. May be tank mixed with Prowl or Command for annual grass control. Valor SX, Valor XLT, Envive, Surveil or Gangster should not be mixed with alachlor, metolachlor, S-metolachlor, or dimethenamid-P and applied preemergence. Combinations of Valor SX, Valor XLT, Envive, Surveil or Gangster plus alachlor, metolachlor, S-metolachlor, or dimethenamid-P can be applied 14 or more days ahead of planting. Do not apply Valor XLT within 14 days before or after application of organophosphate insecticide or any variety that is not DuPont BOLT, STS or STS/RR due to injury potential.
flumioxazin, MOA 14 + chlorimuron, MOA 2 (Valor XLT) 40.3 WDG (Rowel FX) 40.3 WDG	3 to 5 oz	0.056 to 0.094 + 0.019 to 0.032	
flumioxazin, MOA 14 + chlorimuron, MOA 2 + thifensulfuron, MOA 2 (Envive) 41.3 WDG	2.5 to 4 oz	0.046 to 0.074 + 0.0214 to 0.023 + 0.0045 to 0.007	
flumioxazin, MOA 14 + cloransulam, MOA 2 (Surveil) 48 WG	2.8 to 4.2 oz	0.063 to 0.095 + 0.021 to 0.032	
flumioxazin, MOA 14 + cloransulam, MOA 2 (Gangster, co-pack of Gangster V [51% flumioxazin] and Gangster FR [84% cloransulam])	1.5 to 3 oz + 0.3 to 0.6 oz	0.047 to 0.094 + 0.016 to 0.032	
imazaquin, MOA 2 (Scepter) 70 WDG	2.8 oz	0.123	Controls most broadleaf weeds if adequate rainfall received for activation. A follow-up postemergence herbicide application often needed for adequate sicklepod control. Follow all precautions on the label, including rotational restrictions. For annual grass control, Scepter may be tank mixed with alachlor, Command, dimethenamid-P, pendimethalin, or S-metolachlor. When tank mixing, see respective labels for application rates, weeds controlled, specific application directions, and precautions.
linuron, MOA 7 (Linex) 4 L	0.66 to 3 pt	0.33 to 1.5	Rate depends greatly on soil texture and organic matter content; follow label directions carefully when selecting rates. Do not use on sand or loamy sand soils or any soil with less than 0.5% organic matter. Linuron controls pigweed, lambsquarters, and common ragweed. For annual grass control, linuron may be tank mixed with alachlor, Command, dimethenamid-P, pendimethalin, or S-metolachlor. When tank mixing, see respective labels for application rates, weeds controlled, specific application directions, and precautions. Generic brands may be available.
metribuzin, MOA 5 (Sencor) 75 DF	0.33 to 0.67 pt	0.25 to 0.5	Rate depends greatly on soil texture and organic matter content; follow label directions carefully when selecting rates. Do not use Sencor on sand soils with less than 1% organic matter or on any soil with less than 0.5% organic matter. Some varieties are particularly sensitive to metribuzin; see labels for details. Soybeans may be injured when metribuzin is applied to soil treated with organophosphate insecticides and/or nematicides. Does not adequately control cocklebur or morningglory. Adequate sicklepod control may require a follow-up postemergence herbicide application. For annual grass control, metribuzin may be tank mixed with alachlor, Command, dimethenamid-P, pendimethalin, or S-metolachlor. When tank mixing, see respective labels for application rates, weeds controlled, specific application directions, and precautions.
sulfentrazone, MOA 14 + metribuzin, MOA 5 (Authority MTZ) 45 WDG	12 to 20 oz	0.135 to 0.225 + 0.20 to 0.34	Rate depends upon soil texture and organic matter; see label for application rates. Controls most broadleaf weeds, including Palmer amaranth, morningglory, and cocklebur; sicklepod suppressed. See statement on label concerning sensitive varieties. See label for rotational restrictions.
Preemergence, No-Till or Conventional, Any Cultivar, Annual grasses and broadleaf weeds — packaged herbicide mixtures			
s-metolachlor, MOA 15 + fomesafen, MOA 14 (Prefix) 5.29 L	2 to 3 pt	1.09 to 1.63 + 0.24 to 0.36	Controls most annual grasses (except Texas panicum, seedling johnsongrass, shattercane) and broadleaf weeds, such as pigweed species (including Palmer amaranth), lambsquarters, common ragweed, Florida pusley, smartweed, and nightshade. Does not control sicklepod, and it only suppresses morningglory, cocklebur, and prickly sida.
flumioxazin, MOA 14 + pyroxasulfone, MOA 15 (Fierce) 76 WDG	3 to 3.75 oz	0.06 to 0.079 + 0.0796 to 0.0996	Controls most annual grasses (except Texas panicum, seedling johnsongrass, shattercane) and broadleaf weeds, such as pigweed species (including Palmer amaranth), lambsquarters, common ragweed, Florida pusley, smartweed, and nightshade.
flumioxazin, MOA 14 + pyroxasulfone, MOA 15 + chlorimuron, MOA 2 (Fierce XLT) 62.41 WDG	3.75 to 4.5 oz	0.06 to 0.069 + 0.073 to 0.088 + 0.016 to 0.019	Controls most annual grasses (except Texas panicum, seedling johnsongrass, shattercane) and broadleaf weeds, such as pigweed species (including Palmer amaranth), lambsquarters, common ragweed, Florida pusley, smartweed, and nightshade.
metribuzin, MOA 5 + S-metolachlor, MOA 15 (Boundary) 7.8 L	1 to 2.5 pt	0.19 to 0.47 + 0.79 to 1.96	See comments for metribuzin applied preemergence. Rate depends on soil texture and organic matter content; follow label directions carefully when selecting rate. Do not use on coarse-textured soils with less than 0.5% organic matter. Controls weeds normally controlled by Sencor and Dual Magnum. May be mixed with Command, FirstRate, Prowl, Python, or Scepter. Follow-up postemergence herbicide needed in most cases.
imazethapyr, MOA 2 + saflufenacil, MOA 14 (OpTILL)	2 oz	0.022 + 0.63	Optill is labeled for application up to soybean emergence. Application to emerged soybean can result in significant injury. See label for adjuvant selection.
imazethapyr, MOA 2 + saflufenacil, MOA 14 + pyroxasulfone, MOA 15 (Zidua PRO) 4.09 SC	4.5 oz + 6 fl oz	0.0169 to 0.0225 + 0.0468 to 0.062 + 0.08 to 0.107	Zidua PRO is labeled for burndown application up to preemergence. Application to emerged soybean can result in significant injury. Delay soybean planting 30 days after application on coarse soils where organic matter is ≤ 2.0%. See label for adjuvant selection. See label for tank mixing information.

Table 7-5A. Chemical Weed Control in Soybeans

Herbicide, Mode of Action Code[1], and Formulation	Amount of Formulation Per Acre	Pounds Active Ingredient Per Acre	Precautions and Remarks
Preemergence, No-Till or Conventional, Any Cultivar, Annual grasses and broadleaf weeds — packaged herbicide mixtures (continued)			
sulfentrazone, MOA 14 + s-metolachlor. MOA 15 (Broadaxe) 7 EC (Broadaxe XC) 7 EC	19 to 38.7 fl oz	0.104 to 0.212 + 0.94 to 1.90	Controls most annual grasses (except Texas panicum, seedling johnsongrass, shattercane) and broadleaf weeds, such as pigweed species (including Palmer amaranth), lambsquarters, morningglory, prickly sida smartweed, and nightshade. Also provides control of yellow nutsedge. Does not control sicklepod or cocklebur.
acetochlor, MOA 15 + fomesafen, MOA 14 (Warrant Ultra) 3.45 L	48 to 70 fl oz	1.06 to 1.54 + 0.236 to 0.345	May be applied preplant, at-planting, preemergence or postemergence. May only be applied ONCE per growing season. Controls most annual grasses (except Texas panicum, seedling johnsongrass, shattercane) and broadleaf weeds, such as pigweed species (including Palmer amaranth), lambsquarters, common ragweed, Florida pusley, smartweed, and nightshade. Does not control sicklepod, and it only suppresses morningglory, cocklebur, and prickly sida. See label for rate and timing information.
Preemergence, No-Till or Conventional, Any Cultivar, Annual grasses and broadleaf weeds — packaged herbicide mixtures			
sulfentrazone, MOA 14 + cloransulam-methyl, MOA 2 (Sonic) 78 WG	6.45 to 8 oz	0.25 to 0.31 + 0.032 to 0.04	May be applied preplant incorporated, preplant, or preemergence within 3 days after planting. Apply 6.45 oz where organic matter is less than 3%. May be applied alone or in tank mix combination with other registered herbicides. Controls several annual grasses (except barnyardgrass, Texas panicum, seedling johnsongrass, shattercane) and broadleaf weeds such as pigweed species (including Palmer amaranth), lambsquarters, horseweed, cocklebur, morningglory, smartweed, and nightshade. Also provides control of purple and yellow nutsedge. Does not control sicklepod.
Preemergence, No-Till or Conventional, Any Cultivar, Nutsedge			
dimethenamid-P, MOA 15 (Outlook) 6.0 EC	14 to 21 fl oz	0.66 to 1	Dimethenamid and metolachlor control or suppress only yellow nutsedge. These herbicides are more effective on yellow nutsedge when incorporated. However, these herbicides applied preemergence may provide adequate control of lighter infestations of yellow nutsedge. Neither product controls purple nutsedge. Follow labels carefully for use rates on various soil types. Do not apply Outlook to sand soils if organic matter is less than 3% and depth to groundwater is 30 feet or less. Generic brands containing metolachlor, not S-metolachlor, are available. See previous comments concerning these products. Prefix, which contains s-metolachlor, will suppress or control yellow nutsedge.
S-metolachlor, MOA 15 (Brawl) 7.62 EC (Cinch) 7.64 EC (Dual Magnum) 7.62 EC (Dual II Magnum) 7.64 EC (Medal) 7.62 EC (Medal II) 7.64 EC	1.33 to 2 pt	1.27 to 1.91	
sulfentrazone, MOA 14 + metribuzin, MOA 5 (Authority MTZ) 45 WDG	12 to 20 oz	0.135 to 0.225 + 0.20 to 0.34	See comments under Annual Grasses and Broadleaf Weeds. Controls yellow and purple nutsedge.
sulfentrazone, MOA 14 + s-metolachlor. MOA 15 (Broadaxe) 7 EC (Broadaxe XC) 7 EC	19 to 38.7 fl oz	0.104 to 0.212 + 0.94 to 1.90	See comments under Annual Grasses and Broadleaf Weeds. Controls yellow and purple nutsedge.
Postemergence Overtop; Roundup Ready Cultivars, Annual grasses and broadleaf weeds plus suppression of perennial weeds—**Roundup Ready cultivars Only**			
glyphosate, MOA 9 (numerous brands and formulations)	See label	0.75 to 1.5 (lb a.e.)	**Apply only to Roundup Ready Cultivars. See comments on resistance management in TABLE 7-10.** A preemergence herbicide is highly recommended to control weeds not controlled by glyphosate (such as Florida pusley), to reduce early season weed competition, to broaden the window of application for glyphosate, and to aid in resistance management. Any registered soil-applied herbicide can be used on Roundup Ready soybeans. Glyphosate controls most annual weeds; exceptions include dayflower, hemp sesbania, and Florida pusley. Timely application required for morningglory control. Can be applied from cracking stage throughout flowering. Multiple applications can be made, but do not exceed 2.2 pounds a.e. per acre per year during this period. Total glyphosate use (preplant, in-crop, and preharvest) should not exceed 6 pounds a.e. per acre per year. Glyphosate is available as an isopropylamine salt and a potassium salt. Glyphosate formulations and application rates should be compared on the basis of pounds of glyphosate acid equivalent (a.e.) per gallon and per acre, respectively. Rate in the preceding column is expressed as a.e. See Table 7-10 for glyphosate rate conversions. Adjuvant recommendations vary by glyphosate product: see label of brand used for details. Rate depends upon weed species and size; see labels for details. Timely application is encouraged. The first application should be made 18 to 20 days after planting. Repeat applications can be made if needed. The following products can be mixed with at least some of glyphosate brands. Refer to label of tank mix partner or glyphosate product used for timing of application, weed sizes, and use of adjuvants. **Ultra Blazer** (1 pint): improves control of hemp sesbania, black nightshade, and larger morningglory. Minor antagonism sometimes noted on grasses and pigweed. Use 1.5 pints to control glyphosate-resistant Palmer amaranth up to 4 inches **Classic** (0.25 to 0.33 oz): improves control of hemp sesbania, spreading dayflower, and larger morningglory. **FirstRate** (0.2 to 0.3 fluid ounce): improves control of spreading dayflower, dove weed, and larger morningglory. **Harmony SG** (0.125 ounces/acre): improves control of lambsquarters and velvetleaf. Controls glyphosate-resistant Palmer amaranth unless it is also ALS resistant. Apply before Palmer amaranth exceeds 8 inches Apply after first trifoliate has fully expanded. See label for use of ammonium sulfate. Some soybean injury can be expected. **Reflex, Flexstar** (6 to 12 fluid ounces): improves control of hemp sesbania, black nightshade, and larger morningglory. Antagonism sometimes noted on grasses and pigweed. Must be applied at 16 fluid ounces to control glyphosate-resistant Palmer amaranth 4 inches tall, or at 24 fluid ounces if 6 inches tall. **Resource** (2 to 4 fluid ounces): improves control of larger morningglory. Minor antagonism sometimes observed on pigweed. Apply at 6 to 8 ounces for glyphosate-resistant Palmer amaranth up to 4 inches **Synchrony XP** (0.375 ounce): improves control of lambsquarters, morningglory, and velvetleaf. Controls glyphosate-resistant Palmer amaranth unless it is also ALS resistant. Apply before Palmer amaranth exceeds 4 inches **Storm** (0.75 to 1.5 pints): improves control of hemp sesbania, black nightshade, and larger morningglory. Minor antagonism sometimes noted on grasses and pigweed.
glyphosate, MOA 9 + fomesafen, MOA 14 (Flexstar GT) 3.29 L	3 to 4.5 pt	1 to 1.5 (lb a.e.) + 0.25 to 0.37	Controls grasses and most annual broadleaf weeds. See label for suggested rates according to weed size. Also see label for adjuvant recommendations. Apply with flat fan nozzles. Do not exceed 4.5 pints per acre of Flexstar GT per year. Also do not exceed 0.375 pound a.i. of fomesafen per year from all sources.

Table 7-5A. Chemical Weed Control in Soybeans

Herbicide, Mode of Action Code[1], and Formulation	Amount of Formulation Per Acre	Pounds Active Ingredient Per Acre	Precautions and Remarks
Postemergence Overtop; Roundup Ready Cultivars, Annual grasses and broadleaf weeds plus suppression of perennial weeds—**Roundup Ready cultivars Only** (continued)			
glyphosate isopropylamine salt, MOA 9 + imazethapyr, MOA 2 (Extreme) 2.17 SL	3 pt	0.56 + 0.064	**Apply only to Roundup Ready cultivars.** Apply before soybean bloom and make only one application per year. Add nonionic surfactant at 1 pint per 100 gallons spray solution plus 2.5 pounds of ammonium sulfate or 1 to 2 quarts of UAN. See label for application directions, precautions, and rotational restrictions. May be more effective on yellow nutsedge and morningglory than glyphosate alone.
glyphosate, MOA 9 + acetochlor, MOA 15 (Warrant) 3.0 ME	See label 1.5 qt	0.56 to 0.75 (lb a.e.) + 1.1	**Apply only to Roundup Ready cultivars.** Apply overtop soybean with glyphosate at V2-V3 soybean for best results. Apply prior to R3. Warrant can be directed at V5-V6. Warrant provides residual control only.
Postemergence Overtop; Roundup Ready Cultivars, Volunteer Roundup Ready corn in Roundup Ready soybeans			
glyphosate, MOA 9 (numerous brands and formulations) + clethodim, MOA 1 (Select) 2 EC (Select Max) 0.97 EC	See label + 4 to 8 fl oz 6 to 12 fl oz	0.56 to 0.75 (lb a.e.) + 0.063 to 0.125 0.045 to 0.106	See comments for glyphosate alone. For corn up to 12 inches tall, apply 4 to 6 ounces of Select or 6 ounces of Select Max. For corn up to 24 inches tall, apply 6 to 8 ounces of Select or 9 ounces of Select Max. For corn up to 36 inches, apply 12 ounces of Select Max. Add 2.5 pounds per acre ammonium sulfate or equivalent. If brand of glyphosate used does not contain surfactant, add nonionic surfactant at 0.25 to 0.5% by volume. If applying Select or Select Max alone, see labels for adjuvant recommendations.
glyphosate, MOA 9 (numerous brands and formulations) + fluazifop-p-butyl, MOA 1 (Fusilade DX) 2 EC	See label + 4 to 6 fl oz	0.56 to 0.75 (lb a.e.) + 0.063 to 0.094	See comments for glyphosate alone. Apply 4 ounces Fusilade for corn less than 12 inches Increase rate to 6 ounces for corn up to 24 inches Add any adjuvants suggested on the label of the glyphosate product used. Additionally, add 0.25% by volume of crop oil concentrate. If applying Fusilade alone, see label for adjuvant recommendations.
glyphosate, MOA 9 (numerous brands and formulations) + quizalofop-p-ethyl, MOA 1 (Assure II) 0.88 EC	See label + 5 to 8 fl oz	0.56 to 0.75 (lb a.e.) + 0.034 to 0.055	See comments for glyphosate alone. Apply Assure at 4 ounces to corn up to 12 inches, 5 ounces to corn up to 18 inches, and 8 ounces to corn up to 30 inches If the brand of glyphosate contains adjuvant, add 0.125% nonionic surfactant by volume. If the brand of glyphosate does not contain adjuvant, add surfactant according to the glyphosate label. If applying Assure alone, see label for adjuvant recommendations.
Postemergence Overtop; Liberty Link Cultivars, Annual grasses and broadleaf weeds			
glufosinate-ammonium, MOA 10 (Liberty SL)	29 to 36 fl oz	0.53 to 0.66 (lb a.i.)	**Apply only to Liberty Link cultivars.** Can be applied as single or sequential applications prior to V4 stage of soybean. Do not apply more than 36 ounces as a single application. If applied as a burndown prior to planting, Liberty SL can be applied in season to soybean with a maximum seasonal use of 65 ounces/acre. See product label for possible tank mixtures with other herbicides.
Postemergence Overtop; Any Cultivar, Annual grasses			
clethodim, MOA 1 (Select) 2 EC (Select Max) 0.97 EC	6 to 8 fl oz 9 to 16 fl oz	0.094 to 0.125 0.068 to 0.121	Apply to actively growing grasses not under drought stress. See label for specific rates and weed size to treat. Add crop oil concentrate at 1 quart per acre to Select. To Select Max, add nonionic surfactant at 0.25% volume, crop oil concentrate at 1% by volume, or methylated seed oil at 1% by volume. Do not cultivate for 7 days before or after application. Generic brands are available.
fluazifop-p-butyl, MOA 1 (Fusilade DX) 2 EC	6 to 12 fl oz	0.094 to 0.188	Apply to actively growing grass not under drought stress. Suggested application rate varies by species; see label for application directions, rates, maximum weed sizes to treat, etc. Add either 1% crop oil concentrate (1 gallon per 100 gallons) or 0.25% nonionic surfactant (1 quart per 100 gallons). Do not cultivate for 7 days before or after application.
quizalofop p-ethyl, MOA 1 (Assure II) 0.88 EC	5 to 8 fl oz	0.034 to 0.055	Apply to actively growing grass not under drought stress. Suggested application rate varies by species; see label for application directions, rates, maximum weed sizes to treat, etc. Add 1% (1 gallon per 100 gallons) crop oil concentrate or 0.25% (1 quart per 100 gallons) nonionic surfactant. Do not cultivate for 7 days before or after application. Generic brands are available.
sethoxydim, MOA 1 (Poast) 1.5 EC (Poast Plus) 1 EC	16 fl oz 24 fl oz	0.19	Apply to actively growing grass not under drought stress. Consult label for maximum grass size to treat, application directions, etc. Add 2 pints per acre of crop oil concentrate. Do not cultivate for 7 days before or after application.
Postemergence Overtop; Any Cultivar, Annual broadleaf weeds			
acifluorfen, MOA 14 (Ultra Blazer) 2 SL	0.5 to 1.5 pt	0.13 to 0.38	See label for weeds controlled, recommended rates for specific weeds, and maximum weed size to treat. Label recommends nonionic surfactant at 1 to 2 pints per 100 gallons spray solution. For broader spectrum control, acifluorfen may be tank mixed with Basagran, Classic, FirstRate, Pursuit, Raptor, Resource, Scepter, Synchrony, or 2,4-DB. When tank mixing, see respective labels for application rates, weeds controlled, maximum weed size to treat, specific application directions, and precautions.
acifluorfen, MOA 14 + bentazon, MOA 6 (Storm) 4 SL	1.5 pt	0.25 + 0.5	See label for weeds controlled and maximum weed size to treat. Add 1 pint per acre of crop oil concentrate or nonionic surfactant at 1 quart per 100 gallons. For broader spectrum control, Storm may be tank mixed with Basagran, Classic, FirstRate, Pursuit, Raptor, Resource, or Scepter. See respective labels for application rates, weeds controlled, maximum weed size to treat, specific application directions, and precautions.
bentazon, MOA 6 (Basagran) 4 SL	1 to 2 pt	0.5 to 1	See label for weeds controlled, recommended rates for specific weeds, and maximum weed size to treat. Add 1.25% by volume (not to exceed 2 pints per acre) of crop oil concentrate when treating for lambsquarters, common ragweed, or hemp sesbania. If velvetleaf is primary target, add 0.5 to 1.0 gallon per acre of liquid nitrogen instead of crop oil. For broader spectrum control, Basagran may be tank mixed with Classic, Cobra, FirstRate, Flexstar, Pursuit, Raptor, Reflex, Resource, Scepter, Storm, Ultra Blazer, or 2,4-DB. When tank mixing, see respective labels for application rates, weeds controlled, maximum weed size to treat, specific application directions, and precautions.
cloransulam-methyl, MOA 2 (FirstRate) 84 WDG	0.3 oz	0.016	Controls cocklebur, jimsonweed, morningglory, ragweed, smartweed, velvetleaf, spreading dayflower, dove weed, and small horseweed. See label for recommended weed size to treat. FirstRate will usually control sicklepod in the cotyledonary to first leaf stage; larger sicklepod will not be controlled. Add either nonionic surfactant at 1 to 2 pints per 100 gallons or crop oil concentrate at 1.2 gallons per 100 gallons. If velvetleaf is the target weed, also add 2.5 gallons 30% UAN per 100 gallons. FirstRate can be applied twice per season. For broader spectrum control, FirstRate may be tank mixed with Basagran, Classic, Cobra, Flexstar, Pursuit, Raptor, Reflex, Resource, Storm, Synchrony, or Ultra Blazer.

Table 7-5A. Chemical Weed Control in Soybeans

Herbicide, Mode of Action Code[1], and Formulation	Amount of Formulation Per Acre	Pounds Active Ingredient Per Acre	Precautions and Remarks
Postemergence Overtop; Any Cultivar, Annual broadleaf weeds (continued)			
chlorimuron ethyl, MOA 2 (Classic) 25 WDG	0.5 to 0.75 oz	0.008 to 0.012	See label for weeds controlled, recommended rates for specific weeds, maximum weed size to treat, rotational restrictions, and sprayer cleanup. Add 0.25% by volume (1 quart per 100 gallons) of nonionic surfactant. Under hot, dry conditions, 1% crop oil concentrate may be used instead of surfactant; crop oil increases potential for injury. See label for specific adjuvant recommendations when treating velvetleaf. For broader spectrum control, Classic may be tank mixed with Basagran, Ultra Blazer, Cobra, FirstRate, Flexstar, Harmony GT, Reflex, Resource, Storm, or Ultra Blazer. When tank mixing, see respective labels for application rates, weeds controlled, maximum weed size to treat, specific application directions, and precautions.
chlorimuron ethyl, MOA 2 + thifensulfuron methyl, MOA 2 (Synchrony XP) 28.4 WDG	0.375 to 1.125 oz	0.005 to 0.015 + 0.0016 to 0.0049	**For non-BOLT, STS cultivars, use only 0.375 ounce rate.** Rate can be increased to 1.125 ounces on BOLT, STS cultivars. See label for weeds controlled, maximum weed size to treat, and rotational restrictions. Add crop oil concentrate at 1% by volume except when tank mixing with a product whose label precludes use of crop oil concentrate; in that case, use nonionic surfactant at 0.25% by volume. Under dry conditions, adding 2 quarts per acre of UAN may enhance control. Synchrony may be tank mixed with Cobra, FirstRate, Flexstar, Harmony GT, Reflex, Resource, or Ultra Blazer.
flumiclorac pentyl ester, MOA 14 (Resource) 0.86 EC	4 to 8 fl oz	0.027 to 0.054	Suggested for use where velvetleaf is a problem. Excellent control of velvetleaf. Also controls small lambsquarters, pigweed species, prickly sida, and common ragweed. See label for weeds controlled and recommended weed size for treatment. Add 1 quart per acre of crop oil concentrate. Resource may be tank mixed with Basagran, Classic, Cobra, FirstRate, Flexstar, Harmony GT, Pursuit, Raptor, Reflex, Scepter, Storm, Synchrony, or Ultra Blazer.
fomesafen, MOA 14 (Flexstar) 1.88 SL (Reflex) 2 SL (Dawn) 2 SL (Rhythm) 1.88 SL	1 to 1.5 pt	0.25 to 0.38 0.24 to 0.35 0.24 to 0.35 0.25 to 0.38	See labels for weeds controlled, recommended rates for specific weeds, maximum weed size to treat, and rotational restrictions. Add 1% crop oil concentrate by volume (4 quarts per 100 gallons) or 0.25% nonionic surfactant (1 quart per 100 gallons). For broader spectrum control, Reflex or Flexstar may be tank mixed with Basagran, Classic, FirstRate, Harmony GT, Pursuit, Raptor, Resource, Scepter, Synchrony, or 2,4-DB. When tank mixing, see respective labels for application rates, weeds controlled, maximum weed size to treat, specific application directions, and precautions. Flexstar is somewhat more active than Reflex and can be more effective on lambsquarters, prickly sida, spurred anoda, and velvetleaf. Foliar burn on the crop may also be greater with Flexstar under conditions of high moisture and high temperatures. See label for tax mix partners with Dawn and Rhythm.
s-metolachlor, MOA 15 + fomesafen, MOA 14 (Prefix) 5.29 L	2 to 2.33 pt	1.09 to 2.27 + 0.24 to 0.28	Apply from cracking to third trifoliate of soybean. Add 0.25% by volume of nonionic surfactant; do not use crop oil. Do not exceed 3 pints per acre of Prefix per year, and do not exceed 0.375 pound a.i. of fomesafen from all sources combined.
fomesafen, MOA 14 + fluthiacet methyl, MOA 14 (Marvel) 3 L	5 to 7.25 oz	0.117 to 0.17	Apply from preplant through full flowering (prior to R3). Add 0.25% by volume of nonionic surfactant, 0.5% by volume crop oil concentrate, or 0.5% by volume methylated seed oil. Crop oil or methylated seed oil are recommended under dry conditions. Do not exceed 0.375 pound a.i. of fomesafen from all sources combined. See label for tankmix partners.
imazamox, MOA 2 (Raptor) 1S L	5 fl oz	0.04	Controls many common broadleaf weeds. Foxtails, fall panicum, broadleaf signalgrass, seedling johnsongrass, and shattercane usually adequately controlled. Does not control sicklepod. May not adequately control ragweed, prickly sida, or Palmer amaranth. May tank mix with Basagran, FirstRate, Flexstar, Reflex, Resource, Storm, or Ultra Blazer for improved control of ragweed and Palmer amaranth; tank mixes may reduce grass control. Suppresses yellow and purple nutsedge. Add either crop oil concentrate at 2 pints per acre or nonionic surfactant at 1 quart per 100 gallons. See label concerning addition of nitrogen-containing fertilizer.
imazaquin, MOA 2 (Scepter) 70 WDG	1.4 to 2.8 oz	0.063 to 0.125	See label for weeds controlled, recommended rates for specific weeds, maximum weed size to treat, and rotational restrictions. Primarily for control of cocklebur and pigweed. Add 0.25% by volume (1 quart per 100 gallons) nonionic surfactant. Alternatively, a crop oil concentrate can be used at the rate recommended on crop oil label. For broader spectrum control, Scepter may be tank mixed with Basagran, Cobra, Flexstar, Reflex, Resource, Storm, or Ultra Blazer. When tank mixing, see respective labels for application rates, weeds controlled, maximum weed size to treat, specific application directions, and precautions.
imazethapyr, MOA 2 (Pursuit) 70 WDG	1.44 oz	0.063	See label for weeds controlled, recommended rates for specific weeds, maximum weed size to treat, and rotational restrictions. Also suppresses johnsongrass, broadleaf signalgrass, and foxtails. Add 0.25% by volume (1 quart per 100 gallons) nonionic surfactant or 1.5 to 2 pints per acre of crop oil concentrate. For broader spectrum control, Pursuit may be tank mixed with Basagran, Cobra, FirstRate, Flexstar, Harmony GT, Reflex, Resource, Storm, or Ultra Blazer. When tank mixing, see respective labels for application rates, weeds controlled, maximum weed size to treat, specific application directions, and precautions.
lactofen, MOA 14 (Cobra) 2 EC	6 to 12.5 fl oz	0.094 to 0.2	See label for weeds controlled, recommended rates, weed size to treat, and recommended adjuvants. At higher rates, Cobra usually causes excessive foliar burn on soybeans. Lower rates tank mixed with other herbicides may be of some value in specific situations. Cobra may be tank mixed with Basagran, Classic, FirstRate, Pursuit, Resource, Scepter, Synchrony, or 2,4-DB. See labels for weeds controlled and specific use directions.
thifensulfuron methyl, MOA 2 (Harmony SG) 50 WDG	0.125 oz	0.004	See label for weeds controlled, maximum weed size to treat, and sprayer cleanup. Add 0.125% to 0.25% by volume (1 to 2 pints per 100 gallons) of nonionic surfactant when applying Harmony SG alone or 0.125% in tank mixes. Under dry or cool conditions, a crop oil concentrate may be used; see label for details. In addition to surfactant or crop oil, Harmony SG label specifies use of an ammonium nitrogen fertilizer. This is usually of value only when treating for velvetleaf. For broader spectrum control, Harmony SG may be tank mixed with Classic, Flexstar, Pursuit, Reflex, Resource, or Synchrony. When tank mixing, see respective labels for application rates, weeds controlled, maximum weed size to treat, specific application directions, and precautions.
pyraflufen-ethyl, MOA 14 (ET) 1 SL	0.5 to 2.0 fl oz	0.003 to 0.015 (lb a.i.)	ET can be applied from emergence to V6 soybean to suppress small broadleaf weeds. Some leaf speckling can occur but is transient. Do not apply crop oil concentrate. See label for adjuvant and spray volume recommendations. Research with ET is limited in North Carolina.

Table 7-5A. Chemical Weed Control in Soybeans

Herbicide, Mode of Action Code[1], and Formulation	Amount of Formulation Per Acre	Pounds Active Ingredient Per Acre	Precautions and Remarks
Postemergence Overtop; Any Cultivar, Annual grasses and broadleaf weeds—tank mixtures			
quizalofop p-ethyl, MOA 1 (Assure II) + Basagran, Classic, Cobra, FirstRate, Flexstar, Harmony SG, Pursuit, Raptor, Reflex, Scepter, Storm, Synchrony STS, or Ultra Blazer (See TABLE 7-10 for MOAs)	See labels	See labels	The listed two-way tank mixes are covered on one or more of the respective labels. Consult the labels of the products to be used for specific application rates, directions, precautions, and adjuvant usage. Formulations and active ingredients of the various products can be found elsewhere in this publication. A number of three-way tank mixes (not listed here) also are registered. While mixing postemergence grass and broadleaf herbicides is convenient and saves time and trips across the field, best results often are obtained when the grass and broadleaf herbicides are applied separately. Antagonism of the grass herbicide (reduced grass control) often occurs when the grass herbicide is mixed with a broadleaf herbicide. Antagonism is more likely to occur under marginal spraying conditions, such as large grasses and dry weather. Some of the broadleaf herbicides also are more antagonistic than others. The antagonism may be partially or completely overcome by increasing the rate of the grass herbicide; labels for some of the grass herbicides suggest increased rates when tank mixing. The adjuvants needed for good activity of the grass herbicide also may enhance crop injury from the broadleaf herbicide; follow label directions carefully for use of adjuvants. Tank mixing should be considered only when the optimum timing for application of the grass and broadleaf herbicides coincides. Tank mixes generally should not be used when treating for rhizome johnsongrass or bermudagrass. If sequential applications are made, the recommended waiting interval between application of the grass and broadleaf herbicides varies depending upon the herbicides used and the order in which they are applied. See the labels for specific recommendations. However, the following are general guidelines: 1) If the grass herbicide is applied first, the broadleaf herbicide can be applied 24 hr later; 2) if Basagran or Resource is applied first, the grass herbicide can be applied 24 hr later; 3) if Classic, FirstRate, Harmony SG, or Synchrony STS is applied first, wait at least 3 days before applying the grass herbicide; 4) if Pursuit, Raptor, or Scepter is applied first, wait at least 5 days before applying the grass herbicide; and 5) if Cobra, Flexstar, Reflex, Storm, or Ultra Blazer is applied first, delay application of the grass herbicide until the grass resumes active growth with development of new leaves.
fenoxaprop-ethyl, MOA 1 (Fusilade DX) + Basagran, Classic, Cobra, Flexstar, Pursuit, Raptor, Reflex, Scepter, Storm, Synchrony STS, or Ultra Blazer (See TABLE 7-10 for MOAs)			
sethoxydim, MOA 1 (Poast or Poast Plus) + Basagran, Classic, Cobra, FirstRate, Flexstar, Pursuit, Reflex, Resource, Scepter, Storm, Synchrony STS, or Ultra Blazer (See TABLE 7-10 for MOAs)			
clethodim, MOA 1 (Select or Select Max) + Basagran, Classic, Cobra, FirstRate, Flexstar, Pursuit, Raptor, Reflex, Resource, Storm, Synchrony STS, or Ultra Blazer (See TABLE 7-10 for MOAs)			
Postemergence Overtop; Any Cultivar, Annual broadleaf weeds, Rhizome johnsongrass			
clethodim, MOA 1 (Select) 2 EC (Select Max) 0.97 EC	8 to 16 fl oz 12 to 32 fl oz	0.125 to 0.25 0.091 to 0.24	Apply when johnsongrass is 12 to 24 inches tall. If needed, make second application of 6 to 8 oz of Select or 9 to 24 fluid ounces of Select Max when regrowth is 6 to 18 inches Add crop oil concentrate at 1 quart per acre to Select. To Select Max, add nonionic surfactant at 0.25% by volume, crop oil concentrate at 1% by volume, or methylated seed oil at 1% by volume. Generic brands available.
fluazifop p-butyl, MOA 1 (Fusilade DX) 2 EC	12 fl oz	0.19	Apply when johnsongrass is 8 to 18 inches tall and before boot stage. Add either a nonionic surfactant at 0.25% by volume (1 quart per 100 gallons) or a crop oil concentrate at 1% by volume (1 gallon per 100 gallons). If needed, make second application of 8 fluid ounces when regrowth is 6 to 12 inches
quizalofop p-ethyl, MOA 1 (Assure II) 0.88 EC	10 fl oz	0.07	Apply when johnsongrass is 10 to 24 inches tall. If needed, make second application of 7 fluid ounces per acre when regrowth is 6 to 10 inches Add either crop oil concentrate at 1% (1 gallon per 100 gallons) or nonionic surfactant at 0.25% (1 quart per 100 gallons). Generic brands available.
sethoxydim, MOA 1 (Poast) 1.5 EC (Poast Plus) 1 EC	24 fl oz 36 fl oz	0.28	Apply to actively growing johnsongrass 20 to 25 inches tall. Add 2 pints per acre of crop oil concentrate. A second application of 16 ounces of Poast or 24 ounces of Poast Plus may be made when regrowth is 12 inches.
Postemergence Overtop; Any Cultivar, Bermudagrass			
clethodim, MOA 1 (Select) 2 EC (Select Max) 0.97 EC	8 to 16 fl oz 12 to 32 fl oz	0.125 to 0.25 0.091 to 0.24	Apply before bermudagrass runners exceed 6 inches If needed, make second application of 8 to 16 ounces of Select or 12 to 32 fluid ounces of Select Max when regrowth is less than 6 inches Add crop oil concentrate at 1 quart per acre to Select. To Select Max, add nonionic surfactant at 0.25% by volume, crop oil concentrate at 1% by volume, or methylated seed oil at 1% by volume. Generic brands available.
fluazifop p-butyl, MOA 1 (Fusilade DX) 2 EC	12 fl oz	0.19	Apply when bermudagrass runners are 4 to 8 inches If regrowth occurs, apply 8 fluid ounces when regrowth is 4 to 8 inches Add crop oil concentrate at 1% by volume (1 gallon per 100 gallons) or nonionic surfactant at 0.25% by volume (1 quart per 100 gallons).
quizalofop p-ethyl, MOA 1 (Assure II) 0.88 EC	10 fl oz	0.07	Apply when bermudagrass is 3 inches tall or has up to 6 inches runners. If regrowth occurs, make second application of 7 fluid ounces/acre when runners are 6 inches Add either crop oil concentrate at 1% (1 gallon per 100 gallons) or nonionic surfactant at 0.25% (1 quart per 100 gallons). Generic brands available.
sethoxydim, MOA 1 (Poast) 1.5 EC (Poast Plus) 1 EC	24 fl oz 36 fl oz	0.28	Apply to actively growing bermudagrass before runners exceed 6 inches Add 2 pints per acre of crop oil concentrate. A second application of 16 ounces of Poast or 24 ounces Poast Plus may be made when regrowth is 4 inches.
Postemergence Overtop; Any Cultivar, Nutsedge			
bentazon, MOA 6 (Basagran) 4 SL	1.5 to 2 pt	0.75 to 1	For yellow nutsedge only; Basagran does not control purple nutsedge. Apply when yellow nutsedge is 6 to 8 inches tall. Add 2 pints per acre of crop oil concentrate. If needed, make second application of same rate 7 to 10 days later.
chlorimuron ethyl, MOA 2 (Classic) 25 WDG	0.5 to 0.75 oz	0.008 to 0.012	Controls yellow nutsedge; suppresses purple nutsedge. Apply when yellow nutsedge is 2 to 4 inches tall. Add surfactant according to label directions.
chlorimuron ethyl, MOA 2 + thifensulfuron methyl, MOA 2 (Synchrony STS SP) 42 WDG	0.5 oz	0.01 + 0.003	Controls yellow nutsedge; suppresses purple nutsedge. Apply when yellow nutsedge is 2 to 3 inches tall. Add crop oil concentrate according to label directions. **Apply only to BOLT, STS soybean.**
imazethapyr, MOA 2 (Pursuit) 70 WDG	1.44 oz	0.063	Apply when nutsedge is 1 to 3 inches tall. Add surfactant or crop oil according to label directions. Pursuit is more effective on purple nutsedge than on yellow nutsedge.

Table 7-5A. Chemical Weed Control in Soybeans

Herbicide, Mode of Action Code[1], and Formulation	Amount of Formulation Per Acre	Pounds Active Ingredient Per Acre	Precautions and Remarks
Late Postemergence Overtop, Salvage Treatment; Any Cultivar, Cocklebur and morningglory			
2,4-DB, MOA 4 (various brands) 2 SL 1.75 SL 75 WP	1 pt 1.1 pt 0.33 lb	0.25	Spray overtop soybeans from 1 week before bloom up to midbloom. This treatment may be used when needed as an aid to control cocklebur and morningglory and as a supplement to but not a replacement for early postemergence treatments. Salvage treatment only—substantial crop injury may occur. Do not add surfactant or crop oil. Not suggested for use on soybeans to be saved for seed.
Postemergence Directed; Any Cultivar, Small grasses and broadleaf weeds			
metribuzin, MOA 5 (Sencor) 75 WDG	0.33 to 0.67 lb	0.25 to 0.5	Apply only as a directed spray to soybeans at least 8 inches tall. Do not spray higher than 2 to 3 inches on the soybean stem. Add surfactant according to label directions. Do not use if soil has been wet for 2 to 3 days. Do not use on sandy soils or any soil having less than 0.5% organic matter. Some varieties of soybeans are sensitive to metribuzin; see label for details.
Postemergence Directed; Any Cultivar, Cocklebur and morningglory			
2,4-DB, MOA 4 (various brands) 2 SL 1.75 SL 75 WP	13 fl oz 15 fl oz 4.3 oz	0.2	Soybeans must be at least 8 inches tall. Contact no more than the lower third of soybean plant. Follow other precautions on label.
Postemergence with Wiper Applicators; Any Cultivar, Certain weeds taller than crop, especially grasses			
glyphosate, MOA 9 (numerous brands and formulations)	Not applicable; see label	Not applicable; see label	Apply glyphosate above crop with wiper-type applicator. Follow label directions carefully. Do not let glyphosate contact crop plants. Johnsongrass and tall annual grasses such as fall panicum are very susceptible; broadleaf weeds are less susceptible. Use only as supplement to a good early season weed management program.
Postemergence with Hooded Sprayer; Any Cultivar, Annual and perennial grasses and broadleaf weeds			
glyphosate, MOA 9 (numerous brands and formulations)	See label	0.38 to 1.13 (lb a.e.)	Glyphosate is available as an isopropylamine salt and a potassium salt. Glyphosate formulations and application rates should be compared on the basis of pounds of glyphosate acid equivalent (a.e.) per gallon and per acre, respectively. Rate in the preceding column is expressed as a.e. See TABLE 7-10 for glyphosate rate conversions. Glyphosate rate depends upon weed species and size; see labels for specific rates. Higher rates can be used for perennial weeds; see labels for details. Keep hoods as close to the ground as possible. Contact of spray with foliage of non-Roundup Ready soybeans will cause severe injury.
Harvest Aid; Any Cultivar, Annual and perennial weeds			
glyphosate, MOA 9 (numerous brands and formulations)	See label	0.75 to 2.25 (lb a.e.)	Glyphosate is available as an isopropylamine salt and a potassium salt. Glyphosate formulations and application rates should be compared on the basis of pounds of glyphosate acid equivalent (a.e.) per gallon and per acre, respectively. Rate in the preceding column is expressed as a.e. See TABLE 7-10 for glyphosate rate conversions. See labels for weeds controlled, maximum weed size to treat, and specific application rates for various species. Apply after pods have set and lost all green color. Apply at least 7 days before harvest. Can be applied by ground or air. Do not apply to soybeans grown for seed.
sodium chlorate (Defol 5) 5 L	4.8 qt	6	Apply when soybeans are mature and ready for harvest. Apply 7 to 10 days before anticipated harvest date. Apply in a minimum of 20 gallons of water per acre. Will control most grass and broadleaf weed species and desiccate soybean prior to harvest. Do not graze or harvest treated soybean for forage.
Harvest Aid; Any Cultivar, Annual grasses and broadleaf weeds			
paraquat, MOA 22 (Gramoxone Inteon) 2 SL	8 to 16 fl oz	0.13 to 0.25	Apply when pods are fully developed and at least one-half of leaves have dropped and leaves left on plants are turning yellow. Can be applied by ground or air. Generic brands containing 3 pounds active per gallon are available. This product would be used at 5.3 to 10.7 fluid ounces.
Harvest Aid; Any Cultivar, Annual broadleaf weeds			
carfentrazone, MOA 14 (Aim) 2 EC	1.5 fl oz	0.023	Desiccates morningglory, pigweed, and cocklebur. Apply 3 or more days ahead of harvest. Add 1 gallon crop oil concentrate per 100 gallons spray solution. Thorough coverage is essential; use a minimum of 20 GPA by ground equipment. May be applied by air. May tank mix with Gramoxone. Tank mixes with Gramoxone should be applied 15 days ahead of harvest.
saflufenacil, MOA 14 (Sharpen) 2.85 SC	1.0 to 2.0 fl oz	0.022 to 0.044	Apply when soybeans have reached physiological maturity, no green color on pods and seeds. Apply 3 or more days ahead of harvest, allow up to 10 days for optimum desiccation. See label for specific directions on indeterminant and determinant varieties. Add 1 gallon methylated seed oil per 100 gallons spray solution plus ammonium-based adjuvant system (AMS or UAN) for optimum desiccation. Do not apply to soybeans grown for seed.

[1] Mode of Action (MOA) code developed by the Weed Science Society of America. See Table 7-10, Herbicide Resistance Management, for details.

Weed Response to Preplant Incorporated and Preemergence Herbicides in Soybeans

W. J. Everman, Crop and Soil Sciences Department

Ratings are based upon average to good soil and weather conditions for herbicide performance and upon proper application rate, technique, and timing.

Table 7-5B. Weed Response to Preplant Incorporated and Preemergence Herbicides in Soybeans

Species	Prowl or Treflan PPI	Sonalan PPI	Authority MTZ PRE	Command PRE	Dual Magnum, Dual II Magnum PRE	Envive PRE	Fierce	Fierce XLT	Intrro or Micro-Tech PRE	Linex PRE	Outlook PRE	Prefix PRE	Prowl PRE	Python PRE	Reflex PRE	Scepter PRE	Sencor PRE	Valor SX PRE	Valor XLT PRE	Zidua/Zidua SC
Bermudagrass	N	N	N	PF	N	N	N	N	N	N	N	N	N	N	N	N	N	N	N	N
Broadleaf signalgrass	G	G	F	E	G	N	E	E	FG	P	FG	G	P	N	FG	PF	PF	N	N	E
Crabgrass	E	E	F	E	E	N	E	E	E	FG	E	E	F	P	FG	NP	F	N	N	E
Fall panicum	G	G	P	E	E	N	E	E	E	F	E	E	E	PF	N	NP	NP	N	N	E
Foxtails	E	E	F	E	E	N	E	E	E	FG	E	E	F	P	—	FG	NP	N	N	E
Goosegrass	E	E	FG	E	E	N	E	E	E	FG	E	E	E	PF	P	NP	F	N	N	E
Johnsongrass, Seedling	G	G	—	G	PF	N	F	F	PF	NP	PF	PF	PF	N	—	FG	PF	N	N	PF
Johnsongrass, Rhizome	P	P	N	N	N	N	N	N	N	N	N	N	N	N	—	N	N	N	N	N
Shattercane	G	G	—	F	P	N	G	G	P	N	P	P	PF	N	—	F	N	N	N	P
Texas panicum	G	G	PF	F	PF	N	F	F	PF	PF	PF	F	PF	N	F	NP	N	N	N	F
Nutsedge, Yellow	N	N	E	N	FG³	N	F	F	P	N	F	GE	N	N	GE	PF	N	N	N	F
Nutsedge, Purple	N	N	E	N	N	N	N	N	N	N	N	—	N	N	N	NP	N	N	N	N
Balloonvine	N	N	—	G	N	—	G	G	N	F	N	N	N	P	—	F	G	—	—	N
Eastern black nightshade	N	F	—	P	F	E	E	E	FG	NP	F	—	P	PF	—	PF	N	E	E	F
Burcucumber¹	N	N	—	NP	N	—	N	N	N	N	N	N	N	P	—	PF	P	—	—	N
Cocklebur	N	N	G	F	N	FG	P	P	N	N	N	G	N	G	G	E	PF	P	FG	N
Cowpea	N	N	—	N	N	—	—	—	N	N	N	—	N	NP	—	N	PF	—	—	N
Crotalaria	N	N	—	N	N	—	—	—	N	N	N	—	N	—	—	—	NP	—	—	N
Florida beggarweed	N	N	G	FG	F	E	E	E	F	F	F	F	N	F	P	F	G	E	E	E
Florida pusley	E	E	—	P	G	GE	GE	GE	G	G	G	G	E	G	G	GE	G	GE	GE	G
Hemp sesbania	N	N	GE	N	N	G	G	G	N	F	N	P	N	N	P	N	GE	G	G	N
Jimsonweed	N	N	E	G	E	E	E	E	N	—	N	N	G	—	FG	G	G	E	E	F
Lambsquarters	G	G	E	G	G	F	E	E	E	F	GE	FG	E	G	E	E	G	E	E	FG
Morningglory	P	P	E	P²	N	G	G	G	N	P	N	PF	N	F	PF	G	P	G	G	N
Palmer amaranth	G	G	G	N	FG	E	E	E	GE	PF	G	E	PF	G⁴	E	GE⁴	G	E	E	E
Pigweed, Redroot and Smooth	G	G	E	N	G	E	E	E	GE	E	GE	E	FG	E	E	E	E	E	E	E
Prickly sida	N	N	GE	E	P	E	E	E	P	F	P	—	N	E	—	G	E	E	E	P
Ragweed, Common	N	N	GE	G	PF	G	G	G	PF	G	N	G	N	FG	G	G	G	G	G	F
Ragweed, Giant	N	N	—	PF	N	F	F	F	N	P	N	G	N	PF	G	G	P	F	F	N
Sicklepod	N	N	G	P	NP	F	P	F	PF	NP	NP	P	N	G	P	FG	G	P	F	N
Smartweed	N	N	E	E	N	—	F	—	N	F	N	—	N	G	—	G	G	F	—	F
Spurred anoda	N	N	G	E	N	E	E	E	N	P	N	E	N	—	FG	P	G	E	E	N
Tropic croton	N	N	—	E	E	E	E	E	PF	N	FG	N	—	FG	NP	FG	E	E	E	N
Velvetleaf	N	N	GE	E	N	G	F	G	P	N	N	GE	—	N	—	F	G	F	G	F

¹ Multiple flushes of germination; one application of any herbicide will seldom be adequate.
² Fair on pitted morningglory.
³ Good on yellow nutsedge when incorporated.
⁴ Palmer amaranth resistant to ALS inhibitors is common in NC. This ALS-inhibiting herbicide will perform poorly on resistant biotypes.

Key:
E = excellent control, 90% or better
G = good control, 80% to 90%
F = fair control, 50% to 80%
P = poor control, 25% to 50%
N = no control, less than 25%

Weed Response to Postemergence Herbicides in Soybeans

W. J. Everman, Crop and Soil Sciences Department

Ratings based upon average to good soil and weather conditions for herbicide performance and upon proper application rate, technique, and timing.

Table 7-5C. Weed Response to Postemergence Herbicides in Soybeans

Species	Assure II	Fusilade	Poast	Select, Select Max	Basagran	Classic	Cobra	FirstRate	Flexstar	Flexstar GT	Glyphosate[1]	Harmony SG	Pursuit	Raptor	Reflex	Resource	Scepter	Storm	Synchrony STS[2]	Ultra Blazer	Liberty
Bermudagrass	G	G	FG	G	N	N	N	N	N	—	G[3]	N	N	N	N	N	N	N	N	N	N
Broadleaf signalgrass	GE	GE	E	E	N	N	N	N	N	E	E	N	G	G	N	N	N	NP	N	NP	G
Crabgrass	G	G	GE	GE	N	N	N	N	N	E	E	N	PF	PF	N	N	N	N	N	N	FG
Fall panicum	E	E	E	E	N	N	P	N	N	E	E	N	F	G	N	N	N	P	N	P	G
Foxtails	E	E	E	E	N	N	P	N	N	E	E	N	G	G	N	N	N	P	N	P	G
Goosegrass	GE	GE	GE	GE	N	N	N	N	N	E	E	N	NP	NP	N	N	N	N	N	N	P
Johnsongrass, Seedling	E	E	E	E	N	NP	P	N	N	E	E	N	GE	GE	N	N	N	P	NP	P	G
Johnsongrass, Rhizome	E	GE	G	GE	N	N	N	N	N	E	E	N	G[5]	G[5]	N	N	N	N	N	N	F
Shattercane	E	E	E	E	N	N	P	N	N	E	E	N	G	G	N	N	N	P	N	P	—
Texas panicum	G	G	E	E	N	N	N	N	N	E	E	N	PF	PF	N	N	N	N	N	N	G
Nutsedge, Purple	N	N	N	N	NP	PF	N	PF	N	G	G	N	G	—	N	N	N	N	PF	N	P
Nutsedge, Yellow	N	N	N	N	G[3]	G	N	PF	F	—	FG[4]	N	FG	FG	F	N	P	F	G	N	P
Balloonvine	N	N	N	N	P	FG	GE	P	G	E	G	—	P	—	G	P	P	G	FG	GE	—
Eastern black nightshade	N	N	N	N	P	F	G	N	G	E	FG	N	G	G	G	P	P	G	F	G	—
Burcucumber[6]	N	N	N	N	P	G	G	F	FG	G	F	G	PF	PF	FG	F	P	F	G	FG	—
Cocklebur	N	N	N	N	E	E	GE	E	E	E	E	FG	E	E	E	G	E	E	E	G	E
Cowpea	N	N	N	N	N	GE	F	P	PF	E	E	—	N	—	P	—	N	P	GE	PF	G
Crotalaria	N	N	N	N	P	G	G	—	G	E	G	—	N	—	G	—	N	E	G	E	—
Florida beggarweed	N	N	N	N	N	E	FG	FG	P	G	G	—	N	—	P	P	NP	P	E	PF	G
Hemp sesbania	N	N	N	N	P	E	G	PF	E	E	PF	—	N	—	E	P	N	GE	E	E	—
Jimsonweed	N	N	N	N	E	E	GE	E	E	E	E	F	GE	E	E	G	NP	E	E	E	E
Lambsquarters	N	N	N	N	FG	N	P	PN	F	E	E	E	PF	G	PF	G	P	G	E	G	E
Morningglory	N	N	N	N	P	G	G	E	GE	E	FG[8]	FG	FG	FG	GE	FG	P	GE	G	GE	E
Palmer amaranth	N	N	N	N	N	F[11]	G	P	G	E[10]	E[10]	GE[11]	GE[11]	—	G	FG	G[11]	G	E[11]	G	FG
Pigweed, Redroot or Smooth	N	N	N	N	N	G	E	P	GE	E	E	E	E	E	GE	G	E	E	E	E	G
Prickly sida	N	N	N	N	G	N	G	P	F	G	G	N	P	G	NP	N	PF	FG	P	N	G
Ragweed, Common	N	N	N	N	G[9]	G	E	E	GE	E	E	F	PF	F	GE	G	F	E	G	E	E
Ragweed, Giant	N	N	N	N	GE	FG	G	GE	E	E	G	P	F	F	E	P	P	GE	FG	GE	E
Sicklepod	N	N	N	N	N	G	NP	F[7]	P	E	P	N	P	N	FG	NP	G	NP	E		
Smartweed	N	N	N	N	E	E	F	E	G	E	G	E	GE	G	G	P	FG	E	E	GE	GE
Spurred anoda	N	N	N	N	G	F	F	F	F	E	N	F	F	P	P	NP	F	F	P	P	
Tropic croton	N	N	N	N	F	NP	G	P	G	E	N	N	G	P	G	N	G	NP	G	G
Velvetleaf	N	N	N	N	G	F	G	G	F	E	G	F	E	P	E	NP	FG	G	PF	—	

1. Apply to Roundup Ready (glyphosate-resistant) cultivars only.
2. Apply only to BOLT, STS cultivars.
3. Assumes two applications.
4. Yellow nutsedge control is good with two applications of glyphosate.
5. Follow-up treatment with a postemergence grass herbicide may be necessary.
6. Multiple flushes of germination; one application of any herbicide will seldom be adequate.
7. FirstRate is good on sicklepod if applied at cotyledonary to first leaf stage.
8. With good timing and a follow-up application as needed, morningglory control can be good.
9. Assumes addition of crop oil concentrate.
10. Palmer amaranth resistant to glyphosate is common in NC. Glyphosate will perform poorly on resistant biotypes.
11. Palmer amaranth resistant to ALS-inhibiting herbicides is common in NC. ALS-inhibiting herbicides will perform poorly on resistant biotypes.

Key:
E = excellent control, 90% or better
G = good control, 80% to 90%
F = fair control, 50% to 80%
P = poor control, 25% to 50%
N = no control, less than 25%

Chemical Weed Control in Sunflowers

D. L. Jordan, Crop and Soil Sciences Department

NOTE: A mode of action code has been added to the Herbicide and Formulation column of this table. Use MOA codes for herbicide resistance management. See Table 7-10, Herbicide Resistance Management, for details.

Table 7-6. Chemical Weed Control in Sunflowers

Herbicide, Mode of Action Code, and Formulation	Amount of Formulation Per Acre	Pounds Active Ingredient Per Acre	Precautions and Remarks
Preplant Foliar, Burndown of weeds and cover crops before planting			
glyphosate, MOA 9 (numerous brands and formulations)	See label	0.75 (lb a.e.)	Apply before or after planting but before sunflowers emerge. See labels for suggested weed sizes to treat and for application rates. Add nonionic surfactant or crop oil concentrate according to Gramoxone label. The need for an adjuvant with glyphosate depends upon the brand used; see the label of the brand used.
paraquat, MOA 22 (Gramoxone Inteon) 2 SL	2 to 4 pt	0.5 to 1	
Preplant Incorporated, Annual grasses and small-seeded broadleaf weeds			
ethalfluralin, MOA 3 (Sonalan HFP) 3 EC	1.5 to 3 pt	0.56 to 1.13	Controls common annual grasses plus pigweed and lambsquarters. Incorporate into top 2 to 3 inches of seedbed. See label for application rate based upon soil texture. Generic brands of pendimethalin and trifluralin are available.
pendimethalin, MOA 3 (Prowl) 3.3 EC (Prowl H$_2$O) 3.8 L	1.2 to 3.6 pt 1.5 to 3.0 pt	0.5 to 1.5 0.71 to 1.43	
trifluralin, MOA 3 (Treflan HFP) 4 EC	1 to 2 pt	0.5 to 1	
Preplant Incorporated, Annual grasses, pigweed, and yellow nutsedge			
S-metolachlor, MOA 15 (Brawl) 7.62 EC (Dual Magnum) 7.62 EC	1 to 2 pt	0.95 to 1.91	Controls common annual grasses except Texas panicum, seedling johnsongrass, and shattercane. Also controls pigweed. At higher rates, controls yellow nutsedge. Incorporate into top 2 to 3 inches of seedbed. See label for application rate based upon soil texture.
Preemergence, Annual grasses and small-seeded broadleaf weeds			
pendimethalin, MOA 3 (Prowl H$_2$O) 3.8 L (Prowl) 3.3 EC	1.5 to 3 pt 1.2 to 3.6 pt	0.71 to 1.43 0.5 to 1.5	See above comments for pendimethalin applied preplant incorporated. Pendimethalin is more consistently effective when incorporated.
S-metolachlor, MOA 15 (Brawl) 7.62 EC (Dual Magnum) 7.62 EC	1 to 2 pt	0.95 to 1.91	See above comments for Dual Magnum applied preplant incorporated. Dual Magnum is more consistently effective on yellow nutsedge when incorporated.
Preemergence, Annual broadleaf weeds and nutsedge			
sulfentrazone, MOA 14 (Spartan) 4 F	3 to 8 oz	0.094 to 0.25	Controls nutsedge and most common annual broadleaf weeds. Only fair control of cocklebur, and no control of ragweed or sicklepod. May tank mix with other registered preemergence herbicides for annual grass control. Adjust Spartan and Spartan Charge application rate according to soil texture and organic matter as specified on label. May also be shallowly incorporated in the top 2 inches of seed bed. Do not plant corn, sweet potatoes, or cotton for 10, 12, and 12 months, respectively, after Spartan or Spartan Charge application. See label for rotational restrictions on other crops.
sulfentrazone, MOA 14 + carfentrazone, MOA 14 (Spartan Charge) 0.35 + 3.15 F	3.75 to 10.2 fl oz		
Postemergence, Annual grasses			
clethodim, MOA 1 (Select) 2 EC (Select Max) 0.97 EC	6 to 8 fl oz 9 to 16 fl oz	0.094 to 0.125 0.068 to 0.121	Apply to actively growing grasses not under drought stress. See label for grass size, application rates, and directions. Add 2 pints of crop oil concentrate to Select or Poast. To Select Max, add one of the following: nonionic surfactant at 0.25% by volume; crop oil concentrate at 1.0% by volume; or methylated seed oil at 1% by volume. Other formulations of clethodim are available.
sethoxydim, MOA 1 (Poast) 1.5 EC	16 fl oz	0.19	
Postemergence, Bermudagrass			
clethodim, MOA 1 (Select) 2 EC (Select Max) 0.97 EC	6 to 8 fl oz 12 to 32 fl oz	0.094 to 0.125 0.091 to 0.24	Apply before bermudagrass runners exceed 6 inches If needed, make second application of 8 to 16 fluid ounces of Select or 12 to 32 fluid ounces of Select Max when regrowth is less than 6 inches Add crop oil concentrate to Select at 1 quart per acre. To Select Max, add nonionic surfactant at 0.25% by volume, crop oil concentrate at 1.0% by volume, or methylated seed oil at 1% by volume. Other formulations of clethodim are available.
sethoxydim, MOA 1 (Poast) 1.5 EC	24 fl oz	0.28	Apply before bermudagrass runners exceed 6 inches If needed, make second application of 16 fluid ounces per acre when regrowth is less than 4 inches Add crop oil concentrate at 1 quart per acre.
clethodim, MOA 1 (Select) 2 EC (Select Max) 0.97 EC	8 to 16 fl oz 12 to 32 fl oz	0.125 to 0.25 0.091 to 0.24	Apply when johnsongrass is 12 to 24 inches tall. If needed, make second application of Select at 6 to 8 fluid ounces or 9 to 24 fluid ounces of Select Max when regrowth is 6 to 18 inches Add crop oil concentrate to Select at 1 quart per acre. To Select Max, add nonionic surfactant at 0.25% by volume, crop oil concentrate at 1.0% by volume, or methylated seed oil at 1% by volume. Other formulations of clethodim are available.
sethoxydim, MOA 1 (Poast) 1.5 EC	24 fl ounces	0.28	Apply when johnsongrass is 20 to 25 inches tall. If needed, make second application of 16 fluid ounces per acre when regrowth is 12 inches Add crop oil concentrate at 1 quart per acre.
quizalofop, MOA 1 (Assure II) 0.88 EC	5 to 12 oz	0.034 to 0.069	Apply to rhizome johnsongrass 10 to 24 inches tall and repeat when regrowth is 6 to 10 inches. Apply to bermudagrass with runners up to 6 inches and repeat when regrowth is present. Apply with crop oil concentrate or nonionic surfactant based on label recommendations. Do not apply more than 18 oz per season and do not apply after seed set.
Postemergence, Annual grasses and broadleaf weeds: Clearfield Cultivars Only			
imazamox, MOA 2 (Beyond) 1 L	4 fl oz	0.031	**APPLY ONLY TO CLEARFIELD CULTIVARS**. Beyond will severely injure or kill non-Clearfield cultivars. Apply to sunflower in the 2- to 8-leaf stage when broadleaf weeds are 3 inches or less. Add crop oil concentrate or nonionic surfactant plus nitrogen according to label directions. Controls many common broadleaf weeds plus some annual grasses. See label for weeds controlled.
Postemergence with Hooded Sprayer, Annual broadleaf weeds, annual and perennial grasses			
glyphosate, MOA 9 (numerous brands and formulations)	See label	0.75 (lb a.e.)	Glyphosate is available as an isopropylamine salt and a potassium salt. Glyphosate formulations and application rates should be compared on the basis of pounds of glyphosate acid equivalent (a.e.) per gallon and per acre, respectively. Rate in the preceding column is expressed as a.e. Apply to row middles using hooded or shielded sprayer that allows no contact of sunflowers by spray solution.

Table 7-6. Chemical Weed Control in Sunflowers

Herbicide, Mode of Action Code, and Formulation	Amount of Formulation Per Acre	Pounds Active Ingredient Per Acre	Precautions and Remarks
Postemergence with Hooded Sprayer, Annual broadleaf weeds, annual and perennial grasses (continued)			
carfentrazone, MOA 14 (Aim) 2 EC	1.0 to 2.0 oz	0.016 to 0.032	Apply using hooded or shielded sprayers and avoid contact with sunflower foliage. Apply with 1 quart of nonionic surfactant per 100 gallons water. Very effective in controlling morning glory

Chemical Weed Control in Tobacco

M. C. Vann, L. R. Fisher, M. D. Inman, and D. S. Whitley, Crop and Soil Sciences Department

NOTE: A mode of action code has been added to the Herbicide and Formulation column of this table. Use MOA codes for herbicide resistance management. See Table 7-10, Herbicide Resistance Management, for details.

Table 7-7A. Chemical Weed Control in Tobacco

Herbicide, Mode of Action Code,[1] and Formulation	Amount of Formulation Per Acre	Pounds Active Ingredient Per Acre	Precautions and Remarks
Flue-Cured Field (before transplanting), Most annual grasses and some broadleaf weeds plus nutsedge suppression			
pebulate, MOA 8 (Tillam) 6 EC	2.7 qt	4.0	Apply to soil surface before bedding and immediately incorporate according to label instructions. Transplant as soon as possible. Early season stunting may occur under unfavorable growing conditions. Does not control cocklebur, morningglory, ragweed, or perennial weeds. Cultivate tobacco at least twice. See label for tank mixes with other pesticides.
Flue-Cured Field (before transplanting), Some annual grasses and some broadleaf weeds			
napropamide, MOA 17 (Devrinol) 2 XT (Devrinol) 50 DF	2 to 4 qt 2 to 4 lb (broadcast, see label for band application)	1.0 to 2.0 1.0 to 2.0	Lower rates usually adequate for most soils. Apply to soil surface and incorporate according to label instructions. Some early season stunting may occur under unfavorable growing conditions. Does not control cocklebur, morningglory, or perennial weeds. Gives some suppression of ragweed. NOTE: Do not seed crops not specified on label for 12 months after application.
Flue-Cured Field (before transplanting), Most annual grasses and some broadleaf weeds			
pendimethalin, MOA 3 (Prowl) 3.3 EC (Prowl) H₂O (Helena-Pendimethalin)	2.4 to 3.0 pt 2.0 to 2.5 pt 2.4 to 3.0 pt	1.0 to 1.25 0.95 to 1.19 1.0 to 1.25	Can be applied up to 60 days before transplanting. Apply before bedding and incorporate into soil according to label instructions. Some early season stunting may occur under unfavorable growing conditions. Lower application rates should be used on coarser soil types with low organic matter content. Does not control cocklebur, morningglory, ragweed, or perennial weeds.
Flue-Cured Field (before transplanting), Annual grasses and some broadleaf weeds			
clomazone, MOA 13 (Command) 3 ME (Willowood Clomazone) 3ME	2 to 2.67 pt 2 to 2.67 pt	0.75 to 1.0 0.75 to 1.0	Excellent annual grass control plus control of certain broadleaf weeds, such as prickly sida, jimsonweed, tropic croton, smartweed, and common ragweed. Partial control of cocklebur; does not control pigweed, sicklepod, or morningglory. Some whitening of lower leaves may occur, but plants should recover. Do not plant small grains or alfalfa in the fall or following spring after Command application. Apply no more than once per season.
Flue-Cured Field (before transplanting), Broadleaf weeds, nutsedges, and some grasses			
sulfentrazone, MOA 14 (Spartan) 4F (Willowood Sulfentrazone) 4SC (Helm Sulfentrazone) 4F (Shutdown) 4.16	4.5 to 12 fl oz 4.5 to 12 fl oz 4.5 to 12 fl oz 4.5 to 11.8 fl oz	0.14 to 0.38 0.14 to 0.38 0.14 to 0.38 0.15 to 0.38	Excellent control of pigweed, morningglories, and nutsedges. Application rate is based on soil type and organic matter. See appropriate label for rate determination and application methods. Early season stunting may occur especially when incorporated. Rainfall or irrigation needed within 7 to 10 days of application for maximum weed control, particularly when surface applied. Observe rotational crop guidelines on label. Most formulations can be tank mixed with clomazone for improved control of grass and ragweed species, verify each label independently.
sulfentrazone + carfentrazone, 0.35 + 3.15 MOA 14 + 14 (Spartan Charge)	5.7 to 15.2 fl oz	0.16 to 0.41 lb	
Flue-Cured Field (after transplanting), Most annual grasses and some broadleaf weeds			
napropamide, MOA 17 (Devrinol) 2 XT (Devrinol) 50 DF	2 to 4 qt 2 to 4 lb (broadcast, see label for band application)	1.0 to 2.0 1.0 to 2.0	Apply overtop immediately after transplanting tobacco. See remarks for Devrinol under "Before Transplanting." NOTE: Do not seed crops not specified on label for 12 months after application. Small grain seeded for cover crop in fall may be stunted. Do not use small grain for food or feed.
Flue-Cured Field (after transplanting)			
clomazone, MOA 13 (Command) 3 FME (Willowood Clomazone) 3ME	2 to 2.67 pt 2 to 2.67 pt	0.75 to 1.0 0.75 to 1.0	Excellent annual grass control plus control of certain broadleaf weeds, such as prickly sida, jimsonweed, tropic croton, smartweed, and common ragweed. Partial control of cocklebur; does not control pigweed, sicklepod, or morningglory. Make a single broadcast application in a minimum of 20 gallons of water. Apply no more than once per season. Apply over the top of tobacco plants immediately or up to 7 days after transplanting but prior to emergence of weeds. Some whitening of lower leaves may occur, but plants should recover. Do not plant small grains or alfalfa in the fall or following spring after Command application.
Flue-Cured Field (after transplanting), Postemergence control of annual grasses			
sethoxydim, MOA 1 (Poast) 1.5 EC	1 to 1.5 pt	0.19 to 0.28	Apply to actively growing grass not under drought stress. Apply in 5 to 20 gallons of spray at 40 to 60 psi. Add 2 pints of crop oil concentrate per acre. Do not apply within 42 days of harvest. Do not apply more than 4 pints per acre per season. Complete coverage of grass required for control.
Flue-Cured Field (after transplanting), Postemergence control of some broadleaf weeds			
Carfentrazone MOA 14 (Aim) 2 EC	0.8 to 1.5 oz	0.0125 to 0.024	Apply using SHIELDED SPRAYER or HOODED SPRAYER to emerged, actively growing weeds PRIOR TO LAYBY. Do not apply when conditions favor drift. MUST PREVENT CONTACT OF SPRAY SOLUTION WITH TOBACCO PLANT. See Label for further instruction.

Table 7-7A. Chemical Weed Control in Tobacco

Herbicide, Mode of Action Code,[1] and Formulation	Amount of Formulation Per Acre	Pounds Active Ingredient Per Acre	Precautions and Remarks
Flue-Cured Lay-by, Most annual grasses and some broadleaf weeds			
napropamide, MOA 17 (Devrinol), 2 XT (Devrinol), 50 DF	2 to 4 qt 2 to 4 lb (band, see label for band application)	1 to 2	Apply in a band to row middles immediately after last cultivation. Lower rates usually adequate for most tobacco soils. Incorporate lightly or sprinkler irrigate, if no rainfall within 3 days after application. Do not apply more than a total of 4 lb of Devrinol per acre in a season. See remarks for Devrinol under "Before Transplanting" and "After Transplanting."
pendimethalin, MOA 3 (Prowl) 3.3 EC (Prowl) H_2O (Helena-Pendimethalin)	1.8 to 2.4 pt 1.5 to 2.0 pt 1.8 to 2.4 pt	0.75 to 1.0 0.71 to 0.95 0.75 to 1.0	Apply to row middles immediately after last cultivation. Avoid contact with tobacco leaves. Use higher rate on medium- or fine-textured soils where grass infestation is heavy or if no herbicide was used previously. Rainfall or irrigation is needed within 7 days. Does not control emerged weeds.
Flue-Cured after first harvest, Postemergence control of some broadleaf weeds			
carfentrazone MOA 14 (Aim) 2 EC	0.8 to 1.5 oz	0.0125 to 0.024	Apply AFTER FIRST HARVEST for control of actively growing, emerged weeds. Position nozzles 3 to 4 inches above the soil and directed underneath the crop canopy. Do not apply when conditions favor drift. MUST PREVENT CONTACT OF SPRAY SOLUTION WITH TOBACCO PLANT. See label for further instruction.
Burley Field (before transplanting), Most annual grasses and some broadleaf weeds plus nutsedge suppression			
pebulate, MOA 8 (Tillam) 6 EC	2.7 qt	4.0	See remarks for Tillam under Flue-Cured. Tank mix suppresses hairy galinsoga under ideal conditions (ample rainfall after application).
Burley Field (before transplanting), Some annual grasses and broadleaf weeds			
napropamide, MOA 17 (Devrinol), 2 XT (Devrinol), 50 DF	2 to 4 qt 2 to 4 lb (broadcast, see label for band application)	1.0 to 2.0 1.0 to 2.0	See remarks for Devrinol under Flue-Cured. Suppresses hairy galinsoga under ideal conditions (ample rainfall after application).
Burley Field (before transplanting), Most annual grasses and some broadleaf weeds			
pendimethalin, MOA 3 (Prowl) 3.3 EC (Prowl) H_2O (Helena-Pendimethalin)	2.4 to 3.0 pt 2.0 to 2.5 pt 2.4 to 3.0 pt	1.0 to 1.25 0.95 to 1.19 1.0 to 1.25	See remarks for Prowl under Flue-Cured. The higher labelled rates may be needed for soils where burley tobacco is grown in N.C. due to higher organic matter content. Add hairy galinsoga to list of weeds not controlled.
Burley Field (before transplanting), Annual grasses and some broadleaf weeds			
clomazone, MOA 13 (Command 3 ME) 3 FME (Willowood Clomazone) 3ME	2 to 2.67 pt 2 to 2.67 pt	0.75 to 1 0.75 to 1.0	Excellent annual grass control plus control of certain broadleaf weeds, such as prickly sida, jimsonweed, tropic croton, smartweed, and common ragweed. Partial control of cocklebur; does not control pigweed, sicklepod, or morningglory. Some whitening of lower leaves may occur, but plants should recover. Do not plant small grains or alfalfa in the fall or following spring after Command application. Apply no more than once per season.
Burley Field (before transplanting), Broadleaf weeds, nutsedges, and some grasses			
sulfentrazone, MOA 14 (Spartan) 4F (Willowood Sulfentrazone) 4SC (Helm Sulfentrazone) 4F (Shutdown) 4.16	4.5 to 12 fl oz 4.5 to 12 fl oz 4.5 to 12 fl oz 4.5 to 11.8 fl oz	0.14 to 0.38 0.14 to 0.38 0.14 to 0.38 0.15 to 0.38	Excellent control of pigweed, morningglories, and nutsedges. Application rate is based on soil type and organic matter. See appropriate label for rate determination and application methods. Early season stunting may occur especially when incorporated. Rainfall or irrigation needed within 7 to 10 days of application for maximum weed control, particularly when surface applied. Observe rotational crop guidelines on label. Most formulations can be tank mixed with clomazone for improved control of grass and ragweed species, verify each label independently.
sulfentrazone + carfentrazone, 0.35 + 3.15 MOA 14 + 14 (Spartan Charge)	5.7 to 15.2 fl oz	0.16 to 0.41	
Burley Field (after transplanting), Most annual grasses and some broadleaf weeds			
napropamide, MOA 17 (Devrinol), 2 XT (Devrinol) 50 DF	2 to 4 qt 2 to 4 lb (broadcast, see label for band application)	1.0 to 2.0	Apply overtop immediately after transplanting tobacco. See remarks for Devrinol under "After Transplanting" in flue-cured section. Suppresses hairy galinsoga under ideal conditions (ample rainfall after application).
Burley Field (after transplanting), Annual grasses and some broadleaf weeds			
clomazone, MOA 13 (Command) 3 ME (Willowood Clomazone) 3ME	2 to 2.67 pt 2 to 2.67 pt	0.75 to 1 0.75 to 1.0	Excellent annual grass control plus control of certain broadleaf weeds, such as prickly sida, jimsonweed, tropic croton, smartweed, and common ragweed. Partial control of cocklebur; does not control pigweed, sicklepod, or morningglory. Make a single broadcast application in a minimum of 20 gallons of water. Apply no more than once per season. Apply over the top of tobacco plants immediately, or up to 7 days after transplanting, but prior to emergence of weeds. Some whitening of lower leaves may occur, but plants should recover. Do not plant small grains or alfalfa in the fall or following spring after applying Command.
Burley Field (after transplanting), Postemergence control of annual grasses			
sethoxydim, MOA 1 (Poast) 1.5 EC	1 to 1.5 pt	0.19 to 0.28	Apply to actively growing grass not under drought stress. Apply in 5 to 20 gallons of spray at 40 to 60 psi. Add 2 pints of crop oil concentrate per acre. Do not apply within 42 days of harvest. Do not apply more than 4 pints per acre per season. Complete coverage of grass required for control.
Burley Field (after transplanting), Postemergence control of some broadleaf weeds			
carfentrazone MOA 14 (Aim) 2 EC	0.8 to 1.5 oz	0.0125 to 0.024	Apply using SHIELDED SPRAYER or HOODED SPRAYER to emerged, actively growing weeds PRIOR TO LAYBY. Do not apply when conditions favor drift. MUST PREVENT CONTACT OF SPRAY SOLUTION WITH TOBACCO PLANT. See Label for further instruction.

Table 7-7A. Chemical Weed Control in Tobacco

Herbicide, Mode of Action Code,[1] and Formulation	Amount of Formulation Per Acre	Pounds Active Ingredient Per Acre	Precautions and Remarks
Burley Lay-by, Most annual grasses and some broadleaf weeds			
napropamide, MOA 17 (Devrinol), 2 XT (Devrinol) 50 DF	2 to 4 qt 2 to 4 lb (band, see label for band application)	1 to 2	Apply in a band to row middles immediately after last cultivation. Lower rates usually adequate for most tobacco soils. Incorporate lightly or sprinkler irrigate if no rainfall within 3 days after application. Do not apply more than a total of 4 pounds of Devrinol 50 WP per acre in a season. See remarks for Devrinol under "After Transplanting" in flue-cured section. Suppresses hairy galinsoga under ideal conditions (ample rainfall after application).
pendimethalin, MOA 3 (Prowl) 3.3 EC (Prowl) H$_2$O (Helena-Pendimethalin)	1.8 to 2.4 pt 1.5 to 2.0 pt 1.8 to 2.4 pt	0.75 to 1.0 0.71 to 0.95 0.75 to 1.0	Apply to row middles immediately after last cultivation. Avoid contact with tobacco leaves. Use higher rate on medium- or fine-textured soils where grass infestation is heavy or if no herbicide was used previously. Rainfall or irrigation is needed within 7 days. Does not control emerged weeds.

[1] Mode of Action (MOA) code developed by the Weed Science Society of America. See Table 7-10, Herbicide Resistance Management, for details.

Weed Response to Herbicides in Tobacco

L. R. Fisher, M. C. Vann, M. D. Inman, and D. S. Whitley, Crop and Soil Sciences Department

Ratings based upon average to good soil and weather conditions for herbicide performance and upon proper application rate, technique, and timing.

Table 7-7B. Weed Response to Herbicides in Tobacco

Species	Command	Devrinol	Poast	Prowl	Spartan	Tillam	Aim
Barnyardgrass	E	GE	E	GE	F	GE	N
Bermudagrass	PF	P	FG	P	P	P	N
Broadleaf signalgrass	E	G	E	G	F	P	N
Crabgrass	E	E	GE	E	F	E	N
Crowfootgrass	E	E	FG	E	F	E	N
Fall panicum	E	G	E	GE	—	G	N
Foxtails	E	E	E	E	F	E	N
Goosegrass	E	E	GE	E	F	G	N
Sandbur	G	—	FG	G	—	G	P
Seedling johnsongrass	G	F	E	G	—	G	N
Texas panicum	G	—	E	G	F	P	N
Nutsedge	P	P	N	P	E	FG	N
Cocklebur	F	P	N	P	FG	P	G
Common purslane	FG	E	N	P	G	G	G
Ragweed							
Common	G	F	N	P	P	P	N
Giant	PF	PF	N	P	—	P	N
Hairy galinsoga	G	PF	N	P	G	P	P
Jimsonweed	G	P	N	P	—	P	G
Lambsquarters	G	G	N	G	E	G	G
Morningglory	P	P	N	P	E	P	E
Pigweed	P	G	N	G	E	G	E
Prickly sida	E	P	N	P	G	P	P
Sicklepod	P	P	N	P	P	P	P
Smartweed	G	P	N	P	E	P	G

KEY

E = excellent control, 90% or better

G = good control, 80% to 90%

F = fair control, 50% to 80%

P = poor control, 25% to 50%

N = no control, less than 25%

Chemical Weed Control in Wheat, Barley, Oats, Rye, and Triticale

W. J. Everman, Crop and Soil Sciences Department

NOTE: A mode of action code has been added to the Herbicide and Formulation column of this table. Use MOA codes for herbicide resistance management. See Table 7-10, Herbicide Resistance Management, for details.

Table 7-8A. Chemical Weed Control in Wheat, Barley, Oats, Rye, and Triticale

Herbicide, Mode of Action Code[1], and Formulation	Amount of Formulation per Acre	Pounds Active Ingredient per Acre	Precautions and Remarks
Wheat Preplant No-Till, Emerged annual broadleaf and grass weeds, volunteer corn, top-kill of perennials			
paraquat, MOA 22 (Gramoxone Inteon) 2 SL (Gramoxone Max) 3 SL	2 to 4 pt	0.5 to 1	Rate depends upon weed size; see label. Apply before crop emerges. Add nonionic surfactant at 1 pint per 100 gallons spray solution or crop oil concentrate at 1 gallon per 100 gallons spray solution. See application directions on label. May be tank mixed with Hoelon. Generic formulations of paraquat containing 3 pounds active per gallon are available. Apply these products at 1.3 to 2.7 pints.
Wheat Preplant No-Till, Emerged annual broadleaf and grass weeds, control or suppression of perennial weeds			
glyphosate, MOA 9 (numerous brands and formulations)	See label	0.38 to 1.13 (lb a.e.)	Glyphosate is available as an isopropylamine salt and a potassium salt. Glyphosate formulations and application rates should be compared on the basis of pounds of glyphosate acid equivalent (a.e.) per gallon and per acre, respectively. Rate in the preceding column is expressed as a.e. See Table 7-10 for glyphosate rate conversions. Rate depends upon weed species and size; see label. Apply before crop emergence. Adjuvant recommendations vary by glyphosate brand; see label of brand used for details. Select or Select Max may be mixed with glyphosate to control volunteer Roundup Ready corn. For corn up to 12 inches tall, apply 4 to 6 fluid ounces of Select or 6 fluid ounces of Select Max. For corn up to 24 inches increase Select rate to 6 to 8 fluid ounces or Select Max rate to 9 fluid ounces. Select or Select Max must be applied at least 30 days ahead of wheat planting. Valor SX at 1 to 2 ounces per acre will suppress ryegrass and bluegrass and controls several broadleaf weeds. Do not till after application and apply at least 30 days prior to planting.
Wheat Preplant No-Till or Preemergence, Broadleaf weeds			
saflufenacil, MOA 14 (Sharpen) 3.42	1.0 to 2.0 fl oz	0.027 to 0.054	See label for broadleaf weeds controlled. Sharpen does not control grasses. Apply with ammonium sulfate (1 to 2 gallons/100 gallons) and methylated seed oil (1 gallons/100 gallons). Do not apply if wheat has germinated. See label for tank mixtures.
flumioxazin, MOA 14 (Valor SX) 51 WDG + paraquat, MOA 22 (Gramoxone Inteon) 2.5 SL OR glyphosate, MOA 9 (numerous brands and formulations)	2.0 oz 2 to 4 pt See label	0.063 0.5 to 1 0.38 to 1.13 (lb a.e.)	Use only on no-till or minimum till fields where stubble from the previous crop has not been incorporated. Valor SX must be applied at least 7 days ahead of planting wheat. Residual control of broadleaf weeds and Italian ryegrass. Apply with nonionic surfactant at 1 quart/100 gallons. Can be applied with nitrogen carriers. Do no perform tillage after application. Carefully follow label directions for sprayer cleaning after each day's use.
thifensulfuron, MOA 2 + tribenuron, MOA 2 (FirstShot SG with TotalSol) 50 WDG	0.5 to 0.8 oz	0.008 to 0.013 + 0.008 to 0.013	FirstShot does not control grasses. May be applied up to planting to control emerged broadleaf weeds. Add nonionic surfactant at 2 to 4 pints per 100 gallons spray solution or crop oil concentrate or methylated seed oil at 1 gallon per 100 gallons spray solution. In addition, add nitrogen fertilizer at a rate of 2 quarts per acre or ammonium sulfate at 2 pounds per acre. See label for tank mixtures.
Wheat Preplant No-Till or Preemergence, Emerged annual broadleaf weeds			
pyraflufen-ethyl, MOA 14 (ET) 1 SL	0.5 to 2.0 fl oz	0.003 to 0.015 (lb a.i.)	ET can be used for limited suppression of small emerged summer annual and winter weeds. See label for adjuvant and spray volume recommendations. Research with ET is limited in North Carolina.
Wheat Preemergence, Italian ryegrass and annual broadleaf weeds			
chlorsulfuron, MOA 2 + metsulfuron methyl, MOA 2 (Finesse) 75 WDG	0.5 oz	0.0195 + 0.0039	Ryegrass control is variable; expect only suppression. May stunt wheat on sandy soils. Suggested primarily for fields with Hoelon-resistant ryegrass. Also controls most annual broadleaf weeds. Do not use where a later application of Osprey or PowerFlex is anticipated. Plant only STS soybeans following wheat harvest. May cause severe injury in non-STS soybeans.
pyroxasulfone, MOA 15 + carfentrazone, MOA 14 (Anthem Flex) 4 SC	2.0 to 4.5 fl oz	0.058 to 0.13 + 0.004 to 0.009	Apply to wheat as a preemergence treatment from 2.0 to 4.5 fluid ounces depending on soil type. Read label and adjust rates for soil texture. Under extended periods of dry weather, adequate weed control may not be achieved, 0.5 inch of rainfall may be necessary for activation and optimal weed control. Do not apply to broadcast seeded wheat. Do not apply preemergence if 0.25 inch or more rain is expected within 48 hours of application. Plant wheat a minimum of 1 inch deep, but not over 1.5 inches deep.
Wheat Spike Stage, Italian ryegrass			
flufenacet, MOA 15 + metribuzin, MOA 5 (Axiom) 68 WDG	4 to 10 oz	0.136 to 0.034 + 0.34 to 0.085	Apply to wheat in the spike stage. Preemergence application can cause severe injury on coarse-textured soils. Application rate depends on soil type; see label. In general, North Carolina research has shown best results with 6 to 7 ounces on coarse soils and 8 to 9 ounces on medium and heavy soils. If rainfall is received timely, Axiom controls ryegrass well. It also controls chickweed, henbit, and wild radish.
pyroxasulfone, MOA15 (Zidua) 85 WG (Zidua SC) 4.17 SC	0.7 to 2.0 oz/A 1.25 to 4 fl oz/A	0.037 to 0.106	Apply to wheat when 80% of germinated wheat seeds have a shoot at least 1/2-inch long until wheat spiking. May be applied as a broadcast spray to wheat at spiking up to the fourth-tiller. Will not control emerged weeds. Sequential application may be applied, but do not exceed a total of 2.25 ounces/acre Zidua (fl oz Zidua SC) in a season.
pyroxasulfone, MOA 15 + carfentrazone, MOA 14 (Anthem Flex) 4 SC	2.5 to 4.5 fl oz	0.073 to 0.13 + 0.005 to 0.009	Apply to wheat when 80% of germinated wheat seeds have a shoot at least 1/2-inch long until wheat spiking. May be applied as a broadcast spray to wheat at spiking up to the fourth-tiller. Early postemergence broadcast applications will provide suppression of emerged chickweed and henbit. Do not apply more than a maximum cumulative amount of 4.55 fluid ounces per acre in one cropping season.
flumioxazin, MOA 14 + pyroxasulfone, MOA 15 (Fierce) 76 WDG	1.5 oz	0.07	Apply to wheat when 95% of wheat is in the spike to 2-leaf stage of growth for control of ryegrass and wild radish. Wheat seed must be planted between 1 and 1.5 inches deep or injury may occur. Do not apply to fields where wheat seed has been broadcast and shallow incorporated. Do not tank mix with any adjuvant, fertilizer, or pest control product or severe injury will occur. Rainfall of at least 0.5 inch within 10 days after application is necessary for activation. Avoid applications to heavy sand and low organic matter areas or excessive injury may occur following heavy rainfall.

Table 7-8A. Chemical Weed Control in Wheat, Barley, Oats, Rye, and Triticale

Herbicide, Mode of Action Code[1], and Formulation	Amount of Formulation per Acre	Pounds Active Ingredient per Acre	Precautions and Remarks
Wheat Postemergence, Italian ryegrass			
diclofop-methyl, MOA 1 (Hoelon) 3 EC	1.33 to 2.67 pt	0.5 to 1	Apply when ryegrass is in the one- to five-leaf stage. See label for specific rates depending upon weed size. Application to smaller ryegrass is more effective. Make only one application per season. Do not tank mix with broadleaf herbicides or use liquid nitrogen as the carrier. Apply before the first wheat node (joint) develops. May add 1 to 2 pints per acre of crop oil concentrate under dry conditions or when ryegrass is large. In most cases, crop oil is not necessary. See precautions on label concerning temperatures. Biotypes of ryegrass resistant to Hoelon are common within the state. Hoelon should not be applied to fields where resistance is suspected.
mesosulfuron, MOA 2 (Osprey) 4.5 WDG	4.75 oz	0.013	Apply when ryegrass is in one-leaf to two-tiller stage. Add adjuvant as directed on label. In North Carolina, nonionic surfactant at 1 to 2 quarts per 100 gallons spray solution plus 1 to 2 quarts of 30% liquid nitrogen per acre is preferred. See label for broadleaf weeds controlled. For additional control, Osprey may be mixed with Harmony Extra. Do not tank mix with 2,4-D or dicamba. Do not apply using liquid nitrogen as the carrier. Do not topdress wheat within 14 days of Osprey application. In fields with Hoelon-susceptible ryegrass, it is recommended that Osprey or PowerFlex and Hoelon or Axial be used on an alternating basis (i.e., rotated) as part of a resistance management strategy. See comments for Axial. Ryegrass with multiple resistance to Axial, PowerFlex, Hoelon, Osprey, and PowerFlex occurs in North Carolina.
pinoxaden, MOA 1 (Axial XL) 0.42 EC	16.4 fl oz	0.054	Apply to wheat with two or more leaves when ryegrass has one to five leaves on the main stem. More effective when applied to smaller ryegrass. May be tank mixed with Harmony Extra. When mixing, add Harmony Extra first, then Axial. No adjuvants are necessary. May be applied in water-nitrogen mixtures containing up to 50% liquid nitrogen by volume. Add water to tank, then add Axial. Mix thoroughly, and then add the nitrogen. Axial and Hoelon have the same mode of action. Ryegrass resistant to Hoelon may be cross-resistant to Axial, although in some cases Axial will control Hoelon-resistant ryegrass. If Hoelon resistance is expected, trial use of Axial on limited acreage is suggested to determine if the biotype is susceptible to Axial.
pyroxsulam, MOA 2 (PowerFlex HL) 13 WDG	2.0 oz	0.0375	Can be applied to wheat from the 3-leaf stage until jointing. Apply after the majority of the ryegrass has emerged but before it exceeds the 2-tiller stage. Add nonionic surfactant at 1 to 2 quarts/100 gallons spray solution. See label for broadleaf weeds controlled. For additional control, PowerFlex may be mixed with Harmony Extra. Do not mix with dicamba or with amine formulations of 2,4-D or MCPA. Can be applied in water-nitrogen mixtures containing up to 50% liquid nitrogen by volume, or a maximum of 30 pounds/acre. If applying in liquid nitrogen, reduce surfactant rate to 1 pint/100 gallons. Rainfast in 4 hours. Do not apply to wet foliage. Current labeling specifies a 5-month rotation for soybeans. Limited research in North Carolina has shown no problems with soybeans double-cropped behind PowerFlex-treated wheat. Corn, cotton, or peanuts can be planted 9 months after application. See comments under mesosulfuron (Osprey) and pinoxaden (Axial) concerning resistance management.
Wheat Postemergence, Cheat			
pyroxsulam, MOA 2 (PowerFlex) 7.5 WDG	3.5 oz	0.0164	See comments for PowerFlex under Italian Ryegrass.
Wheat Postemergence, Annual bluegrass			
mesosulfuron, MOA 2 (Osprey) 4.5 WDG	4.75 oz	0.013	See comments for Osprey under Italian Ryegrass. Apply to bluegrass from the 1-leaf to 2-tiller stage. Application when plants are about the size of a quarter coin has worked well.
Wheat Postemergence, Wild garlic, curly dock, and most winter annual broadleaf weeds except cornflower and vetch			
thifensulfuron-methyl, MOA 2 + tribenuron-methyl, MOA 2 (Harmony Extra SG with TotalSol) 50 WDG	0.45 to 0.9 oz	0.0094 to 0.0188 + 0.0047 to 0.0094	Apply after the two-leaf stage of wheat but before flag leaf is visible. Use 0.45 to 0.6 ounce for most winter annual weeds. Use 0.75 to 0.9 ounce for wild garlic and wild radish. Wild garlic should be less than 12 inches tall and should have 2 to 4 inches of new growth. Control is enhanced when application is made during warm temperatures (50 degrees F or more) to actively growing garlic plants. Add 1 quart of nonionic surfactant per 100 gallons of spray solution. Liquid nitrogen may be used as the carrier. May tank mix Harmony Extra with 0.125 to 0.375 pound active ingredient of 2,4-D for improved control of wild radish. Follow mixing instructions on the label when using nitrogen as the carrier or when mixing with 2,4-D. Reduce surfactant rate according to label instructions when using nitrogen as the carrier or when mixing with 2,4-D. Do not tank mix with Hoelon. May be tank mixed with Axial or Osprey.
Wheat Postemergence, Most winter annual broadleaf weeds except chickweed, henbit, and knawel			
2,4-D amine, MOA 4 (various brands) 3.8 SL	1 pt	0.48	Apply after wheat is fully tillered (usually 4 to 8 inches tall; stages 4 and 5 on Feekes scale) but before jointing. Spraying wheat too young or after jointing can cause deformed heads, reduced yields, and uneven ripening. Better results are obtained when day-time temperatures are above 50 degrees F. Increase the rate of 2,4-D by 50% to control corn cockle. Liquid nitrogen may be used as the carrier for 2,4-D. Ester formulations can be added directly to the nitrogen. If using amine formulation, premix in water (1 part 2,4-D to 4 parts water) and add mixture to nitrogen with strong agitation. Amine formulations give less burn than ester formulations in nitrogen.
2,4-D ester, MOA 4 (various brands) 3.8 SL	1 pt	0.48	
2,4-D ester, MOA 4 (various brands) 5.7 SL	0.67 pt	0.48	
2,4-D acid/ester, MOA 4 (Weedone 638) 2.8 SL	1 pt	0.35	
Wheat Postemergence, Most winter annual broadleaf weeds			
dicamba, MOA 4 (Banvel) 4 SL (Clarity) 4 SL	0.25 pt	0.125	Apply after wheat is fully tillered but before jointing. Better results will be obtained if applied when daytime temperatures are above 50 degrees F. Liquid nitrogen may be used as the carrier.
dicamba, MOA 4 (Banvel) 4 SL (Clarity) 4 SL + 2,4-D amine, MOA 4 (various brands) 3.8 SL OR 2,4-D ester, MOA 4 (various brands) 3.8 SL	0.25 pt 0.25 pt + 0.5 to 0.75 pt or 0.5 to 0.75 pt	0.125 + 0.24 to 0.36 or 0.24 to 0.36	Apply after wheat is fully tillered (usually 4 to 8 inches tall; stages 4 and 5 on Feekes scale) but before jointing. Compared to dicamba alone, tank mixture is more effective on buttercup, cornflower, field pennycress, Virginia pepperweed, shepherdspurse, wild mustard, and wild radish. Use this tank mix only if both herbicides are necessary for weed control. Tank mix may injure wheat.
halauxifen-methly, MOA 2 + florasulam, MOA 4 (Quelex) 20WG	0.75 oz	0.009	Apply from 2-leaf to flag leaf stage when weeds are actively growing in the 2 to 4 leaf stage. May be tank mixed with other herbicides labeled for use on wheat. Do not apply within 60 days of crop harvest. Add 1.6 to 4 pt of nonionic surfactant per 100 gallons of spray solution, or a crop oil concentrate or methylated seed oil at 4 to 8 pt per 100 gallons. Liquid nitrogen may be used as the carrier, see label for guidelines.
pyraflufen-methyl, MOA 14 (ET) 1 SL	0.5 to 2.0 fl oz	0.003 to 0.015 (lb a.i.)	ET can be used to suppress small annual winter weeds. Although application is registered to flag leaf appearance, coverage of small weeds is necessary but difficult when wheat is tall. See label for adjuvant and spray volume and carrier recommendations. Research with ET is limited in North Carolina.

Table 7-8A. Chemical Weed Control in Wheat, Barley, Oats, Rye, and Triticale

Herbicide, Mode of Action Code[1], and Formulation	Amount of Formulation per Acre	Pounds Active Ingredient per Acre	Precautions and Remarks
Wheat Preharvest, Annual broadleaf and grass weeds, suppression of perennial weeds			
glyphosate, MOA 9 (numerous brands and formulations)	See label	0.75 (lb a.e.)	Apply after hard dough stage of grain (30% or less grain moisture) and at least 7 days before harvest. Do not apply to wheat grown for seed. Glyphosate is available as an isopropylamine salt and a potassium salt. Glyphosate formulations and application rates should be compared on the basis of pounds of glyphosate acid equivalent (a.e.) per gallon solution and per acre, respectively. Rate in the preceding column is expressed as a.e. See Table 7-10 for glyphosate rate conversions.
Wheat Preharvest, Annual broadleaf weeds			
2,4-D amine, MOA 4 (various brands) 3.8 SL	1 to 2 pt	0.48 to 0.95	Apply when grain is in hard dough stage or later. Do not allow drift to sensitive crops (be especially careful with ester formulations). Amine formulations strongly encouraged if sensitive crops are nearby, especially cotton and tobacco.
2,4-D ester, MOA 4 (various brands) 3.8 SL	1 to 2 pt	0.48 to 0.95	
2,4-D ester, MOA 4 (various brands) 5.7 SL	0.67 to 1.3 pt	0.48 to 0.95	
2,4-D acid/ester, MOA 4 (Weedone 638) 2.8 SL	1 to 2 pt	0.35 to 0.7	
Barley Preplant No-Till, Emerged annual broadleaf and grass weeds, volunteer corn, top-kill of perennials			
paraquat, MOA 22 (Gramoxone Inteon) 2 SL	2 to 4 pt	0.5 to 1	Rate depends upon weed size; see label. Apply before crop emerges. Add nonionic surfactant at 1 pint per 100 gallons spray solution or crop oil concentrate at 1 gallon per 100 gallons spray solution. See application directions on label. Generic formulations of paraquat containing 3 pounds active per gallon are available. Apply these products at 1.3 to 2.7 pints.
Barley Preplant No-Till, Emerged annual broadleaf and grass weeds, control or suppression of perennial weeds			
glyphosate, MOA 9 (numerous brands and formulations)	See label	0.56 to 1.13 (lb a.e.)	Glyphosate is available as an isopropylamine salt and a potassium salt. Glyphosate formulations and application rates should be compared on the basis of pounds of glyphosate acid equivalent (a.e.) per gallon and per acre, respectively. Rate in the preceding column is expressed as a.e. See Table 7-10 for glyphosate rate conversions. Rate depends upon weed species and size; see labels. Apply before crop emergence. Adjuvant recommendations vary by glyphosate brand; see label of brand used for details. Select or Select Max may be mixed with glyphosate to control volunteer Roundup Ready corn. For corn up to 12 inches tall, apply 4 to 6 fluid ounces of Select or 6 fluid ounces of Select Max. For corn up to 24 inches increase Select rate to 6 to 8 fluid ounces or Select Max rate to 9 fluid ounces. Select or Select Max must be applied at least 30 days ahead of barley planting.
thifensulfuron, MOA 2 + tribenuron, MOA 2 (FirstShot SG with TotalSol) 50 WDG	0.5 to 0.8 oz	0.008 to 0.013 + 0.008 to 0.013	FirstShot does not control grasses. May be applied up to planting to control emerged broadleaf weeds. Add nonionic surfactant at 2 to 4 pints per 100 gallons spray solution or crop oil concentrate or methylated seed oil at 1 gallon per 100 gallons spray solution. In addition, add nitrogen fertilizer at a rate of 2 quarts per acre or ammonium sulfate at 2 pounds per acre. See label for tank mixtures.
Barley Postemergence, Italian ryegrass			
diclofop-methyl, MOA 1 (Hoelon) 3 EC	1.33 to 2.67 pt	0.5 to 1	Apply when ryegrass is in the 1- to 4-leaf stage and after tiller initiation but prior to jointing of barley. Make only one application per year. Do not tank mix with broadleaf herbicides or use liquid nitrogen as the carrier. Do not add crop oil. Apply only to the following varieties of barley: Anson, Boone, Henry, Milton, Molly Bloom, Mulligan, Nomini, Pennco, Starling, Sussex, and Wysor. Cold (lower than 40 degrees F) and/or prolonged wet conditions increase barley sensitivity to Hoelon. Biotypes of ryegrass resistant to Hoelon are becoming more common within the state. Hoelon should not be applied to fields where resistance is suspected.
pinoxaden, MOA 1 (Axial XL) 0.42 EC	16.4 fl oz	0.054	Apply to barley with 2 or more leaves when ryegrass has one to five leaves on the main stem. More effective when applied to smaller ryegrass. May be tank mixed with Harmony Extra. When mixing, add Harmony Extra first, then Axial. May be applied in water/nitrogen mixtures containing up to 50% liquid nitrogen by volume. Add water to tank, then add Axial. Mix thoroughly, and add the nitrogen. Axial and Hoelon have the same mode of action. Ryegrass resistant to Hoelon may be cross-resistant to Axial, although in some cases Axial will control Hoelon-resistant ryegrass. If Hoelon resistance is expected, trial use of Axial on limited acreage is suggested to determine if the ryegrass biotype is susceptible to Axial.
Barley Postemergence, Wild garlic, curly dock, and most winter annual broadleaf weeds except cornflower and vetch			
thifensulfuron-methyl, MOA 2 + tribenuron-methyl, MOA 2 (Harmony Extra SG with TotalSol) 50 WDG	0.45 to 0.9 oz	0.0094 to 0.0188 + 0.0047 to 0.0094	Apply after the 2-leaf stage of barley but before flag leaf is visible. Use 0.45 to 0.6 ounce for most winter annual weeds. Use 0.75 to 0.9 ounce for wild garlic and wild radish. Wild garlic should be less than 12 inches tall and should have 2 to 4 inches of new growth. Control is enhanced when application is made during warm temperatures (50 degrees F or more) to actively growing garlic plants. Add 1 quart of nonionic surfactant per 100 gallons of spray solution. Liquid nitrogen may be used as the carrier. May tank mix Harmony Extra with 0.125 to 0.375 pound a.i. of 2,4-D for improved control of wild radish. Follow mixing instructions on the label when using nitrogen as the carrier or when mixing with 2,4-D. Reduce surfactant rate according to label instructions when using nitrogen as the carrier or when mixing with 2,4-D. Do not tank mix with Hoelon.
Barley Postemergence, Most winter annual broadleaf weeds except chickweed, henbit, and knawel			
2,4-D amine, MOA 4 (various brands) 3.8 SL	1 pt	0.48	Apply after barley is fully tillered but before jointing. Spraying barley too young or after jointing can cause deformed heads, reduced yields, and uneven ripening. Better results are obtained when day-time temperatures are above 50 degrees F. Increase the rate of 2,4-D by 50% to control corn cockle. Liquid nitrogen may be used as the carrier for 2,4-D. Ester formulations can be added directly to the nitrogen. If using amine formulation, premix in water (1 part 2,4-D to 4 parts water) and add mixture to nitrogen with strong agitation. Amine formulations give less burn than ester formulations in nitrogen.
2,4-D ester, MOA 4 (various brands) 3.8 SL	1 pt	0.48	
2,4-D ester, MOA 4 (various brands) 5.7 SL	0.67 pt	0.48	
2,4-D acid/ester, MOA 4 (Weedone 638) 2.8 SL	1 pt	0.35	
Barley Postemergence, Most winter annual broadleaf weeds			
dicamba, MOA 4 (Banvel) 4 SL (Clarity) 4 SL	0.25 pt	0.125	Apply before jointing stage of growth. Risk of crop injury is least if applied after winter dormancy and before grain begins to joint. Better results will be obtained if applied when daytime temperatures are above 50 degrees F. Liquid nitrogen may be used as the carrier.

Table 7-8A. Chemical Weed Control in Wheat, Barley, Oats, Rye, and Triticale

Herbicide, Mode of Action Code[1], and Formulation	Amount of Formulation per Acre	Pounds Active Ingredient per Acre	Precautions and Remarks
Barley Postemergence, Most winter annual broadleaf weeds (continued)			
dicamba, MOA 4 (Banvel) 4 SL (Clarity) 4 SL + 2,4-D amine, MOA 4 (various brands) 3.8 SL OR 2,4-D ester, MOA 4 (various brands) 3.8 SL	0.25 pt 0.25 pt + 0.5 pt or 0.5 pt	0.125 + 0.24 or 0.24	Apply after barley is fully tillered but before jointing. Compared to dicamba alone, tank mixture is more effective on buttercup, cornflower, field pennycress, Virginia pepper-weed, shepherdspurse, and wild mustard. Use this tank mix only if both herbicides are necessary for weed control. Tank mix may injure barley.
halauxifen-methly, MOA 2 + florasulam, MOA 4 (Quelex) 20WG	0.75 oz	0.009	Apply from 2-leaf to flag leaf stage when weeds are actively growing in the 2 to 4 leaf stage. May be tank mixed with other herbicides labeled for use on barley. Do not apply within 60 days of crop harvest. Add 1.6 to 4 pt of nonionic surfactant per 100 gallons of spray solution, or a crop oil concentrate or methylated seed oil at 4 to 8 pt per 100 gallons. Liquid nitrogen may be used as the carrier, see label for guidelines.
Barley Preharvest, Annual broadleaf weeds			
2,4-D amine, MOA 4 (various brands) 3.8 SL 2,4-D ester, MOA 4 (various brands) 3.8 SL 2,4-D ester, MOA 4 (various brands) 5.7 SL 2,4-D acid/ester, MOA 4 (Weedone 638) 2.8 SL	1 to 2 pt 1 to 2 pt 0.67 to 1.3 pt 1 to 2 pt	0.48 to 0.95 0.48 to 0.95 0.48 to 0.95 0.35 to 0.7	Apply when grain is in hard dough stage or later. Do not allow drift to sensitive crops (be especially careful with ester formulations). Amine formulations strongly encouraged if sensitive crops are nearby, especially cotton or tobacco.
Barley Preharvest, Annual broadleaf and grass weeds, suppression of perennial weeds			
glyphosate, MOA 9 (numerous brands and formulations)	See label	0.75 (lb a.e.)	Apply after hard dough stage of grain (and 20% or less moisture) and at least 7 days before harvest. Do not apply to barley grown for seed. Glyphosate is available as an isopropylamine salt and a potassium salt. Compare glyphosate formulations and application rates on the basis of pounds of glyphosate acid equivalent (a.e.) per gallon and per acre, respectively. Rate in the preceding column is expressed as a.e. See Table 7-10 for glyphosate rate conversions.
Oats Preplant No-Till, Emerged annual broadleaf and grass weeds, volunteer corn, control or suppression of perennial weeds			
glyphosate, MOA 9 (numerous brands and formulations)	See label	0.56 to 1.13 (lb a.e.)	Glyphosate is available as an isopropylamine salt and a potassium salt. Glyphosate formulations and application rates should be compared on the basis of pounds of glyphosate acid equivalent (a.e.) per gallon and per acre, respectively. Rate in the preceding column is expressed as a.e. See Table 7-10 for glyphosate rate conversions. Rate depends upon weed species and size; see labels. Apply before crop emergence. Adjuvant recommendations vary by glyphosate brand; see label of brand used for details. Select or Select Max may be mixed with glyphosate to control volunteer Roundup Ready corn. For corn up to 12 inches tall, apply 4 to 6 fluid ounces of Select or 6 fluid ounces of Select Max. For corn up to 24 inches, increase Select rate to 6 to 8 fluid ounces or Select Max rate to 9 fluid ounces. Select or Select Max must be applied at least 30 days ahead of planting oats.
Oats Postemergence, Wild garlic, curly dock, and most winter annual broadleaf weeds except cornflower and vetch			
thifensulfuron-methyl, MOA 2 + tribenuron-methyl, MOA 2 (Harmony Extra SG with TotalSol) 50 WDG	0.45 to 0.6 oz	0.0094 to 0.0125 + 0.0047 to 0.0063	Apply after the two-leaf stage of oats but before flag leaf is visible. Wild garlic should be less than 12 inches tall and should have 2 to 4 inches of new growth. Control is enhanced when application is made during warm temperatures (50 degrees F or more) to actively growing garlic plants. Add 1 quart of nonionic surfactant per 100 gallons of spray solution. Liquid nitrogen may be used as the carrier. May tank mix Harmony Extra with 0.125 to 0.375 pound a.i. of 2,4-D for improved control of wild radish. Follow mixing instructions on the label when using nitrogen as the carrier or when mixing with 2,4-D. Reduce surfactant rate according to label instructions when using nitrogen as the carrier or when mixing with 2,4-D. Oats are more sensitive to 2,4-D than wheat.
Oats Postemergence, Most winter annual broadleaf weeds except chickweed, henbit, and knawel			
2,4-D amine, MOA 4 (various brands) 3.8 SL	1 pt	0.48	Apply after oats are fully tillered but before jointing. Spraying oats too young or after jointing can cause deformed heads, reduced yields, and uneven ripening. Also, oats are less tolerant of 2,4-D than wheat. Better results are obtained when daytime temperatures are above 50 degrees F. Liquid nitrogen may be used as the carrier for 2,4-D. Premix in water (1 part 2,4-D to 4 parts water) and add mixture to nitrogen with strong agitation.
Oats Postemergence, Most winter annual broadleaf weeds			
dicamba, MOA 4 (Banvel) 4 SL (Clarity) 4 SL	0.25 pt 0.25 pt	0.125	Apply before jointing stage of growth. Risk of crop injury is least if applied after winter dormancy and before grain begins to joint. Better results will be obtained if applied when daytime temperatures are above 50 degrees F. Liquid nitrogen may be used as the carrier.
Oats Preharvest, Annual broadleaf weeds			
2,4-D amine, MOA 4 (various brands) 3.8 SL 2,4-D ester, MOA 4 (various brands) 3.8 SL	1 to 2 pt 1 to 2 pt	0.48 to 0.95	Apply when grain is in hard dough stage or later. Do not allow drift to sensitive crops, especially cotton and tobacco. Amine formulations strongly encouraged if sensitive crops are nearby, especially cotton or tobacco.
Rye Preplant No-Till, Emerged annual broadleaf and grass weeds, control or suppression of perennial weeds			
glyphosate, MOA 9 (numerous brands and formulations)	See label	0.56 to 1.13 (lb a.e.)	Glyphosate is available as an isopropylamine salt and a potassium salt. Glyphosate formulations and application rates should be compared on the basis of pounds of glyphosate acid equivalent (a.e.) per gallon and per acre, respectively. Rate in the preceding column is expressed as a.e. See Table 7-10 for glyphosate rate conversions. Rate depends upon weed species and size; see labels. Apply before crop emergence. Adjuvant recommendations vary by glyphosate brand; see label of brand used for details. Select or Select Max may be mixed with glyphosate to control volunteer Roundup Ready corn. For corn up to 12 inches tall, apply 4 to 6 fluid ounces of Select or 6 fluid ounces of Select Max. For corn up to 24 inches, increase Select rate to 6 to 8 fluid ounces or Select Max rate to 9 fluid ounces. Select or Select Max must be applied at least 30 days ahead of planting rye.

Table 7-8A. Chemical Weed Control in Wheat, Barley, Oats, Rye, and Triticale

Herbicide, Mode of Action Code[1], and Formulation	Amount of Formulation per Acre	Pounds Active Ingredient per Acre	Precautions and Remarks
Rye Postemergence, Most winter annual broadleaf weeds except chickweed, henbit, and knawel			
2,4-D amine, MOA 4 (various brands) 3.8 SL	1 pt	0.48	Apply after rye is fully tillered but before jointing. Spraying rye too young or after jointing can cause deformed heads, reduced yields, and uneven ripening. Better results are obtained when daytime temperatures are above 50°F. Increase the rate of 2,4-D by 50% to control corn cockle. Liquid nitrogen may be used as the carrier for 2,4-D. Ester formulations can be added directly to the nitrogen. If using amine formulation, premix in water (1 part 2,4-D to 4 parts water) and add mixture to nitrogen with strong agitation. Amine formulations give less burn than ester formulations in nitrogen.
2,4-D ester, MOA 4 (various brands) 3.8 SL	1 pt		
2,4-D ester, MOA 4 (various brands) 5.7 SL	0.67 pt		
Rye Preharvest, Annual broadleaf weeds			
2,4-D ester, MOA 4 (various brands) 3.8 SL	1 to 2 pt	0.48 to 0.95	Apply when grain is in hard dough stage or later. Do not allow drift to sensitive crops, especially cotton and tobacco. Amine formulations are strongly encouraged if sensitive crops are nearby, especially cotton or tobacco.
2,4-D ester, MOA 4 (various brands) 5.7 SL	0.67 to 1.3 pt		
Triticale Preplant No-Till, Emerged annual broadleaf and grass weeds, control or suppression of perennial weeds			
glyphosate, MOA 9 (numerous brands and formulations)	See label	0.56 to 1.13 (lb a.e.)	Glyphosate is available as an isopropylamine salt and a potassium salt. Glyphosate formulations and application rates should be compared on the basis of pounds of glyphosate acid equivalent (a.e.) per gallon and per acre, respectively. Rate in the preceding column is expressed as a.e. See Table 7-10 for glyphosate rate conversions. Rate depends upon weed species and size; see labels. Apply before crop emergence. Adjuvant recommendations vary by glyphosate brand; see label of brand used for details. Select or Select Max may be mixed with glyphosate to control volunteer Roundup Ready corn. For corn up to 12 inches tall, apply 4 to 6 fluid ounces of Select or 6 fluid ounces of Select Max. For corn up to 24 inches, increase Select rate to 6 to 8 fluid ounces or Select Max rate to 9 fluid ounces. Select or Select Max must be applied at least 30 days ahead of planting triticale.
thifensulfuron, MOA 2 + tribenuron, MOA 2 (FirstShot SG with TotalSol) 50 WDG	0.5 to 0.8 oz	0.008 to 0.013 + 0.008 to 0.013	FirstShot does not control grasses. May be applied up to planting to control emerged broadleaf weeds. Add nonionic surfactant at 2 to 4 pints per 100 gallons spray solution or crop oil concentrate or methylated seed oil at 1 gallon per 100 gallons spray solution. In addition, add nitrogen fertilizer at a rate of 2 quarts per acre or ammonium sulfate at 2 pounds per acre. See label for tank mixtures.
Triticale Preemergence, Italian ryegrass and annual broadleaf weeds			
chlorsulfuron, MOA 2 + metsulfuron methyl, MOA 2 (Finesse) 75 WDG	0.5 oz	0.0195 + 0.0039	See comments under Wheat, Preemergence.
Triticale Postemergence, Wild garlic and annual broadleaf weeds			
thifensulfuron-methyl, MOA 2 + tribenuron-methyl, MOA 2 (Harmony Extra SG with TotalSol) 50 WDG	0.45 to 0.9 oz	0.0094 to 0.0188 + 0.0047 to 0.0094	Apply after 2-leaf stage of triticale but before flag leaf is visible. See comments for Harmony Extra under Wheat-Postemergence.
Triticale Postemergence, Annual broadleaf weeds			
2,4-D, MOA 4 (Amine 4 2,4-D) 3.8 SL (Weedar 64) 3.8 L	1 pt	0.48	See comments for 2,4-D under Wheat- Postemergence.
dicamba, MOA 4 (Banvel) 4 SL (Clarity) 4 SL	0.25 oz	0.125	See comments for dicamba under Wheat- Postemergence.
halauxifen-methly, MOA 2 + florasulam, MOA 4 (Quelex) 20WG	0.75 oz	0.009	See comments for dicamba under Wheat- Postemergence.

[1] Mode of Action (MOA) code developed by the Weed Science Society of America. See Table 7-10, Herbicide Resistance Management, for details.

Weed Response to Herbicides in Small Grains

W. J. Everman, Crop and Soil Sciences Department

Ratings based upon average to good soil and weather conditions for herbicide performance and upon proper application rate, technique, and timing.

Table 7-8B. Weed Response to Herbicides in Small Grains

Species	Anthem Flex	Axial	Axiom	Banvel/ Clarity	Fierce	Finesse[2]	2,4-D	Harmony Extra	Hoelon	Osprey	PowerFlex[7] HL	Valor SX	Zidua/ Zidua SC
Annual bluegrass	E[4]	N	G[4]	N	E[4]	N	N	N	N	G		F[4]	E[4]
Annual ryegrass	E[4]	GE[6]	G[4]	N	E[4]	F	N	N	E[1]	E[5]	E[5]	GE4	E[4]
Buttercup		N		F	-	G	G	G	N			-	-
Chickweed, common	F[4]	N	G	G	E[4]	G	P	G	N	FG		E[4]	F[4]
Cornflower	-	N		FG	-	F	G	P	N	P			-
Curly dock	N	N	N	F	N		P	E	N	P		N	N
Cutleaf eveningprimrose	-	N		G	E[4]		E	G	N	P		E4	-
Field pennycress	-	N		F	E[4]	G	G	G	N			E[4]	-
Henbit	F[4]	N	GE[4]	F	E[4]	G	P	G	N	G		E[4]	F[4]
Knawel	-	N		G	-		P	G	N			-	-
Shepherd's-purse	-	N		FG	E[4]	G	GE	E	N			E[4]	-
Swinecress	-	N		G	-		G	E	N	E		-	-
Vetch	-	N		E	-		G	P	N	N		-	-
Virginia pepperweed	-	N		F	G[4]		E	G	N			G[4]	-
Wild garlic	N	N	N	F	N	P	F	E	N	P		N[4]	N
Wild mustard	P[4]	N	G[4]	F	E[4]	G	GE	G	N	E[3]	GE	E[4]	P[4]
Wild radish	P[4]	N	G[4]	F	E[4]	G	GE	G	N	E[3]	GE	E[4]	P[4]

[1] A biotype of ryegrass resistant to Hoelon is common in North Carolina, especially in the piedmont. If resistance is suspected, do not apply Hoelon.
[2] Applied preemergence.
[3] Rating assumes mustard or radish is 1 to 2 inches Osprey is less effective on larger plants.
[4] Assumes adequate rainfall for activation prior to weed emergence.
[5] A biotype of ryegrass resistant to Osprey and PowerFlex has been found in the southern piedmont.
[6] May not control Hoelon-resistant ryegrass. See comments under "Wheat-Postemergence."
[7] Inadequate research has been conducted in North Carolina to determine response of most broadleaf weeds to PowerFlex. The label claims control of a number of broadleaf species, including Carolina geranium, chickweed, hairy bittercress, field pennycress, shepherdspurse, buttercup, Virginia pepperweed, and vetch.

Key:

E = excellent control, 90% or better

G = good control, 80% to 90%

F = fair control, 50% to 80%

P = poor control, 25% to 50%

N = no control, less than 25%

Glyphosate Formulations

C. W. Cahoon, Assistant Professor and Extension Weed Specialist, Crop and Soil Sciences Department

Table 7-9. Glyphosate Formulations

				Equivalent rates	
Formulation salt	% formulated salt by weight	Concentration (lb formulated salt/gal)	lb acid equivalent (a.e.) / gal	lb a.e./acre	fl oz product/acre
Diammonium	34.0	3.6	3.0	0.375 0.560 0.750 1.125	16 24 32 48
Dimethylamine	50.2	5.07	4.0	0.375 0.560 0.750 1.125	12 18 24 36
Isopropylamine	41.0	4.0	3.0	0.375 0.560 0.750 1.125	16 24 32 48
	50.2	5.0	3.75	0.375 0.560 0.750 1.125	12.8 19.1 25.6 38.4
	53.8	5.5	4.0	0.375 0.560 0.750 1.125	12 18 24 36
Isopropylamine + Monoammonium	37.54 + 3.42	3.64 + 0.33	2.7 + 0.3	0.375 0.560 0.750 1.125	16 24 32 48
Isopropylamine + Potassium	30.94 + 22.99	3.33 + 2.5	2.5 + 2.0	0.375 0.560 0.750 1.125	10.7 15.9 21.3 32
Potassium	44.9	5.0	4.17	0.375 0.560 0.750 1.125	11.5 17.2 23 34.5
	48.8	5.5	4.5	0.375 0.560 0.750 1.125	10.7 15.9 21.3 32
	52.3	6.0	5.0	0.375 0.560 0.750 1.125	9.6 14.3 19.2 28.8

Herbicide Resistance Management

W. J. Everman and F. H. Yelverton, Crop and Soil Sciences Department

Herbicide resistance is becoming a serious problem in North Carolina and across the country. Herbicide resistance is not a new phenomenon in North Carolina. Goosegrass resistant to dinitroaniline herbicides was first reported in North Carolina in the 1970s. Since then, smooth pigweed and common lambsquarters resistant to triazines, cocklebur resistant to organoarsenicals and ALS inhibitors, Palmer amaranth resistant to ALS inhibitors, Italian ryegrass resistant to ACCase inhibitors and ALS inhibitors, and common ragweed resistant to ALS inhibitors have been observed. Of greatest concern is weed resistance to glyphosate. Horseweed (2003), Palmer amaranth (2005), common ragweed (2006), and Italian ryegrass (2009). resistant to glyphosate have been found. Resistance of common lambsquarters to glyphosate is suspected. Weed resistance to glyphosate is a highly significant concern in light of the extensive use of glyphosate for burndown in conservation tillage systems and in Roundup Ready cotton, soybeans, and corn.

Crop rotation, along with appropriate herbicide rotation, should be employed to the extent possible. Additionally, cultivation, where feasible, can be very helpful in herbicide resistance management. However, the most important component of a resistance management strategy is rotation of herbicide modes of action and use of multiple herbicide modes of action within each crop.

Mode of action describes the process whereby an herbicide kills susceptible plants. Table 7-10 lists the mode of action, along with the chemical family, of all the herbicides likely to be used on agronomic and horticultural crops in North Carolina. Each herbicide mode of action is assigned a numerical code for ease of use. Wherever possible, at least two modes of action should be used within each crop. This can be accomplished by preemergence herbicide applications followed by postemergence applications and by tank mixtures of herbicides with two or more modes of action. Also, within a rotation, one should try to avoid dependence on herbicides with the same mode of action in all crops in the rotation. For example, in a corn and soybean rotation, it is best not to use an ALS inhibitor (#2) in each crop. Alternatively, if an ALS inhibitor is used in each crop, herbicides with other modes of action should also be included. Similarly, it would be best to not rely exclusively on glyphosate in both crops.

In Roundup Ready corn and soybeans, it is recommended that glyphosate plus herbicides with at least one other mode of action be used. In Roundup Ready cotton, it is recommended that at least two modes of actions, in addition to glyphosate, be used.

Table 7-10A. Herbicide Modes of Action

Brand Names	Active Ingredient(s)	Chemical Family	Mode of Action[1]
AAtrex	atrazine	triazine	5
Accent Q	nicosulfuron	sulfonylurea	2
Acumen	pendimethalin	dinitroaniline	3
Aim	carfentrazone	triazolinone	14
Alachlor	alachlor	chloroacetamide	15
Alanap	naptalam	phthalamate semicarbazone	19
Armezon	topramezone	triketone	27
Arrow	clethodim	cyclohexanedione	1
Atrazine	atrazine	triazine	5
Assure II	quizalofop	aryloxyphenoxy-propionate	1
Authority Assist	sulfentrazone + imazethapyr	triazolinone + imidazolinone	14 + 2
Authority First	sulfentrazone + cloransulam	triazolinone + triazolopyrimidine	14 + 2
Authority MTZ	sulfentrazone + metribuzin	triazolinone + triazinone	14 + 5
Axial	pinoxaden	phenylpyrazoline	1
Axiom	flufenacet + metribuzin	oxyacetamide + triazinone	15 + 5
Balance FLEXX	Isoxaflutole	Isoxazole	27
Banvel	dicamba	benzoic acid	4
Banvel-K + Atrazine	dicamba + atrazine	benzoic acid + triazine	4 + 5
Basagran	bentazon	benzothiadiazinone	6
Beyond	imazamox	imidazolinone	2
Bicep II Magnum	s-metolachlor + atrazine	chloroacetamide + triazine	15 + 5
Blazer	acifluorfen	diphenylether	14
Boundary	s-metolachlor + metribuzin	chloroacetamide + triazinone	15 + 5
Brawl, Brawl II	s-metolachlor	chloroacetamide	15
Brawl II ATZ	s-metolachlor + atrazine	chloroacetamide + triazine	15 + 5
Breakfree ATZ	s-metolachlor + atrazine	chloroacetamide + triazine	15 + 5
Break-Up	pronamide	benzamide	3
Buctril	bromoxynil	nitrile	6
Bullet	alachlor + atrazine	chloroacetamide + triazine	15 + 5
Butoxone	2,4-DB	phenoxy-carboxylic acid	4
Butyrac	2,4-DB	phenoxy-carboxylic acid	4
Cadre	imazapic	imidazolinone	2
Callisto	mesotrione	triketone	27
Camix	mesotrione + s-metolachlor	triketone + chloroacetamide	27 + 15
Canopy	metribuzin + chlorimuron	triazine + sulfonylurea	5 + 2
Canopy EX	chlorimuron + tribenuron	sulfonylurea + sulfonylurea	2 + 2
Canopy XL	sulfentrazone + chlorimuron	diphenylether + sulfonylurea	14 + 2
Caparol	prometryn	triazine	5
Capreno	tembotrione + thiencarbazone	benzoyl pyrazole + triazolone	27 + 2
Celebrity, Celebrity Plus	nicosulfuron + dicamba	sulfonylurea + benzoic acid	2 + 4

Table 7-10A. Herbicide Modes of Action

Brand Names	Active Ingredient(s)	Chemical Family	Mode of Action[1]
Charger Basic	s-metolachlor	chloroacetamide	15
Cinch	s-metolachlor	chloroacetamide	15
Cinch ATZ	s-metolachlor + atrazine	chloroacetamide + triazine	15 + 5
Clarity	dicamba	benzoic acid	4
Classic	chlorimuron	sulfonylurea	2
Clethodim	clethodim	cyclohexanedione	1
Clopyr AG	clopyralid	pyridine carboxylic acid	4
Cobra	lactofen	diphenylether	14
Command	clomazone	isoxazolidinone	13
Confidence	acetochlor	chloroacetamide	15
Confidence Xtra	acetochlor + atrazine	chloroacetamide + triazine	15 + 5
Corvus	Isoxaflutole + thiencarbazone-methyl	isoxazole + triazolone	27 + 2
Cotoran	fluometuron	urea	7
Cotton-Pro	prometryne	triazine	5
Curbit	ethalfluralin	dinitroaniline	3
Dacthal	DCPA	benzoic acid	3
Dawn	fomesafen	diphenyl ether	14
Define	flufenacet	oxyacetamide	15
Degree	acetochlor	chloroacetamide	15
Degree Xtra	acetochlor + atrazine	chloroacetamide + triazine	15 + 5
Devrinol	napropamide	acetamide	15
Diablo	dicamba	benzoic acid	4
Dicamba DMA Salt	dicamba	benzoic acid	4
Direx	diuron	urea	7
Distinct	dicamba + diflufenzopyr	benzoic acid + semicarbazone	4 + 19
Diuron	diuron	urea	7
Double Team	acetochlor + atrazine	chloroacetamide + triazine	15 + 5
DSMA, numerous brands	DSMA	organoarsenical	17
Dual II, Dual II Magnum	s-metolachlor	chloroacetamide	15
Envive	flumioxazin + chlorimuron + thifensulfuron	n-phenylphthalimide + triazolopyrimidine + sulfonylurea	14 + 2 + 2
Envoke	trifloxysulfuron	sulfonylurea	2
Eptam	EPTC	thiocarbamate	8
Equip	formasulfuron + iodosulfuron	sulfonylurea	2 + 2
Eradicane	EPTC	thiocarbamate	8
Establish	dimethenamid	chloroacetamide	15
Establish ATZ	dimethenamid + atrazine	chloroacetamide + triazine	15 + 5
ET	pyraflufen ethyl	phenylpyrazole	14
Evik	ametryne	triazine	5
Expert	glyphosate + s-metolachlor + atrazine	glycine + chloroacetamide + triazine	9 + 15 + 5
Express	tribenuron	sulfonylurea	2
Extreme	glyphosate + imazethapyr	glycine + imidazolinone	9 + 2
Fierce	flumioxazin + pyroxasulfone	n-phenylphthalimide + chloroacetamide	14 + 15
Finesse	chlorsulfuron + metsulfuron	sulfonylurea + sulfonylurea	2 + 2
Firestorm	paraquat	bipyridylium	22
Firstrate	cloransulam	triazolopyrimidine	2
Flexstar	fomesafen	diphenylether	14
Fultime	acetochlor + atrazine	chloroacetamide + triazine	15 + 5
Fusilade DX	fluazifop	aryloxyphenoxy-propionate	1
Fusion	fluazifop + fenoxaprop	aryloxyphenoxy-propionate + aryloxyphenoxy-propionate	1 + 1
Galigan	oxyfluorfen	diphenylether	14
Gangster	flumioxazin + cloransulam	n-phenylphthalimide + triazolopyrimidine	14 + 2
Guardsman Max	dimethenamide + atrazine	chloroacetamide + triazine	15 + 5
Glyphosate (numerous brands)	glyphosate	glycine	9
Goal	oxyflurofen	diphenylether	14
Goal Tender	oxyfluorfen	diphenylether	14
Gramoxone (Inteon)	paraquat	bipyridylium	22
Halex GT	s-metolachlor + glyphosate + mesotrione	chloroacetamide + glycine + triketone	15 + 9 + 27
Harmony Extra	thifensulfuron + tribenuron	sulfonylurea + sulfonylurea	2 + 2
Harmony GT, Harmony SG	thifensulfuron	sulfonylurea	2
Harness	acetochlor	chloroacetamide	15
Harness Xtra	acetochlor + atrazine	chloroacetamide + triazine	15 + 5
Hoelon	diclofop	aryloxyphenoxy-propionate	1
Ignite, Ignite 280	glufosinate	phosphinic acid	10
Impact	topramezone	triketone	27
Intrro	alachlor	chloroacetamide	15
Karmex	diuron	urea	7

Table 7-10A. Herbicide Modes of Action

Brand Names	Active Ingredient(s)	Chemical Family	Mode of Action[1]
Kerb	pronamide	benzamide	3
Keystone NXT	acetochlor + atrazine	chloroacetamide + triazine	15 + 5
Lariat	alachlor + atrazine	chloroacetamide + triazine	15 + 5
Laudis	tembotrione	benzoyl pyrazole	27
Layby Pro	diuron + linuron	urea + urea	7 + 7
Leadoff	rimsulfuron + thifensulfuron	sulfonylurea	2 + 2
Lexar	mesotrione + s-metolachlor + atrazine	triketone + chloroacetamide + triazine	27+15+5
Liberty	glufosinate	phosphinic acid	10
Liberty ATZ	glufosinate + atrazine	phosphinic acid + triazine	10 + 5
Lightning	imazethapyr + imazapyr	imidazolinone + imidazolinone	2 + 2
Linex	linuron	urea	7
Lorox	linuron	urea	7
Lumax	mesotrione + s-metolachlor + atrazine	triketone + chloroacetamide + atrazine	27+15+5
Marksman	dicamba + atrazine	benzoic acid + triazine	4 + 5
Matrix	rimsulfuron	sulfonylurea	2
Medal, Medal II	s-metolachlor	chloroacetamide	15
Medal II AT	s-metolachlor + atrazine	chloroacetamide + triazine	15 + 5
Me-Too-Lachlor	metolachlor	chloroacetamide	15
Metribuzin	metribuzin	triazinone	5
Metri DF	metribuzin	triazinone	5
Micro-Tech	alachlor	chloroacetamide	15
Moxy	bromoxynil	nitrile	6
MSMA (numerous brands)	MSMA	organoarsenical	17
OpTILL	imazethapyr + saflufenacil	imidazolinone + pyrimidinedione	2 + 14
Option	foramsulfuron	sulfonylurea	2
Osprey	mesosulfuron	sulfonylurea	2
Outlook	dimethenamid	chloroacetamide	15
OxiFlo	oxyfluorfen	diphenylether	14
Parallel, Parallel PCS	metolachlor	chloroacetamide	15
Parallel Plus	metolachlor + atrazine	chloroacetamide + atrazine	15 + 5
Parazone	paraquat	bipyridylium	22
Parrlay	metolachlor	chloroacetamide	15
Peak	prosulfuron	sulfonylurea	2
Pendant	pendimethalin	dinitroaniline	3
Pendimax	pendimethalin	dinitroaniline	3
Permit	halosulfuron	sulfonylurea	2
Poast, Poast Plus	sethoxydim	cyclohexanedione	1
PowerFlex	pyroxsulam	triazolopyrimidine	2
Prefar	bensulide	phosphorodithioate	8
Prefix	s-metolachlor + fomesafen	chloroacetamide + diphenylether	15 + 14
Princep	simazine	triazine	5
Prometryn	prometryn	triazine	5
Prowl, Prowl H$_2$O	pendimethalin	dinitroaniline	3
Pursuit	imazethapyr	imidazolinone	2
Pyramin	pyrazon	pyridazinone	6
Python	flumetsulam	triazolopyrimidine	2
Raptor	imazamox	imidazolinone	2
Realm Q	rimsulfuron + mesotrione	sulfonylurea + triketone	2 + 27
Reflex	fomesafen	diphenylether	14
Resicore	acetochlor + mesotrione + clopyralid	chloroacetamide + triketone + pyridine carboxylic acid	15 + 27 + 4
Resolve	rimsulfuron	sulfonylurea	2
Resolve Q	rimsulfuron + thifensulfuron	sulfonylurea + sulfonylurea	2 + 2
Resource	flumiclorac-pentyl	n-phenylphthalimide	14
Ro-Neet	cycloate	thiocarbamate	8
Roundup (and other brands)	glyphosate	glycine	9
Rhythm	fomesafen	diphenyl ether	14
Sandea	halosulfuron	sulfonylurea	2
Scepter	imazaquin	imidazolinone	2
Select, Select Max	clethodim	cyclohexanedione	1
Sencor	metribuzin	triazinone	5
Sequence	glyphosate + s-metolachlor	glycine + chloroacetamide	9 + 15
Sharpen	saflufenacil	pyrimidinedione	14
Simazine	simazine	triazine	5
Sim-Trol	simazine	triazine	5
Sinbar	terbacil	uracil	5
Sonalan	ethalfluralin	dinitroaniline	3

Table 7-10A. Herbicide Modes of Action

Brand Names	Active Ingredient(s)	Chemical Family	Mode of Action[1]
Sonic	sulfentrazone + cloransulam	triazolinone + triazolopyrimidine	14 + 2
Spartan	sulfentrazone	triazolinone	14
Spartan Charge	sulfentrazone + carfentrazone	triazolinone + triazolinone	14 + 14
Spin-Aid	phenmedipham	phenyl-carbamate	6
Squadron	imazaquin + pendimethalin	imidazolinone + dinitroaniline	2 + 3
Stalwart, Stalwart C	metolachlor	chloroacetamide	15
Stalwart Xtra	metolachlor + atrazine	chloroacetamide + triazine	15 + 5
Staple	pyrithiobac	pyrimidinyl(thio)benzoate	2
Status	dicamba + diflufenzopyr	benzoic acid + semicarbazone	4 + 19
Steadfast Q	nicosulfuron + rimsulfuron	sulfonylurea + sulfonylurea	2 + 2
Steadfast ATZ	nicosulfuron + rimsulfuron + atrazine	sulfonylurea + sulfonylurea + triazine	2 + 2 + 5
Stealth	pendimethalin	dinitroaniline	3
Sterling	dicamba	benzoic acid	4
Stinger	clopyralid	pyridine carboxylic acid	4
Storm	acifluorfen + bentazon	diphenylether + benzothiadiazinone	14 + 6
Strategy	ethalfluralin + clomazone	dinitroaniline + isoxazolidinone	3 + 13
Strongarm	diclosulam	triazolopyrimidine	2
Suprend	prometryn + trifloxysulfuron	triazine + sulfonylurea	5 + 2
Sutan+	butylate	thiocarbamate	8
SureStart	acetochlor + flumetsulam + clopyralid	chloroacetamide + triazolopyrimidine + pyridine carboxylic acid	15 + 2 + 4
Surpass	acetochlor	chloroacetamide	15
Synchrony XP	chlorimuron + thifensulfuron	sulfonylurea + sulfonylurea	2 + 2
Targa	quizalofop	aryloxyphenoxy-propionate	1
Tillam	pebulate	thiocarbamate	8
TopNotch	acetochlor	chloroacetamide	15
Treflan	trifluralin	dinitroaniline	3
Triangle	metolachlor + atrazine	chloroacetamide + triazine	15 + 5
Trifluralin	trifluralin	dinitroaniline	3
Trigger	clethodim	cyclohexanedione	1
Trilin	trifluralin	dinitroaniline	3
Trizmet II	atrazine + metolachlor	triazine + chloroacetamide	5 + 15
Trust	trifluralin	dinitroaniline	3
Ultra Blazer	acifluorfen	diphenylether	14
Valor SX	flumioxazin	n-phenylphthalimide	14
Valor XLT	flumioxazin + chlorimuron	n-phenylphthalimide + sulfonylurea	14 + 2
Vision	dicamba	benzoic acid	4
Volley	acetochlor	chloroacetamide	15
Volley ATZ	acetochlor + atrazine	chloroacetamide + triazine	15 + 5
Volunteer	clethodim	cyclohexanedione	1
Warrant	acetachlor	chloroacetamide	15
Weedmaster	2,4-D + dicamba	phenoxy-carboxylic acid + benzoic acid	4 + 4
Yukon	halosulfuron + dicamba	sulfonylurea + benzoic acid	2 + 4
Zidua	pyroxasulfone	chloroacetamide	14
2,4-D (numerous brands)	2,4-D	phenoxy-carboxylic acid	4
2,4-DB (numerous brands)	2,4-DB	phenoxy-carboxylic acid	4

Mode of Action Code Key:

1	ACCase inhibition
2	ALS inhibition
3	Microtubule assembly inhibition
4	Synthetic auxin
5	Photosystem II inhibition, different binding behavior than groups 6 and 7
6	Photosystem II inhibition, different binding behavior than groups 5 and 7
7	Photosystem II inhibition, different binding behavior than groups 5 and 6
8	Inhibition of lipid synthesis - not ACCase inhibition
9	EPSP synthase inhibition
10	Glutamine synthase inhibition
12	Inhibition of carotenoid biosynthesis at PDS
13	Inhibition of carotenoid biosynthesis, unknown target
14	PPO inhibition
15	Inhibition of very long-chain fatty acids
17	Unknown mode of action
19	Auxin transport inhibition
22	Photosystem I electron diversion
27	Inhibition of HPPD

Herbicide Modes of Action for Hay Crops, Pastures, Lawns and Turf

Table 7-10B. Herbicide Modes of Action for Hay Crops, Pastures, and Lawns and Turf

Brands listed were registered for sale and use in North Carolina in 2015 according to http://www.kellysolutions.com/nc/searchbychem.asp. Active ingredients enclosed within parentheses are prepackaged herbicides.

Active Ingredient(s)	Brand Names	Chemical Family	Mode of Action[1]
(2, 4-D + mecoprop + dichlorprop)	Spoiler, Triamine	phenoxycarboxylic acid + phenoxyalkanoic acid + chlorinated phenoxy	4 + 4 + 4
2,4-D amine	(4 SL): various trade names	phenoxy-carboxylic acid	4
(2,4-D + aminopyralid)	GrazonNext HL	phenoxy-carboxylic acid + pyridinecarboxylic acid	4 + 4
(2,4-D ester + carfentrazone-ethyl)	Rage D-Tech	phenoxy-carboxylic acid + triazolinone	4 + 14
(2,4-D + clopyralid + dicamba)	Millennium Ultra 2	phenoxycarboxylic acid + pyridinecarboxylic acid + benzoic acid	4 + 4 + 4
(2,4-D amine + cloypyralid)	Curtail	phenoxy-carboxylic acid + pyridinecarboxylic acid	4 + 4
(2,4-D + dicamba)	Brash, Range Star, Rifle-D, Weedmaster	phenoxy-carboxylic acid + benzoic acid	4 + 4
(2,4-D + fluroxypyr + dicamba)	E-2, Escalade 2	phenoxy + pyridinecarboxylic acid + benzoic acid	4 + 4 + 4
(2,4-D + glyphosate)	Campaign	phenoxy-carboxylic acid + glycine	4 + 9
(2,4-D + mecoprop + dicamba)	3-D, Three-Way Selective Herbicide, Triplet SF, various Trimec formulations	phenoxycarboxylic acid + phenoxyalkanoic acid + benzoic acid	4 + 4 + 4
(2,4-D + picloram)	Grazon P+D, Trooper P+D	phenoxy-carboxylic acid + pyridinecarboxylic acid	4 + 4
(2,4-D + triclopyr)	Candor, Crossbow	phenoxy-carboxylic acid + pyridinecarboxylic acid	4 + 4
2,4-DB	(1.75 EC): 2,4-DB 175 (2 EC): 2,4-DB 200, Butyrac 200	phenoxy-carboxylic acid	4
amicarbazone	Xonerate	triazolinone	5
aminopyralid	Milestone	pyradinecarboxylic acid	4
(aminopyralid + metsulfuron methyl)	Chaparral	pyradinecarboxylic acid + sulfonylurea	4 + 2
asulam	Asulox	carbamate	18
atrazine	(4 L): AAtrex 4 L, Atrazine 4 L (90 DF, 90 WDG): Aatrex Nine-O, Atrazine 90 WDG, Atrazine 90 DF	triazine	5
benefin	Lebanon Balan 2.5 G, The Andersons Crabgrass Preventer with 2.5 Balan	dinitroaniline	3
(benefin + trifluralin)	Fertilizer with Team Pro 0.86%	dinitroaniline + dinitroaniline	3 + 3
bensulide	(4 EC): Bensumec 4 LF (8.5 G): Weedgrass Preventer (12.5 G): Pre-San Granular	organophosphorus	8
(bensulide + oxadiazon)	Goosegrass / Crabgrass Control	organophosphorus + oxadiazole	8 + 14
bentazon	Basagran Sedge Control, Basagran T/O, Bentazon 4, Lescogran	benzothiadiazole	6
bispyribac-sodium	Velocity	pyrimidinyloxybenzoic acid	2
bromoxynil	(2 EC): Broclean, Maestro 2 EC (4 EC): Buctril 4 EC	nitrile	6
carfentrazone-ethyl	Aim EC, QuickSilver T&O	triazinone	14
(carfentrazone + 2,4-D ester + mecoprop + dicamba)	(2.2 EC): Speed Zone (0.81 EC): Speed Zone Southern	triazinone + phenoxycarboxylic acid + phenoxyalkanoic acid + benzoic acid	14 + 4 + 4 + 4
(carfentrazone + MCPA + mecoprop + dicamba)	Power Zone	triazinone + phenoxy + phenoxyalkanoic acid + benzoic acid	14 + 4 + 4 + 4
(carfentrazone + quinclorac)	SquareOne	triazinone + quinoline carboxylic acid	14 + (27 + 4)
chlorsulfuron	Alligare Chlorsulfuron 75, Telar XP	sulfonylurea	2
clethodim	(2 EC): Arrow 2 EC, Avatar S2, Cleanse 2 EC, Clethodim 2 LC, Clethodim 2 E, Dakota, Section 2 EC, Shadow, Tide Clethodim 2 EC, Select 2 EC, Volunteer, Willowood Clethodim 2 EC (0.97 EC): Envoy Plus, TapOut	cyclohexanedione	1
cloypyralid	Lontrel T&O	pyridinecarboxylic acid	4
dicamba / diglycolamine	Banvel, Clarity, Clash, Detonate, Rifle, Strut, Topeka, Vanquish	benzoic acid	4
diclofop-methyl	Illoxan	aryloxyphenoxy propionate	1
(diflufenzopyr-sodium + dicamba)	Overdrive	semicarbazone + benzoic acid	19 + 4
dimethenamid	Tower	chloroacetamide	15
(dimethenamid + pendimethalin)	Freehand	chloroacetamide + dinitroaniline	15 + 3
diquat	Diquash, Diquat., Diquat SPC, Harvester, Priceto Diquat 2 L, Reward LS, Tribune	bipyridylium	22
diuron	Direx 4 L, Diuron 4 L	phenylurea	7
dithiopyr	(2 EW, 2 L): Armortech CGC 2 L, Dimension 2 EW, Dithiopyr 2 L (40 WP): Armortech CGC 40, Dimension Ultra, Dithiopyr 40 WSP	pyridine	4
EPTC	Eptam 7-E	thiocarbamate	8
ethofumesate	(1.5 EC): Progress (4 EC): Phoenix Thrasher, PoaConstrictor, Progress SC	benzofuranes	8
fenoxaprop	Acclaim Extra	aryloxyphenoxy propionate	1
flazasulfuron	Katana	sulfonylurea	2

Table 7-10B. Herbicide Modes of Action for Hay Crops, Pastures, and Lawns and Turf

Brands listed were registered for sale and use in North Carolina in 2015 according to http://www.kellysolutions.com/nc/searchbychem.asp. Active ingredients enclosed within parentheses are prepackaged herbicides.

Active Ingredient(s)	Brand Names	Chemical Family	Mode of Action[1]
florasulam	Defendor	triazolopyrimidine	2
fluazifop	Fusilade II	aryloxyphenoxy propionate	1
flumioxazin	Sureguard	N-phenylphthalimide	14
foramsulfuron	Revolver	sulfonylurea	2
glufosinate	Finale	organophosphorus	10
glyphosate – Roundup formulations	(4 SL, 5 SL, 5.4 SL, 5.5 SL): various trade names	glycine	9
glyphosate – Touchdown formulations	(3 SL): Touchdown Pro (4.17 SL): Touchdown Total (5 SL): Touchdown HiTech		
(glyphosate + diquat)	Razor Burn, Roundup QuikPRO	glycine + bipyridylium	9 + 22
halosulfuron	HiYield Nutsedge Control, Nutgrass Killer II, Profine 75, Sandea	sulfonylurea	2
imazapic	Imazapic 2 SL, Impose, Panoramic, Plateau	imadazolinone	2
imazaquin	Image 70 DG	imidazolinone	2
imazethapyr	Pursuit, Slay, Thunder	Imidazolinone	2
imazosulfuron	Celero	sulfonylurea	2
indaziflam	(20 WSP): Specticle 20 WSP (0.0224 G): Specticle G (0.622F): Specticle Flo	benzamide	21
(indaziflam + diquat bromide + glyphosate)	Specticle Total	benzamide + bipyridylium + glycine	21 + 22 + 9
isoxaben	Gallery, Isoxaben	benzamide	21
(MCPA + mecoprop + dicamba)	Ortho Weed B Gon Pro Southern, Tri-Power	mcpa + phenoxyalkanoic acid + benzoic acid	4 + 4 + 4
(MCPA amine + fluroxypyr ester + dicamba)	Change Up	phenoxy + pyridinecarboxylic acid + benzoic acid	4 + 4 + 4
(MCPA amine + fluroxypyr ester + triclopyr amine)	Battleship III	phenoxy + pyridinecarboxylic acid + pyridinecarboxylic acid	4 + 4 + 4
(MCPA amine + triclopyr amine + dicamba)	Horsepower, Lesco Eliminate	phenoxy + pyridinecarboxylic acid + benzoic acid	4 + 4 + 4
(MCPA ester + triclopyr ester + dicamba)	Cool Power, Lesco Three-Way Ester II, Monterey Spurge Power	phenoxy + pyridinecarboxylic acid + benzoic acid	4 + 4 + 4
mecoprop	MCPP-p4 Amine, Mecomec 2.5, Mecomec 4	phenoxyalkanoic acid	4
mesotrione	Tenacity	benzoylcyclohexanedione	27
metolachlor	Pennant Magnum	chloroacetamide	15
metribuzin	(75 DF): Dimetric DF, Glory, Metribuzin 75, Tricor DF 4 L, 4 F): Glory 4L, Tricor 4 F	triazinone	5
metsulfuron methyl	Accurate, Amtide MSM 60 DF, Amtide MSM Turf, Manor, MSM 60 DF, MSM Turf, Ortho Weed B Gon Pro St. Augustine, Plotter, Purestand, Rometsol	sulfonylurea	2
(metsulfuron + 2,4-D + dicamba)	Cimarron Max	sulfonylurea + phenoxy-carboxylic acid + benzoic acid	2 + 4 + 4
(metsulfuron + chlorsulfuron)	Chisom, Cimarron Plus	sulfonylurea + sulfonylurea	2 + 2
(metsulfuron methyl + rimsulfuron)	Negate	sulfonylurea + sulfonylurea	2 + 2
monosodium methylarsonate	(6 SL): Target 6 Plus, Drexel MSMA 6 Plus (6.6 SL): Target 6.6 Plus, Drexel MSMA 6.6 Plus	organic arsenical	17
napropamide	Devrinol 50 DF	acetamide	15
(nicosulfuron + metsulfuron methyl)	Pastora	sulfonylurea + sulfonylurea	2 + 2
oryzalin	(4 AS, 4 L): Monterey Weed Impede 4 AS, Oryzalin 4 AS, Phoenix Harrier 4 L, Surflan AS (85WDG): Surflan WDG	dinitroaniline	3
oxadiazon	(2 G): Oxadiazon 2 G, Ronstar G (50 WP): Oxadiazon 50 WSB, Ronstar 50 WSB (3.17 SC): Oxadiazon SC, Phoenix Starfighter L, Ronstar Flo	oxadiazole	14
(oxadiazon + prodiamine)	Pro-mate Ronstar + Barricade 1.2 G, Regalstar II	oxadiazole + dinitroaniline	14 + 3
paraquat	(2 SL): Cyclone SL 2.0, Gramoxone Inteon, Gramoxone SL (3 SL): Bonedry, Firestorm, Helmquat 3 SL, Parazone 3 SL	bipyridylium	22
pendimethalin	(3.8 CS): HydroCap, Pendulum AquaCap, Pre-M AquaCap, Prowl H2O, Satellite HydroCap (3.3 EC): Drexel Pin-Dee 3.3 T&O (2 G): Pendulum 2 G (0.86 G): fertilizers – Pendimethalin, Pre-M, Propendi, Pro-mate Pendi (1.29 G): Step 1 Crabgrass Preventer, Turf Builder with Halts	dinitroaniline	3
penoxsulam	(0.31 L): Sapphire (0.03 G): Harrels Fert. with Penoxsulam, lesco LockUp, Pennington Seed LockUp (014 G): Harrels Fert. with Penoxsulam	triazolopyrimidine	2
(picloram + fluroxypyr)	Surmount	pyradinecarboxylic acid + pyradinyloxyacetic acid	4 + 4
prodiamine	(65 WDG): Armortech Kade, Barricade, Cavalcade, Halts Pro, Phoenix Knighthawk, ProClipse, Prodiamine, Quali-Pro Prodiamine, RegalKade, Resolute, Stonewall (4 FL): Barricade, Evade, Resolute (0.5 G): RegalKade, Signature Crabgrass Preventer, Turf Pride	dinitroaniline	3

Table 7-10B. Herbicide Modes of Action for Hay Crops, Pastures, and Lawns and Turf

Brands listed were registered for sale and use in North Carolina in 2015 according to http://www.kellysolutions.com/nc/searchbychem.asp. Active ingredients enclosed within parentheses are prepackaged herbicides.

Active Ingredient(s)	Brand Names	Chemical Family	Mode of Action[1]
pronamide	(50 WP): Kerb 50-W (3.3 SC): Kerb SC T&O, Willowood Pronamide 3.3SC	benzamide	3
pyraflufen ethyl	Octane 2% SC	phenylpyrazole	14
quinclorac	(75 DF): Armortech Quinclorac Pro, Quinclorac, Quinclorac SPC, Quinstar (1.5 SL): Drive XLR8, Quinclorac 1.5 L	quinoline carboxylic acid	(27 + 4)
(quinclorac + mecoprop + dicamba)	Onetime	quinoline carboxylic acid + phenoxyalkanoic acid + benzoic acid	(27 + 4) + 4 + 4
(quinclorac + sulfentrazone + 2,4-D amine + dicamba)	Q4 Plus	quinoline carboxylic acid + triazinone + phenoxycarboxylic acid + benzoic acid	(27 + 4) + 14 + 4 + 4
rimsulfuron	Rimsulfuron	sulfonylurea	2
sethoxydim	(1.5 EC): Poast (1 EC): Poast Plus, Segment, Sethoxydim SPC	cyclohexanedione	1
siduron	Tupersan	phenylurea	7
simazine	(4 L): Princep Liquid, Simazine, Sim-Trol (90 DF, 90 WDG): Simazine, Sim-Trol, Princep Caliber 90	triazine	5
sulfentrazone	Dismiss Turf	triazinone	14
(sulfentrazone + 2,4-D + mecoprop + dicamba)	Surge	triazinone + phenoxycarboxylic acid + phenoxyalkanoic acid + benzoic acid	14 + 4 + 4 + 4
(sulfentrazone + imazethapyr)	Dismiss South	triazinone + imidazolinone	14 + 2
(sulfentrazone + metsulfuron-methyl)	Blindside	triazinone + sulfonylurea	14 + 2
(sulfentrazone + prodiamine)	Echelon	triazinone + dinitroaniline	14 + 3
(sulfentrazone + quinclorac)	Solitare	triazinone + quinoline carboxylic acid	14 + (27 + 4)
sulfosulfuron	Certainty, Outrider	sulfonylurea	2
tebuthiuron	(20 P): Spike 20 P, Alligare 20 P (80 DF, 80 WG): Spike 80 DF, Alligare 80 WG	thiadiazolyurea	7
terbacil	Sinbar 80 WDG	uracil	5
(thiencarbazone + foramsulfuron + halosulfuron)	Tribute Total	triazolinone + sulfonylurea + sulfonylurea	14 + 2 + 2
(thiencarbazone + iodosulfuron + dicamba)	Celsius WG	triazolinone + sulfonylurea + benzoic acid	14 + 2 + 4
(thiencarbazone + iodosulfuron + foramsulfuron)	Derigo	triazolinone + sulfonylurea + sulfonylurea	14 + 2 + 2
topramezone	Pylex	benzoylpyrazole	27
(triclopyr)	Monterey Turflon Ester, Remedy Ultra, Turflon Ester Ultra	pyradinecarboxylic acid	4
(triclopyr + cloypyralid)	Confront, 2-D	pyradinecarboxylic acid + pyradinecarboxylic acid	4 + 4
(triclopyr + fluroxypyr)	PastureGard HL	pyradinecarboxylic acid + pyradinyloxyacetic acid	4 + 4
(triclopyr + sulfentrazone + 2,4-D + dicamba)	Tzone SE	pyridinecarboxylic acid + triazinone + phenoxycarboxylic acid + benzoic acid	4 + 14 + 4 + 4
trifloxysulfuron	Monument	sulfonylurea	2
trifluralin	(10 G): Treflan TR-10, Trifluralin 10 G (4 EC): Treflan HFP, Trifluralin 4 EC	dinitroaniline	3

Chemical Weed Control in Clary Sage

R. B. Batts, Horticultural Science Department

Table 7-11. Chemical Weed Control in Clary Sage

Herbicide, Mode of Action Code[1], and Formulation	Amount of Formulation Per Acre	Pounds Active Ingredient Per Acre	Precautions and Remarks
Preplant Injection, Weeds, diseases, and nematodes.			
metam sodium (Vapam HL) 42% (Sectagon-42) 42%	37.5 to 75 gal 15 to 75 gal	15.7 to 31.5 6.3 to 31.5	Rates are dependent on soil types and weeds known to be present. Apply when soil moisture is at field capacity. Read label thoroughly for regulatory, safety and application instructions. Interval to planting is usually 14 to 21 days, but can be as long as 30 days in certain environments.
metam potassium (K-PAM HL, Sectagon-K54) 54%	30-60 gal	16.2 to 32.4	

Table 7-11. Chemical Weed Control in Clary Sage

Herbicide, Mode of Action Code[1], and Formulation	Amount of Formulation Per Acre	Pounds Active Ingredient Per Acre	Precautions and Remarks
Preplant and preemergence, Annual and perennial grass and broadleaf weeds. Stale bed application.			
flumioxazin, MOA 14 (Valor SX) 51 WDG	1 to 3 oz	0.03 to 0.095	Apply preplant, 30 days or more before planting. Application rate varies based on preplant interval. Treated area must receive a minimum of 0.5 inches of rainfall/irrigation between application and planting. Sequential preplant applications are prohibited. Some soil texture restrictions apply. See label for details.
glyphosate, MOA 9 (various brands) 4 SL (various brands) 5 SL (Roundup Weather Max) 5.5 L	1 to 3 pt 0.8 to 2.4 pt 11 to 32 oz	0.5 to 1.5	Apply to emerged weeds before crop emergence. Perennial weeds may require higher rates of glyphosate. Consult the manufacturer's label for rates for specific weeds. Certain glyphosate formulations require the addition of surfactant. Adding nonionic surfactant to glyphosate formulation formulated with nonionic surfactant may result in reduced weed control.
d-limonene, MOA unclassified (Avenger AG) 55%	14 to 20%	—	
pelargonic acid, MOA unclassified (Scythe) 4.2 EC	3 to 10%	—	Apply broadcast as 1:6 ratio (14%) of Greenmatch: water in a minimum of 60 gallons per acre. For spot application on difficult to control weeds, use 1:4 ratio (20%). For best results, apply to dry foliage with temperatures above 50 degrees F. Do not exceed 8.5 gallons Greenmatch per acre per application.
Ammonium nonanoate, MOA unclassified (Axxe) 3.3	6.15%	—	Label recommends application in 75 to 200 gallons of solution per acre when applied alone. When tank mixing with other herbicides, delivery rates can be reduced to 10 to75 gallons per acre. See label for recommended ratio to apply on different types of weeds and other instructions. Rate is dependent on weed type and size at application.
Postemergence, residual weed control.			
flumioxazin, MOA 14 (Valor SX) 51 WDG	1 to 2	0.03 to 0.06	Apply to crop that is a minimum of 4 inches in diameter. Do not apply after crop begins to enter reproductive stage. Do not add adjuvant, fertilizer or other pest control products. Do not use on clary sage grown for food or feed purposes.
Postemergence, Cutleaf eveningprimrose and certain other broadleaf weeds.			
linuron, MOA 7 (Linex) 4 L	1 to 1.5 pt	0.5 to 0.75	Crop must be a minimum of 4 inches in diameter for application. Do not use on clary sage grown for food or feed purposes. Do not use on sands or loamy sands or on soils with less than 1% organic matter. Temporary yellowing or stunting of crop may occur.
Postemergence, Henbit and other winter annual broadleaf weeds.			
oxyfluorfen, MOA 14 (Goal 2XL) 2EC (Galigan, Oxystar, others) 2 E (GoalTender) 4E (Galigan H_2O) 4EC	0.5 to 1 pt 0.5 to 1 pt 0.25 to 0.5 pt 0.25 to 0.5 pt	0.12 to 0.25	Apply to 2-4 leaf henbit. Additional applications may be needed for subsequent henbit emergence. Do not apply more than 6 pints of Goal 2XL or Galigan per acre per year. Do not apply more than 3 pints of GoalTender or Galigan H_2O per acre per year. Clary sage may exhibit phytotoxicity on leaf margins after application, but recovery should occur quickly.
Postemergence, Emerged weeds.			
paraquat MOA 22 (Parazone) 3SL	3 to 32 oz	0.07 to 0.75	Application rate varies based on size and physiological status of the crop. See product label for specific rates and timings. Do not exceed 1.125 pounds per growing season. Do not use clary sage for food or feed.
paraquat MOA 22 (Parazone) 3SL	See comments	See comments	**Rope wick/carpet roller applicator.** Mix 1 part Parazone with 3-4 parts water to prepare a 25-33% solution. Add nonionic surfactant at 0.25% v/v, or 1 quart per 100 gallons. Do not allow mixture to contact crop. Do not apply Parazone in this manner within 30 days of harvest.
carfentrazone, MOA 14 (Aim) 2 EC (Aim) 1.9 EW	Up to 2 oz Up to 2 oz	Up to 0.031 Up to 0.031	**Hooded/shielded application to row middles. Do not allow crop contact. Do not apply more than 6.1 per acre per season. See label for details.**
pelargonic acid, MOA unclassified (Scythe) 4.2 EC	3 to 10%	—	**Hooded/shielded application to row middles. Do not allow crop contact.** Label recommends application in 75 to 200 gallons of solution per acre when applied alone. When tank mixing with other herbicides, delivery rates can be reduced to 10 to 75 gallons per acre. See label for recommended ratio to apply on different types of weeds and other instructions.
ammonium nonanoate, MOA unclassified (Axxe) 3.3	6.15%	—	Hooded/shielded application to row middles. Do not allow crop contact. Rate is dependent on weed type and size at application.
Postemergence, Emerged annual and perennial grasses			
clethodim, MOA 1 (Arrow, Clethodim, others) 2EC	6 to 8 oz	0.09 to 0.125	Arrow and Clethodim require addition of crop oil concentrate to spray mixture. See label for precautions regarding crop oil concentrates. Do not apply within 14 days of harvest.
(Select Max) 1 EC	9 to 32 oz	0.07 to 0.25	
(Intensity One, Tapout) 1 EC	9 to 16 oz	0.07 to 0.125	Select Max allows the use of a nonionic surfactant, methylated seed oil, or crop oil concentrate in the mixture. Label suggests different rate ranges for annual and perennial grasses. See label for details. Do not apply within 14 days of harvest. Label recommends the inclusion of nonionic surfactant at 0.25% v/v. Do not apply more than 64 fluid ounces per acre per crop. Do not apply within 14 days of harvest.

[1] Mode of Action (MOA) code developed by the Weed Science Society of America. See Table 7-10, Herbicide Resistance Management, for details.

Chemical Weed Control in Small Fruit Crops

W. E. Mitchem and K. M. Jennings, Horticultural Science Department

NOTE: A mode of action code has been added to the Herbicide and Formulation column of this table. Use MOA codes for herbicide resistance management. See Table 7-10, Herbicide Modes of Action, for details.

Table 7-12A. Chemical Weed Control in Small Fruit Crops

Timing/Targeted Weeds	Herbicide, Mode of Action Code[1] and Formulation	Amount of Formulation Per Acre	Pounds Active Ingredient Per Acre	Precautions and Remarks
Blackberries				
PREPLANT Annual and perennial weeds	glyphosate, MOA 9 (various brands) 4 SL (various brands) 5 SL (Roundup WeatherMax) 5.5 L	1 to 3 pt 0.8 to 2.4 pt 11 to 32 oz	0.5 to 1.5	Apply to emerged weeds at least 30 days before crop transplanting. Perennial weeds may require higher rates of glyphosate. Consult the manufacturer's label for rates for specific weeds. See label for further instructions.
PREEMERGENCE Annual grass and broadleaf weeds	indaziflam, MOA 29 (Alion) 1.67 SC	3.5 to 5 fl. oz	0.045 to 0.13	Use in plantings established 1 year or longer exhibiting good growth and vigor. Apply **ONLY** as a dormant application between late fall and early spring prior to bud swell. Two applications may be applied so long as there are at least 90 days between applications. DO NOT use on caneberries grown in sand or soils having a gravel content more than 20%. Total use rate cannot exceed more than 7 fl. oz/A in soils having < 1% OM or 10 fl. oz/A in soils having ≥ 1% OM. DO NOT allow spray to contact green stems, flowers, fruit, or foliage or unacceptable injury may occur. Tank mix with paraquat for non-selective POST weed control.
PREEMERGENCE Annual grass and small seeded broadleaf weeds	napropamide, MOA 15 (Devrinol) 50 WDG	8 lb	4	Apply to soil surface free of weeds and plant residue. Rainfall or overhead irrigation is needed within 1 to 2 days of application for effective herbicide performance. In new plantings allow soil to settle after transplanting before applying Devrinol.
	oryzalin, MOA 3 (Oryzalin or Surflan) 4 AS	2 to 6	2 to 6	Oryzalin may be tank mixed with paraquat for non-selective POST weed control. Total use rate cannot exceed 12 quarts/acre per year. Sequential applications may be used along as there is 2.5 months between applications. Tank mix with simazine for broad spectrum PRE weed control.
PREEMERGENCE and POSTEMERGENCE Annual broadleaf weeds	mesotrione, MOA 27 (Callisto) 4L	3 to 6 oz	0.094 to 0.185	May be applied as split applications of 3 oz per acre followed by 3 oz per acre. If 2 applications are made, there must be at least 14 days between applications. Do not apply more than 6 ounces per acre per year. Do not apply after the onset of bloom stage or illegal residues may occur. The addition of crop oil concentrate at a rate of 1% v/v (1 gallon per 100 gallons of spray solution) is recommended for POST weed control.
PREEMERGENCE Annual broadleaf and grass weeds. Nutsedge suppression	norflurazon, MOA 12 (Solicam) 80 DF	1.25 to 5	1 to 4	Apply to dormant caneberries established at least 1 year. Tank mix with paraquat for control of emerged weeds. The addition of simazine will expand spectrum of PRE control. Solicam has a 60 day PHI. In areas prone to soil movement injury to ground cover planted in row middles can occur.
PREEMERGENCE and POSTEMERGENCE Annual broadleaf and grass weeds	terbacil, MOA 5 (Sinbar) 80 WDG	1 to 2	0.8 to 1.6	Apply as a directed spray in early fall, winter, or spring before fruit set. Do not allow contact with desirable foliage. Sinbar has a 70 day PHI. Do not use on loamy sand and sandy soils or on soils having less than 1% organic matter. Do not use on eroded areas where subsoil is exposed. Tank mix with paraquat for non-selective POST weed control. Sinbar has POST emergence activity on some weeds like horseweed.
PREEMERGENCE Broadleaf and some grass weeds	simazine, MOA 5 (Princep or Simazine) 4L (Princep or Simazine) 90 DF	1 to 4 qt 1.1 to 4.4 lb	1 to 4	Use half rate on caneberries established less than 6 months. Rate is soil texture dependent therefore higher rates cannot be used sandy loam, loamy sand, or sand soils. See label for soil texture precautions. Apply as a single application in spring or as a split application of 2 quarts/acre in spring followed by 2 quarts/acre in the fall. Do not apply when fruit is present or illegal residues may occur. Tank mix with oryzalin or norflurazon for expanded residual control of annual grasses like crabgrass and goosegrass. Simazine may be applied in combination with paraquat for non-selective POST weed control.
PREEMERGENCE Annual and perennial broadleaf weeds as well as grass weeds	dichlobenil, MOA 20 (Casoron) 4G (Casoron) CS	100 lb 1.4 to 2.8 gal	4 2 to 4	Granular formulation should be applied in January or February. Rainfall or snow is needed for activation. Warm temperatures increase herbicide loss due to volatilization. The liquid formulation should be used during warmer temperatures (up to 70 degrees F) as it is less likely to volatilize. As with the granular formulation, it needs to be activated with rainfall. The liquid formulation may be tank mixed with other herbicides like paraquat.
PREEMERGENCE + POSTEMERGENCE Controls annual broadleaf and some grass weeds	Sulfentrazone + Carfentrazone, MOA 14 (Zeus Prime XC)	8 to 15 fl. oz	0.22 to 0.41	Use in blackberry plantings that have been establish at least 2 years. Zeus Prime may be applied twice per year so long as use rate does not exceed 15 fluid ounces per acre on a broadcast basis. Rainfall is needed within 2 weeks of application to ensure herbicide activation. Applications made after bud break must be made using hooded application equipment. There must be at least 60 days between sequential applications. Zeus Prime has a 3 day PHI. Tank mix with non-selective POST herbicides for broad spectrum control of emerged weeds. The addition of oryzalin is necessary for expanded residual control of annual grass weeds. **Zeus will control yellow nutsedge.**
PREEMERGENCE + POSTEMERGENCE Controls annual broadleaf and some grass weeds (continued)	rimsulfuron, MOA 2 (Solida) 25 WG (Matrix) 25 WG	4 oz	0.063	Rimsulfuron has POST and PRE activity on broadleaf and some grass weeds. For broad spectrum residual control of annual grass weeds, tank mix rimsulfuron with oryzalin or Sinbar. For nonselective POST weed control, tank mix rimsulfuron with paraquat. Do not treat blackberry plantings established less than 1 year. Rainfall for herbicide activation is necessary within 2 to 3 weeks of application. Do not apply within 21 days of harvest. The pH of spray solution should be in the range of 4 to 8. Rimsulfuron may be applied as a sequential application so long as total use rate does not exceed 4 ounces per acre per year and application is made in band to less than 50% of orchard floor. Allow at least 30 days between sequential applications. To reduce the risk of primocane injury, apply prior to primocane emergence or wait until primocanes are 3 feet tall. If primocanes are emerged at the time of application, chlorosis and stunting is likely but in most instances those primocanes recover.
	Flumioxazin, MOA 14 (Chateau) 51 SW	6 oz	0.188	Apply as a directed to spray. Use ONLY a single application per year. Chateau has a 7 day PHI for caneberries. Tank mix with paraquat for non-selective POST weed control.

Table 7-12A. Chemical Weed Control in Small Fruit Crops

Timing/Targeted Weeds	Herbicide, Mode of Action Code[1] and Formulation	Amount of Formulation Per Acre	Pounds Active Ingredient Per Acre	Precautions and Remarks
Blackberries (continued)				
POSTEMERGENCE Annual broadleaf weeds like pigweed, morningglory, lambsquarter, purslane, nightshade, tropical spiderwort, and smartweed	carfentrazone, MOA 14 (Aim) 2 EC	0.8 to 2 oz	0.013 to 0.03	Do not allow spray solution to contact desirable vegetation, foliage, flowers, or fruit. Every precaution should be taken to avoid herbicide injury related to herbicide drift. Use rate should not exceed 25 ounces/acre per year and there must be at least a 14 day interval between applications. The addition of a non-ionic surfactant at 0.25% v/v (1 quart per 100 gallons of spray solution) or crop oil concentrate at 1 to 2% v/v (1 to 2 gallons per 100 gallons of spray solution is necessary for optimum herbicide performance. Aim has a 15 day PHI. Aim may be used to suppress primocane emergence. See label for instructions and rate information relative to primocane suppression.
POSTEMERGENCE Non-selective broadleaf and grass weed control	paraquat, MOA 22 (Firestorm, Parazone, Paraquat Concentrate) 3 SL (Gramoxone) 2 SL	1.3 to 2.7 pt 2 to 4 pt	0.5 to 1.0	Do not allow herbicide to directly contact desirable foliage or green canes. Young plants must be shielded, or severe crop injury or death will result. The addition of a non-ionic surfactant at 0.25% v/v (1 quart/100 gallons of spray solution) is necessary for adequate control. Paraquat can be tank mixed with PRE herbicides. DO NOT make more than 5 applications per year.
POSTEMERGENCE Annual and perennial grass weeds	clethodim, MOA 1 (Select Max or Intensity One) 1 EC (Select, Arrow, Volunteer, Clethodim) 2 EC	9 to 16 oz 6 to 8 oz	0.07 to 0.125	Low rates are for annual grass weeds. Use higher rates and sequential applications for perennial grass (bermudagrass or johnsongrass) control. Select Max and Intensity One need to be applied in combination with a non-ionic surfactant. Select and other generic formulations of clethodim require the addition of either a non-ionic surfactant or a crop oil concentrate. See label for specific information related to spray adjuvants. **The Select Max formulation is labeled for bearing caneberries and has a 7 day PHI. ALL other clethodim formulations are registered for use in non-bearing caneberries ONLY.**
	fluazifop, MOA 1 (Fusilade DX) 2 EC	12 to 24 oz	0.19 to 0.38	Sequential applications will be necessary for perennial grass control. The addition of a non-ionic surfactant (1 qt/100 gal of spray solution) or crop oil concentrate (1 gallon/100 gallons of spray solution) is necessary. **Use in non-bearing caneberry plantings only.**
	sethoxydim, MOA 1 (Poast) 2 EC	1 to 2.5 pt	0.25 to 0.63	Sequential application will be necessary for perennial grass control. The addition of a non-ionic surfactant (1 qt/100 gal of spray solution) or crop oil concentrate (1 gal/100 gal of spray solution) is necessary for optimum herbicide performance. Poast has a 50 day PHI.
Blueberries				
PREPLANT Annual and perennial weeds	glyphosate, MOA 9 (various brands) 4 SL (various brands) 5 SL (Roundup WeatherMax) 5.5 L	1 to 3 pt 0.8 to 2.4 pt 11 to 32 oz	0.5 to 1.5	Apply to emerged weeds at least 30 days before crop transplanting. Perennial weeds may require higher rates of glyphosate. Consult the manufacturer's label for rates for specific weeds. See label for further instruction.
PREEMERGENCE Annual weeds and some perennial (goldenrod) weeds	hexazinone, MOA 5 (Velpar) 2 SL (Velpar) 75 DF	0.5 to 1 gal 1.3 to 2.6 lb	1 to 2	Apply as a directed spray to soil and weeds before blueberry leaf emergence but at least 90 days before harvest. Use lower rates on poorly drained or sandy soils. Bushes must be established for at least 3 years.
PREEMERGENCE Annual weeds (crabgrass, chickweed) and some perennial (dogfennel) weeds	dichlobenil, MOA 20 (Casoron) 4 G (Casoron CS) 1.4 L	100 to 150 lb 1.4 to 2.8 gal	4 to 6 2 to 3.92	Apply in the early winter, no later than mid-February, to plants that have been established 1 year or longer (Casoron CS formulation). Casoron 4G may be used in blueberry planted at least 4 weeks earlier.
PREEMERGENCE Annual broadleaf weeds (Maryland meadowbeauty, pigweed spp., morningglory spp.) and some annual grasses (large crabgrass)	flumioxazin, MOA 14 (Chateau) 51 WDG	6 to 12 oz	0.19 to 0.375	Do not apply to blueberries established less than 2 years unless they are protected from spray contact by nonporous wrap, grow tubes, or waxed containers. Do not apply more than 12 ounces per acre during a 12-month period. If a sequential application is applied, it must occur no earlier than 60 days after the first application. Do not apply more than 6 ounces per application to bushes less than 3 years old on soils having a sand plus gravel content greater than 80%. Apply at the base of the bush. Chateau should be tank-mixed with a registered burndown herbicide to control emerged weeds. Residual weed control will be reduced if vegetation prevents Chateau from reaching soil surface. PHI = 7 days.
PREEMERGENCE Directed to the soil surface Annual grass and broadleaf weeds, yellow nutsedge suppression	S-metolachlor, MOA 15 (Dual Magnum) 7.62 EC	0.67 to 1.33 qt	0.64 to 1.27	This is a Section 24(c) North Carolina Special Local Need Label. Growers must obtain label prior to making Dual Magnum applications. Growers must obtain label at www.farmassist.com. Avoid contact with crop foliage, or crop injury may occur. See label for rate range based on soil type. Blueberry plants less than 1 year may be more sensitive.
PREEMERGENCE Annual broadleaf weeds (morningglory, chickweed) and some annual grasses (large crabgrass)	diuron, MOA 7 (Karmex) 80 DF (Direx) 4 L	1.5 to 2 lb 1.2 to 1.6 qt	1.2 to 1.6	Use only in fields that have been established for at least 1 year. Apply as a band treatment at the base of bushes. The addition of a surfactant will kill many small emerged weeds. May be applied in the spring and again in the fall after harvest.
PREEMERGENCE Annual broadleaf weeds (pigweed, common purslane) and some annual grasses (crabgrass spp., fall panicum)	simazine, MOA 5 (Princep Caliber 90) 90 WDG (Princep) 4 L	2.2 to 4.4 lb 2 to 4 qt	2 to 4	Apply half the maximum annual application in the spring before buds break and weeds emerge, and half after harvest. Do not apply more than 1 pound a.i. simazine on newly planted blueberries. Apply in minimum of 40 gallons of water per acre.

Table 7-12A. Chemical Weed Control in Small Fruit Crops

Timing/Targeted Weeds	Herbicide, Mode of Action Code[1] and Formulation	Amount of Formulation Per Acre	Pounds Active Ingredient Per Acre	Precautions and Remarks
Blueberries (continued)				
PREEMERGENCE Directed Underneath Bushes, Broadleaf weeds and some annual grasses	rimsulfuron, MOA 2 (Matrix) 25 WG (Solida) 25 WG	4 oz	0.063	Application after bud break may cause temporary chlorosis and/or stunting of leaves contacted by the spray solution. Rimsulfuron has POST and PRE activity on broadleaf and some grass weeds. For broad spectrum residual control, tank mix rimsulfuron with oryzalin. For nonselective POST weed control, tank mix rimsulfuron with glyphosate, paraquat, or Rely. Do not treat blueberries established less than 1 year. The best residual control is obtained when at least 0.5 inches of rain or overhead irrigation comes within the first week after application. The pH of spray solution should be in the range of 4 to 8. Do not use on soils classified as a Sand. Rimsulfuron may be applied as a sequential application so long as total use rate does not exceed 4 ounces per acre per year and application is made in band to less than 50% of floor. Do not apply within 21 days of first harvest.
PREEMERGENCE Annual broadleaf weeds (chickweed, red sorrel from seed) and some annual and perennial grasses	pronamide, MOA 3 (Kerb) 3.3 SC	2.5 to 5 pt	1 to 2	Apply in the fall or winter. May be applied to newly planted bushes as long as roots are well established. Do not apply more than 2 lb ai or make more than one application per year.
PREEMERGENCE Most annual broadleaf and grass weeds plus many perennials	terbacil, MOA 5 (Sinbar) 80 WDG	0.5 to 2 lb	0.4 to 1.6	Apply as directed spray in early spring or in fall after harvest. May be applied before weeds emerge or shortly after they emerge. Use only in plantings established 1 year or longer. Do not use on sandy soils with less than 3% organic matter. This herbicide can be very active but injurious on blueberries. See label for further information.
PREEMERGENCE Annual broadleaf weeds	mesotrione, MOA 27 (Callisto) 4L	3 to 6 oz	0.094 to 0.185	May be applied as a split application of 3 ounces per acre followed by 3 ounces per acre no less than 14 days apart. Do not apply more than 6 oz per acre per year. Do not apply after the onset of bloom stage or illegal residues may occur.
PREEMERGENCE Broadleaf weeds including corn spurry, common cocklebur, dayflower, horseweed, smartweed, wild mustard, wild radish, and pigweed	halosulfuron, MOA 2 (Sandea) 75 DG	0.5 to 1 oz	0.024 to 0.047	Do not apply to plants established less than 1 year. Apply as a directed treatment to avoid contact with the crop. Occasional injury may occur. For nutsedge control, apply Sandea postemergence to the nutsedge (see Sandea listed below for postemergence control). PHI = 14 days
PREEMERGENCE Annual grasses and small seeded broadleaf weeds	napropamide, MOA 15 (Devrinol, Devrinol DF-XT) 50 DF (Devrinol) 10 G (Devrinol 2-XT)	8 lb 40 lb 8 qt	4	Apply to weed-free soil surface. Enough irrigation or rainfall to wet the soil to a depth of 4 inches is necessary within 24 hours of application. Apply as a directed spray to the base of the blueberry plant. May be used on first-year plantings. If using Devrinol DF-XT or Devrinol 2-XT, time between application and irrigation is extended to 48 hours.
PREEMERGENCE Will suppress yellow and purple nutsedge	norflurazon, MOA 12 (Solicam) 78.6 WDG	2.5 to 5 lb	2 to 4	Apply as a directed spray from fall to early spring when the crop is dormant and before weeds emerge. Make only 1 application per year. Blueberries must be established 6 months prior to Solicam use. Application of Solicam may result in temporary bleaching or chlorosis of the leaves. Preharvest interval is 60 days.
PREEMERGENCE Annual broadleaf (morningglory, pigweed spp., common purslane) and annual grasses (crabgrass spp., barnyardgrass)	oryzalin, MOA 3 (Oryzalin or Surflan) 4 AS	2 to 4 qt	2 to 4	This treatment may be used on first year plants.
PREEMERGENCE Broadleaf, grasses and sedge weeds	sulfentrazone, MOA 14 (Zeus XC) 4L	12 oz	0.375	Apply to crop that has been growing for at least 3 years and is in good condition. Avoid direct or indirect spray contact to foliage and green bark. Apply as a uniform broadcast soil application in a minimum of 10 gallons of water per acre. For best control, Zeus XC should be applied when there are no weeds present. If weeds are present, tank mix with a postemergence herbicide to eliminate emerged weeds. Do not apply more than 0.375 lb ai per acre per 12-month period. Appropriate soil moisture at time of application or at least ½ inch of rainfall or irrigation within two weeks after application is required to achieve desired level of weed control. Preharvest interval is 3 days.
PREEMERGENCE Broadleaf weeds	Isoxaben, MOA 21 (Trellis) 0.75	1.33 lb	1.0	APPLY ONLY ON ESTABLISHED NON-BEARING PLANTS. Apply as a uniform broadcast soil application in a minimum of 10 gallons of water per acre. For best control, Trellis should be applied when there are no weeds present. Appropriate soil moisture at time of application or at least ½ inch of rainfall or irrigation within 21 days after application is required to achieve desired level of weed control.
PREEMERGENCE Most annual grasses and some broadleaf weeds	pendimethalin, MOA 3 (Prowl H_2O) 3.8 AS	2 to 6.3 qt	1.9 to 5.96	Do not exceed a total of 6.3 qt per acre per year in blueberries. Do not apply over the top of bushes with leaves, shoots, or buds. Contact by the spray mixture may cause injury. PHI = 30 days.
	glyphosate, MOA 9 (various brands) 4 SL (various brands) 5 SL (Roundup WeatherMax) 5.5 L	1 to 3 pt 0.8 to 2.4 pt 11 to 32 oz	0.5 to 1.5	DO NOT SPRAY GREEN CANES, BARK, OR FOLIAGE. Apply as a directed shielded spray to base of established plants. Do not apply within 14 days of harvest. Wiper applications may also be used. Perennial weeds may require higher rates of glyphosate. Certain glyphosate formulations require the addition of a surfactant. See label for specific rates for herbicide and surfactant.
POSTEMERGENCE NON-SELECTIVE Contact kill of all green foliage	paraquat, MOA 22 (Firestorm, Parazone) 3 SL (Gramoxone Inteon, Gramoxone SL, Gramoxone SL 2.0) 2 SL	1.3 to 2.7 pt 2 to 4 pt	0.56 to 1	Apply as a directed spray to weeds before new canes emerge. Avoid paraquat contact with new canes, as injury will occur. Use of paraquat in rabbiteye blueberry can increase incidence of stem blight if herbicide contacts green stems. Rabbiteye producers should consider using other non-selective herbicides.
POSTEMERGENCE	pelargonic acid, MOA 27 (Scythe) 4.2 SL	3 to 10% soln.		Apply as a directed or shielded spray.

Table 7-12A. Chemical Weed Control in Small Fruit Crops

Timing/Targeted Weeds	Herbicide, Mode of Action Code[1] and Formulation	Amount of Formulation Per Acre	Pounds Active Ingredient Per Acre	Precautions and Remarks
Blueberries (continued)				
POSTEMERGENCE Broadleaf weeds up to 4 inches tall or 3 inches in diameter	carfentrazone-ethyl, MOA 14 Aim 2EC	1 to 2 oz	0.015 to 0.031	Apply as a hooded spray with application equipment designed to prevent spray deposit on green stems, leaf tissue, flowers, or fruit. Use in established fields only; do not use on newly set plants. May be used alone or tank-mixed with other herbicides. May be a good option for sodded middles as it does not control grasses. Add crop oil concentrate at 1% by volume (1 gal/100 gal of spray solution) or a nonionic surfactant at 0.25% by volume (1 qt/100 gal of spray solution).
POSTEMERGENCE Yellow nutsedge and some broadleaf weeds	bentazon, MOA 6 (Basagran) 4SL	1.5 to 2 pt	0.75 to 1	NONBEARING ONLY. For yellow nutsedge control, 2 applications may be needed. Apply when plants are 6 to 8 inches tall. If needed, make a second application at the same rate 7 to 10 days later. Add crop oil concentrate to the spray solution at a rate of 2 pints in 20 to 50 gallons of water per acre.
POSTEMERGENCE Yellow and purple nutsedge, pigweed, common ragweed, wild radish, wild mustard velvetleaf, smartweed, common cocklebur, dayflower, rice flatsedge	halosulfuron, MOA 2 (Sandea) 75 DG	0.75 to 1 oz	0.036 to 0.047	Do not apply to plants established less than 1 year. Apply as a directed treatment to avoid contact with the crop. Occasional injury may occur. See label for further instructions regarding nutsedge control. PHI = 14 days.
POSTEMERGENCE Annual and perennial grasses	clethodim, MOA 1 (Arrow, Select, and others) 2 EC	6 to 8 oz	0.09 to 0.125	Select formulation is for use on nonbearing crop only (within 1 year of harvest). Select Max formulation may be applied as a directed spray to nonbearing and bearing crop. Select Max formulation requires the use of a nonionic surfactant rather than crop oil concentrate. PHI for Select Max is 14 days.
	(Select Max and others) 1 EC	9 to 16 oz	0.07 to 0.125	
	fluazifop, MOA 1 (Fusilade DX) 2 EC	16 to 24 oz	0.25 to 0.38	USE ON NONBEARING CROP ONLY. Postemergence grass control. Check label for specific rates and timings. Do not apply within 1 year of the first harvest. Use of a crop oil or surfactant will be necessary. Sequential applications are necessary for adequate control of perennial grasses.
	sethoxydim, MOA 1 (Poast) 1.5 EC	1.5 to 2.5 pt	0.3 to 0.5	Check label for specific rates and timings. Use a crop oil at a rate of 1 qt per acre. May be used on bearing blueberries. PHI = 30 days.
Grapes				
PREEMERGENCE Directed Underneath Vines, Annuals and many perennials	dichlobenil, MOA 20 (Casoron) 4 G	100 to 150 lb	4 to 6	Do not apply Casoron 4G within 4 weeks of transplanting. Apply in January and February. High rate is necessary for perennial weed control. Casoron CS may only be used in vineyards established at least 1 year.
	(Casoron CS) 1.4 CS	1.4 to 2.8 gal	2 to 4	
PREEMERGENCE Directed Underneath Vines, Annual broadleaf and grass weeds	simazine, MOA 5 (Princep Simazine) 90 WDG	2.2 to 4.4 lb	2 to 4	Apply before germination of annual weeds. Do not apply in vineyards less than 3 years old. Tank mix with glyphosate, paraquat, or glufosinate for POST weed control. Tank mixing simazine with oryzalin or Prowl H2O will improve residual control of annual grasses and certain broadleaf weeds.
	(Princep Simazine) 4 L	2 to 4 qt		
	flumioxazin, MOA 14 (Chateau) 51 SW (Tuscany) 51 WDG (Tuscany) 4 SC	6 to 12 oz 6 to 12 fl oz	0.19 to 0.375	Apply as a directed spray using hooded or shielded application equipment. The trunks of grape vines established less than 2 years must be shielded from contact with spray solution using grow tubes. Do not apply after flowering unless using hooded or shielded application equipment and applicator can insure spray material does not contact fruit or desirable foliage. DO NOT tank mix with glyphosate when applying Chateau after bud break due to increased injury potential. Do not apply more than 6 ounces per acre per application to vines less than 3 years old on soils having a sand plus gravel content greater than 80%. DO NOT apply sequential applications closer than 30 days apart. Total use rate cannot exceed 24 ounces/acre per year. Chateau has a 60 day PHI.
	diuron, MOA 7 (Diuron, Karmex XP)	2 to 3 lb	1.6 to 2.4	Apply before germination of annual weeds. Vineyards must be at least 3 years old. Higher rate may be used on soils with greater than 2% organic matter and high clay content. Do not use on sandy loam or coarser soils. Tank mix with glyphosate, paraquat, or glufosinate for POST weed control. Applications in vineyards having less than 2% organic matter may cause injury if heavy rainfall occurs soon after application. This risk is assumed by user.
	indaziflam, MOA 29 (Alion) 1.67 SC	3.5 to 5 fl. oz	.045 to .065	Use in vineyards established 3 years or longer and on vines having good growth and vigor. Grapes must have a 6" barrier between the soil surface and a major portion of the vine's root system. DO NOT use on grapes planted in sand soils. Rate is soil texture dependent. See label for details. Total use rate cannot exceed 5 fl. oz/A in a 12-month period. When making more than 1 application per year allow at least 90 days between applications. Tank mix with glyphosate, glufosinate, or paraquat for non-selective POST weed control.
PREEMERGENCE Directed Underneath Vines, Annual grasses and small-seeded broadleaf weeds	oryzalin, MOA 3 (Oryzalin, Surflan) 4 AS	2 to 4 qt	2 to 4	Apply once soil has settled after transplanting. Multiple applications per year are permitted; see label for details. Apply in combination with Trellis in newly planted vineyards for improved control of broadleaf weeds. Sequential applications may be used so long as total use rate does not exceed 12 quarts per acre per year. Allow 2.5 months between applications. In established planting oryzalin may be tank mixed with simazine for broad spectrum residual control of annual weeds. Apply in combination with paraquat, glyphosate, or glufosinate for non-selective POST weed control.
	pendimethalin, MOA (Prowl) H_2O 4E	2 to 6.3 qt	2 to 6	In newly planted grapes allow soil to settle after transplanting before applying Prowl. Use only during dormancy (prior to bud swell) when applying around newly planted and 1 year old vines. In bearing vineyards apply any time after harvest, during winter dormancy, and spring. Use rate cannot exceed 6.3 quarts/acre per year. Prowl H_2O has a 90-day PHI. Tank mix with simazine or rimsulfuron for expanded residual control of broadleaf weeds. Prowl may be applied in combination with paraquat, glyphosate, or glufosinate for non-selective POST weed control.

Table 7-12A. Chemical Weed Control in Small Fruit Crops

Timing/Targeted Weeds	Herbicide, Mode of Action Code[1] and Formulation	Amount of Formulation Per Acre	Pounds Active Ingredient Per Acre	Precautions and Remarks
Grapes (continued)				
PREEMERGENCE Directed Underneath Vines, Annual grasses and broadleaf weeds	oryzalin, MOA 3 (Oryzalin, Surflan) 4 AS + simazine, MOA 5 (Princep) 90 WDG (Princep) 4 L	2 to 4 qt + 2.2 to 4.4 lb 2 to 4 qt	2 to 4 + 2 to 4	Tank mix for use before weed emergence. See comments for oryzalin and simazine.
PREEMERGENCE + POSTEMERGENCE Directed Underneath Vines for control of broadleaf weeds and some annual grasses	Sulfentrazone + Carfentrazone (Zeus Prime XC)	8 to 15 fl. oz	0.22 to 0.41	Use in vineyards that have been establish at least 2 years. When applied in a band treating less than 50% of the vineyard floor, Zeus Prime may be applied twice per year so long as use rate does not exceed 15 fluid ounces per acre on a broadcast basis. Rainfall is needed within 2 weeks of application to ensure herbicide activation. Applications made after bud break must be made using hooded application equipment. There must be at least 60 days between sequential applications. Zeus Prime has a 3 day PHI. Tank mix with non-selective POST herbicides for broad spectrum control of emerged weeds. The addition of oryzalin is necessary for expanded residual control of annual grass weeds. **Zeus will control yellow nutsedge.**
PREEMERGENCE Directed Underneath Vines, Broadleaf weeds and some annual grasses	rimsulfuron, MOA 2 (Matrix) 25 WG (Pruvin) 25 WG (Solida) 25 WG	4 oz	0.063	Rimsulfuron has POST and PRE activity on broadleaf and some grass weeds. For broad spectrum residual control, tank mix rimsulfuron with oryzalin, Prowl H_2O, or diuron. For nonselective POST weed control, tank mix rimsulfuron with glyphosate, paraquat, or Rely. Do not treat vineyards established less than 1 year. Rainfall for herbicide activation is necessary within 2 to 3 weeks of application. Do not apply within 14 days of harvest. The pH of spray solution should be in the range of 4 to 8. Rimsulfuron may be applied as a sequential application so long as total use rate does not exceed 4 ounces/acre per year and application is made in band to less than 50% of vineyard floor.
POSTEMERGENCE Directed Underneath Vines, Non-selective weed control	paraquat, MOA 22 (Firestorm) 3 SL (Paraquat Concentrate) 3 SL (Parazone) 3 SL (Gramoxone) 2 SL	1.7 to 2.7 pt 2.5 to 4 pt	0.66 to 1	Apply in 20 gallons per acre spray mix when grass and weeds are 1 to 6 inches high and succulent for best results. Direct spray with low pressure to avoid contact with foliage or bark less than 1 year old. Add a nonionic surfactant at a rate of 32 ounces per 100 gallons of spray solution. May be used for sucker suppression. See label for details.
	glyphosate, MOA 9 various brands and formulations)	See label	1 to 2	DO NOT SPRAY GREEN BARK OR FOLIAGE. Apply preplant or as a directed spray to base of established vines. Do not treat within 14 days of harvest. Wiper applications may also be used. Perennial weeds may require higher rates of glyphosate. See label for specific rates. Do not apply in late summer or fall. Some formulations may require the addition of a surfactant.
	glufosinate, MOA 10 (Cheetah, Reckon, Rely, Lifeline, Surmise) 2.34 SL	48 to 82 oz	0.88 to 1.5	Apply as a directed spray to emerged weeds in a minimum of 20 gallons water per acre with a minimum of 30 psi spray pressure when weeds are 1 to 6 inches high. For spot application, use 1.7 ounces per gal of water and spray to wet but not runoff; however, spot spray solution should not contact vines or injury can occur. Do not allow spray to contact desirable foliage or green bark. Do not apply within 14 days of harvest. See label for specific rates. Do not make more than 3 applications per year. The addition of a spray grade ammonium sulfate will enhance Rely 280 activity on difficult to control weeds. The use of additional surfactants or crop oil is not needed and/or may increase potential for crop injury.
POSTEMERGENCE, Directed Underneath Vines, Annual broadleaf weeds	carfentrazone-ethyl, MOA 14 (Aim) 2 EC	0.5 to 1.6 oz	0.008 to 0.025	Apply as a directed spray or as a hooded spray. DO NOT allow spray solution to contact green tissue, leaves, flowers, or fruit. Aim may be used alone, or tank mixed with other herbicides; see label for tank mixing instructions. Aim controls cocklebur, pigweed, nightshade, velvetleaf, carpetweed, and spreading dayflower. Do not apply within 3 days of harvest. Apply in minimum spray volume of 20 GPA. Apply in combination with crop oil concentrate at 1% v/v (1 gal/100 gal of spray solution) or a nonionic surfactant at 0.25% v/v (1 qt/100 gal of spray solution). Do not use on newly transplanted vines.
POSTEMERGENCE, Directed Underneath Vines, Annual and perennial grasses	clethodim, MOA 1 (Arrow, Select, others) 2 EC (Select Max and others) 1 EC	6 to 8 oz 9 to 16 oz	0.07 to 0.125	USE ON NONBEARING CROP ONLY. Postemergence grass control. Very effective in controlling bluegrass. Do not apply within 1 year of harvest. See label for all other instructions. Sequential applications necessary for adequate control of perennial grass weeds. Always apply 80% active ingredient nonionic surfactant at a rate of 0.25% volume per volume (1 pt/50 gal of spray).
	fluazifop, MOA 1 (Fusilade DX) 2 EC	16 to 24 oz	0.25 to 0.38	Sequential applications will be necessary for perennial grass control. Check label for rates and timings for specific weeds. Do not apply within 50 days of harvest. Use of a crop oil or surfactant will be necessary.
	sethoxydim, MOA 1 (Poast) 1.5 EC	1.5 to 2.5 pt	0.3 to 0.5	Postemergence grass control. Check label for rates and timings for specific grasses. Use a crop oil at a rate of 1 quart per acre. Do not apply within 50 days of harvest. Sequential applications necessary for adequate control of perennial grass weeds.
Strawberries (matted row)				
PREEMERGENCE Most annual grasses and small-seeded broadleaf weeds	DCPA, MOA 3 (Dacthal) W-75 (Dacthal) 6 F	8 to 12 lb 8 to 12 pt	6 to 9	Apply over the top of newly planted transplants or in fall or early spring for preemergence weed control. Do not apply after first bloom through harvest.
	napropamide, MOA 15 (Devrinol, Devrinol DF-XT) 50 DF (Devrinol) 2 EC	4 to 8 lb 8 qt	2 to 4	Apply to established plants before weed emergence anytime, except the interval between bloom and harvest. Apply to established plantings in fall to early spring prior to bloom. See label for notes on irrigation requirement.

Chapter VII—2019 N.C. Agricultural Chemicals Manual

Table 7-12A. Chemical Weed Control in Small Fruit Crops

Timing/Targeted Weeds	Herbicide, Mode of Action Code[1] and Formulation	Amount of Formulation Per Acre	Pounds Active Ingredient Per Acre	Precautions and Remarks
Strawberries (matted row) (continued)				
PREEMERGENCE Most annual broadleaf weeds and grass weeds	terbacil, MOA 5 (Sinbar) 80 WDG	2 to 6 oz	0.1 to 0.3	For planting year: apply 2 to 3 ounces of Sinbar per acre after transplanting but before new runner plants start to root. If strawberry transplants are allowed to develop new foliage prior to Sinbar application, apply 0.5 to 1 inch irrigation or rainfall immediately after application. For control of winter weeds: apply 2 to 6 ounces Sinbar per acre in late summer or early fall. If crop is not dormant, the application must be followed immediately by 0.5 to 1 inch irrigation or rainfall. To extend weed control through harvest of the following year, apply 2 to 4 ounces Sinbar per acre just prior to mulching in the late fall. For harvest years: after postharvest renovation and before new growth begins in midsummer, apply 4 to 6 ounces of Sinbar per acre. To extend weed control through harvest of the following year, apply 4 to 6 ounces of Sinbar per acre just prior to mulching in late fall. Do not apply within 110 days of harvest. See label for more information.
PREEMERGENCE Henbit, chickweed, cutleaf evening primrose, wild radish	flumioxazin, MOA 14 (Chateau) 51 SW	3 oz	0.096	**Crop row.** Apply to **dormant** strawberries for the preemergence control of weeds. Crop oil concentrate at 1% v/v or nonionic surfactant at 0.25% v/v may be added to help control small emerged broadleaf weeds. **Row middle.** Use a hooded or shielded applicator. DO NOT apply over strawberries. Apply prior to weed emergence. Crop spotting may occur if adjuvant is used. DO NOT APPLY AFTER FRUIT SET.
PREEMERGENCE Yellow nutsedge, purple nutsedge, corn spurry, yellow woodsorrel, henbit, chickweed	sulfentrazone, MOA 14 (Spartan) 4F	4 to 8 oz		Please refer to label for soil type restrictions.
POSTEMERGENCE Broadleaf weeds including vetch, clover, dock, sowthistle, and thistle	clopyralid, MOA 4 (Stinger) 3 EC	Spring: 0.67 pt Post harvest 0.33 to 0.67 pt	0.125 to 0.25	Apply postemergence in spring or postharvest to control emerged broadleaf weeds in established strawberries. Do not tankmix with surfactant or other pesticides. PHI = 30 days.
POSTEMERGENCE Annual broadleaf weeds	2,4-D amine, 4, MOA 1 (Amine 4 2,4-D Weed Killer) 4 SL	2 to 3 pt	1 to 1.5	Apply to well-established strawberries after harvest and before runners form or when crop is dormant. Not more than 2 treatments per year. Do not apply during bud, flower, or fruit stage. Timing is very critical to avoid damage. Do not apply unless possible injury to crop is acceptable.
	acifluorfen, MOA 14 (UltraBlazer) 2 L	0.5 to 1.5 pt	0.125 to 0.375	**Crop row.** Apply after last harvest or following bed renovation. A second application can be made in late fall or early spring when plants are dormant. **Row middle.** May be applied up to 1.5 pints/acre. PHI = 120 days.
POSTEMERGENCE NON-SELECTIVE Contact kill of all green foliage	paraquat, MOA 22 (Firestorm) 3 SL (Parazone) 3 SL (Gramoxone Inteon) 2 L	1.3 pt 2 pt	0.5 to 1	For control of emerged broadleaf and grass weeds, use shields and direct spray between the rows to prevent contact with strawberry foliage. Use a nonionic surfactant at a rate of 16 to 32 ounces per 100 gal spray mix or 1 gallon approved crop oil concentrate per 100 gal spray mix. Do not apply within 21 days of harvest.
POSTEMERGENCE Annual and perennial grasses only	clethodim, MOA 1 (Arrow, Clethodim, Intensity, Select) 2 EC (Intensity One, Select Max) 1 EC	6 to 8 oz 9 to 16 oz	0.094 to 0.125 0.07 to 0.125	Apply postemergence for control of emerged grasses in strawberries. With Arrow and Select, add 1 gallon crop oil concentrate per 100 gallons spray mix. With Select Max, add 0.25% nonionic surfactant, 1 quart per 100 gallons spray mix. Very effective in controlling annual bluegrass. Apply to actively growing grasses not under drought stress. Do not apply within 4 days of harvest.
	fluazifop, MOA 1 (Fusilade DX) 2 EC	16 to 24 oz	0.25 to 0.38	USE ON NONBEARING CROP ONLY. Postemergence grass control. Check label for rates and timings for specific weeds. Do not apply within 1 year of the first harvest. Use of a crop oil or surfactant will be necessary.
	sethoxydim, MOA 1 (Poast) 1.5 EC	1 to 1.5 pt	0.2 to 0.3	Apply to emerged grasses. Consult manufacturer's label for specific rates and best times to treat. Add 1 quart of crop oil concentrate per acre. Do not apply on days that are unusually hot and humid. Do not apply within 7 days of harvest.
Strawberries (plasticulture) Preplant				
PREEMERGENCE Yellow and purple nutsedge, broadleaf and grass weeds	EPTC, MOA 8 (Eptam) 7E	3.5 to 7 pt	3 to 6	No research has been conducted in NC. For best control of nutsedge, soil must have enough moisture for tuber sprouting. Allow 10 to 14 days for nutsedge tuber sprouting to occur, then lightly till to destroy shoots and dry the soil surface. Apply and incorporate Eptam 7E to prevent volatilization, immediately incorporate into soil to a depth of approximately 2 to 4 inches. If possible, use a leveling device behind the incorporating equipment to leave soil surface as smooth as possible. Field traffic, excessive rainfall, or irrigation and other soil disturbances will reduce the level of nutsedge suppression. To avoid injury to following crops, irrigating at least 30 days prior to planting is recommended. Do not plant crops not on Eptam 7E label for 45 days after application.
PREEMERGENCE Broadleaf weeds	acifluorfen, MOA 14 (Ultra Blazer) 2 L	0.5 to 1.5 pt	0.125 to 0.375	**Crop row.** Make 1 banded application before laying plastic mulch and after final land preparation, and prior to transplanting the crop. For best results, avoid soil disturbance during laying of plastic and planting of crop. **Row middles between plastic mulch rows.** Apply as a direct-shielded application to strawberry row middles between mulched beds. DO NOT ALLOW ULTRA BLAZER TO CONTACT STRAWBERRY PLANTS. Limited research has been conducted with Ultra Blazer in North Carolina.

Table 7-12A. Chemical Weed Control in Small Fruit Crops

Timing/Targeted Weeds	Herbicide, Mode of Action Code[1] and Formulation	Amount of Formulation Per Acre	Pounds Active Ingredient Per Acre	Precautions and Remarks
Strawberries (plasticulture) Preplant (continued)				
PREEMERGENCE Annual grasses and broadleaf weeds	napropamide, MOA 15 (Devrinol, Devrinol 2-XT) 2 EC (Devrinol, Devrinol DF-XT) 50 DF	8 qt 8 lb	4	Devrinol applied to bed before laying the plastic has potential to injure strawberry plants. For plantbed treatment, preplant incorporate to weed-free soil before laying plastic mulch. Soil should be well worked yet moist enough to permit a thorough incorporation to a depth of 2 inches incorporated within 24 hours of application before laying of plastic mulch. If weed pressure is from small-seeded annuals, apply Devrinol to the surface of the bed immediately in front of the laying of the plastic mulch. If soil is dry, water or sprinkler irrigate with sufficient water to wet to a depth of 2 to 4 inches before covering with plastic mulch. Lay the plastic mulch over the treated soil on the same day as the Devrinol application.
PREEMERGENCE Broadleaf weeds including Carolina geranium and cutleaf eveningprimrose, and a few annual grasses	oxyfluorfen, MOA 14 (Goal) 2 XL	up to 2 pt	up to 0.5	Apply to the soil surface of pre-formed beds at least 30 days prior to transplanting crop for control of many broadleaf weeds that will emerge from hole near crop. While incorporation is not necessary, it may result in less crop injury. Soil disturbance after application will reduce weed control. Plastic mulch can be applied any time after applying Goal, but best results are likely if it is applied soon after Goal.
PREEMERGENCE Annual broadleaf weeds including cutleaf evening primrose, henbit, chickweed, horseweed, wild radish and some annual grasses	flumioxazin, MOA 14 (Chateau SW) 51 WDG	3 oz	0.096	**Crop row.** Apply a minimum of 30 days prior to transplanting and prior to plastic mulch being laid. **Row middles between plastic mulch rows.** Apply only to row middles. DO NOT APPLY over top of strawberries. Apply prior to weed emergence and prior to fruit set. Crop spotting may occur if an adjuvant is added. Application after fruit set may result in spotting of fruit and should be avoided. Do not allow spray drift to come in contact with fruit or foliage.
PREEMERGENCE Yellow nutsedge, purple nutsedge, corn spurry, yellow woodsorrel, henbit, chickweed	sulfentrazone, MOA 14 (Spartan) 4F	4 to 8 oz	0.125 to 0.25	Please refer to label for soil type restrictions.
POSTEMERGENCE Annual and perennial grasses only	sethoxydim, MOA 1 (Poast) 1.5 EC	1 to 2.5 pt	0.2 to 0.5	Apply as a postemergence application to kill emerged grasses. Most effective on actively growing grasses. See label for specific rates and best times to treat. Add 1 quart per acre of crop oil concentrate to spray solution. Very effective control of ryegrass but will not control sedges. Also, effective on volunteer small grains (wheat, etc.). Do not tank mix with other pesticides. PHI = 7 days.
	clethodim, MOA 1 (Arrow, Clethodim, Intensity, Select) 2 EC (Intensity One, Select Max) 1 EC	6 to 8 oz 9 to 16 oz	0.094 to 0.125 0.07 to 0.125	Apply as a postemergence application to kill emerged grasses. Use high rate and sequential applications for perennial grasses (bermudagrass or johnsongrass). With Arrow, Clethodim, Intensity, and Select, add 1 gallon crop oil concentrate per 100 gallons spray mix. With Intensity One and Select Max, add 0.25% nonionic surfactant, 1 quart per 100 gallons spray mix. Very effective in controlling annual bluegrass. Apply to actively growing grasses not under drought stress. PHI = 4 days.
POSTEMERGENCE Broadleaf weeds including vetch, clover, dock, sowthistle, and thistle	clopyralid, MOA 4 (Stinger) 3 EC	0.33 to 0.5 pt	0.125 to 0.187	**Crop row.** Apply postemergence over crop for postemergence control. Do not use with other pesticides or surfactants. Do not apply within 30 days of harvest. DO NOT compost treated vegetation if compost will be used on sensitive plants
Strawberries (plasticulture) Row Middles				
PREEMERGENCE Annual grasses and small-seeded broadleaf weeds	napropamide, MOA 15 (Devrinol, Devrinol DF-XT) 50 DF (Devrinol, Devrinol 2-XT) 2EC	8 lb 8 qt	4	Apply as a banded preemergence treatment to the middles between plastic before weed emergence. Tank mixture with paraquat will provide pre- and postemergence weed control. Rainfall or irrigation within 24 hours after Devrinol application is needed for optimum control. Effective on volunteer small grains (wheat, etc.) if applied before emergence.
PREEMERGENCE Annual grasses and small-seeded broadleaf weeds	pendimethalin, MOA 3 (Prowl H_2O) 3.8 EC	1.5 pt	0.72	Avoid contact with strawberry plant. See label for more information. PHI = 35 days.
POSTEMERGENCE Annual broadleaf weeds	carfentrazone-ethyl, MOA 14 (Aim) 1.9 EW or 2 EC	up to 2 oz	up to 0.031	Apply post directed using hooded sprayers for control of emerged weeds. If crop is contacted, burning of contacted area will occur. Most effective on weeds less than 4 inches tall or rosettes less than 3 inches across. Use a crop oil concentrate at up to 1 gallon per 100 gallons solution or a nonionic surfactant at 2 pints per 100 gallons of spray solution. Coverage is essential for good weed control. Does not control grass weeds. Can be tank mixed with other registered herbicides. PHI = 0 days.
POSTEMERGENCE Broadleaf weeds including vetch, clover, dock, sowthistle, thistle	clopyralid, MOA 4 (Stinger) 3 EC	0.33 to 0.67 pt	0.125 to 0.25	Do not tank mix with other pesticides. Do not include an adjuvant. PHI = 30 days.
POSTEMERGENCE NON-SELECTIVE Contact kill of green foliage	paraquat, MOA 22 (Firestorm, Parazone) 3 SL (Gramoxone Inteon, Gramoxone SL, Gramoxone SL 2.0) 2 SL	1.3 pt 2 pt	0.5	Apply as a banded treatment using shields to the middles between plastic to kill emerged weeds. To avoid injury, do not allow spray to contact strawberry plants. Add a nonionic surfactant at a rate of 16 to 32 ounces per 100 gallons or 1 gallon approved crop oil concentrate per 100 gallons spray solution. Do not apply more than 3 times per season. PHI = 21 days.
POSTEMERGENCE NON-SELECTIVE Most emerged weeds	glyphosate, MOA 9 (Roundup WeatherMax) 5.5L	11 to 22 oz	0.5 to 0.94	Apply as a hooded spray in row middles or shielded spray in row middles or wiper applications in row middles or post harvest. To prevent severe injury to crop, do not let herbicide contact foliage, green shoots or stems, exposed roots, or fruit of crop. PHI = 14 days.
POSTEMERGENCE	pelargonic acid, MOA 27 (Scythe) 4.2 SL	3 to 10% soln.		Apply with hooded or shielded sprayer for weed control in row middles. See label for directions.

Table 7-12A. Chemical Weed Control in Small Fruit Crops

Timing/Targeted Weeds	Herbicide, Mode of Action Code[1] and Formulation	Amount of Formulation Per Acre	Pounds Active Ingredient Per Acre	Precautions and Remarks
Strawberries (plasticulture) Row Middles (continued)				
POSTEMERGENCE Annual and perennial grasses only	sethoxydim, MOA 1 (Poast) 1.5 EC	1 to 2.5 pt	0.2 to 0.5	Apply as a postemergence application to kill emerged grasses. Most effective on actively growing grasses. See label for specific rates and best times to treat. Add 1 quart per acre of crop oil concentrate to spray solution. Very effective control of ryegrass but will not control sedges. Also, effective on volunteer small grains (wheat, etc.). PHI = 7 days.
	clethodim, MOA 1 (Arrow, Clethodim, Intensity, Select) 2 EC (Intensity One, Select Max) 1 EC	6 to 8 oz 9 to 16 oz	0.094 to 0.125 0.07 to 0.125	Apply as a postemergence application to kill emerged grasses. With Arrow, Clethodim, Intensity, and Select, add 1 gallon crop oil concentrate per 100 gallons spray mix. With Intensity One and Select Max, add 0.25% nonionic surfactant, 1 quart per 100 gallons spray mix. Very effective in controlling annual bluegrass. Apply to actively growing grasses not under drought stress. PHI = 4 days.

[1] Mode of action (MOA) code developed by the Weed Science Society of America. See TABLE 7-10, Herbicide Modes of Action, for details.

Chemical Weed Control in Tree Fruit Crops

W. E. Mitchem, Horticultural Science Department

NOTE: A mode of action code has been added to the Herbicide and Formulation column of this table. Use MOA codes for herbicide resistance management. See Table 7-10, Herbicide Resistance Management, for details.

Table 7-12B. Chemical Weed Control in Fruit Crops—Tree Fruits

Weed	Herbicide, Mode of Action,* and Formulation	Amount of Formulation Per Acre	Pounds Active Ingredient Per Acre	Precautions and Remarks
Apples, Preemergence Directed Underneath Tree				
Annual and perennial grass and broadleaf weeds	dichlobenil, MOA 20 (Casoron) 4 G (Casoron) 1.4 CS	100 to 150 lb 1.4 to 2.8 gal	4 to 6 2 to 3.92	For best results apply Casoron 4G in January or February. In order to prevent loss to volatilization, Casoron CS should be applied when temperatures are less than 70 degrees F. Casoron CS may be tank mixed with glyphosate or other herbicides registered for use in apples. Casoron 4G can be used in newly planted orchards once trees have been planted for 4 weeks. Casoron CS should only be used in established orchards 1 year after transplanting.
Annual grasses and broadleaf weeds	diuron, MOA 7 (Diuron, Karmex DF) 80 WDG (Direx, Diuron) 4L	2 to 4 lb 1.6 to 3.2 qt	1.6 to 3.2	Apply in spring (March thru May) to trees established in the orchard for at least 1 year. Best results occur if rainfall occurs within 2 weeks of application. DO NOT treat varieties grafted on full-dwarf rootstocks. When using sequential applications allow at least 90 days between applications and total use rate cannot exceed 4 pounds/acre per year.
	flumioxazin, MOA 14 (Chateau) 51 SW (Tuscany) 51 WG (Tuscany) 4 SC	6 to 12 oz 6 to 12 fl oz	0.19 to 0.38	Flumioxazin is for newly planted and established orchards. Shield the trunks of trees established less than 1 year from contact with spray solution. Tank mix with glyphosate, glufosinate, or paraquat for POST weed control. After budbreak, only tank mix with glufosinate or paraquat. Do not apply more than 6 ounces per acre to trees planted less than 3 years in soil having a sand plus gravel content more than 80%. Sequential applications are very effective. Do not apply within 60 days of harvest. Flumioxazin may only be applied after final harvest and no later than pink flower bud in bearing orchards. Do not use more than 24 ounces of flumioxazin in a 12-month period. When applying after bud break in non-bearing orchards use hooded application equipment.
	indaziflam, MOA 29 (Alion) 1.67 SC	3.5 to 6.5 oz	0.046 to 0.085	Use in orchards established 3 years or more. See label for details pertaining to replants in established orchards. Allow at least 90 days between applications. Use rate cannot exceed 3.5 fluid ounces per acre per application on soils having less than 1% organic matter. On soils with an organic matter content from 1 to 3%, no more than 5 fluid ounces/acre can be applied in a single application and the total use rate for the year cannot exceed 8.5 fluid ounces/acre. In order to apply more than 5 fluid ounces/acre in a single application soil organic matter must be >3%. Do not use on soils that have a 20% or greater gravel content. Do not use in orchards with open channels or cracks in soil - Alion has a 14-day PHI. Tank mix with glyphosate, glufosinate, or paraquat for nonselective POST weed control.
Annual grasses and small seeded broadleaf weeds	norflurazon, MOA 12 (Solicam) 80 WDG	2.5 to 5 lb	2 to 4	Can be tank mixed with Karmex, Goal, paraquat, Prowl, glyphosate, Princep, rimsulfuron, or oryzalin. Do not apply to newly transplanted trees until ground has settled. Rate is soil texture dependent. See label for details. PHI is 60 days. Multiple applications can be made per season so long as cumulative rate does not exceed maximum use rate for soil texture and crop.
	oryzalin, MOA 3 (Oryzalin or Surflan) 4 AS	2 to 6 qt	2 to 6	Allow soil to settle around newly transplanted trees before application. Rainfall (1/2 to 1") with 14 days of application is necessary for optimum herbicide performance. Oryzalin may be tank mixed with Goal, paraquat, glyphosate, Solicam, and simazine. May be applied sequentially. See label for details. Sequential applications may be used so long as total use rate does not exceed 12 quarts per acre per year. Allow 2.5 months between applications.
	pendimethalin, MOA 3 (Prowl H$_2$O) 4 AS	2 to 4.2 qt	2 to 4	Most effective when adequate rainfall or irrigation is received within 7 days after application. Do not apply to newly transplanted trees until ground has settled. Rainfall within 7 to 10 days of application is necessary for herbicide activation. Tank mix with paraquat for POST weed control. 60-day preharvest interval (PHI). May be applied as a sequential application as long as rate does not exceed 4.2 quarts/acre per year. Allow 30 days between applications

Table 7-12B. Chemical Weed Control in Fruit Crops—Tree Fruits

Weed	Herbicide, Mode of Action,* and Formulation	Amount of Formulation Per Acre	Pounds Active Ingredient Per Acre	Precautions and Remarks
Apples, Preemergence Directed Underneath Tree (continued)				
Broadleaf weeds and some annual grasses	rimsulfuron, MOA 2 (Matrix) 25 WG (Solida) 25 WG (Pruvin) 25 WG	4 oz	0.063	For broad spectrum residual control, tank mix with diuron, Sinbar, oryzalin or Prowl H_2O. For nonselective POST control, apply in combination with glyphosate or paraquat. Rimsulfuron does have POST activity on certain broadleaf weeds (see label for list). Rimsulfuron will control emerged horseweed less than 3 inches tall when applied in combination with a non-ionic surfactant and a spray grade ammonium sulfate (2 pound/acre). **DO NOT** treat orchards established less than 1 year. Rainfall within 2 to 3 weeks of application is necessary for herbicide activation. Spray solutions having a pH of less than 4.0 or greater than 8.0 will result in herbicide degradation. Rimsulfuron has a 7-day PHI for apples. Rimsulfuron may be applied as a sequential application so long as total use rate does not exceed 4 ounces/acre per year and application is made in a band on less than 50% of orchard floor. Allow at least 30 days between applications.
	Penoxsulam + oxyfluorfen, MOA 2 and 14 (Pindar GT) 4	1.5 to 3 pt	0.75 to 1.5	Pindar GT can be applied after harvest up until bud swell in the fall to late winter/early spring time frame. **DO NOT** use after bud break until completion of final harvest or in orchards established less than 4 years. **DO NOT** use in soils that contain less than 20% clay or greater than 70% sand. For non-selective POST control tank mix with glufosinate, glyphosate, or paraquat. For expand residual control of annual grass weeds tank mix with oryzalin or Solicam.
Annual broadleaf and grass weeds	simazine, MOA 5 (Princep, Simazine) 4 L 90 WDG	2 to 4 qt 2.2 to 4.4 lb	2 to 4	Apply preemergence to trees that have been established 1 year or more. Apply with glyphosate, paraquat, or glufosinate for postemergence weed control. PHI for simazine is 150 days. Tank mixing simazine with oryzalin, Solicam, or Prowl H_2O will expand residual control of annual grasses and certain broadleaf weeds.
Annual broadleaf weeds and yellow nutsedge	sulfentrazone + carfentrazone-ethyl, MOA 14 (Zeus Prime XC) 3.5 XC	7.7 to 15.1 oz	0.21 to 0.41	Use only in orchards established 3 years or longer. Avoid contact with green tissue or bark. Sequential applications of Zeus Prime can be applied when directed as a banded application (50% band or less of orchard floor) so long as total use rate does not exceed 15.1 fl. oz/A on a broadcast basis within a year and the second application is not applied within 60 days of the initial application. Zeus Prime has a 14 day PHI. For optimum residual control of annual grass weeds tank mix with oryzalin, pendimethalin, or norflurazon. Tank mix glyphosate, glufosinate, or paraquat for non-selective POST weed control.
Broadleaf and grass weeds	terbacil, MOA 5 (Sinbar) 80 WDG	2 to 4 lb	1.6 to 3.2	Use only on trees that have been established 3 years or more. Tank mixing Sinbar with Karmex allows Sinbar to be used in orchards established 1 year or longer. Rate varies with soil organic matter. See label for details. Apply no more than 3 lb unless soil organic matter is greater than 2%. Do not use on sand or loamy sand soils. Do not use on soils having less than 1% organic matter. Sinbar has a 60-day PHI. Tank mix with glyphosate, paraquat or glufosinate for non-selective postemergence weed control.
Most annual broadleaf weeds and grass weeds in **NEWLY PLANTED NON-BEARING ORCHARDS**	terbacil, MOA 5 (Sinbar) 80 WDG	0.5 to 1.0 lb	0.4 to 0.8	Apply once adequate rainfall has occurred allowing the soil to settle after transplanting. Apply no more than 1 pound per acre per year. For best results, apply 0.5 pound in spring followed by another 0.5 pound when control from initial application fails. Do not use on soils coarser than sandy loams of soils with less than 1% organic matter. Tank mix with paraquat for non-selective postemergence weed control.
Apples, Preemergence Tank Mixes				
Many annuals and perennial grass and broadleaf weeds	diuron, MOA 7 (Diuron or Karmex XP) 80 WDG (Direx) 4 L + terbacil, MOA 5 (Sinbar) 80 WDG	1 to 2 lb 1.6 to 3.2 qt + 1 to 2 lb	0.8 to 1.6 0.5 to 1 + 0.8 to 1.6	DO NOT treat varieties grafted on full-dwarf rootstocks. Use only on trees established in orchard for 1 year. See labels for details. See labels for soil texture and organic matter restrictions.
Annual grasses and broadleaf weeds	norflurazon, MOA 12 (Solicam) 80 WDG + simazine, MOA 5 (Princep, Simazine) 4 L 90 WDG	2.5 to 5 lb + 2 to 4 qt 2.2 to 4.4 lb	2 to 4 + 2 to 4	See labels for details. Apply in combination with paraquat, glyphosate, or glufosinate for postemergence control.
	oryzalin, MOA 3 (Oryzalin, Surflan) 4 AS + simazine, MOA 5 (Princep, Simazine) 4 L 90 WDG	2 to 4 qt + 2 to 4 qt 2.2 to 4.4 lb	2 to 4 + 2 to 4	See labels for details. Apply in combination with paraquat, glyphosate, or glufosinate for postemergence control.
	rimsulfuron, MOA 2 (Matrix) 25 WG (Solida) 25 WG (Pruvin) 25 WG + terbacil, MOA 5 (Sinbar) 80 WDG	2 oz + 1 to 2 lb	0.031 + 0.8 to 1.6	See each product label for use precautions. Tank mix with glyphosate, glufosinate, or paraquat for non-selective POST weed control. See label for soil texture and organic matter restrictions.
	norflurazon, MOA 12 (Solicam) 80 WDG + diuron, MOA 7 (Diuron or Karmex XP) 80 WDG	2.5 to 5 lb + 2 to 4 lb	2 to 4 + 1.6 to 3.2	Use only on trees established 1 year or more. Do not treat varieties grafted on full-dwarf rootstocks. See label for details. Apply in combination with glyphosate, paraquat, or glufosinate for postemergence control.
	rimsulfuron, MOA 2 (Matrix) 25 WG (Solida) 25 WG (Pruvin) 25 WG + oryzalin, MOA 3 (Oryzalin, Surflan) 4 AS	4 oz + 2 to 4 qt	0.0625 + 2 to 4	Tank mix with glyphosate, glufosinate, or paraquat for non-selective POST weed control.

Table 7-12B. Chemical Weed Control in Fruit Crops—Tree Fruits

Weed	Herbicide, Mode of Action,* and Formulation	Amount of Formulation Per Acre	Pounds Active Ingredient Per Acre	Precautions and Remarks
Apples, Preemergence Tank Mixes (continued)				
	rimsulfuron, MOA 2 (Matrix) 25 WG (Solida) 25 WG (Pruvin) 25 WG + diuron, MOA 7 (Diuron or Karmex XP) 80 WDG	4 oz + 2 to 4 lb	0.0625 + 1.6 to 3.2	Tank mix with glyphosate, glufosinate, or paraquat for non-selective POST weed control.
Annual broadleaf and grass weeds as well as yellow nutsedge	sulfentrazone + carfentrazone-ethyl, MOA 14 (Zeus Prime) 3.5 XC + oryzalin, MOA 3 (Oryzalin, Surflan) 4 AX	7.7 to 15.1 oz	0.21 to 0.41	
Apples, Postemergence, Directed Underneath Tree				
Broadleaf weeds including morningglory, pigweed, dayflower, lambsquarters, and prickly lettuce	carfentrazone-ethyl, MOA 14 (Aim) 2 EC	0.5 to 2 oz	0.008 to 0.032	Apply alone or tank mixed with other herbicides. Apply in a minimum spray volume of 20 gpa. Applications can be made with boom equipment, hooded sprayers, or shielded sprayers. Do not allow Aim to contact green bark, desirable foliage, flowers, or fruit. Contact with fruit or foliage will result in spotting and leaf necrosis. The trunks of trees less than 2 years old must be protected from Aim. Do not apply within 3 days of harvest. Sequential applications may be used so long as there is 14 days between applications and total use rate for year does not exceed 7.9 ounces/acre. Best results are obtained when applied to weeds in the 2- to 3-leaf stage. Apply in combination with a nonionic surfactant (1 quart/100 gallons of spray solution) or crop oil concentrate (1 gallon/100 gallons of spray solution).
Broadleaf weeds including perennials like blackberry, horsenettle, poison ivy, Virginia creeper, and white clover.	fluroxypyr, MOA 4 (Starane Ultra) 2.8	0.7 to 1.4 pt	0.35 to 0.70	DO NOT apply during bloom or to trees less than 4 years old. Make only 1 application per year. Starane may be tank mixed with other herbicides registered for use on apples. Starane has a 14-day PHI. **This use is not listed on the Starane label because it is on a supplemental label.**
Broadleaf weeds including horseweed, morningglory, pigweed, ragweed, smartweed and purslane	saflufenacil, MOA 14 (Treevix) 70 WG	1 oz	0.044	Do not apply more than 3 ounces/acre per year. Allow at least 21 days between applications. Treevix has a 0 day PHI. Treevix may be tank mixed with glyphosate, glufosinate, Poast, and oxyfluorfen. Treevix provides excellent control of horseweed, purslane, morningglory species, ragweed, and smartweed. The addition of methylated seed oil at 1% v/v (1 gallon per 100 gallons of spray solution) plus ammonium sulfate at 8.5 to 17 lbs/100 gallons of spray solution is required for optimum herbicide performance.
Most annual broadleaf and grass weeds plus many perennials	glufosinate, MOA 10 (Cheetah, Lifeline, Reckon, Rely, Surmise) 2.34 L	48 to 82 oz	0.88 to 1.5	DO NOT SPRAY GREEN BARK OR FOLIAGE. Glufosinate should not be used on trees within 1 year of transplanting. Apply in a minimum of 20 gallons of water per acre as a directed spray under trees. Repeat applications may be necessary for control of perennial weeds. Glufosinate can be tank mixed with diuron, Sinbar, Solicam, oryzalin, Devrinol, Goal, rimsulfuron, and simazine. Glufosinate has a 14 day PHI. DO NOT allow spot spray applications to directly contact tree or suckers. The addition of ammonium sulfate will enhance glufosinate activity on difficult to control species; however, the addition of surfactants and crop oil will increase risk of crop injury.
	glyphosate, MOA 9 (various brands and formulations)	See label	1 to 2	DO NOT SPRAY GREEN BARK OR FOLIAGE. Trees are more susceptible to injury from midsummer until dormant. Repeat applications may be necessary for control of perennial weeds. Can be tank mixed with Goal, Karmex, simazine, Solicam, and oryzalin. Check label for specifics. Generic glyphosate formulations may require the addition of a surfactant at 0.5% by volume (2 quart per 100 gallons of spray solution). See label for spray additive information and for detailed restriction information.
Yellow and purple nutsedge, horsenettle, and pokeweed	halosulfuron, MOA 2 (Sandea) 75 WDG	0.5 to 1 oz	0.023 to 0.047	Apply halosulfuron to actively growing weeds. Do not apply to apple trees established less than 1 year. Do not apply more than 2 ounces/acre per 12-month period. Avoid herbicide contact with tree foliage. Addition of a nonionic surfactant is necessary for optimum herbicide performance. Sequential applications may be more effective on yellow nutsedge than one application. When using sequential applications use at least 0.75 ounces/acre per application. Halosulfuron may be tank mixed with glyphosate for broad spectrum POST weed control.
Broadleaf and some small annual grass weeds	paraquat, MOA 22 (Gramoxone) SL 2 SL (Firestorm) 3 SL (Paraquat Concentrate) 3 SL (Parazone) 3 SL	2.5 to 4 pt 1.7 to 2.7 pt	0.6 to 1	Apply when grass and weeds are 1 to 6 inches high and succulent for best results. Direct spray with low pressure to avoid contact with tree foliage or bark less than 1 year old. Young trees must be shielded to prevent spray contact with bark. Add surfactant at 0.25% by volume (2 pints per 100 gallons). Paraquat may be tank mixed with Goal, Karmex, simazine, Sinbar, Solicam, and oryzalin. Paraquat is a restricted use pesticide.
Broadleaf weeds	2,4-D amine, MOA 4 (various manufacturers and brands) 3.8 SL	1 to 3 pt	0.95 to 1.4	Apply any time during the growing season to actively growing broadleaf weeds except during apple bloom. Trees must be at least 1 year old. Do not apply more than 2 applications per crop cycle (75-day interval between applications) or within 14 days of harvest. Some formulations limit rate to 2 pints per acre. See label for details.
	2,4-D choline, MOA 4 (Embed) 3.8 SL	1 to 4 pt	0.48 to 1.9	Embed offers the stability and reduced drift technology associated with 2,4-D choline. It may be used in orchards established 1 year or longer. DO NOT apply during bloom or use in orchards having sand soils. Embed has a 14 day PHI for pome fruit. No more than 8 pts of Embed can be applied within a 12-month period. If making more than one application allow 75 days between applications. Embed may be tank mixed with glyphosate, glufosinate, as well as various PRE herbicides.

Table 7-12B. Chemical Weed Control in Fruit Crops—Tree Fruits

Weed	Herbicide, Mode of Action,* and Formulation	Amount of Formulation Per Acre	Pounds Active Ingredient Per Acre	Precautions and Remarks
Apples, Postemergence, Directed Underneath Tree (continued)				
Broadleaf weeds including clover, horseweed, thistle, dandelion and mugwort	clopyralid, MOA 4 (Stinger) 3EC	0.33 to 0.66 pt	0.125 to 0.25	Use on trees established that have been established in the orchard at least 1 year or longer. No more than 2 applications of Stinger can be applied per crop year and total use rate within that time frame cannot exceed 2/3 pt per acre. DO NOT apply within 30 days of harvest. DO NOT apply during bloom and avoid contact with foliage, fruit, and tree trunk.
Grasses	clethodim, MOA 1 (Arrow, Clethodim, Intensity, or Select) 2 EC (Select Max or Intensity One) 1 EC	6 to 8 oz 12 to 16 oz	0.094 to 0.125	Apply to actively growing grasses not under stress. See label for rate and optimum grass size to treat. Multiple applications may be necessary to control perennial grass weeds. When using the 2 EC formulations, add crop oil concentrate at 1% by volume (1 gallon per 100 gallons of spray solution). When using the 1 EC formulations, add nonionic surfactant at 0.25% by volume (1 qt per 100 gallons). **Select Max has a 14 day PHI for use in apples. All other clethodim formulations can be used in NON-BEARING apples ONLY!**
	fluazifop, MOA 1 (Fusilade DX) 2 EC	12 to 24 oz	0.125 to 0.38	**NONBEARING TREES ONLY.** Apply to actively growing grasses not under stress. See label for rate and optimum grass size to treat. Multiple applications may be necessary to control perennial grass weeds. Add crop oil at 1% by volume (1 gallon per 100 gallons).
Grasses (continued)	sethoxydim, MOA 1 (Poast) 1.5 EC	1.0 to 2.5 pt	0.19 to 0.47	Apply to actively growing grasses not under stress. See label for rate and optimum grass size to treat. Multiple applications may be necessary to control perennial grass weeds. Add Dash adjuvant at 1 pint per acre or crop oil concentrate at 1 quart per acre. Do not apply within 14 days of harvest. Do not apply more than 7.5 pints per acre per year.
Apples, Ground Cover Suppression				
Suppression of fescue, orchardgrass, and bluegrass	glyphosate, MOA 9 (various brands and formulations)	Rates and application time vary for each grass species. See label for details.		Mow one time in spring. Apply 3 to 4 days after mowing. **Caution:** This treatment will normally discolor the grass. **DO NOT** apply after seedhead emergence. See label for details.
Peaches, Preemergence Directed Underneath Tree				
Annual grasses and some broadleaf weeds	diuron, MOA 7 (Direx, Diuron) 4 L (Diuron, Karmex XP) 80 DF	1.6 to 2.2 qt 2 to 2.75 lb	1.6 to 2.2	Apply in spring to trees at least 3 years old. Rate is soil texture dependent. May be tank mixed with Sinbar, Solicam, glyphosate, or paraquat. Karmex DF, Karmex XP, and Direx 4L have a 20-day PHI. Other formulations of diuron have a 90-day PHI.
Annual grasses and some broadleaf weeds (continued)	flumioxazin, MOA 14 (Chateau) 51 SW (Tuscany) 51 WG (Tuscany) 4 SC	6 to 12 oz 6 to 12 fl oz	0.19 to 0.38	Chateau is for newly planted and established orchards. Shield or protect trees established less than 1 year from contact with spray solution. Tank mix with paraquat for non-selective POST weed control. Do not apply more than 6 ounces per acre to trees planted less than 3 years in soil having a sand plus gravel content more than 80%. Sequential applications are very effective. Due to the potential for crop injury, Chateau should not be applied in bearing orchards after budbreak until after final harvest. Do not apply within 60 days of harvest. Do not use more than 24 ounces per acre per year. In non-bearing orchards Chateau may be applied after bud break however application equipment should be hooded.
	indaziflam, MOA 29 (Alion) 1.67 SC	3.5 to 6.5 oz	0.046 to 0.085	Use in orchards established 3 years or longer. See label for details regarding the management of replants in established orchards. Use rate cannot exceed 3.5 fluid ounces/acre per application on soils having less than 1% organic matter. On soils with an organic matter content from 1 to 3%, no more than 5 fluid ounces/acre can be applied in a single application and the total use rate for the year cannot exceed 8.5 fluid ounces/acre. In order to apply more than 5 fluid ounces/acre in a single application soil organic matter must be >3%. Do not use on soils with 20% or more gravel content. Allow at least 90 days between applications. Research has shown Alion applied in the fall followed by a late spring application will provide summer long control of annual broadleaf and grass weeds. Do not treat soil around trees with cracks or channels, or with depressions. Tank mix Alion with glyphosate or paraquat for nonselective POST weed control. Alion has a 14-day PHI.
	norflurazon, MOA 12 (Solicam) 80 WDG	2.5 to 5 lb	2 to 4	Can be tank mixed with Karmex, Goal, glyphosate, paraquat, Prowl, rimsulfuron, simazine, Sinbar, or oryzalin. Rate is soil texture dependent. See label for details. Do not apply within 6 months of transplanting. PHI is 60 days. Multiple applications can be made per season so long as total use rate does not exceed maximum use rate for soil texture and crop.
	oryzalin, MOA 3 (Oryzalin or Surflan) 4 AS	2 to 6 qt	2 to 6	Allow soil to settle around newly transplanted trees before application. Oryzalin may be tank mixed with glyphosate, paraquat, simazine, and Solicam. Sequential applications permitted as long as there is 2.5 months between the applications. See label for details. In newly planted orchards may be tank mixed with Gallery for broad spectrum preemergence control. Sequential applications may be used if total use rate does not exceed 12 quarts per acre per year.
	pendimethalin, MOA 3 (Prowl H_2O) 4 AS	2 to 4 qt	2 to 4	Most effective when adequate rainfall or irrigation is received within 7 days after application. Do not apply to newly transplanted trees until ground has settled around roots. Apply with paraquat to control emerged weeds. Prowl H_2O has a 60-day preharvest interval (PHI). Pendimethalin may be applied as sequential applications so long as total amount used does not exceed 4.2 qt/A per year. Allow at least 30 days between applications.
Broadleaf and some grass weeds	penoxsulam + oxyfluorfen, MOA 2 and 14 (Pindar GT) 4	1.5 to 3 pt	0.75 to 1.5	Pindar GT can be applied after harvest up until bud swell in the fall to late winter/early spring time frame. **DO NOT** use after bud break until completion of final harvest or in orchards established less than 4 years. **DO NOT** use in soils that contain less than 20% clay or greater than 70% sand. For non-selective POST control tank mix with glufosinate, glyphosate, or paraquat. For expand residual control of annual grass weeds tank mix with oryzalin or Solicam.

Table 7-12B. Chemical Weed Control in Fruit Crops—Tree Fruits

Weed	Herbicide, Mode of Action,* and Formulation	Amount of Formulation Per Acre	Pounds Active Ingredient Per Acre	Precautions and Remarks
Peaches, Preemergence Directed Underneath Tree (continued)				
Broadleaf and some grass weeds (continued)	rimsulfuron, MOA 2 (Matrix) 25 WG (Solida) 25 WG (Pruvin) 25 WG	4 oz	0.063	For broad spectrum PRE control, tank mix with diuron, Sinbar, oryzalin or Prowl H_2O. For nonselective POST control, apply with glyphosate or paraquat. Rimsulfuron does have POST activity on certain broadleaf weeds (see label for list). Rimsulfuron will control emerged horseweed less than 3 inches tall when applied in combination with a non-ionic surfactant and a spray grade ammonium sulfate (2 pounds/acre). Do NOT treat orchards established less than 1 year. Rainfall within 2 to 3 weeks of application is necessary for herbicide activation. Spray solutions having a pH lower than 4.0 or higher than 8.0 will result in herbicide degradation. Rimsulfuron has a 14-day PHI for stone fruit, and sequential applications can be made so long as total use rate does not exceed 4 ounces/acre per year and application is made in a band on less than 50% of orchard floor. Allow at least 30 days between applications.
Broadleaf and grass weed control for NEWLY PLANTED NON-BEARING ORCHARDS	terbacil, MOA 5 (Sinbar) 80 WDG	0.5 to 1.0 lb	0.4 to 0.8	Apply once soil has settled after transplanting. Apply no more than 1 lb per acre per year. For best results, apply 0.5 lb in the spring followed by another 0.5 lb when control from initial application fails. Do not apply on soils coarser than sandy loam. Do not use on soils having less than 1% organic matter.
Annual broadleaf and grass weeds	simazine, MOA 5 (Princep, Simazine) 4 L 90 WDG	1.6 to 4 qt 1.8 to 4.4 lb	1.6 to 4	Apply in early spring before weed emergence. Use only on trees established 1 year or more. Do not use on sand or loamy sand soils. Tank mixing simazine with oryzalin, Prowl H_2O, or Solicam will improve residual control of annual grasses and certain broadleaf weeds.
Annual broadleaf and grass weeds plus many perennial grasses	terbacil, MOA 5 (Sinbar) 80 WDG	2 to 4 lb	1.6 to 3.2	Use on trees established 3 years, however when tank mixed with Karmex XP or another diuron containing herbicide Sinbar may be applied in orchards established 1 year or longer. Sinbar may only be used on soils with at least 1% organic matter. Unless soil organic matter is greater than 2% do not exceed 3 pounds/acre. Do not use on sand or loamy sand soils. Sinbar is an excellent choice for tank mixing with diuron or rimsulfuron for extended broad spectrum residual control of those products. Sinbar has a 60-day PHI.
Peaches, Preemergence Tank Mixes				
Many annual and perennial grasses and broadleaf weeds	diuron, MOA 7 (Diuron or Karmex XP) 80 DF + terbacil, MOA 5 (Sinbar) 80 WDG	1 to 2 lb + 1 to 2 lb	0.8 to 1.6 + 0.8 to 1.6	Use only under trees established in the orchard for at least 1 year. Apply to soils having at least 1% organic matter. See label for details.
	oryzalin, MOA 3 (Oryzalin or Surflan) 4 AS + simazine, MOA 5 (Princep, Simazine) 4 L 90 WDG	2 to 4 qt + 1.6 to 4 qt 1.75 to 4.4 lb	2 to 4 + 1.6 to 4	Tank mix for use before weed emergence. Tree must be established at least 1 year.
Many annual and perennial grasses and broadleaf weeds (continued)	norflurazon, MOA 12 (Solicam) 80 WDG + simazine, MOA 5 (Princep, Simazine) 4 L 90 WDG	2.5 to 5 lb + 2 to 4 qt 2.2 to 4.4 lb	2 to 4 + 2 to 4	See labels for details.
	rimsulfuron, MOA 2 (Matrix) 25 WG (Solida) 25 WG (Pruvin) 25 WG + terbacil, MOA 5 (Sinbar) 80 WDG	2 oz + 1 to 2 lb	0.063 + 0.8 to 1.6	See labels for use precautions and details.
	rimsulfuron, MOA 2 (Matrix) 25 WG (Solida) 25 WG (Pruvin) 25 WG + diuron, MOA 7 (Diuron or Karmex XP) 80 WDG	4 oz + 2 to 4 lb	0.0625 + 1.6 to 3.2	Tank mix with glyphosate, glufosinate, or paraquat for non-selective POST weed control.
	norflurazon, MOA 12 (Solicam) 80 WDG + diuron, MOA 7 (Diuron or Karmex XP) 80 DF	2.5 to 5 lb + 2 to 4 lb	2 to 4 + 1.6 to 3.2	See labels for details. Trees must be established at least 3 years.
	rimsulfuron, MOA 2 (Matrix) 25 WG (Solida) 25 WG (Pruvin) 25 WG + oryzalin, MOA 3 (Oryzalin, Surflan) 4 AS	4 oz + 2 to 4 qt	0.063 + 2 to 4	Tank mix with glyphosate or paraquat for non-selective POST weed control.

Table 7-12B. Chemical Weed Control in Fruit Crops—Tree Fruits

Weed	Herbicide, Mode of Action,* and Formulation	Amount of Formulation Per Acre	Pounds Active Ingredient Per Acre	Precautions and Remarks
Peaches, Postemergence, Directed Underneath Tree				
Broadleaf weeds including morningglory, pigweed, lambsquarters, cocklebur, smartweed, and dayflower	carfentrazone-ethyl, MOA 14 (Aim) 2 EC	0.5 to 1.6 oz	0.008 to 0.025	Apply alone or tank mixed with other herbicides. Apply in a minimum spray volume of 20 gpa. Applications can be made with boom equipment, hooded sprayers, or shielded sprayers. Do not allow Aim to contact green bark, desirable foliage, flowers, or fruit of the crop. Contact with fruit or foliage will result in spotting and leaf necrosis. The trunks of trees established less than 2 years must be protected. Do not apply within 3 days of harvest. Best results are obtained when applied to weeds in the 2- to 3-leaf stage. Sequential applications may be used so long as there is at least 14 days between applications and total use rate for year does not exceed 7.9 ounces/acre per year. Apply in combination with a nonionic surfactant (1 quart per 100 gallons of spray solution) or crop oil concentrate (1 gallon per 100 gallons of spray solution).
Kill all green foliage on contact	paraquat, MOA 22 (Gramoxone SL) 2 SL (Firestorm) 3 SL (Paraquat Concentrate) 3 SL (Parazone) 3 SL	2.5 to 4 pt 1.7 to 2.7 pt	0.66 to 1	Apply when grass and weeds are 1 to 6 inches high and succulent for best results. Direct spray with low pressure to avoid contact with tree foliage or bark. Add surfactant at 0.25% by volume (2 pints per 100 gallons) for best results. Paraquat may be tank mixed with Goal, Karmex, simazine, Sinbar, Solicam, and oryzalin. Paraquat is a restricted use pesticide. Newly planted trees can be severely injured by paraquat, so use a shield or wrap to protect the tree from spray. Do not make more than 3 applications per year. Paraquat has a 14-day PHI. Paraquat has a 14-day PHI for peach and 28-day PHI for nectarine.
Non-selective weed control	glyphosate, MOA 9 (various brands and formulations)	See label	1	Do not apply in orchards established less than 2 years. Applications must be made with shielded sprayer. Low hanging limbs and suckers must be removed at least 10 days prior to application. DO NOT use glyphosate 90 days past bloom. DO NOT allow glyphosate to contact foliage or bark; EXTREME care must be taken to prevent injury. See label for details. Some glyphosate formulations may require the addition of a surfactant.
Most annual broadleaf and grass weeds plus many perennials	glufosinate, MOA 10 (Cheetah, Lifeline, Reckon, Rely, Surmise) 2.34 L	48 to 82 oz	0.88 to 1.5	DO NOT SPRAY GREEN BARK, UNCALLOUSED BARK OR DESIRABLE FOLIAGE UNLESS TREES ARE PROTECTED. Glufosinate should not be used on trees within 1 year of transplanting. Apply in a minimum of 20 gallons of water per acre as a directed spray under trees. Repeat applications may be necessary for control of perennial weeds. Glufosinate can be tank mixed with diuron, Sinbar, Solicam, oryzalin, Devrinol, Goal, rimsulfuron, and simazine. Glufosinate has a 14 day PHI. DO NOT apply more than 164 fl oz per acre within a 12 month period. There must be at least 28 days between applications. Glufosinate formulations contain surfactant therefore additional nonionic surfactants or crop oils are not necessary and may increase potential for injury.
Grasses	clethodim, MOA 1 (Arrow, Clethodim, Intensity, or Select) 2 EC (Select Max or Intensity One) 1 EC	6 to 8 oz 12 to 16 oz	0.094 to 0.125	Apply to actively growing grasses not under stress. See label for rate and optimum grass size to treat. Multiple applications may be necessary to control perennial grass weeds. When using 2 EC formulation chemicals, add crop oil concentrate at 1% by volume (1 gallon per 100 gallons). When using 1 EC formulations, use a nonionic surfactant at 0.25% by volume rather than crop oil. **Select Max has a 14-day PHI for peach. Unless otherwise stated on label, all other clethodim products are for non-bearing orchards ONLY.**
	fluazifop, MOA 1 (Fusilade DX) 2 EC	8 to 24 oz	0.125 to 0.38	Apply to actively growing grasses not under stress. See label for rate and optimum grass size to treat. Multiple applications may be necessary to control perennial grass weeds. Add crop oil at 1% by volume (1 gallon per 100 gallons). Do not apply within 14 days or harvest. Do not apply more than 72 fl ounces per acre per year.
	sethoxydim, MOA 1 (Poast) 1.5 EC	1.0 to 2.5 pt	0.19 to 0.47	Apply to annual grasses up to 12 inches tall. For perennial grasses apply early in the growth cycle at the high use rate. Multiple applications may be necessary for perennial grass weeds. Add Dash adjuvant at 1 pint per acre or crop oil concentrate at 1 quart per acre. Do not apply within 25 days of harvest. Do not apply more than 5 pints per acre per year.
Broadleaf weeds	2,4-D amine, MOA 4 (Weedar 64) (various brands) 3.8 SL	1 to 3 pt	0.95 to 1.4	Do not apply within 40 days of harvest. Do not apply more than twice a year and allow 75 days between applications. Trees must be at least 1 year old. Use when trees are dormant. Some formulations limit rate to 2 pints per acre. See label for details.
	2,4-D choline, MOA 4 (Embed) 3.8 SL	1 to 4 pt	0.48 to 1.9	Embed offers the stability and reduced drift technology associated with 2,4-D choline. It may be used in orchards established 1 year or longer. DO NOT apply during bloom or use in orchards having sand soils. Embed has a 40 day PHI for peaches. No more than 8 pts of Embed can be applied within a 12-month period. If making more than 1 application allow 75 days between applications. Embed may be tank mixed with glyphosate, glufosinate, as well as various PRE herbicides.
Broadleaf weeds including clover, horseweed, dock, thistle and mugwort	clopyralid, MOA 4 (Stinger) 3 EC	0.33 to 0.66 pt	0.125 to 0.25	Multiple applications may be used as long as total amount does not exceed maximum rate. Use at least 10 GPA of spray solution. Stinger may be tank mixed with preemergence herbicides. Do not apply within 30 days of harvest. Do not apply more than twice. Total use rate cannot exceed 2/3 pint per acre per crop year.
Pecans, Preemergence, Directed Underneath Tree				
Broadleaf weeds and annual grasses	diuron, MOA 7 (Diuron or Karmex DF) 80 WDG (Direx) 4 L	2 to 4 lb 1.6 to 3.2 qt	1.6 to 3.2	Do not apply to trees less than 3 years old. Rate is soil texture dependent. Do not use on soils with less than 0.5% organic matter.
Annual grass and some broadleaf weeds	norflurazon, MOA 12 (Solicam) 80 WDG	2.5 to 5 lb	2 to 4	Do not apply when nuts are on the ground. Rate is soil texture dependent. See label for details. Do not apply within 6 months of planting. PHI is 60 days. Multiple applications can be made per season so long as total use rate does not exceed maximum use rate for soil texture and crop.
	oryzalin, MOA 3 (Oryzalin or Surflan) 4 AS	2 to 6 qt	2 to 4	Allow soil to settle around newly transplanted trees before application. Oryzalin may be tank mixed with Goal, glyphosate, paraquat, simazine, and Solicam. Sequential application permitted. See label for details. Sequential applications may be used so long as total use rate does not exceed 12 qt per acre per year and there is a minimum of 2.5 months between applications.

Table 7-12B. Chemical Weed Control in Fruit Crops—Tree Fruits

Weed	Herbicide, Mode of Action,* and Formulation	Amount of Formulation Per Acre	Pounds Active Ingredient Per Acre	Precautions and Remarks
Pecans, Preemergence, Directed Underneath Tree (continued)				
Annual grass and some broadleaf weeds (continued)	pendimethalin, MOA 3 (Prowl H$_2$O) 4 AS	2 to 6 qt	2 to 6	Most effective when adequate rainfall or irrigation is received within 7 days after application. Do not apply to newly transplanted trees until ground has settled around roots. Apply with paraquat to control emerged weeds. Prowl H2O has a 60 day PHI. Pendimethalin may be applied in sequential applications so long as total use rate does not exceed maximum of application rate on the label and there is at least 30 days between applications.
Annual broadleaf weeds and grass weeds	simazine, MOA 5 (Princep, Simazine) 4 L 90 WDG	2 to 4 qt 2.2 to 4.4 lb	2 to 4	Apply preemergence to weeds under trees that have been established 2 years or more. Do not apply when nuts are on the ground. Do not use on sand or loamy sand soils. Tank mixing simazine with oryzalin, Solicam, or Prowl H2O will improve residual control of annual grasses and certain broadleaf weeds.
	flumioxazin, MOA 14 (Chateau) 51 SW (Tuscany) 51 WDG (Tuscany) 4 SC	6 to 12 oz 6 to 12 fl oz	0.19 to 0.38	Chateau may be applied in newly planted and established orchards. Trees established less than 1 year must be shielded from contact with spray solution to prevent injury. Do not apply more than 6 ounces per acre to trees planted less than 3 years in soil having a sand plus gravel content more than 80%. Sequential applications are very effective; however, allow 60 days between applications. Do not apply after bud break through final harvest unless using shielded application equipment. When applying Chateau after bud break DO NOT tank mix with glyphosate or 2,4-D amine. When tank mixed with glyphosate and/or 2,4-D amine the potential for drift increases. Chateau has a 60-day PHI. Use rate cannot exceed 24 ounces per acre in a 12-month period.
Annual broadleaf weeds and grass weeds (continued)	indaziflam, MOA 29 (Alion) 1.67 SC	3.5 to 6.5 oz	0.045 to 0.085	Use in orchards established 3 years or longer. See label for details regarding the management of replants in established orchards. Do not use on soils having a 20% or greater gravel content. Use rate cannot exceed 3.5 fluid ounces per acre per application on soils having less than 1% organic matter. On soils with an organic matter content from 1 to 3%, no more than 5 fluid ounces/acre can be applied in a single application and the total use rate for the year cannot exceed 6.5 fluid ounces/acre. In order to apply more than 5 fluid ounces/acre in a single application soil organic matter must be >3%. Allow at least 90 days between applications. Do not treat soil around trees with cracks or channels, or with depressions. Tank mix Alion with glyphosate, glufosinate, or paraquat for nonselective POST weed control. Alion has a 14-day PHI.
	rimsulfuron, MOA 2 (Matrix) 25 WG (Solida) 25 WG (Pruvin) 25 WG	4 oz	0.063	For broad spectrum PRE control, tank mix with, diuron, oryzalin or Prowl H$_2$O. For nonselective POST control, apply in combination with glyphosate or paraquat. Rimsulfuron does have POST activity on certain broadleaf weeds (see label for list). Rimsulfuron will control emerged horseweed less than 3 inches tall when applied in combination with a non-ionic surfactant and a spray grade ammonium sulfate (2 pounds per acre). Do NOT treat orchards established less than 1 year. Rainfall within 2 to 3 weeks of application is necessary for herbicide activation. Spray solutions having a pH lower than 4.0 or higher than 8.0 will result in herbicide degradation. Rimsulfuron has a 14-day PHI for Pecan. Rimsulfuron may be applied as a sequential application so long as total use rate does not exceed 4 ounces/acre per year and application is made in a band on less than 50% of orchard floor. Allow at least 30 days between applications.
Pecans, Preemergence, Tank Mix Options				
Annual broadleaf weeds and grass weeds	diuron, MOA 7 (Diuron or Karmex XP) 80 WDG + norflurazon, MOA 12 (Solicam) 80 WDG	2 to 4 lb + 2.5 to 5 lb	1.6 to 3.2 + 2 to 4	Trees must be established in the orchard for 3 years.
	norflurazon, MOA 12 (Solicam) 80 F + simazine, MOA 5 (Princep, Simazine) 4 L 90 WDG	2.5 to 5 lb + 2 to 4 qt 2.2 to 4.4 lb	2 to 4 + 2 to 4	Trees must be established for at least 2 years. See labels for details.
	oryzalin, MOA 3 (Oryzalin, Surflan) 4 AS + simazine, MOA 5 (Princep, Simazine) 4 L 90 WDG	2 to 4 qt + 2 to 4 qt 2.2 to 4.4 lb	2 to 4 + 2 to 4	See label for details.
Pecans, Postemergence, Directed Underneath Tree				
Broadleaf weeds including morningglory, pigweed, lambsquarters, cocklebur, smartweed, and dayflower	carfentrazone-ethyl, MOA 14 (Aim) 2 EC	0.5 to 2 oz	0.008 to 0.031	Apply alone or tank mixed with other herbicides. Apply in a minimum spray volume of 20 gpa. Applications can be made with boom equipment, hooded sprayers, or shielded sprayers. Do not allow Aim to contact green bark, desirable foliage, flowers, or fruit of the crop. Contact with fruit or foliage will result in spotting and leaf necrosis. The trunks of trees less than 2 years old must be protected from direct contact with Aim. Do not apply within 3 days of harvest. Sequential applications may be used so long as total use rate does not exceed 7.9 ounces/acre per year and there is 14 days between applications. Best results are obtained when applied to weeds in the 2- to 3-leaf stage. Apply in combination with a nonionic surfactant (1 quart per 100 gallons of spray solution) or crop oil concentrate (1 gallon per 100 gallons of spray solution).
Most annual broadleaf and grass weeds plus many perennials	glufosinate, MOA 10 (Cheetah, Lifeline, Reckon, and Rely, Surmise) 2.34L	48 to 82 oz	0.88 to 1.5	DO NOT SPRAY GREEN BARK OR FOLIAGE. Glufosinate should not be used on trees within 1 year of transplanting. Do not make more than 3 applications per year. Apply in a minimum of 20 gallons of water per acre as a directed spray under trees. Repeat applications may be necessary for control of perennial weeds. Glufosinate can be tank mixed with diuron, Solicam, Surflan, Devrinol, Goal, rimsulfuron, and simazine. Do not apply within 14 days of harvest. The addition of ammonium sulfate will enhance glufosinate activity on difficult to control species, however the addition of non-ionic surfactants or crop oil will increase the risk of crop injury.

Table 7-12B. Chemical Weed Control in Fruit Crops—Tree Fruits

Weed	Herbicide, Mode of Action,* and Formulation	Amount of Formulation Per Acre	Pounds Active Ingredient Per Acre	Precautions and Remarks
Pecans, Postemergence, Directed Underneath Tree (continued)				
Most annual broadleaf and grass weeds plus many perennials (continued)	glyphosate, MOA 9 (various brands and formulations)	See label	1 to 2	DO NOT SPRAY GREEN BARK OR FOLIAGE. Repeat applications may be necessary for control of perennial weeds. Tank mix with Goal, Karmex, simazine, Solicam, and Surflan. Check label for details. Generic glyphosate formulations may require the addition of surfactant at 0.5% by volume (2 qt per 100 gal). See label to determine if surfactant is needed for the formulation you use.
Yellow and purple nutsedge, horsenettle, pokeweed and other broadleaf weeds.	halosulfuron, MOA 2 (Sandea) 75 WDG	0.66 to 1.33 oz	0.032 to 0.063	Use on trees established in orchard at least 12 months. Avoid contacting bark or foliage or severe injury or death may occur. The addition of 0.25% surfactant (1 quart per 100 gallons of spray solution) will be necessary for adequate control. Do not make more than two applications per year. Use no more than 1 ounces per acre on soils classified as sand, loamy sand, or sandy loam. Sandea has a 1-day PHI. User assumes risk when treating trees recovering from certain stress conditions. Sandea may be tank mixed with glyphosate to control weeds other than nutsedge.
Annual broadleaf and grass weeds	paraquat, MOA 22 (Firestorm) 3 SL (Paraquat Concentrate) 3 SL (Parazone) 3 SL (Gramoxone) 2 SL	1.75 to 2.7 pt 2.5 to 4.0 pt	0.66 to 1	Apply when grass and weeds are 1 to 6 inches high and succulent for best results. Direct spray with low pressure to avoid contact with tree foliage or bark less than 1 year old. Add surfactant at 0.25% by volume (2 pints per 100 gallon) or 1% crop oil concentrate (1 gallon per 100 gallons) for best results. Paraquat may be tank mixed with Goal, Karmex, simazine, Solicam, and oryzalin. Paraquat is a restricted use pesticide.
Broadleaf weeds	2,4-D amine, MOA 4 (various brands) 3.8 SL	2 to 3 pt	0.95 to 1.4	Apply anytime during the growing season to actively growing broadleaf weeds except during bloom. Do not apply more than 2 applications per year. Allow at least 30 days between sequential applications. Do not use within 60 days of harvest. Do not apply to trees less than 1 year old. Some formulations may limit use rate to 2 pints per acre. Refer to product label for details.
	2,4-D amine + flumioxazin, MOA 4 and 14 (Panther D) SL	3 pt	1.5 + 0.1	Panther D is premix of 2, 4-D amine and flumioxazin. The amount of flumioxazin in this formulation only enhances the POST activity of the 2, 4-D amine and will provide very limited residual control. Orchards must be at least 1 year old and trees in vigorous condition. Panther D has a 60 day PHI for pecans. DO NOT apply within 2 weeks either side of bloom. DO NOT allow contact with foliage, fruit, stems, trunks or exposed roots.
Grasses	clethodim, MOA 1 (Arrow, Intensity, or Select) 2 EC (Select Max or Intensity One) 1 EC	6 to 8 oz 12 to 16 oz	0.094 to 0.125	**NONBEARING TREES ONLY.** Apply to actively growing grasses not under stress. See label for rate and optimum grass size to treat. Multiple applications may be necessary to control perennial grass weeds. For 2 EC formulation chemicals, add crop oil concentrate at 1% by volume (1 gallon per 100 gallons). For 1 EC formulation chemicals, a nonionic surfactant at 0.25% by volume may be used rather than crop oil. **The Select Max formulation has a use in bearing pecans allowing it's use up to within 14 days of harvest.**
	fluazifop, MOA 1 (Fusilade DX) 2 EC	8 to 24 oz	0.125 to 0.38	Postemergence grass control. Annuals up to 12 inches tall and 6 to 10 inches new growth on perennials. Multiple applications may be necessary to control perennial grass weeds. Add crop oil at 1% by volume (1 gallon per 100 gallons). Limited to 72 ounces per year. Do not apply within 30 days of harvest.
	sethoxydim, MOA 1 (Poast) 1.5 EC	1.0 to 2.5 pt	0.19 to 0.47	Apply to annual grasses up to 12 inches tall. For perennial grasses, apply early in the growth cycle at the high use rate. Multiple applications may be necessary to control perennial grass weeds. Add Dash adjuvant at 1 pint per acre or crop oil at 1 quart per acre. Do not apply within 15 days of harvest. Do not apply more than 10 pints per year.
Pecans, Ground Cover Suppression				
Groundcover suppression in row middles	glyphosate, MOA 9 (various brands) 4 SL (various brands) 5 SL	Rate and application times different for grass species. See label.		See label directions specific for each grass species. **DO NOT** apply after seedhead emergence. See label for details.
	(Roundup WeatherMax) 5.5 SL	See label	See label	

* Mode of action (MOA) code developed by the Weed Science Society of America. See Table 7-10, Herbicide Resistance Management, for details.

Chemical Weed Control in Hay Crops and Pastures

F. H. Yelverton and T. W. Gannon, Crop and Soil Sciences Department

Note: A mode of action code has been added to the Herbicide and Formulation column of this table. Use MOA codes for herbicide resistance management. See Table 7-10B, Herbicide Modes of Action for Hay Crops, Lawns and Turf for details concerning active ingredients, brand names, chemical families, and modes of action.

Table 7-13. Chemical Weed Control in Hay Crops and Pastures

Weed	Herbicide, Mode of Action, and Formulation	Amount of Formulation per Acre	Pounds Active Ingredient per Acre	Precautions and Remarks
Alfalfa, Birdsfoot Trefoil, Clovers, Lespedeza Preplant				
Certain annual grass weeds, broadleaf weeds, and nutsedge species	EPTC, MOA 8 (7 EC)	3.5 pt	3	Use on clay and clay loam soils of piedmont. Incorporate into soil immediately after application. See label for directions. Temporary crop stunting may occur if conditions for germination and growth are not optimum. Do not use if grain or grass crop is to be planted with the legume.
Alfalfa, Preplant or Preemergence				
Various grass and broadleaf weeds	paraquat, MOA 22 (2 SL) (3 SL)	2.5 to 4 pt 1.67 to 2.67 pt	0.625 to 1	Apply prior to crop emergence. Add nonionic surfactant at 1 to 2 pints per 100 gallons.
Alfalfa, Established Preemergence				
Crabgrass, foxtails, and other annual grasses	trifluralin, MOA 3 (10 G) (4 EC)	20 lb 2 qt	2	A single rainfall of 0.5 inch or more after application is required to activate trifluralin. Apply 2 quarts trifluralin HFP if chemigation or water incorporated.
Alfalfa, Seedling				
Seedling broadleaf weeds, such as burcucumber, cocklebur, jimsonweed, lambsquarters, velvetleaf, Virginia pepperweed, shepherd's-purse, wild radish, and species of morningglory, mustard, nightshade, pigweed, ragweed, and smartweed	bromoxynil, MOA 6 (2 EC) (4 EC)	1 to 1.5 pt 0.5 to 0.75 pt	0.25 to 0.375	Apply in fall or spring to seedling alfalfa with a minimum of 4 trifoliate leaves and to weeds not greater than 4 leaf stage, 2 inches in height, or 1 inch in diameter, whichever comes first. Unacceptable crop injury can occur 3 days after application if temperatures exceed 70 degrees F. For chemigation only, apply to alfalfa with 2 trifoliate leaves at 2 pints/A to most susceptible weeds not greater than 8 leaf stage, 4 inches in height, or 2 inches in diameter, whichever comes first. Unacceptable crop injury can occur 3 days after application if temperatures exceed 85 degrees F. Bromoxynil can be tank mixed with 2,4-DB 200 or imazethapyr 2 AS. Do not apply in warm humid conditions or to alfalfa under any kind of stress. Do not add surfactant unless specified. Do not cut for feed or graze spring-treated alfalfa until 30 days after treatment. Wait until spring, or 60 days after treatment, for winter-treated alfalfa.
Alfalfa, Birdsfoot Trefoil, Clovers, Seedling				
Certain broadleaf weeds such as cocklebur, lambsquarters, morningglory, pigweed, ragweed, smartweed, curly dock, shepherdspurse, and wild mustard	2,4-DB, MOA 4 (2 EC)	2 to 6 pt	0.5 to 1.5	Apply postemergence when weeds are less than 3 in. tall and legume has at least 2 to 4 trifoliate leaves. Do not graze or feed seedling legume crops to livestock within 60 days after application.
Alfalfa, Clover, Birdsfoot Trefoil, Crown Vetch, Established and Seedling				
Ryegrass species, annual bluegrass, perennial bluegrass, orchardgrass, chickweed, and volunteer grain	pronamide, MOA 3 (3.3 SC) (50-W)	1.25 to 5 pt 1 to 4 lb	0.516 to 2.06 0.5 to 2	Use preemergence or postemergence to the weeds only on established legume plantings or on new plantings after the legume has reached the trifoliate leaf stage or beyond. Controls henbit, shepherdspurse, wild radish and wild mustard with preemergence applications. Apply from Oct. 15 to Jan. 15. Optimum herbicidal activity occurs when applications are made under cool temperatures (55 degrees F or less) and followed by rainfall or irrigation. Do not graze or harvest for forage or dehydration within 120 days of treatment.
Alfalfa, Established and Seedling				
Lambsquarters, pigweed, ragweed, morningglory, and smartweed	2,4-DB, MOA 4 (2 EC)	2 to 6 pt	0.5 to 1.5	Apply postemergence when weeds are less than 3 inches tall. Do not graze established alfalfa or cut for hay within 30 days after application.
Crabgrass, foxtails, seedling johnsongrass and certain broadleaf weeds such as chickweed, cocklebur, henbit, jimsonweed, morningglory, wild mustard, nightshade, pepperweed, pigweed, ragweed, smartweed, spurge, and Russian thistle	imazethapyr, MOA 2 (2 AS)	3 to 6 fl oz	0.048 to 0.095	Apply postemergence when seedling alfalfa or clover is in the second trifoliate stage or larger. Can also be applied postemergence to established alfalfa or clover in the fall, in the spring to dormant or semi-dormant alfalfa or clover, or between cuttings. Application should be made before significant alfalfa or clover growth or regrowth to allow herbicide to reach target weeds. Use 80% active nonionic surfactant at 1 quart per 100 gallons of water or a crop oil concentrate at 1 quart per acre. Weeds should be 1 to 3 inches in height. Imazethapyr will reduce growth of perennial grasses (fescue, etc.) that are present in the stand. See label for weeds controlled and other precautions.
Annual bluegrass, barnyardgrass, crabgrass, crowfootgrass, foxtail species, goosegrass, Italian ryegrass, seedling johnsongrass, fall panicum, Texas panicum, sandbur, signalgrass, and certain broadleaf weeds such as palmer amaranth, common chickweed, henbit, lambsquarters, pigweed species, and smartweed	pendimethalin, MOA 3 (3.8 CS)	1.1 to 4.2 qt	1.045 to 4	Use on alfalfa grown for forage, hay, or seed. Apply 1.1 to 4.2 quarts/acre prior to weed emergence in fall after last cut, during winter dormancy, in the spring, or between cuttings before alfalfa reaches 6 inches when grown for forage or hay. Apply same rates for alfalfa grown for seed production when dormant or before alfalfa exceeds 10 inches after first or second cut. Use drop nozzles to minimize foliar contact. Apply 1.1 to 2.1 pints/acre to seedling alfalfa in second trifoliate stage before 6 inches of growth. Do not harvest alfalfa forage or hay less than 28 days after applying 2.1 quarts/acre or less than 50 days after applying more than 2.1 quarts/acre. Do not harvest alfalfa seed less than 90 days after application. Some stunting and chlorosis of alfalfa may occur after postemergence applications.

Table 7-13. Chemical Weed Control in Hay Crops and Pastures

Weed	Herbicide, Mode of Action, and Formulation	Amount of Formulation per Acre	Pounds Active Ingredient per Acre	Precautions and Remarks
Alfalfa, Established and Seedling (continued)				
Annual and perennial grasses	clethodim, MOA 1 (2 EC)	8 to 16 fl oz	0.125 to 0.250	Apply postemergence for annual grasses in seedling alfalfa at 6 to 8 fluid ounces per acre or in established alfalfa at 8 ounces per acre. Apply postemergence for bermudagrass and rhizome johnsongrass at 8 to 16 fluid ounces per acre. Add a crop oil concentrate at 1 quart per acre. Can be applied at any stage of alfalfa growth. Apply to actively growing grasses not under drought stress. Be sure grasses have leaves present for contact by the spray solution. Do not apply within 15 days of grazing, feeding, or harvesting (cutting) alfalfa for forage or hay. Select may be tank mixed with 2,4-DB or imazethapyr. When tank mixing, see respective labels for application rates, weeds controlled, maximum weed size to treat, specific application directions, and precautions.
	sethoxydim, MOA 1 (1.5 EC) (1 EC)	1 to 2.5 pt 1.5 to 3.5 pt	0.19 to 0.47	Apply postemergence for annual grasses at 0.19 pound a.i. (the lower rate) per acre and for bermudagrass and johnsongrass at 0.47 pound a.i. (the higher rate) per acre. Add 2 pints crop oil concentrate per acre. Use 10 to 20 gallons of spray solution per acre. Can be applied at any stage of alfalfa growth. Do not apply to weedy grasses or alfalfa under stress. Be sure grasses have leaves present for contact by the spray solution. Do not apply within 7 days of grazing, feeding, or cutting for (undried) forage, or within 14 days of cutting alfalfa for (dry) hay.
Alfalfa, Established Dormant				
Winter annual weeds, such as chickweed, henbit, bittercress, pepperweed, shepherds purse, yellow rocket, and ryegrass	metribuzin, MOA 5 (75 DF) (4 L, 4 F)	0.67 lb 0.5 to 2 pt	0.5 0.25 to 1	Good results have been obtained in N.C. when the herbicide was applied from Nov. 20 to Dec. 20. Do not graze or harvest within 28 days after application. In alfalfa-grass mixtures, it will provide partial reduction of forage grass stands.
	paraquat, MOA 22 (2 SL) (3 SL)	0.75 to 2 pt 0.7 to 1.3 pt	0.1875 to 0.5 0.263 to 0.488	Apply up to 1.25 pints per acre on fall-seeded newly established stands less than 1 year old. Apply up to 2 pints per acre on established stands. Tank mix with metribuzin will improve vegetation control. Apply late fall to winter after last fall cutting and before first spring cutting. Alfalfa must be dormant to avoid injury. There is a 60-day grazing or preharvest interval. Add 1 to 2 pints per acre nonionic surfactant per 100 gallons.
	terbacil, MOA 5 (80 WDG)	0.5 to 1.5 lb	0.4 to 1.2	Apply to established alfalfa stands at least 1 year old before or shortly after weed growth begins. Weeds have been controlled with an application from mid-November through February. Do not use terbacil on alfalfa-grass mixtures on sand or loamy sand soils or on soils containing less than 1% organic matter.
Alfalfa, Between Cuttings (even first year alfalfa)				
Various grass and broadleaf weeds	paraquat, MOA 22 (2 SL) (3 SL)	1 pt 0.7 pt	0.25 0.263	Apply immediately after hay or silage removal but no more than 5 days after a cutting. Apply up to 2 times for first year alfalfa and 3 times for established alfalfa. There is a 30 day grazing or preharvest interval. Add 1 to 2 pints per acre nonionic surfactant per 100 gallons.
Lespedeza, Preplant				
Certain annual grass and broadleaf weeds	EPTC, MOA 8 (7 EC)	3.5 pt	3	See remarks under alfalfa.
No-Till Alfalfa or No-Till Pasture Reseeding				
Complete kill of existing sod	glyphosate, MOA 9 (4 SL) (5.5 SL)	1 to 5 qt 2.75 pt	1 to 5 1.89	Broadcast spray 10 to 14 days before planting. Provides better control of perennial weeds. Check label for rate according to weeds present.
	paraquat, MOA 22 (2 SL) (3 SL)	2 to 4 pt 1.67 to 2.67 pt	0.5 to 1 0.625 to 1	Broadcast spray in 20 to 30 gallons of water per acre. Make 2 applications if needed. If spraying following hay harvest, allow enough regrowth to provide leaf area to absorb the herbicide. Add 1 pint of a nonionic surfactant per 100 gallons of water.
No-Till Pasture Reseeding with Grasses or Clover				
Suppression of existing sod and undesirable emerged broadleaf and grass weeds to permit pasture reseeding	paraquat, MOA 22 (2 SL) (3 SL)	1 to 2 pt 0.7 to 1.3 pt	0.25 to 0.5 0.263 to 0.488	Rates are per sprayed acre. Usually band sprayed for planting clover into existing grass sod. Apply before or at time of seeding. Pasture should not exceed 3 inches in height at time of treatment. Add 1 to 2 pints of a nonionic surfactant per 100 gallons of water. Spray bermudagrass or bahiagrass sod in late summer to early fall.
Pastures, Clover, Velvetbean, Lespedeza, Lupine, Sainfoin, Trefoil, Vetch, Crown Vetch, Milk Vetch				
Various grass and broadleaf weeds	paraquat, MOA 22 (2 SL) (3 SL)	0.75 to 2 pt 0.7 to 1.3 pt	0.1875 to 0.5 0.263 to 0.488	Apply up to 1.2 pints per acre to fall-seeded newly established stands less than1 year old. Apply up to 2 pints per acre to established stands. Apply when dormant in late fall or winter after last cutting and before first spring cutting. Do not apply if regrowth is >2 inches. Make only 1 application per season. There is a 60 day grazing or preharvest interval. Add 1 to 2 pints per acre nonionic surfactant per 100 gallons.
Pastures, Ladino Clover, Orchardgrass, Fescue, and other grasses				
Curly dock, ragweed, bitterweed, pigweed, dandelion, and other broadleaf weeds	2,4-D amine, MOA 4 (4 SL)	1 to 2 pt	0.5 to 1	Spray when weeds are 4 to 8 inches tall and before heading. Clover may be stunted, and growth retarded 3 to 6 weeks. Use lower rate in warm, wet weather. For wild garlic, apply late February or early March. Repeat for 2 years. **Do not graze dairy animals on treated areas within 7 days after application.** Remove meat animals from treated areas for 3 days before slaughter. Withdrawal is not necessary if more than 2 weeks have elapsed since treatment. Do not cut treated grass for hay within 30 days after application.
Wild garlic	2,4-D amine, MOA 4 (4 SL)	1 qt	1	
Perennial Grasses, Rangeland, Permanent Grass Pastures				
Many annual and perennial grass and broadleaf weeds and nutsedge species (goosegrass not controlled)	imazapic, MOA 2 (2 SL)	4 to 12 fl oz	0.0625 to 0.1875	Apply to common and coastal bermudagrass varieties. Jiggs bermudagrass is more sensitive than other types. Expect 30 to 45 days of bermudagrass suppression. Do not apply 1) to drought stressed bermudagrass, 2) during spring transition, 3) to newly aerated fields for 30 days, 4) to newly sprigged or seeded bermudagrass, or 5) to World Feeder bermudagrass varieties. To speed bermudagrass recovery, apply with nitrogen fertilizer and do not add a spray adjuvant. If spray carrier is water, add a nonionic surfactant at 0.25% by volume or a methylated seed oil at 1.5 to 2 pints per acre. Imazapic also controls winter weeds when applied to dormant bermudagrass, and can be mixed with glyphosate at this time. There is a 7-day hay restriction.

Table 7-13. Chemical Weed Control in Hay Crops and Pastures

Weed	Herbicide, Mode of Action, and Formulation	Amount of Formulation per Acre	Pounds Active Ingredient per Acre	Precautions and Remarks
Perennial Grasses, Rangeland, Permanent Grass Pastures				
Johnsongrass, kyllinga species, purple and yellow nutsedge	sulfosulfuron, MOA 2 (75 WG)	1.33 oz	0.0625	Apply to established bermudagrass and bahiagrass pastures. A second application can be made 40 days after initial application if needed, but do not exceed 2.66 ounces per acre per year. Apply a nonionic surfactant at 0.25% v/v. There are no grazing restrictions. Do not harvest for hay within 14 days of application. Johnsongrass is best controlled if sulfosulfuron is applied at 18 to 24 inches and up to heading stage.
Amaranth and pigweed species, ladysthumb, wild mustard, wild radish, maypop passionflower, pokeweed, ragweed species, hemp sesbania, velvetleaf, purple and yellow nutsedge	halosulfuron, MOA 2 (75 DF)	0.67 to 1.33 oz	0.0314 to 0.0628	Listed weeds are controlled postemergence if small and actively growing at application. Check label for weeds that are controlled preemergence. The following adjuvants can be used with halosulfuron; nonionic surfactant applied at 0.25 to 0.5% v/v, crop oil concentrate or methylated seed oil at 1% v/v, high quality granular spray grade ammonium sulfate at 2 to 4 pounds per acre or liquid ammonium sulfate at equivalent nitrogen rate of 2 to 4 lb per acre. If nutsedge populations resprout or reemerge, a second spot application not to exceed 0.75 ounce per treated acre in these areas can be made. There are no grazing or slaughter restrictions. There is a 37 day pre-harvest interval.
Emerged annual, biennial, and perennial broadleaf weeds and certain woody species	[picloram + 2,4-D], MOA 4 + 4 (0.54 + 2 lb/gal SL)	1 to 8 pt	0.3175 to 2.54	Due to crop sensitivity, [picloram + 2,4-D] should not be used in cotton- or tobacco-growing regions of the state. Do not graze lactating dairy animals for 7 days after application. Do not harvest grass for hay for 30 days after application. Meat animals must be withdrawn from treated forage at least 3 days before slaughter. There are no other grazing restrictions for non-lactating dairy animals or other livestock. Newly seeded grasses may be injured. Check label for livestock transfer restrictions due to possible urine and manure contamination with picloram. Check label for all other restrictions.
	[picloram + fluroxypyr], MOA 4 + 4 (1.19 + 0.96 lb/gal EC)	1.5 to 6 pt	0.40 to 1.60	Due to crop sensitivity, [picloram + fluroxypyr] should not be used in cotton- or tobacco-growing regions of the state. Do not allow lactating dairy animals to graze or consume harvested forage within 14 days after application. There are no grazing restrictions for nonlactating dairy animals or other livestock. Do not harvest hay within 7 days after application. Withdraw meat animals from treated forage at least 3 days before slaughter. Newly seeded or sprigged grasses may be injured. Check label for livestock transfer restrictions due to possible urine and manure contamination with picloram. Check label for all other restrictions.
Mustard, radish, cocklebur, vetch, and other susceptible broadleaf weeds	2,4-D amine, MOA 4 (4 SL)	1 to 2 pt	0.5 to 1	Do not spray in seedling stages or just before heading. Apply after the perennial grass seedlings have reached the 2- to 4-leaf stage. For wild garlic apply February or March. Repeat for 2 years. **Do not graze dairy animals on treated areas within 7 days after application.** Remove meat animals from treated areas for 3 days before slaughter. Withdrawal is not necessary if more than 2 weeks have elapsed since treatment. Do not cut treated grass for hay within 30 days after application.
Wild garlic	2,4-D amine, MOA 4 (4 SL)	3 qt	3	
Many broadleaf weeds including certain ones resistant to 2,4-D	dicamba, MOA 4 (4 SL)	0.5 to 2 pt	0.25 to 1	Rate dependent on weed species and size. See label for specific rates and precautions concerning grazing.
Many broadleaf weeds including dogfennel, thistles, and horsenettle	dicamba (4 SL) + 2,4-D amine (4 SL) (a tank mix) MOA 4 + 4	0.5 pt + 1.5 pt	0.25 + 0.75	The tank-mix combination will control a greater number of broadleaf weeds than either herbicide alone. Observe each label for restrictions on grazing and cutting for hay. For 1 pint of dicamba or 2 pints of [2,4-D amine + dicamba], do not graze lactating dairy animals for 7 days or harvest hay for 37 days. No grazing restrictions for other livestock; however, meat animals must be removed 30 days before slaughter.
	[dicamba + 2,4-D amine], MOA 4 + 4 (a premix) (1 + 2.87 lb/gal SL)	1 to 2 pt	0.13 to 0.25 + 0.36 to 0.72	
	[2,4-D +triclopyr], MOA 4 + 4 (2 + 1 lb/gal EC)	1 to 6 qt	0.5 to 3 + 0.25 to 1.5	Rate depends on weeds to be controlled. Woody plant control requires 6 quarts or more. Consult label for specific rates. Withdraw livestock from treated forage at least 3 days before slaughter during the year of treatment. Do not graze lactating dairy animals on treated areas for 14 days following treatment. Do not harvest grass for hay from treated areas for 1 year following treatment for lactating dairy animals. Wait 7 days for other livestock.
	metsulfuron methyl, MOA 2 (60 WDG)	0.1 to 1 oz	0.0038 to 0.038	Bermudagrass, bluegrass, orchardgrass, bromegrass, and timothy are tolerant. Metsulfuron methyl may cause stunting and seedhead suppression of tall fescue. Therefore, do not exceed 0.4 ounce product per acre. Pensacola bahiagrass controlled at 0.3 ounce product per acre in established bermudagrass. Also controls wild garlic. Alfalfa, clover, and ryegrass are highly sensitive. Use a nonionic surfactant of at least 60% active ingredient with a hydrophilic/lipophilic balance greater than 12 at 1 to 2 pints per 100 gallons of spray solution. Use a COC of 80% quality or MSO with 15% surfactant emulsifiers at 1 gallon per 100 gallons spray solution. Metsulfuron methyl has no grazing restrictions.
	[metsulfuron methyl + chlorsulfuron], MOA 2 + 2 (48% + 15% WDG)	0.125 to 1.25 oz	0.005 to 0.05	Tolerant to native grasses, such as bluestems, blue grama, and buffalograss, as well as bermudagrass, bluegrass, orchardgrass, bromegrass (but not Matua), and fescue. To minimize fescue injury, do not exceed 0.5 ounce per acre and use a nonionic surfactant unless liquid nitrogen is the carrier. Apply to bermudagrass 2 months after establishment and fescue 24 months after establishment. Generally, treat actively growing weeds less than 4 inches tall or 4 inches in diameter. However, this product also provides preemergence control. Unless otherwise recommended, use a nonionic surfactant at 0.25% v/v or crop oil concentrate at 1% v/v. There are no grazing or hay harvest restrictions.
Many annual and perennial broadleaf weeds and woody brush	[metsulfuron methyl + dicamba + 2,4-D amine], MOA 2 + 4 + 4 (a 2-part product) [60%+(1+2.87) lb/gal]	0.25 to 1 oz + 1 to 4 pt	0.009375 to 0.038 + 0.48 to 1.94	Observe same precautions as metsulfuron methyl except for the following changes or additions: Tall fescue: do not exceed 0.25 ounce per acre Part A + 1 pint per acre Part B. Nonlactating meat animals: remove 30 days prior to slaughter. Lactating dairy animals: 7-day grazing and 37-day hay restriction.

Table 7-13. Chemical Weed Control in Hay Crops and Pastures

Weed	Herbicide, Mode of Action, and Formulation	Amount of Formulation per Acre	Pounds Active Ingredient per Acre	Precautions and Remarks
Perennial Grasses, Rangeland, Permanent Grass Pastures (continued)				
Henbit, common chickweed, mustards, buttercup, Carolina geranium, pigweed species, common lambsquarters, and other susceptible broadleaf weeds	chlorsulfuron, MOA 2 (75 DF)	0.25 to 1 oz	0.0117 to 0.047	Treat perennial weeds in bud to bloom stage or fall rosette stage. There are no grazing or hay restrictions with rates up to 1.33 ounces per acre. For bermudagrass and orchardgrass, apply a maximum of 1 ounce per acre. Apply up to 0.5 ounce per acre on tall fescue. Spot treat with 1.33 ounces per acre if grass injury can be tolerated. Use a high quality spray adjuvant for improved postemergence control, but do not use LI-700 or other acidifying spray adjuvants.
Buttercup species, cocklebur, henbit, horsenettle, horseweed, ragweed, thistles, and other susceptible broadleaf weeds	aminopyralid, MOA 4 (2 SL)	3 to 7 fl oz	0.04688 to 0.10938	Due to crop sensitivity, use extreme caution around sensitive crops, including but not limited to alfalfa, cotton, potatoes, soybeans, tobacco, and other broadleaf or vegetable crops, fruit trees, or ornamental plants. Do not use aminopyralid-treated plant residues, including hay or straw from treated areas, or manure from animals that have grazed treated areas in compost or mulch that will be in contact with susceptible broadleaf plants. Hay treated within the preceding 18 months can only be used on the farm or ranch where the product was applied. There are no other restrictions on grazing or hay harvest following aminopyralid applications. Check product label for list of precautions.
	[aminopyralid + 2,4-D amine], MOA 4 + 4 (0.41 + 3.33 lb/gal L)	1.2 to 2.1 pt	0.56 to 0.975	Do not use on areas where loss of desirable broadleaf forage plants (legumes) cannot be tolerated. Do not use hay, straw, or manure from farm animals that have grazed forage or eaten hay harvested from treated areas within previous 3 days in compost or mulch that will be in contact with susceptible broadleaf plants. Wait 7 days after application to harvest forage for hay. Wait 30 days to make second application. Do not transfer grazing animals from treated areas to sensitive broadleaf crop areas without allowing for 3 days grazing on nontreated areas. Hay treated within the preceding 18 months can only be used on the farm or ranch where the product was applied.
	[aminopyralid + metsulfuron methyl], MOA 4 + 2 (71.58 WG)	1 to 3.3 oz	0.0447 to 0.1476	[Aminopyralid + metsulfuron methyl] is effective on Pensacola bahiagrass. At higher rates, [aminopyralid + metsulfuron methyl] may stunt tall fescue, cause yellowing, or cause seedhead suppression. Follow label precautions to minimize these symptoms. Include 1% crop oil concentrate, 0.25% non-ionic surfactant, 0.5% methylated seed oil, or 2 quarts/acre urea ammonium nitrate. Can spot spray less than 50% of an acre with up to 6.6 ounces/acre. There are no grazing or hay harvest restrictions. Do not use on grasses grown for seed. Do not overseed 4 months after treatment. Aminopyralid Precautions: Do not transfer grazing animals from treated areas to sensitive broadleaf crop areas without allowing for 3 days grazing on nontreated areas. Do not use hay, straw, or manure from animals that have grazed forage or eaten hay from treated areas within previous 3 days in compost or mulch that will contact susceptible broadleaf plants. Do not spread manure on land used for growing susceptible broadleaf crops if animals have consumed treated hay within 3 previous days. Conduct a field bioassay before planting a broadleaf crop. Hay treated within the preceding 18 months can only be used on the farm or ranch where the product was applied.
Catchweed bedstraw, common lambsquarters, mustard spp., nightshade spp., amaranthus spp., velvetleaf, bittercress, shepherds-purse, annual sowthistle, corn spurry, Russian thistle, redstem filaree	carfentrazone ethyl, MOA 14 (2 EC)	0.5 to 2 fl oz	0.0078125 to 0.03125	Use in grasses grown for forage, fodder, hay, seed, and sod. There are no grazing or hay restrictions. Add a nonionic surfactant at 0.25% v/v or crop oil concentrate or methylated seed oil at 1% v/v with or without a high quality sprayable liquid N fertilizer at 2 to 4% v/v or ammonium sulfate at 2 to 4 pounds per acre. Apply to weeds up to 4 inches tall. In overseeded pastures, Aim can be applied to barley, millet, oats, rye, teosinte, triticale, and wheat from prior to planting up to joint stage. Tank mix 2,4-d amine or ester for extended broadleaf weed control but don't harvest for forage within 7 days of application.
Catchweed bedstraw, common lambsquarters, mustard spp., nightshade spp., amaranthus spp., velvetleaf, bittercress, shepherds-purse, annual sowthistle, corn spurry, Russian thistle, redstem filaree, and many other weeds susceptible to 2,4-D	[carfentrazone ethyl + 2,4-D ester], MOA 14 + 4 (0.13 + 5.92 lb/gal EC)	0.5 to 2 pt	0.378125 to 1.5125	Use in grasses grown for forage, fodder, hay, seed, and sod. Restrictions after application include: 3-day slaughter; 7-day dairy grazing; and 30-day hay harvest. Add a nonionic surfactant at 0.25% v/v or crop oil concentrate or methylated seed oil at 1.5 to 2% v/v with or without a high quality sprayable liquid N fertilizer at 2 to 4% v/v or ammonium sulfate at 2 to 4 pounds per acre. Apply to weeds up to 6 inches tall. In overseeded pastures, 8 to 16 ounces of [carfentrazone ethyl + 2,4-D ester] can be applied to barley, oats, rye, and wheat from 3 tiller stage up to joint stage. Use only a nonionic surfactant with or without a fertilizer solution as described above in small grains. Do not graze dairy or meat animals being finished for slaughter for 14 days following application. Do not feed treated straw to livestock.
Buttercup species, cocklebur, dogfennel, henbit, horsenettle, horseweed, select amaranth and pigweed species, other susceptible broadleaf weeds.	[diflufenzopyr sodium + dicamba], MOA 19 + 4 (21.3% + 55% DF)	4 to 8 oz	0.05 + 0.125 to 0.1 + 0.25	Use caution around sensitive broadleaf crops, including but not limited to alfalfa, clover, and lespedeza. There are no hay harvesting or grazing restrictions for pasture and rangelands treated with [diflufenzopyr sodium + dicamba]; however, there is a 30-day crop rotation or planting restriction. [Diflufenzopyr + dicamba] is for use on established grasses; do not apply to newly seeded grasses or to small grains grown for pasture.
	quinclorac, MOA 27 + 4 (1.5 lb/gal)	12 to 64 fl oz plus 2 pt crop oil concentrate or 1.5 pt methylated seed oil	1.5 lb per gallon	May be used for postemergence weed control in cool- and warm-season pastures including pastures grown for hay. Facet L also provides short residual weed control (length of control depends on species and environmental conditions. Facet L controls grassy weeds including barnyardgrass, crabgrass spp., foxtail spp., and broadleaf signalgrass and broadleaf weeds including clover spp., eclipta, jointvetch spp., select morningglory spp., and hemp sesbania. Facet L is rainfast 6 hours after application. Do not cut treated area for hay within 7 days after application. There is no waiting period for grazing following application.
	saflufenacil, MOA 14 (2.85 lb/gal)	1.0 to 2.0 fl oz	2.85 lb per gallon	Provides preemergence and early postemergence broadleaf weed control in perennial cool-season and warm-season forage grasses grown for forage, silage, or hay production. Do not apply to annual forage stands (including forage sorghum, sudangrass, etc.) Sharpen does not control grass weeds. Sharpen is labeled at 1.0 to 2.0 fl oz/a during dormant and periods of active growth. Do not apply more than 1.0 oz/a to forage bermudagrass after greenup. Add a methylated seed oil (1% vol/vol or 1.5 pt/a) for optimum control. Do not use a nonionic surfactant.

Table 7-13. Chemical Weed Control in Hay Crops and Pastures

Weed	Herbicide, Mode of Action, and Formulation	Amount of Formulation per Acre	Pounds Active Ingredient per Acre	Precautions and Remarks
Perennial Grasses, Rangeland, Permanent Grass Pastures (continued)				
Many broadleaf weeds including most legumes, cocklebur, curly dock, horseweed, jimsonweed, lambsquarters, prickly lettuce, mustards, nightshades, redroot pigweed, plantains, wild radish, sicklepod, sowthistles, and thistles	[2,4-D amine + cloypyralid], MOA 4 + 4 (2 +0.38 lb/gal)	2 to 4 qt	1.19 to 2.38	Apply to established grass pastures. Allow 7 days grazing on nontreated pasture before livestock transfer to sensitive broadleaf crop areas. Do not use plant residues or manure from treated areas for composting or mulching near sensitive plants. Avoid movement of treated soil and minimize spray drift when possible. There is a 14-day grazing restriction for lactating cattle, a 30-day restriction for haying, and a 7-day restriction for slaughter. Addition of surfactant is usually not necessary.
Many woody plants, such as poplar, sumac, sassafras, wax myrtle, oaks, red maple, locust, eastern persimmon, and broadleaf weeds such as poison oak, poison ivy, blackberry, clover, curly dock, multiflora rose, lespedeza, mustard, plantain, and vetch	triclopyr, MOA 4 (4 EC)	1 to 3 pt	0.5 to 1.5	Apply to established grass pastures. Rate depends on weed species to be controlled. See label for rates for specific weeds and dilution rates for woody plant control. For lactating dairy animals do not graze until next growing season. There are no grazing restrictions for other livestock. Portions of grazed area may be treated if comprised of no more than 10 percent of the total grazable area. There is a 3 day slaughter and 14 day hay harvest restriction. Addition of 2,4-D at 1 pint per acre enhances weed spectrum. A surfactant or oil-based carrier is recommended, but liquid nitrogen carrier may be used as well.
	[triclopyr + fluroxypyr], MOA 4 + 4 (3 + 1 lb/gal EC)	0.75 to 1.5 pt	0.38 to 0.75	Apply to established grass pastures. In general, apply 0.75 to 1 pint per acre for annual broadleaf weeds, 1 to 1.5 pints per acre for biennial and perennial broadleaf weeds, and 1 to 4 pints per acre for woody plants. Do not graze or harvest green forage for lactating dairy animals during the same growing season. There are 14-day haying and 3-day slaughter restrictions. A nonionic surfactant at 1 to 2 quarts per 100 gallons of spray solution may improve control of drought-stressed weeds.
Multiflora rose	dicamba, MOA 4 (4 SL)	1 to 2 gal	4 to 8	Mix 1 gallon of dicamba herbicide with 99 gallons of water. Add a nonionic surfactant at the rate of 2 quarts per 100 gallons of spray solution to improve wetting. Apply 100 to 200 gallons of spray solution per acre. Completely wet foliage and stems, allowing spray solution to run down the stem. Apply at full vegetative stage before bloom. For spot treatment, mix 3 tbsp of dicamba per 1 gallon of water, or directly apply concentrate to root area with a spotgun applicator. Use 1 fluid ounce of dicamba per 10 feet of canopy diameter and apply before bud-break in spring. Do not graze dairy animals for 60 days after treatment. There is no waiting period between treatment and grazing for animals other than dairy animals.
	glyphosate, MOA 9 (4 SL, 5.5 SL)	1% solution		Apply 1% solution of glyphosate in water with a handgun. Spray foliage completely. Apply after full bloom until August 1. Do not graze livestock for 10 days following treatment.
	metsulfuron methyl, MOA 2 (60 WDG)	0.5 oz	0.019	Broadcast application: treat in spring when multiflora rose is fully leafed but less than 3 ft tall. Spot application: treat spring through summer at 1 ounce per 100 gallons of spray solution. Add surfactant. Metsulfuron methyl has no grazing restrictions.
	tebuthiuron, MOA 7 (20 P) (80 DF)	20 lb 2.5	4 2	Apply in the spring just before growth begins or during periods of vigorous growth, late spring or early summer (March 15 to June 1). Best to apply to individual clumps on basis of area (sq ft) covered. Spread uniformly over the plant roots. Check label carefully for restrictions and precautions on use. Do not contaminate water used for irrigation.
	[2,4-D + triclopyr], MOA 4+4 (2 + 1 lb/gal EC)	1.5 gal	4.5	Mix 1 to 1.5 gallons of [2,4-D + triclopyr] with water to make 100 gallons of total spray solution. Spray to give thorough coverage of foliage, wetting leaves and stems to the drip point. The best time for treatment is during early to mid-flowering stage. See restrictions for [2,4-D + triclopyr] on grazing and hay harvest listed previously.
Bermudagrass, Newly Sprigged				
Annual and perennial grass and broadleaf weeds	diuron, MOA 7 (4 L)	0.8 to 2.4 qt	0.8 to 2.4	Apply after planting, before bermudagrass or weed emergence. Apply 0.4 to 0.8 quart per acre if annual weeds are up to 4 inches tall with an NIS at 2 quarts per 100 gallons of water. If bermudagrass has emerged at treatment, expect temporary burn. Plant sprigs 2 inches deep to reduce crop injury potential. Do not graze or feed livestock within 70 days of application.
Bermudagrass, Dormant				
Emerged winter annual broadleaf and grass weeds	metsulfuron methyl, MOA 2 (60 WDG)	0.1 to 0.3 oz	0.0038 to 0.011	Controls many winter annual broadleaf weeds.
	paraquat, MOA 22 (2 SL) (3 SL)	1 to 2 pt 0.7 to 1.3 pt	0.25 to 0.5 0.263 to 0.488	Apply in February or March in dormant bermudagrass and Coastal bermudagrass pastures. Add 1 pint of a nonionic surfactant per 100 gallons of water. Do not mow for hay until 40 days after treatment.
Barnyardgrass, crabgrass, crowfootgrass, foxtail species, goosegrass, seedling johnsongrass, fall panicum, Texas panicum, sandbur, signalgrass, and certain broadleaf weeds such as palmer amaranth, lambsquarters, pigweed species, and smartweed	pendimethalin, MOA 3 (3.8 CS)	1.1 to 4.2 qt	1.045 to 4	Apply to established bermudagrass pasture and hay fields in winter dormancy. Apply as a single full rate or in two split applications with a half rate at the onset of winter dormancy and another half rate prior to spring greenup. Do not harvest for forage or graze until 45 days after treatment. Do not harvest for hay until 60 days after harvest. Observe a plant back interval of 270 days after treatment. Use of pendimethalin on rangeland is prohibited.

Table 7-13. Chemical Weed Control in Hay Crops and Pastures

Weed	Herbicide, Mode of Action, and Formulation	Amount of Formulation per Acre	Pounds Active Ingredient per Acre	Precautions and Remarks
Bermudagrass, Coastal				
Various annual grass and broadleaf weeds	glyphosate MOA 9 (4 SL) (5.5 SL)	1 pt 0.727 pt	0.5	May be applied to coastal bermudagrass prior to spring growth or immediately after first cutting. Cannot be applied prior to spring growth and immediately after first cutting in the same year. Remove domestic livestock from area before making applications. When applying prior to spring growth, apply in late winter or early spring but before new coastal bermudagrass growth begins in spring. Applications to new growth can damage bermudagrass. Wait 60 days after making this application before grazing or harvesting the treated area. When applied after first cutting, apply after the first bermudagrass cutting when the bermudagrass has not yet begun to grow. Applications made after regrowth can damage the bermudagrass. Wait 28 days after making this application before grazing or harvesting the treated area.
Suppresses large crabgrass, goosegrass. Controls barnyardgrass, broadleaf signalgrass, foxtail species, johnsongrass up to 18 inches, panicum species, Italian ryegrass, sandbur, volunteer cereals, Pensacola bahiagrass, wild garlic, many broadleaf weeds such as bitter sneezeweed, buttercup, geranium, chickweed, curly dock, dandelion, dog fennel, henbit, horseweed, jimsonweed, lambs-quarters, morningglory, pigweed, plantain, smartweed, wild mustard.	[nicosulfuron + metsulfuron methyl], MOA 2 + 2 (56.2% + 15% WG)	1 to 1.5 oz	0.0445 to 0.0668	There are no grazing or hay restrictions. [Nicosulfuron + metsulfuron methyl] provides pre and post broadleaf weed control and only post grass weed control. Apply to established bermudagrass. Temporary crop injury may occur if treated on new growth more than 2 inches or after 7 days following harvest. Do not exceed 2.5 ounce/acre per year if sequential applications are needed; 0.25% non-ionic surfactant is preferred but can apply with 1% crop oil concentrate or 2 quarts/acre urea ammonium nitrate. Check label for acceptable tank mix partners. [Nicosulfuron + metsulfuron methyl] will control crabgrass and sandbur up to 2 inches and goosegrass up to 2 tillers. FIFRA Section 2 (ee) allows glyphosate to be tankmixed with [Nicosulfuron + metsulfuron methyl] for improved control or suppression of crabgrass, sandbur, foxtail, rescuegrass, Japanese brome, little barley, and ryegrass. Apply 2.5 to 4.1 ounce a.i. per acre glyphosate. Expect temporary yellowing or stunting of bermudagrass. Add nonionic surfactant at 0.25% v/v.
Hybrid Bermudagrass, (Coastal, Tifton 44)				
Young annual broadleaf weeds	2,4-D amine, MOA 4 (4 SL)	1 qt	1	Apply after sprigging. Gives little preemergence weed control. Later applications may be needed to control broadleaf weeds.
Sorghum-Sudan Hybrids, Preemergence				
Annual broadleaf and grass weeds	atrazine, MOA 5 (4 L) (90 DF, 90 WDG)	3.2 to 4 pt 1.8 to 2.2 lb	1.6 to 2	Use only on silt-loam, clay loam, and clay soils with more than 1% organic matter. Use lower rates on soils 1% to 1.5% organic matter and higher rates on soils having more than 1.5% organic matter. On highly erodible soils (as defined by SCS), if conservation tillage is practiced, leaving at least 30% of soil covered with plant residues at planting, apply a maximum of 2 pounds active per acre as broadcast spray. If soil coverage with plant residue is less than 30% at planting, a maximum of 1.6 pounds active per acre may be applied. On soils not highly erodible, apply 2 pounds active per acre as a broadcast spray.

Chemical Weed Control in Lawns and Turf

F. H. Yelverton and T. W. Gannon, Crop and Soil Sciences Department

Note: A mode of action code has been added to the Herbicide and Formulation column of this table. Use MOA codes for herbicide resistance management. See Table 7-10B, Herbicide Modes of Action for Hay Crops, Lawns and Turf for details concerning active ingredients, brand names, chemical families and modes of action.

Several of the preemergence herbicides are available on fertilizer carriers for homeowner application.

Table 7-14. Chemical Weed Control in Lawns and Turf

Herbicide and Formulation	Amount of Formulation Per 1,000 sq ft	Amount of Formulation per Acre	Pounds Active Ingredient per Acre	Precautions and Remarks
Preemergence Control, Smooth and Large Crabgrass, Goosegrass, Foxtails, other annual grasses				
benefin, MOA 3 (2.5 G)	2.75 lb	120 lb	3	Safe to apply to all established turfgrass except bentgrass. Do not apply in the spring to lawns seeded the previous fall or to golf course greens. Do not use on newly sprigged turfgrasses.
[benefin + trifluralin], MOA 3 + 3 (0.86 G)	8 lb	349 lb	3	Use on lawns and golf course fairways of bahiagrass, bentgrass, bermudagrass, centipedegrass, fescue, perennial ryegrass, St. Augustinegrass, and zoysiagrass.
bensulide, MOA 8 (4 EC) (8.5 G, 12.5 G)		Varies, several concentrations available	10	May be applied to all established turfgrass and dichondra, residential lawns, and golf course greens and tees. Limit 2 applications per year to greens and tees. Do not use on newly sprigged turfgrasses. Not effective for goosegrass control.
[bensulide + oxadiazon], MOA 8 + 14 (6.56 G)	2.6 lb	116 lb	6 + 1.5	Controls crabgrass and goosegrass. Use on established bermudagrass, zoysiagrass, tall fescue, bentgrass, perennial bluegrass, or perennial ryegrass fairways and tees. Use also on bermudagrass and bentgrass greens.
dithiopyr, MOA 4 (2 EW, 2 L) (40 WP)	0.75 fl oz 0.46 oz	1 qt 20 oz	0.5	May be applied to most all cool-season and warm-season turfgrasses except colonial bentgrass. See label for injury precautions regarding certain varieties. Also controls pre-tillered crabgrass. Split applications recommended in southern and coastal regions of the state (0.25 pound a.i. at 8-week intervals). Timely irrigation or rainfall is critical for activation.
indaziflam, MOA 21 (20 WSP)	0.057 to 0.115 oz	2.5 to 5 oz	0.03125 to 0.0625	Use only on established turf (1 year after seeding) such as bermudagrass, zoysiagrass, centipedegrass, St. Augustinegrass, seashore paspalum, and bahiagrass. Labeled for commercial and residential lawns, golf courses (roughs, tees, fairways), sod farms, athletic fields, parks and cemeteries. Use a minimum of 2.5 ounces per acre for crabgrass, annual bluegrass and broadleaf weed control and a minimum of 3.75 ounces per acre for goosegrass, annual sedge and kyllinga species control. Apply up to 2.5 ounces per acre on centipedegrass and St. Augustinegrass due to tolerance concerns. For all other tolerant turfgrasses, do not exceed 5 ounces per acre in a single application or 7.1 ounces per acre within a calendar year. There is an 8 month overseeding restriction following a 2.5 ounces per acre application. Can sprig 2 months following application, or if sprigged first, wait 4 months before spraying. Can sod 4 months following application, or if sodded first wait 2 months after rooting before spraying.
(0.622F)	0.69 to 0.23 fl oz	3 to 10 fl oz	0.01458 to 0.0486	Use up to 6 fluid ounces per acre on common bermudagrass, centipedegrass and St. Augustinegrass and 10 fluid ounces per acre on hybrid bermudagrass, zoysiagrass and bahiagrass established 16 months in areas such as golf course roughs and fairways, residential and commercial turf, sod farms, athletic fields, parks and cemeteries. 10 fluid ounces per acre needed for annual sedge and kyllinga species control. Don't exceed 18.5 fluid ounces per acre per year. Do not vertical mow 1 month before or after application. Irrigate within 2 days of treatment for maximum benefit. Check label for split or multiple application rates and timings. Delay overseeding 10 months if 4.5 to 6 fluid ounces used and 12 months if 6 to 9 fluid ounces used. For sod production, only apply to bermudagrass, zoysiagrass or bahiagrass. Apply if 80% ribbon coverage and before 4 months prior to harvest. Wait 6 month after treatment if sodding bare ground. Apply to actively growing sod established for 3 months.
(0.0224 G)	2.9 to 4.6 lb	125 to 200 lb	0.028 to 0.045	Use on same warm season turf species established at least 16 months and sites as above. Do not exceed 400 pounds product per year. Allow a 15 feet buffer from cool season turf areas. Do not apply upslope from cool season turf.
metolachlor, MOA 15 (7.62 EC)	0.96 fl oz	2.6 pt	2.48	Apply to established bermudagrass, centipedegrass, St. Augustinegrass, bahiagrass, and zoysiagrass. Can apply up to 4.2 pints per acre per year to same area used for commercial sod production.
napropamide, MOA 15 (50 DF)	1.5 to 2.2 oz	4 to 6 lb	2 to 3	Use in established bahiagrass, bermudagrass, centipedegrass, St. Augustinegrass, and tall fescue.
oryzalin, MOA 3 (4 AS, 4 L)	1.5 fl oz	2 qt	2	Use on established bahiagrass, centipedegrass, tall fescue, St. Augustinegrass, zoysiagrass, and bermudagrass except greens and tees. A total of 3 quart per acre may be used if application is split by applying 1.5 quarts per acre followed by 1.5 quarts per acre 8 to 10 weeks later. Follow label directions. Do not apply in the spring or summer to tall fescue reseeded the previous fall.
oryzalin, MOA 3 (85 WDG)	0.64 to 0.88 oz	1.75 to 2.4 lb	1.4875 to 2.04	Observe same turf tolerances and tall fescue precautions as above. Successful preemergence activity should occur if activated by 0.5 inch of water within 21 days of application. Apply 2.4 pounds per acre as a single application or 1.75 pounds per acre in sequential applications spaced 12 weeks apart.
oxadiazon, MOA 14 (2 G)	2.3 to 4.6 lb	100 to 200 lb	2 to 4	Use in established perennial bluegrass, perennial ryegrass, bentgrass, bermudagrass, tall fescue, zoysiagrass, and St. Augustinegrass. Red fescue is not tolerant. Do not apply to dichondra, centipedegrass, putting greens or tees, or to newly seeded areas. Do not apply to bentgrass mowed at less than 3/8 inch. Do not apply to wet turf. Rainfall or irrigation after application will improve weed control activity. May be applied when sprigging bermudagrass and zoysiagrass. Do not apply to home lawns.

Table 7-14. Chemical Weed Control in Lawns and Turf

Herbicide and Formulation	Amount of Formulation Per 1,000 sq ft	Amount of Formulation per Acre	Pounds Active Ingredient per Acre	Precautions and Remarks
Preemergence Control, Smooth and Large Crabgrass, Goosegrass, Foxtails, other annual grasses (continued)				
oxadiazon, MOA 14 (50 WP)	1.5 to 2.2 oz	4 to 6 lb	2 to 3	Use in dormant, established bermudagrass, St. Augustinegrass, and zoysiagrass in fairways and parks. Should be applied at least 2 to 3 weeks before greenup of turf. May be applied when sprigging bermudagrass and zoysiagrass. Do not use on home lawns.
oxadiazon, MOA 14 (3.17 SC)	1.85 to 2.8 fl oz	2.52 to 3.81 qt	2 to 3	Use in dormant, established bermudagrass, St. Augustinegrass, and zoysiagrass in fairways and parks. May apply 2 lb a.i. per acre when sprigging bermudagrass. Apply at least 2 to 3 weeks before greenup of turf. Do not use on home lawns.
[oxadiazon + prodiamine], MOA 14 + 3 (1.2 G)	4.5 lb	200 lb	2 + 0.4	Use on turf, golf courses (excluding putting greens) of established bermudagrass, zoysiagrass, St. Augustinegrass, ryegrass, centipedegrass, bentgrass, bluegrass, and tall fescue. Contains 38% N. Apply to dry foliage.
pendimethalin, MOA 3 (2 G) (0.86 G) (1.29 G)	1.72 to 3.44 lb 2.67 to 5.34 lb 2.67 lb	75 to 150 lb 116 to 232 lb 116 lb	1.5 to 3 1 to 2 1.5	Use on established bahiagrass, bermudagrass, centipedegrass, fine fescue, Kentucky bluegrass, perennial ryegrass, St. Augustinegrass, tall fescue, and zoysiagrass. Do not use on winter-overseeded grasses. Wait 4 months after treatment to seed or sod. Do not apply to newly seeded turf until after the 4th mowing. Do not apply to newly sprigged turf until 5 months establishment.
pendimethalin, MOA 3 (3.8 CS)	1.15 to 2.3 fl oz	3.1 to 6.3 pt	1.5 to 3	Use on noncropland as well as established nonresidential and residential turf areas mowed at least 4 times consisting of bahiagrass, bermudagrass, buffalograss, centipedegrass, St. Augustinegrass, zoysiagrass, Kentucky bluegrass, perennial ryegrass, bentgrass, established *Poa annua* (0.5 inch height or taller), fine fescue, and tall fescue. Do not use on bentgrass or *Poa annua* greens and tees. If lower rate is applied initially, repeat in 6-8 weeks for extended control. Do not reseed or overseed into treated turfgrass for 3 months, or sprig turfgrass for 5 months following application. Do not exceed 4.2 pints per acre on residential and sod farm turfgrass.
[pendimethalin + dimethenamid], MOA 3 + 15 (1.75 G)	2.3 to 4.6 lb	100 to 200	1.75 to 3.5	Use on residential, commercial, recreational, sod farm and golf course turf, excluding greens. Tolerant turf species include bermudagrass, centipedegrass, St. Augustinegrass, seashore paspalum and zoysiagrass. For extended control, make sequential applications within 5 to 8 weeks not to exceed 400 pounds per acre. Irrigate within 24 hours of application for optimum control. Following application, wait 3 months to overseed, reseed or sprig. If sprigged first, wait 2 months for root establishment to treat. On new sod, mow at least twice before application. On new seedlings, mow at least 4 times before application. Wait 2 weeks after aerification or verticutting before applying.
prodiamine, MOA 3 (65 WG) (4 FL)	0.185 to 0.83 oz 0.23 to 1.1 fl oz	0.5 to 2.3 lb 0.625 to 3 pt	0.325 to 1.5 0.3125 to 1.5	May be used on established bahiagrass, bermudagrass, centipedegrass, St. Augustinegrass, zoysia, tall fescue, creeping red fescue, perennial bluegrass and ryegrass, and creeping bentgrass. Do not apply to greens. May apply when sprigging or plugging bermudagrass, up to 0.8 pound product per acre.
prodiamine, MOA 3 (0.5 G)	1.5 to 6.9 lb	64 to 300 lb	0.32 to 1.5	See precautions for prodiamine 65 WG and 4 FL above except may be used on established turf only. Do not apply more than 150 pounds per acre per application. Do not make more than two applications per calendar year. Wait at least 60 days after initial application before making a second application. Prodiamine is coated on a 32-3-12 dry fertilizer carrier.
siduron, MOA 7 (50 WP)	7.3 oz	20 lb	10	Use only on bluegrass, fescue, perennial ryegrass, and certain bentgrasses (check label). Can be used at the rate of 8 pounds of formulation when seeding bentgrass, bluegrass, fescue, and ryegrass. Can be used in newly sprigged or established zoysia. Do not use on bermudagrass, carpetgrass, centipedegrass.
Preemergence Control, Goosegrass				
dimethenamid, MOA 15 (6 L)	0.48 to 0.73 fl oz	21 to 32 fl oz	1 to 1.5	Use on residential, commercial, recreational, sod farm and golf course turf, excluding greens. Apply 21 ounces to established bentgrass, bluegrass species, fescue species and perennial ryegrass maintained at 0.5 inch cut, but expect yellowing and stand reduction. Apply 32 ounces to bahiagrass, bermudagrass species, centipedegrass, St. Augustinegrass, seashore paspalum and zoysiagrass. For extended control, make sequential applications within 5 to 8 weeks at 32 fluid ounces per acre rate. Irrigate within 24 hours of application for optimum control. Following application, wait 6 weeks to overseed or reseed, wait 2 months to sprig, wait 2 mowings for new sod, and wait 4 mowings for newly seeded turf.
				For use in bermudagrass and seashore paspalum. Use 0.5 to 0.75 ounces per acre plus a methylated seed oil (MSO) at 1.5 pints per acre. Apply only to established Bermuda grass and seashore paspulum.
Preemergence Control, Annual Bluegrass (*Poa annua*)				
[benefin + trifluralin], MOA 3 + 3 (0.86 G)	4 to 8 lb	174 to 349 lb	1.5 to 3	Apply during late summer before *Poa annua* germinates. Do not apply to turf areas that are to be overseeded.
bensulide, MOA 8 (4 EC) (8.5 G, 12.5 G)		several concentrations available	12.5	See section on preemergence control of crabgrass and goosegrass or product labels for turfgrass tolerance, precautions and remarks for the listed preemergence annual bluegrass herbicides.
dithiopyr, MOA 4 (2 EW, 2 L) (40 WP)	0.75 fl oz 0.46 oz	1 qt 20 oz	0.5	Timely irrigation or rainfall is critical for activation.
indaziflam, MOA 21 (20 WSP) (0.622 F) (0.0224 G)	0.057 to 0.115 oz 0.138 to 0.23 fl oz 2.9 to 4.6 lb	2.5 to 5 oz 6 to 10 fl oz 125 to 200 lb	0.031 to 0.063 0.029 to 0.049 0.028 to 0.045	
metolachlor, MOA 15 (7.62 EC)	0.48 to 0.96 fl oz	1.3 to 2.6 pt	1.24 to 2.48	

Table 7-14. Chemical Weed Control in Lawns and Turf

Herbicide and Formulation	Amount of Formulation Per 1,000 sq ft	Amount of Formulation per Acre	Pounds Active Ingredient per Acre	Precautions and Remarks
Preemergence Control, Annual Bluegrass (*Poa annua*) (continued)				
napropamide, MOA 15 (50 DF)	1.5 to 2.25 oz	4 to 6 lb	2 to 3	
oryzalin, MOA 3 (4 AS)	1.1 fl oz	1.5 qt	1.5	
(85 WDG)	0.64 to 0.88 oz	1.75 to 2.4 lb	1.4875 to 2.04	Apply full rate unless potentially thin turfgrass cover is a problem caused by dense poa infestation.
oxadiazon, MOA 14 (2 G)	2.3 to 4.6 lb	100 to 200 lb	2 to 4	
pendimethalin, MOA 3 (2 G) (0.86 G) (1.29 G) (3.8 CS)	1.72 to 3.44 lb 2.67 to 5.34 lb 2.67 lb 1.15 to 1.55 fl oz	75 to 150 lb 116 to 232 lb 116 lb 3.1 to 4.2 pt	1.5 to 3 1 to 2 1.5 1.5 to 2	
(pendimethalin + dimethenamid), MOA 3 + 15 (1.75 G)	2.3 to 4.6 lb	100 to 200	1.75 to 3.5	
prodiamine, MOA 3 (65 WG) (4 FL)	0.185 to 0.83 oz 0.23 to 1.1 fl oz	0.5 to 2.3 lb 0.625 to 3 pt	0.325 to 1.5 0.3125 to 1.5	
pronamide, MOA 3 (3.3 SC)	0.46 to 1.29 fl oz	1.25 to 3.5 pt	0.5 to 1.5	Not for home use. Can be applied from Sept. 15 to Feb. 1 for preemergence or postemergence annual bluegrass control in bermudagrass, zoysiagrass, centipedegrass, and St. Augustinegrass grown for sod, nonresidential or industrial sites, golf course turf, and stadium or professional athletic fields. 1.25 to 2.5 pints per acre provides preemergence to pre tiller stage control. 2 to 2.5 pints per acre provides postemergence control from early tiller to early seedhead stage. 2.5 to 3.5 pints per acre for postemergence control at seedhead stage. Henbit and chickweed species controlled at preemergence timings. Can be used for removal of overseeded grasses; do not overseed if it is desired to maintain a stand. Do not overseed treated area within 90 days of treatment. Injury symptoms from postemergence applications can take up to 5 weeks to develop.
Preemergence Control, Annual Bluegrass in Overseeded Bermudagrass				
benefin, MOA 3 (2.5 G)	2.75 lb	120 lb	3	Apply in late summer before *Poa annua* germinates. Perennial ryegrass can be overseeded 6 weeks after benefin is applied.
dithiopyr, MOA 4 (2 EW, 2 L) (40 WP)	0.75 fl oz 0.46 oz	1 qt 20 oz	0.5	Apply in late summer before *Poa annua* germinates. Perennial ryegrass can be overseeded 6 to 8 weeks after application. Apply only on well-established bermudagrass. Do not reapply in fall or winter after overseeding unless injury can be tolerated.
prodiamine, MOA 3 (65 WG)	0.213 to 0.367 oz	0.58 to 1 lb	0.37 to 0.65	Use on golf courses (excluding putting greens) when overseeding with perennial ryegrass at a minimum seeding rate of 350 pounds per acre. Apply 8 to 10 weeks before overseeding and expect 70% or greater control. For best potential control, use higher rate and shorter time interval before overseeding. However, this could increase ryegrass seedling mortality or temporarily reduce root growth.
Preemergence and Postemergence Control, Annual Bluegrass				
ethofumesate, MOA 8 (1.5 EC)	2 fl oz	2.67 qt	1	For control of annual bluegrass in dormant bermudagrass overseeded with perennial ryegrass or in established perennial ryegrass turf. Rates are per application. The first application should be 30 to 45 days after overseeding with perennial ryegrass. The second application should be 21 to 28 days later. Do not apply ethofumesate to overseeded bermudagrass after Jan. 1 in N.C.
ethofumesate, MOA 8 (4 SC)	0.55 to 1.47 fl oz	1.5 to 4 pt	0.75 to 2	Must be professionally applied to residential and nonresidential turf including golf courses and sod farms. May be applied to established perennial ryegrass, Kentucky bluegrass, creeping bentgrass, tall fescue, St. Augustinegrass, and dormant bermudagrass. Do not apply to putting greens. Delay application at least 8 weeks after a pgr application. Fall annual bluegrass control best during period of maximum germination. Spring applications most effective following fall applications. For overseeded bermudagrass, apply 1 to 2 weeks after perennial ryegrass emergence and repeat at 21- to 28-day intervals. Do not apply to bermudagrass 4 weeks prior to breaking winter dormancy.
Postemergence Control and Seedhead Suppression, Annual Bluegrass in Overseeded Bermudagrass Fairways, Tees				
bispyribac-sodium, MOA 2 (17.6 SG)	0.046 to 0.138 oz	2 to 6 oz	0.021875 to 0.065625	Do not apply to putting greens, ryegrass mowed to less than 0.375 inch, or non-overseeded bermudagrass. Apply between Feb. 1 and March 15. Make first application when annual bluegrass begins flowering. If actively flowering, use the low rate and re-treat in 28 to 35 days. If not actively flowering, use the low rate and retreat in 14 to 21 days with the low rate. Do not apply if air temperature is less than 50 degrees F within 3 days after application. Check label for further special instructions.
amicarbazone, MOA 5 (70 WG)	0.023 to 0.23 oz	1 to 10 oz	0.044 to 0.44	Also tolerant to 6 month established turfgrasses such as bahiagrass, centipedegrass, seashore paspalum, St. Augustinegrass, zoysiagrass, bentgrass, Kentucky bluegrass, perennial ryegrass, fine and tall fescue. Labeled for use on golf course, sod farm, residential, commercial, athletic field and roadside turf. Bentgrass tees: 1 ounce/acre at 7 day intervals for 4 applications. Bentgrass roughs and fairways: 2 to 3 ounces/acre for 14 to 21 day intervals for 2 applications. Cool season turf: 2 to 4 ounces/acre for 14 to 21 day intervals for 2 applications. Warm season turf: 3 to 10 ounces/acre for 14 to 21 day intervals for 2 applications not to exceed 10 ounces/acre per year. Allow 4 weeks before cutting or lifting sod. Allow 1 week before overseeding winter grasses.

Table 7-14. Chemical Weed Control in Lawns and Turf

Herbicide and Formulation	Amount of Formulation Per 1,000 sq ft	Amount of Formulation per Acre	Pounds Active Ingredient per Acre	Precautions and Remarks
Postemergence Control, Annual Bluegrass, Overseeded Perennial Ryegrass, Tall Fescue, *Poa trivialis*				
flazasulfuron, MOA 2 (25 DG)	0.011 to 0.069 oz	0.5 to 3 oz	0.0078 to 0.0469	For use on well-established bermudagrass, zoysiagrass, centipedegrass, and seashore paspalum grown turf including golf courses (including fairways, roughs, greens (bermudagrass and seashore paspalum only), tees, collars and approaches), industrial parks, tank-sod- and seed farms, cemeteries, athletic field and commercial lawns. Residential turf applications are limited to spot applications. Apply a maximum of 1.5 ounces per acre on fully green centipedegrass and seashore paspalum. 3 ounces per acre needed for annual bluegrass control and best if applied in spring. 0.5 to 1.5 ounces per acre will control perennial and Italian ryegrass. For clumpy ryegrass, use 1.5 to 3 ounces per acre. 1.5 ounces per acre needed for tall fescue control. 2.25 to 3 ounces per acre needed for *poa trivialis* control. Include a nonionic surfactant at 0.25% by volume.
foramsulfuron, MOA 2 (0.19 SC)	0.2 to 0.6 fl oz	8.8 to 26.2 fl oz	0.013 to 0.039	For use on bermudagrass and zoysiagrass grown on home lawns, golf courses and sod farms. Do not use on warm season turfgrass collars surrounding bentgrass greens. May be applied up to 1 week prior to overseeding. Do not apply within 2 weeks of bermudagrass sprigging. Apply in 25 to 60 gallons water per acre. Rainfast after 2 hours. Surfactant not required.
rimsulfuron, MOA 2 (25 DF)	0.011 to 0.092 oz	0.5 to 4 oz	0.0078 to 0.0625	May be applied to bermudagrass, zoysiagrass and centipedegrass on professionally managed sports facilities at professional and collegiate levels, golf courses, sod farms, roadsides, industrial and commercial lawns. For annual bluegrass control, apply November through December and again February through March if needed at 2 ounces per acre. May be applied 10 to 14 days prior to overseeding. For overseeded removal, apply 2 ounces per acre 3 to 4 weeks before desired removal date, and repeat 3 weeks later if needed. For weed control along roadsides, apply 4 ounces per acre if single application only. A nonionic surfactant at 0.25% by volume or an oil adjuvant such as crop oil concentrate and modified seed oil at 1% by volume are required. Do not apply to cool-season turfgrasses, residential lawns or newly sprigged/sodded bermudagrass.
[metsulfuron + rimsulfuron], MOA 2 + 2 (37 WG)	0.0344352 oz	1.5 oz	0.0346875	Use on well-established bermudagrass and zoysiagrass grown on nonresidential turf including golf courses, sod farms, industrial and commercial lawns, and professionally managed college and professional sports fields. Overseeding can occur 2 months after application. Include a nonionic surfactant at 0.25% by volume.
sulfosulfuron, MOA 2 (75 DG)	0.017 to 0.046 oz	0.75 to 2 oz	0.035 to 0.09375	May be applied to certain ornamental native grasses and also bermudagrass species, zoysiagrass, centipedegrass, St. Augustinegrass, and kikuyugrass grown on sod farms, golf courses (excluding greens), commercial and residential turf that is highly managed, and other noncrop areas. Use 1.5 to 2 ounces per acre for fall annual bluegrass control 7 to 10 days before overseeding. Use 0.75 to 1.25 ounces per acre for fall or winter control in nonoverseeded bermudagrass, and reapply if needed but not before 21 days after initial application. For tall fescue control, two applications may be required at 4- to 10-week intervals. Perennial ryegrass control not as complete as with foramsulfuron, rimsulfuron, or trifloxysulfuron. Use a nonionic surfactant at 0.25% by volume. Do not exceed 2.66 ounces per acre per year.
trifloxysulfuron, MOA 2 (75 WG)	0.0023 to 0.0129 oz	0.1 to 0.56 oz	0.0047 to 0.0263	May be applied to residential bermudagrass and zoysiagrass and also on golf courses, sod farms, and other nonresidential turf areas. A nonionic surfactant at 0.25 to 0.5% by volume is recommended. Temporary discoloration may occur if used with MSO or COC. May be applied 3 weeks prior to overseeding. Use rates of 0.1 to 0.3 ounces per acre to remove overseeded perennial ryegrass and *Poa trivialis* to aid bermudagrass spring transition. Labeled turf species can be seeded or sprigged into treated areas 4 weeks after application.
Preemergence and Postemergence Control, Annual Bluegrass and certain winter annual broadleaf weeds				
atrazine, MOA 5 (4 L) (90 DF, 90 WG)	0.75 to 1.5 fl oz 0.025 to 0.05 lb	1 to 2 qt 1.1 to 2.2 lb	1 to 2	Use on centipedegrass, St. Augustinegrass, and dormant bermudagrass. Apply Nov. 15 to Dec. 31. Follow label directions.
simazine, MOA 5 (90 WDG, 90 DF) (4 L)	0.4 to 0.8 oz 0.75 to 1.5 fl oz	1.1 to 2.2 lb 1 to 2 qt	1 to 2	Use on bermudagrass, centipedegrass, St. Augustinegrass, and zoysiagrass. See label for instructions on newly sprigged turfgrass or on hybrid bermudagrass. Apply Nov. 15 to Dec. 15. Follow label directions.
Preemergence Control, Certain Broadleaf Weeds				
isoxaben, MOA 21 (75 DF, 75 WG)	0.25 to 0.5 oz	0.66 to 1.33 lb	0.5 to 1	All established turfgrasses are tolerant. However, do not apply to putting greens or turfgrass grown for seed. Check label for specific weeds controlled.
pendimethalin, MOA 3 (3.8 CS)	1.15 to 1.55 fl oz	3.1 to 4.2 pt	1.5 to 2	See section on preemergence control of crabgrass or product label for turfgrass tolerance. Provides preemergence control of summer broadleaf weeds, such as prostrate spurge, prostrate knotweed, and purslane species, as well as winter broadleaf weeds, such as yellow woodsorrel, hop clover, cudweed species, common chickweed, lawn burweed, henbit, and corn speedwell when applied before expected germination.
Preemergence and Postemergence Control Crabgrass, Goosegrass, Other Annual Grasses, Broadleaf Weeds, Sedges				
mesotrione, MOA 27 (4 SC)	0.092 to 0.183 fl oz	4 to 8 fl oz	0.125 to 0.25	Use on residential turf, golf courses (not greens) and sod farms for pre- and postemergence weed control. Tolerant turfgrasses include St. Augustinegrass, centipedegrass, tall fescue, fine fescue, Kentucky bluegrass, and perennial ryegrass. Add a nonionic surfactant and repeat application after 2 to 3 weeks for improved postemergence control. Tank mix with prodiamine 65 WG for extended preemergence grassy weed control. Can be applied at seeding to all tolerant grasses except fine fescue. After turf germination, wait 4 weeks or until turf has been mowed twice before making a postemergence application. Also controls henbit, chickweed, dandelion, white clover, Florida betony, Florida pusley, ground ivy, oxalis, wild violet, creeping bentgrass, and yellow nutsedge.

Table 7-14. Chemical Weed Control in Lawns and Turf

Herbicide and Formulation	Amount of Formulation Per 1,000 sq ft	Amount of Formulation per Acre	Pounds Active Ingredient per Acre	Precautions and Remarks
Preemergence and Postemergence Control Crabgrass, Goosegrass, Other Annual Grasses, Broadleaf Weeds, Sedges (continued)				
[sulfentrazone + prodiamine], MOA 14 + 3 (4 SC)	0.184 to 0.826 fl oz	0.5 to 2.25 pt	0.25 to 1.125	For use in residential and institutional lawns, athletic fields, sod farms, golf course fairways and roughs, utility rights-of-way, roadsides, railways, and industrial areas. Apply to turf following a second mowing if a good root system has been established. Apply up to 12 fluid ounces per acre to bentgrass at 0.5 inch or higher, fine fescue, and perennial ryegrass. Apply 18 to 24 fluid ounces per acre to perennial bluegrass, tall fescue, and all warm season grasses except St. Augustinegrass (do not apply) and bermudagrass (apply 18 to 36 fluid ounces per acre). For sod production, apply 6 months after establishment, and do not harvest within 3 months. Do not apply with adjuvants or surfactants. [Sulfentrazone + prodiamine should not be applied to cool-season turf with N-containing fertilizers unless some short-term discoloration is tolerable.
Postemergence Control, Crabgrass, Goosegrass				
fenoxaprop, MOA 1 (0.57 EC)	0.3 to 0.9 fl oz	0.8 to 2.4 pt	0.057 to 0.174	Use only on perennial ryegrass, fine fescue, tall fescue, Kentucky bluegrass, and zoysiagrass. Reduced vigor or discoloration can occur. Rate depends upon leaf number or tillers of grass weeds and turf tolerance. Check label. A second application may be applied after 14 days.
	0.08 fl oz	3.5 fl oz	0.016	Apply only to established Penncross bentgrass maintained at a minimum cutting height of at least 0.25 inch. Bentgrass should be established for one growing season. Do not apply to greens. Applications should be made at a minimum of 21-day intervals, beginning in the spring when grassy weeds first emerge and are not larger than two-leaf. Repeat applications throughout the summer as new infestations of one- to two-leaf grassy weeds occur. See label for restrictions.
metribuzin, MOA 5 (75 DF)	0.12 to 0.24 oz	0.33 to 0.67 lb	0.25 to 0.5	Recommended for application by commercial applicators only on established bermudagrass turf (such as parks, athletic fields, golf course fairways, cemeteries, and sod farms) that has a mowing height of 0.5 inch or greater. Apply when turf is vigorously growing and not under stress. Repeat if necessary, in 7 to 10 days. Do not make more than two applications per season. Do not apply to greens, tees, or aprons.
sethoxydim, MOA 1 (1 EC)	0.8 to 1.38 fl oz	2.25 to 3.75 pt	0.28 to 0.47	Use in seedling and established centipedegrass and fine fescues. Apply 2.25 pint to grasses up to 6 inches and 3.75 pints to grasses up to 12 inches if turf is tolerant. Does not control yellow and purple nutsedge, annual bluegrass or broadleaf weeds. Apply no sooner than 3 weeks after spring greenup of centipedegrass. Apply before crabgrass becomes extensively tillered. Delay all treatments until newly planted centipedegrass has 3 inches of new stolon growth. Do not mow within 7 days before or after application. Two applications 3 weeks apart will suppress bahiagrass. Additives or adjuvants not required.
Postemergence Control, Smooth and Large Crabgrass, Barnyardgrass, White and Hop Clover, Common Dandelion, Dollarweed, Foxtails				
quinclorac, MOA (27 + 4) (75 DF) (1.5 SL)	0.367 oz / 1.45 fl oz	1 lb / 2 qt	0.75	For use in residential and nonresidential turf that is established or newly seeded, overseeded, or sprigged. Refer to label for specific varieties. Apply to common and hybrid bermudagrass, Kentucky bluegrass, annual bluegrass, buffalograss, tall fescue, annual and perennial ryegrass, creeping bentgrass, and zoysiagrass. Can also be applied to fine fescue but must be in a blend. Some discoloration of hybrid bermudagrass, creeping bentgrass or fine fescue may occur. Do not apply to bahiagrass, centipedegrass, St. Augustinegrass, or dichondra. Do not use on golf course greens or collars. The addition of methylated seed oil (1.5 pints per acre or 0.55 ounces per 1,000 square feet) or a crop oil concentrate (2 pint per acre or 0.73 ounces per 1,000 square feet) is required for control. Application to weeds under stress will result in poor control. Irrigation 24 hours prior to application is recommended if drought conditions exist. Some ornamental plants are sensitive to quinclorac. See label for further precautions.
Postemergence Control, Smooth and Large Crabgrass, Barnyardgrass, Foxtails, and many broadleaf weeds				
[quinclorac + sulfentrazone + 2,4-D amine + dicamba], MOA (27 + 4) + 14 + 4 + 4 (1.79 L)	1.8 to 3 fl oz	5 to 8 pt	1.12 to 1.79	For use in fully dormant bermudagrass as well as actively growing bermudagrass after spring greenup but use only 5 to 7 pints per acre. Also labeled in fully dormant zoysiagrass s well as cool-season turf including annual bluegrass and ryegrass, perennial bluegrass and ryegrass, and fescue species. Do not apply to bahiagrass, bentgrass (creeping, Seaside, Colonial), centipedegrass, St. Augustinegrass, carpetgrass, and golf course greens, tees, and collars. May be applied to home lawns. Apply to seedling grasses after second or third mowing, or 28 days after emergence. Wait 3 to 4 weeks after sodding, sprigging, or plugging operations to apply. Wait 4 weeks after application to seed.
[quinclorac + mecoprop + dicamba], MOA (27 + 4) + 4 + 4 (2.45 SL)	0.5 to 1.45 fl oz	0.68 to 2 qt	0.4165 to 1.225	For use in warm- and cool-season residential and non-residential turf, including but not limited to commercial property, parks, roadsides, schools, athletic fields, cemeteries, and golf courses. May be applied to species of bermudagrass, bluegrass, fescue, and ryegrass as well as creeping bentgrass, seashore paspalum, and zoysiagrass. Use with methylated seed oil at 1.5 pints per acre. Allow 28 days of seedling or sprig growth before application. If treating first, allow 28 days before seeding or sprigging. Do not apply to golf course collars or greens or to turf grown for sod. Use low rate in 2 split applications when treating creeping bentgrass.
[carfentrazone + quinclorac], MOA 14 + (27 + 4) (75 WG)	0.184 to 0.413 oz	8 to 18 oz	0.35 to 0.79	Can use up to 12 ounces per acre 7 days after emergence from seed or sod installment on bluegrass and fescue species and perennial ryegrass; 18 ounces per acre can be used 7 days after seed, sod or sprig operations on bermudagrass species, centipedegrass and seashore paspalum. Wait 14 days after emergence for zoysiagrass. May apply to residential, commercial, and institutional lawns, athletic fields, sod farms, and golf course fairways and roughs. Adjuvants not required but may help on mature weeds.

Table 7-14. Chemical Weed Control in Lawns and Turf

Herbicide and Formulation	Amount of Formulation Per 1,000 sq ft	Amount of Formulation per Acre	Pounds Active Ingredient per Acre	Precautions and Remarks
Postemergence Control, Smooth and Large Crabgrass, Barnyardgrass, Foxtails, and many broadleaf weeds (continued)				
[sulfentrazone + quinclorac], MOA 14 + (27 + 4) (75 WG)	0.367 to 0.735 oz	1 to 2 lb	0.75 to 1.5	Use up to 21 ounces per acre on well-established tall fescue, Kentucky bluegrass and perennial ryegrass; up to 32 ounces per acre on well-established bermudagrass, centipedegrass, zoysiagrass and seashore paspalum. May be applied to residential, commercial, and institutional lawns, athletic fields, sod farms, and golf course fairways and roughs. After treatment, wait at least 1 month before reseeding, overseeding (use slit seeder for best results), or sprigging. Wait at least 3 months for sod establishment and do not spray within 3 months of harvest. Controls goosegrass in the 1 to 4 leaf stage. Yellow nutsedge and kyllinga species are also controlled. Do not apply with a spray adjuvant.
[fenoxaprop + fluroxypyr + dicamba], MOA 1 + 4 + 4 (0.75 EC)	1.3 to 1.5 fl oz	3.5 to 4 pt	0.33 to 0.375	Tolerant turfgrass species include zoysiagrass, Kentucky bluegrass, perennial ryegrass, fine and tall fescue. May be applied to golf courses excluding greens and tees, athletic fields, commercial and residential turf. Sod farm use is not permitted. Best grass weed control will be achieved when treated from 1 leaf to 4 tiller stage. Do not apply more than 15 pints per acre per year. Do not reapply within 14 days of an application. Surfactant not required. Spot treat using 0.6 to 1 fluid ounces per 1 gallon water.
Postemergence Control, Large Crabgrass, Carpetgrass, Bull Paspalum, Bahiagrass, Foxtails, and many broadleaf weeds, including Chamberbitter, Corn Speedwell, Dichondra, Dollarweed, Dovewood, Florida Betony, Florida Pusley, Lespedeza, Oxalis, Spurge, Virginia Buttonweed, Kyllinga				
[thiencarbazone-methyl + iodosulfuron-methyl + dicamba], MOA 14 + 2 + 4 (68 WG)	0.057 to 0.113 oz	2.5 to 4.9 oz	0.106 to 0.208	For use by licensed applicators in residential and commercial lawns, golf courses (excluding greens), sports fields, parks, recreational areas, roadsides, school grounds, and sod farms. Provides up to 60 days residual control. Use on bermudagrass, zoysiagrass, centipedegrass, and St Augustinegrass. Apply maximum 7.4 ounces per acre per season. Safe to use at high temperatures. Ryegrass can be overseeded 2 weeks after application. Apply 30 days prior to seeding bermudagrass or zoysiagrass. Wait 2 weeks after bermudagrass seedling emergence or sprigging operation before applying. For zoysiagrass, wait 3 weeks after seedling emergence before applying. A nonionic surfactant or methylated seed oil at 0.25% v/v is required for optimum control.
Postemergence Control or Suppression of Summer Weeds Such as Crabgrass Species, Goosegrass, Dallisgrass, Virginia Buttonweed, Dovewood, Florida Pusley, Nutsedge and Kyllinga Species; Winter Weeds Such as poa annua, poa trivialis, Tall Fescue, Henbit, Corn Speedwell, and Species of Ryegrass, Chickweed, and Clover				
[thiencarbazone-methyl + foramsulfuron + halosulfuron], MOA 14 + 2 + 2 (60.5 WG)	0.0735 oz	3.2 oz	0.121	Apply to well-established residential and commercial bermudagrass and zoysiagrass (Emerald, Meyer, Zeon) lawns, golf courses (excluding greens), athletic fields, sod farms, roadsides, parks, cemeteries and recreational areas. Do not exceed 3.2 ounces per acre per application or 6.4 ounces per acre yearly. Use 0.25 to 0.5% by volume nonionic surfactant or 0.5 to 1% by volume methylated seed oil. After application, wait 12 weeks to overseed ryegrass or bermudagrass. Wait 1 month after bermudagrass seedling emergence and 2 weeks after sprigging or sodding bermudagrass before treating. Temporary stunting and yellowing may last up to 2 weeks, but turf will recover. Crabgrass and goosegrass are controlled up to 2 tiller stage.
Postemergence Control, Goosegrass				
diclofop-methyl, MOA 1 (3 EC)	0.75 to 1 fl oz	32 to 43 fl oz	0.75 to 1	Apply in established bermudagrass. Rate depends on number of goosegrass leaves from one to four leaves. Check label for specific rates.
foramsulfuron, MOA 2 (0.19 SC)	0.39 fl oz	17 fl oz	0.025	For use on bermudagrass and zoysiagrass grown on home lawns, golf courses and sod farms. See precautions listed under annual bluegrass section. For goosegrass control, apply 17 fl ounces per acre on plants up to 2 tillers followed by 17 fluid ounces per acre 2 weeks later.
sulfentrazone, MOA 14 (4 SC)	0.275 fl oz	0.75 pt	0.375	May be applied to home lawns. For use on creeping bentgrass, tall and fine fescue, perennial ryegrass, Kentucky bluegrass, and all warm-season turf species except St. Augustinegrass. See precautions listed under purple and yellow nutsedge section. For goosegrass control, apply 0.75 pint per acre on plants up to 2 tillers.
				For use in bermudagrass and seashore paspalum. Use 0.5 to 0.75 ounces per acre plus a methylated seed oil (MSO) at 1.5 pints per acre. Apply only to established bermudagrass and seashore paspulum.
Postemergence Control, Bahiagrass, Crabgrass, Dallisgrass, Goosegrass, Nutsedge, Annual Sedges, Sandbur				
MSMA, MOA 17 (6 SL, 6.6 SL)		several concentrations	1.82 to 4.5	MSMA is only registered for golf course, sod farm, and highway right-of-way use. Bermudagrass, bluegrass and zoysiagrass are tolerant. Injury may result on bentgrass, fescue and also St. Augustinegrass grown for commercial sod production only. Do not use on carpetgrass or centipedegrass. MSMA restrictions: For existing golf courses, spot treat (100 square feet per spot) not to exceed 25% of total acreage. For new courses, make 1 broadcast application per year. For sod farms, make 1 to 2 broadcast applications per year and maintain 25 feet buffer around permanent water bodies. For highway rights of way, make 2 broadcast applications and maintain 100 feet buffer around permanent water bodies.
Postemergence Control, Crabgrass, Goosegrass, Sandbur, Dallisgrass				
MSMA, MOA 17 (6 SL, 6.6 SL) + metribuzin, MOA 5 (75 DF)		several concentrations + 0.17 to 0.33 lb	1.5 to 2 + 0.125 to 0.25	See remarks for MSMA and metribuzin. The combination improves goosegrass control. Should be applied to bermudagrass only.
Postemergence Control, Crabgrass, Goosegrass, Sandbur				
asulam, MOA 18 (3.34 SL)	1.8 fl oz	5 pt	2	Use only on St. Augustinegrass and Tifway 419 turf. On golf courses, use only on fairways and roughs.

Table 7-14. Chemical Weed Control in Lawns and Turf

Herbicide and Formulation	Amount of Formulation Per 1,000 sq ft	Amount of Formulation per Acre	Pounds Active Ingredient per Acre	Precautions and Remarks
Postemergence Control, Crabgrass and Foxtail Species, Goosegrass, Broadleaf Signalgrass, Japanese Stiltgrass				
Postemergence Suppression, Creeping Bentgrass, Common Bermudagrass, Dallisgrass, Nimblewill				
topramezone, MOA 27 (2.8 L)	0.023 to 0.034 fl oz	1 to 1.5 fl oz	0.021875 to 0.0328125	Labeled for broadcast treatment use in residential and athletic field turf, as well as in nonresidential turf sites including sod farms, golf courses (excluding greens and collars), parks, roadsides, cemeteries, and commercial properties. Tolerant turf species include Kentucky bluegrass, tall and fine fescue, perennial ryegrass, and centipedegrass at seeding and then anytime beyond 28 days after seeding. Add crop oil concentrate or methylated seed oil for enhanced control at 0.5 to 1% by volume. Don't apply greater than 2 fluid ounces per acre per application or 4 fluid ounces per acre per year. Bleaching intensity of susceptible weeds reduced, and broadleaf weed spectrum increased if tankmixed with quinclorac, [quinclorac + mecoprop + dicamba] or triclopyr. For suppression of above-listed weeds, add triclopyr at 1 pound ae per acre and make either 2 or 3 applications at 3 to 4 week intervals depending on topramezone rate. Creeping bentgrass is marginally tolerant to topramezone at 0.25 fluid ounces per acre. Test on a small area before large-scale use. Sequential applications may be required to achieve desired level of weed control.
Postemergence Control, Yellow Nutsedge, Annual Sedge				
bentazon, MOA 6 (4 SL)	0.75 to 1.5 fl oz	1 to 2 qt	1 to 2	For control of yellow nutsedge in established bluegrass, fescues, bentgrass, ryegrass, bermudagrass, bahiagrass, St. Augustinegrass, centipedegrass, and zoysiagrass. Apply to yellow nutsedge when actively growing under good soil moisture conditions. Additional applications may be made at intervals of 10 to 14 days until nutsedge is controlled.
Postemergence Control, Purple and Yellow Nutsedge, Kyllinga Species				
flazasulfuron, MOA 2 (25 DG)	0.034 to 0.069 oz	1.5 to 3 oz	0.023 to 0.0469	For use on well-established bermudagrass, zoysiagrass, centipedegrass and seashore paspalum grown on nonresidential turf including golf course fairways, roughs and tees, and industrial parks, tank-sod- and seed farms, cemeteries, athletic field and commercial lawns. Apply a maximum of 1.5 ounces per acre on fully green centipedegrass and seashore paspalum. 3 ounces per acre needed for perennial nutsedge and some annual sedge species control. Repeat applications in 2 to 6 weeks when nutsedge or sedge growth is evident. 1.5 to 2.25 ounces per acre will control kyllinga species. Maintain a 25 feet nontreated border beside susceptible turf species. Can overseed in 2 weeks if applied up to 1.5 ounces per acre. Wait 4 weeks if applied more than 1.5 ounces per acre. Include a nonionic surfactant at 0.25% by volume.
imazaquin, MOA 2 (70 DG)	0.128 to 0.256 oz	0.357 to 0.714 lb	0.25 to 0.5	Use on bermudagrass, centipedegrass, St. Augustinegrass, and zoysiagrass. Do not apply during spring greenup. Temporary yellowing may occur. Add a nonionic surfactant at 2 pt per 100 gal of spray solution. Addition of MSMA at 1.5 lb active per acre will improve sedge control in MSMA tolerant turfgrasses.
imazosulfuron, MOA 2 (75 WG)	0.184 to 0.322 oz	8 to 14 oz	0.38 to 0.66	May be applied to established (two mowings) residential and commercial bermudagrass, zoysiagrass, centipedegrass, St. Augustinegrass, creeping bentgrass, Kentucky bluegrass, perennial ryegrass, tall fescue, and fine fescue. Do not apply to putting greens. Reapply 3 weeks after initial application when using the 8 ounces per acre rate. Reapply as needed 3 weeks after initial application when using rates above 8 ounces per acre. Wait 4 weeks to seed or sod after application. Use an 80% active nonionic surfactant at 0.25% by volume. For spot treatment, add 0.25 to 0.33 oz in 1 to 2 gallons of water per 1000 square feet. Add 2 teaspoons nonionic surfactant per gallon.
halosulfuron, MOA 2 (75 WDG)	0.9 g	0.67 to 1.33 oz	0.031 to 0.062	May be applied to established residential and commercial bermudagrass, bahiagrass, zoysiagrass, centipedegrass, St. Augustinegrass, creeping bentgrass, Kentucky bluegrass, perennial ryegrass, tall fescue, and fine fescue. Apply broadcast when sedges have reached the 3- to 8-leaf stage. Use lower rate for light infestations and higher rate for heavy infestations. A second treatment will usually be required 6 to 10 weeks after the initial treatment. Use an 80% active nonionic surfactant at 2 quarts per 100 gallons of spray solution (0.5% by volume). Do not exceed 1 to 2 pints of surfactant per acre. Do not apply to putting greens. Halosulfuron only suppresses green kyllinga.
MSMA, MOA 17 (6 SL, 6.6 SL)		several concentrations	2 to 3	See remarks for MSMA above. Will require at least 2 applications 7 to 10 days apart.
sulfosulfuron, MOA 2 (75 DG)	0.017 to 0.029 oz	0.75 to 1.25 oz	0.035 to 0.059	May be applied to certain ornamental native grasses and also bermudagrass species, zoysiagrass, centipedegrass, St. Augustinegrass, and kikuyugrass grown on sod farms, golf courses (excluding greens), commercial and residential turf that is highly managed, and other noncrop areas. Use 0.75 to 1.25 ounces per acre, and repeat in 4 to 10 weeks if needed. Use a nonionic surfactant at 0.25% by volume.
trifloxysulfuron, MOA 2 (75 WG)	0.0023 to 0.0129 oz	0.1 to 0.56 oz	0.0047 to 0.0263	May be applied to residential bermudagrass and zoysiagrass and on golf courses, sod farms, and other nonresidential turf areas. A nonionic surfactant at 0.25 to 0.5% by volume is recommended. Temporary discoloration may occur if used with MSO or COC. Use rates of 0.33 to 0.56 ounces per acre for sedge and kyllinga species control. Labeled turf species can be seeded or sprigged into treated areas 4 weeks after application. Repeat application may be needed in 4 to 6 weeks.
Postemergence Control, Purple and Yellow Nutsedge, Kyllinga Species, and various broadleaf weeds				
sulfentrazone, MOA 14 (4 SC)	0.092 to 0.275 fl oz	0.25 to 0.75 pt	0.125 to 0.375	May be applied to home lawns. For use on creeping bentgrass, tall and fine fescue, perennial ryegrass, Kentucky bluegrass, and all warm-season turf species except St. Augustinegrass. Wait 3 months to seed, overseed, or sprig unless overseeding bermudagrass with perennial ryegrass, which only requires a 4- to 6-week waiting period after application. Apply to seedling grasses after second mowing and to new sod 6 months after establishment.

Table 7-14. Chemical Weed Control in Lawns and Turf

Herbicide and Formulation	Amount of Formulation Per 1,000 sq ft	Amount of Formulation per Acre	Pounds Active Ingredient per Acre	Precautions and Remarks
Postemergence Control, Purple and Yellow Nutsedge, Kyllinga Species, and various broadleaf weeds (continued)				
[sulfentrazone + imazethapyr], MOA 14 + 2 (4 SC)	0.22 to 0.33 fl oz	9.5 to 14.4 fl oz	0.29 to 0.45	May be applied to home lawns, athletic fields, sod farms, golf course fairways and roughs, and various non-crop sites. For use on bahiagrass, bermudagrass, centipedegrass, and zoysiagrass. Do not apply to soils classified as sand with less than 1% organic matter. Do not reseed, overseed, or sprig within 1 month of application. Expect slight perennial ryegrass injury if overseeded 2 to 4 weeks after application. Allow 3 month sod establishment before treatment.
[sulfentrazone + metsulfuron], MOA 14 + 2 (66 WG)	0.075 to 0.23 oz	3.25 to 10 oz	0.134 to 0.413	May be applied to established residential, commercial and institutional lawns, athletic fields, sod farms, and golf course fairways and roughs. Use up to 6.5 ounces per acre on Kentucky bluegrass and tall fescue and 10 ounces per acre on bermudagrass, centipedegrass, St. Augustinegrass and zoysiagrass. Do not reseed, overseed, or sprig within 1 month of application. Expect slight perennial ryegrass injury if overseeded 6 to 8 weeks after application. Allow 3 months sod establishment before treatment. No adjuvant needed.
Postemergence Control, Bahiagrass, Crabgrass, Yellow and Purple Nutsedge, Annual Sedge, Kyllinga Species				
imazapic, MOA 2 (2 AS)	0.092 to 0.184 fl oz	4 to 8 fl oz	0.063 to 0.125	For use on unimproved centipedegrass after complete greenup only. Not for use in home lawns. Do not use on other turfgrass species. A repeat application may be needed on tough to control perennial weeds such as bahiagrass. The highest labeled rate may discolor centipedegrass by causing a red color.
Postemergence Control, Dandelion, Carpetweed, Carolina Cranesbill, Curly Dock, Plantain, Dichondra, Shepherds-Purse, Yellow Rocket				
2,4-D amine, MOA 4 (4 SL)	3 to 4 tsp	1.5 to 2 pt	0.75 to 1	Cut rate one-half for bentgrass, carpetgrass, centipedegrass, and St. Augustinegrass. Spray when weeds are young and actively growing. To reduce danger of injury to flowers and ornamentals by spray drift, use low pressure and do not spray on windy days.
Postemergence Control, Common Chickweed, Mouseear Chickweed, Creeping Charlie or Ground Ivy, Dandelion, Lespedeza, Black Medic, Spotted Spurge, Hop or White Clover				
mecoprop, MOA 4 (1.9 L) (1.16 L) (1.74 L)	1 to 1.5 fl oz 1.5 to 2.25 fl oz 0.75 to 1.5 fl oz	2.7 to 4 pt 4 to 6 pt 2 to 4 pt	0.64 to 0.95 0.58 to 0.87 0.43 to 0.87	Observe same precaution as for 2,4-D. May be used on bentgrass, carpetgrass, centipedegrass, St. Augustinegrass, and other turf grasses.
Postemergence Control, Chickweed, White Clover, Dandelion, Curly Dock, Hawkweed, Henbit, Knotweed, Red Sorrel, Knawel, Spurweed, Spotted Spurge, Wild Strawberry, Yarrow				
dicamba, MOA 4 (4 SL)	1 to 2 tsp	0.5 to 1 pt	0.25 to 0.5	Apply as foliar spray to growing weeds. Prevent injury to ornamentals. Avoid rooting zone of shallow-rooted trees and shrubs.
diglycolamine, MOA 4 (4 SL)	1 to 4.5 tsp	0.5 to 2 pt	0.25 to 1	Do not exceed 1 pint per acre on bentgrass, carpetgrass, buffalograss, and St. Augustinegrass. Apply to newly seeded grasses after the second mowing. Do not exceed 0.25 pint per acre on extended sensitive plant roots on sandy soils and 0.5 pint per acre on clay soils.
Postemergence Control, All Weeds Listed Under 2,4-D Amine, MCPP, Dicamba, and Diglycolamine Sections				
[2,4-D amine + MCPP + dicamba], MOA 4 + 4 + 4 (various formulations)	See individual label	See individual label	See individual label	Check individual labels for specific rates, instructions and precautions. Generally, 1) apply to grass seedlings after second mowing; 2) apply to sodded, sprigged, or plugged turf 3 to 4 weeks after operations; and 3) wait 3 to 4 weeks after application to seed. Many products labeled for tall fescue, perennial ryegrass, perennial bluegrass, bermudagrass, and St. Augustinegrass. Some products labeled for bentgrass putting greens, bahiagrass, zoysiagrass, and centipedegrass. Some products labeled for home use when applied by a commercial applicator.
[2,4-D amine + MCPP + dichlorprop], MOA 4+ 4 + 4 (4.11 L) (2.48 L)	0.62 to 1.47 fl oz 0.64 to 1.47 fl oz	1.7 to 4 pt 1.75 to 4 pt	0.873 to 2.06 0.543 to 1.24	
[MCPA + MCPP + dicamba], MOA 4+4+4 (4 L)	0.7 to 1.5 fl oz	2.5 to 4.1 pt	1.25 to 2.05	
Postemergence Control, Curly Dock, Broadleaf Dock, Galinsoga, Nightshade, Clover (Red, Hop, White, Sweet), Goldenrod, Musk Thistle, Speedwells, Common Vetch, Hairy Buttercup, Broadleaf Plantain				
clopyralid, MOA 4 (3 EC)	0.1 to 0.5 fl oz	0.25 to 1.33 pt	0.09 to 0.5	Do not apply to home lawns. May be used on bentgrass, Kentucky bluegrass, creeping, red, chewings, sheep and tall fescue, perennial ryegrass, bermudagrass, bahiagrass, buffalograss, centipedegrass, zoysiagrass, and St. Augustinegrass. Do not apply to putting greens and tees. Should be applied in a minimum of 20 gallons of water per acre. Surfactants are not necessary. Do not apply to exposed roots of certain trees and shrubs (legumes such as acacia, locust, mimosa, redbud, or mesquite) or *Tilia* spp. Do not use treated clippings for mulching and compost during the growing season of application.
All Weeds Listed Under 2,4-D Amine, Clopyralid, Dicamba, and Diglycolamine Sections				
[2,4-D amine + clopyralid + dicamba], MOA 4 + 4 + 4 (3.56 L)	0.55 to 1.1 fl oz	1.5 to 3 pt	0.67 to 1.34	Do not apply to home lawns. Use on perennial bluegrass, ryegrass, and fescue species, bentgrass (excluding greens and tees), bermudagrass, zoysiagrass, and bahiagrass. Do not apply to seedling grasses until well established. Wait 3 to 4 weeks after application to seed.

Table 7-14. Chemical Weed Control in Lawns and Turf

Herbicide and Formulation	Amount of Formulation Per 1,000 sq ft	Amount of Formulation per Acre	Pounds Active Ingredient per Acre	Precautions and Remarks
Postemergence Control, Virginia Buttonweed, Chickweed Species, White Clover, Dandelion, Henbit, Ground Ivy, Prostrate Knotweed, Matchweed, Black Medic, Plantain Species, Common Woodsorrel				
[2,4-D amine + fluroxypyr + dicamba], MOA 4 + 4 + 4 (4 SL)	0.36 to 1.1 fl oz	1 to 3 pt	0.5 to 1.5	Use on perennial bluegrass and ryegrass, tall fescue, creeping bentgrass (excluding greens and tees), bermudagrass species, bahiagrass, zoysiagrass, and St. Augustinegrass in residential, industrial, and institutional lawns, parks, cemeteries, athletic fields, golf courses, and sod farms. Use on St. Augustinegrass sod farms only. Apply 1 to 2 pints per acre on creeping bentgrass and 1.5 to 1.8 pints per acre on warm season turf grown for sod. Apply 2 to 3 pints per acre to all other turf areas. For non-turf areas, rate can be increased to 2 to 5 pints per acre. Application can be made to grass seedlings after second mowing and to newly sodded, sprigged, or plugged grasses 3 to 4 weeks after operations.
[MCPA amine + fluroxypyr ester + dicamba], MOA 4 + 4 + 4 (4.8 SL)	0.73 to 1.1 fl oz	2 to 3 pt	1.2 to 1.8	Same turf tolerances and uses as [2,4-D amine + fluroxypyr + dicamba] in addition to centipedegrass. Only spot treat St. Augustinegrass when temperature exceeds 80 degrees F. Do not apply more than two applications per year totaling 3 pints per acre. For non-turf areas, rate can be increased to 2 to 5 pints per acre. Application can be made to grass seedlings after second mowing and to newly sodded, sprigged, or plugged grasses 3 to 4 weeks after operations. Sod farm rates include 1.25 pints per acre for creeping bentgrass, 2 to 3 pints per acre for all other cool season grasses listed on label and 1.5 to 1.8 pints per acre for all warm season grasses listed on label.
Postemergence Control, Winter and Summer Annual Broadleaf Weeds				
bentazon + atrazine, MOA 6 + 5 Create by tank mixing			0.5 to 0.75 + 0.5 to 0.75	Apply to bermudagrass, centipedegrass, St. Augustinegrass, and zoysiagrass. Check individual labels for weeds controlled and weed size for proper application.
Postemergence Control, Black Medic, White, Hop Clover, Buckhorn Plantain, Common Chickweed, Mouseear Chickweed, Henbit, Spurweed (Lawn Burweed), Broadleaf Plantain, Dandelion, False Dandelion, Lespedeza, Prostrate Spurge, Wild Violet				
[triclopyr + cloypyralid], MOA 4 + 4 (3 SL)	0.37 to 0.74 fl oz	1 to 2 pt	0.28 to 0.56 + 0.09 to 0.19	Do not apply to home lawns. May be used on centipedegrass, bermudagrass, zoysiagrass, tall fescue, creeping red fescue, chewing fescue, Kentucky bluegrass, perennial ryegrass. Repeat treatment may be necessary for prostrate spurge and wild violet. Quali-Pro formulation: maintain 0.5 inch height for warm season turf. Do not apply to bermudagrass sod farms. Wait 3 weeks to reseed. Do not use grass clippings for compost or mulch.
[MCPA ester + triclopyr ester + dicamba], MOA 4 + 4 + 4 (3.6 EC)	0.91 to 1.29 fl oz	2.5 to 3.5 pt	1.125 to 1.575	May be applied to home lawns by a commercial applicator. Not for use on turf grown for resale or other commercial use as sod or seed production. Use on perennial bluegrass, ryegrass, fescue species, bentgrass (excluding greens and tees), bermudagrass, zoysiagrass, and bahiagrass. Do not apply to seedling grasses until well established. Wait 3 to 4 weeks after application to seed.
[MCPA amine + triclopyr amine + dicamba], MOA 4 + 4 + 4 (4.56 L)	0.73 to 1.1 fl oz	2 to 3 pt	1.14 to 1.71	
[MCPA amine + fluroxypyr ester + triclopyr amine], MOA 4+4+4 (3.41 L)	0.37 to 1.47 fl oz	1 to 4 pt	0.42625 to 1.705	Apply by a commercial applicator to residential, industrial, and institutional lawns, sod farms, parks, cemeteries, athletic fields, roadsides, and golf courses excluding greens and tees. May apply to bentgrass, Kentucky bluegrass, perennial ryegrass, fescue species, bahiagrass, bermudagrass, centipedegrass, and zoysiagrass. Do not spray on warm season turf less than 0.5 inch and do not exceed 3 pints per acre. Generally apply 3 to 4 pints per acre except on fairway bentgrass, which can only tolerate 2 pints per acre. Wait 3 to 4 weeks after application to reseed. Check label for spray adjuvant recommendation.
Postemergence Control, Plantain, Chickweed, Dandelion, Purslane, and Thistle Species, Ground Ivy, Lawn Burweed, Henbit, Corn Speedwell, Spotted Spurge				
carfentrazone-ethyl, MOA 14 (1.9 EW)	0.0126 to 0.048 fl oz	0.55 to 2.1 fl oz	0.008 to 0.031	May be applied to bahiagrass, bermudagrass, buffalograss, centipedegrass, St. Augustinegrass, zoysiagrass, Kentucky bluegrass, tall fescue, fine fescue, perennial ryegrass, and bentgrass. To expand the weed spectrum and extend control of the weeds listed here and, on the label, carfentrazone-ethyl can be tank mixed with the entire range of phenoxy products—amines, esters, and other salts—and is also compatible with dicamba, atrazine, glyphosate, glufosinate, clopyralid, triclopyr, and MSMA. When applied alone, add 0.12 to 0.25% nonionic surfactant.
Postemergence Control, White Clover, Dandelion, Ground Ivy, Spurges, Plantains, Chickweeds, Henbit, Lawn Burweed, Woodsorrels, Dollarweed, Poison Ivy, Poison Oak, Corn Speedwell, Wild Strawberry, Wild Violet, Virginia Pepperweed, Shepherd's Purse				
[carfentrazone + 2,4-D ester + MCPP + dicamba], MOA 14 + 4 + 4 + 4 (2.2 EC)	0.75 to 1.8 fl oz	2 to 5 pt	0.55 to 1.375	May be used on annual and perennial bluegrass, annual and perennial ryegrass, tall and fine fescue, creeping and colonial bentgrass, common and hybrid bermudagrass, and zoysiagrass. For use in ornamental turf, golf courses, lawns, sod farms, cemeteries, and parks. Optimum results when applied when temperatures are between 45 and 75 degrees F but may be applied up to 90 degrees F. Lower rates may be used in cooler weather. Rainfast within 3 hr and may reseed after 2 weeks. May apply 3 to 4 wks after sodding, sprigging, or plugging. Also may be used on bahiagrass, buffalograss, St. Augustinegrass, centipedegrass, seashore paspalum, and kikuyugrass. May reseed after 1 week.
[carfentrazone + 2,4-D ester + MCPP + dicamba], MOA 14 + 4 + 4 + 4 (0.81 EC)	0.55 to 2.2 fl oz	1.5 to 6 pt	0.1519 to 0.6075	
[carfentrazone + MCPA ester + MCPP + dicamba], MOA 14 + 4 + 4 + 4 (2.91 EC)	0.75 to 2.2 fl oz	2 to 6 pt	0.7275 to 2.1825	Same precautions and turf uses as [carfentrazone + 2,4-D ester + MCPP + dicamba] 2.2 EC except cannot be applied to creeping and colonial bentgrass.
penoxsulam + sulfentrazone + 2,4-D + dicamba	1.0 to 2.2 fl oz	2.7 to 6 pt	0.271 to 0.602	May be used on established Kentucky bluegrass, annual bluegrass, perennial ryegrass, annual ryegrass, tall fescue, common and hybrid bermudagrass, zoysiagrass, centipedegrass, and seashore paspalum. Do not apply more than 2.7 pints per acre on fescue or perennial ryegrass unless turf injury can be tolerated. Do not apply more than 3.5 pints per acre on St. Augustinegrass unless injury, discoloration, stunting, and thinning can be tolerated.

Table 7-14. Chemical Weed Control in Lawns and Turf

Herbicide and Formulation	Amount of Formulation Per 1,000 sq ft	Amount of Formulation per Acre	Pounds Active Ingredient per Acre	Precautions and Remarks
Postemergence Control, White Clover, Dandelion, Ground Ivy, Spurges, Plantains, Chickweeds, Henbit, Lawn Burweed, Woodsorrels, Dollarweed, Poison Ivy, Poison Oak, Corn Speedwell, Wild Strawberry, Wild Violet, Virginia Pepperweed, Shepherd's Purse (continued)				
[sulfentrazone + 2,4-D amine + MCPP + dicamba], MOA 14 + 4 + 4 + 4 (2.18 SL)	0.92 to 1.84 fl oz	2.5 to 5 pt	0.68 to 1.36	Apply 2.5 to 3.25 pints per acre on warm season turf including bermudagrass species, zoysiagrass, bahiagrass, and buffalograss. Apply 3.25 to 4 pints per acre on cool season turf including species of bluegrass, ryegrass, fescue, and bentgrass (excluding greens and tees). 4 to 5 pints per acre needed to control corn speedwell and wild violet. Turf areas include residential, ornamental, institutional, and sod farms. Apply to grass seedlings after second mowing. Apply to sodded, sprigged, or plugged areas 3 to 4 weeks after operations. Treated areas may be reseeded 3 weeks after application.
[triclopyr ester + sulfentrazone + 2,4-D ester + dicamba], MOA 4 + 14 + 4 + 4 (2.51 EC)	0.75 to 1.5 fl oz	2 to 4 pt	0.628 to 1.26	Apply 2 to 2.25 pints per acre on fully dormant bermudagrass, zoysiagrass, and bahiagrass. Apply 3.25 to 4 pints per acre on annual and perennial bluegrass and ryegrass, and tall, red, and fine fescue. Rainfast within 3 hours. Approved turf areas include residential, ornamental, institutional, noncropland, and sod farms. Apply to grass seedlings after the second or third mowing. Apply to sodded, sprigged, or plugged areas 3 to 4 weeks after operations. Treated areas may be reseeded 3 weeks after application.
Postemergence Control, Chickweed, Clover, Plantain and Dandelion Species, Florida Betony, Dollarweed, Ground Ivy, Lespedeza, and Yellow Woodsorrel				
florasulam, MOA 2 (0.42 SC)	0.09 fl oz	4 fl oz	0.013125	Can be used on all established major warm and cool season turfgrass species in residential lawns, golf courses (excluding putting greens), sports fields, sod farms and commercial turf areas. Controls Carolina geranium, species of chickweed, clover and dandelion, vetch, dollarweed and common groundsel. Do not exceed 3 applications or 12 fluid ounces per acre per year. Apply to newly seeded or sprigged turf after third mowing or when tillering and secondary root development has occurred. Wait 4 weeks to reseed. When used alone, add a nonionic surfactant at 0.2% by volume.
penoxsulam, MOA 2 (0.014 G) (0.03 G)	3.4 to 10.3 lb 1.7 to 4.6 lb	150 to 450 lb 75 to 200 lb	0.02 to 0.06	May be applied to residential and commercial lawns, golf courses (excluding greens and tees), parks, athletic fields, and sod farms. Use on turf that has been mowed at least 3 times or sprigs that have developed secondary root systems. Apply up to 75 pounds per acre of 0.03 G or 150 pounds per acre of 0.014 G to perennial ryegrass and tall fescue. Apply up to 150 pounds per acre of 0.03 G or 300 pounds per acre of 0.014 G to bentgrass, Kentucky bluegrass, and fine fescue. Apply up to 200 pounds per acre of 0.03 G or 450 pounds per acre of 0.014 G to bermudagrass, centipedegrass, zoysiagrass, and St. Augustinegrass. Do not apply to dormant centipedegrass. Reapply at 4 weeks if needed but do not exceed 300 pounds per acre of 0.03 G or 650 pounds per acre of 0.014 G per season. After treatment, wait 3 to 4 weeks to reseed.
(0.31 L)	0.092 to 0.55 fl oz	0.25 to 1.5 pt	0.01 to 0.058	Same statement as above concerning turf uses and reseeding intervals. Bermudagrass and kikuyugrass are the only warm season grasses labeled for use. Apply up to 1 pint per acre on bentgrass, 1.5 pint per acre on bermudagrass and kikuyugrass and 2 pints per acre on tall fescue and perennial ryegrass. Do not apply greater than 2.3 pints per acre per year. Surfactant not required.
Carpetweed, Chickweed, Dandelion, Curly Dock, Cutleaf Eveningprimrose, Henbit, Knotweed, Common Mallow, Poison Ivy, and Annual Sowthistle				
pyraflufen ethyl, MOA 14 (0.177 SC)	0.016 to 0.092 fl oz	0.7 to 4 fl oz	0.000938 to 0.0055	Used in established sod farm and ornamental turf by commercial applicators and professional landscapers only. Turf can be newly seeded, sodded, or sprigged as long as it is established and not under stress. Tolerant turfgrasses include bermudagrass, centipedegrass, St. Augustinegrass, zoysiagrass, tall fescue, perennial ryegrass, perennial bluegrass, and creeping bentgrass (not greens or tees). Apply 1 to 4 fluid ounces alone to 3- to 6-inch tall weeds. For larger weeds and broader spectrum control, apply 0.75 to 1.5 fluid ounces and tank mix with 2,4-D, mecoprop, dicamba, chloroprop, MCPA, triclopyr, or fluroxypyr.
Postemergence Control, Bahiagrass, Perennial Ryegrass, Wild Garlic, Spurweed, Henbit, Miscellaneous Other Broadleaf Weeds				
metsulfuron, MOA 2 (60 WDG)	0.003 to 0.02 oz	0.125 to 1 oz	0.005 to 0.038	May be applied to established bermudagrass, zoysiagrass (Meyer or Emerald), St. Augustinegrass, Kentucky bluegrass or fine fescue. Do not apply to turf less than 1 year old. Do not exceed 0.5 ounces per acre on centipedegrass, fine fescue, or Kentucky bluegrass. See label for a complete list of weeds controlled. The addition of 0.25% nonionic surfactant will enhance control. May be used for removal of perennial ryegrass from overseeded warm-season turf species. For bahiagrass control, use 0.25 to 0.75 ounces per acre after spring greenup but before seedhead development. A repeat treatment may be necessary in 4 to 6 weeks.
metsulfuron (Patriot) 60 WDG	0.007 to 0.046 oz	0.33 to 2 oz	0.012 to 0.075	Apply to unimproved industrial turf only. Use maximum of 0.5 ounce per acre for fescue and bluegrass and 2 ounces per acre for bermudagrass.
[metsulfuron + rimsulfuron], MOA 2 + 2 (37 WG)	0.0344352 oz	1.5 oz	0.0346875	See comments under postemergence annual bluegrass control. For bahiagrass control, a repeat treatment may be necessary 4 to 6 weeks after initial application.
Postemergence Control, Wild Garlic, Wild Onion				
imazaquin, MOA 2 (70 DG)	0.128 to 0.256 oz	0.357 to 0.714 lb	0.25 to 0.5	Use on bermudagrass, centipedegrass, St. Augustinegrass, and zoysiagrass. Do not apply during spring greenup. Temporary yellowing may occur. Add a nonionic surfactant at 2 pints per 100 gallons of spray solution.
2,4-D amine, MOA 4 (4 SL)	2.2 fl oz	3 qt	3	Apply in fall when garlic is young and actively growing. Add a wetting agent to keep spray from bouncing off garlic leaves. Repeat treatment for 2 years. Avoid spray drift which can injure susceptible plants. Use on bluegrass, fescue, bermudagrass, or zoysia. For more susceptible grasses, uses spot treatment below.
		Spot treatment		One tbsp of 1% 2,4-D solution per garlic clump or use pressurized applicator. Apply December to April. Use as spot treatment for widely scattered clumps in small areas. Avoid excessive spraying as turfgrass injury may result.

Table 7-14. Chemical Weed Control in Lawns and Turf

Herbicide and Formulation	Amount of Formulation Per 1,000 sq ft	Amount of Formulation per Acre	Pounds Active Ingredient per Acre	Precautions and Remarks
Postemergence Control of Various Grass and Broadleaf Weeds in Unimproved Turf and Other Noncrop Areas				
glyphosate, MOA 9 (5.5 SL) (5 SL) (4 SL)	0.14 to 1.01 0.12 to 0.87 fl oz 0.75 to 2.94 fl oz	0.375 to 2.75 pt 0.3125 to 2.375 pt 1 to 4 qt	0.26 to 1.89 0.2 to 1.48 0.5 to 4	Check specific labels for correct rates. Apply to dormant or actively growing well established bermudagrass and bahiagrass. Bahiagrass growth will be suppressed if treated after spring greenup and before seedhead formation. Treat winter annual weeds when less than 6 inches tall. Higher rates are needed for more mature plants. Apply in 10 to 40 gallons of water per acre and use an NIS at 2 quarts per 100 gallons of spray solution.
[glyphosate + 2,4-D amine], MOA 9 + 4 (1.2 + 1.9 lb/gal SL)	0.55 to 1.47 fl oz	1.5 to 4 pt	0.58 to 1.55	Apply in 15 to 30 gallons of water per acre. May be applied to highly maintained dormant bermudagrass at 2 to 4 pt per acre. In low maintenance bermudagrass, sulfometuron can be added at 0.25 to 1 ounce per acre when dormant or actively growing. Apply 2 to 4 pints per acre on dormant bahiagrass and 1.5 to 2 pints per acre on actively growing bahiagrass. Tank mix with sulfometuron if needed. Check label for sulfometuron rates. Tall fescue applications can be made in the spring or summer at 2 to 3 pints per acre with or without sulfometuron. Spray tall fescue at 4 to 6 inches tall and before seedhead emergence to minimize injury.
sulfosulfuron, MOA 2 (75 WG)	0.017 to 0.046 oz	0.75 to 2 oz	0.035 to 0.094	May be used in well-established dormant and actively growing bermudagrass and bahiagrass. Wait 30 days to re-treat if needed; do not exceed 2.66 ounces per acre per year. If treating weeds postemergence, use an NIS at 2 quarts per 100 gallons spray solution unless tank mixed with glyphosate. Sulfosulfuron can be tank mixed with [glyphosate + 2,4-D amine], metsulfuron, sulfometuron, and chlorsulfuron, but check label for proper turf species and timing. Expect temporary injury or discoloration with tank mix partners. For well-established tall fescue, do not exceed 1 ounce per acre per year, and do not tank mix. Effective on johnsongrass.
[thiencarbazone-methyl + iodosulfuron-methyl + foramsulfuron], MOA 14 + 2 + 2 (36.4 WDG)	0.069 to 0.138 oz	3 to 6 oz	0.068 to 0.137	May be applied to unimproved bermudagrass, zoysiagrass, centipedegrass and bare ground sites on private, public and military land for control of many annual and perennial broadleaf and grass weeds. Check label for complete weed listing, rate needed and recommended adjuvant. Repeat application in 4 to 6 weeks if weed regrowth occurs not to exceed 6 ounces product per year. Spot treatment (spray-to-wet) rate is 3 to 6 ounces product per 25- to 100-gallon solution. Nonionic surfactant is generally recommended at 0.25 to 0.5%. Use 0.5 to 1% methylated seed oil for difficult to control broadleaf weeds and perennial grasses. Increased control may be achieved with 1.5 to 3 pounds per acre ammonium sulfate in high humidity climates or 1.5 to 2 quarts per acre urea ammonium nitrate in low humidity climates. Do not use an organosilicone surfactant.
Postemergence Control in Dormant Warm Season Turf Annual Bluegrass, Various Other Winter Annual Weeds				
diquat, MOA 22 (2 SL)	0.4 to 0.75 fl oz	1 to 2 pt	0.25 to 0.5	Apply in 20 to 100 gallons spray mix as a broadcast application. Add 1 to 2 pints of a nonionic surfactant per 100 gallons of solution. Bermudagrass must be dormant. More than one application may be needed.
flumioxazin, MOA 14 (51 WG)	0.1837 to 0.2755 oz	8 to 12 oz	0.255 to 0.3825	Use on completely dormant bermudagrass turf including residential and commercial lawns, golf courses (excluding greens), sod farms, roadsides, athletic fields, parks and schools. Add 0.25% by volume nonionic surfactant for postemergence applications. Provides preemergence control of annual grasses such as crabgrass, goosegrass, foxtail species, barnyardgrass and annual bluegrass. Does control annual bluegrass postemergence along with many common winter annual broadleaf weeds such as chickweed species, henbit, Carolina geranium and hairy bittercress. Allow a 15 feet buffer zone when applying upslope from bentgrass greens or bermudagrass greens overseeded with *Poa trivialis*. To limit potential lateral movement, do not apply to saturated soil.
glyphosate - Roundup, MOA 9 (4 SL) (5 SL) (5.5 SL)	0.37 fl oz 0.29 fl oz 0.27 fl oz	1 pt 0.8 pt 0.73 pt	0.5	Check specific labels for correct rates. Apply in 5 to 40 gallons water per acre with 0.5% by volume of a nonionic surfactant. Application to actively growing annual bluegrass must be made before initiation of bermudagrass greenup in the spring.
glyphosate - Touchdown, MOA 9 (3 LC) (4.17 LC) (5 LC)	0.18 to 1.47 fl oz 0.13 to 1.06 fl oz 0.11 to 0.88 fl oz	0.5 to 4 pt 0.36 to 2.88 pt 0.3 to 2.4 pt	0.1875 to 1.5	Apply to dormant bermudagrass and bahiagrass before spring greenup. Apply in 10 to 40 gallons water per acre. Will control winter annual weeds up to 6 inches tall and 4- to 6-leaf tall fescue. Use a 75% active ingredient nonionic surfactant at 0.25% by volume or dry ammonium sulfate at 0.5% by weight.
[glyphosate + diquat], MOA 9 + 22 (76 WG) (4.21 SL)	0.11 to 0.37 fl oz 0.18 to 0.62 fl oz	5 to 16 oz 8 to 27 fl oz	0.24 to 0.78 0.26 to 0.89	Apply to dormant bermudagrass and bahiagrass not grown for research, sale, or other commercial uses, such as sod, seed production. Apply in 10 to 80 gallons water per acre. Rates greater than 9 ounces per acre of 76 WG product or 15 fluid ounces per acre of 4.21 SL product may cause injury or delay greenup in highly maintained areas. Controls tall fescue.
metribuzin, MOA 5 (75 WDF)	0.25 oz	0.67 lb	0.5	For application by commercial applicators to dormant bermudagrass turf. Broadcast spray before greenup of turf. Do not apply to greens, tees, or aprons. Controls common chickweed, corn speedwell, henbit, parsley-piert, and spurweed.
Suppression/Control, Bermudagrass				
fenoxaprop, MOA 1 (0.57 EC)	0.46 fl oz	1.25 pt	0.089	Use on Kentucky bluegrass, perennial ryegrass, fine and tall fescue, and zoysiagrass. Apply June 1, July 1, Aug. 1, Sept. 1, repeat for 2 years. Can be tankmixed with 1 pt per acre triclopyr following the same schedule as above. Apply June 1 and Aug. 1 for 2 years if tank mixed with 1 quart per acre triclopyr. Zoysia may show discoloration but should recover in 10 to 14 days following tankmix applications.

Table 7-14. Chemical Weed Control in Lawns and Turf

Herbicide and Formulation	Amount of Formulation Per 1,000 sq ft	Amount of Formulation per Acre	Pounds Active Ingredient per Acre	Precautions and Remarks
Suppression/Control, Bermudagrass (continued)				
fluazifop, MOA 1 (2 EC)	0.05 to 0.14 fl oz	2 to 6 fl oz	0.03 to 0.09	Use on tall fescue or zoysia. For fescue, apply 5 to 6 ounces per acre during warm weather in early spring when bermudagrass is breaking dormancy; repeat in fall when bermudagrass is preparing for dormancy. For zoysia, apply 4 ounces per acre on June 1, Aug. 1; repeat for 2 years. Can tank-mix with 1 quart per acre triclopyr following schedule above. Zoysia or tall fescue may show slight discoloration but should recover in 10 to 14 days. Add a nonionic surfactant at 0.25% v/v. Apply in a minimum of 30 gallons of water per acre.
siduron, MOA 7 (50 WP)	0.5 to 1 lb	21.78 to 43.56 lb	10.88 to 21.78	Apply as 8- to 12-inch band treatment with a single nozzle sprayer along putting green perimeter to suppress bermudagrass stolon encroachment. Initiate in March or April, and continue subsequent applications at 4- to 5-week intervals.
triclopyr, MOA 4 (4 EC)	0.73 fl oz	1 qt	1.0	Use on perennial bluegrass, perennial ryegrass, tall fescue or ornamental turf including sod farms and golf courses. Do not apply to zoysia unless injury can be tolerated. Apply June 1, July 1, Aug. 1, Sept. 1, repeat for 2 years. Can be tank-mixed with fenoxaprop or fluazifop at rates, timings listed above. New low-odor formulation uses methylated seed oil solvents instead of petroleum distillates.
Postemergence Control Bermudagrass				
clethodim, MOA 1 (0.97 EC)	0.4 to 0.8 fl oz	17 to 34 fl oz	0.125 to 0.25	For use on sod farms only. Do not apply to centipedegrass being grown for seed. Do not apply until 3 weeks after full greenup of centipedegrass in spring. Do not mow for 1 week before and after application. The addition of a nonionic surfactant at 0.25 % solution (1 pint per 50 gallons water) or a crop oil concentrate at 1% solution (2 quart per 50 gallons water) is necessary for control. A repeat application usually 3 to 4 weeks after the first application will be required for bermudagrass control. Use higher rates for more established bermudagrass. Do not apply more than 68 ounces of clethodim per acre per year. Some discoloration of centipedegrass will occur at higher rates.
Preplant Control or Lawn Renovation — Emerged Annual and Perennial Grass and Broadleaf Weeds				
glyphosate - Roundup, MOA 9 (4 SL) (5 SL) (5.5 SL)	0.75 to 3 fl oz 0.54 to 2.17 fl oz 0.54 to 2.14 fl oz	1 to 4 qt 0.8 to 3.2 qt 0.73 to 2.91 qt	1 to 4	Where existing vegetation is growing in a field or unmowed situation, apply to actively growing weeds at the stages according to label. Where existing vegetation is growing under mowed turfgrass management, apply after omitting at least one regular mowing to allow sufficient growth for good interception of the spray. Tillage or renovation techniques such as vertical mowing, coring, or slicing should be delayed for 7 days after application. Desirable turfgrass may be established following treatment.
glyphosate - Touchdown, MOA 9 (3 LC) (4.17 LC) (5 LC)	0.18 to 1.47 fl oz 0.13 to 1.06 fl oz 0.11 to 0.88 fl oz	0.5 to 4 pt 0.36 to 2.88 pt 0.3 to 2.4 pt	0.1875 to 1.5	Same remarks as glyphosate, above. In addition, use a 75% active ingredient nonionic surfactant at 0.25% by volume or dry ammonium sulfate at 0.5% by weight.
[glyphosate + diquat], MOA 9 + 22 (76 WG) (4.21 SL)	1.65 to 4.5 oz 2.75 to 5.5 fl oz	4.5 to 12.25 lb 3.75 to 7.5 qt	3.4 to 9.3 3.95 to 7.89	Generally use the 75 WG product at 4.5 pounds per acre on annuals, 9 pounds per acre on perennials, and 12.25 pounds per acre on dusty or stressed plants, dense stands, or difficult-to-control perennials. Generally use the 4.21 SL product at 3.75 quarts per acre on annuals and 7.5 quarts per acre on perennials. Do not use on turf grown for research, for sale, or for commercial uses, such as sod or seed production. Do not use if renovating bermudagrass or kikuyugrass sods. Delay tillage for 7 days after application.
[indaziflam + diquat dibromide + glyphosate], MOA 21 + 22 + 9 (1.958 SL)	1 pt	5.44 gal	10.66	For nonselective preemergence and postemergence control in noncrop areas. Reapply 4 months after initial application if needed not to exceed 1 quart per 1000 square feet per year. Apply 1 pint in 1 gallon of water to cover 1000 square feet. Do not seed for 12 months after application.
Trimming and Edging and Control of Emerged Weeds				
diquat, MOA 22 (2 SL)	0.4 to 0.75 fl oz	1 to 2 pt	0.25 to 0.5	Add nonionic surfactant at 0.25 ounce per gallon of water. Water volumes above 15 gal per acre should be used. For spot sprays, use 0.3 to 0.75 fluid ounce per gallon.
glufosinate, MOA 10 (1 SL)	2.2 to 4.4 fl oz	3 to 6 qt	0.75 to 1.5	Rate depends on weed to be controlled and stage of growth. Consult label. For spot or directed spray use 1.5 to 4 fluid ounces per gallon of water.
glyphosate + diquat, MOA 9 + 22 (76 WG) (4.21 SL)	1.65 to 4.5 oz 2.75 to 5.5 fl oz	4.5 to 12.25 lb 3.75 to 7.5 qt	3.4 to 9.3 3.95 to 7.89	May be used in general noncrop areas. Do not use on plants grown for sale or other commercial uses, such as seed production. See rate comments in lawn renovation section. For spray to wet treatments, apply the 76 WG product at 1.2 ounces per gal of water for annuals and 1.5 ounces per gal of water for perennials. Apply the 4.21 SL product at 2 fluid ounces per gallon of water for annuals and 2.5 fluid ounces per gal water for perennials. For directed spot treatment of perennials using hand-held low volume equipment, apply 4 to 8 ounces per gallon of water.

Chemical Weed Control in Ornamentals

J. C. Neal, Horticultural Science Department

Table 7-15. Chemical Weed Control in Ornamentals

More detailed information about herbicides labeled for use in nursery crops and landscape plantings is available at the NC State Extension portal: "Weed Management in Nurseries, Landscapes, and Christmas Trees" (weeds.ces.ncsu.edu).

Weed	Herbicide and Formulation	Mode of Action	Amount Formulation Per Acre	Pounds Active Ingredient Per Acre	Precautions and Remarks
Preplant to all Ornamentals					
Most annuals and perennials	dazomet (Basamid granular)	27	218 to 525 lb	216 to 520	Preplant soil fumigant. Dazomet is a restricted-use pesticide. Follow label directions and restrictions. Incorporate to 8 inches deep. Drench with water or cover with plastic for best results. Planting must be delayed until fumigant has dissipated from the soil. This may take a month or more.
	diquat dibromide (Reward) 2 L	22	1 to 2 qt	0.5 to 1	Non-selective, contact-type, postemergence control of seedling weeds. A nonionic surfactant should be added to the spray solution. Apply for full coverage and thorough weed contact. Retreatment will be necessary for established weeds. May be used in landscapes, nurseries and greenhouses.
	glufosinate (Finale) 1 L	10	3 to 6 qt	0.75 to 1.5	Non-selective postemergence control of weeds, including many glyphosate-resistant weeds. Thorough coverage is essential. Apply in a minimum of 20 gallons of water per acre. No residual control. Repeat applications may be necessary for control of perennial weeds.
	glyphosate (Roundup Pro, Touchdown Pro, and many others)	9	1 to 5 qt	1 to 5	Non-selective, systemic herbicide. Apply to emerged weeds prior to planting ornamentals.
	paraquat (Gramoxone Extra) 2.5 L	22	2 to 3 pt	0.6 to 0.9	Non-selective, contact herbicide. Apply when grass and weeds are 1 to 6 inches high and succulent for best results. Direct spray with low pressure to avoid contact with foliage or bark of crop less than 1 year old. Add nonionic surfactant, 0.25% by volume (2 pints per 100 gallons of water). Not for use in landscapes or greenhouses.
Annual grasses and broadleaf weeds (preemergence) See label for susceptible species	benefin + oryzalin (XL) 1 + 1 G	3	200 to 300 lb	2 to 3 + 2 to 3	Apply preemergence to weeds. May be applied in spring and fall to ornamental plants.
	dichlobenil (Casoron) 4 G	20	100 to 150 lb	4 to 6	Do not use on fir, hemlock, *Ilex crenata, I. rotunda*, or *I. vomitoria*. Do not use more than 6 pounds per acre on azalea, rhododendron, boxwood, holly, euonymus, forsythia, leucothoe, ivy, lilac, heather, or any plantings less than 1 year old. Do not use in seedbeds, cutting, or transplant beds. Do not apply until 4 weeks after transplanting any plants. Apply in winter. http://content.ces.ncsu.edu/casoron-dichlobenil
	dithiopyr (Dimension) 2 EW 40 WP	3	2 pt / 20 oz	0.5	Preemergence control of annual grasses and some small seeded broadleaf weeds in turf, landscape plantings, and nurseries. Use as a directed application around ornamental plants unless specified otherwise on the product label. See label for tolerant species and restrictions. http://content.ces.ncsu.edu/dimension-dithiopyr
	dimethenamid-p (Tower) 6 EC	15	21 to 32 oz	1 to 1.5	Preemergence control of annual sedges, annual grasses and many annual broadleaf weeds in woody landscape plantings, field and container nurseries. Suppression of yellow nutsedge. Avoid foliar treatments over the top of early spring growth flushes as injury to ornamental plants can occur. http://content.ces.ncsu.edu/tower-dimethenamid-p
	dimethenamid-p + pendimethalin (Freehand) 1.75 (0.75 + 1) G	15 + 3	100 to 200 lb	1.75 to 3.5	Preemergence control of annual grasses and many broadleaf weeds from seed as well as suppression of yellow nutsedge in container and field nurseries, and woody landscape plantings. http://content.ces.ncsu.edu/freehand-dimethenamid-p-pendimethalin
	flumioxazin (Broadstar) G	14	150	0.375	Preemergence control of most annual broadleaf and annual grasses in container and field-grown woody nursery crops. Not for use in landscape plantings. See label for species and precautions. Do not apply to wet foliage or newly potted liners. http://content.ces.ncsu.edu/broadstar-flumioxazin
	flumioxazin (Sureguard) DG (Sureguard) 4SC	14	8 to 12 oz / 8 to 12 fl oz	0.25 to 0.375	Preemergence control of most annual broadleaf and annual grasses, and early postemergence control of seedling broadleaf weeds in field and container-grown woody nursery crops and certain landscape plantings. See label for species and precautions. http://content.ces.ncsu.edu/sureguard-flumioxazin
	imazaquin (Image) 1.5 LC	2	1 to 1.3 qt	0.4 to 0.5	Apply as a directed spray away from rooting zone. Labeled for over-the-top sprays on a few species.
	Indaziflam (Marengo) 0.0224% G (Marengo) 0.622 SC	29	100 to 200 lb / 7.5 to 15.5 fl oz	0.022 to 0.044 / 0.036 to 0.075	Preemergence control of many annual weeds in container or field grown nursery crops. Not labeled for use in landscape plantings. See label for specific species. http://content.ces.ncsu.edu/marengo-indaziflam Preemergence weed control in field grown nursery crops. Use as a directed spray. Also for use as a ground treatment in container nurseries and greenhouses (and similar covered structures). When treating inside covered houses, structures should have no crop present at the time of treatment. See label for full details.
	Indaziflam (Specticle) 0.0224% G (Specticle) 0.622 SC	29	100 to 200 lb / 6 to 12 fl oz	0.022 to 0.044 / 0.029 to 0.058	Preemergence control of many annual weeds in established woody landscape plantings. Do not use in areas planted to or to be planted to bedding plants. Labeled for use on a limited number of herbaceous perennials. The GR formulation is less injurious to ornamental plants than the SC formulation. The SC formulation should be applied as a directed spray, avoiding contact with the foliage of ornamental plants. Check labels for details. http://content.ces.ncsu.edu/marengo-indaziflam
	isoxaben (Gallery) 75 DF (Gallery) 4.16 SC	21	0.66 to 1.33 lb / 16 to 31 fl oz	0.5 to 1	Use as preemergence control of broadleaf weeds in many field and container grown ornamentals, turf, and landscape plantings. Generally used in combination with another herbicide for broader spectrum weed control. http://content.ces.ncsu.edu/gallery-isoxaben
	isoxaben + prodiamine (Gemini) 3.7 SC (1.5 + 2.2) (Gemini) 0.65 GR (0.25 + 0.5)	21 + 3	43.5 to 87 fl oz / 100 to 200 lb	1.25 (0.5 + 0.75) to 2.5 (1 + 1.5)	Preemergence control of weeds in field- and container-grown ornamentals and landscape plantings. See label for specific species.

Table 7-15. Chemical Weed Control in Ornamentals

More detailed information about herbicides labeled for use in nursery crops and landscape plantings is available at the NC State Extension portal: "Weed Management in Nurseries, Landscapes, and Christmas Trees" (weeds.ces.ncsu.edu).

Weed	Herbicide and Formulation	Mode of Action	Amount Formulation Per Acre	Pounds Active Ingredient Per Acre	Precautions and Remarks
Post-plant Preemergence Weed Control (control) (continued)					
Annual grasses and broadleaf weeds (preemergence) See label for susceptible species (continued)	isoxaben + trifluralin (Snapshot TG) 2.5 G	21 + 3	100 to 200 lb	2.5 to 5	Preemergence control of weeds in field- and container-grown ornamentals and landscape plantings. See label for specific species. https://content.ces.ncsu.edu/snapshot-tg-isoxaben-trifluralin
	S-metolachlor (Pennant Magnum) 7.62 EC	15	1.3 to 2.6 pt	1.2 to 2.5	Preemergence control of annual grasses, annual sedges, and some annual broadleaf weeds including doveweed, as well as suppression of yellow nutsedge. Apply to soil surface immediately after planting. Avoid foliar treatments over the top of early spring growth flushes as injury to ornamental plants can occur. https://content.ces.ncsu.edu/pennant-magnum-s-metolachlor
	napropamide (Devrinol) 50 DF	15	8 to 12 lb	4 to 6	Apply preemergence to weeds and as a directed spray in ornamentals. Can be used in field or container nurseries or landscape plantings. If broadcast over top of ornamentals, irrigate soon after application to reduce risk of foliar injury. https://content.ces.ncsu.edu/devrinol-napropamide
	oryzalin (Surflan) 4 AS	3	2 to 4 qt	2 to 4	Preemergence to weeds in field or container nurseries or landscape plantings. Apply only to established plantings. Do not use in seedbeds or transplant beds. Not recommended for use on soils containing more than 3% organic matter. Use higher rate for longer term control. Do not apply on hemlock. http://content.ces.ncsu.edu/surflan-oryzalin
	oxadiazon (Ronstar) 2 G	14	100 to 200 lb	2 to 4	Apply preemergence to weeds. Can be used on container- and field-grown ornamentals. Repeat applications are labeled for some species. Injury has been observed on ajuga, liriope, mondo, and fig. Granules may burn tender foliage of several species if irrigation is not used to wash them off. Caution: Plants that trap granules in leaf axil can be injured.
	oxyfluorfen (Goal) 2 XL (Galigan) 2E	14	5 to 10 pt	1 to 2	Preemergence and postemergence control of many broadleaf and grass weeds in conifers and dormant deciduous trees. Do not apply when conifers have young tender growth. Lower rates are used in conifer seedbeds and for postemergence treatments. https://content.ces.ncsu.edu/goal-goaltender-oxyfluorfen
	oxyfluorfen + oryzalin (Rout or Double O) 3 (2+1) G	14 + 3	100 lb	3 (2 + 1)	Apply preemergence to weeds. Can be used on container and field-grown ornamentals. Repeat applications are labeled. Injury is to be expected to herbaceous plants or to plants with leaf orientation that might trap granules. Check label for genera of plants on which it can be used.
	oxyfluorfen + oxadiazon (Regal OO) 3 (2 + 1) G	14 + 14	100 lb	3 (2 + 1)	Apply preemergence to weeds. May be used on container- or field-grown woody ornamentals, including liner production. Injury is to be expected to herbaceous plants or to plants with leaf orientation that might trap granules. Check label for genera of plants on which it can be used. https://content.ces.ncsu.edu/regal-o-o-oxyfluorfen-oxadiazon
	oxyfluorfen + pendimethalin (Ornamental Herbicide 2) 3 (2+1) G	14 + 3	100 lb	3 (2 + 1)	Apply preemergence to weeds. Can be used on container- and field-grown ornamentals. Repeat applications are labeled. Injury is to be expected to herbaceous plants or to plants with leaf orientation that might trap granules. Check label for genera of plants on which it can be used. https://content.ces.ncsu.edu/regal-o-o-oxyfluorfen-oxadiazon
	oxyfluorfen + prodiamine (Biathlon) 2.75 G	14 + 3	100 lb	2.75 (2 + 0.75)	Apply preemergence to weeds. Can be used on container- and field-grown ornamentals. Repeat applications are labeled. Injury is to be expected to herbaceous plants or to plants with leaf orientation that might trap granules. Check label for genera of plants on which it can be used. https://content.ces.ncsu.edu/biathlon-oxyfluorfen-prodiamine
	oxyfluorfen + trifluralin (HGH 75) 5 (2 + 3) G	14 + 3	100 lb	5 (2 + 3)	Preemergence control of annual grasses and broadleaf weeds in container- and field-grown woody ornamentals and landscape beds. Injury is to be expected to herbaceous ornamentals or plants with leaf orientation that might trap granules. https://content.ces.ncsu.edu/hgh-75-oxyfluorfen-trifluralin
	pendimethalin MOA 3 (Corral, Pendulum) several formulations	3	See label	2 to 4	Preemergence control of annual grasses and some broadleaf weeds in turf, landscape plantings, container and field-grown nursery crops, and Christmas trees. Pendulum Aqua Cap is labeled only for turf and landscape uses. See labels for details. https://content.ces.ncsu.edu/pendulum-aquacap-corral-pendimethalin
	prodiamine (Barricade) 65 WG, 4 FL (Regalkade) 0.5 G	3	1 to 1.15 lb 21 to 48 oz 150 lb	0.65 to 0.75	Apply preemergence to weeds. Labeled for use in turf, landscape plantings, and nurseries. See label for tolerant species and restrictions. https://content.ces.ncsu.edu/barricade-prodiamine-regalkade-g-prodiamine
	pronamide (Kerb) 50 WP	15	2 to 4 lb	1 to 2	Pre and postemergence control of cool-season grasses and some annual broadleaf weeds from seed. Apply in late winter just before rain or snowfall. Not recommended for soils that are high in muck or peat. Check label for use restrictions.
	simazine (Princep) 4 L	5	2 to 3 qt	2 to 3	Apply preemergence to weeds in field nurseries and Christmas trees. Injury has occurred on azaleas, Japanese holly, euonymus, lilac, privet, pittosporum, mock orange, hemlock, boxwood, and several other broadleaf species. High rates will injure Fraser fir. https://content.ces.ncsu.edu/princep-simazine-simazine
	trifluralin (Preen) 1.47G (Treflan) 5 G	3	136 to 272 lb 80 lb	2 to 4 4	Preemergence to weeds. Irrigate after application. May injure some azalea cultivars.
Post-Plant, Postemergence Selective Grass Control					
Annual and perennial grasses (postemergence) See label for tolerant species	clethodim (Envoy and others)	1	8 to 34 fl oz	0.06 to 0.25	Postemergent grass control. Annuals 2 to 6 inches tall, perennials at 4 to 12 inches new growth. Add nonionic surfactant at 0.25% v/v (2 pints per 100 gallons) to final spray. https://content.ces.ncsu.edu/envoy-plus-clethodim
	fenoxaprop-P (Acclaim Extra) .57EC	1	13 to 39 oz	0.06 to 0.17	Apply to emerged grass using at least 40 gpa. Can be used overtop of many flowers and woody ornamentals. Check label. Injury has been observed on Bar Harbor juniper, philodendron, Salvia, Podocarpus, and Pittosporum when sprayed with this product. https://content.ces.ncsu.edu/acclaim-extra-fenoxaprop-p

Table 7-15. Chemical Weed Control in Ornamentals

More detailed information about herbicides labeled for use in nursery crops and landscape plantings is available at the NC State Extension portal: "Weed Management in Nurseries, Landscapes, and Christmas Trees" (weeds.ces.ncsu.edu).

Weed	Herbicide and Formulation	Mode of Action	Amount Formulation Per Acre	Pounds Active Ingredient Per Acre	Precautions and Remarks
Post-Plant, Postemergence Selective Grass Control (continued)					
Annual and perennial grasses (postemergence) See label for tolerant species (continued)	fluazifop-P (Fusilade II) 2 EC	1	2 to 3 pt	0.25 to 0.4	Postemergence grass control. Annuals not over 2 to 8 inches tall, perennials at 4 to 12 inches new growth. Consult label for tolerant species. Use nonionic surfactant and no oil. https://content.ces.ncsu.edu/fusilade-ii-fluazifop-p-butyl
	sethoxydim (Segment, Sethoxydim) 1 EC	1	36 to 60 oz	0.3 to 0.5	Postemergence grass control. Annuals up to 12 inches tall and 6 to 10 inches new growth on perennials. https://content.ces.ncsu.edu/segment-sethoxydim
Post-Plant, Postemergence Weed Control					
Annual grasses and broadleaf weeds (postemergence)	2,4-D amine (Weedar 64)	4	2 to 8 pt	0.95 to 3.8	For postemergence broadleaf weed control in Christmas tree production. Not for use in commercial nurseries or landscape plantings. Use as a directed spray prior to bud break. Avoid contact with the desirable plants. Applications after conifer bud-break carry a much greater risk of crop injury from spray drift or misapplication than do applications made before bud break.
	asulam (Asulox) 3.34 L	18	77 to 128 oz	2 to 7	Apply postemergence to weeds in many conifers.
	bentazon (Basagran TO) 4 L	6	1.5 to 2 pt	0.75 to 1	Postemergent directed to many established ornamentals for yellow nutsedge and seedling broadleaf weed control.
	clopyralid (Lontrel) 3 L	4	4 to 11 oz	0.09 to 0.25	Postemergence control of legume and many aster weeds. Can be used as a directed spray around on several field-grown woody ornamentals. Can be applied overtop of actively growing conifers transplanted 1 year or more. Apply when weeds are young and actively growing. https://content.ces.ncsu.edu/lontrel-clopyralid
	dichlobenil (Casoron) 4 G	20	100 to 150 lb	4 to 6	Pre and Postemergence control of many annual and perennial weeds. Do not use on fir, hemlock, Ilex crenata, I. rotunda, or I. vomitoria. Do not use more than 6 pounds per acre on azalea, rhododendron, boxwood, holly, euonymus, forsythia, leucothoe, ivy, lilac, heather, or any plantings less than 1 year old. Do not use in seedbeds, cutting, or transplant beds. Do not apply until 4 weeks after transplanting any plants. Winter applications are best. https://content.ces.ncsu.edu/casoron-dichlobenil
	diquat dibromide (Reward) 2 L	22	1 to 2 qt	0.5 to 1	Non-selective, contact-type, postemergence control of seedling weeds. A nonionic surfactant should be added to the spray solution. Apply for full coverage and thorough weed contact. Retreatment will be necessary for established weeds. May be used in landscapes, nurseries and greenhouses.
	flumioxazin (Sureguard) DG 4SC	14	8 to 12 oz 8 to 12 fl oz	0.375	Early-postemergence and residual control of many seedling annual broadleaf weeds in field and container-grown woody nursery crops and certain landscape plantings. See label for species and precautions. https://content.ces.ncsu.edu/sureguard-flumioxazin
	glufosinate (Finale) 1 L	10	3 to 6 qt	0.75 to 1.5	Non-selective postemergence control of weeds. Thorough coverage is essential. Apply in a minimum of 20 gallons of water per acre. No residual control. Repeat applications may be necessary for control of perennial weeds.
	glyphosate (Many formulations are available including Roundup-Pro Max, Touchdown Pro, and others)	9	1 to 5 qt	1 to 5	Non-selective postemergence, systemic control of weeds. DO NOT SPRAY GREEN BARK OR FOLIAGE of crop. Exercise extreme caution in applications near small plants. Use of a shielded sprayer can increase crop safety. Apply in 20 to 30 gallons of water per acre as a directed spray under shrubs or trees. No residual control. Repeat applications may be necessary for control of perennial weeds. https://content.ces.ncsu.edu/glyphosate-1
	imazaquin (Image) 1.5 LC	2	1 to 1.3 qt	0.4 to 0.5	Apply as a directed spray away from rooting zone. May injure landscape plants through foliar or root absorption. Read the label carefully before use.
	oxyfluorfen (Goal) 2 EC	14	1 to 2 pt	0.25 to 0.5	Pre- and postemergence control. Apply 1 to 2 pints of Goal 2 EC per acre as a postemergence application on some conifers. Add 0.25% (v/v) nonionic surfactant. https://content.ces.ncsu.edu/goal-goaltender-oxyfluorfen
	paraquat (Gramoxone Extra) 2.8 L	22	2 to 3 pt	0.6 to 0.9	Non-selective postemergence control of weeds. Apply when grass and weeds are 1 to 6 inches high and succulent for best results. Direct spray with low pressure to avoid contact with foliage or bark of crop less than 1 year old. Add wetting agent to make 0.25% (2 pints per 100 gallons) by volume of spray for best results. Not for use in landscapes.
	pelargonic acid (Scythe)	27	3 to 10% by volume	na	Non-selective, contact-type control of seedling broadleaf and grass weeds. Use as a directed spray avoiding contact with foliage and stems of desirable plants. Thorough spray coverage is required. Use the lower concentration for small, succulent seedling weeds. Higher concentrations are needed for larger weeds. Repeated applications are generally required. May be used in landscapes, nurseries and greenhouses.
	triclopyr (Garlon 3A)	4	2 to 5 pt	0.75 to 1.75 lb ae	Postemergence control of broadleaf weeds among some conifers grown as Christmas trees. Labeled for directed applications avoiding contact with desirable plants.
Sedges (postemergence)	bentazon (Basagran T/O) 4 L	6	1.5 to 2 pt	0.75 to 1	Postemergent directed spray. For best results add 1 quart per acre crop oil concentrate. For yellow nutsedge and annual sedges; does not control purple nutsedge. May be used in nurseries or landscape plantings.
	halosulfuron (Sledgehammer) 75 DF	2	0.67 to 1.33 oz	0.031 to 0.062	Early postemergence control of yellow and purple nutsedge. Use only as a directed spray around established woody plants. Add 0.25% nonionic surfactant. For use in landscape plantings only. Not labeled for use in nurseries. https://content.ces.ncsu.edu/sedgehammer-halosulfuron
	imazaquin (Image) 1.5 LC	2	1 to 1.3 pt	0.4 to 0.5	Pre- or early postemergence control of purple nutsedge. Use as a directed spray. Labeled as an over the top spray on a few species. Add a nonionic surfactant. For use in landscape plantings only. Read label carefully before use.
	sulfentrazone (Dismiss) 4L	14	6 oz	0.188	Postemergence suppression of yellow and purple nutsedge, as well as several broadleaf weeds. Apply as a directed spray avoiding contact with foliage of desirable plants. Reapply when regrowth of treated weeds is observed. Do not exceed 12 ounces per acre per year. May be used in landscape plantings, as well as field and container nurseries. https://content.ces.ncsu.edu/dismiss-sulfentrazone

Chapter VII—2019 N.C. Agricultural Chemicals Manual

Chemical Weed Control in Vegetable Crops

K. M. Jennings, S. Chaudhari, and D. W. Monks, Horticultural Science

NOTE: A mode of action code (MOA) has been added to the Herbicide and Formulation column in this table. Use MOA codes for herbicide resistance.

Table 7-16. Chemical Weed Control in Vegetable Crops

Weed	Herbicide, Mode of Action Code* and Formulation	Amount of Formulation Per Acre	Pounds Active Ingredient Per Acre	Precautions and Remarks
Asparagus (seeded and new crown plantings), Preemergence				
Contact kill of all green foliage, stale bed application	paraquat, MOA 22 (Firestorm, Parazone) 3 SL (Gramoxone SL) 2 SL	1.7 to 2.7 pt 2.5 to 4 pt	0.6 to 1	Apply to emerged weeds before crop emergence as a broadcast or band treatment over a preformed row. Row should be formed several days ahead of planting and treating to allow maximum weed emergence. Use a nonionic surfactant at a rate of 16 to 32 ounces per 100 gallons spray mix or 1 gallon approved crop oil concentrate per 100 gallons spray mix. No more than 3 applications per year.
Annual and perennial grass and broadleaf weeds, stale bed application	glyphosate, MOA 9 (numerous brands and formulations)	See labels	See labels	Apply to emerged weeds before crop emergence. Row should be formed several days ahead of planting and treating to allow maximum weed emergence. Perennial weeds may require higher rates. The need for an adjuvant depends on brand used.
Annual grasses and small-seeded broadleaf weeds	linuron, MOA 7 (Lorox DF) 50 WDG	1 to 2 lb	0.5 to 1	**Preemergence application.** Plant seed 0.5 inch deep in coarse soils. Apply to soil surface. See label for further instruction. **Postemergence application.** Apply when ferns are 6 to 18 inches tall. Make one or two applications, but do not exceed 2 pounds active ingredient total per acre. Do not use fertilizer and surfactant or crop oil, as injury will occur. Use the lower rate on coarse soils. Not recommended on sand or loamy sand soils. Do not apply within 1 day of harvest.
Annual grasses and certain broadleaf weeds	pendimethalin, MOA 3 (Prowl H_2O) 3.8 AS	8.2 pt	3.9	Newly planted crown asparagus only. Do not apply to newly seeded asparagus. Newly planted crowns must be covered with at least 2 to 4 inches of soil prior to application. **Do not apply Prowl H20 at more than 2.4 pints per acre in sandy soils. See label for more information.**
Asparagus (seeded and new crown plantings), Postemergence				
Annual and perennial grasses	clethodim, MOA 1 (Intensity One, Select Max) 1 EC (Arrow) 2 EC	9 to 16 oz 6 to 8 oz	0.07 to 0.125 0.094 to 0.125	Apply to emerged grasses. Consult the manufacturer's label for best times to treat specific grasses. For Select Max, add 2 pints nonionic surfactant per 100 gallons spray mixture. With sethoxydim, add 1 quart crop oil concentrate per acre. Adding crop oil to Poast may increase the likelihood of crop injury at high air temperatures. With fluazifop, add 1 quart of nonionic surfactant or 1 gallon crop oil concentrate per 100 gallons of spray mix. PHI is 1 day.
	fluazifop, MOA 1 (Fusilade DX) 2 EC	6 to 16 oz	0.1 to 0.25	
	sethoxydim, MOA 1 (Poast) 1.5 EC	1.5 to 2.5 pt	0.3 to 0.5	
Asparagus (established - at least 2 years old) Preemergence				
Annual grasses and small-seeded broadleaf weeds	linuron, MOA 7 (Lorox DF) 50 WDG	1 to 2 lb	0.5 to 1	Apply before spear emergence or immediately after a cutting. Do not use a surfactant or fertilizer solution in spray mixture. Use the lower rates on coarse soils. Not recommended for sand or loamy sand soils. Repeat applications may be made but do not exceed 4 pounds per acre per year.
	napropamide, MOA 15 (Devrinol DF-XT) 50 DF (Devrinol 2-XT) 2 EC	8 lb 2 gal		Apply to the soil surface in spring before weed and spear emergence. Do not exceed 8 pounds per acre per year. See XT labels for information regarding delay in irrigation event.
	trifluralin, MOA 3 (Trilin, Treflan HFP, Treflan) 4 EC	1 to 4 pt	0.5 to 2	In winter or early spring, apply to dormant asparagus after ferns are removed but before spear emergence, or apply after harvest in late spring or early summer. In a calendar year, the maximum rate is 2 pints per acre for coarse soils, 3 pints on medium soils and 4 pints on fine soils. See label for further restrictions on rates for soil types.
Annual grasses and certain broadleaf weeds	pendimethalin, MOA 3 (Prowl H_2O) 3.8 AS	8.2 pt	3.9	Apply at least 14 days prior to the first spear harvest or after seasonal harvest complete. Do not apply over the top of emerged spears as severe injury may occur. **Do not apply Prowl H20 at more than 2.4 pints per acre in sandy soils.**
Annual broadleaf and grass weeds	diuron, MOA 7 (Karmex) 80 DF (Direx) 4 L	1 to 4 lb 0.8 to 3.2 qt	0.8 to 3.2	Apply in spring before spear emergence but no *earlier* than 4 weeks before spear emergence. A second application may be made immediately after last harvest. **For the majority of N.C. plantings, a 1 to 2 pounds per acre dosage of 80 DF or 0.8 to 1.6 quarts rate of Direx should be used.** Diuron also controls small emerged weeds but less effectively.
	flumioxazin, MOA 14 (Chateau) 51 SW	6 oz	0.188	Apply only to dormant asparagus no sooner than 14 days before spears emerge or after the last harvest. Do not apply more than 6 ounces per acre during a single growing season. Provides residual weed control. Can be tank mixed with paraquat for control of emerged weeds. Apply in a minimum of 15 gallons spray mix per acre. Add a nonionic surfactant at 1 quart per 100 gallons of spray mix. A spray-grade nitrogen source (either ammonium sulfate at 2 to 2.5 pounds per acre or 28 to 32 percent nitrogen solutions at 1 to 2 quarts per acre) may be added to increase herbicidal activity.
	metribuzin, MOA 5 (Metribuzin) 75 WDG (TriCor DF) 75 WDG	1.3 to 2.67 lb 1.3 to 2.67 lb	1 to 2	Make a single application to small emerged weeds and the soil surface in early spring before spear emergence or following last cutting. Do not apply within 14 days of harvest or after spear emergence. For the majority of N.C. plantings, the low rate should be used. Do not make postharvest applications until after the last harvest of spears. Tricor DF may be used in sprinkler irrigation. A split application can be used. See label for rates.
	terbacil, MOA 5 (Sinbar) 80 WDG	see label for rate	see label for rate	Apply in spring before weed emergence and spear emergence or immediately after last clean-cut harvest. Use the lower rate on sandy soils and the higher rate on silty or clay soils. Do not use on soils containing less than 1% organic matter nor on gravelly soils or eroded areas where subsoil or roots are exposed. Do not harvest within 5 days after application. See label about rotation restrictions.
Grasses and broadleaf weeds	norflurazon, MOA 12 (Solicam) 80 DF	2.5 to 5 lb	2 to 4	Either rainfall or irrigation necessary for herbicide activation within 4 weeks after application. Solicam DF rates depend on type of soil. Solicam may be tank mixed with other herbicide registered for use in asparagus. See label for rates and tank mix information. PHI 14 days

Table 7-16. Chemical Weed Control in Vegetable Crops

Weed	Herbicide, Mode of Action Code* and Formulation	Amount of Formulation Per Acre	Pounds Active Ingredient Per Acre	Precautions and Remarks
Asparagus (established - at least 2 years old) Preemergence (continued)				
Annual broadleaf weeds	mesotrione, MOA 27 (Callisto) 4 L	3 to 7.7 fl oz	0.093 to 0.25	**Preemergence application: apply as a spring application prior to spear emergence, after final harvest, or both.** For optimum preemergence weed control callisto must be applied after fern mowing, disking or other tillage operations but before to spear emergence. **Directed or semidirected application:** Apply after final harvest with care to minimize contact with any standing asparagus spears to avoid crop injury. Do not make more than two applications per year or apply more than 7.7 ounces per acre per year.
Asparagus (established - at least 2 years old), Postemergence				
Broadleaf weeds including trumpetcreeper	2,4-D, MOA 4 (Amine 4 and various other brands) 3.8 SL	1.5 to 2 qt	1.5 to 2	Apply in spring before spear emergence or immediately following a clean cutting. Make no more than two applications during the harvest season and these should be spaced at least 1 month apart. Postharvest sprays should be directed under ferns, avoiding contact with ferns, stems, or emerging spears. Add a nonionic surfactant at a rate of 1 quart per 100 gallons spray mix. **Do not apply if sensitive crops are planted nearby or if conditions favor drift. PHI 3 days**
Broadleaf weeds including trumpetcreeper, annual sowthistle, black mustard, nettleleaf goosefoot, and wild radish	dicamba, diglycolamine salt, MOA 4 (Clarity) 4 L	8 to 16 oz	0.25 to 0.5	Apply to emerged and actively growing weeds in 40 to 60 gallons of diluted spray per treated acre immediately after cutting in the field but at least 24 hours before the next cutting. If spray contacts emerged spears, twisting of spears may occur. Discard twisted spears. See label for more information. **Follow precautions on label concerning drift to sensitive crops. PHI = 1 day**
Contact kill of emerged annual weeds, suppression of emerged perennial weeds, and contact kill of volunteer ferns	paraquat, MOA 22 (Firestorm, Parazone) 3 SL (Gramoxone SL) 2 SL	1.7 to 2.7 pt 2.5 to 4 pt	0.6 to 1	Apply to control emerged weeds (including volunteer ferns). Apply in a minimum of 20 gallons spray mix per acre to control weeds before spears emerge or after last harvest. Do not apply within 6 days of harvest. Use a nonionic surfactant at a rate of 1 quart per 100 gallons spray mix or 1 gallon approved crop oil concentrate per 100 gallons spray mix.
Volunteer ferns (seedling) and certain broadleaf weeds	linuron, MOA 7 (Lorox DF) 50 WDG	2 lb	1	Apply before cutting season or immediately after. Do not apply within 1 day of harvest. Lorox will also control emerged annual broadleaf weeds that are up to 3 inches in height. Lorox can also be applied as a directed spray to the base of the ferns.
Annual and perennial grass and broadleaf weeds Established volunteer ferns	glyphosate, MOA 9 (numerous brands and formulations)	See labels	See labels	Apply to emerged weeds up to 1 week before spear emergence or immediately after last cutting has removed all above-ground parts or as a directed spray under mature fern. Avoid contact with the stem to reduce risk of injury. Perennial weeds may require higher rates of glyphosate. For spot treatment, apply immediately after cutting but prior to emergence of new spears. Certain glyphosate formulations may require the addition of a surfactant. Adding nonionic surfactant to glyphosate formulated with nonionic surfactant may result in reduced weed control.
Yellow and purple nutsedge, non-ALS resistant pigweed, cocklebur, ragweed, wild radish	halosulfuron, MOA 2 (Profine 75) 75 DF (Sandea) 75 DF	0.5 to 1.5 oz	0.024 to 0.072	**Postemergence and post-transplant.** Apply before or during harvesting season. **Do not use nonionic surfactant or crop oil or unacceptable crop injury may occur.** Without the addition of a nonionic surfactant, postemergence weed control may be reduced. Do not exceed 1 ounce per acre per year. Do not harvest within 24 hours of application. **Postharvest.** Apply after final harvest with drop nozzles to limit contact with crop. Contact with the fern may result in temporary yellowing. Add a nonionic surfactant at 1 quart per 100 gallons of spray mixture. Under heavy nutsedge pressure, split applications will be more effective; see label for details. Do not exceed 1 ounce per acre per year.
Annual and perennial grasses	clethodim, MOA 1 (Intensity One, Select Max) 1 EC (Arrow) 2 EC	9 to 16 oz 6 to 8 oz	0.07 to 0.125 0.094 to 0.125	For Select Max, add 2 pints nonionic surfactant per 100 gallons spray mixture. DO NOT USE CLETHODIM WITHIN 1 DAY OF HARVEST.
	fluazifop, MOA 1 (Fusilade DX) 2 EC	6 to 16 oz	0.1 to 0.25	Apply to emerged grasses. Consult the manufacturer's label for best times to treat specific grasses. With sethoxydim, add 1 qt crop oil concentrate per acre. With fluazifop, add 1 quart nonionic surfactant or 1 gallon crop oil concentrate per 100 gallons of spray mix. Adding crop oil to Poast may increase the likelihood of crop injury at high air temperature. DO NOT USE FLUAZIFOP OR SETHOXYDIM WITHIN 1 DAY OF HARVEST.
	sethoxydim, MOA 1 (Poast) 1.5 EC	1.5 to 2.5 pt	0.3 to 0.5	
Beans, Preplant and Preemergence				
Contact kill of all green foliage, stale bed application	paraquat, MOA 22 (Firestorm, Parazone) 3 SL (Gramoxone SL) 2 SL	1.3 to 2.7 pt 2 to 4 pt	0.5 to 1	**Lima or snap beans only.** Apply in a minimum of 10 gallons spray mix per acre to emerged weeds before crop emergence as a broadcast or band treatment over a preformed row. Use sufficient water to give thorough coverage. Row should be formed several days ahead of planting and treating to allow maximum weed emergence. Use a nonionic surfactant at a rate of 16 to 32 ounces per 100 gallons spray mix or 1 gallon approved crop oil concentrate per 100 gallons spray mix.
Most broadleaf weeds less than 4 inches tall or rosettes less than 3 inches in diameter; does not control grasses	carfentrazone-ethyl, MOA 14 (Aim) 1.9 EW or 2 EC	up to 2 oz	up to 0.031	Legume vegetable group (Group 6) such as but not limited to edamame, kidney bean, lima bean, pinto bean, snap bean, soybean, and wax bean. Apply prior to or no later than one day after planting. Use a nonionic surfactant or crop oil with Aim. See label for rate. Coverage is essential for good weed control. Can be tank mixed with other registered burndown herbicides.
Annual and perennial grass and broadleaf weeds, stale bed application	glyphosate, MOA 9 (numerous brands and formulations)	See labels	See labels	**Various beans are covered.** Apply to emerged weeds before crop emergence. Perennial weeds may require higher rates of glyphosate. Consult the manufacturer's label for rates for specific weeds. Certain glyphosate formulations may require the addition of a surfactant. Adding nonionic surfactant to glyphosate formulated with nonionic surfactant may result in reduced weed control. See label for details.
Annual grasses and small-seeded broadleaf weeds	ethalfluralin, MOA 3 (Sonalan HFP) 3 EC	1.5 to 3 pt	0.6 to 1.1	**Dry beans only.** See label for specific bean. Apply preplant and incorporate into the soil 2 to 3 inches deep using a rototiller or tandem disk. If groundcherry or nightshade is a problem, the rate range can be increased to 3 to 4.5 pints per acre. For broader spectrum control, Sonalan may be tank mixed with Eptam or Dual. Read the combination product labels for directions, cautions, and limitations before use.
	dimethenamid, MOA 15 (Outlook) 6.0 EC	12 to 18 oz	0.55 to 0.85	**Dry beans only.** See label for specific bean. Apply preplant incorporated, preemergence to the soil surface after planting, or early postemergence (first to third trifoliate stage). Dry beans may be harvested 70 or more days after Outlook application. For soils having 3% or greater organic matter, see label for rate. See label for further instructions, including those for tank mixtures.

Table 7-16. Chemical Weed Control in Vegetable Crops

Weed	Herbicide, Mode of Action Code* and Formulation	Amount of Formulation Per Acre	Pounds Active Ingredient Per Acre	Precautions and Remarks
Beans, Preplant and Preemergence (continued)				
Annual grasses and small-seeded broadleaf weeds (continued)	trifluralin, MOA 3 (Treflan, Trifluralin, Trifluralin HF, other brands) 4 EC	1 to 1.5 pt	0.5 to 0.75	**Dry, lima, or snap beans only.** See label for specific bean. Apply preplant and incorporate into the soil 2 to 3 inches deep within 8 hr. Incorporate with a power-driven rototiller or by cross disking.
	pendimethalin, MOA 3 (Prowl H_2O) 3.8 AS	1.5 to 3 pt	0.75 to 1.5	**Edible beans: dry, lima, or snap beans, and certain others.** See label for specific bean. Apply preplant and incorporate into the soil 2 to 3 inches using a power-driven rototiller or by cross disking. **Do not apply after seeding.**
	S-metolachlor, MOA 15 (Brawl, Dual Magnum, Medal) 7.62 EC (Brawl II, Dual II Magnum, Medal II) 7.64 EC	1 to 2 pt	0.95 to 1.91	**Dry, lima, or snap beans, and certain others.** See label for specific bean, and specific rate based on soil texture. Apply preplant incorporated or preemergence to the soil surface after planting.
Annual grasses and broadleaf weeds	clomazone, MOA 13 (Command) 3ME	0.4 to 0.67 pt	0.15 to 0.25	**Snap beans (succulent) only.** Apply to the soil surface immediately after seeding. Offers weak control of pigweed. See label for further instructions. Do not apply within 45 days of harvest. Limited research has been done on this product in this crop in North Carolina.
Yellow and purple nutsedge, grasses and some small-seeded broadleaf weeds	EPTC, MOA 8 (Eptam) 7 EC	2.25 to 3.5 pt	2 to 3	**Dry or snap beans only.** See label for specific bean. Apply preplant and incorporate immediately to a depth of 3 inches or may be applied at lay-by as a directed application before bean pods start to form to control late season weeds. See label for instructions on incorporation. May be tank mixed with Prowl. Do not use on black-eyed beans, lima beans, or other flat-podded beans except Romano.
Many broadleaf weeds	fomesafen, MOA 14 (Reflex 2 EC)	1 to 1.5 pt	0.25 to 0.375	**Dry bean and snap beans only.** Apply preplant surface or preemergence. Total use per year cannot exceed 1.5 pints per acre. See label for further instructions and precautions.
Yellow and purple nutsedge, common cocklebur, and other broadleaf weeds	halosulfuron-methyl, MOA 2 (Profine 75) 75 DG (Sandea) 75 DG	0.5 to 0.75 oz	0.024 to 0.036	**Dry beans and succulent snap beans including lima beans only.** Apply after seeding but prior to cracking. Do not apply more than 0.67 ounce product per acre to dry bean. Data are lacking on runner-type snap beans. See label for other instructions.
Broadleaf weeds including morningglory, pigweed, smartweed, and purslane	imazethapyr, MOA 2 (Pursuit) 2 EC	1.5 oz	0.023	**Dry beans and lima beans only.** See label for specific bean. Apply preemergence or preplant incorporated. Pursuit should be applied with a registered preemergence grass herbicide. **Snap beans only.** Apply preemergence or preplant incorporated. For preplant incorporated application, apply within 1 week of planting. May be used with a registered grass herbicide. Reduced crop growth, quality, yield, and/or delayed crop maturation may result. Do not apply within 30 days of harvest of snapbeans.
Beans, Postemergence				
Annual broadleaf weeds and yellow nutsedge	bentazon, MOA 6 (Basagran) 4 SL	1 to 2 pt	0.5 to 1	**Dry, lima, or snap beans only.** Apply overtop of beans and weeds when beans have one to two expanded trifoliate leaves. Two applications spaced 7 to 10 days apart may be made for nutsedge control. Do not apply more than 2 quarts per season or within 30 days of harvest. Use of crop oil as an adjuvant will improve weed control but will likely increase crop injury. See label regarding crop oil concentrate use. Do not within 30 days of harvest.
Many broadleaf weeds	fomesafen, MOA 14 (Reflex 2 EC)	0.75 to 1 pt	0.188 to 0.25	**Dry or snap beans only.** See label for specific bean. Apply postemergence to dry beans or snap beans that have at least one expanded trifoliate leaf. Include a nonionic surfactant at 1 quart per 100 gallons spray mixture. Total use per year cannot exceed 1.5 pints per acre. Do not apply within 45 days of dry bean harvest or 30 days of snap bean harvest. **Postemergence application of fomesafen can cause significant injury to the crop.** See label for further information.
Most broadleaf weeds less than 4 inches tall or rosettes less than 3 inches in diameter. Does not control grasses	carfentrazone-ethyl, MOA 14 (Aim) 1.9 EW or 2 EC	up to 2 oz	up to 0.031	**Edible beans: edamame, kidney bean, lima bean, pinto bean, snap bean and wax bean only.** Apply post-directed using hooded sprayers for control of emerged weeds. If crop is contacted, burning of contacted area will occur. Use a nonionic surfactant or crop oil with Aim. See label for rate. Coverage is essential for good weed control. Can be tank mixed with other registered herbicides.
Yellow and purple nutsedge	EPTC, MOA 8 (Eptam) 7 EC	3.5 pt	3	**Green or dry beans only.** See label for specific bean. Do not use on lima bean or pea. Apply and incorporate at last cultivation as a directed spray to soil at the base of crop plants before pods start to form.
Yellow and purple nutsedge, common cocklebur, and other broadleaf weeds	halosulfuron-methyl, MOA 2 (Profine 75) 75 DG (Sandea) 75 DG	0.5 to 0.67 oz	0.024 to 0.031	**Succulent snap beans, including lima beans.** Apply after crop has reached 2-to 3-trifoliate leaf stage but prior to flowering. Postemergence application may cause significant but temporary stunting and may delay crop maturation. Use directed spray to limit crop injury. Do not apply within 30 days of harvest. See label for further precautions. Data lacking on runner-type snap beans.
Annual broadleaf weeds, including morningglory, pigweed, smartweed, and purslane	imazethapyr, MOA 2 (Pursuit) 2.EC	1.5 to 3 oz	0.023 to 0.047	**Dry beans and snap beans only.** See label for specific bean. Use only 1.5 ounces EC formulation on snap bean and up to 3 ounces on dry beans. Apply postemergence to 1- to 3-inches weeds (one to four leaves) when beans have at least one fully expanded trifoliate leaf. Add nonionic surfactant at 2 pints per 100 gallons of spray mixture with all postemergence applications. For snap beans, allow at least 30 days between application and harvest. For dry bean, do not apply within 60 days of harvest. See label for instructions on use.
Most emerged weeds	glyphosate, MOA 9 (Roundup PowerMax) 5.5 L (Roundup WeatherMax) 5.5 L	11 to 22 oz	0.5 to 0.94	**Row middles only.** See label for specific bean. Apply as a hooded spray in row middles, as shielded spray in row middles, as wiper applications in row middles, or postharvest. Spot treatment is allowed in some bean crops. To avoid severe injury to crop, do not allow herbicide to contact foliage, green shoots, stems, exposed, roots, or fruit of crop. Do not apply within 14 days of harvest.
Annual and perennial grasses	sethoxydim, MOA 1 (Poast) 1.5 EC	1 to 1.5 pt	0.2 to 0.3	**Dry or succulent beans only.** See label for specific bean. For succulent beans, products with quizalofop are limited to snap beans. Apply to emerged grasses. Consult manufacturer's label for specific rates and best times to treat. With sethoxydim, add 1 quart of crop oil concentrate per acre. With quizalofop, add 1 qal oil concentrate or 1 quart nonionic surfactant per 100 gallons spray. Adding crop oil to Poast may increase the likelihood of crop injury at high air temperatures. Do not apply on days that are unusually hot and humid. Do not apply within 15 days and 30 days of harvest for succulent and dry beans, respectively.
	quizalofop p-ethyl, MOA 1 (Assure II) 0.88 EC (Targa) 0.88 EC	6 to 12 oz	0.04 to 0.08	

Table 7-16. Chemical Weed Control in Vegetable Crops

Weed	Herbicide, Mode of Action Code* and Formulation	Amount of Formulation Per Acre	Pounds Active Ingredient Per Acre	Precautions and Remarks
Beans, Postemergence (continued)				
Annual and perennial grasses (continued)	clethodim, MOA 1 (Arrow, Clethodim, Intensity, Select) 2 EC (Intensity One, Select Max) 1 EC	6 to 16 oz 9 to 16 oz	0.094 to 0.25 0.07 to 0.125	**Dry or succulent beans.** See label for specific bean. Select is registered for dry beans only. Apply postemergence for control of emerged grasses. See label for specific rate for crop. For Arrow, Clethodim, or Select, add a crop oil concentrate at 1 quart per acre. For Select Max or Intensity One, add 2 pints nonionic surfactant per 100 gallons spray mixture. Adding crop oil may increase the likelihood of crop injury at high air temperatures. Very effective in controlling annual bluegrass. Apply to actively growing grasses not under drought stress. See label for minimum time from application to harvest.
Beets (Garden or Table), Preplant				
Most emerged weeds except for resistant pigweed, primrose, or spiderwort	glyphosate, MOA 9 (numerous brands and formulations)	See labels	See labels	**Garden beets only.** Apply to emerged weeds before seeding or after seeding but before crop emergence. Perennial weeds may require higher rates. Certain glyphosate formulations may require the addition of a surfactant. Adding nonionic surfactant to glyphosate formulated with nonionic surfactant may result in reduced weed control.
Emerged broadleaf weeds	pyraflufen, MOA 14 (ET Herbicide) 0.208 EC	0.5 to 2 fl oz	0.0008 to 0.003	**Garden beets only.** Apply as a preplant burndown treatment in a minimum of 10 gallons per acre. **Addition of a crop oil concentrate at 1 to 2% is recommended for optimum weed control.** See label for additional information.
Emerged broadleaf and grass weeds	pelargonic acid, MOA 27 (Scythe) 4.2 EC	3 to 10% v/v		Apply as a preplant burndown treatment or prior to crop emergence from seed.
Beets (Garden or Table), Preemergence				
Annual grasses (crabgrass spp., foxtail spp., barnyardgrass, annual ryegrass, annual bluegrass) and broadleaf weeds (Lamium spp., lambsquarters, common purslane, redroot pigweed, shepherdspurse)	cyclohexylethylthio-carbamate, MOA 3 (Ro-Neet) 6E	0.5 to 0.67 gal	3 to 4	Use on mineral soils only. Use higher dosage rate on heavier soils. Read label for further instructions.
Annual broadleaf and grass weeds including common chickweed, common purslane, nightshade spp.	ethofumesate, MOA 16 (Ethotron) 4 SC	See label	See label	Apply preplant, preemergence, or postemergence to the beets and prior to weed germination. Application rate is soil type dependent. The use of higher than specified rates may cause beet injury and/or carry over problems. See label for more information about planting restrictions.
Beets (Garden or Table), Postemergence				
Broadleaf weeds including sowthistle clover, cocklebur, jimsonweed, and ragweed	clopyralid, MOA 4 (Stinger) 3EC	0.25 to 0.5 pt	0.093 to 0.187	Apply to beets having 2 to 8 leaves when weeds are small and actively growing. Will control most legumes. Do not apply within 30 days of harvest. Do not apply more than 0.5 pint per acre per year. See label for information regarding rotational restrictions. The PHI is 30 days.
Broadleaf weeds including wild mustard, common lambsquarters, common chickweed, purslane suppression	phenmedipham, MOA 6 (Spin-Aid) 1.3 EC	1.5 to 3 pt	0.25 to 0.5	**Red garden beets only.** Apply to red garden beets in the 2 to 6 leaf stage. Rate is dependent on crop stage. See label for specific rate. Best control occurs when applied to seeds in cotyledon to 2 leaf stage. Minor crop stunting may be observed for approximately 10 days. Do not add spray adjuvant. Do not apply within 60 days of harvest.
Broadleaf weeds including wild mustard, shepherd's and velvetleaf	triflusulfuron methyl, MOA 2 (Upbeet) 50 DF	0.5 oz	0.0156	**Garden beets.** Apply when beets are at the 2 to 4 leaf stage. Additional applications may be made at the 4 to 6, and 6 to 8 leaf stages. See label for information on adjuvants. Total amount of Upbeet must not exceed 1.5 ounces per acre per growing season. Do not apply within 30 days of harvest.
Annual and perennial grasses	sethoxydim, MOA 1 (Poast) 1.5 EC	1 to 1.5 pt	0.2 to 0.3	Apply to emerged grasses. Consult manufacturer's label for specific rates and best times to treat. Add 1 quart of crop oil concentrate per acre. Adding crop oil to Poast may increase the likelihood of crop injury at high air temperatures. Do not apply Poast on days that are unusually hot and humid. Do not apply within 60 days of harvest.
	clethodim, MOA 1 (Arrow, Clethodim, Intensity, Select) 2 EC (Select Max, Intensity One) 1 EC	6 to 8 oz 9 to 16 oz	0.094 to 0.125 0.07 to 0.125	Apply postemergence for annual grasses at 6 to 8 ounces per acre or bermudagrass and johnsongrass at 8 ounces per acre. For Arrow, Clethodim, or Select, add a crop oil concentrate at 1% per spray volume. For Select Max, add 2 pints nonionic surfactant per 100 gallons spray mixture. Adding crop oil may increase the likelihood of crop injury at high air temperatures. Very effective in controlling annual bluegrass. Apply to actively growing grasses not under drought stress. Do not apply within 30 days of harvest.
Beets (Garden or Table), Row Middles Only				
Most emerged weeds except resistant pigweed	glyphosate, MOA 9 (numerous brands and formulations)	See labels	See labels	Apply as a hooded spray in row middles, as shielded spray in row middles, as wiper applications in row middles, or postharvest. To avoid severe injury to crop, do not allow herbicide to contact foliage, green shoots, stems, exposed, roots, or fruit of crop. The need for an adjuvant depends on brand used. Do not apply within 14 days of harvest.
Annual broadleaf weeds including morningglory, spiderwort, and very small pigweed	carfentrazone-ethyl, MOA 14 (Aim) 1.9 EW or 2 EC	up to 2 oz	up to 0.031	Apply post-directed using hooded sprayers for control of emerged weeds. If crop is contacted, burning of contacted area will occur. Use a crop oil concentrate or a nonionic surfactant with Aim. See label for directions. Coverage is essential for good weed control. Can be tank mixed with other registered herbicides.
Broccoli – See Cole Crops				
Cabbage – See Cole Crops				
Cantaloupes (Muskmelons), Preplant and Preemergence				
Suppression or control of most annual grasses and broadleaf weeds, full rate required for nutsedge control	metam sodium (Vapam HL) 42%	37.5 to 75 gal	15.7 to 31.5	Rates are dependent on soil type and weeds present. Apply when soil moisture is at field capacity (100 to 125%). Apply through soil injection using a rotary tiller or inject with knives no more than 4 inches apart; follow immediately with a roller to smooth and compact the soil surface or with mulch. May apply through drip irrigation prior to planting a second crop on mulch. Plant back interval is often 14 to 21 days and can be 30 days in some environments. See label for all restrictions and additional information.

Table 7-16. Chemical Weed Control in Vegetable Crops

Weed	Herbicide, Mode of Action Code* and Formulation	Amount of Formulation Per Acre	Pounds Active Ingredient Per Acre	Precautions and Remarks
Cantaloupes (Muskmelons), Preplant and Preemergence (continued)				
Most broadleaf weeds less than 4 inches tall or rosettes less than 3 inches in diameter; does not control grasses	carfentrazone-ethyl, MOA 14 (Aim) 1.9 EW or 2 EC	up to 2 oz	up to 0.031	**Transplant crop: apply no later than one day before transplanting crop.** **Seeded crop: apply no later than 7 days before seeding crop.** Use a crop oil at up to 1 gallon per 100 gallons of spray solution or a nonionic surfactant at 2 pints per 100 gallons of spray solution. Coverage is essential for good weed control. Can be tank mixed with other registered burndown herbicides.
Contact kill of all green foliage, stale bed application	paraquat, MOA 22 (Firestorm, Parazone) 3 SL (Gramoxone SL) 2 SL	1.3 to 2.7 pt 2 to 4 pt	0.5 to 1	Apply in a minimum of 10 gallons spray mix per acre to emerged weeds before crop emerges or before transplanting as a broadcast or band treatment over a preformed row. Seedbeds or plant beds should be formed as far ahead of treatment as possible to allow maximum weed emergence. Use a nonionic surfactant at a rate of 16 to 32 ounces per 100 gallons spray mix or 1 gallon approved crop oil concentrate per 100 gallons spray mix.
Annual and perennial grass and broadleaf weeds, stale bed application	glyphosate, MOA 9 (numerous brands and formulations	See labels	See labels	Apply to emerged weeds at least 3 days before seeding or transplanting. Perennial weeds may require higher rates of glyphosate. Consult manufacturer's label for rates for specific weeds. When applying Roundup before transplanting crops into plastic mulch, carefully remove residues of this product from the plastic prior to transplanting. To prevent crop injury, residues can be removed by 0.5 inch rainfall or by applying water via a sprinkler system. Certain glyphosate formulations may require the addition of a surfactant. Adding nonionic surfactant to glyphosate formulated with nonionic surfactant may result in reduced weed control.
Morningglory and small pigweed < 1 inch	pyraflufen ethyl, MOA 14 (ET Herbicide) 0.208 L	1 to 2 oz	0.016 to 0.0032	Bareground. Wait 1 day following preplant burndown application before planting. Plasticulture. May apply over mulch; however, a single 0.5 inch irrigation/rain event plus a 7 day waiting period is needed before transplanting.
Emerged broadleaf and grass weeds	pelargonic acid, MOA 27 (Scythe) 4.2 EC	3 to 10% v/v		Apply before crop emergence and control emerged weeds. There is no residual activity. May be tank mixed with soil residual compounds. See label for instruction. May also be used as a banded spray between row middles. Use a shielded sprayer directed to the row middles to reduce drift to the crop.
Annual grasses and small-seeded broadleaf weeds	bensulide, MOA 8 (Prefar) 4 EC	5 to 6 qt	5 to 6	Registered for cucurbit vegetable group (Crop grouping 9). Apply preplant and incorporate into the soil 1 to 2 inches (1 inch incorporation is optimum) with a rototiller or tandem disk, or apply preemergence after seeding and follow with irrigation. Check replant restrictions for small grains and other crops on label.
Annual grasses and broadleaf weeds; weak on pigweed and morningglory	clomazone, MOA 13 (Command) 3 ME	0.4 to 0.67 pt	0.15 to 0.25	Apply immediately after seeding or just prior to transplanting with transplanted crop. Roots of transplants must be below the chemical barrier when planting. See label for further instruction.
Annual grasses and some small-seeded broadleaf weeds	ethalfluralin, MOA 3 (Curbit) 3 EC	3 to 4.5 pt	1.1 to 1.7	Apply postplant to seeded crop prior to crop emergence or as a banded spray between rows after crop emergence or transplanting. See label for timing. Shallow cultivation, irrigation, or rainfall within 5 days needed for good weed control. Do not use under mulches, row covers, or hot caps. Under conditions of unusually cold or wet soil and air temperatures, crop stunting and injury may occur. Crop injury can occur if seeding depth is too shallow.
Annual grasses and broadleaf weeds	ethalfluralin, MOA 3 + clomazone, MOA 13 (Strategy) 2.1 L	2 to 6 pt	0.4 to 1.2 + 0.125 to 0.375	Apply to the soil surface immediately after seeding crop for preemergence control of weeds. **Do not apply prior to planting crop. Do not soil incorporate.** May also be used as a **banded** treatment **between** rows after crop emergence or transplanting. Do not apply over or under plastic mulch. See label for application rate according to soil type and crop restriction instructions.
Yellow and purple nutsedge and broadleaf weeds	halosulfuron-methyl, MOA 2 (Profine 75) 75 DG (Sandea) 75 DG	0.5 to 0.75 oz	0.024 to 0.036	Apply after seeding or prior to transplanting crop. For transplanted crop, do not transplant until 7 days after application. Rate can be increased to 1 ounce of product per acre to middles between rows. Do not apply within 57 days of harvest.
Cantaloupes (Muskmelons), Postemergence				
Annual grasses and some small-seeded broadleaf weeds	DCPA, MOA 3 (Dacthal) W-75 (Dacthal) 6 F	8 to 10 lb 6 to 10 pt	4.5 to 7.5	**Not labeled for transplanted crop.** To improve preemergence control of late emerging weeds. Apply only when crop has 4 to 5 true leaves, is well-established, and growing conditions are favorable. Will not control emerged weeds. Incorporation not recommended.
	trifluralin, MOA 3 (Treflan HFP, Trifluralin, Trifluralin HF) 4EC	1 to 2 pt	0.5 to 0.75	**Apply as a directed spray to soil between rows after crop emergence when crop plants have reached** three to four true leaf stage of growth. Avoid contacting foliage as slight crop injury may occur. Set incorporation equipment to move treated soil around base of crop plants. Do not apply within 30 days of harvest. Will not control emerged weeds.
Yellow and purple nutsedge and broadleaf weeds including cocklebur, galinsoga, smartweed, ragweed, wild radish, and pigweed	halosulfuron-methyl, MOA 2 (Profine 75) 75 DG (Sandea) 75 DG	0.5 to 0.75 oz	0.024 to 0.036	Apply postemergence only after the crop has reached 3 to 5 true leaves but before first female flowers appear. Do not apply sooner than 14 days after transplanting. Use nonionic surfactant at 1 quart per 100 gallons of spray solution with all postemergence applications. Avoid over-the-top applications during late summer when temperature and humidity are high. Do not apply within 57 days of harvest.
Most broadleaf weeds less than 4 inches tall or rosettes less than 3 inches in diameter; does not control grasses	carfentrazone-ethyl, MOA 14 (Aim) 1.9 EW or 2 EC	up to 2 oz	up to 0.031	Apply post-directed using hooded sprayers for control of emerged weeds. If crop is contacted, burning of contacted area will occur. Use a crop oil concentrate or a nonionic surfactant with Aim. See label for directions. Coverage is essential for good weed control. Can be tank mixed with other registered herbicides. PHI = 0 days.
Most emerged weeds	glyphosate, MOA 9 (numerous brands and formulations)	See labels	See labels	Apply as a hooded spray in row middles, as shielded spray in row middles, as wiper applications in row middles, or postharvest. To avoid severe injury to crop, do not allow herbicide to contact foliage, green shoots, stems, exposed, roots, or fruit of crop. Do not apply within 14 days of harvest.

Table 7-16. Chemical Weed Control in Vegetable Crops

Weed	Herbicide, Mode of Action Code* and Formulation	Amount of Formulation Per Acre	Pounds Active Ingredient Per Acre	Precautions and Remarks
Cantaloupes (Muskmelons), Postemergence				
Annual and perennial grasses only	sethoxydim, MOA 1 (Poast) 1.5 EC	1 to 1.5 pt	0.2 to 0.3	Apply to emerged grasses. Consult manufacturer's label for specific rates and best times to treat. Add 1 quart of crop oil concentrate per acre. Adding crop oil to Poast may increase the likelihood of crop injury at high air temperatures. Do not apply Poast on days that are unusually hot and humid. Do not apply within 3 days of harvest.
	clethodim, MOA 1 (Arrow, Clethodim, Intensity, Select) 2 EC (Intensity One, Select Max) 1 EC	6 to 8 oz 9 to 16 oz	0.094 to 0.125 0.07 to 0.125	Apply postemergence for control of grass in cantaloupes (muskmelons). For Arrow, Clethodim, or Select, add 1 gallon crop oil concentrate per 100 gallons spray mix. For Select Max, add 2 pints nonionic surfactant per 100 gallons spray mixture. Adding crop oil may increase the likelihood of crop injury at high air temperatures. Very effective in controlling annual bluegrass. Apply to actively growing grasses not under drought stress. Do not apply within 14 days of harvest.
Broadleaf, grass and nutsedge	imazosulfuron, MOA 2 (League) 0.5 DF	4 to 6.4 oz	0.19 to 0.3	**ROW MIDDLE APPLICATION ONLY.** Apply anytime during the cropping season (up to 48 days prior to harvest), as long as the melons are well-established and at least 5 inches wide. Avoid contact with the melon crop. In plasticulture, prevent the spray from contacting the plastic. Consult label for further instructions. PHI = 48 days.
Carrots, Preplant				
Contact kill of all green foliage, stale bed application	paraquat, MOA 22 (Firestorm, Parazone) 3 SL (Gramoxone SL) 2 SL	1.3 to 2.7 pt 2 to 4 pt	0.5 to 1	Apply to emerged weeds before crop emergence as a broadcast or band treatment over a preformed row. Use sufficient water to give thorough coverage. Row should be formed several days ahead of planting and treating to allow maximum weed emergence. Use a nonionic surfactant at a rate of 16 to 32 ounces per 100 gallons spray mix or 1 gallon approved crop oil concentrate per 100 gallons spray mix.
Most emerged weeds except resistant pigweed	glyphosate, MOA 9 (numerous brands and formulations)	See labels	See labels	Apply to emerged weeds before seeding or crop emergence. Perennial weeds may require higher rates. Certain glyphosate formulations require the addition of surfactant. Adding nonionic surfactant to glyphosate formulated with nonionic surfactant may result in reduced weed control.
Emerged broadleaf and grass weeds	pelargonic acid, MOA 27 (Scythe) 4.2 EC	3 to 10% v/v		Apply as a preplant burndown or prior to emergence of plants from seed. There is no residual activity. May be tank mixed with soil residual compounds. See label for instruction. May also be used as a banded spray between row middles. Use a shielded sprayer directed to the row middles to reduce drift to the crop.
Carrots, Preplant incorporated (PPI) or Preemergence (PRE)				
Annual grasses and small-seeded broadleaf weeds	trifluralin, MOA 3 (Treflan, Trifluralin) 4 EC	1 to 2 pt	0.5 to 1	Apply preplant and incorporate into the soil 2 to 3 inches within 8 hr. Use lower rate on coarse soils with less than 2% organic matter.
Broadleaf and grass weeds	pendimethalin, MOA 3 (Prowl H_2O) 3.8 AS	2 pt	0.95	Apply postplant within 2 days after planting but prior to crop emergence. See label for instruction on layby treatment. PHI interval is 60 days.
Broadleaf and grass weeds	prometryn, MOA 5 (Caparol) 4L	2 to 4 pt	1 to 2	Apply as preemergence and or postemergence over the top to carrot. Make Post application through the six leaf stage of carrot. See label for application rate and crop rotation restrictions. PHI is 30 days.
Carrots, Postemergence				
Annual grasses and broadleaf weeds	linuron, MOA 7 (Lorox DF) 50 WDG	1.5 to 3 lb	0.75 to 1.5	Apply as a broadcast spray after carrots are at least 3 inches tall. If applied earlier crop injury may occur. Avoid spraying after three or more cloudy days. Repeat applications may be made, but do not exceed 4 pounds of Lorox DF per acre per season. Do not use a surfactant or crop oil. Carrot varieties vary in their resistance, therefore determine tolerance to Lorox DF before adoption as a field practice to prevent potential crop injury. See label for further directions. PHI = 14 days.
Annual broadleaf weeds and some grasses	metribuzin, MOA 5 (Dimetric, Metribuzin, TriCor DF) 75 WDG (Metri, TriCor 4F) 4 F	0.33 lb 0.5 pt	0.25 0.25	Apply overtop when weeds are less than 1 inch tall and carrots have 5 to 6 true leaves. A second application may be made after a time interval of at least 3 weeks. Do not apply unless 3 sunny days precede application. Do not apply within 3 days of other pesticide applications. Preharvest interval is 60 days.
Annual and perennial grasses	clethodim, MOA 1 (Arrow, Clethodim, Intensity, Select) 2 EC (Select Max) 1 EC	6 to 8 oz 9 to 16 oz	0.094 to 0.125 0.07 to 0.125	Apply to actively growing grasses not under drought stress. With Arrow, Clethodim, or Select, add 1 gallon crop oil concentrate per 100 gallons spray mix. With Select Max, add 2 pints nonionic surfactant per 100 gallons of spray mixture. Adding crop oil may increase the likelihood of crop injury at high air temperatures. Do not mix with other pesticides. Very effective in controlling annual bluegrass. Do not apply within 30 days of harvest.
	fluazifop, MOA 1 (Fusilade DX) 2 EC	6 to 16 oz	0.1 to 0.25	Apply to actively growing grasses not under drought stress. Up to 48 ounces of Fusilade DX may be applied per year. See label for rates for specific weeds. Add 1 gallon crop oil concentrate or 1 quart nonionic surfactant per 100 gallons spray mix. Do not mix with other pesticides. Do not apply within 45 days of harvest.
	sethoxydim, MOA 1 (Poast) 1.5 EC	1 to 1.5 pt	0.2 to 0.3	Apply to actively growing grasses not under drought stress. Consult manufacturer's label for specific rate and best times to treat. Add 1 quart of crop oil concentrate per acre. Adding crop oil may increase the likelihood of crop injury at high air temperatures. Do not apply on days that are unusually hot and humid. Do not apply with other pesticides. Do not apply within 30 days of harvest.
Carrots, Row Middles				
Most broadleaf weeds less than 4 inches tall or rosettes less than 3 inches in diameter; does not control grasses	carfentrazone-ethyl, MOA 14 (Aim) 1.9 EW or 2 EC	up to 2 ounces	up to 0.031	Apply as a hooded spray in row middles for control of emerged weeds. If crop is contacted, burning of contacted area will occur. Use a crop oil concentrate or a nonionic surfactant with Aim. See label for directions. Coverage is essential for good weed control. Can be tank mixed with other registered herbicides.
Most emerged weeds except resistant pigweed	glyphosate, MOA 9 (numerous brands and formulations)	See labels	See labels	Apply as a hooded spray in row middles, as shielded spray in row middles, as wiper applications in row middles, or postharvest. To avoid severe injury to crop, do not allow herbicide to contact foliage, green shoots or stems, exposed roots, or fruit of crop. Do not apply within 14 days of harvest.
Cauliflower – See Cole Crops				

Table 7-16. Chemical Weed Control in Vegetable Crops

Weed	Herbicide, Mode of Action Code* and Formulation	Amount of Formulation Per Acre	Pounds Active Ingredient Per Acre	Precautions and Remarks
Celery, Preplant				
Annual and perennial grass and broadleaf weeds, stale bed application	glyphosate, MOA 9 (numerous brands and formulations)	See labels	See labels	Apply to emerged weeds before crop emergence. Perennial weeds may require higher rates. Consult the manufacturer's label for rates for specific weeds. Certain glyphosate formulations require the addition of surfactant. Adding nonionic surfactant to glyphosate formulated with nonionic surfactant may result in reduced weed control.
Cutleaf evening primrose, Carolina geranium, henbit, and a few grasses	oxyfluorfen, MOA 14 (Goaltender) 4 F (Goal 2 XL) 2 EC	up to 1 pt up to 2 pt	up to 0.5	**Transplants only.** Apply to soil surface of pre-formed beds at least 30 days prior to transplanting. No research has been conducted in North Carolina, therefore, try on a limited number of acres first.
Emerged broadleaf and grass weeds	pelargonic acid, MOA 27 (Scythe) 4.2 EC	3 to 10% v/v		Apply as a preplant burndown. There is no residual activity. May be tank mixed with soil residual compounds. See label for instruction. May also be used as a banded spray between row middles. Use a shielded sprayer directed to the row middles to reduce drift to the crop.
Celery, Preplant incorporate (PPI) or Preemergence (PRE)				
Annual grasses and small-seeded broadleaf weeds	trifluralin, MOA 3 (Treflan, Treflan HFP, Trifluralin) 4 EC	1 to 2 pt	0.5 to 1 lb	Apply incorporated to direct seeded or transplant celery before planting, at planting, or immediately after planting. Incorporate within 8 hours of application. Use lower rate on coarse soils with less than 2% organic matter.
	Bensulide (Prefar) 4-E	5 to 6 qt	5 to 6	**Transplants only.** Apply PRE after planting, irrigate immediately. See label for rotation restrictions.
Celery, Postemergence				
Annual broadleaf and grass weeds	linuron, MOA 7 (Lorox DF) 50 WDG	1.5 to 3 lb	0.75 to 1.5	Apply after celery is transplanted and established but before celery is 8 inches tall. Grasses should be less than 2 inches in height, and broadleaf weeds should be less than 6 inches tall. Do not tank mix with other products including surfactant or crop oil. Avoid spraying after 3 or more cloudy days or when temperature exceeds 85 F. Not recommended for sands or loamy sand soil. Preharvest interval is 45 days.
Annual and perennial grasses only	clethodim, MOA 1 (Arrow, Clethodim, Intensity, Select) 2 EC (Select Max) 1 EC	6 to 8 oz 9 to 16 oz	0.094 to 0.125 0.07 to 0.125	Apply to actively growing grasses not under drought stress. With Arrow, Clethodim, or Select, add 1 gallon crop oil concentrate per 100 gallons spray mix. With Select Max, add 2 pints of nonionic surfactant per 100 gallons spray mixture. Very effective in controlling annual bluegrass. Apply to actively growing grasses not under drought stress. Adding crop oil may increase the likelihood of crop injury at high air temperature. Do not apply within 30 days of harvest.
	sethoxydim, MOA 1 (Poast) 1.5 EC	1 to 1.5 pt	0.2 to 0.3	Apply to actively growing grasses not under drought stress. Consult label for specific rates and best times to treat. Add 1 quart of crop oil concentrate per acre. Adding crop oil to Poast may increase the likelihood of crop injury at high air temperatures. Do not apply Poast on unusually hot and humid days. Do not apply within 30 days of harvest.
Most broadleaf weeds less than 4 inches tall or rosettes less than 3 inches in diameter; does not control grasses	carfentrazone-ethyl, MOA 14 (Aim) 1.9 EW or 2 EC	up to 2 oz	up to 0.031	Apply post-directed using hooded sprayers for control of emerged weeds. If crop is contacted, burning of contacted area will occur. Use a crop oil concentrate or a nonionic surfactant with Aim. See label for directions. Coverage is essential for good weed control. Can be tank mixed with other registered herbicides.
Most emerged weeds	glyphosate, MOA 9 (numerous brands and formulations)	See labels	See labels	**Row middles only.** Apply as a hooded spray in row middles, as shielded spray in row middles, as wiper applications in row middles, or postharvest. To avoid severe injury to crop, do not allow herbicide to contact foliage, green shoots, stems, exposed, roots, or fruit of crop. Do not apply within 14 days of harvest.
Cole Crops: Broccoli, Cabbage, Cauliflower — Preplant and Preemergence				
Contact kill of all green foliage, stale bed application	paraquat, MOA 22 (Firestorm, Parazone) 3 SL (Gramoxone SL) 2 SL	1.3 to 2.7 pt 2 to 4 pt	0.5 to 1	Apply in minimum of 10 gallons spray mix per acre to emerged weeds before crop emergence or transplanting as a broadcast or band treatment over a preformed row. Use sufficient water for thorough coverage. Row should be formed several days ahead of planting and treated to allow maximum weed emergence. Use nonionic surfactant at rate of 16 to 32 ounces per 100 gallons spray mix or 1 gallon approved crop oil concentrate per 100 gallons spray mix.
Most broadleaf weeds less than 4 inches tall or rosettes less than 3 inches in diameter; does not control grasses	carfentrazone-ethyl, MOA 14 (Aim) 2 EC	up to 2 oz	up to 0.031	Apply no later than one day before transplanting or seven days before seeding. See label for rate for crop oil or nonionic surfactant. Coverage is essential for good weed control. See label for more information.
Annual and perennial grass and broadleaf weeds, stale bed application	glyphosate, MOA 9 (numerous brands and formulations)	See labels	See labels	Apply to emerged weeds before crop emergence or before transplanting. Perennial weeds may require higher rates of glyphosate. Consult the manufacturer's label for rates for specific weeds. When applying Roundup before transplanting crops into plastic mulch, care must be taken to remove residues of this product from the plastic prior to transplanting. To prevent crop injury, residues can be removed by 0.5 inch natural rainfall or by applying water via a sprinkler system. Certain glyphosate formulations may require the addition of a surfactant. Adding nonionic surfactant to glyphosate formulated with nonionic surfactant may result in reduced weed control.
Emerged broadleaf and grass weeds	pelargonic acid, MOA 27 (Scythe) 4.2 EC	3 to 10% v/v		Also labeled for collards, kale, mustard/turnip greens. Apply as a preplant burndown or prior to emergence of plants from seed. There is no residual activity. May be tank mixed with soil residual compounds. May also be used as a banded spray between row middles. Use a shielded sprayer directed to the row middles to reduce drift to the crop.
Annual grasses and small-seeded broadleaf weeds	bensulide, MOA 8 (Prefar) 4 EC	5 to 6 qt	5 to 6	Also labeled for Chinese broccoli, broccoli rabe, Chinese cabbage (bok choy, Napa), Chinese mustard cabbage (gai choy), and kohlrabi. Apply preplant or preemergence after planting. With preemergence application, irrigate immediately after application. See label for more directions.
	trifluralin, MOA 3 (Treflan HFP, Trifluralin, Trifluralin HF) 4 EC	1 to 1.5 pt	0.5 to 0.75	Also labeled for Brussels sprouts. Apply and incorporate prior to transplanting. **Caution**: If soil conditions are cool and wet, reduced stands and stunting may occur. Direct seeded cole crops exhibit marginal tolerance to higher than recommended rates.
	DCPA, MOA 3 (Dacthal) W-75 (Dacthal) 6 F	8 to 10 lb 8 to 10 pt	6 to 7.5	Also labeled for Brussels sprouts and all other Brassica (cole) leafy vegetables in this crop group. Apply immediately after seeding or transplanting. May also be incorporated.

Table 7-16. Chemical Weed Control in Vegetable Crops

Weed	Herbicide, Mode of Action Code* and Formulation	Amount of Formulation Per Acre	Pounds Active Ingredient Per Acre	Precautions and Remarks
Cole Crops: Broccoli, Cabbage, Cauliflower — Preplant and Preemergence (continued)				
Annual grasses and broadleaf weeds; weak on pigweed	clomazone, MOA 13 (Command) 3ME	0.67 pt	0.25	**Direct seeded cabbage only.** Apply to the soil surface immediately after seeding. Offers weak control of pigweed. See label for further instructions. Limited research has been done on this product in this crop in North Carolina.
		0.67 to 1.3 pt	0.25 to 0.50	**Transplanted cabbage only.** Apply broadcast to the soil prior to transplanting cabbage. See label for further instructions. Limited research has been conducted with this product on this crop in North Carolina.
Hairy galinsoga, common lambsquarters, redroot pigweed, and Palmer amaranth	sulfentrazone, MOA 14 (Spartan) 4 F	2.25 to 4.5 oz	0.07 to 0.14	**Cabbage only (Transplanted Processing only).** May be applied 60 days prior to planting up to planting time. Application rate depends on soil type.
Annual grasses and small-seeded broadleaf weeds, including galinsoga, common ragweed, and smartweed	napropamide, MOA 15 (Devrinol DF) 50 DF (Devrinol DF-XT) 50 DF (Devrinol 2-XT) 2 EC	4 lb 4 lb 4 qt	2 2 2	Includes Brussels sprouts. Apply to weed-free soil just after seeding or transplanting as a surface application. Light cultivations, rainfall, or irrigation will be necessary within 24 hours to activate this chemical.
Many broadleaf weeds, including galinsoga, common ragweed, and smartweed	oxyfluorfen, MOA 14 (Goal 2 XL, Galigan) 2 EC (GoalTender) 4 E	1 to 2 pt 0.5 to 1 pt	0.25 to 0.5	**Transplants only.** Surface apply before transplanting. Do not incorporate or knock the bed off after application. *Do not spray over the top of transplants.* Oxyfluorfen is weak on grasses. Expect to see some temporary crop injury.
Cole Crops: Broccoli, Cabbage, Cauliflower — Postemergence				
Annual grasses, small seeded broadleaf weeds and nutsedge	S-metolachlor, MOA 15 (Dual Magnum) 7.62 EC	0.5 to 1 pt	0.5 to 1	A section 24(c) North Carolina Label must be obtained at www.farmassist.com prior to use. Label includes Chinese cabbage. **Transplants.** After transplanting in bareground, irrigate to seal soil around root ball; about 10 days after sealing soil apply Dual Magnum overtop. If applying in mulched systems, apply 10 days after transplanting. **Seeded.** Apply overtop after crop reaches 3 inches. **Row middle.** May be applied at a rate up to 1.25 pt/A.
Grass and broadleaf weeds	pendimethalin, MOA 3 (Prowl H2O) 3.8 AS	2.1 pt	0.97	Apply postemergence-directed-spray on the soil, beneath plants, and between rows. Do not spray foliage or stems because crop injury can occur. PHI for broccoli is 60 days and for other crops 70 days.
Broadleaf weeds including sowthistle, clover, cocklebur, jimsonweed, and ragweed	clopyralid, MOA 4 (Stinger) 3 EC	0.25 to 0.5 pt	0.09 to 0.187	Labeled for broccoli, cabbage, cauliflower, broccoli rabe, brussels sprouts, cavalo broccoli, Chinese cabbage (bok choy), Chinese broccoli, Chinese mustard, and Chinese cabbage (Napa). Apply to crop when weeds are small and actively growing. Will control most legumes. Do not apply within 30 days of harvest.
Most emerged weeds	glyphosate, MOA 9 (numerous brands and formulations)	See labels	See labels	**Row middles only.** Apply as a hooded spray in row middles, as shielded spray in row middles, as wiper applications in row middles, or postharvest. To avoid severe injury to crop, do not allow herbicide to contact foliage, green shoots, stems, exposed, roots, or fruit of crop. Do not apply within 14 days of harvest.
Most broadleaf weeds less than 4 inches tall or rosettes less than 3 inches in diameter, does not control grasses	carfentrazone-ethyl, MOA 14 (Aim) 1.9 EW or 2 EC	up to 2 oz	up to 0.031	Apply post directed using hooded sprayers for control of emerged weeds. If crop is contacted, burning of contacted area will occur. Use crop oil concentrate at up to 1 gallon per 100 gallons solution or a nonionic surfactant at 2 pints per 100 gallons of spray solution. Coverage is essential for good weed control. Can be tank mixed with other registered herbicides.
Annual and perennial grasses only	clethodim, MOA 1 (Arrow, Clethodim, Intensity, Select) 2 EC (Select Max) 1 EC	6 to 8 oz 9 to 16 oz	0.094 to 0.125 0.07 to 0.125	Apply to emerged grasses. Consult manufacturer's label for specific rates and best times to treat. For sethoxydim, add 1 quart of crop oil concentrate per acre. For Arrow, Clethodim, or Select, add crop oil concentrate at 1 gallon per 100 gallons of spray solution. For Select Max, add 2 pints nonionic surfactant per 100 gallons of spray mixture. Adding crop oil to Poast or Select may increase the likelihood of crop injury at high air temperature. Do not apply Poast or Select plus crop oil on days that are unusually hot and humid. Do not apply within 30 days of harvest.
	sethoxydim, MOA 1 (Poast) 1.5 EC	1 to 1.5 pt	0.2 to 0.3	
Corn (sweet), Preplant Burndown				
Most broadleaf weeds less than 4 inches tall or rosettes less than 3 inches in diameter; does not control grasses	carfentrazone-ethyl, MOA 14 (Aim) 1.9 EW or 2 EC	up to 2 oz	up to 0.031	Apply prior to planting or within 24 hours after planting. Use a crop oil concentrate or a nonionic surfactant with Aim. For optimum performance, make applications to actively growing weeds up to 4 inches high or rosettes less than 3 inches across. Coverage is essential for good weed control. Optimum broad spectrum control of annual and perennial weeds requires a tank mix with burndown herbicides such as glyphosate, paraquat or 2,4-D. Must be applied prior to the preharvest interval of 14 leaf collars. See label for directions.
Contact kill of all green foliage, stale bed and minimum tillage application	paraquat, MOA 22 (Firestorm, Parazone) 3 SL (Gramoxone SL) 2 SL	1.5 to 2.7 pt 2.4 to 4 pt	0.6 to 1	Apply in a minimum of 20 gallons spray mix per acre to emerged weeds before crop emergence as a broadcast or band treatment over a preformed row. Seedbeds should be formed several days ahead of planting and treating to allow maximum weed emergence. Plant with a minimum of soil movement for best results. Use a nonionic surfactant at a rate of 16 to 32 ounces per 100 gallons spray mix or 1 gallon approved crop oil concentrate per 100 gallons spray mix. May be tank mixed with preemergence sweet corn herbicides and herbicide combinations. See section on Corn (sweet), Preemergence. Check label for directions and specific rates.
Annual and perennial grass and broadleaf weeds, stale bed application	glyphosate, MOA 9 (numerous brands and formulations)	See labels	See labels	Apply to emerged weeds before crop emergence. Do not feed crop residue to livestock for 8 weeks following treatment. Perennial weeds may require higher rates of glyphosate. Consult manufacturer's label for rates for specific weeds. Check label for directions. Certain glyphosate formulations require addition of surfactant. Adding nonionic surfactant to glyphosate formulated with nonionic surfactant may result in reduced weed control. Glyphosate-resistant horseweed (marestail) is now common in eastern North Carolina counties. If horseweed is present at planting time, a tank mixture of paraquat and atrazine is suggested.

Table 7-16. Chemical Weed Control in Vegetable Crops

Weed	Herbicide, Mode of Action Code* and Formulation	Amount of Formulation Per Acre	Pounds Active Ingredient Per Acre	Precautions and Remarks
Corn (sweet), Preplant Burndown (continued)				
Broadleaf weeds	2,4-D amine 4, MOA 4 (various brands)	1 to 2 pt	0.5 to 1	May be tank mixed with glyphosate for broad spectrum weed control including glyphosate-resistant horseweed (marestail). See label for planting restrictions if applied prior to planting.
Corn (sweet), Preemergence				
Most annual grass weeds, including fall panicum, broadleaf signalgrass, and small-seeded broadleaf weeds	alachlor, MOA 15 (Micro-Tech) 4 FME	2 to 4 qt	2 to 4	Apply to soil surface immediately after planting. Higher rates will improve control of ragweed and lambsquarter. May be tank mixed with atrazine, glyphosate, or simazine. Various other brands are available. Check label for directions.
	dimethenamid, MOA 15 (Outlook) 6.0 EC	12 to 21 oz	0.56 to 1.0	Apply to soil surface immediately after planting. May be tank mixed with atrazine, glyphosate, or paraquat. See label for other herbicides that may be tank mixed to broaden weed control spectrum.
	metolachlor, MOA 15 (Me-Too-Lachlor II) 7.8 EC (Parallel) 7.8 EC	1 to 2 pt	0.98 to 1.98	See comments for s-metolachlor products. Products containing s-metolachlor are more active on weeds per unit of formulated product than those containing metolachlor. See label for all instructions.
	S-metolachlor, MOA 15 (Brawl II, Dual II Magnum, Medal II) 7.64 EC	1 to 2 pt	0.95 to 1.91	Apply to soil surface immediately after planting. May be tank mixed with atrazine, glyphosate, or simazine. Check label for directions. Rate is soil-texture and organic-matter dependent. See label for details.
	pyroxasulfone, MOA 15 (Zidua) 85 WG	1.5 to 4 oz	0.0796 to 0.213	Rate ranges based on soil texture. See label for specific rate relating to your fields. Sweet corn seed must be planted a minimum of 1-inch deep. Provides suppression of Texas panicum, seedling johnsongrass, and shattercane. Do not harvest sweet corn ears for human consumption less than 37 days after application of this herbicide. See label regarding tank mixtures for broader spectrum control and/or control of emerged weeds. Research with Zidua in sweet corn is limited in North Carolina.
Most annual broadleaf and grass weeds	atrazine, MOA 5 (various brands) 4 F (various brands) 90 WDG	1 to 2 qt 1.1 to 2.2 lb	1 to 2	Apply to soil surface immediately after planting. Shallow cultivations will improve control. Check label for restrictions on rotational crops. See label for reduced rate if soil coverage with plant residue is less than 30% at planting. Does not control fall panicum or smooth crabgrass. May be tank mixed with metolachlor, alachlor, glyphosate, paraquat, bentazon, or simazine. Check label for directions.
	alachlor, MOA 15 + atrazine, MOA 5 (Bullet or Lariat) 4 F	2.5 to 4.5 qt	1.56 to 2.81 + 0.94 to 1.69	Apply to soil surface immediately after planting. Soil texture and organic matter influence application rate. See label for further instruction.
	dimethenamid, MOA 15 + atrazine, MOA 5 (Guardsman Max) 5 F	2.4 to 4.6 pt	0.51 to 1 + 1 to 1.9	Apply to soil surface immediately after planting. Does not control Texas panicum, seedling johnsongrass, or shattercane adequately. Adjust rate for soil texture and organic matter according to label. See label for reduced rate if soil coverage with plant residue is less than 30% at planting. See labels for comments on rotational crops. See label for additional instructions.
	S-metolachlor, MOA 15 + atrazine, MOA 5 (Bicep II Magnum) 5.5 F	1.3 to 2.6 qt	0.78 to 1.56 + 1 to 2	Apply to soil surface immediately after planting. Does not adequately control Texas panicum, seedling johnsongrass, or shattercane. May not adequately control cocklebur, morningglory, or sicklepod. Cultivation or other herbicides may be needed. See label for rates based on soil texture and organic matter and for information on setback requirements from streams and lakes. See label for reduced rate if soil coverage with plant residue is less than 30% at planting and for comments on rotational crops.
Grass and broadleaf weeds	pendimethalin, MOA 3 (Prowl H2O) 3.8 AS	2 to 4 pt	1 to 2	Apply preemergence before crop germinates, or postemergence until sweet corn is 20 to 24 inches tall or in the V8 growth stage. Do not apply in reduced, minimum or no-till sweet corn. See label for use rates according to organic matter. See label for additional information. See label for tank mix options.
Broadleaf and grass weeds	Simazine, MOA 5 (Princep) 4L	2 qt	2	Apply preemergence before weeds and crop emerge. See label for tank mix options. PHI is 45 days.
Corn (sweet), Postemergence				
Most annual broadleaf and grass weeds	atrazine, MOA 5 (various brands) 4 L (various brands) 90 WDG	2 qt 2.2 lb	2	Apply overtop before weeds exceed 1.5 inches in height. See label for additional information in controlling larger weeds. See label for amount of oil concentrate to add to spray mix. See label on setback requirements from streams and lakes.
Annual grasses and broadleaf weeds	dimethenamid, MOA 15 (Outlook) 6.0 EC + atrazine, MOA 5 (AAtrex) 4 F or 90 WDG	8 to 21 oz + See label for rate	0.375 to 1 + See label for rate	Apply overtop corn before crop reaches 12 inches tall and before weeds exceed the two-leaf stage. Larger weeds will not be controlled. Good residual control of annual grass and broadleaf weeds. Do not apply within 50 days of sweet corn ear harvest. Do not apply to corn 12 inches or taller. Also available as the commercial products Guardsman.
	S-metolachlor, MOA 15 (Dual II Magnum) 7.64 EC + atrazine, MOA 5 (AAtrex) 4 F (AAtrex) 90 WDG	1 to 1.67 pt + 1 to 2 qt 1.3 to 2.2 lb	0.95 to 1.58 + 1 to 2	Apply overtop corn (5 inches or less) before weeds exceed the two-leaf stage. Larger weeds will not be controlled. Do not apply within 30 days of sweet corn ear harvest. Good residual control of annual grass and broadleaf weeds. Also available as Bicep II or Bicep II Magnum
Cocklebur, common ragweed, jimsonweed, Pennsylvania smartweed, velvetleaf, yellow nutsedge, and morningglory	bentazon, MOA 6 (Basagran) 4 SL	1 to 2 pt	0.5 to 1	Apply early postemergence overtop when weeds are small and corn has one to five leaves. See label for rates according to weed size and special directions for annual morningglory and yellow nutsedge control. Use a crop oil at a rate of 1 quart per acre.
Many broadleaf weeds	mesotrione, MOA 27 (Callisto) 4 EC	3 oz	0.094	Apply overtop corn 30 inches or less or 8 leaves or less to control emerged broadleaf weeds. Use nonionic surfactant at 2 pints per 100 gallons of spray solution. DO NOT add VAN or AMS adjuvants when making post application in sweetcorn or severe injury will occur. Most effective on small weeds, however, if weeds are greater than 5 inches or for improved control of certain weeds, certain atrazine formulations may be mixed with this herbicide. See label for further information. Do not apply within 45 days of harvest.

Table 7-16. Chemical Weed Control in Vegetable Crops

Weed	Herbicide, Mode of Action Code* and Formulation	Amount of Formulation Per Acre	Pounds Active Ingredient Per Acre	Precautions and Remarks
Corn (sweet), Postemergence				
Annual broadleaf weeds and some grasses	tembotrione, MOA 27 (Laudis) 3.5 L	3 fl oz	0.082	Can be applied overtop or with drop nozzles to sweet corn from emergence up to V7 stage. Controls most broadleaf weeds. Does not control sicklepod or prickly sida and only suppresses morningglory. Controls or suppresses some grasses. See label for weeds controlled and recommended size for treatment. Herbicide sensitivity in all hybrids and inbreds of sweet corn has not been tested. See label for information on adjuvant use. May be tankmixed with atrazine to increase weed spectrum and consistency of control. If tankmixed with atrazine do not apply if corn is 12 inches tall or greater. See label for further restrictions and instructions.
	Topramezone, MOA 27 (Impact) 2.8 L	0.75 fl oz	0.016	Can be applied overtop or with drop nozzles to sweet corn from emergence until 45 days prior to harvest. Does not control sicklepod and only suppresses morningglory. Controls or suppresses some grasses. See label for weeds controlled and recommended size for treatment. This product has not been tested on all inbred line for tolerance. See label for information on adjuvant use. See label for further restrictions and instructions. Do not apply within 45 days of sweet corn ear harvest.
Velvetleaf, spreading dayflower, morningglory species, and redroot pigweed. Will not control grasses	fluthiacet-methyl, MOA 14 (Cadet) 0.91 L	0.6 to 0.9 oz	0.0042 to 0.06	**Processing sweet corn only.** Apply to small weeds, generally about 2 inches tall. Will control large velvetleaf up to 36 inches. See label for information on adjuvant use. See label for further restrictions and instructions.
Annual broadleaf weeds	fluthiacet-methyl, MOA 14 + mesotrione, MOA 27 (Solstice)	2.5 to 3.15 fl oz	0.004 to 0.0053 0.074 to 0.0931	Apply up to the V8 growth stage (or 30 inches tall). See label for crop rotation restrictions. Do not include nitrogen based adjuvants (UAN or AMS) when making postemergence application or severe injury will occur. Use nonionic surfactant at 1 qt per 100 gallons of spray. Do not apply within 40 days of sweet corn ear harvest. See label for further instructions.
Velvetleaf, pigweed, nightshade, morningglory, common lambsquarters	carfentrazone-ethyl, MOA 14 (Aim) 2.0 EC	0.5 oz	0.008	Apply postemergence to actively growing weeds less than 4 inches high (rosettes less than 3 inches across) up to the 14-leaf collar stage of corn. Rates above 0.5 oz will aid in controlling larger weeds and certain weeds (see label for specific rate). Directed sprays will lessen the chance of crop injury and allow later application. Coverage of weeds is essential for control. Use nonionic surfactant (2 pints per 100 gallons of spray) with all applications. Under dry conditions, the use of crop oil concentrate may improve weed control. Mix with atrazine to improve control of many broadleaf weeds. Limited information is available concerning the use of this product in sweet corn. Do not apply more than 2 ounces per acre per season. Do not apply within 3 days of sweet corn ear harvest.
Broadleaf weeds including sowthistle, clover, cocklebur, jimsonweed, ragweed, Jerusalem artichoke, and thistle	clopyralid, MOA 4 (Stinger) 3 EC	0.25 to 0.67 pt	0.095 to 0.25	**Processing sweet corn only.** Apply to sweet corn when weeds are small (less than 5-leaf stage) and actively growing. Addition of surfactants, crop oils, or other adjuvants is not usually necessary when using Stinger. Use of adjuvants may reduce selectivity to the crop. Do not apply to sweet corn over 18 inches tall. Will control most legumes. Do not apply within 30 days of harvest.
Cocklebur, passionflower (maypop), pigweed, pokeweed, ragweed, smartweed (Pennsylvania), velvetleaf	halosulfuron-methyl, MOA 2 (Profine 75, Sandea) 75 WDG	0.67 oz	0.032	Apply over the top or with drop nozzles to sweet corn from spike to lay-by for control of emerged weeds. Add nonionic surfactant at 1 to 2 quarts per 100 gallons of spray solution. See label for all instructions and restrictions. Do not apply within 30 days of harvest.
Cocklebur, pigweed, lambsquarters, morningglory, sicklepod, and many other annual broadleaf weeds	2,4-D amine, MOA 4 (various brands) 3.8 SL	0.5 to 1 pt	0.24 to 0.48	Use 0.5 pt of 2,4-D overtop when corn is 4 to 5 inches tall and weeds are small. Increase rate to 1 pt as corn reaches 8 inches. Use drop nozzles and direct spray toward base if corn is over 8 inches tall. Do not cultivate for about 10 days after spraying as corn may be brittle. Reduce rate of 2,4-D if extremely hot and soil is wet. For better sicklepod and horsenettle control, add a nonionic surfactant when using a directed spray at a rate of 1 quart per 100 gallons spray solution. Do not apply within 45 days of sweet corn harvest.
Annual grasses and broadleaf weeds	paraquat, MOA 22 (Firestorm, Parazone) 3 SL (Gramoxone SL) 2 SL	0.7 to 1.3 pt 1 to 2 pt	0.25 to 0.5	DO NOT SPRAY OVERTOP OF CORN OR SEVERE INJURY WILL OCCUR. Make a post directed application in a minimum of 20 gallons spray mix per acre to emerged weeds when the smallest corn is **at least 10 inches tall.** Use nonionic surfactant at a rate of 16 to 32 ounces per 100 gallons spray mix or 1 gallon approved crop oil concentrate per 100 gallons spray mix. Use of a hooded or shielded sprayer will reduce crop injury.
Certain grasses, including barnyardgrass, foxtails, Texas panicum, and johnsongrass; and broadleaf weeds, including burcucumber, jimsonweed, pigweed, pokeweed, and smartweeds	nicosulfuron, MOA 2 (Accent) 75 WDG	0.67 oz	0.031	Apply to sweet corn up to 12 inches tall or up to and including 5 leaf collars. For corn 12 to 18 inches tall, apply only with drop nozzles. Sweet corn hybrids vary in their sensitivity to Accent. Do not apply to Merit sweet corn variety. Contact company representative for information on other local hybrids that have been evaluated with Accent. Accent may be applied to corn previously treated with Fortress, Aztec, or Force, or non-organophosphate soil insecticides regardless of soil type. See label for more information on use of soil insecticides with Accent. Label prohibits application of Accent to corn previously treated with Counter insecticide, and also indicates that applying Accent to corn previously treated with Counter 20 CR, Lorsban, or Thimet may result in unacceptable crop injury, especially on soils with less than 4% organic matter. See label for information on use of adjuvants.
Cucumbers, Preplant and Preemergence				
Suppression or control of most annual grasses and broadleaf weeds, full rate required for nutsedge control	metam sodium (Vapam HL) 42%	37.5 to 75 gal	15.7 to 31.5	Rates are dependent on soil type and weeds present. Apply when soil moisture is at field capacity (100 to 125%). Apply through soil injection using a rotary tiller or inject with knives no more than 4 inches apart; follow immediately with a roller to smooth and compact the soil surface or with mulch. May apply through drip irrigation prior to planting a second crop on mulch. Plant back interval is often 14 to 21 days and can be 30 days in some environments. See label for all restrictions and additional information.
Emerged broadleaf and grass weeds	pelargonic acid, MOA 27 (Scythe) 4.2 EC	3 to 10% v/v		Apply before crop emergence and control emerged weeds. There is no residual activity. May be tank mixed with soil residual compounds. See label for further instructions. May also be used as a banded spray between row middles. Use a shielded sprayer directed to the row middles to reduce drift to the crop.

Table 7-16. Chemical Weed Control in Vegetable Crops

Weed	Herbicide, Mode of Action Code* and Formulation	Amount of Formulation Per Acre	Pounds Active Ingredient Per Acre	Precautions and Remarks
Cucumbers, Preplant and Preemergence				
Most broadleaf weeds less than 4 inches tall or rosettes less than 3 inches in diameter, does not control grasses.	carfentrazone-ethyl, MOA 14 (Aim) 1.9 EW or 2 EC	up to 2 oz	up to 0.031	Aim 1.9 EW is registered for application in transplant production systems only. Aim 2 EC is registered in seeded and transplant production systems. Apply no later than one day before transplanting or no later than 7 days before seeding crop. See label for information about application timing. Use a crop oil at up to 1 gallon per 100 gallons of spray solution or a nonionic surfactant at 2 pints per 100 gallons of spray solution. Coverage is essential for good weed control. Can be tank mixed with other registered burndown herbicides.
Contact kill of all green foliage, stale bed application	paraquat, MOA 22 (Firestorm, Parazone) 3 SL	1.3 to 2.7 pt	0.5 to 1	Apply in a minimum of 10 gallons spray mix per acre to emerged weeds before crop emergence as a broadcast or band treatment over a preformed row. Use sufficient water to give thorough coverage. Row should be formed several days ahead of planting and treating to allow maximum weed emergence. Use a nonionic surfactant at a rate of 16 to 32 ounces per 100 gallons spray mix or 1 gallon approved crop oil concentrate per 100 gallons spray mix.
	(Gramoxone SL) 2 SL	2 to 4 pt		
Annual and perennial grass and broadleaf weeds, stale bed application	glyphosate, MOA 9 (numerous brands and formulations)	See labels	See labels	Apply to emerged weeds at least 3 days before seeding or transplanting. Perennial weeds may require higher rates of glyphosate. Consult the manufacturer's label for rates for specific weeds. When applying Roundup before transplanting crops into plastic mulch, care must be taken to remove residues of this product from the plastic prior to transplanting. To prevent crop injury, residues can be removed by 0.5 inch natural rainfall or by applying water via a sprinkler system. Certain glyphosate formulations require the addition of surfactant. Adding nonionic surfactant to glyphosate formulated with nonionic surfactant may result in reduced weed control.
Morningglory and small pigweed < 1 inch	pyraflufen ethyl, MOA 14 (ET Herbicide) 0.208 L	1 to 2 oz	0.016 to 0.0032	Bareground. Wait 1 day following preplant burndown application before planting. Plasticulture. May apply over mulch; however, a single 0.5 inch irrigation/rain event plus a 7 day waiting period is needed before transplanting.
Annual grasses and small-seeded broadleaf weeds	bensulide, MOA 8 (Prefar) 4 EC	5 to 6 qt	5 to 6	Registered for cucurbit vegetable group (Crop grouping 9). Apply preplant and incorporate into the soil 1 to 2 inches (1 inch incorporation is optimum) with a rototiller or tandem disk, or apply to the soil surface after seeding and follow with irrigation within 36 hours of application. Check replant restrictions for small grains on label.
Annual grasses and some small-seeded broadleaf weeds; weak on pigweed	clomazone, MOA 13 (Command) 3 ME	0.4 to 1 pt	0.15 to 0.375	Apply immediately after seeding. See label for further information.
Annual grasses and some small-seeded broadleaf weeds	ethalfluralin, MOA 3 (Curbit) 3 EC	3 to 4.5 pt	1.1 to 1.7	Apply postplant to seeded crop prior to crop emergence or as a banded spray between rows after crop emergence or transplanting. See label for timing. Shallow cultivation, irrigation, or rainfall within 5 days is needed for good weed control. Do not use under mulches, row covers, or hot caps. Under conditions of unusually cold or wet soil and air temperatures, crop stunting or injury may occur. Crop injury can occur if seeding depth is too shallow.
Annual grasses and broadleaf weeds	ethalfluralin, MOA 3 + clomazone, MOA 13 (Strategy) 2.1 L	2 to 6 pt	0.4 to 1.2 + 0.125 to 0.375	Apply to the soil surface immediately after crop seeding for preemergence control of weeds. **Do not apply prior to planting crop. Do not soil incorporate.** May also be used as a **banded** treatment **between** rows after crop emergence or transplanting. Do not apply over or under plastic mulch.
Yellow and purple nutsedge and broadleaf weeds	halosulfuron-methyl, MOA 2 (Profine 75) 75 DG (Sandea) 75 DG	0.5 to 0.75 oz	0.024 to 0.036	Apply after seeding but before soil cracking or prior to transplanting crop. For transplanting, do not transplant until 7 days after application. For seeded or transplanting cucumbers in plasticulture, do not plant within 7 days of Sandea application. Rate can be increased to 1 ounce of product per acre to middles between rows.
Cucumbers, Postemergence				
Annual grasses and small-seeded broadleaf weeds	trifluralin, MOA 3 (Treflan HFP, Trifluraline, Trifluralin HF) 4EC	1 to 2 pt	0.5 to 0.75	**Will not control emerged weeds. Row middles only.** To improve preemergence control of late emerging weeds. **Apply as a directed spray to soil between rows after crop emergence when crop plants have reached** three to four true leaf stage of growth. Avoid contacting crop foliage as slight crop injury may occur. Set incorporation equipment to move treated soil around base of crop plants. Do not apply within 30 days of harvest.
Yellow and purple nutsedge and broadleaf weeds, including cocklebur, galinsoga, smartweed, ragweed, wild radish and pigweed	halosulfuron-methyl, MOA 2 (Profine 75, Sandea) 75 DG	0.5 to 0.75 oz	0.024 to 0.036	Apply postemergence only after the crop has reached 3 to 5 true leaves but before first female flowers appear. Do not apply sooner than 14 days after transplanting. Use nonionic surfactant at 1 quart per 100 gallons of spray solution with all postemergence applications. Do not apply within 14 days of harvest.
Most broadleaf weeds less than 4 inches tall or rosettes less than 3 inches in diameter; does not control grasses	carfentrazone-ethyl, MOA 14 (Aim) 1.9 EW or 2 EC	up to 2 oz	up to 0.031	Apply post-directed using hooded sprayers for control of emerged weeds. If crop is contacted, burning of contacted area will occur. Use crop oil concentrate at up to 1 gallon per 100 gallons solution or a nonionic surfactant at 2 pints per 100 gallons of spray solution. Coverage is essential for good weed control. Can be tank mixed with other registered herbicides.
Most emerged weeds	glyphosate, MOA 9 (numerous brands and formulations)	See labels	See labels	**Row middles only.** Apply as a hooded spray in row middles, as shielded spray in row middles, as wiper applications in row middles, or postharvest. To avoid severe injury to crop, do not allow herbicide to contact foliage, green shoots, stems, exposed, roots, or fruit of crop. Do not apply within 14 days of harvest.
Annual and perennial grasses only	sethoxydim, MOA 1 (Poast) 1.5 EC	1 to 1.5 pt	0.2 to 0.3	Apply to emerged grasses. Consult manufacturer's label for specific rates and best times to treat. Add 1 quart of crop oil concentrate per acre. Adding crop oil to Poast may increase the likelihood of crop injury at high air temperatures. Do not apply Poast on days that are unusually hot and humid. Do not apply within 14 days of harvest.
	clethodim, MOA 1 (Arrow, Clethodim, Intensity, Select) 2 EC (Select Max, Intensity One) 1 EC	6 to 8 oz 9 to 16 oz	0.094 to 0.125 0.07 to 0.125	Control of emerged grasses. For Arrow, Clethodim, and Select, add 1 gallon crop oil concentrate per 100 gallons spray mix. For Select Max and Intensity One, add 2 pints nonionic surfactant per 100 gallons spray mixture. Adding crop oil may increase the likelihood of crop injury at high air temperatures. Very effective in controlling annual bluegrass. Apply to actively growing grasses not under drought stress. Do not apply within 14 days of harvest.

Table 7-16. Chemical Weed Control in Vegetable Crops

Weed	Herbicide, Mode of Action Code* and Formulation	Amount of Formulation Per Acre	Pounds Active Ingredient Per Acre	Precautions and Remarks
Eggplant, Preplant				
Suppression or control of most annual grasses and broadleaf weeds, full rate required for nutsedge control	metam sodium (Vapam HL) 42%	37.5 to 75 gal	15.7 to 31.5	Rates are dependent on soil type and weeds present. Apply when soil moisture is at field capacity (100 to 125%). Apply through soil injection using a rotary tiller or inject with knives no more than 4 inches apart; follow immediately with a roller to smooth and compact the soil surface or with mulch. May apply through drip irrigation prior to planting a second crop on mulch. Plant back interval is often 14 to 21 days and can be 30 days in some environments. See label for all restrictions and additional information. Chloropicrin (150 pounds per acre broadcast) will also be needed when laying first crop mulch to control nutsedge.
Contact kill of all green foliage, stale bed application	paraquat, MOA 22 (Firestorm) 3 SL (Parazone) 3 SL (Gramoxone SL) 2 SL	1.3 to 2.7 pt 2 to 4 pt	0.5 to 1	Apply in a minimum of 10 gallons spray mix per acre to emerged weeds before transplanting as a broadcast or band treatment over a preformed row. Use sufficient water to give thorough coverage. Row should be formed several days ahead of planting and treating to allow maximum weed emergence. Use a nonionic surfactant at a rate of 16 to 32 ounces per 100 gallons spray mix or 1 gallon approved crop oil concentrate per 100 gallons spray mix.
Most broadleaf weeds less than 4 inches tall or rosettes less than 3 inches in diameter; does not control grasses	carfentrazone-ethyl, MOA 14 (Aim) 1.9 EW or 2 EC	up to 2 oz	up to 0.031	**Aim 1.9 EW is registered for application in transplant production systems only. Aim 2 EC is registered in seeded and transplant production systems.** Apply no later than one day before transplanting crop (Aim 1.9 EW or Aim 2 EC) or no later than 7 days before seeding crop (Aim 2 EC only). See label for information about application timing. Use a crop oil at up to 1 gallon per 100 gallons of spray solution or a nonionic surfactant at 2 pints per 100 gallons of spray solution. Coverage is essential for good weed control. Can be tank mixed with other registered burndown herbicides.
Annual and perennial grass and broadleaf weeds, stale bed application	glyphosate, MOA 9 (numerous brands and formulations)	See labels	See labels	Apply to emerged weeds at least 3 days before seeding or transplanting. Perennial weeds may require higher rates of glyphosate. Consult the manufacturer's label for rates for specific weeds. When applying Roundup before transplanting crops into plastic mulch, care must be taken to remove residues of this product from the plastic prior to transplanting. To prevent crop injury, residues can be removed by 0.5 inch natural rainfall or by applying water via a sprinkler system. Certain glyphosate formulations require the addition of surfactant. Adding nonionic surfactant to glyphosate formulated with nonionic surfactant may result in reduced weed control.
Eggplant, Preemergence				
Annual grasses and small-seeded broadleaf weeds	bensulide, MOA 8 (Prefar) 4 EC	5 to 6 qt	5 to 6	Apply preplant incorporated (1 inch incorporation is optimum) or preemergence after planting. With preemergence application, irrigate immediately after application. See label for more directions.
Annual grasses and some broadleaf weeds	napropamide, MOA 15 (Devrinol, Devrinol DF-XT) 50 DF (Devrinol 2-XT) 2 EC	2 to 4 lb 2 to 4 qt	1 to 2	**Transplanted eggplant only.** Apply preplant and incorporate into soil 1 to 2 inches using a rototiller or tandem disk. Shallow cultivations or irrigation will improve control. See label for small grains replanting restrictions. May also be applied in the **row middles** between plastic covered beds. See label for more information. See XT labels for information regarding delay in irrigation event.
Grass and broadleaf weeds including black nightshade and hairy nightshade	pendimethalin, MOA 3 (Prowl H2O) 3.8 AS	1.0 to 1.5 pt	0.48 to 0.72	**Transplanted eggplant only.** Apply to row middles or under the plastic. Do not exceed 3 pt/A per year. 70 day PHI.
Eggplant, Postemergence				
Annual grasses and small-seeded broadleaf weeds	DCPA, MOA 3 (Dacthal) W-75 (Dacthal) 6 F	6 to 10 lb 6 to 10 pt	4.5 to 7.5	Application confined to a period of 4 to 6 weeks after transplanting. To improve preemergence control of late emerging weeds. Apply to weed-free soil over the top of transplants.
Annual and perennial grasses only	sethoxydim, MOA 1 (Poast) 1.5 EC	1 to 1.5 pt	0.2 to 0.3	Apply to emerged grasses. Consult manufacturer's label for specific rates and best times to treat. Add 1 quart of crop oil concentrate per acre. Adding crop oil to Poast may increase the likelihood of crop injury at high air temperatures. Do not apply Poast on days that are unusually hot and humid. Do not apply within 20 days of harvest.
	clethodim, MOA 1 (Arrow, Clethodim, Intensity, Select) 2 EC (Select Max, Intensity One) 1 EC	6 to 8 oz 9 to 16 oz	0.094 to 0.125 0.07 to 0.125	Apply postemergence for control of grasses. With Arrow, Clethodim, or Select, add 1 gallon crop oil concentrate per 100 gallons spray mix. With Select Max, add 2 pints of nonionic surfactant per 100 gallons spray mixture. Adding crop oil may increase the likelihood of crop injury at high air temperature. Very effective in controlling annual bluegrass. Apply to actively growing grasses not under drought stress. Do not apply within 20 days of harvest.
Eggplant, Row Middles				
Most broadleaf weeds less than 4 inch tall or rosettes less than 3 inches in diameter; does not control grasses	carfentrazone-ethyl, MOA 14 (Aim) 1.9 EW or 2 EC	up to 2 oz	up to 0.031	Apply post-directed using hooded sprayers for control of emerged weeds. If crop is contacted, burning of contacted area will occur. Use crop oil concentrate at up to 1 gallon per 100 gallons solution or a nonionic surfactant at 2 pints per 100 gallons of spray solution. Coverage is essential for good weed control. Can be tank mixed with other registered herbicides. PHI 0 days.
Most emerged weeds	glyphosate, MOA 9 ((numerous brands and formulations)	See labels	See labels	Apply as a hooded spray in row middles, as shielded spray in row middles, as wiper applications in row middles, or postharvest. To avoid severe injury to crop, do not allow herbicide to contact foliage, green shoots or stems, exposed roots, or fruit of crop. Do not apply within 14 days of harvest.
Yellow and purple nutsedge and broadleaf weeds	halosulfuron-methyl, MOA 2 (Profine 75) 75 DG (Sandea) 75 DG	0.5 to 1 oz	0.024 to 0.048	Apply between rows as a postemergence spray. Do not allow spray to contact crop or plastic mulch. Early season application will give postemergence and preemergence control. For postemergence applications, use nonionic surfactant at 1 quart per 100 gallons of spray solution. Preharvest interval is 30 days.
Contact kill of all green foliage	paraquat, MOA 22 (Firestorm, Parazone) 3SL (Gramoxone SL) 2 SL	1.3 pt 2 pt	0.5	Apply in 10 gallons spray mix as a shielded spray to emerged weeds between rows of eggplant. Use a nonionic surfactant at a rate of 16 to 32 ounces per 100 gallons spray mix or 1 gallon approved crop oil concentrate per 100 gallons spray mix. Do not allow spray to contact crop or injury will result.

Table 7-16. Chemical Weed Control in Vegetable Crops

Weed	Herbicide, Mode of Action Code* and Formulation	Amount of Formulation Per Acre	Pounds Active Ingredient Per Acre	Precautions and Remarks
Garlic, Preplant or Preemergence				
Annual and perennial grass and broadleaf weeds	glyphosate, MOA 9 (numerous brands and formulations)	See labels	See labels	Stale bed application. Apply to emerged weeds at least 3 days before planting. Perennial weeds may require higher rates of glyphosate. Consult the manufacturer's label for rates for specific weeds. Certain glyphosate formulations require the addition of surfactant. Adding nonionic surfactant to glyphosate formulated with nonionic surfactant may result in reduced weed control.
	paraquat, MOA 22 (Firestorm, Parazone) 3 SL (Gramoxone SL) 2 SL	1.7 to 2.7 pt 2.5 to 4 pt	0.6 to 1	Apply in a minimum of 20 gallons spray mix per acre to emerged weeds before crop emergence as a broadcast or band treatment over a preformed row. Row should be formed several days ahead of planting and treating to allow maximum weed emergence. Use a nonionic surfactant at a rate of 16 to 32 ounces per 100 gallons spray mix or 1 gallon approved crop oil concentrate per 100 gallons spray mix. Do not apply within 60 days of harvest.
Most broadleaf weeds less than 4 inches tall or rosettes less than 3 inches in diameter; does not control grasses	carfentrazone-ethyl, MOA 14 (Aim) 2.0 EC	Up to 2oz	0.031	Apply no later than 30 days before planting. See label for specific Aim rate relating to weed species and proper adjuvant and rate. Coverage is essential for good weed control. Can be tank mixed with other registered burndown herbicides.
Emerged broadleaf weeds	pyraflufen, MOA 14 (ET Herbicide) 0.208 EC	0.5 to 2 fl oz	0.0008 to 0.003	Apply as a preplant burndown treatment in a minimum of 10 gallon solution per acre. See label for information on use of adjuvant.
Emerged broadleaf and grass weeds	pelargonic acid, MOA 27 (Scythe) 4.2 EC	3 to 10% v/v		Apply as a preplant burndown treatment or use in row middles using shielded sprayer.
Annual grasses and small-seeded broadleaf weeds	bensulide, MOA 8 (Prefar) 4 EC	5 to 6 qt	5 to 6	Apply preplant incorporated (1 inch incorporation is optimum) or preemergence after planting. With preemergence application, irrigate immediately after application. See label for more directions.
Annual broadleaf weeds Including henbit, purslane, pigweed, primrose, smartweed, and many others; controls small emerged weeds as well	oxyfluorfen, MOA 14 (Galigan, Goal 2XL) 2 E	1 to 2 pt	0.25 to 0.5	**Transplanted garlic only.** For use on a fallow bed. Garlic may be planted immediately following application of 1 pt of product. For rates above 1 pt do not plant within 30 days. PHI 60 days.
Annual broadleaf and grass weeds	ethofumesate, MOA 16 (Ethotron) 4 SC	16 to 32 oz	0.5 to 1	Can use as preplant, preemergence or postemergence application. The use of higher than specified rates may cause injury and/or carry over problems. Rainfall of at least 0.5 inch is needed for activation. See label for more information.
Garlic, Preemergence				
Annual grasses and small-seeded broadleaf weeds	dimethenamid-P, MOA 15 (Outlook) 6 EC	12 to 21 oz	0.6 to 1	For preemergence weed control. Apply after crop has reached 2 true leaves until a minimum of 30 days before harvest. If applications are made to transplanted crop, DO NOT apply until transplants are in the ground and soil has settled around transplants with several days to recover.
	flumioxazin, MOA 14 (Chateau) 51 SW	6 oz	0.188	For preemergence weed control. Apply prior to garlic and weed emergence. Application should be made within 3 days after planting garlic.
	pendimethalin, MOA 3 (Prowl) 3.3 EC (Prowl H$_2$O) 3.8 AS	1.2 to 3.6 pt 1.5 to 3 pt	0.5 to 1.5 0.75 to 1.5	For preemergence weed control. Apply preemergence after planting but prior to weed and crop emergence, or postemergence to garlic in the 1 to 5 true leaf stage. Prowl can be applied sequentially by applying preemergence followed by a postemergence application. Preharvest interval is 45 days.
Garlic, Postemergence				
Annual grasses and small-seeded broadleaf weeds	dimethenamid-P, MOA 15 (Outlook) 6 EC	12 to 21 oz	0.6 to 1	For preemergence weed control applied after crop has reached 2 true leaves until a minimum of 30 days before harvest. If applications are made to transplanted crop, DO NOT apply until transplants are in the ground and soil has settled around transplants with several days to recover.
Most annual broadleaf weeds	oxyfluorfen, MOA 14 (Galigan) 2 E (Goal 2 XL) 2 EC (GoalTender) 4 E	0.5 pt 0.5 pt 0.25 pt	0.12	**Transplanted dry bulb only.** May be used as a postemergence spray to both the weeds and crop after the garlic has at least two fully developed true leaves. Some injury to garlic may result. Injury will be more severe if the chemical is applied during cool, wet weather. Weeds should be in the 2- to 4-leaf stage for best results. Preharvest interval is 60 days.
Annual and perennial grasses only	clethodim, MOA 1 (Arrow, Clethodim, Intensity, Select) 2 EC (Select Max, Intensity One) 1 EC	6 to 16 oz 9 to 32 oz	0.09 to 0.25 0.07 to 0.25	Apply to emerged grasses. Consult manufacturer's label for specific rates and best times to treat. With Arrow, Clethodim, or Select, add 1 gallon crop oil concentrate per 100 gallons spray mix. With Select Max, add 2 pints of nonionic surfactant per 100 gallons spray mixture. Adding crop oil may increase the likelihood of crop injury at high air temperatures. Do not apply Arrow, Clethodim, or Select on unusually hot and humid days. Do not apply within 45 days of harvest. Very effective in controlling annual bluegrass.
	fluazifop, MOA 1 (Fusilade DX) 2 EC	6 to 16 oz	0.1 to 0.25	Apply to emerged grasses. Consult label for specific rates and best times to treat. Add 1 gallon crop oil concentrate or 1 quart nonionic surfactant per 100 gallons spray mix. Do not apply on unusually hot and humid days. Do not apply within 45 days of harvest.
	sethoxydim, MOA 1 (Poast) 1.5 EC	1 pt	0.2	Apply to emerged grasses. Consult manufacturer's label for specific rates and best times to treat. Add 1 quart of crop oil concentrate per acre. Adding crop oil to Poast may increase the likelihood of crop injury at high air temperatures. Do not apply Poast on days that are unusually hot and humid. Do not apply within 30 days of harvest.
Garlic, Row Middles				
Most emerged weeds	glyphosate, MOA 9 (numerous brands and formulations)	See labels	See labels	**Row middles only.** Apply as a hooded spray in row middles, as shielded spray in row middles, as wiper applications in row middles, or post harvest. To avoid severe injury to crop, do not allow herbicide to contact foliage, green shoots, stems, exposed, roots, or fruit of crop. Do not apply within 14 days of harvest.
Most broadleaf weeds less than 4 inches tall or rosettes less than 3 inches in diameter; does not control grasses	carfentrazone-ethyl, MOA 14 (Aim) 1.9 EW or 2 EC	up to 2 oz	up to 0.031	Apply post-directed using hooded sprayers for control of emerged weeds. If crop is contacted, burning of contacted area will occur. Use a nonionic surfactant or crop oil with Aim. See label for rate. Coverage is essential for good weed control. Can be tank mixed with other registered herbicides.

Table 7-16. Chemical Weed Control in Vegetable Crops

Weed	Herbicide, Mode of Action Code* and Formulation	Amount of Formulation Per Acre	Pounds Active Ingredient Per Acre	Precautions and Remarks
Greens (Collard, kale, mustard, and turnip greens or roots), Preplant				
Emerged broadleaf and grass weeds	pelargonic acid, MOA 27 (Scythe) 4.2 EC	3 to 10% v/v		Apply as preplant burndown emerged weeds. See label for instruction. May also be used as a banded spray between row middles. Use a shielded sprayer directed to the row middles to reduce drift to the crop.
Contact kill of all green foliage, stale bed application	paraquat, MOA 22 (Firestorm, Parazone) 3 SL (Gramoxone SL) 2 SL	1.3 to 2.7 pt 2 to 4 pt	0.5 to 1	**Collard and turnip only.** Apply in a minimum of 10 gallons spray mix per acre to emerged weeds before crop emergence or transplanting as a broadcast or band treatment over a preformed row. Use sufficient water to give thorough coverage. Row should be formed several days ahead of planting and treating to allow maximum weed emergence. Use a nonionic surfactant at a rate of 16 to 32 ounces per 100 gallons spray mix or 1 gallon approved crop oil concentrate per 100 gallons spray mix.
Annual and perennial grass and broadleaf weeds, stale bed application	glyphosate, MOA 9 (numerous brands and formulations)	See labels	See labels	Apply to emerged weeds before crop emergence. Do not feed crop residue to livestock for 8 weeks following treatment. Perennial weeds may require higher rates of glyphosate. Consult the manufacturer's label for rates for specific weeds. Certain glyphosate formulations require the addition of surfactant. Adding nonionic surfactant to glyphosate formulated with nonionic surfactant may result in reduced weed control.
Annual grasses and small-seeded broadleaf weeds	trifluralin, MOA 3 (Treflan) 4 EC	0.75 to 1.5 pt	0.5 to 0.75	**Greens.** Collard, kale, mustard and turnip (fresh or processing). Apply preplant and incorporate into the soil 2 to 3 inches within 8 hr using a rototiller or tandem disk. **Turnip root.** TREFLAN HFP ONLY. A Section 24 (c) North Carolina Local Need Label must be obtained prior to use. Apply Treflan HFP as a preplant soil incorporated treatment.
	bensulide, MOA 8 (Prefar) 4 EC	5 to 6 qt	5 to 6	Brassica (cole) leafy vegetable group. Not labeled for turnip. Apply preplant or preemergence after planting. With preemergence application, irrigate immediately after application. See label for more directions.
	DCPA, MOA 3 (Dacthal) W-75 (Dacthal) 6 F	6 to 10 lb 6 to 10 pt	4.5 to 7.5	Also labeled for broccoli raab (raab, raab salad), and hanover salad. Apply immediately after seeding. May also be incorporated.
Greens, Postemergence				
Broadleaf weeds including sowthistle clover, cocklebur, jimsonweed, and ragweed	clopyralid, MOA 4 (Stinger) 3 EC	0.25 to 0.5 pt	0.11 to 0.187	**Kale, collards, mustard, turnip, mizuna, mustard spinach, and rape.** Apply to crop when weeds are small and actively growing. Will control most legumes. For kale, collards, mustard, and turnip (roots), do not apply within 30 days of harvest. For turnip tops, do not apply within 15 days of harvest. Mustard green injury observed in some research trials.
Annual and perennial grasses only	clethodim, MOA 1 (Arrow, Clethodim, Intensity, Select) 2 EC (Select Max) 1 EC	6 to 8 oz 9 to 16 oz	0.094 to 0.125 0.07 to 0.125	Apply postemergence for grass control. With Arrow, Clethodim, or Select, add 1 gallon crop oil concentrate per 100 gallons spray mix. With Select Max, add 2 pints nonionic surfactant per 100 gallons spray mixture. Adding crop oil may increase the likelihood of crop injury at high air temperatures. Very effective in controlling annual bluegrass. Apply to actively growing grasses not under drought stress. Do not apply within 14 days of harvest of green crops. Do not apply within 30 days of harvest of turnips grown for roots.
Annual and perennial grasses only (continued)	sethoxydim, MOA 1 (Poast) 1.5 EC	1 to 1.5 pt	0.2 to 0.3	ALSO LABELED FOR RAPE GREENS. Do not apply within 14 days of harvest of turnip and 30 days of harvest of other greens. Apply to emerged grasses. Consult manufacturer's label for specific rates and best times to treat. Add 1 quart crop oil concentrate per acre. Adding crop oil to Poast may increase the likelihood of crop injury at high air temperatures. Do not apply on unusually hot and humid days.
Greens, Row middles				
Most broadleaf weeds less than 4 inches tall or rosettes less than 3 inches in diameter; does not control grasses	carfentrazone-ethyl, MOA 14 (Aim) 1.9 EW or 2 EC	up to 2 oz	up to 0.031	Apply post-directed using hooded sprayers for control of emerged weeds. If crop is contacted, burning of contacted area will occur. Use a nonionic surfactant or crop oil with Aim. See label for rate. Coverage is essential for good weed control. Can be tank mixed with other registered herbicides.
Most emerged weeds	glyphosate, MOA 9 (numerous brands and formulations)	See labels	See labels	**Not labeled for turnip greens.** Apply as a hooded spray in row middles, as shielded spray in row middles, as wiper applications in row middles, or postharvest. To avoid severe injury to crop, do not allow herbicide to contact foliage, green shoots or stems, exposed roots, or fruit of crop. Do not apply within 14 days of harvest.
Lettuce, Preplant				
Suppression or control of most annual grasses and broadleaf weeds, full rate required for nutsedge control	metam sodium (Vapam HL) 42%	37.5 to 75 gal	15.7 to 31.5	Rates are dependent on soil type and weeds present. Apply when soil moisture is at field capacity (100 to 125%). Apply through soil injection using a rotary tiller or inject with knives no more than 4 inches apart; follow immediately with a roller to smooth and compact the soil surface or with mulch. May apply through drip irrigation prior to planting a second crop on mulch. Plant back interval is often 14 to 21 days and can be 30 days in some environments. See label for all restrictions and additional information.
Contact kill of all green foliage, stale bed application	paraquat, MOA 22 (Firestorm, Parazone) 3 SL (Gramoxone SL) 2 SL	1.3 to 2.7 pt 2 to 4 pt	0.5 to 1	Apply in a minimum of 10 gallons spray mix per acre to emerged weeds before crop emerges as a broadcast or band treatment over a preformed row. Row should be formed several days ahead of planting and treating to allow maximum weed emergence. Use a nonionic surfactant at a rate of 16 to 32 ounces per 100 gallons spray solution or 1 gallon approved crop oil concentrate per 100 gallons spray mix.
Annual and perennial grass and broadleaf weeds, stale bed application	glyphosate, MOA 9 (numerous brands and formulations)	See labels	See labels	Apply to emerged weeds before crop emergence. Do not feed crop residue to livestock for 8 weeks following treatment. Perennial weeds may require higher rates of glyphosate. Consult the manufacturer's label for rates for specific weeds. Certain glyphosate formulations require the addition of surfactant. Adding nonionic surfactant to glyphosate formulated with nonionic surfactant may result in reduced weed control.
Lettuce, Preplant or preemergence				
Annual grasses and small-seeded broadleaf weeds	benefin, MOA 3 (Balan) 60 WDG	2 to 2.5 lb	1.2 to 1.5	Apply preplant and incorporate 2 to 3 inches deep with a rototiller or tandem disk before seeding or transplanting.
	bensulide, MOA 8 (Prefar) 4 EC	5 to 6 qt	5 to 6	Apply preplant incorporated (1 inch incorporation is optimum) or preemergence after planting. With preemergence application, irrigate immediately after application. See label for more directions.

Table 7-16. Chemical Weed Control in Vegetable Crops

Weed	Herbicide, Mode of Action Code* and Formulation	Amount of Formulation Per Acre	Pounds Active Ingredient Per Acre	Precautions and Remarks
Lettuce, Preplant or preemergence				
Most annual grasses and broadleaf weeds	pronamide, MOA 3 (Kerb) 50 WP (Kerb) 3.3 SC	2 to 4 lb 1.25 to 5 pt	0.5 to 2 lb	Kerb 3.3 SC has supplemental label allowing application on leaf lettuce as well as head lettuce. Also labeled in endive, escarole or radicchio greens. Can be used preplant or preemergence. Application can also be made postemergence to head lettuce but should be made before weed germination if possible or before weeds are beyond the two-leaf stage. Moisture is necessary to activate. Do not apply within 55 days of harvest. Make only one application per crop. Consult label for planting restrictions for rotational crops.
Lettuce, Postemergence				
Annual and perennial grasses only	sethoxydim, MOA 1 (Poast) 1.5 EC	1 to 1.5 pt	0.2 to 0.3	Apply to emerged grasses. Arrow, Clethodim, and Select are only registered for leaf lettuce. Consult manufacturer's label for specific rates and best times to treat. For sethoxydim, add 1 quart of crop oil concentrate per acre. Use of Poast or clethodim with crop oil may increase the likelihood of crop injury at high air temperatures. For Arrow, Clethodim, or Select, add 1 gallon crop oil concentrate per 100 gallons spray solution. With Select Max, add 2 pints nonionic surfactant per 100 gallons spray mixture. Do not apply on unusually hot and humid days. Do not apply sethoxydim within 30 days of harvest on head lettuce or within 15 days of harvest on leaf lettuce. For clethodim, do not apply within 14 days of harvest.
	clethodim, MOA 1 (Arrow, Clethodim, Intensity, Select) 2 EC	6 to 8 oz	0.09 to 0.125	
	(Select Max, Intensity One) 1 EC	9 to 16 oz	0.07 to 0.125	
Most emerged weeds	glyphosate, MOA 9 (numerous brands and formulations)	See labels	See labels	**Row middles only.** Apply as a hooded spray in row middles, as shielded spray in row middles, as wiper applications in row middles, or postharvest. To avoid severe injury to crop, do not allow herbicide to contact foliage, green shoots, stems, exposed, roots, or fruit of crop. Do not apply within 14 days of harvest.
Most broadleaf weeds less than 4 inches tall or rosettes less than 3 inches in diameter; does not control grasses	carfentrazone-ethyl, MOA 14 (Aim) 1.9 EW or 2 EC	up to 2 oz	up to 0.031	Apply post-directed using hooded sprayers for control of emerged weeds. If crop is contacted, burning of contacted area will occur. Use a nonionic surfactant or crop oil with Aim. See label for rate. Coverage is essential for good weed control. Can be tank mixed with other registered herbicides.
Okra, Preplant and Preemergence				
Annual and perennial grass and broadleaf weeds, stale bed application	glyphosate, MOA 9 (numerous brands and formulations)	See labels	See labels	Apply to emerged weeds before crop emergence. Perennial weeds may require higher rates. Consult the manufacturer's label for rates for specific weeds. Certain glyphosate formulations require the addition of surfactant. Adding nonionic surfactant to glyphosate formulated with nonionic surfactant may result in reduced weed control.
Contact kill of green foliage	paraquat, MOA 22 (Gramoxone) 2 SL	2 to 4 pt	0.5 to 1	Apply to emerged weeds before planting or up to 1 day after planting.
Most broadleaf weeds less than 4 inches tall or rosettes less than 3 inches in diameter; does not control grasses	carfentrazone-ethyl, MOA 14 (Aim) 1.9 EW or 2 EC	up to 2 oz	up to 0.031	Apply no later than 1 day before transplanting crop. Use a nonionic surfactant or crop oil with Aim. See label for rate. Coverage is essential for good weed control. Can be tank mixed with other registered burndown herbicides. PHI 0 days.
Annual grasses and small-seeded broadleaf weeds	trifluralin, MOA 3 (Treflan, Treflan HFP, Trifluralin, Trilin) 4 EC	1 to 2 pt	0.5 to 1	Apply preplant and incorporate into the soil 2 to 3 inches within 8 hr using a rototiller or tandem disk.
Okra, Preplant and Preemergence (continued)				
Broadleaf and grass weeds	prometryn, MOA 5 (Caparol) 4L	1.5 to 3 pt	0.75 to 1	Apply preemergence and or post-directed to okra. Do not exceed 3 pints per acre of Caparol per season. See label for crop rotation restrictions. PHI is 14 days.
Annual broadleaf weeds including pigweed spp.	mesotrione, MOA 27 (Callisto) 4 L	6 oz	0.19	May be applied as row middle or hooded POST-directed application but not both. For preemergence row middle application, apply as a banded application to the row middles prior to weed emergence. Leave 1 foot of untreated area over the okra row or 6 inches on each side of the planted row. Do not apply Callisto directly over the planted row or severe injury may occur. Injury risk is greatest on coarse textured soils (sand, sandy loam or loamy sand).
Okra, Postemergence				
Annual and perennial grasses only	sethoxydim, MOA 1 (Poast) 1.5 EC	1 to 1.5 pt	0.3 lb	Apply to actively growing grasses not under drought stress. Do not apply on days that are unusually hot and humid. Do not apply within 14 days of harvest.
Okra, Row middles				
Most broadleaf weeds less than 4 inches tall or rosettes less than 3 inches in diameter; does not control grasses	carfentrazone-ethyl, MOA 14 (Aim) 1.9 EW or 2 EC	up to 2 oz	up to 0.031	Apply post-directed using hooded sprayers for control of emerged weeds. If crop is contacted, burning of contacted area will occur. Use a nonionic surfactant or crop oil with Aim. See label for rate. Coverage is essential for good weed control. Can be tank mixed with other registered herbicides. PHI 0 days.
Contact kill of green foliage	paraquat, MOA 22 (Gramoxone) 2 SL	2 pt	0.5	Spray must not contact okra plants. Hooded sprayers must be used. Two applications can be made, allow 14 day interval between the two applications.
Most emerged weeds	glyphosate, MOA 9 (numerous brands and formulations)	See labels	See labels	**Row middles only.** Apply as a hooded spray in row middles, as shielded spray in row middles, as wiper applications in row middles, or postharvest. To avoid severe injury to crop, do not allow herbicide to contact foliage, green shoots or stems, exposed roots, or fruit of crop. Do not apply within 14 days of harvest.
Annual broadleaf weeds including pigweed 3 inches or less	mesotrione, MOA 27 (Callisto) 4 L	3 oz	0.093	May be applied as a row middle or hooded POST-directed application but not both. For postemergence hooded application, okra must be at least 3 inches tall. Minimize amount of Callisto that contacts okra foliage or crop injury will occur. PHI 28 days.
Yellow and purple nutsedge and broadleaf weeds	halosulfuron-methyl, MOA 2 (Sandea) 75 DG	0.5 to 1 oz	0.024 to 0.048	Apply to row middles as a postemergence shielded or hooded spray to avoid contact of herbicide with planted crop. In plasticulture, do not allow spray to contact plastic. Do not apply more than 2 ounces per acre per 12-month period. PHI 30 days.

Table 7-16. Chemical Weed Control in Vegetable Crops

Weed	Herbicide, Mode of Action Code* and Formulation	Amount of Formulation Per Acre	Pounds Active Ingredient Per Acre	Precautions and Remarks
Onions, Preplant and Preemergence				
Suppression or control of most annual grasses and broadleaf weeds, full rate required for nutsedge control	metam sodium (Vapam HL) 42%	37.5 to 75 gal	15.7 to 31.5	**Dry bulb and green onion.** Rates are dependent on soil type and weeds present. Apply when soil moisture is at field capacity (100 to 125%). Apply through soil injection using a rotary tiller or inject with knives no more than 4 inches apart; follow immediately with a roller to smooth and compact the soil surface or with mulch. May apply through drip irrigation prior to planting a second crop on mulch. Plant back interval is often 14 to 21 days and can be 30 days in some environments. See label for all restrictions and additional information.
Contact kill of all green foliage, stale bed application	paraquat, MOA 22 (Firestorm, Parazone) 3 SL (Gramoxone SL 2.0) 2 SL	1.7 to 2.7 pt 2.5 to 4 pt	0.65 to 1	**Seeded onion only.** Apply in a minimum of 10 gallons spray mix per acre to emerged weeds before crop emergence or transplanting as a broadcast or band treatment over a preformed row. Row should be formed several days ahead of planting and treating to allow maximum weed emergence. Use a nonionic surfactant at a rate of 16 to 32 ounces per 100 gallons spray mix or 1 gallon approved crop oil concentrate per 100 gallons spray mix. PHI 60 days.
Annual and perennial grass and broadleaf weeds	glyphosate, MOA 9 (numerous brands and formulations)	See labels	See labels	Apply to emerged weeds before crop emergence. Perennial weeds may require higher rates of glyphosate. Consult the manufacturer's label for rates for specific weeds. Use on direct seeded onions only. Certain glyphosate formulations require the addition of surfactant. Adding nonionic surfactant to glyphosate formulated with nonionic surfactant may result in reduced weed control.
Most broadleaf weeds less than 4 inches tall or rosettes less than 3 inches in diameter; does not control grasses	carfentrazone-ethyl, MOA 14 (Aim) 2.0 EC	up to 2 oz	0.031	Apply no later than 30 days before planting. See label for specific Aim rate relating to weed species and proper adjuvant and rate. Coverage is essential for good weed control. Can be tank mixed with other registered burndown herbicides.
Annual grasses and small-seeded broadleaf weeds	bensulide, MOA 8 (Prefar) 4 E	5 to 6 qt	5 to 6	**Dry bulb only.** Apply preplant incorporated (1 inch incorporation is optimum) or preemergence after planting. With preemergence application, irrigate immediately after application. See label for more directions and for rotation restrictions.
	DCPA, MOA 3 (Dacthal) W-75 (Dacthal) 6 F	8 to 10 lb 8 to 10 pt	6 to 7.5	**Dry bulb and green.** Apply immediately after seeding or transplanting and/or at layby. See label for timing layby treatments.
Annual broadleaf weeds	oxyfluorfen, MOA 14 (Galigan) 2 E (Goal 2 XL) 2 EC (GoalTender) 4 E	1 to 2 pt 1 to 2 pt 1 pt	0.25 to 0.5 0.25 to 0.5 0.5	**Transplanted dry bulb only.** Apply as a single application immediately (within 2 days) after transplanting for preemergence control of weeds. Injury can occur if applications are made during cool, wet weather or prior to the full development of the true leaves. See label for rates and instructions for use. Do not apply within 45 days of harvest.
Most annual grasses and some broadleaf weeds	pendimethalin, MOA 3 (Prowl) 3.3 EC (Prowl) 3.8 AS	See label	See label	**Dry bulb and green onion (chives, leeks, spring onions, scallions, Japanese bunching onions, green shallots, and green eschalots). Prowl 3.3 EC is not registered for green onion.** For preemergence weed control. Apply when onions have two to nine true leaves (dry bulb) and two to three leaves (green onion) but prior to weed emergence. For green onion the soil must be a muck soil or be a mineral soil with at least 3% organic matter. See label for additional information on rate depending on soil type. PHI for dry bulb onion is 45 days. PHI for green onion is 30 days.
	dimethenamid-P, MOA 15 (Outlook) 6 EC	12 to 21 oz	0.6 to 1	**Dry bulb and green onion (leeks, spring onions or scallions, Japanese bunching onions, green shallots or eschalots).** For preemergence weed control. Apply after crop has reached 2 true leaves until a minimum of 30 days before harvest. If applications are made to transplanted crop, DO NOT apply until transplants are in the ground and soil has settled around transplants with several days to recover. See label for further details.
Onions, Postemergence				
Annual grasses, small seeded broadleaf weeds and yellow nutsedge	S-metolachlor, MOA 15 (Dual Magnum) 7.62 EC	0.5 to 1 pt	0.47 to 0.96	A Section 24(c) North Carolina Label must be obtained at www.farmassist.com prior to use. **Seeded (green or dry bulb).** Do not apply before 4 leaf stage. From the 4-6 leaf stage may apply 0.5 pt/A; rate can be increased to 1 pt/A after the 6 leaf stage. **Transplant (dry bulb only).** Transplant, irrigate to seal soil around the root ball, and then apply Dual Magnum within 48 hours of planting and sealing soil around onions.
Most annual broadleaf weeds	oxyfluorfen, MOA 14 (Galigan) 2 E (Goal 2 XL) 2 EC (GoalTender) 4 E	0.5 pt 0.5 pt 0.25 pt	0.12	**Dry bulb only.** May be used as a postemergence spray to both the weeds and crop after the onions have at least two fully developed true leaves. Some injury to onions may result. Injury will be more severe if the chemical is applied during cool, wet weather. Weeds should be in the two- to fourleaf stage for best results. Do not make more than four applications per year. Do not apply within 45 days of harvest.
Common lambsquarters, common chickweed, common purslane, black nightshade, ladysthumb, Pennsylvania smartweed, redroot pigweed, and some annual grasses	ethofumesate, MOA 16 (Nortron) 4 SC	16 to 32 oz	0.5 to 1	Apply at planting or just after planting prior to weed emergence. Can be used postemergence at 16 oz per acre. See label for more information. Rainfall of at least 0.5 inch is needed for activation.
Most broadleaf weeds less than 4 inches tall or rosettes less than 3 inches in diameter; does not control grasses	carfentrazone-ethyl, MOA 14 (Aim) 1.9 EW or 2 EC	up to 2 oz	up to 0.031	Apply post-directed using hooded sprayers for control of emerged weeds. If crop is contacted, burning of contacted area will occur. Use a nonionic surfactant or crop oil with Aim. See label for rate. Coverage is essential for good weed control. Can be tank mixed with other registered herbicides.
Most emerged weeds	glyphosate, MOA 9 (numerous brands and formulations)	See labels	See labels	**Row middles only.** Apply as a hooded spray in row middles, as shielded spray in row middles, as wiper applications in row middles, or postharvest. To avoid severe injury to crop, do not allow herbicide to contact foliage, green shoots, stems, exposed, roots, or fruit of crop. Do not apply within 14 days of harvest.

Table 7-16. Chemical Weed Control in Vegetable Crops

Weed	Herbicide, Mode of Action Code* and Formulation	Amount of Formulation Per Acre	Pounds Active Ingredient Per Acre	Precautions and Remarks
Onions, Postemergence (continued)				
Annual and perennial grasses only	fluazifop, MOA 1 (Fusilade DX) 2 EC	6 to 16 oz	0.1 to 0.25	**Dry bulb only.** Apply to emerged grasses. Consult manufacturer's label for specific rates and best times to treat. Add 1 gallon crop oil concentrate or 1 quart nonionic surfactant per 100 gallons spray mix. Do not apply on days that are unusually hot and humid. Do not apply within 45 days of harvest.
	sethoxydim, MOA 1 (Poast) 1.5 EC	1 to 1.5 pt	0.2 to 0.3	**Dry bulb and green.** Apply to emerged grasses. Consult manufacturer's label for specific rates and best times to treat. Add 1 quart of crop oil concentrate per acre. Adding crop oil to Poast may increase the likelihood of crop injury at high air temperatures. Do not apply Poast on days that are unusually hot and humid. Do not apply within 30 days of harvest.
	clethodim, MOA 1 (Arrow, Clethodim, Intensity, Select) 2 EC	6 to16 oz	0.09 to 0.25	**Dry bulb only.** Apply to emerged grasses. Consult label for specific rates and best times to treat. With Arrow, Clethodim, or Select, add 1 gallon crop oil concentrate per 100 gallons spray mix. With Select Max or Intensity One, add 2 pints nonionic surfactant per 100 gallons spray mixture. Adding crop oil may increase the likelihood of crop injury at high air temperatures. Do not apply Select on unusually hot and humid days. Do not apply within 45 days of dry bulb onion harvest. Intensity One may be applied to dry bulb onions or green onions (Leeks, scallions or spring onions, Japanese bunching onion, shallots or eschalots). Do not exceed 16 ounces of Intensity One per acre on green onions. Do not apply Intensity One herbicide within 14 days of green onion harvest.
	(Intensity One, Select Max) 1 EC	9 to 32 oz	0.07 to 0.25	
Peas, Green, Preplant and Preemergence				
Contact kill of all green foliage, stale bed application	paraquat, MOA 22 (Firestorm, Parazone) 3 SL	1.3 to 2.7 pt	0.5 to 1	Apply in minimum of 10 gallons spray mix per acre to emerged weeds before crop emergence as broadcast or band treatment over preformed row. Use sufficient water for thorough coverage. Row should be formed several days before planting and treating to allow maximum weed emergence. Use nonionic surfactant at a rate of 16 to 32 ounces per 100 gallons spray mix or 1 gallon approved crop oil concentrate per 100 gallons spray mix.
	(Gramoxone SL) 2 SL	2 to 4 pt		
Most broadleaf weeds less than 4 inches tall or rosettes less than 3 inches in diameter; does not control grasses	carfentrazone-ethyl, MOA 14 (Aim) 1.9 EW or 2 EC	up to 2 oz	up to 0.031	Apply prior to planting or emergence of crop. Use a nonionic surfactant or crop oil with Aim. See label for rate. Coverage is essential for good weed control. Can be tank mixed with other registered burndown herbicides.
Annual and perennial grass and broadleaf weeds	glyphosate, MOA 9 (numerous brands and formulations)	See labels	See labels	Apply to emerged weeds before crop emergence. Do not feed crop residue to livestock for 8 weeks following treatment. Perennial weeds may require higher rates of glyphosate. Consult the manufacturer's label for rates for specific weeds. Certain glyphosate formulations require the addition of a surfactant. Adding nonionic surfactant to glyphosate formulated with nonionic surfactant may result in reduced weed control.
Annual grasses and small-seeded broadleaf weeds	pendimethalin, MOA 3 (Prowl H$_2$O) 3.8 AS	1.5 to 3 pt	0.75 to 1.5	**Southern peas (cowpeas) and snap beans only.** Apply preplant and incorporate into the soil 2 to 3 inches using a power driven rototiller or by cross disking. **Do not apply after seeding. Do not apply when air temperature is below 45 degrees F.**
Broadleaf weeds	saflufenacil, MOA 14 (Sharpen) 3.42 SL	1 oz	0.027	**Dry field pea, edible pea (sugar snap, English pea, garden pea, green pea, marrowfat pea) and chickpea only:** Apply as a preplant/ preemergence burndown of small actively growing broadleaf weeds. Can also be used preplant incorporated or preemergence in edible pea. See label for directions. Do not apply more than 2 fluid ounces per acre per season.
Annual grasses and small-seeded broadleaf weeds	trifluralin, MOA 3 (Treflan, Trifluralin, Trifluralin HF, other brands) 4 EC	1 to 1.5 pt	0.5 to 0.75	**English peas only.** Apply preplant and incorporate to a depth of 2 to 3 inches within 8 hr with a rototiller or tandem disk.
Annual grasses and broadleaf weeds; weak on pigweed	clomazone, MOA 13 (Command) 3ME	1.3 pt	0.5	Apply to the soil surface immediately after seeding. See label for further instruction. Limited research has been done on this product in this crop in North Carolina.
Annual grasses, small-seeded broadleaf weeds, and suppression of yellow nutsedge	S-metolachlor, MOA 15 (Brawl, Dual Magnum, Medal) 7.62 EC (Brawl II, Dual II Magnum, Medal II) 7.64 EC	1 to 2 pt	0.95 to 1.91	Apply to soil surface immediately after seeding. Shallow cultivations will improve control. See label for specific rate.
Annual broadleaf weeds including morningglory, pigweed, smartweed, and purslane	imazethapyr, MOA 2 (Pursuit) 2 EC	up to 3 oz	Up to 0.047	**English peas only.** Apply preplant incorporated or to soil surface immediately after planting. See label for more details.
Peas, Green, Postemergence				
Annual broadleaf weeds and yellow nutsedge	bentazon, MOA 6 (Basagran) 4 SL	1 to 2 pt	0.5 to 1	Apply overtop of peas when weeds are small and peas have at least three pairs of leaves (four nodes). **Do not add crop oil concentrate to spray mix.** Do not apply within 10 days of harvest. Do not apply when peas are in bloom.
Most broadleaf weeds less than 4 inches tall or rosettes less than 3 inches in diameter; does not control grasses	carfentrazone-ethyl, MOA 14 (Aim) 1.9 EW or 2 EC	up to 2 oz	up to 0.031	Apply post-directed using hooded sprayers for control of emerged weeds. If crop is contacted, burning of contacted area will occur. Use a nonionic surfactant or crop oil with Aim. See Label for rate. Coverage is essential for good weed control. Can be tank mixed with other registered herbicides.
Most emerged weeds	glyphosate, MOA 9 (numerous brands and formulations)	See labels	See labels 0.5 to 0.94	**Row middles only.** Apply as a hooded spray in row middles, as shielded spray in row middles, as wiper applications in row middles, or post harvest. To avoid severe injury to crop, do not allow herbicide to contact foliage, green shoots or stems, exposed roots, or fruit of crop. Do not apply within 14 days of harvest.
Annual and perennial grasses	sethoxydim, MOA 1 (Poast) 1.5 EC	1 to 1.5 pt	0.2 to 0.3	Apply to emerged grasses. Consult manufacturer's label for specific rates and best times to treat. With sethoxydim, add 1 quart of crop oil concentrate per acre. Adding crop oil to Poast or Assure II may increase the likelihood of crop injury at high air temperatures. With quizalofop, add 1 gallon oil concentrate or 1 quart nonionic surfactant per 100 gallons spray. Do not apply Poast or Assure II on days that are unusually hot and humid. Do not apply sethoxydim within 15 days or Assure II or Targa within 30 days of harvest.
	quizalofop p-ethyl, MOA 1 (Assure II) 0.88 EC (Targa) 0.88 EC	6 to 12 oz	0.04 to 0.08	

Table 7-16. Chemical Weed Control in Vegetable Crops

Weed	Herbicide, Mode of Action Code* and Formulation	Amount of Formulation Per Acre	Pounds Active Ingredient Per Acre	Precautions and Remarks
Peas, Green, Postemergence (continued)				
Annual broadleaf weeds including morningglory, pigweed, smartweed, and purslane	imazethapyr, MOA 2 (Pursuit) 2 EC	up to 3 oz	Up to 0.047	**See label for pea type.** Apply postemergence to 1- to 3-inch weeds (one to four leaves) when peas are at least 3 inches high but prior to five nodes and before flowering. Add nonionic surfactant at 2 pints per 100 gallons of spray mix. See label for crop rotation restrictions. PHI 30 days.
Broadleaf and grass weeds	imazamox, MOA 2 (Raptor) 1 SL	4 fl oz	0.31	**Dry peas only:** Apply postemergence before bloom stage but after dry peas have at least 3 pairs of leaves. See label for further information.
Peas, Southern (cowpeas, blackeyed peas), Preplant or Preemergence				
Contact kill of all green foliage, stale bed application	paraquat, MOA 22 (Firestorm, Parazone) 3 SL	1.3 to 2.7 pt	0.5 to 1	Apply in a minimum of 20 gallons spray solution to emerged weeds before crop emergence as a broadcast or band treatment over a preformed row. Use sufficient water to give thorough coverage. Row should be formed several days ahead of planting and treating to allow maximum weed emergence. Use a nonionic surfactant at a rate of 16 to 32 ounces per 100 gallons spray mix or 1 gallon approved crop oil concentrate per 100 gal spray mix.
	(Gramoxone SL) 2 SL	2 to 4 pt		
Most broadleaf weeds less than 4 inches tall or rosettes less than 3 inches in diameter; does not control grasses	carfentrazone-ethyl, MOA 14 (Aim) 1.9 EW or 2 EC	Up to 2 oz	Up to 0.031	Apply prior to planting or emergence of crop. Use a nonionic surfactant or crop oil with Aim. See label for rate. Coverage is essential for good weed control. Can be tank mixed with other registered burndown herbicides.
Annual and perennial grass and broadleaf weeds, stale bed application.	glyphosate, MOA 9 (numerous brands and formulations)	See labels	See labels	Apply to emerged weeds before crop emergence. Do not feed crop residue to livestock for 8 weeks following treatment. Perennial weeds may require higher rates of glyphosate. Consult the manufacturer's label for rates for specific weeds. Certain glyphosate formulations require the addition of a surfactant. Adding nonionic surfactant to glyphosate formulated with nonionic surfactant may result in reduced weed control.
Annual grasses and small-seeded broadleaf weeds	pendimethalin, MOA 3 (Prowl H_2O) 3.8 AS	1.5 to 3 pt	0.75 to 1.5	**Not labeled for blackeyed peas.** Apply preplant and incorporate into the soil 2 to 3 inches using a power driven rototiller or by cross disking. **Do not apply after seeding**.
	trifluralin, MOA 3 (Treflan HFP, Trifluralin, Trifluralin HF) 4 EC	1 to 2 pt	0.5 to 1	Apply preplant and incorporate into the soil 2 to 3 inches deep within 8 hr with a rototiller or tandem disk.
Annual grasses and broadleaf weeds	clomazone, MOA 13 (Command) 3ME	0.4 to 0.67 pt	0.15 to 0.25	**Succulent southern peas only.** Apply to the soil surface immediately after seeding. Offers weak control of pigweed. See label for further instruction. Limited research has been done on this product in this crop in North Carolina.
Annual grasses, small-seeded broadleaf weeds, and suppression of yellow nutsedge	S-metolachlor, MOA 15 (Brawl, Dual Magnum, Medal) 7.62 EC (Brawl II, Dual II Magnum, Medal II) 7.64 EC	1 to 2 pt	0.95 to 1.91	Apply to soil surface immediately after planting. Shallow cultivations will improve control. May also be soil incorporated before planting.
Annual grasses and broadleaf weeds including morningglory, pigweed, smartweed, and purslane	imazethapyr, MOA 2 (Pursuit) 2 EC	Up to 4 oz	Up to 0.063	Apply preemergence or preplant incorporated. See label for rate for specific pea species.
Peas, Southern, Postemergence				
Annual broadleaf weeds and yellow nutsedge	bentazon, MOA 6 (Basagran) 4 SL	1 to 2 pt	0.5 to 1	Apply overtop of peas when weeds are small and peas have at least three pairs of leaves (four nodes). **Do not add crop oil concentrate to spray mix.** See label for weeds controlled with Basagran. Do not apply within 30 days of harvest. Do not apply when peas are in bloom.
Annual broadleaf weeds including morningglory, pigweed, smartweed, and purslane	imazethapyr, MOA 2 (Pursuit) 2 EC	Up to 4 oz	Up to 0.063	**Southern peas and certain dry peas.** Apply postemergence to 1- to 3-inch weeds (one to four leaves) when peas are at least 3 inches in height but prior to five nodes and flowering. Add nonionic surfactant at 2 pints per 100 gallons of spray mixture with all postemergence applications. Do not apply within 30 days of harvest. See label for rate for specific pea species.
Most broadleaf weeds less than 4 inches tall or rosettes less than 3 inches in diameter; does not control grasses	carfentrazone-ethyl, MOA 14 (Aim) 1.9 EW or 2 EC	up to 2 oz	up to 0.031	Apply post-directed using hooded sprayers for control of emerged weeds. If crop is contacted, burning of contacted area will occur. Use a nonionic surfactant or crop oil with Aim. See label for rate. Coverage is essential for good weed control. Can be tank mixed with other registered herbicides.
Most emerged weeds	glyphosate, MOA 9 (numerous brands and formulations)	See labels	See labels	**Row middles only.** Apply as a hooded spray in row middles, as shielded spray in row middles, as wiper applications in row middles, or post harvest. To avoid severe injury to crop, do not allow herbicide to contact foliage, green shoots, stems, exposed, roots, or fruit of crop. Do not apply within 14 days of harvest.
Annual and perennial grasses	quizalofop p-ethyl, MOA 1 (Assure II, Targa) 0.88 EC	6 to 12 oz	0.04 to 0.08	Apply to emerged grasses. Consult manufacturer's label for specific rates and best times to treat. With sethoxydim, add 1 quart of crop oil concentrate per acre. With quizalofop, add 1 gallon oil concentrate or 1 quart nonionic surfactant per 100 gallons spray. Adding crop oil to Assure II or Poast may increase the likelihood of crop injury at high air temperatures. Do not apply Assure II or Poast on days that are unusually hot and humid. With sethoxydim, do not apply within 15 days and 30 days of harvest for succulent and dry peas, respectively. With quizalofop, do not apply within 60 days of harvest of dry Southern peas, or within 30 days of harvest of succulent Southern peas.
	sethoxydim, MOA 1 (Poast) 1.5 EC	1 to 1.5 pt	0.2 to 0.3	
	clethodim, MOA 1 (Intensity One, Select Max) 1 EC	9 to 16 oz	0.07 to 0.125 lb	For Select Max or Intensity One, add 2 pints nonionic surfactant per 100 gallons spray mixture. Apply before bloom. Do not make more than one application per acre per season. Do not apply clethodim within 21 days of harvest.

Table 7-16. Chemical Weed Control in Vegetable Crops

Weed	Herbicide, Mode of Action Code* and Formulation	Amount of Formulation Per Acre	Pounds Active Ingredient Per Acre	Precautions and Remarks
Peppers, Preplant				
Suppression or control of most annual grasses and broadleaf weeds, full rate required for nutsedge control	metam sodium (Vapam HL) 42%	37.5 to 75 gal	15.7 to 31.5	Rates are dependent on soil type and weeds present. Apply when soil moisture is at field capacity (100 to 125%). Apply through soil injection using a rotary tiller or inject with knives no more than 4 inches apart; follow immediately with a roller to smooth and compact the soil surface or with mulch. May apply through drip irrigation prior to planting second crop on mulch however adhere to label guidelines on crop plant back interval. Plant back interval is often 14 to 21 days and can be 30 days in some environments. See label for all restrictions and additional information. Chloropicrin (150lb/A broadcast) will also be needed when laying first crop mulch to control nutsedge.
Contact kill of all green foliage, stale bed application	paraquat, MOA 22 (Firestorm, Parazone) 3 SL (Gramoxone SL) 2 SL	1.3 to 2.7 pt 2 to 4 pt	0.5 to 1	Apply in a minimum of 10 gallons of spray mix per acre to emerged weeds before transplanting as a broadcast or band treatment over a preformed row. Row should be formed several days ahead of planting and treating to allow maximum weed emergence. Use a nonionic surfactant at a rate of 16 to 32 ounces per 100 gallons spray mix or 1 gallon approved crop oil concentrate per 100 gallons spray mix.
Most broadleaf weeds less than 4 inches tall or rosettes less than 3 inches in diameter; does not control grasses	carfentrazone-ethyl, MOA 14 (Aim) 1.9 EW or 2 EC	up to 2 oz	up to 0.031	**Transplanted crop. Apply no later than 1 day before transplanting crop.** **Seeded crop. Apply no later than 7 days before planting seeded crop.** Use a nonionic surfactant or crop oil. See label for rate. Coverage of weed is essential for good weed control. Can be tank mixed with other registered burndown herbicides.
Annual and perennial grass and broadleaf weeds, stale bed application	glyphosate, MOA 9 (numerous brands and formulations)	See labels	See labels	Apply to emerged weeds at least 3 days before seeding or transplanting. When applying Roundup before transplanting crops into plastic mulch, care must be taken to remove residues of this product from the plastic prior to transplanting. To prevent crop injury, residues can be removed by 0.5 inch natural rainfall or by applying water via a sprinkler system. Perennial weeds may require higher rates of glyphosate. Consult the manufacturer's label for specific weeds. Certain glyphosate formulations require the addition of a surfactant. Adding nonionic surfactant to glyphosate formulated with nonionic surfactant may result in reduced weed control.
Broadleaf weeds including Carolina geranium and cutleaf eveningprimrose and a few annual grasses	oxyfluorfen, MOA 14 (Goal) 2XL (GoalTender) 4 F	up to 2 pt up to 1 pt	0.5 lb	**Plasticulture only.** Apply to soil surface of pre-formed beds at least 30 days prior to transplanting crop. While incorporation is not necessary, it may result in less crop injury. Plastic mulch can be applied any time after application, but best results are likely if applied soon after application.
Palmer amaranth, redroot pigweed, smooth pigweed, galinsoga sp., black nightshade, Eastern black nightshade, common purslane, partial control of yellow nutsedge	fomesafen, MOA 14 (Reflex) 2 EC	1 to 1.5 pt	0.25 to 0.375	This is a Section 24(c) special local needs label for transplanted pepper in North Carolina. Growers must obtain the label at Farmassist.com prior to making an application of Reflex. **See label for further instructions.** **Plasticulture In-row Application for Transplanted Pepper.** Apply after final bed formation and the drip tape is laid but prior to laying plastic mulch. Avoid soil disturbance after application. Unless restricted by other products such as fumigants, pepper may be transplanted immediately following the application of Reflex and the application of the mulch. **Bareground for Transplanted Pepper.** Apply pretransplant up to 7 days prior to transplanting pepper. Weed control will be reduced if soil is disturbed after application. During the transplanting operation make sure the soil in the transplant hole settles flush or above the surrounding soil surface. Avoid cultural practices that may concentrate Reflex-treated soil around the transplant root ball. An overhead irrigation or rainfall event between Reflex herbicide application and transplanting will ensure herbicide activation and will likely reduce the potential for crop injury due to splashing. **Plasticulture Row Middle Application.** Apply to row middles with a hooded or shielded sprayer. Avoid drift of herbicide on mulch. If drift occurs, 0.5 inch of rain or irrigation must occur prior to transplanting. **Carryover is a large concern; see label for more information.**
Annual grasses and small-seeded broadleaf weeds	clomazone, MOA 13 (Command) 3 ME	0.67 to 2.67 pt	0.25 to 1	**Not labeled for banana pepper.** Apply preplant before transplanting. Weak on pigweed. See label for instructions on use.
	napropamide, MOA 15 (Devrinol, Devrinol DF-XT) 50 DF (Devrinol, Devrinol 2-XT) 2 EC	2 to 4 lb 2 to 4 qt	1 to 2	**Bare ground:** Apply preplant and incorporate into the soil 1 to 2 inches as soon as possible with a rototiller or tandem disk. Can be used on direct-seeded or transplanted peppers. See label for instructions on use. **Plasticulture:** Apply to a weed-free soil before laying plastic mulch. Soil should be well worked yet moist enough to permit a thorough incorporation to a depth of 2 inches. Mechanically incorporate or irrigate within 24 hours after application. If weed pressure is from small seeded annuals, apply to the surface of the bed immediately in front of the laying of plastic mulch. If soil is dry, water or sprinkle irrigate with sufficient water to wet to a depth of 2 to 4 inches before covering with plastic mulch. **Between rows:** Apply to a weed free soil surface between the rows (bareground or plastic mulch). Mechanically incorporate or irrigate Devrinol into the soil to a depth of 1 to 2 inches within 24 hours of application. See XT labels for information regarding delay in irrigation event.
Annual grasses and small-seeded broadleaf weeds (continued)	pendimethalin, MOA 3 (Prowl H$_2$O) 3.8	1 to 3 pt	0.5 to 1.5	May be applied in chili pepper, cooking pepper, pimento, Jalapeno, and sweet pepper. Do not apply more than 3 pints per acre per season. See label for specific use rate for your soil type. Avoid direct contact with pepper foliage or stems. Do not apply within 70 days of harvest. See label for further instructions and precautions. **Between rows.** Can be applied as a post-directed spray on the soil at the base of the plant beneath plants and between rows. **In-row.** May be applied as a broadcast preplant incorporated surface application prior to transplanting peppers. Limited research has been conducted in NC.
	trifluralin, MOA 3 (Treflan, Treflan HFP, Trifluralin HF) 4 EC	1 to 2 pt	0.5 to 1	Apply pretransplant, and incorporate to a depth of 2 to 3 inches within 8 hr with a rototiller or tandem disk. Do not apply after transplanting.
Peppers, Preplant and Preemergence				
Annual grasses and small seeded broadleaf weeds	bensulide, MOA 8 (Prefar) 4 EC	5 to 6 qt	5 to 6	Apply preplant incorporated (1 inch incorporation is optimum) or preemergence. With preemergence application, irrigate immediately after application. See label for directions.

Table 7-16. Chemical Weed Control in Vegetable Crops

Weed	Herbicide, Mode of Action Code* and Formulation	Amount of Formulation Per Acre	Pounds Active Ingredient Per Acre	Precautions and Remarks
Peppers, Preplant or Postemergence (continued)				
Annual grass and broadleaf weeds, yellow nutsedge suppression	S-metolachlor, MOA 15 (Dual Magnum) 7.62 EC	8 to 12 oz	0.47 to 0.7	**Bell pepper transplants only.** This is a Section 24(c) North Carolina Special Local Needs Label. Growers must obtain label from www.farmassist.com prior to making Dual Magnum applications. Option 1: Apply 8 to 12 ounces to the soil surface of pre-formed beds prior to laying plastic. Insure the plastic laying process does not incorporate or disturb the treated bed. Option 2: Apply 12 ounces overtop of bell pepper between 1 and 3 weeks after planting. Does not control emerged weeds. Limited data are available for NC. Do not apply more than 12 ounces per acre as it is likely that injury will occur including decreased crop vigor. Read label for further instructions.
Peppers, Postemergence				
Annual and perennial grasses only	sethoxydim, MOA 1 (Poast) 1.5 EC	1 to 1.5 pt	0.2 to 0.3	Apply to emerged grasses. Consult manufacturer's label for specific rates and best times to treat. Add 1 quart of crop oil concentrate per acre. Adding crop oil to Poast may increase the likelihood of crop injury at high air temperatures. Do not apply Poast on days that are unusually hot and humid. Do not apply within 7 days of harvest.
	clethodim, MOA 1 (Arrow, Clethodim, Intensity, Select) 2 EC (Select Max, Intensity One) 1 EC	6 to 8 oz 9 to 16 oz	0.094 to 0.125 0.07 to 0.125	Apply postemergence to control grasses. With Arrow, Clethodim, or Select, add 1 gallon crop oil concentrate per 100 gallons spray mix. With Select Max, add 2 pints nonionic surfactant per 100 gallons spray mixture. Adding crop oil may increase likelihood of crop injury at high air temperatures. Very effective in controlling annual bluegrass. Apply to actively growing grasses not under drought stress. Do not apply within 20 days of harvest.
Broadleaf, grass (suppression only), and yellow nutsedge	imazosulfuron, MOA 2 (League) 0.5 DF	4 to 6.4 oz	0.19 to 0.3	**Pepper (Bell and non-bell).** Apply to pepper plants that are well established and at least 10 inches tall. Apply directed to the base of the plants stem, no higher than 2 inches from the soil surface and do not contact fruit. Consult label for approved surfactants and crop rotation restrictions. PHI 21 days.
Peppers, Row Middles				
Most broadleaf weeds less than 4 inches tall or rosettes less than 3 inches in diameter; does not control grasses	carfentrazone-ethyl, MOA 14 (Aim) 1.9 EW or 2 EC	up to 2 oz	up to 0.031	Apply post-directed using hooded sprayer for control of emerged weeds. If crop is contacted, burning of contacted area will occur. Use a nonionic surfactant or crop oil with Aim. See label for rate. Coverage is essential for good weed control. Can be tank mixed with other registered herbicides. PHI 0 days.
Most emerged weeds	glyphosate, MOA 9 (numerous brands and formulations)	See labels	See labels	Apply as a hooded spray in row middles, as shielded spray in row middles, as wiper applications in row middles, or postharvest. To avoid severe injury to crop, do not allow herbicide to contact foliage, green shoots, stems, exposed, roots, or fruit of crop. Do not apply within 14 days of harvest.
Yellow and purple nutsedge and broadleaf weeds	halosulfuron-methyl, MOA 2 (Profine 75, Sandea) 75 DG	0.5 to 1 oz	0.024 to 0.048 lb	Apply to row middles as a postemergence spray. In plasticulture, do not allow spray to contact plastic. Early season application will give postemergence and preemergence control. Do not apply within 30 days of harvest. For postemergence applications, use nonionic surfactant at 1 quart per 100 gallons of spray solution.
Contact kill of all green foliage	paraquat, MOA 22 (Firestorm, Parazone) 3 SL (Gramoxone SL) 2 SL	1.3 pt 2 pt	0.5	Apply in a minimum of 20 gallons spray mix per acre as a shielded spray to emerged weeds between rows of peppers. Use a nonionic surfactant at a rate of 16 ounces per 100 gallons spray mix. Do not apply more than three times per season.
Potatoes, Irish, Preplant and Preemergence				
Contact kill of all green foliage, stale bed application	paraquat, MOA 22 (Firestorm, Parazone) 3 SL (Gramoxone SL) 2 SL	0.7 to 1.3 pt 1 to 2 pt	0.25 to 0.5	Apply in a minimum of 20 gallons spray mix per acre to emerged weeds up to ground cracking before crop emergence. May be used instead of the drag-off operation to kill emerged weeds before the application of preemergence herbicides. Use a nonionic surfactant at a rate of 16 to 32 ounces per 100 gallons spray mix or 1 gallon approved crop oil concentrate per 100 gallons spray mix.
Most broadleaf weeds less than 4 inches tall or rosettes less than 3 inches in diameter; does not control grasses	carfentrazone-ethyl, MOA 14 (Aim) 1.9 EW or 2 EC	up to 2 oz	up to 0.031	Apply prior to planting, or within 1 day after planting crop. Use a nonionic surfactant or crop oil with Aim. See label for rate. Coverage of weed is essential for good weed control. Can be tank mixed with other registered burndown herbicides.
Annual and perennial grass and broadleaf weeds, stale bed application	glyphosate, MOA 9 (numerous brands and formulations)	See labels	See labels	Apply to emerged weeds before crop emergence. Do not feed crop residue to livestock for 8 weeks following treatment. Perennial weeds may require higher rates of glyphosate. Consult the manufacturer's label for rates for specific weeds. Certain glyphosate formulations require the addition of a surfactant. Adding nonionic surfactant to glyphosate formulated with nonionic surfactant may result in reduced weed control.
	trifluralin, MOA 3 (Treflan) 4 EC (Trifluralin) 4 L	1 to 1.5 pt	0.5 to 0.75	Apply and incorporate after planting but before emergence, immediately following dragoff, or after potato plants have fully emerged. Rate is soil texture dependent.
Annual grasses and small-seeded broadleaf weeds	pendimethalin, MOA 3 (Prowl) 3.3 EC (Prowl H_2O) 3.8 AS	1.8 to 3.6 pt 1.5 to 3 pt	0.75 to 1.5 0.75 to 1.5	Apply just after planting or drag-off to weed-free soil before crop emerges or from emergence until crop reaches 6 inches tall.
Annual grasses and small-seeded broadleaf weeds, plus yellow nutsedge suppression	S-metolachlor, MOA 15 (Brawl, Dual Magnum, Medal) 7.62 EC (Brawl II, Dual II Magnum, Medal II) 7.64 EC	1 to 2 pt	0.95 to 1.91	Apply just after planting or drag-off to weed-free soil before crop emerges. Dual Magnum can also be applied at lay-by for control of late season weeds. Do not apply within 60 days after the at-planting to drag-off application, or within 40 days after a lay-by application. See label for further instruction.
	dimethenamid-P, MOA 15 (Outlook) 6 EC	12 to 21 oz	0.6 to 1	Apply just after planting or drag-off to weed-free soil before crop emerges. See label for further instruction. PHI 40 days.

Table 7-16. Chemical Weed Control in Vegetable Crops

Weed	Herbicide, Mode of Action Code* and Formulation	Amount of Formulation Per Acre	Pounds Active Ingredient Per Acre	Precautions and Remarks
Potatoes, Irish, Preplant and Preemergence (continued)				
Annual grasses, most broadleaf weeds, plus yellow and purple nutsedge suppression	EPTC, MOA 8 (Eptam) 7 EC	3.5 pt	3	Apply preplant and incorporate into the soil 2 to 3 inches with a rototiller or tandem disk. The variety "Superior" has been shown to be sensitive to Eptam. See label for specific methods of incorporation. For late season preemergence nutsedge control, apply and incorporate as a directed spray to the soil on both sides of the crop row. See label for more detail.
Most annual broadleaf weeds and some annual grasses	flumioxazin, MOA 14 (Chateau) 51 SW	1.5 oz	0.047	Apply immediately after hilling. A minimum of 2 inches of soil must cover the vegetative portion of the potato plant at the time of application of Chateau. DO NOT apply to emerged potatoes. DO NOT incorporate Chateau or weed control will be reduced. Can be tank mixed with burndown herbicides if weeds present at application. See label for further instructions.
	linuron, MOA 7 (Lorox DF) 50 WDG (Linex) 4L	1.5 to 3 lb 1.5 to 3 pt	0.75 to 1.5	Apply just after planting or drag-off or hilling but before crop emerges. If emerged weeds are present, add 1 pint surfactant for each 25 gallons spray mixture. Weeds may be up to 3 inches tall at time of application.
	metribuzin, MOA 5 (TriCor DF, Dimetric DF, and other trade names) 75 WDG	0.33 to 1.33 lb	0.23 to 1	Apply just after planting or drag-off but before crop emerges. Weeds may be emerged at time of application. On sand soils or sensitive varieties, do not exceed 0.67 pound per acre. See label for list of sensitive varieties.
	rimsulfuron, MOA 2 (Matrix, Pruvin) 25 WDG	1 to 1.5 oz	0.016 to 0.023	Apply after drag-off or hilling but before potatoes and weeds emerge. If emerged weeds are present, add surfactant. See label for specific rate. Can be tank mixed with Eptam, Prowl, Sencor, Lorox, or Dual Magnum. See label for further instructions.
Broadleaf, weeds and yellow nutsedge	fomesafen, MOA 14 (Reflex) 2 EC	1 pt	0.25	Apply as a preemergence after planting but prior to potato emergence. Do not apply as s pre-plant incorporated application or emerged potato plant severe crop injury may occur. Do not exceed Reflex at 1 pint per acre per season. PHI 70 days.
Broadleaf weeds and nutsedge	imazosulfuron, MOA 2 (League) 0.5 DF	4 to 6.4 oz	0.19 to 0.3	Apply as a preemergence (4 to 6.4 oz per acre) after crop has been planted but prior to emergence or immediately after hilling. Postemergence application (3.2 to 4 oz per acre) may be made after crop has emerged if weeds are less than 3 inches in height. Do not apply League more than 6.4 oz per acre per season. Consult label for sequential application program and crop rotation restrictions. PHI 45 days.
Potatoes, Irish, Postemergence				
Most annual broadleaf weeds and some annual grasses	metribuzin, MOA 5 (TriCor DF, Dimetric DF and other trade names) 75 WDG	0.33 to 0.67 lb	0.25 to 0.5	Do not use on early maturing smooth-skinned white or red-skinned varieties. Apply only if there have been at least three successive days of sunny weather before application. Treat before weeds are 1 inch tall. Treatment may cause some chlorosis or minor necrosis. Do not apply within 60 days of harvest.
	rimsulfuron, MOA 2 (Matrix, Pruvin) 25 WDG	1 to 1.5 oz	0.016 to 0.023	Apply to young actively growing weeds after crop emergence. More effective on small weeds. Add nonionic surfactant at 1 to 2 pints per 100 gallons water. Can be tank mixed with Eptam or Sencor or some foliar fungicides. See label for further instructions. PHI 60 days.
Most broadleaf weeds less than 4 inches tall or rosettes less than 3 inches in diameter; does not control grasses	carfentrazone-ethyl, MOA 14 (Aim) 1.9 EW or 2 EC	up to 2 oz	up to 0.031	Apply post-directed using hooded sprayers for control of emerged weeds. If crop is contacted, burning of contacted area will occur. Use a nonionic surfactant or crop oil with Aim. See label for rate. Coverage is essential for good weed control. Can be tank mixed with other registered herbicides. Do not apply within 7 days of harvest.
Most emerged weeds	glyphosate, MOA 9 ((numerous brands and formulations)	See labels	See labels	**Row middles only.** Apply as a hooded spray in row middles, as shielded spray in row middles, as wiper applications in row middles, or postharvest. To avoid severe injury to crop, do not allow herbicide to contact foliage, green shoots, stems, exposed, roots, or fruit of crop. Do not apply within 14 days of harvest.
Annual and perennial grasses only	clethodim, MOA 1 (Arrow, Clethodim, Intensity. Select 2 EC (Intensity One, Select Max) 1 EC	6 to 8 oz 9 to 32 oz	0.094 to 0.125 0.07 to 0.25	Apply postemergence for control of grasses. With Arrow, Clethodim, Intensity or Select, add 1 quart crop of oil concentrate per acre. With Intensity One or Select Max, nonionic surfactant of 2 pints per 100 gallons spray mixture can be used instead of crop oil concentrate. Adding crop oil may increase the likelihood of crop injury at high air temperatures. Very effective in controlling annual bluegrass. Apply to actively growing grasses not under drought stress. Do not apply within 30 days of harvest.
	sethoxydim, MOA 1 (Poast) 1.5 EC	1 to 1.5 pt	0.2 to 0.3	Apply to emerged grasses. Consult manufacturer's label for specific rates and best times to treat. Add 1 quart of crop oil concentrate per acre. Adding crop oil to Poast may increase the likelihood of crop injury at high air temperatures. Do not apply on days that are unusually hot and humid. Do not apply within 30 days of harvest.
Pumpkins, Preplant and Preemergence				
Contact kill of all green foliage, stale bed application	paraquat, MOA 22 (Firestorm, Parazone) 3 SL (Gramoxone SL) 2 SL	1.3 to 2.7 pt 2 to 4 pt	0.5 to 1	Apply in a minimum of 20 gallons spray mix per acre to emerged weeds before crop emergence or transplanting as a band or broadcast treatment over a preformed row. Use sufficient water to give thorough coverage. Row should be formed several days ahead of planting or treating to allow maximum weed emergence. Use a nonionic surfactant at a rate of 16 to 32 ounces per 100 gallons spray solution or 1 gallon approved crop oil concentrate per 100 gallons spray mix.
Most broadleaf weeds less than 4 inches tall or rosettes less than 3 inches in diameter; does not control grasses	carfentrazone-ethyl, MOA 14 (Aim) 1.9 EW or 2 EC	up to 2 oz	up to 0.031	**Not registered for use on seeded crop.** Apply prior to transplanting crop. Use a nonionic surfactant or crop oil with Aim. See label for rate. Coverage is essential for good weed control. Can be tank mixed with other registered burndown herbicides.
Annual and perennial grass and broadleaf weeds, stale bed application	glyphosate, MOA 9 (numerous brands and formulations)	See labels	See labels	Apply to emerged weeds at least 3 days before seeding or transplanting. Perennial weeds may require higher rates of glyphosate. Consult the manufacturer's label for rates for specific weeds. Certain glyphosate formulations require the addition of a surfactant. Adding nonionic surfactant to glyphosate formulated with nonionic surfactant may result in reduced weed control.

Table 7-16. Chemical Weed Control in Vegetable Crops

Weed	Herbicide, Mode of Action Code* and Formulation	Amount of Formulation Per Acre	Pounds Active Ingredient Per Acre	Precautions and Remarks
Pumpkins, Preplant and Preemergence (continued)				
Morningglory and small pigweed < 1 inch	pyraflufen ethyl, MOA 14 (ET Herbicide) 0.208 L	1 to 2 oz	0.016 to 0.0032	**Bareground.** Wait 1 day following preplant burndown application before planting. **Plasticulture.** May apply over mulch; however, a single 0.5 inch irrigation/rain event plus a 7 day waiting period is needed before transplanting.
Palmer amaranth, redroot pigweed, smooth pigweed, galinsoga sp., black nightshade, Eastern black nightshade, common purslane, partial control of yellow nutsedge	fomesafen, MOA 14 (Reflex) 2 EC	8 to 10 oz	0.13 to 0.16	A Section 24(c) North Carolina Local Need Label must be obtained at www.farmassist.com prior to this use. **Bareground transplants.** Prepare land for planting; apply Reflex; lightly irrigate to activate herbicide and move it into soil; and then prepare plant holes and plant.
Annual grasses and some small-seeded broadleaf weeds	bensulide, MOA 8 (Prefar) 4 EC	5 to 6 qt	5 to 6	Registered for cucurbit vegetable group (Crop grouping 9). Apply preplant and incorporate into the soil 1 to 2 inches (1 inch incorporation is optimum) with a rototiller or tandem disk, or apply to the soil surface after seeding and follow with irrigation. Check replant restrictions for small grains on label. See label for use rate if Prefar 4 EC is used.
	ethalfluralin, MOA 3 (Curbit) 3 EC	3 to 4.5 pt	1.1 to 1.7	Apply postplant to seeded crop prior to crop emergence, or as a banded spray between rows after crop emergence or transplanting. See label for timing. Shallow cultivation, irrigation, or rainfall within 5 days is needed for good weed control. Do not use under mulches, row covers, or hot caps. Under conditions of unusually cold or wet soil and air temperatures, crop stunting or injury may occur. Crop injury can occur if seeding depth is too shallow.
	ethalfluralin, MOA 3 + clomazone, MOA 13 (Strategy) 2.1 L	2 to 6 pt	0.4 to 1.2 + 0.125 to 0.375	Apply to the soil surface immediately after crop seeding for preemergence control of weeds. **Do not apply prior to planting the crop. Do not soil incorporate.** May also be used as a **banded** treatment **between** rows after crop emergence or transplanting.
Yellow and purple nutsedge suppression, non-ALS resistant pigweed, wild radish, and ragweed	halosulfuron-methyl, MOA 2 (Profine 75) 75 DG (Sandea) 75 DG	0.5 to 0.75 oz	0.024 to 0.036 lb	**Direct-seeded pumpkin or winter squash:** Apply after seeding but prior to soil cracking. **Transplanted pumpkin and winter squash:** Apply 7 days prior to transplanting. See label for specific rate. See label for crop rotational restrictions and other information. **Post Transplant in pumpkin and winter squash:** Can be applied as an over-the-top application, a directed spray application, or with crop shields. Apply to transplants that are established, actively growing and in the 3 to 5 true leaf stages or no sooner than 14 days after transplanting unless local conditions demonstrate safety at an earlier interval, but before first female flowers appear. **Row middle/furrow applications in direct-seeded and transplant pumpkin and winter squash:** Apply between rows of direct-seeded or transplanted crop while avoiding contact of the herbicide with the planted crop. If plastic is used on the planted row, adjust equipment to keep the application off the plastic. Reduce rate and spray volume in proportion to area actually sprayed. Rate can be increased to 1 oz per acre if needed for row middle/furrow applications.
Annual broadleaf, grass and yellow nutsedge	S-metolachlor, MOA 15 (Brawl™, Dual Magnum)	1 to 1.33 pt	0.95 to 1.26	Apply as interrow or interhill application. Leave a 1 foot untreated area over the seeded row (6 in. on either side of the row). **Application made as a broadcast spray over the planted row or hill or directly to crop foliage will increase the risk of injury to the crop.** Apply before weeds emerge. See label for further instructions.
Pumpkins, Postemergence				
Yellow and purple nutsedge suppression, non-ALS resistant pigweed, wild radish, and ragweed	halosulfuron-methyl, MOA 2 (Profine 75) 75 DG (Sandea) 75 DG	0.5 to 0.75 oz	0.024 to 0.036 lb	**Direct-seeded pumpkin and winter squash:** Apply after crop has reached the 2 to 5 true leaf stage, preferably 4 to 5 true leaves, but before first female flowers appear. Do not apply within 30 days of harvest. **Post Transplant in pumpkin and winter squash:** Can be applied as an over-the-top application, a directed spray application, or with crop shields. Apply to transplants that are established, actively growing and in the 3 to 5 true leaf stages or no sooner than 14 days after transplanting unless local conditions demonstrate safety at an earlier interval, but before first female flowers appear. Do not apply within 30 days of harvest. **Row middle/furrow applications in direct-seeded and transplant pumpkin or winter squash:** Apply between rows of direct-seeded or transplanted crop while avoiding contact of the herbicide with the planted crop. If plastic is used on the planted row, adjust equipment to keep the application off the plastic. Reduce rate and spray volume in proportion to area actually sprayed. Rate can be increased to 1 oz per acre if needed for row middle/furrow applications. Do not apply within 30 days of harvest.
Annual and perennial grasses only	clethodim, MOA 1 (Arrow, Clethodim, Intensity, Select) 2 EC (Intensity One, Select Max) 1 EC (Select Max) 1 EC	6 to 8 oz 9 to 16 oz	0.094 to 0.125 0.07 to 0.125	Apply postemergence for control of grasses. With Arrow, Clethodim, or Select, add 1 gallon crop oil concentrate per 100 gallons spray mix. With Select Max or Intensity One, add 2 pints of nonionic surfactant per 100 gallons spray mixture. Adding crop oil concentrate may increase the likelihood of crop injury at high air temperatures. Very effective in controlling annual bluegrass. Apply to actively growing grasses not under drought stress. Do not apply within 14 days of harvest.
	sethoxydim, MOA 1 (Poast) 1.5 EC	1 to 1.5 pt	0.2 to 0.3	Apply to emerged grasses. Consult manufacturer's label for specific rates and best times to treat. Add 1 quart of crop oil concentrate per acre. Crop oil may increase the likelihood of crop injury at high temperatures. Do not apply Poast on days that are unusually hot and humid. Do not apply within 14 days of harvest.
Pumpkins, Row Middles				
Annual grasses and some small-seeded broadleaf weeds	trifluralin, MOA 3 (Treflan) 4 EC (Treflan HFP) 4 EC	1 to 1.5 pt	0.5 to 0.75	**Row middles.** To improve preemergence control of late emerging weeds. Apply after emergence when crop plants have reached the three to four true leaf stage of growth. Apply as a directed spray to soil between the rows. Avoid contacting foliage as slight crop injury may occur. Set incorporation equipment to move treated soil around base of crop plants. Do not apply within 30 days of harvest.

Table 7-16. Chemical Weed Control in Vegetable Crops

Weed	Herbicide, Mode of Action Code* and Formulation	Amount of Formulation Per Acre	Pounds Active Ingredient Per Acre	Precautions and Remarks
Pumpkins, Row Middles (continued)				
Most broadleaf weeds less than 4 inches tall or rosettes less than 3 inches in diameter; does not control grasses	carfentrazone-ethyl, MOA 14 (Aim) 1.9 EW or 2 EC	up to 2 oz	up to 0.031	Apply post-directed using hooded sprayers for control of emerged weeds. If crop is contacted, burning of contacted area will occur. Use a nonionic surfactant or crop oil with Aim. See label for rate. Coverage is essential for good weed control. Can be tank mixed with other registered herbicides. PHI 0 days.
Most emerged weeds	glyphosate, MOA 9 (numerous brands and formulations)	See labels	See labels	**Row middles.** Apply as a hooded spray in row middles, as shielded spray in row middles, as wiper applications in row middles, or post harvest. To avoid severe injury to crop, do not allow herbicide to contact foliage, green shoots, stems, exposed, roots, or fruit of crop. Do not apply within 14 days of harvest.
Yellow and purple nutsedge and broadleaf weeds	halosulfuron-methyl, MOA 2 (Sandea, Profine 75) 75 DG	0.5 to 1 oz	0.024 to 0.048 lb	**Row middles.** Apply to row middles as a postemergence spray. In plasticulture, do not allow spray to contact plastic. Early season application will give postemergence and preemergence control. Do not apply within 30 days of harvest. For postemergence applications, use nonionic surfactant at 1 quart per 100 gallons of spray solution.
Radish, Preplant				
Annual and perennial grass and broadleaf weeds	glyphosate, MOA 9 (numerous brands and formulations)	See labels	See labels	Apply to emerged weeds before planting. Perennial weeds may require higher rates of glyphosate. Consult the manufacturer's label for rates for specific weeds. Certain glyphosate formulations may require addition of a surfactant. Adding nonionic surfactant to glyphosate formulated with nonionic surfactant may result in reduced weed control.
Annual grasses and broadleaf weeds	trifluralin, MOA 3 (Treflan, Treflan HFP, Trifluralin, Trifluralin HF) 4 EC	1 to 1.5 pt	0.5 to 0.75	Apply preplant and incorporate immediately after application for preemergence weed control. Low rate should be used on coarse-textured soil.
Radish, Postemergence				
Most broadleaf weeds less than 4 inches tall or rosettes less than 3 inches in diameter; does not control grasses	carfentrazone-ethyl, MOA 14 (Aim) 1.9 EW or 2 EC	up to 2 oz	up to 0.031	Apply post-directed using hooded sprayers for control of emerged weeds. If crop is contacted, burning of contacted area will occur. Use a nonionic surfactant or crop oil with Aim. See label for rate. Coverage is essential for good weed control. Can be tank mixed with other registered herbicides.
Annual and perennial grasses	clethodim, MOA 1 (Arrow, Clethodim, Intensity, Select) 2 EC	6 to 8 oz	0.94 to 0.125	Apply postemergence to emerged grasses. See label for rates for specific grasses. With Arrow, Clethodim, or Select, add crop oil concentrate at 1 gallon per 100 gallons of spray solution. With Select Max, add nonionic surfactant at 2 pints per 100 gallons spray mixture. Do not spray within 15 days of harvest.
	(Select Max, Intensity One) 1 EC	9 to 16 oz	0.07 to 0.125	
Spinach, Preemergence				
Annual and perennial grass and broadleaf weeds, stale bed application	glyphosate, MOA 9 (numerous brands and formulations)	See labels	See labels	Apply to emerged weeds before crop emergence. Do not feed residue to livestock for 8 weeks. Perennial weeds may require higher rates of glyphosate. Consult the manufacturer's label for rates for specific weeds. Certain glyphosate formulations require the addition of a surfactant. Adding nonionic surfactant to glyphosate formulated with nonionic surfactant may result in reduced weed control.
Annual grasses (crabgrass spp., foxtail spp., barnyardgrass, annual ryegrass, annual bluegrass) and broadleaf weeds (Lamium spp., lambsquarters, common purslane, redroot pigweed, shepherdspurse)	cyclohexylethylthio-carbamate, MOA 3 (Ro-Neet) 6E	2 qt	3	Use on sandy mineral soils only. Read label for further instructions.
Annual grass and broadleaf weeds, Palmer amaranth, yellow nutsedge suppression	S-metolachlor, MOA 15 (Dual Magnum) 7.62 EC	0.33 to 0.67 pt	0.32 to 0.65	This is a section 24(c) North Carolina Special Local Need Label. Growers must obtain label from www.farmassist.com. Apply over top of sweetpotatoes after transplanting but prior to weed emergence. Do not apply preplant. Do not incorporate after application. Injury potential is greatest when applied to sands or loamy sands especially if a heavy rainfall event occurs following application. See label for further information.
Spinach, Postemergence				
Broadleaf weeds including sowthistle clover, cocklebur, jimsonweed, and ragweed	clopyralid, MOA 4 (Stinger) 3 EC	0.17 to 0.33 pt	0.0625 to 0.125 lb	Apply to spinach in the 2- to 5-leaf stage when weeds are small and actively growing. Will control most legumes. See label for more precautions. Do not apply within 21 days of harvest.
Broadleaf weeds	phenmedipham, MOA 6 (Spin-aid) 1.3 EC	3 to 6 pt	0.5 to 1	**For processing spinach only.** Do not use when expected high temperatures will be above 75 degrees F. For best results, spray when weeds are in the two-leaf stage. Use the 6 pints rate only on well-established crops that are not under stress. Do not apply within 21 days of harvest. Spinach plants must have more than six true leaves.
Annual and perennial grasses only	sethoxydim, MOA 1 (Poast) 1.5 EC	1 to 1.5 pt	0.2 to 0.3	Apply to emerged grasses. Consult manufacturer's label for specific rates and best times to treat. For sethoxydim, add 1 quart of crop oil concentrate per acre. For Arrow, Clethodim, or Select, add 1 gallon of crop oil concentrate per 100 gallons spray solution. For Select Max, add nonionic surfactant at 2 pints per 100 gallons of spray mixture. Adding crop oil to Poast or Select may increase the likelihood of crop injury at high air temperatures. Do not apply Poast, Arrow, Clethodim, or Select on days that are unusually hot and humid. Do not apply sethoxydim within 15 days of harvest or clethodim within 14 days of harvest.
	clethodim, MOA 1 (Arrow, Clethodim, Intensity, Select) 2 EC	6 to 8 oz	0.094 to 0.125	
	(Intensity One, Select Max) 1 EC	9 to 16 oz	0.07 to 0.125	
Most broadleaf weeds less than 4 inches tall or rosettes less than 3 inches in diameter; does not control grasses	carfentrazone-ethyl, MOA 14 (Aim) 1.9 EW or 2 EC	up to 2 oz	up to 0.031	Apply post-directed using hooded sprayers for control of emerged weeds. If crop is contacted, burning of contacted area will occur. Use a nonionic surfactant or crop oil with Aim. See label for rate. Coverage is essential for good weed control. Can be tank mixed with other registered herbicides.

Table 7-16. Chemical Weed Control in Vegetable Crops

Weed	Herbicide, Mode of Action Code* and Formulation	Amount of Formulation Per Acre	Pounds Active Ingredient Per Acre	Precautions and Remarks
Spinach, Postemergence (continued)				
Most emerged weeds	glyphosate, MOA 9 (numerous brands and formulations)	See labels	See labels	**Row middles only.** Apply as a hooded spray in row middles, as shielded spray in row middles, as wiper applications in row middles, or post harvest. To avoid severe injury to crop, do not allow herbicide to contact foliage, green shoots, stems, exposed, roots, or fruit of crop. Do not apply within 14 days of harvest.
Squash, Preplant and Preemergence				
Suppression or control of most annual grasses and broadleaf weeds, full rate required for nutsedge control	metam sodium (Vapam HL) 42%	37.5 to 75 gal	15.7 to 31.5	Rates are dependent on soil type and weeds present. Apply when soil moisture is at field capacity (100 to 125%). Apply through soil injection using a rotary tiller or inject with knives no more than 4 inches apart; follow immediately with a roller to smooth and compact the soil surface or with mulch. May apply through drip irrigation prior to planting a second crop on mulch. Plant back interval is often 14 to 21 days and can be 30 days in some environments. See label for all restrictions and additional information.
Emerged broadleaf and grass weeds	pelargonic acid, MOA 27 (Scythe) 4.2 EC	3 to 10% v/v		Apply before crop emergence and control emerged weeds. There is no residual activity. May be tank mixed with soil residual compounds. See label for instruction. May also be used as a banded spray between row middles. Use a shielded sprayer directed to the row middles to reduce drift to the crop.
Contact kill of all green foliage, stale bed application	paraquat, MOA 22 (Firestorm, Parazone) 3 SL (Gramoxone SL) 2 SL	1.3 to 2.7 pt 2 to 4 pt	0.5 to 1	Apply in a minimum of 10 gallons spray mix per acre to emerged weeds before transplanting or crop emergence as a band or broadcast treatment over a preformed row. Use sufficient water to give thorough coverage. Row should be formed several days ahead of planting or treating to allow maximum weed emergence. Use a nonionic surfactant at a rate of 16 to 32 ounces per 100 gallons spray mix or 1 gallon approved crop oil concentrate per 100 gallons spray mix.
Most broadleaf weeds less than 4 inches tall or rosettes less than 3 inches in diameter; does not control grasses	carfentrazone-ethyl, MOA 14 (Aim) 1.9 EW or 2 EC	up to 2 oz	up to 0.031	Not registered for seeded crop. Apply prior to transplanting crop. Use a nonionic surfactant or crop oil with Aim. See label for rate. Coverage is essential for good weed control. Can be tank mixed with other registered burndown herbicides.
Annual and perennial grass and broadleaf weeds, stale bed application	glyphosate, MOA 9 (numerous brands and formulations)	See labels	See labels	Apply to emerged weeds at least 3 days before seeding or transplanting. When applying Roundup before transplanting crops into plastic mulch, care must be taken to remove residues of this product from the plastic prior to transplanting. To prevent crop injury, residues can be removed by 0.5 inch natural rainfall or by applying water via a sprinkler system. Perennial weeds may require higher rates of glyphosate. Consult the manufacturer's label for rates for specific weeds. Certain glyphosate formulations require the addition of a surfactant. Adding nonionic surfactant to glyphosate formulated with nonionic surfactant may result in reduced weed control.
Morningglory and small pigweed < 1 inch	Pyraflufen ethyl, MOA 14 (ET Herbicide) 0.208 L	1 to 2 oz	0.0016 to 0.0032	**Bareground.** Wait 1 day following preplant burndown application before planting. **Plasticulture.** May apply over mulch, however, a single 0.5 inch irrigation/rain event plus a 7 day waiting period is needed before transplanting.
Annual grasses and small-seeded broadleaf weeds	bensulide, MOA 8 (Prefar) 4 EC	5 to 6 qt	5 to 6	Registered for cucurbit vegetable group (Crop grouping 9). Apply preplant and incorporate into the soil 1 to 2 inches (1 inch incorporation is optimum) with a rototiller or tandem disk, or apply to the soil surface after seeding and follow by irrigation. Check replant restrictions for small grains on label.
	ethalfluralin, MOA 3 (Curbit) 3 EC	1.5 to 2 pt	0.56 to .75	**For squash grown on bare ground only.** Apply to the soil surface immediately after seeding. Seed must be covered with soil to prevent crop injury. For coarse-textured soils, use lowest rate of rate range. Shallow cultivation, irrigation, or rainfall within 5 days is needed for good weed control. If weather is unusually cold or soil wet and cold, crop stunting or injury may occur. Crop injury can also occur if seeding depth is too shallow. See label for further precautions and instruction.
		3 to 4.5 pt	1.1 to 1.7	For squash grown on plastic only. Apply to soil surface between the rows of black plastic immediately after seeding or transplanting. **Do not use under mulches, row covers, or hot caps.** Do not apply prior to planting or over plastic. See label for further instruction.
Annual grasses and broadleaf weeds	ethalfluralin, MOA 3 + clomazone, 13 (Strategy) 2.1 L	2 to 6 pt	0.4 to 1.2 + 0.125 to 0.375	Apply to the soil surface immediately after crop seeding for preemergence control of weeds. **Do not apply prior to planting crop. Do not soil incorporate.** May also be used as a **banded** treatment **between** rows after crop emergence or transplanting.
Suppression of annual grasses and broadleaf weeds; weak on pigweed and morningglory	clomazone, MOA 13 (Command) 3 ME	0.67 to 1.3 pt	0.25 to 0.48	Apply immediately after seeding or prior to transplanting. Seeds and roots of transplants must be below the chemical barriers when planting. Command should only be applied between rows when squash is grown on plastic. Some cultivars may be sensitive to Command (see label). Use lower rates on coarse soils. Higher rates can be used on winter squashes. See label about rotation restrictions.
Yellow and purple nutsedge and broadleaf weeds	halosulfuron-methyl, MOA 2 (Sandea, Profine 75) 75 DG	0.5 to 1 oz	0.024 to 0.048	**Row middles only.** Apply to row middles as preemergence spray. In plasticulture, do not allow spray to contact plastic. Early season application will give postemergence and preemergence control. Do not apply within 30 days of harvest. For postemergence applications, use nonionic surfactant at 1 quart per 100 gallons of spray solution. **WINTER SQUASH: Direct seeded:** Apply 0.5 to 0.75 ounce per acre Sandea as preemergence after planting, but before soil cracking **OR** postemergence after crop has reached 2 to 5 true leaf stage, but before first female flowers appear. **Transplanted:** Apply 0.5 to 0.75 ounce per acre Sandea prior to transplant. Transplanting should not be made sooner than 7 days after application. May be applied after crop emergence over the top of the crop when plants reach the four to five true leaf stage but before first female flowers appear. See label for further instructions.
Squash, Postemergence				
Annual grasses and small-seeded broadleaf weeds	trifluralin, MOA 3 (Treflan) 4 EC (Treflan HFP) 4 EC	1 to 1.5 pt	0.5 to 0.75	**Row middles only.** To improve preemergence control of late emerging weeds. **Apply as a directed spray to soil between rows after crop emergence when crop plants have reached** three to four true leaf stage of growth. Avoid contacting foliage as slight crop injury may occur. Set incorporation equipment to move treated soil around base of crop plants. Do not apply within 30 days of harvest. Will not control emerged weeds.

Table 7-16. Chemical Weed Control in Vegetable Crops

Weed	Herbicide, Mode of Action Code* and Formulation	Amount of Formulation Per Acre	Pounds Active Ingredient Per Acre	Precautions and Remarks
Squash, Postemergence (continued)				
Most broadleaf weeds less than 4 inches tall or rosettes less than 3 inches in diameter; does not control grasses	carfentrazone-ethyl, MOA 14 (Aim) 1.9 EW or 2 EC	up to 2 oz	up to 0.031	Apply post-directed using hooded sprayers for control of emerged weeds. If crop is contacted, burning of contacted area will occur. Use a nonionic surfactant or crop oil with Aim. See label for rate. Coverage is essential for good weed control. Can be tank mixed with other registered herbicides. PHI 0 days.
Most emerged weeds	glyphosate, MOA 9 (numerous brands and formulations)	See labels	See labels	**Row middles only.** Apply as a hooded spray in row middles, as shielded spray in row middles, as wiper applications in row middles, or post harvest. To avoid severe injury to crop, do not allow herbicide to contact foliage, green shoots, stems, exposed, roots, or fruit of crop. Do not apply within 14 days of harvest.
Yellow and purple nutsedge and broadleaf weeds	halosulfuron-methyl, MOA 2 (Sandea, Profine 75) 75 DG	0.5 to 1 oz	0.024 to 0.048 lb	**Row middles only.** Apply to row middles as postemergence spray. In plasticulture, do not allow spray to contact plastic. Early season application will give postemergence and preemergence control. Do not apply within 30 days of harvest. For postemergence applications, use nonionic surfactant at 1 quart per 100 gallons of spray solution.
Annual and perennial grasses only	clethodim, MOA 1 (Arrow, Clethodim, Intensity, Select) 2 EC (Intensity One, Select Max) 1 EC	6 to 8 oz 9 to 16 oz	0.094 to 0.125 0.07 to 0.125	Apply postemergence for control of grasses. With Arrow, Clethodim, or Select, add 1 gallon crop oil concentrate per 100 gallons spray mix. With Select Max or Intensity One, add 2 pints nonionic surfactant per 100 gallons spray mixture. Adding crop oil may increase likelihood of crop injury at high air temperatures. Very effective control of annual bluegrass. Apply to actively growing grasses not under drought stress. Do not apply within 14 days of harvest.
	sethoxydim, MOA 1 (Poast)1.5 EC	1 to 1.5 pt	0.2 to 0.3	Apply to emerged grasses. Consult manufacturer's label for specific rates and best times to treat. Add 1 quart of crop oil concentrate per acre. Adding crop oil to Poast may increase the likelihood of crop injury at high air temperatures. Do not apply Poast on days that are unusually hot and humid. Do not apply within 14 days of harvest.
Sweetpotato, Preplant				
Annual and perennial grass and broadleaf weeds, stale seed bed application	glyphosate, MOA 9 (numerous brands and formulations)	See labels	See labels	Apply to emerged weeds before transplanting. Perennial weeds may require higher glyphosate rates. Consult label for rates for specific weeds. Certain glyphosate formulations may require the addition of a surfactant. Adding nonionic surfactant to glyphosate formulated with nonionic surfactant may result in reduced weed control.
Suppression or control of most annual grasses and broadleaf weeds, full rate required for nutsedge control	metam sodium (Vapam HL) 42%	37.5 to 75 gal	15.7 to 31.5	Rates are dependent on soil type and weeds present. Plant back interval is often 14 to 21 days and can be 30 days in some environments. See label for all restrictions and additional information.
Annual broadleaf weeds including Palmer amaranth and other pigweeds, smartweed, morningglory, wild mustard, wild radish, common purslane, common lambsquarters	flumioxazin, MOA 14 (Valor SX, Chateau) 51 WDG	3 oz	0.094	Apply prior to transplanting crop for control of many annual broadleaf weeds and annual sedges. Do not incorporate. Movement of soil during transplanting should not occur or reduced weed control may result. Do not apply postemergence or serious crop injury will occur. Do not use on transplant propagation beds. See label for further instructions.
Sweetpotato, Preemergence				
Annual grass and broadleaf weeds, Palmer amaranth, yellow nutsedge suppression	S-metolachlor, MOA 15 (Dual Magnum) 7.62 EC	0.75 pt	0.7 to 0.96	This is a Section 24(c) Special Local Needs Label. Growers must obtain label from www.farmassist.com Apply over top of sweetpotatoes after transplanting but prior to weed emergence. Do not apply preplant. Do not incorporate after application. Injury potential is greatest when applied to sands or loamy sands especially if a heavy rainfall event occurs following application. See label for further information.
Annual grasses such as large crabgrass and broadleaf weeds including velvetleaf, purslane, prickly sida	clomazone, MOA 13 (Command) 3 ME	up to 2 pt	up to 0.75	Apply preplant or after transplanting prior to weed emergence for preemergence control. Weak on pigweed. The label allows up to 4 pt per acre. See label for other instructions and precautions.
Annual grasses including large crabgrass and broadleaf weeds including purslane, Florida pusley, common lambsquarters	DCPA, MOA 3 (Dacthal) W-75 (Dacthal) 6 F	8 to 10 lb 8 to 10 pt	6 to 7.5	Apply to the soil surface immediately after transplanting. May also be applied at layby for preemergence weed control late in the growing season. Do not apply in plant beds or crop injury will occur.
Annual grasses including crabgrass, foxtail, goosegrass, fall panicum and broadleaf weeds including pigweed, Florida pusley, purslane	napropamide, MOA 15 (Devrinol, Devrinol DF-XT) 50 DF (Devrinol, Devrinol 2-XT) 2 EC	2 to 4 lb 2 to 4 qt	1 to 2	**Plant Beds.** Apply to the soil surface after sweetpotato roots are covered with soil but prior to soil cracking and sweetpotato plant emergence. Does not control emerged weeds. Check label for more information. **Production Fields.** Apply to the soil surface immediately after transplanting. If rainfall does not occur within 24 hr, shallow incorporate or irrigate with sufficient water to wet the soil to a depth of 2 to 4 inches Check label for more information. See XT labels for information regarding delay in irrigation event.

Chapter VII—2019 N.C. Agricultural Chemicals Manual

Table 7-16. Chemical Weed Control in Vegetable Crops

Weed	Herbicide, Mode of Action Code* and Formulation	Amount of Formulation Per Acre	Pounds Active Ingredient Per Acre	Precautions and Remarks
Sweetpotato, Postemergence				
Annual and perennial grasses only	clethodim, MOA 1 (Arrow, Clethodim, Intensity, Select) 2 EC (Select Max, Intensity One) 1 EC	6 to 16 oz 9 to 32 oz	0.094 to 0.25 0.07 to 0.25	Apply to actively growing grasses not under drought stress. For Arrow, Clethodim, or Select, add 1 gallon crop oil concentrate per 100 gallons spray mix. For Select Max, add 2 pints nonionic surfactant per 100 gallons spray mixture. Adding crop oil may increase the likelihood of crop injury at high air temperatures. Very effective in controlling annual bluegrass. Do not apply within 30 days of harvest.
	fluazifop, MOA 1 (Fusilade DX) 2 EC	6 to 16 oz	0.1 to 0.25	Apply to actively growing grasses not under drought stress. Consult manufacturer's label for specific rates and best times to treat. Add 1 gallon crop oil concentrate or 1 quart nonionic surfactant per 100 gallons spray mix. Do not apply Fusilade on days that are unusually hot and humid. Do not apply within 14 days of harvest.
	sethoxydim, MOA 1 (Poast) 1.5 EC	1 to 1.5 pt	0.2 to 0.3	Apply to actively growing grasses not under drought stress. Add 1 quart of crop oil concentrate per acre. Adding crop oil to Poast may increase the likelihood of crop injury at high air temperatures. Do not apply Poast on days that are unusually hot and humid. Do not apply within 30 days of harvest.
Sweetpotato, Row Middles				
Most broadleaf weeds less than 4 inches tall or rosettes less than 3 inches in diameter; does not control grasses	carfentrazone-ethyl, MOA 14 (Aim) 1.9 EW or 2 EC	up to 2 oz	up to 0.031	Apply post-directed using hooded sprayers for control of emerged weeds. If crop is contacted, burning of contacted area will occur. Use a nonionic surfactant or crop oil with Aim. See label for rate. Coverage is essential for good weed control. Can be tank mixed with other registered herbicides.
Most emerged weeds	glyphosate, MOA 9 ((numerous brands and formulations)	See labels	See labels	Apply as a hooded spray in row middles, as shielded spray in row middles, as wiper applications in row middles, or postharvest. To avoid severe injury to crop, do not allow herbicide to contact foliage, green shoots, stems, exposed, roots, or fruit of crop. May cause cracking of sweetpotato storage roots if spray solution comes in contact with sweetpotato foliage. Do not apply within 14 days of harvest.
Tomato, Preplant				
Suppression or control of most annual grasses and broadleaf weeds, full rate required for nutsedge control	metam sodium (Vapam HL) 42%	37.5 to 75 gal	15.7 to 31.5	Rates are dependent on soil type and weeds present. Apply when soil moisture is at field capacity (100 to 125%). Apply through soil injection using a rotary tiller or inject with knives no more than 4 inches apart; follow immediately with a roller to smooth and compact the soil surface or with mulch. May apply through drip irrigation prior to planting a second crop on mulch; however, adhere to label guidelines on crop plant back interval. Plant back interval is often 14 to 21 days and can be 30 days in some environments. See label for all restrictions and additional information. Chloropicrin (150 pounds per acre broadcast) will also be needed when laying first crop mulch to control nutsedge.
Most broadleaf weeds less than 4 inches tall or rosettes less than 3 inches in diameter; does not control grasses	carfentrazone-ethyl, MOA 14 (Aim) 1.9 EW or 2 EC	up to 2 oz	up to 0.031	**Transplanted crop: Apply no later than 1 day before transplanting.** **Seeded crop (Aim 2EC only): Apply no later than 7 days before planting seeded crop.** Use a nonionic surfactant or crop oil with Aim. See label for rate. Coverage is essential for good weed control. Can be tank mixed with other registered burndown herbicides.
Contact kill of all green foliage, stale bed application	paraquat, MOA 22 (Firestorm, Parazone) 3 SL (Gramoxone SL) 2 SL	1.3 to 2.7 pt 2 to 4 pt	0.5 to 1	Apply to emerged weeds in a minimum of 20 gallons spray mix per acre before crop emergence as a broadcast or band treatment over a preformed row. Row should be formed several days ahead of planting and treating to allow maximum weed emergence. Use a nonionic surfactant at a rate of 16 to 32 ounces per 100 gallons spray mix or 1 gallon approved crop oil concentrate per 100 gallons spray mix.
Broadleaf weeds including Carolina geranium and cutleaf eveningprimrose and a few annual grasses	oxyfluorfen, MOA 14 (Goal) 2 XL	up to 2 pt	0.5 lb	**Plasticulture only.** Apply to soil surface of pre-formed beds at least 30 days prior to transplanting crop. While incorporation is not necessary, it may result in less crop injury. Plastic mulch can be applied any time after application, but best results are likely if applied soon after application.
Annual grasses and small-seeded broadleaf weeds including common lambsquarters, pigweed, carpetweed, and common purslane	napropamide, MOA 15 (Devrinol, Devrinol DF-XT) 50 DF (Devrinol, Devrinol 2-XT) 2 EC	2 to 4 lb 2 to 4 qt	1 to 2	**Bare ground:** Apply preplant and incorporate into the soil 1 to 2 inches as soon as possible with a rototiller or tandem disk. Can be used on direct-seeded or transplanted tomatoes. See label for instructions on use. **Plasticulture:** Apply to a weed-free soil before laying plastic mulch. Soil should be well worked yet moist enough to permit a thorough incorporation to a depth of 2 inches. Mechanically incorporate or irrigate within 24 hours after application. If weed pressure is from small seeded annuals, apply to the surface of the bed immediately in front of the laying of plastic mulch. If soil is dry, water or sprinkle irrigate with sufficient water to wet to a depth of 2 to 4 inches before covering with plastic mulch. **Between rows:** Apply to a weed free soil surface between the rows (bareground or plastic mulch). Mechanically incorporate or irrigate Devrinol into the soil to a depth of 1 to 2 inches within 24 hours of application. See XT labels for information regarding delay in irrigation event.
	pendimethalin, MOA 3 (Prowl H$_2$O) 3.8 CS	1 to 3 pt	0.5 to 1.5	**Plasticulture In-row:** May be applied as a preplant surface application or a preplant incorporated application prior to transplanting tomato. Limited research has been conducted in NC. **Bareground In-row:** May be applied as a broadcast preplant surface application or preplant incorporated application prior to transplanting tomato. **Post-directed spray:** May be applied as a post-directed spray on the soil at the base of the plant, beneath plants, and between rows. Avoid direct contact with tomato foliage or stems. Do not apply over the top of tomato. Do not apply within 21 days of harvest. Do not apply more than 3 pt per acre per season. See label for specific use rate for your soil type. Emerged weeds will not be controlled. See label for further instructions and precautions.
	trifluralin, MOA 3 (Treflan HFP, Trifluralin, Trifluralin HF, various other trade names,) 4 EC	1 pt	0.5	**Transplant tomato.** Apply pretransplant and incorporate into the soil 2 to 3 inches within 8 hr using a rototiller or tandem disk. Can be applied post-plant as a directed spray to soil between the rows and beneath plants and then incorporated. Not suggested for mulch systems.

Table 7-16. Chemical Weed Control in Vegetable Crops

Weed	Herbicide, Mode of Action Code* and Formulation	Amount of Formulation Per Acre	Pounds Active Ingredient Per Acre	Precautions and Remarks
Tomato, Preplant (continued)				
Yellow and purple nutsedge and broadleaf weeds including non-ALS resistant pigweed, wild radish, common ragweed, suppression of purslane	halosulfuron-methyl, MOA 2 (Profine, Sandea) 75 DG	0.5 to 1 oz	0.024 to 0.048	For pretransplant application under plastic mulch, apply to pre-formed bed just prior to plastic mulch application and delay transplanting at least 7 days. Can be applied for pretransplant application in bareground tomato. Early season application will give postemergence and preemergence control. The 1 ounce rate is for preemergence and postemergence control in row middles only. For postemergence applications, use nonionic surfactant at 1 quart per 100 gallons of spray solution. Do not apply within 30 days of harvest.
Yellow nutsedge, annual grasses, and broadleaf weeds including pigweed, Palmer amaranth, Florida pusley, Hairy galinsoga, Eastern black nightshade, and carpetweed	S-metolachlor, MOA 15 (Brawl, Dual Magnum) 7.62 EC	1 to 2 pt	0.95 to 1.50	Apply preplant or postdirected to transplants after the first settling rain or irrigation. In plasticulture, apply to preformed beds just prior to applying plastic mulch. Lower rates of rate range for S-metolachlor are safest to tomato. May also be used to treat row middles in bedded tomato. Minimize contact with crop. Do not apply within 90 days of harvest. Also registered for use in row middles, and in seeded crop. See label for further instructions.
Palmer amaranth, redroot pigweed, smooth pigweed, Galinsoga sp., black nightshade, Eastern black nightshade, common purslane, partial control of yellow nutsedge	fomesafen, MOA 14 (Reflex) 2 EC	1 to 1.5 pt	0.25 to 0.375	This is a Section 24(c) special local needs label for transplanted tomato in North Carolina. Growers must obtain the label at Farmassist.com prior to making an application of Reflex. **See label for further instructions.** **Plasticulture In-row Application for Transplanted Tomato.** Apply after final bed formation and the drip tape is laid but prior to laying plastic mulch. Avoid soil disturbance after application. Unless restricted by other products such as fumigants, tomato may be transplanted immediately following the application of Reflex and the application of the mulch. **Bareground for Transplanted Tomato.** Apply pretransplant up to 7 days prior to transplanting tomato. Weed control will be reduced if soil is disturbed after application. During the transplanting operation, make sure the soil in the transplant hole settles flush or above the surrounding soil surface. Avoid cultural practices that may concentrate Reflex-treated soil around the transplant root ball. An overhead irrigation or rainfall event between Reflex herbicide application and transplanting will ensure herbicide activation and will likely reduce the potential for crop injury due to splashing. **Plasticulture Row Middle Application.** Apply to row middles with a hooded or shielded sprayer. Avoid drift of herbicide on mulch. If drift occurs, 0.5 inch of rain or irrigation must occur prior to transplanting. **Carryover is a large concern; see label for more information.**
Annual grasses and broadleaf weeds including jimsonweed, common ragweed, smartweed, and velvetleaf	metribuzin, MOA 5 (TriCor DF, Metribuzin) 75 WDG (Metri) 4 F	0.33 to 0.67 lb 0.5 to 1 pt	0.25 to 0.5	Apply to soil surface and incorporate 2 to 4 inches deep before transplanting. See label for instructions. Place tomato transplant roots below herbicide layer to avoid injury. Use lower rate when on sands or if it is cool/wet.
Broadleaf weeds including Carolina geranium and cutleaf eveningprimrose and a few annual grasses	oxyfluorfen, MOA 14 (Goal) 2XL (GoalTender) 4 F	up to 2 pt up to 1 pt	0.5 lb	**Plasticulture (fallow beds) only.** Apply to soil surface of pre-formed beds at least 30 days prior to transplanting crop. While incorporation is not necessary, it may result in less crop injury. Plastic mulch can be applied any time after application, but best results are likely if applied soon after application.
Broadleaf, grass (suppression),yellow nutsedge (PRE or POST), purple nutsedge (POST only)	imazosulfuron, MOA 2 (League) 0.5 DF	4 to 6.4 oz	0.19 to 0.3	Apply to planting beds before plastic is laid. Tomato may be transplanted 1 day after application. Postemergence application may be made on direct-seeded plants that are well established (4 to 5 leaf stage) and in transplanted tomato 3 to 5 days after transplanting. 10 inches tall, **if a pretransplant application was not made.** Consult label for approved surfactants and crop rotation restrictions. PHI 21 days.
Common lambsquarters, ivyleaf morningglory, redroot pigweed, and yellow nutsedge	sulfentrazone, MOA 14 (Spartan, Helm, Sulfentrazone) 4 F	2.25 to 8 oz	0.07 to 0.25	**Transplant.** Pretransplant application. Application rate depends on soil type.
Tomato, Postemergence				
Annual grasses and small-seeded broadleaf weeds	DCPA, MOA 3 (Dacthal) W-75 (Dacthal) 6 F	6 to 10 lb 6 to 10 pt	4.5 to 7.5	Apply over the top of transplants only between 4 to 6 weeks after transplanting to improve preemergence control of late emerging weeds. Will not control emerged weeds.
Yellow and purple nutsedge and broadleaf weeds	halosulfuron-methyl, MOA 2 (Sandea, Profine 75) 75 DG	0.5 to 1 oz	0.024 to 0.048 lb	Apply no sooner than 14 days after transplanting. For postemergence applications, use nonionic surfactant at 1 quart per 100 gallons of spray solution. Some weeds, such as nutsedge, may require two applications of Sandea; if a second application is needed, spot-treat only weed-infested areas. Do not apply within 30 days of harvest. See label for further instructions.
Annual grasses and broadleaf weeds, including cocklebur, common ragweed, smartweed, velvetleaf, jimsonweed, yellow nutsedge, and morningglory	metribuzin, MOA 5 (TriCor DF, Metribuzin) 75 WDG	0.33 to 1.33 lb	0.25 to 1	Use either as a broadcast or directed spray but do not exceed 0.5 pound a.i. with a broadcast spray. Do not apply within 7 days of harvest. Do not exceed 1 pound a.i. per year. Do not apply as a broadcast spray unless 3 sunny days precede application.
Most broadleaf weeds including wild radish, common purslane, redroot and smooth pigweed	rimsulfuron, MOA 2 (Matrix) 25 WDG (Pruvin) 25 WDG	1 to 2 oz	0.25 to 0.5 oz	Apply in tomatoes after the crop has at least two true leaves and weeds are small (1 inch or less) and actively growing. Add nonionic surfactant at 1 quart per 100 gallons of spray solution. Do not apply within 45 days of tomato harvest. See label for further instruction.
Yellow nutsedge, morningglory, common cocklebur, common lambsquarters, and other broadleaf weeds	trifloxysulfuron-sodium, MOA 2 (Envoke) 75 DG	0.1 to 0.2 oz	0.0047 to 0.0094	Apply post-directed to tomato grown on plastic for control of nutsedge and certain broadleaf weeds. Crop should be transplanted at least 14 days prior to application. The application should be made prior to fruit set and at least 45 days prior to harvest. Use nonionic surfactant at 1 quart per 100 gallons spray solution with all applications.

Table 7-16. Chemical Weed Control in Vegetable Crops

Weed	Herbicide, Mode of Action Code* and Formulation	Amount of Formulation Per Acre	Pounds Active Ingredient Per Acre	Precautions and Remarks
Tomato, Postemergence (continued)				
Annual and perennial grasses only	clethodim, MOA 1 (Arrow, Clethodim, Intensity, Select) 2 EC	6 to 16 fl oz	0.094 to 0.25	Apply to actively growing grasses not suffering from drought stress. With Arrow, Clethodim, or Select, add a crop oil concentrate at 1% by volume (1 gallon per 100 gallons spray mix). With Select Max, add 2 pints of nonionic surfactant per 100 gallons spray mixture. Adding crop oil may increase the likelihood of crop injury at high air temperatures. Do not apply on unusually hot and humid days. Very effective in controlling annual bluegrass. Do not apply within 20 days of harvest.
	(Select Max, Intensity One) 1 EC	9 to 32 oz	0.07 to 0.25	
	sethoxydim, MOA 1 (Poast) 1.5 EC	1 to 1.5 pt	0.2 to 0.3	Apply to actively growing grasses not under drought stress. Add 1 quart of crop oil concentrate per acre. Adding crop oil to Poast may increase the likelihood of crop injury at high air temperatures. Do not apply Poast on days that are unusually hot and humid. Do not apply within 20 days of harvest.
Tomato, Row Middles				
Yellow nutsedge, morningglory, common cocklebur, common lambsquarters, and other broadleaf weeds	trifloxysulfuron-sodium, MOA 2 (Envoke) 75 DG	0.1 to 0.2 oz	0.0047 to 0.0094	Crop should be transplanted at least 14 days prior to application. Use nonionic surfactant at 1 quart per 100 gallons spray solution with all applications. The application should be made prior to fruit set and at least 45 days prior to harvest. See label for information on registered tank mixes. Tank mixtures with Select or Poast may reduce grass control. See label for more information.
Yellow and purple nutsedge and broadleaf weeds	halosulfuron-methyl, MOA 2 (Profine, Sandea) 75 DG)	0.5 to 1 oz	0.024 to 0.048 lb	For postemergence applications, use nonionic surfactant at 1 quart per 100 gallons of spray solution. Some weeds, such as nutsedge, may require two applications of Sandea; if a second application is needed, spot-treat only weed-infested areas. Do not apply within 30 days of harvest. See label for further instructions.
Most broadleaf weeds less than 4 inches tall or rosettes less than 3 inches in diameter; does not control grasses	carfentrazone-ethyl, MOA 14 (Aim) 1.9 EW or 2 EC	up to 2 oz	up to 0.031	Apply post-directed using hooded sprayers for control of emerged weeds. If crop is contacted, burning of contacted area will occur. .Use a nonionic surfactant or crop oil with Aim. See label for rate. Coverage is essential for good weed control. Can be tank mixed with other registered herbicides. PHI 0 days.
Annual grasses and small-seeded broadleaf weeds	napropamide, MOA 15 (Devrinol, Devrinol DF-XT) 50 DF	2 to 4 lb	1 to 2	**Plasticulture:** Apply to a weed-free soil surface. Apply within 24 hours of rainfall, or mechanically incorporate or irrigate into the soil to a depth of 1 to 2 inches
	(Devrinol, Devrinol 2-XT) 2 EC	2 to 4 qt		
	pendimethalin, MOA 3 (Prowl H$_2$O) 3.8	1 to 3 pt	0.5 to 1.5	Post-directed spray on the soil at the base of the plant, beneath plants and between rows. Avoid direct contact with tomato foliage or stems. Do not apply more than 3 pints per acre per season. See label for specific use rate for your soil type. Emerged weeds will not be controlled. Avoid direct contact with tomato foliage or stems. Do not apply within 21 days of harvest. See label for further instructions and precautions.
Contact kill of all green foliage	paraquat, MOA 22 (Firestorm, Parazone) 3 SL (Gramoxone SL) 2 SL	1.3 pt 2 pt	0.5 to 1	Apply for control of emerged weeds between rows of tomatoes. Do not allow spray to contract crop or injury will occur. Do not make more than 3 applications per season. Do not apply within 30 days of harvest.
Watermelon, Preplant				
Suppression or control of most annual grasses and broadleaf weeds, full rate required for nutsedge control	metam sodium (Vapam HL) 42%	37.5 to 75 gal	15.7 to 31.5	Rates are dependent on soil type and weeds present. Apply when soil moisture is at field capacity (100 to 125%). Apply through soil injection using a rotary tiller or inject with knives no more than 4 inches apart; follow immediately with a roller to smooth and compact the soil surface or with mulch. May apply through drip irrigation prior to planting a second crop on mulch. Plant back interval is often 14 to 21 days and can be 30 days in some environments. See label for all restrictions and additional information.
Contact kill of all green foliage, stale bed application	paraquat, MOA 22 (Firestorm, Parazone) 3 SL (Gramoxone SL) 2 SL	1.3 to 2.7 pt 2 to 4 pt	0.5 to 1	Apply in a minimum of 10 gallons spray mix per acre to emerged weeds before crop emergence or transplanting as a broadcast or band treatment over a preformed row. Row should be formed several days ahead of planting and treating to allow maximum weed emergence. Plant with a minimum of soil movement for best results. Use a nonionic surfactant at a rate of 16 to 32 ounces per 100 gallons spray mix or 1 gallon approved crop oil concentrate per 100 gallons spray mix.
Most broadleaf weeds less than 4 inches tall or rosettes less than 3 inches in diameter; does not control grasses	carfentrazone-ethyl, MOA 14 (Aim) 1.9 EW or 2 EC	up to 2 oz	up to 0.031	**Transplants only.** Apply prior to transplanting of crop. Use a nonionic surfactant or crop oil with Aim. See label for rate. Coverage is essential for good weed control. Can be tank mixed with other registered burndown herbicides.
Annual and perennial grass and broadleaf weeds, stale bed application	glyphosate, MOA 9 (numerous brands and formulations)	See labels	See labels	Apply to emerged weeds at least 3 days before seeding or transplanting. When applying Roundup before transplanting crops into plastic mulch, care must be taken to remove residues of this product from the plastic prior to transplanting. To prevent crop injury, residues can be removed by 0.5 inch natural rainfall or by applying water via a sprinkler system. Perennial weeds may require higher rates of glyphosate. Consult the manufacturer's label for rates for specific weeds. Certain glyphosate formulations require the addition of a surfactant. Adding nonionic surfactant to glyphosate formulated with nonionic surfactant may result in reduced weed control.
Morningglory and small pigweed < 1 inch	pyraflufen ethyl, MOA 14 (ET Herbicide) 0.208 L	1 to 2 oz	0.0016 to 0.0032	**Bareground.** Wait 1 day following preplant burndown application before planting. **Plasticulture.** May apply over mulch; however, a single 0.5 inch irrigation/rain event plus a 7 day waiting period is needed before transplanting.
Annual grasses	bensulide, MOA 8 (Prefar) 4 E	5 to 6 qt	5 to 6	Registered for cucurbit vegetable group (Crop grouping 9). Apply preplant and incorporate into the soil 1 to 2 inches (1 inch incorporation is optimum) with a rototiller or tandem disk, or apply to the soil surface after seeding and follow with irrigation. Check replant restrictions for small grains on label.

Table 7-16. Chemical Weed Control in Vegetable Crops

Weed	Herbicide, Mode of Action Code* and Formulation	Amount of Formulation Per Acre	Pounds Active Ingredient Per Acre	Precautions and Remarks
Watermelon, Preplant or Preemergence				
Palmer amaranth, redroot pigweed, smooth pigweed, galinsoga sp., black nightshade, Eastern black nightshade, common purslane, partial control of yellow nutsedge	Fomesafen, MOA 14 (Reflex) 2 EC	10 to 12 oz	0.16 to 0.19	This is a Section 24(c) Special Local Needs Label for watermelon in North Carolina. Growers must obtain the label at www.farmassist.com prior to making an application of Reflex. **See label for further instructions.** **Mulch transplants or seeds.** May apply under mulch as long as plastic laying process does not disturb treated soil; thus, do not apply prior to laying drip or forming bed. May apply over mulch as long as the mulch is washed with 0.5" rainfall/irrigation in a single event prior to punching holes and planting; bed formation must allow herbicide to wash off the mulch and not concentrate in low areas on the mulch. **Baregound transplant.** Prepare land for planting; apply Reflex; lightly irrigate to activate herbicide and move it into soil; and then prepare plant holes and plant. **Baregound seeded.** Apply within 1 day of planting; lightly irrigate after application but at least 36 hours prior to emergence. **Row middles.** Must apply prior to crop emergence or transplanting. May use up to 16 oz/A in watermelon. **Carryover is a large concern; see label for more information.**
Watermelon, Preemergence				
Annual grasses and broadleaf weeds	clomazone, MOA 13 (Command) 3 ME	0.4 to 0.67 pt	0.15 to 0.25	Apply immediately after seeding, or just prior to transplanting. Roots of transplants must be below the chemical barrier when planting. Offers weak control of pigweed. See label for further instructions.
Annual grasses and some small-seeded broadleaf weeds	ethalfluralin, MOA 3 (Curbit) 3 EC	3 to 4.5 pt	1.1 to 1.7	Apply postplant to seeded crop prior to crop emergence, or as a banded spray between rows after crop emergence or transplanting. See label for timing. Shallow cultivation, irrigation, or rainfall within 5 days is needed for good weed control. Do not use under mulches, row covers, or hot caps. Under conditions of unusually cold or wet soil and air temperatures, crop stunting or injury may occur. Crop injury can occur if seeding depth is too shallow.
Annual grasses and broadleaf weeds	ethalfluralin, MOA 3 + clomazone, MOA 13 (Strategy) 2.1 L	2 to 6 pt	0.4 to 1.2 + 0.125 to 0.375	Apply to soil surface immediately after crop seeding for preemergence control of weeds. **Do not apply prior to planting. Do not incorporate. Do not apply under mulch.** May also be used as a banded treatment between rows after crop emergence or transplanting.
Broadleaf weeds	terbacil, MOA 5 (Sinbar) 80 WP	2 to 4 oz	0.1 to 0.2	Apply after seeding but before crop emerges, or prior to transplanting crop. With plasticulture, Sinbar may be applied preemergence under plastic mulch or to row middles. May be applied over plastic mulch prior to transplanting, or prior to punching holes into the plastic mulch for transplanting. Sinbar must be washed off the surface of the plastic mulch with a minimum of 0.5 inch of rainfall or irritation prior to punching transplant holes or transplanting watermelon. Do not apply within 70 days of harvest. See label for further instructions.
Yellow and purple nutsedge suppression, pigweed and ragweed control	halosulfuron-methyl, MOA 2 (Profine 75, Sandea) 75 DG	0.5 to 0.75 oz 0.5 to 1 oz	0.024 to 0.036 0.024 to 0.048	**Baregound.** Apply after seeding but before cracking or prior to transplanting crop. **Plasticulture.** Application may be made to preformed beds prior to laying plastic. If application is made prior to planting, wait 7 days after application to seed or transplant. Stunting may occur but should be short lived with no negative effects on yield or maturity in favorable growing conditions. **See label for information on rotation and other restrictions.** **Row middles only.** Apply to row middles as a preemergence spray. In plasticulture, do not allow spray to contact plastic. Early season application will give postemergence and preemergence control. Do not apply within 57 days of harvest. For postemergence applications, use nonionic surfactant at 1 quart per 100 gallons of spray solution.
Watermelon, Postemergence				
Annual grasses and some small seeded broadleaf weeds	DCPA, MOA 3 (Dacthal) W-75 (Dacthal) 6 F	8 to 10 lb 8 to 10 pt	6 to 7.5	**Not labeled for transplanted crop.** To improve preemergence control of late emerging weeds, apply only when crop has 4 to 5 true leaves, is well-established, and growing conditions are favorable. Will not control emerged weeds. Incorporation not recommended.
Annual and perennial grasses only	clethodim, MOA 1 (Arrow, Clethodim, Intensity, Select) 2 EC (Intensity One, Select Max) 1 EC	6 to 8 oz 9 to 16 oz	0.094 to 0.125 0.07 to 0.125	Apply postemergence for control of grasses. With Arrow, Clethodim, or Select, add 1 gallon crop oil concentrate per 100 gallons spray mix. With Select Max, add 2 pints nonionic surfactant per 100 gallons spray mixture. Adding crop oil may increase the likelihood of crop injury at high air temperatures. Very effective in controlling annual bluegrass. Apply to actively growing grasses not under drought stress. Do not apply within 14 days of harvest.
	sethoxydim, MOA 1 (Poast) 1.5 EC	1 to 1.5 pt	0.2 to 0.3	Apply to emerged grasses. Consult manufacturer's label for specific rates and best times to treat. Add 1 quart of crop oil concentrate per acre. Adding crop oil to Poast may increase the likelihood of crop injury at high air temperatures. Do not apply Poast on days that are unusually hot and humid. Do not apply within 14 days of harvest.
Watermelon, Row Middles				
Annual grasses and some small-seeded broadleaf weeds	trifluralin, MOA 3 (Treflan HFP, Trifluralin, Trifluralin HF) 4 EC	1 to 2 pt	0.5 to 0.75	To improve preemergence control of late emerging weeds. Apply after emergence when crop plants have reached the three to four true leaf stage of growth. Apply as a directed spray to soil between the rows. Avoid contacting foliage as slight crop injury may occur. Set incorporation equipment to move treated soil around base of crop plants. Do not apply within 60 days of harvest. Will not control emerged weeds.
Broadleaf weeds	terbacil, MOA 5 (Sinbar) 80 WP	2 to 4 oz	0.1 to 0.2	With plasticulture, Sinbar may be applied to row middles. Do not apply within 70 days of harvest. See label for further instructions.
Most broadleaf weeds less than 4 inches tall or rosettes less than 3 inches in diameter; does not control grasses	carfentrazone-ethyl, MOA 14 (Aim) 1.9 EW or 2 EC	up to 2 oz	up to 0.031	Apply post-directed using hooded sprayers for control of emerged weeds. If crop is contacted, burning of contacted area will occur. Use a nonionic surfactant or crop oil with Aim. See label for rate. Coverage is essential for good weed control. Can be tank mixed with other registered herbicides.

Table 7-16. Chemical Weed Control in Vegetable Crops

Weed	Herbicide, Mode of Action Code* and Formulation	Amount of Formulation Per Acre	Pounds Active Ingredient Per Acre	Precautions and Remarks
Watermelon, Row Middles (continued)				
Most emerged weeds	glyphosate, MOA 9 (numerous brands and formulations)	See labels	See labels	Apply as a hooded spray in row middles, as shielded spray in row middles, as wiper applications in row middles, or postharvest. To avoid severe injury to crop, do not allow herbicide to contact foliage, green shoots, stems, exposed, roots, or fruit of crop. Do not apply within 14 days of harvest.
Yellow and purple nutsedge and broadleaf weeds	halosulfuron-methyl, MOA 2 (Profine 75, Sandea) 75 DG	0.5 to 1 oz	0.024 to 0.048	Apply to row middles as a postemergence spray. In plasticulture, do not allow spray to contact plastic. Early season application will give postemergence and preemergence control. For postemergence applications, use nonionic surfactant at 1 quart per 100 gallons of spray solution. Do not apply within 57 days of harvest.
Broadleaf weeds and yellow and purple nutsedge	Imazosulfuron, MOA 2 (League) 0.5 DF	4 to o6.4 oz	0.19 to 0.3	Apply any time during the cropping season (up to 48 days before harvest), as long as the melons are well established and at least 5 inches wide. Avoid contact with the melon. In plasticulture prevent the spray from contacting the plastic. Consult label for further instructions. PHI = 48 days.

* Mode of action (MOA) code developed by the Weed Science Society of America.

Chemical Weed Control in Forest Stands

Compiled by C. Lambert and M. Megalos, NCSU Forestry and Environmental Resources

With professional assistance by:
David Spak, Ph.D., Bayer U.S.
Jeff Tapley, Nutrien Ag Solutions

NOTE: A mode of action code has been added to the Herbicide and Formulation column of this table. Use MOA codes for herbicide resistance management.

Table 7-17. Pine Release

Type Application / Plants Controlled	Herbicide and Formulation	Amount of Formulation Per Acre	Uses, Remarks, and Precautions
Foliar Spray			
Many broadleaf weeds, grasses, and woody species (black locust, cherry, oaks, persimmon, maple, sassafras, sumac, sweetgum, yellow-poplar)	Glyphosate, MOA 9 (Rodeo or Aquaneat) 5.4L OR Other brands 1.125 to 1.875 qt 12 to 18 oz	0.75 qt Rate may vary	Use at end of first growing season for first-year plantings after planted pine seedlings buds have hardened off. May be tank mixed with Oust XP and/or Arsenal or Arsenal AC or Chopper Gen2 for many pines. See labels for mix details.
		Use late summer or early fall after conifers have hardened off (second and subsequent growing seasons).	
		Herbaceous release (early season) for loblolly and longleaf pines. May be tank mixed with Oust XP and/or Arsenal or Arsenal AC for many species. See labels for mix details.	
Woody plants (oaks, sweetgum, elm, and sumac)	Hexazinone, MOA 5 (Velpar L) 2EC OR (Velpar DF) 75%	4 to 6 pt (first-year plantings) 4 to 8 pt (established trees) 1.33 to 1.8 lb (first-year plantings) 1.33 to 2.66 (established trees)	Use spring to early summer. All rates depend on soil type. Use lower rates on coarser soils. See label.
Many grasses, broadleaf weeds, vines, brambles, woody brush, and trees	Imazapyr, MOA 2 Arsenal AC (53.1%) or Arsenal 2NS (27.8%) or Chopper Gen2 (26.7%) or Polaris AC Complete)	12 to 20 oz or 24 to 40 oz or 24 to 80 oz or 48 to 80 oz for loblolly 12 to 20 oz for loblolly	See label for rates for specific species of pine. May be tank mixed with Rodeo (only loblolly after late summer hardening off), Oust XP and, for loblolly only, Oust Extra.

Table 7-17. Pine Release

Type Application / Plants Controlled	Herbicide and Formulation	Amount of Formulation Per Acre	Uses, Remarks, and Precautions
Many grasses, broadleaf weeds, vines, brambles, woody brush, and trees	Imazapyr, MOA 2 (63.2%) or Arsenal AC (53.1%) or Arsenal 2NS (27.8%) or Chopper Gen2 (26.7%) + Metsulfuron methyl, MOA 2 (9.5%) (Lineage Clearstand)	9 to 16 oz	Labeled for loblolly and slash pines. Can use up to 19 oz for mid rotation release of pines.
Herbaceous plants (crabgrass, dog fennel, fescue, willowweed [fireweed], goldenrod, horseweed, Kentucky bluegrass, yellow nutsedge, panicums [broadleaf, fall, narrow], pokeweed, ragweed, white snakeroot, and yellow sweetclover)	Sulfometuron methyl, MOA 2 (Oust XP 75% or Spyder)	2 to 8 oz	Rates depend on soil type and species. Labeled for loblolly, slash, longleaf, Virginia, and white pines. May be tank mixed with Velpar and/or Arsenal for some species. See labels. Do not exceed 8 oz per acre per year.
Many annual grasses and broadleaf weeds	Hexazinone, MOA 5 63.2% + Sulfometuron methyl, MOA 2 11.8% (Oustar)	10 to 19 oz (first-year control) + 12 to 24 oz (after first year)	Use spring to early summer. All rates depend on soil type. For loblolly, slash, and longleaf pine only. See labels.
Certain hardwoods, weeds, and grasses	Sulfometuron methyl, MOA 2 (15%) + Metsulfuron methyl, MOA 2 (56.25%) (Oust Extra or Spyder Extra)	2.66 to 4 oz	Loblolly and slash only. May be tank mixed with Arsenal AC for loblolly only. Use lower rates on coarser soils.
Herbaceous weeds	Imazapyr, MOA 2, (Arsenal 2NS (27.8%) or Arsenal AC (53.1%) or Chopper Gen2 (26.7%) or Polaris AC Complete)	8 to 20 oz 4 to 10 oz 8 to 20 oz 4 to 10 oz	May be tank mixed with Oust see label for rates for specific species of pine.
Broadleaf herbaceous plants such as bedstraw, hemp dogbane, sericea lespedeza, cocklebur, coffeeweed, common and giant ragweed, dog fennel, grape, marestail, morning glory, goldenrod, blackberry, ironweed, wild carrot, common mullein, mustard, vetch, and prickly pear cactus.	Fluroxypyr MOA 4, (Vista XRT OR Comet 1.5lb/g fluroxypyr)	6 to 23 fl oz .67 to 2.3 pts	Do not apply as an over-the-top broadcast treatment during active terminal growth (from initiation of budbreak/growth flush until seasonal terminal growth has hardened off and overwintering buds have formed.) Directed spray applications may be made to pines during periods of active growth, but care should be taken to avoid spray contact with actively growing foliage.
Broadleaf herbaceous plants such as ragweed, coffeeweed, clover, marestail, knapweed, partridge pea, sicklepod, smartweed, thistle, and morning glory	Clopyralid MOA 4, (Transline or Clean Slate)	8 to 21 fl oz	Labeled for use on first year pine seedlings and older trees. Will not control henbit or chickweed.
Herbaceous weeds (ragweed, marestail, morning glory, sicklepod, pigweed, and blackberry.) See label for complete list of susceptible broadleaf weeds	Aminopyralid MOA 4 (Milestone)	5 to 7 fl oz	Labeled for use for release of longleaf pine only. Do not apply over the top of other desirable pine species. May be tank mixed with approved products for use over longleaf. Applications can be performed over the top of longleaf still in the "grass stage.". Newly planted seedlings should be at least three months old before application and not under stress. Read and follow label recommendations. Do not add surfactant/adjuvant.
Many broadleaf weeds and warm season grasses	Topramezone, MOA 27 (Frequency)	4 to 8 fl oz	Labeled for use for grass and broadleaf weed release for loblolly and longleaf pine. Can be tank mixed with Arsenal AC for increased weed control. See label for rates. Excellent control of bermudagrass and other warm season grasses. DO NOT add surfactant/adjuvant when applying to longleaf pine.
Preemergent/Foliar			
Most annual grasses and many broadleaf weeds	Pendimethalin, MOA 3 (Pendulum) 2 G OR (Pendulum) 3.3 EC OR Pendulum AquaCap (38.7%)	100 to 200 lb 2.4 to 4.8 qt 2.1 to 4.2 qt	Apply at time of planting or to established trees. Planting slit must be closed to avoid root contact. Labeled for loblolly, white and Virginia pines and a number of hardwood species. See label for details. See label for details.
Spot-gun			
Woody plants (cherry, blackgum, dogwood, elm, hawthorn, hickory, oaks, maple, sweetgum, and sumac)	Hexazinone, MOA 5 (Velpar L) 2EC OR (Velpar DF) 75%	3 to 10 qt 1.3 to 5.3 lb	Apply in calibrated spots on grid pattern. Rate per acre and grid pattern depend on soil texture and species composition. Apply with exact delivery handgun. Apply late winter to early summer. Do not apply spots within 36 inches or directly upslope from seedlings. Poor results may occur if site is burned 3 to 6 months before treatment or on stump sprouts less than 1 year old.

Forest Site Preparation, Stand Conversion, Timber Stand Improvement

Compiled by C. Lambert and M. Megalos, NCSU Forestry and Environmental Resources

Table 7-18. Forest Site Preparation, Stand Conversion, Timber Stand Improvement

Type Application	Plants Controlled	Herbicide and Formulation	Amount of Formulation Per Acre	Uses, Remarks, and Precautions
Spot-gun (continued)				
Foliar Spray	For control of many broadleaf weeds, annual and perennial grasses, brush, vines and brambles	Imazapyr, MOA 2 (63.2%) + metsulfuron methyl, MOA 2 (9.5%) (Lineage Clearstand)	8 to 25 oz for loblolly and slash pine	Can tank mix with KRENITE-S at 4 to 6 quarts for control of natural pines.
		OR Tank Mix Polaris AC Complete + Patriot	10 to 24 oz + 1 to 2 oz	
		OR Imazapyr, MOA 2 Arsenal AC (53.1%) or Arsenal 2NS (27.8%) or Chopper Gen2 (26.7%) + metsulfuron methyl, MOA 2 (9.5%)	20 to 40 oz or 40 to 80 oz or 32 to 64 oz See label	For loblolly and slash pine. See label for details.
Foliar Spray	For control of natural pines	Ammonium salt of fosamine, MOA 17 (41.2%) (Krenite-S)	4 to 6 qts	Can tank mixed with most other site prep products. Need highest label rate for control of Virginia pine.
		Detail	2 oz/A	See label for further instructions. Detail should be used in combination with 4 qt/A of glyphosate.
Foliar Spray	Control of woody brush including but not limited to loblolly pine, oak species, legumes, maple, cherry, poplar, blackberry, etc.	Glyphosate, MOA 9 (Accord XRT II) 5.07 lb/gal + Aminopyralid, MOA 4 (Milestone) +	6 to 7 qts + 7 fl oz +	Recommended for use from June – July 31. Offers broad spectrum hardwood and natural pine control.
		Imazapyr, MOA 2 Arsenal AC (53.1%) or Arsenal 2NS (27.8%) or Chopper Gen2 (26.7%) or Polaris AC Complete)	12 to 40 oz or 24 to 80 oz or 24 to 64 oz or 12 to 40 oz	See label for details.
Foliar Spray	Control of woody brush including but not limited to southern pines, oak species, legumes, maple, cherry, blackberry, poplar, gallberry, wax myrtle, yaupon, bay spp. etc.	Glyphosate, MOA 9 (Accord XRT II) 5.07 lb/gal + Aminopyralid, MOA 4 (Milestone) + Triclopyr ester, MOA 4 (Forestry Garlon XRT) +	5 qts + 7 fl oz + 21 fl oz +	Recommended for use from Aug. 1 to Oct. 15th. Offers broad spectrum hardwood and natural pine control.
		Imazapyr, MOA 2 (Arsenal 2NS (27.8%) or Arsenal AC (53.1%) or Chopper Gen2 (26.7%)or Polaris AC Complete)	16 to 40 oz or 24 to 80 oz or 24 to 64 oz or 16 to 40 oz	See label for details.
Foliar Spray	Control of woody brush including but not limited to southern and Virginia pine, oak spp., cherry, maple, poplar, legumes, wax myrtle, blackberry, etc.	Glyphosate (Accord XRT II 5.07 lb/gal or Aquaneat) + Saflufenacil MOA14 (Detail) + Triclopyr ester, MOA 4 (Forestry Garlon XRT) +	6 qts + 2 fl oz + 21 fl oz +	Recommended for use from June 1 to Sept. 15. Recommended when VA pine are part of the target species.
		Imazapyr, MOA 2 Arsenal AC (53.1%) or Arsenal 2NS (27.8%) or Chopper Gen2 (26.7%) or Polaris AC Complete	12 to 40 oz or 24 to 80 oz or 24 to 64 oz or 12 to 40 oz	See label for details.
Foliar Spray	Woody brush, trees, vines, grasses, and broadleaf weeds (Alder, berries [blackberry, dewberry, raspberry], elderberry, honeysuckle, maples [red, sugar], oaks [red, Northern pine, white], multi flora rose, poison ivy, poison oak, trumpet creeper, willow)	Glyphosate, MOA 9 (Accord XRT II) 5.07lb/gal OR Razor Pro OR Other brands	2% solution as a high volume spray using a spray to wet basis Rates may vary	Apply over actively growing plants when foliage has fully developed. See labels for appropriate rates for weeds and woody brush. See labels.
Foliar Spray	Kudzu, wisteria	Aminopyralid, MOA 4 (Milestone)	7 fl oz	0.5 ounce/gallon of water for spot treatments (assuming 20 gpa). See label for full details. Do not apply over the top of susceptible tree species.
	Kudzu	Clopyralid, MOA 4 (Transline) Clean Slate	21 fl oz	Application June to October to actively growing plants

Table 7-18. Forest Site Preparation, Stand Conversion, Timber Stand Improvement

Type Application	Plants Controlled	Herbicide and Formulation	Amount of Formulation Per Acre	Uses, Remarks, and Precautions
Directed Foliar Spray	Woody vines and brush	Aminopyralid, MOA 4 + Triclopyr Amine, MOA 4 (Capstone)	5 to 7%	Excellent control of woody vines such as kudzu, wisteria, blackberry, honeysuckle, and small seedling brush. For example, cherry, maple, southern pine, oak spp., and legumes, etc. Does not hurt most grass species.
	Over 50 species of woody brush, broadleaf perennials, etc.	2,4-D, 2,4-DP (Patron 170)	1.5% to 4.25%	Selective to most grasses control of over 50 broadleaf perennial and woody species including poison ivy, kudzu, blackberry, dewberry, honeysuckle, locust, and many more.
Cut Stumps	Sweetgum, poplar, oak, sycamore. Suppression of dogwood, blackgum, hickory, and red maple	Glyphosate, MOA 9 (Rodeo or Aquaneat) 5.4L OR Other brands	Spray or brush 50 to 100% solution to freshly cut stumps Rates may vary	Best used mid growing season. Avoid spring sapflow. Treat stumps immediately after cutting. Treat entire stump surface for small trees and just cambium for large trees. See labels.
Cut Stump and Basal treatment	Works on more than 90 woody brush and vine species. See label for a complete list	Triclopyr ester, MOA 4 (Garlon 4 Ultra) or Relegate or Triclopyr ester (Pathfinder II)	20 to 25% Garlon 4 Ultra + 75 to 80% basal oil or use Pathfinder II as a ready-to-use	Treat the lower 12 to 15 inches of the target stem down to the ground line making sure to treat completely around the stem. Treat to wet not to the point of runoff. See label for full details. May be used on smooth bark species up to 6 inches in basal diameter and up to 4 inches in basal diameter for rough bark species. Can be performed all year except when the bark is wet and any time after cutting.
Foliar Spray	Annual bluegrass, ash, aspens (bigtooth, trembling), asters, balsam poplar, barnyardgrass, bentgrass, birch, common groundsel, common ragweed, elksedge, elm, false dandelion, fleabane, flowering dogwood, foxtail, hawthorn, hazel, hickory, oaks, oxeye daisy, Pennsylvania smartweed, pinegrass, red maple, sourwood sweetgum, velvetgrass, wild carrot, wild cherry, and willows. Treatment provides partial control of Canada thistle (suppression only), catsear, crabgrass, curly dock, dandelion, willowweed (fireweed) fescue, goldenrod, heath aster, honeysuckle, horseweed, orchardgrass, and perennial grasses (quackgrass and ryegrass)	Hexazinone, MOA 5 (Velpar L) 2 EC OR (Velpar DF) (75%)	4 to 10 qt 2 1/3 to 6 2/3 lb	Labeled for longleaf pine, shortleaf pine, slash pine, Virginia pine, loblolly pine, Scotch pine, and white spruce. All rates depend on soil type. Use lower rates on coarser soils. Read and follow label directions.
Spotgun	Many woody plants (black cherry, blackgum dogwood, elm, hawthorn, hickory, oaks, maple, sweetgum, and sumac)	Hexazinone, MOA 5 (Velpar L) 2 EC OR (Velpar DF) (75%)	4 to 10 qt Mix 2 2/3 lb/gal of water. Apply suspension at 3 to 10 qt/acre, using grid pattern. See label for grid spacing.	Use an exact-delivery handgun applicator and calibrate for precise delivery of the undiluted product. Selection of the rate per acre and "grid" pattern will depend on soil texture and woody plant species composition. Use the higher rates on fine-textured soils and when the major component of the hardwoods is a difficult-to-control species such as blackgum, dogwood, hickory, or red maple. Use the lower rates on medium-to-coarse textured soils where elm, cherry, oak, and sweetgum are dominant. Results may be unsatisfactory where stump sprouts of less than one year's growth predominate. Labeled for loblolly, slash, shortleaf, and longleaf pines. Suspensions of DF require intermittent agitation.
Broadcast Granular	Many woody plants (American elder, balsam poplar, birch, black cherry, blackgum, hickory, boxelder, brambles [blackberry, dewberry, and raspberry], cherry, chokeberry, cottonwood, dogwood Eastern red cedar, elm, green ash, hawthorn, hornbeam, mulberry, multiflora rose, Norway maple, red maple, oaks [black, blackjack, bluejack, post, southern red, Turkey, water, and white oak], Russian olive, sumac, sweetgum, white ash, wild plum, willow, sourwood, and pine)	Hexazinone, MOA 5 (Velpar ULW) 75% granule (no longer commercially available)	2.5 to 6.3 lb	May be applied broadcast or to individual plant stems. Rainfall is required to dissolve the granules and move the herbicide into the root zone. Apply in spring when hardwood leaves are 50% developed but no later than June 15. Results are best on coarse-to-medium textured soils. Not recommended for poorly drained or marsh sites. Labeled for loblolly, longleaf, shortleaf, slash, and Virginia pine.
Foliar Spray	Many grasses, broadleaf annual, and perennial weeds, vines, brambles, woody brush, and trees	Imazapyr, MOA 2 Arsenal AC (53.1%) or Arsenal (27.8%) or Chopper Gen2 (26.7%) or Polaris AC Complete)	12 to 40 oz or 24 to 80 oz or 24 to 64 oz or 12 to 40 oz Add 0.5% v/v nonionic surfactant to mix.	Broadcast spray during growing season to prepare site for many conifers. May be tank mixed with Accord XRT II, Oust, Oust Extra (loblolly only), and others. See labels. Not labeled for hardwood site preparation.

Table 7-18. Forest Site Preparation, Stand Conversion, Timber Stand Improvement

Type Application	Plants Controlled	Herbicide and Formulation	Amount of Formulation Per Acre	Uses, Remarks, and Precautions
Tree Injection, Frill, or Hack 'N' Squirt	Brush species listed above	Imazapyr, MOA 2 Arsenal AC (53.1%) or Arsenal 2NS (27.8%) or Chopper Gen2 (26.7%) or Polaris AC Complete)	25% solution in water	Use hatchet or tree injector and space sloping cuts evenly around the trunk. Apply 1 ml/3 in. DBH. Best used mid growing season but dormant applications may be used. Avoid spring sapflow. Nearby desirable hardwoods may be injured or killed. See label for details
		OR Aminopyralid, MOA 4 + triclopyr amine, MOA 4 (Capstone) OR	Use undiluted and apply 1 ml of undiluted product into each cut	Use hatchet or machete and space cuts no more than 2 inches apart and at a 45-degree angle. Make cuts around the tree trunk at a convenient height so that the cuts overlap slightly and make a continuous circle around the trunk. Avoid treatment during periods of heavy sap flow such as the spring.
		Triclopyr choline, MOA 4 (Vastlan)	40% solution mixed with water	Use hatchet or machete and space cuts no more than 2 inches apart and at a 45-degree angle. Make cuts around the tree trunk at a convenient height so that the cuts overlap slightly and make a continuous circle around the trunk. Avoid treatment during periods of heavy sap flow such as the spring.
	Sweetgum, poplar, oak, sycamore. Suppression of dogwood, blackgum, hickory, and red maple	Glyphosate, MOA 9 (Rodeo) 5.4 lb/gal or Aquaneat	1 ml/2 to 3 in. of trunk diameter, or use continuous frill and a diluted mix.	Use hatchet or tree injector and space sloping cuts evenly around the trunk. Best used mid growing season. Avoid spring sapflow.
		OR Imazapyr, MOA 2 Arsenal AC (53.1%) or Arsenal 2NS (27.8%) or Chopper Gen2 (26.7%)	1 mL of solution in each cut site	See label for details.
		OR Other brands	Rates for other brands may vary	See labels.
Cut Stumps	Brush species listed above	Imazapyr, MOA 2 Arsenal AC (53.1%) or Arsenal 2NS (27.8%) or Chopper Gen2 (26.7%) or Polaris AC Complete)	25% solution in water 6 oz/gal water on freshly cut stump surface.	Arsenal AC Dilute solution mix 4-6 ounces in 1 gallon of water. Concentrate solution mix 1 quart with 1 pint of water. See label for details. Arsenal (27.8%) Dilute solution mix 8-12 ounces in 1 gallon of water. Concentrate solution mix 2 quarts with 1 pint of water. See label for details. Chopper Gen2 Dilute solution mix 8-12 ounces in 1 gallon of water. Concentrate solution mix 2 quarts with 1 pint of water. See label for details. Best used during growing season; avoid spring sapflow. Treat stumps promptly after cutting. Nearby desirable hardwoods may be killed or injured.
		OR Aminopyralid, MOA 4 + triclopyr amine, MOA 4 (Capstone)	Use 100% solution.	Best used immediately after cut. Avoid spring sap flow. Spray or brush cambium layer immediately after cut.
		OR Triclopyr ester, MOA 4 (Garlon 4 Ultra 4lb/gal or Relegate, contains no petroleum distillates)	Use 20 to 25% solution mixed with 75 to 80% basal oil.	Can be used all year except when bark is wet. Treat the cambium layer and root collar down to the ground line. Treat to wet and not to the point of run-off. Can be used up to several weeks after cutting.
Foliar Spray	Many species of brush and broadleaf weeds, including: BRUSH: Ash, aspen, birches, brambles, cedar, cherry, dogwood, elms, gooseberry, honeylocust, multiflora rose, oak, poplar, shortleaf pine, sumac, wild plum, willows, sycamore, spruce, blackberry, black locust, buckbrush, cottonwood, honeysuckle. WEEDS: bodotraw, burdock, chicory, dock, kudzu, morning glory, poison ivy, poison oak, thistles, trumpet vine	Imazapyr, MOA 2 (Chopper GEN2 or Polaris SP) + Glyphosate, MOA 9 (Accord XRT II)	32 to 64 oz + 2 qts	Apply during growing season to prepare site for many conifers. A seed oil mix that is 12 to 50% of volume is recommended. Use 5 to 40 gallons mix/acre for good coverage. Not labeled for hardwood site preparation. Tank mixing Accord XRT II with Chopper treatments will enhance brown-out of target vegetation thus enabling landowner to more effectively burn.
	Most grasses, broadleaf annual and perennial weeds, vines, brambles, hardwoods and pines	Imazapyr, MOA 2 Arsenal AC (53.1%) or Arsenal 2NS (27.8%) or Chopper Gen2 (26.7%) or Polaris SP) + Glyphosate, MOA 9 (Accord XRT II)) + Aminopyralid, MOA 4 (Milestone)	16 to 40 oz or 24 to 80 oz or 24 to 64 oz or 32 to 64 oz + 6 to 7.5 qts + 7 fl oz	Effective broad-spectrum site preparation for hardwoods and natural pine.

Table 7-18. Forest Site Preparation, Stand Conversion, Timber Stand Improvement

Type Application	Plants Controlled	Herbicide and Formulation	Amount of Formulation Per Acre	Uses, Remarks, and Precautions
Basal Bark	Controls over 90 woody plants and vines	Triclopyr, MOA 4 (Pathfinder II)	100% solution ready to use formulation of Garlon 4 Ultra	Spray or wet lower 12 to 15 inches of stems. Best used on rough bark species no more than 4 inches diameter and smooth bark species no more than 6 inches diameter. Perform year round except when bark is wet. Spray to wet but not to the point of run-off.
		OR Triclopyr, MOA 4 (Garlon 4 Ultra or Relegate)	20 to 25% solution mixed with 75-80% basal oil	
Foliar Spray	Many woody species and broadleaf weeds, grasses	Glufosinate-ammonium, MOA 10 (Derringer F) 1 lb a.i./gal	2 to 6 qt	May be used to prepare site for planting hardwoods and conifers. Use higher rate for woody or heavy dense brush. See label for specific details. May be tank mixed with Arsenal AC and Chopper for conifer site preparation.
Foliar Spray	Alder, birch, blackberry, black cherry, blackjack oak, black locust, currant, fir, gooseberry, hemlock, honeysuckle, oaks, pine, poison ivy, poison oak, poplar, red elm, red maple, serviceberry, spruce, sycamore, tulip poplar, willow, winged elm	Dicamba, MOA 4 + 2,4-D, MOA 4 (Veteran 720) 2.9 lb a.i./gal	1 gal	Apply 1 gallon in 20 to 100 gallons water per acre depending upon density of brush. Do not apply more than 1 gallon per acre per year.
Foliar Spray	Many brush species (alder, ash, basswood, beech, birch, blackberry, black gum, cedar, cherry, cottonwood, dogwood, elm, grape, hawthorn, hemlock, hickory, honeysuckle, hornbeam, poison ivy, kudzu, locust, maple, oaks, persimmon, pines, poplar, sassafras, sumac, sweetgum, sycamore, willow) Controls many woody plants including yaupon, gallberry, wax myrtle, Baccharis spp., and many bay species.	Triclopyr (choline), MOA 4 (Vastlan) 4 lb a.i./gal OR Triclopyr (ester), MOA 4 (Forestry Garlon XRT) 6.3 lb/gal	0.5 to 1.5 gal 21 fl. oz. to 1 gal	Ground applications: Mix product in water (20 to 100 gallons; enough for good coverage) + 0.5% v/v nonionic surfactant. May also be tank mixed with other products. See label for details. Use higher rates for woody plant control.
Tree Injection, Frill, Hack 'N' Squirt	Same as above	Triclopyr (choline), MOA 4 (Vastlan) 4 lb a.i./gal	Mix 40:60 ratio with water to make a 40% solution	Inject 0.5 ml undiluted or 1 ml diluted/inch tree diameter. Space cuts evenly. Best used during growing season.
Cut Stump	Same as above	Triclopyr (choline), MOA 4 (Vastlan) 4 lb a.i./gal	Use 40% solution mixed with water	Spray or paint fresh cut stumps immediately after cut. Best used during the growing season but avoid period of spring sap flow.
Cut Surface Treatment	Same as above	Dicamba, MOA 4 (Banvel CST) 1 S	See label	Inject 1 ml undiluted at 1- to 2-inch intervals. Spray or paint frill or girdle OR spray or paint cut stumps with undiluted herbicide
Basal Bark or Dormant Stem Treatment	Same as above	Dicamba, MOA 4 + 2,4-D(ester), MOA 4 (Banvel 520) 2.9 lb total a.i./gal	1 to 5 gal	For hydraulic sprayers use 1 to 3 gallons Banvel 520 per 100 gallons water or oil to water emulsion per acre. For backpack sprayers mix 8 to 16 gallons Banvel per 100 gallons water or oil to water emulsion and apply 30 gallons mix per acre. **BASAL BARK**—Apply year around to lower 1.5 to 2 feet of stem, allowing runoff. **DORMANT STEM**—Apply anytime brush is dormant by thoroughly wetting stems to point of runoff.
Foliar Spray	**WOODY**: Alder, ash, beech, birch, blackberry, blackgum, cherry, cottonwood, dogwood, elderberry, elm, hawthorn, hickory, hornbeam, locust, maples, mulberry, oaks, persimmon, pine, poison oak, poplar, sassafras, sumac, sweetbay, magnolia, sweetgum, sycamore, tulip poplar, willow, winged elm **ANNUAL AND PERENNIAL BROADLEAF WEEDS**: Bindweed, burdock, Canada thistle, chicory, curly dock, dandelion, field bindweed, lambsquarter, plantain, ragweed, smartweed, tansy ragwort, vetch, wild lettuce	Triclopyr-(choline), MOA 4 (Vastlan) 4 lb/gal	0.3 to 1.5 gal	**Aerial Application**—Mix 0.75-1 gallon herbicide with 0.25 to 1 pint agricultural surfactant. For broader spectrum control, tank mix with other suitable herbicides. Apply in 10 to 30 gal water per acre. **Ground High-Volume Application**—Mix 0.5 to 1 gallon herbicide in water to make 100 gallons solution + 8-16 oz adjuvant. For broader spectrum control, tank mix with other suitable herbicides. **Ground Low-Volume Directed Spot Application**—Mix 0.5-2% solution (0.64 – 2.5 fl oz herbicide/gallon of water) with 1 % (1.28 fl oz surfactant/gal of water)
Tree Injection, Frill, Hack 'N' Squirt	See woody plants above	Aminopyralid, MOA 4 + triclopyr amine, MOA 4 (Capstone) OR Picloram, MOA 4 + 2,4-D, MOA 4 (Pathway) 1.25 lb a.i./gal	Apply as ready to use mixtures (100% solution)	Inject Capstone undiluted making cuts no more than 2 inches apart. Use 1 ml/inch of Pathway undiluted. Space injections evenly. Best used during the summer or fall.

Table 7-18. Forest Site Preparation, Stand Conversion, Timber Stand Improvement

Type Application	Plants Controlled	Herbicide and Formulation	Amount of Formulation Per Acre	Uses, Remarks, and Precautions
Frill or Girdle	See woody plants above	Same as above	Use undiluted	Make a single hack girdle or "frill" of overlapping cuts through the bark completely around the tree. Spray or paint the cuts with undiluted product using enough volume to wet the treated areas.
Cut Stump	See woody plants above	Aminopyralid, MOA 4 + triclopyr amine, MOA 4 (Capstone) OR Picloram, MOA 4 + 2,4-D, MOA 4 (Pathway) 1.25 lb a.i./gal	Apply as ready to use mixtures (100% solution)	Spray or paint cambium of freshly cut stumps with undiluted herbicide solution.
Foliar Spray	Same as above	Triclopyr (ester), MOA 4 (Forestry Garlon XRT) 6.3 lb a.i./gal OR 4 lb/gal triclopyr (Relegate)	21 fl oz to 1 gal 32 oz to 1.5 gal	Use higher rates to control woody species. Apply in enough water for good coverage, usually 20 to 40 gallons per acre. A nonionic surfactant (0.5 to 1.0%) is recommended. Tank mixes may provide additional benefits. Will not control most native grasses. Complete coverage of target vegetation is necessary for best results.
Basal Bark	Same as above	Same as above OR Triclopyr, MOA 4 (Pathfinder II) 0.75 lb a.i./gal	13% Forestry Garlon XRT + 87% basal oil Apply as a ready to use formulation (100% solution)	Spray 13% solution of Forestry Garlon XRT in oil penetrant (diesel or vegetable oil) to lower 12 to 18 inches of tree trunk to point of runoff. Most effective on trees less than 6 inches in diameter with smooth or thin bark. Apply to dry bark in winter or summer. Avoid spring sap flow
Foliar	Many annual and biennial broadleaf weeds and woody brush	Picloram, MOA 4 (Tordon K) 2 lb a.i./gal	0.25 to 4 qt	Restricted use chemical. See label for specific rates for given situations, tank mixes, and volumes per acre.
Foliar	Many annual and biennial broadleaf and woody brush species	Diglycolamine salt of dicamba, MOA 4 (Vanquish) 4 lb a.i./gal	0.5 to 4 pt	Use lower rates for annual and biennial weeds and heavy rates for woody brush. Many tank mixes are available. Apply in 15 or more gallons of water/acre with a nonionic surfactant (0.5 to 1% v/v).
Tree Injection, Frill, Hack 'N' Squirt	Many species of hardwood trees and brush	Same as above	See label	Apply undiluted product using injector or hatchet with one cut per inch tree diameter. Or apply diluted mixture (1:1-3 parts water) to overlapping cuts (frill).
Cut Stump	Same as above	Same as above	See label	Dilute 1 to 1 with water. Spray or paint the freshly cut stump cambium.
Dormant Stem	Many species of hardwood trees and brush. Controls woody vines and conifers	Triclopyr ester (Garlon 4 Ultra) 4 lb/gal	2 to 3 gallons of Garlon 4 Ultra + 3 gal of Crop oil per 100 gallons of water	Spray upper and lower stems of dormant brush to wet. See label for more details.
Cut Surface and Basal Spray	Many species of hardwood trees and brush	Triclopyr ester, MOA 4 + 2,4-D, MOA 4 (Crossbow or Candor)	See label	Read label carefully as rates and dilutions vary widely depending on tree species to control and type of application.

Aquatic Weed Control

R. J. Richardson, Crop and Soil Sciences Department, and K. D. Getsinger, U.S. Army Engineer Research and Development Center, Vicksburg, MS, and Adjunct Professor, Crop and Soil Sciences Department, NC State University

Several options, including hand removal, cultural, mechanical, biological, and chemical control techniques are available for the management of aquatic weeds. The applicator should choose the most efficacious, environmentally acceptable and cost-effective alternative that is available for a particular weed problem. The site-specific management strategy to use in a given situation will depend on the intended use of the body of water, fish, and wildlife populations that may be impacted, type of environment in which the weed problem occurs, and the particular weed species of concern. Before selecting your management strategy, **be sure to have the weed(s) of concern identified by a qualified individual.**

Assistance in weed identification is available from the Cooperative Extension center in your county. Additional information on management techniques also may be obtained from the local Extension center; ask for AG437, *Weed Management in Small Ponds*; AG-438, *Weed Control in Irrigation Water Supplies;* and AG-449, *Hydrilla: A Rapidly Spreading Aquatic Weed in North Carolina*. Information on pond construction, stocking, and general pond management may be found in AG-424, *Pond Management Guide*. Additional information may be found on the Aquatic Weed Management Web site: go.ncsu.edu/aquatic-weed-management

For the purpose of description and management, aquatic weeds may be grouped either on the basis of their botanical relationships or on the basis of their growth habits. Most plants in each group are managed similarly, with some exceptions.

Table 7-19A. Aquatic Plant Groups — Grouping of Aquatic Plants on the Basis of Botanical Relationships

Category and Description	Examples
1. *Algae* — These plants may be either microscopic or visible to the naked eye, exist as single cells or occur in clusters or filaments containing many cells, and may be either free floating (planktonic) or attached to the soil, rocks, or vegetation. Filamentous algae may be unbranched, slightly or highly branched, or net-like. Some planktonic algae are mobile. Certain types of algae (macroalgae) may be large, very coarse, and resemble submersed vascular plants. Most algae (except macroalgae) usually require magnification to be identified accurately. Algae do not contain vascular (water conducting) tissues, consequently all chemicals used for algae control have only contact activity. Algae reproduce by cell division, fragmentation, and sexually by spores.	**Filamentous Algae** Bluegreens or Cyanobacteria Giant *Lyngbya* Green algae *Oedogonium* *Hydrodictyon* (water net) *Spirogyra* *Pithophora* **Planktonic Algae** Bluegreens or Cyanobacteria *Lyngbya* *Anabaena* *Oscillatoria* *Microcystis* Euglenoids (*Euglena*) **Macroalgae** Muskgrass (*Chara*) Stonewort (*Nitella*)
2. *Mosses* — These plants are visible to the naked eye and resemble delicate, leafy submersed plants. The mosses lack vascular tissues or roots, but usually are attached to the soil. Mosses reproduce sexually by spore production.	*Fontinalis* *Sphagnum* (peat moss)
3. *Ferns* — These plants are visible to the naked eye, either free floating or rooted to the bottom, occasionally forming loosely consolidated floating mats. Ferns have vascular tissues and reproduce by vegetative propagation and sexually by spores.	Giant salvinia (*Salvinia molesta*) Mosquito fern (*Azolla* spp.) Water clover (*Marsilea quadrifolia*) Water spangles (*Salvinia minima*)
4. *Vascular flowering plants* — These plants may be rooted or unrooted, free floating, submersed, floating-leaved, or emergent. Most reproduce vegetatively by means of rhizomes, stolons, and various other vegetative perennating structures including turions and tubers. Most also produce flowers and may set seeds. This group has a vascular system that shows varying degrees of development from rudimentary in the case of the duckweeds and submersed species to very complex and highly developed in emergent plants and includes annual and perennial herbaceous forms and several woody species.	Bald cypress (*Taxodium distichum*) Bladderwort (*Utricularia* spp.) Bulrushes (*Scirpus* spp.) Cattail (*Typha* spp.) Duckweed (*Lemna* spp. and *Spirodela* spp.) Hydrilla (*Hydrilla verticillata*) Naiads (*Najas* spp.) Pondweeds (*Potamogeton* spp.) Rushes (*Juncus* spp.) Spikerushes (*Eleocharis* spp.) Waterhyacinth (*Eichhornia crassipes*) Watermilfoils (*Myriophyllum* spp.)
1. *Submersed plants* — Plants in this group grow beneath the surface of the water and may be rooted to the bottom or free floating, with or without roots. Flowers usually are produced above the surface of the water and occasionally may be supported by specialized floatation structures. Some species will produce emergent floral spikes that extend several inches above the surface of the water and are covered with bracts that resemble leaves. Submersed plants usually have poorly developed vascular systems and very limited structural tissue and depend on the water's buoyancy for support. Filamentous algae and macroalgae also could be considered submersed plants.	American elodea (*Elodea canadensis* and *E. nuttallii*) Bladderwort (*Utricularia* spp.) Brazilian elodea (*Egeria densa*) Brittle naiad (*Najas minor*) Coontail (*Ceratophyllum demersum*) Creeping rush (*Juncus repens*) Eurasian watermilfoil (*Myriophyllum spicatum*) Fanwort (*Cabomba caroliniana*) Hydrilla (*Hydrilla verticillata*) Parrotfeather (*Myriophyllum aquaticum*) Pondweeds (*Potamogeton* spp.) Proliferating spikerush (*Eleocharis baldwinii*) Southern naiad (*Najas guadalupensis*) Variable-leaf milfoil (*Myriophyllum heterophyllum*) Widgeongrass (*Ruppia maritima*) Wild celery (*Vallisneria americana*)

Table 7-19B. Aquatic Weed Groups — Grouping of Aquatic Plants on the Basis of Growth Habit

NOTE: Some species have growth habits that overlap and may be listed more than once.

Category and Description	Examples
1. *Submersed plants* — Plants in this group grow beneath the surface of the water and may be rooted to the bottom or free floating, with or without roots. Flowers usually are produced above the surface of the water and occasionally may be supported by specialized floatation structures. Some species will produce emergent floral spikes that extend several inches above the surface of the water and are covered with bracts that resemble leaves. Submersed plants usually have poorly developed vascular systems and very limited structural tissue and depend on the water's buoyancy for support. Filamentous algae and macroalgae also could be considered submersed plants.	American elodea (*Elodea canadensis* and *E. nuttallii*) Bladderwort (*Utricularia* spp.) Brazilian elodea (*Egeria densa*) Brittle naiad (*Najas minor*) Coontail (*Ceratophyllum demersum*) Creeping rush (*Juncus repens*) Eurasian watermilfoil (*Myriophyllum spicatum*) Fanwort (*Cabomba caroliniana*) Hydrilla (*Hydrilla verticillata*) Parrotfeather (*Myriophyllum aquaticum*) Pondweeds (*Potamogeton* spp.) Proliferating spikerush (*Eleocharis baldwinii*) Southern naiad (*Najas guadalupensis*) Variable-leaf milfoil (*Myriophyllum heterophyllum*) Widgeongrass (*Ruppia maritima*) Wild celery (*Vallisneria americana*)
2. *Free-floating plants* — Plants in this group float on the surface of the water and may lie flat on the water or be raised well above the surface. These plants, with the exception of the duckweeds, watermeal, and mosquito ferns, have well-developed vascular systems and substantial supportive tissues. Most form true roots. Flowers extend above the surface of the water in the flowering plants.	Duckweeds (*Lemna* spp. and *Spirodela* spp.) Floating heart (*Nymphoides aquatica*) Frogbit (*Limnobium spongia*) Giant salvinia (*Salvinia molesta*) Mosquito fern (*Azolla caroliniana*) Waterhyacinth (*Eichhornia crassipes*) Waterlettuce (*Pistia stratiotes*) Watermeal (*Wolffia* spp.)
3. *Floating leaf plants* — These plants are rooted in the bottom and have their leaves attached to long, tough stems that extend to the surface from depths up to 6 ft or more. The leaves float directly on the surface of the water. Mature leaves of some species may push well above the surface into an emergent position. Most of these plants have extensive root and rhizome systems and well-developed vascular systems and supportive tissues. Flowers float just above the surface or are extended well above the surface on a tough stem. A few nonvascular representatives.	American lotus (*Nelumbo lutea*) Fragrant waterlily (*Nymphaea odorata*) Illinois pondweed (*Potamogeton illinoiensis*) Spatterdock (*Nuphar luteum*) Water clover (*Marsilea quadrifolia*) Watershield (*Brasenia schreberi*)
4. *Emergent plants* — These plants grow rooted in the bottom with their leaves and green stems extending well above the surface of the water. A few species also may form floating mats. All have extensive root and rhizome systems and well-developed vascular systems and supportive tissues. Reproduction occurs vegetatively by rhizomes and stolons; floating mat-forming species also reproduce readily by stem fragmentation. Most flower prolifically and form many seeds.	**Broadleaf Species** Arrow arum (*Peltandra virginica*) Arrowhead (*Sagittaria* spp.) Asian spiderwort (*Murdannia keisak*) Frogbit (*Limnobium spongia*) Lizard's tail (*Saururus cernuus*) Pickerelweed (*Pontederia cordata*) Smartweeds (*Polygonum* spp.) **Mat-forming Broadleaf Species** Alligatorweed (*Alternanthera philoxeroides*) Creeping waterprimrose (*Ludwigia hexapetala*) Water pennywort (*Hydrocotyle* spp.) Water willow (*Justicia americana*) **Sedges, Rushes, Spikerushes, and Grasses** Bulrush (*Scirpus* spp.) Cattail (*Typha* spp.) Common reed (*Phragmites australis*) Flat sedge (*Carex* spp.) Foursquare (*Eleocharis quadrangulata*) Maidencane (*Panicum hemitomon*) Rushes (*Juncus* spp.) Sedge (*Cyperus* spp.) Soft rush (*Juncus effusus*) Softstem bulrush (*Scirpus validus*) Southern wildrice (*Zizaniopsis miliacea*) Spikerushes (*Eleocharis* spp.) Threesquare bulrush (*Scirpus americanus*) Torpedograss (*Panicum repens*) Water paspalum (*Paspalum repens*) Woolgrass (*Scirpus cyperinus*) **Other Common Species** Bur-reed (*Sparganium americanum*) Scouring rush (*Equisetum hymale*)
5. *Woody plants* — These are obligate, aquatic species of trees usually growing totally flooded or in saturated soils, but occasionally occur in upland areas (usually planted there). Some form systems of "knees" to provide aeration for the root systems. They are deciduous, dropping leaves in the autumn, and are rarely if ever vegetative during winter months.	Bald cypress (*Taxodium distichum*) Pond cypress (*Taxodium ascendens*) Tupelo (*Nyssa aquatica*)

Biological Control of Aquatic Weeds with Triploid Grass Carp

While the triploid, sterile grass carp is a cost-effective control method, it is best suited for use in small ponds, where submersed aquatic plants are not required for fish and wildlife habitat. Grass carp are effective on most submersed weeds. They generally are less effective on algae and weeds in the floating and emergent groups. Refer to the chart below for information on the relative effectiveness of grass carp for different weeds.

Grass carp are normally stocked at 15 fish per acre in small ponds. In larger ponds, they are usually stocked at 15 to 20 fish per vegetated acre. Large fish (minimum of 8 to 10 inches long) should be stocked to prevent loss due to predation by large bass and wading birds. If the surface of the pond is completely covered with vegetation, some limited herbicide application or mechanical removal of weeds from a portion of the pond will be necessary before stocking to allow oxygen to reach the underlying water. Grass carp may be stocked at any time of the growing season, but best results are usually obtained by a late summer or fall stocking.

No permit is required to purchase up to 150 triploid grass carp for stocking a private pond. At a stocking rate of 15 fish per acre of water, 150 triploid grass carp are adequate to control vegetation in a 10-acre pond. A permit from the Wildlife Resources Commission is required for larger stockings. Grass carp may be purchased from a licensed distributor. For a list of North Carolina vendors, see http://www.ncagr.com/aquacult/grasscarp.htm. Permits, a list of certified distributors, and additional information on stocking of triploid grass carp may be obtained from the Wildlife Resources Commission, Chief of Inland Fisheries, 1721 Mail Service Center, Raleigh, NC 27699-1721, or call at (919) 707-0220.

Table 7-20. Biological Control of Aquatic Weeds with Triploid Grass Carp

Weed	Relative Effectiveness	Comments
Algae Filamentous (green and bluegreen) and planktonic	Poor	High stocking rates (60 to 75 or more fish per acre) with small fish (4 to 6 inches size) are required to achieve temporary control; control usually decreases as fish grow larger and are unable to feed on the algae.
Macroalgae Chara and Nitella	Good to Excellent	Chara usually is beneficial to fish and wildlife.
Floating and Floating-Leaved Weeds Duckweeds, watermeal	Poor	Small fish at very high stocking rates (see filamentous algae above) may give control; larger fish at normal stocking rates usually are not effective.
Water ferns (Azolla and Salvinia)	Fair to Poor	
Alligatorweed, water lilies, water primrose, lotus, watershield, spadderdock, waterhyacinth	Poor	Grass carp may feed lightly on weeds in this group, but control is usually unacceptable.
Emergent and Marginal Weeds Cattails, rushes, common reed, bulrushes, pickerelweed, pennywort, arrowhead	Poor	Grass carp may feed lightly on weeds in this group, but control is usually unacceptable.
Submersed Weeds	Good to Excellent	Most rooted and free-floating submersed weeds in ponds are readily controlled with triploid grass carp; control may be poorer on the watermilfoils, particularly Eurasian waterfoil.

Chemical Control of Aquatic Plants

Table 7-21. Chemical Control of Aquatic Plants

Herbicide, Formulation, and Mode of Action Code	Amount of Formulation	Active Ingredient Rate or Concentration	Precautions and Remarks[2]
Algae, blue-green			
copper sulfate (various)	See label	0.5 to 1 ppm	Apply crystals or powder at early stage of growth by any method to give rapid and uniform dispersion. For best results, apply on a clear day. Do not apply to muddy water. **Warning: Copper is toxic to fish.** Copper products formulated with a chelating agent (copper complex) have a greater margin of safety to fish.
sodium carbonate peroxyhydrate (various)	See label	0.3 to 1.7 ppm	Apply with 8 to 10 hours of daylight remaining. Do not reapply within 48 hours.
Algae, filamentous and planktonic			
copper complex (various)	0.6 gal/acre ft	0.2 ppm	Dilute with water in ratio of at least 9-to-1 and apply uniformly. For best results, apply on a clear day and break up floating mats of filamentous algae before treatment. **Warning: Copper is toxic to fish.**
copper sulfate (various)	See label	0.5 to 1 ppm	Same as under Algae, blue-green. For best results break up floating mats of filamentous algae before treatment. **Warning: Copper is toxic to fish.** Copper products formulated with a chelating agent (copper complex) have a greater margin of safety to fish.
diquat (Reward) 2 lb/gal MOA 22	See label	0.18 to 0.37 ppm	For certain filamentous algae—*Pithophora* spp. and *Spirogyra* spp. Check label for application instructions. For best results, break up floating mats before treatment.
Algae, macro, chara, nitella			
copper complex (Cutrine-Plus Granular) 3.7 G (Cutrine-Plus) 0.9 lb/gal (K-Tea) 0.8 lb/gal	60 lb/surface acre 1.2 gal/acre ft 1.7 to 3.4 gal/acre ft	2.2 lb/acre 0.4 ppm 0.5 to 1.0 ppm	Distribute granular formulation evenly over infested area when plants are young. If chara is in water less than 3 ft deep or growth is near the surface, the liquid formulation may be used. Dilute with water in ratio of at least 9-to-1 and apply uniformly. **Warning: Copper is toxic to fish.**
Algae, Pithophora and cladophora			
flumioxazin (Clipper) 51% MOA 14	6 to 12 oz/A	3 to 6 ai/A or 100 to 400 ppb	Early morning applications may be more effective when water pH tends to be lower. If vegetation is dense, treat in sections to avoid reducing dissolved oxygen. Water pH greater than 7.5 will reduce effectiveness.
Floating Weeds (except watermeal)			
2,4-D amine (various) MOA 4	See label	2 to 4 lb/acre	Thorough wetting of foliage is essential. Apply with 100 gallons of water per acre. Use low pressure, large nozzle, and spray thickener.

Table 7-21. Chemical Control of Aquatic Plants

Herbicide, Formulation, and Mode of Action Code	Amount of Formulation	Active Ingredient Rate or Concentration	Precautions and Remarks[2]
Floating Weeds (except watermeal) (continued)			
bispyribac (Tradewind) 80% MOA 2	1 to 2 oz/A	0.8 to 1.6 oz ai/A	Controls duckweed, mosquito fern, salvinia, water hyacinth, water lettuce, and water pennywort. Apply with at least 30 gpa water volume. Include aquatic-approved adjuvant.
carfentrazone (Stingray) 1.9 lb/gal, MOA 14	3.4 to 13.5 fl oz/acre	0.05 to 0.2 lb/acre	Controls water lettuce, waterhyacinth, salvia, duckweed, mosquito fern, and water spinach. Rates vary according to target species. Methylated seed oil or nonionic surfactant (aquatic-approved) recommended.
diquat (Reward) 2 lb/gal MOA 22	0.5 to 0.75 gal/surface acre	1 to 1.5 lb/acre	Weeds controlled: pennywort, salvinia, waterhyacinth, waterlettuce. Apply in a spray volume of 150 to 200 gallons of water per acre plus aquatic-approved nonionic surfactant.
	1 gal/surface acre	2 lb/acre	For duckweed control, apply in a spray volume of 50 to 150 gallons of water per acre. Take care to cover all plants on water and damp marginal areas. Will require retreatment. An aquatic-approved nonionic surfactant at 0.5% by volume may be used.
florpyrauxifen (Procellacor)	1 to 2 prescription dose units (PDU)	1 to 2 prescription dose units (PDU)	Controls several emergent species including alligatorweed. Labelled rates provided as PDUs. Product only available to approved aquatic applicators. Addition of aquatic-approved nonionic surfactant is recommended.
glyphosate (various) MOA 9	See label	See label	For control of waterlilies, spadderdock, and lotus, apply as foliar spray on a calm day when there is little to no wave action. Vegetation must be on or above the surface for treatment to be effective. A nonionic surfactant approved for aquatic use is required with some formulations. If applying from a boat, take care not to create waves that may wash the herbicide off floating leaves. Will not control small floating plants, such as azolla, duckweed, or watermeal.
imazamox (Clearcast) 1 lb/gal, MOA 2	32 to 64 fl oz/acre	0.25 to 0.5 lb ai/acre 50 to 150 ppb	See label for specific weeds controlled. An aquatic-approved nonionic surfactant or methylated seed oil is recommended for foliar applications. Spot treatments may be made with up to 5% solution by volume.
imazapyr (Habitat) MOA 2	1 to 4 pt/acre	0.25 to 1.5 lb/acre	Rates vary according to target species. Retreatment of some plants may be required. An aquatic-approved nonionic surfactant is recommended. Will not control small floating plants, such as azolla, duckweed, or watermeal.
penoxsulam (Galleon) 2 lb/gal, MOA 2	2 to 5.6 fl oz/acre	0.03 to 0.09 lb/acre 5 to 150 ppb	An aquatic-approved nonionic surfactant is recommended for foliar applications.
triclopyr (Renovate 3) MOA 4	0.5 to 2 gal/acre	1.5 to 6 lb/acre	Rates vary according to target species. Addition of aquatic-approved nonionic surfactant is recommended.
topramezone (Oasis) 29.7%	up to 16 fl oz/acre	up to 0.35 lb/acre	Use of an aquatic-approved surfactant is recommended for foliar applications. Check label for specific irrigation restrictions.
Floating Weeds (Watermeal and others)			
flumioxazin (Clipper) 51% MOA 14	6 to 12 oz/acre	3 to 6 ai/A or 100 to 400 ppb	Early morning applications may be more effective when water pH tends to be lower. If vegetation is dense, treat in sections to avoid reducing dissolved oxygen. Water pH greater than 7.5 will reduce effectiveness. A follow-up application will likely be needed for long-term watermeal control. Application with diquat, flumioxazin, or fluridone may provide enhanced watermeal control.
fluridone (Sonar) 4 AS MOA 12	Ponds: 0.16 to 1.5 qt/acre	0.16 to 1 lb/acre 45 to 90 ppb	Product amount will depend on average depth of water body. Do not apply when there is substantial outflow from the pond. Do not apply as a spot treatment. See label for specific weeds controlled. For watermeal, use 45 to 90 ppb. Other floating species may be controlled with lower rates. Do not use treated water for irrigation for 7 to 30 days. See label for irrigation precautions. **Warning: 30 days may be insufficient restriction if pond water will be used to irrigate very sensitive crops, such as tobacco, tomatoes, or peppers.**
Emergent, Marginal, and Ditchbank Weeds			
2,4-D amine (various) MOA 4	See label	2 to 4 lb/acre	Thorough wetting of foliage is essential. Apply in 100 to 400 gallons of water per acre. Use low pressure, large nozzle and spray thickener. An aquatic-approved adjuvant may improve efficacy.
2,4-D granular (Navigate) 20 G (2,4-D Gran 20) 20 G MOA 4	150 to 200 lb/surface acre	30 to 40 lb/acre	Weeds controlled: arrowhead, bulrush, creeping waterprimrose, pickerelweed, smartweed, spadderdock, waterchestnut, waterlily, watershield. Rate depends upon species and depth of water. Check label. Apply early, when weeds are actively growing, with a rotary seeder. Spadderdock may require retreatment.
bispyribac (Tradewind) 80% MOA 2	1 to 2 oz/A	0.8 to 1.6 oz ai/A	Controls alligatorweed and parrotfeather. Apply with at least 30 gpa water volume. Include aquatic-approved adjuvant.
carfentrazone (Stingray) 1.9 lb/gal, MOA 14	6.7 to 13.5 fl oz/acre	0.2 lb/acre	Suppresses alligatorweed and waterprimrose. Methylated seed oil or nonionic surfactant (aquatic-approved) recommended.
diquat (Reward) 2 lb/gal (Weedtrine) 0.4 lb/gal MOA 22	1 gal/surface acre	2 lb/acre	For control of cattails in ponds or lakes. For top kill, apply in 100 gal of water per acre with 0.25% to 0.5% nonionic surfactant. Apply before flowering for best results. Retreat as needed.
florpyrauxifen (Procellacor)	1 to 2 prescription dose units (PDU)	1 to 2 prescription dose units (PDU)	Controls several emergent species including alligatorweed. Labelled rates provided as PDUs. Product only available to approved aquatic applicators. Addition of aquatic-approved nonionic surfactant is recommended.
flumioxazin (Clipper) 51% MOA 14	6 to 12 oz/A	3 to 6 ai/A or 100 to 400 ppb	Early morning applications may be more effective when water pH tends to be lowest. If vegetation is dense, treat in sections to avoid reducing dissolved oxygen. Ensure adequate coverage of dense vegetation or a follow-up application may be necessary. Addition of aquatic-approved nonionic surfactant is recommended.
glyphosate (various) MOA 9	See label	See label	Rates vary according to target species. Retreatment of alligatorweed is necessary. Aquatic-approved nonionic surfactant is recommended. Note: The use of very hard water or water containing high concentrations of iron to prepare spray solutions may result in reduced efficacy of glyphosate.
imazamox (Clearcast) 1 lb/gal MOA 2	32 to 64 fl oz/acre	0.25 to 0.5 lb ai/acre 50 to 500 ppb	See label for specific weeds controlled. An aquatic-approved nonionic surfactant or methylated seed oil is recommended for foliar applications. Spot treatments may be made with up to 5% solution by volume. Rates vary according to target species. Retreatment of some plants may be required. An aquatic-approved nonionic surfactant is recommended.
imazapyr (Habitat) MOA 2	1 to 6 pt/acre	0.25 to 1.5 lb/acre	Rates vary according to target species. Retreatment of some plants may be required. An aquatic-approved nonionic surfactant is recommended.
penoxsulam (Galleon) 2 lb/gal, MOA 2	2 to 5.6 fl oz/acre	0.03 to 0.09 lb/acre 5 to 500 ppb	See label for specific weeds controlled and application details. An aquatic-approved nonionic surfactant is recommended.
triclopyr (Renovate 3) MOA 4	0.5 to 2 gal/acre	1.5 to 6 lb/acre	Rates vary according to target species. Addition of an aquatic-approved nonionic surfactant is recommended.
topramezone (Oasis) 29.7%	up to 16 fl oz/acre	up to 0.35 lb/acre	Use of an aquatic-approved surfactant is recommended for all foliar applications. Check label for specific irrigation restrictions.

Table 7-21. Chemical Control of Aquatic Plants

Herbicide, Formulation, and Mode of Action Code	Amount of Formulation	Active Ingredient Rate or Concentration	Precautions and Remarks[2]
Submersed Weeds[3]			
2,4-D granular (Navigate) 20 G, MOA 4	100 to 200 lb/ surface acre	20 to 40 lb/acre	Controls milfoils and certain other submersed species. Rate depends upon weed to be controlled and depth of water. Check labels for species and rates. Apply uniformly with a rotary spreader.
bispyribac (Tradewind) 80% MOA 2	See label	10 to 45 ppb	Controls hydrilla, sago pondweed, and Eurasian watermilfoil. Do not apply in areas of high water flow or water diffusion. Refer to label for specific details on application rate based on water volume.
carfentrazone (Stingray) 1.9 lb/gal MOA 4	0.286 to 5.75 gal/acre	200 ppb	Controls Eurasian watermilfoil. Apply in spring or early summer as a subsurface application or with an appropriate adjuvant to ensure sinking and mixing of the spray mix. Early morning applications may be more effective when water pH tends to be lowest. Water pH greater than 7.5 will reduce effectiveness.
diquat (Reward) 2 lb/gal MOA 22	1 to 2 gal/ surface acre	2 to 4 lb/acre	Weeds controlled: bladderwort, coontail, elodea, naiads, pond weeds. Apply early in season by pouring directly into water in strips 40 feet apart. Later in season, as weeds reach surface, pour in strips 20 feet apart or inject a dilute solution. Not effective in turbid or muddy water. If vegetation is dense, treat in sections to avoid reducing dissolved oxygen.
endothall (Aquathol K) 4.2 lb/gal (Aquathol Super K) 63 G	0.3 to 2.6 gal/acre ft 2.2 to 17.6 lb/acre ft	0.5 to 5 ppm	Weeds controlled: bass weed, bur reed, coontail, hydrilla (Aquathol K only), pondweeds, watermilfoil, water star grass. Rate depends upon weed species and type of treatment. Spot or marginal treatments require higher rates. Aquathol Granular is especially useful for spot or marginal treatments.
florpyrauxifen (Procellacor)	1 to 5 prescription dose units (PDU)	1 to 5 prescription dose units (PDU)	Controls several submersed species including hydrilla and milfoils. Labelled rates provided as PDUs. Product only available to approved aquatic applicators.
flumioxazin (Clipper) 51% MOA 14	See label	100 to 400 ppb	Early morning applications may be more effective when water pH tends to be lowest. If vegetation is dense, treat in sections to avoid reducing dissolved oxygen. Water pH greater than 7.5 will reduce effectiveness.
fluridone (Sonar) AS MOA 12	Ponds: 0.16 to 1 qt/acre Lakes: 0.2 to 4 qt/acre Canals: 2 qt/acre	0.16 to 1 lb/acre 0.2 to 4 lb/acre 2 lb/acre	Do not use water for irrigation for 7 to 30 days. See label for specific irrigation precautions. Application to canals should be made only if water flow can be restricted. **Warning: 30 days may be insufficient restriction if applied to small ponds and pond water will be used to irrigate very sensitive crops, such as tobacco, tomatoes, or peppers.**
(Sonar SRP) MOA 12	Ponds: 3.2 to 30 lb/acre Lakes: 4 to 80 lb/acre Canals: 40 lb/acre Rivers: 40 lb/acre	0.16 to 1.5 lb/acre 0.2 to 4 lb/acre 2 lb/acre 2 lb/acre	
imazamox (Clearcast) 1 lb/gal, MOA 2	See label	50 to 500 ppb	Rates vary according to target species and depth to be treated. See label for specific weeds controlled and application details.
penoxsulam (Galleon) 2 lb/gal, MOA 2	See label	5 to 150 ppb	Rates vary according to target species and depth to be treated. See label for specific weeds controlled and application details.
triclopyr (Renovate 3 or OTF), MOA 4	See label	1.5 to 6 lb/acre 0.5 to 2.5 ppm	Controls milfoils and certain other submersed species. Rates vary according to target species and depth to be treated. See label for specific weeds controlled and application details.
topramezone (Oasis) 29.7%	up to 16 fl oz/acre	up to 0.35 lb/acre	Rates vary according to target species and depth to be treated. See label for specific weeds controlled and application details. Check label for specific irrigation restrictions.

[1] Mode of Action (MOA) code developed by the Weed Science Society of America. Cooper compounds, endothall. and sodium carbonate peroxyhydrate have not been assigned codes.

[2] Also see comments for specific herbicides under "Table 7-25. Labeled Sites and Restrictions."

[3] Grass carp give cost-effective control on the majority of the weeds in this group and should be given consideration *before* using herbicides. See text at beginning of this section under *Biological Control of Aquatic Weeds with Triploid Grass Carp*. A permit is required to purchase more than 150 grass carp or for stocking in impoundments larger than 10 acres. Grass carp usually are **not effective** on filamentous algae, duckweed, watermeal, or any of the plants in the floating or emergent groups.

Table 7-22. Waiting Period (in Days) Before Using Water After Application of Herbicides for Aquatic Weed Control

Herbicide	Irrigation[1]	Fish Consumption	Watering Livestock	Swimming
2,4-D (various formulations and manufacturers)	Water use restrictions vary by formulation and manufacturer. In general, if water is used for irrigating sensitive crops, 2,4-D should not be used. Turfgrasses are generally tolerant to low concentrations of 2,4-D. Also, many 2,4-D formulations are NOT labelled for aquatic use. Read the label before purchasing and/or use.			
Bispyribac (Tradewind)	Do not irrigate until concentrations are < 1 ppb	No restrictions	Do not water livestock until concentrations are ≤ 1 ppb	No restrictions
carfentrazone (Stingray)	1 to 14[2]	No restrictions	0 to 1	No restrictions
copper (Copper sulfate pentahydrate, including Bluestone and EarthTec; and complexed copper formulations, including Algae-Pro, Captain, Clearigate, Cutrine-Plus, Cutrine-Plus Granular, K-Tea, Komeen, etc.)	No restrictions	No restrictions	No restrictions	No restrictions
diquat (Reward)	3 to 5[3]	No restrictions	1	No restrictions
endothall (Aquathol K) (Aquathol Super K) (Hydrothol 191) (Hydrothol 191 granular)	No restrictions for many situations. See label for specific restrictions	No restrictions	7 to 25	No restrictions
Florpyrauxifen (Procellacor)	Do not use treated water for irrigation unless allowed by product label	No restrictions	Do not allow livestock to drink treated water unless allowed by product label	No restrictions
Flumioxazin (Clipper)	0 to 5[3]	No restrictions	No restrictions	No restrictions
fluridone (Sonar 4AS, Sonar SRP)	7 to 30[3]	No restrictions	No restrictions	No restrictions
Glyphosate (AquaMaster, Aqua Neat, Rodeo, Touchdown Pro)	No restrictions	No restrictions	No restrictions	No restrictions
imazamox (Clearcast)	0+[3]	No restrictions	No restrictions	No restrictions
Imazapyr (Habitat)	120	No restrictions	No restrictions	No restrictions
penoxsulam (Galleon)	Do not irrigate food crops until residues ≤ 1 ppb	No restrictions	No restrictions	No restrictions
sodium carbonate peroxhydrate (GreenClean Pro, Pak 27)	No restrictions	No restrictions	No restrictions	No restrictions
topramezone (Oasis)	See label for specific irrigation restrictions	No restrictions	No restrictions	No restrictions
triclopyr (Renovate 3, Renovate OTF)	120 0 to established grass	No restrictions	Next growing season for lactating dairy animals	No restrictions

[1] Irrigation restrictions may be removed for specific products if a laboratory assay of treated water meets a standard as stated on the product label.
[2] Do not use treated water for irrigation in commercial nurseries or greenhouses.
[3] Refer to product label for specific restrictions.

Table 7-23. Effectiveness of Herbicides and Triploid Grass Carp for Control of Common Aquatic Weeds in N.C.

Weeds	2,4-D	bispyribac	carfentrazone	copper compounds	diquat	diquat + copper	endothall Aquathol	endothall Hydrothol	florpyrauxifen	flumioxazin	fluridone	glyphosate	imazamox	imazapyr	peroxide compounds	penoxsulam	triclopyr	triploid grass carp
Algae																		
Planktonic	NR	ID	NR	G	P	G	NR	P	NR	ID	NR	NR	NR	NR	G	NR	NR	NR
Filamentous	NR	ID	NR	G	E	E	NR	E	NR	G	NR	NR	NR	NR	ID	NR	NR	P
Chara / Nitella	NR	ID	ID	G	G	E	NR	G	NR	P	NR	NR	NR	NR	ID	NR	NR	E
Floating Plants																		
Azolla (mosquito fern)	NR	G	F	F	E	E	NR	NR	G	ID	E	NR	ID	NR	NR	G	NR	P
Duckweed	P	G	G	P	G	G	NR	NR	NR	E	E	NR	NR	NR	NR	G	P	P
Frogbit	F	ID	ID	NR	E	E	NR	NR	ID	G	NR	P	E	E	NR	ID	G	P
Salvinia, common	NR	G	G	P	E	E	NR	NR	G	E	G	E	ID	ID	NR	NR	NR	P
Salvinia, giant	NR	G	G	P	E	E	F	NR	NR	F	E	G	P	G	NR	E	NR	P
Waterhyacinth	E	G	G	NR	G	G	NR	E	P	F	G	E	G	NR	NR	E	E	P
Watermeal	NR	NR	NR	NR	P	P	NR	NR	NR	G	G	NR	NR	NR	NR	P	NR	P
Water lettuce	NR	G	G	NR	G	G	G	G	NR	E	NR	E	G	E	NR	E	NR	P

Table 7-23. Effectiveness of Herbicides and Triploid Grass Carp for Control of Common Aquatic Weeds in N.C.

Weeds	2,4-D	bispyribac	carfentrazone	copper compounds	diquat	diquat +copper	endothall Aquathol	endothall Hydrothol	florpyrauxifen	flumioxazin	fluridone	glyphosate	imazamox	imazapyr	peroxide compounds	penoxsulam	triclopyr	triploid grass carp
Emersed Plants
Alligatorweed	P	G	F	NR	NR	NR	NR	NR	G	F	F	G	G	G	NR	G	G	P
American lotus	G	ID	NR	NR	NR	NR	NR	NR	ID	ID	G	E	F	G	NR	ID	G	P
Cattail	F	ID	NR	NR	F	F	NR	NR	NR	P	G	E	G-E	E	NR	ID	F	P
Creeping waterprimrose	E	ID	F	NR	NR	NR	NR	NR	G	ID	F	E	F	E	NR	ID	E	P
Floating hearts	P	ID	NR	NR	F	F	E	E	G	F	F	G	G	G	NR	F	P	P
Fragrant waterlily	G	ID	NR	NR	NR	NR	NR	NR	E	ID	G	E	G	E	NR	ID	G	P
Grass species	NR	ID	NR	NR	F	F	NR	NR	NR	NR	F	E	F	E	NR	ID	NR	P
Parrotfeather	E	G	F	NR	NR	NR	NR	NR	E	F	NR	F	G	E	NR	G	E	NR
Phragmites (Common reed)	NR	ID	NR	NR	NR	NR	NR	NR	NR	P	NR	G	F-G	E	NR	NR	F	P
Pickerelweed	G	ID	NR	NR	NR	NR	NR	NR	E	ID	NR	F	E	E	NR	ID	G	P
Rush	NR	ID	NR	NR	NR	NR	NR	NR	NR	ID	NR	G	ID	G	NR	ID	F	P
Spatterdock	G	ID	NR	NR	NR	NR	NR	NR	P	ID	G	E	G	E	NR	ID	F	P
Smartweeds	F	ID	NR	NR	F	F	NR	NR	G	ID	F	G	G	G	NR	F	G	P
Waterpennywort	G	G	NR	NR	F	F	NR	NR	ID	G	G	E	E	E	NR	F	G	P
Watershield	E	ID	NR	NR	F	F	NR	NR	G	ID	F	E	G	G	NR	ID	E	P
Submersed Plants
Bladderwort	P	ID	ID	NR	F	F	P	P	F	ID	E	NR	F-G	NR	NR	ID	P	E
Cabomba	NR	ID	ID	NR	F	F	F	F	F	G	F	NR	F	NR	NR	ID	NR	F
Coontail	G	ID	ID	NR	E	E	E	E	F	G	E	NR	NR	NR	NR	ID	G	E
Egeria (Brazilian elodea)	NR	ID	ID	F	E	E	P	P	F	ID	E	NR	ID	NR	NR	G	NR	E
Eurasian watermilfoil	E	G	G	NR	G	G	E	NR	E	G	E	NR	F	NR	NR	G	E	P
Hydrilla, monoecious	NR	G	ID	F	G	E	E	E	E	G	E	NR	F	NR	NR	G	NR	E
Naiad, brittle	NR	ID	ID	G	E	E	E	E	ID	G	E	NR	ID	NR	NR	F	NR	E
Naiad, Southern	NR	ID	ID	G	P	G	P	P	ID	G	G	NR	ID	NR	NR	F	NR	E
Parrotfeather	E	G	ID	NR	G	G	E	E	E	G	E	NR	F	NR	NR	G	E	F
Pondweed species	NR	G	ID	NR	E	E	E	E	P-F	G	E	NR	G	NR	NR	G	NR	E
Proliferating spikerush	NR	ID	ID	NR	NR	NR	NR	NR	F	P	F	NR	F	NR	NR	F	NR	E
Variable leaf milfoil	E	ID	G	NR	E	E	E	E	E	E	G	NR	NR	NR	NR	NR	E	P

Key: NR = Not Recommended; P = Poor; G=Good ; ID = Insufficient Data; F = Fair; E = Excellent

Pond Dyes

Pond dyes may be used to prevent the growth of filamentous algae and submersed macrophyte vegetation. Pond dyes are not herbicides and do not directly kill aquatic plants. They function by blocking light penetration to the bottom of the pond. As a result, these products are most effective when applied very early in the growing season.

The use of a pond dye in aquacultural ponds usually is not recommended, as they tend to inhibit phytoplankton productivity that is needed to produce oxygen and provide food for zooplankton, which are the major food of fry and the smaller juvenile fishes. Application rates usually are about one part per million or 1 gallon per acre for a pond averaging 4 feet deep (i.e., 1 gallon per 4 acre-feet of water) for algae and most submersed weeds. For hydrilla, the rate needs to be doubled, due to its ability to grow at very low light levels. Several of the available pond dyes are registered by the USEPA for aquatic weed control. Pond dyes *should not be applied to drinking water supplies or to streams or any body of water where there is any substantial outflow.*

Table 7-24. Pond Dyes

Examples of Pond Dyes	USEPA Registered
Admiral Liquid	Yes
Aquashade	Yes

Table 7-25. Labeled Sites and Restrictions

Herbicide and Formulation	Labeled Sites	Restrictions (others may apply)
2, 4-D amine (Weedar 64) 3.8 lb a.i./gal Other formulations	potable water reservoirs, farm and fish ponds, lakes, golf course water hazards, fish hatcheries	Delay the use of treated waters for irrigation and domestic purposes for 3 weeks after application unless an assay indicates that chemical water concentrates are below the minimum amount as specified on the product label. Do not treat irrigation ditches where water will be used for overhead irrigation of susceptible crops. Refer to specific product label for restrictions.
2,4-D granular (Navigate) 20 G	ponds and lakes	Do not apply to water used for irrigation, agricultural sprays, watering dairy animals, or domestic water supplies.
bispyribac (Tradewind) 80%	bayous, canals, fresh water ponds, lakes, marshes, and reservoirs	Do not irrigate until water concentrations are less than 1 ppb. Do not treat water used for crawfish production.
copper-complex (Cutrine-Plus) 0.9 lb/gal (Cutrine-Plus) 3.7 G (K-Tea) 0.8 lb/gal **copper sulfate**	potable water reservoirs, farm and fish ponds, lakes, golf course water hazards, fish hatcheries	No restrictions on use of treated water. Check tolerance of crop to copper applied in irrigation water. Trout are very susceptible to copper. Toxicity to other fish increases with decreasing hardness of water.
carfentrazone (Stingray) 1.9 lb/gal	ponds, lakes, reservoirs, marshes, wetlands, drainage ditches, canals, streams, rivers, etc.	Irrigation: Do not use treated water in commercial nurseries or greenhouses. Field crops may be irrigated after 1 day if less than 20% of surface area was treated, or after 14 days if treatment was 20% or more of surface area or until an assay indicates that chemical water concentrates are below a minimum amount as specified on the product label. Treated water may be used for turf irrigation with no restriction if less than 20% of the total water body was treated. A 14-day restriction applies for larger area treatments. Do not apply within 0.25 miles an active potable water intake (upstream only in flowing waters), or turn intake off for at least 24 hours as specified on product label. Do not drink or water livestock for 1 day if 20% or more of total surface area was treated. Applicators must be licensed or certified by the state.
diquat (Reward) 2 lb/gal	lakes, still ponds, ditches, laterals, waterways	Apply only to still water and/or public waters. Do not apply to turbid waters. Do not use treated water for irrigation of food crops, preparation of agricultural sprays, or for drinking for 5 days after application. Turf and nonfood crops may be irrigated 3 days after treatment. Do not use water for livestock for 1 day after treatment. Water use restrictions may be removed if an approved assay is conducted and water concentration is less than the maximum contaminant level as specified on product label.
dyes (Admiral Liquid) (Aquashade)	ponds and lakes with little to no outflow	No not apply to water bodies not under direct control of user. Do not apply to water that will be used for human consumption.
endothall (Aquathol K) 4.23 lb/gal (Aquathol Super K granular) 63% (Hydrothol 191) 2 lb a.i./gal (Hydrothol granular) 11.2%	drainage canals, lakes, ponds	Restrictions up to 25 days may apply to waters used for domestic uses, irrigation, or watering livestock. Setback distance of at least 600 feet from functioning potable water intakes may also apply. Refer to specific product label for current restrictions on domestic use, irrigation, livestock use, and setback distance. Hydrothol formulations may kill fish when rates exceed 0.3 ppm. Check label for drinking water restrictions. Fish may be killed by rates exceeding 0.3 ppm. Irrigation and animal consumption restrictions of 7 to 25 days, depending on rate.
florpyrauxifen (Procellacor)	ponds, lakes, reservoirs, drainage ditches, etc.	Irrigation and livestock watering are generally restricted unless specifically allowed by product label. Do not apply to salt or brackish water. Prevent contact to or drift on sensitive species. Do not use with organosilicone surfactants.
flumioxazin (Clipper) 51%	bayous, canals, fresh water ponds, lakes, marshes, and reservoirs	Do not irrigate from treated water for at least 5 days. Do not treat water used for crawfish production.
fluridone (Sonar 4 AS or SRP)	lakes, ponds, canals	Treated ponds may not be used for irrigation for 7 to 30 days. See label for irrigation precautions.[1]
glyphosate (AquaMaster) 5.4 lb a.i./gal (AquaNeat) 5.4 lb a.i./gal (Rodeo) 5.4 lb a.i./gal (Touchdown Pro) 3 lb a.e./gal Other formulations	all bodies of fresh water and all types of aquatic sites	Do not apply within 0.5 mile of an active potable water intake (upstream only in flowing waters), or turn intake off for at least 48 hours as specified on product label. Refer to specific product label for restrictions.
imazamox (Clearcast)	in and around aquatic and noncropland sites	Irrigation: Do not apply to water to be used for irrigation of greenhouse or nursery plants. Do not irrigate from still or quiescent water bodies within 24 hours of application. Do not irrigate if concentrations exceed 50 ppb.
imazapyr (Habitat)	in and around standing and flowing waters, including estuarine and marine sites	Irrigation: Do not use treated water for 120 days following application or until an assay indicates that chemical water concentrations are below a minimum amount as specified on the product label. Do not apply within 0.5 mile of an active potable water intake (upstream only in flowing waters), or turn intake off for at least 48 hours as specified on product label. Do not apply to fast-moving waters. Do not apply to irrigation ditches or canals within 1 mile of an active irrigation water intake unless the irrigation restrictions can be observed. Applicators must be licensed or certified by the state.
penoxsulam (Galleon)	in and around quiescent water bodies and exposed sediments of de-watered areas	Do not apply to flowing water. Irrigation: Do not apply to water to be used for irrigation of greenhouse or nursery plants. Do not irrigate established food crops, other than rice, if concentrations exceed 1 ppb. Do not irrigate established rice if concentrations in treated water exceed 30 ppb. No restrictions on use of treated water for turf irrigation, if concentrations are less than 30 ppb. Consult SePRO for other situations/commodities.
sodium carbonate peroxyhydrate (GreenClean Pro) (PAK 27)	ponds, lakes, lagoons, canals, ditches, etc.	Do not apply to treated, finished drinking water reservoirs.
triclopyr (Renovate 3) 3 lb/gal (Renovate OTF) 10 G	quiescent and slow-moving waters; non-irrigation canals	Irrigation: Do not use treated water for 120 days following application or until treated water has a non-detectable triclopyr level by an assay as specified on the product label. No restriction on irrigation of established grass. Applications around potable water intakes must observe minimum setback distances and/or minimum water concentrations as specified on the product label. Do not apply directly to or allow to come in direct contact with grapes, tobacco, vegetable crops, flowers, and other desirable broadleaf plants. Do not apply to estuarine or marine sites; do not apply directly to un-impounded rivers or streams; and do not apply to irrigation ditches or canals. Do not allow lactating dairy animals to graze treated areas until the next growing season after application unless spot-treatment was applied to less than 10% of total grazable area. Animals for slaughter must be removed from the treated area for at least 3 days.

[1] Water use restrictions for irrigation vary with formulation. See label for precautions. A 30-day restriction may be insufficient if applied to small ponds intended for irrigation of very sensitive crops, such as tobacco, tomatoes, or peppers.

Chemical Control of Specific Weeds

F. H. Yelverton and T. W. Gannon, Crop and Soil Sciences; D. W. Monks, Horticultural Science

Table 7-26. Chemical Control of Specific Weeds

Herbicide and Formulation	Amount of Formulation	Time of Application	Precautions and Remarks
Artichoke, Betony or Florida Betony			
dichlobenil (Casoron) 4 G (Gordon's Barrier) 4 G (Proteam Zapzit) 4 G	100 to 250 lb/acre	Any time. Fall or winter best.	*Do not use on cropland.* See Precautions and Remarks under **Weed Control in Woody Ornamentals.** In order to comply with state noxious weed regulations, contact your local plant protection specialist or NCDA&CS weed specialist at 919-707-3749.
mesotrione (Tenacity) 4 SC	4 to 8 fl oz/acre	Apply to young, actively growing weeds.	Use a nonionic surfactant at 0.25% v/v. Reapply in 3 weeks. Controls Florida betony.
Bamboo (Cane)			
glyphosate (Accord) 4 SL (Roundup PRO) 4 SL and various other trade names	2% solution sprayed to wet	Midsummer to early fall while actively growing.	For large canes; first cut canes and allow to regrow to 4 to 6 feet in height. Avoid drift to desirable vegetation. Use Accord for forestry and utility rights-of-way sites.
	1 part herbicide + 2 parts water, applied with sponge	Whenever new shoots are in the "husk" stage (before leaves open) and are 12 feet to 24 inches in height.	Wear rubber gloves. Wipe entire shoot with a sponge dampened with the herbicide. Sponge should not be dripping wet. Do not allow contact with desirable vegetation; avoid dripping onto grass.
Bermudagrass			
glyphosate + fluazifop (Roundup PRO) 4 SL + (Fusilade DX) 2 EC	2 to 0.75 qt/acre	See Precautions and Remarks	Apply when bermudagrass is actively growing. Repeat applications when bermudagrass regrows. Wait 30 days after last application to seed, sprig, or sod new bermudagrass.
Dazomet (Basamid)	218 to 525 lb/acre	See Precautions and Remarks	Restricted Use Pesticide. May be used to control many annual and perennial weed species including bermudagrass. Apply to properly prepared soil and incorporate physically or with water. Consult and follow label directions.
Berries (*Rubus* spp.)			
glyphosate (Accord) 4 SL (Roundup PRO) 4 SL and various other trade names	3 to 4 qt/acre as broadcast spray or 1% to 1.5% solution with handheld equipment	After bloom stage	Use higher rate for plants that have reached the woody stage of growth. Best results are obtained when sprayed in late summer after berries are formed. Use Accord for forestry and utility rights-of-way sites. Roundup PRO should be used for industrial and other noncropland areas. Use Roundup ULTRA on agricultural areas.
triclopyr amine (Garlon 3 A) 3 SL	1 gal/100 gal water	After leaves fully expand in spring and before leaf color change in fall	See comments under kudzu.
Florida Betony			
atrazine (AAtrex 4L) 4L	1 to 2 qt/A	Fall and/or winter	May be used on centipedegrass, St. Augustinegrass and dormant bermudagrass. Do not apply after Dec. 31 on bermudagrass unless delay in greenup is acceptable.
triclopyr + clopyralid (Confront) 3 SL	1 to 2 pt/A	Fall and/or winter	Do not apply to home lawns. May be used on centipedegrass, bermudagrass, and zoysiagrass. Do not apply to bermudagrass during transition. Repeat applications may be required.
MCPA amine + triclopyr amine + dicamba (Horsepower) 4.56 L	2 to 3 pt/A	Fall and/or winter	May be applied to home lawns by a commercial applicator. Not for use on turf grown for resale or other commercial uses such as sod or seed production. May be used on bermudagrass, zoysiagrass and bahiagrass.
MCPA ester + triclopyr ester + dicamba (Coolpower) 3.6 L	2.5 to 3.5 pt/A	Fall and/or winter	May be applied to home lawns by a commercial applicator. Not for use on turf grown for resale or other commercial uses such as sod or seed production. May be used on bermudagrass, zoysiagrass and bahiagrass.
Honeysuckle			
dicamba (Banvel) 4 SL	1 gal/100 gal water	When actively growing, prior to bloom.	Using hand-held equipment, spray to wet leaves. Add a nonionic surfactant at the rate of 2 quarts per100 gallons of finished spray solution to improve wetting. Keep spray off desired plants. Do not spray in rooting zone of desired plants.
2,4-D amine (various) 4 SL	1.5 qt/50 gal water		
glyphosate (Accord) 4 SL (Roundup PRO) 4 SL and various other trade names	3 to 4 qt/acre as broadcast spray or 1% to 1.5% solution with handheld equipment	When plants are actively growing at or beyond the bloom stage of growth.	Use the higher rate for plants that have reached the woody stage of growth. Ensure thorough spray coverage with hand-held equipment. Use Accord for forestry and utility rights-of-way sites. Roundup PRO should be used for industrial and other noncropland areas. Use Roundup ULTRA on agricultural areas.
Kudzu			
aminopyralid (Milestone) 2 SL	7 oz/A	Apply to young, actively growing plants	May be used in permanent grass pastures, rangeland, noncrop areas, nonirrigation ditch banks and other natural areas. Due to crop sensitivity, use extreme caution around sensitive crops including but not limited to alfalfa, cotton, potatoes, soybeans, tobacco and other broadleaf or vegetable crops, fruit trees, or ornamental plants. Do not use aminopyralid-treated plant residues, including hay or straw from treated areas, or manure from animals that have grazed treated areas in compost or mulch that will be in contact with susceptible broadleaf plants. There are no restrictions on grazing or hay harvest following aminopyralid applications. Check and follow label directions for completed list and precautions.
clopyralid (Transline) 3 SL	1 to 3 qt/100 gal	During active growth.	Spray to wet leaves. Do not apply more than 1.337 pints per acre.
dicamba (Banvel) 4 SL	1 gal/100 gal water	When actively growing, before bloom.	Using hand-held equipment, spray to wet leaves. Keep spray off desired plants. Do not spray in rooting zones of desired plants. Add a nonionic surfactant at the rate of 2 quarts per 100 gallons of finished spray solution to improve wetting.
(Vanquish) 4 SL	0.5 gal/100 gal water	Dormant season; just prior to budbreak (early to mid-March).	Do not spray in rooting zones of desired plants. Regrowth should be sprayed in mid- to late summer with glyphosate or clopyralid.
fosamine (Krenite S) 4 SL	1.5 to 3 gal/acre	August through September.	Spray to wet leaves thoroughly. Good coverage is necessary.
glyphosate (Accord) 4 SL (Roundup PRO) 4 SL and various other trade names	4 qt/acre as broadcast spray or 2% solution with handheld equipment	When actively growing at or beyond bloom stage of growth.	Repeat applications are necessary to maintain control. Ensure thorough spray coverage with hand-held equipment. Apply before frost in the fall. Use Accord for forestry and utility rights-of-way sites. Roundup PRO should be used for industrial and other noncropland areas. Use Roundup ULTRA on agricultural areas.

Table 7-26. Chemical Control of Specific Weeds

Herbicide and Formulation	Amount of Formulation	Time of Application	Precautions and Remarks
Kudzu (continued)			
metsulfuron methyl (Escort) 60 WDG	3 to 4 oz/acre	When actively growing.	Add 1 quart nonionic surfactant per 100 gallons of water plus a drift control agent. Do not apply by air.
sulfometuron methyl (Oust) 75 WDG	12 oz/acre per 100 gal water	When actively growing.	Add 1 quart nonionic surfactant per 100 gallons of water plus a drift control agent. Do not apply by air.
triclopyr amine (Garlon 3 A) 3 SL	2 gal/100 gal water	During mid-season when plants are actively growing.	Spray to wet leaves thoroughly. Do not allow to drift; product is very toxic to tobacco, soybeans, many other broadleaf crops, trees, and ornamentals. Most grasses very resistant.
Mugwort (*Artemisia vulgaris*)			
dichlobenil (Casoron) 4 G (Gordon's Barrier) 4 G (Proteam Zapzit) 4 G	100 to 250 lb/acre	Any time. Fall or winter best.	*Do not use on cropland.* See Precautions and Remarks under **Weed Control in Woody Ornamentals.**
EPTC (Eptam 7-E) 7 EC	6.75 pt/acre	Spring or fall.	Plow area before treatment. Incorporate chemical into soil immediately after application. Rototilling is the preferred method, but deep cross-disking is satisfactory. Treatment most effective when soil is moist but not wet. Under normal conditions the herbicide will be dissipated in 8 to 12 weeks. Tilling the soil several times before planting will help dissipate chemical from soil.
Multiflora Rose (See publication AG-536 for more details)			
2,4-D + triclopyr (Crossbow) 3 EC	1 to 1.5 gal in 100 gal water (handgun application) or a 1% to 1.5% solution for smaller amounts	At fall vegetative stage prior to full bloom	Spray to wet all leaves and green stems to drip point. Use low spraying pressure to prevent drift. For best results, apply when plants are actively growing during the early to mid-flowering stage.
	Undiluted herbicide		Small plants may be controlled by a thinline basal application of undiluted herbicide across all stems at a height where the stems are less than 0.5 inch in diameter. Apply approximately 20 ml of undiluted product per bush. Treat when bark is dry, and rain is not forecast. For bushes with more than 3 or 4 stems, coverage of each stem may be difficult; basal bark or dormant stem applications may be more effective (see these sections under **Chemical Control of Woody Vegetation**). Warning: *Restrictions on grazing or harvesting of green forage*: Do not graze lactating dairy animals or harvest green forage for 14 days following treatment with 2 gallons per acre or less; with treatment rates greater than 2 gallons per acre, do not graze or harvest green forage until the following growing season. For other livestock, no grazing restrictions apply at rates under 2 gallons per acre. Above 2 gallons per acre, do not graze or harvest green forage from treated areas for 14 days after treatment. *Restrictions on haying (harvesting of dried forage)*: For lactating dairy animals, do not harvest hay until the next growing season. For other livestock, do not harvest hay for 7 days after treatment at rates under 2 gallons per acre. Above 2 gallons per acre, do not harvest hay for 14 days after treatment. *Slaughter restrictions*: Withdraw livestock from grazing treated grass or treated hay at least 3 days before slaughter. This restriction applies to grazing during the season following treatment or hay harvested during the season following treatment.
dicamba (Banvel) 4 SL	1 gal/100 gal water	At full vegetative stage, prior to bloom.	Can be used in pastures and noncropland. A maximum of 200 gallons of spray solution can be used per acre. Spray with a handgun and completely wet foliage and stems, allowing spray solution to run down the stems. Add a nonionic surfactant at the rate of 2 quarts/100 gallons of finished spray solution to improve wetting. Do not graze dairy animals for 60 days. There is no waiting period between treatment and grazing beef cattle or other livestock. Do not spray desired plants or in rooting zone of desired plants. Follow-up treatments may be necessary in subsequent years.
fosamine (Krenite S) 4 SL	1.5 to 3 gal/100 gal water	Apply to foliage during the 2-month period before fall leaf coloration.	Thoroughly and uniformly cover plants without drenching. Use 1 to 2 quarts per 100 gallons of a penetrating type, oil-based surfactant to improve activity. Use in noncropland, fence lines, etc.
glyphosate (Accord) 4 SL (Roundup PRO) 4 SL and various other trade names	1 gal/100 gal water	Apply to foliage after full bloom until August 1.	Use handgun and thoroughly cover bush. May be used in noncropland and pasture. Do not graze livestock for 10 days following treatment. Use Accord for forestry and utility rights-of-way sites. Roundup PRO should be used for industrial and other noncropland areas. Use Roundup ULTRA on agricultural areas.
metsulfuron methyl (Escort) 60 WDG	0.5 to 1 oz/acre	Early spring after bushes are fully leaved out.	Escort is not labelled for pastures; Cimarron may be used to control roses in pastures. For effective broadcast treatments, rose bushes should not be taller than 3 feet. For spot treatment, apply as foliar spray to runoff, and do not exceed 75 gallons total spray per acre. Use 1 pint to 1 quart surfactant per 100 gallons spray.
metsulfuron methyl (Cimarron) 60 WDG	0.3 oz/acre for broadcast; 1 oz per 100 gal water for spot treatment		
tebuthiuron (Spike) 20 P	20 lb/acre	Apply after ground thaws in spring before or soon after leaf flush. May require 2 yr for kill of large canes.	Broadcast over root zone. Do not apply near desirable trees or shrubs. Ground may be bare for 3 to 5 years where applied.
(Spike) 80 WP	5 lb/acre		Apply in water with a backpack sprayer as a band at the base of bushes or lace overtop of bushes. Same precautions as above for Spike 20 P.
Nutsedge			
2,4-D amine (various brands) 4 SL	1 qt/acre	Early in growing season following a thorough disking. Repeat at 3-week intervals for 3 treatments.	Corn crop can be produced while using 2,4-D. Apply preemergence rate. Follow 3 to 4 weeks later with rate suggested for corn and repeat. Rate suggested can be used following tobacco harvest.
dichlobenil (Casoron)	See section on **Weed Control in Woody Ornamentals**. Do not use on row crop land.		
glyphosate (Accord) 4 SL (Roundup PRO) 4 SL and various other trade names	3 qt/acre	See remarks.	Apply when plants are in flower or when new nutlets can be found at rhizome tips. Tillage two weeks after application will improve control. Repeat treatment will be required for long-term control. Use Accord for forestry and utility rights-of-way sites. Roundup PRO should be used for industrial and other noncropland areas. Use Roundup ULTRA on agricultural areas.

Table 7-26. Chemical Control of Specific Weeds

Herbicide and Formulation	Amount of Formulation	Time of Application	Precautions and Remarks
Nutsedge (continued)			
bentazon (Basagran) 4 SL	Follow label directions	Follow label directions	See weed control for specific field, turf, or vegetable crop for nutsedge suppression. Bentazon and S-metolachlor control yellow nutsedge only.
chlorimuron (Classic) 25 WDG	Follow label directions	Follow label directions	See weed control for specific field, turf, or vegetable crop for nutsedge suppression.
EPTC (Eptam 7-E) 7 EC	Follow label directions	Follow label directions	See weed control for specific field, turf, or vegetable crop for nutsedge suppression.
halosulfuron (Permit, Sandea, Sedgehammer) 75 DF	Follow label directions	Follow label directions	See weed control for specific field, turf, or vegetable crop for nutsedge suppression.
imazaquin (Image) 70 DG	Follow label directions	Follow label directions	See weed control for specific field, turf, or vegetable crop for nutsedge suppression.
imazethapyr (Pursuit) 2 AS	Follow label directions	Follow label directions	See weed control for specific field, turf, or vegetable crop for nutsedge suppression.
S-metolachlor (Dual Magnum, Dual II Magnum)	Follow label directions	Follow label directions.	See weed control for specific field, turf, or vegetable crop for nutsedge suppression. Bentazon and S-metolachlor control yellow nutsedge only.
sulfosulfuron (Certainty) 75 WG	1.25 oz/acre	Apply in May for June.	Add a nonionic surfactant at 0.25% v/v. Make a second application in 6 to 10 weeks, if needed. Certainty and Monument control yellow nutsedge, purple nutsedge, and kyllinga species.
trifloxysulfuron (Monument) 75 WG	0.45 to 0.56 oz/acre	Apply in May for June.	Add a nonionic surfactant at 0.25% v/v. Make a second application in 6 to 10 weeks, if needed. Certainty and Monument control yellow nutsedge, purple nutsedge, and kyllinga species.
Poison Ivy and Poison Oak			
2,4-D amine (various brands) 4 SL	2 qt/100 gal water	Apply in late spring or early summer when the plants are growing rapidly.	Apply only to plant material to be killed. Apply as a wetting spray. Avoid drift. Repeat in 6 to 8 weeks if needed. Use Accord for forestry and utility rights-of-way sites. Roundup PRO should be used for industrial and other noncropland areas. Use Roundup ULTRA on agricultural areas.
dicamba (Banvel) 4 SL	1 gal/100 gal water	At full vegetative stage, before bloom.	See comments for honeysuckle.
glyphosate (Accord) 4 SL (Roundup PRO) 4 SL and various other trade names	4 to 5 qt/acre as a broadcast spray or 2% solution with handheld equipment	After leaves fully expand in the spring and before leaf color changes in the fall.	
triclopyr amine (Garlon 3 A) 3 SL	1 gal/100 gal water	After leaves fully expand in spring and before leaf color changes in fall.	See comments for Kudzu.
Tree of Heaven *(Ailanthus altissima)*			
metsulfuron methyl (Escort) 60 WDG	1 oz/100 gal water for high-volume treatment 1 oz/20 gal water for low-volume treatment	Mid-summer	For right-of-way use only. Escort is not labeled for pastures.

Chemical Control of Woody Plants

R. J. Richardson, F. H. Yelverton, and T. W. Gannon, Crop and Soil Sciences Department

Table 7-27. Chemical Control of Woody Plants

Herbicide and Formulation	Amount of Formulation	Precautions and Remarks
Foliage Treatment, most woody species: ash, red maple, and persimmon generally resistant. Rhododendron resistant		
2,4-D amine (various brands) 4 SL, MOA 4	2 gallons in 100 gallons water	Use amine formulations to reduce vapor drift hazard. Use low spraying pressure to prevent spray drift. Wet foliage and stems thoroughly. Most effective results obtained by spraying within 6 weeks after plants have reached full-leaf stage. This treatment used primarily on trees or brush less than 6 feet tall. Only certain brands of 2,4-D can be used on ditch banks or near other bodies of water; check labels.
2,4-D low volatile ester or oil-soluble amine (various brands and concentrations)	varies	Use as invert emulsion to reduce drift hazards See remarks for 2,4-D amine.
2,4-D + triclopyr (Crossbow) EC 2.0 + 1.0 pound/gallon, MOA 4	1 to 1.5 gallons in 100 gallons water (handgun application) 1.5 to 4 gallons in water to deliver 15 to 30 gallons total spray/acre	Spray to wet all leaves and green stems to drip point. Use low spraying pressure to prevent drift. For best results, apply when plants are actively growing after full leaf in spring to early summer. This treatment is used primarily on trees and brush less than 6 feet tall. For application via boom or other broadcast spray equipment. For aerial application (helicopter only), use Nalcotrol to prevent drift. See label for specific information. *Warning: Restrictions on grazing or harvesting of green forage:* Do not graze lactating dairy animals or harvest green forage for 14 days following treatment with 2 gallons per acre or less; with treatment rates greater than 2 gallons per acre, do not graze or harvest green forage until the following growing season. For other livestock, no grazing restrictions apply at rates under 2 gallons per acre. Above 2 gallons per acre, do not graze or harvest green forage from treated areas for 14 days after treatment. *Restrictions on haying (harvesting of dried forage):* For lactating dairy animals, do not harvest hay until the next growing season. For other livestock, do not harvest hay for 7 days after treatment at rates under 2 gallons per acre. Above 2 gallons per acre, do not harvest hay for 14 days after treatment. *Slaughter restrictions:* Withdraw livestock from grazing treated grass or treated hay at least 3 days before slaughter. This restriction applies to grazing during the season following treatment or hay harvested during the season following treatment.
fosamine (Krenite S) 4 SL	1.5 to 3 gallons in 100 gallons water	Apply to foliage during the 2-month period prior to fall leaf coloration. Thoroughly and uniformly cover plants without drenching. Add surfactant WK at the rate of 1 quart per 100 gallons of spray. Surfactant WK is not needed with Krenite S. Rate and gallonage depend on plant size and species to be controlled. Check label. Use in noncropland, fence lines, etc.
dicamba (Banvel) 4 SL	1 gallon in 100 gallons	Apply when leaves are fully developed. Spray with a handgun to completely wet foliage, and allow spray to run down the stem. Add a nonionic surfactant at the rate of 2 quarts per 100 gallons of finished spray solution to improve wetting. Retreatment may be required, but do not exceed 2 gallons per treated acre during one growing season. Keep spray off desired plants. Do not spray in rooting zone of desired plants.
triclopyr (Garlon 3A) 3 SL (Garlon) 4.4 EC	2 to 3 gallons in 100 gallons water 1 to 3 gallons in 100 gallons water	Spray to thoroughly wet leaves, stems, and root collars. Can be mixed with other woody plant herbicides. See label. Avoid drift.
Foliage Treatment, woody brush and trees		
2,4-D amine (DMA 4 IVM) 3.8 SL, MOA 4	2 to 8 pints/acre	Apply when weeds are small and actively growing before bud stage. Biennial and perennial species are best controlled in seedling to rosette stage before flower stalks appear.
dicamba (Vanquish) 4 SL, MOA 4	0.5 to 4 pints in 25 to 200 gallons water	For low volume applications, apply 3 to 5% v/v rate. Check product label for tank mix partners for woody brush and vines.
glyphosate (Accord Concentrate) 5.4 SL, MOA 9	5 to 8% solution	If brush has been mowed or trees cut, wait until regrowth reaches recommended stage before treating. Apply as a low volume directed spray on at least 50% of the targeted foliage using a lateral zigzag motion from top to bottom. Spray to wet, not runoff. Add NIS at 2 quarts per 100 gallons of spray solution.
metsulfuron methyl (Escort XP) 60 DF, MOA 2	0.33 to 4 ounces/acre in 10 to 50 gallons water	For industrial, noncrop sites on young, actively growing weeds and brush. High volume ground application: mix 0.5 to 3 ounces per 100 gallons spray solution, and apply at 100 to 400 gallons per acre. Low volume and ultra-low volume ground applications: mix 4 to 8 ounces per 100 gallons spray solution, and apply at 10 to 50 gallons per acre.
triclopyr (Remedy) 4 EC, MOA 4	2 pints in 10 gallons water/acre	Treat after rapid growth period in spring when leaf tissue is fully expanded, and terminal growth has slowed. During drought or for hard-to-control weeds, add 2 to 3 quarts of 2,4-D low volatile ester to spray solution.
triclopyr + fluroxypyr (PastureGard) 2 EC 1.5 + 0.5 pounds/gallon, MOA 4	3 to 8 pints/acre	Broadcast applications: treat in late spring through summer when leaves are fully expanded, and terminal growth has slowed. If brush has been mowed, allow 9 to 12 months of regrowth before treating. NIS or liquid fertilizer at 1 to 2 quarts per 100 gallons of spray solution may improve control. High volume foliar treatment of individual plants: apply 1 to 2 gallons of PastureGard plus 1 quart NIS per 100 gallons of spray solution.
Foliage Treatment, black locust, honey locust, mimosa, redbud, and wisteria		
aminopyralid (Milestone VM) 2 SL, MOA 4	4 to 7 fluid ounces/Acre	Treat when weeds are actively growing. Include a non-ionic surfactant. Avoid mowing for 14 days after application.
Foliage Treatment, numerous woody species		
aminopyralid + triclopyr (Milestone VM Plus) 1.1 SL, MOA 4	6 to 9 pints/acre	Treat when weeds are actively growing. Include a non-ionic surfactant.
Foliage Treatment, most vegetation		
imazapyr (various) 2 SL, MOA 2	0.5 to 5% v/v 0.6 to 6.4 fluid ounces/gallon	Most effective with 1% methylated seed oil.
Basal Stem Treatment, most woody species; black locust resistant		
2,4-D low volatile ester (various brands) 4 SL, MOA 4	2 gallons in 100 gallons high quality mineral oil	Spray lower 12 inches of stem or trunk and let some solution run into ground. May be used any time of year, but is much more effective during dormant season. One growing season required before plants die completely. This treatment used primarily on plants less than 6 inches in diameter. Root suckering species may be resistant. *Both dormant stem and basal treatments useful to farmers and landowners because during winter there is less hazard to crops and more labor probably available.* Do not use around the home or ditch banks.
triclopyr (Garlon) 4.4 EC, MOA 4	1 to 3 gallons in 100 gallons high quality mineral oil	

Table 7-27. Chemical Control of Woody Plants

Herbicide and Formulation	Amount of Formulation	Precautions and Remarks
Basal Stem Treatment, most woody species; black locust resistant (continued)		
2,4-D + triclopyr (Crossbow) EC 2.0 + 1.0 pound/gallon, MOA 4	4 gallons in high quality mineral oil to make 100 gallons spray	Spray basal portions of trees or brush to a height of 15 to 20 inches from the ground. Thoroughly wet all basal bark areas, including crown and ground sprouts and ground area at base of stems or trunk. For trees larger than 6 to 8 inches diameter, use stump treatment. Winter and early spring treatments give best results. See warning for livestock and haying usage for Crossbow listed above under "Most Woody Species."
Basal Stem Treatment, most woody species		
imazapyr (Stalker) 2 SL, MOA 2	8 to 12 fluid ounces in 1 gallon high quality mineral oil	Treat lower 18 inches of stem. May be used on stems up to 4 inches DBH. Do not apply to point of dripping or puddling.
Basal Stem Treatment, woody brush and trees		
2,4-D amine (DMA 4 IVM) 3.8 SL, MOA 4	8 qt in 100 gal water or 2.6 fl oz in 1 gal water	Thoroughly wet the base and root collar of all stems until the spray accumulates around the root collar at the ground line. Wetting the stems will aid in control.
triclopyr (Remedy) 4 EC, MOA 4	2 gallons in 98 gallons high quality mineral oil	Spray basal 15 to 20 inches of plant to point of runoff at soil surface.
triclopyr + fluroxypyr (PastureGard) 2 EC 1.5 + 0.5 pounds/gallon, MOA 4	50% PastureGard + 50% high quality mineral oil	Apply at any time to stems less than 6 inches in diameter except when snow or water prevents spraying to ground line. Use solid cone or flat fan nozzles at low pressure. Spray to wet but not runoff.
Dormant Stem Treatment, most woody species		
2,4-D + triclopyr (Crossbow) EC 2.0 + 1.0 lb/gal, MOA 4	1 to 4 gallons in high quality mineral oil to make 100 gallons spray	Thoroughly wet upper and lower stems, including root collar and any ground sprouts. Treat when brush is dormant, and the bark is dry, but not when snow or water prevents spraying to ground line. Best results occur with late-winter to early spring applications. Brush over 8 feet in height is difficult to control with this method. See warning for livestock and haying usage for Crossbow listed above under "Most Woody Species."
Dormant Stem Treatment, woody brush and trees		
triclopyr (Remedy) 4 EC, MOA 4	3 to 6 quarts in high quality mineral oil to make 100 gallons spray	Treat any time brush is dormant and most foliage has dropped. Use 20 to 40 psi with knapsack or power spraying equipment. Do not apply if snow or water prevents spraying to ground line. Wet stems to point of runoff and ground below the plant for root suckering species, such as sumac, sassafras, or locust.
Stump Treatment To Prevent Regrowth, most woody species		
2,4-D low volatile ester (various brands) 4 SL, MOA 4	3 gallons in 100 gallons high quality mineral oil	Soak freshly cut stumps with spray solution to prevent sprouting, or use AMS crystals on stump. Hasten decay of stump by covering with layers of soil and a nitrogen fertilizer. Keep moist.
2,4-D + triclopyr (Crossbow) EC 2.0 + 1.0 lb/gal, MOA 4	4 gallons in high quality mineral oil to make 100 gallons spray	Cut down trees and treat stumps, including the freshly cut surface, bark, crown, and ground sprouts. Winter and early spring treatments (before growth begins) give best results.
dicamba (Banvel) 4 SL, MOA 4	16.5 gal in 100 gal water	Spray or paint freshly cut surface with the solution. Area adjacent to bark should be thoroughly wet.
Stump Treatment To Prevent Regrowth, Woody brush and trees		
2,4-D amine (DMA 4 IVM) 3.8 SL, MOA 4	8 qt in 100 gal water or 2.6 fl oz in 1 gal water	Apply as soon as possible after cutting trees. Thoroughly soak entire stump including cut surface, bark, and exposed roots.
dicamba (Vanquish) 4 SL, MOA 4	1 gal in 1 to 3 gal water	NIS or oil may be added to enhance control. Make application within 30 minutes of cutting. Area adjacent to the bark should be thoroughly wet.
triclopyr (Remedy) 4 EC, MOA 4	20 to 30 gallons in high quality mineral oil to make 100 gallons spray	Treat with a backpack or knapsack sprayer using low pressure and a solid cone or flat fan nozzle. Spray stump sides and outer portion of cut surface but not to point of runoff. Apply anytime except when snow or water prevent spraying to ground line.
Stump Treatment To Prevent Regrowth, Woody brush and trees (continued)		
triclopyr + fluroxypyr (PastureGard) 2 EC 1.5 + 0.5 lb/gal, MOA 4	50% PastureGard + 50% high quality mineral oil	Apply to freshly cut stumps using solid cone or flat fan nozzles at low pressure. Wet stump sides, root collar, and outer portion of cut surface but not to point of runoff. Apply anytime except when snow or water prevent spraying to ground line.
Stump Treatment To Prevent Regrowth, woody species, such as alder, dogwood, hickory, maple, oak, poplar, sweet gum, sycamore, and willow		
glyphosate (Accord Concentrate) 5.4 SL, MOA 9	50 to 100% solution	Treat freshly cut stumps or resprouts. Apply to freshly cut stumps immediately after cutting or reduced performance may occur.
Stump Treatment, numerous wood species		
aminopyralid + triclopyr (Milestone VM Plus) 1.1 SL, MOA 4	apply undiluted	Apply as soon as possible after cutting stems.
Stump Treatment, most woody species		
imazapyr (Stalker) 2 SL, MOA 2	8 to 16 ounces in 1 gallon high quality mineral oil	Apply as soon as possible after cutting stems.
Soil Treatment Beneath Woody Plants, most woody species		
hexazinone (Velpar L) 2 SL, MOA 5	2 to 4 gallons in 100 gallons water	Apply as a coarse spray, using a handgun applicator. Direct spray beneath plants to be controlled. Apply during the period between late winter and early summer. Do not apply in vicinity of desirable plants.
bromacil (Hyvar X-L) 2 SL, MOA 5	varies	Apply as a coarse spray, using a handgun applicator. Use at least 200 gallons of spray per acre. Direct spray beneath plants to be controlled just before or during the period of active growth. Do not apply in vicinity of desirable plants. Rates depend on species to be controlled. Check label.
tebuthiuron (Spike) 20 P, MOA 7	5 to 30 pounds/acre	Rates depend on species to be controlled. Check label for specific rates. Apply when ground is not frozen. Do not apply to the root zone of desirable trees or shrubs or where runoff can carry the herbicide to desired plants.

Total Vegetation Control in Noncropland

R. J. Richardson, F. H. Yelverton, L. S. Warren Jr., and T. W. Gannon, Crop and Soil Sciences Department

Controlling all weeds for extended periods is expensive; it is practical only where complete vegetation control is desirable and soil erosion is not an important factor. Such areas are around signposts and buildings, along highways and railroads, under guardrails and fences, and in parking lots. Do not use any of the treatments where adjacent trees, ornamentals, or crops might be affected. Roots of nearby desirable plants, especially trees and shrubs, may grow into an area that has been treated and be killed.

Effective rates for control vary with the weed species, degree of infestation, soil type, and environmental conditions. The lower rates are generally applied to annuals, biennials, shallow-rooted perennials, and seedling perennials, whereas the higher rates are applied to established, deep-rooted, and other hard-to-kill perennials. For specific details, read and follow directions on the label.

Table 7-28. Total Vegetation Control in Noncropland

Herbicide and Formulation	Amount of Formulation	Precautions and Remarks
bromacil (Hyvar X) 80 WP (Hyvar X-L) 2 SL	3 to 30 pounds/acre 1.5 to 12 gallons/acre	For handgun sprayer, use at least 200 gallons of water per acre. For treating small areas, use a hand sprayer or sprinkling can. For retreatment apply 2 to 6 pounds of active material per acre when annual weeds and grasses reappear on sites where weed growth has been controlled. Rates depend on weeds to be controlled. Check label.
bromacil + diuron (Krovar I) 80 WDG	4 to 30 pounds/acre	Rates depend on weeds to be controlled. Check label. For retreatment use 4 to 6 pounds per acre.
diuron (Karmex) 80 WDG	5 to 15 pounds/acre	For most annual weeds.
	20 to 60 pounds/acre	For perennial weeds. Addition of paraquat at the rate of 0.5 pound per acre plus nonionic surfactant will provide quick kill of existing vegetation and allow lower rates of diuron to be used.
imazapic + glyphosate (Journey) 0.75 + 1.5 lb/gal AS	0.33 to 1 quart/acre	For broadcast or spot treatment of annual and perennial grass and broadleaf weeds and vine species. May be applied preemergence or postemergence but postemergence is preferred. Depending on weed species, use a 5 to 20% active MSO at 1 to 2 pints per acre or a NIS at least 60% active with HLB ratio between 12 and 17 at 0.25% v/v. Nitrogen-based liquid fertilizers may be added but not substituted for spray adjuvants. For extended residual control, tank mix with Arsenal, Endurance, Escort, Karmex, Krova, Oust, Pendulum, Roundup Pro, Sahara, Tordon, Vanquish, or 2,4-D. Spot treat with 0.8 to 17 ounces per gallon + 1% v/v MSO.
glyphosate (Roundup PRO) 4 SL (Roundup ULTRA) 4 SL	4 quarts/100 gallons water or 2 to 3 ounces/gallon water/1,000 square feet	For industrial and nonagricultural uses in hand-held, high-volume equipment, mix 4 quarts of Roundup PRO in 100 gallons of water and spray to wet. For farmstead weed control, use 1% to 2% solution of Roundup ULTRA. Use a 2% solution for perennial weeds.
imazapyr (Arsenal) 2 SL	0.25 to 0.5 gallons/acre	See label instructions.
imazapic + glyphosate (Journey) 0.75 + 1.5 lb/gal AS	0.33 to 1 quart/acre	For broadcast or spot treatment of annual and perennial grass and broadleaf weeds and vine species. May be applied preemergence or postemergence but postemergence is preferred. Depending on weed species, use a 5 to 20% active MSO at 1 to 2 pints per acre or a NIS at least 60% active with HLB ration between 12 and 17 at 0.25% v/v. Nitrogen-based liquid fertilizers may be added but not substituted for spray adjuvants. For extended residual control, tank mix with Arsenal, Endurance, Escort, Karmex, Krovan, Oust, Pendulum, Roundup Pro, Sahara, Tordon, Vanquish, or 2,4-D. Spot treat with 0.8 to 17 ounces per gallon + 1% v/v MSO.
oryzalin + glyphosate (Surflan AS Specialty) 4 AS + (Roundup PRO) 4 SL	4 to 6 quart/acre + 4 to 6 quart/acre	For safer total vegetation control in vicinity of desirable vegetation. Rates depend upon size and weeds to be controlled. See Surflan label. Apply in 100 gallons water per acre.
prometon (Pramitol 25 E) 2 EC	5 to 30 gallons/acre	For annuals, use 5 to 7.5 gallons per acre. For perennials, use 20 to 30 gallons per acre. Apply in 50 to 100 gallons of water. For faster knockdown of established weeds and grasses, apply in 100 to 200 gallons oil. For maintenance application in following seasons, reduce rate in half.
(Pramitol) 5 PS	0.5 to 2 pounds/100 square feet	Pellets containing 5% prometon, 0.5% simazine, 40% sodium chlorate, and 50% sodium metaborate. For maintenance use 1 pound per 100 square feet.
sulfometuron methyl + chlorsulfuron (Landmark MP) 50 + 25 DG (Landmark II MP) 56.25 + 18.75 DG	4.5 to 9 ounces/acre 2.66 to 10 ounces/acre	Controls many annual and perennial grass and broadleaf weeds on terrestrial noncrop sites, including public, private, and military lands. Do not apply to recreation areas or paved surfaces. Can be applied to areas where temporary surface water has collected. Treat weeds preemergence or early postemergence when actively germinating or growing.

VIII — PLANT GROWTH REGULATORS

Growth Regulators for Cotton	406
Guide for use of Defoliants on Cotton	406
Harvest Aids, Preharvest Desiccants, and Postharvest Desiccants	407
Growth-Regulating Chemicals for Apples	408
Growth Regulators for Floricultural Crops in Greenhouses	411
Growth Regulators for Woody Ornamental Crops	438
Sucker Control for Flue-Cured Tobacco	438
Yellowing Agents for Flue-Cured Tobacco	440
Sucker Control for Burley Tobacco	440
Growth Regulators for Fruiting Vegetables	441
Growth Regulators for Peanuts	441
Growth Regulators for Turfgrasses	442

Growth Regulators for Cotton

K. L. Edmisten and G. D. Collins, Crop and Soil Sciences Department

Table 8-1. Growth Regulators for Cotton

Chemical and Formulation	Amount of Formulation Per Acre	Remarks
To suppress excessive vegetative growth		
mepiquat chloride (various brands) 0.35 lb/gal	0.5 to 1 pt	Base rate and timing on field conditions. Most consistent results occur when applied to cotton that is at least 24 inches tall at early bloom (5 to 6 white blooms per 25 feet of row). A follow-up application may be warranted 10 to 14 days later if excessive growth continues. When early season conditions promote excessive prebloom growth, mepiquat chloride may be applied at the rate of 0.125 to 0.25 pint beginning at match head square (first square 0.125 to 0.25 inch in diameter). Repeat applications can be made when renewed growth occurs. Follow label directions. Mepiquat chloride will consistently suppress vegetative growth. Other benefits may include easier scouting, better insecticide coverage, less boll rot, earlier maturity, easier defoliation, and more efficient use of ground sprayers and low drum pickers. Yield response is inconsistent.
mepiquat pentaborate (Pentia)	0.82 lb/gal 0.5 to 1 pt	
cyclanilid 1.84 lb/gal + mepiquat chloride 0.736 lb/gal (Stance)	2 to 3 fl oz	Use the 2 fluid ounces rate if applied at match-head square and the 2.5 to 3 ounces rate on applications following match-head square.
To stimulate boll opening and enhance defoliation		
ethephon (various brands) 6 lb/gal	0.66 to 1.33 qt	Use higher rate during cool weather. Prep boll opener will accelerate boll opening and enhance the activity of defoliants. It will not stimulate boll maturity. Micronaire may be reduced if Prep is applied to cotton that is less than 60 percent open.

For more information, see Extension publication AG-417, *Cotton Information,* which is available at your local Cooperative Extension center.

Guide for use of Defoliants on Cotton

K. L. Edmisten and G. D. Collins, Crop and Soil Sciences Department

Apply defoliants when at least 60%, preferably 70 to 75%, of the bolls are open and the remaining bolls expected to be harvested are mature. A boll is mature enough for defoliation when it is too hard to be squeezed between thumb and fingers, when it is too hard to be sliced with a sharp knife, and when the seed coats turn light brown. Apply defoliants in a volume of 15 to 25 GPA by ground or at least 5 GPA by air. Consult the label for recommended rates when tank mixed with other harvest aids and for maximum season-long rates.

Table 8-2. Guide for use of Defoliants on Cotton

Chemical Name	Brand Name and Formulation	Amount of Formulation Per Acre	Remarks
carfentrazone	Aim (2 lb/gal)	1 to 1.5 oz	Aim, like other herbicidal defoliants, can cause desiccation; however, the desiccation appears to be more transitory than that achieved with some of the other herbicidal defoliants. Aim appears to desiccate mature morningglories very well, and it does not seem like the addition of ethephon-type products is needed to improve morningglory desiccation. The label recommends a minimum of 10 gallons per acre for ground applications and the use of oil concentrate at 1% by volume (1 gallon per 100 gallons of spray solution). Do not use crop oil concentrate with Aim in mixtures with CottonQuik or FirstPick. Aim can be tank mixed with other defoliant products if boll opening or regrowth control is desired. Lower rates may be needed in defoliation mixtures.
ethephon + cyclanilide	Finish (6 lb/gal) (various brands)	1.33 to 2 pt	Use higher rates in cool weather. Finish is defoliant and boll opener. Finish also provides some regrowth control. Terminal regrowth control is stronger than basal regrowth control. Finish will provide acceptable regrowth control in many situations. In situations where extended regrowth control is needed (in the 20- to 28-day range), Roundup (in conventional cotton only) or thidiazuron would provide more acceptable regrowth control. Finish performance may benefit from the addition of a low rate of a standard defoliant in situations where cotton is actively growing with juvenile growth. Use compatibility agent when mixing with Def 6 or Folex.
glyphosate	Roundup (various brands)	24 to 32 oz	Based on limited trials in N.C. concerning varieties that do not convey tolerance to glyphosate, Roundup provides effective regrowth control. Roundup provides very little defoliation and should be used in combination with a defoliant. Experience in N.C. is limited to tank mixes with 1.5 pints of Def or Folex per acre. Although Roundup prevents regrowth, it will not defoliate juvenile growth that may occur prior to defoliation. Roundup also provides some weed control when defoliating weedy cotton. Roundup will not provide regrowth control on Roundup Ready cotton.

Table 8-2. Guide for use of Defoliants on Cotton

Chemical Name	Brand Name and Formulation	Amount of Formulation Per Acre	Remarks
pyraflufen-ethyl	ET (0.2 lb/gal)	1.5 to 2.5 oz	Like all herbicidal defoliants, ET can cause desiccation, although the desiccation tends to be transitory compared to some herbicidal defoliants. ET should help with desiccation of morningglory and other weeds listed on the herbicide label. ET can be mixed with other harvest-aid materials to provide boll opening or regrowth control. Consult the label for additive recommendations. In general, the use of a crop oil concentrate at 0.5% by volume is recommended. Do not use crop oil concentrate with ET in mixtures with CottonQuik or FirstPick. Use lower rates in warm weather.
flumiclorac-pentyl	Resource (0.86 lb/gal)	4 to 6 fl oz	Resource is an herbicidal defoliant that, like all herbicidal defoliants, can cause desiccation, although the desiccation tends to be more transitory than with some herbicidal defoliants. Under ideal defoliation conditions (warm, sunny days), add a NIS at 1 quart per 100 gallons of spray solution. Under dry or cool weather, use a methylated seed oil (MSO) or organosilicone spray solution. Apply in a minimum of 10 gallons per acre for ground applications and a minimum of 5 gallons per acre for aerial applications. Do not use flood jet or air induction nozzles. Resource can be tank mixed with other products if boll opening or regrowth control is desired. Resource only needs a 1 hour rain-free period. Preharvest interval (PHI) is 7 days.
fluthiacetmethyl	Blizzard (0.91 lb/gal)	0.5 to 0.66 fl oz	Blizzard is a PPO-inhibitor herbicidal-type defoliant. Experience with Blizzard has been limited in North Carolina. It can be tank mixed with ethephon-based products. Similar to other PPO-inhibitor defoliants, Blizzard should be very useful in desiccating juvenile foliage and as a second application prior to harvest. Add a crop oil concentrate or surfactant to tank mixes containing Blizzard according to label directions.
thidiazuron	Dropp (50% WP) Free Fall (50% WP)	0.2 to 0.4 lb	Dropp is very effective at regrowth suppression. However, the activity of Dropp is reduced to a greater extent by cool weather than the activity of Def 6, Folex. Dropp requires the longest rain-free period of any of the defoliants—24 hours. Follow label directions and safety precautions.
thidiazuron	Dropp SC Freefall SC (4 lb/gal) (various brands)	1.6 to 6.4 oz	The addition of crop oil concentrate may improve performance in cooler weather.
thidiazuron + diuron	Ginstar (1 lb/gal of thidiazuron plus 0.5 lb/gal of diuron) (various brands)	0.4 to 1 pt	Ginstar is a mixture of Dropp and Diuron with an emulsifier that makes it more likely to desiccate than Dropp Ultra. Do not use Ginstar if temperatures are over 90 degrees F. The addition of other defoliants is not recommended except for Prep or generic ethephon products. Ginstar rates should not exceed 10 ounces per acre in N.C. except under very adverse conditions or cold weather.
urea sulfate and ethephon	CottonQuik or FirstPick	1.5 to 3.5 qt	Use higher rates in cool weather. The 2 quarts rate is recommended for most situations. Use a low rate of a standard defoliant with FirstPick unless the cotton is well cut-out with no juvenile growth.
tribufos	Def 6 or Folex (6 lb/gal)	0.5 to 1.5 pt	Use lower range of rates if crop is well matured as indicated by yellow- or red-tinged foliage or when temperatures are warm, especially when tankmixing with boll openers and thidiazuron. Use higher recommended rates when plants are still green and actively growing or in dry weather or cool weather; needs a 1-hour rain-free period. Follow label directions and safety precautions.

Harvest Aids, Preharvest Desiccants, and Postharvest Desiccants

K. M. Jennings, Horticultural Science Department

All crops. Maximum coverage of plants is essential for adequate chemical desiccation. Dense vine growth and/or heavy weed growth necessitates use of maximum permitted rates and high spray gallonage per acre. Observe instructions for use of wetting agents.

Potato desiccation. Vine destruction in North Carolina has usually been accomplished by mechanical methods immediately before digging. Several chemicals are also available for preharvest destruction. All of these chemicals must be used in advance of digging. Vine killing with chemicals often aids skin set, depending upon maturity of the crop. Inhibition of heat sprouts has also been observed with chemical vine killing. Browning of the vascular ring of potato tubers sometimes occurs after using vine kills.

Table 8-3. Harvest Aids and Preharvest Desiccants

Crop Chemical and Formulation	Amount of Formulation per Acre	Pounds Active Ingredient per Acre	Precautions and Remarks
Bulb Vegetable[1]			
pelargonic acid (Scythe) 4.2 L	7 to 10% solution	0.3 to 0.5	Preharvest interval is 24 hours. See label for further instructions.
Dry Bean, Southern Pea, Lentil, Guar Bean[2]			
paraquat (Firestorm) 3 SL (Gramoxone SL) 2 SL (Gramoxone SL 2.0) 2 SL	0.8 to 1.3 pt 1.2 to 2 pt 1.2 to 2.0 pt	0.3 to 0.5	Apply paraquat for weed and bean desiccation. Apply when the crop is mature and **at least 80%** of the pods are yellowing and mostly ripe with no more than 40% (bush-type beans) or 30% (vine-type bean and lentil) of the leaves still green. Add nonionic surfactant. Do not harvest or graze for at least 7 days after application. Make ground application in at least 20 gallons of water per acre. See label for further instructions.
sodium chlorate (Defol 750) 7.5 L (Defol 5) 5 L	3.2 qt 4.8 qt	6	Apply approximately 7 to 10 days before harvest. Do not graze livestock on treated fields or feed treated fodder or forage to livestock. See label for further instructions.

Table 8-3. Harvest Aids and Preharvest Desiccants

Crop Chemical and Formulation	Amount of Formulation per Acre	Pounds Active Ingredient per Acre	Precautions and Remarks
Pepper, Chili			
sodium chlorate (Defol 750) 7.5 L (Defol 5) 5 L	1.6 to 4 qt 4 to 10 qt	3 to 7.5	**Processing only.** Apply 10 days before harvest. Consult with processor before applying. See label for further instructions.
Potato			
diquat (Reglone) 2 L	1 to 2 pt	0.25 to 0.5	**Preharvest desiccation of potato vines.** Apply at least 7 days before harvest in 20 gallons of water per acre with ground equipment or 5 gallons of water per acre with aerial equipment. If vine growth is very dense, make a second application 5 days after the first application. Do not apply to drought-stressed potatoes.
glufosinate (Rely 280) 2.34 SL	1.3 pt	0.38	**Desiccation of potato vines.** Use sufficient water for thorough coverage of potato vines. For best results, apply at the beginning of the natural senescence of potato vines. See label for further instructions.
sodium chlorate (Defol 750) 7.5 L (Defol 5) 5 L	3.2 qt 4.8 qt	6	Apply 10 days before harvest. DO NOT apply under conditions of extreme heat during the middle of the day.
pelargonic acid (Scythe) 4.2 L	7 to 10% solution		Preharvest interval is 24 hours. See label for further instructions.
pyraflufen (ET) 0.208 L (ETX) 0.335 L	2.75 to 5.5 oz 1.7 to 3.4 oz	0.045 to 0.009	Apply in early stage of crop senescence. Two applications may be necessary for complete desiccation. Preharvest interval is 7 days. See label for further information.
Root and Tuber Vegetable[3]			
pelargonic acid (Scythe) 4.2 L	7 to 10% solution		Preharvest interval is 24 hours. See label for further instructions.
Sunflower			
paraquat (Firestorm) 3 SL (Gramoxone SL) 2 SL (Gramoxone SL 2.0) 2 SL	0.8 to 1.3 pt 1.2 to 2 pt 1.2 to 2 pt	0.3 to 0.5	**Preharvest desiccation.** Apply when sunflower seeds reach maturity (when seed moisture is 35% or lower). For many varieties, this corresponds to the time when the back of the heads are yellow and bracts are turning brown. Do not graze treated areas or feed treated forage to livestock. Do not apply within 7 days of harvest.
sodium chlorate (Defol 750) 7.5 L (Defol 5) 5 L	3.2 qt 4.8 qt	6	Apply 7 or more days before harvest of mature sunflower. Do not graze livestock on treated fields or feed treated forage to livestock within 14 days of application. See label for further instructions.
Tomato			
paraquat (Firestorm) 3 SL (Gramoxone SL) 2 L (Gramoxone SL 2.0) 2 SL	1.6 to 2.5 pt 2.4 to 3.75 pt 2.4 to 3.75 pt	0.6 to 0.94	**Postharvest desiccation of crop and weeds.** Add a nonionic surfactant at 1 pint per 100 gallons spray solution. Apply in a minimum of 40 gallons spray solution. See label for further information.

[1] Garlic, leek, onion, and shallot.

[2] Sweet lupin, white sweet lupin, white lupin, grain lupin, adzuki beans, asparagus beans, black beans, broad beans, field beans, garbanzo beans, kidney beans, lablab beans lima beans, moth beans, mung beans, navy beans, pinto beans, rice beans, tepary beans, urd beans, guar beans, blackeyed peas, chickpeas, cowpeas, crowder peas, southern peas, and catjang.

[3] Artichoke, beet, carrot, ginger, ginseng, horseradish, parsnip, potato, rutabaga, sweetpotato, and turnip.

Growth-Regulating Chemicals for Apples

T. M. Kon and M. L. Parker, Horticultural Science Department

Table 8-4. Growth-Regulating Chemicals for Apples

Material/ Purpose	Amount of Formulation per 100 Gallons	Minimum Interval (Days) Between Application and Harvest	Remarks and Precautions
To improve shape and increase fruit weight of responsive apple cultivars, such as Delicious, Gala, or Ginger Gold			
Perlan or Promalin	1 to 2 pts	—	Apply between king bloom opening and full-bloom stage as a fine mist. Can be applied as a single spray at 1 or 2 pints per 100 gallons or as two split applications at 1 pint per 100 gallons. Do not exceed the maximum rate of 2 pints per acre for the combined sprays. Do not apply later than full bloom because only late blossoms will be affected and early blossoms may be thinned. Do not apply when air temperatures are lower than 40 degrees F or higher than 90 degrees F. For optimum results, have the water pH near neutral (pH 7) and always below 8.5.
To increase lateral bud break and shoot growth and improve branch angle on nonbearing and nursery trees			
Perlan/Promalin ± nonionic surfactant	0.25 to 1 pt/5 gal ± 0.2 to 0.3% (v/v) or 1.25 fl. oz./5 gal	—	Apply to previous season's leader growth with thorough coverage when new terminal growth is 1 to 3 inches long or nursery trees have reached the height where branching is required. Apply using a hand held sprayer and ensure thorough wetting of the foliage and bark. May also be applied as a latex paint mixture with a brush or sponge at 0.2 to 0.33 pints per pint of latex paint. Do not apply the -latex paint mixture after bud break.
To increase branching of nursery stock and young (nonbearing) trees, to improve branch angles, stimulate bud break and improve tree structure			
MaxCel or Exilis 9.5 SC	128 fl. oz./40 gal of water 16-32 fl oz/50 gal of water		Make the first of 3 to 4 applications to new growth at the height where branching is desired and apply at every 10 to 12 inches of new growth. Do not tank mix with streptomycin or apply streptomycin on the same day.

Table 8-4. Growth-Regulating Chemicals for Apples

Material/ Purpose	Amount of Formulation per 100 Gallons	Minimum Interval (Days) Between Application and Harvest	Remarks and Precautions
To promote lateral branching on current season's terminal growth			
Perlan/Promalin or MaxCel	4 fl. oz./5 gal (125 ppm) 3.2 fl. oz./gal (500 ppm)		Apply to every 8 to 10 inches of new terminal growth in conjunction with removal of at least half of each immature terminal leaf. Do not damage the growing point (Summer Knipping). Apply 2 to 3 times with a single nozzle directed to the shoot tip, beginning when the leader has reached the height at which lateral branches are required. Apply at 7 to 10 day intervals.
To increase fruit set after frost			
Perlan or Promalin	16 to 32 fl. oz.	—	Apply within 24 hours after a frost event when the majority of the crop is between early bloom and full bloom. Apply in 75 to 150 gallons of water per acre. Do not apply to frozen foliage, blossoms or developing fruit. Allow trees to completely thaw prior to application. Do not use a surfactant. This is a rescue treatment and should only be used if significant crop loss is anticipated. Parthenocarpic fruit have reduced storage potential and may be misshapen.
To increase the duration of floral receptivity and subsequent fruit set in the event of poor pollination conditions during bloom			
aminoethoxyvinylglycine or AVG (ReTain)	Apply one 333-q pouch (50 g a.i.)	7	Apply one pouch of ReTain per acre, as a single application from pink to full bloom. Applications made prior to pink or after full bloom will significantly reduce efficacy of the treatment. Do not apply after petal fall.
To decrease June drop on trees with light bloom			
prohexadione-calcium (Apogee or Kudos 27.5 WDG)	10 to 12 oz.	45	Apply 10 to 12 oz per 100 gallons when shoots are 1 to 3 inches long (i.e. during bloom).
To suppress russet formation on Golden Delicious and other russet sensitive varieties			
Novagib 10L or ProVide 10SG	20 to 33 fl. oz. 2.1 to 3.5 oz.	—	Recommended application interval is at petal fall (PF), then 10, 20, and 30 days after petal fall. There is an 80 fluid ounces per acre limit per season for Novagib 10L. Apply as a fine mist. **Do not** apply until runoff. Make 2 to 4 applications. Always make the first ProVide application at the beginning of petal fall and continue at 7 to 10 day intervals. Direct 85% of the spray volume to the upper two-thirds of the tree.
To reduce the occurrence of Stayman fruit cracking			
Novagib 10L or ProVide 10SG	32 to 64 fl. oz. 1.8 to 3.5 oz.		Apply in 3 to 6 consecutive applications at 14 to 21 day intervals beginning 2 to 3 weeks before fruit cracking is expected to occur. Apply at 50% of Tree Row Volume. **Do not** treat for cracking suppression apples that have received ProVide 10SG applications to suppress russet during the same growing season. Always apply when conditions favor slow drying.
To thin Golden Delicious fruit on trees where a heavy fruit set has occurred and to aid in encouraging annual bearing			
naphthalene-acetic acid Fruitone L, PoMaxa or Refine + surfactant (to increase thinning activity)	2 lb (1 lb a.i.) 2 pt (1 lb a.i.) + 5 ppm	— 3 2	Until experience is gained in your orchard, use all concentrations of all thinning chemicals only on a trial basis on a few trees. Optimum thinning with NAA is achieved when fruit diameter is 5 to 10 mm. Under optimum growing conditions, the best size for thinning usually occurs between 15 to 21 days after full bloom. Applying NAA too early may result in fruit persistence rather than removal. Applications are most effective at temperatures from 70 degrees F to 75 degrees F. Do not exceed 161 fluid ounces of Fruitone L, PoMaxa, or Refine per acre per season.
To thin heavily set non-spur Red Delicious and Rome and to aid in encouraging annual bearing			
carbaryl Carbaryl 50WP or Carbaryl 4L/Sevin XLR	1 to 2 lb 1 to 2 pt	3	Apply at 8 to 12 mm average fruit diameter for optimum thinning. Applying carbaryl to Golden Delicious may cause russet.
To thin heavily set spur strains of Red Delicious and to aid in encouraging annual bearing			
carbaryl Carbaryl 50WP or Carbaryl 4L/Sevin XLR + naphthalene-acetic acid (Fruitone L, PoMaxa, or Refine)	2 lb (1 lb a.i.) 2 pt (1 lb a.i.) + 5 ppm	— 3 2	Apply at 8 to 10 mm average fruit diameter. Application of NAA at larger fruit size to Red Delicious may cause small fruit (i.e. pygmies, nubbins, or mummies) to persist on the tree through harvest. Apply as a dilute spray to runoff. Do not use NAA to thin Fuji, since this cultivar is prone to pygmy fruit formation.
carbaryl Carbaryl 50WP or Carbaryl 4L/Sevin XLR + ethephon (Ethrel, Ethephon 2, Motivate, etc.)	2 lb (1 lb a.i.) 2 pt (1 lb a.i.) + 1.5 pt	3 7	Apply at 12 to 15 mm average fruit diameter. Apply only as a dilute spray to runoff (for use only on spur strains of Delicious). Applying ethephon when daily maximum temperatures exceed 80 degrees F may result in excessive thinning. Do not apply ethephon if daily maximum temperature exceeds 90 degrees F.
To defruit trees too young or too small to begin bearing or where fruit load needs to be eliminated (all varieties)			
naphthalene-acetic acid (Fruitone L, PoMaxa, or Refine) + carbaryl Carbaryl 50WP or Carbaryl 4L/Sevin XLR + ethephon (Ethrel, Ethephone 2, Motivate, etc.)	2 to 4 fl. oz. (5 to 10 ppm) + 2 lb (1 lb a.i.) 2 pt (1 lb a.i.) + 2 pt	2 3 7	Apply dilute at 8 to 10 mm average fruit diameter. Caution: This will temporarily suppress vegetative growth.

Table 8-4. Growth-Regulating Chemicals for Apples

Material/ Purpose	Amount of Formulation per 100 Gallons	Minimum Interval (Days) Between Application and Harvest	Remarks and Precautions
Post-bloom thinning agent for cultivars such as Gala, Fuji, Spur Red Delicious			
N-(phenylmethyl)-1H-purine-6-amine Exilis 9.5 SC	9.6 to 25.6 fl. oz.	86	Make 1 to 2 applications of 75 to 200 ppm when king fruit are 5 to 10 mm in diameter. Do not apply more than 296 fluid ounces of Exilis 9.5 SC per acre per season. Do not apply Exilis 9.5 SC if the temperature is below 60 degrees F.
or MaxCel +	48 to 128 fl. oz.	86	Make 1 to 2 applications of 75 to 200 ppm when king fruit are 5 to 15 mm in diameter. Applications will be most effective when the maximum temperature is above 65 degrees F on the day of application, and the following 2 to 3 days. Do not apply more than 308 fluid ounces of MaxCel per acre per season
Carbaryl 50 WP or Carbaryl 4L/Sevin XLR	2 lb (1 lb a.i.) 2 pt (1 lb a.i.)	3	
To thin Stayman, Rome, McIntosh, Jonathan, or Gala			
carbaryl Carbaryl 50WP or Carbaryl 4L/Sevin XLR +	2 lb (1 lb. a.i) 2 pt (1 lb a.i.)	3	Apply as a dilute spray when fruit are 9 to 11 mm average fruit diameter. Do not use NAA to thin Fuji since this cultivar is prone to pygmy fruit formation.
naphthalene-acetic acid (Fruitone L, PoMaxa, or Refine)	5 to 10 ppm	2	
or carbaryl Carbaryl 50 WP or Carbaryl 4L/Sevin XLR +	1 to 2 lb (0.5 to 1 lb a.i.) 1 to 2 pt (0.5 to 1 lb a.i.)	3	
naphthalene-acetic acid (Fruitone L, PoMaxa, or Refine)	+ 2.5 to 5 ppm	2	
To increase return bloom for the following season, especially on heavily cropped trees			
naphthalene-acetic acid (Fruitone L, PoMaxa, or Refine + surfactant)	5 to 20 ppm or 2.5 to 10 ppm + 1 pt surfactant	2	After the chemical fruit thinning activity window is past (typically 6 to 8 weeks after petal fall) use biweekly applications of naphthalene-acetic acid (NAA) in the next 3 to 4 cover sprays. (Typically two applications in June and two applications in July for Southeastern U.S. apple-growing areas). These NAA applications may be tank mixed with routine pesticide cover sprays. Even when used at low rates, reduced fruit quality such as early ripening or water core or leaf drop can result on sensitive varieties such as Early McIntosh or other early summer varieties.
To reduce vegetative growth and to reduce later season tree canopy volume and density for improving pesticide efficiency			
prohexadione-calcium (Apogee or Kudos 27.5 WDG)	3 to 6 oz. biweekly or 8 to 12 oz. monthly	45	Apply as a sequential application every other week beginning when new shoots are 1 inch long using 3 to 6 ounces in 100 gallons per acre or apply as a sequential monthly application beginning when new shoots are 1 inch long using 8 to 12 ounces in 100 gallons. Do not tank mix Apogee with calcium nutrient sprays; but Apogee/Kudos can be tank mixed in pesticide cover sprays. Do not apply more than 99 ounces of Apogee/Kudos per acre per season. Addition of a nonionic surfactant can improve coverage. Use an acidifier if the pH of the spray water is greater than 7.0. If calcium is present in the spray water (i.e. hard water), it can deactivate prohexadione-calcium and a reduction in growth control may occur. In this case, add a water conditioner, such as ammonium sulfate. Do not mix with calcium or boron sprays.
To aid in preventing preharvest fruit drop			
naphthalene-acetic acid (Fruitone L, PoMaxa, or Refine)	4 to 8 fl. oz. per acre	2	Apply 10 to 20 ppm NAA at anticipated start of fruit drop. One application will normally prevent fruit drop for 7 to 10 days. If necessary, repeat applications can be made at weekly intervals. Apply only when temperature is 70 degrees F or higher for maximum effectiveness. Higher application rates (>10 ppm NAA) may cause fruit softening at temperatures above 85 degrees F. Use sufficient water to ensure adequate coverage of fruit.
To aid in delaying the onset of preharvest fruit drop before loosening begins			
naphthalene-acetic acid (Fruitone L, PoMaxa, or Refine)	5 ppm applied as weekly sprays beginning 4 weeks before anticipated harvest	2	Preloading with the low rate of NAA is a more effective drop control program than waiting to use a higher rate as the fruit begins to loosen. Preloading may also increase return bloom of Golden Delicious and Red Delicious.
To delay preharvest fruit drop and fruit maturity to allow time for added fruit size increase			
aminoethoxyvinylglycine or AVG (ReTain)	Apply one 333-g pouch (50 g a.i.) per acre for most cultivars	7	Apply 28 days before anticipated harvest in most years. This is the most effective stop-drop control program; however, fruit maturity and harvest date will be significantly delayed. Do not apply ReTain to plants under stress. Avoid ReTain application during the heat of the day. Maintain solution pH between 6 and 8. Do not apply ReTain if rain is expected within 8 hours of application. Delay ReTain application until 2 to 3 weeks before normal harvest in hot years. For optimal response, use ReTain with a 100% organosilicone surfactant at a final surfactant concentration of 0.05 to 0.1% (v/v) in the spray tank. To prevent possible spotting, use the 0.05% (v/v) concentration when high temperature (in excess of 86 degrees F) weather conditions prevail or are anticipated. On sensitive cultivars such as Gala, the amount of a.i. per acre can be reduced to 25 g to obtain a response similar to that of 50 g on less sensitive varieties. To further delay preharvest drop and delay the onset of maturity, up to 2 pouches of ReTain can be applied per acre. Very good drop control and maintenance of fruit firmness on the tree is achieved by combining ReTain with 10 to 20 ppm NAA (4 to 8 fluid ounces Fruitone L or PoMaxa per 100 gallons) applied 2 weeks before harvest. Higher application rates of NAA may cause fruit softening at temperatures above 85 degrees F.
To control suckers from the ground around the trunk of apple trees			
naphthalene-acetic acid (NAA), ethyl ester (Tre-Hold sprout inhibitor A-112)	Apply a 1 % (v/v) solution	—	Apply a 1 % (v/v) solution during the dormant season prior to the green tip stage or during the summer pruning season when new sucker growth is 4 to 12 inches long. Apply as a low-pressure, large droplet, directed spray with hand-held equipment. A thorough application, giving complete wetting and coverage, is necessary for good results. Do not allow spray to drift onto tree foliage or fruiting spurs. For best results, cut off woody sucker growth at ground level during the dormant season. Do not apply during the period from bloom to 4 weeks after bloom.

Table 8-4. Growth-Regulating Chemicals for Apples

Material/ Purpose	Amount of Formulation per 100 Gallons	Minimum Interval (Days) Between Application and Harvest	Remarks and Precautions
To control water sprout regrowth around pruning cuts			
naphthalene-acetic acid (NAA), ethyl ester (Tre-Hold sprout inhibitor A-112)	Use 0.5 % (v/v) on newly planted trees. Use 1.0 % (v/v) on established plantings	—	Apply with a cloth or brush as a localized application to the pruning cut and surrounding bark any time after pruning and before growth starts in the spring. Do not spray Tre-Hold up in the trees. Do not allow Tre-Hold to contact buds or fruiting spurs. Tre-Hold use in the tree is not recommended when green growth is present. One to 4 pints of light-colored latex (water based) paint may be added per gallon to mark completed application. Thorough application, giving complete wetting and coverage, is necessary for good results.
To aid in controlling scald on stored apples			
diphenylamine (DPA)	1,000 ppm; amount per 100 gallons depends on the a.i. in the formulated product. Consult the product label.	—	Apply as a dip or spray to harvested fruit. For maximum effect, uncooled fruit should be sprayed or dipped as soon after harvest as possible. The longer treatment is delayed after harvest, the less its effectiveness. Fruit should be thoroughly covered; however, fruit should not be dipped for longer than 30 seconds to prevent excess residue. Best control is obtained when both the fruit and the solution are at room temperature.
To maintain apple flesh firmness, reduce scald, and maintain fruit acidity.			
1-methylcyclopropene (1-MCP) (SmartFresh or SmartTabs)	Sufficient to achieve a final concentration of 1 ppm of the a.i.	—	SmartFresh/SmartTabs is a postharvest treatment that is introduced into the atmosphere of an airtight facility or container in which the fruit is held for 24 hours. Fruit must be treated within 3 to 5 days of harvest. The amount of product used depends upon the volume (cubic feet) of the treatment container. After treatment, the fruit is held in regular cold storage. Product must be purchased directly from Agro Fresh. Online orders: www.agrofreshstore.com, Phone orders: 1-877-537-3135.

Growth Regulators for Floricultural Crops in Greenhouses

B. E. Whipker, Horticultural Science Department

This table lists labeled rates of plant growth regulators (PGR) for greenhouse crops. These rates are guidelines based on labeled recommendations, research at North Carolina State University, and recommendations by suppliers. Read the label for a complete listing of precautions. The degree of control can vary by a number of factors, including: plant type, cultivar, stage of development, fertilization program, growing temperatures, and crop spacing. When using a PGR for the first time, it is good to test the rate on a few plants prior to spraying the entire crop. Keep accurate records and adjust rates for your location.

General Recommendations: Plug culture and flat culture have different recommended rates. The rates in this table include recommendations for both plug (lower rates) and flat culture (higher rates). Apply ALL foliar sprays of plant growth regulators using 0.5 gallons per 100 square feet of bench area.

Table 8-5. Growth Regulators for Floricultural Crops in Greenhouses

CROP	PURPOSE	CHEMICAL	RATE*	PRECAUTIONS AND REMARKS
ABUTILON	To control plant growth	Citadel/Cycocel	750 to 1,500 ppm spray	
		Dazide/B-Nine	2,500 ppm spray	Rate for use on plugs.
		Piccolo/Piccolo 10 XC/	5 ppm spray	Can be applied once plant fills the pot, 2 to 3 weeks after transplanting.
		Bonzi/Paczol		
	To increase branching	Florel/Collate	250 to 500 ppm spray	Applied 2 weeks after transplanting. Follow with a pinch if needed.
ACHILLEA	To control plant growth	Dazide/B-Nine	2,500 ppm spray	One or 2 sprays may be needed to keep plants more compact.
		Piccolo/Piccolo 10 XC/	0.5 to 1 ppm drench	Apply to moderately moist substrate.
		Bonzi/Paczol/Downsize		
ACHMELLA OLERAEA	To control plant growth	Piccolo/Piccolo 10 XC/	15 ppm spray	Apply 2 weeks after transplant. Repeat a week later or a week after pinch if needed.
		Bonzi/Paczol		
AGASTACHE	To control plant growth	Citadel+Dazide/Cycocel+B-Nine	3,000 ppm + 1,500 ppm spray	Rates for compact genetics needing slight growth control.
AGERATUM	To control plant growth	Abide/A-Rest	7 to 26 ppm spray	
		Dazide/B-Nine	2,500 to 5,000 ppm spray	One or 2 sprays may be needed to keep plants more compact.
		Piccolo/Piccolo 10 XC/	15 to 45 ppm spray	High rates of Piccolo 10 XC may delay flowering. Late applications and overdosing may cause slow growth on transplantation. This can be avoided by using multiple applications of 25% to 50% of the specified rate and monitoring plant growth.
		Bonzi/Paczol		
		Citadel/Chlormequat E-Pro/	800 to 1,500 ppm spray	
		Cyclocel		
		Concise/Sumagic	2 to 30 ppm spray	Cultivar response rates vary. Use lower rates to hold plants.
		Topflor	20 to 60 ppm spray	Based on NC State University trials. Adjust rates for other locations.
AGERATUM, Plugs	To control plant growth	Piccolo/Piccolo 10 XC/	5 to 10 ppm spray	Timing of application should normally begin at the 1 to 2 true leaf stage.
		Bonzi/Paczol		

Table 8-5. Growth Regulators for Floricultural Crops in Greenhouses

CROP	PURPOSE	CHEMICAL	RATE*	PRECAUTIONS AND REMARKS
ALCEA ROSEA	To control plant growth	Piccolo/Piccolo 10 XC/	30 to 50 ppm spray	
		Bonzi/Paczol		
		Piccolo/Piccolo 10 XC/	0.12 to 0.24 mg a.i. (1 to 2 ppm) drench for a 6-in. pot; apply 4 fl. oz./6-in. pot	
		Bonzi/Paczol/Downsize		
ALTERNANTHERA (Joseph's coat)	To control plant growth	Abide/A-Rest	25 to 132 ppm spray	
			0.25 to 0.5 mg a.i. (2 to 4 ppm) for a 6-in. pot (1 to 2 fl. oz./gal of drench solution: apply 4 fl. oz./6-in. pot)	Drench volumes and mg a.i. vary with pot size.
		Citadel/Chlormequat E-Pro/	Spray	Apply only if needed. Not recommended on some cultivars due to potential phytotoxicity.
		Cyclocel		
		Dazide/B-Nine	5,000 ppm spray	
		Florel/Collate	500 ppm spray	To keep plants more compact. Based on Texas A&M University trials.
		Piccolo/Piccolo 10 XC/	30 to 45 ppm spray	Rate for Alternanthera dentata.
		Bonzi/Paczol/Downsize	4 ppm drench	To keep plants more compact. Apply to moderately moist substrate
ALYSSUM	To control plant growth	Piccolo/Piccolo 10 XC/	40 to 60 ppm spray	
		Bonzi/Paczol		
		Concise/Sumagic	5 to 25 ppm spray	
		Dazide/B-Nine	2,500 ppm spray	
ALYSSUM, Plugs	To control plant growth	Piccolo/Piccolo 10 XC/	10 to 20 ppm spray	Timing of application should normally begin at the 1 to 2 true leaf stage.
		Bonzi/Paczol		
AMARYLLIS	To control plant growth	Piccolo/Piccolo 10 XC/	23.66 mg a.i. (200 ppm) drench for a 6-in. pot (6.4 fl. oz./gal. of drench solution; apply 4 fl. oz./6-in. pot)	Drench volumes and mg a.i. vary with pot size.
		Bonzi/Paczol	100 ppm bulb soak	
ANAGALLIS	To control plant growth	Piccolo/Piccolo 10 XC/	0.5 ppm drench	To keep plants more compact. Apply to moderately moist substrate.
		Bonzi/Paczol/Downsize		
ANEMONE	To control plant growth	Piccolo/Piccolo 10 XC/	2 ppm drench	Rates for Mona Lisa series. Apply about 6 weeks after transplant when the foliage has covered the pot and the first visible flower bud is showing. Rates up to 4 ppm can be used after conducting your own trial. Apply one week earlier during warm weather if needed.
		Bonzi/Paczol/Downsize		
ANGELONIA	To control plant growth	Citadel + Dazide/Cycocel + B-Nine	1,500 to 3,000 ppm Dazide/B-Nine + 750 to 1,000 ppm Citadel/Cycocel applied as a tank-mix spray	At planting, soft pinch to promote lateral shoot development.
		Citadel/Cycocel	1,500 ppm spray	
		Concise/Sumagic	10 to 20 ppm spray	Based on NC State University trials.
		Dazide/B-Nine	3,000 ppm spray	
		Florel/Collate	Spray	Not recommended.
		Topflor	45 to 60 ppm spray	Based on NC State University trials.
AQUILEGIA	To control plant growth	Dazide/B-Nine	3,000 to 5,000 ppm spray	
ARGYRANTHEMUM	To control plant growth	Citadel/Cycocel	750 to 1,500 ppm spray	
		Citadel+Dazide/Cycocel+B-Nine	750 to 1,000 ppm + 1,000 to 2000 ppm spray	Rates for compact genetics needing slight growth control.
		Concise/Sumagic	3 to 40 ppm spray	Based on NC State University trials conducted during late spring. Trial rates of 3 to 5 ppm for compact genetics.
		Piccolo/Piccolo 10 XC/	5 to 10 ppm spray	Rates for compact genetics needing slight growth control.
		Bonzi/Paczol	1 to 5 ppm drench	Rates for compact genetics needing slight growth control.
		Dazide/B-Nine	1,500 to 2,500 ppm spray	
		Topflor	50 to 75 ppm spray	Based on NC State University trials conducted during late spring. Slight phytotoxicity occurred with rates greater than 40 ppm, but damage was quickly hidden by new leaf growth.
	To induce basal branching	Collate/Florel	500 ppm spray	Apply one week after establishment.
ASCLEPIAS	To control plant growth	Piccolo/Piccolo 10 XC/	30 to 60 ppm spray	
		Bonzi/Paczol		

Table 8-5. Growth Regulators for Floricultural Crops in Greenhouses

CROP	PURPOSE	CHEMICAL	RATE*	PRECAUTIONS AND REMARKS
ASTER NOVI-BELGII (Perennial)	To control plant growth	Concise/Sumagic	80 to 160 ppm spray	
		Dazide/B-Nine	1,500 to 5,000 ppm spray	
		Piccolo/Piccolo 10 XC/	160 ppm spray	Use lower rates of 5 to 10 ppm later in the season.
		Bonzi/Paczol	12 to 16 ppm drench	
ASTER, Bedding Plant (Callistephus chinensis)	To control plant growth	Abide/A-Rest	7 to 26 ppm spray	
		Dazide/B-Nine	2,500 to 5,000 ppm spray	
ASTER, Cut (Callistephus chinensis)	To promote stem elongation and break dormancy	Florgib/ProGibb T&O	50 to 100 ppm spray	Make one to three applications during the early vegetative period at 2- to 3-week intervals. Apply when plants are 2 to 6 in. tall.
ASTERISCUS MARITIMUS (Compact Gold Coin)	To control plant growth	Dazide/B-Nine	750 to 1,500 ppm spray	
		Citadel/Cycocel	800 to 1,500 ppm spray	
		None	None	Plants grown with good light and optimal growing conditions generally do not need PGRs.
ASTILBE	To control plant growth	Concise/Sumagic	25 ppm drench	Apply just prior to flower stem elongation.
		Dazide/B-Nine	5,000 ppm spray	1 or 2 sprays can be used to keep plants more compact. Begin once flower stalks show color. 1 to 2-week delay in flowering possible.
		Piccolo/Piccolo 10 XC/ Bonzi/Paczol	30 ppm drench	Apply just prior to flower stem elongation.
AZALEA	To control plant growth	Abide/A-Rest	26 ppm spray	
		Concise	5 to 15 ppm spray	Apply as a uniform spray at a volume of 1.5 qt. per 100 sq. ft. of bench area approximately 4 to 6 weeks after the final pinch. Shorter-growing cultivars (Gloria, Solitaire): use 10 ppm. If a second application is required 2 to 3 weeks later, use 5 to 10 ppm. Taller-growing cultivars (Prize): use 10 ppm. If a second application is required 2 to 3 weeks later, use 10 to 15 ppm.
	To promote flower initiation	Dazide/B-Nine	1,500 to 2,500 ppm spray	Apply solution when new growth from final pinch is 1 to 2 in. long.
		Citadel/Chlormequat E-Pro/ Cyclocel	1,000 to 4,000 ppm spray	Optimum rates are generally between 1,000 and 2,000 ppm. Two to six multiple sprays may be needed. Make
	To prevent flower bud initiation during vegetative growth	GibGro	130 to 850 ppm spray	Apply two to three sprays at 2- to 3-week intervals.
		Florgib/ProGibb T&O	100 to 750 ppm spray	Apply a first application beginning 2 to 3 weeks after pinching. Weekly applications can continue for 1 to 2 additional weeks, for a maximum of three total applications.
	For partial or full substitution of cold treatment	GibGro	265 to 1,055 ppm spray	Spray timing, concentration and number of applications vary with cultivar, as well as intended degree of cold substitution. Consult label for exact recommendations. Not labeled for California.
		Florgib/ProGibb T&O	250 to 500 ppm spray	Spray timing, concentration and number of applications vary with cultivar, as well as intended degree of cold substitution. Consult label for exact recommendations.
	To promote lateral shoot growth on vegetative plants	Off-Shoot-O	Use a 3 to 5% solution in greenhouses; use a 5 to 7% solution outdoors. Apply as a foliar spray.	Efficacy is related to relative humidity and temperature. Spray a few plants to check activity prior to treating the entire crop; effect should be visible in about 1 hr. Be certain chemical covers shoot tip. Ineffective if microscopic flower buds are present.
	To increase lateral branching	Augeo	3,125 to 6,250 ppm spray	
		Florel/Collate	2,500 to 5,000 ppm spray	
	To control plant growth, reduce bypass shoot elongation and promote flower bud initiation	Piccolo/Piccolo 10 XC/ Bonzi/Paczol	100 to 200 ppm spray	To control plant growth and promote flower bud initiation, apply after final shaping when new growth is 1.5 to 2 in. long. To reduce bypass shoot development, apply after bud set when bypass shoots are barely visible, or about 5 to 7 weeks prior to cooling.
		Piccolo/Piccolo 10 XC/ Bonzi/Paczol/Downsize	0.59 to 1.77 mg a.i. (5 to 15 ppm) drench for a 6-in. pot; apply 4 fl. oz./6-in. pot)	Drench volumes mg a.i. vary with pot size.
	To control plant growth	Concise/Sumagic	10 to 15 ppm spray	Apply at 1.5 qt per 100 sq. ft. of bench area.
BACOPA (SUTERA)	To control plant growth	Dazide/B-Nine	750 to 1,500 ppm spray	At planting, soft pinch to promote lateral shoot development. Initially try with lower rate.
		Piccolo	4 to 8 ppm liner root soak	Irrigation of the liners occurred within 24 hours prior to application, which results in a moderately dry substrate (the stage the plants would be watered but not wilted). Soak for a minimum of 30 to 60 seconds. Transplant after 3-hour waiting period. Rates based on Michigan State University trials.
		Piccolo/Piccolo 10 XC/ Bonzi/Paczol	1 to 2 ppm drench	
		Florel/Collate	150 to 200 ppm spray	Early spray will increase branching and reduce early flowering.
	To increase lateral branching	Florel/Collate	150 to 200 ppm spray	

Table 8-5. Growth Regulators for Floricultural Crops in Greenhouses

CROP	PURPOSE	CHEMICAL	RATE*	PRECAUTIONS AND REMARKS
BEDDING PLANTS (Not specifically listed in this table)	To control plant growth	Abide/A-Rest	6 to 66 ppm spray; use 15 ppm spray as a base rate and adjust as needed	
			0.06 to 0.12 mg a.i. drench for a 4-in. pot; apply 2 fl. oz./4-in. pot)	Drench volumes and mg a.i. vary with pot size.
		Citadel + Dazide/Cycocel + B-Nine	800 to 5,000 ppm + 1,000 to 1,500 ppm Cycocel applied as a tank-mix spray	Use the highest rate of Cycocel that doesn't cause excessive leaf yellowing, and then adjust the B-Nine/Dazide rate up and down within the labeled range to attain the desired level of height control.
		Piccolo/Bonzi/Paczol	5 to 90 ppm spray. Use 30 ppm spray as a base rate and adjust as needed.	Conduct trials on a small number of plants, adjusting the rates as needed for desired final plant height and duration of height control. Not recommended for use on fibrous begonia or vinca.
		Piccolo/Bonzi/Paczol/Downsize	0.118 mg a.i. drench for a 6-in. pot; apply 4 fl. oz./6-in. pot)	Drench applications are recommended only for bedding plants in 6-in. or larger containers. Not recommended for use on fibrous begonia or vinca.
		Citadel/Cycocel	800 to 1,500 ppm spray	Conduct trials on a small number of plants, adjusting the rates as needed for desired final plant height and duration of height control.
	To control plant growth	Concise/Sumagic	1 to 50 ppm spray	Conduct trials on a small number of plants, adjusting the rates as needed for desired final plant height and duration of height control. Apply spray as elongation begins (plant height about 2 to 4 in.).
			0.1 to 2 ppm drench	
		Piccolo 10 XC	15 to 30 ppm spray	General starting point for conducting trials for plants not specifically on the label. Use lowest rate in the Northern Belt Region and the upper rate in the Sunbelt Region.
			1 ppm drench	General starting point for conducting trials for plants not specifically on the label.
	To promote plant growth and overcome over-application of gibberellin-inhibiting PGRs	Florgib/ProGibb T&O	1 to 25 ppm spray	Conduct trials on a small number of plants initially using 1 ppm unless previous experience warrants higher use rates. Following assessment of plant response, and if desired results were not evident, reapplication or an increase in rate may be warranted. Consult the label for additional precautions.
		Fresco/Fascination	1 to 25 ppm spray	Conduct trials on a small number of plants initially using 1 ppm unless previous experience warrants higher use rates. Following assessment of plant response, and if desired results were not evident, reapplication or an increase in rate may be warranted. The most common rates for use are 3 to 5 ppm. SEE LABEL FOR ADDITIONAL PRECAUTIONS BEFORE USE.
	To induce lateral or basal branching	Configure	50 to 500 ppm spray	The supplemental label allows legal use on greenhouse-grown plants not specifically listed on the original label. See label for trialing suggestions and precautions.
BEDDING PLANT PLUGS (Not specifically listed in this table)	To control plant growth	Abide/A-Rest	3 to 35 ppm spray	
			Drench plug flats with a 0.5 to 1 ppm solution	For uniform application, use a subirrigation delivery system. Plug trays should not be excessively dry prior to the subirrigation treatment. Plants should develop one to two true leaves prior to first application.
		Dazide/B-Nine	1,500 to 2,500 ppm spray	Conduct trials on a small number of plants, adjusting the rate as needed for desired final plant height and duration of height control. Can be used at the beginning of the true first leaf stage through the finishing stage.
		Citadel + Dazide/Cycocel + B-Nine	800 to 5,000 ppm Dazide/B-Nine + 1,000 to 1,500 ppm Citadel/Cycocel applied as a tank-mix spray	Use the highest rate of Citadel/Cycocel that doesn't cause excessive leaf yellowing and then adjust the B-Nine/Dazide rate up and down within the labeled range to attain desired level of height control.
		Piccolo/Piccolo 10 XC/	1 to 20 ppm spray. Use 5 ppm spray as a base rate and adjust as needed.	Conduct trials on a small number of plants, adjusting the rate as needed for desired final plant height and duration of height control. Plants should develop one to two true leaves prior to first application.
		Bonzi/Paczol		
		Citadel/Cycocel	400 to 1,500 ppm spray	Conduct trials on a small number of plants. Start with lower rates and adjust the rates as needed for desired final plant height and duration of height control.
		Concise/Sumagic	0.5 to 10 ppm spray	Conduct trials on a small number of plants, adjusting the rates as needed for desired final plant height and duration of height control. Plugs can be especially sensitive to Concise/Sumagic.

Table 8-5. Growth Regulators for Floricultural Crops in Greenhouses

CROP	PURPOSE	CHEMICAL	RATE*	PRECAUTIONS AND REMARKS
BEGONIA, Hiemalis (Elatior)	To control plant growth	Citadel/Cycocel	500 to 1,000 ppm spray	Applied 1 week after short days begin in summer or when short days begin in winter. Late applications can result in insufficient flower stalk elongation.
	To increase lateral branching	Augeo	781 to 1,562 ppm spray	
BEGONIA, Seed (Wax)	To control plant growth	Abide/A-Rest	3 to 15 ppm spray	Use lower half of rate range for plugs and upper range for finishing plants.
		Dazide/B-Nine	2,500 to 5,000 ppm spray	
		Florel/Collate	500 ppm spray	Apply to increase lateral branching, prevent flower initiation and development, and inhibit internode elongation.
		Concise/Sumagic	Sprays	Not registered for use. Can result in excessive control.
		Piccolo/Piccolo 10 XC/ Bonzi/Paczol	Sprays	Not registered for use. Can result in excessive control.
		Topflor	Sprays	Not registered for use. Can result in excessive control.
		Citadel/Cycocel	500 ppm spray	
		Citadel + Dazide/Cycocel + B-Nine	1,000 to 1,250 ppm Dazide/B-Nine +800 to 1,250 ppm Citadel/Cycocel applied as a tank-mix spray	
BEGONIA, Tuberous	To control plant growth	Citadel/Cycocel	250 to 500 ppm spray	Rate can be used on Stage 4 plugs or beginning 2 weeks after transplanting.
		Citadel/Cycocel	1,000 ppm spray	Rate for actively growing plants.
		Dazide/B-Nine	2,500 ppm spray	Rate for actively growing plants.
BEGONIA, Vegetative	To control plant growth	Citadel/Cycocel	750 to 1,000 ppm spray	
BEGONIA, Vegetative (Dragon Wing)	To control plant growth	Piccolo/Piccolo 10 XC/ Bonzi/Paczol	3 to 5 ppm spray	For 4-in. pots, apply a weekly 3 ppm spray starting 2 weeks after transplanting for 3 weeks. For 6-in. pots, use 5 ppm starting 2 weeks after transplant. A second and third application may be useful.
BELLIS	To control plant growth	Dazide/B-Nine	2,500 ppm spray	If needed.
		Concise/Sumagic	5 ppm spray	If needed.
BIDENS	To control plant growth	Dazide/B-Nine	1,500 to 2,500 ppm spray	At planting, soft pinch to promote lateral shoot development.
		Concise/Sumagic	1 to 5 ppm spray	Rates for genetics needing slight growth control.
			0.25 ppm drench	Rates for genetics needing slight growth control.
	To increase lateral branching	Florel/Collate	300 to 500 ppm spray	
BOUGAINVILLEA	To control plant growth	Abide/A-Rest	50 ppm drench	
		Piccolo/Piccolo 10 XC/ Bonzi/Paczol	25 to 100 ppm drench	
	To increase lateral branching	Augeo	400 to 1,600 ppm spray	Cultivar response rates vary. Conduct your own trials to determine suitability and appropriate timing.
			1,600 ppm drench	Cultivar response rates vary. Conduct your own trials to determine suitability and appropriate timing.
BRACHYSCOME	To control plant growth	Florel/Collate	500 to 1,000 ppm spray	To keep plants more compact. Based on Texas A&M University trials.
		Dazide/B-Nine	2,500 to 5,000 ppm spray	
BRACTEANTHA, BRACTEATA	To control plant growth	Dazide/B-Nine	2,500 ppm spray	
		Piccolo/Bonzi/Paczol	20 to 30 ppm spray	
			1 ppm drench	
		Concise/Sumagic	10 to 20 ppm spray	
	To increase lateral branching	Florel/Collate	300 to 500 ppm	
BROMELIAD	To promote flower initiation	Florel/Collate	2,471 ppm spray	Cultivar response rates vary. Conduct your own trials to determine suitability and appropriate timing.
BROWALLIA	To control plant growth	Dazide/B-Nine	2,500 to 5,000 ppm spray	

Table 8-5. Growth Regulators for Floricultural Crops in Greenhouses

CROP	PURPOSE	CHEMICAL	RATE*	PRECAUTIONS AND REMARKS
BULB CROPS (Not specifically listed in this table)	To control plant growth	Abide/A-Rest	25 to 50 ppm spray	
			0.25 mg a.i. (2 ppm) drench for a 6-in. pot; apply 4 fl. oz./6-in. pot)	Drench volumes and mg a.i. vary with pot size.
		Piccolo/Piccolo 10 XC/	100 ppm spray	Conduct trials on a small number of plants, adjusting the rate as needed for desired final plant height and duration of height control.
		Bonzi/Paczol	1.183 mg a.i. (10 ppm) drench for a 6-in. pot; apply 4 fl. oz./6-in. pot)	Drench volumes and mg a.i. vary with pot size.
			20 ppm bulb soak	Soak for 15 min. Conduct trials on a small number of bulbs, adjusting the rate and soaking period (up to 1 hour) as needed for desired final plant height.
		Concise/Sumagic	2.5 to 20 ppm spray	Conduct trials on a small number of plants, adjusting the rate as needed for desired final plant height and length of height control.
			1 to 3 ppm drench	Drench volumes and mg a.i. vary with pot size. Application should be made when newly emerged shoots are 1 to 2 in. tall.
			1 to 10 ppm bulb soak	Soak for 1 to 5 min. Conduct trials on a small number of bulbs, adjusting the rate and soaking period as needed for desired final plant height.
	To promote plant growth and overcome over-application of gibberellin-inhibiting PGRs.	Fascination	1 to 25 ppm spray	Conduct trials on a small number of plants initially using 1 ppm, unless previous experience warrants higher use rates. Following assessment of plant response, and if desired results were not evident, reapplication or an increase in rate may be warranted. The most common rates for use are 3 to 5 ppm. SEE LABEL FOR ADDITIONAL PRECAUTIONS BEFORE USE.
CALADIUM	To control plant growth	Dazide/B-Nine	2,500 to 5,000 ppm spray	
		Piccolo/Bonzi/Paczol/Downsize	100 to 200 ppm spray (3.2 to 6.4 fl oz/gal)	Make first spray application when plants are 2 to 4 in. tall.
			0.24 to 1.77 mg a.i. (5 to 15 ppm) drench for a 6-in. pot; apply 4 fl. oz./6-in. pot)	Make first application when plants are 1 to 2 in. tall. Drench volumes and mg a.i. vary with pot size.
		Piccolo/Bonzi/Paczol	60 ppm tuber soak	Soak tubers for 30 min. prior to planting.
		Piccolo 10 XC	100 to 200 ppm spray	Spray applications of Piccolo 10 XC are the least desirable method for controlling bulb plant height and must be applied sequentially to maximize uniformity of the crop. Begin spray applications when plants reach a height of 2 to 4 in.
			2 to 16 ppm drench	Drench volume varies with pot size. Begin drench applications when plants reach a height of 1 to 2 in.
		Topflor	0.5 to 2 mg a.i. drench for a 6-in. pot	Based on NC State University trials. Adjust rates for other locations. Use lower rates for less vigorous cultivars.
CALCEOLARIA	To control plant growth	Citadel/Cycocel	400 to 800 ppm spray	Used to control internode length. Apply 400 ppm when flower buds are 1-in. in diameter. Repeat 2 weeks later if needed.
		Dazide/B-Nine	1,000 to 1,500 ppm spray	Used to control internode length.
CALENDULA	To control plant growth	Dazide/B-Nine	2,500 to 5,000 ppm spray	Can be used when the visible flower bud is pea sized. Rates of 3,500 ppm be used 4 to 5 weeks after germination (when 3 to 4 mature leaves formed).
			2,500 to 5,000 ppm spray	Plugs: Use 2,500 ppm with Stage 1 and 5,000 ppm with Stages 2 or 3.
		Concise/Sumagic	1 ppm spray	Plugs: Use at Stages 2 or 3.
		Piccolo/Piccolo 10 XC/	4 ppm spray	Plugs: Use at Stages 2 or 3.
		Bonzi/Paczol		
CALIBRACHOA	To control plant growth	Dazide/B-Nine	2,500 to 5,000 ppm spray	At planting, soft pinch to promote lateral shoot development. Multiple applications may be required.
		Citadel + Dazide/Cycocel + B-Nine	2,500 ppm Dazide + 500 to 1,500 ppm Citadel applied as a tank-mix spray	
		Concise/Sumagic	10 to 25 ppm spray	Try lower rate initially. Apply 2 weeks after transplanting.
		Piccolo/Piccolo 10 XC/	3 to 50 ppm spray	Use rates of 3 to 5 ppm for compact genetics needing slight growth control.
		Bonzi/Paczol		
		Piccolo/Piccolo 10 XC/	3 to 8 ppm drench	Rates for compact genetics needing slight growth control. Begin with 1 to 2 ppm to determine suitable rates.
		Bonzi/Paczol/Downsize		
		Florel/Collate	300 to 500 ppm spray	Early spray will increase branching and reduce early flowering.
		Topflor	5 to 10 ppm spray	

Table 8-5. Growth Regulators for Floricultural Crops in Greenhouses

CROP	PURPOSE	CHEMICAL	RATE*	PRECAUTIONS AND REMARKS
CALLA LILY (Zantedeschia)	To control plant growth	Piccolo/Piccolo 10 XC/	0.59 to 1.77 mg a.i. (5 to 15 ppm) drench for a 6-in. pot; apply 4 fl. oz./6-in. pot	Make first application when plants are 1 to 2 in. tall. Drench volumes and mg a.i. vary with pot size.
		Bonzi/Paczol/Downsize		
		Piccolo/Bonzi/Paczol	20 ppm rhizome/tuber soak	Soak the rhizomes/tubers for 15 min. prior to planting.
		Concise/Sumagic	1 to 2 mg a.i. drench (8.45 to 16.9 ppm); apply 4 fl. oz./6-in. pot)	Optimal rate based on NC State University trials. Adjust rate for plant vigor. Drench volumes and mg a.i. vary with pot size.
		Topflor	1 to 2.25 mg a.i drench for a 6-in. pot	Based on NC State University trials. Adjust rates for other locations.
	To promote flowering	Florgib/ProGibb T&O	500 ppm rhizome/tuber soak	Soak the rhizomes or tubers for 10 min. prior to planting. See label for details.
CAMPANULA	To control plant growth	Dazide/B-Nine	2,500 to 5,000 ppm spray	Use at visible bud.
		Topflor	10 to 30 ppm spray	Use at visible bud.
CANNA LILY	To control plant growth	Topflor	50 to 80 ppm spray	
CELOSIA	To control plant growth	Abide/A-Rest	7 to 26 ppm spray	
		Dazide/B-Nine	2,500 to 5,000 ppm spray	
		Piccolo/Piccolo 10 XC/ Bonzi/Paczol	15 to 45 ppm spray	
		Citadel/Chlormequat E-Pro/ Cyclocel	800 to 1,500 ppm spray	
		Concise/Sumagic	10 to 20 ppm spray	
		Topflor	10 to 40 ppm spray	Based on NC State University trials. Adjust rates for other locations.
CELOSIA, Plugs	To control plant growth	Piccolo/Piccolo 10 XC/	5 to 10 ppm spray	Timing of application should normally begin at the 1 to 2 true leaf stage.
		Bonzi/Paczol		
CENTAUREA	To control plant growth	Abide/A-Rest	10 to 15 ppm spray	
		Dazide/B-Nine	2,500 to 5,000 ppm spray	
CENTRADENIA HYBRID	To control plant growth	None	None	Plants grown with good light and optimal growing conditions generally do not need PGRs.
CHRISTMAS CACTUS (Schlumbergera spp.)	To increase branching under vegetative conditions	Configure	100 ppm spray	After planting when new vegetative growth begins, uniformly apply 1 to 2 quarts of finished spray solution to 100 sq. ft. of area.
	To increase the number of flower buds under reproductive conditions	Configure	100 to 200 ppm spray	Apply as a uniform foliar spray after the start of short days following leveling, or when flower buds become visible. See the label for specific guidelines based on lighted or natural-season growth plants.
CHRYSANTHEMUM, Cut	To reduce "neck" stretching	Dazide/B-Nine	2,500 ppm spray	Spray upper foliage 5 weeks after start of short-day treatment.
	To elongate peduncles of pompom-type mums	Florgib/ProGibb T&O	25 to 60 ppm spray	Use a single application 4 to 5 weeks after initiation of short days. Direct spray solution towards the flower buds. See label for precautions.
CHRYSANTHEMUM, Perennial	To control plant growth	Piccolo/Piccolo 10 XC/	50 to 200 ppm spray	
		Bonzi/Paczol		
		Piccolo/Piccolo 10 XC/ Bonzi/Paczol/Downsize	0.12 to 0.48 mg a.i. (1 to 4 ppm) drench for a 6-in. pot; apply 4 fl. oz./6-in. pot	

Table 8-5. Growth Regulators for Floricultural Crops in Greenhouses

CROP	PURPOSE	CHEMICAL	RATE*	PRECAUTIONS AND REMARKS
CHRYSANTHEMUM, Potted	To control plant growth	Abide/A-Rest	25 to 50 ppm spray	
			0.25 to 0.5 mg a.i. drench for a 6-in. pot; apply 4 fl. oz./6-in. pot	Apply when plants are 2 to 6 in. in height (about 2 weeks after pinch). Drench rates and application volumes vary with pot size.
		Dazide/B-Nine	1,000 ppm preplant foliar dip	Rooted cuttings can be dipped in solution to thoroughly wet leaves and stems and then potted. Allow foliage to dry before watering in. For unrooted cuttings, dip stems in solution, remove to flat, cover to prevent dehydration and hold overnight under cool conditions. Stick the next day.
			2,500 to 5,000 ppm spray	Spray when new growth from pinch is 1 to 2 in. long. Some varieties may require another application 3 weeks later.
		Piccolo/Piccolo 10 XC/ Bonzi/Paczol	50 to 200 ppm spray	Applications should begin when axillary shoots are 2 to 3 in. long. Sprays can be applied earlier to vigorous cultivars if additional control is desired. Sequential applications of lower rates generally provide more uniformly shaped plants than single-spray applications. Uniform application of both sprays and drenches is critical for uniform crop development.
		Piccolo/Piccolo 10 XC/ Bonzi/Paczol/Downsize	0.118 to 0.473 mg a.i. (1 to 4 ppm) drench for a 6-in. pot; apply 4 fl. oz./6-in. pot	Drench volumes and mg a.i. vary with pot size. Begin when the axillary shoots are to 2 to 3 in. long. Uniform application is required.
		Concise	5 to 10 ppm dip treatment on cuttings	Apply when the lateral shoots are 1.5 to 2.0 in. tall (about 7 to 14 days after pinching). Test for cultivar sensitivity. Multiple applications of the lower label rate may elicit a more satisfactory response and/or increasing the spray volume from 2 qts/100 sq. ft. to 3 qts/100 sq. ft. For Florida only: use a foliar spray concentration between 5 to 10 ppm (1.3 to 2.56 fl. oz./gal). For medium to tall cultivars, increase the spray volume to 3 qts/100 sq. ft.
			2.5 to 10 ppm spray	Apply as a dip treatment on unrooted cuttings followed by a foliar spray in the low rate range. On rooted cuttings, use a solution of 2.5 ppm or less, followed by a foliar spray in the low rate range.
	To control plant growth	Concise/Sumagic	2.5 to 10 ppm spray	
		Topflor	7.5 to 25 ppm spray	Based on NC State University trials. Adjust rates for other locations. Use lower rates for less vigorous cultivars.
CHRYSANTHEMUM, Garden	To control plant growth	Concise	5 to 10 ppm dip treatment on cuttings	Apply when the lateral shoots are 1.5 to 2.0 in. tall (about 7 to 14 days after pinching). Test for cultivar sensitivity. Multiple applications of the lower label rate may elicit a more satisfactory response and/or increasing the spray volume from 2 qts/100 sq. ft. to 3 qts/100 sq. ft. For Florida only: use a foliar spray concentration between 5 to 10 ppm (1.3 to 2.56 fl. oz./gal). For medium to tall cultivars, increase the spray volume to 3 qts/100 sq. ft.
		Concise/Sumagic	2.5 to 10 ppm spray	
	To increase lateral branching	Florel/Collate	500 ppm spray	Florel and Collate applications will provide some growth retardant effects and delay flowering. Read the label for restrictions on timing of applications.
CHRYSOCEPHALUM APICULATUM	To control plant growth	Dazide/B-Nine	2,500 ppm spray	Plants pinched and grown with good light and optimal growing conditions generally do not need PGRs.
CLARKIA (Godetia)	To control plant growth	Concise/Sumagic	15 to 25 ppm drench	Trial rates for cultivar response. Rates based on older cultivars.
		Dazide/B-Nine	3,000 ppm foliar spray	Trial rates for cultivar response. Rates based on older cultivars.
		Piccolo/Piccolo 10 XC/ Bonzi/Paczol	20 to 30 ppm drench	Trial rates for cultivar response. Rates based on older cultivars.
CLEMATIS	To control plant growth	Abide/A-Rest	25 to 132 ppm spray	
			0.25 to 0.5 mg a.i. (2 to 4 ppm) drench for a 6-in. pot; apply 4 fl. oz./6-in. pot	Drench volumes and mg a.i. vary with pot size.
CLEOME	To control plant growth	Abide/A-Rest	7 to 26 ppm spray	
		Citadel/Chlormequat E-Pro/ Cyclocel	800 to 1,500 ppm spray	
		Dazide/B-Nine	4,000 to 5,000 ppm spray	Multiple applications may be required. Make them at 7- to 10-day intervals.
		Piccolo/Piccolo 10 XC/ Bonzi/Paczol	20 to 30 ppm spray	Multiple applications may be required. Make them at 7- to 10-day intervals.
CLERODENDRUM	To control plant growth	Abide/A-Rest	50 ppm spray	
			0.9 mg a.i. drench	
		Piccolo/Piccolo 10 XC/ Bonzi/Paczol	100 ppm drench	
			0.5 mg a.i. drench	
	To increase lateral branching	Augeo	1,042 to 2,083 ppm spray	

Table 8-5. Growth Regulators for Floricultural Crops in Greenhouses

CROP	PURPOSE	CHEMICAL	RATE*	PRECAUTIONS AND REMARKS
COLEUS PLUGS, Seed	To control plant growth	Piccolo/Piccolo 10 XC/	5 to 10 ppm spray	Timing of application should normally begin at the 1 to 2 true leaf stage.
		Bonzi/Paczol		
COLEUS, Seed	To control plant growth	Dazide/B-Nine	2,500 to 5,000 ppm spray	
		Piccolo/Piccolo 10 XC/	15 to 30 ppm spray	
		Bonzi/Paczol		
		Citadel/Chlormequat E-Pro/	400 to 3,000 ppm spray	
		Cyclocel		
		Concise/Sumagic	10 to 20 ppm spray	
		Topflor	20 to 40 ppm spray	Based on NC State University trials. Adjust rates for other locations.
COLEUS, Vegetative	To control plant growth	Citadel + Dazide/Cycocel + B-Nine	2,500 to 4,000 ppm + 1,000 to 1,500 ppm Cycocel applied as a tank-mix spray	See General Recommendations. Scheduling the crop to avoid excessive stretch is the most effective means of controlling growth.
		Piccolo/Piccolo 10 XC/	5 to 30 ppm spray	
		Bonzi/Paczol	1 to 2 ppm drench	
		Citadel/Chlormequat E-Pro/	800 to 1,500 ppm spray	
		Cyclocel		
		Concise/Sumagic	5 to 20 ppm spray	Use rates of 5 to 10 ppm for compact genetics needing slight growth control.
		Collate/Florel	500 ppm spray	
COLUMBINE	To control plant growth	Abide/A-Rest	65 to 132 ppm spray	
			0.25 to 0.5 mg a.i. (2 to 4 ppm) drench for a 6-in. pot; apply 4 fl. oz./6-in. pot	Drench volumes and mg a.i. vary with pot size.
CONEFLOWER (Echinacea spp.)	To control plant growth	Concise/Sumagic	30 to 40 ppm spray	
	To increase branching	Configure	300 to 900 ppm spray	Apply after plant establishment and resumption of growth (i.e., approximately 2 weeks after potting). Apply in a uniform spray volume of 2 qts/100 sq. ft. of area. Application timing and rate may vary with cultivar.
CONSOLIDA (Larkspur)	To control plant growth	Abide/A-Rest	35 to 132 ppm spray	
			0.25 to 0.5 mg a.i. drench for a 6-in. pot (1 to 2 fl. oz./gal of drench solution; apply 4 fl. oz./6-in. pot)	Drench volumes and mg a.i. vary with pot size.
		Concise/Sumagic	5 ppm drench	
		Dazide/B-Nine	2,500 to 5,000 ppm spray	
		Piccolo/Piccolo 10 XC/	30 to 60 ppm spray	
		Bonzi/Paczol		
CONSOLIDA, Cut (Larkspur)	To promote growth and stem elongation	Florgib/ProGibb T&O	50 to 100 ppm spray	Apply when plants are 4- to 8-in. tall. Apply at 2- to 3-week intervals. See label for precautions.
COREOPSIS	To control plant growth	Concise/Sumagic	2 to 4 ppm spray	Rates for compact genetics needing slight growth control.
		Piccolo/Piccolo 10 XC/	3 to 100 ppm spray	Use rates of 3 to 6 ppm for compact genetics needing slight growth control.
		Bonzi/Paczol		
		Piccolo/Piccolo 10 XC/	0.59 to 1.18 mg a.i. (5 to 10 ppm) drench for a 6-in. pot; apply 4 fl. oz./6-in. pot	Rates for vigorous genetics needing moderate growth control.
		Bonzi/Paczol/Downsize		
		Topflor	2 to 4 ppm spray	Rates for compact genetics needing slight growth control.
CORNFLOWER (Centaurea)	To control plant growth	Abide/A-Rest	7 to 26 ppm spray	
		Dazide/B-Nine	2,500 to 5,000 ppm spray	
COSMOS	To control plant growth	Dazide/B-Nine	2,500 to 5,000 ppm spray	
CROSSANDRA	To control plant growth	Dazide/B-Nine	2,500 to 5,000 ppm spray	Apply after pinch when new growth is 2-in. long.
		Piccolo/Piccolo 10 XC/	50 ppm spray	Apply 2 weeks after pinch.
		Bonzi/Paczol		
CUPHEA	To control plant growth	Dazide/B-Nine	1,500 to 2,500 ppm spray	PGRs not required on compact cultivars.
		Piccolo/Piccolo 10 XC/	1 to 5 ppm spray	Initially, test on a few plants to determine rate for optimum control. Cuphea is sensitive to excessive rates.
		Bonzi/Paczol		
		Piccolo/Piccolo 10 XC/	0.25 to 2 ppm drench	Use rates of 0.25 to 0.5 ppm for compact genetics needing slight growth control. Use 2 ppm for vigorous cultivars grown in the south.
		Bonzi/Paczol/Downsize		

Table 8-5. Growth Regulators for Floricultural Crops in Greenhouses

CROP	PURPOSE	CHEMICAL	RATE*	PRECAUTIONS AND REMARKS
DAFFODIL	To control plant growth	Piccolo/Piccolo 10 XC/ Bonzi/Paczol/Downsize	2.37 to 4.73 mg a.i. (20 to 40 ppm) drench for a 6-in. pot; apply 4 fl. oz./6-in. pot	See CALADIUM.
		Piccolo/Piccolo 10 XC/ Bonzi/Paczol	80 ppm bulb soak	Soak bulbs for 1 hr. prior to planting. Ten-minute soaks of 400 ppm provided excellent results in NC State University trials.
		Florel/Collate	2,000 ppm spray	Controls plant height and stem topple. Apply when shoots are 3 to 4 in. tall. See label for cultivar differences in rates.
DAHLIA, Bedding Plant	To control plant growth	Abide/A-Rest	7 to 26 ppm spray	
		Dazide/B-Nine	2,500 to 5,000 ppm spray	
		Citadel + Dazide/Cycocel + B-Nine	2,500 to 4,000 ppm + 1,000 to 1,500 ppm Cycocel applied as a tank-mix spray	
		Piccolo/Piccolo 10 XC/ Bonzi/Paczol	15 to 45 ppm spray	
	To control plant growth	Citadel/Chlormequat E-Pro/ Cyclocel	800 to 1,500 ppm spray	
		Concise/Sumagic	10 to 20 ppm spray	
DAHLIA PLUGS, Bedding Plant	To control plant growth	Piccolo/Piccolo 10 XC/ Bonzi/Paczol	5 to 10 ppm spray	Timing of application should normally begin at the 1 to 2 true leaf stage.
DAHLIA, Tuberous	To control plant growth	Abide/A-Rest	0.25 to 0.5 mg a.i. (2 to 4 ppm) drench for a 6-in. pot; apply 4 fl. oz./6-in. pot	Drench volumes and mg a.i. vary with pot size.
		Piccolo/Piccolo 10 XC/ Bonzi/Paczol/Downsize	1.18 to 4.73 mg a.i. (10 to 40 ppm) drench for a 6-in. pot; apply 4 fl. oz./6-in. pot	
		Piccolo/Piccolo 10 XC/ Bonzi/Paczol	Greater than 40 ppm tuber soak	Soak tubers for 20 min. prior to planting.
		Concise/Sumagic	0.25 to 0.5 mg a.i. drench (2.1 to 4.2 ppm); apply 4 fl. oz./6-in. pot	Optimal rate based on NC State University trials. Adjust rate for plant vigor. Drench volumes and mg a.i. vary with pot size.
		Topflor	0.25 to 2 mg a.i. (2.1 to 16.9 ppm) drench for a 6-in. pot	Based on NC State University trials. Adjust rates for other locations. Use lower rates for less vigorous cultivars.
DELPHINIUM	To control plant growth	Abide/A-Rest	35 to 132 ppm spray	
			0.25 to 0.5 mg a.i. drench for a 6-in. pot (1 to 2 fl. oz./gal of drench solution; apply 4 fl. oz./6-in. pot)	Drench volumes and mg a.i. vary with pot size.
		Concise/Sumagic	5 ppm drench	
		Dazide/B-Nine	2,500 to 5,000 ppm spray	
		Piccolo/Piccolo 10 XC/ Bonzi/Paczol	30 to 60 ppm spray	
DELPHINIUM, Cut	To promote plant growth and stem elongation	Florgib/ProGibb T&O	50 to 100 ppm spray	Apply when plants are 4 to 8 in. tall. More than one application is possible at 2- to 3-week intervals. See label for precautions.
DIANTHUS, Bedding Plant	To control plant growth	Abide/A-Rest	7 to 26 ppm spray	
		Dazide/B-Nine	2,500 to 5,000 ppm spray	
		Piccolo/Piccolo 10 XC/ Bonzi/Paczol	5 to 60 ppm spray	Cultivar response rates vary. Conduct your own trials to determine suitability and appropriate timing. Some series recommend the use of 5 to 8 ppm sprays.
		Citadel/Chlormequat E-Pro/ Cyclocel	800 to 1,500 ppm spray	
		Concise/Sumagic	3 to 5 ppm spray	
DIANTHUS PLUGS, Bedding plant	To control plant growth	Piccolo/Piccolo 10 XC/ Bonzi/Paczol	10 to 20 ppm spray	Timing of application should normally begin at the 1 to 2 true leaf stage.
DIANTHUS, Cut	To promote plant growth and stem elongation	Florgib/ProGibb T&O	50 to 100 ppm spray	Apply when plants are 4 to 8 in. tall. More than one application is possible at 2- to 3-week intervals. See label for precautions.
DIANTHUS, Pot	To control plant growth	Concise/Sumagic	15 ppm spray	
		Piccolo/Piccolo 10 XC/ Bonzi/Paczol	15 ppm spray	

Table 8-5. Growth Regulators for Floricultural Crops in Greenhouses

CROP	PURPOSE	CHEMICAL	RATE*	PRECAUTIONS AND REMARKS
DIASCIA Hybrid	To control plant growth	Dazide/B-Nine	1,250 to 5,000 ppm	At planting, soft pinch to promote lateral shoot development. Use higher rates on vigorous cultivars.
		Concise/Sumagic	5 to 15 ppm spray	Use lower rates to ensure taller flower spikes.
		Florel/Collate	200 to 500 ppm spray	Use 2 weeks after pinch.
		Piccolo/Piccolo 10 XC/	30 ppm spray	
		Bonzi/Paczol	1 to 2 ppm drench	
DIASCIA, Seed	To control plant growth	Abide/A-Rest	20 ppm spray	Start application 7 to 10 days after transplant. Repeat 7 days later.
		Concise/Sumagic	5 to 10 ppm spray	To hold plants under warm conditions. Use caution, plants very responsive.
		Dazide/B-Nine	3,000 to 5,000 ppm spray	Start application 7 to 10 days after transplant.
		Piccolo/Piccolo 10 XC/ Bonzi/Paczol	10 to 20 ppm spray	To hold plants under warm conditions. Use caution, plants very responsive.
DICENTRA SPECTABILIS (Bleeding Heart)	To control plant growth	Abide/A-Rest	65 to 132 ppm spray	
			0.25 to 0.5 mg a.i. drench for a 6-in. pot; apply 4 fl. oz./6-in. pot)	Drench volumes and mg a.i. vary with pot size.
		Dazide/B-Nine	2,500 to 5,000 ppm spray	Apply as new sprouts emerge from the pot. Repeat if needed due to non-uniform emergence.
DICHONDRA ARGENTEA	To control plant growth	Citadel+Dazide/Cycocel+B-Nine	1,000 ppm + 5,000 ppm spray	Also increases branching and improves silver color.
		Dazide/B-Nine	5,000 ppm spray	Also increases branching and improves silver color. Apply 2 weeks after transplanting.
DIGITALIS	To control plant growth	Piccolo/Piccolo 10 XC/ Bonzi/Paczol	80 to 160 ppm spray	
		Piccolo/Piccolo 10 XC/ Bonzi/Paczol/Downsize	0.24 to 0.48 mg a.i. (2 to 4 ppm) drench for a 6-in. pot; apply 4 fl. oz./6-in. pot	
DOROTHEANTHUS BELLIDIFORMIS	To control plant growth	None	None	Plants pinched and grown with good light and optimal growing conditions generally do not need PGRs.
DRACAENA	To control plant growth	Abide/A-Rest	25 to 132 ppm spray	
			0.25 to 0.5 mg a.i. (2 to 4 ppm) drench for a 6-in. pot; apply 4 fl. oz./6-in. pot	Drench volumes and mg a.i. vary with pot size.
DUSTY MILLER *(Senecio cineraria)*	To control plant growth	Dazide/B-Nine	2,500 to 5,000 ppm spray	
		Concise/Sumagic	30 ppm spray	
EASTER LILY (See Lily, Easter)				
ECHEVERIA spp	To induce offsets and induce flower development	Configure	100 to 400 ppm spray	Based on NC State University trials when applied 2 weeks after potting. A slight increase in offsets occurred along with the induction of flowering.
EGGPLANT	To control plant growth	Sumagic	2 to 10 ppm spray	See label for application suggestions and precautions. Make initial foliar applications when 2 to 4 true leaves are present. Apply uniformly as a foliar spray using 2 qt/100 sq. ft. Sequential applications at lower recommended rates will generally provide more growth control than a single high rate application. First-time users should apply the lowest recommended rate in order to determine optimal rate for individual cultivars under local environmental conditions. If additional growth control is required, a sequential spray application at the lowest recommended rate should be made 7 to 14 days after the initial application. If multiple applications are made to the transplants, the total amount of Sumagic applied may not exceed that from a single application of a 10 ppm spray. The final application may not occur later than 14 days after the 2 to 4 true leaf stage.
ERYSIMUM	To control plant growth	None	None	Plants grown with good light and optimal growing conditions generally do not need PGRs.
EUPATORIUM	To control plant growth	Piccolo/Piccolo 10 XC/ Bonzi/Paczol	>240 ppm spray	
		Piccolo/Piccolo 10 XC/ Bonzi/Paczol/Downsize	0.96 to 1.18 mg a.i. (8 to 10 ppm) drench for a 6-in. pot; apply 4 fl. oz./6-in. pot	
***EUPHORBIA HYPERICIFOLIA* HYBRID**	To control plant growth	Dazide/B-Nine	2,500 ppm spray	Plant growth slow early on. Apply PGRs if control is needed.
		Citadel+Dazide/Cycocel+B-Nine	750 ppm + 2,500 ppm spray	
		Florel/Collate	Spray	Not recommended.
		Piccolo/Piccolo 10 XC/ Bonzi/Paczol	0.5 to 2 ppm drench	Can be applied 3 to 4 weeks before finish, using the lower rate in the North and higher rate in the South.
EVOLVULUS	To control plant growth	None	None	Plants grown with good light and optimal growing conditions generally do not need PGRs.

Chapter VIII — 2019 N.C. Agricultural Chemicals Manual

Table 8-5. Growth Regulators for Floricultural Crops in Greenhouses

CROP	PURPOSE	CHEMICAL	RATE*	PRECAUTIONS AND REMARKS
EXACUM	To control plant growth	Dazide/B-Nine	2,500 to 5,000 ppm spray	
		Piccolo/Piccolo 10 XC/	75 ppm spray	
		Bonzi/Paczol	0.25 to 0.75 mg a.i. drench for a 6-in. pot	
		Topflor	25 to 50 ppm spray	Based on NC State University trials. Adjust rates for other locations.
			0.01 to 0.03 mg a.i. (0.08 to 0.25 ppm) drench for a 6-in. pot	Based on NC State University trials. Adjust rates for other locations. Exacum is very responsive to Topflor drenches, so start trials with lower rates.
FATSHEDERA	To control plant growth	Abide/A-Rest	65 to 132 ppm spray	
			0.25 to 0.5 mg a.i. drench for a 6-in. pot; apply 4 fl. oz./6-in. pot	Drench volumes and mg a.i. vary with pot size.
FELICIA	To control plant growth	Citadel+Dazide/Cycocel+B-Nine	1,000 to 1,500 ppm + 2,500 to 4,000 ppm	Pinch plant as needed to improve shape.
		Citadel/Cycocel	1,500 ppm spray	Applied to pinched plants.
FLOWERING/ FOLIAGE PLANTS, Herbaceous Species (Not specifically listed in this table)	To control plant growth	Abide/A-Rest	20 to 50 ppm spray	Recommended starting rate for an Abide/A-Rest spray on a new herbaceous flowering or foliage species is 33 ppm (16 fl. oz./gal).
			0.125 to 0.25 mg a.i. (1 to 2 ppm) drench for a 6-in. pot; apply 4 fl. oz./6-in. pot	Drench volumes and mg a.i. vary with pot size.
		Piccolo/Piccolo 10 XC/	30 ppm spray	Conduct trials on a small number of plants, adjusting the rate as needed for desired final plant height and length of height control.
		Bonzi/Paczol		
		Piccolo/Piccolo 10 XC/	0.118 mg a.i. (1 ppm) drench for a 6-in. pot; apply 4 fl. oz./ 6-in. pot	Drench volumes and mg a.i. vary with pot size. Conduct trials on a small number of plants.
		Bonzi/Paczol/Downsize		
		Citadel/Cycocel	800 to 3,000 ppm spray	Optimum rate depends on species, desired amount of height control and environmental conditions. The suggested initial rate for small-scale trials is 1,250 ppm. Example: herbaceous species known to respond to Cycocel are—Achimenes, Aster, Astilbe, Begonia (hiemalis), Begonia (tuberous), Calceolaria, Carnation, Chrysanthemum, Columbine, Easter lily, *Gynura aurantiaca*, Ivy, Kalanchoe, *Lilium spp.*, Morning glory, Pachystachys, *Pilea spp.*, Pentas, *Salvia spp.*, Schefflera, *Sedum spp.* and Sunflower.
			2,000 to 4,000 ppm drench	Drench volumes vary with pot size. See label for recommended volumes. Herbaceous species known to respond to Cycocel are listed above.
		Concise/Sumagic	5 to 40 ppm spray	Conduct trials on a small number of plants, adjusting the rate as needed for desired final plant height and length of height control.
			0.1 to 1 ppm drench	Drench volumes and mg a.i. vary with pot size.
	To promote plant growth and overcome over-applications of gibberellin-inhibiting PGRs	Florgib/ProGibb T&O	1 to 25 ppm spray	Conduct trials on a small number of plants initially using 1 ppm, unless previous experience warrants higher use rates. Following assessment of plant response, and if desired results are not evident, reapplication or an increase in rates may be warranted. Consult the label for additional precautions.
		Fresco/Fascination	1 to 25 ppm spray	Conduct trials on a small number of plants initially using 1 ppm, unless previous experience warrants higher use rates. Following assessment of plant response, and if desired results were not evident, reapplication or an increase in rate may be warranted. The most common rates for use are 3 to 5 ppm. SEE LABEL FOR ADDITIONAL PRECAUTIONS BEFORE USE.
	To induce lateral or basal branching	Configure	50 to 500 ppm spray	The supplemental label allows legal use on greenhouse grown plants not specifically listed on the original label. See label for trialing suggestions and precautions.

Table 8-5. Growth Regulators for Floricultural Crops in Greenhouses

CROP	PURPOSE	CHEMICAL	RATE*	PRECAUTIONS AND REMARKS
FLOWERING/ FOLIAGE PLANTS, Woody Species (Not specifically listed in this table)	To control plant growth	Abide/A-Rest	50 ppm spray	
			0.25 mg a.i. (2 ppm) drench for a 6-in. pot; apply 4 fl. oz./ 6-in. pot	Drench volumes and mg a.i. vary with pot size.
		Dazide/B-Nine	2,500 to 7,500 ppm spray	Two or more applications may be necessary if new growth begins to stretch or for enhanced coloration.
		Piccolo/Piccolo 10 XC/ Bonzi/Paczol	50 ppm spray	Conduct trials on a small number of plants, adjusting the rate as needed for desired final plant height and length of height control.
		Piccolo/Piccolo 10 XC/ Bonzi/Paczol/Downsize	0.237 mg a.i. drench for a 6-in. pot; apply 4 fl. oz./6-in. pot	Drench volumes and mg a.i. vary with pot size.
		Citadel/Cycocel	800 to 3,000 ppm spray	Optimum rate depends on species, desired amount of height control and environmental conditions. The suggested initial rate for small-scale trials is 1,250 ppm. Example: woody species known to respond to Cycocel are—Barleria cristata, Bougainvillea, Camellia, Gardenia, Fuchsia, Hollies, Hydrangea, Lantana, Pseuderanthemum lactifolia, Rhododendron and Roses (potted).
			2,000 to 4,000 ppm drench	Drench volumes vary with pot size. See label for recommended volumes. Woody species known to respond to Cycocel are listed above.
		Concise/Sumagic	20 to 50 ppm spray	Conduct trials on a small number of plants, adjusting the rate as needed for desired final plant height and length of height control.
			0.5 to 2 ppm drench	Drench volumes and mg a.i. vary with pot size.
FREESIA	To control plant growth	Abide/A-Rest	100 to 200 ppm corm soak	Soak corms in the solution for 1 hour before planting. Cultivar response varies, so conduct your own trials.
		Piccolo/Piccolo 10 XC/ Bonzi/Paczol/Downsize	0.22 to 0.48 mg a.i. (2 to 4 ppm) drench for a 6-in. pot; apply 4 fl. oz./6-in. pot	To increase lateral branching.
		Piccolo/Piccolo 10 XC/ Bonzi/Paczol	50 to 200 ppm corm soak	Soak corms in the solution for 1 hour before planting. Cultivar response varies, so conduct your own trials.
FUCHSIA	To control plant growth	Abide/A-Rest	25 to 75 ppm spray	May also increase flowering.
		Dazide/B-Nine	1,250 to 2,500 ppm spray	
		Piccolo/Piccolo 10 XC/ Bonzi/Paczol	5 to 10 ppm spray	Make applications prior to visible bud to avoid delay.
		Concise/Sumagic	2 to 5 ppm spray	Make applications prior to visible bud to avoid delay.
	To increase lateral branching	Augeo	781 to 2,343 ppm spray	
		Florel/Collate	500 ppm spray	Florel and Collate applications will provide some growth retardant effects and delay flowering. Read the label for restrictions on timing of applications.
	To promote stem elongation for topiary	Florgib/ProGibb T&O	200 to 400 ppm spray	For use on upright growing cultivars used for topiary. Weekly sprays can be used, maximum 3 applications.
GARDENIA	To control plant growth	Abide/A-Rest	50 ppm spray	
			0.25 mg a.i. (2 ppm) drench for a 6-in. pot; apply 4 fl. oz./ 6-in. pot	Drench volumes and mg a.i. vary with pot size.
		Dazide/B-Nine	5,000 ppm spray	Spray when plants are at two-thirds final market size.
		Piccolo/Piccolo 10 XC/ Bonzi/Paczol	12 ppm drench	Flower delay possible. Apply prior to floral initiation (short days) or 6 weeks after pinching.
		Topflor	100 to 200 ppm spray	Apply prior to floral initiation (short days) or 6 weeks after pinching.
	To increase lateral branching	Augeo	2,343 to 4,687 ppm spray	
GAURA	To control plant growth	Dazide/B-Nine	3,000 to 4,000 ppm spray	
		Piccolo/Piccolo 10 XC/ Bonzi/Paczol	30 to 50 ppm spray	
		Piccolo/Piccolo 10 XC/ Bonzi/Paczol/Downsize	3.54 mg a.i. (30 ppm) drench for a 6-in. pot; apply 4 fl. oz./ 6-in. pot	
		Concise/Sumagic	10 to 30 ppm spray	
GAZANIA	To control plant growth	Citadel/Chlormequat E-Pro/ Cyclocel	1,500 ppm spray	Make applications prior to visible bud to avoid delay.
		Dazide/B-Nine	2,500 ppm spray	Make applications prior to visible bud to avoid delay.

Table 8-5. Growth Regulators for Floricultural Crops in Greenhouses

CROP	PURPOSE	CHEMICAL	RATE*	PRECAUTIONS AND REMARKS
GERANIUM	To control plant growth	Abide/A-Rest	26 to 66 ppm spray	See AGERATUM.
		Piccolo/Bonzi/Paczol	5 to 30 ppm spray	Apply to zonal geraniums when new growth is 1.5 to 2 in. long. Apply to seed geraniums approximately 2 to 4 weeks after transplanting.
		Concise	3 to 8 ppm spray	Use lower rates for less vigorous plants and higher rates for more vigorous growing plants. Flower delay on some cultivars can occur when using rates >6 ppm.
		Citadel/Chlormequat E-Pro/ Cyclocel	800 to 1,500 ppm spray	Make first application 2 to 4 weeks after planting plugs or rooted cuttings (after stems have started elongating). Multiple applications may be needed.
		Piccolo 10 XC	10 to 30 ppm spray	See Piccolo remarks for GERANIUM. Early applications may require lower rates to avoid overdosing. Piccolo 10 XC will reduce late stretch when applied as the flower stems begins to elongate.
		Concise/Sumagic	3 to 6 ppm spray for cutting geraniums and 2 to 4 ppm spray for seed geraniums	
		Topflor	15 to 25 ppm spray	Apply to zonal geraniums when new growth is 1.5 to 2 in. long.
	To promote earlier flowering in seed geraniums	Citadel/Chlormequat E-Pro/ Cyclocel	1,500 ppm spray	Make two applications at 35 and 42 days after seeding. Treated plants should flower earlier and be more compact and more well-branched than untreated plants.
		Florgib/ProGibb	5 to 15 ppm spray (0.02 to 0.06 fl. oz./gal)	Make a single foliar application when first flower bud set is noted. Spray the entire plant until runoff. See label for precautions.
	To increase flower number and size in cutting geranium	Florgib/ProGibb T&O	1 to 5 ppm spray	Make a single foliar application when first flower bud set is noted. Spray the entire plant until runoff. See label for precautions.
	To increase lateral branching	Florel/Collate	300 to 500 ppm spray	Labeled for zonal and ivy geraniums. Use the lower concentration for ivy geraniums. Florel and Collate will also provide some growth retardant effect and delay flowering. Read the label for restrictions on timing of applications.
GERANIUM, IVY	To control plant growth	Citadel/Chlormequat E-Pro/ Cyclocel	750 to 1,500 ppm spray	
	To increase branching	Augeo	1,562 ppm spray	Labeled for ivy geraniums only.
		Florel/Collate	200 to 300 ppm spray	
GERANIUM, Seed	To promote earlier flowering	Citadel	1,500 ppm spray	See label. Make two spray applications at 35 and 42 days after seeding. Plants flower quicker, are compact and have increased lateral breaks.
	To control plant growth	Concise	2 to 4 ppm spray	Apply when plant height is approximately 4 in. tall.
GERBERA DAISY	To control plant growth	Abide/A-Rest	25 to 132 ppm spray	Do not apply when flower stems are visible.
			0.25 to 0.5 mg a.i. drench for a 6-in. pot; apply 4 fl. oz./6-in. pot	Drench volumes and mg a.i. vary with pot size. Do not apply when flower stems are visible.
		Dazide/B-Nine	1,200 to 5,000 ppm spray	Do not apply when flower stems are visible. Apply lower rate at 10 to 14 interval if needed.
GLADIOLUS	To control plant growth	Abide/A-Rest	1.5 mg drench per 0.5 gal. pot	For container-grown plants.
		Piccolo/Piccolo 10 XC/ Bonzi/Paczol	2.5 to 5.0 mg drench per 0.5 gal. pot	For container-grown plants.
GLOXINIA *(Sinningia speciosa)*	To control peduncle length	Dazide/B-Nine	1,250 ppm spray	PGRs may not be required on compact cultivars. Make first application when the leaves reach the side of the pot. A repeat application can be made 7 to 10 days later if needed. Flower streaking can develop if PGR applied when the buds show color. Phytotoxicity may occur at rates >1,250 ppm.
		Piccolo/Piccolo 10 XC/	30 ppm spray	Can be applied when buds grow above the foliage.
		Bonzi/Paczol	4 to 8 ppm drenches	For elongation control late in the season (10 weeks after transplant).
GOMPHRENA	To control plant growth	Dazide/B-Nine	2,500 to 5,000 ppm spray	
		Citadel/Chlormequat E-Pro/ Cyclocel	800 to 1,5,00 ppm spray	
GOODENIA	To control plant growth	None	None	Plants grown with good light and optimal growing conditions generally do not need PGRs.
GRAPE IVY	To increase lateral branching	Augeo	781 to 1,562 ppm spray	
GROUNDCHERRY	To control plant growth	Concise/Sumagic	2 to 10 ppm spray	See precautions listed with EGGPLANT.
GYPSOPHILA	To accelerate plant growth, increase stem and flower number and increase flower uniformity	Florgib/ProGibb T&O	150 to 500 ppm spray	Make 3 to 4 foliar applications after 4 weeks of new growth has occurred after pinching. Use 2-week intervals between sprays. See label for precautions.
HELENIUM AMARUM	To control plant growth	Dazide/B-Nine	5,000 ppm spray	Apply after plant established (2 weeks after transplant).

Table 8-5. Growth Regulators for Floricultural Crops in Greenhouses

CROP	PURPOSE	CHEMICAL	RATE*	PRECAUTIONS AND REMARKS
HELICHRYSUM PETIOLARE/ H. ITALICUM (Licorice plant)	To control plant growth	Piccolo/Piccolo 10 XC/ Bonzi/Paczol	1 ppm drench	Plants grown with good light and optimal growing conditions generally do not need PGRs.
	To increase lateral branching	Florel/Collate	300 to 500 ppm spray	Make first application after 2 weeks. Repeat in 2 weeks if needed (with larger pots).
HELICONIA	To control plant growth	Piccolo/Piccolo 10 XC/	15 to 30 ppm spray	Apply when axillary shoots are 4 to 6-in. high after removal of primary shoot (2 to 3 months after planting). Cultivar variation possible, so conduct your own trials to determine optimal rates.
		Bonzi/Paczol	0.375 mg a.i. drench /6-in. pot	Apply when axillary shoots are 4 to 6-in. high after removal of primary shoot (2 to 3 months after planting). Cultivar variation possible, so conduct your own trials to determine optimal rates.
HELIOTROPIUM ARBORESCENS	To control plant growth	Citadel/Chlormequat E-Pro/ Cyclocel	500 ppm spray	Rate for compact genetics needing slight growth control.
		Citadel+Dazide/Cycocel+B-Nine	750 to 1,000 ppm + 1,500 to 3,000 ppm spray	Rate for compact genetics needing slight growth control.
HIBISCUS MOSCHEUTOS	To control plant growth	Citadel/Cycocel	1000 ppm foliar spray	Multiple applications may be required.
		Concise/Sumagic	15 ppm foliar spray	
HIBISCUS ROSA-SINENSIS	To control plant growth	Dazide/B-Nine	2,500 to 5,000 ppm spray	
		Piccolo/Piccolo 10 XC/ Bonzi/Paczol	5 to 150 ppm spray	Application should be made when laterals are 1 to 4 in. long. Single applications control lateral growth for 3 to 6 weeks.
		Concise	10 ppm spray	Apply within 7 days after pruning. Make additional applications as necessary to obtain desired results. Florida only: Use a foliar spray concentration between 5 to 10 ppm and apply a uniform spray volume of 3 qts/100 sq. ft.
		Citadel/Chlormequat E-Pro/ Cyclocel	200 to 600 ppm spray	Multiple applications starting prior to first pinch are recommended. See label for additional precautions. Avoid applications after flower buds are visible.
		Concise/Sumagic	0.025 to 0.2 mg a.i. drench per pot	
HOLLY	To control plant growth	Abide/A-Rest	50 ppm spray	
			0.25 mg a.i. (2 ppm) drench for a 6-in. pot; apply 4 fl. oz./6-in. pot	Drench volumes and mg a.i. vary with pot size.
HOLLYHOCK	To control plant growth	Piccolo/Bonzi/Paczol	30 to 50 ppm spray	
		Concise/Sumagic	5 to 40 ppm spray	
HOSTA	To promote lateral growth on finished plants	Configure	1,000 to 3,000 ppm spray	Apply in a uniform spray volume. Application is most effective when plants are fully established prior to application (i.e. at least 3 to 4 weeks after potting), when there is evidence of surface root development but before flower initiation.
	To increase production of offsets for propagation	Configure	1,000 to 3,000 ppm spray	Apply in a uniform spray volume to fully established, actively growing stock plants. Repeat the application at 30-day intervals during the growing season. Offsets may be harvested at any time. Treatment effects may vary by Hosta cultivar and may respond differently to a given rate. Multiple applications at 30-day intervals using lower rates may be more effective than a single application at a higher rate. Conduct trials on a small number of plants under actual use conditions to establish the proper use rates and timings.
HYACINTH	To reduce stem topple	Florel/Collate	1,000 ppm spray	To reduce stem topple at time of full flower, apply foliar spray before florets have opened.
	To control plant growth	Piccolo/Bonzi/Paczol	100 ppm bulb soak	Ten-minute soaks provided excellent results in NC State University trials. Cultivar response varied.
		Concise/Sumagic	20 to 40 ppm bulb soak	Two to ten-minute preplant soaks provided excellent results in NC State University trials. Cultivar response varied.
		Topflor	0.5 to 1 mg a.i. (4.2 to 8.45 ppm) drench for a 6-in. pot	Based on NC State University trials. Adjust rates for other locations.
			10 to 25 ppm bulb soak	Two to ten-minute preplant soaks provided excellent results in NC State University trials. Cultivar response varied.
HYBRID LILY (See Lily, Hybrid)				
HYDRANGEA	To control plant growth	Abide/A-Rest	50 ppm spray	
			0.25 mg a.i. (2 ppm) drench for a 6-in. pot; apply 4 fl. oz./6-in. pot	Drench volumes and mg a.i. vary with pot size.
		Dazide/B-Nine	1,250 to 7,500 ppm spray	Use lower rate in spring when 4 to 5 pairs of leaves are visible and new growth is starting to unfold, but not later than 4 weeks after initiation of forcing. Use higher rate for summer when regrowth after pinching is 1 to 2 in. long.
		Topflor	100 to 200 ppm spray	

Table 8-5. Growth Regulators for Floricultural Crops in Greenhouses

CROP	PURPOSE	CHEMICAL	RATE*	PRECAUTIONS AND REMARKS
HYPOESTES	To control plant growth	Chlormequat E-Pro	800 to 1,500 ppm spray	Initially apply after second set of leaves have developed. If needed, reapply 2 weeks later.
		Citadel/Cycocel	400 to 1,500 ppm spray	Initially apply after second set of leaves have developed. If needed, reapply 2 weeks later.
		Dazide/B-Nine	1,000 ppm spray	Initially apply after second set of leaves have developed. If needed, reapply 2 weeks later.
IMPATIENS, Seed	To control plant growth	Abide/A-Rest	10 to 44 ppm spray	
		Piccolo/Piccolo 10 XC/ Bonzi/Paczol	10 to 45 ppm spray	
		Concise/Sumagic	5 to 10 ppm spray	
		Topflor	20 to 60 ppm spray	Based on NC State University trials. Adjust rates for other locations.
	To increase branching	Florel/Collate	100 to 300 ppm spray	Use if better branching needed.
IMPATIENS PLUGS, Seed	To control plant growth	Piccolo/Piccolo 10 XC/ Bonzi/Paczol	0.5 to 10 ppm spray (0.015 to 0.32 fl. oz./gal)	Timing of application should normally begin at the 1 to 2 true leaf stage.
IMPATIENS, Vegetative	To control plant growth	Piccolo/Bonzi/Paczol	2 to 15 ppm spray	Cultivars' response to PGRs varies, so test a few plants to determine rate for optimum control.
			0.5 to 1 ppm drench	Drench volumes and mg a.i. vary with pot size. See label for recommended volumes.
		Florel/Collate	100 to 300 ppm spray	Will improve branching.
IMPATIENS, Seashell-type	To control plant growth	Piccolo/Bonzi/Paczol	5 to 8 ppm spray	Apply when plants have reached 75% of finished height. Don't apply to plants under stress. Recommendations based on Michigan trials.
IOCHROMA	To control plant growth	Dazide/B-Nine	5,000 ppm spray	
		Piccolo/Piccolo 10 XC/ Bonzi/Paczol	2 ppm spray	
IPOMOEA	To control plant growth	Concise/Sumagic	10 to 25 ppm spray	Not needed if optimal scheduling is used. If needed, apply when plants have reached 75% of finished growth. Recommendations based on NC State University trials.
		Dazide/B-Nine	2,500 ppm spray	Apply as needed.
		Florel/Collate	500 to 1,000 ppm spray	Will improve branching and control growth.
		Piccolo/Piccolo 10 XC/ Bonzi/Paczol	8 ppm drench	Applied to plugs prior to transplanting.
IRESINE HYBRID	To control plant growth	Citadel+Dazide/Cycocel+B-Nine	1,000 to 1,500 ppm + 2,500 to 4,000 ppm spray	
		Piccolo/Piccolo 10 XC/ Bonzi/Paczol	5 to 10 ppm spray	
		Piccolo/Piccolo10XC/Bonzi/ Paczol/Downsize	1 to 3 ppm drench	
JACOBINIA (Pink)	To control plant growth	Piccolo/Piccolo 10 XC/ Bonzi/Paczol	5 to 10 ppm spray	
		Piccolo/Piccolo 10 XC/ Bonzi/Paczol/Downsize	0.06 to 0.12 mg a.i. (0.5 to 1 ppm) drench for a 6-in. pot; apply 4 fl. oz./6-in. pot	
JERUSALEM CHERRY (*Solanum pseudocapsicum*)	To control plant growth	Citadel/Chlormequat E-Pro	800 to 1,500 ppm spray	
		Citadel/Cycocel	400 to 1,500 ppm spray	
	To promote stem elongation for topiary	Florgib/ProGibb T&O	250 ppm spray	For plants grown in 6-in. pots and with 4- to 6-in. of growth, apply 2 foliar sprays 10 days apart to promote stem elongation for topiary plants. Stake plants to support stem.

Table 8-5. Growth Regulators for Floricultural Crops in Greenhouses

CROP	PURPOSE	CHEMICAL	RATE*	PRECAUTIONS AND REMARKS
KALANCHOE	To control plant growth	Abide/A-Rest	50 ppm spray	Apply when axillary growth begins and repeat 20 to 30 days after short days begin. Trial to determine optimal rates and timing for your location.
		Dazide/B-Nine	2,500 to 5,000 ppm spray	Rates and timing vary with the season and cultivar. Applications typically begin 2 weeks after pinching. Apply sprays every 7 days in the summer, 10 to 15 days in the spring and fall, and 14 to 21 days in the winter. Trial to determine optimal rates and timing for your location.
		Piccolo/Piccolo 10 XC/ Bonzi/Paczol	2 to 4 ppm spray	Trial to determine optimal rates and timing for your location.
	To increase lateral branching	Augeo	1,042 to 2,343 ppm spray	
	To control peduncle length	Dazide/B-Nine	1,200 to 5,000 ppm spray	Phytotoxicity possible if B-Nine/Dazide accumulates in cupped areas of certain cupped-leafed varieties.
LACHENALIA sp.	To control plant growth	Concise/Sumagic	20 ppm corm soaks	Rates based on trials at Cornell University.
		Piccolo/Piccolo 10 XC/	100 to 200 ppm spray	Rates based on trials at Cornell University.
		Bonzi/Paczol	1 to 2 mg a.i./pot drench	Rates based on trials at Cornell University.
LAMIUM	To control plant growth	Concise/Sumagic	5 ppm spray	
		Piccolo/Piccolo 10 XC/	30 ppm spray	
		Bonzi/Paczol	1 ppm drench	
	To increase lateral branching	Collate/Florel	500 ppm spray	Improves branching and produces compact growth.
LANTANA	To control plant growth	Citadel + Dazide/Cycocel + B-Nine	2,500 to 5,000 ppm + 1,000 to 1,500 ppm Cycocel applied as a tank-mix spray	Cultivar response varies.
		Piccolo/Bonzi/Paczol	20 to 40 ppm spray	
		Concise/Sumagic	10 to 20 ppm spray	
	To increase lateral branching	Augeo	781 to 1,562 ppm spray	
		Florel/Collate	500 ppm spray	Florel and Collate applications will provide some growth retardant effects and delay flowering. Read the label for restrictions on timing of applications.
LAURENTIA AXILLARIS	To control plant growth	Abide/A-Rest	2 to 4 ppm spray	
		Dazide/B-Nine	2,500 ppm spray	
		Piccolo/Piccolo 10 XC/ Bonzi/Paczol	1 to 2 ppm drench	
LIATRIS	To control plant growth	Abide/A-Rest	25 to 132 ppm spray	
			0.25 to 0.5 mg a.i. (2 to 4 ppm) drench for a 6-in. pot; apply 4 fl. oz./6-in. pot)	Drench volumes and mg a.i. vary with pot size.
		Dazide/B-Nine	2,500 to 5,000 ppm spray	
LILY, Easter	To control plant growth	Abide/A-Rest	30 to 132 ppm spray. Use 50 ppm spray as a base rate and adjust as needed.	Apply when newly developing shoots are 2 to 3 in. long; a second application when shoots average 6 in. long may be needed.
			0.25 to 0.5 mg a.i. (2 to 4 ppm) drench for a 6-in. pot; apply 4 fl. oz./6-in. pot	Single drench should be applied when shoots average 3 to 5 in. long. Drench volumes and mg a.i. vary with pot size.
	To control plant growth	Concise	3 to 15 ppm spray	Apply when shoots average 3 in. tall. It is best to make only one foliar application per crop.
			0.03 to 0.06 mg a.i. (0.23 to 0.5 ppm) drench for a 6-in. pot; apply 4 fl. oz./6-in. pot	Apply when shoots average 3 in. tall. Use lower rates on cultivars such as Nellie White and higher rates for Ace. For Florida only: use a solution concentration of between 0.05 to 0.12 mg a.i. (0.4 to 1.0 ppm) drench for a 6-in. pot (0.11 to 0.26 fl. oz./gal of drench solution, apply 4 fl. oz./6-in. pot).
		Concise/Sumagic	3 to 15 ppm spray	Apply when shoots average 3 in. tall.
			0.03 to 0.06 mg a.i. (0.25 to 0.5 ppm) drench for a 6-in. pot; apply 4 fl. oz./6-in. pot	Drench volumes and mg a.i. vary with pot size.
	To prevent leaf yellowing	Fresco/Fascination	5 to 10 ppm spray	Apply early season (7 to 10 days PRIOR to visible bud stage) and mid-season (7 to 10 days AFTER visible bud stage). Apply spray only to lower leaves to minimize stem elongation. See label.
	To prevent leaf yellowing and prolong flowering	Fresco/Fascination	100 ppm spray	Apply late season (when first bud reaches at least 3 in. in length) and no more than 14 days prior to placement in a cooler or shipping. Apply to foliar and flower buds. See label.

Chapter VIII — 2019 N.C. Agricultural Chemicals Manual

Table 8-5. Growth Regulators for Floricultural Crops in Greenhouses

CROP	PURPOSE	CHEMICAL	RATE*	PRECAUTIONS AND REMARKS
LILY, Hybrid	To control plant growth	Piccolo/Piccolo 10 XC/ Bonzi/Paczol	200 to 500 ppm spray	See CALADIUM.
		Piccolo/Bonzi/Paczol	5 to 30 ppm bulb soak	Soak bulbs in the solution for 15 min. prior to planting.
		Piccolo/Piccolo 10 XC/ Bonzi/Paczol/Downsize	0.25 to 0.5 mg a.i. (4 to 30 ppm) drench for a 6-in. pot; apply 4 fl. oz./6-in. pot	Single drench should be applied when shoots average 3 to 5 in. long. Drench volumes and mg a.i. vary with pot size and cultivar.
		Concise	2.5 to 20 ppm spray	Conduct a trial to determine optimal rates for each cultivar and adjust the rate as needed. Spray when shoots average 3 in. tall. If a second application is needed or a split application is made, it should be applied when the shoots average 6 in. tall. Usually two applications of foliar sprays at a lower rate are more effective than one application at a higher rate. Avoid applications after visible bud stage.
			1 to 3 ppm drench	Drench volume varies with pot size. Applications should be made when newly emerged shoots are 1 to 2 in. tall.
			1 to 10 ppm bulb soak	Treatment soak time should range from 1 to 5 minutes. Soak time will vary depending on bulb size, cultivar, and final desired height. Lower rates may require longer soak times (5 to 10 minutes) than higher rates (1 minute).
		Concise/Sumagic	3 to 15 ppm spray	Apply when shoots average 3 in. tall.
			0.03 to 0.06 mg a.i. (0.25 to 0.5 ppm) drench for a 6-in. pot; apply 4 fl. oz./6-in. pot	Drench volumes and mg a.i. vary with pot size.
		Topflor	0.25 to 0.5 mg a.i. (2.1 to 4.2 ppm) drench for a 6-in. pot	Based on NC State University trials. Adjust rates for other locations and plant response.
	To prevent leaf yellowing	Fresco/Fascination	5 to 10 ppm spray	Apply early season (7 to 10 days PRIOR to visible bud stage) and mid-season (7 to 10 days AFTER visible bud stage). Apply spray only to lower leaves to minimize stem elongation. See label.
	To prevent leaf yellowing and prolong flowering	Fresco/Fascination	100 ppm spray	Apply late season (when first bud reaches at least 3 in. in length) and no more than 14 days prior to placement in a cooler or shipping. Apply to foliar and flower buds. See label.
LILY, Oriental	To control plant growth	Piccolo/Bonzi/Paczol	100 to 200 ppm bulb soak	Ten-minute preplant soaks provided excellent results in NC State University trials. Cultivar response varied.
		Concise	2.5 to 10 ppm spray	See Concise label comments for Hybrid lilies.
			1 to 10 ppm bulb soak	See Concise label comments for Hybrid lilies.
		Concise/Sumagic	1 to 10 ppm bulb soak	See Concise label comments for Hybrid lilies. Ten-minute preplant soaks of 5 ppm provided excellent results in NC State University trials. Cultivar response varied.
		Piccolo 10 XC	200 to 500 ppm spray	Begin spray applications when plants reach a height of 2 to 4 inches.
			4 to 30 ppm drench	Drench volume varies with pot size. Begin drench applications when plants reach a height of 1 to 2 inches.
		Topflor	0.5 mg a.i. drench (4.2 ppm); apply 4 fl. oz./6-in. pot	Optimal rate based on NC State University trials. Adjust rate for plant vigor. Drench volumes and mg a.i. vary with pot size.
			25 ppm bulb soak	Ten-minute preplant soaks provided excellent results in NC State University trials. Cultivar response varied.
	To prevent leaf yellowing	Fresco/Fascination	100 ppm spray	Apply early season (7 to 10 days PRIOR or AFTER visible bud stage). Apply spray only to lower leaves to minimize stem elongation. See label.
	To prevent leaf yellowing and prolong flowering	Fresco/Fascination	100 ppm spray	Apply late season (no more than 14 days prior to placement in a cooler or shipping). Apply to foliar and flower buds. See label.
LINARIA HYBRIDA (Baby snapdragon)	To control plant growth	Dazide + Citadel/B-Nine + Cycocel	2,500 ppm Dazide/B-Nine + 300 to 500 ppm Citadel/Cycocel applied as a tank-mix spray	Controlled plant growth, but didn't strengthen stems, as well as paclobutrazol sprays.
		Piccolo/Piccolo 10 XC/ Bonzi/Paczol	10 to 30 ppm spray	Use 10 ppm 1 week after transplant. Make a second application of 20 to 30 ppm once the secondary shoots are 2-in. long. Strengthened stems and improved flower coloration.
LINER DIPS	To control plant growth	Piccolo	0.5 to 8 ppm preplant liner dip	See label: for detailed recommendations for chemical application techniques, adjusting rates for northern or southern locations, and the specific rates for achieving the desired level of activity.
LIPSTICK VINE	To increase lateral branching	Augeo	521 to 1,042 ppm spray	

Table 8-5. Growth Regulators for Floricultural Crops in Greenhouses

CROP	PURPOSE	CHEMICAL	RATE*	PRECAUTIONS AND REMARKS
LISIANTHUS (Eustoma)	To control plant growth	Abide/A-Rest	0.5 mg a.i. drench	Cultivar response varies.
		Concise/Sumagic	5 to 10 ppm spray	Cultivar response varies.
		Dazide/B-Nine	2,500 to 5,000 ppm spray	Cultivar response varies.
		Piccolo/Piccolo 10 XC/ Bonzi/Paczol	4 to 16 ppm drench	Cultivar response varies.
LOBELIA	To control plant growth	Dazide/B-Nine	1,500 to 2,500 ppm spray	
		Concise/Sumagic	1 to 10 ppm spray	
		Piccolo/Piccolo 10 XC/	4 ppm spray	
		Bonzi/Paczol	1 ppm drench	Can be used 3 to 5 weeks before sale to control stretch.
LOBULARIA	To control plant growth	Piccolo	4 to 8 ppm liner root soak	See BACOPA. Rate based on North Carolina State University trials with Snow Princess.
			75 to 100 ppm spray	Sprays less effective than preplant liner soaks or substrate drenches. Rate based on North Carolina State University trials with Snow Princess.
			2 to 4 ppm drench	Drench volume varies with pot size. Rate based on North Carolina State University trials with Snow Princess.
	To control plant growth	Concise	0.5 to 1 ppm liner root soak	See BACOPA. Rate based on North Carolina State University trials with Snow Princess.
			20 to 25 ppm spray	Sprays less effective than preplant liner soaks or substrate drenches. Rate based on North Carolina State University trials with Snow Princess.
			1 to 2 ppm drench	Drench volume varies with pot size. Rate based on North Carolina State University trials with Snow Princess.
		Topflor	10 ppm spray	
Lophsopermum (Lofus)	To control plant growth and improve branching	Collate/Florel	250 to 500 ppm spray	Cultural requirements vary with the cultivar grown. Many cultivars only require high light, optimal growing conditions and regular pinching to control growth. Use a PGR if needed. Multiple applications may be needed in warmer climates. Avoid applications within 8 weeks of sale to ensure flowering is not delayed.
MANDEVILLA SANDERI (Dipladenia)	To control plant growth	None	None	Cultural requirements vary with the cultivar grown. Many cultivars only require high light, optimal growing conditions, and regular pinching to control growth.
		Dazide/B-Nine	2,500 to 3,500 ppm spray	Use a PGR if needed. Multiple applications may be needed in warmer climates.
		Dazide/B-Nine + Citadel/Cyclocel	1,000 to 1,500 ppm Dazide/B-Nine + 750 ppm Citadel/Cycocel spray	Use a PGR if needed. Multiple applications may be needed in warmer climates.
MARIGOLD	To control plant growth	Abide/A-Rest	13 to 33 ppm spray	
		Dazide/B-Nine	2,500 to 5,000 ppm spray	
		Piccolo/Piccolo 10 XC/ Bonzi/Paczol	15 to 60 ppm spray	See remarks for AGERATUM. Use 15 to 30 ppm for French type and 30 to 60 ppm for African type (apply at an early stage of plant growth for African type with good stem coverage, especially for vigorous varieties).
		Citadel/Chlormequat E-Pro/ Cyclocel	800 to 1,500 ppm spray	
	To control plant growth	Concise/Sumagic	10 to 20 ppm spray	
		Topflor	20 to 60 ppm spray	Based on NC State University trials. Adjust rates for other locations.
MARIGOLD, Plugs	To control plant growth	Piccolo/Piccolo 10 XC/ Bonzi/Paczol	5 to 20 ppm spray	Timing of application should normally begin at the 1 to 2 true leaf stage. Use 5 to 10 ppm for French types and 10 to 20 ppm for African types.
MATTHIOLA, Bedding Plant (Stock)	To control plant growth	Dazide + Citadel/B-Nine + Cycocel	800 to 5,000 ppm Dazide/B-Nine + 1,000 to 1,500 ppm Citadel/Cycocel applied as a tank-mix spray	
MATTHIOLA, Cut (Stock)	To promote growth and stem elongation	Florgib/ProGibb T&O	50 to 100 ppm spray	Apply when plants are 4 to 8 in. tall. Apply at 2- to 3-week intervals. See label for precautions.
MELAMPODIUM	To control plant growth	Dazide/B-Nine	2,500 ppm spray	Use when plants reach 75% of marketable size to tone.
MIMULUS	To control plant growth	Dazide/B-Nine	2,500 ppm spray	Use if needed. Delay in flowering possible with multiple applications.
MONARDA	To control plant growth	Piccolo/Piccolo 10 XC/ Bonzi/Paczol	60 to 160 ppm spray	
		Piccolo/Piccolo 10 XC/ Bonzi/Paczol/Downsize	>0.48 mg a.i. (>4 ppm) drench for a 6-in. pot; apply 4 fl. oz./6-in. pot	
		Concise/Sumagic	15 to 30 ppm spray	

Table 8-5. Growth Regulators for Floricultural Crops in Greenhouses

CROP	PURPOSE	CHEMICAL	RATE*	PRECAUTIONS AND REMARKS
MONSTERA	To control plant growth	Abide/A-Rest	25 to 132 ppm spray	
			0.25 to 0.5 mg a.i. (2 to 4 ppm) drench for a 6-in. pot; apply 4 fl. oz./6-in. pot	Drench volumes and mg a.i. vary with pot size.
MONTBRETIA	To control plant growth	Piccolo/Bonzi/Paczol	20 to 30 ppm corm soak	Soak corms in the solution for 15 min. prior to planting.
NARSiSSUS	To control plant growth	Florel/Collate	500 to 2,000 ppm spray	For types requiring a vernalization period (Narcissus hybrids), apply when new leaves reach 3 to 4 in. of height. For paperwhite narcissus (Narcissus tazetta), apply 2,000 ppm when the new leaves are 3- to 4-in. tall. Cultivar response varies, so conduct your own trial to determine suitable concentrations. Results based on Cornell University trials.
NASTURTIUM	To control plant growth	Citadel/Chlormequat E-Pro/ Cyclocel	800 to 1,500 ppm spray	Use only on non-food plants.
NEMESIA	To control plant growth	Dazide/B-Nine	2,500 to 5,000 ppm spray	Use on compact varieties to tone and hold crop.
		Piccolo/Bonzi/Paczol	10 to 20 ppm spray	Based on NC State University trials.
		Collate/Florel	250 to 500 ppm spray	Make final application 4 to 6 weeks before sale.
		Concise/Sumagic	3 to 30 ppm spray	In NC State University trials, 5 ppm worked well on Vanilla Sachet.
		Topflor	2.5 to 5 ppm spray	Recommendation based on NC State University trials with Vanilla Sachet.
NEPHTYTIS, Green and Green Gold	To control plant growth	Abide/A-Rest	25 to 132 ppm spray	
			0.25 to 0.5 mg a.i. (2 to 4 ppm) drench for a 6-in. pot; apply 4 fl. oz./6-in. pot	Drench volumes and mg a.i. vary with pot size.
NEW GUINEA IMPATIENS	To control plant growth	Piccolo/Piccolo 10 XC/ Bonzi/Paczol	0.25 to 15 ppm spray	Apply 2 to 4 weeks after transplanting. Cultivars' response to PGRs varies greatly. Test a few plants to determine rate for optimal control.
		Piccolo/Bonzi/Paczol	0.25 to 2 ppm drench	Drench volumes vary with pot size. See label for recommendations. Cultivars response to PGRs varies greatly. Test a few plants to determine rate for optimal control.
		Florel/Collate	100 to 300 ppm spray	To increase lateral branching and reduce premature flowering, don't apply within 8 weeks of desired flower date.
		Topflor	5 to 15 ppm spray	Apply 2 to 4 weeks after transplanting. Cultivars' response to PGRs varies greatly. Test a few plants to determine rate for optimal control.
NEW GUINEA IMPATIENS, Plugs	To control plant growth	Piccolo 10 XC	0.25 to 5 ppm spray	See Piccolo remarks for AGERATUM, Plugs.
NICOTIANA	To control plant growth	Dazide/B-Nine	2,500 to 5,000 ppm spray	Higher initial rates can be used after the plant becomes established. Use lower rate with multiple applications at 3-week interval.
NOLANA PARADOXA	To control plant growth	Florel/Collate	500 ppm spray	To keep plants more compact. Based on Texas A&M University trials.
OENOTHERA	To control plant growth	Concise/Sumagic	5 to 10 ppm spray	Apply if needed.
ORNAMENTAL CABBAGE and KALE (Non-food)	To control plant growth	Dazide/B-Nine	2,500 to 5,000 ppm spray	Use the higher rates for more vigorous types/cultivars. Multiple applications may be needed. Recommendation based on North Carolina conditions.
		Concise/Sumagic	2.5 to 8 ppm spray	Use higher rates for more vigorous cultivars. Cultivar response can vary. Recommendation based on North Carolina conditions.
ORNAMENTAL PEPPERS (Capsicum) (Non-food)	To control plant growth	Piccolo/Bonzi/Paczol	20 ppm foliar spray	Recommendation based on North Carolina conditions for a moderately vigorous cultivar.
		Concise/Sumagic	5 to 15 ppm spray	
ORNAMENTAL VEGETABLES (Non-food)	To control plant growth	Dazide/B-Nine	2,500 to 5,000 ppm spray	Use the higher rates for more vigorous types/cultivars like kale Red Bor. Multiple applications may be needed. Recommendation based on North Carolina conditions.
		Concise/Sumagic	10 to 25 ppm spray	Use higher rates for more vigorous cultivars. Recommendation based on North Carolina conditions.
ORNITHOGALUM	To increase stem length	Florgib/ProGibb T&O	100 ppm dip	Soak the bulbs for 20 minutes prior to potting.

Chapter VIII—2019 N.C. Agricultural Chemicals Manual

Table 8-5. Growth Regulators for Floricultural Crops in Greenhouses

CROP	PURPOSE	CHEMICAL	RATE*	PRECAUTIONS AND REMARKS
OSTEOSPERMUM	To control plant growth	Citadel/Cycocel	750 to 1,500 ppm spray	Two applications may be required. Two applications of 1,500 ppm (with the first applied at the start and the second at the end of the vernalization period) provided excellent results in NC State University trials.
			1,500 to 3,000 ppm drench	Drench volumes vary with pot size. See label for recommended volumes.
		Concise/Sumagic	8 ppm spray	Recommendation based on European trials on a cultivar with prostrate growth. Rates less than 24 ppm were not effective in NC State University trials.
			0.25 to 2 ppm drench; apply 3 fl. oz./5-in pot	One application of 1 to 2 ppm (at the start of vernalization) or two applications of 1 ppm (at the start of vernalization) and 0.5 ppm (at the end of the vernalization period) provided excellent results in NC State University trials for 4.5-in. production.
		Dazide/B-Nine	2,500 to 4,000 ppm spray	Can be applied 3 or 4 times (weekly) after pinch.
		Dazide + Citadel/B-Nine + Cycocel	1,500 to 3,000 ppm Dazide/B-Nine + 1,000 to 1,500 ppm Citadel/Cycocel applied as a tank-mix spray	Multiple sprays required. Stop applications after visible bud to avoid flower delay and smaller flowers. Not effective in NC State University trials.
		Piccolo	4 to 8 ppm liner root soak	See BACOPA. Rate based on Michigan State University trials.
		Piccolo/Bonzi/Paczol	27 to 54 ppm drench (8 to 16 mg a.i.) during production	Drench volumes vary with pot size. See label for recommended volumes. (based on NC State University trials)
			2 to 3 ppm drench (0.236 to 0.35 mg a.i.) for holding plants	
		Piccolo/Piccolo 10 XC/ Bonzi/Paczol	15 to 30 ppm spray	
		Topflor	20 to 60 ppm spray	
			1 to 2 ppm drench; apply 3 fl. oz./5-in pot	One application of 1 to 2 ppm (at the start of vernalization) or two applications of 1 ppm (at the start of vernalization) and 0.5 ppm (at the end of the vernalization period) provided excellent results in NC State University trials for 4.5-in. production.
OTACANTHUS	To control plant growth	Dazide/B-Nine	2,500 ppm spray	Make first application when new growth appears after pinching. A second application may be used if a second pinch is planned.
OTOMERIA	To control plant growth	Dazide/B-Nine	1,700 ppm spray	Apply 1 to 3 times if needed to tone the plant.
OXALLIS	To control plant growth	Abide/A-Rest	33 ppm spray	To limit petiole stretch.
		Concise/Sumagic	0.1 mg a.i. /4.5-in. pot drench	
		Piccolo/Piccolo 10 XC/	1 to 4 ppm sprays	Rates for *O. regnellii*.
		Bonzi/Paczol	1 to 10 ppm preplant dip	Dip for 5 minutes. Rates for *O. regnellii*.
PANSY	To control plant growth	Abide/A-Rest	3 to 15 ppm spray	See AGERATUM.
		Piccolo/Piccolo 10 XC/ Bonzi/Paczol	5 to 15 ppm spray	Apply when plants are 2 in. in diameter. Use higher rates for higher temperatures and more vigorous cultivars. Late applications may delay flowering.
		Concise/Sumagic	1 to 6 ppm spray	Apply when plants are 3 to 4 in. tall. Use higher rates for higher temperatures and more vigorous cultivars. Late applications may delay flowering.
		Topflor	2.5 to 7.5 ppm spray	Based on NC State University trials. Adjust rates for other locations. Pansies are very responsive to Topflor, so start trials with lower rates.
PANSY PLUGS	To control plant growth	Piccolo/Piccolo 10 XC/ Bonzi/Paczol	1 to 5 ppm spray	Timing of application should normally begin at the 1 to 2 true leaf stage. Pansies are sensitive as plugs, so determine optimal rates.
PENNISETUM GLAUCUM	To control plant growth	Collate/Florel	500 ppm spray	Apply first application 4 weeks after sowing or 1 week after transplant. If needed, a second application can be made 10 to 14 days later. Promotes side shoot production more than providing height control.
		Piccolo/Piccolo 10 XC/	6 to 8 ppm drench	For direct-sown seed, apply palcobutrazol 4 weeks after sowing. A second application possible 10 days later, if needed.
		Bonzi/Paczol	3 to 5 ppm drench	For plugs, apply 1 week after transplant.
***PENNISETUM SETACEUM* 'Rubrum'**	To control plant growth	Concise/Sumagic	5 ppm spray	First application can be made 21 days after transplanting. Repeat if needed 14 days later.
PENSTEMON HARTWEGII	To control plant growth	Citadel+Dazide/Cycocel+B-Nine	1,000 ppm + 2,500 ppm spray	Rates for moderately vigorous cultivars. Up to 2 sprays may be needed.
		Concise/Sumagic	5 to 10 ppm spray	Rates for moderately vigorous cultivars. Up to 2 sprays may be needed.
		Dazide/B-Nine	2,500 ppm spray	Rates for moderately vigorous cultivars. Up to 2 sprays may be needed.
		Florel/Collate	Spray	Not recommended because of flower delay.
PENTAS	To control plant growth	Abide/A-Rest	2 to 4 ppm spray	
		Citadel/Cycocel	1,000 to 1,500 ppm spray	
		Dazide/B-Nine	2,500 to 5,000 ppm spray	
		Piccolo/Piccolo 10 XC/ Bonzi/Paczol	2 to 3 ppm spray	

PLANT GROWTH REGULATORS

Table 8-5. Growth Regulators for Floricultural Crops in Greenhouses

CROP	PURPOSE	CHEMICAL	RATE*	PRECAUTIONS AND REMARKS
PEPINO	To control plant growth	Sumagic	2 to 10 ppm spray	See precautions listed with EGGPLANT.
PEPPER	To control plant growth	Sumagic	2 to 10 ppm spray	See precautions listed with EGGPLANT.
PERENNIALS (Not specifically listed in this table)	To induce lateral or basal branching	Configure	50 to 500 ppm spray	The supplemental label allows legal use on greenhouse grown plants not specifically listed on the original label. See label for trialing suggestions and precautions.
PERICALLIS (Cineraria)	To control plant growth	Dazide/B-Nine	2,000 ppm spray	Apply every 14 days, if needed.
PERILLA	To control plant growth	Concise/Sumagic	3 to 5 ppm spray	Apply if needed.
		Dazide/B-Nine	2,000 to 4,000 ppm spray	Apply 1 to 3 times as needed.
		Dazide + Citadel/B-Nine + Cycocel	2,500 to 4,000 ppm + 1,000 to 1,500 ppm Citadel/Cycocel applied as a tank-mix spray	
		Piccolo/Bonzi/Paczol	10 to 20 ppm spray	
PETUNIA, Seed	To control plant growth	Abide/A-Rest	10 to 26 ppm spray	See AGERATUM.
		Dazide/B-Nine	2,500 to 5,000 ppm spray	
		Piccolo/Piccolo 10 XC/ Bonzi/Paczol	15 to 45 ppm spray	
		Concise/Sumagic	25 to 50 ppm spray	
		Topflor	20 to 60 ppm spray	Based on NC State University trials. Adjust rates for other locations.
PETUNIA PLUGS, Seed	To control plant growth	Piccolo/Piccolo 10 XC/ Bonzi/Paczol	5 to 10 ppm spray	Timing of application should normally begin at the 1 to 2 true leaf stage.
PETUNIA, Vegetative	To control plant growth	Abide/A-Rest	10 to 26 ppm spray	Multiple applications may be required.
		Dazide/B-Nine	2,500 to 5,000 ppm spray	
		Dazide/B-Nine + Bonzi/Piccolo/Paczol	2,500 ppm spray + 40 ppm Bonzi/Piccolo/Paczol applied as a tank-mix spray	Recommendation based on NC State University trials.
		Dazide/B-Nine + Topflor	2,500 ppm spray + 15 to 30 ppm Topflor applied as a tank-mix spray	Recommendation based on NC State University trials.
		Piccolo/Bonzi/Paczol	5 to 45 ppm spray	An application at 2 to 4 ppm can be made 1 to 2 weeks after transplanting, followed by a 20 to 30 ppm spray 2 to 3 weeks later. Cultivars' responses to PGRs vary. Test a few plants to determine rate for optimal control. Finished plants can be maintained and have prolonged shelf life when 5 to 10 ppm sprays are applied on full-grown, mature plants. Recommendations based on Michigan conditions.
		Concise/Sumagic	20 to 50 ppm spray	20 ppm worked well in NC State University trials.
		Piccolo	12 ppm liner root soak	See BACOPA. Rate based on Michigan State University trials with petunia multiflora prostrate Wave Purple.
		Topflor	15 to 60 ppm spray	Recommendation based on NC State University trials.
	To increase lateral branching	Florel/Collate	300 to 500 ppm spray	
PHALAENOPSIS Orchids	To increase flower number and earlier flowering	Configure	200 to 400 ppm spray	Apply Configure 1 week after the start of forcing (cooling). Cultivar response varies. Some cultivars are sensitive to Configure and distorted flower stalks may form, so conduct your own trials to determine suitability. Recommendation based on Michigan State University trials.
	To control inflorescence length	Concise/Sumagic	100 to 200 ppm spray	Apply when the flower spike length is 1 in. (3 cm).
		Piccolo/Piccolo 10 XC/ Bonzi/Paczol	250 ppm spray	Apply when the flower spike length is 1 in. (3 cm).
PHILODENDRON	To control plant growth/vine control	Abide/A-Rest	25 to 132 ppm spray	
			0.25 to 0.5 mg a.i. (2 to 4 ppm) drench for a 6-in. pot; apply 4 fl. oz./6-in. pot	Drench volumes and mg a.i. vary with pot size.
		Citadel/Cycocel	3,000 ppm spray	
		Dazide/B-Nine	2,500 to 7,500 ppm spray	
PHLOX DRUMMONDII	To control plant growth	Dazide/B-Nine	2,500 to 5,000 ppm spray	
PHLOX MACULATA, (Hybrid)	To control plant growth	Concise/Sumagic	5 to 10 ppm spray	
		Dazide/B-Nine	2,500 to 5,000 ppm spray	
		Topflor	10 to 15 ppm spray	
PILEA	To control plant growth	Abide/A-Rest	25 to 132 ppm spray	
			0.25 to 0.5 mg a.i. (2 to 4 ppm) drench for a 6-in. pot; apply 4 fl. oz./6-in. pot	Drench volumes and mg a.i. vary with pot size.

Chapter VIII—2019 N.C. Agricultural Chemicals Manual

Table 8-5. Growth Regulators for Floricultural Crops in Greenhouses

CROP	PURPOSE	CHEMICAL	RATE*	PRECAUTIONS AND REMARKS
PLATYCODON	To control plant growth	Abide/A-Rest	100 ppm spray	PGRs usually not required.
		Dazide/B-Nine	1,500 to 5,000 ppm spray	PGRs usually not required. High rates have been reported to cause edge burn.
PLECTRANTHUS	To control plant growth	Dazide + Citadel/B-Nine + Cycocel	1,500 to 2,500 ppm + 750 to 1,000 ppm Citadel/Cycocel applied as a tank-mix spray	Cultivars' responses to PGRs vary. Test a few plants to determine rate for optimal control. See label.
		Piccolo/Bonzi/Paczol	5 to 20 ppm spray	Cultivars' responses to PGRs vary.
PLUMBAGO AURICULATA	To control plant growth	Collate/Florel	1,000 ppm spray	Pinching plants help improve the overall form. In addition, to further enhance secondary shoots, apply PGR 1 week before pinch.
POINSETTIA	To control plant growth	Abide/A-Rest	0.06 to 0.25 mg a.i. (2 to 4 ppm) drench for a 6-in. pot; apply 4 fl. oz./6-in. pot	Drench volume and mg a.i. vary with pot size. Start with lower rates.
		Dazide/B-Nine	2,000 to 3,000 ppm spray	Not effective in NC State University studies.
		Dazide + Citadel/B-Nine + Cycocel	800 to 2,500 ppm + 1,000 to 1,500 ppm Citadel/Cycocel applied as a tank-mix spray	Use the higher rates of this tank-mix spray on stock plants and for finishing crops in very warm regions. Outside of very warm areas, use the lower rates. Late applications can delay flowering and reduce bract size.
		Piccolo/Bonzi/Paczol	10 to 30 ppm spray	Use higher rates of 15 to 45 ppm in southern Florida. Applications to slower-growing cultivars in cool climates should begin when axillary shoots are 2 to 3 in. long. For vigorous growing cultivars in warm climates, applications should begin when axillary shoots are 1.5 to 3 in. long. See label for other precautions.
		Piccolo/Bonzi/Paczol/Downsize	0.237 to 0.473 mg a.i. (0.25 to 3 ppm) drench for a 6-in. pot; apply 4 fl. oz./6-in. pot	Drenches generally have less of an effect on bract size than sprays. Drench volume and mg a.i. vary with pot size. Start with lower rates.
		Concise/Sumagic	2.5 to 10 ppm spray	Apply when the lateral shoots are 1.5 to 2.5 in. tall (about 10 to 14 days after pinching). Test for cultivar sensitivity. Multiple applications of the lower label rate may elicit short days. For Florida only: use a foliar spray concentration between 10 to 15 ppm (2.5 to 3.8 fl. oz./gal) and do not apply after October 25.
		Citadel/Chlormequat E-Pro/	800 to 1,500 ppm spray	For natural season crops in N.C., don't apply Cycocel after mid-October to November 1. Late applications can reduce bract size and delay flowering.
		Cyclocel	3,000 to 4,000 ppm drench	Drench volume varies with pot size. Consult the label for recommended volumes.
		Topflor	2.5 to 80 ppm spray	Use lower rates for less vigorous cultivars. SEE LABEL FOR ADDITIONAL RATE RECOMMENDATIONS.
			0.03 to 0.5 mg a.i. (0.25 to 4.2 ppm) drench for a 6-in. pot	
	To promote plant growth	Fascination	3 ppm spray	Use an early-season application during vegetative growth prior to the start of short days and flower initiation if promoting vegetative growth. SEE LABEL FOR ADDITIONAL PRECAUTIONS BEFORE USE.
		Fresco/Fascination	3 to 10 ppm spray	Use a late-season application to promote bract expansion. SEE LABEL FOR ADDITIONAL PRECAUTIONS BEFORE USE.
POINSETTIA, Tree	To control plant growth	Concise	2 to 3 ppm drench for a 6-in. pot	For use in Florida only: Apply when the lateral shoots are 1.5 to 2.5 in. tall (about 10 to 14 days after pinching). Test for cultivar sensitivity. Do not apply after October 25.
PORPHYROCOMA POHLIANA (Brazilian Fireworks)	To improve foliage color and for earlier flowering	Piccolo/Piccolo 10 XC/	3 to 5 ppm spray	Height control generally not needed and rates above 5 ppm can cause leaf puckering.
		Bonzi/Paczol		
PORTULACA OLERACEA	To control plant growth	Abide/A-Rest	7 to 26 ppm spray	
		Concise/Sumagic	15 to 30 ppm spray	
		Piccolo/Piccolo 10 XC/ Bonzi/Paczol	5 ppm drench	Apply 7 days after transplant. May replace the need to pinch.
		Topflor	30 ppm spray	Apply 7 days after transplant. Repeat 2 weeks later, if needed.
	To increase lateral branching	Citadel/Cyclocel	5,000 ppm spray	Apply 5 to 6 days after pinching to improve branching of cuttings.
		Collate/Florel	300 to 500 ppm spray	Recommendations based on Michigan conditions. Defoliation can occur with rates greater than 300 ppm.
POTHOS	To control plant growth	Abide/A-Rest	25 to 132 ppm spray	
			0.25 to 0.5 mg a.i. (2 to 4 ppm) drench for a 6-in. pot; apply 4 fl. oz./6-in. pot	Drench volumes and mg a.i. vary with pot size.
		Dazide/B-Nine	2,500 to 7,500 ppm spray	
		Piccolo/Piccolo 10 XC/ Bonzi/Paczol	4 to 6 mg a.i. drench for an 8-in. pot; apply 10 fl. oz./8-in. pot	
PRIMULA ACAULIS	To control plant growth	Dazide/B-Nine	1,000 to 2,500 ppm spray	PGRs usually not required.
PRIMULA OBCONICA	To control plant growth	Dazide/B-Nine	5,000 ppm spray	PGRs usually not required.

Table 8-5. Growth Regulators for Floricultural Crops in Greenhouses

CROP	PURPOSE	CHEMICAL	RATE*	PRECAUTIONS AND REMARKS
PURPLE CONEFLOWER	To control plant growth	Concise/Sumagic	30 to 40 ppm spray	
PURPLE PASSION	To control plant growth	Abide/A-Rest	26 to 132 ppm spray	
			0.25 to 0.5 mg a.i. (2 to 4 ppm) drench for a 6-in. pot; apply 4 fl. oz./6-in. pot	Drench volumes and mg a.i. vary with pot size.
RANUNCULUS	To control peduncle length	Dazide/B-Nine	2,500 to 5,000 ppm spray	Make first application after 4 weeks. Repeat at lower rate every 2 weeks if needed. 3 to 4 applications may be needed. Conduct trials to determine optimal concentrations and timing.
ROSE, Pot	To control plant growth	Concise/Sumagic	0.1 to 0.2 mg a.i./pot drenches	Usually only a single application is made.
		Piccolo/Piccolo 10 XC/	16 to 25 ppm sprays	Begin applications after the final pinch. Make the first one in 14 to 21 days. Repeat weekly if needed. Discontinue applications after visible bud.
		Bonzi/Paczol		
SALVIA, Annual	To control plant growth	Abide/A-Rest	10 to 26 ppm spray	
		Dazide/B-Nine	2,500 to 5,000 ppm spray	
		Piccolo/Piccolo 10 XC/	20 to 60 ppm spray	
		Bonzi/Paczol		
		Citadel/Chlormequat E-Pro/ Cyclocel	800 to 1,500 ppm spray	
		Concise/Sumagic	5 to 10 ppm spray	
		Topflor	20 to 80 ppm spray	Based on NC State University trials. Adjust rates for other locations.
SALVIA PLUGS, Annual	To control plant growth	Piccolo/Piccolo 10 XC/	5 to 10 ppm spray	Timing of application should normally begin at the 1 to 2 true leaf stage.
		Bonzi/Paczol		
SALVIA FARINACEA	To control plant growth	Citadel+Dazide/Cycocel+B-Nine	1,000 ppm + 2,500 ppm spray	Apply if growth control is needed.
		Florel/Collate	Spray	Not recommended because of flower delay.
SALVIA GUARANITICA	To control plant growth	Citadel+Dazide/Cycocel+B-Nine	1,000 to 1,500 ppm + 2,000 to 3,500 ppm spray	
SALVIA HYBRID	To control plant growth	Dazide/B-Nine	1,500 to 2,500 ppm spray	
		Piccolo/Piccolo 10 XC/	0.5 to 1 ppm drench	
		Bonzi/Paczol/Downsize		
SALVIA LONGISPICATA x FARINACEA	To control plant growth	Dazide/B-Nine	2,500 to 3,000 ppm spray	
SALVIA PATENS	To control plant growth	Citadel+Dazide/Cycocel+B-Nine	1,000 ppm + 2,500 ppm spray	
		Piccolo/Piccolo 10 XC/	1 ppm drench	Trial rate before use.
		Bonzi/Paczol/Downsize		
SALVIA, Perennial	To control plant growth	Piccolo/Piccolo 10 XC/	40 to 60 ppm spray	
		Bonzi/Paczol		
SALVIA, Vegetative	To control plant growth	Dazide/B-Nine	1,000 to 2,000 ppm spray	Multiple applications may be needed to tone crop.
		Dazide + Citadel/B-Nine + Cycocel	2,000 to 3,500 ppm + 1,000 to 1,500 ppm Citadel/Cycocel applied as a tank-mix spray	
SANVITALIA	To control plant growth	Dazide/B-Nine	1,200 to 5,000 ppm spray	Use to tone plants. Cultivars' response to PGRs varies. Test a few plants to determine rate for optimal control.
SCAEVOLA AEMULA	To control plant growth	Concise/Sumagic	30 ppm spray	Based on NC State University trials, 30 ppm worked well. Adjust rates to other locations; test on a few plants to determine rate for optimal control.
			0.125 ppm drench (0.011 mg a.i.) for a 5-in. pot; apply 3 fl. oz./5-in. pot	Drench volumes vary with pot size. See label for recommended volumes. Scaevola is very responsive to Concise/Sumagic drenches. Test on a few plants to determine rate for optimal control. Recommendations based on NC State University trials.
		Dazide/B-Nine	2,500 ppm spray	
		Piccolo/Bonzi/Paczol	20 to 40 ppm spray	
			1 to 3 ppm drench (0.12 to 0.35 mg a.i.)	Drench volumes vary with pot size. See label for recommended volumes. Cultivars' response to PGRs varies. Start with lowest rate in your trials. Scaevolas are very responsible to paclobutrazol.
		Topflor	45 to 60 ppm spray	Recommendations based on NC State University trials.
			0.79 to 2.25 ppm drench (0.075 to 0.2 mg a.i.)	Drench volumes will vary with pot size. See label for recommended volumes. Scaevola is very responsive to Topflor. Test the lower rates on a few plants. Recommendations based on NC State University trials.
			2 to 4 ppm liner dip	Scaevola is very responsive to Topflor. Test the lower rates on a few plants. Recommendations based on NC State University trials.
	To increase lateral branching	Florel/Collate	300 to 500 ppm spray	Apply early, typically 2 to 3 weeks after pinching. Late applications can delay flowering.

Table 8-5. Growth Regulators for Floricultural Crops in Greenhouses

CROP	PURPOSE	CHEMICAL	RATE*	PRECAUTIONS AND REMARKS
SCHEFFLERA	To control plant growth	Abide/A-Rest	25 to 132 ppm spray	
			0.25 to 0.5 mg a.i. (2 to 4 ppm) drench for a 6-in. pot; apply 4 fl. oz./6-in. pot	Drench volumes and mg a.i. may vary with pot size.
		Dazide/B-Nine	2,500 to 7,500 ppm spray	
	To increase lateral branching	Augeo	3,125 ppm spray	Labeled for Schefflera arboricola only.
SCHIZANTHUS	To control plant growth	Abide/A-Rest	1 to 2 ppm spray	
		Dazide/B-Nine	1,500 to 3,000 ppm spray	
SCOPARIA	To control plant growth	Dazide/B-Nine	1,000 to 2,500 ppm spray	Use to tone plants if needed.
SCUTELLARIA JAVANICA (Skullcap)	To control plant growth	Dazide + Citadel/B-Nine + Cycocel	2,500 ppm Dazide/B-Nine + 1,000 ppm Citadel/Cycocel tank mix spray	Begin applications 2 to 3 weeks after transplanting. Repeat as needed every 2 weeks.
SEMPERVIVUM spp.	To induce offsets	Configure	100 to 400 ppm spray	Based on NC State University trials when applied 2 weeks after potting. For retail sales, 400 ppm produced the most offsets. For stock plant production, 100 to 200 ppm provided a balance between an increase in offset number and a larger offset size.
SHASTA DAISY	To control plant growth	Concise/Sumagic	15 to 30 ppm spray	
SHRIMP PLANT	To control plant growth	Abide/A-Rest	25 to 50 ppm spray	Apply after plants established.
		Dazide/B-Nine	1,000 ppm	Apply after plants established.
	To increase lateral branching	Augeo	781 to 1,562 ppm spray	
SNAPDRAGON, Seed (ANTIRRHINUM)	To control plant growth	Abide/A-Rest	10 to 26 ppm spray	
		Concise/Sumagic	25 to 50 ppm spray	
		Dazide + Citadel/B-Nine + Cycocel	800 to 1,000 ppm Dazide/B-Nine + 800 to 1,000 ppm Citadel/Cycocel applied as a tank-mix spray	
		Piccolo/Piccolo 10 XC/ Bonzi/Paczol	30 to 90 ppm spray	Apply at an early stage of plant growth with good stem coverage, especially for vigorous varieties.
SNAPDRAGON PLUGS, Seed (ANTIRRHINUM)	To control plant growth	Piccolo/Piccolo 10 XC/ Bonzi/Paczol	10 to 20 ppm spray	Timing of application should normally begin at the 1 to 2 true leaf stage.
SNAPDRAGON, Vegetative (ANTIRRHINUM)	To control plant growth	Piccolo/Bonzi/Paczol	30 to 60 ppm spray	
		Concise/Sumagic	20 to 45 ppm spray	
	To control plant growth and peduncle stretch	Dazide/B-Nine	1,500 ppm spray	Use during periods of high temperatures.
SPATHIPHYLLUM	To induce flowering	GibGro	265 ppm spray	Apply one full-coverage spray during non-seasonal bloom period (June through January). Some cultivars exhibit distorted blooms, increased petiole length and narrow leaves.
	To accelerate bloom and increase flower number	Florgib/ProGibb T&O	150 to 250 ppm spray	Use a single application approximately 9 to 12 weeks prior to expected sale date. Spray to the point of runoff and thoroughly wet all growing points.
STATICE, Cut (Limonium)	To promote plant growth and stem elongation	Florgib/ProGibb T&O	50 to 100 ppm spray	Apply when plants are 4 to 8 in. tall. Other applications can be made at 2- to 3-week intervals. See label.
	For earlier flowering and increased flowering	Florgib/ProGibb T&O	400 to 500 ppm spray	Give each plant 0.33 fl. oz. (10 ml) of solution. Use when plants are 10 in. or more in diameter (approximately 90 to 100 days after sowing). See label.
STEPHANOTIS, Pot	To tone plant growth	Dazide + Citadel/B-Nine + Cycocel	100 ppm + 100 ppm spray	Controls vine elongation and shortens days until flowering.
STOKESIA	To control plant growth	Piccolo/Piccolo 10 XC/ Bonzi/Paczol	40 to 80 ppm spray	
STREPTOCARPUS	To control plant growth	Abide/A-Rest	10 to 50 ppm spray	Rate based on Louisiana State University trial.
		Dazide/B-Nine	1,500 to 2,500 ppm spray	Supplier rate recommendation.
		Topflor	5 to 20 ppm spray	
	To delay premature bloom and promote additional plant growth	Collate	250 to 1000 ppm spray	Optimal rates varied significantly by cultivar. Conduct your own trials to determine optimal rates for each Streptocarpus series and specific cultivar. Results based on Iowa State University trial.
STROBILANTHES DYERIANUS (Persian Shield)	To control plant growth	Dazide/B-Nine	2,500 to 5,000 ppm spray	
		Piccolo/Piccolo 10 XC/ Bonzi/Paczol	30 ppm spray	

Table 8-5. Growth Regulators for Floricultural Crops in Greenhouses

CROP	PURPOSE	CHEMICAL	RATE*	PRECAUTIONS AND REMARKS
SUNFLOWER	To control plant growth	Citadel/Chlormequat E-Pro/ Cyclocel	800 to 1,500 ppm spray	
		Piccolo/Bonzi/Paczol	2 to 4 mg a.i. drench; apply 4 fl. oz./6-in. pot	Optimal rate based on NC State University trials. Adjust rate for plant vigor. Drench volumes and mg a.i. vary with pot size.
		Concise/Sumagic	16 to 32 ppm sprays	Optimal rate based on NC State University trials. Adjust rate for plant vigor.
		Topflor	30 to 50 ppm spray	
			1 to 2 mg a.i. (8.45 to 16.9 ppm) drench for a 6-in. pot	
TALINUM PANICULATUM	To control plant growth	Dazide/B-Nine	2,500 to 3,500 ppm spray	For toning the crop. Apply once after transplanting.
TECOMA STANS	To control plant growth	Dazide + Citadel/B-Nine + Cycocel	2,500 ppm Dazide/B-Nine + 1,000 ppm Citadel/Cycocel tank mix spray	Begin applications 2 to 3 weeks after transplanting. Repeat as needed every 2 weeks.
THUNBERGIA ALATA	To control stem elongation/plant growth	Dazide + Citadel/B-Nine + Cycocel	2,500 ppm Dazide/B-Nine + 1,000 ppm Citadel/Cycocel tank mix spray	Apply to cuttings in propagation.
TIBOUCHINA	To control plant growth	Dazide/B-Nine	2,500 ppm spray	
TOMATILLO	To control plant growth	Sumagic	2 to 10 ppm spray	See precautions listed with EGGPLANT.
TOMATO	To control plant growth	Sumagic	2 to 10 ppm spray	See precautions listed with EGGPLANT.
TORENIA FOURNIERI	To control plant growth	Concise/Sumagic	5 to 15 ppm spray	Apply if growth control is needed.
		Dazide/B-Nine	1,500 to 2,500 ppm spray	Apply if growth control is needed.
TORENIA spp.	To control plant growth	Dazide/B-Nine	1,500 ppm spray	Apply if growth control is needed
		Florel/Collate	Avoid use	Florel and Collate significantly delay flowering.
TROPICAL PLANTS (Not specifically listed in this table)	To induce lateral or basal branching	Configure	50 to 500 ppm spray	The supplemental label allows legal use on greenhouse-grown plants not specifically listed on the original label. See label for trialing suggestions and precautions.
TULIP	To control plant growth	Abide/A-Rest	0.125 to 0.5 mg a.i. (1 to 4 ppm) drench for a 6-in. pot; apply 4 fl. oz./6-in. pot	Drench volumes and mg a.i. vary with pot size.
		Piccolo/Piccolo 10 XC/ Bonzi/Paczol/Downsize	0.591 to 4.732 mg a.i. (5 to 40 ppm) drench for a 6-in. pot; apply 4 fl. oz./6-in. pot	Drench volumes and mg a.i. vary with pot size. Apply drenches 1 to 5 days after forcing begins.
		Piccolo/Bonzi/Paczol	2 to 5 ppm bulb soak	Soak bulbs for 1 hr. prior to planting. Ten-minute soaks of 50 ppm (1.6 oz./gal.) provided excellent results in NC State University trials. Cultivar response varied.
		Concise/Sumagic	10 ppm bulb soak	Ten-minute preplant soaks provided excellent results in NC State University trials. Cultivar response varied.
		Topflor	0.5 to 1 mg a.i. (4.2 to 8.45 ppm) drench for a 6-in. pot	Based on NC State University trials. Adjust rates for other locations.
			80 to 100 ppm spray	
			10 to 40 ppm bulb soak	Ten-minute preplant soaks provided excellent results in NC State University trials. Cultivar response varied.
VERBENA, Annual	To control plant growth	Dazide/B-Nine	2,500 to 5,000 ppm spray	
		Piccolo/Piccolo 10 XC/ Bonzi/Paczol	15 to 30 ppm spray	
		Citadel/Chlormequat E-Pro/ Cyclocel	800 to 1,500 ppm spray	Begin applications 7 days after pinching. Repeat as needed every 2 weeks.
		Concise/Sumagic	15 to 30 ppm spray	
	To increase lateral branching	Augeo	521 to 1,042 ppm spray	
		Florel/Collate	500 ppm spray	Florel and Collate applications will provide some growth retardant effects and delay flowering. Read the label for restrictions on timing of applications.
VERBENA PLUGS, Annual	To control plant growth	Piccolo/Piccolo 10 XC/ Bonzi/Paczol	5 to 10 ppm spray	Timing of application should normally begin at the 1 to 2 true leaf stage.
VERBENA, Perennial	To control plant growth	Piccolo/Piccolo 10 XC/ Bonzi/Paczol	120 to 160 ppm spray	
		Piccolo/Piccolo 10 XC/ Bonzi/Paczol/Downsize	>0.36 mg a.i. (>3 ppm) drench for a 6-in. pot; apply 4 fl. oz./6-in. pot	

Table 8-5. Growth Regulators for Floricultural Crops in Greenhouses

CROP	PURPOSE	CHEMICAL	RATE*	PRECAUTIONS AND REMARKS
VERBENA, Vegetative	To control plant growth	Dazide + Citadel/B-Nine + Cycocel	2,000 to 3,500 ppm Dazide/B-Nine + 750 to 1,000 ppm Citadel/Cycocel applied as a tank-mix spray	See General Recommendations.
		Piccolo	8 to 12 ppm liner root soak	See BACOPA. Rate based on Michigan State University trials.
		Citadel/Chlormequat E-Pro/ Cyclocel	1,500 to 2,000 ppm spray	
		Concise/Sumagic	5 to 10 ppm spray	Apply as needed.
		Dazide/B-Nine	1,500 to 2,500 ppm spray	Do not apply within 2 weeks of a Florel or Collate application.
		Florel/Collate	250 to 300 ppm spray	Make last application 8 weeks before sale.
VERONICA	To control plant growth	Piccolo/Piccolo 10 XC/ Bonzi/Paczol	20 to 40 ppm spray	
		Concise/Sumagic	20 to 40 ppm spray	
VINCA (Catharanthus)	To control plant growth	Abide/A-Rest	5 to 18 ppm spray	
		Dazide/B-Nine	2,500 to 5,000 ppm spray	
		Citadel/Chlormequat E-Pro/ Cyclocel	800 to 1,500 ppm spray	
		Concise/Sumagic	1 to 3 ppm spray	Apply after plants reach a height of 4 in.
		Topflor	2.5 to 7.5 ppm spray	Based on NC State University trials. Adjust rates for other locations. Vinca is very responsive to Topflor, so start trials with lower rates.
VINCA VINE *(Vinca spp.)*	To increase lateral branching	Florel/Collate	500 ppm spray	Florel and Collate applications will provide some growth retardant effects and delay flowering. Read the label for restrictions on timing of applications.
VIOLA	To control plant growth	Concise/Sumagic	1 to 5 ppm spray	
WANDERING JEW	To control plant growth	Abide/A-Rest	26 to 132 ppm spray	
WOODY LANDSCAPE PLANT (Not specifically listed in this table)	To control plant growth	Abide/A-Rest	50 ppm spray	
			0.25 mg a.i. (2 ppm) drench for a 6-in. pot; apply 4 fl. oz./6-in. pot	Drench volumes and mg a.i. vary with pot size.
		Piccolo/Piccolo 10 XC/ Bonzi/Paczol	100 ppm spray	See BEDDING PLANTS.
		Piccolo/Piccolo 10 XC/ Bonzi/Paczol/Downsize	0.47 mg a.i. (4 ppm) drench for a 6-in. pot; apply 4 fl. oz/6-in. pot	
		Concise/Sumagic	10 to 50 ppm spray	
			1 to 2 ppm drench	
ZINNIA	To control plant growth	Abide/A-Rest	7 to 26 ppm spray	
		Citadel/Chlormequat E-Pro/ Cyclocel	800 to 1,500 ppm spray	
		Concise/Sumagic	5 to 25 ppm spray	
		Dazide/B-Nine	2,500 to 5,000 ppm spray	Multiple applications may be required. Use higher rates for summer crops.
		Piccolo/Piccolo 10 XC/ Bonzi/Paczol	15 to 45 ppm spray	
ZINNIA PLUGS	To control plant growth	Piccolo/Piccolo 10 XC/ Bonzi/Paczol	4 to 10 ppm spray	Timing of application should normally begin at the 1 to 2 true leaf stage.

Growth Regulators for Woody Ornamental Crops

A. V. LeBude, Horticultural Science Department

Table 8-6. Growth Regulators for Woody Ornamental Crops

Purpose	Chemical	Rate	Precautions and Remarks
Azalea			
To produce compact plants.	succinic acid (daminozide) (B-Nine) 85% WSP	9 oz/gal water	Spray foliage to runoff. Apply during early part of July. Follow label instructions.
Woody Ornamentals			
To stimulate rooting of cuttings.	K-IAA (Rhizopon A Water Soluble Tablet) 0.0018 oz (50 mg) K-IAA	Variable	Concentration will depend on species and also on time of year rooting is to take place. Follow label instructions.
	IBA (C-mone) 1% to 2%	1% to 2%	
	K-IBA (Rhizopon AA Water Soluble Tablet) 0.0018 oz (50 mg) K-IBA	Variable	
	K-IBA (C-mone K) 1%	1%	
	K-IBA + K-NAA (C-mone K+) 1.5% L	1.5%	
	IBA + NAA (Dip 'N Grow) 1.5% L	1.5%	
	IBA (Hormex) 0.1% to 4.5% D	0.1% to 4.5%	
	IBA (Hormodin) 0.1% to 0.8% D	0.1% to 0.8%	
	IBA + thiram (Hormo-Root) 0.1% to 2% IBA + 15% thiram D	0.1% to 2% + thiram 15%	
	IBA (Rhizopon AA) 0.1% to 0.8% D	0.1 to 0.8%	
	K-NAA (Rhizopon B Water Soluble Tablet) 0.0009 oz (25 mg) K-NAA	Variable	
	NAM + thiram (Rootone) 4.24% D	4.24%	
	IBA + NAA (Wood's Rooting Compound) 1.54% L	1.54%	
To promote lateral branching and produce compact plants in various species and suppress flowering and fruit formation in various species.	dikegulac sodium (Atrimmec) 20% L; (Augeo) (Pinscher) 18.5%	Depending on species, size, vigor of plants, and specific use	Amount used and concentration will vary depending on a number of factors. Follow label directions.
To control height on a wide variety of woody landscape plants (container grown in greenhouses or shadehouses or in landscapes) using both spray or drench applications.	paclobutrazol (Bonzi) (Piccolo) 0.4% L; (Profile 2 SC) 21.8%;	variable	Amount used and concentration will vary depending on a number of factors. Follow label directions.
To reduce or eliminate undesirable fruit development on many ornamental trees and shrubs such as apples, cottonwood, crabapples, elm, flowering pear, horse chestnut, maples, oaks, pines, sour orange, sweetgum, and sycamore.	ethephon (Florel) 3.9% L; (Ethephon 2 SL) 21.7%	1 qt/10 gal water (3 oz/gal)	Timing is extremely critical. Application must be made prior to fruit set; so apply at the full bloom stage in sufficient water to wet (do not spray to run off). Follow label directions.
To retard regrowth of most trees (hickory, red oak, silver maple), shrubs (viburnum, glossy abelia), and vines (ajuga, periwinkle, English ivy).	chlorflurenol (Maintain CF 125)12.5% EC, A	0.33 pt to 8 pt/100 gal water	Concentration will depend on particular species. Apply after new flush of growth or after pruning and new leaves have fully expanded. Take care to confine the use of this material to the particular area treated.
To promote lateral shoot growth on vegetative plants of azalea, cotoneaster, juniper, taxus.	methyl decanoate/octanoate (Off-Shoot-O) 45% EC	2 to 5 oz/qt water	Amount will vary with genera. Follow label for specific conditions.
To retard regrowth of most trees, (sycamore, sweetgum, willow), shrubs (pyracantha, privet), and ivy.	maleic hydrazide (Royal Slo-Gro) 21.7% SOL	1.33 gal/100 gal water	Prune plant to desired height. After regrowth occurs (as new leaves expand), apply to drip point. Uniform coverage is important for desired results.
To reduce terminal growth on shrubs (not trees) by shortening internode length. May increase quality of plants by darkening leaf color and thickening leaves and stems.	flurprimidol (Cutless) 0.33% G; (TopFlor) 0.17% G and 0.38%	Depending on height and mass of woody stems, foliage volume, species	Prune plants to desired height. Apply any time of year to top of soil or substrate and water thoroughly.

Sucker Control for Flue-Cured Tobacco

M. C. Vann, L. R. Fisher, M. D. Inman, and D. S. Whitley, Crop and Soil Sciences Department

Table 8-7. Sucker Control for Flue-Cured Tobacco

Type Chemical and Formulation	Purpose	Amount of Formulation Per Acre	Precautions and Remarks
Contact Type			
C_8–C_{10} fatty alcohol (various brands) 6.01 lb/gal	Normal sucker control	2 or 2.5 gal (4% or 5%)	Apply in 48 gallons of water per acre (4% solution) to plants in button stage with second application 3 to 5 days later at any time of day, except when plants are wet or temperature exceeds 90 degrees F or plants are wilted. Use two TG-3 nozzle tips plus a TG-5 in the center or equivalents per row with approximately 20 psi operated from 12 to 16 inches above the top of the button or stalk at 2.5 to 3 mph. Rate of second application may be increased to 2.5 gallons in 47.5 gallons of water (5% solution) unless crop is tender. Will not control suckers more than 1 inch long. Excess nitrogen increases the chance of leaf drop.
C_{10} fatty alcohol 5.72 lb/gal	Normal sucker control	1.5 gal (3%)	Apply in 48.5 gallons water per acre (3% solution) for both applications. Follow application instructions above for C_8–C_{10} alcohol.

Table 8-7. Sucker Control for Flue-Cured Tobacco

Type Chemical and Formulation	Purpose	Amount of Formulation Per Acre	Precautions and Remarks
C_8-C_{10} fatty alcohol 6 .01 lb/gal	Control of late-season sucker regrowth	2.5 gal (5%)	Apply 3 to 4 weeks after MH application if suckers begin to grow. Apply in 47.5 gallons of water (5% solution) per acre. Follow same directions as above. Will not control suckers more than 1 inch long. Do not make more than three applications of a contact per crop per season.
Systemic Type			
Maleic hydrazide [MH] Liquids, various brands 1.5 lb/gal 2.25 lb/gal	Normal sucker control	1.5 gal (1 qt/1,000 plants) 1 gal (1 qt/1,500 plants)	Rate varies with plant population. 1.5 gallons of the 1.5 pounds per gallons material assumes 6,000 plants per acre. For plant populations other than 6,000, adjust rate accordingly. Apply to plants 5 to 7 days after the last contact application. Apply in the morning, using 30 to 50 gallons of water per acre, two to three cone nozzle tips per row, and 40 to 60 psi. Effectiveness will be reduced if applied to wet plants or those that are drought stressed or wilted from too much rainfall or high temperatures. Do not make more than one application per season. Should wash-off occur within 6 hours, a single repeat application may be made. **Do not apply at higher than suggested rates or within 7 days before harvest in order to minimize mh residues.**
60% Water-Soluble Products Fair 80 SP or Sucker Stuff 60 WS		3.75 lb	Rate for 6,000 plants per acre. Adjust rate accordingly for other plant populations.
Royal MH-30 SG		4 to 5 lb	
Contact Local-Systemic Type			
flumetralin (Prime +, Flupro or Drexalin Plus) 1.2 lb/gal	Normal sucker control, power sprayer	2 qt	Mix in 49 gallons of water per acre and apply like a contact at elongated button to early flower stage with three nozzles per row (TG-3, TG-5, TG-3) at 20 psi. Remove suckers longer than 1 inch within 24 hours before application and remove missed suckers as observed later. Excess spray to the point of rundown on the soil increases the risk of carryover residues, which may stunt early growth of next crop, including tobacco if a dinitroaniline herbicide is also used. **Do not apply these products through any type of irrigation system and apply only once per season.** Rainfall within 2 hours after application may reduce effectiveness. Follow WSP requirements and other precautions and restrictions listed on product labels.
flumetralin (Prime +, Flupro or Drexalin Plus) 1.2 lb/gal	Hand application	1.2 to 2.4 qt (2.5 oz/gal water)	Mix in desired amount of water at rates shown in parenthesis and apply mixture as a coarse spray or drench to top of stalk. Apply about 0.5 ounces of mixture per plant after topping and removing suckers longer than 1 inch, but do not exceed 25 to 30 gallons per acre. See remarks above for power sprayer application and follow precautions, restrictions, and WPS requirements shown on product labels.
flumetralin (Prime +, Flupro or Drexalin Plus) 1.2 lb/gal	Control of late season sucker regrowth	2 qt	Apply only if control with MH is beginning to break down. Mix in 49 gallons water per acre and apply like a contact at 20 to 25 psi 3 to 4 weeks after MH application; will not control suckers longer than 1 inch. **To reduce the risk of soil residue carryover, do not use for late-season control if used earlier in the season.**
Systemic Type + Contact Local-Systemic Type			
maleic hydrazide (MH) + flumetralin (Prime +, Flupro or Drexalin Plus)	Normal sucker control	Full rate MH + 2 qt	See precautions and remarks for MH to determine "full rate" of MH. Mix in sufficient water to total 50 gallons per acre and apply 5 to 7 days after the last contact or when MH alone is normally applied. Apply like a contact, using three nozzles (TG-3, TG-5, TG-3) per row at approximately 20 psi. Follow precautions and restrictions on labels. **Do not apply at higher than suggested rates or within 7 days before harvest in order to minimize MH residues.**
Contact-Systemic Type			
C_{10} fatty alcohol + MH (FST-7 or Leven-38) 4 lb/gal	Normal sucker control	3 gal	Apply in 47 gallons water to plants in early flower stage (1 week after button) any time during the day except when plants are wet or temperatures exceed 90 degrees F or plants are wilted. Use three nozzles per row with tips that deliver a coarse spray and desired rate when operated at 20 psi. Operate sprayer at a speed of 2.5 to 3 miles per hour and spray 50 gallons of diluted emulsion per acre. Use a semi-coarse spray covering the top 1/3 to 1/2 of the plant and allowing the liquid to run down the stalk to the bottom of each plant. **Do not apply at higher than suggested rates or within 7 days before harvest in order to minimize MH residues.** Effectiveness will be reduced if applied to plants that are drought-stressed or wilted from too much rainfall or high temperatures.
Contact + Contact-Local Systemic			
C_8-C_{10} fatty alcohol + flumetralin (Plucker-Plus)	Normal Sucker Control	2.5 gal	Refer to the discussion above on C_8-C_{10} fatty alcohol and flumetralin Apply in 47.5 gallons of water per acre at normal timing for flumetralin application. Remove suckers longer than 1 inch within 24 hours before application and remove missed suckers as observed later. Excess spray to the point of rundown on the soil increases the risk of carryover residues, which may stunt early growth of next crop, including tobacco if dinitroaniline herbicide is also used. Do not apply these products through any type of irrigation system. Rainfall within 2 hours after application may reduce effectiveness. Follow WPS requirements and other precautions and restrictions listed on product labels. Do not make more than two applications per season so not to exceed maximum rate of flumetralin.

Yellowing Agents for Flue-Cured Tobacco

M. C. Vann, L. R. Fisher, M. D. Inman, and D. S. Whitley Crop and Soil Sciences Department

Table 8-8. Yellowing Agents for Flue-Cured Tobacco

To Increase the Rate of Yellowing

Chemical	Amount of Formulation Per Acre	Pounds Active Ingredient Per Acre	Precautions and Remarks
ethephon (Prep), (Super Boll), (Mature XL), or (Ethephon 6)	1.33 to 2.67 pt	1 to 2 lb	Use after second or third priming when remaining leaves are physiologically mature. Determine if tobacco is ready to spray by treating several representative plants at several locations with test kit (or prepare test spray by mixing 1 teaspoon of product in 1 quart of water). If test leaves begin to yellow in 24 to 72 hours, apply product to tobacco in 40 to 60 gallons water per acre as a fine spray mist (40 to 60 psi). Effectiveness may be reduced by application on cool, cloudy days, poor spray coverage, or rain within 4 hours after application. Harvest leaves within 24 to 48 hours or when they reach the desired degree of yellowness; prolonged delay in harvest may result in yield and quality loss or leaf drop. Therefore, do not spray more acreage than can be harvested before major rain is anticipated. **Do not use surfactants.**
(Oskie)	2.67 to 5.33 pt	1 to 2 lb	

Sucker Control for Burley Tobacco

M. C. Vann, L. R. Fisher, M. D. Inman, and D. S. Whitley, Crop and Soil Sciences Department

Table 8-9. Sucker Control for Burley Tobacco

Chemicals and Formulations	Amount of Formulation Per Acre	Precautions and Remarks
Contact Type		
C_6–C_{10} fatty alcohol (various brands) 6.01 lb/gal	1.5 to 2 gal[1] (3% to 4%)	Apply in button to early flower stage as coarse, low pressure (approximately 20 psi) spray directed downward on plant tops. Leaf burn may occur with high application rates and pressure, especially on tender or wilted plants when temperature exceeds 90 degrees F. High rates or reapplication may contribute to leaf drop. Application before dew dries will reduce effectiveness.
Systemic Type (maleic hydrazide [MH])		
Liquids, various brands 1.5 lb/gal 2.25 lb/gal	1.5 to 2 gal 1 to 1.33 gal	For all systemic products, apply to upper 1/3 to 1/4 of plant in 20 to 50 gallons water per acre at approximately 20 psi after topping to 8-inch leaf. Effectiveness reduced when applied to drought-stressed or wilted plants, or before dew has dried. Apply a single repeat application ONLY if wash-off occurs within 6 hours. For water-soluble products, see rate information below and read labels carefully for mixing instructions.
60% Water-Soluble Products Fair 80 SP or Sucker Stuff 60 WS	3.75 lb	
Royal MH-30 SG	4 to 5 lb	Rate for 6,000 plants per acre. Adjust rate accordingly for other plant populations.
Contact + Systemic Mixture		
(FST-7 or Leven-38) C_{10} fatty alcohol + maleic hydrazide (MH)	9 qt[1]	Apply downward on plant tops as coarse, low-pressure (approximately 20 psi) spray after topping down to 8-inch leaf. Follow precautions given above and label restrictions for both contact- and systemic-type chemicals. High rates or reapplication after wash-off may contribute to leaf drop and increase MH residues on cured tobacco.
Contact, Local-Systemic Type		
flumetralin (Prime +, Flupro or Drexalin Plus)	1 gal[1]	Apply downward on plant tops as coarse, low-pressure (approximately 20 psi) spray after topping down to 8-inch leaf. Suckers longer than 1 inch should be removed at application time and missed suckers removed when seen. Apply only once per plant per season. Excessive spray volume that causes downstalk runoff on soil increases the chance of soil residue carryover, which may stunt the growth of small grains and corn or cause early season stunting of the next tobacco crop when a dinitroanaline herbicide is also used. Rainfall within 2 hours may reduce effectiveness.
butralin (Butralin)	3 to 4 qt*	
Systemic + Contact, Local-Systemic		
maleic hydrazide (MH) + flumetralin (Prime +, Flupro or Drexalin Plus)	1/2 to full rate MH + 2 qt flumetralin[1]	Apply as tank mix downward on topped plants as coarse, low pressure (approximately 20 psi) spray at time recommended for MH application. Follow precautions given above and label restrictions for both systemic and contact, local-systemic-type chemicals. The 3/4 rate of MH (1.5 gallons for most products) tank mixed with Prime+ has given satisfactory sucker control on vigorous crops and/or those harvested more than 3 weeks after application.
maleic hydrazide (MH) + butralin (Butralin)	Full rate MH +2 qt Butralin	
Contact + Contact-Local Systemic		
C_6-C_{10} fatty alcohol + flumetralin (Plucker-Plus)	2.5 gal	Refer to the discussion above on C_6-C_{10} fatty alcohol and flumetralin. Apply in 47.5 gallons of water per acre at normal timing for flumetralin application. Remove suckers longer than 1 inch within 24 hours before application and remove missed suckers as observed later. Excess spray to the point of rundown on the soil increases the risk of carryover residues, which may stunt early growth of next crop, including tobacco if dinitroanaline herbicide is also used. Do not apply these products through any type of irrigation system. Rainfall within 2 hours after application may reduce effectiveness. Follow WPS requirements and other precautions and restrictions listed on product labels. Do not make more than two applications per season so not to exceed maximum rate of flumetralin.

[1] Mix in sufficient water to total 50 gallons spray per acre.

Growth Regulators for Fruiting Vegetables

C. C. Gunter, Horticultural Science Department

Table 8-10. Growth Regulators for Fruiting Vegetables

Crop	Chemical	Amount of Formulation	Precautions and Remarks
Ethephon 2 — To Stimulate Uniform Ripening for One Harvest			
Processing Tomatoes Early and midseason crops or warm conditions	Ethephon 2 Ethephon 2 SL (Ethrel)	1.25 to 3.25 pt/A	Apply to cover foliage and fruit uniformly when 5% to 15% of the fruit is pink and red. Apply in a minimum of 20 gallons per acre for ground application (10 gallons for aerial). Under warm temperatures (above 85 degrees F) rates as low as 1.25 pints/acre can be effective. Thorough coverage is essential. Observe treated fields closely and harvest fruit at proper maturity.
Processing Tomatoes Late season or coastal crops or cool conditions	Ethephon 2 Ethephon 2 SL (Ethrel)	3.25 to 6.5 pt/A	Apply to cover foliage and fruit uniformly when 5% to 15% of the fruit is pink and red. Apply in a minimum of 20 gallons per acre for ground application (10 gallons for aerial). Use the higher rate when nighttime temperatures are cool (below 65 degrees F) or vegetative growth is dense. Thorough coverage is essential. Observe treated fields closely and harvest fruit at proper maturity.
Pepper	Ethephon 2 Ethephon 2 SL (Ethrel)	1.25 to 4 pt/A	Apply to bell peppers when 10% have red or chocolate coloration; chili and pimento peppers when 10-30% have red to chocolate coloration. Application should not be made until enough fruit exists for sufficient yield. Product will not ripen immature, green fruit. Rates between 1.25 and 2 pints/acre should be applied in 20 gallons/acre and 3 to 4 pints/acre in 40 gallons/acre. Use of spray volumes less than 40 gallons/acre in hot dry weather may result in foliage burn. Applications should not be made at temperatures of 100 degrees F or greater. In addition, use the higher rate when nighttime temperatures are cool (below 65 degrees F) or vegetative growth is dense. Thorough coverage is essential. Observe treated fields closely and harvest fruit at proper maturity. Maximum of 4 pints/acre per year.
Sumagic — For Commercial Greenhouse, Lathhouse, and Shadehouse Use Only			
Eggplant Groundcherry Pepino Pepper Tomatillo Tomato	Sumagic (Uniconizol-P)	0.52 to 2.6 fl oz/Gal or 16 to 76 ML/Gal or 2 to 10 PPM	Apply uniformly as a foliar spray at a volume of 2 quarts/100 square feet. Initial foliar applications when 2-4 true leaves are present. Sequential applications of lower rates provide more growth control than a single high rate application. First-time user should apply the lowest recommended rate in order to determine optimal rate for individual cultivars under local environmental conditions. If additional growth regulation is required, a sequential spray application at the lowest recommended rate should be made at 7-14 days after initial application. Total uniconizol-P applied may not exceed that from a single application of a 10 ppm spray concentration at 2 quarts/100 feet. Final application may not occur later than 14 days after the 2-4 true leaf stage.

Growth Regulators for Peanuts

D. L. Jordan, Crop and Soil Sciences Department

Table 8-11. Growth Regulators for Peanuts

Chemical and Formulation	Amount of Formulation per Acre	Remarks
To Suppress Excessive Vegetative Growth and Reduce Pod Shed		
prohexadione calcium (Apogee) 27.5% WDG (Kudos) 27.5% WDG	7.25 + 7.25 oz	Apply Apogee or Kudos when 50% of vines from adjacent rows are touching. Follow with a repeat application 2 to 3 weeks later. Two applications are needed in most circumstances. Although a third application may be made, it is generally discouraged because of expense and possible adverse impact on peanut. Do not exceed 21.75 ounces per acre per year. The preharvest interval is 25 days. Do not apply more than 2 applications of Apogee or Kudos within 6 weeks. Apply in a minimum of 20 GPA. Always apply Apogee and Kudos with a nitrogen source. Apply one pound of spray grade ammonium sulfate for every pound of Kudos. Apply one pound of spray grade ammonium sulfate per acre or one pint of UAN (28%, 30% or 32% nitrogen solution) per acre with Apogee. Apply Apogee or Kudos with either nonionic surfactant or crop oil concentrate depending on product labels when mixing with other products.

For further information, see Extension publication *Peanut Information* (AG-331). Copies are available from your local Cooperative Extension center.

Growth Regulators for Turfgrasses

F. H. Yelverton, R. Cooper, and T. W. Gannon, Crop and Soil Sciences Department

Table 8-12. Growth Regulators for Turfgrasses

Brand	Amount of Formulation Per Acre	Pounds Active Ingredient Per Acre	Precautions and Remarks
Cool Season Grasses—Well-Maintained Turf: Seedhead and Foliar Suppression			
mefluidide (Embark) 0.2	5 pt/15 to 150 gal water	0.125	See Embark 2-S for low-maintenance cool-season turf. Follow label directions and precautions.
trinexapac-ethyl (Governor) 0.17 G	30 to 258 lb	0.05 to 0.44	Apply 30 to 41 pounds per acre to greens, 53 to 152 pounds per acre to fairways less than 0.5 inch cut, and 152 to 258 pounds per acre to residential and commercial turf. Do not exceed 2.5 pounds active ingredient per acre per year. These rates should provide 50% turf growth suppression for 4 weeks with minimal yellowing.
(Primo Maxx) 1 MEC or (T-Nex) 1 AQ (Primo WSB) 25 WP	6 to 44 fl oz 2.75 to 21.8 oz	0.085 to 0.34 0.085 to 0.34	Application rate varies with turfgrass species and height of cut. Apply to actively growing, nonstressed turf. More growth suppression occurs at lower mowing heights. See label for specific rate and other directions and precautions. Repeat applications can be made, but do not exceed a total of 21.4 pints per acre per year of Primo Maxx or a total of 174 ounces per acre per year of Primo WSB. Do not exceed a total of 19 pints per acre per year of T-Nex. Refer to the respective Primo label for guidelines regarding mowing prior to and following application. Mix with 0.5 to 4 gallons of water per 1,000 sq ft (20 to 174 gallons per acre). Primo can be applied to putting greens. See label for instructions.
Cool Season Grasses—Well-Maintained Turf: Foliar Suppression			
ethephon (Ethephon or Proxy) 2 SL	1.7 gal	3.4	May be applied to Kentucky bluegrass, perennial ryegrass, bentgrass, and tall and fine fescues. Apply in 22 to 174 gallons of water per acre. Do not use a surfactant. Plant growth regulator effect will not be seen until 7 to 10 days after application. May be reapplied to Kentucky bluegrass and perennial ryegrass at 7-week intervals. Repeat applications to bentgrass and tall and fine fescue may be made at 4-week intervals.
flurprimidol (Cutless 50 W) 50 WP	0.75 to 3 lb/50 to 200 gal water	0.37 to 1.5	Rates depend upon grass species and cultivar. Apply to bentgrass, Kentucky bluegrass, and perennial ryegrass in late spring-early summer and/or late summer-early fall. Time the second application to occur at least 3 months before expected winter dormancy. Do not apply to putting greens. Do not exceed 1.5 pounds per acre per application on coarse-textured soils. Treated areas should receive 0.5 inch of irrigation within 24 hours after application. Resume mowing 3 to 5 days after application.
flurprimidol + trinexapac-ethyl (Legacy) 1.51 SL	5 to 22 fl oz	0.059 to 0.26	Tolerant species include bentgrass greens and fairways, Kentucky bluegrass, and perennial ryegrass. Do not use on turf grown for sale or other commercial use as sod or seed production. Do not seed 3 weeks before or 3 weeks after application. Wait 6 to 8 weeks after sprigging or laying sod before applying. Use only 5 to 8 fluid ounces per acre on bentgrass greens. Repeat applications at 2- to 6-week intervals until 4 weeks before the onset of inactive growth.
paclobutrazol (TGR Turf Enhancer 2 SC or Trimmit 2 SC) 2 SC	1 to 2 pt /43 to 200 gal water	0.25 to 0.5	Apply in spring after greenup and after turf has been mowed once or twice. Apply at least 1 month before onset of high temperatures. In late summer-early fall, apply at least 1 month before anticipated first killing frost. Apply with 0.5 to 0.9 pound nitrogen per 1000 sq ft of a nonburning fertilizer. Apply 0.25 inch of water within 24 hours after application to remove product from foliage and onto soil surface. See label for special rates and directions for applications to bentgrass, putting greens, and overseeded bermudagrass. Repeat applications within the same growing season may be made but refer to label for instructions. Do not apply more than three times annually. Do not use on areas containing greater than 70% poa annua. Do not seed within 6 weeks prior to or 2 weeks after applications.
prohexadione calcium (Anuew) 27.5 WG	1.8 to 29.1 oz	0.031 to 0.5	Apply to golf course fairways, tees, greens and roughs and also athletic fields, residential and commercial lawns, sod farms, parks, cemeteries and roadsides. Apply 1.8 to 7.25 ounces per acre on bentgrass greens and tees at 1 to 2 wk intervals and 7.25 to 14.5 ounces per acre on bentgrass fairways and roughs at 2 to 4 wk intervals. Apply 14.5 to 21.8 ounces per acre on perennial bluegrass and 21.8 to 29.1 ounces per acre on perennial ryegrass at 2 to 4 wk intervals. For cool season grass sod production, apply 7.25 to 29.1 ounces per acre at 2 to 4 wk intervals. Use a spray volume of 1 to 2 gallons of water per 1000 sq ft. A nonionic surfactant may improve product performance. Do not irrigate for 4 hours after application or mow until 1 day after application.
trinexapac-ethyl (Primo Maxx) 1 MEC or (T-Nex) 1 AQ (Primo WSB) 25 WP	6 to 22 fl oz 2.75 to 10.9 oz	0.085 to 0.17 0.085 to 0.17	Application rates are for mowing heights of less than or equal to 0.5 inch Apply to actively growing, non-stressed turf. Rate varies with turfgrass species. See label for specific rate and other directions and precautions. Repeat applications can be made but do not exceed a total of 21.4 pints per acre per year of Primo Maxx or a total of 174 ounces per acre per year of Primo WSB. Do not exceed a total of 19 pints per acre per year of T-Nex. Refer to the respective Primo label for guidelines regarding mowing prior to and following applications. Mix with 0.5 to 4 gallons of water per 1,000 sq ft (20 to 174 gallons per acre). Primo can be applied to putting greens. See label for instructions.
Cool Season Grasses—Low-Maintenance Turf: Seedhead and Foliar Suppression			
chlorsulfuron (Telar DF) 75 DF + mefluidide (Embark 2-S) 2S	0.25 oz + 0.5 pt	0.012 + 0.125	For growth and seedhead suppression in fescue/bluegrass stands. Apply up until seedhead emergence. Do not apply Telar DF to turf less than 1 year old. Grass seed may be planted in treated areas 6 months after treatment but cultivation is recommended. For broadcast applications, do not exceed 0.5 ounces Telar DF per acre within a 12-month period. Telar DF alone can also be used for weed control in bahiagrass, bermudagrass, fescue, and bluegrass.
glyphosate (Touchdown Pro) 3 LC	4 to 8 fl oz/10 to 40 gal water	0.09375 to 0.1875	Touchdown Pro may be used on turf described in "GENERAL USE AREAS" section of the label. 4 to 5 ounces will suppress annual grasses, such as ryegrass, wild barley, and wild oats, growing in turf areas. 6 ounces will suppress Kentucky bluegrass and serve as a mowing substitute. 8 ounces will suppress fine fescue and tall fescue and serve as a mowing substitute. A nonionic surfactant containing at least 75% active ingredient at 0.25% v/v (1 quart per 100 gallons) or ammonium sulfate at 0.5% by weight (4.25 to 17 pounds per 100 gallons) may be added.
imazethapyr + imazapyr (Event) 1.46 lb/gal	8 to 10 fl oz	0.09 to 0.11	Apply to tall fescue, perennial ryegrass, and bluegrass only. Apply after the turf is at 100% greenup and has at least 2 inches of vertical growth. The addition of a nonionic surfactant containing at least 80% active ingredient at 0.25% v/v of the spray (2 pints per 100 gallons of spray mixture) is required. Do not use on newly established stands less than 1 year old or on highly managed turf. Do not reseed before 3 months after application. See label for herbicide tank mix options. Follow label directions and precautions.

Table 8-12. Growth Regulators for Turfgrasses

Brand	Amount of Formulation Per Acre	Pounds Active Ingredient Per Acre	Precautions and Remarks
Cool Season Grasses—Low-Maintenance Turf: Seedhead and Foliar Suppression (continued)			
maleic hydrazide (Retard) 2.25 lb/gal (Royal Slo-Gro) 1.5 lb/gal (Liquid Growth Retardant) 0.6 lb/gal	1.3 gal/50 gal water 2 gal/30 to 50 gal water 5 gal/45 gal water	3	Treat in the spring when the grass is actively growing but before seedhead appears. Applications made after seedhead appears will suppress subsequent seedheads. Do not apply to turf less than 3 years old, and do not reseed within 3 days after application. Treated turf may appear less dense and temporarily discolored. Optimum results may not be obtained if rainfall or overhead irrigation occurs within 12 hours following application. Remove excess grass clippings and fallen leaves before application. Do not add a surfactant. Follow label directions and precautions.
mefluidide (Embark 2-S) 2 S	1.5 to 2 pt/15 to 150 gal water	0.38 to 0.5	Apply after uniform spring greenup until approximately 2 weeks before seedheads appear. Do not apply to turf within 4 growing months after seeding, and do not reseed within 3 days after application. Treated turf may appear less dense and temporarily discolored. Optimum results may not be obtained if rainfall or overhead irrigation occurs within 8 hours following application. Remove excess clippings and fallen leaves before application. Adding 1 to 2 quarts of nonionic surfactant per 100 gallons of spray solution may enhance suppression; however, discoloration may also be increased. Follow label directions and precautions.
metsulfuron methyl (Escort XP) 60 DF	0.25 to 0.5 oz	0.009 to 0.018	Apply to well-established tall fescue and perennial bluegrass turf. Can tank mix with 0.125 to 0.25 pints per acre of Embark to improve pgr performance. Treat after 2 to 3 inches of new growth but before seed stalk formation. Temporary discoloration may occur. Do not use on stressed turf.
Warm Season Grasses— Well-Maintained Turf: Seedhead and Foliar Suppression			
trinexapac-ethyl (Governor) 0.17 G	12 to 258 lb	0.02 to 0.44	Apply 12 to 41 pounds per acre to greens, 30 to 77 pounds per acre to fairways less than 0.5 inch cut, and 41 to 258 pounds per acre to residential and commercial turf. Do not exceed 2.5 pounds active ingredient per acre per year. These rates should provide 50% turf growth suppression for 4 weeks with minimal yellowing.
(Primo Maxx) 1 MEC or (T-Nex) 1 AQ (Primo WSB) 25 WP	2.7 to 88 fl oz 1.35 to 43.6 oz	0.085 to 0.68 0.085 to 0.68	Application rate varies with turfgrass species and height of cut. Apply to actively growing, nonstressed turf. More growth suppression occurs at lower mowing heights. See label for specific rate and other directions and precautions. Repeat applications can be made but do not exceed a total of 21.4 pints per acre per year of Primo Maxx or 174 ounces per acre per year of Primo WSB. Do not exceed a total of 19 pints per acre per year of T-Nex. Refer to the respective Primo label for guidelines regarding mowing prior to and following application. Mix with 0.5 to 4 gallons of water per 1,000 sq ft (20 to 174 gallons per acre). Primo can be applied to putting greens. See label for directions.
mefluidide (Embark) 0.2	10 pt/15 to 150 gal water	0.25	For St. Augustinegrass. See Embark 2-S for low-maintenance warm season turf. Follow label directions and precautions.
Warm Season Grasses— Well-Maintained Turf: Foliar Suppression			
flurprimidol (Cutless 50 W) 50 WP	0.75 to 3 lb/50 to 200 gal water	0.37 to 1.5	Rates depend upon grass species and cultivar. Apply to Tifway, Tifgreen, common bermudagrass, or zoysiagrass. Treated areas should receive 0.5 inch of irrigation within 24 hours of application. Resume mowing. Overseed 2 to 3 weeks after fall application with a desired perennial ryegrass.
flurprimidol + trinexapac-ethyl (Legacy) 1.51 SL	8 to 15 fl oz	0.094 to 0.177	Tolerant species include Tifway and Tifsport bermudagrass, zoysiagrass, and seashore paspalum. Do not use on turf grown for sale or other commercial use as sod or seed production. Do not seed 3 weeks before or 3 weeks after application. Wait 6 to 8 weeks after sprigging or laying sod before applying. Repeat applications at 2- to 6-week intervals until 4 weeks before winter dormancy.
paclobutrazol (TGR Turf Enhancer 2 SC or Trimmit 2 SC) 2 SC	2 to 3 pt/43 to 200 gal water	0.5 to 0.75	Use any time when established hybrid bermudagrass and St. Augustinegrass are green, are actively growing, and have recovered from dormancy (filled in fully following winter). Apply with 0.5 to 0.9 pound nitrogen per 1,000 sq ft of a nonburning fertilizer. Apply 0.25 inch of water within 24 hours after application to remove product from foliage and onto soil surface. A repeat application within the same growing season may be made, but not sooner than 8 weeks following initial application. Do not apply more than 3 times annually. Do not use on areas containing greater than 70% *poa annua*. Refer to label to determine bermudagrass and St. Augustine cultivar response relating to sensitivity, growth, and color response. Do not seed within 6 weeks prior to or 2 weeks after application.
prohexadione calcium (Anuew) 27.5 WG	7.25 to 43.6 oz	0.125 to 0.75	Apply to golf course fairways, tees, greens and roughs and also athletic fields, residential and commercial lawns, sod farms, parks, cemeteries and roadsides. Apply 7.25 to 14.5 ounces per acre on hybrid bermudagrass greens and tees at 1 to 2 wk intervals and 29.1 to 43.6 ounces per acre on hybrid bermudagrass fairways and roughs at 2 to 4 wk intervals. For warm season grass sod production, apply 14.5 to 43.6 ounces per acre at 2 to 4 wk intervals. Use a spray volume of 1 to 2 gallons of water per 1000 sq ft. A nonionic surfactant may improve product performance. Do not irrigate for 4 hours after application or mow until 1 day after application.
trinexapac-ethyl (Primo Maxx) 1 MEC or (T-Nex) 1 AQ (Primo WSB) 25 WP	2.7 to 13 fl oz 1.35 to 6.5 oz	0.042 to 0.085 0.042 to 0.085	Application rates are for mowing heights of less than or equal to 0.5 inch Apply to actively growing, non-stressed turf. Rate varies with turfgrass species. See label for specific rate and other directions and precautions. Repeat applications can be made but do not exceed a total of 21.4 pints per acre per year of Primo Maxx or a total of 174 ounces per acre per year of Primo WSB. Do not exceed a total of 19 pints per acre per year of T-Nex. Refer to the respective Primo label for guidelines regarding mowing prior to and following applications. Mix with 0.5 to 4 gallons of water per 1,000 sq ft (20 to 174 gallons per acre). Primo can be applied to putting greens. See label for directions.
Warm Season Grasses—Low-Maintenance Turf: Seedhead and Foliar Suppression			
glyphosate (Roundup Pro) 4 lb/gal	6 fl oz/10 to 25 gal water	0.2	Apply to bahiagrass only. Apply after full greenup of the bahiagrass (about late May) and make only one application per year. Do not apply to turf less than 3 years old. Treated turf may appear less dense and temporarily discolored. Optimum results may not be obtained if rainfall or overhead irrigation occurs within 6 hours following application. This is a nonselective herbicide. If application exceeds the above recommended rates, it can result in permanent loss of turf.
(Touchdown Pro) 3 LC	0.375 to 4 pt/10 to 40 gal water	0.14 to 1.5	Touchdown Pro may be used on dormant or actively growing bermudagrass and bahiagrass turf described in "GENERAL USE AREAS" section of label. May be tank mixed with 0.25 to 2 ounces of Oust for residual weed control. Check label for correct rates. Touchdown Pro will control winter annual weeds less than 6 inches tall and also 4-to 6-leaf tall fescue in dormant turf. Use only on well-established bermudagrass. Injury may occur, but regrowth will occur under moist conditions. Bahiagrass vegetative growth and seedheads may be suppressed approximately 45 days when applied 1 to 2 wk after spring greenup and before seedhead emergence. A second application at 45 days will extend suppression to approximately 120 days.

Table 8-12. Growth Regulators for Turfgrasses

Brand	Amount of Formulation Per Acre	Pounds Active Ingredient Per Acre	Precautions and Remarks
Warm Season Grasses—Low-Maintenance Turf: Seedhead and Foliar Suppression			
imazapic (Plateau) 2 ASU	2 fl oz	0.031	Only government entities may buy Plateau. Used for bahiagrass seedhead suppression. Apply to bahiagrass in spring after full greenup but approximately 3 to 4 weeks prior to expected seedhead emergence or 7 to 10 days after mowing. Do not apply to wetlands. Add a surfactant according to label directions. Bahiagrass may appear less dense and discolored following application.
imazapic (Panoramic) 2 SL	2 to 3 fl oz	0.031	May be used for seedhead suppression of bahiagrass or tall fescue turf areas including industrial turf, golf courses, and non-residential areas. Apply 2-3 ounces/A for tall fescue seedhead suppression prior to seedhead emergence. Apply 2 ounces/acre after bahiagrass greenup but prior to emergence. Temporary turf discoloration may occur.
imazapic + glyphosate (Journey) 2.25 AS	11 to 32 fl oz	0.19 to 0.56	Use in noncrop areas. Temporary turf discoloration may occur. Apply 4 to 8 fluid ounces per acre on a small area first to determine rate needed for desired results. Do not use with methylated seed oil. Do not apply to drought-stressed turf. Apply after full turf greenup.
imazethapyr + imazapyr (Event) 1.46 lb/gal	8 to 10 fl oz	0.09 to 0.11	Apply to bahiagrass only. Apply after the turf is at 100% greenup and has at least 2 inches of vertical growth. The addition of a nonionic surfactant containing at least 80% active ingredient at 0.25% v/v of the spray (2 pints per 100 gallons of spray mixture) is required. Do not use on newly established stands less than 1 year old or on highly managed turf. Do not reseed before 3 months after application. See label for herbicide tank mix options. Follow label directions and precautions.
maleic hydrazide (Retard) 2.5 lb/gal (Royal Slo-Gro) 1.5 lb/gal (Liquid Growth Retardant) 0.6 lb/gal	1.3 gal/50 gal water 2 gal/30 to 50 gal water 5 gal/45 gal water	3 3 3	Apply to bahiagrass only. Apply in late spring but before seedheads appear. Applications made after seedhead appearance will suppress subsequent seedheads. Do not apply to turf less than 3 years old and do not reseed within 3 days after application. Treated turf may appear less dense and temporarily discolored. Optimum results may not be obtained if rainfall or overhead irrigation occurs within 12 hours following application. Remove excess grass clippings and leaves before application. Do not add a surfactant. Follow label directions and precautions. A repeat application may be needed 6 weeks after initial application.
mefluidide (Embark 2-S) 2 S	2 qt/15 to 150 gal water	1	Apply to bermudagrass only. Apply in late spring until about 2 weeks before seedhead appearance. Do not apply to turf within 4 growing months after seeding, and do not reseed within 3 days after application. Treated turf may appear less dense and temporarily discolored. Optimum results may not be obtained if rainfall or overhead irrigation occurs within 8 hours following application. Remove excess grass clippings and leaves before application. Adding 1 to 2 quarts of a nonionic surfactant per 100 gallons of spray solution may enhance suppression; however, discoloration may also be increased. Follow label directions and precautions.
sulfometuron methyl (Oust) 75 DG	0.5 oz/30 to 50 gal water	0.02 lb	Apply to bahiagrass in late spring or early summer before seedheads appear. Do not apply to wetlands or where runoff water may flow onto agricultural lands or forests. Injury of desirable trees may result if applications are made near plants or where their roots extend or may be subjected to runoff from treated areas. Do not apply to turf less than 3 years old. Treated turf may appear less dense and temporarily discolored. Do not add a surfactant. Follow label directions and precautions.
sulfometuron methyl + chlorsulfuron (Landmark MP) 50 + 25 DG (Landmark II MP) 56.25 + 18.75 DG	0.9 oz 1.0 oz	0.042 0.047	For established bermudagrass and centipede-improved turf. Temporarily suppresses foliar and seedhead growth while controlling many grass and broadleaf weeds. Apply 30 days after breaking dormancy or either late fall or early winter. Landmark MP may discolor or cause top kill of desired turf species. Do not apply to turf less than 1 year old. Annual retreatments may reduce turf vigor.
sulfometuron methyl + metsulfuron methyl (Oust Extra) 56.25 + 15 DG	0.5 to 2 oz	0.022 to 0.088	For use on well-established, unimproved bermudagrass and centipedegrass. Apply 30 days after breaking dormancy. Can also be applied in late fall or early winter depending on weed presence. Oust Extra can be tank mixed with 3 to 4 pounds active ingredient per acre MSMA on bermudagrass during the summer. Do not add a surfactant.
Annual Bluegrass: Suppression			
flurprimidol (Cutless 50 W) 50 WP	0.25 to 0.5 lb/50 to 100 gal water	0.12 to 0.25	Apply to actively growing bentgrass putting greens in spring after third or fourth mowing or in the fall. Repeat, if necessary, at 3- to 6-week intervals, not to exceed 2 pounds per acre per growing season. Delay overseeding 2 weeks after application. Make final fall application 8 weeks before onset of winter dormancy.
	1 to 1.5 lb/50 to 200 gal water	0.5 to 0.75	Apply to bentgrass, Kentucky bluegrass, and perennial ryegrass in late spring-early summer and/or late summer-early fall. Time the second application to occur at least 3 months before expected winter dormancy. Management practices that encourage vigorous growth of perennial turfgrasses following application will enhance conversion. *Poa annua* discoloration will be visible 7 to 10 days after treatment and last for 3 to 6 weeks. Do not apply to putting greens. Treated areas should receive 0.5 inch of irrigation within 24 hours after application. Resume mowing 3 to 5 days after application.
flurprimidol + trinexapac-ethyl (Legacy) 1.51 SL	5 to 30 fl oz	0.059 to 0.354	Use in cool season turfgrasses, such as bentgrass greens and fairways, Kentucky bluegrass, and perennial ryegrass. Repeat applications at 2- to 6-week intervals. Annual bluegrass suppression is gradual and could take several growing seasons. Start treatments in early spring and continue through early fall.
maleic hydrazide (Retard) 2.25 lb/gal (Royal Slo-Gro) 1.5 lb/gal (Liquid Growth Retardant) 0.6 lb/gal	1 qt/30 to 40 gal water 2 qt/30 to 40 gal water 1.25 gal/30 to 40 gal water	0.56 0.75 0.75	Treat after two normal mowings but before seedhead appears. Applications made after seedhead appears will suppress subsequent seedheads. Do not apply to golf greens. Do not apply to turf less than 3 years old, and do not reseed within 3 days after application. Treated turf may appear less dense and temporarily discolored. Optimum results may not be obtained if rainfall or overhead irrigation occurs within 12 hours following application. Remove excess grass clippings and fallen leaves before application. Do not add a surfactant. Follow label directions and precautions for use on fairways.
mefluidide (Embark 2-S) 2 S (Embark) 0.2	0.5 pt/15 to 150 gal water 2 to 5 pt/15 to 150 gal water	0.125 0.05 to 0.125	Apply after uniform greenup but before first appearance of seedheads. Do not apply to turf within 4 growing months after seeding, and do not reseed within 3 days after application. Treated turf may appear less dense and temporarily discolored. Optimum results may not be obtained if rainfall or overhead irrigation occurs within 8 hours following application. Remove excess grass clippings and leaves before application. Adding 1 to 2 quart of a nonionic surfactant per 100 gallons of spray solution enhances suppression; however, discoloration may also be increased. Follow label directions and precautions for use of fairways and tees.

Table 8-12. Growth Regulators for Turfgrasses

Brand	Amount of Formulation Per Acre	Pounds Active Ingredient Per Acre	Precautions and Remarks
Annual Bluegrass: Suppression (continued)			
paclobutrazol (31-3-9 Fertilizer with TGR *Poa annua* Control 0.42%)	128 lb	0.5	Apply only to bentgrass, Kentucky bluegrass, perennial ryegrass fairways, or bentgrass greens with less than a 70% *poa annua* infestation. Follow label directions and precautions. Note: This product supplies 0.9 pound N per 1,000 sq ft.
prohexadione calcium (Anuew) 27.5 WG	0.9 to 1.75 oz	0.015 to 0.03	Apply to overseeded hybrid bermudagrass during periods of active *poa annua* growth at 3 to 4 wk intervals.
(15-0-29 High K Fertilizer with TGR *Poa annua* Control 0.34%)	98 lb to 146 lb	0.33 to 0.5	Apply only to bentgrass, zoysiagrass, Kentucky bluegrass, and Kentucky bluegrass/perennial ryegrass fairways, tees, and roughs, as well as bentgrass greens with less than 70% *poa annua* infestation. Note: This product supplies 0.5 pound N per 1,000 sq ft.
(TGR Turf Enhancer 2 SC or Trimmit 2 SC) 2 SC	6.4 to 48 fl oz/43 to 200 gal water	0.1 to 0.75	Apply on hybrid bermudagrass, bentgrass, perennial ryegrass, and Kentucky bluegrass/perennial ryegrass fairways, tees, and roughs. Can also be applied to bentgrass putting greens. Apply in spring after greenup or regrowth has begun and after mowing once or twice. Apply with a nonburning fertilizer. Apply 0.25 inch of water within 24 hours after application to remove product from foliage and onto soil surface. See label for rates and other directions for applications to bentgrass putting greens and overseeded bermudagrass. Do not apply more than 3 times annually. Do not use on areas containing more than 70% *poa annua*. For bentgrass putting greens, do not apply more than 0.25 pound active ingredient per acre per application.
ethephon (Proxy) 2 SL	1.7 gal	3.4	May be used to suppress annual bluegrass seedheads and growth of other cool season turfgrasses including golf course greens, fairways, tees, and roughs. Do not use an adjuvant. Do not apply to stressed turfgrass or where excessive thatch is present. Scalping may occur on bentgrass surfaces after application. Consult label for repeat application intervals.
Overseeded Bermudagrass Turf: Foliar Suppression			
flurprimidol (Cutless 50 W) 50 WP	0.75 to 3 lb/50 to 200 gal water	0.37 to 1.5	Rates depend upon grass species and cultivar. Apply to zoysiagrass, Tifway, Tifgreen, and common bermudagrass in late spring-early summer and/or late summer-early fall. Time the second application to occur 8 to 10 weeks before expected winter dormancy. Do not apply to putting greens. Do not exceed 1.5 pound per acre per application on coarse-textured soils. Treated areas should receive 0.5 inch of irrigation within 24 hours after application. Resume mowing 3 to 5 days after application.
flurprimidol + trinexapac-ethyl (Legacy) 1.51 SL	5 to 30 fl oz	0.059 to 0.354	Use in cool season turfgrasses, such as bentgrass greens and fairways, Kentucky bluegrass, and perennial ryegrass. Repeat applications at 2- to 6-week intervals. Annual bluegrass suppression is gradual and could take several growing seasons. Start treatments in early spring and continue through early fall.
maleic hydrazide (Royal Slo-Gro) 1.5 lb/gal (Liquid Growth Retardant) 0.6 lb/gal	1.5 gal/50 gal water 3.3 gal/50 gal water	2.25	Apply in late September or early October to inhibit bermudagrass growth and allow winter overseeding to establish. Overseed no sooner than 48 hours after application. Follow label directions and precautions for use on greens and fairways.
paclobutrazol (TGR Turf Enhancer 2 SC or Trimmit 2 SC) 2 SC	6.4 to 16 fl oz/43 to 200 gal water	0.1 to 0.25	Apply any time after overseeded turf has successfully established itself. Do not apply after March 15 to avoid delay in bermudagrass green-up. Apply with 0.25 to 0.5 pound N per 1,000 sq ft of a nonburning fertilizer. Apply 0.25 inch of water within 24 hours after application to remove product from foliage and onto soil surface. Repeat applications can be made but *do not apply* more than 3 times annually. Do not use on areas containing more than 70% *poa annua*. Do not seed within 6 weeks prior to or 2 weeks after application. Do not apply to 'Tifdwarf' putting greens.
prohexadione calcium (Anuew) 27.5 WG	0.9 to 1.75 oz	0.015 to 0.03	To enhance overseeding establishment, apply to hybrid bermudagrass 3 to 5 days prior to seeding. Delay verticutting, spiking or scalping for 1 to 2 days after application.
trinexapac-ethyl (Governor) 0.17 G	129 to 165 lb	0.22 to 0.28	Apply before verticutting, scalping, or spiking the bermudagrass. Apply 1 to 5 days before overseeding. To minimize yellowing, use iron at recommended rates or available nitrogen at 0.2 to 0.5 pound per 1,000 square feet.
(Primo Maxx) 1 MEC or (T-Nex) 1 AQ (Primo WSB) 25 WP	6 to 44 fl oz 2.75 to 21.8 oz	0.08 to 0.34 0.08 to 0.34	Application rate varies with turfgrass species and height of cut. Apply to actively growing, nonstressed turf. More growth suppression occurs at lower mowing heights. See label for specific rate and other directions and precautions. Repeat applications can be made but do not exceed a total of 21.4 pints per acre per year of Primo Maxx or a total of 174 ounces per acre per year of Primo WSB. Do not exceed 19 pints per acre per year of T-Nex. Refer to the respective Primo label for guidelines regarding mowing prior to and following application. Mix with 0.5 to 4 gallons of water per 1,000 sq ft (20 to 174 gallons per acre). Primo can be applied to putting greens. See label for directions.
Lawn Edging			
maleic hydrazide (Retard) 2.25 lb/gal (Royal Slo-Gro) 1.5 lb/gal (Liquid Growth Retardant) 0.6 lb/gal	1.33 gal/100 gal water 2 gal/100 gal water 6.67 gal/100 gal water	3 3 4	Apply in spring to a 6-inch band along sidewalks. Consult instructions on applicator for delivery dosage.
mefluidide (Embark) 0.2	1.36 gal/174 gal water	0.27	For Kentucky bluegrass, tall fescue, chewings fescue, red fescue, perennial ryegrass, and St. Augustinegrass. For bermudagrass, use 5.45 gallons in 174 gallons water. Apply in 6- to 12-inch bands. Avoid overlapping.
trinexapac-ethyl (Governor) 0.17 G	100 to 259 lb	0.17 to 0.44	Do not exceed 2.5 pounds active ingredient per acre per year. These rates should provide 50% turf growth suppression for 4 weeks with minimal yellowing.
(Primo Maxx) 1 MEC (T-Nex) 1 AQ (Primo WSB) 25 WP			Apply 0.75 to 2 ounces per 1,000 linear feet of Primo Maxx or T-Nex, or 0.4 to 2 ounces per 1,000 linear ft of Primo WSB. Apply to actively growing, nonstressed turf. Apply along perimeter of lawns, sidewalks, curbs, parking lots, driveways, flower beds, or fences. Apply in an 8- to 12-inch band along the perimeter of the lawn to reduce growth of turf into adjacent areas. Application rate varies with turf species. Follow label directions for repeat applications and other precautions.

IX — ANIMAL DAMAGE CONTROL

Animal Damage Control 448
Description of Potential Animal Pests 450
Animal Control Suggestions 453
Rodenticides 457
Fish Control 458

Animal Damage Control

C. S. DePerno, Department of Forestry and Environmental Resources

General Practices

Before undertaking any control measures, read the North Carolina law regarding Wildlife Killed for Depredations or Accidentally (15 NCAC 10B.0106).

Damage caused by rodents, other mammals, and birds can be prevented or reduced by one or more of the following practices.

1. Removing or minimizing available food, water, and shelter. This includes general cleanup, rodent-proofing, and bird-proofing.
2. Using repellents.
3. Capturing offenders and releasing them away from the problem area when permitted.
4. Eliminating the offenders by trapping, shooting, or poisoning.

North Carolina legislation restricts the use of pesticides on wild birds and mammals to species that are not protected by the N.C. Wildlife Resources Commission and that have been classified as pests by the North Carolina Pesticide Board.

Currently, five rodent and eight bird species have been classified as unprotected pests, and pesticides may be used in controlling these species subject to specific restrictions of the Pesticide Board. These species are: Norway rat, black rat (roof rat), house mouse, pine vole and meadow vole, common grackle, boat-tailed grackle, brown-headed cowbird, red-winged blackbird, starling, English sparrow, pigeon, and gulls (near airports).

Federal, state, and local laws on animal control are rather strict and subject to change. Contact authorities at the N.C. Wildlife Resources Commission (1701 Mail Service Center, Raleigh, NC 27606; 919-707-0062); or State Director, USDA-APHIS, Wildlife Services (6213-E Angus Dr., Raleigh, NC 27617; 866-487-3297 or 919-786-4480) to clarify the current legal status of a particular control practice.

Report any rabid animals or suspected rabid animals to the local health department.

Traps or chemicals mentioned in this section can usually be purchased at hardware or farm supply stores. If not locally available, check with your county N.C. Cooperative Extension agent.

Wildlife Killed for Depredations or Accidentally, 15A N.C. Administrative Code 10B .0106

a) *Depredation Permit*

1) **Endangered or Threatened Species.** No permit shall be issued to take any endangered or threatened species of wildlife listed under 15A NCAC 10I by reason of depredations to property. An individual may take an endangered or threatened species in immediate defense of his own life or of the lives of others without a permit. Any endangered or threatened species that may constitute a demonstrable but nonimmediate threat to human safety shall be reported to a federal or state wildlife enforcement officer, who, upon verification of the report, may take or remove the specimen as provided by 15A NCAC 10I .0102.

2) **Other Wildlife Species.** Except as provided in subparagraph (1) of this paragraph, the executive director or an agent of the Wildlife Resources Commission may, upon application of a landholder and after such investigation of the circumstances as required, issue a permit to such landholder to take any species of wildlife that is or has been damaging or destroying landholder's property provided there is evidence of substantial property damage. No permit may be issued for the taking of any migratory birds and other federally protected animals unless a corresponding valid U.S. Fish and Wildlife Service depredation permit has been issued. The permit shall name the species allowed to be taken and, at the discretion of the Executive Director or an agent, may contain limitations as to age, sex or any other condition within the species so named. The permit may be used only by the landholder or another person named on the permit.

Wildlife Damage Control Agents. Upon satisfactory completion of Wildlife Resources Commission-approved training and satisfactory demonstration of a knowledge of wildlife laws and safe, humane wildlife handling techniques, an individual may apply to the Wildlife Resources Commission to become a Wildlife Damage Control Agent (WDCA). Those persons approved as agents by the commission may then issue depredation permits to landholders and list themselves as a second party to provide the control service. WDCAs may not issue depredation permits for big game animals, bats, or species listed as endangered, threatened, or of special concern under Rules 10I .0103, .0104, and .0105 of this chapter. WDCAs must report to the Wildlife Resources Commission the number and disposition of animals taken, by county, annually. Records must be available for inspection by a Wildlife Enforcement officer at any time during normal business hours. WDCA status may be revoked at any time by the executive director when there is evidence of violations of wildlife laws, failure to report, or inhumane treatment of animals by the WDCA. WDCAs may not charge for the permit but may charge for their investigations and control services. To maintain a knowledge of current laws, rules, and techniques, WDCAs must renew their agent status every three years by showing proof of having attended at least one Wildlife Commission-approved training course provided for the purpose of reviewing and updating information on wildlife laws and safe, humane wildlife handling techniques within the previous 12 months.

b) *Term of Permit*

Each depredation permit issued by the executive director or an agent shall have entered thereon a date or time of expiration after which date or time the same shall become invalid for any purpose, except as evidence of lawful possession of any wildlife that may be retained thereunder.

c) *Manner of Taking*

1) **Taking without a Permit.** Wildlife taken without a permit while committing depredations to property may, during the open season on the species, be taken by the landholder by any lawful method. During the closed season such depredating wildlife may be taken without a permit only by the use of firearms.

2) **Taking with a Permit.** Wildlife taken under a depredation permit may be taken only by the method or methods specifically authorized by the permit. When trapping is authorized, to limit the taking to the intended purpose, the permit may specify a reasonable distance from the property sought to be protected, according to the particular circumstances, within which the traps must be set. The executive director or agent may also state in a permit authorizing trapping whether or not bait may be used and the type of bait, if any, that is authorized. In addition to any trapping restrictions that may be contained in the permit, the method of trapping must be in accordance with the requirements and restrictions imposed by G.S. 113-291.6 and other local laws passed by the N.C. General Assembly. No depredation permit shall authorize the use of poisons or pesticides in taking wildlife except in accordance with the provisions of the North Carolina Pesticide Law of 1971, the Structural Pest Control Act of 1955, and Article 22A of Chapter 113 of the General Statutes of North Carolina. No depredation permit shall authorize the taking of wildlife by any method by any landholder upon the lands of another.

3) **Intentional Wounding**. It is unlawful for any landholder, with or without a depredation permit, intentionally to wound a wild animal in a manner so as not to cause its immediate death as suddenly and humanely as the circumstances permit.

d) *Disposition of Wildlife Taken*

1) **Generally**. Except as provided by the succeeding subparagraphs of this paragraph, any wildlife killed accidentally or without a permit while committing depredations shall be buried or otherwise disposed of in a safe and sanitary manner on the property. Wildlife killed under a depredation permit may be transported to an alternate disposal site if desired. Anyone in possession of carcasses of animals being transported under a depredation permit must have the depredation permit in their possession. Except as provided by the succeeding subparagraphs of (d)(2) through (6) of this rule, all wildlife killed under a depredation permit must be buried or otherwise disposed of in a safe and sanitary manner.

2) **Deer**. The edible portions of up to five deer may be retained by the landholder for consumption but must not be transported from the property where the depredations took place without a valid depredation permit. An enforcement officer, if so requested by the permittee, shall provide the permittee a written authorization for the use by a charitable organization of the edible portions of the carcass. The nonedible portions of the carcass, including head, hide, feet, and antlers, shall be disposed of as specified in subparagraph (1) of this paragraph or turned over to a wildlife enforcement officer for disposition. When a deer is accidentally killed on a road or highway by reason of collision with a motor vehicle, the law enforcement officer who investigates the accident shall, upon request of the operator of the vehicle, provide such operator a written permit authorizing him to possess and transport the carcass of such deer for his or her personal and lawful use, including delivery of such carcass to a second person for his or her private use or the use by a charitable organization upon endorsement of such permit to such person or organization by name and when no money or other consideration of value is received for such delivery or endorsement.

3) **Fox**. Any fox killed accidentally by a dog or dogs, motor vehicle, or otherwise shall be disposed of as provided by subparagraph (1) of this paragraph. Any fox killed under a depredation permit may be disposed of in the same manner or, upon compliance with the fur tagging requirements of 15A NCAC 10B .0400, the carcass or pelt thereof may be sold to a licensed fur dealer. Any live fox taken under a depredation permit may be sold to a licensed controlled hunting preserve for fox in accordance with G.S. 113-273(g).

4) **Furbearing Animals**. The carcass or pelt of any furbearing animal killed during the open season for taking such furbearing animal either accidentally or for control of depredations to property, whether with or without a permit, may be sold to a licensed fur dealer provided that the person offering such carcass or pelt for sale has a valid hunting or trapping license, provided further that, bobcats and otters may only be sold upon compliance with any required fur tagging requirement set forth in 15A NCAC 10B .0400.

5) **Animals Taken Alive**. Wild animals in the order Carnivora and beaver shall be humanely euthanized either at the site of capture or at an appropriate facility designed to humanely handle the euthanasia or released on the property where captured. Animals transported or held for euthanasia must be euthanized within 12 hours of capture. Anyone in possession of live animals being transported for relocation or euthanasia under a depredation permit must have the depredation permit in their possession.

6) **Wild Birds or Animals Killed Accidentally with Motor Vehicles or Found Dead**. A person killing a wild bird or wild animal accidentally with a motor vehicle or finding a dead wild bird or wild animal that was killed accidentally may possess that wild bird or wild animal for a period not to exceed 10 days for the purpose of delivering it to a licensed taxidermist for preparation. The licensed taxidermist may accept the wild bird or wild animal after determining that the animal was killed accidentally. The taxidermist shall certify and record the circumstances of acquisition as determined by his or her inquiry. Licensed taxidermists shall keep accurate records of each wildlife specimen received as required by rule 10H .1003 of this chapter. Upon delivery of the finished taxidermy product to the person presenting the animal, the taxidermist shall give the person a receipt in the form required by the Wildlife Resources Commission indicating the species, date of delivery, circumstances of initial acquisition, and any other information that may be required on the form. A copy of this receipt shall be filed with the Wildlife Resources Commission within 10 days of the date of delivery of the mounted specimen. The receipt shall serve as the nontransferable permit for continued possession of the mounted specimen and shall be retained by the person for as long as the mounted specimen is kept. Mounted specimens possessed pursuant to this rule may not be sold, and if such specimens are transferred by gift or inheritance, the new owner must apply for a new permit and must submit the written receipt originally obtained from the taxidermist to document the legality of possession. This provision does not allow possession of accidentally killed raptors; migratory birds; species listed as endangered, threatened, or of special concern under Rules 10I .0103, .0104, and .0105 of this chapter; bear or wild turkey.

e) *Reporting Requirements*

Any landholder who kills a deer, bear, or wild turkey under a currently valid depredation permit shall report such kill on the form provided with the permit and mail the form immediately upon the expiration date to the Wildlife Resources Commission. The killing and method of disposition of every game animal and game bird, every furbearing animal, and every nongame animal or nongame bird for which there is no open season, when killed for committing depredations to property, without a permit, shall be reported to the Wildlife Resources Commission within 24 hours following the time of such killing, except that when the carcass or pelt of a fox killed under a depredation permit or of a furbearing animal killed with or without a permit is lawfully sold to a licensed fur dealer in this state, the fur dealer is required to report the source of acquisition, and no report is required of the seller.

If you have questions, contact:

N.C. Wildlife Resources Commission
NC State University Centennial Campus
1751 Varsity Drive
Raleigh, NC 27606
1-919-707-0010

Chapter IX — 2019 N.C. Agricultural Chemicals Manual

Description of Potential Animal Pests

C. S. DePerno, Department of Forestry and Environmental Resources

Table 9-1. Description of Potential Animal Pests

Animal	Description	Indicators or Signs of Presence	Range and Habits	Bait or Control
Bat*	Seventeen species occur in North Carolina. Mouselike body 3 to 5.5 inches long, including tail, wingspread 6 to 16 inches. Prominent ears.	Strong odor, noise, or movement, and squeaking. Brown stains (rubs) near openings or vents in roof. Bats seen entering or leaving premises at dusk or dawn.	Statewide. Roosts in caves, hollow trees, attics, and walls by day. Nocturnal, insectivorous. Some species hibernate in attics, but most are migratory.	See Animal Control Suggestions.
Beaver*	Very large rodent with brown to reddish brown fur, large orange incisors, and a large, flat, scaly tail. Weight up to record of 120 pounds, length to 48 inches, including an 11-inch scaly tail. Webbed hind feet to 7 inches long.	Girdled and felled trees, mud and stick dams, trails to terrestrial feeding sites. Dens may be lodge style of sticks and mud in middle of pond or on bank.	Statewide. Eats wide variety of herbaceous and woody plant material. Most active from dusk till dawn. Needs constant supply of water; ponded areas may cover a hundred or more acres in flat terrain. Potential for significant timber loss in bottomland drainages.	Protect gardens with electric fence, individual trees with wire mesh and pond drains with a wire cage. Remove local populations by trapping. No depredation permit needed.
Bobcat*	Brown and buff mixed with black and gray. Weight 15 to 35 pounds. Total length 24 to 40 inches. Short tail. Bobbing motion noticed while running.	Will cache food; vegetation scratched over remains of kill. Round, 2-inch footprints without claw marks. Droppings contain bone fragments and mouse and/or rabbit hair.	Statewide. Prefers thickets, river bottoms, swamps, and brushy areas. Seldom seen. Feeds primarily on rodents, rabbits, poultry, birds, amphibians, and rarely on livestock and game.	Use a No. 2 or 3 steel foot trap Recreational hunting and trapping allowed, check NC Wildlife Resources Commission laws and regulations for details.
Chipmunk*	Small, striped, squirrel-like rodent; grayish or reddish brown.	1.5- to 2-inch open burrows with few signs of digging.	Piedmont and western part of state. May dig flower bulbs or seeds. Occasionally damages truck crops. Hibernates.	Use a small box-type live trap or kill trap (wooden-base rat snap) placed under a board or bucket at holes or near runways.
Cotton Rat* (field rat)	Up to 12 inches long, including 5-inch tail; dark brown, grizzled appearance. Scaly and nearly hairless tail. Hindfoot to 1.3 inches long.	Trails in grass, nests of cut grass usually above ground. Burrows with several 1.5- to 2-inch openings.	Statewide. Lives in dense cover, field edges, and wet sites. May damage alfalfa, cotton, and row crops.	Use a small box-type live trap or kill trap (wooden-base rat snap) placed under a board or bucket at holes or near runways.
Cowbird*†	Robin-sized. Males glossy black with brown head. Females plain gray with pale throat.	Damage to corn occurs when grackles expose ears.	Statewide, chiefly in eastern half. Forms large winter roosts with red-winged blackbirds and starlings.	See Animal Control Suggestions.
Coyote*	Medium-dog size, 20 to 50 pounds Gray or reddish-gray hair. Ears pointed, muzzle long and narrow, and a bushy tail. Length from tip of nose to end of tail up to 48 inches.	Hindfoot diameter is 2 to 2.75 inches with claw marks. Scat is doglike and often contains hair and bone. Dens at bases of trees in cultivated fields and pastures, brush covered slopes, sandy ridges and caves.	Statewide. All habitats. Eats vegetables, crops, rodents, fruits, poultry, game, and livestock. Will scavenge.	Shoot, trap, and animal husbandry practices.
Crow*† (American and Fish)	Abundant in the state. Solid black; about 17 inches long with long, rounded wings. Larger than blackbird.	Roosting in trees along planted fields. Uprooted seedlings; mature corn opened and shattered.	Statewide, except for high mountain peaks.	See Animal Control Suggestions.
House Sparrow	Small, reddish brown; upper parts streaked with black. Gray underparts. Males have black throat and gray cap.	Soiling of feed lots and window ledges.	Statewide. Feeds and roosts in flocks of 100 to 300 birds. Frequents feed lots; roosts on ledges of buildings and in trees, vines, and shrubs.	See Animal Control Suggestions. No depredation permit needed.
Field Mouse,* Deer Mouse, White-Footed Mouse	Up to 6 inches; tan above, light below; tail furred and bicolored. Large ears and eyes.	0.75-inch footprints and tail drag marks. Seed dug when no sprout visible. Sprouts dug from side, not pulled. Holes in bare, freshly planted fields and sparse cover.	Statewide. Often follows other rodent burrows.	Trap with small snap or mouse-size box traps or use glue boards within residences. See RODENTICIDE section.
Flying Squirrel*	Head and body 6 inches with 4-inch tail. Unique folds of loose skin on sides of body allow gliding. Thick, soft fur is gray.	Seldom seen; active at night. Eats seeds, nuts, insects, and bird eggs.	Statewide. Lives in holes in trees, buildings.	Invades attics where movements and food storage usually require removal by trapping. Northern flying squirrels are listed as endangered throughout North Carolina. See Animal Control Suggestions.
Grackle*† (Common and boat-tailed)	Slightly larger than robin; long, keel-shaped tail. Iridescent, glossy black.	Sprouting corn pulled up. Mature corn ears opened and shattered.	Statewide, chiefly in eastern half. Forms large winter roosts with red-winged blackbirds, cowbirds, and starlings.	See Animal Control Suggestions
Gray Fox*	Small dog-sized, to 15 pounds and measures 32 to 45 inches long from nose to tip of tail. Bushy tail is black-tipped with a black central stripe. Grizzled gray to reddish color.	Small, dainty, doglike tracks, 1 to 2 inches wide with claw marks. Large burrows under rocks or in hollow logs or dens in piles of sawdust. Climbs trees readily.	Statewide. Prefers brush, timbered, and swampy areas. Mostly nocturnal. Opportunistic carnivore. Eats rodents, squirrels, birds, eggs, turtles, insects, fruits, poultry, and game.	Occasional rabies carrier. Report any rabid animals or suspected rabid animals to local health department or animal control.
Gray Squirrel*	Up to 18 inches long; excellent tree climber; bushy tail; grayish color. Tail 8.5 inches long.	Twigs clipped; fruit and nuts of large trees eaten; footprints 2 to 2.5 inches.	Lives close to trees and shrubs; eats nuts and fruits and occasionally girdles bark; may move into attics.	See Animal Control Suggestions.
Groundhog* (Woodchuck)	A large, grizzled brown, burrowing rodent. Weight 4 to 15 pounds. Thick bodied, short tail, short legs, coarse fur.	Large burrows with earth mounded and scattered at entrance in fence rows, hedgerows, meadows, hayfields, and edges of woods.	Mountains through piedmont. Multiple entrances to den. Mainly vegetarian. Hibernates during the winter.	Use body grip or box-type live trap. Shoot.
Gulls*†	Robust birds with long, pointed wings and webbed feet. Tails usually square.	Large flocks near dumps, cultivated fields, lakes, and rivers.	May be seen occasionally in the piedmont but more common near the coast.	See Animal Control Suggestions.

Table 9-1. Description of Potential Animal Pests

Animal	Description	Indicators or Signs of Presence	Range and Habits	Bait or Control
House Mouse	Up to 5 inches; long, nearly hairless tail; body about same color top and bottom.	Feed sacks cut; holes 1 inch and less; footprints less than 0.5 inch. Nests on finely shredded material. Black-colored droppings the size of grains of rice.	Statewide. Lives in walls, furniture, feed sacks, and other enclosures. Can live without water for a long time.	Trap with small snap or mouse-size box traps or use glue boards within residences. For poison, see RODENTICIDE section.
House Rat (Norway)	Up to 18 inches long; tail length less than head and body; dark brown color; hindfoot 1.5 inches or more.	Feed sacks cut; gnawed holes; droppings up to 0.75 inch, footprints 1.5 inches; underground burrows nearby.	Statewide. Good climber. Lives at ground level in burrows, litter, walls, and other protective enclosures.	Use environmental control. Minimize food, shelter, and water. Set snap traps and rat-size glue boards. For poison, see RODENTICIDE section. No depredation permit needed.
Meadow Vole* (Meadow Mouse)	3.5 to 5 inches long body and head; tail longer than length of hind leg; dark brown color. Tail bicolored. Lighter belly. Rather long, grizzled fur.	Small, shallow burrows and trails in grass. Trees or shrubs gnawed from ground level up under mulch or snow cover.	Statewide. Damage occurs to trees and shrubs from slightly below ground level and up. Lives in areas of dense cover.	Environmental control through close mowing or clean cultivation. Trap with mouse traps placed in runs and covered with tar paper or large flower pot. A depredation permit is required for any trapping. For poison, see RODENTICIDE section.
Mole* (Eastern)	4 to 6 inches; short tail and pointed nose. Very short, paddlelike front feet; eyes and ears not visible. Very soft fur.	Extensive ridges, sometimes with mounds 1 to 2 inches high with lumpy appearance; ridges pushed up instead of dug out; no gnawings.	Statewide. Lives underground; makes tunnels, sometimes small mounds; feeds on worms, and insects, but not plants.	Use spear-type traps for best control. Ground insecticides will lessen food supply and give some control.
Mole (Star-nosed)*	6 to 7 inches, large scaled feet and a long, thick tail with concentric rings of short, coarse hair. Has a distinctive flesh-colored ring of retractable tentacles around its nose.	Extensive ridges, sometimes with mounds 1 to 2 inches high with lumpy appearance; ridges pushed up instead of dug out; no gnawings.	Limited distribution in the central Piedmont and western mountains.	Considered "non-game" and protected. Star-nosed moles cannot be killed without requesting and receiving a permit from the North Carolina Wildlife Resources Commission (NCWRC).
Muskrat*	Dark or reddish-brown, ratlike rodent; 20 to 24 inches total length; tail 8 to 10 inches, hairless, scaly, flattened from side to side. Weight 2 to 4 pounds.	Houses of rushes, etc. or burrows in banks, with entrances below water level. Trails at water's edge.	Statewide. Eats aquatic plants; occasionally animal material and crops. Burrows into earthen dams.	See Animal Control Suggestions.
Nutria*	Large, ratlike, brownish rodent. The most noticeable characteristics are its size and a long, round tail that is scantily haired. Weight 15 to 18 pounds.	Bank burrows or platform nests anchored to marsh vegetation. Worn trails, eaten-away marsh vegetation. Trails at water's edge.	Northeastern coastal area of state. An introduced species. Feeds on rushes, grasses, seeds, roots, and coarse vegetation. Usually consumes food on shore in evening.	Use No. 1.5 foot trap, box trap, or body-grip trap.
Opossum*	Generally light gray. Average length 33 inches, including a 12-inch rat-like prehensile tail. Female has pouch for carrying young (marsupial). Weight up to 15 pounds.	Omnivorous; particularly detrimental to poultry. Badly mauled poultry typical of opossum kill.	Statewide. Prefers living by streams or swamps in wooded areas. Lives in hollow logs, rocks, crevices, etc. Frequently found in towns and cities, killed by vehicles.	Easily caught in box-style live traps.
Pigeon	Crow-sized. Slate blue background, two black wing bars. Greenish purple sheen to neck feathers.	Soiling of feed lots and window sills.	Statewide. Feed, roost, and loaf in flocks, except in breeding season.	See Animal Control Suggestions. No depredation permit needed.
Pine Vole* (Orchard Rat)	Body and head about 3 to 4.25 inches long with tail shorter than length of hind leg; dark brown with reddish tinge.	Soft, spongy ground. Extensive underground burrows. Plants severed or girdled underground or dying without visible cause. Quarter-sized holes in lawns without soil mounds.	Statewide. Lives underground. Feeds on roots and bases of plants below ground level. Damage prevalent on bulbs, ornamental shrubs, and apple trees. Often follows mole runs and other burrows.	Close mowing and clean cultivation deter voles. Set mouse traps in runs, and cover them with tar paper or large flower pot. Depredation permit required for trapping. For poison, see RODENTICIDE.
Rabbit* (Cottontail)	Up to 15 inches long, including 2-inch fluffy tail. 2.5 to 4 pounds. Long ears.	Presence indicated by round fibrous pellets. Gnawing and clipping of vegetation at 45 degree angle.	Statewide. Feeds on all parts of plants. Prefers succulent foods-young shoots and vegetables.	See Animal Control Suggestions.
Raccoon*	Small dog-size; black face mask; color rings on tail; grizzled gray color. Good climber. Weight to 35 pounds.	2- to 3-inch handlike tracks; seeds dug, leaving 2-inch holes. Ripening corn opened and eaten. May den in structures, including attics and chimneys.	Statewide. Prefers wooded areas around water; eats fish, waterfowl, poultry, eggs, grain, and fruits.	In urban settings, remove any excess pet food immediately after feeding (dog, cat, etc.) or feed pets indoors. See Animal Control Suggestions. Occasional rabies carrier. Report rabid animals or suspected rabid animals to local health department.
Red Fox*	Medium dog-size to 15 pounds and measures 36 to 41 inches long from nose to tip of tail. Primarily reddish with white-tipped tail and white throat. However, 3-color phases (red, yellow, black) may occur.	Triangular foot prints with claw marks. Dens in burrows located in cleared or semicleared areas, under stumps, and vegetated slopes.	Statewide. Prefers fairly open country; agricultural areas and forest edges. Mostly nocturnal. Eats rodents, birds, fruits, poultry, livestock, and game.	Occasional rabies carrier. Report rabid animals or suspected rabid animals to local health department.
Red-Winged Blackbird *†	Robin-size. Male black with red-wing patch; female brown with heavily striped underside.	Corn husks opened during soft dough stage. Soft kernels fed upon. Fungi and molds may develop on corn.	Statewide, chiefly eastern half. Forms large winter roosts with grackles, cowbirds, and starlings. Migrates north in early spring. During late summer and early fall, roosts without blackbirds and starlings.	See Animal Control Suggestions.

Table 9-1. Description of Potential Animal Pests

Animal	Description	Indicators or Signs of Presence	Range and Habits	Bait or Control
Roof Rat (Black or Brown)	Up to 17 inches long; tail longer than head and body; dark brown, lighter belly. Hindfoot 1.5 inches or less.	1.5-inch holes in feed sacks; droppings to 0.5 inches; large, gnawed holes; 1- to 1.5-inch tracks. Smears and swing marks.	Statewide, especially coastal area. Excellent climber; lives above ground in walls, litter, etc.	Similar to house rat. See RODENTICIDE section. No depredation permit needed.
Skunk*	House cat size; black with white markings; bushy tail; bad smell. Poor climber.	Rooting in lawns or other fertile spots that have been pushed up instead of dug; tracks usually dainty, from 1 to 3 inches	Statewide. Roots for insects; literally upends some areas; hunts mice and eggs around farm buildings. Common rabies carrier.	See Animal Control Suggestions.
Snakes (37 species in N.C.)*	Beneficial for their predation on rodents. Six species are venomous. Cottonmouths, copperheads, and three species of rattlesnakes have a pit between eyes and nose with vertical eye pupil. Coral snakes have round pupils and yellow, red, and black rings encircling body. Yellow and red bands touch and the nose area is black.	Droppings with white ends similar to that of birds, shed skins.	Coral Snake (rare) found in sandhills and coastal plain. Cottonmouth Moccasin found in sandhills and coastal plain. Copperhead found statewide. Other-nonvenomous species are represented throughout state. Venomous species often nocturnal during summer months. All hibernate.	See Animal Control Suggestions.
Starling	Robin-size. Large yellow beak; gold-flecked, iridescent blue-black plumage. Delta shape of wing in flight. Short tail.	Soiling of feed lots and window ledges. Nests in holes.	Statewide, chiefly in eastern half. Feeds, roosts, and loafs in flocks during nonbreeding season.	See Animal Control Suggestions. No depredation permit needed.
White-Tailed Deer*	Length 65 to 70 inches; height at shoulder 30 to 40 inches; weight 75 to 250 pounds. Reddish brown hair; slender, long-legged. Males have antlers.	Tracks are best sign. Twigs and leaves clipped off raggedly about 24 inches above ground. Raisin-sized, tear-shaped pellets in groups of 10 to 25.	Statewide. Found in all habitats from rural farming communities to urban residential subdivisions. Active any time of the day. Primarily feeds on agricultural crops, grasses, fruits, mushrooms, acorns.	For population control, see Animal Control Suggestions.
Woodpecker*†	Birds with sharply pointed beaks used for chipping and digging. Stiff tail and gripping toes.	Drumming sounds on trees or buildings. Some peck holes in siding for nest cavity. Cedar-sided houses frequently damaged.	Statewide. Undulating flight patterns.	Woodpeckers are federally protected. Prevent access by tacking netting, used to prevent bird damage to blackberries, to eves and to side of house. They may be discouraged by using frightening devices, such as rubber snakes, pie pans, metal streamers, or other materials that either flash or scare. Attaching bags of suet to trees may provide food, and thus prevent birds from drilling on siding. If the problem persists, contact State Director, USDA-APHIS-Wildlife Services Agency, 6213-E Angus Drive, Raleigh, NC 27613, (919) 786-4480.

* Covered by N.C. law regarding Wildlife Killed for Depredations or Accidentally (10B.0106).

† Covered by Migratory Bird Treaty Act (50 CFR 21).

Animal Control Suggestions

C. S. DePerno, Department of Forestry and Environmental Resources

Table 9-2. Animal Control Suggestions

Problem Areas	Repellent Formulation	Application Methods	Rates or Procedures	Mode of Action	Remarks
Bat — Gray bats and Indiana bats, Virginia Big-eared bats, and Rafinesque's Big-eared bats are classified as endangered/threatened species.					
Attics	Flood lights in attic. Other repellents, such as mothballs, are usually not successful.		Illumination works best on repelling colonies that have just begun.		All bats in North Carolina are beneficial insectivores. Removal or displacement of large colonies should be carried out under the supervision and guidance of federal or state agencies and licensed pest control operators. Exclusion of large natal colonies during April through August should NOT occur until after the young have left the structure. Use tongs or heavy leather gloves to pick up dead or dying bats. Bats occasionally carry rabies. If bitten or scratched, wash vigorously with soap and water and see a doctor. Collect the bat without crushing its skull, place it on ice, and take it to the county or city health department for analysis.
	Ultrasound devices in attic.		Ultrasound is a new technique that has not been evaluated and may not work.		
		Bat proofing is the only permanent solution.	Seal all but a few openings with hardware cloth or steel wool for smaller openings. Wait a few days then seal remaining openings half hour after dark.		
Cowbird — Covered by Migratory Bird Treaty Act (50 CFR 21).					
All problem areas		Scare tactics.	Propane exploders, shell crackers, and electronic sound devices.		Use combination of scare devices; vary times and locations. Be persistent.
Feedlots, Standing field corn *(NOT SWEET CORN)*	4-Aminopyridine 1.0% (Avitrol)	Prebait with untreated grain, then replace with treated grain. Carefully observe cowbirds' habits to establish feeding habits and locations to prevent ingestion by desirable or non-target birds.	Rates vary with extent of bird infestation.	Blocks potassium channels resulting in increased cholinergic nervous system activity.	Restricted label. For use by or under the supervision of government agencies or pest control operators. Follow label instructions carefully.
Structures, nesting and roosting sites	4-Aminopyridine 0.5% (Avitrol)			Blocks potassium channels resulting in increased cholinergic nervous system activity.	
Crow — Covered by Migratory Bird Treaty Act (50 CFR 21).					
Newly planted and standing crops		Scare tactics.	Propane exploders and shell crackers.		Use combination of scare devices; vary times and locations. Be persistent.
Newly planted corn fields	4-Aminopyridine 1.0% (Avitrol)	Prebait with untreated corn, then replace with treated corn. Be careful to place treated corn in the crows' feeding area to prevent ingestion by desirable or non-target birds.	Rates vary with extent of bird infestation.	Blocks potassium channels resulting in increased cholinergic nervous system activity.	Restricted label. For use by or under the supervision of government agencies or pest control operators. Follow label instructions carefully.
House Sparrow					
Structures, nesting and roosting sites		Scare tactics.	Propane exploders and shell crackers.		Use combination of scare devices; vary times and locations. Be persistent.
		Roost barrier (Nixalite).	Attach to ledges.		Long lasting way to prevent birds from roosting.
	Polybutenes 93%	Paste.	Spread on ledges.	Surface repellent that results in an unpleasant sticky sensation.	Follow label instructions carefully.
	Polybutenes 49%	Liquid spray.	Spray on shrubs and rafters.	Surface repellent that results in an unpleasant sticky sensation.	Follow label instructions carefully.
	4-Aminopyridine 0.5% (Avitrol)	Prebait with untreated grain, then replace with treated grain. Carefully observe sparrows' habits to establish feeding habits and locations to prevent ingestion by desirable or non-target birds.	Rates vary with extent of bird infestation.	Blocks potassium channels resulting in increased cholinergic nervous system activity.	Restricted label. For use by or under the supervision of government agencies or pest control operators. Follow label instructions carefully.
	Methyl anthranilate 26.4%	Liquid Spray	Spray on problem areas	Targets taste senses	Follow label instructions carefully.

Table 9-2. Animal Control Suggestions

Problem Areas	Repellent Formulation	Application Methods	Rates or Procedures	Mode of Action	Remarks
Grackle — Covered by Migratory Bird Treaty Act (50 CFR 21).					
All problem areas		Scare tactics	Propane exploders, shell crackers, and electronic sound devices.		Use combination of scare devices; vary times and locations. Be persistent.
Standing field corn (NOT SWEET CORN)	4-Aminopyridine .03% (Avitrol)	Prebait with untreated grain then replace with treated grain. Carefully observe grackles' habits to establish feeding habits and locations to prevent ingestion by desirable or non-target birds. Ground and air application permitted.	Rates vary with extent of bird infestation.	Blocks potassium channels resulting in increased cholinergic nervous system activity.	Restricted label. For use by or under the supervision of government agencies or pest control operators. Follow label instructions carefully.
Gulls — Covered by Migratory Bird Treaty Act (50 CFR 21).					
On or near airport runways, endangering air traffic; normally fall and winter flocks cause problems.		Scare tactics	Propane exploders and shell crackers.		Use combination of scare devices; vary times and locations. Be persistent.
	Polybutenes 49%	Liquid spray.	Spray on landing sites.	Surface repellent that results in an unpleasant sticky sensation.	Follow label instructions carefully.
Muskrat — Covered by 15 NCAC 10B.0106. A depredation permit is needed prior to setting a trap for this species outside of the legal trapping season. Where shooting is recommended, a depredation permit is also needed unless the animal is in the act of destroying property.					
Water control structures and dams		Trapping	Use 1.5 foot trap, #110 body-grip trap, single-catch box trap, or live catch muskrat colony trap. Place traps in muskrat trails and at feeding sites. Bait live-catch traps with carrot or apple.		Check traps daily.
Moles					
Burrows	Castor Oil 14.85%	Place in burrows		Surface repellent	
Pigeon					
Structures, nesting and roosting sites		Scare tactics	Propane exploders and shell crackers.		Use combination of scare devices; vary times and locations. Be persistent.
		Roost barrier (Nixalite)	Attach to ledges.		Long lasting way to prevent birds from roosting.
	Polybutenes 93% & 93.5%	Paste	Spread on ledges.	Surface repellent that results in an unpleasant sticky sensation.	Follow label instructions carefully.
	Polybutenes 49%	Liquid spray	Spray on shrubs and rafters.	Surface repellent that results in an unpleasant sticky sensation.	
	4-Aminopyridine 0.5% (Avitrol)	Prebait with untreated corn that looks the same as treated corn. Carefully observe pigeons' habits to establish proper feeding locations to prevent ingestion by desirable or non-target birds. Place bait in high locations and on high ledges.	Rates vary with extent of pigeon infestation.	Blocks potassium channels resulting in increased cholinergic nervous system activity.	Restricted label. For use by or under the supervision of government agencies or pest control operators. Follow label instructions carefully.
	Methyl anthranilate 20% & 26.4%	Liquid Spray	Spray on problem areas.	Targets taste senses.	Follow label instructions carefully.
In and around farm buildings; pipe yards; loading docks; building tops; inside other buildings; bridges		Shooting			Discharging firearms within city or town limits is typically illegal.

Table 9-2. Animal Control Suggestions

Problem Areas	Repellent Formulation	Application Methods	Rates or Procedures	Mode of Action	Remarks
Rabbit (cottontail) — Covered by 15 NCAC 10B.0106. A depredation permit is needed prior to setting a trap for this species outside of the legal box-trapping season. Where shooting is recommended, a depredation permit is also needed outside of the legal hunting season unless the animal is in the act of destroying property.					
Ornamentals, garden shrubs, vegetables, and trees	Ammonium soap 15% liquid (Hinder)	Paint or spray on plants	Rates vary according to crop or plant. See label directions.	Repellent resulting in an unpleasant odor/taste.	May be used on home gardens. Follow label instructions carefully.
	Dried blood 15% Tobacco dust 70% (Rabbit and Dog Chaser)	Ring soil around plants, trees, and shrubs with several inch-wide bands.	Use every 2 to 3 weeks.	Repellent resulting in an unpleasant odor/taste.	Follow label instructions carefully.
		Trapping	Bait box traps with apple.		Can be box trapped during trapping season with trapping license; other times, depredation permit required.
		Fencing	2-inch mesh or smaller fencing dug in ground 2 inches, 2 feet high all around garden.		
		Shooting			Permitted with a hunting license during the legal season, and when animal is in the act of causing damage. Discharging firearms within city or town limits is typically illegal.
	Capsaicin 0.0001%	Spray on plants	Use every 2-3 weeks.	Repellent	Follow label instructions carefully.
Raccoon — Covered by 15 NCAC 10B.0106. A depredation permit is needed prior to setting a trap for this species outside of the legal trapping season. Where shooting is recommended, a depredation permit is also needed outside of the legal hunting season unless the animal is in the act of destroying property.					
Gardens and fields; Attics and other structures		Trapping	Live trap with box-style trap. Bait with sweet roll or corn-on-the-cob. Place bait at rear of trap to prevent stealing by raccoon. Secure trap to tree or stake.		Check traps daily. Raccoons can be vicious when trapped. Contact local state agencies (NCWRC) and licensed pest control operators for advice.
			Set a steel foot-hold trap with dirt hole.		
		Shooting			Discharging firearms within city or town limits is typically illegal.
	Capsaicin 0.256% Black Pepper, Oils Piperidine 1.48%	Spread along active areas.	Use every 2-3 weeks.	Repellent	Follow label instructions carefully.
	Meat meal 99% Red pepper 1%	Spread along active areas.	Use every 2-3 weeks.	Repellent	Follow label instructions carefully.
Chimneys		Exclusion			All chimneys should be protected from animal entry by installation of a wire chimney cap.
Red-Winged Blackbird					
All problem areas		Scare tactics	Propane exploders, shell crackers, and electronic sound devices.		Use combination of scare devices; vary times and locations. Be persistent.
Feedlots, standing field corn *(NOT SWEET CORN)*	4-Aminopyridine 1.0% (Avitrol)	Prebait with untreated grain then replace with treated grain. Carefully observe birds' habits to establish feeding habits and locations to prevent ingestion by desirable or non-target birds.	Rates vary with extent of bird infestation.	Blocks potassium channels resulting in increased cholinergic nervous system activity.	Restricted label. For use by or under the supervision of government agencies or pest control operators. Follow label instructions carefully.
Structures, nesting and roosting sites	4-Aminopyridine 0.5% (Avitrol)			Blocks potassium channels resulting in increased cholinergic nervous system activity.	Restricted label. For use by or under the supervision of government agencies or pest control operators. Follow label instructions carefully.
Skunk — Covered by 15 NCAC 10B.0106. A depredation permit is needed prior to setting a trap for this species outside of the legal trapping season. Where shooting is recommended, a depredation permit is also needed outside of the legal hunting season unless the animal is in the act of destroying property.					
Houses and barns			Exclusion.		Place flour at crawl space entrance to determine if skunks have left. Once gone, seal the area. To get rid of skunk odor, use a garden sprayer to mist strong smelling areas with a solution of 1 gallon vinegar with 1 cup liquid cleanser. Bury clothes. Report rabid or suspicious animals to local health department or animal control.
	Capsaicin 0.032% Black Pepper, Oils 0.48% Piperidine 0.185%	Spread along active areas.	Use every 2-3 weeks.	Repellent	Follow label instructions carefully.

Table 9-2. Animal Control Suggestions

Problem Areas	Repellent Formulation	Application Methods	Rates or Procedures	Mode of Action	Remarks
Skunk (continued)					
Gardens		Trapping	Live trap with box-style trap. Bait with fish or fish-flavored cat food. Covering trap with a tarp will reduce the chance of spray.		Check traps daily and release non-target animals.
		Shooting			In agricultural settings, producers may shoot a skunk in the act of destroying property; however, it is strongly advised to contact the local Wildlife Enforcement officer or Wildlife Biologist with the N.C. Wildlife Resources Commission to obtain a Wildlife Depredation Permit. Discharging firearms within city or town limits is typically illegal.
Snakes					
Gardens and houses		Environmental	Remove debris, trash, woodpiles, thick vegetation, and food sources.		No methods are totally effective. Take sensible precautions. Learn the difference between venomous and nonvenomous snakes. Snakes are very beneficial. No chemical controls available.
		Snake-proofing	Screen or fill any openings into house. Check door sills and where pipes enter.		
		Fencing	18- to 36-inch high, 0.25 inch mesh buried several inches below surface. Keep a 2- to 3-feet strip outside fence free of vegetation.		
		Rodent glue boards	Staple several glue boards to a piece of plywood and place next to foundation under house.		Remove snakes without harming them by pouring mineral spirits or vegetable oil on the snake and glue board. Always use glue boards inside residences, never outdoors.
	Napthelene (7%) Sulfur (28%)	Spread along area to protect.	Variable	Repellent	Impacts sensory receptors.
Squirrel (gray and flying) — Covered by 15 NCAC 10B.0106. A depredation permit is needed prior to setting a trap for this species. Where shooting is recommended, a depredation permit is also needed outside of the legal hunting season unless the animal is in the act of destroying property.					
Houses, shrubs, and trees		Trapping	Suitable size live (box-style) trap. Bait with peanuts, walnuts, peanut butter, or black oil sunflower seeds. Prebait for several days with trap doors in open position.		Check live traps daily. Release non-target animals immediately. Release squirrels at least 10 miles away. If squirrels are in the house, repair building after last squirrel is removed.
		Shooting			Discharging firearms within city or town limits is typically illegal.
Starling					
All problem areas		Scare tactics.	Propane exploders, shell crackers, and electronic sound devices.		Use combination of scare devices; vary times and locations. Be persistent.
Feedlots, standing field corn (NOT SWEET CORN)	4-Aminopyridine 1.0% (Avitrol)	Prebait with untreated grain, then replace with treated grain. Carefully observe starlings to establish feeding habits and locations to prevent ingestion by desirable or non-target birds.	Rates vary with extent of bird infestation.	Blocks potassium channels resulting in increased cholinergic nervous system activity.	Restricted label. For use by or under the supervision of government agencies or pest control operators. Follow label instructions carefully.
Structures, nesting and roosting sites	4-Aminopyridine 0.5% (Avitrol)	Mix with shelled corn in hopper before planting.	One packet per bushel of shelled corn.	Blocks potassium channels resulting in increased cholinergic nervous system activity.	Same as Feedlots.
	Polybutenes 93%	Paste.	Spread on ledges.	Surface repellent that results in an unpleasant sticky sensation.	Follow label instructions carefully.
	Methyl anthranilate 20% & 26.4%	Liquid spray.	Spray on problem areas.	Targets taste senses.	Follow label instructions carefully.

Table 9-2. Animal Control Suggestions

Problem Areas	Repellent Formulation	Application Methods	Rates or Procedures	Mode of Action	Remarks
White-Tailed Deer — Covered by 15 NCAC 10B.0106. Where shooting is recommended, a depredation permit is needed outside of the legal hunting season unless the animal is in the act of destroying property. Contact your District Biologist with the N.C. Wildlife Resources Commission for information.					
Ornamentals, gardens, field crops, and forest regeneration		Scare tactics	Propane exploders, shell crackers, and human hair hung in bags around the area.		Results may vary. Deer quickly become habituated. Many devices are expensive, require maintenance, and are dangerous.
	Soap	Hang small bars of soap from branches, using 20 pound-test fishing lines.	One bar per shrub, several per tree, replace weekly.	Repellent resulting in an unpleasant odor/taste.	Aromatic soap may be effective. However, results may vary and deer may become habituated.
	Ammonium Soap 15% Liquid (Hinder)	Painting or spraying	Rates vary according to crop or plant. See label directions.	Repellent resulting in an unpleasant odor/taste.	May be used on home gardens. Follow label instructions carefully.
	Add putrefied egg solid containing the fungicide Thiram (Big Game Repellant, Deer Away, Hinder)	Spray all parts of plant within reach of animal; repeat as needed to protect new growth. Spray twigs and trunk of dormant trees to 5 feet.	Instructions for use vary depending on formulation or brand.	Repellent resulting in an unpleasant odor/taste.	**Do not** treat vegetables, fruits, or plant parts to be eaten. Spray on dry day. Clean sprayer immediately. Follow instructions carefully. Results will vary and applications may need to be numerous and often.
		Fencing	Electric fence, high-tensile wire fence, 8 feet high woven wire, mesh fences		Must be properly maintained. Initially very expensive, however, over the long run may be cost effective.
		Shooting			In agricultural settings, producers may shoot a deer in the act of destroying property; however, it is strongly advised to contact the local Wildlife Enforcement officer or Wildlife Biologist with the N.C. Wildlife Resources Commission to obtain a Wildlife Depredation Permit. Discharging firearms within city or town limits is typically illegal.
	Capsaicin 0.0002%	Spray on plants	Use every 2-3 weeks	Repellent	Follow label instructions carefully.

Rodenticides
C. S. DePerno, Department of Forestry and Environmental Resources

Table 9-3. Rodenticides

Compound	For Use On	Formulation	Food-Handling Establishments Inside	Food-Handling Establishments Outside	Farm Use[1]	Precautions and Remarks
Aluminum Phosphide	Norway rats, Gophers, Mice, Moles	Commercially prepared baits 57-60%[7]	Yes	Yes	Yes	For retail sale to and use by pest control operators only. (Not for home use.)
Brodifacoum	Norway rats, Roof rats, House mice	Commercially prepared baits .005%[4]	Yes	Yes	Yes	Active against Warfarin-resistant rats. For use in homes, industrial buildings, and commercial buildings. Keep away from humans, domestic animals, and pets. For retail sale to and use by pest control operators only. For use in and around the periphery of homes, industrial, commercial, and public buildings in urban areas. Do not place bait in areas where there is a possibility of contaminating food or surfaces that come in direct contact with food.
Bromadiolone	Norway rats, Roof rats, House mice	Commercially prepared baits .005%[4]	Yes	No	Yes	For use in and around the periphery of homes, industrial, commercial, and public buildings in urban areas.
Bromethalin	Norway rats, Roof rats, House mice	Commercially prepared baits 0.1%[5]	No	No	Yes	Keep away from humans, domestic animals, pets, and wildlife.
Bromethalin	Moles, Eastern moles	Commercially prepared baits 0.025%	No	No	Yes	Keep away from humans, domestic animals, pets, and wildlife.
Cholecalciferol (Rampage)	Norway rats, House mice	Commercially prepared bait 0.075%[6]	No	No	Yes	Keep away from humans, domestic animals, pets, and wildlife.
Chlorophacinone[2]	Norway rats, Roof rats, House mice	Commercially prepared baits .005%[4]	Yes	Yes	Yes	Restricted Use; for retail sale to and use by pest control operators only. Keep away from humans, domestic animals, pets, and wildlife.
		Tracking powder 0.2%[4]	Yes	No		For use as a tracking powder in prescribed and specific operations.
	Norway rats, Roof rats, House mice, Meadow voles,[3] Pine voles[3]	Commercially prepared baits .005%[4]		Yes	Yes	Registered for home use. Follow label directions.

Table 9-3. Rodenticides

Compound	For Use On	Formulation	Food-Handling Establishments		Farm Use[1]	Precautions and Remarks
			Inside	Outside		
Diphacinone[2]	Norway rats, Norway rats, Roof rats, House mice	Commercially prepared baits .005%[4]	Yes	Yes	Yes	Keep away from humans, domestic animals, pets, and wildlife.
	Meadow voles,[3] Pine voles[3]	Commercially prepared baits .005%[4]			No	Not registered for home use. Follow label directions.
Warfarin[2]	Norway rats, Roof rats, House mice, Meadow voles,[3] Pine voles[3]	Powder concentrate 0.5%[4] Commercially prepared baits .025%[4]	Yes	Yes	Yes	Place baits in area inaccessible to children, pets, wildlife, and domestic animals or in tamper-proof bait boxes.
Zinc Phosphide	Norway rats, Roof rats, House mice, Moles, Pocket gophers	Commercially prepared baits 2%[7]	No	Yes		For retail sale to and use by pest control operators only. (Not for home use.)
		Tracking powder 10%[7]	Yes	No		For retail sale to and use by pest control operators only. (Not for home use.)
		Commercially or farm-prepared baits 2%[7]			Yes	For retail sale to and use by pest control operators only. (Not for home use.)
	Meadow voles,[3] Pine voles[3]	Commercially prepared grain baits 2%[7]			Yes	Broadcast application is ineffective against pine voles. For retail sale to and use by pest control operators only. (Not for home use.)

[1] Additional instructions on control of rodents on farms may be found in *Poultry Science and Technology Guide No. 4*, which is available at your county Cooperative Extension center.

[2] Requires multiple feedings to kill.

[3] Apply rodenticides for mole and vole control only after harvest or during the dormant period. Do not pick or use any crops after rodenticide application. Do not allow animals to graze in treated orchards.

[4] Mode of action: anticoagulant.

[5] Mode of action: oxidative phosphorylation of central nervous system.

[6] Mode of action: calcium mobilizer.

[7] Mode of action: phosphine gas.

Fish Control

C. S. DePerno, Department of Forestry and Environmental Resources

Address inquiries about fish control to the county wildlife law enforcement officer or to other representatives of the Boating and Inland Fisheries Division of the North Carolina Wildlife Resources Commission.

In overpopulated waters, reducing numbers may improve fish growth. Sometimes it is desirable to eliminate a fish population so that a more preferred mix of species can be introduced. All population control measures in public waters must, by law, be undertaken only by the North Carolina Wildlife Resources Commission. Within limits of available personnel, the Wildlife Resources Commission provides advice on materials and methods for fish control and help in determining if waters are public or private.

X — DISEASE CONTROL

Foliar Fungicides for Wheat Leaf Disease Control	460
Seed Treatment for Wheat Foliar Disease Control	462
Nematode Control in Corn	462
Fungicides for Control of Corn Foliar Diseases	462
Cotton Disease Control	464
Peanut Disease Control	464
Peanut Disease Management Calendar	468
Soybean Disease Control	468
Tobacco Disease Control	473
Turfgrass Disease Control	475
Nematicides for Turf	498
Floral, Nursery, and Landscape Diseases	498
Fungicides and Bactericides for Disease Control of Greenhouse Floriculture Crops	498
Disease Control for Forest, Christmas, and Ornamental Trees	512
Commercial Landscape and Nursery Crops Disease Control	519
Disease Control for Commercial Vegetables	533
Disease Control by Crop	534
Asparagus	534
Basil	535
Bean	535
Brassicas (Broccoli, Brussel Sprout, Cabbage, Cauliflower)	539
Corn, Sweet	544
Cucurbits (Cucumber, Cantaloupe, Melon, Pumpkin, Squash, Watermelon)	545
Example Spray Program for Foliar Disease Control in Watermelon Production	552
Eggplant	553
Garlic	555
Hop	556
Leafy Brassica Greens (Collard, Kale, Mustard, Rape, Salad Greens Turnip Greens)	557
Jerusalem Artichoke (Sunchoke)	559
Lettuce and Endive	560
Okra	562
Onion	563
Parsley	568
Pea	569
Pepper	571
Potato, Irish	576
Root Vegetables (Except Sugar Beet) Beet, Carrot, Parsnip, Radish, Turnip	581
Spinach	583
Sweetpotato	585
Tomatillo	589
Tomato	590
Nematode Control in Vegetable Crops	600
Fumigants	600
Relative Efficacy of Currently Registered Fumigants or Fumigant Combinations for Managing Soilborne Nematodes, Diseases, and Weeds in Plasticulture Strawberries	600
Management of Soilborne Nematodes with Non-Fumigant Nematicides	601
Greenhouse Disease Control	602
Seed Treatments	607
Sanitation	612
Various and Alternative Fungicides	613
Fungicide Resistance Management	618

Foliar Fungicides for Wheat Leaf Disease Control

L. D. Thiessen, Entomology and Plant Pathology

Anyone using any agricultural chemical should refer to the current chemical label, which contains information about the safe and effective use of the chemical, before using the chemical.

Table 10-1A. Foliar Fungicides for Wheat Leaf Disease Control

Disease Fungicide Type and (FRAC Code)	Fungicide[1]	Amount of Formulation Per Acre	Remarks[2]
Powdery mildew, Leaf Rust			
Triazoles (3)	metconazole (Caramba)	10 to 14 oz	For Powdery Mildew, apply fungicide only when mildew covers 5% to 10% of area of upper leaves. For leaf rust, apply fungicide only when disease covers 1% to 3% of total leaf area. Do not apply after head emergence (Feekes Growth Stage 10.5). Make no more than one application of tebuconazole per year. Apply Caramba immediately after flag leaf emergence for optimum control of diseases other than Fusarium head blight.
	propiconazole (Propimax, Tilt 3.6 EC)	4 fl oz	
	prothioconazole (Proline)	4.3 to 5.0 oz	
Combinations	fluapyroxad 2.8% pyraclostrobin 18.7% propiconazole 11.7% (Nexicor)	7 to 13 fl oz	For Powdery Mildew, apply fungicide only when mildew covers 5% to 10% of area of upper leaves. For leaf rust, apply fungicide only when disease covers 1% to 3% of total leaf area. Do not apply after head emergence (Feekes Growth Stage 10.5). Do not apply if head scab is anticipated to become a problem.
	propiconazole 11.7% + azoxystrobin 7.0% (Quilt)	10.5 to 14 fl oz	
	prothioconazole 10.8% + trifloxystrobin 32.3% (StrategoYLD)	4.0 to 4.65 oz	
	propiconazole 11.7% + azoxystrobin 13.5% (QuiltXcel)	10.5 to14 fl oz	
	cyproconazole 7.2% + picoxystrobin 32.3% (Aproach Prima)	3.4 to 6.8 fl oz	
Strobilurins (11)	azoxystrobin (Quadris 2.08 F)	6.2 to 10.8 fl oz	
	pyraclostrobin (Headline 2.09 EC)	6 to 9 fl oz	
	picoxystrobin 22.5% (Aproach)	6.0 to 12.0 fl oz	
Stagonospora Leaf and Glume Blotch, Tan Spot, Helminthosporium Leaf Spot			
Multi-site action (M3)	mancozeb (various brands) 4 F 80 WP 75 DF	1.6 qt 2 lb 2 lb	If 25% of the indicator leaves have one or more lesions, then a fungicide application is indicated. Indicator leaves are: Feekes Growth Stage 6 to 8: Flag - 4 and Flag - 5 Feekes Growth Stage 8 to10: Flag - 3 Feekes Growth Stage 10 to 10.51: Flag - 2 Feekes Growth Stage 10.52 to 11: Flag - 1 Do not apply mancozeb after late heading (Feekes Growth Stage 10.5) or Tilt after flag leaf emergence (Feekes Growth Stage 8).
Strobilurins (11)	pyraclostrobin (Headline 2.09 EC)	6 to 9 fl oz	For Stagonospora, if 25% of the indicator leaves have one or more lesions, then a fungicide application is indicated. Indicator leaves are: Feekes Growth Stage 6 to 8: Flag - 4 and Flag - 5 Feekes Growth Stage 8 to 10: Flag - 3 Feekes Growth Stage 10 to 10.51: Flag - 2 Feekes Growth Stage 10.52 to 11: Flag - 1
	azoxystrobin (Quadris 2.08 F)	6.2 to 10.8 fl oz	
	picoxystrobin 22.5% (Aproach)	6.0 to 12.0 fl oz	
Combinations of Strobilurins and Triazoles (3, 11)	trifloxystrobin 32.3% + prothioconazole 10.8% (Stratego Yld)	4.0 to 4.65 oz	
	metconazole 7.4% + pyraclostrobin 12.0% (Twinline)	7 to 9 fl oz	
	cyproconazole 7.2 % + picoxystrobin 32.3% (Approach Prima)	3.4 to 6.8 fl oz	Do not apply if head scab is anticipated to become a problem.
Head Scab			
Triazoles (3)[2]	tebuconazole (generic brands)	4 fl oz	Specifically, forward and backward mounted nozzles, or nozzles that have two-directional spray, should be used. Spraying at 45 degrees down from horizontal has been shown to be most effective. Operate nozzles within the spray pressure directions suggested by the manufacturer. Do not make more than one application of tebuconazole per year
	Tebuconazole 19.0% + prothioconazole 19.0% (Prosaro 421 SC)	6.5 to 8.2 fl oz	
	metconazole (Caramba)	13.5 to 17 oz	Do not apply Caramba within 30 days of harvest.
	prothioconazole (Proline)	5.0 to 5.7 oz	Do not apply Proline or Prosaro within 30 days of harvest or after full flower (Feekes 10.52).

[1] Fungicides are more likely to be profitable when the yield potential is 50 bushels/acre or more.

[2] Triazole fungicides are generally more effective in control of powdery mildew, while the strobilurins are generally more effective against leaf rust and Stagonospora. Some triazoles can suppress but not eliminate head scab, whereas strobilurins should not be used if there is concern about head scab.

Fungicide Efficacy for Control of Wheat Diseases (Updated 2017)
L. D. Thiessen, Entomology and Plant Pathology

The North Central Regional Committee on Management of Small Grain Diseases (NCERA-184) has developed the following information on fungicide efficacy for control of certain foliar diseases of wheat for use by the grain production industry in the U.S. Efficacy ratings for each fungicide listed in the table were determined by field testing the materials over multiple years and locations by the members of the committee. Efficacy is based on proper application timing to achieve optimum effectiveness of the fungicide as determined by labeled instructions and overall level of disease in the field at the time of application. Differences in efficacy among fungicide products were determined by direct comparisons among products in field tests and are based on a single application of the labeled rate as listed in the table. Table includes most widely marketed products and is not intended to be a list of all labeled products.

Table 10-1B. Efficacy of Fungicides for Wheat Disease Control Based on Appropriate Application Timing

Class	Active ingredient	Product	Rate/A (fl. oz)	Powdery mildew	Stagonospora leaf/glume blotch	Septoria leaf blotch	Tan spot	Stripe rust	Leaf rust	Stem rust	Head scab	Harvest Restriction
Strobilurin	Picoxystrobin 22.5%	Aproach SC	6.0 – 12	G[1]	VG	VG[2]	VG	E	VG	VG	NL	Feekes 10.5
Strobilurin	Fluoxastrobin 40.3%	Evito 480 SC	2.0 – 4.0	G	--	--	VG	--	VG	--	NL	Feekes 10.5 and 40 days
Strobilurin	Pyraclostrobin 23.6%	Headline SC	6.0 – 9.0	G	VG	VG[2]	E	E[3]	E	E	G	NL	Feekes 10.5
Triazole	Metconazole 8.6%	Caramba 0.75 SL	10.0 – 17.0	VG	VG	--	VG	E	E	E	G	30 days
Triazole	Tebuconazole 38.7%	Folicur 3.6 F[5]	4.0	NL	NL	NL	NL	E	E	E	F	30 days
Triazole	Prothioconazole 41%	Proline 480 SC	5.0 – 5.7	--	VG	VG	VG	VG	VG	VG	G	30 days
Triazole	Prothioconazole 19% Tebuconazole 38.7%	Prosaro 421 SC	6.5 – 8.2	VG	VG	VG	VG	E	E	E	G	30 days
Triazole	Propiconazole	Tilt 3.6 EC[4,5]	4.0	VG	VG	VG	VG	VG	VG	VG	P	Feekes 10.5.4
Mixed modes of action[5]	Tebuconazole 22.6% Trifloxystrobin 22.6%	Absolute Maxx SC	5.0	G	VG	VG	VG	VG	E	VG	NL	35 days
Mixed modes of action[5]	Cyproconazole 7.17% Picoxystrobin 17.94%	Aproach Prima SC	3.4 – 6.8	VG	VG	VG	VG	E	VG	--	NR	45 days
Mixed modes of action[5]	Fluoxastrobin 14.8% Flutriafol 19.3%	Fortix	4.0 – 6.0	--	--	VG	VG	E	VG	--	NL	Feekes 10.5 and 40 days
Mixed modes of action[5]	Fluapyroxad 2.8% Pyraclostrobin 18.7% Propiconazole 11.7%	Nexicor EC	7.0 – 13.0	G	VG	VG	E	E	E	VG	NL	Feekes 10.5
Mixed modes of action[5]	Fluxapyroxad 14.3% Pyraclostrobin 28.6%	Praxair	4.0 – 8.0	G	VG	VG	E	VG	VG	G	NL	Feekes 10.5
Mixed modes of action[5]	Propiconazole 11.7% Azoxystrobin 13.5%	Quilt Xcel 2.2 SE[5]	10.5 – 14.0	VG	VG	VG	VG	E	E	VG	NL	Feekes 10.5.4
Mixed modes of action[5]	Prothioconazole 10.8% Trifloxystrobin 32.3%	Stratego YLD	4.0	G	VG	VG	VG	VG	VG	VG	NL	Feekes 10.5 and 35 days
Mixed modes of action[5] (continued)	Benzovindiflupyr 2.9% Azoxystrobin 10.5%	Trivapro SE	9.4 – 13.7	VG	VG	VG	VG	E	E	VG	NL	Feekes 10.5.4 and 14 days
Mixed modes of action[5] (continued)	Metconazole 7.4% Pyraclostrobin 12%	TwinLine 1.75 EC[3]	7.0 – 9.0	G	VG	VG	E	E	E	GE	NL	Feekes 10.5

[1] Efficacy categories: NL=Not Labeled and Not Recommended; P = Poor; F = Fair; G = Good; VG = Very Good; E=Excellent; ND = Insufficient data to make statement about efficacy of this product.
[2] Product efficacy may be reduced in areas with fungal populations that are resistant to strobilurin fungicides.
[3] Efficacy may be significantly reduced if solo strobilurin products are applied after stripe rust infection has occurred.
[4] Application of products containing strobilurin active ingredients may result in elevated levels of the mycotoxin Deoxynivalenol (DON) in grain damaged by head scab.
[5] Products with mixed modes of action generally combine triazole and strobilurin active ingredients. Nexicor, Priaxor and Trivapro include carboxamide active ingredients.

Seed Treatment for Wheat Foliar Disease Control

L. D. Thiessen, Entomology and Plant Pathology

Anyone using any agricultural chemical should refer to the current chemical label, which contains information about the safe and effective use of the chemical, before using the chemical.

Table 10-2. Seed Treatment for Wheat Foliar Disease Control

Disease and Fungicide	Amount of Formulation per cwt	Remarks
Seed-Borne Stagonospora Nodorum Blotch, Damping Off		
difenoconazole (3) + mefenoxam (4) (Dividend XL)	5 to 10 fl oz	Seed treatments are moderately effective for control of seed-borne SNB.
clothianidin + metalaxyl (4) + metconazole (3) (NipsItSuite)	5 to 7.5 fl oz	
prothioconazole + tebuconazole + metalaxyl (Raxil, Pro MD)	0.6 to 0.2 fl oz	
Barley Yellow Dwarf Virus		
imidacloprid (Gaucho 600 F) (Gaucho XT)	0.8 to 2.4 fl oz 3.4 fl oz	
thiamethoxam (Cruiser 5FS)	0.75 to 1.33 fl oz	
clothianidin (NipsIt)	0.75 to 1.79 fl oz/100 lbs seed	

Nematode Control in Corn

L. D. Thiessen, Entomology and Plant Pathology

Anyone using any agricultural chemical should refer to the current chemical label, which contains information about the safe and effective use of the chemical, before using the chemical.

Table 10-3. Nematode Control in Corn[1]

Material and Formulation	Amount of Formulation Per 1,000 Feet	Amount of Formulation Per Acre (36-inch rows)	Remarks
terbufos (Counter) 20 G	5 to 6 ounces	5.0 pounds	Apply in furrow. Do not exceed 6.5 pounds per acre of Counter 20 CR.
abamectin (Avicta)	0.15 mg per seed	Seed treatment	Seed treatment
tioxazafen (Acceleron NemaStrike ST)	0.5 to mg ai/seed	Seed treatment	Seed treatment
clothianidin 40.30% + *Bacillus firmus* I-1582 8.10% (Ponco/Votivo)	0.25 to 0.50 mg per seed	Seed treatment	Seed treatment

[1] Efficacy of chemical treatments is dependent upon nematode present and population density. For best control, sample fields in fall (by the end of November).

Fungicides for Control of Corn Foliar Diseases

L. D. Thiessen, Entomology and Plant Pathology

Anyone using any agricultural chemical should refer to the current chemical label, which contains information about the safe and effective use of the chemical, before using the chemical.

Table 10-4A. Fungicides for Control of Corn Foliar Diseases (Northern and Southern Blight, Gray Leaf Spot, and Rusts)[1]

Fungicide Type and FRAC Code	Fungicide	Rate Per Acre Formulated (fluid ounces per acre)	Remarks
Strobilurins (11)	azoxystrobin (Quadris)	6.0 to 9.0	See label for restrictions.
	pyraclostrobin (Headline)	6.0 to 12.0	Application should be with 20 gallons of water/acre for adequate coverage with ground application, or with 5 gallons of water/acre with aerial applications.
	picoxystrobin (Aproach)	6.0 to 12.0	
Triazoles (3)	propiconazole (Tilt)[1]	2.0 to 4.0	See label for restrictions. Application should be with 20 gallons of water/acre for adequate coverage with ground application, or with 5 gallons of water/acre with aerial applications.
	tetraconazole (Domark)	4.0 to 6.0	See label for restrictions. Application should be with 5 gallons of water/acre for adequate coverage with ground application, or with 2 gallons of water/acre with aerial applications.
	prothioconazole (Proline)	5.7	Apply at R1 to R3 as a preventative or curative spray and 14 to 21 days later if disease pressure is high. Do not apply after R5 or within 21 days of harvest.
Combinations of Strobilurins and Triazoles (3,11)	azoxystrobin + propiconazole (Quilt)[1]	7.0 to 14.0	See label for restrictions.
	trifloxystrobin + propiconazole (Stratego)[1]	10.0	Application should be with 20 gallons of water/acre for adequate coverage with ground application, or with 5 gallons of water/acre with aerial applications. Alternate with another non-Group 11 fungicide if making more than one application per season.
	prothioconazole + trifloxystrobin (Stratego YLD)	4.0 to 5.0	Do not apply more than 10 ounces per year. Do not apply within 14 days of harvest.
	azoxystrobin + propiconazole (QuiltXcel)[1]	10.5 to 14	Do not apply more than 84 ounces per year. Do not apply within 30 days of harvest.
	pyraclostrobin + metconazole (Headline AMP)	10.0 to 14.4	Do not apply more than 57.6 ounces/acre per season. See label for pre-harvest interval.
	fluoxastrobin + flutriafol (Fortix)	4.0 to 6.0	Field corn only, not labeled for sweet corn.
	cyproconazoe + picoxystrobin (Aproach Prima)	3.4 to 6.8	Apply at R1 to R3 as a preventative or curative spray and 14 to 21 days later if disease pressure is high. Do not apply after R5 or within 21 days of harvest.
	tetraconazole + azoxystrobin (Affiance)	10.0 to 17.0	Apply at R1 to R3 as a preventative or curative spray and 14 to 21 days later if disease pressure is high. Do not apply after R5 or within 21 days of harvest.

Table 10-4A. Fungicides for Control of Corn Foliar Diseases (Northern and Southern Blight, Gray Leaf Spot, and Rusts)[1]

Fungicide Type and FRAC Code	Fungicide	Rate Per Acre Formulated (fluid ounces per acre)	Remarks
Combinations of dicarboximides and Strobilurins (7, 11)	Fluxapyroxad + Pyraclostrobin (Priaxor)	4.0 to 8.0	Apply at R1 to R3 as a preventative or curative spray and 14 to 21 days later if disease pressure is high. Do not apply after R5 or within 21 days of harvest.

[1] Fungicides often significantly increase yields only in intensive production systems with high plant populations and adequate moisture.

Fungicide Efficacy for Control of Corn Diseases
L. D. Thiessen, Entomology and Plant Pathology

The Corn Disease Working Group (CDWG) has developed the following information on fungicide efficacy for control of major corn diseases in the United States. Efficacy ratings for each fungicide listed in the table were determined by field testing the materials over multiple years and locations by the members of the committee. Efficacy ratings are based upon level of disease control achieved by product, and are not necessarily reflective of yield increases obtained from product application. Efficacy depends upon proper application timing, rate, and application method to achieve optimum effectiveness of the fungicide as determined by labeled instructions and overall level of disease in the field at the time of application. Differences in efficacy among fungicide products were determined by direct comparisons among products in field tests and are based on a single application of the labeled rate as listed in the table. **Table includes systemic fungicides available that have been tested over multiple years and locations. The table is not intended to be a list of all labeled products.**[1]

Efficacy categories: NR = Not Recommended; P = Poor; F = Fair; G = Good; VG = Very Good; E = Excellent; NL = Not Labeled for use against this disease; U = Unknown efficacy or insufficient data to rank product.

Table 10-4B. Fungicide Efficacy for Control of Corn Diseases

Class	Active ingredient (%)	Product/ Trade name	Rate/A (fl oz)	Anthracnose leaf blight	Common rust	Eyespot	Gray leaf spot	Northern leaf blight	Southern rust	Harvest Restriction[2]
QoI Strobilurins Group 11	Azoxystrobin 22.9%	Quadris 2.08 SC Multiple Generics	6.0 – 15.5	VG	E	VG	E	G	G	7 days
	Pyraclostrobin 23.6%	Headline 2.09 EC/SC	6.0 – 12.0	VG	E	E	E	VG	VG	7 days
	Picoxystrobin	Aproach 2.08 SC	3.0 – 12.0	VG	VG-E	VG	F-VG	VG	G	7 days
DMI Triazoles Group 3	Propiconazole 41.8%	Tilt 3.6 EC Multiple Generics	2.0 – 4.0	NL	VG	E	G	G	F-G	30 days
	Prothioconazole 41.0%	Proline 480 SC	5.7	U	VG	E	U	VG	G	14 days
	Tebuconazole 38.7%	Folicur 3.6 F Multiple Generics	4.0 – 6.0	NL	U	NL	U	VG	F-G	36 days
	Tetraconazole 20.5%	Domark 230 ME Multiple Generics	4.0 – 6.0	U	U	U	E	VG	G	R3 (milk)
Mixed mode of action	Azoxystrobin 13.5% Propiconazole 11.7%	Quilt Xcel 2.2 SE Multiple Generics	10.5 – 14.0	VG	VG-E	VG-E	E	VG	VG	30 days
	Benzovindiflupyr 10.27% Azoxystrobin 13.5% Propiconazole 11.7%	Trivapro A 0.83 + Trivapro B 2.2 SE	A = 4.0 B = 10.5	U	U	U	E	VG	E	7 days (A) 30 days (B)
	Cyproconazole 7.17% Picoxystrobin 17.94%	Aproach Prima 2.34 SC	3.4 – 6.8	U	U	U	E	VG	G-VG	30 days
	Flutriafol 19.3% Fluoxastrobin 14.84%	Fortix 3.22 SC Preemptor 3.22 SC	4.0 – 6.0	U	U	U	E	VG-E	VG	30 days
	Pyraclostrobin 28.58% Fluxapyroxad 14.33%	Priaxor 4.17 SC	4.0 – 8.0	U	VG	U	VG	VG-E	G	21 days
	Pyraclostrobin 13.6% Metconazole 5.1%	Headline AMP 1.68 SC	10.0 – 14.4	U	E	E	E	VG	G-VG	20 days
	Trifloxystrobin 32.3% Prothioconazole 10.8%	Stratego YLD 4.18 SC	4.0 – 5.0	VG	E	VG	E	VG	G-VG	14 days
	Tetraconazole 7.48% Azoxystrobin 9.35%	Affiance 1.5 SC	10.0 – 14.0	U	U	U	F-VG	U	G	7 days

[1] Additional fungicides are labeled for disease on corn, including contact fungicides such as chlorothalonil. Certain fungicides may be available for diseases not listed in the table, including Gibberella and Fusarium ear rot. Applications of Proline 480 SC for use on ear rots requires a FIFRA Section 2(ee) and is only approved for use in Illinois, Indiana, Iowa, Louisiana, Maryland, Michigan, Mississippi, North Dakota, Ohio, Pennsylvania, and Virginia.

[2] Harvest restrictions are listed for field corn harvested for grain. Restrictions may vary for other types of corn (sweet, seed, or popcorn, etc.), and corn for other uses such as forage or fodder.

Many products have specific use restrictions about the amount of active ingredient that can be applied within a period of time or the amount of sequential applications that can occur. Please read and follow all specific use restrictions prior to fungicide use. This information is provided only as a guide. It is the responsibility of the pesticide applicator by law to read and follow all current label directions. Reference to products in this publication is not intended to be an endorsement to the exclusion of others that may be similar. Persons using such products assume responsibility for their use in accordance with current directions of the manufacturer. Members or participants in the CDWG assume no liability resulting from the use of these products.

Cotton Disease Control

L. D. Thiessen, Entomology and Plant Pathology

Anyone using any agricultural chemical should refer to the current chemical label, which contains information about the safe and effective use of the chemical, before using the chemical.

Table 10-5A. Nematode Control on Cotton

Nematodes Nematicide	Amount of Formulation Per Acre	Precautions and Remarks
Root-Knot, Columbia Lance, Sting, Reniform		
1,3-dichloropropene (Telone II)	3 to 6 gallons	Inject 1 to 2 weeks before planting 8 to 12 inches deep.
sodium methyldithiocarbamate (Vapam HL)	6 to 12 gallons	Inject 2 to 3 weeks before planting at 6 to 12 inches deep.
Reniform, Root-Knot		
abamectin (Avicta)	0.15 mg per seed	Seed treatment.
sodium methyldithiocarbamate (Vapam HL)	6 to 12 gallons	Inject 2 to 3 weeks before planting at 6 to 12 inches deep.
thiodicarb (Aeris) + Bacillus fermis (Poncho/Votivo)	NA	Seed treatment

[1] Seed treatments are most effective under low to moderate nematode populations. Accurate assessment of nematode populations by soil sampling in fall will indicate if seed treatments will be effective for managing nematodes economically in a specific field.

Table 10-5B. Disease Control on Cotton

Fungicide Type and FRAC Code	Fungicide	Rate of Formulation	Remarks
Damping Off			
Seed Treatments[1]	Trifloxystrobin + Triadimenol + Metalaxyl (Trilex Advanced)	1.6 fl oz/100 lbs seed	Efficacy on *Fusarium, Rhizoctonia,* and *Pythium* spp.
	Azoxystrobin + Fludioxonil + Mefenoxam + Difenconazole (Seed Shield)	4.0 fl oz/100 lbs seed	Efficacy on *Fusarium, Rhizoctonia,* and *Pythium* spp.
	Mefenoxam (Apron XL)	0.32 to 0.64 fl oz/100 lbs seed	Efficacy on *Pythium* and *Phytophthora* spp.
In-furrow Fungicides	Mefenoxam (Ridomil Gold SL)	0.075 to 0.15 oz/1000 row ft	Efficacy on *Pythium* and *Phytophthora* spp.
	Pyraclostrobin (Headline SC)	0.1 to 0.8 fl oz/1000 row ft	Efficacy on *Rhizoctonia solani*
	Fluxapyroxad + Pyraclostrobin (Priaxor)	0.1 to 0.6 fl oz/1000 row ft	Efficacy on *Rhizoctonia solani* and *Fusarium* spp.
Foliar Diseases			
Strobilurins (11)	Azoxystrobin (Quadris)	6 to 9 fl oz/acre	45 day pre-harvest interval. Max 27 fl oz/acre/season
	Pyraclostrobin (Headline SC)	4 to 8 fl oz/acre	30 day pre-harvest interval. Two applications allowed
Triazoles (3)	Flutriafol (TopGuard)	7 to 14 fl oz/acre	30 day pre-harvest interval. Three applications allowed.
Combinations (7, 11)	Fluxapyroxad + Pyraclostrobin (Priaxor)	4 to 8 fl oz/acre	30 day pre-harvest interval. Three applications allowed.

[1] This is not an exhaustive list, but represents several products for seed treatments available in cotton seed production that have differential efficacy on damping off pathogens.

Peanut Disease Control

B. B. Shew, Entomology and Plant Pathology

Most peanut disease control chemicals leave residues on peanut vines that make them unsuitable for hay. Check each label before using the material if you intend to feed hay to livestock.

Note: certain products labeled on peanut in the United States contain active ingredients (propiconazole, phosphites, or phosphorous acid) that were not acceptable to export markets in 2019. Products containing these ingredients are not listed in the tables below. Check with local North Carolina Cooperative Extension personnel and your buyer for the latest information on the current status of these ingredients and products.

Table 10-6A. Peanut Disease Control

Disease or Diseases Controlled Pesticide Formulation (FRAC Group Number)	Amount of Formulation Per Acre	Application Schedule	Minimum Days to Harvest	Precautions and Remarks
Aspergillus Crown Rot (Aspergillus); see also seedling diseases				
azoxystrobin (Abound, various brands)[1] 2.08 F (11)	0.4 to 0.8 fl oz/1,000 ft of row	At planting	NA	Apply as in-furrow spray with 3 to 5 gallons water.
Black Root Rot (CBR) (Cylindrocladium); see also seedling diseases				
metam sodium 42% (various brands) 4.25 F	7.5 gal (36-inch rows) or 6.61 fl oz/100 ft of row	At least 2 weeks before planting or longer if cool and/or wet	NA	Inject 8 to 10 inches below seed placement. Apply only when soil temperature at 3-inch depth is between 60 and 90 degrees F. If wet and/or cold weather occurs following fumigation, the waiting period should be extended. Soil aeration helps reduce residual chemical. When in doubt use a bioassay such as the lettuce seed germination test to determine if safe to plant. Buffer zones, fumigant management plans, and other restrictions on metam sodium must be followed. See your local Extension center for details.

Table 10-6A. Peanut Disease Control

Disease or Diseases Controlled Pesticide Formulation (FRAC Group Number)	Amount of Formulation Per Acre	Application Schedule	Minimum Days to Harvest	Precautions and Remarks
Black Root Rot (CBR) (Cylindrocladium); see also seedling diseases (continued)				
prothioconazole (Proline) 480 SC (3)	5.7 oz per acre (36-inch rows) or 0.4 fl oz/1,000 ft of row	At planting or at full emergence	NA	Apply as in-furrow spray or banded at full emergence for suppression of CBR. Not a substitute for fumigation in fields with a history of more than 10% CBR and rotations of less than 4 years. Use with a CBR-resistant cultivar.
prothioconazole + fluopyram (Propulse) 3.3 SC (3 + 7)	13.7 fl oz	At planting	NA	Apply as in-furrow spray for suppression of CBR and nematodes. Not a substitute for fumigation in fields with a history of more than 10% CBR and rotations of less than 4 years. Use with a CBR resistant cultivar.
Early Leafspot (Cercospora)				
cupric hydroxide (Kocide, various brands and formulations) [2] (M1)	Various; see label	Begin applications at very early pod (R3). Repeat applications every 7 to 14 days.	0	Use nozzles that give a cone-shaped spray pattern. Use 12 to 24 gallons of water for spray materials applied by ground sprayers. Use at least 5 gallons of water for materials applied by air. **Calendar program:** 5 or 6 applications suggested. Begin applications at very early pod (R3). Repeat applications at 7- to 14-day intervals. **Not suitable for use with the leaf spot advisory.**
basic copper sulfate (various brands and formulations) [2] (M1)	Various; see label	See above	0	See above
mancozeb and copper hydroxide (Mankocide) [2] 61.1 DF (M3)	2 to 2.6 lb	See above	14	See above
mancozeb (Manzate, Koverall, various brands and formulations) [2] (M3) M45 F45 75 WDG 80 WP	1 to 2 lb .8 to 1.6 qt 1 to 2 lb 1 to 2 lb	See above	14	See above
Sulfur (various brands and formulations) [2] (M2)	Various; see label	See above	0	See above
Early Leafspot (Cercospora); Late Leafspot (Cercosporidium); Web Blotch (Ascochyta)				
chlorothalonil (Bravo, Echo, various brands) (M5) 720, 6 F 82.5 WDG 90 DF 500	1 to 1.5 pt .9 to 1.36 lb .875 to 1.25 lb 1.5 to 2.25 pt	Begin applications at very early pod (R3). Repeat applications every 14 days or according to daily weather based advisories. Begin 14-day program if web blotch is found.	14	Use nozzles that give a cone-shaped spray pattern. Use 12 to 24 gallons of water for spray materials applied by ground sprayers. Use at least 5 gallons of water for materials applied by air. **14-day program:** 5 or 6 applications suggested. Begin applications at very early pod (R3). Repeat applications at 10- to 14-day intervals. **Advisory:** Begin applications at very early pod (R3). Repeat applications when weather conditions become favorable as determined by peanut leaf spot advisories. This schedule requires strict adherence to the program guidelines and usually results in fewer fungicide applications than the 14-day schedule. Contact your local Extension center for details. Leafspot advisories are most effective if used with long rotations, resistant varieties, and high rates of effective fungicides. Repeated applications of chlorothalonil can make spider mites and Sclerotinia blight more difficult to control.
boscalid (Endura) [2,3] 70 WDG (7)	10 oz	Make up to 2 or 3 applications in mid-season as part of a full-season, 14-day, or advisory program	14	See above. Primarily controls web blotch. Alternate with another fungicide or mix with 0.75 to 1 pint chlorothalonil to improve leaf spot control. Also controls Sclerotinia blight; see below.
thiophanate methyl (Topsin, Topsin M) (1) 4.5 F 70 WSB	10 fl oz .5 lb	14 day or advisory beginning at R3	14	See above. **Do not** apply alone. Always mix with another leaf spot fungicide.
dodine (Elast) 400 F (U12)	1.5 pt	Make no more than 3 applications as part of a full-season, 14-day, or advisory program	14	See above
cyproconazole (Alto) 100 SL (3)	5.5 fl oz	Apply up to 2 times in a 14-day or advisory program beginning at R3	30	See above. Mixing with chlorothalonil is recommended.
tetraconazole (Eminent VP, Domark) (3) 125 SL 230ME	6 to 13 fl oz 5.25 to 6.9 fl oz	Apply up to 2 times in a 14-day or advisory program beginning at R3	14	See above. Mix or alternate with another fungicide to reduce the risk of fungicide resistance
Pydiflumetofen (Miravis)[3] SC (7)	3.4 fl oz	Apply up to 2 times in a 14-day or advisory program beginning at R3	14	See above. Can be effective up against leaf spots for up to 28 days; if using an extended interval, apply a fungicide that is effective against stem rot after 14 days or tank mix with a stem rot fungicide. Do not make more than 2 applications of an unmixed group 7 fungicide.
Early Leaf Spot (Cercospora); Late Leaf Spot (Cercosporidium); Web Blotch (Ascochyta); Limb Rot (Rhizoctonia)				
tebuconazole + trifloxystrobin (Absolute) 500 3C (3 + 11)	3.5 to 7 fl oz	14 day or advisory beginning at R3	14	See above. Use highest rate for soil-borne pathogens. No more than 3 applications per season. **Resistance management:** Site-specific fungicides (groups 3, 7, and 11) should be mixed or rotated with a fungicide from a different group to minimize the risk of fungus resistance development.
Early Leaf Spot (Cercospora); Late Leaf Spot (Cercosporidium); Web Blotch (Ascochyta); Limb Rot (Rhizoctonia); Stem Rot (Sclerotium rolfsii); Pod Rot (Sclerotium rolfsii, Rhizoctonia)				
prothioconazole + tebuconazole (Provost Opti) [4] 433 SC (3+3)	7 to 10.7 fl oz	Make up to 2 to 4 applications in mid-season as part of a full-season, 14-day, or advisory program.	14	See Early Leaf Spot, Late Leaf Spot, and Web Blotch above. For best control of limb and pod rot, do not use a surfactant. Do not apply more than 3 times in a 5-spray program or after the first week in September. **Resistance management:** Site-specific fungicides (groups 3, 7, and 11) should be mixed or rotated with a fungicide from a different group to minimize the risk of fungus resistance development.

Table 10-6A. Peanut Disease Control

Disease or Diseases Controlled Pesticide Formulation (FRAC Group Number)	Amount of Formulation Per Acre	Application Schedule	Minimum Days to Harvest	Precautions and Remarks
Early Leaf Spot (Cercospora); Late Leaf Spot (Cercosporidium); Web Blotch (Ascochyta); Limb Rot (Rhizoctonia); Stem Rot (Sclerotium rolfsii); Pod Rot (Sclerotium rolfsii, Rhizoctonia) (continued)				
metconazole (Quash) 50 WDG (3)	2.5 to 4 oz	Make up to 2 to 4 applications in mid-season as part of a full-season, 14-day, or advisory program.	14	See above
prothioconazole + fluopyram (Propulse) 3.3 SC (3 + 7)	13.7 fl oz	Make up to 2 applications per season as part of a full-season, 14-day or advisory program.	14	See Early Leaf Spot, Late Leaf Spot, and Web Blotch above. Application for leaf spot control requires a FIFRA Section 2(ee) label.
penthiopyrad (Fontelis)[3] 1.67 SC (7)	16 to 24 fl oz	Make up to 3 applications per season as part of a full-season, 14-day, or advisory program	14	See Early Leaf Spot, Late Leaf Spot, and Web Blotch above. FRAC guidelines recommend no more than 2 applications of an unmixed group 7 fungicide in a 5-spray program. Use higher rates for web blotch control. Also suppresses Sclerotinia blight; see below.
azoxystrobin (Abound; various brands)[1,4] 2.08 F (11)	12.0 to 24.6 fl oz	Make up to 2 applications per season as part of a full-season, 14-day, or advisory program. Use higher rates for limb rot and stem rot control.	14	See above. Use in mid-season for best control of soil-borne pathogens. Use no more than 2 applications in a 5 spray program. **Resistance management**: Site-specific fungicides (groups 3, 7, and 11) should be mixed or rotated with a fungicide from a different group to minimize the risk of developing fungal resistance. Some populations of leaf spot fungi are not controlled by group 11 fungicides.
chlorothalonil + tebuconazole (Muscle ADV)	32 fl oz	Make up to 3 applications per season as a part of a full-season, 14-day, or advisory program.	14	See above. Use in mid-season for best control of soil-borne pathogens. Chlorothalonil rate is equivalent to 1 pt/A of a 720 formulation.
pyraclostrobin (Headline)[1] 2.09 EC, 2.08 SC (11)	6 to 15 fl oz	Make up to 2 applications per season as part of a full-season, 14-day, or advisory program. Use higher rates for limb rot and stem rot control.	14	See above **Resistance management**: Site-specific fungicides (groups 3, 7, and 11) should be mixed or rotated with a fungicide from a different group to minimize the risk of developing fungal resistance. Some populations of leaf spot fungi are not controlled by group 11 fungicides.
fluoxastrobin (Exito, Aftershock)[1] 480 SC (11)	5.7 fl oz	Make up to 2 applications per season as part of a full-season, 14-day, or advisory program.	14	See above **Resistance management**: Site-specific fungicides (groups 3, 7, and 11) should be mixed or rotated with a fungicide from a different group to minimize the risk of developing fungal resistance. Some populations of leaf spot fungi are not controlled by group 11 fungicides.
azoxystrobin + tebuconazole (various generic) (11 + 3)	15.5 fl oz	Make up to 2 to 4 applications in mid-season as part of a full-season, 14-day, or advisory program.	14	See above. **CAUTION:** Check labels. Most products contain 1.0 lb/A azoxystrobin per gallon, which is about one-half the amount found in 2.08 F formulations of azoxystrobin. Consider mixing with another fungicide to improve leaf spot control.
fluoxastrobin + tebuconazole (Evito T) (11+3)	11.2 fl oz	Make up to 2 to 4 applications in mid-season as part of a full-season, 14-day, or advisory program.	14	See above
azoxystrobin + benzovindiflupyr 45 DF (Elatus) (11 + 7)	7.3 to 9.5 fl oz	Make up to 2 or 3 applications per season as part of a full-season, 14-day, or advisory program. Use higher rates for limb rot and stem rot control.		See above. Extended spray intervals (up to 21 days) may be possible at the highest rate. **Resistance management**: Site-specific fungicides (groups 3, 7, and 11) should be mixed or rotated with a fungicide from a different group to minimize the risk of developing fungal resistance.
fluxapyroxad + pyraclostrobin 4.17 SC (Priaxor) (7+11)	4 to 8 fl oz	Use 1 to 3 times per season as part of a full-season, 14-day, or advisory program. Use higher rates for limb rot and stem rot control.	14	See above
flutriafol + azoxystrobin (Topguard EQ)[2] 4.3 SC (3+11)	5 to 7 fl oz	Use up to 2 times per season as part of a full-season, 14-day, or advisory program. Use higher rates for limb rot and stem rot control.	14	See above. Mix or alternate with another fungicide to reduce the risk of fungicide resistance. Mix or alternate with another leaf spot fungicide to improve leaf spot control.
Stem Rot (white mold, Southern blight, Sclerotium rolfsii); Limb Rot (Rhizoctonia); Pod Rot (Sclerotium rolfsii, Rhizoctonia)				
tebuconazole (various brands) (3) 3.6F 20AQ 75DF	7.2 fl oz 15.4 fl oz 4.3 oz	Following leafspot advisories, make 1 to 3 applications in mid-season. May provide some control of foliar diseases.	14	Effective against stem rot. Not effective against many populations of leaf spot fungi. **Always** mix with chlorothalonil or another fungicide (other than group 3) that is effective against leaf spots.
flutolanil (Convoy)[3] 3.8 SC (7)	10 to 16 fl oz or 20 -32 fl oz (see remarks)	Following leafspot advisories, make 1 to 3 applications in mid-season. **Does not control foliar diseases.**	40	Apply up to 16 fluid ounces per acre at 2-week intervals or up to 32 fluid ounces per acre at 3- to 4-week intervals. Do not apply more than a combined total of 64 fluid ounces in a single growing season. See label for detailed information on rates. Wheat may be planted 30 days after last application; do not plant other small grains within 5 months of last application. See label for other plant-back restrictions.
Nematodes—Fumigants				
1-3 dichloropropene 97.5% (Telone II) 93.6% (Telone EC)	Depends on application method; see label for details	At least 2 weeks before planting	NA	Inject 8 to 10 inches below the soil surface. Very effective against all nematodes. Does not control soil-borne fungi. Regulations require handler training and impose buffer zones and other restrictions on fumigant use. See the label and your local Extension center for details.
1-3 dichloropropene + chloropicrin 81.2% + 16.5% (Telone C-17) 63.4% + 34.7% (Telone C-35) 60.8% + 33.3% (InLine)	Depends on application method; see label for details	At least 2 weeks before planting	NA	Inject 8 to 10 inches below the soil surface. Very effective against all nematodes. Regulations require handler training and impose buffer zones and other restrictions on fumigant use. See the label your local Extension center for details.

Chapter X — 2019 N.C. Agricultural Chemicals Manual

Table 10-6A. Peanut Disease Control

Disease or Diseases Controlled Pesticide Formulation (FRAC Group Number)	Amount of Formulation Per Acre	Application Schedule	Minimum Days to Harvest	Precautions and Remarks
Nematodes—Fumigants (continued)				
metam sodium 42% (various brands) [5] 4.25 F	7.5 gal	At least 2 weeks before planting	NA	Inject 8 to 10 inches below the soil surface. If wet and/or cold weather occurs following fumigation, the waiting period should be extended. Soil aeration helps reduce residual chemical. When in doubt use a bioassay such as the lettuce seed germination test to determine if safe to plant. Moderately effective against Northern root knot nematode (M. hapla). Not very effective on peanut root knot nematode (M. arenaria). Buffer zones and other restrictions on metam sodium use are required. See your local Extension center for details.
Nematodes—Nonfumigant				
fluopyram + imidacloprid (Velum Total) 3.67 SC (7 + insecticide group 4A)	18 fl oz	At planting	30	Apply in-furrow at planting, directed on or below the seed. Also controls thrips, leaf hoppers, and aphids (see Chapter 5). May provide early season control of leaf spots and stem rot. Do not exceed 0.5 pound a.i./a imidacloprid for all seed, in-furrow, and foliar applications. See label for plant-back restrictions.
prothioconazole + fluopyram (Propulse SC) 3.3 SC (3 + 7)	13.7 fl oz	Apply approx. 45 days after planting	NA	Apply in a minimum of 15 gal of water per acre and follow with 0.1 to 0.25 inches of irrigation. Use after application of a nematicide at planting. Application requires possession of a FIFRA Section 2(ee) label.
Seed and Seedling Rot; Pythium Pod Rot				
mefenoxam + azoxystrobin (Uniform) 390 SE (4 + 11)	.34 fl oz/1,000 ft of row	At planting	75	Apply as an in-furrow spray at planting. Only one application per season.
azoxystrobin[1] (Abound 2.08 F; various brands) (11)	.4 to .8 fl oz/1,000 ft of row	At planting	14	Apply as an in-furrow spray at planting; counts as a group 11 application for resistance management purposes.
mefenoxam (Ridomil Gold GR; various brands) (4) (Ridomil Gold SL; various brands) (4)	Per 1,000 ft of row: 6.5 oz .25 pt	At planting	75	Apply in-furrow or as a 7-inch band over row at planting.
mefenoxam (Ridomil Gold 2.5 GR; various brands) (4) (Ridomil Gold SL; various brands) (4)	Per 1,000 ft of row (GR): 13 oz SL per Acre: .5 to 1 pt	Early pegging	75	Apply in an 8- to 12-inch band. Do not apply to wet foliage as foliar toxicity may result. Use with other fungicides for late-season control of stem rot (*Sclerotium rolfsii*) and Rhizoctonia stem and pod rot (*Rhizoctonia* spp.).
Seedling Diseases—Seed Treatments				
azoxystrobin + fludioxonil + mefenoxam (Dynasty PD) [4] (11 + 12 + 4)	4 oz/100 lb seed	Seedling diseases: Apply to conditioned, untreated seed. Commercial application strongly recommended.	NA	Peanuts can be replanted immediately. Do not plant other crops within 45 days of planting treated seed.
thiamethoxam + mefenoxam + fludioxonil + azoxystrobin (CruiserMaxx Peanuts) [4] (MOA 4A + 11 + 4)	3 to 4 oz/100 lb seed	See above	NA	Peanuts can be replanted immediately. See label for additional information about plant-back restrictions. Do not make any soil or foliar application of products containing thiamethoxam to crops grown from seed treated with CruiserMaxx Peanuts. Also controls some early season insects; see Chapter 5 for more information.
carboxin + ipconazole + metalaxyl (Rancona V PD) (3 + 4 + 7)	4 oz/100 lb seed	See above	NA	Dust formulation.
Sclerotinia Blight				
fluazinam (Omega) 500 F (29)	1 to 1.5 pt	1 to 3 applications according to weather-based advisory field history, and scouting.	30	If favorable conditions persist, reapply at 21 to 30 day intervals. Do not apply more than a combined total of 4 pints in a single growing season. Contact your local Extension center for details on weather-based Sclerotinia advisories.
boscalid (Endura) [2,3] 70 WG (7)	8 to 10 oz	See above	14	If favorable conditions persist, reapply at 14- to 21-day intervals. Make no more than 2 consecutive applications per season. Contact your local Extension center for details on weather-based Sclerotinia advisories. Also controls or suppresses leaf spots and web blotch.
penthiopyrad (Fontelis) [3] 1.67 SC (7)	24 fl oz	See above	14	Suppression only. Apply at 2-week intervals or according to advisory. FRAC guidelines recommend no more than 2 applications of an unmixed group 7 fungicide in a 5-spray foliar disease control program. Do not apply more than 72 fl oz per season. Use on cultivars that have some Sclerotinia blight resistance, for example, Bailey. Also controls or suppresses leaf spots, web blotch, southern stem rot, and Rhizoctonia limb and pod rot; see above.
fluopyram + prothioconazole (Propulse) (7 + 3)	13.7 fl oz	See above	14	Suppression only. Apply at 2-week intervals or according to advisory. Use on cultivars that have some Sclerotinia blight resistance, for example, Bailey. Also controls southern stem rot and Rhizoctonia limb rot; see above.
Iprodione 4F (Rovral, various brands) (2)	32 fl oz	See above	10	Suppression only. Apply at 2- to 3-week intervals or according to advisory. Apply at low pressure for a spray volume of at least 40 gal per acre. Do not apply more than 72 fl oz per season. Use on cultivars that have some Sclerotinia blight resistance, for example, Bailey.

[1] QOI (group 11) fungicide. Do not apply group 11 fungicides more than 2 times in sequence or more than 3 times per season. Some populations of leaf spot fungi are not controlled by group 11 fungicides. See www.FRAC.org for information on fungicide resistance management.

[2] Less effective against leaf spots than many other fungicides; more frequent application may be necessary. If using advisories, alternate or mix with more effective fungicides.

[3] Do not apply unmixed group 7 fungicides more than 2 times in sequence. FRAC guidelines: no more than 2 applications of a group 7 fungicide per 5-spray program; no more than 3 applications per ≥6-spray program. May be alternated with group 11 or group 3 fungicides. See www.FRAC.org for information on fungicide resistance management.

[4] Also suppresses CBR. See label for details.

[5] Probably not as effective as the other fumigants against nematodes.

Further Information: *2019 Peanut Information* and peanut disease control information are available at your local Cooperative Extension center.

Peanut Disease Management Calendar

B. B. Shew, Entomology and Plant Pathology

Table 10-6B. Peanut Disease Management Calendar

Time of Year	Disease	Threshold	Management Tactics
Spring (April–June)	Tomato spotted wilt virus (TSWV)	See TSWV risk index	Plant a resistant cultivar (Bailey, Sullivan, or Wynne); use a high seeding rate or twin rows; plant after May 5 and before May 16; apply an insecticide in furrow. Consider an additional post-emergence insecticide application.
	CBR (Cylindrocladium black rot)	1% to 10% disease in this field last time peanuts were grown	Rotate 2-4 years; avoid soybeans in rotations. Plant a resistant cultivar (Bailey). Consider an in-furrow fungicide application.
		More than 10% disease in this field last time peanuts were grown	Rotate 3-4 years; avoid soybeans in rotations. Plant a resistant cultivar (Bailey) and fumigate before planting.
June–Harvest	Leaf spots, Web blotch, Pepper spot	R3 (beginning pods)	Rotate at least 2 years to any crop other than peanuts. Longer rotations are preferred. Plant a partially resistant cultivar (Bailey, Sullivan, or Wynne). Begin calendar sprays or advisory program. Use nozzles that give a cone-shaped spray pattern. Use 12 to 24 gallons of water for spray materials applied by ground sprayers. Use at least 5 gallons of water for materials applied by air. **14-day program:** 4 to 6 applications suggested. Begin applications at very early pod (R3). The first application can be delayed 2 weeks on Bailey. Repeat applications at 14-day intervals. **Advisory:** Begin applications at very early pod (R3) The first spray can be delayed 10 days in low-risk fields. Repeat applications when weather conditions become favorable as determined by peanut leaf spot advisories. This schedule requires strict adherence to the program guidelines and usually results in fewer fungicide applications than the 14-day schedule. Contact your local Extension center for details. Not recommended for rotations of less than 3 years. Scout fields: if 20% or more of leaflets have spots in the worst part of the field, begin a 14-day spray program.
		20% leaflets with spots in any area of the field	Reduce intervals between sprays when over threshold. Switch to a more effective fungicide if late leaf spot, web blotch, or pepper spot becomes predominant. If using advisory, switch to a 14-day spray schedule.
	Southern stem rot	Mid-July or on demand	Plant a partially resistant cultivar (Bailey). Avoid highly susceptible cultivars. Rotate 3 to 4 years to non-host crops. Use a soil fungicide or a foliar fungicide with efficacy against soil-borne pathogens at least once from Mid-July to mid-August on resistant cultivars, or up to 3 times on susceptible cultivars in fields with a history of disease, or if signs and symptoms of disease are present. See leaf spots above for application information. Using a surfactant, higher volumes of water (15 to 25 gallons per acre), or spraying at night may improve control.
	Sclerotinia blight	In fields with a history of disease, but less than 10% disease: early July or according to advisory	Plant a partially resistant cultivar (Bailey). Avoid highly susceptible cultivars. Rotate 4+ years with non-host crops. Scout every 2 weeks or according to advisory; begin fungicide applications if disease is seen.
		In fields with a history of greater than 10% disease: just before vines close or according to weather-based Sclerotinia advisory	Plant a partially resistant cultivar (Bailey). Avoid highly susceptible cultivars. Begin fungicide applications. Using higher volumes of water (15-25 gallons per acre) may improve control. Rotate 4-plus years with non-host crops.
September–October	CBR, Sclerotinia blight, Southern stem rot	At digging	Make disease maps to decide future rotations, use of resistant varieties, and to pinpoint areas for fumigation and fungicide application.
October–November	Nematodes, All diseases	Sample areas to be planted to peanut the following spring for nematodes and soil fertility. See the NCDA &CS website for nematode and fertility sampling instructions.	Plan rotation and nematicide use based on recommendations. Adjust soil fertility and pH as recommended. Avoid planting in areas with high levels of Zn and/or a history of manure or litter applications.

Soybean Disease Control

L. D. Thiessen, Entomology and Plant Pathology

Please refer to the current chemical label for all directions regarding safe use and the most up-to-date recommendations for application rates, timing, and harvest intervals.

Table 10-7A. Soybean Nematode Control

Nematodes	Nematicide and Formulation	Amount of Formulation [1]	Precautions and Remarks
Root-Knot, Columbia Lance	1,3-dichloropropene (Telone II)	3 to 6 gal/acre	Apply 10 to 14 days prior to planting. Inject at least 12 inches deep. Do not use in dry, wet, or cold soils.
Soybean Cyst Nematode	*Pasteuria nishizawae* (Clariva)	1 to 3 oz/100 lbs of seed	Seed treatment[1].
Root-Knot, Columbia Lance, Lesion, Soybean Cyst, Sting	abamectin (Avicta)	0.15 mg per seed	Seed treatment.
	Bacillus fermis (Poncho/Votivo)	0.13 mg a.i. per seed	Seed treatment.
	Bacillus amyloliquefaciens (Aveo EZ)	0.1 fl oz per 140,000 seeds	
	fluopyram (Ilevo)	0.075 to 0.25 mg a.i. per seed	Seed treatment

[1] Seed treatments may not be effective in high pressure environments. Soil samples should be taken in the fall when nematode populations are highest to determine if seed treatments may be effective in managing nematode populations.

Table 10-7B. Soybean Foliar Disease Control

Fungicide Type and FRAC Code	Fungicide	Rate Per Acre Formulated	Remarks
Frogeye Leaf Spot and Target Spot			
Strobilurins (11)[1]	azoxystrobin (Quadris)	6.0 to 15.5 fl oz	Apply fungicide at R1 to R3 and make a second application 14 to 21 days later with a different mode of action if disease pressure is high. Do not apply after R5 (small bean) or within 14 days of harvest. Higher rates provide longer residual activity and may reduce the need for a second application.
	picoxystrobin (Aproach)	6.0 to 12.0 fl oz	
	pyraclostrobin (Headline)	6.0 to 12.0 fl oz	
Nitriles (M5)	chlorothalonil (various brands)	1.5 to 2.25 pints	Apply fungicide at R1 to R3 and make a second application 14 days later if disease pressure is high. Do not apply within 48 days of harvest.
Thiophanates (1)	thiophanate-methyl (Topsin M 70WP) (Cercobin)	0.5 to 1 lb 10.9 to 21.8 fl oz	Apply fungicide at R1 to R3 and make a second application 14 to 21 days later if disease pressure is high. Do not apply after R5 (small bean). Higher rates provide longer residual activity and may reduce the need for a second application. Thiophanate-methyl is not labeled for Asiatic soybean rust.
Combinations of Strobilurins and Triazoles (3,11)	prothioconazole (10.8 %) + trifloxystrobin (32.3 %) (Stratego YLD)	4 to 4.65 fl oz	Apply at R1 to R3 and 14 to 21 days later if disease pressure is high. Do not apply after R5 or within 21 days of harvest.
	fluoxastrobin (14.84 %) + flutriafol (19.3 %) (Fortix)	4 to 6 fl oz	Apply at R1 to R3 and 14 to 21 days later if disease pressure is high. Do not apply after R5 or within 21 days of harvest.
	azoxystrobin (18.2 %) + difenoconazole (11.4%) (Quadris Top)[1]	8 to 14 fl oz	Apply at R1 to R3 and 14 to 21 days later if rust is expected. Do not apply after R6 or within 14 days of harvest. Make no more than 2 applications of materials containing azoxystrobin or difenoconazole per year soybean per season. Corn or wheat may be planted within 180 days of last application; do not plant other crops with 360 days of last application.
	tetraconazole (7.48) + azoxystrobin (9.35 %) (Affiance)	10.0 to 14.0 fl oz	Apply at R1 to R3 or when conditions are favorable for disease development and make a second application 15 to 21 days after application if disease pressure is high. Do not apply after R5 or within 14 days of harvest.
	cyproconazoe (7.17 %) + picoxystrobin (17.94 %) (Aproach Prima)	5.0 to 6.8 fl oz	Apply fungicide at R1 to R3 and a second application 14 to 21 days later if disease pressure is high. Do not apply after R5 or within 30 days of harvest.
Combinations of dicarboximides and Strobilurins and (7, 11)	fluxapyroxad (14.33 %) + pyraclostrobin (28.58 %) (Priaxor)	4.0 to 8.0 fl oz	Apply at R1 to R3 prior to infection and a second application 7 to 14 days later if disease pressure is high. Do not apply after R5 or within 21 days of harvest.
Asiatic Soybean Rust[2]			
Strobilurins (11)	azoxystrobin (Quadris)	6.2 to 15.5 oz	Apply at R1 to R3 as a preventative spray and 14 to 21 days later if rust is expected. Do not apply after R5 or within 14 days of harvest. Make no more than 2 applications to soybean per season.
	picoxystrobin (Aproach)	6.0 to 12.0 fl oz	Apply at R1 to R3 as a preventative spray and 14 to 21 days later if disease pressure remains high. Do not apply after R5 or within 14 days of harvest. Make no more than 2 applications to soybean per season
	pyraclostrobin (Headline)	6.0 to 12.0 fl oz	Apply at R1 to R3 as a preventative spray and 14 to 21 days later if rust is expected. Do not apply after R5 or within 21 days of harvest. Make no more than 2 applications to soybean per season.
Triazoles (3)	cyproconazole (Alto)	2.75 to 5.5 fl oz	Apply at R1 to R3 and a second application 14 to 28 days later if rust is expected. Use higher rates if rust is present in field. Do not apply after R6 or within 21 days of harvest. Make no more than 2 applications of cyproconazole to soybean per season. Corn or wheat may be planted within 180 days of last application; do not plant other crops within 360 days of last application.
	flutriafol (Topguard)[1]	7 to 14 fl oz	Apply at R1 to R3 as a preventative spray and 21 to 35 days (7 fl oz) later if rust is expected. Do not apply within 21 days of harvest. Make no more than 2 applications of flutriafol to soybean per season. Plant-back restrictions for all crops except soybean is 120 days after last application.
	propiconazole (Tilt, Propimax, Bumper)	4.0 to 6.0 fl oz	Apply at R1 to R3 as a preventative spray and 14 to 21 days later if rust is expected. Use higher rates if rust is present in field. Do not apply after R5 or within 21 days of harvest. Make no more than 2 applications of propiconazole-containing materials to soybean per season.
	tetraconazole (Domark)	4.0 to 5.0 fl oz	Apply at R1 to R5 as a preventative spray if rust is expected. A second application of another fungicide may be required if disease pressure is high. Use higher rates if rust is present in field. Do not apply after R5 or within 22 days of harvest. Make no more than 2 applications of tetraconazole per season. Peanut, soybean and sugar beets may be planted immediately after the last application; small grains (barley, buckwheat, millet, oats, rice, rye, triticale, and wheat) and sugarcane can be planted 45 days after the last application; all other crops can be planted 120 days after the last application.
	prothioconazole (Proline)	2.5 to 3 fl oz	Apply at R1 to R3 as a preventative spray and 14 to 21 days later if rust is expected. Do not apply after R6 or within 30 days of harvest. Make no more than 3 applications of materials containing triazoles per year. Any crop not listed on the label may be planted within 30 days of last application.
Combinations of Strobilurins and Triazoles (3,11)	azoxystrobin (18.2%) + difenoconazole (11.4%) (Quadris Top)	8.0 to 14.0 fl oz	Apply at R1 to R3 as a preventative spray and 14 to 21 days later if rust is expected. Do not apply after R6 or within 30 days of harvest. Make no more than 2 applications of materials containing azoxystrobin or cyproconazole per year. Make no more than 2 applications of cyproconazole to soybean per season. Corn or wheat may be planted within 180 days of last application; do not plant other crops with 360 days of last application.
	azoxystrobin (7 0%) + propiconazole (11.7%) (Quilt)	14.0 to 20.5 fl oz	Apply at R1 to R3 as a preventative application and 14 to 21 days later if rust is expected. Do not apply after R6. Make no more than 2 applications containing propiconazole per year.
	fluoxastrobin (14.84%) + flutriafol (19.3%) (Fortix)	4.5 to 6 fl oz	Apply at R1 to R3 as a preventative spray and 21 to 35 days later if disease pressure is high. Do not apply after R5 or within 30 days of harvest.
	prothioconazole (10.8%)+ trifloxystrobin (32.3%) (Stratego YLD)	4.0 to 4.65 fl oz	Apply at R1 to R3 as a preventative spray and 14 to 21 days later if disease pressure is high. Do not apply after R5 or within 21 days of harvest. Make no more than 3 applications.
	trifloxystrobin (11.4%)+ propiconazole (11.4%) (Stratego)	10.0 fl oz	Apply at R1 to R3 as a preventative spray and 10 to 21 days later if disease pressure is high. Do not apply after R5 or within 21 days of harvest. Make no more than 2 applications of materials containing propiconazole per year.
	cyproconazole (7.2%) + picoxystrobin (32.3%) (Approach Prima)	5.0 to 6.8 fl oz	Apply at R1 to R3 as a preventative spray and 14 to 21 days later if disease pressure is high. Do not apply after R5 or within 21 days of harvest.
	tetraconazole (7.48) + azoxystrobin (9.35 %) (Affiance)	10.0 to 14.0 fl oz	Apply at R1 to R3 as a preventative spray and 14 to 21 days later if disease pressure is high. Do not apply after R5 or within 21 days of harvest.

Table 10-7B. Soybean Foliar Disease Control

Fungicide Type and FRAC Code	Fungicide	Rate Per Acre Formulated	Remarks
Combinations of Dicarboximides and Strobilurins (7, 11)	fluxapyroxad (14.3%) + pyraclostrobin (28.6%) (Priaxor)	4.0 -8.0 fl oz	Apply at R1 to R3 prior to infection and a second application 7 to 14 days later if disease pressure is high. Do not apply after R5 or within 21 days of harvest.
Nitriles (M5)	chlorothalonil (various brands)	1.5 to 2.25 pints	Apply fungicide at R1 to R3 and make a second application 14 days later. Do not apply within 48 days of harvest.

[1] QOI group fungicides. Avoid using group 11 fungicides (Strobilurins) more than one time in sequence to reduce fungicide resistance development. If fungicide-resistant frogeye leaf spot is identified in your region, avoid using group 11 fungicides.

[2] Fungicides are not recommended for Asiatic Soybean Rust unless it has been confirmed within 100 miles of a field.

Relative Fungicide Efficacy for Control of Soybean Seedling Diseases
L. D. Thiessen, Entomology and Plant Pathology

The members of the Identification and Biology of Seedling Pathogens of Soybean project funded by the North Central Soybean Research Program and plant pathologists across the United States have developed the following ratings for how well fungicide seed treatments control seedling diseases of soybeans in the United States. Efficacy ratings for each fungicide active ingredient listed in the table were determined by field-testing the materials over multiple years and locations by the members of this group, and include ratings summarized from national fungicide trials published in Plant Disease Management Reports (and formerly Fungicide and Nematicide Tests) by the American Phytopathological Society at http://www.apsnet.org. Each rating is based on the fungicide's level of disease control, and does not necessarily reflect efficacy of fungicide active ingredient combinations and/or yield increases obtained from applying the active ingredient.

The list includes the most widely marketed products available. It is not intended to be a list of all labeled active ingredients and products. Additional active ingredients may be available, but have not been evaluated in a manner allowing a rating. Products listed are the most common products available as of the release date of the table; all available products may not be listed. Additional active ingredients may be included in some products for insect and nematode control, however; only active ingredients for pathogen control are listed and rated.

Many active ingredients and their products have specific use restrictions. Read and follow all use restrictions before applying any fungicide to seed, or before handling any fungicide-treated seed. This information is provided only as a guide. It is the applicator's and user's legal responsibility to read and follow all current label directions. Reference in this publication to any specific commercial product, process, or service, or the use of any trade, firm, or corporation name is for general informational purposes only and does not constitute an endorsement, recommendation, or certification of any kind by members of the group, or by the North Central Soybean Research Program. Individuals using such products assume responsibility for their use in accordance with current directions of the manufacturer. Efficacy categories: E = Excellent; VG = Very Good; G = Good; F = Fair; P = Poor; NR = Not Recommended; NS = Not Specified on product label; U = Unknown efficacy or insufficient data to rank product. Please note: Efficacy ratings may be dependent on the rate of the fungicide product on seed. Contact your local Extension plant pathologist for recommended fungicide product rate information for your area.

Table 10-7C. Relative Fungicide Efficacy for Control of Soybean Seedling Diseases

Fungicide active ingredient	*Pythium* sp.[1]	*Phytophthora* root rot	*Rhizoctonia* sp.	*Fusarium* sp.[1,3]	Sudden Death Syndrome (SDS) (*Fusarium virguliforme*)	*Phomopsis* sp.
Azoxystrobin	P-G	NS	VG	F-G	NR	P
Carboxin	U	U	G	U	NR	U
Chloroneb	U	P	E	P	NR	P
Ethaboxam	E	E	U	U	U	U
Fludioxonil	NR	NR	G	F-VG	NR	G
Fluopyram	NR	NR	NR	NR	VG	NR
Fluxapyroxad	U	U	E	G	NR	G
Ipconazole	P	NR	F-G	F-E	NR	G
Mefenoxam	E[2]	E	NR	NR	NR	NR
Metalaxyl	E[2]	E	NR	NR	NR	NR
PCNB	NR	NR	G	U	NR	G
Penflufen	NR	NR	G	G	NR	G
Prothioconazole	NR	NR	G	G	NR	G
Pyraclostrobin	P-G	NR	F	F	NR	G
Sedaxane	NR	NR	E	NS	NR	G
Thiabendazole	NR	NR	NS	NS	P	U
Trifloxystrobin	P	P	F-E	F-G	NR	P-F

[1] Products may vary in efficacy against different *Fusarium* and *Pythium* species.

[2] Areas with mefenoxam or metalaxyl insensitive populations may see less efficacy with these products.

[3] Listed seed treatments do not have efficacy against *Fusarium virguliforme*, causal agent of sudden death syndrome.

Relative Fungicide Efficacy for Soybean Foliar Diseases

L. D. Thiessen, Entomology and Plant Pathology

The North Central Regional Committee on Soybean Diseases (NCERA-137) has developed the following information on foliar fungicide efficacy for control of major foliar soybean diseases in the United States. Efficacy ratings for each fungicide listed in the table were determined by field-testing the materials over multiple years and locations by the members of the committee. Efficacy ratings are based upon level of disease control achieved by product, and are not necessarily reflective of yield increases obtained from product application. Efficacy depends upon proper application timing, rate, and application method to achieve optimum effectiveness of the fungicide as determined by labeled instructions and overall level of disease in the field at the time of application. <u>Differences in efficacy among fungicide products were determined by direct comparisons among products in field tests and are based on a single application of the labeled rate as listed in the table, unless otherwise noted.</u> **Table includes systemic fungicides available that have been tested over multiple years and locations. The table is not intended to be a list of all labeled products**[1]. Efficacy categories: NR=Not Recommended; P = Poor; F = Fair; G = Good; VG = Very Good; E = Excellent; NL = Not Labeled for use against this disease; U = Unknown efficacy or insufficient data to rank product efficacy.

Table 10-7D. Relative Fungicide Efficacy for Soybean Foliar Diseases

Class	Active ingredient (%)	Product/ Trade name	Rate/A (fl oz)	Aerial web blight	Anthracnose	Brown spot	Cercospora leaf blight[2]	Frogeye leaf spot[3]	Phomopsis/Diaporthe (Pod and stem blight)	Soybean rust	White mold[4]	Harvest restriction[5]
QoI Strobilurins Group 11	Azoxystrobin 22.9%	Quadris 2.08 SC Multiple Generics	6.0 to 15.5	VG	VG	G	P	P	U	G-VG	P	14 days
	Fluoxastrobin 40.3%	Aftershock 480 SC Evito 480 SC	2.0 to 5.7	VG	G	G	P	P	U	U	NL	R5 (beginning seed) 30 days
	Picoxystrobin	Aproach 2.08 SC	6.0 to 12.0	VG	G	G	P	P	U	G	G-VG[9]	14 days
	Pyraclostrobin 23.6%	Headline 2.09 EC/SC	6.0 to 12.0	VG	VG	G	P	P	U	VG	NL	21 days
DMI Triazoles Group 3	Cyproconazole 8.9%	Alto 100SL	2.75 to 5.5	U	U	VG	F	F	U	VG	NL	30 days
	Flutriafol 11.8%	Topguard 1.04 SC	7.0 to 14.0	U	VG	VG	P-G	VG	U	VG-E	F	21 days
	Propiconazole 41.8%	Tilt 3.6 EC Multiple Generics[7]	4.0 to 6.0	P	VG	G	NL	F	NL	VG	NL	R5 (beginning seed)
	Prothioconazole 41.0%	Proline 480 SC[8]	2.5 to 5.0	NL	NL	NL	NL	G-VG	NL	VG	F	21 days
	Tetraconazole 20.5%	Domark 230 ME	4.0 to 5.0	NL	VG	VG	P-G	G-VG	U	VG-E	F	R5 (beginning seed)
MBC Thiophanates Group 1	Thiophanate-methyl	Topsin-M Multiple Generics	10.0 to 20.0	U	U	U	F	VG	U	G	F	21 days
2,6-dinitro-anilines Group 29	Fluazinam 40%	Omega 500 DF	0.75 to 1.0 pints	NL	NL	NL	NL	NL	NL	NL	G	R3 (beginning pod)
SDHI Carboxamides Group 7	Boscalid 70%	Endura 0.7 DF	3.5 to 11.0	U	NL	VG	U	P	NL	NL	VG	21 days

Table 10-7D. Relative Fungicide Efficacy for Soybean Foliar Diseases

Class	Active ingredient (%)	Product/ Trade name	Rate/A (fl oz)	Aerial web blight	Anthracnose	Brown spot	Cercospora leaf blight[2]	Frogeye leaf spot[3]	Phomopsis/Diaporthe (Pod and stem blight)	Soybean rust	White mold[4]	Harvest restriction[5]
Mixed mode of action	Azoxystrobin 18.2% Difenconazole 11.4%	Quadris Top 2.72 SC	8.0 to 14.0	U	U	G-VG	P-G	VG	U	VG	NL	14 days
	Azoxystrobin 19.8% Difenconazole 19.8%	Quadris Top SBX 3.76 SC	7.0 to 7.5	U	U	U	U	G-VG	U	U	U	14 days
	Azoxystrobin 7.0% Propiconazole 11.7%	Quilt 1.66 SC Multiple Generics	14.0 to 20.5	U	U	G	F	F	U	VG	NL	21 days
	Azoxystrobin 13.5% Propiconazole 11.7%	Quilt Xcel 2.2 SE	10.5 to 21.0	E	VG	G	F	F	U	VG	NL	R6
	Benzovindiflupyr 10.27% Azoxystrobin 13.5% Propiconazole 11.7%	Trivapro A 0.83 + Trivapro B 2.2 SE	A = 4.0 B = 10.5	E	U	VG	U	VG	U	U	NL	14 days R6
	Cyproconazole 7.17% Picoxystrobin 17.94%	Aproach Prima 2.34 SC	5.0 to 6.8	U	U	VG	P-G	G	U	U	NL	14 days
	Flutriafol 19.3% Fluoxastrobin 14.84%	Fortix SC Preemptor SC	4.0 to 6.0	U	U	G	U	G	U	U	U	R5
	Pyraclostrobin 28.58% Fluxapyroxad 14.33%	Priaxor 4.17 SC	4.0 to 8.0	E	VG	E	P-G	P-F	U	VG	P	21 days
	Pyraclostrobin 28.58% Fluxapyroxad 14.33% Tetraconazole 20.50%	Priaxor D 4.17 SC 1.9 SC	4.0 (each component)	U	U	VG	U	G-VG	U	U	P	21 days
	Trifloxystrobin 32.3% Prothioconazole 10.8%	Stratego YLD 4.18 SC[9]	4.0 to 4.65	VG	VG	VG	F	F	U	VG	NL	21 days
	Tetraconazole 7.48% Azoxystrobin 9.35%	Affiance 1.5 SC	10.0-14.0	U	VG	VG	F	G	U	U	U	R5 14 days

[1] Multiple fungicides are labeled for soybean rust only, powdery mildew, and Alternaria leaf spot, including tebuconazole (multiple products) and Laredo (myclobutanil). Contact fungicides such as chlorothalonil may also be labeled for use.

[2] Cercospora leaf blight efficacy relies on accurate application timing, and standard R3 application timings may not provide adequate disease control. Fungicide efficacy may improve with earlier or later applications; however, efficacy has been inconsistent with some products. Fungicides with a solo or mixed QoI or MBC mode of action may not be effective in areas where QoI or MBC resistance has been detected in the fungal population that causes Cercospora leaf blight.

[3] In areas where QoI-fungicide resistant isolates of the frogeye leaf spot pathogen are not present, QoI fungicides may be more effective than indicated in this table.

[4] White mold efficacy is based on R1-R2 application timing, and lower efficacy is obtained at R3 or later application timings, or if disease symptoms are already present at the time of application.

[5] Harvest restrictions are listed for soybean harvested for grain. Restrictions may vary for other types of soybean (edamame, etc.) and soybean for other uses such as forage or fodder.

[6] Multiple generic products containing this mode of action may also be labeled in some states.

[7] Proline has a supplemental label (2ee) for soybean, only for use on white mold in IL, IN, IA, MI, MN, NE, ND, OH, SD, WI. A separate 2ee for NY exists for white mold.

[8] Stratego YLD has a supplemental label (2ee) for white mold on soybean only in IL, IN, IA, MI, MN, NE, ND, OH, SD, WI.

[9] Rating is based on two applications of a 9 fl oz/A rate of Aproach at R1 and R3.

Many products have specific use restrictions about the amount of active ingredient that can be applied within a period of time or the amount of sequential applications that can occur. Please read and follow all specific use restrictions prior to fungicide use. This information is provided only as a guide. It is the responsibility of the pesticide applicator by law to read and follow all current label directions. Reference to products in this publication is not intended to be an endorsement to the exclusion of others that may be similar. Persons using such products assume responsibility for their use in accordance with current directions of the manufacturer. Members or participants in the NCERA-137 group assume no liability resulting from the use of these products.

Chapter X — 2019 N.C. Agricultural Chemicals Manual

Tobacco Disease Control

L. D. Thiessen, Entomology and Plant Pathology

Table 10-8A. Tobacco Disease Control — Nematode Control

Nematicide	Amount of Formulation Per Acre	Waiting Period Before Planting[1] (Days)	Precautions and Remarks
Fumigants			Rates are for in-row injection. Where labeled, broadcast rates are usually 50% to 100% more than in-row rates. Apply fumigants and multi-purpose fumigants at a final depth of 12 to 14 inches. Apply only when the soil temperature is between 55 degrees F and 80 degrees F and soil is moist but not wet. Should soil become wet for an extended time following application, a longer waiting period before transplanting may be necessary to avoid fumigant injury.
dichloropropene (Telone II)	6.0 gal	21	
dichloropropene + chloropicrin (Telone C-35)	12.0 gal	21	
Contact Nematicides[2] fluensulfone (Nimitz)	3.5 to 7 pt	7	Adequate watering in or rain after application is needed.

[1] Read and follow product label directions concerning worker reentry periods.

[2] Contact nematicides may not be effective in high pressure environments. Assess nematode populations in the fall prior to use of nematicides for effective selection of chemistries to control populations.

Table 10-8B. Tobacco Disease Control — Field Blue Mold, Target Spot, and Frog-eye Leafspot Control

Material (FRAC Code)[1]	Rate Per Acre (Formulated)	Method of Application
azoxystrobin (Quadris) (11)	6.0 to 12.0 fl oz	Apply on a 7- to 14-day interval with sufficient water volume for adequate coverage and canopy penetration. May be applied up to day of harvest. Do not tank mix with thiodan. **Application directions for blue mold:** Applications should begin prior to disease development or at first indication of blue mold in the area. If blue mold is present in the field, apply dimethomorph prior to Quadris applications.
Blue Mold Only (FRAC Code)[1]		
acibenzolar-S-methyl (Actigard) (21)	0.5 oz	Begin preventative applications after plants reach a height of 12 inches (Flue-Cured)/18 inches (Burley). Make up to 3 applications on a 10-day schedule. Apply in a minimum of 20 gallons of water per acre.
aluminum tris (O-ethyl phosphate) (Aliette WDG) (33)	2.5 to 4.0 lb	Apply immediately after transplanting and continue on a 7- to 10-day schedule. Begin with a minimum spray volume of 20 gallons per acre, and increase by 20 gallons per acre weekly to a maximum of 100 gallons per acre. The pH of spray solution should not be less than 6.0. No more than 20 pounds per season.
mandipropamid (Revus) (40)	8.0 fl oz	Apply prior to disease development and continue on a 7- to 10- day schedule. Revus can be tank mixed with another fungicide of different FRAC Code. No more than 2 consecutive applications before switching to another mode of action fungicide. No more than 32 fluid ounces per season. Do not apply within 7 days of harvest.
dimethomorph (Forum) (40)	2.0 to 8.0 oz	Increase rate and spray volume as crop size increases. MUST be used in a tank mix with another fungicide (non-Group 40) active against blue mold. Refer to the partner labeling for rates, application method, and restrictions.
mancozeb (Dithane Rainshield) (M3)	2 lb	Use only if there is a threat of metalaxyl-insensitive blue mold. Mix 1.5 to 2 pounds per 100 gallons per acre. Spray weekly for complete coverage. Discontinue when threat of blue mold no longer exists. In flue-cured tobacco, do not spray after first button or within 21 days of harvest. In burley, do not spray within 30 days of harvest.
Mefenoxam[2] (Ridomil Gold) EC, SL (Ultra Flourish) 2 E (4)	0.5 to 1 pt 1 to 2 pt	For mefenoxam-sensitive strains of the blue mold fungus, apply preplant in a minimum of 15 gallons of water per acre. Incorporate in the top 2 to 4 inches of soil and form beds. Use highest rate for burley tobacco. For prolonged control, especially in burley, apply a supplemental soil application of either 0.5 pints per acre Ridomil Gold EC or 1 pint Ultra Flourish 2 E at layby or the last cultivation. Do not make the supplemental application if more than the highest rate was applied preplant.
Presidio (43)	4.0 fl oz	MUST be used in a tank mix with another fungicide with different mode of action active against blue mold. No more than 2 foliar applications per season.

[1] To prevent resistance in pathogens, alternate fungicides from a group with fungicides in another group. Fungicides in the "M" group are generally considered "low risk" with no signs of resistance developing to the majority of fungicides.

[2] Mefenoxam resistant blue mold has been reported in tobacco producing regions; use multiple modes of action in a growing season or avoid mefenoxam.

Table 10-8C. Tobacco Disease Control — Black Shank, Granville Wilt, and Black Root Rot Control

Material	Amount of Formulation Per Acre	Waiting Period Before Planting (Days)	Precautions and Remarks
chloropicrin 98% (Chlor-O-Pic 100, Chloropicrin 100) chloropicrin 85% (Pic Plus Fumigant)	3.0 gal 4.0 gal	21	Rates are for in-row injection. Where labeled, broadcast rates are usually 25% to 100% more than in-row rates. Apply multipurpose fumigants to a depth of 6 to 8 inches and form a high, wide bed immediately. Apply only when the soil temperature is above 55 degrees F and soil is moist but not wet. Should soil become wet for extended time following applications, a longer waiting period before transplanting may be necessary to avoid fumigant injury. Use with Ridomil for black shank control.
dichloropropene + chloropicrin (Telone C-35)	12.0 gal	21	
Black Shank Only (FRAC Code)[1]			
Mefenoxam (Ridomil Gold) EC, SL (Ultra Flourish) 2 E (4)	1 to 3 pt 2 to 6 pt	0 0	Use in combination with crop rotation and resistant varieties where applicable. For prolonged control apply either 1 pint Ridomil Gold EC or SL or 2 pints Ultra Flourish 2 E just before transplanting followed by either 0.5 to 1 pint Gold EC or 1 to 2 pints Ultra Flourish 2 E at first cultivation and at lay-by. Also, control nematodes for best results.
Oxathiapiprolin (Orondis)		0	If this product is in the co-pack form, it must be mixed with the other chemistry in the co-pack.
Presidio (43)	4 fl oz		No more than 2 soil applications per season. Applications cannot be consecutive. Alternate with another fungicide of different mode of action. Apply at first cultivation or layby.

[1] To prevent resistance in pathogens, alternate fungicides within a group with fungicides in another group. Fungicides in the "M" group are generally considered "low risk" with no signs of resistance developing to the majority of fungicides.

Table 10-8D. Tobacco Disease Control — Tobacco Seedling Disease Control

DISEASE Material (FRAC Code)[1]	Rate	Precautions and Remarks
Blue Mold		
aluminum tris (O-ethyl phosphonate) (Aliette WDG) (33)	0.5 lb/50 gal water	Apply 3 gallons of spray solution per 1,000 square feet for small plants. Increase the volume as the plants grow to a maximum of 12 gallons per 1,000 square feet. Apply preventatively or at the first sign of blue mold. Apply every 5 to 7 days, and do not exceed 2 applications. After application, wait 24 hours before applying any material over top. Apply insecticides that require a wash down to the soil prior to Aliette.
Blue Mold; Anthracnose; Damping-Off (Rhizoctonia); Stem Rot (Rhizoctonia); Target Spot		
mancozeb (Dithane Rainshield) (M)	1 lb/100 gal water (outdoor plant bed)	Begin sprays when seedlings are quarter size: For outdoor beds, mix 1 tablespoon per gallon water, apply 3 to 5 gallons per 900 square feet every 5 to 7 days.
	0.5 lb/100 gal water (greenhouse and float-bed systems)	For greenhouse and float systems, mix 1 teaspoon per gal water, apply 3 to 12 gallons per 1,000 square feet every 5 to 7 days. Use low gallonage on small plants and higher gallonage on larger plants. Do not contaminate float water with mancozeb.
Stem Rot (Rhizoctonia); Target Spot		
azoxystrobin (Quadris) (11)	0.14 oz (4ml)/1000 ft² (equal to 6 fl oz/acre)	Use enough water for thorough coverage (recommend 5 gallons per 1,000 square feet or more). Make ONLY ONE application prior to transplanting.
Mosaic		
milk (whole or skim) OR dry skim milk	5 gal/100 sq yd of bed 5 lb in 5 gal water/100 sq yd	Spray plants within 24 hours of pulling.
soap OR milk (skim or whole)		Wash hands with soap or dip hands every 20 minutes while pulling and transplanting to field.
dry skim milk	1 lb in 1 gal water	
Pythium Root Rot		
etridiazole (Terramaster) 4 EC (14)	1.4 oz/100 gal float water	Apply at least 2 to 3 weeks after seeding. Mix thoroughly in the float water. May be used preventively or curatively. A second application may be made, but no later than 8 weeks after seeding.
Wildfire; Angular Leafspot (Burley Tobacco); Blue Mold		
streptomycin sulfate (Agri-Mycin 17) (25)	Spray 200 ppm using 5 gal/100 sq yd, Drench 100 ppm using 10 gal/100 sq yd	Spray or drench when plants are in two-leaf stage and repeat once a week for five sprays. Prepare the solution by mixing 2 (200 ppm) or 1 (100 ppm) teaspoon of streptomycin (17% to 21%) per gallon of water.

[1] To prevent resistance in pathogens, alternate fungicides within a group with fungicides in another group. Fungicides in the "M" group are generally considered "low risk" with no signs of resistance developing to the majority of fungicides.

Table 10-8E. Tobacco Disease Control of Tomato Spotted Wilt (TSWV) — Virus suppression

Material (FRAC Code)[1]	Rate	Precautions and Remarks
acibenzolar (Actigard 50W) (21)	0.5 oz/25,000 to 50,000 plants (sprayed over the top) OR 10 to 25 ppm (added to the float water)	**Waiver of liability must be signed to obtain label.** Apply to trays or flats 5 to 7 days before transplanting. If sprayed over the top, rinse it off into potting soil. Apply only with calibrated boom sprayer to ensure no overlap. If applied to float water, ensure water is circulated uniformly to all tobacco plants. For better results, dilute Actigard in a small volume of water first, and then add this volume to the float water. Use lower rate in areas with moderate TSWV risk and highest in areas with severe TSWV risk.
	0.5 oz/acre	Up to 3 field applications in 10-day increments may be made starting 10 days after the greenhouse application. Begin applications after plants reach a height of 18 inches.
imidacloprid (Admire Pro) (4A)	1.8 oz/1,000 plants	Apply to trays IN THE GREENHOUSE 3 to 5 days prior to transplanting. Mix with water prior to application; do not add wetting agents or defoamers, and do not use in combination with other pesticides. Immediately after application, wash the material off the plants to transfer it to the potting soil. Observe worker protection standards for greenhouse application.
acibenzolar (Actigard 50W) + imidacloprid (21 + 4A)	See above	See comments above for both products. Apply Actigard first, then imidacloprid. Tank mixing has not been determined to be safe.

[1] To prevent resistance in pathogens, alternate fungicides within a group with fungicides in another group. Fungicides in the "M" group are generally considered "low risk" with no signs of resistance developing to the majority of fungicides.

Turfgrass Disease Control

J. P. Kerns and E. L. Butler, Entomology and Plant Pathology Extension

When more than one brand name exists for an agricultural chemical, the brand name that first came onto the market is listed first. Otherwise, brand names are listed in alphabetical order. The order in which brand names are given is not an indication of a recommendation or criticism. Products marked with an asterisk are not labeled for home lawn use.

Table 10-9. Turfgrass Disease Control

Disease	Fungicide and Formulation[1]	Amount of Formulation (oz/1,000 sq ft)[2]	Application Interval (days)[3]
Algae (Cyanobacteria)	chlorothalonil* (Daconil) 82.5 WDG (Daconil Weather Stik, Legend) 6 F (Daconil Zn) 4.16 F	 1.8 to 3.25 2 to 3.6 4 to 5.5 3 to 5 6 to 8	 7 to 14 7 to 14 14 7 to 14 14
	chlorothalonil + acibenzolar-S-methyl (Daconil Action) 6.1 F*	2 to 3.6 4 to 5.4	7 to 14 14
	chlorothalonil + azoxystrobin (Renown) 5.16 SC*	2.5 to 4.5	10 to 14
	chlorothalonil + fluoxastrobin (Fame C) 4.25 SC*	3 to 5.4	7 to 14
	chlorothalonil + thiophanate-methyl (Spectro) 90 WDG*	2 to 5.76	7 to 14
	chlorothalonil + triticonazole (Reserve) 4.79 SC*	3.2 to 5.4	14 to 28
	fluazinam (Secure) 4.17 SC*	0.5	14
	fluazinam + acibenzolar-S-methyl (Secure Action) 4.18 SC*	0.5	14
	fluazinam + tebuconazole (Traction) 3.24 SC*	1.3	14
	fluxapyroxad (Xzemplar) 2.47 SC	0.21 to 0.26	14 to 28
	mancozeb* (Fore) 80 WP (Dithane, Pentathlon) 75 DF (Pentathlon) 4 LF (Protect) 75 WP (Wingman) 75 WP	 6 6 10 6 6	 7 to 14 refer to label refer to label 7 to 14 refer to label
	mancozeb + copper hydroxide (Junction) 60 DF*	2 to 4	7 to 14
Anthracnose (Colletotrichum cereale)	azoxystrobin (Heritage, Strobe) 50 WG (Heritage) 0.8 TL (Heritage) 0.31 G (Strobe) 2 L	 0.2 to 0.4 1 to 2 2 to 4 lbs 0.38 to 0.77	 14 to 28 14 to 28 14 to 28 14 to 28
	azoxystrobin + acibenzolar-S-methyl (Heritage Action) 51 WG*	0.2 to 0.4	14 to 28
	azoxystrobin + chlorothalonil (Renown) 5.16 SC*	2.5 4.5	7 to 10 14 to 21
	azoxystrobin + difenoconazole (Briskway) 2.7 SC*	0.3 to 0.725	14
	azoxystrobin + propiconazole (Headway) 1.4 ME 1.06 G	 1.5 to 3 2 to 4 lbs	 14 to 28 14 to 28
	azoxystrobin + tebuconazole (Strobe T) 2.67 SC*	0.75 to 1.5	14 to 21
	chlorothalonil* (Daconil Ultrex) 82.5 WDG (Daconil Weather Stik, Legend) 6 F (Daconil Zn) 4.16 F (Chlorothalonil 500ZN) 4.17 F (Chlorothalonil 720SFT) 6 F (Chlorothalonil, Chlorostar) 82.5 DF (Pegasus) 6 L (Pegasus) 82.5 DF (Pegasus HPX) 6 F	 2.75 to 5 3 to 3.6 3.6 to 5.5 4.4 to 5 5.3 to 8 3 to 5 7.9 2.12 to 3.5 5.5 2.0 to 3.2 3.6 to 5.5 3.25 to 5 3.6 to 5.5	 7 to 14 7 to 14 14 7 to 14 14 7 to 14 14 7 to 14 14 7 to 14 7 to 14 7 to 14 7 to 14
	chlorothalonil + acibenzolar-S-methyl (Daconil Action) 6.1 F*	3 to 3.6 3.6 to 5.4	7 to 14 14
	chlorothalonil + fluoxastrobin (Fame C) 4.25 SC*	3 to 5.9	14 to 28
	chlorothalonil + iprodione + thiophanate-methyl + tebuconazole (Enclave) 5.3 F*	3 to 4 7 to 8	14 to 21 28
	chlorothalonil + propiconazole (Concert) 4.3 SC*	4.5 to 8.5	7 to 28
	chlorothalonil + propiconazole + fludioxonil (Instrata) 3.59 SC*	2.75 to 6	14 to 28

Table 10-9. Turfgrass Disease Control

Disease	Fungicide and Formulation[1]	Amount of Formulation (oz/1,000 sq ft)[2]	Application Interval (days)[3]
Anthracnose (*Colletotrichum cereale*) (continued)	chlorothalonil + thiophanate-methyl* (Consyst) 67 WDG (Peregrine) 67 WDG (Spectro) 90 WDG (TM/C) 67 WDG	 2 to 8 2 to 8 3.72 to 5.76 2 to 8	 7 to 14 14 7 to 14 14 to 21
	fenarimol (Rubigan) 1 AS*	1.75 to 3.5	30
	fluazinam (Secure) 4.17 SC*	0.5	14
	fluazinam + acibenzolar-S-methyl (Secure Action) 4.18 SC*	0.5	14
	fluazinam + tebuconazole (Traction) 3.24 SC*	1.3	14
	fludioxonil (Medallion) 50 WP	0.25 to 0.5	14
	fluopyram + trifloxystrobin (Exteris Stressgard) 0.27 SC	2.135 to 6	14 to 28
	fluoxastrobin (Fame) 4 SC 0.25 G	 0.18 to 0.36 2.3 to 4.6 lbs	 14 to 28 14 to 28
	fluoxastrobin + myclobutanil (Fame M) 3.9 SC	0.25 to 1	14 to 28
	fluoxastrobin + tebuconazole (Fame T) 4 SC*	0.45 to 0.9	21 to 28
	flutolanil + thiophanate-methyl (SysStar) 80 WDG	2 to 3	14 to 30
	iprodione + thiophanate-methyl* (26/36) 3.8 F (Dovetail) 3.8 F	 2 to 4 1 to 4	 14 to 21 14 to 21
	iprodione + trifloxystrobin (Interface) 2.27 SC*	4 to 7	refer to label
	isofetamid + tebuconazole (Tekken) 1.8 SC*	3	14 to 28
	metconazole (Tourney) 50 WDG	0.28 to 0.37	14 to 21
	mineral oil (Civitas) + proprietary pigment (Civitas Harmonizer)*	(8 to 32) + (1 to 4)	7 to 21
	myclobutanil (Eagle, Myclobutanil, Siskin) 20 EW	1.2	14 to 21
	penthiopyrad (Velista) 50 WG	0.3 to 0.5	14
	phosphorous acid (Jetphiter) 5.41 F	5	7
	polyoxin D (Affirm) 11.3 WDG (Endorse) 2.5 WP	 0.88 4	 7 to 14 7 to 14
	propiconazole (Banner MAXX, Kestrel, Propiconazole, Savvi, Strider) 1 ME	1 to 2	14 to 28
	pyraclostrobin (Insignia) 20 WG 2 SC	 0.5 to 0.9 0.4 to 0.7	 14 to 28 14 to 28
	pyraclostrobin + boscalid (Honor) 28 WG*	0.55 to 1.1	14 to 28
	pyraclostrobin + fluxapyroxad (Lexicon Intrinsic) 4.17 SC	0.34 to 0.47	14 to 28
	pyraclostrobin + triticonazole (Pillar) 0.81 G	3 lbs	14 to 28
	tebuconazole* (Torque) 3.6 F (Mirage, Stressgard) 2 SC (Skylark, Tebuconazole) 3.6 F	 0.6 to 1.1 1 to 2 0.6	 21 14 to 28 28
	thiophanate-methyl (3336) 50 WP or 4 F (3336 Plus) 2 F (SysTec 1998, T-Bird, TM) 85 WDG (3336) 2 G (SysTec 1998, T-Bird, TM) 4.5 L	 2 to 6 2 to 8 0.67 to 1.3 3 to 9 lbs 1 to 2	 14 14 to 28 14 14 14
	triadimefon (Bayleton) 50 WSP	1	30 to 45
	trifloxystrobin (Compass) 50 WDG	0.15 to 0.25	14 to 21
	trifloxystrobin + triadimefon (Armada) 50 WP (Tartan) 2 SC*	 0.6 to 1.2 1 to 2	 14 to 28 14 to 28
	triticonazole (Trinity) 1.7 SC (Triton) 70 WDG (Triton Flo) 3 F	 0.5 to 1 0.15 to 0.225 0.41 to 1.1	 14 to 28 14 to 28 14 to 28
	triticonazole + chlorothalonil (Reserve) 4.79 SC*	3.2 to 5.4	14 to 28

Table 10-9. Turfgrass Disease Control

Disease	Fungicide and Formulation[1]	Amount of Formulation (oz/1,000 sq ft)[2]	Application Interval (days)[3]
Brown Ring Patch (*Rhizoctonia circinata var. circinata*)	azoxystrobin (Heritage) 50 WG 0.8 TL 0.31 G	0.2 to 0.4 1 to 2 2 to 4 lbs	14 to 28 14 to 28 14 to 28
	azoxystrobin + acibenzolar-S-methyl (Heritage Action) 51 WG*	0.2 to 0.4	14 to 28
	azoxystrobin + difenoconazole (Briskway) 2.7 SC*	0.5 to 0.725	14 to 28
	azoxystrobin + propiconazole (Headway) 1.4 ME	1.5 to 3	14 to 28
	fluazinam + tebuconazole (Traction) 3.24 SC*	1.3	21
	fluoxastrobin + myclobutanil (Fame M) 3.9 SC	0.25 to 1	14 to 28
	fluoxastrobin + tebuconazole (Fame T) 4 SC*	0.45 to 0.9	21 to 28
	isofetamid + tebuconazole (Tekken) 1.8 SC*	3	14 to 28
	penthiopyrad (Velista) 50 WG	0.5	14
	polyoxin D (Affirm) 11.3 WDG (Endorse) 2.5 WP	0.88 4	7 to 14 7 to 14
	pyraclostrobin (Insignia) 2 SC	0.7	14 to 28
	pyraclostrobin + fluxapyroxad (Lexicon Intrinsic) 4.17 SC	0.34 to 0.47	14 to 28
	pyraclostrobin + triticonazole (Pillar) 0.81 G	3 lbs	14 to 28
	tebuconazole* (Torque) 3.6 F (Mirage, Stressgard) 2 SC (Skylark, Tebuconazole) 3.6 F	0.6 to 1.1 1 to 2 0.6	21 14 to 28 28
	triticonazole (Trinity) 1.7 SC (Triton FLO) 3 F	1 to 2 0.5 to 1.1	14 to 28 14 to 28
	triticonazole + chlorothalonil (Reserve) 4.79 SC*	3.2 to 5.4	14 to 28
Brown Patch (*Rhizoctonia solani*)	azoxystrobin (Heritage, Strobe) 50 WG (Heritage) 0.8 TL (Heritage) 0.31 G (Strobe) 2 L	0.2 to 0.4 1 to 2 2 to 4 lbs 0.38 to 0.77	14 to 28 14 to 28 14 to 28 14 to 28
	azoxystrobin + acibenzolar-S-methyl (Heritage Action) 51 WG*	0.2 to 0.4	14 to 28
	azoxystrobin + chlorothalonil (Renown) 5.16 SC*	2.5 4.5	14 14 to 21
	azoxystrobin + difenoconazole (Briskway) 2.7 SC*	0.3 to 0.725	14 to 28
	azoxystrobin + propiconazole (Headway) 1.4 ME 1.06 G	0.75 to 3 2 to 4 lbs	14 to 28 14 to 28
	azoxystrobin + tebuconazole (Strobe T) 2.67 SC*	0.75 to 1.5	14 to 21
	chloroneb (Teremec)* 65 SP 2.9 F	3 to 4 5 to 7	7 to 10 7 to 10
	chlorothalonil* (Daconil Ultrex) 82.5 WDG (Daconil Weather Stik, Legend) 6 F (Daconil Zn) 4.16 F (Chlorothalonil 500ZN) 4.17 F (Chlorothalonil 720SFT) 6 F (Chlorothalonil, Chlorostar) 82.5 DF (Pegasus) 6 L (Pegasus) 82.5 DF (Pegasus HPX) 6 F	1.8 to 3.23 3.7 to 5 2 to 3.6 4 to 5.5 3 to 5 6 to 8 3 to 5 7.9 2.12 to 3.5 5.5 1.8 to 3.2 2 to 3.6 1.82 to 3.25 2 to 3.6	7 to 14 14 7 to 14 14 7 to 14 14 7 to 14 14 7 to 14 14 7 to 14 7 to 14 7 to 14 7 to 14
	chlorothalonil + acibenzolar-S-methyl (Daconil Action) 6.1 F*	2 to 3.5 4 to 5.4	7 to 14 14
	chlorothalonil + fluoxastrobin (Fame C) 4.25 SC*	1.5 to 5.9	14 to 28
	chlorothalonil + iprodione + thiophanate-methyl + tebuconazole (Enclave) 5.3 F*	3 to 4 7 to 8	14 to 21 28
	chlorothalonil + propiconazole (Concert) 4.3 SC*	3 to 5.5 5.5 to 8.5	7 to 14 14 to 28
	chlorothalonil + propiconazole + fludioxonil (Instrata) 3.6 SC*	2.75 to 6	14 to 21

Table 10-9. Turfgrass Disease Control

Disease	Fungicide and Formulation[1]	Amount of Formulation (oz/1,000 sq ft)[2]	Application Interval (days)[3]
Brown Patch (*Rhizoctonia solani*) (continued)	chlorothalonil + thiophanate-methyl* (Spectro) 90 WDG (TM/C) 67 WDG	3 to 5.76 2 to 8	14 to 21 14 to 21
	fenarimol (Rubigan) 1 AS*	1.5	7 to 14
	fluazinam (Secure) 4.17 SC*	0.5	14
	fluazinam + acibenzolar-S-methyl (Secure Action) 4.18 SC*	0.5	14
	fluazinam + tebuconazole (Traction) 3.24 SC*	1.3	14
	fludioxonil (Medallion) 50 WP	0.2 to 0.25 0.5	7 14
	fluopyram + trifloxystrobin (Exteris Stressgard) 0.27 SC	2.135 to 6	14 to 28
	fluoxastrobin (Fame) 4 SC 0.25 G	0.09 to 0.36 1.2 to 4.6 lbs	14 to 28 14 to 28
	fluoxastrobin + myclobutanil (Fame M) 3.9 SC	0.25 to 1	14 to 28
	fluoxastrobin + tebuconazole (Fame T) 4 SC*	0.45 to 0.9	21 to 28
	fluxapyroxad (Xzemplar) 2.47 SC	0.21 to 0.26	14 to 21
	flutolanil (Prostar) 70 WP, 70 DG	1.5 to 3	14 to 21
	flutolanil + thiophanate-methyl (SysStar) 80 WDG	2 to 3	14 to 21
	iprodione 26GT, Iprodione Pro, IPro, Raven* 2 F, 2 SC, 2 SE	3 to 4	14 to 28
	iprodione + thiophanate-methyl* (26/36) 3.8 F (Dovetail) 3.8 F	2 to 4 1 to 4	14 to 21 14 to 21
	iprodione + trifloxystrobin (Interface) 2.27 SC*	3 to 5	refer to label
	isofetamid + tebuconazole (Tekken) 1.8 SC*	3	14 to 28
	mancozeb* (Fore) 80 WP (Dithane) 75 DF (Protect) 75 WP	4 4 4	7 10 7 to 14
	mancozeb + copper hydroxide (Junction) 60 DF*	2 to 4	7
	mandestrobin (Pinpoint) 4SC	0.31	14
	metconazole (Tourney) 50 WDG	0.28 to 0.37	14 to 21
	mineral oil (Civitas) + proprietary pigment (Civitas Harmonizer)*	(8 to 32) + (1 to 4)	7 to 21
	myclobutanil (Eagle, Myclobutanil, Siskin) 20 EW	1.2	14
	penthiopyrad (Velista) 50 WG	0.3 to 0.5	14 to 21
	phosphorous acid (Jetphiter) 5.41 F	5	7
	polyoxin D (Affirm) 11.3 WDG (Endorse) 2.5 WP	0.88 4	7 to 14 7 to 14
	propiconazole (Banner MAXX, Kestrel, Propiconazole, Savvi, Strider) 1 ME	1 to 2	14 to 21
	pyraclostrobin (Insignia) 20 WG 2 SC	0.5 to 0.9 0.4 to 0.7	14 to 28 14 to 28
	pyraclostrobin + boscalid (Honor) 28 WG*	0.55 to 1.1	14 to 28
	pyraclostrobin + fluxapyroxad (Lexicon Intrinsic) 4.17 SC	0.34 to 0.47	14 to 28
	pyraclostrobin + triticonazole (Pillar) 0.81 G	3 lbs	14 to 28
	tebuconazole* (Torque) 3.6 F (Mirage, Stressgard) 2 SC (Skylark, Tebuconazole) 3.6 F	0.6 to 1.1 1 to 2 0.6	21 14 to 28 28
	thiram (Spotrete) 4 F*	3.75 to 7.5	3 to 10
	triadimefon (Bayleton) 50 WSP, 4.15 F	0.5 to 1	15 to 30
	trifloxystrobin (Compass) 50 WDG	0.1 to 0.2 0.15 to 0.25	14 21
	trifloxystrobin + triadimefon (Armada) 50 WP (Tartan) 2 SC*	0.6 to 1.2 1 to 2	14 to 28 14 to 28

Table 10-9. Turfgrass Disease Control

Disease	Fungicide and Formulation[1]	Amount of Formulation (oz/1,000 sq ft)[2]	Application Interval (days)[3]
Brown Patch (*Rhizoctonia solani*) (continued)	triticonazole (Trinity) 1.7 SC (Triton) 70 WDG (Triton Flo) 3 F	0.75 to 2 0.15 to 0.3 0.41 to 1.1	14 to 28 14 to 28 14 to 28
	triticonazole + chlorothalonil (Reserve) 4.79 SC*	3.2 to 5.4	14 to 28
	vinclozolin (Curalan, Touche) 50 EG*	1	14 to 28
Copper Spot (*Gloeocercospora sorghi*)	chlorothalonil* (Daconil Ultrex) 82.5 WDG (Daconil Weather Stik, Legend) 6 F (Daconil Zn) 4.16 F (Chlorothalonil 500ZN) 6 F (Chlorothalonil 720SFT) 6 F (Chlorothalonil, Chlorostar) 82.5 DF (Pegasus) 6 L (Pegasus) 82.5 DF (Pegasus HPX) 6 F	3.7 to 5 4 to 5.5 6 to 8 3 to 5 7.9 2.12 to 3.5 5.5 3.2 3.6 to 5.5 3.25 to 5 3.6 to 5.5	14 14 14 7 to 10 14 7 to 10 14 7 to 10 7 to 14 7 to 14 7 to 14
	chlorothalonil + acibenzolar-S-methyl (Daconil Action) 6.1 F*	4 to 5.4	14
	chlorothalonil + azoxystrobin (Renown) 5.16 SC*	2.5	14
	chlorothalonil + fluoxastrobin (Fame C) 4.25 SC*	5.9	14
	chlorothalonil + iprodione + thiophanate-methyl + tebuconazole (Enclave) 5.3 F*	3 to 4 7 to 8	14 to 21 28
	chlorothalonil + propiconazole (Concert) 4.3 SC*	5.5 to 8.5	14
	chlorothalonil + thiophanate-methyl* (Consyst) 67 WDG (Peregrine) 67 WDG (Spectro) 90 WDG (TM/C) 67 WDG	3 to 8 3 to 8 3 to 5.76 3 to 8	7 to 10 14 14 14 to 21
	fenarimol (Rubigan) 1 AS*	0.75 to 1.5	10 to 28
	fluazinam + tebuconazole (Traction) 3.24 SC*	1.3	14
	fluoxastrobin + myclobutanil (Fame M) 3.9 SC	0.25 to 1	14 to 21
	flutolanil + thiophanate-methyl (SysStar) 80 WDG	2 to 3	14 to 21
	iprodione + thiophanate-methyl (26/36) 3.8 F*	2 to 4	14 to 21
	isofetamid + tebuconazole (Tekken) 1.8 SC*	3	14 to 28
	mancozeb* (Fore) 80 WP (Dithane) 75 DF (Pentathlon) 4 LF (Pentathlon) 75 DF (Protect, Wingman) 75 WP	4 to 8 4 to 8 7 to 14 4 to 8 4 to 8	7 to 14 10 7 to 14 7 7 to 14
	mancozeb + copper hydroxide (Junction) 60 DF*	2 to 4	7 to 14
	myclobutanil (Eagle, Myclobutanil, Siskin) 20 EW	1.2	14
	tebuconazole* (Torque) 3.6 F (Skylark, Tebuconazole) 3.6 F	0.6 to 1.1 0.6	refer to label 28
	thiophanate-methyl (3336) 50 WP or 4 F (3336 Plus) 2 F (SysTec 1998, T-Bird, TM) 85 WDG (3336) 2 G (SysTec 1998, T-Bird, TM) 4.5 L	2 to 4 2 to 4 0.67 to 1.3 1.5 to 6 lbs 1 to 2	14 14 to 28 14 14 14
	thiram (Spotrete) 4F*	3.75 to 7.5	3 to 10
	triadimefon (Bayleton) 50 WSP, 4.15 F	0.5 to 1	15 to 30
Dead Spot (*Ophiosphaerella agrostis*)	azoxystrobin + propiconazole (Headway) 1.4 ME 1.06 G	1.5 to 3 2 to 4 lbs	14 14 to 28
	azoxystrobin + tebuconazole (Strobe T) 2.67 SC*	0.75 to 1.5	14
	boscalid* (Emerald) 70 WG	0.18	14
	chlorothalonil + thiophanate-methyl (Spectro) 90 WDG*	3.72 to 5.76	14
	fludioxonil (Medallion) 50 WP	0.3 to 0.5	14
	pyraclostrobin (Insignia) 20 WG 2 SC	0.5 to 0.9 0.4 to 0.7	14 to 28 14 to 28
	pyraclostrobin + boscalid (Honor) 28 WG*	0.55 to 1.1	14 to 28

Table 10-9. Turfgrass Disease Control

Disease	Fungicide and Formulation[1]	Amount of Formulation (oz/1,000 sq ft)[2]	Application Interval (days)[3]
Dead Spot (*Ophiosphaerella agrostis*) (continued)	pyraclostrobin + fluxapyroxad (Lexicon Intrinsic) 4.17 SC	0.34 to 0.47	14 to 28
	pyraclostrobin + triticonazole (Pillar) 0.81 G	3 lbs	14 to 28
	thiophanate-methyl (3336) 50WP or 4 F (3336 Plus) 2 F (3336) 2 G	4 to 6 4 to 6 6 to 9 lbs	14 14 14
Dollar Spot (*Sclerotinia homoeocarpa*)	azoxystrobin + difenoconazole (Briskway) 2.7 SC*	0.3 to 0.725	14 to 21
	azoxystrobin + propiconazole (Headway) 1.4 ME 1.06 G	0.75 to 3 2 to 4 lbs	7 to 28 14 to 28
	azoxystrobin + tebuconazole (Strobe T) 2.67 SC*	0.75 to 1.5	14 to 21
	boscalid* (Emerald) 70 WG	0.13 to 0.18	14 to 28
	chlorothalonil* (Daconil Ultrex) 82.5W DG (Daconil Weather Stik, Legend) 6 F (Daconil Zn) 4.16 F (Chlorothalonil 500ZN) 4.17 F (Chlorothalonil 720SFT) 6 F (Chlorothalonil, Chlorostar) 82.5 DF (Pegasus) 6 L (Pegasus) 82.5 DF (Pegasus HPX) 6 F	1 to 3.25 3.7 to 5 1 to 3.6 4 to 5.5 1.5 to 5 6 to 8 3 to 5 7.9 2.12 to 3.5 5.5 1.8 to 3.2 2 to 3.6 1.82 to 3.25 2 to 3.6	7 to 21 14 to 21 7 to 21 14 to 21 7 to 21 14 7 to 14 14 7 to 14 14 7 to 10 7 to 14 7 to 14 7 to 14
	chlorothalonil + acibenzolar-S-methyl (Daconil Action) 6.1 F*	1 to 3.5 4 to 5.4	7 to 21 14
	chlorothalonil + azoxystrobin (Renown) 5.16 SC*	2.5 to 4.5	7 to 14
	chlorothalonil + fluoxastrobin (Fame C) 4.25 SC*	3 to 5.9	14 to 21
	chlorothalonil + iprodione + thiophanate-methyl + tebuconazole (Enclave) 5.3 F*	3 to 4 7 to 8	14 to 21 28
	chlorothalonil + propiconazole (Concert) 4.3 SC*	1.5 to 3 3 to 5.5 5.5 to 8.5	7 to 10 14 to 21 14 to 28
	chlorothalonil + propiconazole + fludioxonil (Instrata) 3.6 SC*	2.75 to 6	21 to 28
	chlorothalonil + thiophanate-methyl* (Consyst) 67 WDG (Peregrine) 67 WDG (Spectro) 90 WDG (TM/C) 67 WDG	2 to 8 2 to 8 3.72 to 5.76 2 to 8	7 to 21 14 14 to 21 7 to 14
	fenarimol (Rubigan) 1 AS*	0.75 to 1.5	10 to 28
	fluazinam (Secure) 4.17 SC*	0.5	14
	fluazinam + acibenzolar-S-methyl (Secure Action) 4.18 SC*	0.5	14 to 21
	fluazinam + tebuconazole (Traction) 3.24 SC*	1.3	14
	fluopyram + trifloxystrobin (Exteris Stressgard) 0.27 SC	1.5 to 4.135	7 to 28
	fluoxastrobin (Fame) 4 SC 0.25 G	0.18 to 0.36 2.3 to 4.6 lbs	14 to 21 14 to 21
	fluoxastrobin + myclobutanil (Fame M) 3.9 SC	0.25 to 1	14 to 21
	fluoxastrobin + tebuconazole (Fame T) 4 SC*	0.45 to 0.9	21 to 28
	fluxapyroxad (Xzemplar) 2.47 SC	0.16 to 0.26	14 to 28
	flutolanil + thiophanate-methyl (SysStar) 80 WDG	2 to 3	14 to 30
	iprodione (26GT, Iprodione Pro, IPro, Raven) 2 F, 2 SC, 2 SE*	2 to 4	14 to 28
	iprodione + thiophanate-methyl* (26/36) 3.8 F (Dovetail) 3.8 F	2 to 4 1 to 4	14 to 21 14 to 21
	iprodione + trifloxystrobin (Interface) 2.27 SC*	2 to 5	refer to label
	isofetamid (Kabuto 3.33 SC)	0.4 to 0.5	14
	isofetamid + tebuconazole (Tekken) 1.8 SC*	3	14 to 28

Table 10-9. Turfgrass Disease Control

Disease	Fungicide and Formulation[1]	Amount of Formulation (oz/1,000 sq ft)[2]	Application Interval (days)[3]
Dollar Spot (*Sclerotinia homoeocarpa*) (continued)	mancozeb* (Fore) 80 WP (Dithane) 75 DF (Pentathlon) 4 LF (Pentathlon) 75 DF (Protect, Wingman) 75 WP	6 to 8 6 to 8 10 to 14 6 to 8 6 to 8	7 to 14 10 7 to 14 7 7 to 14
	mancozeb + copper hydroxide (Junction) 60 DF*	2 to 4	7 to 14
	mandestrobin (Pinpoint) 4SC	0.17 to 0.31	14 to 21
	metconazole (Tourney) 50 WDG	0.18 to 0.37	14 to 21
	mineral oil (Civitas) + proprietary pigment (Civitas Harmonizer)*	(8 to 32) + (1 to 4)	7 to 21
	myclobutanil (Eagle, Myclobutanil, Siskin) 20 EW	0.5 to 2.4	7 to 28
	penthiopyrad (Velista) 50 WG	0.3 to 0.5	14 to 21
	propiconazole (Banner MAXX, Propiconazole, Savvi, Spectator) 1 ME	0.5 to 2	7 to 28
	pydiflumetofen (Posterity) 1.67 SC*	0.08 to 0.32	14 to 28
	pyraclostrobin (Insignia) 20 WG 2 SC	0.9 0.7	14 14
	pyraclostrobin + boscalid (Honor) 28 WG*	0.83 to 1.1	14 to 21
	pyraclostrobin + fluxapyroxad (Lexicon Intrinsic) 4.17 SC	0.34 to 0.47	14 to 28
	pyraclostrobin + triticonazole (Pillar) 0.81 G	3 lbs	14 to 28
	tebuconazole* (Torque) 3.6 F (Mirage Stressgard) 2 SC (Skylark, Tebuconazole) 3.6 F	0.6 to 1.1 1 to 2 0.6	refer to label 14 to 28 28
	thiophanate-methyl (3336) 50WP or 4 F (3336 Plus) 2 F (SysTec 1998, T-Bird, TM) 85 WDG (3336) 2 G (SysTec 1998, T-Bird, TM) 4.5 L	2 to 4 2 to 4 0.67 to 1.3 1.5 to 6 lbs 1 to 2	14 14 to 28 14 14 14
	thiram (Spotrete) 4 F*	3.75 to 7.5	3 to 10
	triadimefon (Bayleton) 50 WSP, 4.15 F	0.25 to 1	14 to 30
	trifloxystrobin + triadimefon (Armada) 50 WP (Tartan) 2 SC*	0.6 to 1.2 1 to 2	14 to 28 14 to 28
	triticonazole (Trinity) 1.7 SC (Triton) 70 WDG (Triton FLO) 3 F	1 to 2 0.15 to 0.3 0.28 to 1.1	14 to 28 14 to 28 14 to 28
	triticonazole + chlorothalonil (Reserve) 4.79 SC*	3.2 to 4.5	14 to 28
	vinclozolin (Curalan, Touche) 50 EG*	1	21 to 28
Fairy Ring (Basidiomycetes)	azoxystrobin (Heritage, Strobe) 50 WG (Heritage) 0.8 TL (Heritage) 0.31 G	0.4 2 2 to 4 lbs	28 28 14 to 28
	azoxystrobin + acibenzolar-S-methyl (Heritage Action) 51 WG*	0.2 to 0.4	14 to 28
	azoxystrobin + difenoconazole (Briskway) 2.7 SC*	0.5 to 0.725	14 to 28
	azoxystrobin + propiconazole (Headway) 1.4ME 1.06 G	1.5 to 3 2 to 4 lbs	14 to 28 14 to 28
	azoxystrobin + tebuconazole (Strobe T) 2.67 SC*	0.75 to 1.5	28
	chlorothalonil + fluoxastrobin (Fame C) 4.25 SC*	4.5 to 5.9	21 to 28
	fluoxastrobin (Fame) 4 SC 0.25 G	0.28 to 0.36 2.3 to 4.6 lbs	21 to 28 28
	fluoxastrobin + myclobutanil (Fame M) 3.9 SC	0.5 to 1	21 to 28
	fluoxastrobin + tebuconazole (Fame T) 4 SC*	0.45 to 0.9	21 to 28
	flutolanil (Prostar) 70 WP, 70 WDG	2.2 to 4.5	21 to 30
	flutolanil + thiophanate-methyl (SysStar) 80 WDG	3 to 6.12	21 to 28

Table 10-9. Turfgrass Disease Control

Disease	Fungicide and Formulation[1]	Amount of Formulation (oz/1,000 sq ft)[2]	Application Interval (days)[3]
Fairy Ring (Basidiomycetes) (continued)	isofetamid + tebuconazole (Tekken) 1.8 SC*	3	14 to 28
	mandestrobin (Pinpoint) 4SC	0.31	14
	metconazole (Tourney) 50 WDG	0.37	21
	penthiopyrad (Velista) 50 WG	0.5 to 0.7	14 to 28
	polyoxin D (Affirm) 11.3 WDG (Endorse) 2.5 WP	 1 4	 7 7
	pydiflumetofen (Posterity) 1.67 SC*	0.08 to 0.32	21 to 28
	pyraclostrobin (Insignia) 20 WG 2 SC	 0.9 0.7	 28 28
	pyraclostrobin + boscalid (Honor) 28 WG*	1.1	28
	pyraclostrobin + fluxapyroxad (Lexicon Intrinsic) 4.17 SC	0.47	28
	pyraclostrobin + triticonazole (Pillar) 0.81 G	3 lbs	14 to 28
	tebuconazole* (Torque) 3.6 F (Mirage Stressgard) 2 SC	 0.6 to 1.1 1 to 2	 21 28
	triadimefon (Bayleton) 50DF, 4.15 F	1 to 2	14 to 21
Gray Leaf Spot (*Pyricularia grisea*)	azoxystrobin (Heritage, Strobe) 50 WG (Heritage) 0.8 TL (Heritage) 0.31 G (Strobe) 2 L	 0.2 to 0.4 1 to 2 2 to 4 lbs 0.38 to 0.77	 14 to 28 14 to 28 14 to 28 14 to 28
	azoxystrobin + acibenzolar-S-methyl (Heritage Action) 51 WG*	0.2 to 0.4	14 to 28
	azoxystrobin + chlorothalonil (Renown) 5.16 SC*	2.5 to 4.5	10 to 14
	azoxystrobin + difenoconazole (Briskway) 2.7 SC*	0.5 to 0.725	14 to 21
	azoxystrobin + propiconazole (Headway) 1.4 ME 1.06 G	 1.5 to 3 2 to 4 lbs	 14 to 28 14 to 28
	azoxystrobin + tebuconazole (Strobe T) 2.67 SC*	0.75	1.5
	chlorothalonil* (Daconil Ultrex) 82.5 WDG (Daconil Weather Stik, Legend) 6 F (Daconil Zn) 4.16 F (Chlorothalonil 500ZN) 4.17 F (Chlorothalonil 720SFT) 6 F (Chlorothalonil, Chlorostar) 82.5 DF (Pegasus) 6 L (Pegasus) 82.5 DF (Pegasus HPX) 6 F	 1.8 to 3.25 3.7 to 5 2 to 3.6 4 to 5.5 3 to 5 6 to 8 3 to 5 7.9 2.12 to 3.5 5.5 1.8 to 3.2 2 to 3.6 1.82 to 3.25 2 to 3.6	 7 to 21 14 7 to 10 14 7 to 14 14 7 to 10 14 7 to 14 14 7 to 10 7 to 14 7 to 14 7 to 14
	chlorothalonil + acibenzolar-S-methyl (Daconil Action) 6.1 F*	2 to 3.5 4 to 5.4	7 to 10 14
	chlorothalonil + fluoxastrobin (Fame C) 4.25 SC*	3 to 5.9	14 to 28
	chlorothalonil + iprodione + thiophanate-methyl + tebuconazole (Enclave) 5.3 F*	3 to 4 7 to 8	14 to 21 28
	chlorothalonil + propiconazole (Concert) 4.3 SC*	3 to 5.5 5.5 to 8.5	7 to 14 14 to 21
	chlorothalonil + propiconazole + fludioxonil (Instrata) 3.6 SC*	2.75 to 6	10 to 14
	chlorothalonil + thiophanate-methyl* (Consyst) 67 WDG (Peregrine) 67 WDG (Spectro) 90 WDG (TM/C) 67 WDG	 2 to 8 2 to 8 3.72 to 5.76 2 to 8	 7 to 14 14 14 14 to 21
	fluazinam + tebuconazole (Traction) 3.24 SC*	1.3	21
	fludioxonil (Medallion) 50 WP	0.25 to 0.5	14
	fluopyram + trifloxystrobin (Exteris Stressgard) 0.27 SC	2.135 to 6	14 to 28
	fluoxastrobin (Fame) 4 SC 0.25 G	 0.18 to 0.36 2.3 to 4.6 lbs	 14 to 28 14 to 28
	fluoxastrobin + myclobutanil (Fame M) 3.9 SC	0.25 to 1	14 to 28

Table 10-9. Turfgrass Disease Control

Disease	Fungicide and Formulation[1]	Amount of Formulation (oz/1,000 sq ft)[2]	Application Interval (days)[3]
Gray Leaf Spot (*Pyricularia grisea*) (continued)	fluoxastrobin + tebuconazole (Fame T) 4 SC*	0.45 to 0.9	21 to 28
	flutolanil + thiophanate-methyl (SysStar) 80 WDG	2 to 3	14
	isofetamid + tebuconazole (Tekken) 1.8 SC*	3	14 to 28
	mancozeb* (Fore) 80 WP (Dithane) 75 DF (Pentathlon) 4 LF (Pentathlon) 75 DF (Wingman) 75 WP	8 6.4 to 12.8 9 to 14 8 8	14 7 to 14 5 7 7
	metconazole (Tourney) 50 WDG	0.37	14
	mineral oil (Civitas) + proprietary pigment (Civitas Harmonizer)*	(8 to 32) + (1 to 4)	7 to 21
	myclobutanil (Eagle, Siskin) 20 EW	1.2 to 2.4	14
	polyoxin D (Affirm) 11.3 WDG	0.88	7 to 14
	propiconazole (Banner MAXX, Kestrel, Propiconazole, Savvi, Strider) 1 ME	1 to 2	14
	pyraclostrobin (Insignia) 20 WG 2 SC	 0.5 to 0.9 0.4 to 0.7	 14 to 28 14 to 28
	pyraclostrobin + boscalid (Honor) 28 WG*	0.55 to 1.1	14 to 28
	pyraclostrobin + fluxapyroxad (Lexicon Intrinsic) 4.17 SC	0.34 to 0.47	14 to 28
	pyraclostrobin + triticonazole (Pillar) 0.81 G	3 lbs	14 to 28
	tebuconazole* (Torque) 3.6 F (Mirage Stressgard) 2 SC (Skylark, Tebuconazole) 3.6 F	 0.6 to 1.1 1 to 2 0.6	 21 14 to 28 28
	thiophanate-methyl (3336) 50 WP or 4 F (3336 Plus) 2 F (3336) 2 G (SysTec 1998, T-Bird, TM) 85 WDG (SysTec 1998, T-Bird, TM) 4.5 L	 4 to 6 4 to 8 6 to 9 lbs 2.35 to 3.53 3.5 to 5	 14 14 to 28 14 14 14
	triadimefon (Bayleton) 50 WSP, 4.15 F	0.5 to 1	14
	trifloxystrobin (Compass) 50 WDG	0.15 to 0.2 0.25	14 21
	trifloxystrobin + triadimefon (Armada) 50 WP (Tartan) 2 SC*	 0.6 to 1.2 1 to 2	 14 to 28 14 to 28
Helminthosporium Leaf Spot/ Melting Out (*Bipolaris spp.*; *Drechslera spp.*)	azoxystrobin (Heritage, Strobe) 50 WG (Heritage) 0.8 TL (Heritage) 0.31 G (Strobe) 2 L	 0.2 to 0.4 1 to 2 2 to 4 lbs 0.38 to 0.77	 14 to 21 14 to 21 14 to 21 14 to 21
	azoxystrobin + acibenzolar-S-methyl (Heritage Action) 51 WG*	0.2 to 0.4	14 to 21
	azoxystrobin + chlorothalonil (Renown) 5.16 SC*	2.5 to 4.5	14 to 21
	azoxystrobin + difenoconazole (Briskway) 2.7 SC*	0.5 to 0.725	14 to 21
	azoxystrobin + propiconazole (Headway) 1.4 ME 1.06 G	 1.5 to 3 2 to 4 lbs	 14 to 21 14 to 21
	azoxystrobin + tebuconazole (Strobe T) 2.67 SC*	0.75 to 1.5	14 to 21
	chlorothalonil* (Daconil Ultrex) 82.5 WDG (Daconil Weather Stik, Legend) 6 F (Daconil Zn) 4.16 F (Chlorothalonil 500ZN) 4.17 F (Chlorothalonil 720SFT) 6 F (Chlorothalonil, Chlorostar) 82.5 DF (Pegasus) 6 L (Pegasus) 82.5 DF (Pegasus HPX) 6 F	 1.8 to 3.25 3.7 to 5 2 to 3.6 4 to 5.5 3 to 5 6 to 8 3 to 5 7.9 2.12 to 3.5 5.5 1.8 to 3.2 2 to 3.6 1.82 to 3.25 2 to 3.6	 7 to 21 14 to 21 7 to 21 14 7 to 21 14 7 to 10 14 7 to 10 14 7 to 10 7 to 14 7 to 14 7 to 14
	chlorothalonil + acibenzolar-S-methyl (Daconil Action) 6.1 F*	2 to 3.5 4 to 5.4	7 to 21 14
	chlorothalonil + fluoxastrobin (Fame C) 4.25 SC*	3 to 5.9	14 to 21

Table 10-9. Turfgrass Disease Control

Disease	Fungicide and Formulation[1]	Amount of Formulation (oz/1,000 sq ft)[2]	Application Interval (days)[3]
Helminthosporium Leaf Spot/ Melting Out (continued)	chlorothalonil + propiconazole (Concert) 4.3 SC*	3 to 5.5 5.5 to 8.5	7 to 14 14 to 21
	chlorothalonil + propiconazole + fludioxonil (Instrata) 3.6 SC*	2.75 to 6	10 to 21
	chlorothalonil + thiophanate-methyl* (Consyst) 67 WDG (Peregrine) 67 WDG (Spectro) 90 WDG (TM/C) 67 WDG	2 to 8 2 to 8 3.72 to 5.76 2 to 8	7 to 21 14 14 14 to 21
	fluazinam (Secure) 4.17 SC*	0.5	14
	fluazinam + acibenzolar-S-methyl (Secure Action) 4.18 SC*	0.5	14
	fluazinam + tebuconazole (Traction) 3.24 SC*	1.3	14
	fludioxonil (Medallion) 50 WP	0.25 to 0.5	14 to 21
	fluoxastrobin (Fame) 4 SC 0.25 G	0.18 to 0.36 2.3 to 4.6 lbs	14 to 21 14 to 21
	fluoxastrobin + myclobutanil (Fame M) 3.9 SC	0.25 to 1	14 to 28
	fluoxastrobin + tebuconazole (Fame T) 4 SC*	0.45 to 0.9	21 to 28
	flutolanil + thiophanate-methyl (SysStar) 80 WDG	2 to 3	14
	iprodione (26GT, Iprodione Pro, IPro, Raven) 2 F, 2 SC, 2 SE*	3 to 4	14 to 28
	iprodione + thiophanate-methyl * (26/36) 3.8 F (Dovetail) 3.8 F	2 to 4 1 to 4	14 to 21 14 to 21
	iprodione + trifloxystrobin (Interface) 2.27 SC*	3 to 5	refer to label
	mancozeb* (Fore) 80 WP (Dithane) 75 DF (Pentathlon) 4 LF (Pentathlon) 75 DF (Protect, Wingman) 75 WP	4 4 5 to 14 4 4	7 to 14 10 3 to 5 7 7 to 14
	mancozeb + copper hydroxide (Junction) 60 DF*	2 to 4	7 to 14
	mineral oil (Civitas) + proprietary pigment (Civitas Harmonizer)*	(8 to 32) + (1 to 4)	7 to 21
	myclobutanil (Eagle, Myclobutanil, Siskin) 20 EW	1.2	14
	penthiopyrad (Velista) 50 WG	0.3 to 0.5	14
	polyoxin D (Affirm) 11.3 WDG (Endorse) 2.5 WP	0.88 4	7 to 14 7 to 14
	propiconazole (Banner MAXX, Kestrel, Propiconazole, Savvi, Strider) 1 ME	1 to 2	14
	pyraclostrobin (Insignia) 20 WG 2 SC	0.5 to 0.9 0.4 to 0.7	14 to 28 14 to 28
	pyraclostrobin + boscalid (Honor) 28WG*	0.55 to 1.1	14 to 28
	pyraclostrobin + fluxapyroxad (Lexicon Intrinsic) 4.17 SC	0.34 to 0.47	14 to 28
	pyraclostrobin + triticonazole (Pillar) 0.81 G	3 lbs	14 to 28
	thiophanate-methyl (3336) 50 WP or 4 F (3336 Plus) 2 F (3336) 2 G	4 to 6 4 to 8 6 to 9 lbs	14 14 to 28 14
	thiram (Spotrete) 4 F*	3.75 to 7.5	3 to 10
	trifloxystrobin (Compass) 50 WDG	0.1 to 0.15 0.15 to 0.25	14 21 to 28
	trifloxystrobin + triadimefon (Armada) 50 WP (Tartan) 2 SC*	0.6 to 1.2 1 to 2	14 to 28 14 to 28
	triticonazole (Trinity) 1.7 SC (Triton) 70 WDG	0.5 to 2 0.15 to 0.3	14 to 28 14 to 28
	triticonazole + chlorothalonil (Reserve) 4.79 SC*	3.2 to 4.5	14 to 28
	vinclozolin (Curalan, Touche) 50 EG*	1	14 to 28

Table 10-9. Turfgrass Disease Control

Disease	Fungicide and Formulation[1]	Amount of Formulation (oz/1,000 sq ft)[2]	Application Interval (days)[3]
Large Patch (Zoysia Patch) (*Rhizoctonia solani*)	azoxystrobin (Heritage, Strobe) 50 WG (Heritage) 0.8 TL (Heritage) 0.31 G (Strobe) 2 L	0..2 to 0.4 2 2 to 4 lbs 0.38 to 0.77	14 to 28 14 to 28 14 to 28 28
	azoxystrobin + acibenzolar-S-methyl (Heritage Action) 51 WG*	0.2 to 0.4	14 to 28
	azoxystrobin + chlorothalonil (Renown) 5.16 SC*	2.5 4.5	14 14 to 21
	azoxystrobin + difenoconazole (Briskway) 2.7 SC*	0.3 to 0.725	14 to 28
	azoxystrobin + propiconazole (Headway) 1.4 ME 1.06 G	1.5 to 3 2 to 4 lbs	14 to 28 14 to 28
	azoxystrobin + tebuconazole (Strobe T) 2.67 SC*	0.75 to 1.5	14 to 21
	chloroneb* (Teremec) 65 SP (Teremec) 2.9 F	5 9	21 to 28 21 to 28
	chlorothalonil + fluoxastrobin (Fame C) 4.25 SC*	3 to 5.9	14 to 28
	chlorothalonil + iprodione + thiophanate-methyl + tebuconazole (Enclave) 5.3 F*	3 to 4 7 to 8	14 to 21 28
	chlorothalonil + thiophanate-methyl* (Consyst) 67 WDG (Peregrine) 67 WDG	2 to 8 2 to 8	7 to 14 14
	fluazinam (Secure) 4.17 SC*	0.5	14
	fluazinam + acibenzolar-S-methyl (Secure Action) 4.18 SC*	0.5	14
	fluazinam + tebuconazole (Traction) 3.24 SC*	1.3	14
	fluoxastrobin (Fame) 4 SC 0.25	0.28 to 0.36 2.3 to 4.6 lbs	14 to 28 14 to 28
	fluoxastrobin + myclobutanil (Fame M) 3.9 SC	0.5 to 1	21 to 28
	fluxapyroxad (Xzemplar) 2.47 SC	0.21 to 0.26	14 to 28
	flutolanil (Prostar) 70 WP, 70 WDG	2.2	30
	iprodione (26GT, Iprodione Pro, IPro, Raven) 2 F, 2 SC, 2 SE*	4	14 to 21
	iprodione + thiophanate-methyl (26/36) 3.8 F*	2 to 4	14 to 21
	iprodione + trifloxystrobin (Interface) 2.27 SC*	4	14 to 21
	isofetamid + tebuconazole (Tekken) 1.8 SC*	3	14 to 28
	metconazole (Tourney) 50 WDG	0.37	14
	myclobutanil (Eagle, Myclobutanil, Siskin) 20 EW	2.4	28 (fall)
	penthiopyrad (Velista) 50 WG	0.7	14 to 28
	polyoxin D (Affirm) 11.3 WDG (Endorse) 2.5 WP	0.88 4	7 to 14 7 to 14
	propiconazole (Banner MAXX, Kestrel, Propiconazole, Savvi, Strider) 1 ME	3 to 4	early fall
	pyraclostrobin (Insignia) 20 WG 2 SC	0.5 to 0.9 0.4 to 0.7	14 to 28 14 to 28
	pyraclostrobin + boscalid (Honor) 28 WG*	1.1	14 to 28
	pyraclostrobin + fluxapyroxad (Lexicon Intrinsic) 4.17 SC	0.34 to 0.47	14 to 28
	pyraclostrobin + triticonazole (Pillar) 0.81 G	3 lbs	14 to 28
	tebuconazole* (Torque) 3.6 F (Mirage Stressgard) 2 SC (Skylark, Tebuconazole) 3.6 F	0.6 to 1.1 1 to 2 0.6	21 28 28
	thiophanate-methyl (3336) 50WP or 4 F (3336 Plus) 2 F (SysTec 1998, T-Bird, TM) 85 WDG (3336) 2 G (SysTec 1998, T-Bird, TM) 4.5 L	2 to 4 2 to 4 0.67 to 1.3 1.5 to 6 lbs 1 to 2	14 14 to 28 14 14 14
	thiophanate-methyl + flutolanil (SysStar) 80 WDG	2 to 3	14 to 21
	triadimefon (Bayleton) 50 WSP, 4.15 F	1 to 2	fall and spring

Table 10-9. Turfgrass Disease Control

Disease	Fungicide and Formulation[1]	Amount of Formulation (oz/1,000 sq ft)[2]	Application Interval (days)[3]
Large Patch (Zoysia Patch) (*Rhizoctonia solani*) (continued)	triconazole (Trinity) 1.7 SC (Triton) 70 WDG	1 to 2 0.15 to 0.3	14 to 28 14 to 28
	triconazole + chlorothalonil (Reserve) 4.79 SC*	3.2 to 5.4	14 to 28
Leaf and Sheath Spot (*Rhizoctonia zeae, R. oryzae*)	azoxystrobin (Heritage) 0.8 TL 0.31 G	2 2 to 4 lbs	14 to 28 14 to 28
	azoxystrobin + acibenzolar-S-methyl (Heritage Action) 51 WG*	0.2 to 0.4	14 to 28
	azoxystrobin + chlorothalonil (Renown) 5.16 SC*	2.5 4.5	14 14 to 21
	azoxystrobin + difenoconazole (Briskway) 2.7 SC*	0.5 to 0.725	14 to 28
	azoxystrobin + propiconazole (Headway) 1.4 ME 1.06 G	1.5 to 3 2 to 4 lbs	14 to 28 14 to 28
	azoxystrobin + tebuconazole (Strobe T) 2.67 SC*	0.75 to 1.5	14 to 21
	chlorothalonil + propiconazole + fludioxonil (Instrata) 3.59 SC*	2.75 to 6	14 to 21
	chlorothalonil + thiophanate-methyl (Spectro) 90 WDG*	3 to 5.76	14 to 21
	flutolanil (Prostar) 70 WDG	2.2 to 4.5	14 to 21
	penthiopyrad (Velista) 50 WG	0.3 to 0.5	14
	polyoxin D (Affirm) 11.3 WDG	0.88	7 to 14
	pyraclostrobin (Insignia) 20 WG 2 SC	0.5 to 0.9 0.4 to 0.7	14 to 28 14 to 28
	pyraclostrobin + boscalid (Honor) 28 WG*	1.1	14 to 28
	pyraclostrobin + fluxapyroxad (Lexicon Intrinsic) 4.17 SC	0.34 to 0.47	14 to 28
	pyraclostrobin + triticonazole (Pillar) 0.81 G	3 lbs	28
Pink Patch (*Limonomyces roseipelis*)	azoxystrobin (Heritage, Strobe) 50 WG (Heritage) 0.8 TL (Heritage) 0.31 G (Strobe) 2 L	0.2 to 0.4 1 to 2 2 to 4 lbs 0.38 to 0.77	14 to 28 14 to 28 14 to 28 14 to 28
	azoxystrobin + acibenzolar-S-methyl (Heritage Action) 51 WG*	0.2 to 0.4	14 to 28
	azoxystrobin + chlorothalonil (Renown) 5.16 SC*	2.5 to 4.5	14 to 21
	azoxystrobin + difenoconazole (Briskway) 2.7 SC*	0.5 to 0.725	14 to 28
	azoxystrobin + propiconazole (Headway) 1.4 ME 1.06 G	1.5 to 3 2 to 4 lbs	14 to 28 14 to 28
	azoxystrobin + tebuconazole (Strobe T) 2.67 SC*	0.75 to 1.5	14 to 21
	chlorothalonil + fluoxastrobin (Fame C) 4.25 SC*	3 to 5.9	14 to 28
	chlorothalonil + propiconazole (Concert) 4.3 SC*	3 to 5.5 5.5 to 8.5	7 to 14 14 to 21
	fluazinam (Secure) 4.17 SC*	0.5	14
	fluazinam + acibenzolar-S-methyl (Secure Action) 4.18 SC*	0.5	14
	fluazinam + tebuconazole (Traction) 3.24 SC*	1.3	14
	fluopyram + trifloxystrobin (Exteris Stressgard) 0.27 SC	1.5 to 4.135	14 to 28
	fluoxastrobin (Fame) 4 SC 0.25 G	0.18 to 0.36 2.3 to 4.6 lbs	14 to 28 14 to 28
	flutolanil (Prostar) 70 WP, 70 DG	1.5	21 to 28
	flutolanil + thiophanate-methyl (SysStar) 80 WDG	2	21 to 28
	iprodione + trifloxystrobin (Interface) 2.27 SC*	3 to 4	14
	isofetamid + tebuconazole (Tekken) 1.8 SC*	3	14 to 28
	propiconazole (Banner MAXX, Kestrel, Propiconazole, Savvi, Strider) 1 ME	2	14 to 21
	pyraclostrobin (Insignia) 20 WG 2 SC	0.5 to 0.9 0.4 to 0.7	14 to 28 14 to 28
	pyraclostrobin + boscalid (Honor) 28 WG*	0.55 to 1.1	14 to 28
	pyraclostrobin + fluxapyroxad (Lexicon Intrinsic) 4.17 SC	0.34 to 0.47	14 to 28
	pyraclostrobin + triticonazole (Pillar) 0.81 G	3 lbs	14 to 28

Table 10-9. Turfgrass Disease Control

Disease	Fungicide and Formulation[1]	Amount of Formulation (oz/1,000 sq ft)[2]	Application Interval (days)[3]
Pink Patch (continued)	tebuconazole* (Torque) 3.6 F (Mirage Stressgard) 2 SC (Skylark, Tebuconazole) 3.6 F	0.6 to 1.1 1 to 2 0.6	refer to label 14 to 28 28
	trifloxystrobin (Compass) 50 WDG	0.1 to 0.15 0.2 to 0.25	14 21
	trifloxystrobin + triadimefon (Armada) 50 WP (Tartan) 2 SC*	0.6 to 1.2 1 to 2	14 to 28 14 to 28
	triticonazole (Trinity) 1.7 EC	1 to 2	14 to 28
	triticonazole + chlorothalonil (Reserve) 4.79 SC*	3.2 to 4.5	refer to label
	vinclozolin (Curalan, Touche) 50 EG*	1	14 to 28
Pink Snow Mold/Microdochium Patch (*Microdochium nivale*)	azoxystrobin (Heritage, Strobe) 50 WG (Heritage) 0.8 TL (Heritage) 0.31 G (Strobe) 2L	0.2 to 0.4 0.7 2 3.5 4 lbs 7 lbs 0.77 1.35	10 to 28 1 application 10-28 1 application 10 to 28 1 application 14 1 application
	azoxystrobin + acibenzolar-S-methyl (Heritage Action) 51 WG*	0.4	refer to label
	azoxystrobin + chlorothalonil (Renown) 5.16 SC*	2.5 to 4.5	14 to 21
	azoxystrobin + difenoconazole (Briskway) 2.7 SC*	0.5 to 0.725	14 to 28
	azoxystrobin + propiconazole (Headway) 1.4 ME 1.06 G	1.5 to 3 5.25 2 to 4 lbs 5	10 to 28 1 application 14 to 28 1 application
	azoxystrobin + tebuconazole (Strobe T) 2.67 SC*	0.75 to 1.5 2.4	14 to 21 1 application
	chlorothalonil + acibenzolar-S-methyl (Daconil Action) 6.1 F*	5.4	21 to 28
	chlorothalonil + fluoxastrobin (Fame C) 4.25 SC*	3 to 5.9	28
	chlorothalonil + iprodione + thiophanate-methyl + tebuconazole (Enclave) 5.3 F*	7 to 8	28
	chlorothalonil + propiconazole (Concert) 4.3 SC*	8.5	14 to 28
	chlorothalonil + propiconazole + fludioxonil (Instrata) 3.6 SC*	5 to 11	late fall
	chlorothalonil + thiophanate-methyl* (Consyst, Peregrine, TM/C) 67 WDG (Spectro) 90 WDG	6 to 8 3.72 to 5.76	1 application 14
	fenarimol (Rubigan) 1 AS*	8 4	1 application 30 (2 applications)
	fluazinam (Secure) 4.17 SC*	0.5	late fall
	fluazinam + acibenzolar-S-methyl (Secure Action) 4.18 SC*	0.5	late fall
	fluazinam + tebuconazole (Traction) 3.24 SC*	1.3	late fall
	fludioxonil (Medallion) 50 WP	0.25 to 0.5	14
	fluopyram + trifloxystrobin (Exteris Stressgard) 0.27 SC	4.135 to 12.6	10 to 28
	fluoxastrobin (Fame) 4 SC 0.25 G	0.18 to 0.36 2.3 to 4.6 lbs	14 to 28 14 to 28
	fluoxastrobin + myclobutanil (Fame M) 3.9 SC	0.5 to 1	21 to 28
	fluoxastrobin + tebuconazole (Fame T) 4 SC*	0.45 to 0.9	30
	fluxapyroxad (Xzemplar) 2.47 SC	0.26	14 to 28
	flutolanil + thiophanate-methyl (SysStar) 80 WDG	4 to 6.12 2 to 3	1 application 14 to 21
	iprodione (26GT, Iprodione Pro, IPro, Raven) 2 F, 2 SC, 2 SE*	4 to 8	1 to 2 applications
	iprodione + thiophanate-methyl* (26/36) 3.8 F (Dovetail) 3.8 F	2 to 4 1 to 4	14 to 21 14 to 21
	iprodione + trifloxystrobin (Interface) 2.27 SC*	4 to 7	1 application
	isofetamid + tebuconazole (Tekken) 1.8 SC*	3	14 to 28

Table 10-9. Turfgrass Disease Control

Disease	Fungicide and Formulation[1]	Amount of Formulation (oz/1,000 sq ft)[2]	Application Interval (days)[3]
Pink Snow Mold/Microdochium Patch (*Microdochium nivale*) (continued)	mancozeb* (Fore) 80 WP (Dithane, Pentathlon) 75 DF (Pentathlon) 4 LF (Protect) 75 WP	6 to 8 6 to 8 10 to 14 6 to 8	14 to 42 14 to 42 14 to 42 7 to 14
	mancozeb + copper hydroxide (Junction) 60 DF*	2 to 4	14 to 42
	metconazole (Tourney) 50 WDG	0.37 to 0.44	late fall
	mineral oil (Civitas) + proprietary pigment (Civitas Harmonizer)*	(8 to 32) + (1 to 4)	7 to 21
	myclobutanil (Eagle, Myclobutanil, Siskin) 20 EW	1.2 to 2.4	prior to snow cover
	PCNB (various brands) 75 WP 10 G 4 F	3 to 8 80 to 160 12 to 16	28 to 42 prior to snowfall prior to snowfall
	polyoxin D (Affirm) 11.3 WDG (Endorse) 2.5 WP	0.88 4	7 to 14 7 to 14
	propiconazole (Banner MAXX, Kestrel, Propiconazole, Savvi, Strider) 1 ME	2 to 4	fall to early spring
	pydiflumetofen (Posterity) 1.67 SC*	0.08 to 0.16	14 to 28
	pyraclostrobin (Insignia) 20 WG 2 SC	0.5 to 0.9 0.7	14 to 28 14 to 28
	pyraclostrobin + boscalid (Honor) 28 WG*	0.55 to 1.1	14 to 28
	pyraclostrobin + fluxapyroxad (Lexicon Intrinsic) 4.17 SC	0.47	14 to 28
	pyraclostrobin + triticonazole (Pillar) 0.81 G	3 lbs	28
	tebuconazole* (Torque) 3.6 F (Mirage Stressgard) 2 SC (Skylark, Tebuconazole) 3.6 F	0.6 to 1.1 1 to 2 0.6	prior to snowfall 10 to 28 prior to snowfall
	thiram (Spotrete)* 4 F 75 WDG	3 to 12 3 to 8	fall and spring fall and spring
	thiophanate-methyl (3336) 50WP or 4 F (3336 Plus) 2 F (SysTec 1998, T-Bird, TM) 85 WDG (3336) 2 G (SysTec 1998, T-Bird, TM) 4.5 L	2 to 4 2 to 4 0.67 to 1.3 1.5 to 6 lbs 1 to 2	14 14 to 28 14 14 14
	triadimefon (Bayleton) 50 WSP, 4.15 F	1 to 2	60 to 90
	trifloxystrobin (Compass) 50 WDG	0.2 to 0.25	fall to early spring
	trifloxystrobin + triadimefon (Armada) 50 WP (Tartan) 2 SC*	1.2 2	fall to early spring fall to early spring
	triticonazole (Trinity) 1.7 SC (Triton) 70 WDG (Triton Flo) 3 G	0.5 to 2 0.15 to 0.3 0.28 to 1.1	14 to 28 late fall 10 to 14
	triticonazole + chlorothalonil (Reserve) 4.79 SC*	3.2 to 4.5	14 to 28
	vinclozolin (Curalan, Touche) 50 EG*	1	10 to 21
Powdery Mildew (*Blumeria graminis*)	azoxystrobin (Heritage, Strobe) 50 WG (Heritage) 0.8 TL (Heritage) 0.31 G	0.2 to 0.4 1 to 2 2 to 4 lbs	14 to 28 14 to 28 14 to 28
	azoxystrobin + acibenzolar-S-methyl (Heritage Action) 51 WG*	0.2 to 0.4	14 to 28
	azoxystrobin + chlorothalonil (Renown) 5.16 SC*	2.5 to 4.5	14 to 21
	azoxystrobin + difenoconazole (Briskway) 2.7 SC*	0.5 to 0.725	14 to 28
	azoxystrobin + propiconazole (Headway) 1.4 ME 1.06 G	1.5 to 3 2 to 4 lbs	14 to 28 14 to 28
	azoxystrobin + tebuconazole (Strobe T) 2.67 SC*	0.75 to 1.5	14 to 21
	chlorothalonil + fluoxastrobin (Fame C) 4.25 SC*	3 to 5.9	14 to 28
	chlorothalonil + propiconazole (Concert) 4.3 SC*	4.5 to 8.5	14 to 28
	chlorothalonil + thiophanate-methyl* (Consyst) 67 WDG (Peregrine) 67 WDG (Spectro) 90 WDG (TM/C) 67 WDG	2 to 8 2 to 8 3.72 to 5.76 2 to 8	7 to 14 14 14 14 to 21
	fenarimol (Rubigan) 1 AS*	2 to 4	1 application

Table 10-9. Turfgrass Disease Control

Disease	Fungicide and Formulation[1]	Amount of Formulation (oz/1,000 sq ft)[2]	Application Interval (days)[3]
Powdery Mildew (continued)	fluazinam + tebuconazole (Traction) 3.24 SC*	1.3	14
	fluoxastrobin (Fame) 4 SC 0.25 G	 0.18 to 0.36 2.3 to 4.6 lbs	 14 to 28 14 to 28
	fluoxastrobin + myclobutanil (Fame M) 3.9 SC	0.25 to 1	14 to 28
	isofetamid + tebuconazole (Tekken) 1.8 SC*	3	14 to 28
	mancozeb + copper hydroxide (Junction) 60 DF*	2 to 4	7 to 14
	mineral oil (Civitas) + proprietary pigment (Civitas Harmonizer)*	(8 to 32) + (1 to 4)	7 to 21
	myclobutanil (Eagle, Myclobutanil, Siskin) 20 EW	1.2	14 to 28
	penthiopyrad (Velista) 50 WG	0.3 to 0.5	14
	propiconazole (Banner MAXX, Kestrel, Propiconazole, Savvi, Strider) 1 ME	1 to 2	14 to 28
	pyraclostrobin (Insignia) 20 WG 2 SC	 0.5 to 0.9 0.4 to 0.7	 14 to 28 14 to 28
	pyraclostrobin + boscalid (Honor) 28 WG*	0.55 to 1.1	14 to 28
	pyraclostrobin + fluxapyroxad (Lexicon Intrinsic) 4.17 SC	0.34 to 0.47	14 to 28
	pyraclostrobin + triticonazole (Pillar) 0.81 G	3 lbs	14 to 28
	tebuconazole* (Torque) 3.6 F (Skylark, Tebuconazole) 3.6 F	 0.6 to 1.1 0.6	 refer to label 28
	triadimefon (Bayleton) 50 WSP, 4.15 F	0.5 to 1	15 to 30
Pythium Blight (*Pythium aphanidermatum*)	azoxystrobin (Heritage, Strobe) 50 WG (Heritage) 0.8 TL (Heritage) 0.31 G (Strobe) 2 L	 0.4 2 2 to 4 lbs 0.38 to 0.77	 10 to 14 10 to 14 10 to 14 10 to 14
	azoxystrobin + acibenzolar-S-methyl (Heritage Action) 51 WG*	0.2 to 0.4	10 to 14
	azoxystrobin + propiconazole (Headway) 1.4 ME 1.06 G	 3 2 to 4 lbs	 10 to 14 14 to 28
	azoxystrobin + tebuconazole (Strobe T) 2.67 SC*	0.75 to 1.5	10 to 21
	chloroneb* (Teremec) 65 SP (Teremec) 2.9 F	 4 7	 5 to 7 5 to 7
	chlorothalonil + fluoxastrobin (Fame C) 4.25 SC*	3 to 5.9	7 to 14
	cyazofamid (Segway) 3.33 SC*	0.45 to 0.9	14 to 21
	ethazole* (Koban) 30 WP (Terrazole) 35 WP	 2 to 4.5 2 to 4	 10 10 to 14
	fluopicolide + propamocarb (Stellar) 5.7 SC	1.2	14
	fluoxastrobin (Fame) 4 SC 0.25 G	 0.18 to 0.36 2.3 to 4.6 lbs	 7 to 14 14
	fluoxastrobin + myclobutanil (Fame M) 3.9 SC	0.5 to 1	14
	fluoxastrobin + tebuconazole (Fame T) 4 SC*	0.45 to 0.9	21
	fosetyl Al (Signature, Fosetyl-Al) 80 WDG (Signature Xtra Stressgard) 60 WDG* (Autograph) 70 DF* (Viceroy) 70 DF	 4 to 8 2 to 6 4.6 to 9.2 4.6 to 9.1	 14 to 21 7 to 21 14 to 21 14 to 21
	mancozeb* (Fore) 80 WP (Dithane) 75 DF (Pentathlon) 4 LF (Pentathlon) 75 DF (Protect, Wingman) 75 WP	 8 8 14 8 8	 5 to 14 10 5 5 7 to 14
	mancozeb + copper hydroxide (Junction) 60 DF*	2 to 4	5
	mefenoxam (Subdue) 43 WSP (Subdue MAXX, Quell) 2 ME (Subdue) 1 GR (Fenox, Mefenoxam) 2 AQ, 2 EC	 0.28 to 0.56 0.5 to 1 12.5 to 25 0.2 to 1	 10 to 21 10 to 21 10 to 14 10 to 21
	metalaxyl (Vireo) 2 MEC	1 to 2	10 to 21

Table 10-9. Turfgrass Disease Control

Disease	Fungicide and Formulation[1]	Amount of Formulation (oz/1,000 sq ft)[2]	Application Interval (days)[3]
Pythium Blight (*Pythium aphanidermatum*) (continued)	phosphorus acid (Alude, Resyst) 3.3 F (Jetphiter) 5.41 F (Magellan) 4.3 F (Vital) 4.2 F (Vital Sign) 4.2 F	5 to 10 5 4.1 to 8.2 4 to 6 4 to 8	7 to 14 7 14 to 21 14 7 to 14
	potassium phosphite (Appear) 4.1 SC	3 to 4 4 to 6	7 to 14 14
	propamocarb (Banol) 6 S*	1.3 to 4	7 to 21
	pyraclostrobin (Insignia) 20 WG 2 SC	0.9 0.7	14 to 28 10 to 14
	pyraclostrobin + boscalid (Honor) 28 WG*	1.1	10 to 14
	pyraclostrobin + fluxapyroxad (Lexicon Intrinsic) 4.17 SC	0.47	14
	pyraclostrobin + triticonazole (Pillar) 0.81 G	3 lbs	14
Pythium Root Dysfunction (*Pythium volutum*)	azoxystrobin (Heritage) 50 WG 0.8 TL	0.4 2	21 to 28 21 to 28
	azoxystrobin + acibenzolar-S-methyl (Heritage Action) 51 WG*	0.4	21 to 28
	cyazofamid (Segway) 3.33 SC*	0.45 to 0.9	14 to 21
	fluoxastrobin (Fame) 4 SC 0.25 G	0.27 to 0.36 3.6 to 4.6 lbs	14 to 28 14 to 28
	fluoxastrobin + chlorothalonil (Fame C) 4.25 SC *	4.5 to 5.9	14 to 28
	fluoxastrobin + myclobutanil (Fame M) 3.9 SC	0.5 to 1	14 to 28
	pyraclostrobin (Insignia) 20 WG 2 SC	0.9 0.7	14 to 28 14 to 28
	pyraclostrobin + boscalid (Honor) 28WG*	1.1	14 to 28
	pyraclostrobin + fluxapyroxad (Lexicon Intrinsic) 4.17 SC	0.47	14 to 28
	pyraclostrobin + triticonazole (Pillar) 0.81 G	3 lbs	14
Pythium Root Rot (*Pythium* spp.)	azoxystrobin (Heritage, Strobe) 50 WG (Heritage) 0.8 TL (Heritage) 0.31 G (Strobe) 2 L	0.4 2 2 to 4 0.38 to 0.77	10 to 14 10 to 14 10 to 14 10 to 14
	azoxystrobin + propiconazole (Headway) 1.4 ME 1.06 G	3 2 to 4 lbs	10 to 14 14 to 28
	azoxystrobin + tebuconazole (Strobe T) 2.67 SC*	0.75 to 1.5	10 to 21
	chlorothalonil + fluoxastrobin (Fame C) 4.25 SC *	3 to 5.9	7 to 10
	ethazole* (Koban) 30 WP (Terrazole) 35 WP	4.5 2 to 4	10 10 to 14
	fluoxastrobin (Fame) 4 SC 0.25 G	0.18 to 0.36 2.3 to 4.6 lbs	7 to 10 14
	fluoxastrobin + myclobutanil (Fame M) 3.9 SC	0.5 to 1	14
	fluoxastrobin + tebuconazole (Fame T) 4 SC*	0.45 to 0.9	21
	fosetyl Al (Signature, Fosetyl-Al) 80 WDG (Signature Xtra Stressgard) 60 WDG* (Autograph) 70 DF* (Viceroy) 70 DF	4 to 8 2 to 0 4.6 to 9.2 4.6 to 9.1	14 to 21 7 to 21 14 to 21 14 to 21
	phosphorous acid (Vital Sign) 2.4 F (Jetphiter) 5.41 F	6 to 8 3.5 to 5	7 to 14 7 to 28
	potassium phosphite (Appear) 4.1 SC	6 to 8	7 to 14
	propamocarb (Banol) 6 S*	1.3 to 4	7 to 21
Rapid Blight (*Labyrinthula* spp.)	iprodione + trifloxystrobin (Interface) 2.27 SC*	3 to 5	refer to label
	mancozeb (Fore) 80 WP*	8	14
	penthiopyrad (Velista) 50 WG	0.5	14
	pyraclostrobin (Insignia) 20 WG 2 SC	0.5 to 0.9 0.4 to 0.7	14 14 to 28
	pyraclostrobin + boscalid (Honor) 28WG*	0.55 to 1.1	14 to 28
	pyraclostrobin + fluxapyroxad (Lexicon Intrinsic) 4.17 SC	0.34 to 0.47	14

Table 10-9. Turfgrass Disease Control

Disease	Fungicide and Formulation[1]	Amount of Formulation (oz/1,000 sq ft)[2]	Application Interval (days)[3]
Rapid Blight (*Labyrinthula* spp.) (continued)	pyraclostrobin + triticonazole (Pillar) 0.81 G	3 lbs	14 to 28
	trifloxystrobin (Compass) 50 WDG	0.15 to 0.2 0.25	14 21
	trifloxystrobin + triadimefon (Armada) 50 WP	0.6 to 1.2	14 to 28
Red Thread (*Laetisaria fuciformis*)	azoxystrobin (Heritage, Strobe) 50 WG (Heritage) 0.8 TL (Heritage) 0.31 G (Strobe) 2 L	0.2 to 0.4 1 to 2 2 to 4 lbs 0.38 to 0.77	14 to 28 14 to 28 14 to 28 14 to 28
	azoxystrobin + acibenzolar-S-methyl (Heritage Action) 51 WG*	0.2 to 0.4	14 to 28
	azoxystrobin + chlorothalonil (Renown) 5.16 SC*	2.5 to 4.5	14 to 21
	azoxystrobin + difenoconazole (Briskway) 2.7 SC*	0.5 to 0.725	14 to 28
	azoxystrobin + propiconazole (Headway) 1.4 ME 1.06 G	 1.5 to 3 2 to 4 lbs	 14 to 28 14 to 28
	azoxystrobin + tebuconazole (Strobe T) 2.67 SC*	0.75 to 1.5	14 to 21
	chlorothalonil* (Daconil Ultrex) 82.5 WDG (Daconil Weather Stik, Legend) 6 F (Daconil Zn) 4.16 F (Chlorothalonil 500ZN) 4.17 F (Chlorothalonil 720SFT) 6 F (Chlorothalonil, Chlorostar) 82.5 DF (Pegasus) 6 L (Pegasus) 82.5 DF (Pegasus HPX) 6 F	 1.8 to 3.25 3.25 to 5 2 to 5.5 5.5 3 to 5 5.3 to 8 3 to 5 7.9 2.12 to 3.5 5.5 1.8 to 3.2 3.6 to 5.5 3.25 to 5 3.6 to 5.5	 7 to 10 14 7 to 14 14 7 to 10 14 7 to 10 14 7 to 10 14 7 to 10 7 to 14 7 to 14 7 to 14
	chlorothalonil + acibenzolar-S-methyl (Daconil Action) 6.1 F*	2 to 3.5 3.6 to 5.4	7 to 10 14
	chlorothalonil + fluoxastrobin (Fame C) 4.25 SC*	3 to 5.9	14 to 28
	chlorothalonil + iprodione + thiophanate-methyl + tebuconazole (Enclave) 5.3 F*	3 to 4 7 to 8	14 to 21 28
	chlorothalonil + propiconazole (Concert) 4.3 SC*	3 to 5.5 5.5 to 8.5	7 to 14 14 to 21
	chlorothalonil + propiconazole + fludioxonil (Instrata) 3.6 SC*	2.75 to 6	14 to 21
	chlorothalonil + thiophanate-methyl* (Consyst) 67 WDG (Peregrine) 67 WDG (Spectro) 90 WDG (TM/C) 67 WDG	 3 to 8 3 to 8 3.72 to 5.76 3 to 8	 7 to 10 14 14 14 to 21
	fenarimol (Rubigan) 1 AS*	8	30
	fluazinam (Secure) 4.17 SC*	0.5	14
	fluazinam + acibenzolar-S-methyl (Secure Action) 4.18 SC*	0.5	14
	fluazinam + tebuconazole (Traction) 3.24 SC*	1.3	14
	fluopyram + trifloxystrobin (Exteris Stressgard) 0.27 SC	1.5 to 4.135	14 to 28
	fluoxastrobin (Fame) 4 SC 0.25 G	 0.18 to 0.36 2.3 to 4.6 lbs	 14 to 28 14 to 28
	fluoxastrobin + myclobutanil (Fame M) 3.9 SC	0.25 to 1	14 to 28
	fluoxastrobin + tebuconazole (Fame T) 4 SC*	0.45 to 0.9	21 to 28
	flutolanil (Prostar) 70 WP, 70 WDG	1.5	21 to 28
	flutolanil + thiophanate-methyl (SysStar) 80 WDG	2 to 3	14 to 21
	iprodione (iprodione (26GT, Iprodione Pro, IPro, Raven) 2 F, 2 SC, 2 SE*	4	14
	iprodione + thiophanate-methyl (26/36) 3.8 F*	2 to 4	14 to 21
	iprodione + trifloxystrobin (Interface) 2.27 SC*	3 to 4	14
	isofetamid + tebuconazole (Tekken) 1.8 SC*	3	14 to 28
	mancozeb* (Fore) 80 WP (Dithane) 75 DF (Pentathlon) 4 LF (Pentathlon) 75 DF (Protect, Wingman) 75 W	 4 to 8 4 to 8 7 to 14 4 to 8 4 to 8	 7 to 14 10 7 to 14 7 7 to 14

Table 10-9. Turfgrass Disease Control

Disease	Fungicide and Formulation[1]	Amount of Formulation (oz/1,000 sq ft)[2]	Application Interval (days)[3]
Red Thread (*Laetisaria fuciformis*) (continued)	mancozeb + copper hydroxide (Junction) 60 DF*	2 to 4	7 to 14
	metconazole (Tourney) 50 WDG	0.37	14
	mineral oil (Civitas) + proprietary pigment (Civitas Harmonizer)*	(8 to 32) + (1 to 4)	7 to 21
	myclobutanil (Eagle, Myclobutanil) 20 EW	1.2	14 to 21
	penthiopyrad (Velista) 50 WG	0.3 to 0.5	14
	polyoxin D (Affirm) 11.3 WDG (Endorse) 2.5 WP	0.88 4	7 to 14 7 to 14
	Propiconazole (Banner MAXX, Kestrel, Propiconazole, Savvi, Strider) 1 ME	2	14 to 21
	pyraclostrobin (Insignia) 20 WG 2 SC	0.5 to 0.9 0.4 to 0.7	14 to 28 14 to 28
	pyraclostrobin + boscalid (Honor) 28 WG*	0.55 to 1.1	14 to 28
	pyraclostrobin + fluxapyroxad (Lexicon Intrinsic) 4.17 SC	0.34 to 0.47	14
	pyraclostrobin + triticonazole (Pillar) 0.81 G	3 lbs	14 to 28
	tebuconazole* (Torque) 3.6 F (Mirage Stressgard) 2 SC (Skylark, Tebuconazole) 3.6 F	0.6 to 1.1 1 to 2 0.6	refer to label 14 to 28 28
	thiophanate-methyl (3336) 50WP or 4 F (3336 Plus) 2 F (SysTec 1998, T-Bird, TM) 85 WDG (3336) 2 G (SysTec 1998, T-Bird, TM) 4.5 L	2 to 4 2 to 4 0.67 to 1.3 1.5 to 6 lbs 1 to 2	14 14 to 28 14 14 14
	thiram (Spotrete) 4 F*	3.75 to 7.5	3 to 10
	triadimefon (Bayleton) 50 WSP, 4.15 F	0.5 to 1	15 to 30
	trifloxystrobin (Compass) 50 WDG	0.1 to 0.15 0.2 to 0.25	14 21
	trifloxystrobin + triadimefon (Armada) 50 WP (Tartan) 2 SC*	0.6 to 1.2 1 to 2	14 to 28 14 to 28
	triticonazole (Trinity) 1.7 SC (Triton) 70 WDG	0.5 to 1 0.15 to 0.3	14 to 28 14 to 28
	triticonazole + chlorothalonil (Reserve) 4.79 SC*	3.2 to 4.5	refer to label
	vinclozolin (Curalan, Touche) 50 EG*	1	14 to 28
Rust (*Puccinia* ssp.)	azoxystrobin (Heritage, Strobe) 50 WG (Heritage) 0.8 TL (Heritage) 0.31 G	0.2 to 0.4 1 to 2 2 to 4 lbs	14 to 28 14 to 28 14 to 28
	azoxystrobin + acibenzolar-S-methyl (Heritage Action) 51 WG*	0.2 to 0.4	14 to 28
	azoxystrobin + chlorothalonil (Renown) 5.16 SC*	2.5 to 4.5	14 to 21
	azoxystrobin + difenoconazole (Briskway) 2.7 SC*	0.5 to 0.725	14 to 28
	azoxystrobin + propiconazole (Headway) 1.4 ME 1.06 G	1.5 to 3 2 to 4 lbs	14 to 28 14 to 28
	azoxystrobin + tebuconazole (Strobe T) 2.67 SC*	0.75 to 1.5	14 to 21
	chlorothalonil* (Daconil Ultrex) 82.5 WDG (Daconil Weather Stik, Legend) 6 F (Daconil Zn) 4.16 F (Chlorothalonil 500ZN) 6 F (Chlorothalonil 720SFT) 6 F (Chlorothalonil, Chlorostar) 82.5 DF (Pegasus) 6 L (Pegasus) 82.5 DF (Pegasus HPX) 6 F	3.7 to 5 4.0 to 5.5 6 to 8 3 to 5 7.9 2.12 to 3.5 5.5 3.2 3.6 to 5.5 3.25 to 5 3.6 to 5.5	14 14 14 7 to 14 14 7 to 10 14 7 to 14 7 to 14 7 to 14 7 to 14
	chlorothalonil + acibenzolar-S-methyl (Daconil Action) 6.1 F*	4 to 5.4	14
	chlorothalonil + fluoxastrobin (Fame C) 4.25 SC*	3 to 5.9	14 to 28

Table 10-9. Turfgrass Disease Control

Disease	Fungicide and Formulation[1]	Amount of Formulation (oz/1,000 sq ft)[2]	Application Interval (days)[3]
Rust (*Puccinia* ssp.) (continued)	chlorothalonil + propiconazole (Concert) 4.3 SC*	3 to 5.5 4.5 to 8.5	7 to 14 14 to 28
	chlorothalonil + propiconazole + fludioxonil (Instrata) 3.6 SC*	2.75 to 6	14 to 28
	chlorothalonil + thiophanate-methyl* (Consyst) 67 WDG (Peregrine) 67 WDG (Spectro) 90 WDG (TM/C) 67 WDG	 3 to 8 3 to 8 3.72 to 5.76 3 to 8	 7 to 14 14 14 14 to 21
	fluazinam (Secure) 4.17 SC*	0.5	14
	fluazinam + acibenzolar-S-methyl (Secure Action) 4.18 SC*	0.5	14
	fluazinam + tebuconazole (Traction) 3.24 SC*	1.3	14
	fluopyram + trifloxystrobin (Exteris Stressgard) 0.27 SC	1.5 to 4.135	14 to 28
	fluoxastrobin (Fame) 4 SC 0.25 G fluoxastrobin + myclobutanil (Fame M) 3.9 SC	 0.18 to 0.36 2.3 to 4.6 lbs 0.25 to 1	 14 to 28 14 to 28 14 to 28
	iprodione + trifloxystrobin (Interface) 2.27 SC*	3 to 5	refer to label
	isofetamid + tebuconazole (Tekken) 1.8 SC*	3	14 to 28
	mancozeb* (Fore) 80 WP (Dithane) 75 DF (Pentathlon) 4 LF (Pentathlon) 75 DF (Wingman) 75 WP	 4 4 5 to 7 4 4	 7 to 14 10 7 to 10 7 to 10 7 to 10
	mancozeb + copper hydroxide (Junction) 60 DF*	2 to 4	7 to 14
	mandestrobin (Pinpoint) 4SC	0.31	14
	metconazole (Tourney) 50 WDG	0.37	14
	myclobutanil (Eagle, Myclobutanil, Siskin) 20 EW	1.2	14 to 28
	propiconazole (Banner Maxx, Kestrel, Propiconazole, Savvi, Strider) 1 ME	1 to 2	14 to 28
	pyraclostrobin (Insignia) 20 WG 2 SC	 0.5 to 0.9 0.4 to 0.7	 14 to 28 14 to 28
	pyraclostrobin + boscalid (Honor) 28WG*	0.55 to 1.1	14 to 28
	pyraclostrobin + fluxapyroxad (Lexicon Intrinsic) 4.17 SC	0.34 to 0.47	14 to 28
	pyraclostrobin + triticonazole (Pillar) 0.81 G	3 lbs	14 to 28
	tebuconazole* (Torque) 3.6 F (Mirage Stressgard) 2 SC (Skylark, Tebuconazole) 3.6 F	 0.6 to 1.1 1 to 2 0.6	 refer to label 14 to 28 28
	thiophanate-methyl (3336) 50WP or 4 F (3336 Plus) 2 F (T-Bird) 4.5 L (SysTec 1998, T-Bird, TM) 85 WDG	 4 to 6 4 to 8 3.5 to 5 2.35 to 3.53	 14 14 to 28 14 14
	thiram (Spotrete) 4 F*	3.75 to 7.5	3 to 10
	triadimefon (Bayleton) 50 WSP, 4.15 F	0.5 to 1	15 to 30
	trifloxystrobin (Compass) 50 WDG	0.1 to 0.15 0.2 to 0.25	14 21
	trifloxystrobin + triadimefon (Armada) 50 WP (Tartan) 2 SC*	 0.6 to 1.2 1 to 2	 14 to 28 14 to 28
	triticonazole (Trinity) 1.7 SC (Triton) 70 WDG	 0.5 to 1 0.15 to 0.225	 14 to 28 14 to 28
	triticonazole + chlorothalonil (Reserve) 4.79 SC*	3.2 to 4.5	14 to 28
Slime Mold (*Myxomycetes* spp.)	mancozeb (Fore) 80 WP*	4 to 8	7 to 14
	mancozeb + copper hydroxide (Junction) 60 DF*	2 to 4	7 to 14

Table 10-9. Turfgrass Disease Control

Disease	Fungicide and Formulation[1]	Amount of Formulation (oz/1,000 sq ft)[2]	Application Interval (days)[3]
Southern Blight (*Sclerotium rolfsii*)	azoxystrobin (Heritage, Strobe) 50 WG (Heritage) 0.8 TL (Heritage) 0.31 G (Strobe) 2 L	0.2 to 0.4 1 to 2 2 to 4 lbs 0.38 to 0.77	14 to 28 14 to 28 14 to 28 14 to 28
	azoxystrobin + acibenzolar-S-methyl (Heritage Action) 51 WG*	0.2 to 0.4	14 to 28
	azoxystrobin + chlorothalonil (Renown) 5.16 SC*	2.5 to 4.5	14 to 21
	azoxystrobin + difenoconazole (Briskway) 2.7 SC*	0.5 to 0.725	14 to 28
	azoxystrobin + propiconazole (Headway) 1.4 ME 1.06 G	1.5 to 3 2 to 4 lbs	14 to 28 14 to 28
	azoxystrobin + tebuconazole (Strobe T) 2.67 SC*	0.75 to 1.5	14 to 21
	chloroneb* (Teremec) 65 SP	4	5 to 7
	chlorothalonil + fluoxastrobin (Fame C) 4.25 SC*	3 to 5.9	14 to 28
	fluoxastrobin (Fame) 4 SC 0.25 G	0.18 to 0.36 2.3 to 4.6 lbs	14 to 28 14 to 28
	fluoxastrobin + myclobutanil (Fame M) 3.9 SC	0.25 to 1	14 to 28
	fluoxastrobin + tebuconazole (Fame T) 4 SC*	0.45 to 0.9	21 to 28
	flutolanil (Prostar) 70 WP, 70 WDG	1.5	21 to 28
	flutolanil + thiophanate-methyl (SysStar) 80 WDG	2	21 to 28
	triadimefon (Bayleton) 50 WSP, 4.15 F	0.5 to 2	14 to 28
	trifloxystrobin + triadimefon (Armada) 50 WP (Tartan) 2 SC*	0.6 to 1.2 1 to 2	14 14
Spring Dead Spot (*Ophiosphaerella korrae; O. herpotricha; O. narmari*)	azoxystrobin (Heritage, Strobe) 50 WG (Heritage) 0.8 TL (Strobe) 2L	0.4 2 0.38 to 0.77	14 to 28 14 to 28 28
	azoxystrobin + acibenzolar-S-methyl (Heritage Action) 51 WG*	0.2 to 0.4	14 to 28
	azoxystrobin + propiconazole (Headway) 1.4 ME 1.06 G	3 2 to 4 lbs	14 to 28 14 to 28
	azoxystrobin + tebuconazole (Strobe T) 2.67 SC*	1.5	14 to 21
	chlorothalonil + fluoxastrobin (Fame C) 4.25 SC*	5.9	14 to 28
	chlorothalonil + iprodione + thiophanate-methyl + tebuconazole (Enclave) 5.3 F*	3 to 4 7 to 8	14 to 21 28
	fenarimol (Rubigan) 1 AS*	4 6	14 to 30 (2 applications) 1 application
	fluoxastrobin (Fame) 4 SC 0.25 G	0.36 2.3 to 4.6 lbs	14 to 28 14 to 28
	fluoxastrobin + myclobutanil (Fame M) 3.9 SC	0.5 to 1	14 to 28
	fluoxastrobin + tebuconazole (Fame T) 4 SC*	0.45 to 0.9	21 to 28
	isofetamid (Kabuto) 3.33 SC	0.5 to 1.6	14 to 28
	isofetamid + tebuconazole (Tekken) 1.8 SC*	3	14 to 28
	myclobutanil (Eagle, Myclobutanil, Siskin) 20 EW	2.4	28 (fall)
	penthiopyrad (Velista) 50 WG	0.7	28
	propiconazole (Banner MAXX, Kestrel, Propiconazole, Savvi, Strider) 1 ME	4	30
	pydiflumetofen (Posterity) 1.67 SC*	0.16 to 0.32	28
	tebuconazole* (Torque) 3.6 F (Mirage Stressgard) 2 SC (Skylark, Tebuconazole) 3.6 F	0.6 to 1.1 2 0.6	21 28 fall and spring
	thiophanate-methyl (3336) 50WP or 4 F (3336) 2 G	4 to 6 6 to 9 lbs	14 14
Stripe Smut (*Ustilago striiformis*)	chlorothalonil + iprodione + thiophanate-methyl + tebuconazole (Enclave) 5.3 F*	3 to 4 7 to 8	14 to 21 28
	chlorothalonil + propiconazole (Concert) 4.3 SC*	4.5 to 8.5	fall or spring
	fluazinam + tebuconazole (Traction) 3.24 SC*	1.3	one application
	isofetamid + tebuconazole (Tekken) 1.8 SC*	3	14 to 28
	myclobutanil (Eagle, Myclobutanil, Siskin) 20 EW	1.2	14

Table 10-9. Turfgrass Disease Control

Disease	Fungicide and Formulation[1]	Amount of Formulation (oz/1,000 sq ft)[2]	Application Interval (days)[3]
Stripe Smut (*Ustilago striiformis*) (continued)	propiconazole (Banner MAXX, Kestrel, Propiconazole, Savvi, Strider) 1 ME	1 to 2	fall or spring
	tebuconazole* (Torque) 3.6 F (Skylark, Tebuconazole) 3.6 F	 0.6 to 1.1 0.6	 spring spring
	thiophanate-methyl (3336) 50WP or 4 F (3336 Plus) 2 F (3336) 2 G (T-Bird) 4.5 L (SysTec 1998, T-Bird, TM) 85 WDG (SysTec 1998, T-Bird,TM) 4.5 L	 4 to 6 4 to 8 6 to 9 lbs 5 to 10 3 to 3.53 5	 14 14 to 28 14 14 to 21 14 to 21 14 to 21
	triadimefon (Bayleton) 50 WSP	1	refer to label
	trifloxystrobin + triadimefon (Armada) 50 WP (Tartan) 2 SC*	 0.6 1	 refer to label refer to label
Summer Patch (*Magnaporthe poae*)	azoxystrobin (Heritage, Strobe) 50 WG (Heritage) 0.8 TL (Heritage) 0.31 G (Strobe) 2 L	 0.2 to 0.4 1 to 2 2 to 4 lbs 0.38 to 0.77	 14 to 28 14 to 28 14 to 28 14 to 28
	azoxystrobin + acibenzolar-S-methyl (Heritage Action) 51 WG*	0.2 to 0.4	14 to 28
	azoxystrobin + difenoconazole (Briskway) 2.7 SC*	0.5 to 0.725	14 to 28
	azoxystrobin + propiconazole (Headway) 1.4 ME 1.06 G	 1.5 to 3 2 to 4 lbs	 14 to 28 14 to 28
	chlorothalonil + fluoxastrobin (Fame C) 4.25 SC*	3 to 5.9	14 to 28
	chlorothalonil + iprodione + thiophanate-methyl + tebuconazole (Enclave) 5.3 F*	3 to 4 7 to 8	14 to 21 28
	chlorothalonil + propiconazole + fludioxonil (Instrata) 3.6 SC*	6 to 11	14 to 28
	fenarimol (Rubigan) 1 AS*	2 to 4 2 4 to 8	30 (2 applications) 30 (greens) single application
	fluoxastrobin (Fame) 4 SC 0.25 G	 0.18 to 0.36 2.3 to 4.6 lbs	 14 to 28 14 to 28
	fluoxastrobin + myclobutanil (Fame M) 3.9 SC	0.25 to 1	14 to 28
	fluoxastrobin + tebuconazole (Fame T) 4 SC*	0.45 to 0.9	21 to 28
	fludioxonil (Medallion) 50 WP	0.5	14
	fluxapyroxad (Xzemplar) 2.47 SC	0.26	14 to 28
	isofetamid + tebuconazole (Tekken) 1.8 SC*	3	14 to 28
	metconazole (Tourney) 50 WDG	0.37	14
	myclobutanil (Eagle, Myclobutanil, Siskin) 20 EW	1.2 to 2.4	14 to 28
	penthiopyrad (Velista) 50 WG	0.3 to 0.5	14 to 28
	propiconazole (Banner MAXX, Kestrel, Propiconazole, Savvi, Strider) 1 ME	2 4	14 28
	pyraclostrobin (Insignia) 20 WG 2 SC	 0.5 to 0.9 0.4 to 0.7	 14 to 28 14 to 28
	pyraclostrobin + boscalid (Honor) 28 WG*	1.1	14 to 28
	pyraclostrobin + fluxapyroxad (Lexicon Intrinsic) 4.17 SC	0.34 to 0.47	14 to 28
	pyraclostrobin + triticonazole (Pillar) 0.81 G	3 lbs	28
	tebuconazole* (Torque) 3.6 F (Mirage Stressgard) 2 SC (Skylark, Tebuconazole) 3.6 F	 0.6 to 1.1 1 to 2 0.6	 21 14 to 28 28
	thiophanate-methyl (3336) 50WP or 4 F (3336 Plus) 2 F (3336) 2 G (SysTec 1998, T-Bird, TM) 85 WDG (SysTec 1998, T-Bird, TM) 4.5 L	 4 to 6 4 to 8 6 to 9 lbs 3.53 5	 14 to 21 14 to 28 14 to 21 14 14
	triadimefon (Bayleton) 50 WSP, 4.15 F	1 to 2	30
	trifloxystrobin (Compass) 50 WDG	0.2 to 0.25	21 to 28
	trifloxystrobin + triadimefon (Tartan) 2 SC * (Armada) 50 WP	 2 1.2	 21 to 28 21 to 28

Table 10-9. Turfgrass Disease Control

Disease	Fungicide and Formulation[1]	Amount of Formulation (oz/1,000 sq ft)[2]	Application Interval (days)[3]
Summer Patch (*Magnaporthe poae*) (continued)	triticonazole (Trinity) 1.7 SC (Triton) 70 WDG	 1 to 2 0.3 to 0.6	 14 to 28 14 to 28
	triticonazole + chlorothalonil (Reserve) 4.79 SC*	3.2 to 5.4	14 to 28
Take-All Patch (*Gaeumannomyces graminis* var. *avenae*)	azoxystrobin (Heritage, Strobe) 50 WG (Heritage) 0.8 TL (Heritage) 0.31 G (Strobe) 2 L	 0.4 2 2 to 4 lbs 0.38 to 0.77	 28 28 28 28
	azoxystrobin + acibenzolar-S-methyl (Heritage Action) 51 WG*	0.2 to 0.4	28
	azoxystrobin + difenoconazole (Briskway) 2.7 SC*	0.5 to 0.725	28
	azoxystrobin + propiconazole (Headway) 1.4 ME 1.06 G	 3 2 to 4 lbs	 14 to 28 14 to 28
	chlorothalonil + fluoxastrobin (Fame C) 4.25 SC*	5.9	28
	fenarimol (Rubigan) 1 AS*	4 4 to 8	30 (greens) 30 (1 or 2 applications)
	fluoxastrobin (Fame) 4 SC 0.25 G	 0.36 2.3 to 4.6 lbs	 28 28
	fluoxastrobin + myclobutanil (Fame M) 3.9 SC	0.5 to 1	28
	fluoxastrobin + tebuconazole (Fame T) 4 SC*	0.45 to 0.9	28
	isofetamid + tebuconazole (Tekken) 1.8 SC*	3	14 to 28
	mandestrobin (Pinpoint) 4SC	0.31	14
	myclobutanil (Eagle, Myclobutanil, Siskin) 20 EW	2.4	28 (spring/fall)
	propiconazole (Banner MAXX, Kestrel, Propiconazole, Savvi, Strider) 1 ME	2 to 4	spring and fall
	pyraclostrobin (Insignia) 20 WG 2 SC	 0.9 0.7	 28 28
	pyraclostrobin + boscalid (Honor) 28 WG*	1.1	28
	pyraclostrobin + fluxapyroxad (Lexicon Intrinsic) 4.17 SC	0.47	28
	pyraclostrobin + triticonazole (Pillar) 0.81 G	3 lbs	28
	tebuconazole* (Torque) 3.6 F (Mirage Stressgard) 2 SC (Skylark, Tebuconazole) 3.6 F	 0.6 to 1.1 1 to 2 0.6	 fall and spring 14 to 28 fall and spring
	thiophanate-methyl (3336) 50 WP or 4 F (3336 Plus) 2 F (3336) 2 G	 4 to 6 4 to 8 6 to 9 lbs	 14 14 to 28 14
	triadimefon (Bayleton) 50 WSP, 4.15 F	1 to 2	21 to 28
	trifloxystrobin + triadimefon (Armada) 50 WP	1.2	28
	triticonazole (Trinity) 1.7 SC (Triton) 70 WDG	 1 to 2 0.15 to 0.3	 14 to 28 14 to 28
	triticonazole + chlorothalonil (Reserve) 4.79 SC*	3.2 to 5.4	14 to 28
Take-all Root Rot/Bermudagrass Decline (*Gaeumannomyces graminis* var. *graminis*)	azoxystrobin + acibenzolar-S-methyl (Heritage Action) 51 WG*	0.4	28
	azoxystrobin + difenoconazole (Briskway) 2.7 SC*	0.5 to 0.725	14 to 28
	isofetamid + tebuconazole (Tekken) 1.8 SC*	3	14 to 28
	fluxapyroxad + pyraclostrobin (Lexicon Intrinsic) 4.17 SC	0.34 to 0.47	refer to label
	pyraclostrobin (Insignia) 20 WG 2 SC	 0.9 0.7	 refer to label refer to label
	pyraclostrobin + boscalid (Honor) 28 WG*	1.1	refer to label
	pyraclostrobin + triticonazole (Pillar) 0.81G	3 lbs	28
	tebuconazole* (Torque) 3.6 F (Mirage Stressgard) 2SC	 0.6 to 1.1 2	 14 to 28 28

Table 10-9. Turfgrass Disease Control

Disease	Fungicide and Formulation[1]	Amount of Formulation (oz/1,000 sq ft)[2]	Application Interval (days)[3]
Take-all Root Rot/Bermudagrass Decline (*Gaeumannomyces graminis* var. *graminis*) (continued)	thiophanate-methyl (3336) 50 WP or 4 F (3336 Plus) 2 F (3336) 2 G	4 to 6 4 to 8 6 to 9 lbs	14 14 to 28 14
	triadimefon (Bayleton) 50 WSP, 4.15 F	1 to 2	21 to 28
Yellow Patch (*Rhizoctonia cerealis*)	azoxystrobin (Heritage, Strobe) 50 WG (Heritage) 0.8 TL (Heritage) 0.31 G (Strobe) 2 L	0.4 2 2 to 4 lbs 0.38 to 0.77	28 28 14 to 28 28
	azoxystrobin + acibenzolar-S-methyl (Heritage Action) 51 WG*	0.2 to 0.4	14 to 28
	azoxystrobin + chlorothalonil (Renown) 5.16 SC*	2.5 to 4.5	14 to 28
	azoxystrobin + difenoconazole (Briskway) 2.7 SC*	0.5 to 0.725	14 to 28
	azoxystrobin + propiconazole (Headway) 1.4 ME 1.06 G	3 2 to 4 lbs	28 14 to 28
	azoxystrobin + tebuconazole (Strobe T) 2.67 SC*	1.5	14 to 21
	chlorothalonil + fluoxastrobin (Fame C) 4.25 SC*	3 to 5.9	14 to 28
	chlorothalonil + propiconazole + fludioxonil (Instrata) 3.6 SC*	8 to 11	late fall
	chlorothalonil + thiophanate-methyl (Spectro) 90 WDG*	3 to 5.76	14 to 21
	fludioxonil (Medallion) 50 WP	0.5	1 application
	fluopyram + trifloxystrobin (Exteris Stressgard) 0.27 SC	2.135 to 6	21 to 28
	fluoxastrobin (Fame) 4 SC 0.25 G	0.36 2.3 to 4.6 lbs	28 14 to 28
	fluoxastrobin + myclobutanil (Fame M) 3.9 SC	0.25 to 1	28
	fluoxastrobin + tebuconazole (Fame T) 4 SC*	0.45 to 0.9	21 to 28
	flutolanil (Prostar) 70 WP, 70 WDG	1.5	21 to 28
	flutolanil + thiophanate-methyl (SysStar) 80 WDG	2	21 to 28
	isofetamid + tebuconazole (Tekken) 1.8 SC*	3	14 to 28
	metconazole (Tourney) 50 WDG	0.37 to 0.44	late fall
	polyoxin D (Affirm) 11.3 WDG (Endorse) 2.5 WP	0.88 4	7 to 14 7 to 14
	propiconazole (Banner MAXX, Kestrel, Propiconazole, Savvi, Strider) 1 ME	3 to 4	late fall
	tebuconazole (Mirage Stressgard) 2 SC*	1 to 2	21 to 28
	thiophanate-methyl (3336) 50WP or 4 F (3336 Plus) 2 F (3336) 2 G	4 to 6 4 to 8 6 to 9 lbs	14 14 to 28 14
	triticonazole (Triton FLO) 3 F (Trinity) 1.75 SC	0.55 to 1.1 1 to 2	21 to 28 21 to 28
	triticonazole + chlorothalonil (Reserve) 4.79 SC*	3.2 to 5.4	21 to 28
Yellow Tuft (*Sclerophthora macrospora*)	fosetyl Al (Signature, Fosetyl-Al) 80 WDG (Signature Xtra Stressgard) 60 WDG* (Autograph) 70 DF* (Viceroy) 70 DF	4 to 8 2 to 6 4.6 to 9.2 4.6 to 9.1	14 to 21 14 to 21 14 to 21 14 to 21
	mefenoxam (Subdue WSP) 43 WSP (Subdue Maxx, Quell 2 ME (Subdue GR) 1 G (Mefenoxam, Fenox) 2 AQ, 2 EC	0.28 to 0.56 0.5 to 1 12.5 to 25 0.2 to 1	10 to 21 10 to 21 10 to 14 10 to 21
	metalaxyl (Vireo) 2 MEC	1 to 2	10 to 21
	pyraclostrobin (Insignia) 20 WG 2 SC	0.5 to 0.9 0.4 to 0.7	14 to 28 14 to 28
	pyraclostrobin + boscalid (Honor) 28 WG*	0.55 to 1.1	14 to 28
	pyraclostrobin + fluxapyroxad (Lexicon Intrinsic) 4.17 SC	0.34 to 0.47	14 to 28
	pyraclostrobin + triticonazole (Pillar) 0.81 G	3 lbs	14 to 28
Zoysia Patch	See Large Patch		

[1] Other trade names with the same active ingredients are labeled for use on turfgrasses and can be used according to label directions.
[2] Apply fungicides in 2 to 5 gallons of water per 1,000 square feet according to label directions. Use lower rates for preventive and higher rates for curative applications.
[3] Use shorter intervals when conditions are very favorable for disease.
* Products marked with an asterisk are not labeled for home lawn use.

Nematicides for Turf

J. P. Kerns and E. L. Butler, Entomology and Plant Pathology Extension

Table 10-10. Nematicides for Turf

Nematicide and Formulation	Amount of Formulation Per 1,000 sq ft	Precautions and Remarks
abamectin (Avid) 0.15 EC	1.31	For use on golf course greens only. Only abamectin formulated as Avid can be used for nematode control in turf according to a 24(c) label. Apply Avid 0.15 EC as an early curative treatment (after appropriate nematode extraction, identification, and counts). Apply early in the morning while grass is wet with dew or irrigate the area prior to application with 0.1 inch of water. Immediately after application irrigate with 0.1 inch of water to move treatments through the thatch. Do not over irrigate. Apply 3 to 4 consecutive Avid 0.15 EC applications on a 14- to 21-day interval. Avid is labeled only for sting (*Belonolaimus longicaudatus*) and ring (*Macroposthonia* sp.) nematodes.
abamectin (Divanem) 0.7 SC	3.125 to 12.2	For control of turf-parasitic nematodes on golf course greens, tees, and fairways. It is not labeled for use on golf course roughs, residential turf, sports fields, or commercial turf. Divanem should be applied as a seasonal program. Apply Divanem as an early curative treatment after appropriate nematode extraction, identification, and counts. Multiple applications may be required before improvements in turf quality are observed. Maximum annual rate must not exceed 0.27 lb abamectin/calendar year or 50 fl oz Divanem/A/calendar year. Do not apply to turf under heat or moisture stress. Apply in the early morning while grass is wet with dew or irrigate the area prior to application with 0.1 inches of water. Spray onto wet turf. Irrigate with 0.1 to 0.5 inches of water beginning within 1 hour of application to move Divanem through the thatch. For best results, irrigate before the spray droplets have dried on the turf. Apply in 2 gallons of water per 1,000 square feet of turf. Application rate is 3.125 to 6.25 fl oz/A every 14 to 21 days or 6.25 to 12.2 fl oz/A every 21 to 28 days.
Bacillus firmus (Nortica)	10 to 30	Nortica is a biological agent for the protection of plant roots against plant parasitic nematodes on turf, lawns, sod farms, and golf courses. Do not mix with other chemicals or fertilizers during application without first contacting a local Bayer representative. Do not apply for a month after a formaldehyde application. Do not apply within 2 weeks of a fumigant application. Do not combine in the spray tank with pesticides, surfactants or fertilizers if there has been no previous experience or use of the combination to show it is physically compatible, effective, and non-injurious under your use conditions. Refer to product label for further information about mixing compatibilities. Nortica is suitable for application by spraying, drenching, or by drip irrigation. Optimal results are obtained by pre-plant applications (from 2 to 7 days prior to planting) and immediately irrigating after application to a minimum of 3-4 inches. If product is applied prior to planting, maintain moist soil with daily irrigation until planting. Refer to label for further information about application techniques. Make applications every 3 months as necessary and irrigate to a depth of 4 inches.
fluensulfone (Nimitz Pro G)	1.38 to 2.75 lbs	Intended for use by a commercial applicator. For nematode control in bermudagrass, St. Augustinegrass, zoysiagrass, centipedegrass, seashore paspalum, tall fescue, and creeping bentgrass on golf courses, sports fields, commercial turfgrass areas, sod farms and residential turfgrass lawns. Nimitz must be immediately watered in after application with a minimum of ¼ inch water. Do not make applications when soil temperature is below 55 degrees F. Do not exceed 240 pounds of product per acre per calendar year. Do not allow bystanders to enter the treated area until the granules have been watered-in.
fluopyram (Indemnify)	0.195 to 0.39	Intended for use by commercial applicators. For use on golf courses, sod farms, sport fields, residential, institutional, municipal, commercial, and other turfgrass areas. Do not apply more than the maximum annual rate for each specific use from any combination of products containing fluopyram. Do not apply via aerial application. For ground application equipment, apply 2 to 5 gallons of solution per 1000 sq. ft. When using against nematodes, for optimum control irrigate within 24 hours of application to depth of the root zone to be protected. Do not apply more than 17.1 fl oz of Indemnify per acre per year. For residential turf, do not apply more than 15.5 fl oz of Indemnify per acre per year.
furfural (Multiguard Protect)	0.126 to 0.184	For terrestrial (outdoor) non-food use on established turf on golf course tees and greens, practice greens, spot treatment of fairways, roughs, and turf/sod farms. Areas to be treated must be at least 70% of field capacity before application. Apply up to 6 applications using only ground boom sprayers set to release spray at no more than 2 feet above the ground. Use the high rate at the start of the season and under high infestation and/or until acceptable control is achieved every 14 to 28 days. Then use the lower rate as a maintenance application at 14 to 28 day intervals.

Floral, Nursery, and Landscape Diseases
Fungicides and Bactericides for Disease Control of Greenhouse Floriculture Crops

Inga Meadows, Entomology and Plant Pathology

Anyone using any agricultural chemical should refer to the current chemical label, which contains information about the safe and effective use of the chemical, before using the chemical. Consult the product label to ensure that the variety of ornamental plant that you wish to treat is listed on the label. Check for phytotoxicity by making trial applications on a smaller number of plants before you treat an entire crop.

Table 10-11. Disease Control of Annual, Perennial, Bedding, and Flowering Potted Plants in Greenhouses

Disease Pesticide and Formulation	Rate of Formulation	Schedule and Remarks
Bacterial Leaf Spot (*Pseudomonas, Xanthomonas*)		
Bacillus subtilis strain OST 713 (Rhapsody, Cease)	2 to 8 qt/100 gal	Begin applications when conditions favor disease development prior to the onset of disease. Thorough coverage is essential. Repeat at 3- to 10-day intervals. Rhapsody is OMRI listed.
copper hydroxide (Nu-Cop) 50DF (CuPRO 2005) (CuPRO 5000) (Kentan DF)	1 lb/100 gal 0.75 to 2.0 lb/100 gal 1.5 to 2.0 lb/100 gal 1.5 to 2.0 lb/A	Begin at first sign of disease, and repeat at 7- to 14-day intervals. Do not tank mix copper formulations with Aliette. Avoid contact with metal surfaces. Discoloration of blooms may occur on certain plant varieties - check label.
copper hydroxide + mancozeb (Junction)	1.5 to 3.5 lb/A	Begin at first sign of disease, and repeat at 7- to 14-day intervals.
copper diammonia diacetate complex (Copper-Count-N)	1 qt/100 gal	Begin at first sign of disease, and repeat at 7- to 14-day intervals.
copper octanoate (Camelot O)	0.5 to 2.0 gal/A	Begin at first sign of disease, and repeat at 7- to 10-day intervals. Camelot O may cause copper toxicity on some plant species. OMRI approved.
copper oxychloride + copper hydroxide (Badge SC) (Badge X2)	1.5 to 2 pt/A 1.5 to 2 lb/A	Repeat at 7- to 14-day intervals. Badge X2 is OMRI approved. Do not mix with Aliette.

Table 10-11. Disease Control of Annual, Perennial, Bedding, and Flowering Potted Plants in Greenhouses

Disease / Pesticide and Formulation	Rate of Formulation	Schedule and Remarks
Bacterial Leaf Spot *(Pseudomonas, Xanthomonas)* **(continued)**		
copper sulfate, basic (CUPROFIX Ultra 40 Disperss) (Cuproval)	See label	See label. Rates vary depending on host.
cuprous oxide (Nordox 75WG)	0.66 lb/100 gal	Begin at first sign of disease and repeat at 7- to 14-day intervals. OMRI approved.
didecyl dimethyl ammonium chloride (KleenGrow)	0.06 to 0.38 fl. oz. per gallon of water	Apply starting at week 3 or earlier if conditions are favorable for disease. Use a watering device to drench the top and bottom of the leaves and stems, avoiding flowers in full bloom, every 14 days to prevent the spread of spores and the build-up of organic material. Remove severely infected plants and disinfect the area with 1.0 fluid ounce of KleenGrow per gallon of water.
fosetyl-Al (Aliette 80WDG)	2 to 4 lb/100 gal or 2.5 to 5 lb/A	For suppression of Xanthomonas campestris pathovars. Aliette: Do not exceed 400 gallons of spray solution per acre. Do not exceed one application every 7 days. Check label for compatibility with copper and other compounds.
(Viceroy 70DF)	1.4 to 4.6 lb/100 gal	Viceroy: Do not apply more than once every 14 days. Check label for compatibility with copper and other compounds.
mono- and di- potassium salts of phosphorous acid (Alude)	26 to 54 fl oz/100 gal OR 9 to 18 ml/gal	Apply prior to disease development. Spray to thoroughly wet all foliage. Follow labels for repeat application limits.
Black root rot *(Thielaviopsis basicola)*		
dodecyl dimethyl ammonium chloride (KleenGrow)	0.06 to 0.38 fl oz per gallon of water	Apply starting at week 3 or earlier if conditions are favorable for disease. Use a watering device to drench the top and bottom of the leaves and stems, avoiding flowers in full bloom, every 14 days to prevent the spread of spores and the build-up of organic material. Remove severely infected plants and disinfect the area with 1.0 fluid ounce of KleenGrow per gallon of water.
etridiazole + thiophanate-methyl (Banrot 40WP) (Banrot 8G)	See label	See label as rates vary depending on application. Apply in sufficient volume to saturate the soil mixture. Irrigate immediately. Repeat at 4- to 12-week intervals if necessary.
fludioxonil (Medallion 50WDG) (Emblem)	See label	Apply as a preventive drench at seeding or transplanting. Make only one application to seeding crop. If needed, re-treat transplants 21 to 28 days after initial application. Do not apply as a seed or soil drench to impatiens or New Guinea impatiens.
iprodione + thiophanate-methyl (Dovetail, Nufarm TM + IP SPC)	17 to 34 oz/100 gal	After transplanting, apply 1 to 2 pt per sq ft. Repeat at 2- to 4-week intervals. Do not use as a drench on impatiens or pothos. Do not use on *Spathiphyllum*.
polyoxin D zinc salt (Affirm WDG)	0.5 lb/100 gal/acre	Apply as a soil drench every 14 to 28 days.
thiophanate-methyl (Cleary 3336 F, or other formulations)	7.5 to 20 oz/100 gal 8 to 16 fl oz/100 gal	Apply as a heavy spray or drench at the rate of 1/2 to 2 pints per sq ft. Repeat at 4- to 8-week intervals. Apply 8 oz as a drench or directed spray after seeding, or apply 12 to 16 oz after transplanting. Repeat at 21- to 28-day intervals.
(OHP 6672 4.5L, T-bird 4.5L) (Incognito 85 WDG) (T-methyl SPC 50 WSB, OHP 6672 50 WP)	7.5 to 20 fl oz/100 gal 4.8 to 9.6 oz/100 gal 8 to 16 oz/100 gal	Apply as heavy spray or drench at a rate of 1 to 2 pints per sq ft. Repeat at 2- to 4-week intervals. Apply as heavy spray or drench at a rate of ½ to 2 pints per sq ft. Repeat at 4-to 8-week intervals. Apply as heavy spray or drench at a rate of ½ to 2 pints per sq ft. Repeat at 4-to 8-week intervals.
Trichoderma harzianum Rifai strain KRI – AG2 (RootShield WP)	3 to 5 oz/100 gal	May be applied as a soil drench. For cuttings or bare root dip, use 0.25 to 5 lb/5 gal or dip into dry powder.
Trichoderma harzianum Rifai strain T-22. *Trichoderma virens* strain G-41 (Rootshield WP Plus)	3 to 8 oz/100 gal	May be applied as a soil drench. For cuttings or bare root dip, use 0.25 to 1.5 lb/2 gal.
triflumizole (Terraguard SC) (Trionic 4SC)	2 to 8 fl oz/100 gal 2 to 8 fl oz/100 gal	Apply as soil drench at 2- to 4-week intervals. Use higher rate under heavy disease pressure. Do not use on impatiens plugs.
Botrytis Blight		
azoxystrobin (Heritage)	4 to 8 oz/100 gal	Apply every 7 to 21 days prior to infection. Do not exceed 2 oz/100 gal on impatiens, pansy, or viola. Do not make more than 3 sequential applications before switching to a nonstrobilurin fungicide. Do not exceed 24 oz/A
azoxystrobin + benzovindiflupyr (Mural)	4 to 7 oz/100 gal	Apply every 7 to 14 days.
Bacillus subtilis strain OST-713 (Rhapsody, Cease)	2 to 8 qt/100 gal	Repeat at 3- to 10-day intervals. Thorough coverage is essential. Begin applications when conditions favor disease development prior to the onset of disease. Rhapsody is OMRI approved.
Bacillus subtilis strain MBI-600 (Subtilex)	0.4 to 1.2 oz/1000 sq ft	Repeat at 7- to 10-day intervals. Begin applications when conditions favor disease development prior to the onset of disease.
chlorothalonil (Chlorothalonil DF) (Chlorothalonil 500ZN) (Chlorothalonil 720 SFT) (Chlorostar VI, Daconil Weather Stik, Echo 720, Manicure 6F) (Daconil Ultrex) 82.5WDG (Echo) 90DF, (Echo Ultimate) (Exotherm Termil)	1 to 2.5 lb/100 gal 1.9 pt/100 gal 1.37 pt/100 gal 1 3/8 pt/100 gal 1.4 lb/100 gal 1.25 lb/100 gal 1 can/1,000 sq ft	Repeat at 7- to 14-day intervals. Apply to foliage or flowers when plants are dry or nearly dry. Discontinue applications prior to bract formation on poinsettia. Rotate with fenhexamid, iprodione, or fludioxonil. On rose, use 1 pt/100 gal. On rose, use 1 lb/100 gal. On rose, use 0.78 lb/100 gal (Echo 90) DF or 0.9 lb/100 gal (Echo Ultimate). See label for method of application.
chlorothalonil + thiophanate-methyl (Spectro) 90WDG (Consyst WDG, TM/C WDG)	1 to 2 lb/100 gal 0.75 to 1.5 lb/100 gal	Minimum re-treat interval is 7 days. Do not apply to green or variegated pittosporum or schefflera more than once. Repeat every 7 to 10 days as needed during disease period.
copper hydroxide (Nu-Cop) 50DF (CuPRO 2005) (CuPRO 5000) (Kentan DF)	1 lb/100 gal 0.75 to 2.0 lb/100 gal 1.5 to 2.0 lb/100 gal 1.5 to 2.0 lb/A	Begin at first sign of disease, and repeat at 7- to 14-day intervals. Do not tank mix copper formulations with Aliette. Avoid contact with metal surfaces. Discoloration of blooms may occur on certain plant varieties - check label.
copper hydroxide + mancozeb (Junction)	1.5 to 3.5 lb/100 gal	Repeat at 7- to 14-day intervals.

Table 10-11. Disease Control of Annual, Perennial, Bedding, and Flowering Potted Plants in Greenhouses

Disease Pesticide and Formulation	Rate of Formulation	Schedule and Remarks
Botrytis Blight (continued)		
copper, diammonia, diacetate complex (Copper-Count-N)	1 qt/100 gal	Begin at first sign of disease, and repeat at 7- to 14-day intervals.
copper octanoate (Camelot O)	0.5 to 2.0 gal/A	Begin at first sign of disease, and repeat at 7- to 10-day intervals. Camelot O may cause copper toxicity on some plant species. OMRI approved.
copper oxychloride + copper hydroxide (Badge SC) (Badge X2)	1.5 to 2 pt/A 1.5 to 2 lb/A	Repeat at 7- to 14-day intervals. Badge X2 is OMRI approved. Do not mix with Aliette.
copper sulfate, basic (CUPROFIX Ultra 40 Disperss) (Cuproxat)	See label	See label. Rates vary depending on host
cuprous oxide (Nordox 75WG)	0.66 lb/100 gal	Begin at first sign of disease and repeat at 7- to 14-day intervals. OMRI approved
cyprodinil + fludioxonil (Palladium)	4 to 6 oz/100 gal	Spray on a 7- to 14-day interval while conditions are conducive to disease development. After 2 applications, alternate with another fungicide with a different MOA for 2 applications. Cautionary statement on label for applications to Geraniums, Impatiens and New Guinea Impatiens.
dicloran (Botran) 75W	1 lb/150 to 200 gal See label	Apply to stock cuttings or greenhouse plants. Begin when disease is anticipated or first appears. Spray foliage and flowers at 7- to 14-day intervals.
didecyl dimethyl ammonium chloride (KleenGrow)	0.06 to 0.38 fl. oz. per gallon of water	Apply starting at week 3 or earlier if conditions are favorable for disease. Use a watering device to drench the top and bottom of the leaves and stems, avoiding flowers in full bloom, every 14 days to prevent the spread of spores and the build-up of organic material. Remove severely infected plants and disinfect the area with 1.0 fluid ounce of KleenGrow per gallon of water.
fenhexamid (Decree 50WDG)	1 to 1.5 lb/100 gal	RESISTANCE TO THIS CHEMICAL HAS BEEN REPORTED IN BOTRYTIS. Avoid making more than 2 consecutive applications of this product. Treat at 7- to 14-day intervals. Rotate with chlorothalonil, copper, mancozeb, or iprodione. Make trial application before treating poinsettia.
fludioxonil (Medallion 50WDG WSP) (Emblem)	2 to 4 oz/100 gal 2 to 4 fl oz/100 gal	Spray to runoff. Repeat at 7- to 14-day intervals. Use no more than 2 consecutive applications before switching to another fungicide with a different mode of action. Do not use after bract formation on poinsettia. Foliar applications on impatiens, New Guinea impatiens, and some geranium cultivars may cause stunting or chlorosis, especially on young plants.
fluxapyroxad + pyraclostrobin (Orkestra Intrinsic)	8 to 10 fl oz/100 gal	Use prior to disease development on a 7 to 14 day interval.
iprodione (Chipco 26019 N/G) (18 Plus, Iprodione Pro 2SE)	1 to 2.5 lb/100 gal 1 to 2.5 qt/100 gal	Spray to ensure thorough coverage. Repeat at 7- to 14-day intervals. Do not make more than 4 applications per year. Do not apply to *Spathiphyllum*. Do not apply as a soil drench on impatiens or pothos.
iprodione + thiophanate-methyl (26/36) Dovetail, Nufarm TM + IP SPC)	33 to 84 oz/100 gal 17 to 34 fl oz/100 gal	Spray plants to ensure thorough coverage. Do not make more than 4 applications per crop per year. Repeat at 7- to 14-day intervals. Repeat at 10- to 14-day intervals. Do not drench impatiens or pothos. Do not use on *Spathiphyllum*
mancozeb (Dithane 75DF), (Fore 80WP), (Mancozeb DG) (Pentathlon DF) (Pentathlon LF) (Protect DF)	1.5 lb/100 gal 1 to 2 lb/100 gal 0.8 to 1.6 qt/100 gal 1 to 2 lb/100 gal 1 to 2 bags/100 gal	Do not use on French dwarf or signet-type marigolds. Begin at first sign of disease. Repeat at 7- to 10-day intervals. Most effective when applied prior to infection. Not for use on marigold. Begin at first sign of disease. Repeat at 7-to 10-day intervals. Begin at first sign of disease. Repeat at 7-to 10-day intervals. To improve performance, add 2 to 4 oz of an effective spreader-sticker. Apply at 7- to 21-day intervals.
mancozeb + myclobutanil (Clevis, MANhandle)	2 lb/100 gal	Addition of Latron B-1956 will improve performance. Apply at 7- to 10-day intervals.
polyoxin D zinc salt (Affirm WDG)	0.25 to 0.5 lb/100 gal	Apply as a foliar spray every 7 to 10 days. Apply prior to disease development and when conditions are conducive for disease.
propiconazole + chlorothalonil (Concert II)	63 fl oz (4.3 pt)/100 gal	Apply every 14 to 28 days beginning when conditions are favorable for disease. Apply to full coverage to the point of drip. Do not make more than 3 applications.
pyraclostrobin + boscalid (Pageant Intrinsic)	12 to 18 oz/100 gal	Apply prior to disease development. Repeat at 7- to 14-day intervals. Make no more than 2 sequential applications. Do not expose petunia or impatiens in flower or wintercreeper or nine bark to spray or drift as injury may occur.
thiophanate-methyl (Cleary 3336 F, T-methyl SPC 50 WSB) (OHP 6672 4.5F) (OHP 6672 50WP) (Phoenix T-bird 85 WDG) (T-bird 4.5L)	12 to 16 fl oz/100 gal 10 to 14.5 fl oz/100 gal 12 to 16 oz/100 gal 0.4 lb/100 gal 10.75 to 20 fl oz/100 gal	Apply every 7 to 14 days beginning at first signs of disease. Apply every 7 to 14 days beginning at first signs of disease. Apply every 7 to 14 days beginning at first signs of disease. Apply every 10 to 14 days beginning at first signs of disease. Apply every 7 to 14 days beginning at first signs of disease.
thiophanate-methyl + mancozeb (Zyban 79WSB)	see label	Do not use on French mangold or gloxinia. Apply weekly.
trifloxystrobin (Compass O 50WDG)	2 to 4 oz/100 gal	Repeat at 7- to 14-day intervals until the threat of disease is over. Rotate to another nonstrobilurin fungicide after each application of Compass O. Make no more than four foliar applications per crop cycle or season.
triflumizole (Terraguard SC)	4 to 8 oz/100 gal	Make initial application prior to or at first sign of disease. Repeat at 7- to 14-day intervals.
Bulb and Corm Rots (Fusarium, Penicillium)		
didecyl dimethyl ammonium chloride (KleenGrow)	0.06 to 0.38 fl. oz. per gallon of water	Apply starting at week 3 or earlier if conditions are favorable for disease. Use a watering device to drench the top and bottom of the leaves and stems, avoiding flowers in full bloom, every 14 days to prevent the spread of spores and the build-up of organic material. Remove severely infected plants and disinfect the area with 1.0 fluid ounce of KleenGrow per gallon of water
iprodione (Chipco 26019 N/G) (18 Plus, Iprodione Pro 2SE)	2 lb/100 gal 2 qt/100 gal	Dip 5 minutes prior to storage.
didecyl dimethyl ammonium chloride (KleenGrow)	Immerse both crate and bulbs in a solution of 0.15 to 1.5 fl oz KleenGrow per gallon of water for 30 seconds	For control of Botrytis, Rhizoctonia, Fusarium and Penicillium bulb rots and other fungal and bacterial diseases of tulip, narcissus, gladiolus, crocus, dahlia, freesia, iris, lily, daylily, amaryllis, hyacinth, iris, scilla and ornamental onion. Remove from solution and allow to drain prior to planting or storage. Test solution regularly using QAC test strips, (La Motte #2949-BJ or equivalent). Add KleenGrow when necessary to maintain an a.i. concentration between 90 and 900 ppm.
iprodione + thiophanate-methyl (26/36)	66 oz/100 gal	Dip 5 minutes prior to storage. Only labeled for corm rot on Gladiolus.

Table 10-11. Disease Control of Annual, Perennial, Bedding, and Flowering Potted Plants in Greenhouses

Disease Pesticide and Formulation	Rate of Formulation	Schedule and Remarks
Bulb and Corm Rots *(Fusarium, Penicillium)* (continued)		
thiabendazole (Mertect 340-F)	30 fl oz/100 gal	Clean and treat bulbs and corms within 24 to 48 hours of digging. Warm solution prior to dipping. Mix fresh solution per label guidelines. Dip bulbs 15 to 30 minutes and corms 15 minutes for *Fusarium* control, or dip bulbs 10 to 15 minutes for *Penicillium* (blue mold) control.
thiophanate-methyl (Cleary 3336 F) (OHP-6672 50WP) (Incognito 85 WDG) (T-methyl SPC 50 WSB) (OHP 6672 4.5F, T-bird 4.5L0 (T-storm Flowable) (Phoenix T-bird 85 WDG)	12 to 16 fl oz/100 gal foliar spray; 16 to 24 fl oz/100 gal dip 16 to 32 oz/100 gal 9.6 to 19.2 oz/100 gal 16 to 32 oz/100 gal 14.t to 33 fl oz/100 gal 33 fl oz/100 gal 1.4 lb/100 gal	Soak clean bulbs for 15 to 30 minutes in warm (80 to 85 degrees F) solution. Treat bulbs within 48 hours of digging. Dry well before storing.
Cylindrocladium Stem Canker or Root Rot		
chlorothalonil (Chlorothalonil 500ZN) (Chlorothalonil 720 SFT) (Chlorostar VI, Daconil Weather Stik, Echo 720, Manicure 6F) (Daconil Ultrex, Daconil Weatherstik 82.5WDG) (Echo 90DF), (Echo Ultimate) (Exotherm Termil)	1.9 pt/100 gal 1.37 pt/100 gal 1 3/8 pt/100 gal 1.4 lb/100 gal 1.25 lb/100 gal 1 can/1,000 sq ft	Repeat at 7- to 14-day intervals. Apply to foliage or flowers when plants are dry or nearly dry. See label for method of application.
chlorothalonil + thiophanate-methyl (Spectro 90WDG)	1.0 to 2.0 lb/100 gal	For best results use spray mixture the same day it is prepared. Spray uniformly over the area to be treated with a properly calibrated power sprayer, apply as a full coverage spray to run-off when conditions are favorable for disease development.
cyprodinil + fludioxonil (Palladium)	2 to 4 oz/100 gal	For stem diseases, ensure full spray coverage of all stems and inner areas of plants to the soil/media level. Do not apply Palladium to leather leaf fern or other ferns for cutting/harvest.
fludioxonil (Medallion 50WSP) (Emblem)	See label	Completely drench the growing medium. Repeat at 21- to 28-day intervals. Two applications per year when conditions favor disease development are usually adequate for control.
iprodione + thiophanate-methyl (Dovetail, Nufarm TM + IP SPC)	17 to 34 oz/100 gal	After transplanting, apply 1 to 2 pt per sq ft. Repeat at 2- to 4-week intervals. Do not use as a drench on impatiens or pothos. Do not use on *Spathiphyllum*.
pyraclostrobin + boscalid (Pageant Intrinsic)	12 to 18 oz/100 gal	Apply prior to disease development. Completely drench the growing medium. Repeat at 7- to 14-day intervals. Make no more than 2 sequential applications. Do not expose petunia or impatiens in flower or wintercreeper or nine bark to spray, as injury may occur.
thiophanate-methyl (Cleary 3336F) (OHP-6672 50W) (T-Storm Flowable) (Incognito 85WDG) (T-methyl SPC 85WSB) (OHP 6672 4.5F, T-bird 4.5L)	8 to 16 oz/100 gal 8 to 16 oz/100 gal 20 fl oz/100 gal 4.8 to 9.6 oz/100 gal 8 to 16 oz/100 gal 7.5 to 20 fl oz/100 gal	Apply 8 oz after seeding or sticking, or 12 to 16 oz after transplanting as a drench or directed spray at a rate that thoroughly soaks the growing media through the root zone. Repeat every 21 to 28 days. Apply as a drench or heavy spray at a rate of 2 to 2 pints per sq ft. Repeat at 4- to 8-week intervals. Apply as a drench or heavy spray at a rate of 1 to 2 pints per sq ft. Repeat at 2- to 4-week intervals. Apply as a drench or heavy spray at a rate of 1/2 to 2 pints per sq ft. Repeat at 4- to 8-week intervals. Apply as a drench or heavy spray at a rate of 1/2 to 2 pints per sq ft. Repeat at 4- to 8-week intervals. Apply as a drench or heavy spray at a rate of 1/2 to 2 pints per sq ft. Repeat at 4- to 8-week intervals.
Trichoderma harzianum Rifai strain KRL- AG2 (RootShield WP)	3 to 5 oz/100 gal	May be applied as a soil drench. For cuttings or bare root dip, use 0.25 to 5 lb/5 gal or dip into dry powder.
Trichoderma harzianum Rifai strain T-22, *Trichoderma virens* strain G-41 (RootShield WP Plus)	3 to 8 oz/100 gal	May be applied as a soil drench. For cuttings or bare root dip, use 0.25 to 1.5 lb/2 gal.
trifloxystrobin (Compass)	1 to 2 oz/100 gal	Apply as a drench to the upper ½ of the growing media. Begin at time of seeding, again at transplant, and at 21 to 28 days thereafter.
triflumizole (Terraguard SC) (Trionic 4SC)	4 to 8 fl oz/100 gal See label	Can be used as a cutting soak or soil drench. Apply soil drenches at 2 to 4 week intervals as needed. Use higher rate under high disease pressure, which can occur under warmer conditions.
Downy Mildew		
azoxystrobin (Heritage)	1 to 4 oz/100 gal	Apply every 7 to 28 days prior to infection. Do not apply 2 oz rate on less than 14-day intervals.
azoxystrobin + benzovindiflupyr (Mural)	4 to 7 oz/100 gal	Apply every 7 to 14 days.
Bacillus subtilis strain QST 713 (Rhapsody, Cease)	2 to 8 qt/100 gal	Repeat at 3- to 10-day intervals. Thorough coverage is essential. Begin applications when conditions favor disease development, prior to the onset of disease. Rhapsody is OMRI approved.
chlorothalonil + thiophanate-methyl (ConSyst WDG, TM/C WDG)	0.75 to 1 lb/100 gal	Repeat every 7 to 10 days during disease period.
copper hydroxide (Nu-Cop 50DF) (CuPRO 2005) (CuPRO 5000) (Kentan DF)	1 lb/100 gal 0.75 to 2.0 lb/100 gal 1.5 to 2.0 lb/100 gal 1.5 to 2.0 lb/A	Begin at first sign of disease, and repeat at 7- to 14-day intervals. Do not tank mix copper formulations with Aliette. Avoid contact with metal surfaces. Discoloration of blooms may occur on certain plant varieties; check label.
copper oxychloride + copper hydroxide (Badge SC) (Badge X2)	1.5 to 2 pt/A 1.5 to 2 lb/A	Repeat at 7- to 14-day intervals. Badge X2 is OMRI approved. Do not mix with Aliette.
copper diammonia diacetate complex (Copper-Count-N)	1 qt/100 gal	Begin at first sign of disease, and repeat at 7- to 14-day intervals. To avoid phytotoxicity, do not use any copper compound on alyssum.
copper octanoate (Camelot O)	0.5 to 2.0 gal/A	Begin at first sign of disease, and repeat at 7- to 10-day intervals. Camelot O may cause copper toxicity on some plant species. OMRI approved.
cuprous oxide (Nordox 75WG)	0.66 lb/100 gal	Begin at first sign of disease and repeat at 7- to 14-day intervals. OMRI approved.

Chapter X — 2019 N.C. Agricultural Chemicals Manual

Table 10-11. Disease Control of Annual, Perennial, Bedding, and Flowering Potted Plants in Greenhouses

Disease / Pesticide and Formulation	Rate of Formulation	Schedule and Remarks
Downy Mildew (continued)		
cyazofamid (Segway)	2.1 to 3.5 fl oz/100 gal	14- to 21-day intervals using another registered fungicide with a different mode of action. Apply sufficient volume to wet all foliage until runoff (normally 50 to 100 gallons per acre).
didecyl dimethyl ammonium chloride (KleenGrow)	0.06 to 0.38 fl. oz. per gallon of water	Apply starting at week 3 or earlier if conditions are favorable for disease. Use a watering device to drench the top and bottom of the leaves and stems, avoiding flowers in full bloom, every 14 days to prevent the spread of spores and the build-up of organic material. Remove severely infected plants and disinfect the area with 1.0 fluid ounce of KleenGrow per gallon of water.
dimethomorph (Stature SC)	6.12 to 12.25 oz/100 gal	Apply at first sign of disease. Apply to obtain complete coverage of flowers, foliage, and stems. Repeat at 10- to 14-day intervals throughout the production cycle.
dimethomorph + ametoctradin (Orvego)	11 to 14 fl oz/100 gal	Apply on 10- to 14-day intervals using another registered fungicide with a different mode of action. Apply sufficient volume to wet all foliage until runoff (normally 50 to 100 gallons per acre).
fenamidone (Fenstop)	7 to 14 fl oz/100 gal	Apply as a foliar spray until wet. Repeat as necessary on a 28-day schedule. Do not apply more than 2 applications per crop per season.
fluoxastrobin (Disarm G)	see label	Apply in advance of infection periods.
fluopicolide (Adorn)	1 to 4 fl oz/100 gal	Adorn MUST be tank mixed for resistance management with another product that is registered for use against the target disease.
fosetyl-Al (Aliette 80WDG) (Viceroy 70DF)	1.25 to 4 lb/100 gal 1.4 to 4.6 lb/100 gal	For both products, do not apply more than once every 14 days. Aliette: Do not exceed 400 gallons of spray solution per acre. Check label for compatibility with copper and other compounds. Viceroy: Check label for compatibility with copper and other compounds.
mancozeb (Dithane) 75DF, (Fore) 80WSP, (Mancozeb DG) (Pentathlon DF) (Pentathlon LF) (Protect DF)	1.5 lb/100 gal 1 to 2 lb/100 gal 0.8 to 1.6 qt/100 gal 1 to 2 lb/100 gal 1 to 2 bags/100 gal	Repeat at 7- to 10-day intervals. Begin at first sign of disease. Repeat 7- to 10-day intervals. Begin at first sign of disease. Repeat 7- to 10-day intervals. To improve performance, add 2 to 4 oz of an effective spreader-sticker. Repeat at 7- to 21-day intervals.
mancozeb + copper hydroxide (Junction)	1.5 to 3.5 lb/100 gal	Begin applications at first sign of disease and repeat at 7- to 14-day intervals.
mandipropamid (Micora)	4 to 8 fl oz/100 gal	This product can also be used on vegetables sold to the retail market in GH with permanent flooring. Apply prior to disease development. Repeat sprays at 7- to 14-day intervals. Make no more than 2 sequential applications, then rotate to another fungicide with a different MOA.
mefenoxam (Subdue Maxx, Subdue GR)	0.5 to 1 fl oz/100 gal See label for rates	Apply Subdue Maxx as a foliar spray or soil drench treatment. Apply Subdue GR as a soil surface or soil/planting media incorporation treatment.
mono- and di- potassium salts of phosphorous acid (Alude) (Fosphite) (Reliant)	26 to 54 fl oz/100 gal OR 9 to 18 ml/gal 1 to 2 qt/100 gal OR 0.3 to 0.6 fl oz/gal 1.25 to 2.5 qt/100 gal OR 0.5 to 0.125 oz/gal	Apply prior to disease development. Spray to thoroughly wet all foliage. Follow labels for repeat application limits.
oxathiapiprolin (Segovis)	0.6 to 2.4 fl oz/100 gal	Begin foliage applications prior to disease development and continue on 5 to 14 day interval when conditions are conducive for disease development. Do not apply more than 2 consecutive applications before switching to another non-Group U-15 fungicide.
polyoxin D zinc salt (Affirm)	0.25 to 0.5 lb/100 gal	Apply as a foliar spray every 7 to 10 days. Apply prior to disease development and when conditions are conducive for disease.
potassium bicarbonate (MilStop)	1.25 to 5 lb/100 gal	Uniform and complete coverage of foliage is essential for best results. See label for special instructions regarding poinsettia, pansy, and impatiens.
potassium phosphite (Vital)	4 pt/100 gal	Apply prior to disease onset, and repeat at 14-day intervals.
pyraclostrobin (Insignia)	4 to 8 oz/100 gal	Apply prior to onset of disease and repeat every 7 to 14 days. Do not expose flowering impatiens or flowering petunias to Insignia.
pyraclostrobin + boscalid (Pageant Intrinsic)	12 to 18 oz/100 gal	Apply prior to disease development. Repeat at 7- to 10-day intervals. Make no more than 2 sequential applications. Do not expose petunia or impatiens in flower or wintercreeper or nine bark to spray or drift as injury may occur.
thiophanate-methyl + mancozeb (Zyban)	1.5 lb/100 gal	Apply at first sign of disease, and repeat at 7-day intervals.
trifloxystrobin (Compass O 50WDG)	2 to 4 oz/100 gal	Apply as a foliar spray before disease is detected or when conditions are favorable for disease. Repeat at 7- to14-day intervals until threat of disease is over.
Fungal Leaf Spots (consult label for specific fungi controlled)		
azoxystrobin (Heritage)	1 to 8 oz/100 gal See label	Rates may differ depending on disease and host – see label. Repeat at 7- to 28-day intervals. Do not make more than 3 sequential applications before switching to a nonstrobilurin fungicide. Good control of *Alternaria* leaf spot.
azoxystrobin + benzovindiflupyr (Mural)	4 to 7 oz/100 gal	Apply every 7 to 21 days.
Bacillus subtilis strain OST-713(Rhapsody, Cease)	2 to 8 qt/100 gal	Repeat at 3- to 7-day intervals when conditions are favorable for disease and before onset of disease. Thorough coverage is essential. Not effective on *Alternaria*.

Table 10-11. Disease Control of Annual, Perennial, Bedding, and Flowering Potted Plants in Greenhouses

Disease / Pesticide and Formulation	Rate of Formulation	Schedule and Remarks
Fungal Leaf Spots (consult label for specific fungi controlled) (continued)		
chlorothalonil (Chlorothalonil DF) (Chlorothalonil 500ZN) (Chlorothalonil 720 SFT) (Chlorostar VI, Echo 720, Manicure 6 Flowable) (Daconil Ultrex 82.5WDG) (Echo 90DF), (Echo Ultimate)	1 to 2.5 lb/100 gal 1.9 pt/100 gal 1.37 pt/100 gal 1 3/8 pt/100 gal 1.4 lb/100 gal 1.25 lb/100 gal	Repeat at 7- to 14-day intervals. Works well for control of *Alternaria* leaf spot. Discontinue applications prior to bract formation on poinsettias. Applications made during bloom may damage flowers. Apply to plants when both foliage and flowers are dry or nearly dry.
chlorothalonil + thiophanate-methyl (Spectro) 90WDG (Consyst WDG, TM/C WDG)	1 to 2 lb/100 gal 0.75 to 1 lb/100 gal	Apply when foliage and flowers are dry, or nearly dry. Repeat at 7-day intervals. Good control of *Colletotrichum* (anthracnose) and *Alternaria*. Not recommended for Swedish Ivy, Boston Fern, and Easter Cactus. Thorough, uniform coverage is essential for good control. Repeat in 7 to 10 days.
copper hydroxide (Nu-Cop 50DF) (CuPRO 2005) (CuPRO 5000) (Kentan DF)	1 lb/100 gal 0.75 to 2.0 lb/100 gal 1.5 to 2.0 lb/100 gal 1.5 to 2.0 lb/A	Begin at first sign of disease, and repeat at 7- to 14-day intervals. Do not tank mix copper formulations with Aliette. Avoid contact with metal surfaces. Discoloration of blooms may occur on certain plant varieties- check label.
copper hydroxide + mancozeb (Junction)	1.5 to 3.5 lb/A	Repeat at 7- to 21-day intervals. Discoloration of foliage and/or blooms is possible on some varieties of carnation, chrysanthemum, and rose. Do no use on French marigold.
copper diammonia diacetate complex (Copper-Count-N)	1 qt/100 gal	Begin at first sign of disease, and repeat at 7- to 14-day intervals
copper octanoate (Camelot O)	0.5 to 2.0 gal/1A	Begin at first sign of disease, and repeat at 7- to 10-day intervals. Camelot O may cause copper toxicity on some plant species. OMRI approved.
copper oxychloride + copper hydroxide (Badge SC) (Badge X2)	1.5 to 2 pt/A 1.5 to 2 lb/A	Repeat at 7- to 14-day intervals. Badge X2 is OMRI approved. Do not mix with Aliette.
copper sulfate, basic (CUPROFIX Ultra 40 Disperss) (Cuproxat)	See label	See label. Rates vary depending on host.
cuprous oxide (Nordox 75WG)	0.66 lb/100 gal	Begin at first sign of disease and repeat at 7- to 14-day intervals. OMRI approved.
cyprodinil + fludioxonil (Palladium)	2 to 4 oz/100 gal	Spray on a 7- to 14-day interval while conditions are conducive to disease development. After 2 applications, alternate with another fungicide with a different MOA for 2 applications. See cautionary statement on label for applications to Geraniums, Impatiens and New Guinea Impatiens.
didecyl dimethyl ammonium chloride (KleenGrow)	0.06 to 0.38 fl. oz. per gallon of water	Apply starting at week 3 or earlier if conditions are favorable for disease. Use a watering device to drench the top and bottom of the leaves and stems, avoiding flowers in full bloom, every 14 days to prevent the spread of spores and the build-up of organic material. Remove severely infected plants and disinfect the area with 1.0 fluid ounce of KleenGrow per gallon of water.
fludioxonil (Medallion WDG WSP) (Emblem)	1 to 2 oz/100 gal 1 to 2 fl oz/100 gal	Repeat at 7- to 14-day intervals. Good control of *Alternaria* leaf spot.
fluoxastrobin (Disarm O)	1 to 8 fl oz/100 gal	Rate depends on which fungal disease is present; see label.
fluxapyroxad + pyraclostrobin (Orkestra Intrinsic)	4 to 10 fl oz/100 gal	Rate depends on which fungal disease is present; see label.
iprodione (Chipco 26019 N/G) (18 Plus, Iprodione Pro 2SE)	1 to 2.5 lb/100 gal 1 to 2.5 qt/100 gal	Spray plants to ensure thorough coverage. Repeat at 7- to 14-day intervals. Do not make more than 4 applications per crop per year. Do not drench impatiens or pothos. Do not use on *Spathiphyllum*. Good control of *Alternaria* leaf spot.
iprodione + thiophanate-methyl (26/36) (Dovetail, Nufarm TM + IP SPC)	33 to 84 oz/100 gal 17 to 34 fl oz/100 gal	Spray plants to ensure thorough coverage. Do not make more than 4 applications per crop per year. Repeat at 7- to 14-day intervals. Repeat at 10- to 14-day intervals. Do not drench impatiens or pothos. Do not use on *Spathiphyllum*.
mancozeb (Dithane 75DF), (Fore 80WSP), (Mancozeb DG) (Pentathlon DF) (Pentathlon LF) (Protect DF)	1.5 lb/100 gal 1 to 2 lb/100 gal 0.8 to 1.6 qt/100 gal 1 to 2 lb/100 gall	Do not use on French dwarf or signet-type marigolds. Begin at first sign of disease. Repeat at 7- to 10-day intervals. Most effective when applied prior to infection. Begin at first sign of disease. Repeat at 7- to 10-day intervals. Begin at first sign of disease. Repeat at 7-to 10-day intervals. To improve performance, add 2 to 4 oz of an effective spreader-sticker. Apply at 7- to 21-day intervals.
mancozeb + myclobutanil (Clevis, MANhandle)	2 lb/100 gal	Apply at first sign of disease, and repeat at 7- to 10-day intervals. The addition of Latron B-1956 will improve performance. Good control of Alternaria.
myclobutanil (Eagle 20EW) (Eagle 40WP)	6 to 12 fl oz/100 gal 3 to 6 oz/100 gal	Apply at 10- to 14-day intervals, not to exceed 21 days. For chrysanthemums, see label for specific rates.
polyoxin D zinc salt (Affirm)	0.25 to 0.5 lb/100 gal	
potassium bicarbonate (MilStop)	1.25 to 5 lb/100 gal	Uniform and complete coverage of foliage is essential for best results. See label for special instructions regarding poinsettia, pansy, and impatiens.
propiconazole (Banner Maxx II)	See label	Rates vary depending on the disease—see label.
propiconazole + chlorothalonil (Concert II)	22 to 35 fl oz/100 gal	Apply as needed beginning when conditions are favorable for disease. Apply to full coverage to the point of drip. Do not make more than 3 applications.
pyraclostrobin (Insignia)	See label	Apply prior to onset of disease and repeat every 7 to 14 days. Do not expose flowering impatiens or flowering petunias to Insignia.
pyraclostrobin + boscalid (Pageant Intrinsic)	6 to 12 oz/100 gal	Apply prior to disease development. Repeat at 7- to 14-day intervals. Make no more than 2 sequential applications. Do not expose petunia or impatiens in flower or wintercreeper or nine bark to spray or drift as injury may occur.

Table 10-11. Disease Control of Annual, Perennial, Bedding, and Flowering Potted Plants in Greenhouses

Disease Pesticide and Formulation	Rate of Formulation	Schedule and Remarks
Fungal Leaf Spots (consult label for specific fungi controlled) (continued)		
tebuconazole (Torque)	4 to 8 fl oz/100 gal	Begin applications 14 to 21 days prior to when disease is expected, or at very first sign of disease.
thiophanate-methyl		Begin when disease first appears and repeat at 7- to 14-day intervals.
(Cleary 3336 F)	12 to 16 fl oz/100 gal	
(T-methyl SPC 50WSB)	12 to 16 oz/100 gal	Repeat every 7 to 14 days during disease period. Rotations with chlorothalonil can be used.
(Incognito 85WDG)	7.2 to 9.6 fl oz/100 gal	Repeat every 7 to 14 days during disease period. Rotations with chlorothalonil can be used.
(OHP 6672 4.5F)	10 to 14.5 fl oz/100 gal	Repeat every 7 to 14 days during disease period. Rotations with chlorothalonil can be used.
(OHP 6672 50WP)	12 to 16 oz/100 gal	Repeat every 7 to 14 days during disease period. Rotations with chlorothalonil can be used
(Phoenix T-bird 85 WDG)	0.3 to 0.8 lb/100 gal	Repeat every 10 to 14 days during disease period.
(T-bird 4.5L)	10.75 to 20 oz/100 gal	Repeat every 7 to 14 days during disease period.
triadimefon (Strike 25WDG)	2 to 4 oz/100 gal	Apply as needed at first sign of disease. Good control of *Alternaria* leaf spot.
trifloxystrobin (Compass O 50WDG)	1 to 4 oz/100 gal	Repeat at 7- to 14-day intervals until threat of disease is over. Rotate to another nonstrobilurin fungicide after each application. Good control of *Alternaria* leaf spot.
triflumizole (Terraguard SC)	4 to 8 oz/100 gal	Apply at very first sign of disease. Do not use on impatiens plugs, and do not exceed 2 oz/100 gal for impatiens transplants. Repeat at 7- to 14-day intervals.
triticonazole (Trinity 19SC)	4 to 12 fl oz/100 gal	See label as rate varies depending on fungal leaf spot pathogen. Use preventively. Begin applications when conditions favor fungal infection and before disease symptom development. Use of an adjuvant/spreader sticker can aid in control.
***Fusarium* Root and Crown Rot**		
azoxystrobin (Heritage)	1 to 4 oz/100 gal (directed spray) 0.2 to 1 oz/100 gal (drench)	Apply as a directed spray every 7 to 21 days. Apply as a soil drench at 7- to 28-day intervals.
azoxystrobin + benzovindiflupyr (Mural)	5 to 7 oz/100 gal (directed spray) 2 to 3 oz/100 gal (drench)	Apply as a directed spray every 7 to 21 days. Apply 1 to 2 pt per sq ft of solution every 7 to 28 days.
Bacillus subtilis strain QST 713 (Rhapsody, Cease)	4 to 8 oz/100 gal	Apply as a drench directed spray every 21 to 28 days. Apply at or during sticking or transplanting. Rhapsody is OMRI listed.
Bacillus subtilis strain MBI 600 (Subtilex)	0.4 to 0.4 oz/100 gal 0.05 to 0.07 oz/cubic yard of soil or growing media	For greenhouse post-plant application. For pre-plant growing media amendment applications. Apply as a water-based slurry to soil or growing media for preventative control and suppression of root pathogens.
chlorothalonil + thiophanate-methyl (Spectro 90WDG)	1.0 to 2.0 lb/100 gal	For best results use spray mixture the same day it is prepared. Spray uniformly over the area to be treated with a properly calibrated power sprayer, apply as a full coverage spray to run-off when conditions are favorable for disease development.
etridiazole + thiophanate-methyl (Banrot 40WP) (Banrot 8G)	See label	See label as rates vary depending on application. Apply in sufficient volume to saturate the soil mixture. Irrigate immediately. Repeat at 4- to 12-week intervals if necessary. See label for more details.
fludioxonil (Medallion 50WDG) (Emblem)	See label	Apply as a drench at seeding or transplanting. Make only one application to seeding crop. If needed, retreat transplants 21 to 28 days after initial application. See label for incorporation into potting mixture. Do not apply as a seed or soil drench to impatiens or New Guinea impatiens.
fludioxonil + mefenoxam (Hurricane 48WP)	See label	Apply as a pre-potting or growing media drench per label directions. For control of Fusarium, add Medallion Fungicide at a rate of 1.0 oz/gal. Application to impatiens, New Guinea impatiens, pothos, geranium, and Easter lily may cause stunting and/or chlorosis.
fluoxastrobin (Disarm O)	0.15 to 0.6 fl oz/100 gal	Apply in 1 to 2 pt of solution per sq ft surface area (or enough to wet the growing medium).
fluxapyroxad + pyraclostrobin (Orkestra Intrinsic)	8 to 10 fl oz/100 gal	The crown and base of the plant and the soil or growing media surrounding the crown must be thoroughly covered.
iprodione + thiophanate-methyl (Dovetail, Nufarm TM + IP SPC)	17 to 34 fl oz/100 gal	Spray to ensure thorough coverage. After transplanting, apply 1 to 2 pt per sq ft. Repeat at 2- to 4-week intervals. Do not use as a drench on impatiens or pothos. Do not use on *Spathiphyllum*.
pyraclostrobin (Empress Intrinsic)	1 to 6 fl oz/100 gal (see label)	Apply at 1 to 3 fl oz for plants in propagation, rooted cuttings, plugs and seedlings and at 2 to 6 fl oz to all other plants. Do not apply to dry soil media. Apply preventative to disease with sequential at 7 to 28 days after the first application if needed.
pyraclostrobin + boscalid (Pageant Intrinsic)	12 to 18 oz/100 gal	Apply prior to disease development. Completely drench the growing medium. Repeat at 7- to 14-day intervals. Make no more than 2 sequential applications. Do not expose petunia or impatiens in flower or wintercreeper or nine bark to spray, as injury may occur
Streptomyces griseoviridis (Mycostop)	See label	Apply inoculant as a seed dressing, soil drench spray, or transplant dip. Must be applied prior to onset of disease. See label.
thiophanate-methyl		
(Cleary 3336F)	8 to 16 oz/100 gal	Apply 8 oz as a drench or directed spray after seeding or sticking or 12 to 16 oz after transplanting. Repeat at 21- to 28-day intervals.
(OHP 6672 50WP)	8 to 16 oz/100 gal	Apply as a drench or heavy spray at a rate of 1/2 to 2 pints per sq ft. Repeat at 4- to 8-week intervals Apply as
(SysTec 1998 FL)	10 to 20 fl oz/100 gal	a drench or heavy spray at a rate of 1 to 2 pints per sq ft. Repeat at 2- to 4-week intervals.
(T-Storm Flowable)	20 fl oz/100 gal	Apply as a drench or heavy spray at a rate of 1 to 2 pints per sq ft. Repeat at 2- to 4-week intervals.
(Incognito 85WDG)	4.8 to 9.6 oz/100 gal	Apply as a drench or heavy spray at a rate of 1/2 to 2 pints per sq ft. Repeat at 4- to 8-week intervals.
(T-methyl SPC 50WSB)	8 to 16 oz/100 gal	Apply as a drench or heavy spray at a rate of 1/2 to 2 pints per sq ft. Repeat at 4- to 8-week intervals.
(OHP 6672 4.5F, T-bird 4.5L)	7.5 to 20 fl oz/100 gal	Apply as a drench or heavy spray at a rate of 1/2 to 2 pints per sq ft. Repeat at 4- to 8-week intervals.
(Phoenix T-bird 85 WDG)	0.4 to 0.8 lb/100 gal	Apply as a drench or heavy spray at a rate of 1 to 2 pints per sq ft. Repeat at 2- to 4- week intervals.
Trichoderma harzianum Rifai strain KRL-AG2 (RootShield WP)	3 to 5 oz/100 gal	May be applied as a soil drench. For cuttings or bare root dip, use 0.25 to 5 lb/5 gal or dip into dry powder.
Trichoderma harzianum Rifai strain T-22, *Trichoderma virens* strain G-41 (RootShield WP Plus)	3 to 8 oz/100 gal	May be applied as a soil drench. For cuttings or bare root dip, use 0.25 to 1.5 lb/2 gal.
triflumizole (Terraguard 50W)	4 to 8 fl oz/100 gal	Apply soil drenches weekly as needed. Use higher rate under heavy disease pressure.
triticonazole (Trinity 19SC)	8 to 12 fl oz/100 gal	Use preventively. Begin applications when conditions favor fungal infection and before disease symptom development. The crown and base of the plant and the soil or potting medium surrounding the crown must be thoroughly covered.

Table 10-11. Disease Control of Annual, Perennial, Bedding, and Flowering Potted Plants in Greenhouses

Disease / Pesticide and Formulation	Rate of Formulation	Schedule and Remarks
Myrothecium Leaf Blight, Crown, or Petiole Rot		
azoxystrobin (Heritage)	2 to 4 oz/100 gal	Apply every 7 to 21 days. Do not exceed 2 oz per 100 gal on impatiens, pansy, or violas.
azoxystrobin + benzovindiflupyr (Mural)	4 to 7 oz/100 gal	Apply every 7 to 21 days.
chlorothalonil (Chlorostar VI, Daconil Weather Stik, Echo 720, Manicure 6 Flowable)	1 3/8 pt/100 gal	Repeat applications at 7- to 14-day intervals.
(Chlorothalonil DF)	1 to 2.5 lb/100 gal	
(Chlorothalonil 500 ZN)	1.9 pt/100 gal	
(Chlorothalonil 720 SFT)	1.37 pt/100 gal	
(Daconil Ultrex) 82.5WDG	1.4 lb/100 gal	
(Echo Ultimate)	1.25 lb/100 gal	
chlorothalonil + thiophanate-methyl (Spectro 90WDG)	1 to 2 lb/100 gal	Apply when foliage and flowers are dry, or nearly dry. Repeat at 7-day intervals.
cyprodinil + fludioxonil (Palladium)	2 to 4 oz/100 gal	Spray on a 7- to 14-day interval while conditions are conducive to disease development. After 2 applications, alternate with another fungicide with a different MOA for 2 applications. See cautionary statement on label for applications to Geraniums, Impatiens and New Guinea Impatiens.
fludioxonil (Medallion WDG, WSP) (Emblem)	1 to 2 oz/100 gal / 1 to 2 fl oz/100 gal	Spray to runoff at 7- to 14-day intervals. See label for media mix and drench applications. Drench applications to impatiens or New Guinea impatiens may cause stunting and/or chlorosis.
fluoxastrobin (Disarm O)	1 to 4 fl oz/100 gal	
pyraclostrobin + boscalid (Pageant Intrinsic)	8 to 12 oz/100 gal	Apply prior to disease development. Repeat at 7- to 14-day intervals. Make no more than 2 sequential applications. Do not expose petunia or impatiens in flower or wintercreeper or nine bark to spray or drift as injury may occur.
trifloxystrobin (Compass O)	1 to 2 oz/100 gal	Repeat at 7- to 14-day intervals until threat of disease is over.
triflumizole (Terraguard 50W)	4 to 8 oz/100 gal	Apply prior to or at first sign of disease. Repeat at 7- to 14-day intervals. Use higher rates for initial application under disease pressure. Do not use on impatiens plugs. Do not exceed 2 oz per 100 gallons on impatiens transplants.
triticonazole (Trinity 19SC)	8 to 12 fl oz/100 gal	Use preventively. Begin applications when conditions favor fungal infection and before disease symptom development. Use of an adjuvant/spreader sticker can aid in control.
Phytophthora Aerial Shoot Blight		
azoxystrobin (Heritage)	1 to 4 oz/100 gal	Apply every 7 to 28 days. Do not make more than 3 sequential applications before switching to another effective fungicide with a different mode of action.
azoxystrobin + benzovindiflupyr (Mural)	4 to 7 oz/100 gal	Apply every 7 to 14 days.
chlorothalonil + thiophanate-methyl (Spectro 90 WDG)	1 to 2 lb/100 gal	Apply when plants are dry. Spectro has protective and curative action. Repeat at 7-day intervals.
dimethomorph (Stature SC)	12.25 fl oz/100 gal	Begin spraying at first sign of disease. Use a full-coverage spray at 10- to 14-day intervals throughout production cycle.
dimethomorph + ametoctradin (Orvego)	14 fl oz/100 gal	Apply on 10- to 14-day intervals using another registered fungicide with a different mode of action. Apply sufficient volume to wet all foliage until runoff (normally 50 to 100 gallons per acre).
fenamidone (Fenstop)	7 to 14 fl oz/100 gal	Apply as a foliar spray until wet. Repeat as necessary on a 28-day schedule. Do not apply more than 4 applications per crop per season.
fluopicolide (Adorn)	2 to 4 fl oz/100 gal	MUST ALWAYS BE TANK MIXED WITH THE LABELED RATE OF ANOTHER FUNGICIDE WITH A DIFFERENT MODE OF ACTION. Apply before disease development. Use higher rate when treating plants with high potential for disease. Reapply after 14 to 28 days. Do not apply more than 2 applications per cropping cycle.
fluoxastrobin (Disarm 480SC)	1 to 4 fl oz/100 gal	Apply as a crown spray at 7- to 28-day application interval depending on disease pressure
fosetyl-Al (Aliette 80 WDG)	1.25 to 4 lb/100 gal	For both products, do not apply more than once every 14 days. Aliette: Do not exceed 400 gallons of spray solution per acre. Check label for compatibility with copper and other compounds.
(Viceroy 70DF)	1.4 to 4.6/100 gal	Viceroy: Check label for compatibility with copper and other compounds.
mancozeb + myclobutanil (Clevis, MANhandle)	2 lb/100 gal	Apply at first sign of disease, and repeat at 7- to 10-day intervals. The addition of Latron B-1956 will improve performance. Good control of Alternaria.
mandipropamid (Micora)	4 to 8 fl oz/100 gal	This product can also be used on vegetables sold to the retail market in GH with permanent flooring. Apply prior to disease development. Repeat at 7- to 14-day intervals. Make no more than 2 sequential applications before rotating to an alternate MOA.
mono- and di- potassium salts of phosphorous acid (Alude)	26 to 54 fl oz/100 gal (foliar)	See labels for other applications such as root dip, drench, drip irrigation, and overhead irrigation. Apply prior to disease development. Spray to thoroughly wet all foliage. Follow labels for repeat application limits.
(Fosphite)	1 to 2 qt/100 gal (foliar)	
(Reliant)	1 to 2 qt/100 gal (foliar)	
oxathiapiprolin (Segovis)	0.6 to 2.4 fl oz/100 gal	Begin foliar applications prior to disease development and continue on 5- to 14-day interval when conditions are conducive for disease development. Do not apply more than 2 consecutive applications before switching to another non-Group U-15 fungicide.
potassium phosphite (Vital)	see label	Apply preventatively as a soil drench or foliar spray.
pyraclostrobin + boscalid (Pageant Intrinsic)	18 oz/100 gal	Apply prior to disease development. Repeat at 7- to 10-day intervals. Make no more than 2 sequential applications. Do not expose petunia or impatiens in flower or wintercreeper or nine bark to spray or drift as injury may occur.

Table 10-11. Disease Control of Annual, Perennial, Bedding, and Flowering Potted Plants in Greenhouses

Disease Pesticide and Formulation	Rate of Formulation	Schedule and Remarks
Phytophthora or Pythium Root and Crown Rot		
azoxystrobin + benzovindiflupyr (Mural)	3 oz/100 gal	Apply 1 to 2 pints of drench solution per square foot surface area every 7 to 28 days.
Bacillus subtilis strain QST 713 (Rhapsody, Cease)	4 to 8 oz/100 gal	Apply as a drench directed spray every 21 to 28 days. Apply at or during sticking or transplanting. Rhapsody is OMRI listed.
Bacillus subtilis strain MBI 600 (Subtilex)	0.4 to 0.4 oz/100 gal 0.05 to 0.07 oz/cubic yard of soil or growing media	NOT LABELED FOR PHYTOPHTHORA. Apply as a water-based slurry to soil or growing media for preventative control and suppression of root pathogens.
boscalid + pyraclostrobin (Pageant Intrinsic)	12 to 18 oz/100 gal	Thorough coverage and wetting of root zone, crown and base of the plant, and surrounding growing media is necessary for best control.
cyazofamid (Segway)	3.0 to 6.0 fl oz/100 gal	Apply ONLY to ornamentals grown in containers in greenhouses as a soil drench. Make applications on a 14- to 21-day interval using another registered fungicide with a different mode of action. Check label for recommended maximum drench volume based on pot diameter.
didecyl dimethyl ammonium chloride (KleenGrow)	0.06 to 0.38 fl. oz. per gallon of water	Apply starting at week 3 or earlier if conditions are favorable for disease. Use a watering device to drench the top and bottom of the leaves and stems, avoiding flowers in full bloom, every 14 days to prevent the spread of spores and the build-up of organic material. Remove severely infected plants and disinfect the area with 1.0 fluid ounce of KleenGrow per gallon of water. Preventative control only.
dimethomorph (Stature SC)	3.06 to 6.12 oz/50 to 100 gal	Apply when plant roots are well established, or at first sign of disease on 10- to 14-day intervals throughout production cycle. Use enough solution to wet root zone. Avoid watering plants for several hours after application. See label for rates for container-grown perennials and woody ornamentals. **Not effective against Pythium root rot.**
dimethomorph + ametoctradin (Orvego)	11 to 14 fl oz/100 gal	NOT LABELED FOR PYTHIUM. Apply on 10- to 14-day intervals using another registered fungicide with a different mode of action. Apply sufficient volume to wet all foliage until runoff (normally 50 to 100 gallons per acre).
dipotassium phosphonate + dipotassium phosphate (Biophos 1% [v/v])	See label for rates	Apply as a soil drench or foliar spray as a preventive.
etridiazole (Truban 30WP) (Terrazole 35WP)	3 to 10 oz/100 gal 3.5 to 10 oz/100 gal	Apply in sufficient volume to saturate soil. Irrigate immediately. Repeat at 4- to 12-week intervals. Drench 4-inch pot with a minimum of 2 oz and a 6-inch pot with 4 oz. Re-treat at 4- to 12-week intervals. Use higher rates for peat or other high organic potting media.
etridiazole + thiophanate-methyl (Banrot 40W) (Banrot 8G)	See label	See label as rates vary depending on application. Apply in sufficient volume to saturate the soil mixture. Irrigate immediately. Repeat at 4- to 12-week intervals if necessary.
fenamidone (Fenstop)	7 to 14 fl oz/50 to 100 gal/400 sq ft	Apply as a drench using 1 to 2 pints per square foot. Repeat as necessary on a 28-day application schedule. Do not apply more than 4 applications of the maximum rate per crop per season. Higher rate has shown more consistent efficacy in research trials.
fosetyl-Al (Aliette 80WDG)	1.25 to 4 lb/100 gal	For both products, do not apply more than once every 14 days. Aliette: Do not exceed 400 gallons of spray solution per acre. Check label for compatibility with copper and other compounds.
(Viceroy 70DF)	1.4 to 4.6 lb/100 gal	Viceroy: Check label for compatibility with copper and other compounds.
fludioxonil + mefenoxam (Hurricane 48WP)	See label	Apply as a pre-potting or growing media drench per label directions. Application to impatiens, New Guinea impatiens, pothos, geranium, and Easter lily may cause stunting and/or chlorosis.
fluopicolide (Adorn)	1 to 4 fl oz/100 gal	MUST ALWAYS BE TANK MIXED WITH THE LABELED RATE OF ANOTHER FUNGICIDE WITH A DIFFERENT MODE OF ACTION. Apply before disease development. Use higher rate when treating plants with high potential for disease. Reapply after 14 to 28 days. Do not make more than one application per crop on poinsettia- phytotoxicity has been observed with repeat applications.
fluoxastrobin (Disarm O)	0.15 to 0.6 fl oz/100 gal	Apply in 102 pt of solution per sq ft surface area (or enough solution to wet the growing media).
Gliocladium virens strain GL-21 (Soilgard 12G)	see label	Naturally occurring soil fungus. Do not use additional soil fungicides at time of incorporation. Avoid use in media totally void of organic matter. OMRI approved.
mandipropamid (Micora)	4 to 8 fl oz/100 gal	This product can also be used on vegetables sold to the retail market in GH with permanent flooring. Apply prior to disease development. Repeat at 7- to 14-day intervals. Make no more than 2 sequential applications.
mefenoxam (Fenox ME, Mefenoxam 2, Subdue MAXX)	See label for rates	Can be applied as a drench, soil surface spray, or incorporated into the soil mix. Consult label for specific crops and applications.
mono- and di- potassium salts of phosphorous acid (Alude)	26 to 54 fl oz/100 gal (foliar)	See labels for other applications such as root dip, drench, drip irrigation, and overhead irrigation. Apply prior to disease development. Spray to thoroughly wet all foliage. Follow labels for repeat application limits.
(Fosphite)	1 to 2 qt/100 gal (foliar)	
(Reliant)	1 to 2 qt/100 gal (foliar)	
oxathiapiprolin (Segovis)	0.65 to 3.2 fl oz/100 gal	NOT LABELED FOR PYTHIUM. Apply at 5- to 14-day interval preventatively or at first sight of disease symptoms. Do not apply more than 2 consecutive applications before switching to another non-Group U-15 fungicide.
potassium phosphite (Vital)	1 pt/100 gal	Apply preventatively as a soil drench or foliar spray.
propamocarb hydrochloride (Banol, Proplant)	20 to 30 fl oz/100 gal	Apply at seeding or transplanting. See label. Effective for preventing Pythium infections.
pyraclostrobin (Empress Intrinsic)	1 to 3 fl oz/100 gal in propagation and 2 to 6 fl oz/100 gal for all other plants in production	Apply as a preventative drench prior to onset of disease. Can be reapplied 7 to 28 days following the initial application.
Trichoderma harzianum Rifai strain KRL- AG2 (RootShield WP)	3 to 5 oz/100 gal	May be applied as a soil drench. For cuttings or bare root dip, use 0.25 to 5 lb/5 gal or dip into dry powder.

Table 10-11. Disease Control of Annual, Perennial, Bedding, and Flowering Potted Plants in Greenhouses

Disease / Pesticide and Formulation	Rate of Formulation	Schedule and Remarks
Phytophthora or Pythium Root and Crown Rot (continued)		
Trichoderma harzianum Rifai strain T-22, *Trichoderma virens* strain G-41 (RootShield WP Plus)	3 to 8 oz/100 gal	May be applied as a soil drench. For cuttings or bare root dip, use 0.25 to 1.5 lb/2 gal.
trifloxystrobin (Compass O)	1 to 2 oz/100 gal	Apply as drench to wet the upper half of growing media. Start application at time of planting and at 14 to 28 days depending on disease pressure. NOT LABELED FOR PYTHIUM.
Powdery Mildew		
azoxystrobin (Heritage)	1 to 4 oz /100 gal	Spray every 7 to 28 days as needed. To avoid fungicide resistance, do not make more than two sequential applications of Heritage before rotating with non-strobilurin products.
azoxystrobin + benzovindiflupyr (Mural)	4 to 7 oz/100 gal (foliar) 2 to 3 oz/100 gal (drench)	Apply every 7 to 21 days. Do not make more than 2 sequential applications before rotating to another class of fungicide that is not Group 7 or 11. Apply 1 to 2 pints of the solution per square foot surface area every 7 to 28 days.
Bacillus subtilis strain OST-713 (Rhapsody, Cease)	2 to 4 qt/100 gal	Repeat at 3- to 10-day intervals. Thorough coverage is essential. Begin applications when conditions favor disease development prior to the onset of disease. Rhapsody is OMRI listed.
Bacillus subtilis strain MBI 600 (Subtilex)	0.4 to 1.2 oz / 1000 sq ft^2	Repeat at 7- to 10-day intervals. Begin applications when conditions favor disease development prior to the onset of disease.
chlorothalonil (Chlorostar VI, Daconil Weather Stik, Echo 720, Manicure 6 Flowable) (Chlorothalonil DF) (Chlorothalonil 500ZN) (Chlorothalonil 720 SFT) (Daconil Ultrex 82.5WDG) (Echo 90DF), (Echo Ultimate)	1 3/8 pt/100 gal 1 to 2.5 lb/100 gal 1.9 pt/100 gal (1.6 pt/100 gal on rose) 1.37 pt/100 gal 1.4 lb/100 gal 1.25 lb/100 gal	Apply until runoff when flowers and foliage are dry. Repeat at 7- to 14-day intervals. Avoid applications during bloom where flower injury is unacceptable. Discontinue use on poinsettias prior to bract formation.
chlorothalonil + thiophanate-methyl (Spectro 90WDG) (ConSyst WDG, TM/C, WDG)	1 to 2 lb/100 gal 0.75 to 1 lb/100 gal	Apply when foliage and flowers are dry, or nearly dry. Re-treat at 7-day intervals. Repeat at 7- to 10-day intervals.
copper hydroxide (Champ WG) (Nu-Cop 50DF) (CuPRO 2005)	0.5 lb/100 gal 1 lb/100 gal 0.75 to 2.0 lb/100 gal	Begin at first sign of disease, and repeat at 7- to 14-day intervals. Do not tank mix copper formulations with Aliette. Avoid contact with metal surfaces. Discoloration of blooms may occur on certain plant varieties- check label.
copper octanoate (Camelot O)	0.5 to 2.0 gal/100 gal	Begin at first sign of disease, and repeat at 7- to 10-day intervals. Camelot O may cause copper toxicity on some plant species.
cyprodinil + fludioxonil (Palladium)	4 to 6 oz/100 gal	Spray on a 7- to14- day interval while conditions are conducive to disease development. After two applications, alternate with another fungicide with a different MOA for two applications.
didecyl dimethyl ammonium chloride (KleenGrow)	0.06 to 0.38 fl oz/gal	Apply starting at week 3 or earlier if conditions are favorable for disease. Use a watering device to drench the top and bottom of the leaves and stems, avoiding flowers in full bloom, every 14 days to prevent the spread of spores and the build-up of organic material. Remove severely infected plants and disinfect the area with 1.0 fluid ounce of KleenGrow per gallon of water.
fluxapyroxad + pyraclostrobin (Orkestra Intrinsic)	6 to 8 fl oz/100 gal	Use on a 7- to 14-day interval.
fluoxastrobin (Disarm 480SC)	1 to 4 fl oz/100 gal	Use for preventative applications only and apply every 7 to 28 days depending on disease pressure. Do not make more than 2 consecutive applications before switching to a non Group 11 fungicide.
hydrogen peroxide + peroxyacetic acid (ZeroTol 2.0)	125 fl oz/100 gal	Apply in early morning or late evening. Apply consecutive applications until control is achieved, then begin weekly preventative program (see label).
iprodione + thiophanate-methyl (Dovetail, Nufarm TM + IP SPC)	17 to 34 fl oz/100 gal	Spray plants to ensure thorough coverage. Do not make more than 4 applications per crop per year. Repeat at 10- to 14-day intervals. Do not drench impatiens or pothos. Do not use on *Spathiphyllum*.
mancozeb + myclobutanil (Clevis, MANhandle)	2 lb/100 gal	Apply at first sign of disease, and repeat at 7- to 10-day intervals. The addition of Latron B-1956 will improve performance.
metconazole (Tourney)	1 to 4 oz/100 gal	Apply on a 14- to 28-day interval.
myclobutanil (Eagle 20EW) (Eagle 40WSP)	6 to 12 fl oz/100 gal 3 to 6 oz/100 gal	Apply at 10- to 14-day intervals. Use caution if applying to Gerbera daisy as phytotoxicity may occur.
neem oil (Triact 70)	1 gal/100 to 200 gal	Trial first on open blooms. Retreat at 7- to 14-day intervals. Use 1:200 rate as a preventive and 1:100 rate if disease is evident.
phosphorous acid (Rampart)	see label	Apply at two to three week intervals. Do not apply to plants that are under heat or moisture stress. Do not apply to plants treated with copper compounds at less than 20 days intervals.
piperalin (Pipron) LC	4 to 8 fl oz/100 gal	See label for precautions on hydrangea, begonia, and poinsettia. Use high rate if disease is already present.
polyoxin D zinc salt (Affirm)	0.25 to 0.5 lb/100 gal	Apply every 7 to 10 days.
potassium bicarbonate (Armicarb 100) (Kaligreen) (MilStop)	2 to 2.5 lb/100 gal 1 to 3 lb/100 gal 4 to 8 oz/100 gal	Apply every 10- to 14 days. Increase frequency to every 5 - to 7 days under heavy disease pressure. Begin at first sign of disease. Repeat at 7-10 day intervals. See label for precautions for poinsettia, impatiens, and pansy. Repeat at 7- to 14-day intervals.
potassium phosphite (Confine, Fosphite, Rampart)	1 to 2 qt/100 gal	Apply at 2- to 3-week intervals. Do not apply more than 6 times per crop cycle.
propiconazole (Banner Maxx II)	8 to 12 fl oz/100 gal	Apply every 30 days at the first sign of disease. For impatiens, the maximum label rate is 8 fl oz/100 gal.
propiconazole + chlorothalonil (Concert II)	22 to 35 fl oz/100 gal	Apply as needed beginning when conditions are favorable for disease. Apply to full coverage to the point of drip. Do not make more than 3 applications.

Table 10-11. Disease Control of Annual, Perennial, Bedding, and Flowering Potted Plants in Greenhouses

Disease / Pesticide and Formulation	Rate of Formulation	Schedule and Remarks
Powdery Mildew (continued)		
pyraclostrobin + boscalid (Pageant Intrinsic)	6 to 12 oz/100 gal	Apply prior to disease development. Repeat at 7- to 10-day intervals. Make no more than 2 sequential applications. Do not expose petunia or impatiens in flower or wintercreeper or nine bark to spray or drift as injury may occur.
tebuconazole (Torque)	4 to 8 fl oz/100 gal	Begin applications 14 to 21 days prior to when disease is expected, or at very first sign of disease.
thiophanate-methyl (Cleary 3336F) (Incognito 85WDG) (OHP 6672 50WP, T-methyl SPC 50WSB) (OHP 6672 4.5F, T-bird 4.5L) (Phoenix T-bird 85 WDG)	12 to 24 oz/100 gal 4.8 to 9.6 oz/100 gal 8 to 16 oz/100 gal 10 to 20 fl oz/100 gal 0.4 lb/100 gal	Apply when disease first appears and repeat at 7- to 14-day intervals. Apply when disease first appears and repeat at 7 to 14 days. Apply when disease first appears and repeat at 7 to 14 days. Apply when disease first appears and repeat at 7 to 14 days. Apply when disease first appears and repeat at 10 to 14 days
thiophanate-methyl + mancozeb (Zyban 79W)	4 bags/100 gal (24 oz/100 gal)	Repeat at weekly intervals.
triadimefon (Strike) 25WDG	2 to 4 oz/100 gal	Apply as needed at first sign of disease. Repeat at 14- to 21-day intervals. Not effective for powdery mildew control on verbena.
triadimefon + trifloxystrobin (Trigo)	1.2 to 2.4 oz/100 gal	Winter use: 1.2 oz rate; Summer use: 2.4 oz rate.
trifloxystrobin (Compass O) 50WDG	2 to 4 oz/100 gal	Good eradicant. Repeat at 7- to 14-day intervals. Rotate to another fungicide of nonstrobilurin chemistry after each Compass application. Use caution when applying to petunia, violets, and New Guinea impatiens due to possible phytotoxicity.
triflumizole (Terraguard SC)	4 to 16 fl oz/100 gal	Use 16 oz/100 gal for initial applications of existing infections. Use 4 to 8 oz/100 gal for subsequent applications and preventative sprays. Do not exceed 2 oz/100 gal for impatiens transplants. Repeat at 7- to 14-day intervals.
triticonazole (Trinity 19SC)	6 to 12 fl oz/100 gal	Use preventively. Begin applications when conditions favor fungal infection and before disease symptom development. Use of an adjuvant/spreader sticker can aid in control.
Rhizoctonia Aerial Blight		
azoxystrobin (Heritage)	1 to 4 oz/100 gal	Repeat at 7- to 28-day intervals. Do not make more than 3 sequential applications before switching to a nonstrobilurin fungicide.
azoxystrobin + benzovindiflupyr (Mural)	4 to 7 oz/100 gal	Apply every 7 to 14 days. Do not make more than 2 sequential applications before rotating to another class of fungicide that is not Group 7 or 11.
chlorothalonil (Chlorostar VI, Daconil Weather Stik, Echo 720, Manicure 6 Flowable) (Chlorothalonil DF, Chlorothalonil 720 SFT) (Chlorothalonil 500 ZN) (Daconil Ultrex) 82.5WDG (Echo Ultimate)	1 3/8 pt/100 gal See label 1.9 pt/100 gal 1.4 lb/100 gal 1.25 lb/100 gal	Apply when foliage and flowers are dry. Repeat at 7- to 14-day intervals. Apply to hydrangea foliage only. Avoid application during bloom period on plants where flower injury is unacceptable.
chlorothalonil + thiophanate-methyl (Spectro 90WDG)	1 to 2 lb/100 gal	Re-treat at a minimum of 7-day intervals.
copper octanoate (Camelot O)	0.5 to 2.0 gal/A	Begin at first sign of disease, and repeat at 7- to 10-day intervals. Camelot O may cause copper toxicity on some plant species. OMRI approved.
cyprodinil + fludioxonil (Palladium)	2 to 6 oz/100 gal	Spray on a 7- to14-day interval while conditions are conducive to disease development. After 2 applications, alternate with another fungicide with a different MOA for 2 applications. Cautionary statement on label for applications to Geraniums, Impatiens and New Guinea Impatiens.
didecyl dimethyl ammonium chloride (KleenGrow)	0.06 to 0.38 fl. oz. per gallon of water	Apply starting at week 3 or earlier if conditions are favorable for disease. Use a watering device to drench the top and bottom of the leaves and stems, avoiding flowers in full bloom, every 14 days to prevent the spread of spores and the build-up of organic material. Remove severely infected plants and disinfect the area with 1.0 fluid ounce of KleenGrow per gallon of water.
fludioxonil (Medallion, WDG, WSP)	1 to 2 oz/100 gal	Spray to runoff. Repeat at 7- to 14-day intervals until conditions no longer favor disease.
flutolanil (Contrast, Prostar) 70WSP	3 to 12 oz/100 gal	Apply at 14- to 21-day intervals.
iprodione (Chipco 26019 N/G) (18 Plus, Iprodione Pro 2SE)	1 to 2 lb/100 gal 1 to 2.5 qt/100 gal	Spray plants to ensure thorough coverage Repeat at 7- to 14-day intervals. Do not make more than 4 applications per crop per year. Do not use as a soil drench on impatiens or pothos. Do not use on *Spathiphyllum*.
iprodione + thiophanate-methyl (26/36) (Dovetail, Nufarm TM + IP SPC)	33 to 84 oz/100 gal 17 to 34 fl oz/100 gal	Spray plants to ensure thorough coverage. Do not make more than 4 applications per year. Repeat at 7- to 14-day intervals. Repeat at 10- to 14-day intervals. Do not use as a soil drench on impatiens or pothos. Do not use on *Spathiphyllum*.
mancozeb + myclobutanil (Clevis, MANhandle)	2 lb/100 gal	Apply at first sign of disease, and repeat at 7- to 10-day intervals. The addition of Latron B-1956 will improve performance. Good control of Alternaria.
propiconazole (Banner Maxx II)	5 to 8 fl oz/100 gal	Apply as needed beginning when conditions are favorable for disease. Apply to full coverage to the point of drip.
propiconazole + chlorothalonil (Concert II)	22 to 35 fl oz/100 gal	Apply as needed beginning when conditions are favorable for disease. Apply to full coverage to the point of drip. Do not make more than 3 applications.
pyraclostrobin (Empress Intrinsic)	1 to 3 fl oz/100 gal in propagation and 2 to 6 fl oz/100 gal for all other plants in production	Apply as a preventative drench prior to onset of disease. Can be reapplied 7 to 28 days following the initial application.
pyraclostrobin + boscalid (Pageant Intrinsic)	12 to 18 oz/100 gal	Apply prior to disease development. Repeat at 7- to 14 day intervals. Make no more than 2 sequential applications. Do not expose petunia or impatiens in flower or wintercreeper or nine bark to spray or drift as injury may occur.

Table 10-11. Disease Control of Annual, Perennial, Bedding, and Flowering Potted Plants in Greenhouses

Disease Pesticide and Formulation	Rate of Formulation	Schedule and Remarks
Rhizoctonia Aerial Blight (continued)		
thiophanate-methyl (Cleary 3336F) (T-methyl SPC 50WSB), (OHP 6672 50WP) (OHP 6672 4.5F) (T-bird 4.5L)	12 to 16 fl oz/100 gal 12 to 16 fl oz/100 gal 10 to 14.5 fl oz/100 gal 10.75 to 20 fl oz/100 gal	Apply when disease symptoms first appear. Repeat at 7- to 14-day intervals.
thiophanate-methyl + flutolanil (SysStar WDG)	4 to 8 oz/100 gal	For best results apply before disease development.
triflumizole (Terraguard SC)	4 to 8 fl oz/100 gal	Make initial application prior to or at first sign of disease. Use the higher rate under heavy disease pressure. Repeat at 7- to 14-day intervals.
Rhizoctonia Stem and Root Rot		
azoxystrobin (Heritage)	1 to 4 oz/100 gal (spray) 0.2 to 1 oz/100 gal (drench)	Apply as a preventative spray or drench treatment. Repeat at 7 to 21 days. Do not exceed 2 oz/100 gal on impatiens.
azoxystrobin + benzovindiflupyr (Mural)	5 to 7 oz/100 gal (spray) 2 to 3 oz/100 gal (drench)	Apply 1 to 2 pints of drench solution per square foot surface area every 7 to 28 days.
Bacillus subtilis strain QST 713 (Rhapsody, Cease)	4 to 8 oz/100 gal	Apply as a drench directed spray every 21 to 28 days. Apply at or during sticking or transplanting. Rhapsody is OMRI listed.
Bacillus subtilis strain MBI 600 (Subtilex)	0.4 to 0.4 oz/100 gal 0.05 to 0.07 oz/cubic yard of soil or growing media	Apply as a water-based slurry to soil or growing media for preventative control and suppression of root pathogens.
chlorothalonil + thiophanate-methyl (Spectro) 90WDG	1 to 2 lb/100 gal	Spray uniformly over area to be treated. Re-treat at 7-day intervals.
didecyl dimethyl ammonium chloride (KleenGrow)	0.06 to 0.38 fl. oz. per gallon of water	Apply starting at week 3 or earlier if conditions are favorable for disease. Use a watering device to drench the top and bottom of the leaves and stems, avoiding flowers in full bloom, every 14 days to prevent the spread of spores and the build-up of organic material. Remove severely infected plants and disinfect the area with 1.0 fluid ounce of KleenGrow per gallon of water.
etridiazole + thiophanate-methyl (Banrot 40W) (Banrot 8G)	See label	See label as rates vary depending on application. Apply in sufficient volume to saturate the soil mixture. Irrigate immediately. Repeat at 4- to 12-week intervals if necessary.
fludioxonil (Medallion WDG, WSP)) (Emblem)	See label	Apply as a drench at seeding or transplanting. Apply sufficient mix to wet the upper one-half of the growing medium. Make only one application to seedling crop. If needed, re-treat transplants 21 to 28 days after initial application. Do not apply as a seed or soil drench to impatiens or New Guinea impatiens. May cause stunting or chlorosis on some geranium cultivars.
fludioxonil + mefenoxam (Hurricane 48 WP)	See label	Apply as a pre-potting or growing media drench per label directions. Labeled for Rhizoctonia and Phytophthora/Pythium root rots and is best used when both diseases are present or suspected.
fluoxastrobin (Disarm O)	2 to 4 fl oz/100 gal (crown spray) 0.15 to 0.6 fl oz/100 gal (drench)	Crown spray: apply at 7- to 21-day intervals. Drench: Apply in 1 to 2 pt of solution per sq ft surface area. Apply at 14 to 28-day intervals.
flutolanil (Contrast, Prostar 70WSP)	3 to 6 oz/100 gal	Apply drench at 2 oz per 4-inch pot. Repeat 21 to 28 days after initial application. Make no more than 4 applications per year.
Gliocladium virens strain GL-21 (Soilgard 12G)	see label	Naturally occurring soil fungus. Do not use additional soil fungicides at time of incorporation. Avoid use in media totally void of organic matter. OMRI approved.
iprodione (Chipco 26019 N/G) (18 Plus, Iprodione Pro 2SE)	6.5 oz/100 gal 13 fl oz/100 gal	Apply 1 to 2 pints per sq ft at seeding or transplanting. Do not apply as a drench on impatiens or pothos. Repeat every 14 days. Do not make more than 6 applications per year. Do not use on Spathiphyllum.
iprodione + thiophanate-methyl (Dovetail, Nufarm TM + IP SPC)	17 to 34 fl oz/100 gal	Apply 1 to 2 pints per sq ft at seeding or transplanting. Do not apply as a drench on impatiens or pothos. Do not use on *Spathiphyllum*. Repeat every 10 to 14 days.
PCNB (Terraclor 75WP)	4 to 8 oz/100 gal	See label for amount to apply. One repeat application can be made 4 to 6 weeks later, if necessary.
polyoxin D zinc salt (Affirm)	0.25 to 0.5 lb/100 gal	Apply as a drench every 14 to 28 days.
pyraclostrobin (Empress Intrinsic)	1 to 3 fl oz/100 gal in propagation and 2 to 6 fl oz/100 gal for all other plants in production	Apply as a preventative drench prior to onset of disease. Can be reapplied 7 to 28 days following the initial application.
pyraclostrobin + boscalid (Pageant Intrinsic)	12 to 18 oz/100 gal	Apply prior to disease development. Repeat at 7- to 21-day intervals. Make no more than 2 sequential applications. Do not expose petunia or impatiens in flower or wintercreeper or nine bark to spray or drift as injury may occur.
thiophanate-methyl (Cleary 3336 F) (OHP 6672 50WP) (T-Storm Flowable) (Incognito 85WDG) (T-methyl SPC 50WSB) (OHP 6672 4.5F, T-bird 4.5L)	8 to 16 oz/100 gal 8 to 16 oz/100 gal 20 fl oz/100 gal 4.8 to 9.6 oz/100 gal 8 to 16 oz/100 gal 7.5 to 20 fl oz/100 gal	Apply 8 oz as a drench or directed spray after seeding or 12 to 16 oz after transplanting. Repeat at 4- to 8-week intervals. Apply as a drench at a rate of 1/2 to 2 pints per sq ft after transplanting. Repeat at 21- to 28-day intervals. Apply as a drench or heavy spray at a rate of 1 to 2 pints per sq ft after transplanting. Repeat at 2- to 4-week intervals. Apply as a drench or heavy spray at a rate of 1/2 to 2 pints per square foot. Repeat at 4- to 8- week intervals. Apply as a drench or heavy spray at a rate of 1/2 to 2 pints per square foot. Repeat at 4- to 8- week intervals. Apply as a drench or heavy spray at a rate of 1/2 to 2 pints per square foot. Repeat at 4- to 8- week intervals.
thiophanate-methyl + flutolanil (SysStar WDG)	2 to 4 oz/100 gal	Apply according to label directions.
Trichoderma harzianum Rifai strain KRL- AG2 (RootShield WP)	3 to 5 oz/100 gal	May be applied as a soil drench. For cuttings or bare root dip, use 0.25 to 5 lb/5 gal or dip into dry powder.

Table 10-11. Disease Control of Annual, Perennial, Bedding, and Flowering Potted Plants in Greenhouses

Disease Pesticide and Formulation	Rate of Formulation	Schedule and Remarks
Rhizoctonia Stem and Root Rot (continued)		
Trichoderma harzianum Rifai strain T-22, *Trichoderma virens* strain G-41 (RootShield WP Plus)	3 to 8 oz/100 gal	May be applied as a soil drench. For cuttings or bare root dip, use 0.25 to 1.5 lb/2 gal.
trifloxystrobin (Compass O 50WDG)	0.5 oz/100 gal	Apply as a drench to wet upper half of the growing media. Apply at seeding, again at transplanting, and at 21- to 28-day intervals thereafter. May injure petunia, violet, and New Guinea impatiens.
triflumizole (Terraguard SC)	4 to 8 fl oz/100 gal	Apply as soil drench at 2- to 4-week intervals. Use higher rate under heavy disease pressure.
triticonazole (Trinity 19SC)	8 to 12 fl oz/100 gal	Use preventively. Begin applications when conditions favor fungal infection and before disease symptom development. The crown and base of the plant and the soil or potting medium surrounding the crown must be thoroughly covered.
Rusts		
azoxystrobin (Heritage)	1 to 4 oz/100 gal	Apply at 7- to 28-day intervals. Do not make more than 2 sequential applications of Heritage before alternating with a nonstrobilurin fungicide. Not effective for rust control on Hypericum. Rotate with mancozeb or triflumizole.
azoxystrobin + benzovindiflupyr (Mural)	4 to 7 oz/100 gal	Apply every 7 to 14 days. Do not make more than 2 sequential applications before rotating to another class of fungicide that is not Group 7 or 11.
Bacillus subtilis strain QST 713 (Rhapsody, Cease)	2 to 8 qt/100 gal	Repeat at 3- to 10-day intervals. Thorough coverage is essential. Begin applications when conditions favor disease development, prior to the onset of disease. Rhapsody is OMRI listed.
chlorothalonil + thiophanate-methyl (Spectro 90WDG)	1 to 2 lb/100 gal	Apply when foliage and flowers are dry, or nearly dry. Re-treat at a minimum of 7-day intervals.
chlorothalonil (Chlorothalonil DF) (Chlorostar VI, Daconil Weather Stik, Echo 720, Manicure 6 Flowable) (Chlorothalonil DF, Chlorothalonil 720 SFT) (Chlorothalonil 500 ZN) (Daconil Ultrex 82.5WDG) (Echo 90DF), (Echo Ultimate)	1 to 2.5lb/100 gal 1 3/8 pt/100 gal See label 1.9 pt/100 gal 1.4 lb/100 gal 1.25 lb/100 gal	Apply when foliage and flowers are dry. Repeat at 7- to 14-day intervals. Apply to hydrangea foliage only. Avoid application during bloom period on plants where flower injury is unacceptable. Chlorothalonil DF can be applied to impatiens.
flutolanil (ProStar 70WP)	3 to 6 oz/100 gal	Use as foliar application and repeat every 14 to 21 days. Do not exceed 4 applications per growing season. Do not make more than 2 consecutive applications before switching to a non-Group 7 fungicide.
fluoxastrobin (Disarm 480SC)	1 to 4 fl oz/100 gal	Use for preventative applications only and apply every 7 to 28 days depending on disease pressure. Do not make more than 2 consecutive applications before switching to a non Group 11 fungicide.
fluxapyroxad + pyraclostrobin (Orkestra Intrinsic)	6 to 8 fl oz/100 gal	Apply on a 7- to 14-day interval.
mancozeb (Dithane 75DF), (Fore 80WSP), (Mancozeb DG) (Pentathlon DF) (Pentathlon LF) (Protect DF)	1.5 lb/100 gal 1 to 2 lb/100 gal 0.8 to 1.6 qt/100 gal 1 to 2 lb/100 gal	Begin at first sign of disease. Repeat at 7- to 10-day intervals. Begin application at first sign of disease. Repeat at 7- to 10-day intervals. Begin application at first sign of disease. Repeat at 7- to 10-day intervals. Apply at 7- to 21-day intervals.
mancozeb + myclobutanil (Clevis, MANhandle)	2 lb/100 gal	Apply at first sign of disease, and repeat at 7- to 10-day intervals. The addition of Latron B-1956 will improve performance. In a limited number of trials, gave very good to excellent control of rust on geranium (18 oz/100 gal) and snapdragon (1 lb/100 gal).
metconazole (Tourney)	1 to 4 oz/100 gal	Apply on a 14- to 28-day interval.
myclobutanil (Eagle 20EW) (Eagle 40WP)	6 to 12 fl oz/100 gal 3 to 6 oz/100 gal	Apply on a protectant application schedule at 10- to 14-day intervals. See label for rates to control white rust on chrysanthemum.
neem oil (Triact 70)	1 gal/100 to 200 gal	Apply at 7- to 14-day spray intervals. Trial first on open blooms. To control existing disease, apply on a 7-day schedule until disease pressure is eliminated. Not for impatiens, carnation, or hibiscus.
oxycarboxin (Plantvax 75W)	16 to 24 oz/100 gal	Apply at first sign of disease. Repeat at 2-week intervals for a maximum of 2 to 4 applications per season
propiconazole (Banner Maxx II)	5 to 8 fl oz/100 gal	Apply as needed beginning when conditions are favorable for disease. Apply to full coverage to the point of drip.
propiconazole + chlorothalonil (Concert II)	22 to 35 fl oz/100 gal	Apply as needed beginning when conditions are favorable for disease. Apply to full coverage to the point of drip. Do not make more than 3 applications.
pyraclostrobin (Insignia)	See label	Apply prior to onset of disease and repeat every 7 to 14 days. Do not expose flowering impatiens or flowering petunias to Insignia.
pyraclostrobin + boscalid (Pageant Intrinsic)	6 to 12 oz/100 gal	Apply prior to disease development. Repeat at 7- to 10-day intervals. Make no more than 2 sequential applications. Do not expose petunia or impatiens in flower or wintercreeper or nine bark to spray or drift as injury may occur.
thiophanate-methyl (Cleary 3336 F) (T-methyl SPC 50WSB) (OHP 6672 4.5F, T-bird 4.5L) (OHP 6672 50WP)	12 to 16 fl oz/100 gal 12 to 16 oz/100 gal 10.75 to 20 fl oz/100 gal 12 to 16 oz/100 gal	Begin applications at first sign of disease. Repeat at 7- to 14-day intervals.
triadimefon (Strike 25WDG)	2 to 4 oz/100 gal	Spray to the point of drip as needed.

Table 10-11. Disease Control of Annual, Perennial, Bedding, and Flowering Potted Plants in Greenhouses

Disease / Pesticide and Formulation	Rate of Formulation	Schedule and Remarks
Rusts (continued)		
triadimefon + trifloxystrobin (Armada 50WP, Armada 50WDG, Strike Plus)	3 to 9 oz/100 gal	See label for application limits.
trifloxystrobin (Compass O WDG)	2 to 4 oz/100 gal	Apply at 7- to 14-day intervals.
triflumizole (Terraguard 50WP)	2 to 8 fl oz/100 gal	Apply prior to, or at first sign of disease. Repeat at 7- to 14-day intervals.
Scab, Poinsettia (*Sphaceloma*)		
azoxystrobin (Heritage)	1 to 4 oz/100 gal	Apply at 10- to 28-day intervals. Test for phytotoxicity prior to treating entire crop. Do not make more than 3 sequential applications of Heritage before alternating with a nonstrobilurin fungicide.
azoxystrobin + benzovindiflupyr (Mural)	4 to 7 oz/100 gal	Apply every 7 to 14 days. Do not make more than 2 sequential applications before rotating to another class of fungicide that is not Group 7 or 11.
copper sulfate pentahydrate (Phyton 27)	2.0 to 3.5 oz/100 gal	Apply at 7-day intervals.
mancozeb (Dithane 75DF), (Fore 80WSP), (Mancozeb DG)	1.5 lb/100 gal	Spray at first sign of disease. Apply at 7- to 21-day intervals. May leave a residue.
(Protect DF)	1 to 2 lb/100 gal	
mancozeb + myclobutanil (Clevis, MANhandle)	2 lb/100 gal	Apply at first sign of disease. The addition of Latron B-1956 will improve performance.
myclobutanil (Eagle 20EW) (Eagle 40WP)	6 to 12 fl oz/100 gal 3 to 6 oz/100 gal	Retreat at 10- to 14-day intervals.
triticonazole (Trinity 19SC)	6 to 12 fl oz/100 gal	Use preventively. Begin applications when conditions favor fungal infection and before disease symptom development.
Sclerotinia Blight (*Sclerotinia sclerotiorum*)		
azoxystrobin (Heritage)	1 to 4 oz/100 gal (directed spray) 1 oz/100 gal (drench)	Apply as a directed spray every 7 to 21 days. Apply as a drench every 7 to 28 days.
azoxystrobin + benzovindiflupyr (Mural)	5 to 7 oz/100 gal	Apply as a directed spray every 7 to 21 days.
chlorothalonil (Chlorothalonil 500 ZN) (Chlorothalonil 720 SFT) (Chlorostar VI, Daconil Weather Stik, Echo 720, Manicure 6 Flowable) (Daconil Ultrex) 82.5WDG (Echo) 90DF, (Echo Ultimate)	1.9 pt/100 gal 1.37 pt/100 gal 1 3/8 pt/100 gal 1.4 lb/100 gal 1.25 lb/100 gal	Apply at 7- to 14-day intervals. Apply when foliage and flowers are dry, or nearly dry. Applications made during bloom may damage flowers.
chlorothalonil + thiophanate-methyl (Spectro 90WDG)	1 to 2 lb/100 gal	Apply when both foliage and flowers are dry, or nearly dry. Repeat at 7-day intervals. Do not use on Swedish ivy, Boston fern, or Easter cactus.
cyprodinil + fludioxonil (Palladium)	2 to 6 oz/100 gal	Spray on a 7-14 day interval while conditions are conducive to disease development. After 2 applications, alternate with another fungicide with a different MOA for two applications.
fenhexamid (Decree 50WDG)	0.75 to 1.5 lb/100 gal	Apply at 7- to 14-day intervals. Maintain agitation during application. May cause phytotoxicity on poinsettia bracts.
fluoxastrobin (Disarm 480SC)	2 to 4 fl oz/100 gal	Use for preventative applications only and apply every 7 to 21 days depending on disease pressure. Do not make more than 2 consecutive applications before switching to a non Group 11 fungicide.
iprodione + thiophanate-methyl (Dovetail, Nufarm TM + IP SPC)	17 to 34 fl oz/100 gal	Apply 1 to 2 pints per sq ft at seeding or transplanting. Do not apply as a drench on impatiens or pothos. Do not use on *Spathiphyllum*. Repeat every 10 to 14 days.
pyraclostrobin (Empress Intrinsic)	1 to 6 fl oz/100 gal (see label)	Apply at 1 to 3 fl oz for plants in propagation, rooted cuttings, plugs and seedlings and at 2 to 6 fl oz to all other plants. Do not apply to dry soil media. Apply preventative to disease with sequential at 7 to 28 days after the first application if needed.
PCNB (Terraclor) 75WP	See label	Apply as a drench or bulb soak according to label directions.
pyraclostrobin + boscalid (Pageant Intrinsic)	12 to 18 oz/100 gal	Apply prior to disease development. Repeat at 7- to 14-day intervals. Make no more than 2 sequential applications. Do not expose petunia or impatiens in flower or wintercreeper or nine bark to spray or drift as injury may occur.
thiophanate-methyl (OHP 6672 4.5F, T-bird 4.5L) (Cleary 3336 F) (OHP 6672 50WP) (T-Storm Flowable) (Incognito 85WDG) (T-methyl SPC 50WSB) (Phoenix T-bird 85 WDG)	7.5 to 20 fl oz/100 gal 8 to 16 fl oz/100 gal 8 to 16 oz/100 gal 20 fl oz/100 gal 4.8 to 9.6 oz/100 gal 8 to 16 oz/100 gal 0.4 to 0.8 lb/100 gal	Apply as a drench or heavy spray at the rate of 1/2 to 2 pints per sq ft. Apply at 4- to 8- week intervals. Repeat at 21- to 28-day intervals. Repeat at 4- to 8-week intervals. Apply as a drench or heavy spray at the rate of 1 to 2 pints per sq ft. Apply at 2- to 4- week intervals. Apply as a drench or heavy spray at the rate of 1/2 to 2 pints per sq ft. Apply as a drench or heavy spray at the rate of 1/2 to 2 pints per sq ft. Apply as a drench or heavy spray at the rate of 1 to 2 pints per sq ft
triticonazole (Trinity 19SC)	8 to 12 fl oz/100 gal	Use preventively. Begin applications when conditions favor fungal infection and before disease symptom development. The stem areas of the plant must be thoroughly covered using spray to runoff.
Thielaviopsis Root Rot: See Black root rot.		

Disease Control for Forest, Christmas, and Ornamental Trees

Sara M. Villani, Department of Entomology and Plant Pathology

Anyone using any agricultural chemical should refer to the current chemical label, which contains information about the safe and effective use of the chemical, before using the chemical.

Table 10-12. Disease Control for Forest, Christmas, and Ornamental Trees

CROP / Disease	Material	Rate	Method	Schedule	Remarks
Ash					
Anthracnose	thiophanate-methyl (AllBan Flo) (Cleary 3336 F)	10.75 to 20 fl oz/100 gal 12 to 16 fl oz/100 gal	foliar spray	Three to four applications at 14-day intervals.	FRAC GROUP 1 First application at bud break or at first sign of disease.
	chlorothalonil (Daconil Ultrex 82.5 WDG) (Daconil Weather Stick)	1.4 lb/100 gal 1 3/8 pt /100 gal	foliar spray foliar spray	Repeat in 7 to 14 days when conditions favor disease.	
	metconazole (Tourney)	1 to 4 oz/100 gal	foliar spray	Repeat in 14 to 28 days when conditions favor disease.	FRAC GROUP 3
	mancozeb (Dithane 75DF Rainshield)	1 to 2 lb/100 gal	foliar spray	Repeat at 7- to 10-day intervals.	Begin at first sign of disease
	pyraclostrobin (Insignia)	8 to 16 oz/100gal	foliar spray	Repeat at 7- to 14-day intervals as needed.	FRAC GROUP 11 Use for preventative applications only. Do not make more than two consecutive applications before switching to a non Group 11 fungicide.
	pyraclostrobin + boscalid (Pageant Intrinsic)	18 oz./100 gal	foliar spray	Repeat applications 7 to14 days as needed.	FRAC GROUP 11 + 7
	pyraclostrobin + fluxapyroxad (Orkestra)	8 to 10 fl oz/100 gal	foliar spray	Preventative applications every 7 to 14 days.	FRAC GROUP 11 + 7
	tebuconazole (Torque)	4 to 10 fl oz/100 gal	foliar spray	Begin applications 14 to 21 days prior to when disease is expected, or at very first sign of disease.	FRAC GROUP 3
Crabapple					
Fire blight	copper octanoate (Cueva)	2 to 3 qt/A	foliar spray	Repeat applications every 7 to 10 days until approximately 1 month after petal fall.	Can be phytotoxic after bud break. OMRI approved.
	streptomycin (Agri-mycin 17)	50 to 100 ppm	foliar spray	Three to five applications starting at 20% to 30% bloom. Refer to forecasting models.	Spray every 10 to 14 days.
Rusts (cedar apple rust, quince rust, hawthorn rust)	mancozeb (Dithane 75DF, Rainshield)	1 to 2 lb/100 gal	foliar spray	Repeat at 7- to 10-day intervals.	Begin preventatively at tight cluster and continue through petal fall. Monitor throughout summer.
	myclobutanil (Eagle 20EW)	6 to 12 fl oz/A	foliar spray	Repeat at 7- to 10-day intervals.	FRAC GROUP 3 Begin preventative treatments at tight cluster stage and continue through petal fall. Monitor throughout summer.
	propiconazole (Topaz)	See label	foliar spray	See label.	FRAC GROUP 3 Begin preventative treatments at tight cluster stage and continue through petal fall. Monitor throughout summer.
	tebuconazole (Torque)	4 to 8 oz/A	foliar spray	Repeat applications 7 to 10 days as needed.	FRAC GROUP 3 Begin preventative treatments at tight cluster/ pink stage and continue through petal fall. Monitor throughout summer.
Apple scab, powdery mildew	fluxapyroxad + pyraclostrobin (Orkestra)	4 to 8 fl oz/100 gal (scab) 6 to 8 fl oz/100 gal (mildew)	foliar spray	Apply every 7 to 10 days (scab). Apply every 7 to 14 days (mildew).	FRAC GROUPS 7 + 11 Begin preventative treatments starting at tight cluster and continue through petal fall. Do not make more than two consecutive applications before switching to a non Group 7 or non-Group 11 fungicide.
	myclobutanil (Eagle 20EW)	6 to 12 fl oz/100 gal	foliar spray	Apply every 10 to 14 days.	FRAC GROUP 3 Begin preventative treatments starting at tight cluster and continue through petal fall. Monitor for disease throughout summer.
	propiconazole (Banner MAXX II)	2 to 4 fl oz/ 100 gal (scab) 5 to 8 fl oz/100 gal (mildew)	foliar spray	Every 14 to 21 days as needed.	FRAC GROUP 3 Begin preventative treatments starting at tight cluster and continue through petal fall. Monitor for disease throughout summer.
	chlorothalonil + propiconazole (Concert II)	9 to 35 fl oz/100 gal (see label for specific rate instructions)	foliar spray	Every 14 to 21 days.	FRAC GROUPS M5 + 3 See label for additional details

Table 10-12. Disease Control for Forest, Christmas, and Ornamental Trees

CROP Disease	Material	Rate	Method	Schedule	Remarks
Crabapple (continued)					
Apple scab, powdery mildew (continued)	pyraclostrobin + boscalid (Pageant Intrinsic)	6 to 12 oz./100 gal	foliar spray	Repeat applications 7 to 10 days as needed.	FRAC GROUPS 11 + 7
	tebuconazole (Torque)	4 to 10 fl oz/100 gal	foliar spray	Begin applications 14 to 21 days prior to when disease is expected, or at very first sign of disease.	FRAC GROUP 3
Crapemyrtle					
Cercospora leaf spot	azoxystrobin (Heritage)	4 oz/100 gal	foliar spray	Spray every 7 to 28 days.	FRAC GROUP 11
	azoxystrobin + benzovindiflupyr (Mural)	4 to 7 oz/100 gal	Foliar spray	Apply every 7 to 21 days.	FRAC GROUPS 11 + 7 Do not make more than two sequential applications before rotating to another class of fungicide that is not Group 7 or 11.
	fluoxastrobin (Disarm 480SC)	1 to 4 fl oz/100 gal	Foliar spray	Every 7 to 28 days.	FRAC GROUP 11
	fluxapyroxad + pyraclostrobin (Orkestra Intrinsic)	8 to 10 fl oz/100 gal	foliar spray	Apply every 7 to 14 days.	FRAC GROUPS 7 + 11
	pyraclostrobin + boscalid (Pageant Intrinsic)	8 to 12 oz./100 gal	foliar spray	Repeat applications 7 to 14 days as needed.	FRAC GROUPS 11 + 7
	triticonazole (Trinity 19SC)	4 to 8 fl oz/100 gal	foliar spray	Every 7 to 14 days.	FRAC GROUP 3 Use preventively. Begin applications when conditions favor fungal infection and before disease symptom development. Use of an adjuvant/spreader sticker can aid in control.
Powdery mildew	azoxystrobin (Heritage 50)	1 to 4 oz/100 gal	foliar spray	Preventative sprays at 7- to 28-day intervals.	FRAC GROUP 11
	chlorothalonil + propiconazole (Concert II)	22 to 35 fl oz/100 gal	foliar spray	Every 14 to 21 days.	FRAC GROUPS M5 + 3 See label for details
	metconazole (Tourney)	1 to 4 oz/100 gal	foliar spray	Repeat in 14 to 28 days when conditions favor disease.	FRAC GROUP 3
	myclobutanil (Eagle20EW)	6 to 12 fl oz/100 gal	foliar spray	Apply at 10- to 14-day intervals.	FRAC GROUP 3
	potassium bicarbonate (MilStop)	1.25 to 5 lb/100 gal	foliar spray	Apply every 7 to 14 days.	OMRI listed
	propiconazole (Banner Maxx II)	5 to 8 fl oz/100 gal	foliar spray	See label.	FRAC GROUP 3
	pyraclostrobin + boscalid (Pageant Intrinsic)	6 to 12 oz./100 gal	foliar spray	Repeat applications 7 to 10 days as needed.	FRAC GROUPS 11 + 7
	pyraclostrobin + fluxapyroxad (Orkestra Intrinsic)	6 to 8 fl oz/100 gal	foliar spray	Repeat applications 7 to 14 days as needed.	FRAC GROUPS 11 + 7
	sulfur (Micro Sulf)	3 to 10 lb/100 gal	foliar spray	Repeat every 5 to 10 days.	FRAC GROUP M2 Begin with onset of disease
	triadimefon (Bayleton FLO)	See label.	foliar spray	See label.	FRAC GROUP 3
	triticonazole (Trinity 19SC)	6 to 12 fl oz/100 gal	foliar spray	Every 7 to 14 days.	FRAC GROUP 3 Use preventively. Begin applications when conditions favor fungal infection and before disease symptom development. Use of an adjuvant/spreader sticker can aid in control.
Root rot (*Phytophthora* spp.)	cyazofamid (Segway SC)	3 to 6 fl oz/100 gal	soil drench	14 to 21 day intervals using another registered fungicide with a different mode of action.	FRAC GROUP 21
	fosetyl-AL (Aliette WDG)	0.4 to 0.8 lb/100 gal 2.5 to 5 lb/100 gal	drench foliar	Every 30 days as necessary.	FRAC GROUP 33 Thoroughly wet plant and root mass immediately before transplanting.
	mefenoxam (Subdue MAXX)	Seedlings: 1.25 pt/50 gal 2-0 transplants: 2.5 pt/50 gal	directed soil spray over beds	See label for frequency.	FRAC GROUP 4 Apply 0.5 to 1 inch of water after application, if rain is not expected within 3 days.
	phosphorous acid (Alude, Fosphite, Reliant)	See label for rates	Apply as a soil drench or foliar spray as a preventive		FRAC GROUP 33
	pyraclostrobin + boscalid (Pageant Intrinsic)	12 to 18 oz/100 gal	Soil drench	Repeat applications 7 to 21 days as needed.	FRAC GROUPS 11 + 7
	pyraclostrobin + fluxapyroxad (Orkestra Intrinsic)	8 to 10 fl oz/100 gal	soil drench	Repeat applications 7 to 28 days as needed.	FRAC GROUPS 11 + 7
Dogwood					
Anthracnose (*Discula*)	chlorothalonil (Daconil Ultrex)	1.4 lb/100 gal	foliar spray	Repeat every 7 to 14 days when conditions favor disease.	FRAC GROUP M5 Prune out all diseased tissue. Several applications in fall before leaf drop may also be advisable.

Table 10-12. Disease Control for Forest, Christmas, and Ornamental Trees

CROP Disease	Material	Rate	Method	Schedule	Remarks
Dogwood (continued)					
Anthracnose (*Discula*) (continued)	chlorothalonil + propiconazole (Concert II)	9 to 17 fl oz/100 gal 35 fl oz/100 gal	foliar spray	Apply every 14 days. Apply every 28 days.	FRAC GROUPS M5 + 3 See label for additional details.
	chlorothalonil + thiophanate-methyl (Spectro 90WDG)	1 to 2 lb/100 gal	foliar spray	Apply every 7 to 14 days.	FRAC GROUPS M3 + 1 See label for additional details.
	mancozeb (Dithane 75DF Rainshield, Fore 80WP, Protect DF)	1 to 2 lb/100 gal (for Dithane - see labels for other formulations)	foliar spray	Spray every 10 to 14 days from bud break until mid-summer.	FRAC GROUP M3 Prune out all diseased tissue. Several applications in fall before leaf drop may also be advisable.
	metconazole (Tourney)	1 to 4 oz/100 gal	foliar spray	Repeat in 14 to 28 days when conditions favor disease.	FRAC GROUP 3
	myclobutanil (Eagle 20EW)	6 to 12 fl oz/100 gal	foliar spray	Spray every 10 to 14 days.	FRAC GROUP 3 Prune out all diseased tissue. Several applications in fall before leaf drop may also be advisable.
	propiconazole (Banner MAXX II)	2 to 8 fl oz/100 gal	foliar spray	Apply every 14 to 28 days from bud break to mid-summer.	FRAC GROUP 3 Prune out all diseased tissue. Several applications in fall before leaf drop may also be advisable. Apply 8 fl oz/100 gal rate every 28 days, or 2 to 4 fl oz/100 gal rate every 14 days.
	tebuconazole (Torque)	4 to 10 fl oz/100 gal	foliar spray	Begin applications 14 to 21 days prior to when disease is expected, or at very first sign of disease.	FRAC GROUP 3
	thiophanate-methyl + propiconazole (Protocol)	4 to 8 fl oz/100 gal 16 fl oz/100 gal	foliar spray	Apply every 14 days. Apply every 28 days.	FRAC GROUPS 1 + 3 See label for additional application instructions.
	trifloxystrobin (Compass O)	2 to 4 oz/100 gal	foliar spray	Apply every 7 to 14 days.	FRAC GROUP 11 Do not make more than two consecutive applications before switching to a non-Group 11 fungicide.
Powdery mildew	azoxystrobin (Heritage 50 WG)	1 to 4 oz/100 gal	foliar spray	Spray every 7 to 28 days as needed.	FRAC GROUP 11 Do not make more than two sequential applications of Heritage before rotating with nonstrobilurin products to avoid fungicide resistance. See label.
	azoxystrobin + benzovindiflupyr (Mural)	4 to 7 oz/100 gal	foliar spray	Apply every 7 to 21 days.	FRAC GROUPS 11 + 7 Do not make more than 2 sequential applications before rotating to a non-Group 7 or non-Group 11 fungicide.
	chlorothalonil + propiconazole (Concert II)	22 – 35 fl oz/100 gal	foliar spray	Every 14 to 21 days.	FRAC GROUPS M5 + 3 See label for additional details
	metconazole (Tourney)	1 to 4 oz/100 gal	foliar spray	Repeat in 14 to 28 days when conditions favor disease.	FRAC GROUP 3
	myclobutanil (Eagle 20 EW)	6 to 12 fl oz/100 gal	foliar spray	Apply every 10 to 14 days.	FRAC GROUP 3
	petroleum distillate: horticultural oil (Sunspray 11 E)	1 gal/100 gal	foliar spray	Spray at 14-day intervals.	
	potassium bicarbonate (MilStop)	1.25 to 5 lb/A	foliar spray	Apply at 7- to 14-day intervals.	Begin applications at first sign of disease. See label for tank mixing considerations. OMRI listed.
	propiconazole (Banner MAXX II)	5 to 8 fl oz/100 gal	foliar spray	Spray every 21 days in spring.	FRAC GROUP 3
	pyraclostrobin + boscalid (Pageant Intrinsic)	6 to 12 oz/100 gal	foliar spray	Repeat applications 7 to 14 days as needed.	FRAC GROUPS 11 + 7
	sulfur (Micro Sulf)	3 to 10 lb/100 gal	foliar spray	Apply every 5 to 10 days.	FRAC GROUP M2 Begin when disease first appears
	tebuconazole (Torque)	4 to 10 fl oz/100 gal	foliar spray	Begin applications 14 to 21 days prior to when disease is expected, or at very first sign of disease.	
	triadimefon (Bayleton FLO)	5.5 fl oz/275 to 550 gal	foliar spray	Repeat applications as needed.	FRAC GROUP 3
	trifloxystrobin (Compass O)	1 to 2 oz/100 gal	foliar spray	Repeat at 7- to 14-day intervals.	FRAC GROUP 11 Do not make more than two consecutive applications before switching to a non-Group 11 fungicide.
	thiophanate-methyl + chlorothalonil (Spectro 90 WDG)	1 to 2 lb/100 gal	foliar spray	Minimum repeat interval is 7 days.	FRAC GROUPS 1 + M5 Protective and curative activity.
	thiophanate-methyl + propiconazole (Protocol)	10-16 fl oz/100 gal	foliar spray	Apply every 21 days.	FRAC GROUPS 1 + 3
	triflumizole (Terraguard SC)	16 fl oz/100 gal 4 to 8 fl oz/100 gal	foliar spray	Make initial application prior to or at first sign of disease. Repeat at 7- to 14-day intervals.	FRAC GROUP 3 Use 16 fl oz/100 gal rate only for initial application if disease is present. For subsequent and protective applications use lower rates.

Table 10-12. Disease Control for Forest, Christmas, and Ornamental Trees

CROP Disease	Material	Rate	Method	Schedule	Remarks
Dogwood (continued)					
Powdery mildew (continued)	triticonazole (Trinity 19SC)	6 to 12 fl oz/100 gal	foliar spray	Spray every 7 to 14 days.	FRAC GROUP 3 Use preventively. Begin applications when conditions favor fungal infection and before disease symptom development. Use of an adjuvant/spreader sticker can aid in control.
Root rot (*Phytophthora* spp.)	cyazofamid (Segway SC)	3 to 6 fl oz/100 gal	soil drench	14 to 21 day intervals using another registered fungicide with a different mode of action.	FRAC GROUP 21
	fosetyl-AL (Aliette WDG)	2.5 lb/100 gal 2.5 to 5 lb/100 gal	drench foliar	Before transplanting. 30-day minimum interval.	FRAC GROUP 33
	mefenoxam (Subdue MAXX)	1 to 2 fl oz/100 gal 0.5 to 1 oz/100 gal	drench foliar	See label for application frequency.	FRAC GROUP 4 FUNGICIDE RESISTANCE IS POSSIBLE. Do not apply rates of 2 fl oz more often than every 10 weeks.
	oxathiapiprolin (Segovis)	See label	drench	See label.	FRAC GROUP 49 Do not make more than two consecutive applications before switching to a fungicide with a different mode of action.
	phosphorous acid (Fosphite, Reliant)	See label for rates	Apply as a soil drench or foliar spray as a preventive.	See label.	FRAC GROUP 33
	pyraclostrobin + boscalid (Pageant Intrinsic)	12 to 18 oz./100 gal	soil drench	Repeat applications 14 to 21 days as needed.	FRAC GROUP 11 + 7 Apply in tank mix with another effective fungicide.
Spot anthracnose (*Elsinoe*) Septoria leaf spot	azoxystrobin (Heritage)	1 to 4 oz/100 gal	foliar spray	Spray every 7 to 28 days as needed.	Do not make more than three sequential applications of Heritage before rotating with nonstrobilurin products to avoid fungicide resistance. See label.
	azoxystrobin + benzovindiflupyr (Mural)	4 to 7 oz/100 gal	foliar spray	Apply every 7 to 14 days.	Do not make more than 2 sequential applications before rotating to a non-Group 7 or non-Group 11 fungicide.
	chlorothalonil (Daconil Ultrex)	1.4 lb/100 gal	foliar spray	Apply every 7 to 14 days as needed.	
	mancozeb (Dithane 75DF Rainshield, Fore 80 WP, Protect DF)	See label	foliar spray	First spray as buds break in spring. Second as petals fall. Third in midsummer. Fourth when predormant (after flower buds are well formed).	FRAC GROUP M3 See label as rate varies by product.
	myclobutanil (Eagle 20EW)	6 to 12 fl oz/100 gal	foliar spray	Spray every 10 to 14 days.	FRAC GROUP 3
	potassium bicarbonate (MilStop)	1.25 to 5 lb/100 gal	foliar spray	Apply at 7- to 14-day intervals.	Begin applications at first sign of disease. See label for tank mixing considerations. OMRI listed.
	pyraclostrobin + boscalid (Pageant Intrinsic)	8 to 12 oz./100 gal	foliar spray	Repeat applications 7 to 14 days as needed.	FRAC GROUPS 11 + 7
	tebuconazole (Torque)	4 to 10 fl oz/100 gal	foliar spray	Begin applications 14 to 21 days prior to when disease is expected, or at very first sign of disease.	FRAC GROUP 3
Eastern Cedar					
Annosus root rot (*Fomes annosus*)	See PINE				
Phomopsis needle blight	azoxystrobin (Heritage)	1 to 4 oz/100 gal	foliar spray	Spray every 7 to 28 days as needed.	FRAC GROUP 11 Do not make more than 2 sequential applications of Heritage before rotating with nonstrobilurin products to avoid fungicide resistance. See label.
	azoxystrobin + benzovindiflupyr (Mural)	4 to 7 oz/100 gal	foliar spray	Apply every 7 to 21 days.	FRAC GROUP 11 + 7
	pyraclostrobin + fluxapyroxad (Orkestra Intrinsic)	8 to 10 fl oz/100 gal	foliar spray	Apply every 7 to 14 days as needed.	FRAC GROUPS 11 + 7 Apply preventatively when conditions are favorable for disease development
	thiophanate methyl (AllBan Flo) (Cleary 3336 F)	14.5 to 20 fl oz/100 gal 16 to 24 fl oz/100 gal	foliar spray	Beginning when disease appears or during suspected periods of disease incidence, apply every 7 to 14 days.	FRAC GROUP 1
Elm					
Dutch elm disease (*Ophiostoma ulmi*)	All treatments listed must be followed for effective **prevention** of disease on highly valued trees. 1. Sanitation—cut down and destroy diseased trees and dead limbs.				
	2. Elm bark beetle control.				

Table 10-12. Disease Control for Forest, Christmas, and Ornamental Trees

CROP Disease	Material	Rate	Method	Schedule	Remarks
Elm					
Dutch elm disease (*Ophiostoma ulmi*) (continued)	3. SMDC (Vapam) - kills root grafts	1 gal SMDC/ 3 gal water 6 oz/hole	Pour in 1-in. diameter holes 15 in. deep, spaced 6 to 9 in. apart in a line between healthy and diseased trees	Apply with first appearance of disease.	Not closer than 20 ft from healthy tree. Soil temperature above 50 degrees F. Professional applicators only.
	4. Systemic chemical prevention: propiconazole (Alamo) thiabendazole (Arbortect 20S)	See label	root flare injection	See label.	
	5. Therapeutic treatment: propiconazole (Alamo) thiabendazole (Arbortect)	See label		See label.	
Fraser Fir					
Botrytis seedling blight	chlorothalonil (Bravo Ultrex)	1.4 to 2.5 lb/A	foliar	Apply every 7 to 14 days.	Make additional applications to nursery beds when seedlings are 4 inches tall at 7- to 14-day intervals as long as favorable conditions persist.
	metconazole (Tourney)	1 to 4 oz/100 gal	foliar spray	Repeat in 14 to 28 days when conditions favor disease.	
	pyraclostrobin + boscalid (Pageant Intrinsic)	12 to 18 oz/100 gal	foliar spray	Apply every 7 to 14 days.	FRAC GROUPS 11 + 7
	pyraclostrobin + fluxapyroxad (Orkestra Intrinsic)	8 fl oz/100 gal	foliar spray	Apply every 7 to 14 days.	FRAC GROUPS 11 + 7
	thiophanate methyl (Cleary 3336 F)	12 to 16 fl oz/100 gal	foliar	Apply at first sign of disease.	Tank mix combination with chlorothalonil is recommended.
Damping-off, post-plant	mefenoxam (Subdue MAXX) + thiophanate methyl (Cleary 3336 F)	0.25 fl oz + 12 fl oz/100 gal apply 2 pt/sq ft	drench	Apply at first sign of disease.	FRAC GROUPS 4 + 1
	metconazole (Tourney)	1 to 4 oz/100 gal	foliar spray	Repeat in 14 to 28 days when conditions favor disease.	FRAC GROUP 3
Diplodia tip blight, Lophodermium needlecast, Swiss needlecast	azoxystrobin (Heritage)	3.2 to 8 oz/A	foliar spray	Spray every 7 to 28 days.	FRAC GROUP 11 To avoid fungicide resistance, do not make more than two sequential applications before rotating with nonstrobilurin products.
Phytophthora root rot	cyazofamid (Segway SC)	3 to 6 fl oz/100 gal	soil drench	14- to 21-day intervals using another registered fungicide with a different mode of action.	FRAC GROUP 21 Irrigate with at least ½ inch of water if rainfall does not occur within 24 hrs. For container plants, check label for recommended maximum drench volume based on pot diameter.
	fosetyl-AL (Aliette WDG)	0.4 to 0.8 lb/100 gal	drench	30-day minimum interval.	FRAC GROUP 33 Thoroughly wet plant and root mass immediately before transplanting. Field-grown trees in plantations.
		2.5 to 5 lb/100 gal/acre	foliar		
	mefenoxam (Subdue MAXX)	1.5 pt/acre/50 gal/ acre 2.5 pt/acre/50 gal/acre 0.63 to 1.25 gal/50 gal/acre	drench drench directed soil spray	MAXX: May and September.	FRAC GROUP 4 Do not apply to fir growing on bottomlands or poorly drained soils, or near surface water. Seed beds or plug plantings. 2-0 transplants. Field-grown trees in plantations.
	(Subdue GR)	6 to 30 lb/acre 16 to 20 lb/acre 50 to 250 lb/acre	broadcast broadcast broadcast	GR: once in spring and again in fall.	Seed beds. 2-0 transplants. Field-grown trees in plantations. Apply 0.5 to 1 inch water after application if rain is not expected within 3 days.
	oxathiapiprolin (Segovis)	See label	foliar spray	See label.	FRAC GROUP 49 Do not make more than 2 consecutive applications before switching to non-49 fungicide
	phosphorous acid (Fosphite, Reliant)	See label for rates	Soil drench or foliar spray.		FRAC GROUP 33 Apply as a soil drench or foliar spray as a preventive
Hemlock					
Twig rust (*Melampsora farlowii*)	triadimefon (Bayleton FLO)	5.5 fl oz/1.5 to 137.5 gal	foliar spray	Begin at bud break and continue every 14 days until growth stops.	FRAC GROUP 3

Table 10-12. Disease Control for Forest, Christmas, and Ornamental Trees

CROP Disease	Material	Rate	Method	Schedule	Remarks
Leyland Cypress					
Needle blight (*Passalora, Cercosporidium*)	chlorothalonil (Daconil Ultrex)	1.4 lb/100 gal	foliar spray	7- to 10-day intervals.	Begin scouting last year's infection sites for sporulation (tufts of olive green spores) in mid-May to mid-June. At the first sign of sporulation, make 2 applications of a systemic fungicide at 14-day intervals. Follow with 1 or 2 applications of a protectant fungicide (Daconil, Fore, Kocide, Badge, Rainshield) applied at 7-day intervals. In early August, scout trees for new infections. If sporulation is observed, re-treat with 1 application of a systemic fungicide 14 days later with 2 applications of a protectant applied at 7-day intervals.
	copper hydroxide (Kocide 3000)	0.75 to 1.75 lb/A	foliar spray	7- to 30-day intervals.	
	copper hydroxide + copper oxychloride (Badge SC)	3 to 6 pt/A	foliar spray	7- to 14-day intervals.	
	mancozeb (Fore 80 WP, Dithane Rainshield 75DF)	See label for rates	foliar spray	7- to 10-day intervals.	
	myclobutanil (Eagle 20EW)	6 to 12 fl oz/100 gal	foliar spray	10- to 14-day intervals.	
Diplodia tip blight, Lophodermium needlecast, Swiss needlecast	azoxystrobin (Heritage)	3.2 to 8 oz/acre	foliar spray	Spray every 7 to 28 days.	FRAC GROUP 11 To avoid fungicide resistance, do not make more than two sequential applications before rotating with nonstrobilurin products.
	azoxystrobin + benzovindiflupyr (Mural)	7 oz/100 gal	foliar spray	Spray every 7 to 21 days	FRAC GROUP 11 + 7
Longleaf Pine					
Brown spot (*Scirrhia acicola*)	Bordeaux mixture (copper sulfate, lime, and water 8-8-100)	60 gal/acre	foliar spray	Spray at 10- to 14-day interval after emergence of seedlings until July 1.	
Maple					
Anthracnose	azoxystrobin (Heritage)	1 to 4 oz/100 gal	foliar spray	Two to three applications at 7 to 28 days.	FRAC GROUP 11 First application at bud break. Do not make more than three sequential applications of Heritage before rotating with nonstrobilurin products to avoid fungicide resistance. See label.
	chlorothalonil (Echo 720)	1.375 pt/100 gal	foliar spray	Repeat every 7 to 14 days.	FRAC GROUP M5
	pyraclostrobin (Insignia SC Intrinsic)	6.1 to 12.2 fl oz/100 gal	foliar spray	Apply every 7 to 14 days.	FRAC GROUP 11 Do not make more than two sequential applications before switching to a non-Group 11 fungicides
	pyraclostrobin + boscalid (Pageant Intrinsic)	8 to 12 oz./100 gal	foliar spray	Repeat applications 7 to 14 days as needed.	FRAC GROUPS 11 + 7
	pyraclostrobin + fluxapyroxad (Orkestra Intrinsic)	8 to 10 fl oz/100 gal	foliar spray	Repeat applications 7 to 14 days as needed.	FRAC GROUPS 11 + 7
	thiophanate methyl (AllBan Flo)	10.75 to 20 fl oz/100 gal	foliar spray	Three to four applications at 10- to 14-day intervals.	FRAC GROUP 1 Make first application at bud break.
	(Cleary 3336) F	12 to 16 fl oz/100 gal		Apply on 7- to 14-day intervals.	
	triticonazole (Trinity 19SC)	8 to 12 fl oz/100 gal	foliar spray	Spray every 7 to 14 days.	FRAC GROUP 3 Use preventively. Begin applications when conditions favor fungal infection and before disease symptom development.
Oak					
Anthracnose	See MAPLE				
Leaf spot	propiconazole (Banner MAXX II) Bumper ES)	16 fl oz/100 gal 6 fl oz/100 gal	foliar spray	Apply every 14 to 28 days.	FRAC GROUP 3 Apply as needed.
	copper hydroxide + copper oxychloride (Badge SC)	1.5 to 2 pt/A	foliar spray	7- to 14-day intervals.	FRAC GROUP M1 See label for use of higher rates.
	metconazole (Tourney)	1 to 4 oz/100 gal	foliar spray	Repeat in 14 to 28 days when conditions favor disease.	FRAC GROUP 3
	pyraclostrobin (Insignia SC Intrinsic)	1.5 to 6.1 fl oz/100 gal	foliar spray	Apply every 7 to 14 days.	FRAC GROUP 11 Do not make more than two sequential applications before switching to a non-Group 11 fungicides
	pyraclostrobin + boscalid (Pageant Intrinsic)	8 to 12 oz./100 gal	foliar spray	Repeat applications 7 to 14 days as needed.	FRAC GROUPS 11 + 7D
	pyraclostrobin + fluxapyroxad (Orkestra)	8 to 10 fl oz/100 gal	foliar spray	Repeat applications 7 to 14 days as needed.	FRAC GROUPS 11 + 7
	triticonazole (Trinity 19SC)	See label	foliar spray	Spray every 7 to 14 days.	FRAC GROUP 3 Use preventively. Begin applications when conditions favor fungal infection and before disease symptom development. Use of an adjuvant/spreader sticker can aid in control.
Wilt (*Ceratocystis*)	propiconazole (Alamo)	See label	Root flare injection	See label.	FRAC GROUP 3
Pine					
Annosus root rot (*Fomes annosus*)	Borax, dry granular (sodium tetraborate decahydrate)	1 lb/50 sq ft of stump surface OR liberally cover stump surface	Sprinkle liberally on fresh-cut stump	Immediately after felling tree.	To prevent infection from freshly cut stumps.

Table 10-12. Disease Control for Forest, Christmas, and Ornamental Trees

CROP Disease	Material	Rate	Method	Schedule	Remarks
Pine (continued)					
Fusiform rust (*Cronartium fusiforme*)	triadimefon (Bayleton FLO)	5.5 fl oz/71.25 gal	foliar spray	Begin application before infection. Repeat at 2- to 3-week intervals as needed. Use higher rate in high hazard areas.	FRAC GROUP 3
	myclobutanil (Eagle 20EW)	12 to 18 fl oz/A	foliar spray	Apply at 2- to 3-week intervals.	FRAC GROUP 3 Begin applications before infection in early spring. Add spray adjuvant for enhanced disease control.
	prothioconazole (Proline 480SC)	5 fl oz/A	See label	14 to 21 days.	FRAC GROUP 3
Phytophthora root rot	cyazofamid (Segway SC)	3 to 6 fl oz/100 gal	soil drench	14 to 21 day intervals using another registered fungicide with a different mode of action.	FRAC GROUP 21 Irrigate with at least ½ inch of water if rainfall does not occur within 24 hrs. For container plants, check label for recommended maximum drench volume based on pot diameter.
	fosetyl-AL (Aliette WDG)	0.4 to 0.8 lb/100 gal 2.5 to 5 lb/acre	drench foliar	30-day minimum interval.	FRAC GROUP 33
	mefenoxam (Subdue MAXX)	See label	Can be applied as a drench or soil surface spray. Consult label for specific crops and applications.	Every 2 to 3 months.	FRAC GROUP 4 FUNGICIDE RESISTANCE IS POSSIBLE. Do not apply rates of 2 fl oz more often than every 10 weeks.
	phosphorous acid (Fosphite, Reliant)	See label for rates	Apply as a soil drench or foliar spray as a preventive.		FRAC GROUP 33
Scotch and White Pine					
Needle blight (*Lophodermium pinastri*)	mancozeb (Pentathlon DF)	1 to 2 lb/100 gal	foliar spray	Spray every 7 to 10 days August 15 to October 1.	
Sycamore					
Anthracnose	chlorothalonil (Daconil Ultrex)	1.4 lb/100 gal	foliar spray		FRAC GROUP M3 Spray at budswell and repeat at 7- to 14-day intervals during cool, moist weather.
	copper hydroxide (Kocide 3000)	0.75 to 1.25 lb/100 gal	foliar spray	7 to 10 days.	FRAC GROUP M1 Make first application at bud crack and second application 7 to 10 days later at leaf expansion
	copper hydroxide + copper oxychloride (Badge SC)	0.75 to 2.5 pt/100 gal	foliar spray	7 to 10 days.	FRAC GROUP M1 Make first application at bud crack and second application 7 to 10 days later at leaf expansion
	copper sulfate pentahydrate (Phyton 27)	35 fl oz/100 gal	foliar spray	7 to 10 days.	FRAC GROUP M1 Make first application at bud crack and second application 7 to 10 days later at leaf expansion
	metconazole (Tourney)	1 to 4 oz/100 gal	foliar spray	Repeat in 14 to 28 days when conditions favor disease.	FRAC GROUP 3
	propiconazole (Strider)	5 to 8 fl oz/100 gal	foliar spray		FRAC GROUP 3
	pyraclostrobin + boscalid (Pageant Intrinsic)	18 oz./100 gal	foliar spray	Repeat applications 7 to 14 days as needed.	FRAC GROUPS 11 + 7
	thiophanate methyl + mancozeb (Zyban) 70 WSB	4 bags/100 gal	foliar spray		FRAC GROUPS 1 + M3 Spray at budswell and repeat at 7- to 14-day intervals during cool, moist weather.
	thiophanate methyl (Cleary 3336) F	12 to 16 fl oz/100 gal	foliar spray	Apply every 7 to 14 days.	FRAC GROUP 1 Spray at budswell and repeat at 7- to 14-day intervals during cool, moist weather.
	triticonazole (Trinity) 19SC	8 to 12 fl oz/100 gal	foliar spray	Spray every 7 to 14 days.	FRAC GROUP 3 Use preventively. Begin applications when conditions favor fungal infection and before disease symptom development.
Powdery mildew	See Crabapple				

Further Information

All the following Plant Pathology *Information Notes* can be accessed from http://www.ces.ncsu.edu/depts/pp/notes/Ornamental/ornamental_contents.html.
Diseases of Leyland Cypress, Plant Pathology Ornamental Disease Information Note No. 17.
Dutch Elm Disease. Plant Pathology Ornamental Disease Information Note No. 18.
Holly Diseases and Their Control in the Landscape, Plant Pathology Ornamental Disease Information Note No. 7.
Phytophthora Root Rot and Its Control on Established Woody Ornamentals. Plant Pathology Ornamental Disease Information Note No. 13.
Powdery Mildew of Ornamentals and Shade Trees. Plant Pathology Ornamental Disease Information Note No. 4.
Scorch Disease on Shade Trees, Plant Pathology Ornamental Disease Information Note No. 10.
Some Common Pecan Diseases and Their Control in North Carolina. Plant Pathology Ornamental Disease Information Note No. 3.
Some Common Pine Diseases in North Carolina Landscapes and Their Control. Plant Pathology Information Note No. 192.
Dogwood Diseases. Plant Pathology Information Note No. 23.

Chapter X — 2019 N.C. Agricultural Chemicals Manual

Commercial Landscape and Nursery Crops Disease Control

Anyone using any agricultural chemical should refer to the current chemical label, which contains information about the safe and effective use of the chemical, before using the chemical.

Table 10-13. Commercial Landscape and Nursery Crops Disease Control

DISEASE Pesticide and Formulation	Rate of Formulation (per 100 gallons)	Schedule and Remarks
Anthracnose *(Colletotrichum, Gleosporium, Elsinoe, Marssonina, Mycosphaerella,* and others)		
azoxystrobin (Heritage)	1 to 4 oz/100 gal	Repeat every 7 to 28 days. Apply at the first sign of disease. Should not be applied to certain plant species; see label. Do not apply to apple or flowering cherry trees. May cause phytotoxicity on certain crabapple cultivars. See label.
azoxystrobin + benzovindiflupyr (Mural)	4 to 7 oz/100 gal	Repeat every 7 to 14 days. Do not apply to apple or flowering cherry trees. May cause phytotoxicity on certain crabapple cultivars. Do not make more than two consecutive applications before switching to non-Group 7 or a non-Group 11 fungicide.
Bacillus subtilis (Rhapsody, Cease)	2 to 8 qt/100 gal	Begin applications when conditions favor disease development prior to the onset of disease. Thorough coverage is essential. Repeat at 7-day intervals.
chlorothalonil (Daconil Ultrex, Daconil WeatherStik)	1.4 lb/100 gal 1.375 pt/100 gal	Reapply at 7- to 14-day intervals.
chlorothalonil + propiconazole (Concert II)	See label	Apply as a full coverage spray. Reapply at 14- to 21-day intervals. Refer to label for specific rate and application instructions.
chlorothalonil + thiophanate-methyl (Spectro) 90WDG	1 to 2 lb/100 gal	Repeat applications at 7- to 21-day intervals, according to label. See label for maximum seasonal application rules.
copper hydroxide (Nu-Cop) 50DF (CuPRO 2005)	1 lb/100 gal 0.75 to 2.0 lb/100 gal	Begin at first sign of disease, and repeat at 7- to 21-day intervals. Do not tank mix copper formulations with Aliette. Avoid contact with metal surfaces. Discoloration of blooms may occur on certain plant varieties- check label.
copper oxychloride + copper hydroxide (Badge SC) (Badge X2)	1.5-6 pt/A (high) 1.5-2 pt/A (low) 1.5 to 5 lb/A (high) 1.5 to 2 lb/A (low)	High rates should only be used during dormancy. Use lower rate (1.5 to 2 pt/A or 1.5 to 2 lb/A) when new growth is present. Repeat at 7 to 14 day intervals. Badge X2 is OMRI listed. Do not mix with Aliette.
copper hydroxide + mancozeb (Junction)	1.5 to 3.5 lb/100 gal	Phytotoxicity may occur. Reapply at 7- to 14-day intervals.
copper octanoate (Camelot O)	0.5-2.0 gal/100 gal	Begin at first sign of disease, and repeat at 7- to 10-day intervals. Camelot O may cause copper toxicity on some plant species.
copper sulfate pentahydrate (Phyton 27)	See label	See label. To avoid phytotoxicity, do not use any copper compound on alyssum.
cyprodinil + fludioxonil (Palladium)	2 to 4 oz/100 gal	Spray on a 7- to 14-day interval while conditions are conducive to disease development. After two applications, alternate with another fungicide with a different MOA for two applications
mancozeb (Dithane 75 DF Rainshield, Fore 80 WP) (Pentathlon LF) (Protect DF)	See label	Repeat application on 7- to 10-day intervals. Use of a spreader sticker will improve performance. Repeat application on 7- to 10-day intervals. Repeat application on 7- to 21-day intervals. Use of a spreader sticker will improve performance.
metconazole (Tourney)	1 to 4 oz/100 gal	Repeat in 14 to 28 days when conditions favor disease
polyoxin D zinc salt (Affirm)	0.25 to 0.5 lb/100 gal	Repeat every 7 to 10 days when conditions favor disease.
pyraclostrobin (Insignia SC Intrinsic)	6.1 to 12.2 fl oz/100 gal	Repeat every 7 to 14 days when conditions favor disease. Do not make more than two consecutive applications before switching to a non-Group 11 fungicide.
pyraclostrobin + boscalid (Pageant Intrinsic)	18 oz/100 gal	Apply prior to disease development. Repeat at 7- to 14-day intervals. Make no more than two sequential applications. Do not expose petunia or impatiens in flower or wintercreeper or nine bark to spray or drift as injury may occur.
pyraclostrobin + fluxapyroxad (Orkestra Intrinsic)	8 to 10 fl oz/100 gal	Apply prior to disease development. Repeat at 7- to 14-day intervals. Make no more than two sequential applications. See label for sensitive plant species.
tebuconazole (Torque)	4 to 10 fl oz/100 gal	Begin applications 14 to 21 days prior to when disease is expected, or at very first sign of disease.
thiophanate-methyl (AllBan Flo) (Cleary 3336 F) (SysTec 1998 FL)	10.75 to 20 fl oz/100 gal 12 to 16 fl oz/100 gal 20 fl oz/100 gal	Apply as buds break or at first sign of disease. Repeat application on 7- to 14-day intervals.
thiophanate-methyl + mancozeb (Zyban WSB)	24 oz (4 bags)/100 gal	Apply at 7-day intervals.
triticonazole (Trinity 19SC)	8 to 12 fl oz/100 gal	Spray every 7 to 14 days. Use preventively. Begin applications when conditions favor fungal infection and before disease symptom development.

Table 10-13. Commercial Landscape and Nursery Crops Disease Control

DISEASE Pesticide and Formulation	Rate of Formulation (per 100 gallons)	Schedule and Remarks
Bacterial Leaf Spot (*Pseudomonas, Xanthomonas*)		
Bacillus subtilis (Rhapsody, Cease)	2 to 8 qt/100 gal	Begin applications when conditions favor disease development prior to the onset of disease. Thorough coverage is essential. Repeat at 7-day intervals.
copper hydroxide (Nu-Cop 50DF) (CuPRO 2005)	1 lb/100 gal 0.75 to 2.0 lb/100 gal	Begin at first sign of disease, and repeat at 7- to 14-day intervals. Do not tank mix copper formulations with Aliette. Avoid contact with metal surfaces. Discoloration of blooms may occur on certain plant varieties- check label.
copper hydroxide + mancozeb (Junction)	1.5 to 3.5 lb/100 gal	Begin at first sign of disease, and repeat at 7- to 14-day intervals.
copper oxychloride + copper hydroxide (Badge SC) (Badge X2)	1.5-6 pt/A (high) 1.5-2 pt/A (low) 1.5 to 5 lb/A (high) 1.5 to 2 lb/A (low)	High rates should only be used during dormancy. Use lower rate (1.5 to 2 pt/A or 1.5 to 2 lb/A) when new growth is present. Repeat at 7 to 14 day intervals. Badge X2 is OMRI listed. Do not mix with Aliette.
copper octanoate (Camelot O)	0.5-2.0 gal/100 gal	Begin at first sign of disease, and repeat at 7- to 10-day intervals. Camelot O may cause copper toxicity on some plant species.
cuprous oxide (Nordox 75WG)	0.66 lb/100 gal	Begin at first sign of disease and repeat at 7- to 14-day intervals.
copper sulfate pentahydrate (Phyton 27, Phyton 35)	See label	
Black Root Rot (*Thielaviopsis basicola*)		
etridiazole + thiophanate-methyl (Banrot 8G) (Banrot 40WP)	Broadcast 8 to 12 lb/1,000 sq ft 6 to 12 oz/100 gal	After application, rake in or lightly cultivate soil. Apply in sufficient volume to saturate the soil mixture. Irrigate immediately. Repeat at 4- to 12-week intervals if necessary. Protects against *Thielaviopsis* and *Pythium* but is not as effective against *Thielaviopsis* as thiophanate-methyl-only products that have a higher concentration of active ingredient.
fludioxonil (Medallion)	1 to 2 oz/100 gal	Apply as a drench at transplanting as a preventive. If needed, re-treat transplants 21 to 28 days after initial application. Do not apply as a seed or soil drench to impatiens or New Guinea impatiens.
thiophanate-methyl (AllBan Flo) (Cleary 3336 F) (OHP 6672 50WP), (T-Storm 50WSB) (SysTec 1998 FL) (T-Storm Flowable), (OHP 6672 4.5L)	See label	Apply as a heavy spray or drench at the rate of 1/2 to 2 pints per sq ft. Repeat at 4- to 8-week intervals. Apply 8 oz as a drench or directed spray after seeding, or apply 12 to 16 oz after transplanting. Repeat at 21- to 28-day intervals. Apply 1 to 3 pt/sq ft after transplanting to thoroughly soak growing medium. Repeat at 21-to 28-day intervals. Apply as heavy spray or drench at a rate of 1 to 2 pints per sq ft. Repeat at 2- to 4-week intervals. Apply as heavy spray or drench at a rate of 1 to 2 pints per sq ft. Repeat at 2- to 4-week intervals.
triflumizole (Terraguard SC)	2 to 8 fl oz/100 gal	Apply as soil drench at 2- to 4-week intervals. Use higher rate under heavy disease pressure. For use in greenhouses, shadehouses, and nurseries.
Black Rot of Bulb Crops (*Sclerotinia sclerotiorum*)		
PCNB (Terraclor 400) (Revere 10G)	See label	Spread evenly on soil and mix into upper 6 to 7 in. of soil.
thiophanate-methyl (AllBan Flo) (Cleary 3336 F) (OHP 6672 50WP), (T-Storm 50WSB) (SysTec 1998 FL), (OHP 6672 4.5L)	10.75 to 20 oz/100 gal 8 to 16 fl oz/100 gal 12 to 16 oz/100 gal 20 fl oz/100 gal	Apply late Spring or at first sign of disease. Repeat every 7 to 14 days as needed during disease period. Apply 8 oz as a drench or directed spray after seeding, or apply 12 to 16 oz after transplanting. Repeat at 21- to 28-day intervals. Apply late Spring or at first sign of disease. Repeat every 7 to 14 days as needed during disease period. Apply as heavy spray or drench at a rate of 1 to 2 pints per sq ft. Repeat at 2- to 4-week intervals.
Black Spot-Rose (*Diplocarpon rosae*)		
Bacillus subtilis (Rhapsody, Cease)	2 to 8 qt/100 gal	Begin applications when conditions favor disease development, prior to the onset of disease. Thorough coverage is essential. Repeat at 7-day intervals.
azoxystrobin (Heritage)	4 to 8 oz/100 gal	Apply on 7-day interval if disease conditions are favorable. If disease is severe, mix with another registered fungicide for black spot control. Do not make more than two consecutive applications before switching to a non-Group 11 fungicide.
captan (Captec 4L)	1 qt/100 gal	Apply at first sign of disease. Repeat 7- to 10-day intervals.
chlorothalonil (Daconil Ultrex) (Echo 90DF)	1 lb/100 gal 0.875 lb/100 gal	Apply at bud break. Repeat applications at 7- to 14-day intervals. Knock Out and Double Delight roses are sensitive to chlorothalonil.
mancozeb (Protect DF)	1 to 2 lb/100 gal	Apply every 7 to 21 days.
myclobutanil (Eagle 20EW)	6 to 12 fl oz	Apply on 7- to 10-day protectant schedule. Some greenhouse varieties are sensitive to myclobutanil.

Table 10-13. Commercial Landscape and Nursery Crops Disease Control

DISEASE Pesticide and Formulation	Rate of Formulation (per 100 gallons)	Schedule and Remarks
Black Spot-Rose *(Diplocarpon rosae)* (continued)		
paraffinic oil (Organic JMS Stylet Oil)	1 to 2 oz/gal	See label regarding application instructions and phytotoxicity warnings (with captan). OMRI listed.
propiconazole (Banner MAXX II)	5 to 8 fl oz/100 gal	Apply with contact fungicide labeled for black spot.
propiconazole + chlorothalonil (Concert II)	22 to 35 fl oz/100 gal	Apply with contact fungicide labeled for black spot.
pyraclostrobin + fluxapyroxad (Orkestra Intrinsic)	8 fl oz/100 gal	Apply prior to disease development. Repeat at 7- to 14-day intervals. Make no more than two sequential applications. See label for sensitive plant species.
sulfur (Kumulus DF)	3 to 10 lb/A	Apply when disease first appears and continue at 5 to 10 day intervals.
tebuconazole (Torque)	4 to 10 fl oz/100 gal	Begin applications 14 to 21 days prior to when disease is expected, or at very first sign of disease.
thiophanate-methyl (AllBan Flo)	10.75 to 20 oz/100 gal	Apply late Spring or at first sign of disease. Repeat every 7 to 14 days as needed during disease period.
(Cleary 3336 F)	12 to 16 fl oz/100 gal	
(OHP 6672 50WP), (T-Storm 50WSB)	12 to 16 fl oz/100 gal	Apply late Spring or at first sign of disease. Repeat every 7 to 14 days as needed during disease period.
(SysTec 1998 FL), (T-Storm Flowable), (OHP 6672 4.5L)	20 fl oz/100 gal	Apply as heavy spray or drench at a rate of 1 to 2 pints per sq ft. Repeat at 2- to 4-week intervals.
Thiophanate-methyl + chlorothalonil (Spectro 90WDG)	1 to 1.5 lb/100 gal	7- to 14-day reapplication interval.
Thiophanate-methyl + propiconazole (Protocol)	10 to 16 fl oz/100 gal	Apply in tank mixture with contact fungicide registered for black spot.
thiophanate-methyl + iprodione (26/36 Fungicide)	33 to 84 fl oz/100 gal	Do not make more than 4 applications per year.
thiophanate-methyl + mancozeb (Zyban) WSB	24 oz (4 bags)/100 gal	Apply at 7-day intervals.
trifloxystrobin (Compass O)	2 to 4 oz/100 gal	Apply at 7- to 14-day intervals. Do not make more than two consecutive applications before switching to a non-Group 11 fungicide.
Botrytis Blight (Gray Mold)		
Bacillus subtilis (Rhapsody, Cease)	2 to 8 qt/100 gal	Begin applications when conditions favor disease development prior to the onset of disease. Thorough coverage is essential. Repeat at 7-day intervals.
azoxystrobin (Heritage)	4 to 8 oz/100 gal	Apply for suppression only. Repeat at 7- to 21-day intervals. Do not make more than two consecutive applications before switching to a non-Group 11 fungicide
azoxystrobin + benzovindiflupyr (Mural)	4 to 7 oz/100 gal	Apply every 7 to 14 days. Do not apply to apple or flowering cherry trees. May cause phytotoxicity on certain crabapple cultivars. Do not make more than two consecutive applications before switching to non-Group 7 or a non-Group 11 fungicide.
copper oxychloride + copper hydroxide (Badge SC) (Badge X2)	1.5-6 pt/A (high) 1.5-2 pt/A (low) 1.5 to 5 lb/A (high) 1.5 to 2 lb/A (low)	High rates should only be used during dormancy. Use lower rate (1.5 to 2 pt/A or 1.5 to 2 lb/A) when new growth is present. Repeat at 7- to 14-day intervals. May cause discoloration in some azalea varieties. Badge X2 is OMRI listed. Do not mix with Aliette.
chlorothalonil (Daconil Ultrex, Daconil)	1.4 lb/100 gal	Repeat applications at 7- to 14-day intervals.
cyprodinil + fludioxonil (Palladium)	4 to 6 oz/100 gal	Spray on a 7- to 14-day interval while conditions are conducive to disease development. After two applications, alternate with another fungicide with a different MOA for two applications. See cautionary statement on label for applications to Geraniums, Impatiens and New Guinea Impatiens.
fluoxastrobin (Disarm 480SC)	4 to 8 fl oz/100 gal	Apply prior to infection. Repeat at 7- to 21-day intervals. Do not make more than two consecutive applications before switching to a non-Group 11 fungicide.
fludioxonil (Medallion)	2 to 4 oz/100 gal	Spray to runoff at 7- to 14-day intervals. Do not make more than two consecutive applications of Medallion before rotating to another effective product with a different mode of action.
iprodione		Apply at 7- to 14-day intervals. Limit total applications to a maximum of 4 per year.
(Chipco 26GT)	1 to 2.5 qt/100 gal	
(18 Plus, Iprodione Pro 2SE)	1 to 2.5 qt/100 gal	
iprodione + thiophanate methyl (26/36 Fungicide)	33 to 84 fl oz/100 gal	Do not make more than 4 applications per year.
mancozeb (Dithane, Fore, Mancozeb)	1.5 lb/100 gal	Addition of a nonionic surfactant will improve performance. Re-treat at 7- to 10-day intervals.
metconazole (Tourney)	1 to 4 oz/100 gal	Repeat in 14 to 28 days when conditions favor disease
pyraclostrobin + boscalid (Pageant Intrinsic)	8 to 12 oz/100 gal	Apply prior to disease development. Repeat at 7- to 14-day intervals. Make no more than two sequential applications. Do not expose petunia or impatiens in flower or wintercreeper or nine bark to spray or drift as injury may occur.
Cylindrocladium Stem Canker or Root Rot		
chlorothalonil (Chlorothalonil DF) (Chlorothalonil 500ZN) (Chlorothalonil 720 SFT) (Chlorostar VI, Daconil Weather Stik, Echo 720, Manicure 6F) (Daconil Ultrex) 82.5WDG (Echo) 90DF, (Echo Ultimate) (Exotherm Termil)	See label 1.9 pt/100 gal 1.37 pt/100 gal 1 3/8 pt/100 gal 1.4 lb/100 gal 1.25 lb/100 gal 1 can/1,000 sq ft	Repeat at 7- to 14-day intervals. Apply to foliage when plants are dry or nearly dry. See label for method of application.
chlorothalonil + thiophanate-methyl (Spectro) 90WDG	1.0 to 2.15 lb per 100 gallons	For best results use spray mixture the same day it is prepared. Spray uniformly over the area to be treated with a properly calibrated power sprayer, apply as a full coverage spray to run-off when conditions are favorable for disease development.

Table 10-13. Commercial Landscape and Nursery Crops Disease Control

DISEASE Pesticide and Formulation	Rate of Formulation (per 100 gallons)	Schedule and Remarks
Cylindrocladium Stem Canker or Root Rot (continued)		
azoxystrobin + benzovindiflupyr (Mural)	5 to 7 oz/100 gal	Apply every 7 to 14 days. Do not apply to apple or flowering cherry trees. May cause phytotoxicity on certain crabapple cultivars. Do not make more than two consecutive applications before switching to non-Group 7 or a non-Group 11 fungicide.
cyprodinil + fludioxonil (Palladium)	2 to 4 oz/100 gal	For stem diseases ensure full spray coverage of all stems and inner areas of the plant to the soil/media. Apply on a 7- to 14-day interval while conditions are conducive to disease development. After two applications, alternate with another fungicide with a different MOA for two applications. No drench applications on label.
fludioxonil (Medallion) 50WSP	1 to 2 oz/100 gal	Completely drench the growing medium. Repeat at 21- to 28-day intervals. Two applications per year when conditions favor disease development are usually adequate for control.
pyraclostrobin + boscalid (Pageant Intrinsic)	8 to 12 oz/100 gal	Apply prior to disease development. Completely drench the growing medium. Repeat at 7- to 14-day intervals. Make no more than two sequential applications. Do not expose petunia or impatiens in flower or wintercreeper or nine bark to spray, as injury may occur.
thiophanate-methyl (AllBan Flo) (Cleary 3336) F (OHP-6672) 50WP, (T-Storm) 50WSB (SysTec 1998) FL (T-Storm Flowable), (OHP 6672) 4.5L	7.5 to 20 fl oz/100 gal 8 to 16 fl oz/100 gal 12 to 16 oz/100 gal 10 to 20 fl oz/100 gal 20 fl oz/100 gal	Apply as drench or heavy spray at 0.5 to 2 pints per sq ft. Repeat at 4- to 8-week intervals. Apply 8 oz after seeding or sticking, or 12 to 16 oz after transplanting as a drench or directed spray at a rate that thoroughly soaks the growing media through the root zone. Repeat every 21 to 28 days. Apply as a drench or heavy spray at a rate of 1 to 3 pints per sq ft. Repeat at 21- to 28-day intervals. Apply as a drench or heavy spray at a rate of 1 to 2 pints per sq ft. Repeat at 2- to 4-week intervals. Apply as a drench or heavy spray at a rate of 1 to 2 pints per sq ft. Repeat at 2- to 4-week intervals.
triflumizole (Terraguard) 50W	See label	Can be used as a cutting soak or soil drench.
Daylily Leaf Streak (*Aureobasidium microstictum*)		
chlorothalonil (Daconil Ultrex) 82.5WDG	1.4 lb/100 gal	Apply early in the spring as new growth emerges and before disease symptoms appear. Make three to four applications at 14-day intervals.
mancozeb (Dithane) 75 DF, (Fore) 80 WSP (Protect DF)	1.5 lb/100 gal 1 to 2 lb/100 gal	Apply early in the spring as new growth emerges and before disease symptoms appear. Make three to four applications at 14-day intervals.
thiophanate-methyl (Cleary 3336) F	12 to 16 fl oz/100 gal	Apply early in the spring as new growth emerges and before disease symptoms appear. Make three to four applications at 14-day intervals.
Daylily Rust (*Puccinia hemerocallidis*)		
Alternately apply a systemic fungicide from Category 1 with a protective fungicide from Category 2 to protect new foliage as it emerges. Re-treat at 7- to 14-day intervals.		
Category 1 Systemics		
azoxystrobin (Heritage)	1 to 4 oz/100 gal	
flutolanil (Contrast)	3 oz/100 gal	
triadimefon (Bayleton) (Strike) 25 WDG	1 PVA packet/550 to 1,100 gal 4 oz/100 gal	Bayleton cannot be used on plants being grown for sale or other commercial use. Strike is for commercial greenhouse and nursery use only.
trifloxystrobin (Compass)	2 to 4 oz/100 gal	
Category 2 Protectants		
chlorothalonil (Daconil Ultrex)	1.4 lb/100 gal	
mancozeb (Fore)	1.5 lb/100 gal	
Downy Mildew (*Bremia, Pseudoperonospora, Peronospora, Plasmopara* spp.)		
azoxystrobin (Heritage)	1 to 2 oz/100 gal (bedding plants) 2 to 4 oz/100 gal (rose)	Apply every 7 to 14 days prior to infection. Do not apply 2-oz rate on less than 14-day intervals. May damage snapdragons; use 1 oz rate and rotate. Apply every 7 to 21 days on rose during periods of active plant growth and prior to dormancy. Do not apply to apple, flowering cherry, or crabapple. See label.
azoxystrobin + benzovindiflupyr (Mural)	4 to 7 oz/100 gal	Apply every 7 to 14 days. Do not make more than two consecutive applications before switching to a non-Group 7 or non-Group 11 fungicide. Do not apply to apple, flowering cherry, or crabapple. See label.
Bacillus subtilis (Rhapsody, Cease)	2 to 8 qt/100 gal	Repeat at 3- to 10-day intervals. Thorough coverage is essential. Begin applications when conditions favor disease development, prior to the onset of disease.
Bacillus amyloliquefaciens strain D747 (DoubleNickel LC)	0.5 to 6 qt/100 gal	Begin preventative applications at plant emergence and repeat every 3 to 28 days depending on disease pressure. OMRI listed.
copper hydroxide (Champ WG) (Nu-Cop) 50DF (CuPRO 2005)	0.5 lb/100 gal 1 lb/100 gal 0.75 to 2.0 lb/100 gal	Begin at first sign of disease, and repeat at 7- to 14-day intervals. Do not tank mix copper formulations with Aliette. Avoid contact with metal surfaces. Discoloration of blooms may occur on certain plant varieties- check label. To avoid phytotoxicity, do not use any copper compound on alyssum.
copper oxychloride + copper hydroxide (Badge SC) (Badge X2)	1.5-2 pt/A 1.5 to 2 lb/A	Repeat at 7 to 14 day intervals. May cause discoloration in some azalea varieties. See label for other phytotoxicity warnings. Badge X2 is OMRI listed. Do not mix with Aliette.
copper octanoate (Camelot O)	0.5-2.0 gal/100 gal	Begin at first sign of disease, and repeat at 7- to 10-day intervals. Camelot O may cause copper toxicity on some plant species.
copper sulfate pentahydrate (Phyton 27)	See label	See label. To avoid phytotoxicity, do not use any copper compound on alyssum.

Table 10-13. Commercial Landscape and Nursery Crops Disease Control

DISEASE Pesticide and Formulation	Rate of Formulation (per 100 gallons)	Schedule and Remarks
Downy Mildew (Bremia, Pseudoperonospora, Peronospora, Plasmopara spp.) (continued)		
cyazofamid (Segway)	2.1 to 3.5 fl oz/100 gal	Apply on 14- to 21-day intervals using another registered fungicide with a different mode of action. Apply sufficient volume to wet all foliage until runoff (normally 50 to 100 gallons per acre).
dimethomorph (Stature DM)	6.4 to 12.8 oz/100 gal	Apply at first sign of disease. Apply to obtain complete coverage of flowers, foliage, and stems. Repeat at 10- to 14-day intervals throughout the production cycle. For use on greenhouse and nursery-grown ornamentals.
dimethomorph + ametoctradin (Orvego)	11 to 14 fl oz/100 gal	Apply on 10- to 14-day intervals using another registered fungicide with a different mode of action. Apply sufficient volume to wet all foliage until runoff (normally 50 to 100 gallons per acre).
fosetyl-Al (Aliette) 80WDG	1.25 to 4 lb/100 gal (bedding plants) 2.5 lb/100 gal (roses)	Systemic. Apply prior to disease development. Repeat as necessary, but do not make more than one application every 14 days.
fluoxastrobin (Disarm 480SC)	1 to 4 fl oz/100 gal	Apply every 7 to 21 days. Do not make more than two consecutive applications before switching to a non-Group 11 fungicide.
pyraclostrobin (Insignia)	4 to 8 oz/100 gal	Apply every 7 to 14 days prior to disease development. Do not make more than two consecutive applications before switching to a non-Group 11 fungicide.
mancozeb (Dithane) 75 DF (Protect DF)	1.5 lb/100 gal 1 to 2 lb/100 gal	Repeat at 7- to 10-day intervals. Reapply in 7 to 21 days.
mandipropamid (Micora)	4 to 8 fl oz/100 gal	This product can also be used on vegetables sold to the retail market in GH with permanent flooring. Apply prior to disease development. Repeat at 7- to 14-day intervals. Make no more than two sequential applications.
mefenoxam (Subdue MAXX) (Subdue Gr)	0.5 to 1 fl oz/100 gal See label for rates	Apply Subdue MAXX as a foliar spray or soil drench treatment. Apply Subdue GR as a soil surface or soil/planting media incorporation treatment.
oxathiapiprolin (Segovis)	0.6 to 2.4 fl oz/100 gal	Make no more than two consecutive applications before switching to a non-Group 49 fungicide.
pyraclostrobin + boscalid (Pageant)	12 to 18 oz/100 gal	Use preventatively, prior to disease development on a 7- to 10-day schedule. Do not make more than two consecutive applications before switching to a non-Group 7 or non-Group 11 fungicide.
phosphorous acid (Alude, Reliant)	See label	Apply prior to disease development. Spray to thoroughly wet all foliage. Repeat at 14- to 21-day intervals.
potassium phosphite (Vital)	4 pt/100 gal	Apply as a foliar spray prior to disease onset, and repeat at 14-day intervals.
pyraclostrobin + boscalid (Pageant Intrinsic)	8 to 12 oz/100 gal	Apply prior to disease development. Repeat at 7- to 10-day intervals. Make no more than two sequential applications. Do not expose petunia or impatiens in flower or wintercreeper or nine bark to spray or drift as injury may occur.
trifloxystrobin (Compass O) 50WDG	2 to 4 oz/100 gal	Apply as a foliar spray before disease is detected or when conditions are favorable for disease. Repeat at 7- to 14-day intervals until threat of disease is over.
thiophanate-methyl + mancozeb (Zyban) WSB	24 oz (4 bags)/100 gal	Apply at first sign of disease, and repeat at 7-day intervals.
Entomosporium Leaf Blight		
azoxystrobin (Heritage) 50WDG	1 to 4 oz/100 gal	Spray at budbreak, and repeat at 7- to 28-day intervals as needed. To avoid fungicide resistance, make no more than three sequential applications of Heritage before rotating with nonstrobilurin products.
copper oxychloride + copper hydroxide (Badge SC) (Badge X2)	1.5-2 pt/A 1.5 to 2 lb/A	Repeat at 7 to 14 day intervals. May cause discoloration in some azalea varieties. See label for other phytotoxicity warnings. Badge X2 is OMRI listed. Do not mix with Aliette.
chlorothalonil (Daconil Ultrex) 82.5 WDG	1.4 lb/100 gal	Begin applications at budbreak, and continue every 7 to 14 days.
chlorothalonil + propiconazole (Concert II)	22 to 35 fl oz/100 gal	Apply as a full coverage spray. Reapply at 14- to 21-day intervals. Refer to label for specific rate and application instructions.
chlorothalonil + thiophanate-methyl (Spectro 90)	1 to 2 lb/100 gal	Spray at a minimum of 7-day intervals. Apply when foliage is dry.
Iprodione + thiophanate methyl (26/36 Fungicide)	33 to 84 fl oz/100 gal	Do not make more than 4 applications per year.
myclobutanil (Eagle) 20EW (Eagle) 40WP	 6 to 12 fl oz/100 gal 3 to 6 oz/100 gal	Spray every 10 to 14 days.
propiconazole (Banner MAXX II)	5 to 8 fl oz/100 gal	
thiophanate-methyl (3336 F)	12 to 16 oz/100 gal	Apply when disease first appears and repeat every 7 to 14 days.
triadimefon (Bayleton) 50WSP (Strike) 25WDG	1 PVA packet/137.5 to 275 gal 8 to 16 oz/100 gal	In early spring as growth starts, spray every 14 to 21 days until growth is fully expanded. May be phototoxic with repeated applications. Bayleton is not for plants offered for sale or other commercial use.
triticonazole (Trinity) 19SC	12 fl oz/100 gal	Spray every 7 to 14 days. Use preventively. Begin applications when conditions favor fungal infection and before disease symptom development.
Fire blight (Erwinia amylovora) (or see section under fire blight control of crabapple)		
copper hydroxide (Champ WG) (Nu-Cop) 50DF (CuPRO 2005	 0.5 lb/100 gal 1 lb/100 gal 0.75 to 2.0 lb/100 gal	Begin at first sign of disease, and repeat at 7- to 14-day intervals. Do not tank mix copper formulations with Aliette. Avoid contact with metal surfaces. Discoloration of blooms may occur on certain plant varieties- check label.

Table 10-13. Commercial Landscape and Nursery Crops Disease Control

DISEASE Pesticide and Formulation	Rate of Formulation (per 100 gallons)	Schedule and Remarks
Fire blight *(Erwinia amylovora)* **(or see section under fire blight control of crabapple) (continued)**		
copper hydroxide + mancozeb (Junction)	1.5 to 3.5 lb/100 gal	See label for timing of application.
copper sulfate pentahydrate (Phyton 27)	See label	See label. To avoid phytotoxicity, do not use any copper compound on alyssum.
fosetyl-Al (Aliette)	2.5 lb/100 gal	See label for timing of application.
streptomycin sulfate (Agri-mycin 17)	0.5 lb/100 gal	See label.
Flower Blight. *See Petal Blight.*		
Fungal Leaf Spots *(Alternaria, Cercospora, Cylindrosporium, Phyllosticta, Septoria)* **Consult product labels for specific fungi controlled**		
azoxystrobin (Heritage)	1 to 4 oz/100 gal	Repeat every 7 to 28 days. Apply at the first sign of disease as new growth buds out. Do not apply to apple, flowering cherry, or crabapple. See label.
azoxystrobin + benzovindiflupyr (Mural)	4 to 7 oz/100 gal	Apply every 7 to 21 days. Do not make more than two consecutive applications before switching to a non-Group 7 or non-Group 11 fungicide. Do not apply to apple, flowering cherry, or crabapple. See label.
Bacillus subtilis (Rhapsody, Cease)	2 to 8 qt/100 gal	Repeat at 3- to 10-day intervals. Thorough coverage is essential. Begin applications when conditions favor disease development, prior to the onset of disease.
Bacillus amyloliquefaciens strain D747 (DoubleNickel LC)	0.5 to 6 qt/100 gal	Begin preventative applications at plant emergence and repeat every 3 to 28 days depending on disease pressure. OMRI listed.
chlorothalonil (Daconil Ultrex) (Daconil WeatherStik)	 1.4 lb/100 gal 1.375 pt/100 gal	Reapply at 7- to 14-day intervals.
chlorothalonil + propiconazole (Concert II)	9 to 35 fl oz/100 gal	Apply as a full coverage spray. Reapply at 14- to 21-day intervals. Refer to label for specific rate and application instructions.
chlorothalonil + thiophanate-methyl (Spectro) 90WDG	1 to 2 lb/100 gal	Repeat applications at 7- to 21-day intervals depending on plant treated; see label.
copper hydroxide (Champ WG) (Nu-Cop) 50DF (CuPRO 2005)	 0.5 lb/100 gal 1 lb/100 gal 0.75 to 2.0 lb/100 gal	Begin at first sign of disease, and repeat at 7- to 14-day intervals. Do not tank mix copper formulations with Aliette. Avoid contact with metal surfaces. Discoloration of blooms may occur on certain plant varieties- check label.
copper oxychloride + copper hydroxide (Badge SC) (Badge X2)	 1.5 to 2 pt/A 1.5 to 2 lb/A	Repeat at 7- to 14-day intervals. May cause discoloration in some azalea varieties. See label for other phytotoxicity warnings. Badge X2 is OMRI listed. Do not mix with Aliette.
copper hydroxide + mancozeb (Junction)	1.5 to 3.5 lb/100 gal	Reapply at 7- to 14-day intervals.
copper octanoate (Camelot O)	0.5-2.0 gal/100 gal	Begin at first sign of disease, and repeat at 7- to 10-day intervals. Camelot O may cause copper toxicity on some plant species.
copper sulfate pentahydrate (Phyton 27)	See label	
cyprodinil + fludioxonil (Palladium)	2 to 6 oz/100 gal	Spray on a 7- to 14-day interval while conditions are conducive to disease development. After two applications, alternate with another fungicide with a different MOA for two applications. Cautionary statement on label for applications to Geraniums, Impatiens and New Guinea Impatiens.
mancozeb (Dithane) 75 DF, (Fore) 80 WSP, Mancozeb DG) (Pentathlon LF) (Protect DF)	 1.5 lb/100 gal 1.2 qt/100 gal 1 to 2 lb/100 gal	Repeat application at 7- to 10-day intervals. Addition of a spreader sticker will improve performance.
metconazole (Tourney)	1 to 4 oz/100 gal	Repeat in 14 to 28 days when conditions favor disease
myclobutanil (Eagle) 40WP	3 to 6 oz/100 gal	Apply as a protectant every 10 to 14 days.
potassium bicarbonate (MilStop)	1.25 to 5 lb/100 gal	Uniform and complete coverage of foliage is essential for best results. See label for special instructions regarding poinsettia, pansy, and impatiens.
pyraclostrobin (Insignia)	See label	Apply prior to onset of disease and repeat every 7 to 14 days. Do not expose flowering impatiens or flowering petunias to Insignia.
pyraclostrobin + boscalid (Pageant Intrinsic)	8 to 12 oz/100 gal	Apply prior to disease development. Repeat at 7- to 14-day intervals. Make no more than two sequential applications. Do not expose petunia, or impatiens in flower or wintercreeper, or nine bark to spray or drift as injury may occur.
tebuconazole (Torque)	4 to 8 fl oz/100 gal	Begin applications 14 to 21 days prior to when disease is expected, or at very first sign of disease.
thiophanate-methyl (AllBan Flo) (Cleary's 3336 F) (SysTec 1998 FL) (Zyban WSB)	 10 to 14.5 fl oz/100 gal 12 to 16 fl oz/100 gal 20 fl oz/100 gal 24 oz (4 bags)/100 gal	Apply at first sign of disease. Repeat application at 7- to 14-day intervals.
triadimefon (Bayleton, Strike)	See label	
triticonazole (Trinity) 19SC	4 to 8 fl oz/100 gal	Spray every 7 to 14 days. Use preventively. Begin applications when conditions favor fungal infection and before disease symptom development. Use of an adjuvant/spreader sticker can aid in control.

Table 10-13. Commercial Landscape and Nursery Crops Disease Control

DISEASE Pesticide and Formulation	Rate of Formulation (per 100 gallons)	Schedule and Remarks
Fusarium Root and Crown Rot		
azoxystrobin (Heritage)	Directed spray: 1 to 4 oz/100 gal	Repeat every 7 to 21 days.
	Drench: 0.2 to 0.9 oz/100 gal	Apply 1 to 2 pt of solution per sq ft every 7 to 28 days.
cyprodinil + fludioxonil (Palladium)	2 to 4 oz/100 gal	For stem diseases ensure full spray coverage of all stems and inner areas of the plant to the soil/media. Apply on a 7- to 14-day interval while conditions are conducive to disease development. After two applications, alternate with another fungicide with a different MOA for two applications.
fludioxonil (Medallion)	1 to 2 oz/100 gal	Wet entire medium. Reapply at 21- to 28-day intervals. May cause phytotoxicity when applied to impatiens, New Guinea impatiens, and geraniums.
thiophanate-methyl (AllBan Flo) (Cleary's 3336 G)	7.5 to 20 fl oz/100 gal 22 to 30 lb/1,000 sq ft	Apply as drench or heavy spray after transplanting. Repeat at 2- to 4-week intervals. For preventative control, incorporate into media prior to planting or as a broadcast, or make an over the top application after seeding or transplanting. For curative control, apply when disease first appears. Repeat every 21 to 28 days.
(Cleary 3336 F)	Drench: 8 to 16 fl oz/100 gal	Apply after seeding or transplanting at a rate to thoroughly soak growing medium. Repeat every 21 to 28 days.
(SysTec 1998 FL)	10 to 20 fl oz/100 gal	Apply as drench or heavy spray after transplanting. Repeat at 2- to 4-week intervals.
pyraclostrobin (Empress Intrinsic)	1 to 6 fl oz/100 gal (see label)	Apply at 1 to 3 fl oz for plants in propagation, rooted cuttings, plugs, and seedlings and at 2 to 6 fl oz to all other plants. Do not apply to dry soil media. Apply preventative to disease with sequential at 7 to 28 days after the first application if needed.
pyraclostrobin + boscalid (Pageant Intrinsic)	12 to18 oz/100 gal	Apply prior to disease development. Repeat at 7- to 14-day intervals. Make no more than two sequential applications. Do not expose petunia, or impatiens in flower or wintercreeper, or nine bark to spray or drift as injury may occur.
Streptomyces griseoviridis (Mycostop)	See label	Apply inoculant as a seed dressing, soil drench spray, or transplant dip. Must be applied prior to onset of disease. See label.
triflumizole (Terraguard) 50W	4 to 8 oz/100 gal	Apply soil drenches at weekly intervals. Use higher rate under heavy disease pressure.
triticonazole (Trinity) 19SC	8 to 12 fl oz/100 gal	Apply every 7 to 14 days. Use preventively. Begin applications when conditions favor fungal infection and before disease symptom development. The stem areas of the plant must be thoroughly covered using spray to runoff.
Gray Mold. *See* Botrytis Blight.		
Iris Leaf Spot *(Didymellina macrosopora/Mycosphaerella macrospora)*		
azoxystrobin (Heritage)	2 to 4 oz/100 gal	Apply every 7 to 21 days.
azoxystrobin + benzovindiflupyr (Mural)	4 to 7 oz/100 gal	Apply every 7 to 21 days. Do not make more than two consecutive applications before switching to a non-Group 7 or non-Group 11 fungicide.
chlorothalonil (Daconil Ultrex) (Daconil WeatherStik)	 1.4 lb/100 gal 1.375 pt/100 gal	Apply to new growth at 7- to 14-day intervals in spring.
mancozeb (Dithane) 75 DF, (Fore) 80 WSP, Mancozeb DG) (Pentathlon LF)	1.5 lb/100 gal	Addition of a spreader sticker will improve performance.
(Protect DF)	0.8 to 1.6 qt/100 gal	Repeat applications on 7- to 10-day intervals.
	1 to 2 lb/100 gal	Apply at 7- to 21-day intervals.
myclobutanil (Eagle) 20EW (Eagle) 40WP	 12 fl oz/100 gal 6 oz/100 gal	
Pyraclostrobin (Insignia)	2 to 8 oz/100 gal	Apply preventatively on a 7- to 14-day application interval. Do not make more than two consecutive applications before switching to a non-Group 7 or non-Group 11 fungicide.
Thiophanate-methyl (Incognito 85WDG)	7.2 to 9.6 oz/100 gal	Apply every 7 to 14 days prior to or immediately at the onset of disease.
Kabatina Twig Blight		
thiophanate-methyl (Cleary 3336) F	16 to 24 fl oz/100 gal	Apply at 7- to 14-day intervals. Disease not easily controlled. Fall applications may reduce disease the following year.
Leaf and Flower Gall *(Exobasidium* ssp.)		
triadimefon (Bayleton) 50WSP (Strike) 25W	1 PVA packet (11 oz/550 gal) 4 oz/100 gal	Begin applications at bud break, and apply at 10-day intervals. Bayleton cannot be used on plants grown for sale or other commercial use. Strike is for use in commercial nurseries, garden centers, and greenhouses only.
ferbam (Granuflo)	1 to 1.5 lb/100 gal	Apply to plants, flowers, and litter around plants at 3- to 4-day intervals during bloom.
Petal or Flower Blight of Azalea, Rhododendron, or Camellia (*Ovulinia* ssp., *Ciborinia camelliae, Sclerotinia camelliae*)		
mancozeb (Dithane) 75 DF, (Fore) 80 WSP, (Mancozeb DG) (Protect DF)	1.5 lb/100 gal 1 to 2 lb/100 gal	Beginning when flowers start to show color, spray two or three times each week during bloom. Direct spray into flowers and thoroughly spray ground under bushes.
myclobutanil (Eagle) 40WP	 3 to 6 oz/100 gal	Beginning when flowers start to show color, spray every 10 to 14 days.
propiconazole (Banner MAXX, Banner MAXX II)	5 to 8 fl oz/100 gal	Spray every 21 days during bloom.

Table 10-13. Commercial Landscape and Nursery Crops Disease Control

DISEASE Pesticide and Formulation	Rate of Formulation (per 100 gallons)	Schedule and Remarks
Petal or Flower Blight of Azalea, Rhododendron, or Camellia (*Ovulinia* ssp., *Ciborinia camelliae, Sclerotinia camelliae*) (continued)		
tebuconazole (Torque)	4 to 8 fl oz/100 gal	Apply 2 to 3 times per week into the flowers as they open and develop.
thiophanate-methyl (AllBan Flo) (Cleary 3336) F (OHP 6672) 50WP, (T-Storm) 50WSB (SysTec 1998) FL, (T-Storm Flowable), OHP 6672 (4.5L)	10.75 to 20 oz/100 gal 8 to 16 fl oz/100 gal 8 to 16 oz/100 gal 20 fl oz/100 gal	Apply as flowers open. Repeat every 7 to 14 days.
triadimefon (Bayleton) 50WSP (Strike) 25WDG	 1 PVA packet (11 oz/137.5 to 275 gal) 8 to 16 oz/100 gal	Make one application as first flower buds show color. Spray later varieties as they show color at 7- to 14-day intervals. Bayleton cannot be used on plants being grown for sale or other commercial use. Strike is for use in commercial nurseries, garden centers, and greenhouses only.
Phomopsis Twig Blight		
azoxystrobin (Heritage)	1 to 4 oz/100 gal	Apply at the first sign of disease, as new growth buds out. Repeat every 7 to 28 days.
azoxystrobin + benzovindiflupyr (Mural)	4 to 7 oz/100 gal	Apply every 7 to 21 days. Do not make more than two consecutive applications before switching to a non-Group 7 or non-Group 11 fungicide. Do not apply to apple, flowering cherry, or crabapple. See label.
copper oxychloride + copper hydroxide (Badge SC) (Badge X2)	 1.5-2 pt/A 1.5 to 2 lb/A	Repeat at 7 to 14 day intervals. May cause discoloration in some azalea varieties. See label for other phytotoxicity warnings. Badge X2 is OMRI listed. Do not mix with Aliette.
cyprodinil + fludioxonil (Palladium)	2 to 4 oz/100 gal	For stem diseases ensure full spray coverage of all stems and inner areas of the plant to the soil/media. Apply on a 7- to 14-day interval while conditions are conducive to disease development. After two applications, alternate with another fungicide with a different MOA for two applications.
mancozeb (Dithane) 75 DF, (Fore) 80 WSP, (Mancozeb DG) (Pentathlon LF) (Protect DF)	 1.5 lb/100 gal 0.8 to 1.6 qt/100 gal 1 to 2 lb/100 gal	Addition of a spreader sticker will improve performance. Repeat application on 7- to 10-day intervals.
propiconazole (Banner MAXX)	5 to 8 fl oz/100 gal	For junipers, make first application as soon as new growth is observed. Repeat application every 14 to 21 days during period of active plant growth.
pyraclostrobin + boscalid (Pageant Intrinsic)	8 to12 oz/100 gal	Apply prior to disease development. Repeat at 7- to 14-day intervals. Make no more than two sequential applications. Do not expose petunia, or impatiens in flower or wintercreeper, or nine bark to spray or drift as injury may occur.
thiophanate-methyl (AllBan Flo) (Cleary 3336) F (T-Storm) 50WSB (SysTec 1998) FL, (T-Storm Flowable), OHP 6672 (4.5L)	14.5 to 20 fl oz/100 gal 16 to 24 fl oz/100 gal 24 oz/100 gal 20 fl oz /100gal	Repeat at 10- to 14-day intervals. Apply when symptoms first appear. Re-treat every 7 to 14 days as needed during disease period. Apply in spring; repeat every 7 to 10 days.
thiophanate-methyl + mancozeb (Zyban)	6 bags (36 oz/75 gal)	Apply at 7- to 10-day intervals.
Phytophthora and Pythium Root Rot		
cyazofamid (Segway)	3 to 6 fl oz/100 gal	Apply at 14- to 21-day intervals using another registered fungicide with a different mode of action. Irrigate with at least ½ inch of water if rainfall does not occur within 24 hrs. For container plants, check label for recommended maximum drench volume based on pot diameter.
azoxystrobin + benzovindiflupyr (Mural)	3 oz/100 gal	Apply 1 to 2 pints of drench solution per square foot surface area every 7 to 28 days. Do not apply to apple, flowering cherry, or crabapple. See label.
dipotassium phosphonate + dipotassium phosphate (Biophos)	256 fl oz/100 gal	Apply as a soil drench or foliar spray as a preventative.
dimethomorph (Stature DM) 50W	6.4 to 12.8 oz/100 gal	Apply at 10- to 14-day intervals through production cycle. When applied as a drench, use enough solution to wet root zone of the plant. No more than two applications of Stature DM can be applied consecutively in a crop. **Not effective on Pythium root rot.**
dimethomorph + ametoctradin (Orvego)	11 to 14 fl oz	NOT LABELED FOR PYTHIUM. Apply on 10- to 14-day intervals using another registered fungicide with a different mode of action. Apply sufficient volume to wet all foliage until runoff (normally 50 to 100 gallons per acre).
etridiazole (Truban) 30WP (Terrazole) 35WP	 3 to 10 oz/100 gal/400 sq ft 3.5 to 10 oz/100 gal	Apply in sufficient volume to saturate the soil mixture. Water in immediately after application. Repeat at 4- to 12-week intervals.
etridiazole + thiophanate-methyl (Banrot) 40WP (Banrot) 8G	 6 to 12 oz/100 gal See label	Apply in sufficient volume to saturate the soil mixture. Irrigate immediately with additional water equal to at least half the volume of the fungicide drench. Re-treat at 4- to 12-week intervals. For use in nursery crops. See label.

Table 10-13. Commercial Landscape and Nursery Crops Disease Control

DISEASE Pesticide and Formulation	Rate of Formulation (per 100 gallons)	Schedule and Remarks
Phytophthora and Pythium Root Rot (continued)		
fluopicolide (Adorn)	1 to 4 fl oz/100 gal	MUST ALWAYS BE TANK MIXED WITH THE LABELED RATE OF ANOTHER FUNGICIDE WITH A DIFFERENT MODE OF ACTION. Apply before disease development. Use higher rate when treating plants with high potential for disease. Reapply after 14 to 28 days.
fosetyl-Al (Aliette) 80WP	See label	Can be applied as a preventative foliar or drench application. Can be incorporated into the soil for control of Phytophthora species.
mandipropamid (Micora)	4 to 8 fl oz/100 gal	This product can also be used on vegetables sold to the retail market in GH with permanent flooring. Apply prior to disease development. Repeat at 7- to 14-day intervals. Make no more than two sequential applications.
mefenoxam (Fenox ME, Mefenoxam 2, Subdue MAXX, Subdue GR)	See label	Can be applied as a drench or soil surface spray or soil incorporation treatment (granular). Consult label for specific crops and applications. Repeat at 2- to 3-month intervals. Do not apply rates of 1.25 fl oz per 100 gal more often than every 3 months.
oxathiapiprolin (Segovis)	0.65 to 3.2 fl oz/100 gal	NOT LABELED FOR PYTHIUM. Apply on 5- to 14-day interval preventatively or at first sight of disease symptoms. Do not apply more than 2 consecutive applications before switching to another non-Group U-15 fungicide.
phosphorous acid (Alude, Fosphite, Reliant)	See label for rates	Apply as a soil drench or foliar spray as a preventive.
potassium phosphite (Vital)	See label	Apply as a soil drench or foliar spray as a preventive.
propamocarb (Banol)	See label	Do not use for field-grown ornamentals.
trifloxystrobin (Compass O)	1 to 2 oz/100 gal	
Powdery Mildew		
azoxystrobin (Heritage) 50WDG	1 to 4 oz/100 gal	Apply only as a preventive. Spray every 7 to 28 days as needed. To avoid fungicide resistance, make no more than two sequential applications of Heritage before rotating with nonstrobilurin products. Do not apply to apple, flowering cherry, or crabapple. See label.
azoxystrobin + benzovindiflupyr (Mural)	4 to 7 oz/100 gal	Apply every 7 to 21 days. Do not make more than two sequential applications before rotating to another class of fungicide that is not Group 7 or 11. Do not apply to apple, flowering cherry, or crabapple. See label.
Bacillus subtilis (Rhapsody, Cease)	2 to 4 qt/100 gal	Repeat at 7-day intervals. Thorough coverage is essential. Begin applications when conditions favor disease development prior to the onset of disease.
Bacillus amyloliquefaciens strain D747 (DoubleNickel LC)	0.5 to 6 qt/100 gal	Begin preventative applications at plant emergence and repeat every 3 to 28 days depending on disease pressure. OMRI listed.
chlorothalonil (Daconil Ultrex) 82.5 WDG	1.4 lb/100 gal	Spray at 7- to 14-day intervals. Applications made during bloom may damage flowers.
chlorothalonil + propiconazole (Concert II)	22 to 35 fl oz/100 gal	Apply as a full coverage spray. Reapply at 14- to 21-day intervals. Refer to label for specific rate and application instructions.
copper oxychloride + copper hydroxide (Badge SC) (Badge X2)	1.5-2 pt/A 1.5 to 2 lb/A	Repeat at 7 to 14 day intervals. May cause discoloration in some azalea varieties. See label for other phytotoxicity warnings. Badge X2 is OMRI listed. Do not mix with Aliette.
copper octanoate (Camelot O)	0.5-2.0 gal/100 gal	Begin at first sign of disease, and repeat at 7- to 10-day intervals. Camelot O may cause copper toxicity on some plant species.
copper sulfate pentahydrate (Phyton 27)	See label	
cyprodinil + fludioxonil (Palladium)	2 to 6 oz/100 gal	Spray on a 7- to 14- day interval while conditions are conducive to disease development. After two applications, alternate with another fungicide with a different MOA for two applications.
fluoxastrobin (Disarm 480SC)	1 to 4 fl oz/100 gal	Use for preventative applications only and apply every 7 to 28 days depending on disease pressure. Do not make more than two consecutive applications before switching to a non Group 11 fungicide.
metconazole (Tourney)	1 to 4 oz/100 gal	Repeat in 14 to 28 days when conditions favor disease
myclobutanil (Eagle) 20EW (Eagle) 40WP	 8 fl oz/100 gal 3 to 6 oz/100 gal	Apply at 10- to 14-day intervals.
potassium bicarbonate (Kaligreen)	1 to 3 lb/100 gal	Apply at 7- to 10-day intervals.
propiconazole (Banner MAXX, Banner MAXX II)	5 to 12 fl oz/100 gal	See label for appropriate rate and application intervals. For application in field nurseries and landscape plantings.
pyraclostrobin (Insignia)	4 to 8 oz/100 gal	Apply at 7- to 14-day intervals. Do not make more than two consecutive applications before switching to a non Group 11 fungicide.
pyraclostrobin + boscalid (Pageant Intrinsic)	6 to12 oz/100 gal	Apply prior to disease development. Repeat at 7- to 10-day intervals. Make no more than two sequential applications. Do not expose petunia or impatiens in flower or wintercreeper or nine bark to spray or drift as injury may occur.
sulfur (Micro Sulf)	3 to 10 lb/100 gal	Begin when disease first appears and continue at 5 to 10 intervals. See label for phytoxicity considerations.
tebuconazole (Torque)	4 to 8 fl oz/100 gal	Apply every 14 for a total of three applications at the first sign of disease.
thiophanate-methyl (AllBan Flo) (Cleary 3336) F (OHP 6672) 50WP, (T-Storm) 50WSB (SysTec 1998) FL (T-Storm Flowable), OHP 6672 (4.5L)	10 to 20 oz/100 gal 12 to 24 fl oz/100 gal 8 to 16 oz/100 gal 10 fl oz/100 gal 20 fl oz/100 gal	Apply when disease first appears and repeat every 7 to 14 days. Rotations with other effective products are recommended. Apply when disease first appears and repeat every 7 to 14 days. Rotations with other effective products are recommended. Apply when disease first appears and repeat every 7 to 14 days as needed during disease period. Apply as heavy spray or drench at a rate of 1 to 2 pints per sq ft. Repeat at 2- to 4-week intervals. Apply as heavy spray or drench at a rate of 1 to 2 pints per sq ft. Repeat at 2- to 4-week intervals.
thiophanate-methyl + mancozeb (Zyban) WSB	24 oz (4 bags) /100 gal	Apply at first sign of disease. Repeat at 7-day intervals.

Table 10-13. Commercial Landscape and Nursery Crops Disease Control

DISEASE Pesticide and Formulation	Rate of Formulation (per 100 gallons)	Schedule and Remarks
Powdery Mildew (continued)		
triadimefon (Bayleton) 50WSP	1 PVA packet (11 oz/550 to 1,100 gal)	Spray as needed. Bayleton cannot be used on plants for sale or other commercial use. Strike is for greenhouse and nursery use only.
(Strike) 25WDG	2 to 4 oz/100 gal	
trifloxystrobin (Compass O)	2 to 4 oz/100 gal	Apply to point of drip before disease is detected. Apply at 7- to 14-day intervals. Rotate to another nonstrobilurin product after each application.
triticonazole (Trinity) 19SC	6 to 12 fl oz/100 gal	Apply every 7 to 14 days. Use preventively. Begin applications when conditions favor fungal infection and before disease symptom development. Use of an adjuvant/spreader sticker can aid in control.
Rhizoctonia Aerial Blight *(Rhizoctonia solani)*		
chlorothalonil (Daconil Ultrex)	1.4 lb/100 gal	Spray to runoff. Repeat at 7- to 14-day intervals until conditions no longer favor disease.
cyprodinil + fludioxonil (Palladium)	2 to 6 oz/100 gal	Spray on a 7- to 14-day interval while conditions are conducive to disease development. After two applications, alternate with another fungicide with a different MOA for two applications.
fludioxonil (Medallion)	1 to 2 oz/100 gal	Spray to runoff. Repeat at 7- to 14-day intervals until conditions no longer favor disease.
flutolanil (Contrast) 70WSP (Prostar) 70WP	3 to 12 oz/100 gal	Apply at 14- to 21-day intervals.
iprodione (Chipco 26019 N/G) (18 Plus, Iprodione Pro 2SE)	1.0 to 2.5 qt/100 gal 1.0 to 2.5 qt/100 gal	Spray plants to ensure thorough coverage Repeat at 7- to 14-day intervals. Do not make more than four applications per crop per year.
mancozeb (Dithane 75DF)	1 to 2 lb/100 gal	Apple at 7- to 10-day intervals. See label for specific instructions.
myclobutanil (Eagle 20EW)	6 to 12 fl oz/100 gal	Apply at 7- to 10-day intervals. See label for specific instructions.
polyoxin D zinc salt (Endorse 2.5 WP)	1.1 to 2.2 lb/100 gal	Apply as a foliar spray every 7 to 10 days. Apply prior to disease development and when conditions are conducive for disease.
pyraclostrobin + boscalid (Pageant Intrinsic)	12 to18 oz/100 gal	Apply prior to disease development. Repeat at 7- to 14-day intervals. Make no more than two sequential applications. Do not expose petunia or impatiens in flower or wintercreeper or nine bark to spray or drift as injury may occur.
thiophanate-methyl (Cleary 3336) F	12 to 16 fl oz/100 gal	Apply when disease symptoms first appear. Repeat at 7- to 14-day intervals during disease period.
thiophanate-methyl + iprodione (26/36 fungicide)	33-84 fl oz/100 gal	No more than 4 applications per year.
triflumizole (Terraguard) 50W	4 to 8 oz/100 gal	Make initial application prior to or at first sign of disease. Use the higher rate under heavy disease pressure. Repeat at 7- to 14-day intervals.
Rhizoctonia Stem and Root Rot *(Rhizoctonia solani)*		
azoxystrobin (Heritage) 50WDG	directed spray: 1 to 4 oz/100 gal	Apply as a directed spray every 7 to 21 days as needed. To avoid fungicide resistance, make no more than three sequential applications of Heritage before rotating with nonstrobilurin products.
	drench: 0.2 to 0.9 oz/100 gal	Apply 1 to 2 pt of solution per sq ft surface area every 7 to 28 days as a preventative drench treatment. Do not exceed 2 oz/100 gal on impatiens or pansy.
azoxystrobin + benzovindiflupyr (Mural)	3 oz/100 gal	Apply 1 to 2 pints of drench solution per square foot surface area every 7 to 28 days. Should not be applied to some tree fruit varieties. See label.
Bacillus amyloliquefaciens strain D747 (DoubleNickel LC)	0.5 to 4.5 pt/100 gal	Apply as a drench or course spray to growing medium at or immediately before seeding. Repeat applications as needed. OMRI listed.
chlorothalonil + thiophanate-methyl (Spectro) 90WDG	1 to 2 lb/100 gal	Retreat at 7-day intervals. Apply as a spray only. Do not apply more than once to green or variegated Pittosporum due to risk of phytotoxicity. Apply when foliage is dry.
cyprodinil + fludioxonil (Palladium)	2 to 4 oz/100 gal	For stem diseases ensure full spray coverage of all stems and inner areas of the plant to the soil/media. Apply on a 7- to 14-day interval while conditions are conducive to disease development. After two applications, alternate with another fungicide with a different MOA for two applications.
etridiazole + thiophanate-methyl (Banrot) 40W	16 to 12 oz/400 sq ft	Apply in sufficient volume to saturate the soil mixture. Irrigate immediately. Repeat at 4- to 12-week intervals if necessary.
fludioxonil (Medallion) 50W	1 to 2 oz/100 gal	Apply as a drench at seeding or transplanting. Apply sufficient mix to wet the upper one-half of the growing medium. Make only one application to seedling crop. If needed, re-treat transplants 21 to 28 days after initial application. Do not apply as a seed or soil drench to impatiens or New Guinea impatiens. May cause stunting or chlorosis on some geranium cultivars. See label for maximum amounts that can be applied per year.
flutolanil (Contrast) 70WSP (Prostar) 70WP	3 to 6 oz/100 gal	Apply drench according to label. Repeat 21 to 28 days after initial application. Make no more than four applications per year to ornamental plantings.
iprodione (Chipco 26019 N/G) (18 Plus, Iprodione Pro 2SE)	6.5 oz/100 gal 13 fl oz/100 gal	Apply 1 to 2 pints per sq ft at seeding or transplanting. Do not apply as a drench on impatiens or pothos. Repeat every 14 days. Do not make more than six applications per year. Do not use on Spathiphyllum.
metconazole (Tourney)	1 to 4 oz/100 gal	Repeat in 14 to 28 days when conditions favor disease
PCNB (Terraclor) 75WP	4 to 8 oz/100 gal	See label for amount to apply. One repeat application can be made 4 to 6 weeks later, if necessary.
polyoxin D zinc salt (Endorse 2.5 WP)	1.1 to 2.2 lb/100 gal/acre	Apply as a foliar spray every 7 to 10 days. Apply as a drench every 14 to 28 days.
pyraclostrobin + boscalid (Pageant Intrinsic)	12 to18 oz/100 gal	Apply prior to disease development. Repeat at 7- to 14-day intervals. Make no more than two sequential applications. Do not expose petunia or impatiens in flower or wintercreeper or nine bark to spray or drift as injury may occur.
thiophanate-methyl (AllBan Flo) (Cleary 3336) F (OHP-6672) 50WP	7.5 to 20 oz/100 gal 8 to 16 fl oz/100 gal 12 to 16 oz/100 gal	Repeat at 4- to 8-week intervals. Apply as a soil drench or directed spray to thoroughly soak growing media through the root zone after seeding or transplanting at 21- to 28-day interval. Apply 1 to 3 pt/sq ft after transplanting to thoroughly soak growing medium. Repeat every 21 to 28 days.

Table 10-13. Commercial Landscape and Nursery Crops Disease Control

DISEASE Pesticide and Formulation	Rate of Formulation (per 100 gallons)	Schedule and Remarks
Rhizoctonia Stem and Root Rot *(Rhizoctonia solani)* (continued)		
Thiophanate-methyl + iprodione (26/36 fungicide)	See label	See label.
trifloxystrobin (Compass)	0.5 oz/100 gal	Apply as a drench to wet upper half of the growing media. Apply at seeding, transplanting, and at 21- to 28-day intervals thereafter. May injure petunia, violet, and New Guinea impatiens.
triflumizole (Terraguard) 50W	4 to 8 oz/100 gal	Apply as soil drench at 2- to 4-week intervals. Use higher rate under heavy disease pressure.
Rust (also see Daylily Rust)		
azoxystrobin (Heritage)	1 to 4 oz/100 gal commercial rose production: 1.6 to 8 oz	Apply at 7- to 28-day intervals. Do not make more than three sequential applications of Heritage before alternating with a nonstrobilurin fungicide. Should not be applied to certain plant species; see label.
azoxystrobin + benzovindiflupyr (Mural)	4 to 7 oz/100 gal	Apply every 7 to 14 days. Do not make more than two sequential applications before rotating to another class of fungicide that is not Group 7 or 11.
Bacillus amyloliquefaciens strain D747 (DoubleNickel LC)	0.5 to 6 qt/100 gal	Begin preventative applications at plant emergence and repeat every 3 to 28 days depending on disease pressure. OMRI listed.
chlorothalonil (Daconil Ultrex) (Echo) 90DF	 1.4 lb/100 gal; 1 lb for roses 1.4 lb/100 gal; 0.875 oz for roses	Apply when foliage and flowers are dry. Repeat at 7- to 14-day intervals. Apply to hydrangea foliage only. Avoid application during bloom period on plants where flower injury is unacceptable.
chlorothalonil + propiconazole (Concert II)	22 to 35 fl oz/100 gal	Apply as a full coverage spray. Reapply at 14 to 21 day intervals. Refer to label for specific rate and application instructions. Higher rate listed for *Melampsora occidentalis* --use 69 fl oz/100gal.
chlorothalonil + thiophanate-methyl (Spectro) 90	1 to 2 lb/100 gal	Apply when foliage and flowers are dry, or nearly dry. Re-treat at 7-day intervals. Do not exceed 50.6 lb per acre during one season for field-grown ornamentals.
flutolanil (Contrast) 70 WSP	3 to 6 oz/100 gal	Repeat at 14- to 21-day intervals.
fluoxastrobin (Disarm 480SC)	1 to 4 fl oz/100 gal	Use for preventative applications only and apply every 7 to 28 days depending on disease pressure. Do not make more than two consecutive applications before switching to a non Group 11 fungicide.
mancozeb (Dithane, Mancozeb) (Protect DF)	 1.5 lb/100 gal 1 to 2 lb/100 gal	Begin at first sign of disease. Repeat at 7- to 10-day intervals. Apply at 7- to 21-day intervals. To improve performance, add 2 to 4 oz of a spreader-sticker.
metconazole (Tourney)	1 to 4 oz/100 gal	Repeat in 14 to 28 days when conditions favor disease
myclobutanil (Eagle 40WP) (Eagle 20WE)	 3 to 6 oz/100 gal 6 to 12 oz/100 gal	Apply on a protectant application schedule at 10- to 14-day intervals.
neem oil (Triact 70)	1 gal/100 to 200 gal See label	Apply at 7- to 14-day spray intervals. Trial first on open blooms. To control existing disease, apply on a 7-day schedule until disease pressure is eliminated. Not for impatiens, carnation, or hibiscus.
pyraclostrobin (Insignia)	See label	Apply prior to onset of disease and repeat every 7 to 14 days. Do not expose flowering impatiens or flowering petunias to Insignia.
propiconazole (Banner MAXX)	See label	See label. Do not use in greenhouses.
pyraclostrobin + boscalid (Pageant Intrinsic)	12 to18 oz/100 gal	Apply prior to disease development. Repeat at 7- to 14-day intervals. Make no more than two sequential applications. Do not expose petunia or impatiens in flower or wintercreeper or nine bark to spray or drift as injury may occur.
tebuconazole (Torque)	4 to 8 fl oz/100 gal	Apply every 14 days for a total of three applications at the first sign of disease.
thiophanate-methyl (AllBan Flo) (Cleary 3336 F) (OHP 6672 50WP), (T-Storm 50WSB) (OHP 6672 4.5L)	 10.75 to 20 oz/100 gal 12 to 16 fl oz/100 gal 12 to 16 oz/100 gal 20 fl oz/100 gal	For use on crabapples. Do not use treated crabapples for food. Apply late Spring or at first sign of disease. Repeat every 7 to 14 days as needed during disease period. Apply 8 oz as a drench or directed spray after seeding, or apply 12 to 16 oz after transplanting. Repeat at 21- to 28-day intervals. Apply late Spring or at first sign of disease. Repeat every 7 to 14 days as needed during disease period. Apply as heavy spray or drench at a rate of 1 to 2 pints per sq ft. Repeat at 2- to 4-week intervals.
thiophanate-methyl + mancozeb (Zyban) WSB	See label	For use on crabapples.
triadimefon (Bayleton) (Strike)	See label	Spray to the point of drip as needed. See label for spray interval. Bayleton is not for use on plants being grown for sale.
triflumizole (Terraguard) 50W	2 to 8 oz/100 gal	Apply prior to, or at first sign of disease. Repeat at 7- to 14-day intervals.
Scab *(Cladosporium, Fusicladium, Spilocea, Venturia)* For apple scab, see disease under "crabapple" above.		
mancozeb (Dithane, Fore, Mancozeb)	1.5 lb/100 gal	Begin spraying at first sign of disease, and repeat at 7- to 10-day intervals.
azoxystrobin (Heritage)	1 to 4 oz/100 gal	Repeat every 10 to 28 days. Do not make more than two consecutive applications before switching to a non-Group 11 fungicide. PHYTOTOXIC to some crabapple and other tree fruit cultivars. See label
azoxystrobin + benzovindiflupyr (Mural)	4 to 7 oz/100 gal	Repeat every 7 to 14 days. Do not make more than two consecutive applications before switching to a non-Group 11 fungicide. PHYTOTOXIC to some crabapple and other tree fruit cultivars. See label
copper oxychloride + copper hydroxide (Badge SC) (Badge X2)	 1.5-2 pt/A 1.5 to 2 lb/A	Repeat at 7 to 14 day intervals. May cause discoloration in some azalea varieties. See label for other phytoxicity warnings. Badge X2 is OMRI listed. Do not mix with Aliette.
fluoxastrobin + myclobutanil (Disarm M)	3 to 11 fl oz/100 gal	Repeat at 7 to 28 day intervals

Table 10-13. Commercial Landscape and Nursery Crops Disease Control

DISEASE Pesticide and Formulation	Rate of Formulation (per 100 gallons)	Schedule and Remarks
Scab *(Cladosporium, Fusicladium, Spilocea, Ventuna)* For apple scab, see disease under "crabapple" above. (continued)		
metconazole (Tourney)	1 to 4 oz/100 gal	Repeat in 14 to 28 days when conditions favor disease
myclobutanil (Eagle 20EW)	6 to 12 fl oz/100 gal	Spray every 10 to 14 days.
propiconazole (Banner MAXX)	See label	See label. Do not use in greenhouses.
pyraclostrobin (Insignia)	4 to 8 oz/100 gal	Repeat at 7- to 14-day intervals. Do not make more than 2 consecutive applications before switching to a non-Group 11 fungicide.
pyraclostrobin + boscalid (Pageant Intrinsic)	8 to 12 oz/100 gal	Apply prior to disease development. Repeat at 7- to 10-day intervals. Make no more than two sequential applications. Do not expose petunia or impatiens in flower or wintercreeper or nine bark to spray or drift as injury may occur.
tebuconazole (Torque)	4 to 10 fl oz/100 gal	For preventative applications, apply at least 3 times per year, 14 to 21 days apart.
thiophanate-methyl (AllBan Flo)	10.75 to 20 fl oz/100 gal	Spray at bud break. Repeat three to four times at 7- to 14-day intervals.
(Cleary 3336 F)	12 to 16 fl oz/100 gal	
(SysTec 1998 FL)	20 fl oz/100 gal	
thiophanate-methyl + iprodione (26/36 fungicide)	33 to 84 fl oz/100 gal	No more than 4 applications per year.
thiophanate-methyl + mancozeb (Zyban) WSB	24 oz (4 bags)	Repeat at 7-day intervals.
trifloxystrobin (Compass O)	2-4 oz/100 gal	Repeat at 7- to 14-day intervals. Do not make more than 2 consecutive applications before switching to a non-Group 11 fungicide.
triticonazole (Trinity) 19SC	6 to 12 fl oz/100 gal	Apply every 7 to 14 days. Use preventively. Begin applications when conditions favor fungal infection and before disease symptom development.
Sclerotinia Stem Rot *(Sclerotinia sclerotiorum)*		
chlorothalonil (Daconil Ultrex 82.5 WDG)	1.4 lb/100 gal	Repeat at 7- to 14-day intervals when conditions favor disease.
cyprodinil + fludioxonil (Palladium)	2 to 4 oz/100 gal	For stem diseases ensure full spray coverage of all stems and inner areas of the plant to the soil/media. Apply on a 7- to 14-day interval while conditions are conducive to disease development. After two applications, alternate with another fungicide with a different MOA for two applications.
metconazole (Tourney)	1 to 4 oz/100 gal	Repeat in 14 to 28 days when conditions favor disease
PCNB		
(Revere) 10G	20 lb/1,000 sq ft	Apply 1 wk prior to planting; spread on soil surface and mix into soil at a 4-in. depth.
(Terraclor) 75W	6 to 12 fl oz/100 gal	See label for amount to apply. One repeat application may be made 4 to 6 weeks later.
pyraclostrobin + boscalid (Pageant Intrinsic)	12 to 18 oz/100 gal	Apply prior to disease development. Repeat at 7- to 10-day intervals. Make no more than two sequential applications. Do not expose petunia or impatiens in flower or wintercreeper or nine bark to spray or drift as injury may occur.
triticonazole (Trinity) 19SC	8 to 12 fl oz/100 gal	Apply every 7 to 14 days. Use preventively. Begin applications when conditions favor fungal infection and before disease symptom development. The stem areas of the plant must be thoroughly covered using spray to runoff.
Shot Hole *(Blumeriella, Coccomyces* fungal and bacterial)		
copper hydroxide + mancozeb (Junction)	1.5 to 3.5 lb/100 gal	Begin at first sign of disease. Repeat at 7- to 14-day intervals.
mancozeb	1 to 2 lb/100 gal	
(Pentathlon DF)		Begin applications at first sign of disease. Apply at 7- to 10-day intervals.
(Protect DF)		Begin applications at first sign of disease. Apply at 7- to 21-day intervals.
metconazole (Tourney)	1 to 4 oz/100 gal	Apply on a 14- to 28-day interval.
thiophanate-methyl (Cleary 3336) F	12 to 16 fl oz/100 gal	Begin when disease first appears, and repeat every 7 to 14 days.
Southern Stem Blight *(Sclerotium rolfsii)*		
azoxystrobin (Heritage)	directed spray: 1 to 4 oz/100 gal	Apply every 7 to 21 days. Can be used in outdoor nurseries, retail nurseries, residential and commercial landscape areas
	drench: 0.2 to 0.9 oz/100 gal	Apply 1 to 2 pt solution per square foot surface area every 7 to 28 days, prior to infection. Apply to container-grown ornamentals only.
		Do not apply to crabapple, apple, or some tree fruit. See label.
azoxystrobin + benzovindiflupyr (Mural)	5 to 7 oz/100 gal (direct spray)	Repeat every 7 to 21 days. Do not make more than two consecutive applications before switching to a non-Group 11 fungicide. PHYTOTOXIC to some crabapple and other tree fruit cultivars. See label.
	2 to 3 oz/100 gal (drench)	
cyprodinil + fludioxonil (Palladium)	2 to 4 oz/100 gal	For stem diseases ensure full spray coverage of all stems and inner areas of the plant to the soil/media. Apply on a 7- to 14-day interval while conditions are conducive to disease development. After two applications, alternate with another fungicide with a different MOA for two applications.
Fluoxastrobin (Disarm 480SC)	2 to 4 fl oz/100 gal	Crown spray every 7 to 21 days.
flutolanil (Contrast) 70WSP (Prostar) 70WP	3 to 6 oz/100 gal	Drench at 21- to 28-day intervals. Uses 1 pt per sq ft or 2 pt for depths greater than 4 in. See label for container rates. For use in outdoor container and field-grown stock. Make no more than 4 applications per year to nursery ornamental plantings.
PCNB (Terraclor) 75WP	3.25 to 6.5 lb/1,000 sq ft	Apply in sufficient water to ensure uniform ground coverage prior to planting, and thoroughly incorporate to a depth of 6 to 7 in. For use in nursery and landscape plantings.
tebuconazole (Torque)	4 to 8 fl oz/100 gal	Apply every 14 for a total of three applications at the first sign of disease.

Table 10-13. Commercial Landscape and Nursery Crops Disease Control

DISEASE Pesticide and Formulation	Rate of Formulation (per 100 gallons)	Schedule and Remarks
Southern Stem Blight *(Sclerotium rolfsii)* (continued)		
triticonazole (Trinity) 19SC	8 to 12 fl oz/100 gal	Apply every 7 to 14 days. Use preventively. Begin applications when conditions favor fungal infection and before disease symptom development. The stem areas of the plant must be thoroughly covered using spray to runoff.
Volutella Blight		
chlorothalonil		
(Daconil Ultrex) 82.5 WDG	1.4 lb/100 gal	
(Daconil WeatherStik)	1.375 pt/100 gal	Reapply at 7- to 14-day intervals.
copper hydroxide (Champ WG) (Nu-Cop) 50DF (CuPRO 2005)	0.5 lb/100 gal 1 lb/100 gal 0.75 to 2.0 lb/100 gal	Begin at first sign of disease, and repeat at 7- to 14-day intervals. Do not tank mix copper formulations with Aliette. Avoid contact with metal surfaces. Discoloration of blooms may occur on certain plant varieties- check label.
copper hydroxide + mancozeb (Junction)	1.5 to 3.5 lb/100 gal	Begin at first sign of disease, and repeat at 7- to 14-day intervals.
copper oxychloride + copper hydroxide (Badge SC) (Badge X2)	1.5-2 pt/A 1.5 to 2 lb/A	Repeat at 7- to 14-day intervals. May cause discoloration in some azalea varieties. See label for other phytoxicity warnings. Badge X2 is OMRI listed. Do not mix with Aliette.
copper sulfate pentahydrate (Phyton 27, Phyton 35)	See label	See label. To avoid phytotoxicity, do not use any copper compound on alyssum.
mancozeb		
(Dithane) 75 DF, (Fore) 80 WSP, Mancozeb DG	2.0 lb/50 gal/5,000 sq ft of bed	Start at first sign of disease, and apply at 10- to 14-day intervals.
(Pentathlon LF)	0.8 to 1.6 pt/100 gal	Use a drenching spray. Start at first sign of disease, and apply at 10- to 14-day intervals.
(Pentathlon DF)	1 to 2 lb/100 gal	Begin at first sign of disease, and repeat at 7- to 10-day intervals.
(Protect DF)	3 to 4 lb/100 gal/10,000 sq ft of bed	Use a drenching spray. Start at first sign of disease, and apply at least 5 applications at 10- to 14-day intervals.
thiophanate-methyl + mancozeb (Zyban) WSB	24 oz (4 bags)/100 gal	Apply at 7-day intervals while disease is prevalent.

Further Information

Boxwood blight links http://plantpathology.ces.ncsu.edu/pp-ornamentals/
All Plant Pathology *Disease Notes for Ornamentals* can be accessed from http://www.ces.ncsu.edu/depts/pp/notes/Ornamental/ornamental_contents.html.
Holly Diseases and Their Control in the Landscape. Plant Pathology Ornamental Disease Information Note 7, http://www.ces.ncsu.edu/depts/pp/notes/oldnotes/od7.htm
Nematodes and Their Control in Woody Ornamentals in the Landscape. Plant Pathology Information Note 63, http://www.ces.ncsu.edu/depts/pp/notes/oldnotes/no63.htm
Phytophthora Root Rot and Its Control on Established Woody Ornamentals. Plant Pathology Ornamental Disease Information Note 13, http://www.ces.ncsu.edu/depts/pp/notes/oldnotes/odin13/od13.htm
Rose Diseases and Their Control in the Home Garden. Plant Pathology Ornamental Disease Information Note 2, http://www.ces.ncsu.edu/depts/pp/notes/Ornamental/odin002/odin002.htm
Juniper Diseases. Plant Pathology Ornamental Disease Information Note 15, http://www.ces.ncsu.edu/depts/pp/notes/oldnotes/od15.htm
Entomosporium Leaf Spot on Redtip. Plant Pathology Ornamental Disease Information Note 11, http://www.ces.ncsu.edu/depts/pp/notes/Ornamental/odin011/odin011.htm
Rhododendron Diseases. Plant Pathology Ornamental Disease Information Note 12, http://www.ces.ncsu.edu/depts/pp/notes/oldnotes/od12.htm
Azalea Diseases. Plant Pathology Ornamental Disease Information Note 16.

Copies of these publications are available from your local Cooperative Extension center.

Treatments for Sanitizing Tools, Equipment, Cultivation Surfaces, Pots and Flats

Sara M. Villani, Department of Entomology and Plant Pathology

Note: It was not possible to update the information in this table for the 2016 North Carolina Agricultural Chemicals Manual. The information in this table is current as of Jan. 1, 2015. Anyone using any agricultural chemical should refer to the current chemical label, which contains information about the safe and effective use of the chemical, before using the chemical.

All items should be free of organic debris before exposure to the treatments listed below. Sanitizing an entire greenhouse involves physically removing leftover debris and soil as a first step prior to disinfection, as soil and organic residues reduce the effectiveness of disinfectants. There are some commercial cleaners specifically designed for greenhouse use, e.g., Strip-It (best applied by spray, brush, or foam), which is a combination of cleaning and wetting agents formulated to remove algae, dirt, and hard water deposits. High pressure power washing with soap and water is also an option prior to disinfection as listed below.

Table 10-14. Treatments for Sanitizing Tools, Equipment, Cultivation Surfaces, and other Related Items

Material or Treatment	Trade name	Formulation	Remarks	Contact time
alcohol, ethyl and isopropyl (grain, rubbing, wood) (70-100%)	Various commercial brands; Lysol Spray (also includes quaternary ammonium)	Depends on formulation. Read label. Typically full strength for RTU (Ready To Use) formulations.	Evaporates quickly so that adequate contact time may not be achieved; high concentrations of organic matter diminish effectiveness; flammable.	10 min for equipment, pots, flats and surfaces. Tools can be dipped for 10 seconds and allowed to dry. Do not rinse.
alkyl dimethyl benzyl ammonium chloride + alkyl dimethyl ethylbenzyl ammonium chloride	Green-Shield Green-Shield II	See label	Corrosive, Causes irreversible eye damage and skin burns. Pre-clean surfaces/heavily soiled areas prior to use	10 min for hard, non-porous surfaces, pots, flats, cutting tools (see label for other surfaces)
hydrogen peroxide (hydrogen dioxide) and peroxyacetic acid mixture	ZeroTol 2.0; SaniDate 5.0; Oxidate 2.0	2.5 fl oz per gallon of water 0.5 fl oz per gallon of water 0.5 to 1.25 fl oz per gallon of water	Very corrosive; eye/skin irritant. Low odor. Use according to label. Must be stored in cool location.	1-10 min
quaternary ammonium	Physan 20;	Depends on formulation. Typically 1 tablespoon per gallon of water	Effective for non-porous surface sanitation, e.g. floors, walls, benches, pots. Low odor, irritation.	10-15 min Must remain wet for 10 min. Wipe dry with a clean cloth or sponge or allow to air dry.
	KleenGrow	For general disinfection use 0.5 to 1.0 fl oz per gallon of water	Hard, NON-POROUS surfaces use 1.0 fl oz per gal water; Tools, cutters & equipment use 0.5 fl oz per gal water. Apply solution with a cloth, mop, sponge, coarse spray device or by immersion until surfaces are wet. Prepare a fresh solution daily.	Must remain wet for 10 min. Wipe dry with a clean cloth or sponge or allow to air dry.
sodium hypochlorite (8.25%)	Clorox; Commercial bleach;	10%; or a 1:14 ratio of bleach : water	Inactivated by organic matter; fresh solutions should be prepared every 8 hr or more frequently if exposed to sunlight; corrosive to metal; irritating to eyes and skin; Exposure to sunlight reduces efficacy. Keep solution in opaque container.	10-15 min. for equipment, pots, flats and surfaces. Tools can be dipped for 10 seconds and allowed to dry. Do not rinse.
steam	NA	Cover or otherwise seal	For plastic pots and trays, heat center of steamer between 150 degrees F to 160 degrees F;	60 min.
			For less heat-sensitive objects, heat to 180 degrees F.	15 min.
solarization	NA	Place clean items on solid surface, cover tightly with CLEAR plastic	Clear plastic works much better.	140 degrees F, 4 to 8 hr/day for 7 days

Disease Control for Commercial Vegetables

L. M. Quesada-Ocampo, Inga Meadows, and Frank Louws, Entomology and Plant Pathology

This section was prepared as a collaborative effort of vegetable pathology experts in the southeastern United States who yearly update the Southeastern U.S. Vegetable Crop Handbook. Contributors this year included: L. Quesada-Ocampo (North Carolina State University), E. Sikora (Auburn University), A. Keinath (Clemson University), E. Pfeufer (University of Kentucky), B. Dutta (University of Georgia), I. Meadows (North Carolina State University), F. Louws (North Carolina State University), C. Johnson (University of Virginia), R. Singh (Louisiana State University) and R. A. Melanson (Mississippi State University).

Caution: At the time these tables were prepared, the entries were believed to be useful and accurate. However, labels change rapidly, and errors are possible, so the user must follow all directions on the product label. Federal tolerances for fungicides may be canceled or changed at any time.

Information in the following tables must be used in the context of an integrated disease management program. Many diseases are successfully managed by combined strategies—using resistant varieties, crop rotation, deep-turn plowing, sanitation, seed treatments, cultural practices, and fungicides. Always use top-quality seed and plants obtained from reliable sources. Seeds are ordinarily treated by commercial producers for control of decay and damping off diseases.

Preplant fumigation of soils, nematode control chemicals and greenhouse disease control products are provided in separate tables following the crop tables. Efficacy tables will help in selecting appropriate disease control materials for some vegetable crops. These tables are located at the end of each crop table.

Rates: Some foliar rates are based on mixing a specified amount of product in 100 gallons of water and applying the finished spray for complete coverage of foliage just to the point of runoff with high pressure (over 250 psi) drop nozzle sprayers. Actual amount of product and water applied per acre will vary depending on plant size and row spacing. Typically, 25 to 75 gallons (gal) per acre of finished spray are used. Concentrate spray (air blast, aircraft, etc.) rates are based on amount of product per acre. Caution: With concentrate sprays, it is easy to apply too much product. Some fungicides are adversely affected by pH of water; adjust pH of water if specified on label. Some fungicides will cause damage to the plant if applied at temperatures above 90 degrees.

Do not feed treated foliage to livestock unless allowed by the label. Do not reenter fields until sprays have dried; some fungicides may have a reentry requirement of one to several days. Read the label. Do not exceed maximum number of applications on the label. Do not exceed maximum limit of fungicide per acre per application or per year as stated on the label. See label for rotational crops. In all cases, follow directions on the label. The label is the law.

The following online databases provide current product labels and other relevant information:

Database[1]	Web Address
Agrian Label Database	https://home.agrian.com/
Crop Data Management Systems	http://www.cdms.net/Label-Database
EPA Pesticide Product and Label System	https://iaspub.epa.gov/apex/pesticides/f?p=PPLS:1
Greenbook Data Solutions	https://www.greenbook.net/
Kelly Registration Systems[2]	http://www.kellysolutions.com

[1] Additional databases not included in this list may also be available. Please read the database terms of use when obtaining information from a particular website.

[2] Available for AK, AL, AZ, CA, CO, CT, DE, FL, GA, IA, ID, IN, KS, MA, MD, MN, MO, MS, NC, ND, NE, NJ, NV, NY, OK, OH, OR, PA, SC, SD, VA, VT, WA, and WI. Kelly Registration Systems works with State Departments of Agriculture to provide registration and license information.

Disease Control by Crop
Asparagus
E. Sikora, Plant Pathology, Auburn University

Table 10-15. Disease Control Products for Asparagus

Disease	Material	FRAC	Rate of Material	Minimum Days Harvest	Minimum Days Reentry	Method, Schedule, and Remarks
Crown rot	mancozeb (various)	M	See label	See label	See label	Soak crowns 5 minutes in burlap bag with gentle agitation, drain, and plant.
Gray mold (*Botrytis cinerea*)	fenhexamid (Elevate)	17	1.5 lb/A	180	–	Apply at fern stage only. Make up to four applications. Repeat at 7-to 14-day intervals if conditions favor disease development.
Phytophthora crown rot, spear rot	mefenoxam (Ridomil Gold SL)	4	1 pt/A	1	2	Apply over beds after seeding or covering crowns, 30 to 60 days before first cutting, and just before harvest.
	fosetyl-AL (Alliette)	33	5 lb/A	110	0.5	
Rust	myclobutanil (Rally 40W)	3	5 oz/A	180	1	Begin applications to developing ferns after harvest has taken place. Repeat on a schedule not to exceed 14 days. Do not supply to harvestable spears.
	sulfur (various)	M	See label	0	1	
Rust (continued)	tebuconazole (various)	3	4 to 6 fl oz/A	180	0.5	Apply to developing ferns at first sign of rust and repeat on a 14-day interval; no more than 3 applications per season.
	copper oxychloride/ hydroxide (Badge SC)	M	1 to 2.5 pt/A	0	48 hr	Recommended for tank mixture with other registered products. For suppression. Addition of spread/sticker is recommended.
Rust, Cercospora leaf spot	chlorothalonil (various)	M	2 to 4 lb/A	190	0.5	Repeat applications at 14 to 28 day intervals depending on disease pressure. Do not apply more than 12 pints/ acres during each growing season.
	mancozeb (various)	M	See label	See label	See label	Apply to ferns after harvest; spray first appearance of disease at 7- to 10-day intervals. Do not exceed 8 lb product per acre per crop.
Purple spot	azoxystrobin (Quadris)	11	6 to 15.5 fl oz	100	4 hr	Do not apply more than 1 foliar application of Quadris (or other group 11 fungicide) before alternating with a fungicide that has a different mode of action.
	chlorothalonil (various)	M	2 to 4 lb/A	190	0.5	Repeat applications at 14 to 28 day intervals depending on disease pressure. Do not apply more than 12 pints/ acre during each growing season.
	trifloxystrobin (Flint 50 WDG)	11	3 to 4 oz/A	180	12 hr	Make no more than one application before alternating with fungicides that have a different mode of action. Begin applications preventively when conditions are favorable for disease and continue as needed on a 7- to 14-day interval.

Table 10-16. Relative Importance of Alternative Management Practices for Disease Control in Asparagus

Scale E = excellent; G = good; F = fair; P = poor; NC = no control; ND = no data.

Strategy	Rust	Cercospora blight	Stemphylium blight	Fusarium root rot	Phytophthora crown/ spear rot
Avoid overhead irrigation	F	F	F	NC	NC
Crop rotation (5 years or more)	NC	NC	NC	F	P
Clip and bury infected ferns	G	G	G	NC	NC
Destroy infected ferns	E	E	E	NC	NC
Encourage air movement/wider row spacing	P	P	G	NC	NC
Plant in well-drained soil	NC	NC	NC	F	F
Destroy volunteer asparagus	F	NC	NC	NC	NC
Pathogen-free planting material	NC	NC	NC	E	E
Resistant/tolerant cultivars	G	G	NC	G	NC

Basil

L. Quesada-Ocampo, Plant Pathology

Table 10-17. Disease Control Products for Basil

Disease	Material	FRAC	Rate of Material	Minimum Days Harvest	Minimum Days Reentry	Method, Schedule, and Remarks
Damping off (*Pythium*)	mefenoxam (Ridomil Gold SL)	4	1.0 to 2.0 pt/acre	21	2	Limit of 2 soil applications per season.
Leaf spots, fungal (*Botrytis, Alternaria, Fusarium*), powdery mildew	cyprodinil + fludioxonil (Switch) 62.5WG	7	11 to 14 oz/acre	7	0.5	Limit of 56 fl oz per acre per season. Make no more than two consecutive applications before rotating to another effective fungicide with a different mode of action.
	fluopyram (Luna Privilege)	7	4.0 to 6.84 fl oz/acre	3	0.5	Limit of 13.7 fl oz per acre per season. Apply as needed on a 7- to 10-day interval. When disease pressure is severe, use the higher rates and/or shorter intervals.
	fluopyram + trifloxystrobin (Luna Sensation)	7 + 11	5.0 to 7.6 fl oz/acre	7	0.5	Limit of 15.3 fl oz per acre per season. Apply as needed on a 7- to 10-day interval. When disease pressure is severe, use the higher rates and/or shorter intervals.
Downy mildew (*Peronospora belbahrii*)	fluopicolide (Adorn)	43	4 fl oz/acre	1	12 hr	Limit of 12 fl oz per acre per year. Tank mix with a fungicide with a different mode of action.
	fluopicolide (Presidio)	43	4 fl oz/acre	1	12 hr	Limit of 12 fl oz per acre per year. Make no more than 2 sequential applications. Alternate with a fungicide with a different mode of action.
Downy mildew (*Peronospora belbahrii*) (continued)	cyazofamid (Ranman 400SC)	21	2.75 to 3 fl oz/acre	0	0.5	Limit of 27 fl oz per acre per season. Alternate with a fungicide with a different mode of action. May be applied through sprinkler irrigation system.
	mandipropamid (Revus)	40	8 fl oz/acre	1	4 hr	Limit of 32 fl oz per acre per season. Make no more than 2 consecutive applications before rotating to another effective fungicide with a different mode of action.
	phosphorous acid (Confine Extra, K-Phite)	33	1 to 3 qt/20 to 100 gal water/acre	0	4 hr	Do not apply at less than 3-day intervals.
	potassium phosphite (Fosphite, Fungi-phite, Helena Prophyt)	33	1 to 3 qt/100 gal water/acre	0	4 hr	Do not apply at less than 3-day intervals.
Fusarium wilt and *Pythium* and *Rhizoctonia* root rots	phosphorous acid (Confine Extra, K-Phite)	33	1 to 3 qt/20 to 100 gal water/acre	0	4 hr	Do not apply at less than 3-day intervals.
	potassium phosphite (Fosphite, Fungi-phite, Helena Prophyt)	33	1 to 3 qt/100 gal water/acre	0	4 hr	Do not apply at less than 3-day intervals.

Bean

E. Sikora, Plant Pathologist, Auburn University

Table 10-18. Disease Control Products for Bean

Disease	Material	FRAC	Rate of Material	Minimum Days Harvest	Minimum Days Reentry	Method, Schedule, and Remarks
Bean, Snap						
Anthracnose, Botrytis, (*Sclerotinia*)	azoxystrobin (various)	11	6.2 to 15.4 fl oz	0	4 hr	For anthracnose only. Do not apply more than 3 sequential applications.
	boscalid (Endura 70 WG)	7	8 to 11 oz	7	0.5	Many other dried and succulent beans on label.
	chlorothalonil (various)	M	2.7 lb/acre	7	2	Spray first appearance, 11 lb limit per acre per crop, 7-day intervals. **Not for Sclerotinia control.**
	thiophanate-methyl (various)	1	1 to 2 lb/acre	14	1	Spray at 25% bloom; repeat at full bloom. Do not exceed 4 lb product per season.
	fluazinam (Omega 500)	29	0.5 to 0.85 pts	14	3	Apply at 10 to 30% bloom.
	fluxapyroxad + pyraclostrobin (Priaxor)	7 + 11	4.0 to 8.0 fl oz b	7	12 hr	Begin prior to disease development and continue on a 7- to 14-day spray schedule.
Asochyta blight, Botrytis gray mold, white mold	boscalid (Endura 70 WG)	7	8 to 11 oz	7	0.5	
	picoxystrobin (Approach)	7	14 to 30 fl oz/A	0	12 hr	Begin sprays prior to disease development.
Alternaria, Anthracnose, Ascochyta, rust, southern blight, web blight	azoxystrobin + propiconazole (Quilt Xcel; Aframe Plus)	11 + 3	10.5 to 14 oz/A	7	0.5	Apply when conditions are conducive for disease. Up to three applications may be made on 7- to 14-day intervals.
Botrytis gray mold, white mold (*Sclerotinia*)	fludioxonil (various)	12	7 oz	7	0.5	Begin before disease develops and continue on 7-day intervals until conditions no longer favor disease development. Do not apply more than 28 oz per acre. Do not use on cowpeas.
Bacterial blights	fixed copper (various)	M	See labels	1	1	Spray first appearance, 10-day intervals.
Powdery mildew	sulfur (various)	M	See labels	0	1	Spray at first appearance, 10 to 14 day intervals. Avoid days over 90°F.
Cottony leak (*Pythium* spp.)	fenamidone (Reason 500 SC)		5.5 to 8.2 fl oz	3	0.5	Begin applications when conditions become favorable for disease development. Do not make more than one application before alternating to a product with a different mode of action.

Table 10-18. Disease Control Products for Bean

Disease	Material	FRAC	Rate of Material	Minimum Days Harvest	Minimum Days Reentry	Method, Schedule, and Remarks
Bean, Snap (continued)						
Cottony leak, downy mildew, Phytophthora blight	cyazofamid (Ranman)	21	2.75 fl oz	0	0.5	Read label for specific directions for each disease as well as use restrictions.
Rhizoctonia root rot	azoxystrobin (various)	11	0.4 to 0.8 fl oz/1,000 row feet	—	4 hr	Make in-furrow or banded applications shortly after plant emergence.
	myclobutanil (various)	3	4 to 5 oz/acre	0	1	For Rhizoctonia only.
	dichloropropene (Telone C-17) (Telone C-35)	—	10.8 to 17.1 gal/acre 13 to 20.5 gal/acre	—	5	Rate is based on soil type; see label for in-row rates.
	metam-sodium (Vapam 42 HL)	—	37.5 to 75 gal/trt acre	—	—	Rate is based on soil properties and depth of soil to be treated; apply 14 to 21 days before planting.
Rhizoctonia and Fusarium seed rot and damping off	prothioconazole (Redigo 480)	3	0.16 to 0.32 fl oz/100 lbs seed	—	—	For seed rot and damping off caused by Rhizoctonia and Fusarium.
	penflufen + trifloxystrobin (Evergol Xtend)	7 + 11	See label	—	—	For seed rot and damping off caused by Rhizoctonia, Fusarium, Phomopsis, or Botrytis. Seed treatment only.
Rust (Uromyces)	azoxystrobin (various	11	0.4 to 0.8 fl oz/1,000 row feet	—	4 hr	Do not apply more than 3 sequential applications.
	boscalid (Endura 70 WG)	7	8 to 11 oz	7	0.5	Many other dried and succulent beans on label.
	chlorothalonil (various)	M	1.25 to 2.7 lb/acre	7	2	Spray first appearance, 11 lb limit per acre per crop, 7-day intervals.
	myclobutanil (various)	3	4 to 5 oz/acre	0	1	Spray at first appearance.
	pyraclostrobin (various)	11	6 to 9 fl oz		12 hr	Make no more than 2 sequential applications.
	sulfur (various)	M	See label	0	1	Spray at 7 to 10 day intervals.
	tebuconazole (various	3	4 to 6 fl oz/acre	7	0.5	Apply before disease appears when conditions favor rust development and repeat at 14-day intervals; maximum 24 fl oz per season.
White mold (Sclerotinia)	dicloran (Botran 75 W)	14	2.5 to 4 lb/acre	2	0.5	Use low rate for bush varieties and high rate for pole varieties.
Bean, Lima						
Alternaria, Anthracnose, Ascochyta, bean rust, southern blight, web blight (Rhizoctonia)	Azoxystrobin + propiconazole (Quilt Xcel, Aframe Plus)	11 + 3	10.5 to 14 fl oz	7	0.5	Apply when conditions are conducive for disease. Up to 3 applications may be made on a 7- to 14-day interval.
Botrytis, Sclerotinia, leaf spots	azoxystrobin (various)	11	6.2 to 15.4 fl oz/acre	0	4 hr	Leaf spots only; do not make more than 3 sequential applications.
	thiophanate-methyl (various)	1	1.5 to 2 lb/acre	14	1	4 lb limit per acre per crop.
	iprodione (various)	2	1.5 to 2 lb/acre	0	1	DO NOT apply to cowpea. Snap or succulent bean hay must not be fed to livestock. Read label for all restrictions.
	fluazinam (Omega 500)	29	0.5 to 0.85 pts	30	3	Apply at 10 to 30% bloom.
	penthiopyrad (Fontelis)	7	14 to 30 fl oz/acre	0	12 hr	Begin sprays prior to disease development.
	pyraclostrobin (various)	11	6.0 to 9.0 fl oz	21	12 hr	Make no more than 2 sequential applications.
	fluxapyroxad + pyraclostrobin (Priaxor)	7 + 11	4.0 to 8.0 fl oz	21	12 hr	Begin prior to disease development and continue on a 7- to 14-day spray schedule.
Botrytis gray mold, white mold (Sclerotinia)	fludioxonil (various)	12	7 oz	7	0.5	Begin before disease develops and continue on 7-day intervals until conditions no longer favor disease development. Do not apply more than 28 oz per acre. Do not use on cowpeas.
Cottony leak, downy mildew, Phytophthora blight	Cyazofamid (Ranman)	21	2.75 fl oz	0	0.5	Read label for specific directions for each disease as well as use restrictions.
Damping off, Pythium, Rhizoctonia	azoxystrobin (various)	11	0.4 to 0.8 fl oz/1,000 row feet	—	4 hr	Rhizoctonia only. Make in-furrow or banded applications shortly after plant emergence.
	mefenoxam (various)	4	0.5 to 2 pt/trt acre	—	2	For Pythium only. Soil incorporate. See label for row rates. Use proportionally less for band rates.
	azoxystrobin + mefenoxam (Uniform)	11 + 4	0.34 fl oz/1,000 row ft	—	—	Limit of one application per season. In-furrow spray. See label directions.
Rhizoctonia, Fusarium, Phomopsis, Botrytis	penflufen + trifloxystrobin (Evergol Xtend)	7 + 11	See label	—	—	For seed rot and damping off caused by Rhizoctonia, Fusarium, Phomopsis, or Botrytis. Seed treatment only.

Table 10-19. Efficacy of Products for Foliar Disease Control in Beans

Scale: E = excellent; G = good; F = fair; P = poor; NC = no control; ND = no data.

Product[1]	Fungicide group[F]	Preharvest interval (Days)	Aerial Rhizoctonia	Anthracnose	Brown Spot (Pseudomonas)	Cercospora	Common Bacterial Blight	Common Rust	Downy Mildew	Gray Mold (Botrytis)	Halo Blight	Powdery Mildew	Pythium Cottony Leak	Pythium Damping off	Rhizoctonia Sore Shin	Sclerotinia Blight	Southern Blight (S. rolfsii)
azoxystrobin (various)	11	14	G	G	NC	G	NC	E	ND	P	NC	P	F	NC	G	NC	E
azoxystrobin + mefenoxam (Uniform)	11 + 4	—	F	F	NC	G	NC	ND	ND	NC	NC	NC	P	G	G	NC	NC
boscalid (Endura)	7	7 to 21	ND	ND	NC	ND	NC	ND	NC	G	NC	ND	NC	NC	ND	E	P
penthiopyrad (Fontelis)	7	0	ND	ND	NC	ND	NC	ND	NC	G	NC	ND	NC	NC	ND	E	F
dicloran (Botran)	14	—	NC	NC	NC	NC	NC	NC	NC	F	NC	NC	NC	NC	NC	F	NC
fluazinam (Omega 500)	29	30	ND	ND	NC	NC	NC	ND	NC	G	NC	NC	NC	NC	ND	G	F
chlorothalonil (various)	M	7	P	F	NC	G	NC	G	F	NC	NC	P	NC	NC	NC	NC	NC
cyprodonil + fludioxonil (Switch)	9 + 12	7	ND	ND	NC	ND	NC	ND	NC	G	NC	ND	NC	NC	ND	E	P
fixed copper (various)	M	0	NC	P	F	P	F	P	F	P	F	P	NC	NC	NC	NC	NC
iprodione (Rovral)	2	—	P	NC	NC	NC	NC	NC	NC	G	NC	NC	NC	NC	F	G	NC
mefenoxam (Ridomil)	4	—	NC	NC	NC	NC	NC	NC	G	NC	NC	NC	F[R]	G[R]	NC	NC	NC
pyraclostrobin (various)	11	7 to 21	G	G	NC	G	NC	E	ND	P	NC	P	F	NC	F	NC	F
fluxapyroxad + pyraclostrobin (Priaxor)	7 + 11	7 to 21	G	G	NC	G	NC	E	ND	G	NC	P	F	NC	F	E	F
sulfur (various)	M	0	NC	F	NC	F	NC	F	P	P	NC	F	NC	NC	NC	NC	NC
tebuconazole (various)	3	7	NC	NC	NC	F	NC	G	NC	F	NC		NC	NC	P	NC	G
thiophanate-methyl (Topsin M)	1	14 to 28	P	F	NC	G	NC	ND	NC	NC	NC	ND	NC	NC	P	F	NC

[1] Efficacy ratings do not necessarily indicate a labeled use for every disease.

[F] To prevent resistance in pathogens, alternate fungicides within a group with fungicides in another group. Fungicides in the "M" group are generally considered "low risk" with no signs of resistance developing to the majority of fungicides.

[R] Resistance reported in the pathogen.

Table 10-20. Importance of Alternative Management Practices for Disease Control in Beans

Scale: E = excellent; G = good; F = fair; P = poor; NC = no control; ND = no data

Strategy	Anthracnose	Ashy stem blight	Botrytis gray mold	Cercospora	Common bacterial blight and halo blight	Fusarium root rot	Mosaic viruses	Powdery mildew	Pythium damping off	Rhizoctonia root rot	Root knot	Rust (more on pole beans)	Southern blight (*Sclerotium rolfsii*)	White mold (*Sclerotinia*)
Avoid field operations when leaves are wet	E	NC	E	F	E	NC	NC	NC	NC	NC	NC	E	NC	NC
Avoid overhead irrigation	E	NC	E	E	E	NC	NC	NC	P	NC	NC	E	NC	G
Change planting date	F	F	NC	P	F	G	F	P	E	E	P	G (early)	NC	NC
Cover cropping with antagonist	NC	ND	NC	NC	NC	NC	NC	NC	NC	NC	G	NC	NC	NC
Crop rotation	G	P	F	F	G	F	P	P	F	F	G	NC	F	E
Deep plowing	E	F	E	P	E	F	NC	NC	F	F	E	NC	E	E
Destroy crop residue	E	F	E	F	E	NC	NC	NC	P	P	F	F	G	E
Encourage air movement	E	NC	E	F	E	NC	NC	E	P	NC	NC	F	NC	G
Increase between-plant spacing	P	NC	P	F	P	P	P	P	F	F	NC	P	F	G
Increase soil organic matter	NC	F	NC	NC	NC	F	NC	NC	NC	NC	F	NC	NC	NC
Insecticidal oils	NC	NC	NC	NC	NC	NC	F	NC	NC	NC	NC	NC	NC	NC
pH management	NC	NC	NC	NC	NC	F	NC	NC	NC	NC	NC	NC	NC	NC
Plant in well-drained soil	F	F	F	NC	F	E	NC	NC	E	E	NC	NC	P	F
Plant on raised beds	F	P	F	NC	F	E	NC	NC	E	E	NC	NC	P	F
Plastic mulch bed covers	NC	NC	NC	NC	NC	NC	NC	NC	NC	NC	NC	NC	NC	F
Postharvest temperature control	NC	NC	NC	NC	NC	NC	NC	NC	NC	NC	NC	NC	NC	E
Reflective mulch	NC	NC	NC	NC	NC	NC	G	NC	NC	NC	NC	NC	NC	P
Reduce mechanical injury	NC	NC	NC	NC	F	P	NC	NC	NC	NC	NC	NC	P	NC
Rogue diseased plants	NC	NC	P	NC	NC	NC	F	NC	NC	NC	NC	NC	P	F
Row covers	NC	NC	NC	NC	NC	NC	F	NC	NC	NC	NC	NC	NC	NC
Soil solarization	NC	NC	P	NC	NC	F	NC	NC	F	G	F	NC	F	G
Pathogen-free planting material	E	G	NC	F	E	NC	G	NC	NC	NC	NC	NC	NC	NC
Resistant cultivars	E	G	NC	E	E	G	E	E	NC	NC	NC	E	NC	F
Weed control	F	NC	F	NC	F	F	E	F	NC	NC	F	F	P	F

Brassicas (Broccoli, Brussel Sprout, Cabbage, Cauliflower)

A. Keinath, Plant Pathology, Clemson University

Table 10-21. Disease Control Products for Broccoli, Brussel Sprout, Cabbage, and Cauliflower

Disease	Material	FRAC	Rate of Material	Minimum Days Harvest	Minimum Days Reentry	Method, Schedule, and Remarks
Alternaria leaf spot	azoxystrobin + difenoconazole (Quadris Top 2.72 SC)	11 + 3	14 fl oz/acre	1	0.5	Apply prior to disease, but when conditions are favorable, on 7- to 14-day schedule. Alternate to a non-QoI fungicide after 1 application. No more than 4 applications per season.
	boscalid (Endura 70 EG)	7	6 to 9 oz/acre	0	0.5	Begin applications prior to disease development, and continue on a 7- to 14-day interval. Make no more than 2 applications per season.
	cyprodinil + difenoconazole (Inspire Super 2.82 SC)	9 + 3	16 to 20 fl oz/acre	7	0.5	Begin applications prior to disease development, and continue on a 7- to 10-day interval. Make no more than 2 sequential applications before rotating to another effective fungicide with a different mode of action. Do not exceed 80 fl oz per season.
	cyprodinil + fludioxonil (Switch 62.5WG)	9 + 12	11 to 14 oz/acre	7	0.5	Apply when disease first appears, and continue on 7- to 10-day interval. Do not exceed 56 oz of product per acre per year.
	fluxapyroxad + pyraclostrobin (Priaxor 500 SC)	7 + 11	6.0 to 8.2 fl oz/ acre	3	0.5	Make no more than 2 sequential applications before alternating with fungicides that have a different mode of action. Maximum of 3 applications. **Do not apply to turnip greens or roots.**
	triflumizole (Procure 480SC)	3	6 to 8 oz/acre	1	0.5	Apply when disease first appears and continue on 14-day interval. Do not exceed 18 fl oz per season.
Alternaria leaf spot, gray mold	penthiopyrad (Fontelis 1.67 SC)	7	14 to 30 fl oz/acre	0	0.5	Do not exceed 72 fl oz of product per year. Make no more than 2 sequential applications per season before rotating to another effective product with a different mode of action.
Black leg	iprodione (Rovral 4F)	2	2 lb/acre 2 pt/acre	0	—	Apply to base of plant at 2- to 4-leaf stage. A second application may be made up to the harvest date. Do not use as a soil drench. **For broccoli only.**
	fluxapyroxad + pyraclostrobin (Priaxor 500 SC)	7 + 11	6.0 to 8.2 fl oz/ acre	3	0.5	Make no more than two sequential applications before alternating with fungicides that have a different mode of action. Maximum of 3 applications. **Do not apply to turnip greens or roots.**
Black rot, downy mildew	acibenzolar-S-methyl (Actigard 50WG)	P1	0.5 to 1 oz/acre	7	0.5	Begin applications 7 to 10 days after thinning, not to exceed 4 applications per season.
	fixed copper (various)	M1	See label	0	1 to 2	Apply on 7- to 10-day intervals after transplanting or shortly after seeds have emerged. Some reddening on older broccoli leaves and flecking of cabbage wrapper leaves may occur. Check label carefully for recommended rates for each disease on each crop.
Clubroot	cyazofamid (Ranman) 34.5 SC	21	Transplant: 12.9 to 25.75 fl oz/100 gal water Banded: 20 fl oz/acre	0.5	0	Either apply immediately after transplanting with 1.7 fl oz of solution per transplant, or as a banded application with soil incorporation of 6 to 8 inches prior to transplanting. Do not apply more than 39.5 fl oz per acre per season; or 6 (1 soil + 5 foliar) applications per season. Do not make more than 3 consecutive applications without rotating to another fungicide with a different mode of action for 3 subsequent applications.
	fluazinam (Omega 500F)	29	Transplant: 6.45 fl oz/100 gal water Banded: 2.6 pt/acre	50	50	Either apply directly as a drench to transplants or as a banded application with soil incorporation of 6 to 8 inches prior to transplanting. Use of product can delay harvest and cause some stunting without adverse effects on final yields.
Downy mildew	amectoctradin + dimethomorph (Zampro 525 SC)	40 + 45	14 fl oz/acre	0	0.5	Do not make more than 2 sequential applications before alternating to a fungicide with a different mode of action. Addition of an adjuvant may improve performance (see label for specifics).
	cyazofamid (Ranman 400 SC)	21	2.75 fl oz/acre	0	0.5	Begin applications on a 7- to 10-day schedule when disease first appears or weather is conducive. Do not apply more than 39.5 fl oz per acre per season; or 6 (1 soil + 5 foliar) applications per season. Do not make more than 3 consecutive applications without rotating to another fungicide with a different mode of action for 3 subsequent applications.
	fluopicolide (Presidio 4 SC)	43	3 to 4 fl oz/acre	2	0.5	Must be tank mixed with another fungicide with a different mode of action. No more than 2 sequential applications before rotating to another effective product of a different mode of action. Limited to 4 applications, 12 fl oz per acre per season.
	fosetyl-AL (Aliette 80WDG)		2 to 5 lb/acre	3	1	Apply when disease first appears; then repeat on 7- to 21-day intervals. Do not tank mix with copper fungicides. A maximum of 7 applications can be made per season. Also for loose-heading Chinese cabbage, kale, kohlrabi, and greens (collard, mustard, and rape).
	mandipropamid (Revus 2.08 SC)		8 fl oz/acre	0.13 lb/acre	1	0.5	Apply prior to disease development and continue throughout season at 7- to 10-day intervals; maximum 32 fl oz per season.

Table 10-21. Disease Control Products for Broccoli, Brussel Sprout, Cabbage, and Cauliflower

Disease	Material	FRAC	Rate of Material	Minimum Days Harvest	Minimum Days Reentry	Method, Schedule, and Remarks
Downy mildew (continued)	oxathiapiprolin + mandipropamid (Orondis Ultra A + Orondis Ultra B)	49 + 40	2.0 to 4.8 fl oz/acre + 8 fl oz/acre	0	4 hr	**Must tank mix Orondis Ultra and Ultra B before ap- plication.** Apply prior to disease development at 10-day intervals. Make no more than 2 sequential applications before alternating with fungicides that have a different mode of action. Maximum of 4 applications per crop per year of all Orondis products.
	potassium phosphite (various)	33	2 to 4 pt/acre	0	4 hr	Apply when weather is foggy as a preventative. Do not apply to plants under water or temperature stress. Spray solution should have a pH greater than 5.5. Apply in at least 30 gallons water per acre.
Downy mildew, Alternaria leaf spot	azoxystrobin (Quadris 2.08 F)	11	6.2 to 15.5 fl oz/acre	0	4 hr	Do not make more than 2 applications before alternating to a fungicide with a different mode of action. Do not apply more than 92.3 fl oz per acre per season
	chlorothalonil (various)	M5	See label	7	2	Apply after transplanting, seedling emergence, or when conditions favor disease development. Repeat as needed on a 7- to 10-day interval.
	cyprodinil + difenoconazole (Inspire Super 2.82 SC)	9 + 3	16 to 20 fl oz/acre	7	0.5	Begin applications prior to disease development, and continue on a 7- to 10-day interval. Make no more than 2 sequential applications before rotating to another effective fungicide with a different mode of action. Do not exceed 80 fl oz per season.
	fenamidone (Reason 500SC)	11	5.5 to 8.2 fl oz/acre	2	0.5	Begin applications on a 5- to 10-day schedule when disease first appears or weather is conducive. Do not apply more than 24.6 fl oz per acre per season. Do not make more than 1 application without rotating to another fungicide with a different mode of action.
	mancozeb (various)	M3	1.6 to 2.1 lb/acre	10	1	Spray at first appearance of disease and continue on a 7- to 10-day interval. No more than 12.8 lb per acre per season.
	mefenoxam + chlorothalonil (Ridomil Gold/Bravo)	4 + M5	1.5 lb/acre	7	2	Begin applications when conditions favor disease but prior to symptoms. Under severe disease pressure use additional fungicides between 14-day intervals. Do not make more than 4 applications per crop.
	oxathiapiprolin + mandipropamid (Orondis Ultra A + Orondis Ultra B)	49 + M5	2.0 to 4.8 fl oz/acre + 1.5 pt/acre	7	0.5	**Must tank mix Orondis Ultra and Ultra B before ap- plication.** Apply prior to disease development at 10-day intervals. Make no more than 2 sequential applications before alternating with fungicides that have a different mode of action. Maximum of 8 applications at the low rate or 4 applications at the high rate per crop per year of Orondis Opti; if Ultra and Opti are both used, then the maximum is 4 applications.
Powdery mildew	azoxystrobin + difenoconazole (Quadris Top 2.72 SC)	11 + 3	14 fl oz/acre	1	0.5	Apply prior to disease, but when conditions are favorable, on 7- to 14-day schedule. Alternate to a non-QoI fungicide after 1 application. No more than 4 applications per season.
	boscalid (Endura 70 EG)	7	6 to 9 oz/acre	0	0.5	Begin applications prior to disease development, and continue on a 7- to 14-day interval. Make no more than 2 applications per season; disease suppression only.
	cyprodinil + difenoconazole (Inspire Super 2.82 SC)	9 + 3	16 to 20 fl oz/acre	7	0.5	Begin applications prior to disease development and continue on a 7- to 10-day interval. Make no more than 2 sequential applications before rotating to another effective fungicide with a different mode of action. Do not exceed 80 fl oz per season.
	cyprodinil + fludioxonil (Switch 62.5WG)	9 + 12	10 to 12 oz/acre	7	0.5	Apply when disease first appears, and continue on 7- to 10-day intervals. Do not exceed 56 oz of product per acre per year.
	fluxapyroxad + pyraclostrobin (Priaxor 500 SC)	7 + 11	6.0 to 8.2 fl oz/acre	3	0.5	Make no more than 2 sequential applications before alternating with fungicides that have a different mode of action. Maximum of 3 applications. Do not apply to turnip greens or roots.
	penthiopyrad (Fontelis 1.67 SC)	7	14 to 30 fl oz/acre	0	0.5	Do not exceed 72 fl oz of product per year. Make no more than 2 sequential applications per season before rotating to another effective product with a different mode of action.
	sulfur (various)	M2	See label	0	1	Apply when disease first appears; then repeat as needed on 14-day interval. Avoid applying on days over 90 degrees F. Also for use on greens (collard, kale, and mustard), rutabaga, and turnip.
	triflumizole (Procure 480 SC)	3	6 to 8 fl oz/acre	1	0.5	Apply when disease first appears and continue on 14-day interval. Do not exceed 18 fl oz per season.
Pythium damping off, Phytophthora basal stem rot	fluopicolide (Presidio 4 F)	43	3 to 4 fl oz/acre	2	0.5	Apply as a soil drench at transplant. As plants enlarge, apply directly to soil by chemigation on a 7- to 10-day schedule as conditions favor disease, but prior to disease development. No more than 2 sequential applications before rotating to another effective product of a different mode of action. Limited to 4 applications, 12 fl oz per acre per season,
	mefenoxam (Ridomil Gold 4 SL)	4	0.25 to 2 pt/acre	—	2	Apply 1 to 2 pt per acre as a broadcast, preplant application to soil and incorporate in top 2 in. of soil. For Pythium control, use only 0.25 to 0.5 pt per acre.

Table 10-21. Disease Control Products for Broccoli, Brussel Sprout, Cabbage, and Cauliflower

Disease	Material	FRAC	Rate of Material	Minimum Days Harvest	Minimum Days Reentry	Method, Schedule, and Remarks
Pythium damping off, Phytophthora basal stem rot (continued)	metalaxyl (MetaStar 2 E AG)	4	4 to 8 pt/ trt acre	—	2	Preplant incorporated or surface application.
Rhizoctonia bottom rot	boscalid (Endura 70 WP)	7	6 to 9 oz/acre	0	0.5	Begin applications prior to disease development, and continue on a 7- to 14-day interval. Make no more than 2 applications per season; disease suppression only.
Rhizoctonia stem (wire-stem) and root rot	azoxystrobin (Quadris 2.08 SC)	11	5.8 to 8.7 fl oz/acre on 36 in. rows	0	4 hr	Rate is equivalent to 0.4 to 0.6 fl oz per 1000 row feet. Apply at planting as a directed spray to the furrow in a band 7 inches wide. See label for other row spacings.
	boscalid (Endura 70 EG)	7	6 to 9 oz/acre	0	0.5	Begin applications prior to disease development, and continue on a 7- to 14-day interval. Make no more than 2 applications per season.
	penthiopyrad (Fontelis 1.67 SC)	7	16 to 30 fl oz/acre	0	0.5	Do not exceed 72 fl oz of product per year. Make no more than 2 sequential applications per season before rotating to another effective product with a different mode of action.
	Coniothyrium minitans (Contans WG)	—	1 to 4 lb/acre	0	4 hr	**OMRI listed product.** Apply to soil surface and incorporate no deeper than 2 inches. Works best when applied prior to planting or transplanting. Do not apply other fungicides for 3 weeks after applying Contans.
Sclerotinia stem rot (white mold)	boscalid (Endura 70 EG)	7	6 to 9 oz/acre	0	0.5	Begin applications prior to disease development, and continue on a 7- to 14-day interval. Make no more than 2 applications per season.
	penthiopyrad (Fontelis 1.67 SC)	7	16 to 30 fl oz/acre	0	0.5	Do not exceed 72 fl oz of product per year. Make no more than 2 sequential applications per season before rotating to another effective product with a different mode of action.
	Coniothyrium minitans (Contans WG)	—	1 to 4 lb/acre	0	4 hr	**OMRI listed product.** Apply to soil surface and incorporate no deeper than 2 inches. Works best when applied prior to planting or transplanting. Do not apply other fungicides for 3 weeks after applying Contans.

Table 10-22. Efficacy of Products for Disease Control in Brassicas

Scale: E = excellent; G = good; F = fair; P = poor; NC = no control; ND = no data.

Product[1,2]	Crop Group[2]	Fungicide group[F]	Preharvest interval (Days)	Alternaria Leaf Spot	Bacterial Soft Rot	Black Rot	Black Leg	Bottom Rot (*Rhizoctonia*)	Cercospora & Cercosporella	Clubroot	Downy Mildew	Powdery Mildew	Pythium damping-off	Sclerotinia/Raisin Head	Wirestem (*Rhizoctonia*)
acibenzolar-S-methyl (Actigard)	H&S	21	7	NC	ND	F	NC	NC	NC	NC	G	P	ND	ND	NC
ametoctradin + dimethomorph (Zampro)	B	45 + 40	0	NC	NC	NC	NC	NC	NC	NC	E	NC	NC	NC	NC
azoxystrobin (Quadris and others)	B	11	0	E	NC	NC	F	ND	F	NC	G	F	NC	NC	F
azoxystrobin + difenoconazole (Quadris Top)	B	11 + 3	1	E	NC	NC	ND	ND	G	NC	G	F	NC	NC	F
boscalid (Endura)[3]	B	7	0 to 14	G	NC	NC	NC	NC	NC	NC	P	P	NC	F	F
chlorothalonil (Bravo, Echo, Equus, and others)	H&S	M	7	F	NC	NC	NC	P	F	NC	F	F	NC	NC	NC
fixed copper[4]	B	M	0	P	NC	P	NC	NC	P	NC	F	F	NC	NC	NC
cyazofamid (Ranman)	B	21	0	NC	NC	NC	NC	NC	NC	NC	G	NC	NC	NC	NC
cyprodinil + fludioxonil (Switch)	B	9 + 12	7	F	NC	NC	NC	NC	F	NC	NC	F	NC	NC	NC
difenoconazole + cyprodinil (Inspire Super)	B	3 + 9	7	G	NC	NC	ND	NC	G	NC	NC	F	NC	P	NC
dimethomorph (Forum)	B	40	0	NC	NC	NC	NC	NC	NC	NC	G	NC	NC	NC	NC
fenamidone (Reason)	B	11	2	F	NC	NC	NC	NC	F	NC	E	NC	NC	NC	NC
fluopicolide (Presidio)	B	43	2	NC	NC	NC	NC	NC	NC	NC	E	NC	NC	NC	NC
fluazinam (Omega 500)[6]	B	29	20 to 50	NC	NC	NC	NC	NC	NC	F	NC	NC	NC	NC	G
fluxapyroxad + pyraclostrobin (Priaxor)	B	7 + 11	3	G	NC	ND	G	ND	G	NC	F	F	NC	ND	NC
fosteyl-Al[4] (Aliette)		33	3	NC	NC	NC	NC	NC	NC	NC	F	NC	NC	NC	NC
iprodione (Rovral)[5]	H&S	2	—	NC	NC	NC	F	NC	NC	NC	NC	NC	NC	P	P
mancozeb (Manzate, Penncozeb, Dithane)	H&S	M	7	F	NC	NC	NC	NC	F	NC	F	P	NC	NC	NC
mandipropamid (Revus)	B	40	1	NC	NC	NC	NC	NC	NC	NC	E	NC	NC	NC	NC
mefenoxam (Ridomil Gold EC) pre-plant	B	4	—	NC	NC	NC	NC	NC	NC	NC	F	NC	F[R]	NC	NC
mefenoxam + chlorothalonil (Ridomil Gold Bravo)	H&S	4 + M	7	F	NC	NC	NC	P	F	NC	F	F	NC[R]	NC	NC
oxathiapiprolin (Orondis)	H&S	49	7	NC	NC	NC	NC	NC	NC	NC	E	NC	ND	NC	NC
penthiopyrad (Fontelis)	B	7	0	E	NC	NC	ND	NC	ND	NC	NC	G	NC	G	NC
potassium phosphite (various)	B	33	0	NC	NC	NC	NC	NC	NC	NC	G	NC	NC	NC	NC
pyraclostrobin (Cabrio)[3]	B	11	0 to 3	E	NC	NC	ND	NC	E	NC	F	F	NC	NC	P
sulfur (various)	B	M	0	P	NC	NC	NC	NC	P	NC	P	F	NC	NC	NC
tebuconazole (Folicur, Tebuzol, Tegrol)	B	3	7	F	NC	NC	ND	NC	F	NC	NC	ND	NC	NC	NC
triflumizole (Procure)	B	3	1	NC	NC	NC	NC	NC	NC	NC	G	NC	NC	NC	NC

[1] Efficacy ratings do not necessarily indicate a labeled use for every disease.

[2] H&S = fungicides registered only on head and stem brassicas (broccoli, Brussel sprout, cabbage, and cauliflower,). B = fungicides registered on all brassica crops except turnip greens and root turnips; see Tables 10-33 and 10-46 for products registered on turnips. Always refer to product labels prior to use.

[3] Shorter PHI is for head and stem brassicas (broccoli, Brussel sprout, cabbage and cauliflower) and longer PHI is for leafy brassica greens.

[4] Phytotoxicity is seen when fosteyl-Al is tank-mixed with copper. When used in combination with fosteyl-Al or maneb.

[5] Applications of iprodione made for black leg may suppress *Alternaria*, *Sclerotinia*, and wirestem on broccoli.

[6] Use a 20-day PHI for Omega 500 on leafy greens and a 50-day PHI for stem brassicas.

[F] To prevent resistance in pathogens, alternate fungicides within a group with fungicides in another group. Fungicides in the "M" group are generally considered "low risk" with no signs of resistance developing.

[R] Resistance reported in the pathogen.

Table 10-23. Importance of Alternative Management Practices for Disease Control in Brassicas

Scale: E = excellent; G = good; F = fair; P = poor; NC = no control; ND = no data

Strategy	Alternaria leaf spot	Bacterial soft rot	Black rot	Black leg	Bottom rot (*Rhizoctonia*)	Cercospora	Clubroot	Downy mildew	Powdery mildew	Pythium	Sclerotinia head	Wirestem (*Rhizoctonia*)
Avoid field operations when leaves are wet	P	F	G	F	F	P	NC	P	NC	NC	NC	NC
Avoid overhead irrigation	E	E	E	E	F	E	NC	G	P	NC	NC	NC
Change planting date	P	P	NC	NC	P	NC	NC	NC	NC	P	NC	F
Cover cropping with antagonist	NC	NC	NC	NC	NC	NC	P	NC	NC	P	NC	NC
Crop rotation	F	F	G	G	P	F	NC	F	NC	NC	P	P
Deep plowing	F	F	G	G	F	F	NC	F	NC	NC	F	F
Destroy crop residue	F	F	G	G	F	F	NC	F	NC	NC	P	P
Encourage air movement	F	P	P	P	F	F	NC	F	P	NC	F	NC
Increase between-plant spacing	F	P	P	P	F	F	NC	F	NC	P	F	NC
Increase soil organic matter	NC	NC	NC	NC	P	NC	P	NC	NC	NC	NC	P
Hot water seed treatment	P	NC	E	G	NC	NC	NC	NC	NC	NC	NC	NC
pH management	NC	NC	NC	NC	NC	NC	E	NC	NC	NC	NC	NC
Plant in well-drained soil	P	F	P	P	G	P	E	P	NC	F	F	G
Plant on raised beds	NC	F	P	NC	G	NC	E	P	NC	F	F	G
Plastic mulch bed covers	P	NC	NC	NC	F	NC	NC	NC	NC	NC	NC	NC
Postharvest temperature control	NC	E	NC	NC	NC	NC	NC	NC	NC	NC	NC	NC
Reflective mulch	NC	NC	NC	NC	NC	NC	NC	NC	NC	NC	NC	NC
Reduce mechanical injury	NC	E	G	NC	NC	NC	NC	NC	NC	NC	F	P
Rogue diseased plants	P	NC	NC	F	P	NC	NC	NC	NC	NC	NC	NC
Row covers	NC	P	NC	NC	NC	NC	NC	NC	NC	NC	NC	NC
Soil solarization	NC	NC	NC	P	F	NC	NC	NC	NC	P	P	F
Pathogen-free planting material	F	NC	E	E	F	NC	G	NC	NC	NC	P	F
Resistant cultivars	NC	NC	E	NC	NC	NC	P	F	F	NC	NC	P
Weed control	F	NC	F	F	NC	F	F	F	F	NC	F	NC

Cantaloupe — See Cucurbits

Corn, Sweet

A. Keinath, Plant Pathology, Clemson University

Table 10-24. Disease Control Products for Corn, Sweet

Disease	Material	FRAC	Rate of Material	Minimum Days Harvest	Minimum Days Reentry	Method, Schedule, and Remarks
Seedling diseases caused by *Rhizoctonia* and *Penicillium*	pyraclostrobin (Stamina 1.67FC)	11	0.8 to 1.6 fl oz/100 lbs of seed	NA	NA	Seed treatment. Seed treated on farm must be dyed.
Soilborne diseases, *Rhizoctonia* root and stalk rot	fluoxastrobin (Aftershock)	11	0.16 to 0.24 fl oz/ 1000 row feet	7	0.5	May be applied as a banded or in-furrow spray. Consult label for specifics.
	azoxystrobin (various)	11	0.4 to 0.8 fl oz/1000 row feet	7	4 hr	See label for banded or in-furrow sprays. Apply no more than 2.88 qt per crop per acre per season, including soil applications.
Anthracnose, Eyespot, Gray leaf spot (Cercospora leaf spot), Northern corn leaf blight (*Exserohilum* [*Hel- minthosporium*] *turcicum*) Northern corn leaf spot (*Bipolaris zeicola* [*Hel- minthosporium carbonum*]) Southern corn leaf blight (*Bi- polaris* [*Helminthosporium*] *maydis*), Rust, Southern rust	azoxystrobin (various)	11	See labels	7	4 hr	Use lower rate for rust. Make no more than 2 sequential applications before alternating with fungicides that have a different mode of action. Apply no more than 123 fl oz per crop per acre per season. **Not registered for Southern rust.**
	azoxystrobin + propiconazole (Quilt, Quilt XCEL, Avaris)	11 + 3	7 or 10.5 to 14 fl oz/ acre	14	0.5	Use 7 fl oz of Quilt or Avaris for 3 Helminthosporium diseases. Must rotate every application with a non-Group 11 fungicide. Maximum 56 fl oz per acre (4 applications at the high rate) per crop.
	chlorothalonil (various)	M5	See label	14	2	See label. Rates vary depending on the formulation. Spray at first appearance, 4 to 14 day intervals.
	fluoxastrobin (Aftershock)	11	3.8 fl oz/acre	30	0.5	Soil and foliar treatments. Maximum 2 applications per season. Do not apply after early dough stage.
	fluxapyroxad + pyraclostrobin (Priaxor)	7 + 11	4 to 8 fl oz/acre	7	0.5	Do not make more than 2 sequential applications of Priaxor before switching to a fungicide with a different mode of action. May be used with adjuvants (consult label for specifics).
	mancozeb (various)	M3	See label	7	1	Start applications when disease first appears and repeat at 4- to 7-day intervals. **Not registered for anthracnose, eyespot, gray leaf spot, or Southern rust.**
	penthiopyrad (Vertisan)	7	10 to 24 fl oz/acre	7	0.5	No more than 2 sequential applications of the fungicide before switching to a fungicide with another mode of action. **Not registered for anthracnose.**
	propiconazole (various)	3	2 to 4 fl oz/acre	14	0.5	16 fl oz per acre per crop maximum. **Not registered for anthracnose.**
	pyraclostrobin (Headline SC & EC)	11	6 to 12 fl oz/acre	7	0.5	Do not exceed 2 sequential applications or 6 applications of this fungicide or with other group 11 fungicides per crop. **Not registered for eyespot.**
	pyraclostrobin + metconazole (Headline AMP)	11 + 3	10 to 14.4 fl oz/acre	7	0.5	No more than 2 sequential applications before alternating with a different mode of action. Maximum 4 (high rate) or 5 (low rate) applications per crop. **Not registered for eyespot.**
	trifloxystrobin +propiconazole (Stratego)	11 + 3	10 fl oz/acre	14	0.5	Apply Stratego when disease first appears and continue on a 7- to 14-day interval. Alternate applications of Stratego with another product with a different mode of action than Group 11 fungicides. Maximum 3 applications per crop. **Not registered for the 3 Helminthosporium diseases.**
	trifloxystrobin +propiconazole (Stratego YLD)	11 + 3	4 to 5 fl oz/acre	0	0.5	Alternate Stratego YLD sprays with another mode of action than a group 11 fungicide. Maximum 4 (high rate) or 5 (low rate) per crop. **Not registered for the 3 Helminthosporium diseases.**
Brown spot (*Physoderma maydis*)	fluxapyroxad + pyraclostrobin (Priaxor)	7 + 11	4 to 8 fl oz/acre	7	0.5	Do not make more than 2 sequential applications of Priaxor before switching to a fungicide with a different mode of action. Maximum 4 (high rate) or 2 (low rate) applications per crop. Crop damage may occur when an adjuvant is used; read label for specifics.
	penthiopyrad (Vertisan)	7	16 to 24 fl oz/acre	0	0.5	No more than 2 sequential applications of the fungicide before switching to a fungicide with another mode of action.
	pyraclostrobin (Headline SC & EC)	11	6 to 12 fl oz/acre	7	0.5	Do not exceed 2 sequential applications of this fungicide or with other group 11 fungicides.
	pyraclostrobin + metconazole (Headline AMP)	11 + 3	10 to 14.4 fl oz/ acre	7	0.5	No more than 2 sequential applications before alternating with a different mode of action.
Yellow leaf blight (*Peyronellaea zeamaydis* [*Phyllosticta maydis*])	fluxapyroxad + pyraclostrobin (Priaxor)	7 + 11	4 to 8 fl oz/acre	7	0.5	Do not make more than 2 sequential applications of Priaxor before switching to a fungicide with a different mode of action. Maximum 4 (high rate) or 2 (low rate) applications per crop. Crop damage may occur when an adjuvant is used; read label for specifics.

Cucurbits (Cucumber, Cantaloupe, Melon, Pumpkin, Squash, Watermelon)

L. Quesada-Ocampo, Plant Patholog

Table 10-25. Disease Control Products for Cucurbits

Disease	Material	FRAC	Rate of Material	Minimum Days Harvest	Minimum Days Reentry	Method, Schedule, and Remarks
Angular leaf spot	fixed copper (various)	M	See label	See label	See label	See label. Rates vary depending on the formulation. Repeated use may cause leaf yellowing.
Bacterial leaf spot	acibenzolar-*S*-methyl (Actigard 50 WP)	21	0.5 to 1 oz/acre	0	0.5	Apply to healthy, actively growing plants. Do not apply to stressed plants. Apply no more than 8 oz/acre/season.
Bacterial fruit blotch	fixed copper (various)	M	See label	0	0	See label. Rates vary depending on the formulation. Start applications at first bloom; ineffective once fruit reaches full size. Repeated use may cause leaf yellowing.
	acibenzolar-*S*-methyl (Actigard 50 WP)	P1	0.5 to 1 oz/acre	0	0.5	Apply to healthy, actively growing plants. Do not apply to stressed plants. Apply no more than 8 oz/acre/season.
Bacterial wilt	NA	NA	NA	NA	NA	See Insect Control section for Cucumber Beetles.
Belly (fruit) rot, *Rhizoctonia*	azoxystrobin (various)	11	See label	1	4 hr	Make banded application to soil surface or in-furrow application just before seed are covered.
	azoxystrobin + chlorothalonil (Quadris Opti)	11 + M	3.2 pt/acre	1	0.5	Do not apply more than 1 foliar application before alternating with a fungicide with a different mode of action. Do not make more than 4 applications of QoI group 11 fungicides per crop per acre per year.
	difenoconazole + benzovindiflupyr (Aprovia Top)	7 + 3	10.5 to 13.5 fl oz/acre	0	0.5	Make no more than 2 applications before alternating to a fungicide with different active ingredients. Apply no more than 53.6 fl oz/acre/year.
	fluopyram + tebuconazole (Luna Experience) 3.3 F	7 + 3	17 fl oz/acre	7	0.5	Make no more than 2 applications before alternating to a fungicide with different active ingredients.
	thiophanate-methyl (Topsin M 70 WP)	1	0.5 lb/acre	0	0.5	Apply in sufficient water to obtain runoff to soil surface.
Cottony leak (*Pythium* spp.)	metalaxyl (MetaStar) 2 E	4	4 to 8 pt/treated acre	0	2	Soil surface application in 7-in. band.
Damping off (*Pythium* spp.) and fruit rot	mefenoxam (Ridomil Gold 4 SL) (Ultra Flourish 2 EL)	4	1 to 2 pt/acre 2 to 4 pt/acre	0	2	Preplant incorporated (broadcast or band); soil spray (broadcast or band); or injection (drip irrigation).
	metalaxyl (MetaStar 2 E)	4	4 to 8 pt/acre	0	2	Preplant incorporated or surface application.
	propamocarb (Previcur Flex 6 F)	28	12.8 fl oz/100 gal	2	0.5	Rates based on rock wool cube saturation in the greenhouse. See label for use in seed beds, drip system, and soil drench.
Downy mildew	ametoctradin + dimethomorph (Zampro 4.38 SC)	45 + 40	14 oz/acre	0	0.5	Make no more than 2 applications before alternating to a fungicide with different active ingredients. Do not rotate with Forum. Maximum of 3 applications/crop/season.
	azoxystrobin (Quadris 2.08 F)	11	11 to 15.4 fl oz/acre	1	4 hr	Make no more than 1 application before alternating with a fungicide with a different mode of action. Apply no more than 2.88 qt/crop/acre/season. Resistance reported.
	azoxystrobin + chlorothalonil (Quadris Opti)	11 + M	3.2 pt/acre	1	0.5	Do not apply more than 1 foliar application before alternating with a fungicide with a different mode of action. Do not make more than 4 applications of QoI group 11 fungicides/crop/acre/year.
	chlorothalonil (various)	M	See label	See label	See label	See label. Rates vary depending on the formulation. Spray at first appearance and then at 7- to 14-day interval. Avoid late-season application after plants have reached full maturity.
	chlorothalonil + potassium phosphite (Catamaran 5.27 SC)	M + 33	6 pt/acre	0	0.5	Apply no more than 50 pt/acre/season.
	chlorothalonil + zoxamide (Zing!)	M + 22	36 fl oz/acre	0	0.5	May cause sunburn in watermelon fruit, see label for details.
	cyazofamid (Ranman 400 SC)	21	2.1 to 2.75 fl oz/acre	0	0.5	Do not apply more than 6 sprays per crop. Make no more than 3 consecutive applications followed by 3 applications of fungicides from a different resistance management group.
	cymoxanil (Curzate 60 DF)	27	3.2 oz/acre	3	0.5	Use only in combination with labeled rate of protectant fungicide (e.g., mancozeb or chlorothalonil).
	dimethomorph (Forum 4.17SC)	40	6 fl oz/acre	0	0.5	Must be applied as a tank mix with another fungicide with a different mode of action. Do not make more than 2 sequential applications.
	ethaboxam (Elumin)	22	8 fl oz/acre	2	0.5	Make no more than 2 applications before alternating to a fungicide with different active ingredients. Apply no more than 16 fl oz/acre/year.
	famoxadone + cymoxanil (Tanos 50WP)	11 + 27	8 oz/acre	3	0.5	Do not make more than 1 application before alternating with a fungicide that has a different mode of action. Must be tank-mixed with contact fungicide with a different mode of action.

Table 10-25. Disease Control Products for Cucurbits

Disease	Material	FRAC	Rate of Material	Minimum Days Harvest	Minimum Days Reentry	Method, Schedule, and Remarks
Downy mildew (continued)	fenamidone (Reason 500 SC)	11	5.5 fl oz/acre	14	0.5	Begin applications when conditions favor disease development and continue on 5-to 10-day interval. Do not apply more than 22 fl oz per growing season. Alternate with fungicide from different resistance management group and make no more than 4 total applications of Group 11 fungicides per season.
	fixed copper (various)	M	See label	See label	See label	See label. Rates vary depending on the formulation. Repeated use may cause leaf yellowing.
	fluazinam (Omega 500F)	29	0.75 to 1.5 pt/acre	7	0.5	Initiate applications when conditions are favorable for disease development or when disease symptoms first appear. Repeat applications on a 7- to 10-day schedule.
	fluopicolide (Presidio 4F)	43	3 to 4 fl oz/acre	2	0.5	Tank mix with another downy mildew fungicide with a different mode of action.
	fosetyl-AL (Aliette 80 WDG)	33	2 to 5 lb/acre	0.5	0.5	Do not tank mix with copper-containing products. Mixing with surfactants or foliar fertilizers is not recommended.
	mandipropamid (Revus 2.08F)	4 + M	8 fl oz/acre	1	0.5	For disease suppression only. Resistance reported.
	mancozeb (various)	M	See label	See label	See label	See label. Rates vary depending on the formulation. Labeled on all cucurbits.
	mefenoxam + chlorothalonil (Ridomil Gold Bravo, Flouronil 76.5 WP)	4 + M	2 to 3 lb/acre	7	2	Spray at first appearance and repeat at 14 day intervals. Apply full rate of protectant fungicide between applications. Avoid late-season application when plants reach full maturity. Resistance reported.
	oxathiapiprolin + chlorothalonil (Orondis Opti SC)	49 + M	1.7 to 2.5 pt/acre	0	0.5	Limit to 10 pt/acre/year. Limit to 6 foliar applications/acre/year for the same crop. Do not follow soil applications of Orondis with foliar applications of Orondis. Begin foliar applications prior to disease development and continue on a 5- to 14-day interval. Use the higher rates when disease is present.
	oxathiapiprolin + mandipropamid (Orondis Ultra)	49 + 40	5 to 8 fl oz/acre	0	4 hr	Limit to 32 fl oz/acre/year. Limit to 6 foliar applications//acre/year for the same crop. Do not follow soil applications of Orondis with foliar applications of Orondis. Begin foliar applications prior to disease development and continue on a 5- to 14-day interval. Use the higher rates when disease is present.
	propamocarb (Previcur Flex 6 F)	28	1.2 pt/acre	2	0.5	Begin applications before infection; continue on a 7- to 14-day interval. Do not apply more than 6 pints per growing season. Always tank mix with another Downy mildew product.
	pyraclostrobin (Cabrio 20 WG)	11	8 to 12 oz/acre	0	0.5	Make no more than 1 application before alternating to a fungicide with a different mode of action. Resistance reported.
	pyraclostrobin + boscalid (Pristine 38 WG)	11 + 7	12.5 to 18.5 oz/acre	0	1	Make no more than 4 applications per season. Resistance reported.
	trifloxystrobin (Flint 50 WDG)	11	4 oz/acre	0	0.5	Begin applications preventatively and continue as needed alternating applications of Ridomil Gold Bravo on a 7- to14-day interval. Resistance reported.
	zoxamide + mancozeb (Gavel 75 DF)	22 + M	1.5 to 2 lb	5	2	Begin applications when plants are in 2-leaf stage and repeat at 7- to 10-day intervals. Now labeled on all cucurbits. Maximum 8 applications per season.
Fusarium wilt	prothioconazole (Proline 480 SC)	3	5.7 fl oz/acre	7	0.5	1 soil and 2 foliar applications allowed by either ground or chemigation application equipment (including drip irrigation). Do not use in water used for hand transplanting. Not for use in greenhouse/transplant house.
Gummy stem blight, Black rot	prothioconazole (Proline 480 SC)	3	5.7 fl oz/acre	7	0.5	1 soil and 2 foliar applications allowed by either ground or chemigation application equipment (including drip irrigation). Do not use in water used for hand transplanting. Not for use in greenhouse/transplant house.
	tebuconazole (Monsoon 3.6 F)	3	8 oz/acre	7	0.5	Maximum 3 applications per season. Apply as a protective spray at 10- to 14-day intervals. Add a surfactant.
Leaf spots: *Alternaria*, Anthracnose (*Colletotrichum*), *Cercospora*, Gummy stem blight (*Didymella*), Target spot (*Corynespora*)	azoxystrobin (Quadris 2.08 F)	11	11 to 15.4 fl oz/acre	1	4 hr	Make no more than 1 application before alternating with a fungicide with a different mode of action. Apply no more than 2.88 qt/crop/acre/season. Do not use for Gummy stem blight where resistance to group 11(QoI) fungicides exists.
	azoxystrobin + chlorothalonil (Quadris Opti)	11 + M	3.2 pt/acre	1	0.5	Do not apply more than 1 foliar application before alternating with a fungicide with a different mode of action. Do not make more than 4 applications of QoI group 11 fungicides per crop per acre per year.

Chapter X — 2019 N.C. Agricultural Chemicals Manual

Table 10-25. Disease Control Products for Cucurbits

Disease	Material	FRAC	Rate of Material	Minimum Days Harvest	Minimum Days Reentry	Method, Schedule, and Remarks
Leaf spots: *Alternaria*, Anthracnose (*Colletotrichum*), *Cercospora*, Gummy stem blight (*Didymella*), Target spot (*Corynespora*) (continued)	azoxystrobin + difenoconazole (Quadris Top 1.67 SC)	11 + 3	12 to 14 fl oz/acre	1	0.5	Not for Target spot. Make no more than 1 application before alternating with fungicides that have a different mode of action. Apply no more than 56 fl oz/crop/acre/season.
	chlorothalonil (various)	M	See label	See label	See label	See label. Rates vary depending on the formulation.
	chlorothalonil + potassium phosphite (Catamaran 5.27 SC)	M + 33	6 pt/acre	0	0.5	Apply no more than 50 pt/crop/acre/season. Do not apply to watermelon fruit when stress conditions conducive to sunburn occur.
	cyprodinil + fludioxonil (Switch 62.5 WG)	9 + 12	11 to 14 oz/acre	1	0.5	Only for *Alternaria* and Gummy stem blight. Make no more than 2 applications before alternating to a different fungicide. Maximum of 4 to 5 applications at high and low rates.
	difenoconazole + benzovindiflupyr (Aprovia Top)	7 + 3	10.5 to 13.5 fl oz/acre	0	0.5	Make no more than 2 applications before alternating to a fungicide with different active ingredients. Apply no more than 53.6 fl oz/acre/year.
	difenoconazole + cyprodinil (Inspire Super 2.82 SC)	3 + 9	16 to 10 fl oz/acre	7	0.5	Not for Target spot. Make no more than 2 sequential applications before alternating with fungicides that have a different mode of action. Apply no more than 80 fl oz/crop/acre/season.
	famoxadone + cymoxanil (Tanos 50WP)	11 + 27	8 oz/acre	3	0.5	Only for *Alternaria* and Anthracnose; do not make more than 1 application before alternating with a fungicide that has a different mode of action; must be tank-mixed with contact fungicide with a different mode of action
	fenamidone (Reason 500 SC)	11	5.5 fl oz/acre	14	0.5	Begin applications when conditions favor disease development and continue on 5- to 10-day interval. Do not apply more than 22 fl oz per growing season. Alternate with fungicide from different resistance management group and make no more than 4 total applications of Group 11 fungicides per season.
	fixed copper (various)	M	See label	See label	See label	See label. Rates vary depending on the formulation. Repeated use may cause leaf yellowing.
	mancozeb (various)	M	See label	See label	See label	See label. Rates vary depending on the formulation. Labeled on all cucurbits.
	fluopyram + tebuconazole (Luna Experience 3.3 F)	7 + 3	8 to 17 fl oz/acre	7	0.5	Not for *Cercospora* or target spot. Make no more than 2 applications before alternating to a fungicide with different active ingredients. Do not rotate with tebuconazole.
	fluopyram + trifloxystrobin (Luna Sensation 1.67 F)	7 + 11	7.6 fl oz/acre	0	0.5	Make no more than 2 applications before alternating to a fungicide with different active ingredients. Maximum 4 applications per season.
	fluxapyroxad + pyraclostrobin (Merivon 500 SC)	7 + 11	4 to 5.5 oz/acre	0	0.5	Make no more than 2 sequential applications before alternating with fungicides that have a different mode of action. Maximum of 3 applications per crop.
	potassium phosphite + tebuconazole (Viathon)	33 + 3	4 pt/acre	7	0.5	**APPLY ONLY TO WATERMELON.** Maximum 3 applications per crop.
	pyraclostrobin (Cabrio 20 WG)	11	12 to 16 oz/acre	0	0.5	Do not use for Gummy stem blight where resistance to group 11 (QoI) fungicides exists. Make no more than 1 application before alternating with a fungicide with a different mode of action.
	pyraclostrobin + boscalid (Pristine 38 WG)	11 + 7	12.5 to 18.5 oz/acre	0	1	Not for target spot. Do not use for gummy stem blight where resistance to group 7 and group 11 fungicides exists. Use highest rate for anthracnose. Make no more than 4 applications per season.
	thiophanate-methyl (Topsin M 70 WP)	1	0.5 lb/acre	0	0.5	Spray at first appearance and then at 7- to 10-day intervals. Resistance reported in gummy stem blight fungus.
	zoxamide + mancozeb (Gavel 75 DF)	22 + M	1.5 to 2 lb	5	2	*Cercospora* and *Alternaria* only. Begin applications when plants are in 2-leaf stage and repeat at 7- to 10-day intervals. Now labeled on all cucurbits. Maximum 8 applications per season.
Nematodes	fluopyram (Velum Prime)	7	6.5 to 6.84 fl oz/acre	0	0.5	Apply to the root zone. Do not apply more than 13.7 fl oz/acre/year. For follow-up fungicide applications use a product with a different mode of action.
Phytophthora blight	ametoctradin + dimethomorph (Zampro 4.38SC)	45 + 40	14 oz/acre	0	0.5	Make no more than 2 applications before alternating to a fungicide with different active ingredients. Do not rotate with Forum. Maximum of 3 applications per crop per season. Apply at planting as a preventive drench treatment. Addition of a spreading or penetrating adjuvant is recommended
	cyazofamid (Ranman 400 SC)	21	2.75 fl oz/acre	0	0.5	Do not apply more than 6 sprays per crop. Make no more than 3 consecutive applications followed by 3 applications of fungicides from a different resistance management group. Resistant isolates have been found.
	dimethomorph (Forum 4.17SC)	40	6 fl oz/acre	0	0.5	Must be applied as a tank mix with another fungicide with a different mode of action. Do not make more than 2 sequential applications.

Table 10-25. Disease Control Products for Cucurbits

Disease	Material	FRAC	Rate of Material	Minimum Days Harvest	Minimum Days Reentry	Method, Schedule, and Remarks
Phytophthora blight (continued)	ethaboxam (Elumin)	22	8 fl oz/acre	2	0.5	Make no more than 2 applications before alternating to a fungicide with different active ingredients. Apply no more than 16 fl oz/acre/year.
	fluazinam (Omega 500F)	29	0.75 to 1.5 pt/acre	7	0.5	Initiate applications when conditions are favorable for disease development or when disease symptoms first appear. Repeat applications on a 7- to 10-day schedule.
	fluopicolide (Presidio 4F)	43	3 to 4 oz/acre	2	0.5	Tank mix with another *Phytophthora* fungicide with a different mode of action. May be applied through drip irrigation to target crown rot phase.
	mandipropamid (Revus 2.08F)	40	8 fl oz/acre	0	0.5	For disease suppression only; apply as foliar spray with copper-based fungicide.
	oxathiapiprolin + mefenoxam (Orondis Gold 200)	49 + 4	4.8 to 9.6 fl oz/acre	0	4 hr	Limit to 38.6 fl oz/acre/year. Limit to 6 applications per acre per year for the same crop. Do not follow soil applications of Orondis with foliar applications of Orondis. Apply at planting in furrow, by drip, or in transplant water. Use the higher rates for heavier soils, for longer application intervals, or for susceptible varieties.
	oxathiapiprolin + mandipropamid (Orondis Ultra)	49 + 40	5.5 to 8 fl oz/acre	0	4 hr	Limit to 6 applications per acre per year for the same crop. Do not follow soil applications of Orondis with foliar applications of Orondis. Use the higher rates when disease is present.
Plectosporium blight	azoxystrobin + (Quadris 2.08 F)	11	11 to 15.4 fl oz/acre	1	4 hr	Make no more than 1 application before alternating with a fungicide with a different mode of action. Apply no more than 2.88 qt per crop per acre per season and do not make more than 4 applications of Group 11 products.
	azoxystrobin + difenoconazole (Quadris Top 1.67 SC)	11 + 3	12 to 14 fl oz/acre	1	0.5	Make no more than 1 application before alternating with fungicides that have a different mode of action. Apply no more than 56 fl oz per crop per acre per season
	fluxapyroxad + pyraclostrobin (Merivon 500 SC)	7 + 11	4 to 5.5 fl oz/acre	0	0.5	Make no more than 2 sequential applications before alternating with fungicides that have a different mode of action. Maximum of 3 applications per crop.
	trifloxystrobin (Flint 50 WDG)	11	1.5 to 2 oz/acre	0	0.5	Make no more than 1 application before alternating with fungicides that have a different mode of action. Begin applications preventively when conditions are favorable for disease and continue as needed on a 7- to 14-day interval.
	pyraclostrobin (Cabrio 20WG)	11	12 to 16 oz/acre	0	0.5	Make no more than 1 application before alternating to a fungicide with a different mode of action.
Powdery mildew	acibenzolar-S-methyl (Actigard 50 WP)	P1	0.5 to 1 oz/acre	0	0.5	Apply to healthy, actively growing plants. Do not apply to stressed plants. Apply no more than 8 oz/acre/season.
	azoxystrobin + chlorothalonil (Quadris Opti)	11 + M	3.2 pt/acre	1	0.5	Do not apply more than 1 foliar application before alternating with a fungicide with a different mode of action. Do not make more than 4 applications of QoI group 11 fungicides per crop per acre per year.
	azoxystrobin + difenoconazole (Quadris Top 1.67 SC)	11 + 3	12 to 14 fl oz/acre	1	0.5	Make no more than 1 application before alternating with fungicides that have a different mode of action. Apply no more than 56 fl oz/crop/acre/season.
	chlorothalonil (various)	M	See label	See label	See label	Spray at first appearance and then at 7- to 14-day intervals. Avoid late-season application after plants have reached full maturity. Does not control PM on leaf undersides.
	chlorothalonil + potassium phosphite (Catamaran 5.27 SC)	M + 33	6 pt/acre	0	0.5	Apply no more than 50 pt/crop/acre/season. Do not apply to watermelon fruit when stress conditions conducive to sunburn occur.
	difenoconazole + benzovindiflupyr (Aprovia Top)	7 + 3	10.5 to 13.5 fl oz/acre	0	0.5	Make no more than 2 applications before alternating to a fungicide with different active ingredients. Apply no more than 53.6 fl oz/acre/year.
	difenoconazole + cyprodinil (Inspire Super 2.82 SC)	3 + 9	16 to 20 fl oz/acre	7	0.5	Make no more than 2 sequential applications before alternating with fungicides that have a different mode of action. Apply no more than 80 fl oz/crop//acre/season.
	cyprodinil + fludioxonil (Switch 62.5 WG)	9 + 12	11 to 14 oz/acre	1	0.5	Make no more than 2 applications before alternating to a different fungicide. Maximum of 4 to 5 applications at high and low rates. Not for target spot or anthracnose or Cercospora.
	fixed copper (various)	M	See label	See label	See label	See label. Rates vary depending on the formulation. Repeated use may cause leaf yellowing.
	cyflufenamid (Torino 0.85 SC)	U6	3.4 oz/acre	0	4 hr	Do not make more than 2 applications per crop.

Table 10-25. Disease Control Products for Cucurbits

Disease	Material	FRAC	Rate of Material	Minimum Days Harvest	Minimum Days Reentry	Method, Schedule, and Remarks
Powdery mildew (continued)	fluopyram + tebuconazole (Luna Experience 3.3F)	7 + 3	8 to 17 fl oz/acre	7	0.5	Make no more than 2 applications before alternating to a fungicide with different active ingredients. Do not rotate with tebuconazole.
	fluxapyroxad + pyraclostrobin (Merivon 500 SC)	7 + 11	4 to 5.5 fl oz/acre	0	0.5	Make no more than 2 sequential applications before alternating with fungicides that have a different mode of action. Maximum of 3 applications per crop.
	flutianil (Gatten)	U13	6 to 8 fl oz/acre	0	0.5	Not labeled for watermelon. Do not make more than five applications per year.
	metrafenone (Vivando)	U8	15.4 fl oz/acre	0	0.5	Supplemental label expires Dec. 31, 2017. Begin applications prior to disease and continue in a 7- to 10-day interval.
	myclobutanil (Rally 40 WP)	3	2.5 to 5 oz/acre	0	1	Apply no more than 1.5 lb/acre/crop. Observe a 30-day plant-back interval.
	penthiopyrad (Fontelis 1.67 SC)	7	12 to 16 fl oz/acre	1	0.5	Make no more than 2 sequential applications before switching to another fungicide. Do not rotate with Pristine or Luna Experience.
	pyraclostrobin + boscalid (Pristine 38 WG)	11 + 7	12.5 to 18.5 oz/acre	0	1	Make no more than 4 applications per season.
	quinoxyfen (Quintec 2.08 SC)	13	4 to 6 fl oz/acre	3	0.5	Make no more than 2 applications before alternating to a different fungicide. Maximum of 24 fl oz/acre/year. **DO NOT USE ON SUMMER SQUASH or CUCUMBER**; labeled on winter squashes, pumpkins, gourds, melon and watermelon.
	sulfur (various)	M	See label	See label	See label	See label. Rates vary depending on the formulation. Do not use when temperature is over 90 degrees F or on sulfur-sensitive varieties.
	tebuconazole (Monsoon 3.6F)	3	4 to 6 fl oz/acre	7	0.5	Apply before disease appears when conditions favor development and repeat at 10- to 14-day intervals; max 24 fl oz/season.
	triflumizole (Procure 50 WS)	3	4 to 8 oz/acre	0	0.5	Begin applications at vining or first sign of disease and repeat at 7- to 10-day intervals.
Scab	acibenzolar-S-methyl (Actigard 50 WP)	P1	0.5 to 1 oz/acre	0	0.5	Apply to healthy, actively growing plants. Do not apply to stressed plants. Apply no more than 8 oz/acre/season.
	chlorothalonil (various)	M	See label	See label	See label	See label. Rates vary depending on the formulation.
	chlorothalonil + potassium phosphite (Catamaran 5.27 SC)	M + 33	6 pt/acre	0	0.5	Apply no more than 50 pt/crop/acre/season. Do not apply to watermelon fruit when stress conditions conducive to sunburn occur.
Vine decline	fludioxonil (Cannonball)	12	4 to 8 oz/acre	14	0.5	**APPLY ONLY TO MELONS.**

Table 10-26. Efficacy of Products for Disease Control in Cucurbits

Scale: E = excellent; G = good; F = fair; P = poor; NC = no control; ND = no data.

Product[1]	Fungicide group[F]	Preharvest interval (Days)	Alternaria Leaf Blight	Angular Leafspot	Anthracnose	Bacterial Fruit Blotch	Belly Rot	Cercospora Leaf Spot	Cottony Leak	Damping off (Pythium)	Downy Mildew[DM]	Fusarium wilt	Gummy Stem Blight	Phytophthora Blight (foliage and fruit)	Phytophthora Blight (crown and root)	Plectosporium Blight	Powdery Mildew	Target Spot
acibenzolar-S-methyl (Actigard)	P01	0	NC	ND	NC	F	NC	NC	ND	ND	ND	ND	NC	ND	ND	NC	ND	NC
ametoctradin + dimethomorph (Zampro)	45 + 40	0	ND	NC	NC	NC	NC	NC	ND	ND	F	ND	NC	F	F	NC	NC	NC
azoxystrobin[2] (Quadris)	11	1	G	NC	G	NC	F	G	NC	NC	NC[R]	ND	NC[R]	NC	NC	F	NC[R]	G
azoxystrobin + chlorthalonil (Quadris Opti)	11 + M05	0	G	NC	G	NC	F	G	NC	NC	NC[R]	ND	F	NC	NC	F	F	F
azoxystrobin + difenoconazole (Quadris Top)	11 + 3	1	ND	NC	G	NC	ND	ND	ND	ND	F	ND	ND	NC	ND	F	F	ND
boscalid (Endura)	7	0	ND	NC	NC	NC	ND	NC	NC	NC	NC[R]	ND	NC	NC	NC	ND	E[R]	ND
chlorothalonil[5] (various)	M05	0	F	NC	G	NC	NC	G	NC	NC	F	ND	F	NC	NC	F	F	G
cyazofamid (Ranman)	21	0	NC	NC	NC	NC	NC	NC	NC	NC	G	ND	NC	F	NC	NC	NC	NC
cyflufenamid (Torino)	U06	0	NC	NC	NC	NC	NC	NC	NC	NC	NC	ND	NC	NC	NC	NC	G[R]	NC
cymoxanil (Curzate)	27	3	NC	NC	NC	NC	NC	NC	NC	ND	F[R]	ND	NC	F	NC	NC	NC	NC
cyprodinil + fludioxonil (Switch)	9 + 12	1	ND	NC	F	NC	ND	NC	NC	NC	ND	ND	F	NC	NC	F	F	NC
difenoconazole + benzovindiflupyr (Aprovia Top)	3 + 7	0	ND	NC	F	NC	ND	ND	NC	NC	ND	ND	G	NC	NC	ND	ND	ND
difenoconazole + cyprodinil (Inspire Super)	3 + 9	7	F	NC	P	NC	NC	F	NC	NC	NC	ND	F	NC	NC	F	F	ND
dimethomorph (Forum)	40	0	NC	NC	NC	NC	NC	NC	NC	P	ND	ND	NC	P	NC	NC	NC	NC
famoxadone[2] + cymoxanil (Tanos)	11 + 27	3	ND	NC	P	NC	ND	NC	NC	F	ND	ND	NC	ND	NC	NC	NC	NC
fenamidone (Reason)	11	14	F	NC	ND	NC	NC	NC	NC	F[R]	ND	ND	NC	F	NC	NC	NC	NC
fixed copper (various)[P, 5]	M01	1	P	F	P	F	NC	P	NC	NC	P	ND	P	ND	NC	P	P	P
fluopicolide (Presidio)	43	2	NC	NC	NC	NC	NC	NC	NC	P[R]	ND	NC	F	F	NC	NC	NC
fluopyram + tebuconazole (Luna Experience)	7 + 3	7	ND	NC	NC	ND	NC	NC	NC	NC	ND	ND	G	NC	NC	NC	G	NC
fluopyram + trifloxystrobin (Luna Sensation)	7 + 11	0	ND	NC	F	NC	NC	NC	NC	NC	ND	ND	F	NC	NC	NC	F	NC
fluoxastrobin (Evito)	11	1	G	NC	G	NC	F	G	NC	NC	NC[R]	ND	NC[R]	NC	NC	F	NC[R]	F
flutianil (Gatten)	U13	0	NC	NC	NC	NC	NC	NC	NC	NC	NC	ND	NC	NC	NC	NC	G	NC
flutriafol (Rhyme, Topguard)[P]	3	0	ND	NC	NC	NC	NC	NC	NC	NC	NC	ND	F	NC	NC	NC	P	NC
flutriafol + azoxystrobin (Topguard EQ)	3 + 11	0	ND	NC	F	NC	NC	NC	NC	NC	NC	ND	P	NC	NC	P	P	F
fluxapyroxad + pyraclostrobin (Merivon)	7 + 11	0	G	NC	NC	NC	ND	NC	NC	NC	NC	ND	F	NC	NC	F	ND	ND
kresoxim-methyl (Sovran)	11	ND	NC	ND	NC	ND	NC	NC	NC	ND	ND	NC[R]	ND	ND	ND	NC[R]	ND
mancozeb (various)[5]	M03	5	F	NC	G	NC	G	NC	NC	F	ND	F	P	NC	F	P	G
mancozeb + azoxystrobin (Dexter Max)	M03 + 11	5	F	NC	G	NC	G	NC	NC	F	ND	F	P	NC	F	P	G
mancozeb + fixed copper[4] (ManKocide)	M03 + M05	5	P	F	F	F	NC	P	NC	NC	F	ND	NC	P	NC	P	P	F
mandipropamid (Revus)	40	0	NC	NC	NC	NC	NC	NC	NC	NC[R]	ND	NC	F	P	NC	NC	NC
mefenoxam[3, 4] (Ridomil Gold EC, Ultra Flourish)	4	0	NC	NC	NC	NC	NC	NC	F[R]	G[R]	NC	ND	NC	F[R]	F[R]	NC	NC	NC
mefenoxam[2] + chlorothalonil[5] (Ridomil Gold/Bravo, Flouronil)	4 + M05	0	F	NC	F	NC	F	F[R]	F[R]	F[R]	ND	F	F[R]	NC	F	F	F
mefenoxam[2] + copper[5] (Ridomil Gold/Copper)	4 + M01	5	P	P	NC	P	NC	P	F[R]	F[R]	F[R]	ND	NC	F[R]	NC	P	NC	P
mefenoxam[2] + mancozeb[5] (Ridomil Gold MZ)	4 + M03	5	F	NC	F	NC	F	F[R]	F[R]	F[R]	ND	F	F[R]	NC	F	NC	F
metrafenone (Vivando)	U08	0	NC	NC	NC	NC	NC	NC	NC	NC	NC	NC	NC	NC	NC	NC	G	NC
myclobutanil[2] (Rally)	3	0	NC	NC	NC	NC	NC	NC	NC	NC	NC	NC	NC	NC	NC	NC	F	NC
oxathiapiprolin (Orondis Opti,)	49 + M05	0	F	NC	F	NC	ND	F	NC	NC	G	ND	F	G	G	F	P	F
oxathiapiprolin + mandipropamid (Orondis Ultra)	49 + 40	0	ND	NC	ND	NC	ND	ND	NC	NC	G	ND	ND	G	G	ND	ND	ND
oxathiapiprolin + mefenoxam (Orondis Gold)	49 + 4	0	ND	NC	ND	NC	ND	ND	F[R]	NC	G	ND	ND	G	G	ND	ND	ND
penthiopyrad (Fontelis)	7	1	ND	NC	NC	NC	NC	NC	NC	NC	ND	NC[R]	NC	NC	NC	F	NC
Phosphonate[6] (various)	33	0.5	NC	NC	NC	NC	NC	NC	NC	P	ND	NC	NC	NC	F	NC	NC
potassium phosphite + tebuconazole (Viathon)	33 + 3	7	ND	NC	ND	NC	ND	ND	ND	ND	P	ND	F	ND	ND	NC	F	NC
propamocarb (Previcur Flex)	28	2	NC	NC	NC	NC	NC	NC	ND	F[R]	ND	NC	F	NC	NC	NC	NC
prothioconazole (Proline)	3	7	ND	NC	F	NC	F	NC	NC	NC	G	G	NC	NC	ND	F	ND
pyraclostrobin[2] (Cabrio, Pyrac)	11	0	G	NC	G	NC	F	NC	NC	NC[R]	ND	NC[R]	P	NC	G	NC[R]	E
pyraclostrobin[2] + boscalid[2] (Pristine)	11 + 7	0	G	NC	F	NC	G	NC	NC	NC[R]	ND	NC[R]	P	NC	F	F	E
quinoxyfen (Quintec)	13	3	NC	NC	NC	NC	NC	NC	NC	NC	NC	NC	NC	NC	NC	G[R]	NC
sulfur (various)[P, 5]	M	0	NC	NC	NC	NC	NC	NC	NC	NC	NC	NC	NC	NC	NC	F	NC
tebuconazole (various)	3	7	ND	NC	F	NC	NC	F	NC	NC	NC	F	NC	NC	NC	F	NC
tetraconazole (Mettle)	3	0	ND	NC	NC	NC	NC	NC	NC	NC	NC	F	NC	NC	NC	F	NC
thiophanate-methyl[3] (Topsin M)	1	1	F	NC	F	NC	F	F	NC	NC	NC	NC[R]	NC	NC	NC	F	NC[R]	P

Table 10-26. Efficacy of Products for Disease Control in Cucurbits

Scale: E = excellent; G = good; F = fair; P = poor; NC = no control; ND = no data.

Product[1]	Fungicide group[F]	Preharvest interval (Days)	Alternaria Leaf Blight	Angular Leafspot	Anthracnose	Bacterial Fruit Blotch	Belly Rot	Cercospora Leaf Spot	Cottony Leak	Damping off (Pythium)	Downy Mildew[DM]	Fusarium wilt	Gummy Stem Blight	Phytophthora Blight (foliage and fruit)	Phytophthora Blight (crown and root)	Plectosporium Blight	Powdery Mildew	Target Spot
trifloxystrobin[3] (Flint)	11	0	G	NC	G	NC	ND	ND	NC	NC	NC[R]	ND	NC[R]	NC	NC	G	NC[R]	G
triflumizole (Procure)	3	0	NC	NC	NC	NC	NC	NC	NC	NC	ND	NC	NC	NC	NC	NC	F	NC
zoxamide + chlorothalonil (Zing!)	22 + M05	5	F	NC	F	NC	NC	F	NC	NC	F	ND	F	P	NC	F	P	F
zoxamide + mancozeb (Gavel)	22 + M03	5	F	NC	F	NC	NC	F	NC	NC	F	ND	F	P	NC	F	P	F

[1] Efficacy ratings do not necessarily indicate a labeled use for every disease.
[2] Curative activity; locally systemic.
[3] Systemic.
[4] When used in combination with chlorothalonil or mancozeb, gives increased control.
[5] Contact control only; no systemic control.
[6] Check manufacturers label for compatibility with other products.
[P] Can be phytotoxic at temperatures above 90 degrees F; read the label carefully.
[F] To prevent resistance in pathogens, alternate fungicides within a group with fungicides in another group. Fungicides in the "M" group are generally considered "low risk" with no signs of resistance developing to the majority of fungicides.
[R] Resistance reported in the pathogen.
[DM] Ratings based on efficacy and resistance on cucumber.

Table 10-27. Importance of Alternative Management Practices for Disease Control in Cucurbits

Scale: E = excellent; G = good; F = fair; P = poor; NC = no control; ND = no data

Strategy	Alternaria leaf blight	Angular leaf spot	Anthracnose	Bacterial fruit blotch	Bacterial wilt	Belly rot	Cercospora leaf spot	Choanephora fruit rot	Cottony leak	Downy mildew	Fusarium wilt	Gummy stem blight	Mosaic virus	Phytophthora blight	Plectosporium blight	Powdery mildew	Pythium damping off	Root knot	Target spot
Avoid field operations when leaves are wet	P	F	P	F	F	NC	NC	P	NC	P	NC	P	NC	NC	ND	NC	NC	NC	NC
Avoid overhead irrigation	F	F	F	F	P	NC	P	NC	NC	F	NC	F	NC	F	P	P	NC	NC	P
Change planting date from Fall to Spring[1]	G	P	G	P	P	F	G	F	F	G	P	G	F	F	F	F	G	G	G
Cover cropping with antagonist	NC	NC	NC	NC	NC	NC	NC	NC	NC	F	NC	NC	NC	NC	NC	NC	F	NC	NC
Crop rotation with non-host (2 to 3 years)	F	F	F	F	NC	P	F	NC	NC	NC	G	F	NC	F	F	NC	P	F	F
Deep plowing	P	NC	P	NC	NC	P	F	P	NC	NC	F	F	NC	P	P	NC	P	F	P
Destroy crop residue immediately	F	P	F	P	P	P	P	NC	P	F	F	F	P	P	P	F	NC	F	P
Encourage air movement[2]	F	P	F	P	NC	NC	F	F	F	F	NC	F	NC	NC	P	NC	NC	NC	F
Soil organic amendments[3]	ND	NC	ND	NC	NC	P	ND	NC	F	NC	P	ND	NC	P	ND	NC	F	F	ND
Insecticidal/horticultural oils[4]	NC	NC	NC	NC	F	NC	NC	NC	NC	NC	NC	NC	F	NC	NC	F	NC	NC	NC
pH management (soil)	NC	NC	NC	NC	NC	NC	NC	NC	ND	NC	NC	NC	ND	NC	NC	NC	ND	ND	NC
Plant in well-drained soil	NC	NC	NC	NC	NC	F	NC	P	F	NC	NC	NC	NC	F	NC	NC	F	P	NC
Plant on raised beds	NC	NC	NC	NC	NC	P	NC	P	F	NC	F	NC	NC	F	NC	NC	F	P	NC
Plastic mulch bed covers	NC	NC	NC	NC	NC	F	NC	P	F	NC	NC	NC	NC	G	F	NC	F	NC	NC
Postharvest temperature control (fruit)	NC	NC	F	F	NC	F	NC	F	F	NC	NC	NC	NC	F	F	NC	NC	NC	NC
Reflective mulch (additional effect over plastic mulch)	NC	NC	NC	NC	NC	NC	NC	NC	NC	NC	NC	NC	F	NC	NC	NC	NC	NC	NC
Reduce mechanical injury	P	P	P	P	F	P	P	P	P	NC	P	P	P	P	P	P	NC	NC	P
Rogue diseased plants/fruit (home garden)	F	P	P	P	P	NC	P	P	P	P	P	F	F	NC	NC	P	F	P	P
Row covers (insect exclusion)	NC	NC	NC	NC	G	NC	NC	NC	NC	NC	NC	NC	G	NC	NC	NC	NC	NC	NC
Soil solarization (reduce soil inoculum)	P	NC	P	NC	NC	F	P	NC	P	NC	F	P	NC	P	P	NC	F	P	P
Pathogen-free planting material	P	E	F	E	NC	NC	NC	NC	NC	G	E	NC	NC	NC	NC	F	NC	NC	NC
Resistant cultivars[5]		E			E					E	G		E			E			
Destroy volunteer plants	F	F	F	F	NC	F	NC	NC	F	G	F	F	F	NC	F	NC	P	NC	F

[1] Early planting reduces risk.
[2] Air movement can be encouraged by increasing plant spacing, orienting beds with prevailing wind direction and increasing exposure of field to prevailing wind.
[3] Soil organic amendments = cover crops; composted organic wastes.
[4] Insecticidal/Horticultural oil = Sunspray Ultra-Fine Spray Oil (Sun Company, Inc.), JMS Stylet oil; Safe-T-Side (Brandt Consolidated, Inc.); PCC 1223 (United Ag Products).
[5] Resistance available in some cucurbits.

Example Spray Program for Foliar Disease Control in Watermelon Production

A. Keinath, Plant Pathology, Clemson University

This spray program is based on research conducted at the Clemson Coastal Research and Education Center, Charleston, SC, and on a survey of watermelon fields in South Carolina in 2015 and 2016. The most common diseases in both years were gummy stem blight and powdery mildew. The spring program is designed to manage bacterial fruit blotch, bacterial leaf spots, gummy stem blight, powdery mildew, anthracnose, and downy mildew. The fall program is designed to manage gummy stem blight, downy mildew, and anthracnose.

- Protectants (chlorthalonil and mancozeb) are effective against anthracnose all season, but other disease-specific fungicides must be used against gummy stem blight, downy mildew, and powdery mildew. See Disease Control for Cucurbits and Efficacy of Products for Disease Control in Cucurbits.
- Start spraying when vines start to run or no later than when the first blooms (the male ones) open.
- From vine run until mid-May, spray every 10 days. After mid-May or when powdery and downy mildew typically show up in your area, spray every week through harvest regardless of the weather. Weekly sprays are needed to protect watermelon from powdery and downy mildew. Dry weather limits gummy stem blight but promotes powdery mildew; dry weather does not limit downy mildew or anthracnose if they are already present in a field.
- Do not stop spraying until you stop harvesting. Downy and powdery mildew can attack a crop any time it goes more than one week without a fungicide spray. Fungicides with a 7-day PHI are not recommended during the harvest period (usually after week 5); note that mancozeb and Gavel have a 5-day PHI.
- If this spray schedule is used to select fungicides for other cucurbits (vine crops), note that not all fungicides in this spray schedule are labeled on other cucurbits. Luna Experience is registered only on watermelon. Luna Experience is not registered for use in Louisiana. Quintec is not registered on cucumber. To control powdery mildew on other cucurbits or in Louisiana, use Procure in place of Luna fungicides or Quintec.

Table 10-28. Example Spray Program for Foliar Disease Control in Watermelon Production

Spray	Fungicide Program for Spring Watermelon*	Comments on Spring Program	Fungicide Program for Fall Watermelon*
1 (vine run)	mancozeb + fixed copper	For prevention of bacterial leaf spots.	chlorthalonil or Catamaran
2	chlorthalonil or Catamaran	Do not tank mix copper with chlorthalonil.	chlorthalonil
3a**	tebuconazole	If fruit blotch is a concern, add fixed copper.	tebuconazole + **Ranman**
3b**	tebuconazole + Flint	Add Flint if anthracnose fruit rot was found the previous year.	(same as 3a)
4	chlorthalonil (or mancozeb)	If fruit blotch is a concern, substitute mancozeb + fixed copper.	Quadris Top
5a**	mancozeb + Quintec	The protectant switches here from chlorthalonil to mancozeb to avoid injury to fruit on hot, sunny days.	Gavel
5b**	Luna Experience	Use Luna Experience if gummy stem is present.	(same as 5a)
6	Gavel	Note 5-day PHI.	mancozeb
7a**	mancozeb + Torino	Note 5-day PHI on mancozeb.	mancozeb + **Ranman**
7b**	Switch	Use Switch if gummy stem blight is present.	Switch
8	mancozeb + **Ranman**	If downy mildew is seen earlier in the season, apply Ranman as soon as possible.	chlorthalonil
9-12	If more sprays are needed after spray 8 until the last harvest, apply sprays 5 to 8 again.		

* Fungicides that control downy mildew are in bold. Fungicides that control powdery mildew are underlined.

** Option "a" is a lower cost treatment that may be less effective. Option "b" is a more expensive fungicide that also is more effective.

Eggplant

A. Keinath, Plant Pathology, Clemson University

Table 10-29. Disease Control Products for Eggplant

Disease	Material	FRAC	Rate of Material	Minimum Days Harvest	Minimum Days Reentry	Method, Schedule, and Remarks
Anthracnose fruit rot, early blight, gray mold	azoxystrobin (various)	11		0	4 hr	Apply at flowering to manage green fruit rot. Limit of 61.5 fl oz per acre per season. Make no more than **one** application before alternating with fungicides that have a different mode of action. Labeled for anthracnose ONLY.
	boscalid (Endura 70 WG)	7	2.5 to 3.5 oz/acre	0	0.5	Limit of 21 ounces per acre per season. Make no more than 2 sequential applications before alternating with fungicides that have a different mode of action. Labeled for early blight and gray mold ONLY.
	chlorothalonil (various)	M5	1.5 pt/acre	3	1	Limit of 12 pt per acre per season. Labeled for anthracnose and gray mold ONLY.
	difenoconazole + benzovindiflupyr (Aprovia Top 1.62EC)	3 + 7	10.5 to 13.5 fl oz/acre	14	0.5	Make no more than 2 consecutive applications before switching to a non-Group 7 fungicide. Make no more than 5 applications at the low rate or 4 applications at the high rate per year. Labeled for anthracnose ONLY.
	fenamidone (Reason 500SC)	11	5.5 to 8.2 fl oz/acre	14	0.5	Limit of 24.6 fl oz per growing season. Make no more than **one** application before rotating to another effective fungicide with a different mode of action. Labeled for early blight only.
	fluoxastrobin (Aftershock, Evito 280 SC)	11	2 to 5.7 fl oz/acre	3	0.5	Limit of 22.8 fl oz per acre per season. Make no more than **one** application before alternating with fungicides that have a different mode of action. **NOTE: Do not overhead irrigate for 24 hours following a spray application.** Labeled for early blight only.
	penthiopyrad (Fontelis 1.67 SC)	7	16 to 24 fl oz/acre	0	0.5	Limit of 72 fl oz per acre per year. Make no more than 2 consecutive applications before rotating to another effective fungicide with a different mode of action.
	pyraclostrobin (various)	11	8 to 12 oz/acre	0	4 hr	Apply at flowering to manage green fruit rot. Limit of 96 oz per acre per season. Make no more than **one** application before alternating with fungicides that have a different mode of action.
	pyraclostrobin + fluxapyroxad (Priaxor 500 SC)	11 + 7	4.0 to 8.0 fl oz/acre	0	0.5	Limit of 24 fl oz per acre per season. Make no more than 2 consecutive applications before rotating to another effective fungicide with a different mode of action. Labeled for anthracnose and early blight ONLY.
Phomopsis fruit rot	copper (various)	M1	See labels	See labels	2	Make first application at flowering. If disease is present, make additional applications at 7- to 10-day intervals. Do not spray copper when temperatures are above 90 °F.
Phytophthora blight	amectoctradin + dimethomorph (Zampro 525 SC)	45 + 40	14 fl oz/acre	4	0.5	Limit of 3 applications per acre per season. Make no more than 2 sequential applications before rotating to another effective fungicide with a different mode of action.
	cyazofamid (Ranman 400 SC)	21	2.75 fl oz/acre	0	0.5	Limit of 16.5 fluid ounces per acre per season. Apply to the base of the plant at transplanting or in the transplant water. Make no more than three consecutive applications followed by three consecutive applications of another effective fungicide with a different mode of action.
	copper (various)	M1	See label	0	2	Begin applications when conditions first favor disease development and repeat at 3- to 10-day intervals if needed depending on disease severity. Use the higher rates when conditions favor disease. Do not spray copper when temperatures are above 90 degrees F.
	dimethomorph (Acrobat, Forum)	40	6 fl oz/acre	0	0.5	**SUPPRESSION ONLY.** Limit of 30 fl oz per acre per season. Make no more than 2 sequential applications before alternating with fungicides that have a different mode of action. **NOTE: Must tank mix with another fungicide with a different mode of action.**
	famoxadone + cymoxanil (Tanos 50 DF)	11 + 27	8 to 10 oz/acre 4 to 5 oz/acre	3	0.5	**SUPPRESSION ONLY.** Make no more than **one** application before alternating with a fungicide with a different mode of action. **NOTE: Must tank mix with another fungicide with a different mode of action (i.e. copper).**
	fluazinam (Omega 500F)	29	1 to 1.5 pt/acre	30	0.5	Apply as a soil drench at 1.5 pt per acre. For foliar applications use 1 pt per acre. Limit of 9 pt per acre per season.
	fluopicolide (Presidio 4 SC)	43	3 to 4 fl oz/acre	2	0.5	Limit of 4 applications at the low rate or 3 applications at the high rate. Make no more than 2 times sequentially before alternating with fungicides that have a different mode of action. **NOTE: Must be tank-mixed with another mode of action product.**

Table 10-29. Disease Control Products for Eggplant

Disease	Material	FRAC	Rate of Material	Minimum Days Harvest	Minimum Days Reentry	Method, Schedule, and Remarks
Phytophthora blight (continued)	mefenoxam + copper hydroxide (Ridomil Gold/Copper)	4 + M1	2 lb/acre	7	2	See label for an optimal spray program. Limit of 4 applications per crop per year. Do not exceed 0.4 lb a.i. per acre per season of mefenoxam + metalaxyl (MetaStar).
	oxathiapiprolin + mefenoxam (Orondis Gold 200 + Orondis Gold B)	49 + 4	2.4 to 19.2 fl oz/acre + 1 pt/acre	7	4 hr	Make no more than 2 sequential applications before alternating with fungicides that have a different mode of action. Maximum of 3 applications per crop per year. **Must tank mix both products before application.**
	mandipropamid (Revus 2.08 F, Micora)	40	8 fl oz/acre	1	0.5	**SUPPRESSION ONLY.** Limit of 32 fl oz per acre per season. **NOTE: Must tank mix with another fungicide with a different mode of action (i.e. copper).**
Southern blight (*Sclerotium rolfsii*)	fluoxastrobin (Aftershock, Evito 280SC)	11	2 to 5.7 fl oz/acre	3	0.5	Limit of 22.8 fl oz per acre per season. Make no more than **one** application before alternating with fungicides that have a different mode of action. **NOTE: Do not overhead irrigate for 24 hours following a spray application.**
	penthiopyrad (Fontelis 1.67 SC)	7	16 to 24 fl oz/acre	0	0.5	Apply 5 to 10 days after transplanting and again 14 days later. Limit of 2 applications per crop. Follow with a FRAC Group 11 fungicide if additional protection is needed.
	pyraclostrobin (various)	11	12 to 16 oz/acre	0	4 hr	**SUPPRESSION ONLY.** Apply at flowering to manage green fruit rot. Limit of 96 oz per acre per season. Make no more than **one** application before alternating with fungicides that have a different mode of action.
	pyraclostrobin + fluxapyroxad (Priaxor 500 SC)		4.0 to 8.0 fl oz/acre	0	0.5	Limit of 2 applications per season. Best option based on tests on tomato in SC.
Pythium root rot	mefenoxam (various)	4	See label	—	2	**May only be applied at planting.** Apply in a 12- to 16-in. band or in 20 to 50 gallons water per acre in transplant water. Mechanical incorporation or 0.5 to 1 in. irrigation water needed for movement into root zone if rain is not expected. After initial application, 2 supplemental applications (1 pt per treated acre) can be applied.
	metalaxyl (MetaStar 2E)	4	4 to 8 pt/treated acre	7	2	Limit of 12 pt per acre per season. Preplant (soil incorporated), at planting (in water or liquid fertilizer), or as a basil-directed spray after planting. See label for the guidelines for supplemental applications.
Rhizoctonia seedling and root rot	azoxystrobin (various)	11	0.4 to 0.8 fl oz/1,000 row feet	—	4 hr	Make in-furrow or banded applications shortly after plant emergence. Under cool, wet conditions, crop injury from soil directed applications may occur.
	difenoconazole + benzovindiflupyr (Aprovia Top 1.62 EC)	3 + 7	10.5 to 13.5 fl oz/acre	14	0.5	Make no more than 2 consecutive applications before switching to a non-Group 7 fungicide. Make no more than 5 applications at the low rate or 4 applications at the high rate per year.
Verticillium wilt	Polyoxin D (OSO 5%)	19	6.5 to 13 fl oz/acre	0	4 hr	**SUPPRESSION ONLY.** Can be applied using banded or irrigation water applications. Limit of 6 applications at maximum rate per acre per season.

Garlic

E. Pfeufer, Plant Pathology, University of Kentucky

Table 10-30. Disease Control Products for Garlic

Disease	Material	FRAC	Rate of Material	Minimum Days Harvest	Minimum Days Reentry	Method, Schedule, and Remarks
Botrytis blight (*Botrytis* spp.), purple blotch (*Alternaria porri*), downy mildew (*Peronospora destructor*)	azoxystrobin (various)	11	6.2 to 15.4 fl oz/acre	0	4 hr	Use higher rate for downy mildew and *Botrytis*. Do not make more than 2 sequential applications.
	azoxystrobin + difenoconazole (Quadris Top)	11 + 3	14 fl oz/acre	7	0.5	Begin sprays prior to disease onset and spray on a 7- to 14-day schedule. Do not rotate with Group 11 fungicides.
	boscalid (Endura) 70 WG	7	6.8 oz/acre	7	0.5	Not for downy mildew. Do not make more than 2 sequential applications or more than 6 applications per season.
	chlorothalonil (various)	M	See label	7	2	Spray at first appearance; 7 to 14 day intervals.
	chlorothalonil + cymoxanil (Ariston)	M + 27	1.6 to 2.4 pt/acre	7	0.5	Not for Botrytis blight. Apply prior to favorable infection periods; continue on 7- to 9-day interval; alternate with a different mode of action.
	chlorothalonil + zoxamide (Zing!)	M + 22	30 fl oz/acre	7	0.5	Follow protective spray schedule when diseases are in the area; continue on 7-day interval.
	difenoconazole + cyprodinil (Inspire Super)	3 + 9	16 to 20 fl oz/acre	14	0.5	Make no more than 2 applications before alternating with a fungicide with a different mode of action.
	famoxadone + cymoxanil (Tanos)	11 + 27	8 oz/acre	3	0.5	Not for Botrytis.
	fenamidone (Reason)	11	5.5 oz/acre	7	0.5	**Not for Botrytis.**
	fluazinam (Omega 500)	29	1.0 pt/acre	7	1	Initiate sprays when conditions are favorable for disease or at disease onset. Spray on a 7- to 10-day schedule.
	fluxapyroxad + pyraclostrobin (Merivon)	7 + 11	4 to 11 fl oz/acre	7	0.5	Use higher rates for downy mildew suppression. Apply at disease onset; continue on 7- to 14-day schedule. No more than 3 applications per season.
	mefenoxam + chlorothalonil (Ridomil Gold/Bravo)	4 + M	2.5 pt/acre	7	2	Spray at first appearance, 7- to 14-day intervals.
	pyraclostrobin (Cabrio)	11	8 to 12 oz/acre	7	0.5	Not for *Botrytis*. Use highest rate for downy mildew. Make no more than 2 sequential applications and no more than 6 applications per season.
	pyraclostrobin + boscalid (Pristine 38 WG)	11 + 7	10.5 to 18.5 oz/acre	7	1	Use highest rate for suppression only on downy mildew. Make no more than 6 applications per season.
	pyrimethanil (Scala 5F)	9	9 or 18 fl oz/acre	7	0.5	Not for downy mildew. Use lower rate in a tank mix with broad spectrum fungicide and higher rate when applied alone. Do not apply more than 54 fl oz per crop.
Downy mildew (*Peronospora destructor*)	dimethomorph (Forum 50 WP)	40	6.4 oz/acre	0	0.5	Must be applied as a tank mix with another fungicide active against downy mildew; apply every 7 to 10 days. Do not make more than 2 sequential applications.
	mandipropamid (Revus)	40	8.0 fl oz/acre	7	0.5	Apply as a tank mix with another fungicide active against downy mildew. Apply with a silicone-based adjuvant. 7- to 10-day schedule.
	mefenoxam + mancozeb (Ridomil Gold MZ)	4 + M	2.5 lb/acre	7	2	Use with a suitable adjuvant.
	amectoctradin + dimethomorph (Zampro)	45 + 40	14.0 fl oz/acre	0	12 hr	Tank-mix with a broad-spectrum fungicide like chlorothalonil or mancozeb.
White rot (*Sclerotium cupartum*)	azoxystrobin (various)	11	See label	0	4 hr	Do not make more than 2 sequential applications.
	azoxystrobin + chlorothalonil (Quadris Opti)	11 + M	1.6 to 3.2 pt/acre	7	0.5	Make no more than 1 application before alternating with a fungicide with a different mode of action.
	boscalid (Endura)	7	6.8 oz/acre	7	0.5	Apply at planting in a 4- to 6-inch banded spray. Under high disease pressure, apply as a foliar spray.
	iprodione (Rovral 50 WP)	27	4 lb/acre	—	1	Spray cloves as they are being covered by soil (38 to 40 in. bed spacing). One application per year.
	metam-sodium (various)	—gal/acre	37.5 to 75 gal/acre	—	2	Rate is based on soil properties and depth of soil to be treated.

Hop

L. Quesada-Ocampo, Plant Pathology

Table 10-31. Disease Control Products for Hop

Disease	Material	FRAC	Rate of Material	Minimum Days Harvest	Minimum Days Reentry	Method, Schedule, and Remarks
Downy mildew (*Pseudoperonospora humuli*)	fosetyl-Al (Aliette WDG)	33	2.5/acre	24	0.5	Apply as a directed foliar spray. When conditions are warm and humid applications should be made as follows: (1) when shoots are 6-12 inches high; (2) after training when vines are 5-6 feet tall; (3) approximately 3 weeks after the second application; and (4) during bloom. Use sufficient volume of water to insure complete coverage of foliage.
	fixed copper (various)	M	See label	See label	See label	See label. Rates vary depending on the formulation. Repeated use may cause leaf yellowing.
	dimethomorph (Forum)	40	6 fl oz/acre	7	0.5	Begin sprays prior to disease. Minimum interval is 10 days. Maximum 3 applications per season.
	cymoxanil (Curzate 60DF)	27	3.2 oz/acre	7	0.5	Tank mix with a protectant fungicide. Begin applications prior to disease and continue at 10- to 14-day intervals.
	potassium phosphite (various)	33	See label	See label	See label	See label. Rates vary depending on the formulation.
	metalaxyl (Metastar 2E)	4	1 qt/acre	45	2	Apply as a soil drench (1qt/acre in 20 gals) and follow with foliar fixed copper applications. Apply as foliar spray (1 qt/acre in 50 gals) in combination with fixed copper. Do not make more than 3 applications per season.
	ametoctradin + dimethomorph (Zampro)	45 + 40	11 to 14 fl oz/acre	7	0.5	Begin applications prior to disease and continue at 10-day intervals. Do not make more than 3 applications per season.
	famoxadone + cymoxanil (Tanos)	27 + 11	8 oz/acre	7	0.5	Begin applications prior to disease and continue at 6- to 8-day intervals. Do not make more than 6 applications per season.
	fluopicolide (Presidio)	43	4 fl oz/acre	24	12 hr	Limit of 12 fl oz/acre/year. Make no more than 2 sequential applications. Alternate with a fungicide with a different mode of action.
	mefenoxam (Ridomil Gold SL)	4	0.50 pt/acre	45	2	Can apply as soil drench or foliar spray, see label for details. Tank mix with fixed copper.
	mandipropamid (Revus)	40	8 fl oz/acre	7	4 hr	Begin applications prior to disease and continue at 7- to 10-day intervals. Do not make more than 3 applications per season.
	cyazofamid (Ranman 400 SC)	21	2.5 to 2.75 fl oz/acre	3	0.5	Begin applications prior to disease and continue at 7- to 10-day intervals. Do not apply more than 32 fl oz per season.
Powdery mildew (*Sphaerotheca humuli, S. macularis*)	cyflufenamid (Torino 0.85 SC)	U6	3.4 oz/acre	0	4 hr	Do not make more than 2 applications per crop.
	tebuconazole (Folicur 3.6F)	3	4 to 8 fl oz/acre	14	0.5	Begin applications prior to disease and continue at 10- to 14-day intervals. Do not apply more than 16.5 fl oz/acre/season.
	trifloxystrobin (Flint)	11	See label	14	0.5	Several rates available. Begin applications prior to disease and continue at 10- to 14-day intervals.
	metrafenone (Vivando)	U8	15.4 fl oz/acre	3	0.5	Begin applications prior to disease and continue at 7- to 14-day intervals. Do not make more than 2 applications per season.
	quinoxyfen (Quintec)	13	8.2 fl oz/acre	21	0.5	Do not make more than 4 applications per season.
	triflumizole (Procure 400 SC)	3	12 fl oz/acre	7	0.5	Begin applications prior to disease and continue at 14-day intervals. Do not apply more than 36 fl oz/acre/season.
	pyraclostrobin + boscalid (Pristine)	11 + 7	See label	14	0.5	Ground and aerial applications allowed, see label for details. Begin ground applications prior to disease and continue at 10- to 21-day intervals. Do not make more than 3 applications per season.

Leafy Brassica Greens (Collard, Kale, Mustard, Rape, Salad Greens Turnip Greens)

A. Keinath, Plant Pathology, Clemson University

Table 10-32. Disease Control Products for Leafy Brassica Greens

Disease	Material	FRAC	Rate of Material	Minimum Days Harvest	Minimum Days Reentry	Method, Schedule, and Remarks
For turnips harvested for roots, see Remarks and Root Vegetables						
Alternaria leaf spot, Cercospora leaf spot, Anthracnose, White spot, and various foliar diseases (see specific labels)	boscalid (Endura 70 WG)	7	6 to 9 oz/acre	14	0.5	Begin applications prior to disease development, and continue on a 7- to 14-day interval. Make no more than 2 applications per season. **Not labeled for turnip greens or roots.**
	azoxystrobin + difenoconazole (Quadris Top 2.72 SC)	11 + 3	12 to 14 fl oz/acre	1	0.5	Make no more than 1 application before alternating to another fungicide with Group 11 mode of action (NOT Quadris or Cabrio).
	azoxystrobin (various)	11	See label	0	4 hr	Make no more than 2 sequential applications before alternating with fungicides that have a different mode of action. **May be applied to turnip grown for roots.**
	fluxapyroxad + pyraclostrobin (Priaxor 500 SC)	7 + 11	6.0 to 8.2 fl oz/ acre	3	0.5	No more than 2 sequential applications before alternating with fungicides that have a different mode of action. Maximum of 3 applications. **Do not apply to turnip greens or roots.**
	pyraclostrobin (Cabrio 20 EG, Pyrac 2 EC)	11	12 to 16 oz/acre 8 to 12 oz/acre (turnip greens)	3	0.5	Begin applications prior to disease development and continue on a 7- to 10-day interval. No more than 2 sequential applications before alternating to a fungicide with different mode of action.
	tebuconazole 6 F (various)	3	3 to 4 oz/acre	7	0.5	For optimum results use as a preventative treatment. Folicur 3.6 F must have 2 to 4 hours of drying time on foliage for the active ingredient to move systemically into plant tissue before rain or irrigation occurs.
	cyprodonil + fludioxonil (Switch 62.5WG)	9 + 12	11 to 14 oz/acre	7	0.5	Apply when disease first appears, and continue on 7- to 10-day intervals. See label for complete list of greens.
	penthiopyrad (Fontelis 1.67 SC)	7	14 to 30 fl oz/acre	0	0.5	Make no more than 2 sequential applications before alternating with fungicides that have a different mode of action.
	difenoconazole + cyprodinil (Inspire Super 2.82SC)	3 + 9	16 to 20 fl oz/acre	7	0.5	Make no more than 2 sequential applications before alternating to a fungicide with a different mode of action.
Bacterial blight (*Pseudomonas*), Xanthomonas leaf blight	none					Based on field trials in SC, no fungicides, bactericides, or biopesticides are effective against these diseases. Use a 1-yr crop rotation away from all brassicas and early or once-over harvesting if disease appears.
Botrytis gray mold	penthiopyrad (Fontelis 1.67 SC)	7	14 to 30 fl oz/acre	0	0.5	Make no more than 2 sequential applications before alternating with fungicides that have a different mode of action. **May be applied to turnips grown for roots.**
	difenoconazole + cyprodinil (Inspire Super 2.82 SC)	3 + 9	16 to 20 fl oz/acre	7	0.5	Make no more than 2 sequential applications before alternating to a fungicide with a different mode of action.

Table 10-32. Disease Control Products for Leafy Brassica Greens

Disease	Material	FRAC	Rate of Material	Minimum Days Harvest	Minimum Days Reentry	Method, Schedule, and Remarks
Downy Mildew	pyraclostrobin (Cabrio 20 EG)	7	12 to 16 oz/acre	3	0.5	Begin applications prior to disease development and continue on a 7- to 10-day interval. Make no more than 2 sequential applications before alternating to a fungicide with a different mode of action.
	fluopicolide (Presidio 4 SC)	43	3 to 4 fl. oz/acre	2	0.5	Make applications on a 7- to 10-day schedule. Presidio must be tank mixed with another fungicide with a different mode of action. Make no more than 2 sequential applications before rotating to a fungicide with a different mode of action. Apply no more than 12 oz per acre and make no more than 4 applications per season.
	cyazofamid (Ranman 400 SC)	21	2.75 fl. oz/acre	0	0.5	Make applications on a 7- to 10-day schedule. Do not apply more than 39.5 fl oz per acre per crop growing season.
	mandipropamid (Revus 2.08 SC)	40	8.0 fl oz/acre	1	0.5	Begin applications prior to disease development and continue on a 7- to 10-day interval. Make no more than 2 consecutive applications before switching to another effective non-group 40 fungicide. **Not labeled for turnip greens or roots.**
	fenamidone (Reason 500SC)	11	5.5 to 8.2 oz/acre	2	0.5	Begin applications as soon as conditions become favorable for disease development. Applications should be made on a 5- to 10-day interval. Do not make more than one application of Reason 500 SC before alternating with a fungicide from a different resistance management group.
	amectoctradin + dimethomorph (Zampro 525 SC)	45 + 40	14 fl oz/acre	0	0.5	Do not make more than 2 sequential applications before alternating to a fungicide with a different mode of action. Addition of an adjuvant may improve performance (see label for specifics).
	dimethomorph (Forum 4.16 SC)	40	6.4 oz/acre	0	0.5	Must be tank-mixed with another fungicide active against Phytophthora blight. Do not make more than 2 sequential applications before alternating to another effective fungicide with a different mode of action. Do not make more than 5 applications per season. **Do not apply to turnip greens or roots.**
	fosetyl-Al (Aliette 80W DG)	33	2 to 5 lb/acre	3	1	Apply when disease first appears; then repeat on 7- to 21-day intervals. Do not tank mix with copper fungicides. A maximum of 7 applications can be made per season. **Do not apply to turnip greens or roots.**
	potassium phosphite	33	2 to 4 pt/acre	0	4 hr	Apply when weather is foggy as a preventative. Do not apply to plants under water or temperature stress. Spray solution should have a pH greater than 5.5. Apply in at least 30 gallons water per acre.
Powdery mildew	boscalid (Endura 70 WG)	7	6 to 9 oz/acre	14	0.5	Begin applications prior to disease development, and continue on a 7- to 14-day interval. Make no more than 2 applications per season; disease suppression only. **Do not apply to turnip greens or roots.**
	pyraclostrobin (Cabrio 20 EG, Pyrac 2 EC)	11	12 to 16 oz/acre	3	0.5	Begin applications prior to disease development and continue on a 7- to 10-day interval. Make no more than 2 sequential applications before alternating to a fungicide with a different mode of action.
	triflumizole (Procure 480SC)	3	6 to 8 oz/acre	1	0.5	Make no more than 2 sequential applications before rotating with a fungicide with a different mode of action. Do not rotate with Rally or Nova.
	cyprodonil + fludioxonil (Switch 62.5WG)	9 + 12	11 to 14 oz/acre	7	0.5	Apply when disease first appears, and continue on 7- to 10-day intervals. See label for complete list of greens. May be used on turnip where leaves only will be harvested. **Do not apply to turnip grown for roots.**
	penthiopyrad (Fontelis 1.07 SC)	7	14 to 30 fl oz/acre	0	0.5	Make no more than 2 sequential applications before alternating with fungicides that have a different mode of action. **May be applied to turnips grown for roots.**
	difenoconazole + cyprodinil (Inspire Super)	3 + 9	16 to 20 fl oz/acre	7	0.5	Make no more than 2 sequential applications before alternating to a fungicide with a different mode of action.
	fluxapyroxad + pyraclostrobin (Priaxor 500 SC)	7 + 11	6.0 to 8.2 fl oz/ acre	3	0.5	Make no more than 2 sequential applications before alternating with fungicides that have a different mode of action. Maximum of 3 applications. **Do not apply to turnip greens or roots.**
	tebuconazole (various)	3	3 to 4 oz/acre	7	0.5	For optimum results use as a preventative treatment. Folicur 3.6 F must have 2 to 4 hours of drying time on foliage for the active ingredient to move systemically into plant tissue before rain or irrigation occurs. **May be applied to turnip grown for roots.**
Rhizoctonia bottom rot	boscalid (Endura 70 WG)	7	6 to 9 oz/acre	14	0.5	Begin applications prior to disease development, and continue on a 7- to 14-day interval. Make no more than 2 applications per season; disease suppression only. **Do not apply to turnip greens or roots.**

Table 10-32. Disease Control Products for Leafy Brassica Greens

Disease	Material	FRAC	Rate of Material	Minimum Days Harvest	Minimum Days Reentry	Method, Schedule, and Remarks
Sclerotinia stem rot (white mold)	boscalid (Endura 70 WG)	7	6 to 9 oz/acre	14	0.5	Begin applications prior to disease development, and continue on a 7- to 14-day interval. Make no more than 2 applications per season. **Do not apply to turnip greens or roots.**
	penthiopyrad (Fontelis 1.67 SC)	7	16 to 30 fl oz/acre	0	0.5	Do not exceed 72 fl oz of product per year. Make no more than 2 sequential applications per season before rotating to another effective product with a different mode of action.
	Coniothyrium minitans (Contans WG)	—	1 to 4 lb/acre	0	4 hr	**OMRI listed product.** Apply to soil surface and incorporate no deeper than 2 inches. Works best when applied prior to planting or transplanting. Do not apply other fungicides for 3 weeks after applying Contans.
Seedling root rot, basal stem rot (Rhizoctonia)	azoxystrobin (Quadris 2.08 SC)	11	0.4 to 0.8 fl oz per 1000 row feet	0	4 hr	Apply at planting as a directed spray to the furrow in a band 7 inches wide.
White rust	azoxystrobin (Quadris 2.08 SC)	11	6.2 to 15.4 fl oz/ acre	0	4 hr	Make no more than 2 sequential applications.
	fenamidone (Reason 500SC)	11	8.2 oz/acre	2	0.5	Begin applications as soon as conditions become favorable for disease development. Applications should be made on a 5- to 10-day interval. Do not make more than 1 application of Reason 500 SC before alternating with a fungicide from a different resistance management group.
	fluxapyroxad + pyraclostrobin (Priaxor 500 SC)	7 + 11	6.0 to 8.2 fl oz/ acre	3	0.5	Make no more than 2 sequential applications before alternating with fungicides that have a different mode of action. Maximum of 3 applications. **Do not apply to turnip greens or roots.**

Jerusalem Artichoke (Sunchoke)

A. Keinath, Plant Pathology, Clemson University

Table 10-33. Disease Control Products for Jerusalem Artichoke (Sunchoke)

Disease	Material	FRAC	Rate of Material	Minimum Days Harvest	Minimum Days Reentry	Method, Schedule, and Remarks
Pythium damping off	mefenoxam (Ridomil Gold 4 SL) (Ultra Flourish 2 SL)		1 to 2 pt/treated acre 2 to 4 pt/treated acre	1	2	Soil incorporation. See label for row rates.
	fluopicolide (Presidio 4 SC)	43	3 to 4 fl oz/acre	7	0.5	Apply every 10 days if needed. Do not use more than 2 times sequentially and not more than 4 times at the low rate or 3 times at the high rate per acre per season.
Cercospora leaf spot, Powdery mildew, Rust	azoxystrobin (various)	11	See label	14	4 hr	Must rotate every other application with a non-Group 11 fungicide. Maximum of 8 applications per crop per year.
	azoxystrobin + difenoconazole (Quadris Top 2.72 SC)	11 + 3	8 to 14 fl oz/acre	1	0.5	Make no more than 2 applications before alternating to another fungicide with a different mode of action (NOT Quadris).
	difenoconazole + benzovindiflupyr (Aprovia Top 1.62EC)	3 + 7	10.5 to 13.5 fl oz/acre	14	0.5	Make no more than 2 consecutive applications before switching to a non-Group 7 fungicide. Make no more than 3 applications at the low rate or 2 applications at the high rate per year.
Southern blight	azoxystrobin (Quadris 2.08 SC)	11	0.4 to 0.8 fl oz/1000 row ft	14	4 hr	Make 1 application at the high rate before symptoms typically are seen, based on prior year observations.
White mold (*Sclerotinia* basal stalk rot)	boscalid (Endura 70 EG)	7	10 oz/acre	30	0.5	2 applications per crop per season.
	Coniothyrium minitans (Contans WG)	NA	1 to 4 lb/acre	0	4 hr	**OMRI listed product.** Apply to soil surface and incorporate no deeper than 2 inches. Works best when applied prior to planting or transplanting. Do not apply other fungicides for 3 weeks after applying Contans.

Lettuce and Endive
A. Keinath, Plant Pathology, Clemson University

Table 10-34. Disease Control Products for Lettuce and Endive

Disease	Material	FRAC	Rate of Material	Minimum Days Harvest	Minimum Days Reentry	Method, Schedule, and Remarks
Bottom rot, (Rhizoctonia)	azoxystrobin (various)	11	0.4 to 0.8 fl oz/1,000 row ft	—	4 hr	*Rhizoctonia* only. Make in-furrow or banded applications shortly after plant emergence.
Seed decay, Seedling blight, Damping off)	fludioxonil (Spirato 480FS (Maxim 4FS)	12	0.08 to 0.16 fl oz/100 lb of seed	—	12	Used to control diseases of seed such as Aspergillus, Fusarium, and Rhizoctonia among others. Does NOT control Pythium or Phytophthora.
Downy mildew	acibenzolar-*S*-methyl (Actigard 50WG)	P1	0.75 to 1 oz/acre	7	0.5	Do not apply prior to thinning or within 5 days after transplanting. Apply preventatively every 7 to 10 days, not to exceed 4 applications (4 oz) per season.
	ametoctradin + dimethomorph (Zampro 525 SC)	45 + 40	14 fl oz/acre	0	12 hr	Do not make more than 2 sequential applications before alternating to a fungicide with a different mode of action. Addition of an adjuvant may improve performance (see label for specifics). Do not apply more than 42 fl oz/acre/season.
	azoxystrobin (various)	11	6.2 to 15.4 fl oz/acre	0	4 hr	Make no more than 2 sequential applications before alternating with a fungicide with a different mode of action.
	cyazofamid (Ranman 400 SC)	21	2.75 fl oz/acre	0	0.5	Apply on a 7- to 10-day interval when disease first appears or when conditions favorable for disease development. Do not make subsequent applications and limit applications to 6 per year.
	cymoxanil (Curzate)	27	3.2 to 5.0 oz/acre	3	0.5	Curzate is only labeled for lettuce and spinach. Use only in combination with a protectant fungicide. Apply on a 5- to 7-day schedule, not to exceed 30 oz/acre per a 12-month period.
	cymoxanil + famoxadone (Tanos)	27 + 11	8.0 oz	1	0.5	See label for directions.
	dimethomorph (various)	40	6.4 oz/acre	0	0.5	Must be applied as a tank mix with another fungicide active against downy mildew. Do not make more than 2 sequential applications.
	fenamidone (Reason 500SC)	11	5.5 to 8.2 fl oz	2	0.5	Alternate with fungicides with a different mode of action.
	fluopicolide (Presidio)	3 to 4 fl oz/acre		0.09 to 0.125 lb/acre	2	0.5	Tank mix with another downy mildew fungicide with a different mode of action.
	mandipropamid (various)	40	See label	See label	See label	Begin applications as soon as crop and/or environmental conditions become favorable for disease development. Apply on a 7- to 10-day interval depending upon disease conditions.
	oxathiapiprolin + mefenoxam (Orondis Gold 200)	49 + 4	4.8 to 9.6 fl oz/acre	0	4 hr	Limit to 38.6 fl oz/acre/year. Limit to 6 applications/acre/year for the same crop. Do not follow soil applications of Orondis with foliar applications of Orondis. Apply at planting in furrow, by drip or in transplant water. Use the higher rates for heavier soils, for longer application intervals or for susceptible varieties.
	oxathiapiprolin + mandipropamid (Orondis Ultra)	49 + 40	5.5 to 8 fl oz/acre	0	4 hr	Limit to 6 applications/acre/year for the same crop. Do not follow soil applications of Orondis with foliar applications of Orondis. Use the higher rates when disease is present.
	mono- and dipotassium salts of phosphorous acid (Alude, K-Phite)	33	1 to 4 qt in a minimum of 10 gal/acre	0	4 hr	Do not apply at a less than 3 day interval.
	propamocarb (Previcur Flex)	28	2 pt/acre	2	0.5	Previcur Plus is only labeled for head and leaf lettuce. Do not apply more than 8 pt per growing season; begin applications before infection and continue on a 7- to 10-day interval.
	azoxystrobin (various)	11	6.2 to 15.4 fl oz/acre	7	4 hr	Use highest rate for downy mildew. Make no more than 2 sequential applications before alternating with fungicides that have a different mode of action. Apply no more than 2.88 qt/crop/acre/season.
	fixed copper (various)	M1	See label	See label	See label	See label. Rates vary depending on the formulation.
	pyraclostrobin (various)	11	12 to 16 oz/acre	0	0.5	Begin applications prior to disease development and continue on 7- to 14-day intervals.
	fluxapyroxad + pyraclostrobin (Merivon 500 SC)	7 + 11	4 to 11 fl oz/acre	1	0.5	Make no more than 2 sequential applications before alternating with fungicides that have a different mode of action. Suppression only of downy mildew.

Table 10-34. Disease Control Products for Lettuce and Endive

Disease	Material	FRAC	Rate of Material	Minimum Days Harvest	Minimum Days Reentry	Method, Schedule, and Remarks
Downy mildew, leaf spots (continued)	mancozeb (various)	M3	See label	See label	See label	Rates vary depending on the formulation. Spray at first appearance of disease and continue on a 7- to 10-day interval.
Leaf spots	penthiopyrad (Fontelis)	7	14 to 24 fl oz/acre	3	0.5	Begin applications before disease development. DO NOT make more than 2 consecutive applications before switching to a fungicide with a different mode of action.
	flutriafol (Rhyme)	3	5 to 7 oz/acre	7	0.5	Apply preventatively or when conditions are favorable for disease development.
	cyprodinil + fludioxonil (Switch 62.5 WDG)	9 + 12	11 to 14 oz/acre	0	0.5	Switch also has activity against basal rot, Sclerotinia and Gray mold. Alternate with a fungicide with a different mode of action after 2 applications.
Gray mold	dicloran (Botran 5F)	14	See label	14	0.5	Application instructions vary by crop; see label. 2 applications may be applied per season. Do not apply more than 3.2 qt/season.
	penthiopyrad (Fontelis)	7	14 to 24 fl oz/acre	3	0.5	Begin applications before disease development. **DO NOT** make more than 2 consecutive applications before switching to a fungicide with a different mode of action.
	boscalid (Endura)		7 to 9 oz/acre	14	0.5	Begin applications prior to the onset of disease and continue on a 7-day interval.
Seed decay, Seedling blight, damping off	fludioxonil (Spirato 480 FS) (Maxim 4FS)	12	0.08 to 0.16 fl oz/100 lb of seed	—	12	Used to control diseases of seed such as Aspergillus, Fusarium, and Rhizoctonia among others. Does NOT control Pythium or Phytophthora.
Powdery mildew	azoxystrobin (various)	11	6.2 to 15.4 fl oz/acre	0	4 hr	Make no more 2 sequential applications before alternating with fungicides that have a different mode of action.
	fluxapyroxad + pyraclostrobin (Merivon 500 SC)	7 + 11	4 to 11 fl oz/acre	1	0.5	Make no more than 2 sequential applications before alternating with fungicides that have a different mode of action.
	myclobutanil (Rally 40 WSP)	3	5 oz/acre	3	1	For use on lettuce only. Apply when disease first appears and continue on a 14-day interval.
	penthiopyrad (Fontelis)	7	14 to 24 fl oz/acre	0	0.5	Begin applications prior to disease development. DO NOT make more than 2 sequential applications before switching to a fungicide with a different mode of action.
	quinoxyfen (Quintec)	13	6 fl oz	1	1	Alternate with a fungicide with a different mode of action.
	sulfur (various)	M2	See label	See label	See label	Apply at early leaf stage and repeat every 10 to 14 days or as needed. Do not apply if temperatures are expected to exceed 90 degrees F within 3 days of application due to the risk of crop injury.
	triflumizole (various)	3	6 to 8 fl oz/acre	0	0.5	Applications should begin prior to disease development. Repeat on a 14-day schedule. Do not apply more than 18 fl oz/acre/season.
Pythium damping	metalaxyl (various)	4	See label	—	2	Banded over the row, preplant incorporated or injected with liquid fertilizer.
	propamocarb (Previcur Flex)		2 pt/acre	2	0.5	Previcur Plus is only labeled for head and leaf lettuce. Various application methods; see label.
Rust	penthiopyrad (Fontelis)	7	14 to 24 fl oz/acre	3	0.5	Begin applications before disease development. DO NOT make more than 2 sequential applications before switching to a fungicide with a different mode of action.
	sulfur (various)	M2	See label	14	1	Apply at early leaf stage and repeat every 10 to 14 days or as needed. Do not apply if temperatures are expected to exceed 90 degrees F within 3 days of application due to the risk of crop injury.
Sclerotinia	boscalid (Endura)	7	See label	14	0.5	Begin applications prior to onset of disease. Use higher rate when disease pressure is high.
	Coniothyrium minitans (Contans WG)	NA	1 to 4 lb/acre	0	4 hr	OMRI listed product. Apply to soil surface and incorporate no deeper than 2 inches. Works best when applied prior to planting or transplanting. Do not apply other fungicides for 3 weeks after applying Contans.
	dicloran (Botran)	14	See label	14	0.5	Rate depends on specific crop and timing of application. See label.
	fludioxonil (Cannonball WP)	12	7 oz/acre	0	0.5	Ground applications only. Do not apply more than 28 oz/acre/year.
	iprodione (Rovral)	2	1.5 to 2 lb/acre	14	1	Only for use on lettuce. Also effective for bottom rot and Botrytis. Use higher rate when disease pressure is high.
	penthiopyrad (Fontelis)	7	16 to 24 fl oz/acre	3	0.5	Begin applications before disease development. Continue on 7- to 14-day intervals. Do not make more than 2 consecutive applications before switching to a fungicide with a different mode of action.

For Melons – See Cucurbits

Okra

E. Sikora, Plant Pathology, Auburn University

Table 10-35. Disease Control Products for Okra

Disease	Material	FRAC	Rate of Material	Minimum Days Harvest	Minimum Days Reentry	Method, Schedule, and Remarks
Alternaria, gray mold, powdery mildew	cyprodinil; fludioxonil (Switch) 62.5WG	9 + 12	11 to 14 oz/acre	0	0.5	Begin applications before disease development and continue on 7- to 10-day interval. Make no more than 2 consecutive applications before alternating to a fungicide with a different. Do not apply more than 56 oz per acre per season.
Alternaria, gray mold, powdery mildew, Septoria leaf spot, target spot	penthiopyrad (Fontelis)	7	16 to 24 fl oz/acre	0	0.5	Begin applications before disease development and continue on 7- to 14-day interval. Do not exceed more than 72 oz per acre per year.
Anthracnose, bacterial leaf spot, leaf spots, pod spots, powdery mildew	fixed copper (various)	M	See label	0	See label	
Anthracnose, Botrytis leaf mold, powdery mildew, Cercospora leaf spot	chlorothalonil; cymoxanil (Ariston)	M + 27	2 to 4.4 pt/acre	3	0.5	Begin applications before disease development and continue on 7-day interval.
Anthracnose, gray leaf spot, powdery mildew, Cercospora leaf spot	difenoconazole; azoxystrobin (Quadris Top)	3 + 11	8 to 14 fl oz/acre	0	0.5	Begin applications before disease development and continue on 7- to 10-day interval. Make no more than 2 consecutive applications before alternating to a fungicide with a different. Do not apply more than 55 fl oz per acre per season.
	difenoconazole; cyprodinil (Inspire Super)	3 + 9	16 to 20 fl oz/acre	0	0.5	Begin applications before disease development and continue on 7 to 10-day interval. Make no more than 2 consecutive applications before alternating to a fungicide with a different. Do not apply more than 47 fl oz per acre per season.
	fludioxonil (various)	12	5 to 7 fl oz/acre	0	0.5	Begin applications before disease development and continue on 7-day interval.
Anthracnose, gray leaf spot, powdery mildew, Cercospora leaf spot, Rhizoctonia stem rot	difenoconazole; benzovindiflupyr (Aprovia Top)	3 + 7	10.5 to 13.5 oz/acre	0	0.5	Begin applications before disease development and continue on 7- to 10-day interval. Make no more than 2 consecutive applications before alternating to a fungicide with a different mode of action. Refer to label for information on addition of an adjuvant.
Cercospora leaf spot	chlorothalonil (various)	M	1.5 pt/acre	3	0.5	Begin applications when disease is expected. Repeat every 7-to 10-days.
	tebuconazole (various)	3	4 to 6 fl oz/acre	3	0.5	**DO NOT** apply more than 24 fl oz per acre per season.
Downy mildew	mandipropamid (Micora)	40	5.5 to 8 fl oz/acre	—	4 hr	Tank mix Micora with a non-Group 40 fungicide and begin applications prior to disease development. **DO NOT** apply more than 2 applications per crop, or in consecutive applications.
Powdery mildew, anthracnose, Cercospora leaf spot	flutriafol (Topguard)	3	14 fl oz/acre	0	0.5	Apply preventatively or when conditions are favorable for disease development.
Anthracnose, Powdery mildew	azoxystrobin (various)	11	6.0 to 15.5 fl oz/acre	0	4	Do not make more than 2 sequential applications before alternating with a fungicide with a different mode of action. Do not make more than 4 applications strobilurin fungicides per acre per season.
	chlorothalonil (various)	MI	1.5 pt/acre	3	0.5	Begin applications when disease is expected. Repeat every 7 to 10 days.
	myclobutanil (Rally 40WSP)	3	2.5 to 5 oz/acre	0	1	Do not make more than 4 applications per season. Minimum re-treatment interval: 10 to 14 days.
Phytophthora blight	oxathiapiprlin + mandipropamid (Orondis Ultra A + Orondis Ultra B)	49 + 40	2.4 to 4.8 fl oz/acre + 8 fl oz/acre	0	4	Apply at planting, in furrow, by drip, or in transplant water. Disease suppression only. Do not make more than 2 applications before switching to a different mode of action.
Rhizoctonia seedling rot	azoxystrobin (various)	11	0.4 to 0.8 fl oz/1000 row ft	—	4 hr	Make in-furrow or banded applications shortly after plant emergence.

Onion

E. Pfeufer, Plant Pathology, University of Kentucky

Table 10-36. Disease Control Products for Onion

Disease	Material	FRAC	Rate of Material	Minimum Days Harvest	Minimum Days Reentry	Method, Schedule, and Remarks
ONION (green)						
Damping off (*Pythium* spp.)	mefenoxam (Ridomil Gold) 4 SL	4acre	0.5 to 1 pt/trt acre	—	2	See label for low rates. Also for dry onion.
	metalaxyl (various)	4	2 to 4 pt/trt acre	—	2	Preplant incorporated or soil surface spray.
Downy mildew (*Peronospora destructor*)	azoxystrobin (various)	11	9.2 to 15.4 fl oz/acre	0	4 hr	Make no more than 1 application before alternating with a fungicide with a different mode of action. Apply no more than 2.88 qt per crop per acre per season.
	azoxystrobin + chlorothalonil (Quadris Opti)	11 + M	2.4 to 3.6 pt/acre	14	0.5	Make no more than 1 application before alternating with a fungicide with a different mode of action.
	chlorothalonil (various)	M	See label	14	2	Suppression only. Maximum of 3 sprays.
	chlorothalonil + cymoxanil (Ariston)	M + 27	2.0 to 2.4 pt/acre	14	0.5	Apply prior to favorable infection periods; continue on 7- to 9-day interval; alternate with a different mode of action.
	dimethomorph (Forum 50 WP)	40	6.4 oz/acre	0	0.5	Must be applied as a tank mix with another fungicide active against downy mildew. Do not make more than 2 sequential applications.
	fenamidone (Reason 500 SC)	11	5.5 fl oz/acre	7	0.5	Begin applications when conditions favor disease development, and continue on 5- to 10-day interval. Do not apply more than 22 fl oz per growing season. Alternate with fungicide from different resistance group.
	fluxapyroxad + pyraclostrobin (Merivon)	7 + 11	8 to 11 fl oz/ acre	7	0.5	Suppression only. Apply at disease onset; continue on 7- to 14-day schedule. No more than 3 applications per season.
	mandipropamid (Revus 2.08F)	40	8 fl oz/acre	7	0.5	Apply prior to disease development and continue throughout season at 7- to 10-day intervals; maximum 24 fl oz per season.
	amectoctradin + dimethomorph (Zampro)	45 + 40	14.0 fl oz/acre	0	12 hr	Begin applications prior to disease development and continue on a 5- to 7-day spray interval.
	mefenoxam + chlorothalonil (Ridomil Gold/Bravo)	4 + M	2.5 lb/acre	14	2	
	pyraclostrobin (Cabrio)	11	8 to 12 oz/acre	7	0.5	Make no more than 2 sequential applications and no more than 6 applications per season.
	pyraclostrobin + boscalid (Pristine)	11 + 7	18.5 oz/acre	7	1	For suppression only. Make a maximum of 6 applications per season.
Leaf blight (*Botrytis* spp.)	azoxystrobin (various)	11	6.2 to 15.4 fl oz/ acre	7	4 hr	Make no more than 2 sequential applications before alternating with fungicides that have a different mode of action. Apply no more than 2.88 qt per crop per acre per season.
	azoxystrobin + difenoconazole (Quadris Top)	11 + 3	12 to 14 oz/acre	7	0.5	Make no more than 1 application before alternating with a fungicide with a different mode of action.
	azoxystrobin + chlorothalonil (Quadris Opti)	11 + M	1.6 to 3.6 pt/acre	14	2	Make no more than 1 application before alternating with a fungicide with a different mode of action.
	azoxystrobin + propiconazole (various)	11 + 3	14 to 26 fl oz	0	0.5	Make only 1 application before rotating to a non-group 11 fungicide.
	azoxystrobin + tebuconazole (Custodia)	11 + 3	8.6 to 12.9 fl oz	7	0.5	Use higher rate and shorter interval when disease conditions are severe.
	boscalid (Endura 70 WG)	7	6.8 oz/acre	7	0.5	Do not make more than 2 sequential applications or more than 6 applications per season.
	chlorothalonil (various)	M	See label	14	0.5	Spray at first appearance. Maximum of 3 sprays.
	cyprodinil + fludioxonil (Switch)	9 + 12	11 to 14 oz/acre	7	0.5	Do not plant rotational crops other than onions or strawberries for 12 months following the last application.
	dicloran (Botran) 75 W	14	1.5 to 2.7 lb/acre	14	0.5	
	difenoconazole + cyprodinil (Inspire Super)	3 + 9	16 to 20 fl oz/acre	14	0.5	Make no more than 2 applications before alternating with a fungicide with a different mode of action.
	fluopyram + tebuconazole (Luna Experience)	7 + 3	8.0 to 12.8 fl. oz/acre	7	0.5	Observe seasonal application limits for both group 7 and group 3 fungicides.
	fluxapyroxad + pyraclostrobin (Merivon)	7 + 11	4 to 11 fl oz/ acre	7	0.5	Apply at disease onset; continue on 7- to 14-day schedule. No more than 3 applications per season.
	mefenoxam + chlorothalonil (Ridomil Gold/Bravo)	4 + M	2.5 lb/acre	14	2	
	penthiopyrad (Fontelis)	7	16 to 24 fl oz/acre	3	12 hr	Begin sprays prior to disease development and continue on a 7- to 14-day schedule.
	propiconazole (various)	3	4 to 8 oz/acre	0	0.5	Alternate with a different mode of action.
	pyraclostrobin (Cabrio)	11	8 to 12 oz/acre	7	0.5	Make no more than 2 sequential applications and no more than 6 applications per season.
	pyraclostrobin + boscalid (Pristine)	11 + 7	14.5 to 18.5 oz/acre	7	1	Make a maximum of 6 applications per season.

Table 10-36. Disease Control Products for Onion

Disease	Material	FRAC	Rate of Material	Minimum Days Harvest	Minimum Days Reentry	Method, Schedule, and Remarks
ONION (green) (continued)						
Leaf blight (*Botrytis*) (continued)	pyrimethanil (Scala)	9	9 or 18 fl oz/acre	7	0.5	Use lower rate in a tank mix with broad-spectrum fungicide and higher rate when applied alone. Do not apply more than 54 fl oz per crop.
Purple blotch (*Alternaria porri*)	azoxystrobin (various)	11	6.2 to 12.3 fl oz/ acre	7	4 hr	Make no more than 2 sequential applications before alternating with a fungicide with a different mode of action. Apply no more than 2.88 qt per crop per acre per season.
	azoxystrobin + difenoconazole (Quadris Top)	11 + 3	12 to 14 oz/acre	7	0.5	Make no more than 1 application before alternating with a fungicide with a different mode of action.
	azoxystrobin + chlorothalonil (Quadris Opti)		1.6 to 3.2 pt/acre	14	0.5	Make no more than 1 application before alternating with a fungicide with a different mode of action.
	azoxystrobin + propiconazole (various)	11 + 3	14 to 26 fl oz	0	0.5	Make only 1 application before rotating to a non-group 11 fungicide.
	azoxystrobin + tebuconazole (Custodia)	11 + 3	8.6 to 12.9 fl oz	7	0.5	Use higher rate and shorter interval when disease conditions are severe.
	boscalid (Endura) 70WG	7	6.8 oz/acre	7	0.5	Do not make more than 2 sequential applications or more than 6 applications per season.
	chlorothalonil (various)	M	See labels	14	2	Spray at first appearance. Maximum of 3 sprays.
	chlorothalonil + cymoxanil (Ariston)	M + 27	2.0 to 2.4 pt/acre	14	0.5	Apply prior to favorable infection periods; continue on 7- to 9-day interval; alternate with a different mode of action.
	cyprodinil + fludioxonil (Switch)	9 + 12	11 to 14 oz/acre	7	0.5	Do not plant rotational crops other than onions or strawberries for 12 months following the last application.
	difenoconazole + cyprodinil (Inspire Super)	3 + 9	16 to 20 fl oz/acre	14	0.5	Make no more than 2 applications before alternating with a fungicide with a different mode of action.
	fenamidone (Reason)	11	5.5 to 8.2 fl oz	7	0.5	Begin applications when conditions favor disease development, and continue on 5- to 10-day interval. Do not apply more than 22 fl oz per growing season. Alternate with fungicide from different resistance management group.
	mefenoxam + chlorothalonil (Ridomil Gold/Bravo)	4 + M	2.5 lb/acre	14	2	
	penthiopyrad (Fontelis)	7	16 to 24 fl oz/acre	3	0.5	Begin sprays prior to disease development and continue on a 7- to 14-day schedule.
	propiconazole (various)	3	4 to 8 fl oz	0	0.5	Alternate with a different mode of action.
	pyraclostrobin (Cabrio)	11	8 to 12 oz/acre	7	0.5	Make no more than 2 sequential applications and no more than 6 applications per season.
	pyraclostrobin + boscalid (Pristine)	11 + 7	10.5 to 18.5 oz/acre	7	1	Make a maximum of 6 applications per season.
	pyrimethanil (Scala)	9	9 or 18 fl oz/acre	7	0.5	Use lower rate in a tank mix with broad spectrum fungicide and higher rate when applied alone. Do not apply more than 54 fl oz per crop.
Stemphylium leaf blight (*Stemphylium vesicarium*)	azoxystrobin + difenoconazole (Quadris Top)	11 + 3	12 to 14 oz/acre	7	0.5	Make no more than 1 application before alternating with a fungicide with a different mode of action.
	azoxystrobin + propiconazole (Avaris 2XS)	11 + 3	14 to 26 fl oz	0	0.5	Make only 1 application before rotating to a non-group 11 fungicide.
	difenoconazole + cyprodinil (Inspire Super)	3 + 9	16 to 20 fl oz/acre	14	0.5	Make no more than 2 applications before alternating with a fungicide with a different mode of action.
	fluxapyroxad + pyraclostrobin (Merivon)	7 + 11	4 to 11 fl oz/ acre	7	0.5	Apply at disease onset; continue on 7- to 14-day schedule. No more than 3 applications preseason.
	pyraclostrobin + boscalid (Pristine) 38 WG	11 + 7	10.5 to 18.5 oz/acre	7	1	Make no more than 6 applications per season.
ONION (dry)						
Damping off (*Pythium* spp.)	mefenoxam (Ridomil Gold)	4	0.5 to 1 pt/trt acre	—	2	See label for row rates. Also for green onion.
	metalaxyl (various)	4	2 to 4 pt/trt acre	—	2	Preplant incorporated or soil surface spray.
	azoxystrobin + mefenoxam (Uniform)	11 + 4	0.34 fl oz / 1000ft	-	0	In furrow treatment.
Downy mildew	azoxystrobin (various)	11	9.2 to 15.4 fl oz/ acre	0	4 hr	Make no more than 1 application before alternating with a fungicide with a different mode of action. Apply no more than 2.88 qt per crop per acre per season.
	azoxystrobin + chlorothalonil (Quadris Opti)	11 + M	2.4 to 3.2 pt/acre	14	0.5	Make no more than 1 application before alternating with a fungicide with a different mode of action.
	amectoctradin + dimethomorph (Zampro)	45 + 40	14.0 fl oz/acre	0	12 hr	Begin applications prior to disease development and continue on a 5- to 7-day spray interval.
	chlorothalonil + cymoxanil (Ariston)	M + 27	1.6 to 2.4 pt/acre	7	0.5	Apply prior to favorable infection periods; continue on 7- to 9-day interval; alternate with a different mode of action.
	dimethomorph (Forum)	40	6.4 oz/acre	0	0.5	Must be applied as a tank mix with another fungicide active against downy mildew. Do not make more than 2 sequential applications.
	cyazofamid (Ranman)		2.75 to 3.0 oz/acre	0	0.5	Use a surfactant for best results.

Table 10-36. Disease Control Products for Onion

Disease	Material	FRAC	Rate of Material	Minimum Days Harvest	Minimum Days Reentry	Method, Schedule, and Remarks
ONION (green) (continued)						
Downy mildew (continued)	famoxadone + cymoxanil (Tanos)	11 + 27	8.0 oz/acre	3	0.5	Apply preventively on a 5- to 7-day schedule and do not rotate with group 11 fungicides.
	fenamidone (Reason)	11	5.5 fl oz/acre	7	0.5	Use as soon as environmental conditions become favorable.
	fluazinam (Omega 500)		1.0 pt/acre	7	1	Initiate sprays when conditions are favorable for disease or at disease onset. Spray on a 7- to 10-day schedule.
	mandipropamid (Revus)	40	8 fl oz/acre	1	0.5	Apply prior to disease development and continue throughout season at 7- to 10-day intervals; maximum 32 fl oz per season.
	mefenoxam + mancozeb (Ridomil Gold MZ)	4 + M	2.5 lb/trt acre	7	2	Use with a suitable adjuvant.
	pyraclostrobin + boscalid (Pristine)	11 + 7	18.5 oz/acre	7	1	Suppression only. Make no more than 6 applications per season.
Leaf blight (*Botrytis* spp.)	azoxystrobin (various)	11	6.2 to 15.4 fl oz/ acre	7	4 hr	Make no more than 2 sequential applications before alternating with fungicides with different mode of action. Apply no more than 2.88 qt per crop per acre per season.
	azoxystrobin + chlorothalonil (Quadris Opti)	11 + M	1.6 to 3.2 pt/acre	14	0.5	Make no more than 1 application before alternating with a fungicide with a different mode of action.
	penthiopyrad (Fontelis)	7	16 to 24 fl oz/acre	3	12 hr	Begin sprays prior to disease development and continue on a 7- to 14-day schedule.
	cyprodinil + fludioxonil (Switch)	9 + 12	11 to 14 oz/acre	7	0.5	Do not plant rotational crops other than onions or strawberries for 12 months following the last application.
	dicloran (Botran)	14	1.5 to 2.7 lb/acre	14	0.5	Use lower rate in a tank mix with broad-spectrum fungicide and higher rate when applied alone. Do not apply more than 54 fl oz per crop.
	difenoconazole + cyprodinil (Inspire Super)	3 + 9	16 to 20 fl oz/acre	7 to 14	0.5	Make no more than 2 applications before alternating with a fungicide with a different mode of action.
	fixed copper (various)	M	See label			Spray at first appearance, 7- to 10-day intervals. Do not apply to exposed bulbs.
	pyraclostrobin (Cabrio)	11	12 oz/acre	7	0.5	Make no more than 2 sequential applications and no more than 6 applications per season.
	pyrimethanil (Scala)	9	9 or 18 fl oz/acre	7	0.5	Use lower rate in a tank mix with broad-spectrum fungicide and higher rate when applied alone. Do not apply more than 54 fl oz per crop.
Neck rot (*Botrytis* spp.), purple blotch (*Alternaria porri*), downy mildew (*Peronospora destructor*)	azoxystrobin + chlorothalonil (Quadris Opti)	11 + M	1.6 to 3.2 pt/acre	14	0.5	Make no more than 1 application before alternating with a fungicide with a different mode of action.
	azoxystrobin + propiconazole (various)	11 + 3	14 to 26 oz / acre	14	0.5	
	azoxystrobin + tebuconazole (various)	11 + 3	See label	7	0.5	See label for specific rates and application instructions.
	penthiopyrad (Fontelis)	7	16 to 24 fl oz/acre	3	12 hr	Begin sprays prior to disease development and continue on a 7- to 14-day schedule
	chlorothalonil (various)	M	0.9 to 1 lb/acre	7	0.5	Will only suppress neck rot and downy mildew.
	chlorothalonil + zoxamide (Zing)	M + 22	30 fl oz/acre	7	0.5	Follow protective spray schedule when diseases are in the area.
	cyprodinil (Vanguard)	12	10 oz/acre	7	0.5	Suppressive only on neck rot.
	boscalid (Endura)	7	6.8 oz/acre	7	0.5	Not for downy mildew. Do not make more than 2 sequential applications or more than 6 applications per season.
	fixed copper (various)	M	See label	1	1	May reduce bacterial rots.
	fluazinam (Omega 500)	29	1.0 pt/acre	7	1	Initiate sprays when conditions are favorable for disease or at disease onset. Spray on a 7 to 10 day schedule.
	fluopyram + tebuconazole (Luna Experience)	7 + 3	8 to 12.8 oz/acre	7	0.5	Not for downy mildew. Suppresses *Sclerotium* spp.
	fluopyram + pyrimethanil (Luna Tranquility)	7 + 9	16 to 27 oz / acre	7	0.5	Not for downy mildew. Suppresses *Sclerotium* spp.
	fluxapyroxad + pyraclostrobin (Merivon)	7+11	4 to 11 fl oz/ acre	7	0.5	Use higher rates for downy mildew suppression. Apply at disease onset; continue on 7 to 14 day schedule. No more than 3 applications/season.
	iprodione (various)	2	1.5 lb/acre	7	0.5	Not for downy mildew. Apply when conditions are favorable; 14-day intervals.
	mancozeb (various)	M	2 to 3 lb/acre	7	1	Do not exceed 30 lb per acre per crop.
	mefenoxam + chlorothalonil (Ridomil Gold/Bravo)	4 + M	2.5 pt/acre	7	2	

Table 10-36. Disease Control Products for Onion

Disease	Material	FRAC	Rate of Material	Minimum Days Harvest	Minimum Days Reentry	Method, Schedule, and Remarks
Neck rot (*Botrytis* spp.), purple blotch (*Alternaria porri*), downy mildew (*Peronospora destructor*) (continued)	oxathiapiprolin + chlorothalonil	49 + M	1.75 to 2 pt/acre	14	0.5	Observe chlorothalonil season limits.
	propiconazole	3	4 to 8 oz/acre	14	0.5	Not for downy mildew. Alternate with a different mode of action.
	propiconazole					
	pyraclostrobin + boscalid (Pristine)	11 + 7	14.5 to 18.5 oz/acre	7	1	Make no more than 6 applications per season.
	tebuconazole (various)	3	4 to 6 fl oz/acre	7	0.5	Not for downy mildew or Botrytis. Suppresses *Sclerotium* spp.
	tebuconazole + chlorothalonil (Muscle)	3 + M	1.1 to 1.6 pt / acre	7 to 14	0.5	Not for downy mildew or Botrytis.
	tebuconazole + potassium phosphate (Viathon)	3 + 33	2 to 3 pts/acre	7	0.5	
	zoxamide + mancozeb (Zing!)	22 + M	1.5 to 2 lb/acre	7	0.5	Use preventatively.
Pink root (*Phoma* spp.)	metam-sodium (Vapam)	—	37.5 to 75 gal/	—	2	Rate is based on soil properties and depth of soil to be treated.
	dichloropropene (Telone) C-17 C-35	—	10.8 to 17.1 gal/acre 13 to 20.5 gal/acre		5	Rate is based on soil type; see label for in-row rates.
Smut	mancozeb (various)	M	3 lb/29,000 ft row	—	—	
Stemphylium leaf blight	azoxystrobin + difenoconazole (Quadris Top)	11 + 3	14 fl oz/acre	7	0.5	Begin sprays prior to disease onset and spray on a 7- to 14-day schedule. Do not rotate with Group 11 fungicides.
	difenoconazole + cyprodinil (Inspire Super)	3 + 9	16 to 20 fl oz/acre	7	0.5	Make no more than 2 applications before alternating with a fungicide with a different mode of action.
	fluxapyroxad + pyraclostrobin (Merivon)	7 + 11	4 to 11 fl oz/acre	7	0.5	Apply at disease onset; continue on 7- to 14-day schedule. No more than 3 applications per season.
	iprodione (various)	2	1.5 lb/acre 50 to 100 gal/acre	7	0	Start 7-day foliar sprays at first appearance of favorable conditions.
	pyraclostrobin + boscalid (Pristine)	11 + 7	10.5 to 18.5 oz/acre	7	1	Make no more than 6 applications per season.
	fluazinam (Omega 500)	29	1.0 pt/acre	7	2	Initiate sprays when conditions are favorable for disease or at disease onset. Spray on a 7- to 10-day schedule.
	penthiopyrad (Fontelis)	7	10.5 to 18.5 oz/acre	3	0.5	Begin sprays prior to disease development and continue on a 7- to 14-day schedule.
White rot (*Sclerotium cepivorum*)	azoxystrobin + chlorothalonil (Quadris Opti)	11 + M	1.6 to 3.2 pt/acre	14	0.5	Make no more than 1 application before alternating with a fungicide with a different mode of action.
	penthiopyrad (Fontelis)	7	16 to 24 fl oz/acre	3		Begin sprays prior to disease development and continue on a 7- to 14-day schedule.
	fludioxonil (various)	12	7 oz / acre	7	0.5	In furrow treatment only.
	dicloran (Botran) 75 W	14	5.3 lb/acre	14	0.5	Apply 5-inch band over seed row and incorporate in top 1.5 to 3 in. of soil, 1 to 2 weeks before seeding.
	dichloropropene (Telone) C-17 C-35	—	10.8 to 17.1 gal/acre 13 to 20.5 gal/acre	—	5	Rate is based on soil type; see label for in-row rates.
	thiophanate-methyl (various)	1	See label			Spray into open furrow at time of seeding or planting in row.

Table 10-37. Efficacy of Products for Disease Control in Onion

Scale: E = excellent; G = good; F = fair; P = poor; NC = no control; ND = no data.

Product[1]	Fungicide group[F]	Preharvest interval (Days)	Bacterial Streak (*Pseudomonas viridiflava*)	Black Mold (*Aspergillus niger*)	Botrytis Leaf Blight (*B. squamosa*)	Botrytis Neck Rot (*B. allii*)	Damping off (*Pythium spp.*)	Downy Mildew (*P. destructor*)	Fusarium Basal Rot (*F. oxysporum*)	Onion Smut (*Urocystis colchici*)	Center Rot (*Pantoea ananatis*)	Pink Root (*Phoma terrestris*)	Purple Blotch (*Alternaria porri*)	Stemphylium Leaf Blight and Stalk Rot	White Rot (*Sclerotium cepivorum*)
amectoctradin + dimethomporph (Zampro)	40 + 45	0	NC	NC	NC	NC	NC	G	NC	NC	NC	NC	NC	NC	NC
azoxystrobin (various)	11	7	NC	G	F	NC	NC	ND	NC	ND	NC	NC	G	G	ND
azoxystobin + difenoconazole (Quadris Top)	11 + 3	1	NC	NC	F	NC	NC	ND	NC	NC	NC	NC	G	F	NC
boscalid (Endura)	7	7			G								G		G	
chlorothalonil (various)	M	14	NC	NC	F	NC	NC	P	NC	NC	NC	NC	F	F	NC
chlorothalonil + zoxamide (Zing!)	M + 22	7	ND	ND	ND	ND	ND	ND	ND	ND	NDND	ND	ND	ND	ND
chlorothalonil + cymoxanil (Ariston)	M + 27	7	ND	ND	ND	ND	ND	ND	ND	ND	ND	ND	ND	ND	ND
cyprodinil + fludioxonil (Switch)	9 + 12	7	NC	NC	F	ND	NC	NC	NC	NC	NC	NC	F	F	NC
cyprodinil + difenoconazole (Inspire Super)	9 + 3	7	ND	ND	F	ND	ND	ND	ND	ND	ND	ND	ND	G	ND
dichloropropene + chloropicrin, fumigant (Telone C-17)	—	—	NC	NC	NC	NC	P	NC	F	NC	NC	F	NC	NC	F
dimethomorph (Forum)	40	0	NC	NC	NC	NC	NC	F	NC	NC	NC	NC	NC	NC	NC
fenamidone (Reason)	11	7	NC	NC	P	NC	NC	G	NC	NC	NC	NC	P	P	NC
famoxadone/cymoxanil (Tanos)	11 + 27	3	NC	NC	F	NC	NC	P	NC	NC	NC	NC	F	F	NC
fixed copper (various)	M	1	F	NC	F	NC	NC	F	NC	NC	F	NC	F	NC	NC
fluazinam (Omega 500)	29	2	NC	NC	G	NC	NC	G	NC	NC	NC	NC	E	E	NC
fluopyram + pyrimethanil (Luna Tranquility)	7 + 9	7	ND	ND	ND	ND	ND	NDN	ND	ND	ND	ND	ND	E	ND
fluxapyroxad + pyraclostrobin (Merivon)	7 + 11	7	ND	ND	G	ND	ND	ND	ND	ND	ND	ND	G	G	ND
iprodione (Rovral)	2	7	NC	NC	F	P	NC	NC	NC	NC	NC	NC	G	F	F
mancozeb (various)	M	7	NC	NC	F	NC	NC	F	NC	E	NC	NC	F	F	NC
mancozeb + copper (ManKocide)	M + M	7	F	NC	F	NC	NC	F	NC	F	F	NC	F	F	NC
mandipropamid (Revus)	40	7	NC	NC	NC	ND	F	F	NC	NC	NC	NC	NC	NC	NC
mefenoxam (Ridomil Gold EC)	4	7	NC	NC	NC	NC	F	ND	NC	NC	NC	NC	NC	NC	NC
mefenoxam + chlorothalonil (Ridomil Gold Bravo)	4 + M	14	NC	NC	F	NC	P	F	NC	NC	NC	NC	F	F	NC
mefenoxam + copper (Ridomil Gold/ Copper)	4 + M	7	F	NC	NC	NC	P	F	NC	NC	F	NC	NC	NC	NC
mefenoxam + mancozeb (Ridomil Gold MZ)	4 + M	7	NC	NC	F	NC	P	F	NC	F	NC	NC	F	F	NC
metam sodium, fumigant (Vapam)	—	—	NC	NC	NC	NC	F	NC	F	NC	NC	E	NC	NC	F
penthiopyrad (Fontelis)	7		ND	ND	G	ND	ND	ND	ND	ND	ND	ND	G	ND	ND
potassium phosphite + tebuconazole (Viathon)	33 + 3	7	ND	ND	ND	ND	ND	ND	ND	ND	ND	ND	G	ND	ND
pyraclostrobin (Cabrio)	11	7	NC	ND	F	NC	NC	F	NC	NC	NC	NC	G	G	ND
pyraclostrobin + boscalid (Pristine)	11 + 7	7	NC	ND	G	F	NC	F	NC	NC	NC	NC	E	E	ND
pyrimethanil (Scala)	9	7	NC	ND	F	NC	NC	NC	NC	NC	NC	NC	F	F	NC
tebuconazole (various)	3	7	ND	ND	G	ND	ND	ND	ND	NC	ND	ND	G	ND	ND

[1] Efficacy ratings do not necessarily indicate a labeled use for every disease.

[F] To prevent resistance in pathogens, alternate fungicides within a group with fungicides in another group. Fungicides in the "M" group are generally considered "low risk" with no signs of resistance developing to the majority of fungicides.

[R] Resistance reported in the pathogen.

Parsley

A. Keinath, Plant Pathology, Clemson University

Table 10-38. Disease Control Products for Parsley

Disease	Material	FRAC	Rate of Material	Minimum Days Harvest		Method, Schedule, and Remarks
Damping off and root rot (*Pythium, Phytophthora*)	mefenoxam (Ridomil Gold 4 SL) (Ultra Flourish 2 EC)	4	1 to 2 pt/treated acre	0	0.5	Apply preplant incorporated or surface application at planting.
	metalaxyl (MetaStar 2 E)	4	2 to 8 pt/treated acre	0	2	Banded over the row, preplant incorporated, or injected with liquid fertilizer.
Alternaria leaf spot, *Cercospora* leaf spot (Early blight), Powdery mildew, Septoria leaf spot (late blight)	azoxystrobin (various)	11	see label	0	4 hr	Make no more than 2 sequential applications before alternating with fungicides that have a different mode of action. Apply no more than 1.88 lb per crop per acre per season.
	cyprodinil + fludioxonil (Switch 62.5 WG)	9 + 12	11 to 14 oz/acre	0	0.5	Make no more than 2 sequential applications before alternating with fungicides that have a different mode of action for 2 applications. Apply no more than 56 oz per crop per acre per season.
	fixed copper (generic)	M1	See label	0	0	Spray at first disease appearance, 7- to 10- day intervals.
	fluxapyroxad + pyraclostrobin (Merivon 500 SC)	7 + 11	4 to 11 fl oz/ acre	3	0.5	Make no more than 2 sequential applications before alternating with fungicides that have a different mode of action. Maximum of 3 applications per crop.
	penthiopyrad (Fontelis) 1.67 F	7	14 to 24 fl oz	3	0.5	Do not make more than 2 sequential applications. Maximum of 72 fl oz per acre per year.
	propiconazole (various)	3	3 to 4 fl oz/A	14	0.5	Begin at first sign of disease and repeat at 14-day intervals. Make no more than 2 consecutive applications before rotating to another fungicide with a different mode of action.
	pyraclostrobin (Cabrio 20 EG, Pyrac 2 EC)	11	12 to 16 oz/acre	0	0.5	Make no more than 2 sequential applications before alternating with fungicides that have a different mode of action. Apply no more than 64 oz per crop per acre per season.
Web blight and root rot (*Rhizoctonia*)	azoxystrobin (Quadris 2.08 F)	11	0.125 to 0.25 oz/ 1000 row ft (soil application) or 6.0 to 15.5 fl oz/acre (foliar)	0	4 hr	Apply as banded spray to the lower stems and soil surface. Make no more than 2 sequential applications. Apply no more than 1.88 lb per crop per acre per season. Soil applications are included in this maximum.
White mold (Sclerotinia)	cyprodinil + fludioxonil (Switch 62.5 WG)	9 + 12	11 to 14 oz/acre	0	0.5	Make no more than 2 sequential applications before alternating with fungicides that have a different mode of action for 2 applications. Apply no more than 56 oz per crop per acre per season. First application at thinning and second application 2 weeks later.
	penthiopyrad (Fontelis 1.67 F)	7	16 to 30 fl oz	3	0.5	Do not make more than 2 sequential applications. Maximum of 72 fl oz per acre per year.
	Coniothyrium minitans (Contans WG)	NA	1 to 4 lb/acre	0	4 hr	**OMRI listed product.** Apply to soil surface and incorporate no deeper than 2 inches. Works best when applied prior to planting or transplanting. Do not apply other fungicides for 3 weeks after applying Contans.

Table 10-39. Importance of Alternative Management Practices for Disease Control in Parsley

Scale: E = excellent; G = good; F = fair; P = poor; NC = no control; ND = no data.

Strategy	Alternaria leaf spot	Cercospora leaf spot	Powdery mildew	Pythium damping off and root rot	Rhizoctonia damping off and root rot	Root knot (nematode)	Sclerotinia white mold	Septoria blight
Avoid field operations when leaves are wet	G	G	NC	NC	NC	NC	P	G
Avoid overhead irrigation	G	G	NC	NC	NC	NC	G	G
Biofungicide	ND	ND	F	ND	ND	ND	F	ND
Change planting date	NC	NC	NC	NC	E (early)	E (early)	G (late)	NC
Suppressive cover crops	NC	NC	NC	NC	NC	F	NC	NC
Crop rotation with non-host	E	E	NC	P	P	P	F	E
Deep plowing	G	G	P	NC	F	P	F	G
Destroy crop residue	G	G	P	NC	F	P	P	G
Encourage air movement	G	G	P	P	NC	NC	E	G
Flooding (where feasible)	NC	NC	NC	NC	F	G	G	NC
Increase soil organic matter	NC	NC	F	P	P	F	NC	P
Hot water seed treatment	ND	ND	NC	NC	NC	NC	NC	E
Plant in well-drained soil	P	P	NC	E	G	NC	F	P
Plant on raised beds	NC	NC	NC	E	G	NC	F	NC
Plastic mulch bed covers	NC	NC	F	F	F	NC	P	NC
Postharvest temperature control	NC	NC	NC	NC	NC	NC	E	NC
Reduce mechanical injury	NC	NC	NC	NC	P	NC	G	NC
Soil solarization	F	F	NC	P	F	F	P	F
Pathogen-free seed	E	E	P	NC	NC	NC	P	E
Resistant/tolerant cultivars	NC	NC	NC	NC	P	NC	NC	F
Weed control	P	P	F	NC	NC	F	F	P

Pea

E. Sikora, Plant Pathology, Auburn University; A. Keinath, Plant Pathology, Clemson University

Table 10-40. Disease Control Products for Pea

Disease	Material	FRAC	Rate of Material	Minimum Days Harvest	Minimum Days Reentry	Method, Schedule, and Remarks
PEA (English)						
Anthracnose	azoxystrobin (Quadris) 2.08 F	11	6.2 to 15.4 fl oz/acre	0	4 hr	Do not make more than 2 sequential applications.
	penthiopyrad (Fontelis) 1.67 F	7	14 to 30 fl oz	0	0.5	Do not make more than 2 sequential applications. Maximum of 72 fl oz per acre per crop.
	pyraclostrobin + fluxapyroxad (Priaxor) 500 SC	11 + 7	4.0 to 8.0 fl oz/acre	7	12 hr	Do not make more than 2 sequential applications. Maximum of 16 fl oz per acre per crop.
Ascochyta leaf spot and blight	Azoxystrobin (Quadris) 2.08 F	11	6.2 to 15.4 fl oz/ acre	0	4 hr	Do not make more than 2 sequential applications.
	boscalid (Endura) 70 WG	7	8 to 11 oz/ acre	7	0.5	Maximum of 2 applications per crop.
	penthiopyrad (Fontelis) 1.67 F	7	14 to 30 fl oz	0	0.5	Do not make more than 2 sequential applications. Maximum of 72 fl oz per acre per crop.
	pyraclostrobin + fluxapyroxad (Priaxor) 500 SC	11 + 7	4.0 to 8.0 fl oz/acre	7	12 hr	Do not make more than 2 sequential applications. Maximum of 16 fl oz per acre per crop.
Gray mold (*Botrytis*), White mold (*Sclerotinia*)	boscalid (Endura) 70 WG	7	8 to 11 oz/ acre	7	0.5	Maximum of 2 applications per crop.
	penthiopyrad (Fontelis) 1.67 F	7	14 to 30 fl oz	0	0.5	Do not make more than 2 sequential applications. Maximum of 72 fl oz per acre per year.
	pyraclostrobin + fluxapyroxad (Priaxor) 500 SC	11 + 7	4.0 to 8.0 fl oz/acre	7	12 hr	Do not make more than 2 sequential applications. Maximum of 16 fl oz per acre per crop.
White mold (*Sclerotinia*)	*Coniothyrium minitans* (Contans WG)	—	1 to 4 lb/acre	0	4 hr	**OMRI listed product.** Apply to soil surface and incorporate no deeper than 2 inches. Works best when applied prior to planting or transplanting. Do not apply other fungicides for 3 weeks after applying Contans.
Powdery mildew	boscalid (Endura) 70 WG	7	8 to 11 oz/ acre	7	0.5	Maximum of 2 applications per crop.
	fixed copper (various)	M	See label	0	See label	See label.
	penthiopyrad (Fontelis) 1.67 F	7	14 to 30 fl oz	0	0.5	Do not make more than 2 sequential applications. Maximum of 72 fl oz per acre per year.

Table 10-40. Disease Control Products for Pea

Disease	Material	FRAC	Rate of Material	Minimum Days Harvest	Minimum Days Reentry	Method, Schedule, and Remarks
PEA (English) (continued)						
Powdery mildew (continued)	pyraclostrobin + fluxapyroxad (Priaxor) 500 SC	11 + 7	4.0 to 8.0 fl oz/acre	7	12 hr	Do not make more than 2 sequential applications. Maximum of 16 fl oz per acre per crop.
	sulfur (various)	M	See label	0	See label	Spray at first appearance, 10- to 14-day intervals. Do not use sulfur on wet plants or on hot days (in excess of 90 degrees F).
Pythium damping off	mefenoxam (Ridomil Gold) 4 EC	4	0.5 to 1 pt/trt acre	—	2	Incorporate in soil. See label for row rates.
Rhizoctonia root rot	pyraclostrobin + fluxapyroxad (Priaxor) 500 SC	11 + 7	4.0 to 8.0 fl oz/acre	7	12 hr	Do not make more than 2 sequential applications. Maximum of 16 fl oz/acre per crop.
Rust *(Uromyces)*	azoxystrobin (Quadris) 2.08 F	11	6.2 fl oz/acre	0	4 hr	Do not make more than 2 sequential applications.
	penthiopyrad (Fontelis) 1.67 F	7	14 to 30 fl oz	0	0.5	Do not make more than 2 sequential applications. Maximum of 72 fl oz per acre per year.
	pyraclostrobin + fluxapyroxad (Priaxor) 500 SC	11 + 7	4.0 to 8.0 fl oz/acre	7	12 hr	Do not make more than 2 sequential applications. Maximum of 16 fl oz per acre per crop.
PEA (Southern)						
Anthracnose	thiophanate-methyl (various)	1	1 to 1.5 lb/acre	28	0.5	Use no more than 4 lb (2.8 lb a.i.) per acre per year.
Anthracnose, Rust	azoxystrobin (various)	11	2 to 5 oz/acre	14 (dry) 0 (succulent)	4 hr	Make no more than 2 sequential applications before alternating with a fungicide with a different mode of action. Use no more than 1.5 pounds a.i. per acre per season.
Ascochyta blight, Gray mold, White mold	boscalid (Endura) 70 WG	7	8 to 11 oz/acre	21(dry) or 7 (succulent)	0.5	Maximum of 2 applications per season.
Ascochyta blight, Rust, white mold	prothioconazole (various)	3	5.7 fl oz /acre	7	0.5	Maximum of 3 applications per year. Use no more than 17.1 fl oz per acre per year.
Downy mildew, Bacterial blights	fixed copper (various)	M	See label	See label	See label	See label.
Downy mildew, *Cercospora*, Anthracnose, Rust	chlorothalonil (various)	M	1.4 to 2 pt/acre	14	2	Spray early bloom; repeat at 7- to 10-day intervals; for dry beans only.
Alternaria, Anthracnose, Ascochyta, powdery mildew, rust, Cercospora	difenconazole + benzovindiflupyr (Aprovia Top)	3 + 7	10.5 to 11 fl oz	14	0.5	Begin prior to disease development and continue on 14-day schedule.
Alternaria, Anthracnose, Ascochyta, rust, southern blight, web blight	azoxystrobin + propiconazole (various)	3 + 11	10.5 to 14 oz/acre	14 (dry) or 7 (succulent)	0.5	Apply when conditions are conducive for disease. Up to three applications may be made on 7- to 14-day intervals
Alternaria, Anthracnose, Ascochyta, downy mildew, powdery mildew, rust, Cercospora, white mold	picoxystrobin (Approach)	11	6 to 12 fl oz	14	0.5	Do not apply more than 3 sequential applications. For white mold, use higher rates.
	penthiopyrad (Fontelis) 1.67 F	7	14 to 30 fl oz	0	0.5	Do not make more than 2 sequential applications. Maximum of 72 fl oz per acre per year.
Downy mildew, *Cercospora*, Anthracnose, Rust, Powdery mildew	pyraclostrobin (various)	7	6 to 9 fl oz/acre	21	0.5	Make no more than 2 sequential applications before alternating with a fungicide with a different mode of action. Use no more than 18 fl oz per acre per season.
	sulfur (various)	M	See label	0	1	Spray at first appearance; 7- to 10-day interval.
Pythium damping off	mofonoxam (various)	4	0.5 to 1 pt/ treated acre	—	0.5	Broadcast or banded over the row as a soil spray at planting or preplant incorporation into the top 2 inches of soil.
	metalaxyl (various)	4	2 to 4 pt/treated acre	—	2	Broadcast or banded over the row as a soil spray at planting or preplant incorporation into the top 2 inches of soil.
Rhizoctonia root rot	azoxystrobin (various)	11	0.4 to 0.8 fl oz/1,000 row feet	—	4 hr	Make in-furrow or banded application shortly after plant emergence.
	penflufen (Evergol Prime)	7	0.05 to 0.1 fl oz of the product per 100,000 seeds.	—	0.5	Apply using commercial slurry or mist-type seed treatment equipment.
Rhizoctonia and Fusarium seed and seedling decay	fluxapyroxad (various)	7	0.24 to 0.47 fl oz/100 lbs seed	—	—	Seed treatment
Rhizoctonia, and Fusarium seed rot, damping-off, Botrytis seedling blight, Phomopsis seed decay	penflufen + trifloxystrobin (various)	11	Apply 0.25 – 0.5 fl oz/100 lbs seed	—	—	Apply using commercial slurry or mist-type seed treatment equipment.

Table 10-40. Disease Control Products for Pea

Disease	Material	FRAC	Rate of Material	Minimum Days Harvest	Minimum Days Reentry	Method, Schedule, and Remarks
PEA (Southern) (continued)						
White mold (*Sclerotinia*)	*Coniothyrium minitans* (Contans WG)	—	1 to 4 lb/acre	0	4 hr	**OMRI listed product.** Apply to soil surface and incorporate no deeper than 2 inches. Works best when applied prior to planting or transplanting. Do not apply other fungicides for 3 weeks after applying Contans.
	fludioxonil (various)	12	7 oz/acre	7	0.5	Make no more than 2 sequential applications before alternating with a fungicide with a different mode of action for 2 applications. Use no more than 28 oz per acre per year.
Cottony leak (*Pythium* spp.)	fenamidone (Reason 500 C)	11	5.5 to 8.2 fl oz/acre	3	0.5	Begin applications as soon as crop and/or environmental conditions become favorable for disease development. DO NOT use on COWPEA.
Cottony leak, downy mildew, Phytophthora capsici	cyazofamid (Ranman)	21	2.75 fl oz/acre	0	0.5	Application instructions vary by disease; please follow label directions. DO NOT apply to cowpeas used for livestock feed.
Sclerotinia white mold and Botrytis gray mold	fluazinam (Omega 500F)	29	0.5 to 0.85 pt/acre	30	0.5	DO NOT use more than 1.75 pints of per acre. PHI varies by crop; see label restrictions.

Pepper
B. Dutta, Plant Pathology, University of Georgia

Table 10-41. Disease Control Products for Pepper

Disease	Material	FRAC	Rate of Material	Minimum Days Harvest	Minimum Days Reentry	Method, Schedule, and Remarks
Aphid-transmitted viruses: PVY, TEV, WMV, CMV	JMS Stylet-Oil		3 qt/100 gal water	0	Dry	Use in 50 to 200 gallons per acre depending on plant size. Spray weekly when winged aphids first appear.
Anthracnose fruit rot	azoxystrobin (various)	11	See label	0	4 hr	Apply at flowering to manage green fruit rot. Limit of 61.5 fl oz per acre per season. Make no more than **one** application before alternating with fungicides that have a different mode of action.
	azoxystrobin + difenoconazole (Quadris Top)	11 + 3	8 to 14 fl oz/acre	0	0.5	Limit of 55.3 fl oz per acre per season. Make no more than 2 consecutive applications before rotating to another effective fungicide with a different mode of action.
	chlorothalonil (various)	M	See label	7	1	See label. Rates vary depending on the formulation.
	chlorothalonil + cymoxanil (Ariston)	M + 27	2 to 2.44 pt/acre	3	0.5	Limit of 18.1 pt per acre per year.
	difenoconazole + benzovindiflupyr (Aprovia Top)	11 + 3	10.5 to 13.5 fl oz/acre	0	0.5	Limit of 53.6 fl oz per acre per year. Not labeled for greenhouse use. No more than 2 applications of Aprovia top may be applied on a 7-day interval.
	famoxadone + cymoxanil (Tanos)	11 + 27	8 to 10 oz/acre	3	0.5	Make no more than **one** application before alternating with a fungicide with a different mode of action. **NOTE: Must tank mix with another fungicide with a different mode of action (i.e. maneb or copper).**
	fenamidone (Reason)	11	5.5 to 8.2 fl oz/acre	14	0.5	Limit of 24.6 fl oz per growing season. Make no more than **one** application before rotating to another effective fungicide with a different mode of action.
	mancozeb (various)	M	See label	7	1	See label. Rates vary depending on the formulation.
	mancozeb + copper (ManKocide)	M + M	2 to 3 lb/acre	7	2	Limit of 39 lb per acre per season.
	pyraclostrobin (Cabrio) EG	11	8 to 12 oz/acre	0	4 hr	Apply at flowering to manage green fruit rot. Limit of 96 oz per acre per season. Make no more than **one** sequential application before alternating with fungicides that have a different mode of action.
	pyraclostrobin + fluxapyroxad (Priaxor)		4.0 to 8.0 fl oz/acre	0	0.5	**RIPE ROT ONLY.** Limit of 24 fl oz per acre per season. Make no more than 2 consecutive applications before rotating to another effective fungicide with a different mode of action.
	penthiopyrad (Fontelis)	7	24 fl oz/acre	0	0.5	**SUPPRESSION ONLY.** Limit of 72 fl oz per acre per year. Make no more than 2 consecutive applications before rotating to another effective fungicide with a different mode of action.
	trifloxystrobin (Flint)	11	3 to 4 oz/acre	3	0.5	**SUPPRESSION ONLY.** Limit of 16 oz per acre per year. Make no more than **one** application before alternating with fungicides that have a different mode of action.

Table 10-41. Disease Control Products for Pepper

Disease	Material	FRAC	Rate of Material	Minimum Days Harvest	Minimum Days Reentry	Method, Schedule, and Remarks
Bacterial soft rot	famoxadone + cymoxanil (Tanos)	11 + 27	8 to 10 oz/acre	3	0.5	SUPPRESSION ONLY. Make no more than **one** application before alternating with a fungicide with a different mode of action. NOTE: **Must tank mix with another fungicide with a different mode of action** (i.e. maneb or copper).
Bacterial spot (field)	acibenzolar-S-methyl (Actigard 50 WG)	21	0.33 oz to 0.75 oz/acre	14	0.5	**FOR CHILI PEPPERS ONLY EXCEPT IN THE STATE OF GEORGIA.** Begin applications within 1 week of transplanting or emergence. Make up to 6 weekly, consecutive applications.
	fixed copper (various)	M	See label	0	2	See label. Rates vary depending on the formulation. Make first application 7 to 10 days after transplanting. Carefully examine field for disease to determine need for additional applications. If disease is present, make additional applications at 5-day intervals. Applying mancozeb with copper significantly enhances bacterial spot control. Do not spray copper when temperatures are above 90 degrees F.
	famoxadone + cymoxanil (Tanos)	8 to 10 oz/acre	4 to 5 oz/acre	3	0.5	SUPPRESSION ONLY. Make no more than **one** application before alternating with a fungicide with a different mode of action. NOTE: **Must tank mix with another fungicide with a different mode of action** (i.e. maneb or copper).
	mancozeb (various)	M	See label	7	1	See label. Rates vary depending on the formulation.
	mancozeb + copper (ManKocide)	M + M	2 to 3 lb/acre	7	2	Limit of 39 lb per acre per season.
	quinoxyfen (Quintec)	13	6.0 oz/acre	3	0.5	Use 6 oz of product per acre in no less than 30 gallons of water per acre. NOTE: **May only be used to manage bacterial spot in Georgia, Florida, North Carolina, and South Carolina (Section 2(ee)).**
Bacterial spot (transplants)	streptomycin sulfate (Agri-Mycin 17, Firewall, Streptrol)	25	1 lb/100 gal	—	1	**MAY ONLY BE APPLIED TO TRANSPLANTS.** Spray when seedlings are in the 2-leaf stage and continue at 5-day intervals until transplanted into field. NOTE: **Some pathogen strains are resistant to streptomycin sulfate.**
	fixed copper (various)	M	See label	0	2	See label. Rates vary depending on the formulation. Begin applications when conditions first favor disease development and repeat at 3- to 10-day intervals if needed depending on disease severity. Use the higher rates when conditions favor disease. Do not spray copper when temperatures are above 90 degrees F.
Bacterial spot (seed)	sodium hypochlorite (Clorox 5.25%, regular formulation)	—	1 pt + 4 pt water	—	—	Add 1 Tbsp of surfactant (Tween-20 or 80, Silwet) to improve coverage on the seed.
Cercospora leaf spot	azoxystrobin + difenoconazole (Quadris Top) 29.6 SC	11 + 3	8 to 14 fl oz/acre	0	0.5	Limit of 55.3 fl oz per acre per season. Make no more than 2 consecutive applications before rotating to another effective fungicide with a different mode of action. The addition of non-ionic based surfactant or oil concentrate is recommended.
	pyraclostrobin (Cabrio)	11	8 to12 fl oz/acre	0	0.5	Limit of 96 fl oz per acre per season. Do not make more than one application of product before alternating to a labeled fungicide with different mode of action.
	difenoconazole + benzovindiflupyr (Aprovia Top)	7+3	10.5 to 13.5 fl oz/ acre	0	0.5	Limit of 53.6 fl oz per acre per year. Make more than 2 applications before alternating to another fungicide with a non-group 7 mode of action.
	fixed copper (various)	M	See label	0	2	See label. Rates vary depending on the formulation. Begin applications when conditions first favor disease development and repeat at 3- to 10-day intervals if needed depending on disease severity. Use the higher rates when conditions favor disease. Do not spray copper when temperatures are above 90 degrees F.
	mancozeb (various)	M	See label	7	1	See label. Rates vary depending on the formulation.
	mancozeb + copper (ManKocide)	M + M	2 to 3 lb/acre	7	2	Limit of 39 lb per acre per season.
Phytophthora foliar blight and fruit rot (*Phytophthora capsici*) *Phytophthora* or *Pythium* root rot (field)	cyazofamid (Ranman Fungicide)	21	2.75 fl oz/acre	0	0.5	Limit of 16.5 fl oz per acre per season. Apply to the base of the plant at transplanting or in the transplant water. Make no more than 3 consecutive applications followed by 3 consecutive applications of another effective fungicide with a different mode of action.

Table 10-41. Disease Control Products for Pepper

Disease	Material	FRAC	Rate of Material	Minimum Days Harvest	Minimum Days Reentry	Method, Schedule, and Remarks
Phytophthora foliar blight and fruit rot (*Phytophthora capsici*) *Phytophthora* or *Pythium* root rot (field) (continued)	oxathiapiprolin + mefenoxam (Orondis Gold 200 + Orondis Gold B)	49+4	2 to 4.8 fl oz/acre 2.4 to 19.2 fl oz/ acre	0	0.5	Limit of 19.2 fl oz per acre per season. Do not follow soil applications of Orondis Gold 200 with foliar applications of Orondis Opti A or Orondis Ultra A.
	oxathiapiprolin + chlorothalonil (Orondis Opti A + Orondis Opti B)	49 +M	2.0 to 4.8 fl oz/acre 1.5 pt/acre	14	0.5	See labels. For resistance management, do not follow soil applications of Orondis with foliar applications of Orondis Opti A.
	oxathiapiprolin + mandipropamid (Orondis Ultra; premix)	49 +40	5.5 to 8.0 fl oz/acre	See label	4hr	Use higher rate if disease is present. For best results, begin the dis- ease resistance program with an initial treatment at planting or trans- planting with a fungicide registered for its use. Apply Orondis Ultra as a foliar spray in a mixture with copper-based fungicide beginning at first appearance of symptoms.
	fixed copper (various)	M	See label	0	2	See label. Rates vary depending on the formulation. Begin applications when conditions first favor disease development and repeat at 3- to 10-day intervals if needed depending on disease severity. Use the higher rates when conditions favor disease. Do not spray copper when temperatures are above 90 degrees F.
	dimethomorph (Acrobat, Forum)	40	6 fl oz/acre	0	0.5	**SUPPRESSION ONLY.** Limit of 30 fl oz per acre per season. Make no more than 2 sequential before alternating with fungicides that have a different mode of action. **NOTE: Must tank mix with another fungicide with a different mode of action.**
	mancozeb (various)	M	See label	7	1	See label. Rates vary depending on the formulation.
	mancozeb + copper (ManKocide)	M + M	2 to 3 lb/acre	7	2	**SUPPRESSION ONLY.** Limit of 39 lb per acre per season.
	mefenoxam + copper hydroxide (Ridomil Gold/Copper)	4 + M	2 lb/acre	7	2	See label for an optimal spray program. Limit of 4 applications per crop per year. Do not exceed 0.4 lb a.i. per acre per season of mefenoxam + metalaxyl (MetaStar).
	famoxadone + cymoxanil (Tanos)	11 + 27	8 to 10 oz/acre	3	0.5	**SUPPRESSION ONLY.** Make no more than **one** application before alternating with a fungicide with a different mode of action. **NOTE: Must tank mix with another fungicide with a different mode of action (i.e. maneb or copper).**
	fenamidone (Reason) 500SC	11	8.2 fl oz/acre	14	0.5	**SUPPRESSION ONLY.** Limit of 24.6 fl oz per growing season. Make no more than **one** application before rotating to another effective fungicide with a different mode of action.
	fluazinam (Omega) 500F		1 to 1.5 pt/acre	30	0.5	Apply as a soil drench at 1.5 pints per acre. For foliar applications use 1 pt per acre. Limit of 9 pt per acre per season.
	fluopicolide (Presidio)	43	3 to 4 fl oz/acre	2	0.5	Limit of 12 fl oz per acre per season. Make no more than 2 times sequentially before alternating with fungicides that have a different mode of action. **NOTE: Must be tank-mixed with another mode of action product.**
	mandipropamid (Revus, Micora)	40	8 fl oz/acre	1	0.5	**SUPPRESSION ONLY.** Limit of 32 fl oz per acre per season. **NOTE: Must tank mix with another fungicide with a different mode of action (i.e. copper).**
	amectoctradin + dimethomorph (Zampro)	45 + 40	14 fl oz/acre	4	0.5	Limit of 42 fl oz per acre per season. Make no more than 2 sequential applications before rotating to another effective fungicide with a different mode of action.
	mefenoxam (Ridomil Gold, Ultra Flourish)	4	See label	—	2	**MAY ONLY BE APPLIED AT PLANTING.** Apply in a 12- to 16-inch band or in 20 to 50 gallons water per acre in transplant water. Mechanical incorporation or 0.5 to 1 inch irrigation water is needed for movement into root zone if rain is not expected. After initial application, 2 supplemental applications (1 pt per treated acre) can be applied. **NOTE: Strains of *Phytophthora capsici* insensitive to Ridomil Gold have been detected in some North Carolina and Louisiana pepper fields.**
	metalaxyl (MetaStar) 2E	4	4 to 8 pt/treated acre	7	2	Limit of 12 pt per acre per season. Preplant (soil incorporated), at planting (in water or liquid fertilizer), or as a basil-directed spray after planting. See label for the guidelines for supplemental applications.
	oxathiapiprolin + mefenoxam (Orondis Gold 200 + Orondis Gold B)	49 + 4	2.4 to 19.2 fl oz/ acre 1 pt/acre	7	2	See label.

Table 10-41. Disease Control Products for Pepper

Disease	Material	FRAC	Rate of Material	Minimum Days Harvest	Minimum Days Reentry	Method, Schedule, and Remarks
Powdery mildew	azoxystrobin (various)	11	6 to 15.5 fl oz/acre	0	4 hr	Limit of 61.5 fl oz per acre per season. Make no more than **one** application before alternating with fungicides that have a different mode of action.
	azoxystrobin + difenoconazole (Quadris Top) 29.6 SC	11 + 3	8 to 14 fl oz/acre	0	0.5	Limit of 55.3 fl oz per acre per season. Make no more than 2 consecutive applications before rotating to another effective fungicide with a different mode of action.
	chlorothalonil + cymoxanil (Ariston)	M + 27	2 to 2.44 pt/acre	3	0.5	Limit of 18.1 pt per acre per year.
	penthiopyrad (Fontelis)	7	16 to 24 fl oz/acre	0	0.5	Limit of 72 fl oz per acre per year. Make no more than 2 consecutive applications before rotating to another effective fungicide with a different mode of action.
	pyraclostrobin + fluxapyroxad (Priaxor)	11 + 7	6.0 to 8.0 fl oz/acre	0	0.5	Limit of 24 fl oz per acre per season. Make no more than 2 consecutive applications before rotating to another effective fungicide with a different mode of action.
	quinoxyfen (Quintec)	13	4.0 to 6.0 fl oz/acre	3	0.5	Limit of 24 fl oz per acre per year. Make no more than 2 consecutive applications before alternating with fungicides that have a different mode of action. **NOTE:** Under certain environmental conditions leaf spotting or chlorosis may occur after application; discontinue use if symptoms occur.
	sulfur (various)	M	See label			See label. Rates vary depending on the formulation. Apply at first appearance and repeat at 14-day intervals as needed.
	trifloxystrobin (Flint)	11	1.5 to 2 oz/acre	3	0.5	Limit of 16 oz per acre per year. Make no more than **one** application before alternating with fungicides that have a different mode of action.
Southern blight (*Sclerotium rolfsii*)	fluoxastrobin (Aftershock, Evito 280SC)	11	2 to 5.7 fl oz/acre	3	0.5	Limit of 22.8 fl oz per acre per season. Make no more than **one** application before alternating with fungicides that have a different mode of action. **NOTE:** Do not overhead irrigate for 24 hours following a spray application.
	penthiopyrad (Fontelis)	7	16 to 24 fl oz/acre	0	0.5	Limit of 19.2 fl oz per acre per season. Make no more than 2 sequential applications of Fontelis before switching to a fungicide with different mode of action. **For non-bell peppers only**
	PCNB (Blocker 4F) (transplanting)	14	4.5 to 7.5 pt/100 gal; use 0.5 pt of solution per plant.	NA	0.5	Transplanting: Apply at the time of transplanting for Southern blight suppression. The solution should be agitated often to maintain a uniform mixture to assure proper dosage. Limit of 7.5 lb a.i. per acre per season.
	PCNB (Blocker 4F) (in furrow)	14	1.2 to 1.9 gal; apply 10.6 to 16.7 fl oz product per 1000 ft of row	NA	0.5	In furrow: Apply in 8 to 10 gals of water per acre based on 36-inch row spacing. Apply as in-furrow sprays to the open "V" trench just prior to planting. When cultivating, set plows as flat as possible to avoid getting non-treated soil against stems or plants. Limit of 7.5 lb a.i. per acre per season.
	pyraclostrobin (Cabrio) EG	11	12 to 16 oz/acre	0	4 hr	**SUPPRESSION ONLY.** Apply at flowering to manage green fruit rot. Limit of 96 oz per acre per season. Make no more than **one** sequential application before alternating with fungicides that have a different mode of action.
	pyraclostrobin + fluxapyroxad (Priaxor)	11 + 7	4.0 to 8.0 fl oz/acre	0	0.5	**SUPPRESSION ONLY.** Limit of 24 fl oz per acre per season. Make no more than 2 consecutive applications before rotating to another effective fungicide with a different mode of action.
Target spot (*Corynespora cassiicola*)	boscalid (Endura)	7	3.5 oz/acre	0	0.5	Limit of 21 oz per acre per season. Make no more than 2 sequential applications before alternating with fungicides that have a different mode of action.
	cyprodinil + difenoconazole (Inspire Super)	9 + 3	16 to 20 fl oz/acre	0	0.5	Limit of 80 fl oz per acre per season.
	fluoxastrobin (Aftershock, Evito 480SC)	11	2 to 5.7 fl oz/acre	3	0.5	Limit of 22.8 fl oz per acre per season. Make no more than **one** application before alternating with fungicides that have a different mode of action. **NOTE:** Do not overhead irrigate for 24 hours following a spray application.
	penthiopyrad (Fontelis)	7	16 to 24 fl oz/acre	0	0.5	**SUPPRESSION ONLY.** Limit of 72 fl oz per acre per year. Make no more than 2 consecutive applications before rotating to another effective fungicide with a different mode of action.

Table 10-41. Disease Control Products for Pepper

Disease	Material	FRAC	Rate of Material	Minimum Days Harvest	Reentry	Method, Schedule, and Remarks
Target spot (*Corynespora cassiicola*) (continued)	pyraclostrobin (Cabrio) 20EG	11	8 to 12 oz/acre	0	4 hr	Apply at flowering to manage green fruit rot. Limit of 96 oz per acre per season. Make no more than **one** sequential application before alternating with fungicides that have a different mode of action.
	pyraclostrobin + fluxapyroxad (Priaxor)	11 + 7	4.0 to 8.0 fl oz/acre	0	0.5	Limit of 24 fl oz per acre per season. Make no more than 2 consecutive applications before rotating to another effective fungicide with a different mode of action.

Table 10-42. Relative Effectiveness of Various Chemicals for Pepper Disease Control

Scale: E = excellent; G = good; F = fair; P = poor; NC = no control; ND = no data.

Product[1]	Fungicide group	Preharvest interval (Days)	Anthracnose (immature fruit rot)	Bacterial Spot	Phytophthora Blight (root and crown)	Phytophthora Blight (fruit and foliage)	Pythium Damping off	Southern Blight
azoxystrobin (Quadris)	11	0	F	NC	NC	NC	NC	ND
chlorothalonil (various)	M	3	P	NC	NC	P	NC	NC
cyazofamid (Ranman)	21	0	NC	NC	F	G	NC	NC
dimethomorph (Acrobat, Forum)	40	4	NC	NC	NC	P	NC	NC
dimethomorph + amectoctradin (Zampro)	40 + 45	4	NC	NC	F	G	ND	NC
famoxadone + cymoxanil (Tanos)	11 + 27	3	P	NC	NC	P	NC	ND
fixed copper (various)	M	Check label	P	F	NC	F	NC	NC
fluopicolide (Presidio)	43	2	NC	NC	F	G	NC	NC
floxystrobin (Evito)	11	3	NC	NC	NC	NC	NC	ND
fluxapyroxad + pyraclostrobin (Priaxor)	11 + 7	7	F	NC	NC	NC	NC	ND
mancozeb[2] (Dithane, Manzate)	M	5	F	P	P	P	NC	NC
mandipropamid (Revus)	40	1	NC	NC	F	G	NC	NC
mefenoxam[R] (Ridomil Gold EC, Ultra Flourish)	4	0	NC	NC	E	NA	G	NC
mefenoxam[R] + copper (Ridomil Gold + copper)	4 + M	14	P	F	NA	G	NC	NC
methyl salicylate + *Bacillus thuringiensis* subsp. *kurstaki* (Leap)								
oxathiapiprolin (Orondis Gold 200)	49	0	NC	NC	F	G	NC	NC
oxathiapiprolin (Orondis Opti A)	49	0	NC	NC	F	G	NC	NC
penthiopyrad (Fontelis)	7	0	ND	NC	NC	NC	NC	ND
proamocarb (Previcur Flex)	28	5	NC	NC	NC	NC	F	NC
pyraclostrobin (Cabrio)	11	0	G	NC	NC	NC	NC	ND
quinoxyfen (Quintec)	13	3	NC	P	NC	NC	NC	NC
streptomycin sulfate[3] (Agri-mycin, Stretrol, Firewall)	25	Not for field use	NC	F	NC	NC	NC	NC
sulfur (various)	M	0	NC	NC	NC	NC	NC	NC

[1] Efficacy ratings do not necessarily indicate a labeled use for every disease.

[2] Copper tank-mixed with mancozeb enhances the efficacy against bacterial spot.

[3] Streptomycin may only be used on transplants; not registered for field use.

[F] To prevent resistance in pathogens, alternate fungicides within a group with fungicides in another group. Fungicides in the "M" group are generally considered "low risk" with no signs of resistance developing to the majority of fungicides.

[R] Resistance reported in the pathogen.

Table 10-43. Importance of Alternative Management Practices for Disease Control in Pepper

Scale: E = excellent; G = good; F = fair; P = poor; NC = no control; ND = no data.

Strategy	Anthracnose (immature fruit)	Aphid-transmitted viruses (PVX, CMV, TEV, AMV, PVY)	Bacterial soft rot of fruit	Bacterial spot	Blossom-end rot	Phytophthora blight (fruit and foliage)	Phytophthora blight (root and crown)	Pythium damping off	Root-knot nematode	Southern blight	Tomato Spotted Wilt Virus
Avoid field operations when foliage is wet	F	NC	NC	G	NC	F	P	NC	NC	NC	NC
Avoid overhead irrigation	G	NC	F	G	NC	G	G	P	NC	NC	NC
Change planting date within a season	NC	F (early)	NC	F (early)	NC	NC	NC	P (late)	F (early)	P (early)	Varies
Cover cropping with antagonist	NC	NC	NC	NC	NC	NC	NC	NC	F	NC	NC
Rotation with non-host (2 to 3 years)	G	NC	NC	NC	NC	P	P	NC	F	P	NC
Deep plowing	F	NC	NC	NC	NC	NC	NC	NC	P	F	NC
Prompt destruction of crop residue	F	F	NC	P	NC	P	P	NC	F	P	NC
Promote air movement	P	NC	NC	F	NC	P	P	NC	NC	NC	NC
Use of soil organic amendments	NC	NC	NC	NC	NC	P	P	P	F	P	NC
Application of insecticidal/horticultural oils	NC	F	NC	NC	NC	NC	NC	NC	NC	NC	NC
pH management (soil)	NC	NC	NC	NC	F	NC	NC	NC	F	NC	NC
Plant in well-drained soil/raised beds	NC	NC	NC	NC	NC	NC	G	G	NC	NC	NC
Eliminate standing water/saturated areas	NC	NC	NC	NC	NC	NC	G	G	NC	NC	NC
Postharvest temp control (fruit)	NC	NC	G	NC	NC	NC	NC	NC	NC	NC	NC
Use of reflective mulch	NC	F	NC	NC	NC	NC	NC	NC	NC	NC	G
Reduce mechanical injury	NC	NC	NC	NC	NC	NC	NC	NC	NC	NC	NC
Rogue diseased plants/fruit	NC	NC	NC	NC	NC	F	F	NC	NC	NC	NC
Soil solarization	NC	NC	NC	NC	NC	NC	P	NC	F	NC	NC
Use of pathogen-free planting stock	F	NC	NC	G	NC	NC	NC	NC	NC	NC	NC
Use of resistant cultivars	NC	NC	NC	G	F	F	F	NC	G	NC	G
Weed management	P	F	NC	NC	NC	P	P	NC	F	NC	P

Potato, Irish

I. Meadows, Plant Pathology

Table 10-44. Disease Control Products for Potato, Irish

Disease	Material	FRAC	Rate Of Material	Minimum Days Harvest	Minimum Days Reentry	Method, Schedule, and Remarks
Black scurf (*Rhizoctonia solani*) and Silver scurf (*Helminthosporium solani*)	azoxystrobin (various)	11	See label	See label	See label	See labels. Rates may vary depending on the product. Apply in furrow at planting according to label direction. Do not apply more than 1 application without alternating away from fungicides in Group 11.
	azoxystrobin + benzovindiflupyr (Elatus)	11 = 7	0.34 to 0.5 oz/1,000 linear row feet	0.5	—	Limit 9.5 oz/acre per application.
	fludioxonil (Maxim PSP)	12	0.5 lb/100 lb seed pieces	—	0.5	Ensure thorough coverage of each seed piece.
	fludioxonil + mancozeb (Maxim MZ)	12 + M	0.5 lb/100 lb seed pieces	—	0.5	Ensure thorough coverage of each seed piece.
	fludioxonil + thiamethoxam (Cruiser Maxx Potato)	12 + insecticide	0.19 to .27 fl oz/100 lb seed pieces	—	0.5	Rate depends on seeding rate - see label. See label for additional restrictions.

Table 10-44. Disease Control Products for Potato, Irish

Disease	Material	FRAC	Rate Of Material	Minimum Days Harvest	Minimum Days Reentry	Method, Schedule, and Remarks
Black scurf (*Rhizoctonia solani*) and Silver scurf (*Helminthosporium solani*) (continued)	fludioxonil + difenoconazole + sedazane + thiamethoxam (Cruiser Maxx Vibrance Potato)	12 + 3 + 7 + insecticide	0.5 fl oz/100 lb seed pieces	—	0.5	See label for additional restrictions.
	fluopyram (Luna Privilege)	7	5.47 oz/acre (ground); 2.82 oz/acre (aerial)	7	0.5	Use on a 5- to 7-day interval. Do not apply more than 10.95 oz/acre/season for ground application and no more than 8.46 oz/acre/season for aerial application. Do not make more than 2 applications before alternating with a fungicide with a different mode of action. Labeled for **silver scurf only**.
	fluoxastrobin (Aftershock, Evito 480 SC)	11	0.16 to 0.24 fl oz/1,000 ft of row	7	0.5	Apply in furrow at planting according to label directions. Do not apply more than 22.8 fl oz product/ acre/year including seed treatment use. Alternate with fungicide from different resistance management group.
	flutolanil (Moncut) 70DF (Moncut SC)	7	0.71 to 1.1 lb/acre 16.0 to 25.0 fl oz/acre	—	0.5	For **black scurf only**. Apply as an in-furrow spray by directing spray uniformly around and over the seedpiece in a 4 to 8 in band prior to covering with soil.
	flutolanil + mancozeb (MonCoat MZ)	7 + M	0.75 lb to 1.0 lb/100 lb seed piece	—	1	Apply to seedpieces immediately after cutting. Ensure thorough coverage.
	mancozeb (various)	M	See label	—	1	For **black scurf only**.
	penthiopyrad (Vertisan)	7	0.7 to 1.6 fl oz/1,000 ft of row	7	0.5	Maximum rate is 24 fl oz/acre/year. No more than 2 applications before switching to a different mode of action. Provides suppression of **black scurf only**.
	thiophanate-methyl (various)	1	0.5-0.7 fl oz/100 lb seed pieces	—	0.5	
Fusarium seedpiece decay, *Rhizoctonia* stem cancer, *Streptomyces* common scab	fludioxonil (various)	12	See label	—	0.5	Label rates may vary depending on the product.
	fludioxonil + mancozeb (Maxim MZ)	12 + M	0.5 lb/100 lb seed	—	1	Do not use treated seedpieces for feed or food. NOT labeled for *Streptomyces* common scab. See label for treatment instructions.
	mancozeb (various)	M	See label	—	1	Label rates vary depending on the product.
	penthiopyrad (Vertisan)	7	0.7 to 1.6 oz/1,000 ft of row	7	0.5	Maximum rate is 24 fl oz. Labeled for **Rhizoctonia stem canker only**.
Early blight, White mold	azoxystrobin + difenoconazole (Quadris Top)	11 + 3	8 to 14 fl oz/acre	14	0.5	Apply at 7- to 14-day-intervals. Apply no more than 2 sequential applications without alternating with a fungicide with a different mode of action. Limit of 55.3 lb product/acre/year. Limit of 0.46 lb a.i./acre/year of difenoconazole-containing products; limit of 2.0 lbs a.i./acre/year of azoxystrobin-containing products. Labeled for **early blight only**.
	boscalid (Endura)	7	3.5 to 0 oz/acre	10	0.5	For control of *Sclerotinia* White mold, use 5.5 to 10 oz rate and begin applications prior to row closure or at the onset of disease. Make a second application 14 days later if conditions favor disease development. Do not exceed 2 applications per season. For early blight control, use 3.5 to 4.5 oz rate. **Do not exceed 4 applications per season.** Limit of 20.5 oz product/ acre/season. Limit of 2 applications before alternating with a fungicide with a different mode of action.
	fluopyram (Luna Privilege)	7	4.0 to 5.47 oz/acre (ground; 2.82 oz/acre (aerial)	7	0.5	Use on a 5- to 7-day interval. Do not apply more than 10.95 oz/acre/season for ground application and no more than 8.46 oz/acre/season for aerial application. Do not make more than 2 applications before alternating with a fungicide with a different mode of action. For white mold, use 5.47 oz rate.
	fluopyram + pyrimethanil (Luna Tranquility)	7 + 9	11.2 oz/acre	7	0.5	Apply at 7- to 14-day intervals. Do not make more than 2 sequential applications without switching to a fungicide outside of Group 7 or Group 9.
	fluxapyroxad + pyraclostrobin (Priaxor)	7 + 11	4 to 8 oz/acre	7	0.5	Apply at 7- to 14-day intervals. Do not apply more than 24 oz/acre/season including in furrow and foliar uses. Use 6 to 8 oz/acre for SUPPRESSION of white mold. Maximum of 3 applications.
	iprodione (various)	2	See label	14	1	Rates may vary depending on the product.
	metconazole (Quash)	3	2.5 to 4 oz/acre	1	0.5	Limit 16 oz/acre/season. Make no more than 2 applications before changing modes of action. Limit to 4 applications per year. Use the 4 oz rate for white mold.
	metiram + pyraclostrobin (Cabrio Plus)	M + 11	2.0 to 2.9 lb/acre	14	1	Apply at 7- to 14-day intervals. Do not apply more than 17.4 lbs/acre/season. Do not apply more than 2 sequential applications before alternating with a fungicide with a different mode of action. Use at 2.9 lb/acre rate for suppression of white mold.

Table 10-44. Disease Control Products for Potato, Irish

Disease	Material	FRAC	Rate Of Material	Minimum Days Harvest	Minimum Days Reentry	Method, Schedule, and Remarks
Early blight, White mold (continued)	penthiopyrad (Vertisan)	7	10 to 24 oz/acre	7	0.5	Apply at 7- to 14-day intervals. Make no more than 2 applications before alternating with a fungicide with a different mode of action. For SUPPRESSION of white mold, use at 14 to 24 oz/acre. Do not exceed 72 oz/acre/year. Do not apply more than 11.25 oz a.i./acre/year in total from any combination of seed, soil, or foliar applications.
	pyraclostrobin (Headline; Headline SC)	11	6 to 12 fl oz/acre	3	0.5	**DO NOT** exceed more than 6 foliar applications or 72 total ounces of product per acre per season. For early blight, use 6- to 9-oz rate; for SUPPRESSION of white mold, use 6- to 12-oz rate, depending on weather conditions and disease pressure. Do not apply more than 1 time before alternating with a fungicide with a different mode of action.
	pyrimethanil (Scala SC)	9	7 fl oz/acre	7	0.5	Apply at 7- to 14-day intervals. Do not apply more than 35 fl oz/acre/season. For control of **early blight only**.
	thiophanate-methyl (various)	1	See label	See label	0.5	Rates may vary depending on the product.
Late blight, White mold	fluazinam (Omega) 500 F	29	5.5 to 8 fl oz/acre	14	0.5	Begin applications when plants are 6 to 8 in. tall or when conditions favor disease development. Repeat applications at 7- to 10-day intervals. For late blight, use 5.5 fl oz rate. **DO NOT** apply more than 3.5 pints/acre during each growing season.
Early blight, late blight	azoxystrobin (various)	11	See label	14	4 hr	Rates may vary depending on the product. Do not apply more than 1 application without alternating away from fungicides in Group 11.
	azoxystrobin + chlorothalonil (Quadris Opti)	11 + M	1.6 pt/acre	14	0.5	Apply at 5- to 7-day intervals. Do not apply more than one application without alternating away from fungicides in Group 11.
	chlorothalonil (various)	M	See label	7	0.5	See label. Rates vary depending on the formulation.
	chlorothalonil + cymoxanil (Ariston)	M + 27	2 pt/acre	14	0.5	Apply at 5- to 7-day intervals. Do not exceed 17.5 pt of product per acre per year.
	chlorothalonil + zoxamide (Zing!)	M + 22	24 34 fl oz/acre	7	0.5	Apply at 5- to 7-day intervals. Do not make more than 2 sequential applications before alternating with a fungicide that has a different mode of action. Do not make more than 8 applications or per acre per season. Use 30-34 fl oz rate for late blight.
	fixed copper (various)	M1	See label	0	1	See label. Rates vary depending on the formulation.
	cymoxanil + famoxadone (Tanos)	27 + C3	6 to 8 oz/acre	14	0.5	Use rate of 6 fl oz only for early blight. Do not apply more than 48 fl oz/acre/crop season and no more than 72 oz/acre/12 months. Do not make more than one application before alternating with a fungicide with a different mode of action.
	famoxadone + cymoxanil (Tanos)	6 to 8 oz/acre	—	14	1	Begin applications when conditions favor disease development or when disease is present in area. Should be tank mixed with a protectant fungicide (chlorothalonil or mancozeb). **DO NOT** apply more than 48 oz/acre/season.
	dimethomorph (Forum)	40	4 to 6 fl oz/acre	4	1	Must tank mix if using less than 6 fl oz rate; if used alone, use 6 oz rate. **DO NOT** make more than 5 applications per season. Limit 30 fl oz/acre/season.
	fenamidone (Reason 500 SC)	11	5.5 to 8.2 fl oz/acre	14	0.5	Begin applications when conditions favor disease development and continue on 5- to 10-day interval. Do not apply more than 24.6 fl oz per growing season. Alternate with fungicide from different resistance management group.
	fluoxastrobin (Aftershock, Evito 480 SC)	11	2 to 3.8 fl oz/acre	7	0.5	Begin applications when conditions favor disease development, on 7- to 10-day intervals. Do not apply more than once before alternating with fungicides that have a different mode of action. Do not apply more than 22.8 fl oz/acre/season. For late blight, apply at full label rate.
	fluxapyroxad + pyraclostrobin (Priaxor)	7	4 to 8 fl oz/acre	7	0.5	Apply at 7- to 14-day intervals. Do not apply more than 24 oz/acre/season, including in furrow and foliar uses.
	mancozeb + azoxystrobin (Dexter Max)	M + 11	1.6 to 2.1 lb/acre	14	1	Do not exceed 16 lbs product/acre/crop. Season limits apply for azoxystrobin.
	mancozeb + chlorothalonil (Elixir)	M + M	1.8 to 2.4 lb/acre	14	1	Do not apply more than 18.0 lbs product/crop/year.

Table 10-44. Disease Control Products for Potato, Irish

Disease	Material	FRAC	Rate Of Material	Minimum Days Harvest	Minimum Days Reentry	Method, Schedule, and Remarks
Early blight, late blight (continued)	mandipropamid + difenoconazole (Revus Top)	40 + 3	5.5 to 7 fl oz/acre	14	0.5	After 2 applications, switch to a different mode of action. Do not apply more than 28 fl oz/acre/season.
	mefenoxam+ chlorothalonil (Ridomil Gold Bravo SC) 76.5 WP	4 + M	2.5 pints/acre	14	2	See label for limits on application limits per season and application interval.
	mefenoxam+ mancozeb (Ridomil Gold MZ WG)	4 + M	2.5 lb/acre	14	2	Apply at 14-day intervals for up to 3 applications.
	metiram (Polyram 80 DF)	M	1.5 to 2 lb/acre	14	1	Do not apply more than 14 lbs product/crop/year.
	propamocarb hydrochloride (Previcur Flex)	28	0.7 to 1.2 pt/acre	14	0.5	Tank mix with a protectant fungicide such as mancozeb or chlorothalonil. Do not exceed 6 pt of product/acre/season.
	pyraclostrobin (Headline; Headline SC)	11	6 to 12 fl oz/acre	3	1	DO NOT exceed more than 6 foliar applications or 72 total oz of product/acre/season. For early blight, use 6 to 9 oz rate. Do not apply more than 1 time before alternating with a fungicide with a different mode of action.
	pyraclostrobin + chlorothalonil (Cabrio Plus)	11 + M	2.0 to 2.9 lb/acre	14	1	Do not apply more than 2 applications before switching to a different mode of action. Do not exceed 17.4 lbs/acre/season. For late blight, use 12 lb acre rate.
	pyrimethanil (Scala 5F)	9	7 fl oz/acre	7	0.5	Labeled for **early blight only**. Do not apply more than 35 fl oz/crop.
	trifloxystrobin (Gem 500SC)	11	2.9 to 3.8 fl oz/acre	7	0.5	Must tank mix with a non-group 11 fungicide for late blight. Use the 3.8 fl oz rate for late blight. Do not make more than 1 application without switching to a different mode of action. Do not exceed 6 applications or 23 fl oz product/acre/season.
	triphenyltin hydroxide (Super Tin 4L) Super Tin 80WP, Agri-Tin)	30	4 to 6 fl oz/acre 2.5 to 3.75 oz/acre	7	2	For Super Tin 4L, the 3.0 fl oz rate may be used if tank mixed. Add to 3 to 15 gallons of water depending on method of application. Season application limits apply - see label.
	zoxamide + mancozeb (Gavel 75DF)	22 + M	1.5 to 23.0 lb/acre	14	2	Do not make more than 6 applications or apply more than 12 lbs product/acre/season.
Late blight	ametoctradin + dimethomorph (Zampro)	45 + 40	11 to 14 fl oz/acre	4		Do not make more than 2 applications without switching to a different mode of action. Do not exceed 42 fl oz/acre/a season and 3 applications/season.
	cyazofamid (Ranman 400 SC)	21	1.4 to 2.75 fl oz/acre	7	0.5	Do not apply more than 10 sprays per crop. Make no more than 3 consecutive applications and then follow with 3 applications of another mode of action.
	cymoxanil (Curzate) 60 DF	27	3.2 oz/acre	14	1	**USE ONLY WITH A PROTECTANT FUNGICIDE** such as mancozeb or chlorothalonil. No more than 7 applications/crop/year.
	dimethomorph (Forum)	40	4 to 6 fl oz/acre	4	0.5	If applying at less than 6 fl oz rate, must tank mix with a non-group 40 fungicide. Do not exceed 5 applications or 30 fl oz of product/acre/season.
	fluazinam (Omega)	29	5.5 to 8 fl oz/acre	14	0.5	Begin applications when plants are 6 to 8 in. tall or when conditions favor disease development. Repeat applications at 7- to 10-day intervals. For late blight, use the 5.5 fl oz rate. **DO NOT** apply more than 3.5 pt/acre during each growing season.
	mefenoxam + copper hydroxide (Ridomil Gold / Copper)	4 + M	2 lb/acre	14	2	MUST tank mix with a protectant fungicide. Apply at 14-day intervals for up to 3 applications, alternated and followed by the full rate of a protectant.
	mono- and di-potassium salts of phosphorous acid (various)	33	See label	0	4h	Mix with a fungicide labeled for control of late blight. See label for in-furrow application or foliar application rates.
	oxathiapiprolin + chlorothalonil (Orondis Opti A + Orondis Opti B)	49 + M	1.6 to 4.8 fl oz/acre 0.75 to 1.5 pt/acre	7	0.5	Do not make more than 2 sequential applications without switching to a different mode of action and no more than 6 total applications per season. Do not mix soil applications and foliar applications. Apply no more than 27.2 fl oz of Orondis Opti A per season and no more than 15 pt of Orondis Opti B per season. See label for pre-mix.
	oxathiapiprolin + mandipropamid (Orondis Ultra A + Orondis Ultra B))	49 + 40	1.6 to 4.8 fl oz/acre 8.0 fl oz/acre	5 14	4 hr	Do not make more than 2 sequential applications without switching to a different mode of action and no more than 6 total applications per season. Do not mix soil applications and foliar applications. Apply no more than 27.2 fl oz of Orondis Ultra A per season and no more than 32 fl oz of Orondis Opti B per season.
Pink rot, *Pythium* leak, tuber rot	azoxystrobin + mefenoxam (Quadris Ridomil Gold SL)	11 + 4	0.82 fl oz/1,000 ft of row	—	0	Apply as an in-furrow spray in 3 to 15 gallons of water per acre at planting.
	cyazofamid (Ranman 400 SC)	21	1.4 to 2.75 fl oz/acre (foliar) 0.42 fl oz/1,000 ft (in furrow)	7	0.5	For pink rot and *Pythium* leak, apply at the high rate. Do not apply more than 10 sprays per crop or more than 27.5 fl oz/acre/season. Make no more than 3 consecutive applications followed by 3 applications from a different resistance management group.

Table 10-44. Disease Control Products for Potato, Irish

Disease	Material	FRAC	Rate Of Material	Minimum Days Harvest	Minimum Days Reentry	Method, Schedule, and Remarks
Pink rot, *Pythium* leak, tuber rot (continued)	mefenoxam (Ridomil Gold SL) (Ultra Flourish)	4	0.42 fl oz/1,000 ft of row 0.84 fl oz/1,000 ft of row	7	2	See labels for maximum amount of product allowable per season. PHI is based on foliar application for Ultra Flourish.
	mefenoxam + chlorothalonil (Ridomil Gold/Bravo) 76.5 WP	4 + M	2.5 pt/acre	14	2	Apply at flowering and then continue on a 14-day interval. Do not exceed more than 4 applications per crop.
	mefenoxam + copper hydroxide (Ridomil Gold/Copper)	4 + M	2 lb/acre	14	2	Apply at 14-day intervals for up to 3 applications. Alternate with a protectant fungicide.
	mefenoxam + mancozeb (Ridomil Gold MZ)	4 + M	2.5 lb/acre	14	2	Apply at 14-day intervals for up to 4 applications.
	metalaxyl (Metalaxyl 2E AG, MetaStar 2E)	4	12.8 fl oz/acre	14	2	Preplant incorporated or soil surface spray
	mono- and di-potassium salts of phosphorous acid (various)	33	2.5 to 10 pt/acre	0	4h	See label for in furrow application or foliar application rates.
Powdery mildew	azoxystrobin (various)	11	See label	14	4h	See label. Rates may vary depending on the product. Apply in furrow at planting according to label direction. Do not apply more than one application without alternating away from fungicides in Group 11.
	azoxystrobin + chlorothalonil (Quadris Opti)	11 + M	1.6 pt/acre	14	0.5	Do not apply more than 1.5 lb a.i./acre/year of azoxystrobin; do not apply more than 11.25 lb a.i./acre /year of chlorothalonil. Do not make more than 1 application before alternating with a fungicide with a different mode of action. Do not apply this product or other fungicides in Group 11 more than 6 times in a season.
	azoxystrobin + difenoconazole (Quadris Top)	11 + 3	8 to 14 fl oz/acre	14	0.5	Apply at 7- to 14-day intervals. Apply no more than 2 sequential applications without alternating with a fungicide with a different mode of action. Do not apply more than 55.3 lb product/acre/year. Do not apply more than 0.46 pound a.i./acre /year of difenoconazole-containing products; do not apply more than 2.0 lb a.i./acre /year of azoxystrobin-containing products.
	fluopyram + pyrimethanil (Luna Tranquility)	7 + 9	11.2 fl oz/acre	7	0.5	Do not make more than 2 sequential applications without switching to a fungicide outside of Group 7 or Group 9. Limit 54.7 fl oz/acre/season.
	fluxapyroxad + pyraclostrobin (Priaxor Xemium)	7 + 11	6 to 8 fl oz/acre	7	0.5	Limit 3 applications per season and no more than 2 applications before switching to a different mode of action. Do not apply more than 24 fl oz/acre/season including in furrow and foliar uses.
	mancozeb + azoxystrobin (Dexter Max)	M + 11	1.6 to 2.1 lb/acre	14	1	Do not exceed 16 lb product/acre/crop. Season limits apply for azoxystrobin. For suppression of powdery mildew.
	mandipropamid + difenoconazole (Revus Top)	40 + 3	5.5 to 7 fl oz/acre	14	0.5	Begin applications when conditions favor disease development, on 7- to 10-day intervals. Do not apply more than twice before alternating with fungicides that have a different mode of action. Do not apply more than 28 fl oz/acre/season.
	metconazole (Quash)	3	2.5 to 4 oz/acre	1	0.5	Limit 16 oz/acre/season. Make no more than 2 applications before changing modes of action. Limit to 4 applications per year. Use the 4 oz rate for white mold.
	metiram + pyraclostrobin (Cabrio Plus)	M + 11	2.9 lb/acre	14	1	Apply at 7- to 14-day-intervals. Do not apply more than 17.4 lb product/acre/season. Do not apply more than 2 sequential applications before alternating with a fungicide with a different mode of action.
	penthiopyrad (Vertisan)	7	10 to 24 fl oz/acre	7	0.5	Apply at 7- to 14-day intervals. Make no more than 2 applications before alternating with a fungicide with a different mode of action. Do not exceed 72 oz/acre/year. Do not apply more than 11.25 oz a.i./acre/year in total from any combination of seed, soil, or foliar applications.
	pyraclostrobin (Headline; Headline SC)	11	6 to 12 fl oz/acre	3	0.5	DO NOT exceed 6 foliar applications or 72 total fl oz of product/acre/season. Do not apply more than 1 time before alternating with a fungicide with a different mode of action.
	sulfur (various)	M2	See label	—	1	Rates vary among products; see label.

For Pumpkin, Winter Squash, and Summer Squash – See Cucurbits
Radish – See Root Vegetables
Scallion - See Onion, Green Shallot – See Onion, Dry

Chapter X—2019 N.C. Agricultural Chemicals Manual

Root Vegetables (Except Sugar Beet) Beet, Carrot, Parsnip, Radish, Turnip

E. Pfeufer, Plant Pathology, University of Kentucky

Table 10-45. Disease Control Products for Root Vegetables (Except Sugar Beet) — Beet (red, garden or table), Carrot, Parsnip, Radish, Turnip – Harvested for roots only

Disease	Material	FRAC	Rate of Material	Minimum Days Harvest	Minimum Days Reentry	Method, Schedule, and Remarks
Alternaria leaf blight, Cercospora leaf spot	azoxystrobin (various)	11	9.0 to 15.5 fl oz/acre	0	4 hr	No more than 1 application before alternating with a fungicide with a different mode of action. Make no more than 123 fl oz per acre per year.
	azoxystrobin + chlorothalonil (Quadris Opti)	11 + M	2.4 pt / acre	0	0.5	FOR USE ON CARROTS ONLY.
	azoxystrobin + difenoconazole (Quadris Top)	11 + 3	12 to 14 fl oz / acre	7	0.5	FOR USE ON CARROTS ONLY.
	azoxystrobin + propiconazole (various)	11 + 3	14 fl oz	14	0.5	FOR USE ON CARROTS ONLY. No more than 1 application before alternating with a non-Group 11 fungicide. Make no more than 55 fl oz per acre per year.
	boscalid (Endura)	7	4.5 oz/acre	0	0.5	FOR USE ON CARROTS ONLY. Not for Cercospora. Do not make more than 2 consecutive applications or more than 5 applications per season.
	chlorothalonil (various)	M	1.4 to 1.8 lb/acre	—	0.5	FOR USE ON CARROTS ONLY. Spray at first appearance, 7 to 10 day intervals.
	cyprodinil + fludioxonil (Switch)	9 + 12	11 to 14 oz/acre	7	0.5	NOT FOR CERCOSPORA. Apply when disease first appears, and continue on 7- to 10-day intervals if conditions remain favorable for disease development. Do not exceed 56 oz of product per acre per year.
	fixed copper (various)	M	See label	0	1 to 2	FOR USE ON CARROTS AND GARDEN BEETS ONLY.
	fluazinam (Omega)	29	1 pt / acre	7	0.5	FOR USE ON CARROTS ONLY.
	fluopyram + trifloxystrobin (Luna Sensation)	7 + 11	4.0 to 7.6 fl oz/acre	7	0.5	FOR USE ON CARROTS ONLY. Do not make more than 2 consecutive applications before rotating to a labeled non-Group 7 or non-Group 11 fungicide.
	fluopyram + pyrimethanil (Luna Tranquility)	7 + 9	11.2 fl oz/acre	7	0.5	Do not make more than 2 consecutive applications before rotating to a labeled non-Group 7 or non-Group 9 fungicide.
	fluxapyroxad + pyraclostrobin (Merivon)	7 + 11	4 to 5.5 fl oz/acre	7	0.5	Do not make more than 2 consecutive applications before rotating to a labeled non-Group 7 or non-Group 11 fungicide. Make no more than 3 applications per season. Use maximum rate for Cercospora leaf spot.
	iprodione (various)	2	1 to 2 pt/acre	0	1	For use on carrots only. Make no more than 4 applications per season.
	penthiopyrad (Fontelis)	7	16 to 30 fl oz/acre	0	0.5	Make no more than 2 consecutive applications before alternating with a fungicide with a different mode of action. Apply no more than 61 fl oz per acre per year.
	pyraclostrobin (Cabrio)	11	8 to 12 oz/acre	0	0.5	Alternate with a fungicide with a different mode of action.
	pyraclostrobin + boscalid (Pristine)	11 + 7	8 to 10.5 oz/acre	0	0.5	FOR USE ON CARROTS ONLY. Make no more than 2 consecutive applications before alternating with a different mode of action. Use no more than 63 oz or make no more than 6 applications per season.
	sulfur (various)	M	3 to 10 lb/acre	—	1	POWDERY MILDEW ONLY. Spray at first appearance. Avoid applying on days over 90°F.
	trifloxystrobin (Flint) (Gem)	11 oz/acre	2 to 3 oz/acre 1.9 to 2.9 fl oz/acre	7	0.5	NOT FOR RADISHES. Make no more than 1 application before alternating with a fungicide with another mode of action. Make no more than 4 applications of trifloxystrobin or other strobilurin fungicides per season.
Cercospora leaf spot or blight, powdery mildew	tebuconazole (various)	3	4 to 7.2 fl oz/acre	7	0.5	FOR USE ON TURNIP AND GARDEN BEETS ONLY. Repeat applications at 12- to 14-day intervals. Apply no more than 28 fl oz per acre per season.
Phytophthora basal stem rot	mefenoxam (Ridomil Gold) 4 SL (Ultra Flourish) 2EC	4	1 to 2 pt/trt acre 2 to 4 pt/trt/acre	—	2	Apply preplant incorporated into top 2 inches or as a pre-emergent soil spray. Surface spray must be incorporated by rainfall or irrigation.
	metalaxyl (various)	4	4 to 8 pt/trt acre	—	2	
	fenamidone (Reason)	11	8.2 fl oz/acre	14	0.5	Make no more than 1 application before alternating with a mefenoxam-containing fungicide. Apply no more than 24.6 fl oz per growing season. Applied with sprayer or in sprinkler irrigation.
	fluazinam (Omega)	29	1 pt / acre	7	0.5	FOR USE ON CARROTS ONLY.
Pythium root rot, root dieback, cavity spot (Pythium spp.)	azoxystrobin+mefenoxam (Uniform)	11 + 4	0.34 fl oz/1000 row ft	—	0	NOT FOR CARROTS. In-furrow treatment only.
	mefenoxam (Ridomil Gold) 4 SL (Ultra Flourish) 2 EC	4	1 to 2 pt/trt acre 2 to 4 pt/trt acre		2	Apply preplant incorporated into top 2 inches, as a soil spray at planting. Surface spray must be incorporated by rainfall or irrigation.

Table 10-45. Disease Control Products for Root Vegetables (Except Sugar Beet) — Beet (red, garden or table), Carrot, Parsnip, Radish, Turnip – Harvested for roots only

Disease	Material	FRAC	Rate of Material	Minimum Days Harvest	Minimum Days Reentry	Method, Schedule, and Remarks
Pythium root rot, root dieback, cavity spot (*Pythium* spp.) (continued)	metalaxyl (various)	4	4 to 8 pt/trt acre	—	2	Apply preplant incorporated into top 2 inches, as a soil spray at planting. Surface spray must be incorporated by rainfall or irrigation.
	fenamidone (Reason)	11	8.2 fl oz/acre	14	0.5	Make no more than 1 application before alternating with a mefenoxam-containing fungicide. Apply no more than 24.6 fl oz per growing season. Applied with sprayer or in sprinkler irrigation.
	cyazofamid (Ranman)	21	6 fl oz/acre	14	0.5	**FOR USE ON CARROTS ONLY.** May be applied preplant incorporated, as a pre-emergent surface band, or in sprinkler irrigation. Applications can be repeated at 14- day intervals, but must alternate with a Pythium fungicide with a different mode of action.
	fluopicolide (Presidio)	43	3 to 4 fl oz/acre	7	0.5	Can be applied with a sprayer or in sprinkler irrigation. Regardless of method, must be applied in combination with a fungicide with a different mode of action and labeled for that method. No more than 2 consecutive applications before alternating with a Pythium fungicide with a different mode of action. Maximum of 12 fl oz per acre per year. For carrots only, may also be applied preplant incorporated.
Rhizoctonia root canker (*Rhizoctonia* solani)	azoxystrobin (various)	11	0.4 to 0.8 fl oz/1000 row ft	0	4 hr	Make one application, applied either in-furrow at planting, in a 7-inch band over the row prior to or shortly after planting, or in drip irrigation.
Rust (*Puccinia* spp.)	penthiopyrad (Fontelis)	7	16 to 30 fl oz/acre	0	0.5	Make no more than 2 sequential applications before alternating with a fungicide with a different mode of action. Apply no more than 61 fl oz per acre per year.
	sulfur (various)	M	See label	0	1	
White mold (*Sclerotinia* spp.) and gray mold (*Botrytis* spp.)	boscalid (Endura)	7	7.8 oz	0	0.5	**FOR USE ON CARROTS ONLY.** No more than 2 applications before alternating with a fungicide with a different mode of action. Limit of 3 applications per season.
	penthiopyrad (Fontelis)	7	16 to 30 fl oz/acre	0	0.5	Make no more than 2 consecutive applications before alternating with a fungicide with a different mode of action. Apply no more than 61 fl oz per acre per year.
White mold (*Sclerotinia* spp.) and gray mold (*Botrytis* spp.) (postharvest)	thiabendazole (Mertect)		41 fl oz/100 gal	—	0.5	Dip harvested roots 5 to 10 seconds. Do not rinse.
Southern blight	dichloropropene (Telone) C-17 C-35	—	10.8 to 17.1 gal/acre 13 to 20.5 gal/acre	—	5	Fumigate soil in-the-row 3 to 6 weeks before seeding. Rate is based on soil type; see label for in-row rates.
	pyraclostrobin + boscalid (Pristine)	11 + 7	8 to 10.5 oz/acre	0	0.5	**FOR USE ON CARROTS ONLY.** Suppression only. Make no more than 6 applications per season.
	azoxystrobin (various)	11	0.4 to 0.8 fl oz/1000 row ft	0	4 hr	Make 1 application, applied either in-furrow at planting, in a 7-inch band over the row prior to or shortly after planting, or in drip irrigation.
	fluazinam (Omega)	29	1 pt / acre	7	0.5	**FOR USE ON CARROTS ONLY.**
	penthiopyrad (Fontelis)	7	16 to 30 fl oz/acre	0	0.5	Make no more than 2 sequential applications before alternating with a fungicide with a different mode of action. Apply no more than 61 fl oz per acre per year.
White rust (*Albugo* spp.)	azoxystrobin (various)	11	6.0 to 15.5 fl oz/acre	0	4 hr	No more than 1 application before alternating with a fungicide with a different mode of action. Apply no more than 123 fl oz per acre per season.
	pyraclostrobin (Cabrio)	11	8 to 16 oz/acre	0	0.5	Alternate with a fungicide with a different mode of action. Apply no more than 48 oz per acre per season
	mefenoxam + copper hydroxide (Ridomil Gold/ Copper)	4 + M	2 lb/acre	7	1	Spray leaves. Use with preplant Ridomil 2E soil applications. Make 2 to 4 applications if needed on 14-day intervals.

Table 10-46. Importance of Alternative Management Practices for Disease Control in Carrot

Scale: E = excellent; G = good; F = fair; P = poor; NC = no control; ND = no data.

Strategy	Alternaria blight	Cercospora blight	Powdery mildew	Pythium cavity spot	Pythium damping off	Southern blight	Rhizoctonia cavity spot	Sclerotinia postharvest	Botrytis postharvest	Bacterial leaf blight	Root-knot nematode
Avoid field operations when leaves are wet	P	P	NC	NC	NC	NC	NC	NC	NC	F	NC
Avoid overhead irrigation	F	F	NC	NC	NC	NC	NC	F	NC	F	NC
Change planting date	P	P	NC	F	F	F	NC	NC	NC	NC	F
Cover cropping with antagonist	NC	NC	NC	NC	NC	NC	NC	NC	NC	NC	F
Crop rotation	F	F	NC	P	P	P	P	P	NC	F	P
Deep plowing	G	G	P	NC	NC	F	F	F	P	G	NC
Destroy crop residue	E	E	P	NC	NC	NC	P	NC	P	E	P
Encourage air movement	F	F	NC	NC	NC	NC	NC	F	NC	NC	NC
Plant in well-drained soil	NC	NC	NC	G	G	P	F	F	NC	NC	NC
Plant on raised beds	NC	NC	NC	F	F	NC	F	P	NC	NC	NC
Postharvest temperature control	NC	NC	NC	NC	NC	NC	NC	E	E	NC	NC
Reduce mechanical injury	NC	NC	NC	NC	NC	NC	NC	F	G	NC	NC
Destroy volunteer carrots	F	F	P	NC	NC	NC	NC	NC	NC	NC	NC
Pathogen-free planting material	E	E	NC	NC	NC	NC	NC	NC	NC	E	NC
Resistant cultivars	G	G	F	NC	NC	NC	NC	NC	NC	NC	NC

Spinach

E. Pfeufer, Plant Pathology, University of Kentucky

Table 10-47. Disease Control Products for Spinach

Disease	Material	FRAC	Rate of Material	Minimum Days Harvest	Minimum Days Reentry	Method, Schedule, and Remarks
Damping off (*Pythium*)	mefenoxam (Ridomil Gold) (Ultra Flourish)	4	1 to 2 pt/trt acre 2 to 4 pt/trt acre	21	2	Broadcast or banded over the row as a soil spray or preplant incorporation into the top 2 inches of soil.
	metalaxyl (various)	4	4 to 8 pt/trt acre	21	2	Broadcast or banded over the row as a soil spray or preplant incorporation into the top 2 inches of soil.
Seedling blight, damping off, root rot (*Pythium* spp., *Rhizoctonia solani*)()	azoxystrobin + mefenoxam (Uniform)	45 + 40	0.34 fl oz/ 1000 ft of row	—	0	Apply as an in furrow spray in 5 gallons of water per acre prior to covering seed. Make only 1 application per season.
Downy mildew (*Peronospora fainosa* f. sp. *Spinaciae*)	ametoctradin + dimethomorph (Zampro)	45 + 40	14 fl oz/acre	0	0.5	Do not apply with or in rotation with mandipropamid or dimethomorph.
	cymoxanil (Curzate)	27	5 oz /acre	1	0.5	Apply with a protectant fungicide. Apply no more than 30 ounces per acre in a 12-month period.
	dimethomorph (Forum)	40	6 fl oz/acre	0	0.5	Do not apply with or in rotation with mandipropamid.
	mandipropamid (Revus)	40	8 fl oz/acre	1	4 hr	Make no more than 2 consecutive applications before alternating with a fungicide with a different mode of action. Apply no more than 32 fl oz per acre per season. Do not apply with or in rotation with dimethomorph.
Downy mildew (*Peronospora fainosa* f. sp. *Spinaciae*), white rust (*Albugo occidentalis*)	acibenzolar-*S*-methyl (Actigard)	21	0.5 to 0.75 oz/acre	7	0.5	Do not apply to young seedlings or plants stressed due to drought, excessive moisture, cold weather, or herbicide injury.
	famoxadone + cymoxanil (Tanos)	11 + 27	8 to 10 oz/acre	1	0.5	Must be tank-mixed with a contact downy mildew fungicide with a different mode of action. Make no more than 1 application before alternating with a fungicide with a different mode of action. Apply no more than 84 oz per acre per cropping season.
	fluopicolide (Presidio)	43	3 to 4 fl oz/acre	2	0.5	Tank mix with another downy mildew fungicide with a different mode of action. Apply as a foliar spray or in drip irrigation.
	fluxapyroxad + pyraclostrobin (Merivon)	7 + 11	4 to 11 fl oz/acre	1	0.5	**Do not tank-mix Merivon with any pesticides, adjuvants, fertilizers, nutrients, or any other additives.** Do not make more than 2 consecutive applications before rotating to a labeled non-Group 7 or non-Group 11 fungicide. Make no more than 3 applications per season.

Table 10-47. Disease Control Products for Spinach

Disease	Material	FRAC	Rate of Material	Minimum Days Harvest	Minimum Days Reentry	Method, Schedule, and Remarks
Downy mildew, white rust (continued)	fixed copper (various)	M	See label	0	2	Some formulations of copper may cause leaf flecking.
	pyraclostrobin (Cabrio)	11	12 to 16 oz/acre	0	0.5	Make no more than 2 consecutive applications before alternating with a fungicide with a different mode of action. Apply no more than 64 oz per acre per growing season.
	fosetyl-Al (Aliette)	33	2 to 5 lb/acre	3	0.5	Do not mix with surfactants, foliar fertilizers, or products containing copper.
	mefenoxam (Ridomil) 4 SL	4	0.25 pt/acre	21	2	Shank in 21 days after planting or after first cutting. Another application may be shanked in after the next cutting. A total of 2 shank applications may be made on 21-day intervals.
	mefenoxam + copper hydroxide (Ridomil Gold/Copper)	4 + M	2.5 lb/acre	21	2	Spray to foliage. Use with preplant Ridomil Gold soil application.
	metalaxyl (MetaStar) 2 E	4	1 pt/trt acre	21	2	Shank in 21 days after planting. Apply no more than 2 shanked applications on 21-day intervals.
	oxathiapiprolin (Orondis Gold 200)	49	4.8 to 19.2 fl oz/ acre	0	4 hr	**DOWNY MILDEW ONLY.**
White rust	cyazofamid (Ranman)	21	2.1 to 2.75 fl oz/ acre	0	0.5	No more than 5 applications per crop. No more than 3 consecutive applications followed by at least 3 applications of a fungicide with a different mode of action. Do not apply more than 13.75 fl oz per acre per crop per growing season
	fenamidone (Reason)	11	5.5 to 8.2 fl oz/acre	2	0.5	Make no more than 1 application before alternating with a fungicide with a different mode of action. Apply no more than 24.6 fl oz per acre per growing season.
Various leaf spots	azoxystrobin (various)	11	6 to 15.5 fl oz/acre	0	4 hr	Make no more than 2 consecutive applications before alternating with a fungicide with a different mode of action. Apply no more than 92.3 fl oz per acre per season.
	azoxystrobin + flutriafol (Topguard)	11 + 3	6 to 8 fl oz/acre	7	0.5	Use a rotation partner outside of groups 3 and 11.
	cyprodinil + fludioxonil (Switch)	9 + 12	11 to 14 fl oz/acre	0	0.5	Make no more than 1 application before alternating with a fungicide with a different mode of action. Apply no more than 24.6 fl oz per acre per growing season.
	fixed copper (various)	M	See labels	0	2	Some formulations of copper may cause flecking on the leaves.
	flutriafol (Rhyme)	3	5 to 7 fl oz/acre	7	0.5	
	fluopyram + trifloxystrobin (Luna Sensation)	7 + 11	7.6 fl oz/ acre	0	0.5	Use a rotation partner outside of groups 7 and 11.
	pyraclostrobin (Cabrio)	11	12 to 16 oz/acre	0	0.5	Make no more than 2 consecutive applications before alternating with a fungicide with a different mode of action. Apply no more than 64 oz per acre per growing season.
	penthiopyrad (Fontelis)	7	14 to 24 fl oz/acre	3	0.5	Make no more than 2 sequential applications before alternating with a fungicide with a different mode of action. Apply no more than 72 fl oz per acre per year.

For Winter Squash and Summer Squash – See Cucurbits

Sweetpotato

L. Quesada-Ocampo, Plant Pathology

Table 10-48. Disease Control Products for Sweetpotato

Disease	Material	FRAC	Rate of Material	Minimum Days Harvest	Minimum Days Reentry	Method, Schedule, and Remarks
Black rot (*Ceratocystis fimbriata*), scurf (*Monilochaetes infuscans*), and foot rot	thiabendazole (Mertect 340 F)	3	107 fl oz/100 gal	0.5	0.5	Dip seed roots 1 to 2 minutes and plant immediately.
Postharvest Black rot (*Ceratocystis fimbriata*)	thiabendazole (Mertect 340 F)	3	0.42 fl oz per 2,000 lb of roots or 0.42 fl oz/gal	0.5	0.5	SECTION 18 LABEL ONLY IN NORTH CAROLINA. Postharvest treatment of sweetpotato for control of black rot. Limit to 1 application during packing. Mist washed roots on a conveyor line, with tumbling action, before packing with 0.42 fl oz of Mertect to each 2,000 lb of roots in sufficient water for complete coverage. Alternatively, dip the roots for 20 seconds in 0.42 fl oz of Mertect per gal of water. Ensure roots are dry before packing.
Circular spot, Sclerotial blight, *Rhizoctonia* stem canker, *Pythium* root rot	azoxystrobin (Quadris 2.08 F)	11	0.4 to 0.8 fl oz/1,000 row feet	—	4 hr	Make in-furrow or banded applications shortly after transplanting.
	dichloran (Botran 5F)		0.6 qt/7.5 gal (Seed Dip) 5.73 oz in 14 gal/1,000 linear feet of plant bed (Plant bed spray)	—	0.5	Labeled for Southern blight (*Sclerotium rolfsii*). Seed dip: Dip seed sweetpotatoes 10 to 15 seconds in a well-agitated fungicide suspension. Drain sweetpotatoes and bed promptly. Prepare fresh fungicide suspension daily. Plant bed spray: Spray or sprinkle over bedded sweetpotatoes before covering them with soil.
Seed-borne and soilborne fungi that cause decay, damping off or seedling blight	azoxystrobin (Dynasty 0.83 F)	11	0.19 to 0.38 fl oz per 100 lb of propagating roots	—	4 hr	Apply uniformly to seed roots as a water-based slurry.
	fludioxonil (Maxim 4 FS)	12	0.08 to 0.16 fl oz per 100 lb of propagating roots	—	0.5	Apply uniformly to seed roots as a water-based slurry.
Damping off (*Pythium*)	cyazofamid (Ranman 400 SC)	21	6.1 fl oz/acre	7	0.5	Apply at planting. Refer to label for details.
	ethaboxam (Elumin)	22	8 fl oz/acre	—	0.5	Apply in-furrow or as a side dressing over seed piece. Do not make more than 2 applications per year or apply more than 16 fl oz/acre/year.
	fluopicolide (Presidio)	43	3 to 4 fl oz/acre	7	0.5	Must be tank-mixed with a labeled rate of another fungicide active against the target pathogen, but with a different mode of action. Repeat applications at 10-day intervals.
	mefenoxam (Ridomil Gold 4 SL)	4	1 to 2 pt/treated acre	—	2	Incorporate in soil. See label for row rate.
	metalaxyl (MetaStar 2 E)	4	4 to 8 pt/treated acre	7	2	Preplant incorporated or soil surface spray.
Foliar diseases (*Alternaria*)	azoxystrobin (Aframe, generic)	11	6 to 15.5 fl oz/acre	0	4 hr	Limit to 123 fl oz/acre/season. For soilborne disease control, refer to label. Begin foliar applications prior to disease and continue on a 5- to 7-day interval.
	cyprodinil + fludioxonil (Switch 62.5WG)	9 + 12	11 to 14 oz/acre	7	0.5	Begin foliar applications prior to disease and continue on a 7- to 10-day interval.
	difenoconazole + benzovindiflupyr (Aprovia Top)	7 + 3	10.5 to 13.5 fl oz/acre	14	0.5	No more than 2 applications can be made at a 7-day interval, all other applications must be made at a 14-day interval. Apply no more than 27 fl oz/acre/year.
	fenamidone (Reason 500 SC)	11	5.5 to 8.2 fl oz/acre	14	0.5	Begin foliar applications prior to disease and continue on a 5- to 10-day interval.
	fluoxastrobin (Aftershock)	11	2 to 3.8 fl oz/acre	7	0.5	Limit to 22.8 fl oz/acre/year. For soilborne disease control, refer to label. Begin foliar applications prior to disease and continue on a 7- to 10-day interval.
	pyraclostrobin (Cabrio 20 WG)	11	8 to 12 oz/acre	0	0.5	Do not apply more than 48 fl oz/acre/season. Alternate with a fungicide with a different mode of action after each use.
	pyrimethanil (Scala SC)	9	7 fl oz/acre	7	0.5	Begin foliar applications prior to disease and continue on a 7- to 14-day interval.
	trifloxystrobin (Flint Extra)	11	3 to 3.8 oz/acre	7	0.5	Apply on a 7- to 10-day interval as needed. Do not make more than 6 applications per year or apply more than 23 fl oz/acre/year.
Postharvest Fusarium rot	azoxystrobin + fludioxonil + difenoconazole (Stadium)	11 + 12 + 3	1 fl oz/2,000 lb of roots	—	—	Ensure proper coverage, use tumbling, mix the fungicide solution in sufficient water volume. Do not make more than 1 postharvest application.

Table 10-48. Disease Control Products for Sweetpotato

Disease	Material	FRAC	Rate of Material	Minimum Days Harvest	Minimum Days Reentry	Method, Schedule, and Remarks
Mottle necrosis (*Pythium* postharvest)	potassium phosphite (Alude)	33	1 ¼ qt/acre	0	4 hr	Foliar spray at 5- to 14-day intervals depending on disease incidence.
Powdery mildew	azoxystrobin + difenoconazole (Quadris Top)	11 + 3	8 to 14 fl oz/acre	14	0.5	Begin foliar applications prior to disease and continue on a 7- to 14-day interval.
	cyprodinil + fludioxonil (Switch 62.5WG)	9 + 12	11 to 14 oz/acre	7	0.5	Begin foliar applications prior to disease and continue on a 7- to 10-day interval.
	difenoconazole + benzovindiflupyr (Aprovia Top)	7 + 3	10.5 to 13.5 fl oz/acre	14	0.5	No more than 2 applications can be made at a 7-day interval, all other applications must be made at a 14-day interval. Apply no more than 27 fl oz/acre/year.
	fluopyram + pyrimethanil (Luna Tranquility)	7 + 9	11.2 fl oz/acre	7	0.5	Limit to 54.7 fl oz/acre/year. Do not make more than 2 sequential applications of Group 7-containing fungicides. Labeled for Alternaria and Sclerotinia. Apply at 7- or 14-day intervals.
	metconazole (Quash)	3	2.5 to 4 oz/acre	1	0.5	Begin foliar applications prior to disease and continue on a 7- to 10-day interval. Do not apply more than 16 fl oz per year or 4 times per year. Do not make more than 2 sequential applications before alternating with products with different modes of action.
	penthiopyrad (Vertisan)	7	0.7 to 24 fl oz/acre	7	0.5	For soilborne disease control, refer to label. Begin foliar applications prior to disease and continue on a 7- to 14-day interval.
	pyraclostrobin (Cabrio 20 WG)	11	8 to 12 oz/acre	0	0.5	Do not apply more than 48 fl oz/acre/season. Alternate with a fungicide with a different mode of action after each use.
Postharvest Rhizopus soft rot	dicloran (Botran 75 W)	14	1 lb/100gal	—	—	Spray or dip. Dip for 5 to 10 seconds in well-agitated suspension. Add ½ pound Botran to 100 gallons of treating suspension after 500 bushels treated. Do not rinse.
	fludioxonil (Scholar 1.9 SC)	12	16 to 32 fl oz/100 gal	—	—	Dip for approximately 30 seconds in well-agitated solution and allow sweetpotatoes to drain. Add 8 fl oz to 100 gals after 500 bushels are treated. ALTERNATIVELY, mix 16 fl oz in 7 to 25 gals of water, wax/emulsion, or aqueous dilution of wax/oil emulsion. Can also be used to disinfest tanks, refer to label.
Sclerotinia	boscalid (Endura)	7	5.5 to 10 oz/acre	10	0.5	Begin applications prior to disease development and apply again at a 7- to 14-day interval. Do not apply more than 10 fl oz/year. Do not make more than 2 sequential applications before alternating with products with different modes of action.
	Coniothyrium minitans (Contans WG)	—	1 to 4 lb/acre	0	4 hr	**OMRI listed product.** Apply to soil surface and incorporate no deeper than 2 inches. Works best when applied prior to planting or transplanting. Do not apply other fungicides for 3 weeks after applying Contans.
	fluazinam (Omega 500F)	29	5.5 to 8 fl oz/acre	14	0.5	Initiate applications when conditions are favorable for disease development or when disease symptoms first appear. Repeat applications on a 7- to 10-day schedule. Do not apply more than 3.5 pt/year.
	metconazole (Quash)	3	4 oz/acre	1	0.5	Make an application prior to disease development and apply again 14 days later. Do not apply more than 16 fl oz/year or 4 times per year. Do not make more than 2 sequential applications before alternating with products with different modes of action.
Scurf (*Monilochaetes infuscans*)	dicloran (Botran 75 W)	14	1 lb/100 gal	—	—	Seed dip: Dip seed sweetpotatoes 10 to 15 seconds in a well-agitated fungicide suspension. Drain sweetpotatoes and bed promptly. Prepare fresh fungicide suspension daily. Plant bed spray: Spray or sprinkle over bedded sweetpotatoes before covering them with soil.
Southern blight (*Slerotium folfsii*)	dicloran (Botran 75W)	14	1 ll/100 gal	—	—	Seed dip: Dip seed sweetpotatoes 10 to 15 seconds in a well-agitated fungicide suspension. Drain sweetpotatoes and bed promptly. Prepare fresh fungicide suspension daily. Plant bed spray: Spray or sprinkle over bedded sweetpotatoes before covering them with soil.
	difenoconazole + benzovindiflupyr (Aprovia Top)	7 + 3	10.5 to 13.5 fl oz/acre	14	0.5	No more than 2 applications can be made at 7-day interval, all other applications must be made at a 14-day interval. Apply no more than 27 fl oz/acre/year.

Table 10-48. Disease Control Products for Sweetpotato

Disease	Material	FRAC	Rate of Material	Minimum Days Harvest	Minimum Days Reentry	Method, Schedule, and Remarks
White rust	azoxystrobin (Quadris 2.08 F)	11	6.2 to 15.4 fl oz/ acre	7	4 hr	Make no more than 2 sequential applications before alternating with fungicides that have a different mode of action. Apply no more than 2.88 qt/crop/acre/season.
	fenamidone (Reason 500 SC)	11	5.5 to 8.2 fl oz/acre	14	0.5	Begin applications when conditions favor disease development and continue on 5- to 10- day interval. Do not apply more than 16.4 fl oz per growing season. Alternate with a fungicide from different resistance management group.
	pyraclostrobin (Cabrio 20 WG)	11	8 to 16 oz/acre	0	0.5	Do not apply more than 48 oz/acre/season. Alternate with a fungicide with a different mode of action after each use.
Nematodes	fluopyram (Velum Prime)	7	6.0 to 6.84 fl oz/acre	7	0.5	Limit to 0.466 lb/acre/season. Do not make more than 2 sequential applications of Group 7-containing fungicides.

Table 10-49. Efficacy of Products for Disease Control in Sweetpotato

Scale: E = excellent; G = good; F = fair; P = poor; NC = no control; ND = no data.

Product	Nematicide (N) or Fungicide (F)	Alternaria leaf spot	Black rot (*C. fimbriata*)	*Fusarium*	Java black rot (*D. gossypina*)	Nematodes	Pythium	Rhizopus soft rot (*R. stolonifer*)	Southern blight (*S. rolfsii*)	Sclerotinia	Scurf (*M. infuscans*)	Soil rot/Pox (*S. ipomoea*)	Sweetpotato feathery mottle virus
aldicarb (Temik)	N	ND	ND	ND	ND	G	ND	ND	ND	ND	ND	ND	NC
azoxystrobin + fludioxonil + difenoconazole (Stadium)	F	ND	E	ND	ND	ND	ND	E	ND	ND	ND	ND	NC
boscalid (Endura)	F	ND	ND	ND	ND	ND	ND	ND	ND	E	ND	ND	NC
chlorine	F	ND	F	ND	P	ND	ND	F	ND	ND	P	NC	NC
chloropicrin	N, F	ND	P	F	F	F	ND	ND	F	ND	ND	F	NC
Coniothyrium minitans (Contans WG)	F	ND	ND	ND	ND	ND	ND	ND	ND	F	ND	ND	NC
cyazofamid (Ranman)	F	ND	ND	ND	ND	ND	ND	ND	ND	ND	ND	ND	NC
dicloran (Botran 75W)	F	ND	P	ND	P	ND	ND	F	P	G	F	NC	NC
dichloropropene (Telone II)	N	ND	ND	P	ND	G	ND	ND	ND	ND	ND	ND	NC
difenoconazole + benzovindiflupyr (Aprovia Top)	F	ND	ND	ND	ND	ND	E	ND	ND	ND	ND	ND	NC
ethaboxam (Elumin)	F	ND	ND	ND	ND	ND	G	ND	ND	ND	ND	ND	NC
ethoprop (Mocap)	N	ND	ND	ND	ND	P	ND	ND	ND	ND	ND	ND	NC
fludioxonil (Scholar)	F	ND	F	ND	ND	NC	ND	F	NC	ND	ND	NC	NC
fluopicolide (Presidio)	F	ND	ND	ND	ND	ND	G	ND	ND	ND	ND	ND	NC
fluopyram (Velum Prime)	N, F	ND	ND	ND	ND	G	ND	ND	ND	ND	ND	ND	NC
mefenoxam (Ridomil Gold)	F	ND	ND	ND	ND	ND	G	ND	ND	ND	ND	ND	NC
metconazole (Quash)	F	ND	E	ND	ND	ND	ND	ND	ND	ND	ND	ND	NC
metam sodium (Vapam)	N	ND	P	F	ND	F	ND	ND	ND	ND	ND	ND	NC
metalaxyl (Metastar)	F	ND	ND	ND	ND	ND	F	ND	ND	ND	ND	ND	NC
oxamyl (Vydate)	F	ND	ND	ND	ND	F	ND	ND	ND	ND	ND	ND	NC
Pseudomonas syringae (Bio-Save)	F	ND	ND	ND	ND	ND	ND	P	ND	ND	ND	ND	NC
thiabendazole (Mertect 340-F)	F	ND	E	P	F	ND	ND	E	F	ND	P	NC	NC

Table 10-50. Importance of Alternative Management for Disease Control in Sweetpotato

Scale: E = excellent; G = good; F = fair; P = poor; NC = no control; ND = no data.

Strategy	Alternaria leaf spot	Black rot (*C. fimbriata*)	Fusarium	Java black rot (*D. gossypina*)	Nematodes	Pythium	Rhizopus soft rot (*R. stolonifer*)	Sclerotinia	Southern blight	Scurf (*M. infuscans*)	Soil rot/Pox (*S. ipomoea*)	Sweepotato Feathery Mottle Virus
Crop rotation (3 to 4 years)	P	F	F	F	F	P	NC	F	F	P	F	NC
Disease-free planting stock	NC	E	G	G	F	P	NC	NC	P	E	P	G
Resistant cultivars	F	NC	F	F	F[s]	P	F	F	F	P	G	F
Careful handling to reduce mechanical injury	NC	F	F	NC	NC	P	E	F	NC	NC	NC	NC
Cutting plants (in beds) above soil line	NC	G	F	F	G	P	NC	NC	NC	G	P	NC
Soil sample for nematode analysis	NC	NC	NC	NC	E	P	NC	NC	NC	NC	NC	NC
Sanitation (equipment, fields, storage houses)	F	E	P	F	NC	P	E	NC	NC	P	NC	NC
Manage insects that transmit pathogens	NC	P	NC	NC	NC	P	NC	NC	NC	NC	NC	NC
Sulfur added to soil to reduce pH	NC	NC	NC	NC	NC	P	NC	NC	NC	NC	F	NC
Prompt curing and proper storage conditions	NC	E	F	F	NC	P	E	NC	NC	NC	NC	NC
Site selection (drainage)	P	NC	F	F	NC	E	F	G	P	NC	P	NC
Manage insects that cause feeding injuries to roots	NC	G	NC	P	NC	P	P	NC	NC	NC	NC	NC
Avoid harvesting when soils are wet	F	G	NC	F	NC	G	F	F	NC	NC	NC	NC

[s] Resistant cultivars for root knot nematode are susceptible to reniform nematode

Sweetpotato Storage House Sanitation – See Sanitation

Tomatillo

S. Bost, Plant Pathology, University of Tennessee, M. Lewis Ivey, Plant Pathologist, The Ohio State University

Table 10-51. Disease Control Products for Tomatillo

Disease	Material	FRAC	Rate of Material	Minimum Days Harvest	Minimum Days Reentry	Method, Schedule, and Remarks
Early Blight	azoxystrobin (various)	11	5 to 6.2 fl oz/acre	0	4 hr	Limit of 37 fl oz per crop per acre per season. Make no more than **one** application before alternating with fungicides that have a different mode of action.
	azoxystrobin + difenoconazole (Quadris Top 2.72F)	11 + 3	8 fl oz/acre	0	0.5	Limit of 47 fl oz per acre per season. Do not apply until 21 days after transplanting or 35 days after seeding.
	boscalid (Endura 70WDG)	7	2.5 to 3.5 oz/acre	0	0.5	Limit of 21 oz per acre per season. Make no more than 2 sequential applications before alternating with fungicides that have a different mode of action.
	cyprodinil + difenoconazole (Inspire Super 2.82F)	9 + 3	16 to 20 fl oz/acre	0	0.5	Limit of 80 fl oz per acre per season.
	cyprodinil + fludioxonil (Switch 62.5 WG)	9 + 12	11 to 14 oz/acre	0	0.5	Limit of 56 oz per acre per year. After 2 applications, rotate to another fungicide with a different mode of action for two applications.
	difenoconazole + mandipropamid (Revus Top 4.16F)	3 + 40	5.5 to 7 fl oz/acre	1	0.5	Limit of 28 fl oz per acre per season. Make no more than 2 consecutive applications per season before alternating with fungicides that have a different mode of action.
	fenamidone (Reason 500SC)	11	5.5 to 8.2 fl oz/acre	14	0.5	Limit of 24.6 fl oz per growing season. Make no more than **one** application before rotating to another effective fungicide with a different mode of action.
	fluoxastrobin (Aftershock, Evito 480SC 4F)	11	2.0 to 5.7 fl oz/acre	3	0.5	Limit of 22.8 fl oz per acre per season. Make no more than **one** application before alternating with fungicides that have a different mode of action. **NOTE: Do not overhead irrigate for 24 hours following a spray application.**
	penthiopyrad (Fontelis 1.67F)	7	10 to 24 fl oz/acre	0	0.5	Do not exceed 72 fl oz of product per year. Make no more than 2 sequential applications per season before alternating with fungicides that have a different mode of action.
	polyoxin D zinc salt (Ph-D; OSO 5% SC) (OSO)	19	6.2 oz/acre 3.75 to 13.0 fl oz/acre	0	4 hr	Limit of 5 applications per season. Make no more than **one** application before alternating with fungicides that have a different mode of action.
	pyraclostrobin (Cabrio 20% EG)	11	8 to 16 oz/acre	0	4 hr	Limit of 96 oz per acre per season. Make no more than **one** application before alternating with fungicides that have a different mode of action.
	trifloxystrobin (Flint 50WDG)	11	2 to 3 oz/acre	3	0.5	Limit of 16 oz per acre per year. Make no more than **one** application before alternating with fungicides that have a different mode of action.
Powdery mildew	azoxystrobin (various)	11	5 to 6.2 fl oz/acre	0	4 hr	Limit of 37 fl oz per crop per acre per season. Make no more than **one** application before alternating with fungicides that have a different mode of action.
	azoxystrobin + difenoconazole (Quadris Top 2.72F)	11	8 fl oz/acre	0	0.5	Limit of 47 fl oz per acre per season. Do not apply until 21 days after transplanting or 35 days after seeding.
	chlorothalonil (Bravo Weather Stick 6F)	M	1.5 pt/acre	3	0.5	Limit of 12 pt per acre per season.
	chlorothalonil + cymoxanil (Ariston 4.34F)	M + 27	2 to 2.44 pt/acre	3	0.5	Limit of 17.5 pt per acre per year.
	cyprodinil + difenoconazole (Inspire Super 2.82F)	9 + 3	16 to 20 fl oz/acre	0	0.5	Limit of 80 fl oz per acre per season.
	cyprodinil + fludioxonil (Switch 62.5 WG)	9 + 12	11 to 14 oz/acre	0	0.5	Limit 56 oz per acre per year. After 2 applications, rotate to another fungicide with a different mode of action for 2 applications.
	penthiopyrad (Fontelis 1.67F)	7	10 to 24 fl oz/acre	0	0.5	Do not exceed 72 fl oz of product per year. Make no more than 2 consecutive applications per season before rotating to a fungicide with a different mode of action.
	polyoxin D (Ph-D 11.3WDG)	19	6.2 oz/acre	0	4 hr	Limit 5 applications per season. Make no more than **one** application before alternating with fungicides that have a different mode of action.
	pyraclostrobin (Cabrio 20EG)	11	8 to 16 oz/acre	0	4 hr	Limit of 96 oz per acre per season. Make no more than **one** application before alternating with fungicides that have a different mode of action.
	mandipropamid + difenoconazole (Revus Top 4.16F)	40 + 3	5.5 to 7 fl oz/acre	0	0.5	Limit of 28 fl oz per acre per season. Make no more than 2 consecutive applications before alternating with fungicides that have a different mode of action.

Tomato

I. Meadows, Entomology and Plant Pathology

Table 10-52. Disease Control Products for Tomato

Disease	Material	FRAC	Rate of Material	Minimum Days Harvest	Minimum Days Reentry	Method, Schedule, and Remarks
Tomato (transplants produced in a greenhouse or other controlled environment) Treating seed to eliminate plant pathogens on or within the seed is recommended. For a list of seed treatments that are compatible with raw (naked) seed see seed treatment table.						
Bacterial canker	sodium hypochlorite (CPPC Ultra Bleach 2; 6.15%)	NC	1 qt + 4 qt water	NA	0	Wash seed for 40 minutes in solution with continuous agitation; air dry promptly. Use 1 gal of solution per 1 lb seed. **NOTE: Ultra Bleach 2 seed treatment is not compatible with pelleted (coated) seed.**
	streptomycin sulfate (various)	25	1 lb/100 gal	NA	0	Begin application at first true leaf stage: repeat weekly until transplant.
Bacterial spot, Bacterial speck	bacteriophage (AgriPhage)	NC	3 to 8 oz/9,600 sq ft	NA	0	Consult your vegetable Extension Specialist for information on requirements needed to use bacteriophage. Bacteriophages are most effective when applied during or after last watering of the day.
	copper (various)	M1	See label	NA	0	Begin application at first true leaf stage, repeat at 3 to day intervals until transplanting. Alternating with streptomycin sulfate is recommended.
	mancozeb (various)	M3	See label	NA	1	For states east of the Mississippi, use 1.5 to 3 lb product/acre. States west of the Mississippi use 1.5 to 2 lb product/acre. **NOTE: Use a full rate of fixed copper in combination with mancozeb. Mancozeb alone does not control bacteria.**
	streptomycin sulfate (various)	25	1 lb/100 gal	NA	0	Begin application at first true leaf stage, repeat weekly until transplanting. For plant bed use only.
Botrytis (Gray Mold), Botrytis stem canker, Early blight, Powdery mildew	cyprodinil + fludioxonil (Switch 62.5 WG)	9 + 12	11 to 14 oz/acre	NA	0.5	**DO NOT APPLY TO GRAPE OR CHERRY TOMATO.** After 2 applications, switch to a different mode of action for 2 applications.
	fluopyram + pyrimethanil (Luna Tranquility)	7 + 9	11.2 fl oz/acre	NA	0.5	See label for limits on application amounts per season. Do not make more than 2 applications of Group 7 or 9 fungicides without switching to a different mode of action.
	penthiopyrad (Fontelis 1.67 SC)	7	0.5 to 75 fl oz/gal	NA	0.5	Use 1 gallon of spray per 1,360 sq ft. Do not make more than 2 applications before switching to a different mode of action.
Early blight, Gray mold, Late blight	mancozeb (various)	M3	See label	NA	1	For states east of the Mississippi, use 1.5 to 3 lb product/acre. States west of the Mississippi use 1.5 to 2 lb product/acre. **NOTE: Use a full rate of fixed copper in combination with mancozeb if bacteria control is also required.**
Late blight	mandipropamid (Micora)	40	5.5 to 8.0 fl oz/acre (5,000 sq ft)	NA	4h	Apply no more than 2 applications before switching to another mode of action.
	propamocarb (Previcur Flex 6F)	28	0.7 to 1.5 pt/acre	NA	0.5	Can be used as a drench before or after transplanting.
Pythium damping off	cyazofamid (Ranman 400SC)	21	3.0 fl oz/100 gal	NA	0.5	Apply as a soil drench to seedling tray or at the time of transplant.
	propamocarb (Previcur Flex 6F)	28	1.5 pt/acre	NA	0.5	Limit of 7.5 pt/acre/season. Do not apply more than once before alternating with fungicides that have a different mode of action.
Tomato (field)						
Anthracnose	azoxystrobin (various)	11	See label	0	4 hr	See label. Do not make more than 1 application before switching to a fungicide with a different mode of action.
	azoxystrobin + chlorothalonil (Quadris Opti)	11 + M	1.6 pt/acre	0	0.5	Do not make more than 1 application before switching to a fungicide with a different mode of action. Do not apply within 21 days after transplanting or 35 days after seeding.
	azoxystrobin + difenoconazole (Quadris Top)	11 + 3	8 fl oz/acre	0	0.5	Do not make more than 1 application before switching to a fungicide with a different mode of action. Limit 47 fl oz/acre/season. Do not apply with 21 days after transplanting or 35 days after seeding.
	azoxystrobin + flutriafol (Topguard EQ)	11 + 3	4 to 8 fl oz	0	0.5	Do not use adjuvants or EC formulated tank mix partners on fresh market tomatoes. Do not exceed 4 applications per year or 8 fl oz product/acre. Limits on both active ingredients apply.
	chlorothalonil (various)	M	See label	0	0.5	Refer to individual labels for rates and restrictions.
	copper (various)	M	See label	3	2	
	cymoxanil + chlorothalonil (Ariston)	27 + M	1.9 pt/acre	3	0.5	Check copper labels for specific precautions and limitations for mixing with this product.
	difenoconazole + benzovindiflupyr (Aprovia Top)	7 + 3	10.5 to 13.5 fl oz/acre	0	0.5	Do not make more than 2 applications before switching to a non-Group 7 fungicide. See label for application intervals and limits per season.

Table 10-52. Disease Control Products for Tomato

Disease	Material	FRAC	Rate of Material	Minimum Days Harvest	Minimum Days Reentry	Method, Schedule, and Remarks
Tomato (field) (continued)						
Anthracnose (continued)	difenoconazole + cyprodinil (Inspire Super)	3 + 9	16 to 20 fl oz/acre	0	0.5	Limit of 80 fl oz/acre/season. Do not make more than 2 consecutive applications before alternating with fungicides that have a different mode of action. Limits of each a.i. apply - see label.
	fluopyram + trifloxystrobin (Luna Sensation)	7 + 11	7.6 fl oz/acre	3	0.5	Disease suppression ONLY. Do not exceed 5 applications or 27.1 fl oz product/acre/season. Do not make more than 2 applications without switching to a different mode of action.
	flutriafol (Rhyme)	3	5 to 7 fl oz/acre	0	0.5	Do not exceed 4 applications or 28 fl oz product/acre/season.
	fluxapyroxad + pyraclostrobin (Priaxor)	7 + 11	4 to 8 fl oz/acre	7	0.5	Limit of 24 fl oz/acre/season. Do not make more than 2 consecutive applications before alternating with fungicides that have a different mode of action.
	famoxadone + cymoxanil (Tanos)	11 + 27	8 oz/acre	3	0.5	Limit of 72 fl oz/acre/season (12 month cycle). Do not make more than 1 application before alternating to a fungicide with a different mode of action. **NOTE: Must be tanked mixed with a contact fungicide that has a different mode of action.**
	mancozeb (various)	M	See label	5	1	See label for rates.
	mancozeb + azoxystrobin (Dexter Max)	M + 11	0.8 to 1.6 lb/acre	5	1	Do not exceed 12 lb product/acre/season. Do not make more than 1 application before alternating with a fungicide not in group 11. On fresh market tomato do not tank mix with an adjuvant or an EC formulation. Tank mixture with Dimethoate may cause crop injury.
	mandipropamid + difenoconazole (Revus Top)	40 + 3	5.5 to 7 fl oz/acre	1	0.5	Limit of 28 fl oz product/acre/season. Do not make more than 2 consecutive applications before alternating with fungicides that have a different mode of action.
	penthiopyrad (Fontelis)	7	24 fl oz/acre	0	0.5	**Disease suppression only.** Limit of 72 fl oz/acre/season. Do not make more than 2 consecutive applications before alternating with fungicides that have a different mode of action.
	pyraclostrobin (Cabrio EG)	11	8 to 12 oz/acre	0	0.5	Limit of 96 fl oz/acre/season. Do not make more than 2 applications before alternating to a fungicide with a different mode of action.
	trifloxystrobin (Flint 50WDG)	11	3 to 4 oz/acre	3	0.5	**DISEASE SUPPRESSION ONLY.** Limit of 16 fl oz/acre/season. Do not make more than 2 applications before alternating to a fungicide with a different mode of action.
	(Gem 500 SC)		3 to 3.8 fl oz/acre			
Bacterial spot, Bacterial speck	acibenzolar-S-methyl (Actigard 50 WG)	21	0.33 to 0.75 oz/acre	14	0.5	Should only be applied to healthy, actively growing plants. Do not exceed 8 applications per season.
	bacteriophage (AgriPhage)	NC	3 to 8 oz/9,600 sq ft	0	0	Consult your vegetable Extension Specialist for information on requirements needed to use bacteriophage. Bacteriophages are most effective when applied during or after last watering of the day.
	copper (various)	M	See label	0	0	Use a full rate of fixed copper in combination with mancozeb for best results.
	mancozeb (various)	M	See label	0	1	For states east of the Mississippi, use 1.5 to 3 lb product/acre. **NOTE: Use a full rate of fixed copper in combination with mancozeb. Mancozeb alone does not control bacteria.**
Botrytis (gray mold)	boscalid (Endura 70 WG)	7	9 to 12.5 oz/acre	0	0.5	Limit of 25 oz/acre/season. Make no more than 2 sequential applications and no more than 2 per crop year.
	chlorothalonil (various)	M	See label	0	0.5	Refer to individual labels for rates and restrictions.
	chlorothalonil + cymoxanil (Ariston)	M + 27	1.9 pt/acre	3	0.5	Limit of 17.5 pt/acre/season.
	cyprodinil + fludioxonil (Switch 62.5 WG)	9 + 12	11 to 14 oz/acre	0	0.5	Limit of 56 ounces/acre/season. After 2 applications alternate with another fungicide with a different mode of action for 2 applications.
	difenoconazole + cyprodinil (Inspire Super)	3 + 9	16 to 20 fl oz/acre	0	0.5	Limit of 80 fl oz/acre/season. Do not make more than 2 consecutive applications before alternating with fungicides that have a different mode of action.
	fluopyram + trifloxystrobin (Luna Sensation)	7 + 11	7.6 fl oz/acre	3	0.5	Do not exceed 5 applications or 27.1 fl oz/acre/season. Do not make more than 2 applications without switching to a different mode of action.
	fluxapyroxad + pyraclostrobin (Priaxor)	7 + 11	4 to 8 fl oz/acre	7	0.5	**DISEASE SUPPRESSION ONLY.** Limit of 24 fl oz and 3 applications/acre/season. Do not make more than 2 consecutive applications before alternating with fungicides that have a different mode of action.

Table 10-52. Disease Control Products for Tomato

Disease	Material	FRAC	Rate of Material	Minimum Days Harvest	Minimum Days Reentry	Method, Schedule, and Remarks
Tomato (field) (continued)						
Botrytis (gray mold) (continued)	penthiopyrad (Fontelis)	7	16 to 24 fl oz/acre	0	0.5	Limit of 72 fl oz/acre/season. Do not make more than 2 consecutive applications before alternating with fungicides that have a different mode of action.
	pyraclostrobin (Cabrio EG)	11	12 to 16 oz/acre	0	0.5	**Disease suppression only.** No more than 2 applications allowed before switching to a different mode of action. Do not exceed 96 oz/acre/season.
	pyrimethanil (Scala) SC	9	7 fl oz/acre	1	0.5	Limit of 35 fl oz/acre/season.
Buckeye rot	azoxystrobin (various)	11	See Label	0	4 hr	Limit of 37 fl oz/acre/season. Do not make more than 1 application before alternating to a fungicide with a different mode of action. **NOTE: Under high temperatures, Satori in combination with some additives or adjuvants may cause crop injury.**
	azoxystrobin + chlorothalonil (Quadris Opti)	11 + M	1.6 pt/acre	0	0.5	Limit of 5 applications. Do not make more than 1 application before alternating to a fungicide with a different mode of action. Do not apply earlier than 21 days after transplant.
	famoxadone + cymoxanil (Tanos)	11 + 27	8 oz/acre	3	0.5	**DISEASE SUPPRESSION ONLY.** Do not make more than 1 application before alternating to a fungicide with a different mode of action. **NOTE: Must be tanked mixed with a contact fungicide that has a different mode of action.**
	mancozeb + azoxystrobin (Dexter Max)	M + 11	0.8 to 1.6 lb/acre	5	1	Do not exceed 12 lb product/acre/season. Do not make more than 1 application before alternation with a fungicide not in group 11. On fresh market tomato, do not tank mix with an adjuvant or an EC formulation. Tank mixture with Dimethoate may cause crop injury.
	mancozeb + zoxamide (Gavel 75DF)	M + 22	1.5 to 2 lb/acre	5	2	Limit of 8 applications and 16 lb/acre/season east of the Mississippi River.
	mefenoxam + copper hydroxide (Ridomil Gold/ Copper)	4 + M	2 lb/acre	14	2	Tank mix with 0.8 lb a.i. of either maneb or mancozeb. Make up to 3 applications; alternate with full rate of protectant.
	oxathiapiprolin + chlorothalonil (Orondis Opti A + Orondis Opti B)	49 + M	2.0 to 4.8 fl oz/acre 2.0 pt/acre	0	0.5	Do not make more than 2 sequential applications without switching to a different mode of action and no more than 6 total applications per season. Do not mix soil applications and foliar applications. See labels for application limits.
Damping off *(Pythium)*, Root and fruit rots *(Phytophthora)*	fosetyl-Al (Aliette 80 WDG, Linebacker WDG)	33	2.5 to 5 lb/acre	14	0.5	Do not tank mix with copper. So not exceed 20 lb product/season. Not for *Phytophthora* fruit rot. Check label for specific counties in each state where use is prohibited.
	mefenoxam (various)	4	See label	7	2	Apply uniformly to soil at time of planting. Incorporate mechanically if rainfall is not expected before seeds germinate. A second application may be made up to 4 weeks before harvest. See labels for application limits.
	oxathiapiprolin + mefenoxam (Orondis Gold A + Orondis Gold B)	49 + 4	2.4 to 19.2 fl oz/acre 1.0 to 2.0 pt/acre	28	4h	**APPLY ONLY TO THE GROUND; DO NOT APPLY TO FOLIAGE.** Maximum application rate for Root and Fruit rots is 1.0 pt/acre. Limits of a.i. per season apply - see label. Apply any Orondis product to soil OR foliage, but not both. Make no more than 2 sequential applications before switching to a different mode of action.
	propamocarb (Previcur Flex)	28	1.5 pt/acre	5	0.5	Limit of 7.5 pt/acre/season. Do not apply more than once before alternating with fungicides that have a different mode of action. **For *Pythium* (damping-off) only.**
Gray Leaf Spot (*Stemphylium* spp.)	azoxystrobin + difenoconazole (Quadris Top)	11 + 3	8 fl oz/acre	0	0.5	Do not apply until 21 days after transplanting or 35 days after seeding. Limit of 47 fl oz/acre/season. Make no more than 2 consecutive applications before rotating to another effective fungicide with a different mode of action. See label for tank mix cautions.
	chlorothalonil (various)	M	See label	0	0.5	Refer to individual labels for rates and restrictions.
	difenoconazole + benzovindiflupyr (Aprovia Top)	7 + 3	10.5 to 13.5 fl oz/acre	0	0.5	Do not make more than 2 applications before switching to a non-Group 7 fungicide. See label for application intervals and limits per season.
	difenoconazole + cyprodinil (Inspire Super)	3 + 9	16 to 10 fl oz/acre	0	0.5	Limit of 47 fl oz/acre/season. Do not make more than 2 consecutive applications before alternating with fungicides that have a different mode of action.
	fluopyram + trifloxystrobin (Luna Sensation)	7 + 11	7.6 fl oz/acre	3	0.5	Do not exceed 5 applications or 27.1 fl oz/acre/season. Do not make more than 2 applications without switching to a different mode of action.
	fluopyram + pyrimethanil (Luna Tranquility)	7 + 9	11.2 fl oz/acre	1	0.5	See label for limits on application amounts per season. Do not make more than 2 applications of Group 7 or 9 fungicides without switching to a different mode of action.
	mancozeb (various)	M	See label	5	1	See label for limits on application amounts per season.

Table 10-52. Disease Control Products for Tomato

Disease	Material	FRAC	Rate of Material	Minimum Days Harvest	Minimum Days Reentry	Method, Schedule, and Remarks
Tomato (field) (continued)						
Gray Leaf Spot (*Stemphylium* spp.) (continued)	mancozeb + azoxystrobin (Dexter Max)	M + 11	0.8 to 1.6 lb/acre	5	1	Do not exceed 12 lb product/acre/season. Do not make more than 1 application before alternation with a fungicide not in Group 11. On fresh market tomato, do not tank mix with an adjuvant or an EC formulation. Tank mixture with Dimethoate may cause crop injury.
	mancozeb + copper (ManKocide)	M + M	1 to 3 lb/acre	5	2	Limit of 58 pounds/acre/season east of the Mississippi River.
	mancozeb + zoxamide (Gavel 75DF)	M + 22	1.5 to 2 lb/acre	5	2	Limit of 16 lb/acre/season east of the Mississippi River.
	mandipropamid + difenoconazole (Revus Top)	40 + 3	5.5 to 7 fl oz/acre	1	0.5	Limit of 28 fl oz/acre/season. Do not make more than 2 consecutive applications before alternating with fungicides that have a different mode of action.
	trifloxystrobin (Flint) (Gem 500 SC)	11	3 to 4 oz/acre 3.8 fl oz/acre	3	0.5	Limit of 16 oz Flint or 16 fl oz Gem/acre/season. Do not make more than 1 application before alternating to a fungicide with a different mode of action.
Early blight, Powdery mildew, *Septoria* leaf spot and Target spot	azoxystrobin (various)	11	5 to 6.2 fl oz/acre	0	4 hr	Limit of 37 fl oz/acre/season. Do not make more than 1 application before alternating to a fungicide with a different mode of action. **NOTE: Under high temperatures azoxystrobin in combination with some additives or adjuvants may cause crop injury.**
	azoxystrobin + chlorothalonil (Quadris Opti)	11 + M	1.6 pt/acre	0	0.5	Must alternate with a non-FRAC code 11; use of an adjuvant may cause phytotoxicity. Do not make more than 5 applications of a Group 11 fungicide/acre/season.
	azoxystrobin + difenoconazole (Quadris Top)	11 + 3	8 fl oz/acre	0	0.5	Do not apply until 21 days after transplanting or 35 days after seeding. Limit of 47 fl oz/acre/season. Make no more than 2 consecutive applications before rotating to another effective fungicide with a different mode of action.
	azoxystrobin + difenoconazole (Topguard EQ)	11 + 3	4 to 8 fl oz/acre	0	0.5	Do not use adjuvants or EC formulated tank mix partners on fresh market tomatoes. Do not exceed 4 applications per year. Limits on both active ingredients apply.
	chlorothalonil (various)	M	See label	0	0.5	Refer to individual labels for rates and restrictions.
	cyprodinil + fludioxonil (Switch)	9 + 12	11 to 14 oz/acre	0	0.5	Limit of 56 oz/acre/season. After 2 applications alternate with another fungicide with a different mode of action for 2 applications. **For early blight and powdery mildew control only.**
	difenoconazole + benzovindiflupyr (Aprovia Top)	7 + 3	10.5 to 13.5 fl oz/acre	0	0.5	Do not make more than 2 applications before switching to a non-Group 7 fungicide. See label for application intervals and limits per season.
	difenoconazole + cyprodinil (Inspire Super)	3 + 9	16 to 20 fl oz	0	0.5	Limit of 47 fl oz/acre/season. Do not make more than 2 consecutive applications before alternating with fungicides that have a different mode of action.
	famoxadone + cymoxanil (Tanos)	11 + 27	6 to 8 oz/acre	3	0.5	Limit of 72 fl oz/acre/season. Do not make more than 1 application before alternating to a fungicide with a different mode of action. **NOTE: Must be tanked mixed with a contact fungicide that has a different mode of action. For *Septoria* leaf spot and target spot use 8 oz per acre.**
	cymoxanil + chlorothalonil (Ariston)	27 + M	1.9 to 3.0 pt/acre	3	0.5	Check copper labels for specific precautions and limitations for mixing with this product. Limit 17.5 pt/acre/season.
	fenamidone (Reason 500 SC)	11	5.5 to 8.2 fl oz/acre	14	0.5	Limit of 24.6 fl oz/acre/season. Do not apply more than once before alternating with fungicides that have a different mode of action. **NOT labeled for Target spot control.**
	fluopyram + trifloxystrobin (Luna Sensation)	7 + 11	5 to	3	0.5	Do not exceed 5 applications or 27.1 fl oz/acre/season. Do not make more than 2 applications without switching to a different mode of action. Use 7.6 fl oz rate for gray leaf spot and target spot.
	fluopyram + pyrimethanil (Luna Tranquility)	7 + 9	11.2 fl oz/acre	1	0.5	See label for limits on application amounts per season. Do not make more than 2 applications of Group 7 or 9 fungicides without switching to a different mode of action. Disease suppression for powdery mildew only.
	fluoxastrobin (Aftershock, Evito 480 SC)	11	2.0 to 5.7 fl oz/acre	3	0.5	Limit of 22.8 fl oz/acre/season. Do not apply more than once before alternating with fungicides that have a different mode of action. **Controls target spot and early blight only.**
	flutriafol (Rhyme)	3	5 to 7 fl oz/acre	0	0.5	Tank mix with mancozeb for improved early blight control. Do not exceed more than 4 applications or 28 fl oz product/acre/season. **Not labeled for *Septoria* leaf spot control.**
	fluxapyroxad + pyraclostrobin (Priaxor 500SC)	7 + 11	4 to 8 fl oz/acre	0	0.5	Limit of 24 fl oz/acre/season. Do not make more than 2 consecutive applications before alternating with fungicides that have a different mode of action. Use 6 to 8 fl oz rates for control of powdery mildew.
	mancozeb (various)	M3	See label	5	1	**See label.**

Table 10-52. Disease Control Products for Tomato

Disease	Material	FRAC	Rate of Material	Minimum Days Harvest	Minimum Days Reentry	Method, Schedule, and Remarks
Tomato (field) (continued)						
Early blight, Powdery mildew, *Septoria* leaf spot and Target spot (continued)	mancozeb + azoxystrobin (Dexter Max)	M + 11	0.8 to 1.6 lb/acre	5	1	Do not exceed 12 lb product/acre/season. Do not make more than 1 application before alternation with a fungicide not in Group 11. On fresh market tomato, do not tank mix with an adjuvant or an EC formulation. Tank mixture with Dimethoate may cause crop injury. For target spot control east of the Mississippi, use highest rate.
	mancozeb + zoxamide (Gavel 75DF)	M + 22	1.5 to 2 lb/acre	5	2	Limit of 16 lb/acre/season east of the Mississippi River. **Not labeled for target spot.**
	mandipropamid + difenoconazole (Revus Top)	40 + 3	5.5 to 7 fl oz/acre	1	0.5	Limit of 28 fl oz/acre/season. Do not apply more than 2 consecutive applications before alternating with a fungicide that has a different mode of action.
	penthiopyrad (Fontelis 1.67 SC)	7	16 to 24 fl oz/acre	0	0.5	Limit of 72 fl oz/acre/season. Do not make more than 2 consecutive applications before alternating with fungicides that have a different mode of action.
	propamocarb (Previcur Flex)	28	0.7 to 1.5 pt/acre	5	0.5	Limit of 7.5 pt/acre/season. Do not apply more than once before alternating with fungicides that have a different mode of action. Tank mix with a compatible fungicide for optimal early blight control. **For early blight control only.**
	pyraclostrobin (Cabrio EG)	11	8 to 12 oz/acre	0	0.5	Limit of 96 fl oz/acre/season. Do not make more than 2 applications before alternating to a fungicide with a different mode of action.
	pyrimethanil (Scala SC)	9	7 fl oz/acre	1	0.5	Limit of 35 fl oz/acre/season. Use only in a tank mix with another fungicide recommended for early blight. **Labeled for early blight control only.**
	trifloxystrobin (Flint) (Gem 500 SC)	11	See label	3	0.5	Limit of 16 fl oz/acre/season. Do not make more than 1 application before alternating to a fungicide with a different mode of action. **Products are not labeled for target spot management.**
	zinc dimethyldithiocarbamate (Ziram 76 DF)	M	3 to 4 lb/acre	7	2	Limit of 24 lb/acre/season. **Do not use on cherry tomatoes. For early blight and *Septoria* leaf spot only.**
	zoxamide + chlorothalonil (Zing)	22 + M	36 fl oz/acre	5	0.5	Do not use more than 2 sequential applications before alternating to a fungicide with a different mode of action. See label for application limits. **For early blight and *Septoria* leaf spot only.**
Powdery mildew	azoxystrobin + difenoconazole (Quadris Top)	11 + 3	8 fl oz/acre	0	0.5	Do not apply until 21 days after transplanting or 35 days after seeding. Limit of 47 fl oz /acre/season. Make no more than 2 consecutive applications before rotating to another effective fungicide with a different mode of action. See label for tank mix cautions.
	azoxystrobin + flutriafol (Topguard EQ)	11 + 3	4 to 8 fl oz/acre	0	0.5	Do not use adjuvants or EC formulated tank mix partners on fresh market tomatoes. Do not exceed 4 applications per year. Limits on both active ingredients apply.
	flutriafol (Rhyme)	3	5 to 7 fl oz/acre	0	0.5	Tank mix with mancozeb for improved disease control. Use lower rate if tank-mixed. Do not exceed 4 applications per year or more than 28 fl oz/acre/year.
	mancozeb + azoxystrobin (Dexter Max)	M + 11	1.6 lb/acre	5	1	Do not exceed 12 lb/acre/season. Do not make more than 1 application before alternation with a fungicide not in Group 11. On fresh market tomato, do not tank mix with an adjuvant or an EC formulation. Tank mixture with Dimethoate may cause crop injury. For powdery mildew east of the Mississippi, use highest rate.
	myclobutanil (various)	3	See label	1	0	Spray weekly beginning at first sign of disease. Do not apply more than 1.25 lb/acre. Observe a 30-day plant back interval between last application and planting new crop.
	pyraclostrobin (Cabrio EG)	11	8 to 16 oz/acre	0	0.5	Limit of 96 fl oz/acre/season. Do not make more than 2 applications before alternating to a fungicide with a different mode of action.
	tolfenpyrad (Torac)	39	21 fl oz/acre	1	0.5	Do not exceed 42 fl oz/acre/crop. Do not exceed 2 applications per crop cycle and do not exceed 4 applications per year. Provides SUPPRESSION of powdery mildew.
	sulfur (various)	M	See label	See label	1	Follow labels; may cause leaf burn if used under high temperatures.
Late blight	azoxystrobin (various)	11	6.2 fl oz/acre	0	4 hr	Limit of 37 fl oz/acre/season. Do not make more than 1 application before alternating to a fungicide with a different mode of action. **NOTE: Apply at 5- to 7-day intervals for effective late blight management.**
	chlorothalonil (various)	M	See label	0	0.5	Refer to individual labels for rates and restrictions.
	azoxystrobin + chlorothalonil (Quadris Opti)	11 + M	1.6 pt/acre	0	0.5	Must alternate with a non-FRAC code 11; use of an adjuvant may cause phytotoxicity. Do not make more than 5 applications of a Group 11 fungicide/acre/season.

Table 10-52. Disease Control Products for Tomato

Disease	Material	FRAC	Rate of Material	Minimum Days Harvest	Minimum Days Reentry	Method, Schedule, and Remarks
Tomato (field) (continued)						
Late blight (continued)	azoxystrobin + flutriafol (Topguard EQ)	11 + 3	4 to 8 fl oz/acre	0	0.5	Do not use adjuvants or EC formulated tank mix partners on fresh market tomatoes. Do not exceed 4 applications per year. Limits on both active ingredients apply.
	cymoxanil + chlorothalonil (Ariston)	27 + M	1.9 to 3.0 pt/acre	3	0.5	Check copper labels for specific precautions and limitations for mixing with this product.
	cyazofamid (Ranman 400 SC)	21	2.1 to 2.75 fl oz/acre	0	0.5	Limit of 16.5 fl oz/acre/season. Do not make more than 1 application before alternating to a fungicide with a different mode of action.
	cymoxanil (Curzate 60 DF)	27	3.2 to 5 oz/acre	3	0.5	Limit of 30 oz per 12-month period. Use only in combination with a labeled rate of a protectant fungicide. If late blight is present, use 5 oz/acre on a 5-day schedule.
	dimethomorph (Forum 4.18 F)	40	6 fl oz/acre	4	0.5	Limit of 30 fl oz and 5 applications/acre/season. Performance is improved if tank mixed with another fungicide with a different mode of action.
	dimethomorph + ametoctradin (Zampro)	40 + 11	14 fl oz/acre	4	0.5	Limit of 42 fl oz/acre/season. Do not make more than 2 consecutive applications before alternating to a fungicide with a different mode of action. The addition of a spreading or penetrating adjuvant is recommended to improve product performance.
	fenamidone (Reason 500 SC)	11	5.5 to 8.2 fl oz/acre	14	0.5	Limit of 24.6 fl oz/acre/season. Do not apply more than once before alternating with fungicides that have a different mode of action.
	fluopicolide (Presidio 4F)	43	3 to 4 fl oz/acre	2	0.5	Do not make more than 2 consecutive applications before alternating to a fungicide with a different mode of action. Use only in combination with a labeled rate of another fungicide product with a different mode of action.
	fluoxastrobin (Aftershock, Evito 480 SC)	11	5.7 fl oz/acre	3	0.5	**DISEASE SUPPRESSION ONLY.** Limit of 22.8 fl oz/acre/season. Do not apply more than once before alternating with fungicides that have a different mode of action.
	fluxapyroxad + pyraclostrobin (Priaxor 500 SC)	7 + 11	8 fl oz/acre	7	0.5	**DISEASE SUPPRESSION ONLY.** Limit of 24 fl oz/acre/season. Do not make more than 2 consecutive applications before alternating with fungicides that have a different mode of action.
	mancozeb (various)	M	See label	5	1	
	mancozeb + azoxystrobin (Dexter Max)	M + 11	0.8 to 1.6 lb/acre	5	1	Do not exceed 12 lb/acre/season. Do not make more than 1 application before alternation with a fungicide not in Group 11. On fresh market tomato, do not tank mix with an adjuvant or an EC formulation. Tank mixture with Dimethoate may cause crop injury.
	mancozeb + copper hydroxide (ManKocide) (ManKocide) 61 DF	M + M	1 o 3 lb/acre	5	2	Apply at 7- to 10-day intervals.
	mancozeb + zoxamide (Gavel 75DF)	M + 22	1.5 to 2 lb/acre	5	2	Limit of 8 applications and 16 lb/acre/season east of the Mississippi River.
	mandipropamid + difenoconazole (Revus Top)	40 + 3	5.5 to 7 oz/acre	1	0.5	Limit of 28 fl oz/acre/season. Do not apply more than 2 consecutive applications before alternating with a fungicide that has a different mode of action.
	mefenoxam + chlorothalonil (Ridomil Gold Bravo)	4 + M	2.5 pt/acre	5	2	See label for application limits.
	mefenoxam + mancozeb (Ridomil Gold MZ)	4 + M3	2.5 lb/acre	5	2	Do not make more than 3 applications or 7.5 lb/acre/season..
	oxathiapiprolin + chlorothalonil (Orondis Opti A + Orondis Opti B)	49 + M	2.0 to 4.8 fl oz/acre 8.0 fl oz/acre	1	4 hr	Do not make more than 2 sequential applications without switching to a different mode of action and no more than 6 total applications per season. Do not mix soil applications and foliar applications. See label for application limits.
	oxathiapiprolin + mandipropamid (Orondis Ultra A + Orondis Ultra B)	49 + 40	2.0 to 4.8 fl oz/acre 8.0 fl oz/acre	1	4 hr	Do not make more than 2 sequential applications without switching to a different mode of action and no more than 6 total applications per season. Limit applications apply - see label. Do not mix soil applications and foliar applications.
	pyraclostrobin (Cabrio)	11	8 to 16 oz/acre	0	0.5	No more than 2 applications allowed before switching to a different mode of action. Do not exceed 96 oz/acre/season.
	trifloxystrobin (Flint 50WDG) (Gem 500 SC)	11	4 oz/acre 3.8 fl oz/acre	3	0.5	Limit of 16 fl oz/acre/season. Do not make more than one application before alternating with a protectant fungicide. Apply products with 75% of the labeled rate of a protectant fungicide.
	zoxamide + chlorothalonil (Zing!)	22 + M	36 fl oz/acre	5	0.5	Do not use more than 2 sequential applications before alternating to a fungicide with a different mode of action. Do not tank mix with another fungicide if the target pest is only late blight. Tank mix only if a partner product is required to control other diseases.

Table 10-52. Disease Control Products for Tomato

Disease	Material	FRAC	Rate of Material	Minimum Days Harvest	Minimum Days Reentry	Method, Schedule, and Remarks
Tomato (field) (continued)						
Leaf mold (*Fulvia fulva* = *Passalora fulva*)	azoxystrobin + difenoconazole (Quadris Top)	11 + 3	8 fl oz/acre	0	0.5	Do not apply until 21 days after transplanting or 35 days after seeding. Limit of 47 fl oz/acre/season. Make no more than 2 consecutive applications before rotating to another effective fungicide with a different mode of action.
	difenoconazole + benzovindiflupyr (Aprovia Top)	7 + 3	10.5 to 13.5 fl oz/acre	0	0.5	Do not make more than 2 applications before switching to a non-Group 7 fungicide. See label for application intervals and limits per season.
	difenoconazole + cyprodinil (Inspire Super)	3 + 9	16 to 20 fl oz/acre	0	0.5	Limit of 80 fl oz/acre/season. Do not make more than 2 consecutive applications before alternating with fungicides that have a different mode of action.
	difenoconazole + mandipropamid (Revus Top)	3 + 40	5.5 to 7 fl oz/acre	1	0.5	Make no more than 2 consecutive applications before switching to another fungicide with a different mode of action. Application limits apply - see label.
	famoxadone + cymoxanil (Ianos)	11 + 27	8 oz/acre	3	0.5	Do not make more than 1 application before alternating to a fungicide with a different mode of action. **NOTE: Must be tank-mixed with a contact fungicide that has a different mode of action.** See label for application limits.
	mancozeb (various)	M	See label	5	1	
	mancozeb + azoxystrobin (Dexter Max)	M + 11	0.8 to 1.6 lb/acre	5	1	Do not exceed more than 12 lb/acre/season. Do not make more than 1 application before alternation with a fungicide not in Group 11. On fresh market tomato, do not tank mix with an adjuvant or an EC formulation. Tank mixture with Dimethoate may cause crop injury.
	mancozeb + copper hydroxide (ManKocide 61 DF)	M + M	1 to 3 lb/acre	5	2	Apply at 7- to 10-day intervals.
	mancozeb + zoxamide (Gavel 75DF)	M + 22	1.5 to 2 lb/acre	5	2	Limit of 16 lb/acre/season east of the Mississippi River.
Sour rot (*Geotrichum candidum*)	fludioxonil + propiconazole (Chairman)	3 + 12	See label	0	0	Use as a post-harvest dip, drench or high-volume spray to control certain post-harvest rots. See label for details.
	propiconazole (various)	3	See label	0	0	Use as a post-harvest dip, drench or high-volume spray to control certain post-harvest rots. See label for details.
	fludioxonil (Scholar SC)	12	See label	0	0	Use as a post-harvest dip, drench or high-volume spray to control certain post-harvest rots. See label for details.
Southern blight	difenoconazole + benzovindiflupyr (Aprovia Top)	7 + 3	10.5 to 13.5 fl oz/acre	0	0.5	**DISEASE SUPPRESSION ONLY.** Do not make more than 2 applications before switching to a non-Group 7 fungicide. See label for application intervals and limits per season.
	fluoxastrobin (Aftershock, Evito 480 SC)	11	2.0 to 5.7 fl oz/acre	3	0.5	**DISEASE SUPPRESSION ONLY.** Begin applications when conditions favor disease development, on 7- to 10-day intervals. Do not apply more than once before alternating with fungicides that have a different mode of action. Do not apply more than 22.8 fl oz/acre/season.
	fluxapyroxad + pyraclostrobin (Priaxor 500SC)	7 + 11	4 to 8 fl oz/100 gal	7	0.5	**DISEASE SUPPRESSION ONLY.** Limit of 24 fl oz/acre/season. Do not make more than 2 consecutive applications before alternating with fungicides that have a different mode of action.
	PCNB (Blocker 4F) (transplanting) (in furrow)	14	4.5 to 7.5 pt/100 gal; (apply 0.5 pt of solution per plant) 1.2 to 1.875 gal/acre (10.6 to 16 7 fl oz/1,000 ft of row	NA	0.5	Transplanting: Apply at the time of transplanting for Southern blight suppression. In furrow: Apply in 8 to 10 gals of water per acre based on 36-inch row spacing. Limit of 7.5 lb a.i./acre/season.
	penthiopyrad (Fontelis)	7	1 to 1.6 fl oz/1,000 row ft	7	0.5	Apply as a soil drench to seedling tray or at the time of transplant. See label for application limits.
	pyraclostrobin (Cabrio EG)	11	12 to 16 oz/acre	0	4 hr	**DISEASE SUPPRESSION ONLY.** Limit of 96 fl oz/acre/season. Do not make more than 2 applications before alternating to a fungicide with a different mode of action.
Timber rot, white mold or Sclerotinia stem rot	fluxapyroxad + pyraclostrobin (Priaxor)	7 + 11	4 to 8 fl oz/100 gal	7	0.5	**DISEASE SUPPRESSION ONLY.** Limit of 24 fl oz/acre/season. Do not make more than 2 consecutive applications before alternating with fungicides that have a different mode of action. See label for application limits.
	pyraclostrobin (Cabrio EG)	11	12 to 16 oz/acre	0	4 hr	**DISEASE SUPPRESSION ONLY.** Limit of 96 fl oz/acre/season. Do not make more than 2 applications before alternating to a fungicide with a different mode of action.

Table 10-53. Importance of Alternative Management Practices for Disease Control in Tomato

Scale: E = excellent; G = good; F = fair; P = poor; NC = no control; NA = not applicable, ND = no data.

Strategy	Bacterial canker*	Bacterial speck	Bacterial spot	Botrytis	Buckeye rot	Early blight	Late blight	Leaf Mold (greenhouse or open field)	Powdery mildew	Septoria leaf spot	Target Spot (greenhouse or open field)	Tomato spotted wilt virus**
Use of resistant cultivars	NC	P	F	NR	NR	F	G	P	F	NR	P	G
Use of disease-free seed or transplants	G	G	G	NC	NC	NC	NC	F	NC	P	F	NA
Use of seed treatments	G	G	G	NC	NC	P	P	F	NA	P	ND	NA
Use of sanitation practices at the transplant stage	G	G	G	G	NC	NC	NC	F	NC	NC	F	NC
Crop rotation (3-4 years)	F	P	P	NC	F	F	NC	F	NC	P	P	NC
Control of solanaceous weeds	F	NC	NC	F	F	F	F	F	F	F	F	F
Fertility	NC	NC	NC	F	NC	F	NC	ND	NC	NC	ND	NC
Use of cover crops	NC	NC	NC	NC	F	P	NC	ND	NC	NC	ND	NC
Destroy crop residue	F	NC	NC	NC	NC	P	NC	F	NC	F	F	ND
Rogue plants	F	NC	NC	NC	NC	NC	NC	NC	NC	NC	NC	NC
Promote air movement	F	F	F	F	P	P	F	F	P	F	F	NA
Use of plastic or reflective mulches	NC	NC	NC	NC	F	F	NC	NC	NC	F	NC	G
Do not handle plants when wet	G	G	G	NC	NC	P	P	F	NC	P	F	NC
Use of drip irrigation (avoiding overhead irrigation)	F	F	F	F	P	F	F	F	NC	F	F	NC
Use of biological control or biorational products	P	P	F	P	NC	P	P	P	P	NC	P	NC
Use of foliar fungicides/bactericides	F	F	F	F	G	G	G	F	G	G	F	NA
Use of insecticides	NC	NC	NC	NC	NC	NC	NC	NC	NC	NC	NC	F
Soil fumigation	NC	NC	NC	NC	F	P	NC	NC	NC	NC	NC	NC

* Bacterial canker (foliar or systemic) is rarely observed on open field grown tomatoes in Deep South states.

** Tomato spotted wilt virus is transmitted by thrips.

Table 10-54. Efficacy of Products for Disease Control in Tomato

Scale: E = excellent; G = good; F = fair; P = poor; NC = no control; ND = no data.

Product[1]	Fungicide Group[F]	Preharvest Interval (Days)	Bacterial Canker (foliar)	Bacterial Speck	Bacterial Spot	Botrytis Graymold	Buckeye Rot	Early Blight	Late Blight	Leaf mold (*Fulvia fulva*)	Powdery Mildew	Septoria Leaf Spot	Target Spot
azoxystrobin[2] (Quadris)	11	1	NC	NC	NC	NC	ND	E[R]	F	F	G	E	G	F[R]
famoxadone + cymoxanil (Tanos)	11 + 27	3	NC	NC	NC	NC	P	F	F	F	F	ND	F	F[R]
bacteriophage[3] (AgriPhage)	NA	0	NC	P	P	NC	NC	NC	NC	NC	NC	NC	NC
acibenzolar-*S*-methyl[9] (Actigard)	21	14	ND	F	F	NC	NC	NC	NC	NC	NC	NC	F
boscalid (Endura)	7	0	NC	NC	NC	F	NC	G	NC	ND	ND	ND	F
chlorothalonil (various)	M	0	NC	NC	NC	F	P	F	G	G	P	G	F
chlorothalonil + cymoxanil (Ariston)	M + 27	3	NC	NC	NC	F	P	F	G	F	P	F	F
azoxystrobin[2] (Quadris)	11	0	NC	NC	NC	ND	ND	E[R]	F	NC	F	G	P[R]
azoxystrobin + benzovindiflupyr (Mural)	11 + 7	0	NC	NC	NC	ND	ND	G	ND	ND	ND	F	ND
azoxystrobin + difenoconazole (Quadris Top)	11 + 3	0	NC	NC	NC	ND	ND	G	F	ND	G	G	ND
azoxystrobin + flutriafol (Topguard EQ)	11 + 3	0	NC	NC	NC	ND	ND	G	ND	ND	ND	ND	G
acibenzolar-*S*-methyl[9] (Actigard)	21	14	ND	F	F	NC	NC	NC	NC	NC	NC	NC	NC
bacteriophage[3] (AgriPhage)	NG	0	NC	P	P	NC	NC	NC	NC	NC	NC	NC	NC
benzovindiflupyr + difenoconazole (Aprovia Top)	7 + 3	0	NC	NC	NC	ND	ND	G	ND	ND	ND	F	ND
boscalid (Endura)	7	0	NC	NC	NC	P	NC	G	NC	ND	ND	ND	P[R]
chlorothalonil (Bravo, Chloronil, Echo, Equus, Initiate)	M	0	NC	NC	NC	F	P	F	G	F	P	F	F
chlorothalonil + cymoxanil (Ariston)	M + 27	33	NC	NC	NC	F	P	F	G	F	P	F	F
cyazofamid (Ranman)	21	0	NC	NC	NC	NC	NC	NC	F	NC	NC	NC	NC

Table 10-54. Efficacy of Products for Disease Control in Tomato

Scale: E = excellent; G = good; F = fair; P = poor; NC = no control; ND = no data.

Product[1]	Fungicide Group[F]	Preharvest Interval (Days)	Bacterial Canker (foliar)	Bacterial Speck	Bacterial Spot	Botrytis Graymold	Buckeye Rot	Early Blight	Late Blight	Leaf mold (*Fulvia fulva*)	Powdery Mildew	Septoria Leaf Spot	Target Spot
cymoxanil (Curzate)	27	3	NC	NC	NC	NC	P	NC	F	ND	NC	ND	NC
cyprodinil + fludioxonil (Switch)	9 + 12	0	NC	NC	NC	F	NC	F	NC	NC	F	NC	NC
dimethomorph (Forum)	40	4	NC	NC	NC	NC	NC	NC	F	NC	NC	NC	NC
difenoconazole + cyprodinil (Inspire Super)	3 + 9	0	NC	NC	NC	G	NC	G	NC	G	G	F	F
dimethomorph + ametoctradin (Zampro)	40 + 45	4	NC	NC	NC	NC	NC	NC	G	NC	NC	NC	NC
famoxadone + cymoxanil (Tanos)	11 + 27	3	P	P	P	NC	P	F	F	F	ND	F	F[R]
fenamidone (Reason)	11	14	NC	NC	NC	NC	P	F	F	NC	ND	P	NC
fixed copper[4]	M	1	F	F	P[R]	NC	P	F	F	F	P	F	NC
fluopicolide (Presidio)	43	2	NC	NC	NC	NC	P	NC	G	NC	NC	NC	NC
fluopyram + trifloxystrobin (Luna Sensation)	7 + 11	3	NC	NC	NC	NC	ND	G	ND	ND	G	ND	ND
fluopyram + pyrimethanil (Luna Tranquility)	7 + 9	1	NC	NC	NC	ND	ND	G	ND	ND	G	F	ND
flutriafol (Rhyme)	3	0	NC	NC	NC	ND	ND	G	ND	ND	ND	ND	G
fluxapyroxad + pyraclostrobin (Priaxor)	7 + 11	0	NC	NC	NC	P	NC	G	P	ND	G	F	F
mancozeb (Dithane, Koverall, Manzate, Penncozeb)	M	5	NC	NC	P	P	P	F	F	F	NC	F	F
mancozeb + fixed copper (ManKocide)	M + M	5	NC	F	F	NC	NC	F	F	F	NC	F	NC
mancozeb + azoxystrobin (Dexter Max)	M + 11	5	ND	ND	ND	ND	ND	G	ND	ND	ND	ND	ND
mancozeb + zoxamide (Gavel)	M + 22	3	NC	P	P	NC	P	F	F	F	NC	F	NC
mandipropamid + difenoconazole (Revus Top)	40 + 3	1	NC	NC	NC	NC	NC	F	G	F	NC	F	NC
mefenoxam[8] + chlorothalonil (Ridomil Gold Bravo)	4 + M	5	NC	NC	NC	P	E	P	E	F	NC	F	NC
mefenoxam + copper (Ridomil Gold/Copper)	4 + M	14	NC	NC	NC	NC	E	NC	E[R]	NC	NC	NC	NC
mefenoxam + mancozeb (Ridomil Gold MZ)	4 + M	5	NC	NC	NC	NC	E	NC	G[R]	NC	NC	NC	NC
myclobutanil (various)	3	0	NC	NC	NC	NC	NC	NC	NC	NC	G	NC	ND
penthiopyrad (Fontelis)	7	0	NC	NC	NC	F	NC	G	NC	ND	F	F	F
polyoxin D zinc salt (Ph-D; Oso 5% SC)	19	0	ND	ND	ND	F	ND	F	ND	ND	F	ND	F
propamocarb (Previcur Flex)	28	5	NC	NC	NC	NC	NC	P	F	NC	NC	NC	NC
pyraclostrobin (Cabrio)	11	0	NC	NC	NC	P	NC	E[R]	F	NC	E	G	F[R]
pyrimethanil (Scala)	9	1	NC	NC	NC	F	NC	F	NC	ND	ND	ND	F
streptomycin[5] (Agri-Mycin 17, Ag-Steptomycin, Harbour)	25	0	NC	F	F	NC	NC	NC	NC	NC	NC	NC	NC
sulfur[6] (various)	M	7	NC	NC	NC	NC	NC	NC	NC	F	NC	NC	NC
zinc dimethyldithiocarbamate[10] (Ziram)	M	7	NC	NC	NC	NC	NC	F	ND	NC	ND	F	ND

[1] Efficacy ratings do not necessarily indicate a labeled use for every disease.
[2] Contact control only; not systemic.
[3] Biological control product consisting of a virus that attacks pathogenic bacteria.
[4] Fixed coppers include: Basicop, Champ, Champion, Citcop, Copper-Count-N, Kocide, Nu-Cop, Super Cu, Tenn-Cop, Top Cop with Sulfur, and Tri-basic copper sulfate.
[5] Streptomycin may only be used on transplants; not registered for field use.
[6] Sulfur may be phytotoxic; follow label carefully.
[7] Curative activity; not systemic.
[8] Curative activity; systemic.
[9] Systemic activated resistance.
[10] Do not use on cherry tomatoes.
[F] To prevent resistance in pathogens, alternate fungicides within a group with fungicides in another group. Fungicides in the "M" group are generally considered "low risk" with no signs of resistance developing to the majority of fungicides.
[R] Resistance reported in the pathogen.

Table 10-55A. Example Spray Program for Foliar Disease Control in Fresh Market Tomato Production

Week	Chemical[1] (Refer to the label for rates)	Number of Applications of Chemical Per Season
Before Harvest (weeks 1 to 8)		
1	mancozeb + copper + Actigard	mancozeb, 1; Actigard, 1; copper, 1
2	mancozeb + copper	mancozeb, 2; copper, 2
3	mancozeb + Fontelis OR Endura[2] + Actigard	mancozeb, 3; Fontelis, 1; Endura, 1; Actigard, 2
4	mancozeb+ copper	mancozeb, 4; copper, 3
5	mancozeb + Inspire Super + Actigard	mancozeb, 5; Inspire Super, 1; Actigard, 3
6	mancozeb + copper	mancozeb, 6; copper, 4
7	mancozeb + Fontelis OR Endura[2] + Actigard	mancozeb, 7; Fontelis, 2; Endura, 2; Actigard, 4
8	mancozeb + copper	mancozeb, 8; copper 5
During Harvest (weeks 9 to 15)		
9	chlorothalonil + Inspire Super	chlorothalonil, 1; Inspire Super, 2
10	Revus Top OR Presidio OR Ranman	Revus Top, 1; Presidio, 1; Ranman, 1
11	chlorothalonil + Fontelis OR Endura[2]	chlorothalonil, 2; Fontelis, 3; Endura, 3
12	Revus Top OR Presidio OR Ranman	Revus Top, 2; Presidio, 2; Ranman, 2
13	chlorothalonil + Inspire Super	chlorothalonil, 3; Inspire Super, 3
14	Revus Top OR Presidio OR Ranman	Revus Top, 3; Presidio, 3: Ranman, 3
15	chlorothalonil + Fontelis[3] OR Endura[2]	chlorothalonil, 4, Fontelis, 4; Endura, 4
	Finish season with chlorothalonil	

[1] For most products, the label restricts the number of applications or the amount of product applied per season.
[2] Stobilurins were removed from this guide due to widespread resistance in the pathogen that causes early blight. However, a strobilurin may be substituted for Fontelis OR Endura.
[3] This application exceeds the maximum allowable amount per season if the higher rate had been applied.

Table 10-55B. Example Spray Program for Foliar Disease Control in Fresh Market Tomato Production when Early Blight is a Consistent Threat

Week	Chemical (Refer to the label for rates)	Number of Applications of Chemical Per Season[1]
Before Harvest (weeks 1 to 10)		
1	mancozeb + Actigard	mancozeb, 1; Actigard, 1
2	mancozeb + copper	mancozeb, 2: copper, 1
3	Fontelis + Actigard	Fontelis, 1; Actigard, 2
4	mancozeb+ copper	mancozeb, 3; copper, 2
5	Inspire Super + Actigard	Inspire Super, 1; Actigard, 3
6	mancozeb + copper	mancozeb, 4; copper, 3
7	Fontelis + Actigard	Fontelis, 2; Actigard, 4
8	mancozeb + copper	mancozeb, 5; copper 4
9	Inspire Super + Actigard	Inspire Super, 2; Actigard, 5
10	chlorothalonil + copper	chlorothalonil, 1; copper, 5
During Harvest (weeks 11 to 15)		
11	Fontelis	Fontelis, 3
12	chlorothalonil	chlorothalonil, 2
13	Inspire Super	Inspire Super, 3
14	chlorothalonil	chlorothalonil, 3
15	chlorothalonil	chlorothalonil, 4
	Finish season with chlorothalonil	

[1] For most products, the label restricts the number of applications or the amount of product applied per season.
[2] In areas or seasons in which bacterial spot or speck problems are not expected, Actigard and copper can be omitted.

Note: If late blight occurs, appropriate fungicides must be added. Fontelis and Inspire Super do not have any late blight activity.

Turnip Greens – see Greens and Leafy Brassicas
Turnip Roots – see Root Vegetables
For Watermelons – see Cucurbits

Chapter X — 2019 N.C. Agricultural Chemicals Manual

Nematode Control in Vegetable Crops

Crop losses due to nematodes can be avoided or reduced by using the following management tactics.
- Practice crop rotation.
- Plow out and expose roots immediately after the last harvest.
- Plow or disk the field two to four times before planting.
- Use nematode-free planting material.
- Sample soil and have it assayed for nematodes, preferably in the fall. There is a fee for each sample. Ship sample via DHL, FedEx, or UPS to: North Carolina Department of Agriculture and Consumer Services.
- Where warranted, fumigate or use other nematicides according to guidelines listed on the label. (Soil should be warm, well worked, and free from undecomposed plant debris and have adequate moisture for seed germination.)
- For in-row application, insert chisel 6 to 8 inches deep and throw a high, wide bed up over it; do not rework rows after fumigating.
- For broadcast treatments, insert chisels 6 to 8 inches deep, and space chisels 12 inches apart for most fumigants; use 5-inch spacing for Vapam.

Row rates in this section are stated for rows on 40-inch spacing. For other row spacings, multiply the stated acre rate by the appropriate conversion factor to determine the amount of material applied per acre (Do not alter stated amount per 100-foot row). This will be a guide to the amount of material to purchase for the acreage you want to treat.

For example, if 10 gallons per acre are used on 40-inch rows, for 36-inch rows, it will take 11.1 gallons to treat an acre.

CAUTION: Read labels carefully. Some products have restrictive crop rotations.

Your Row Spacing (inches)	Conversion Factor
24	1.67
26	1.54
28	1.43
30	1.33
32	1.25
34	1.18
36	1.11
38	1.05
40	1.00
42	0.952
44	0.909
46	0.870
48	0.833
5 ft	0.667
6 ft	0.556
7 ft	0.476
8 ft	0.417

Fumigants

New labels require extensive risk mitigation measures including fumigant management plans (FMPs), buffer restrictions, worker protection safety standards and other measures. Details are on the labels and see http://www2.epa.gov/soil-fumigants. Some fumigants are registered on multiple crops but with crop- or soil-type-specific rates; others are registered for specific crops and/or in certain states only. Follow all labels carefully.

Relative Efficacy of Currently Registered Fumigants or Fumigant Combinations for Managing Soilborne Nematodes, Diseases, and Weeds in Plasticulture Strawberries

Table 10-56A. Relative Efficacy of Currently Registered Fumigants or Fumigant Combinations for Managing Soilborne Nematodes, Diseases, and Weeds in Plasticulture Strawberries[1]

Product	Rate per Treated Acre[2]		Relative Efficacy[3]			
	Volume (gal)	Weight (lb)	Nematodes	Disease	Nutsedge	Weeds: Annual
Telone II (1,3-dichloropropene; 1,3-D)	15 to 27	153 to 275	E	P	P	P
Telone EC[3]	9 to 24[5]	91 to 242[5]	E	P	P	P
Telone C17 (1,3-D + chloropicrin)	32.4 to 42	343 to 445	E	G	P	P
Telone C35 (1,3-D + chloropicrin)	39 to 50	437 to 560	E	E	P	F
InLine (1,3-D + chloropicrin)[3]	29 to 57.6 (See Label)	325 to 645 (See Label)	E	E	P	G
Pic-Clor 60 (chloropicrin + 1,3-D)	48.6	588	E	E	P	G
Pic-Clor 60 EC[4]	42.6	503	E	E	P	G
Pic-Clor 80	34	440	G	E	P	F
Metam potassium[6]	30 to 62	318 to 657	F	G	P	VG
Metam sodium[6] (MS)	37.5 to 75	379 to 758	F	G	P	VG
Chloropicrin + MS[6]	19.5 to 31.5 + 37.5 to 75	275-444 + 379-758	F	E	F	VG
Chloropicrin	48.6	150 to 360	P	E	ND	ND
Tri-Pic 100EC[4]	8 to 24	100 to 300	P	E	ND	ND
Paladin (dimethyl disulphide)[7]	35.0 to 51.3	310 to 455	VG	VG	VG	G
Paladin PIC-21	41.2 to 60.1	392 to 572	VG	E	VG	G
Paladin EC[3,7]	37.0 to 54.2	326 to 479	VG	VG	VG	G
Dominus (allyl isothiocyanate)[8]	25 to 40[5]	212 to 340[5]	F	G	P	G

[1] Fumigants with lower efficacy against weeds may require a complementary herbicide or hand-weeding program, although use of virtually impermeable film (VIF) or totally impermeable film (TIF) may increase weed control, particularly with chloropicrin + 1,3-D products or Paladin. Refer to the Herbicide Recommendation section of this guide for directions pertaining to herbicide applications. Telone can persist more than 21 days under cool or wet soil conditions.

[2] Rates can sometimes be reduced if products are applied with VIF or TIF.

[3] Efficacy Ratings: The efficacy of a management option is indicated by E = excellent, VG = very good, G = good, F = fair, P = poor, and ND = no data. These ratings are benchmarks; actual performance will vary.

[4] Product is formulated for application through drip lines under a plastic mulch; efficacy is dependent on good distribution of the product in the bed profile.

[5] Labelled rates are per *broadcast-equivalent* acre, NOT per treated acre.

[6] Metam potassium can be Metam KLR, K-Pam, Sectagon K54 or other registered formulations, and should be used in soils with high sodium content. Metam sodium can be Vapam, Sectagon 42, Metam CLR or other registered formulations.

[7] Paladin should be applied with 21% chloropicrin and VIF or TIF to enhance disease control, and has low efficacy on certain small seeded broadleaf weeds and grasses. Paladin may not be registered in all States.

[8] Dominus is registered but there is limited experience with the product through University or independent trials in our region; growers may want to consider this on an experimental basis. Planting interval is 10 days. The active ingredient allyl isothiocyanate is similar to the active ingredient in metam sodium products (methyl isothiocyanate) and is likely to behave in a similar manner with a similar pest control profile.

Chapter X—2019 N.C. Agricultural Chemicals Manual

Management of Soilborne Nematodes with Non-Fumigant Nematicides

Nematodes are best managed through an integrated program (IPM). Key management options may include securing advisory/predictive soil samples, crop rotation, fallow periods, host resistance, soil amendments, flooding, soil solarization, suppressive cover crops and other options. For more details see https://edis.ifas.ufl.edu/cv112.

Table 10-56B. Management of Soilborne Nematodes with Non-Fumigant Nematicides

Vegetable Crop	Product	Application Method	Rate/A	Rate/1k ft	Schedule and Remarks
Bean (snap and lima)	Mocap 15G	Band	13 to 20 lb	0.9 to 1.4 lb	Do not place in-furrow or allow granules to contact seed. Incorporate 2 to 4 inches deep in 12- to 15-inch band, at planting. Use higher rates for higher nematode populations.
	Mocap 15G	Broadcast	40 to 54 lb	NA	Do not place in-furrow or allow granules to contact seed. Incorporate 2 to 4 inches deep no more than 3 days before planting. Use higher rates for higher nematode populations.
Cabbage	Mocap 15G	Band	13 lb	0.9 lb	Do not place in-furrow or allow granules to contact seed. Incorporate 2 to 4 inches deep in 15-inch band, at planting.
	Mocap 15G	Broadcast	34 lb	NA	Do not place in-furrow or allow granules to contact seed. Incorporate 2 to 4 inches deep no more than 1 week before planting.
	Mocap EC	Band	13 lb	2.4 fl.oz.	FOR USE IN CA ONLY. Do not spray in-furrow or allow spray to contact seed. Incorporate 2 to 4 inches deep in 15-inch band, at planting.
	Mocap EC	Broadcast	34 lb	N/A	Do not place in-furrow or allow granules to contact seed. Incorporate 2 to 4 inches deep no more than 1 week before planting.
Corn (field and sweet)	Mocap 15G	Band		0.75 to 1.0 lb	Incorporate 2 to 4 inches deep in 12- to 15-inch band.
	Mocap 15G	Broadcast	40	NA	Incorporate 2 to 4 inches deep no more than 3 days before to at planting.
Cucumber	Mocap 15G	Band	13 lb	2.1 lb	Do not place in-furrow or allow granules to contact seed. Incorporate 2 to 4 inches deep in 12- to 15-inch band (7 ft row spacing) at planting.
Potato	Mocap 15G	Band	20 lb	1.4 lb	For suppression of stubby root nematode populations. Incorporate 2 to 4 inches deep; Band should be 12 to 15 inches wide (36-inch row spacing) at planting; do not apply once seedlings have begun to emerge.
	Mocap 15G	Broadcast	40 to 60 lb	NA	For suppression of moderate to heavy stubby root nematode populations; apply within 2 wk before planting; do not apply once seedlings have begun to emerge.
	Mocap EC	Band	63.9 fl oz	4.4 fl oz	For suppression of stubby root nematode populations. Incorporate 2 to 4 inches deep. Band should be 12 inches wide (36-inch row spacing) at planting or before crop emergence.
	Mocap EC	Broadcast	1 to 1.5 gal	N/A	For suppression of moderate to heavy stubby root nematode populations; apply and immediately incorporate no more than 2 weeks before planting or before crop emergence.
Sweetpotato	Mocap 15G	Band only	20 to 26 lb	1.6 to 2.1 lb	Incorporate 2 to 4 inches deep in centered 12- to 15-band (at least 42-inch row spacing) at bedding.
	Mocap EC	Band only	63.5 to 85.9 fl oz	5.1 to 6.9 fl oz	Incorporate 2 to 4 inches deep in centered 12 to 15 inches band (at least 42-inch row spacing) at bedding.
Carrot	Vydate L	PPI	1 to 2 gal in 20 gpa		Apply within 1 wk of planting or before emergence; thoroughly incorporate at least 2 inches deep in soil.
		Chemigation	1 gal in suffic. water		Before crop emergence; minimum re-treatment interval = 14 days.
		In-Furrow	1 to 2 gal in 20 gpa		
Cucurbit (Crop Group 9) & Fruiting Vegetables (Crop Subgroups 8-10b & 8-10c)	Nimitz	Broadcast soil	3.5 to 5.0 pt in 15 gal	NA	Apply and incorporate 6 to 8 inches deep at least 7 days before transplanting. Irrigate with 0.5 to 1.0 inch water 2 to 5 days after application.
		Banded soil	see label	see label	Table 2 in label specifies rate based on row spacing. Incorporate 6 to 8 inches deep at least 7 days before transplanting. Irrigate with 0.5 to 1.0 inches water 2 to 5 days after application.
		Drip Irrigation			Table 3 in label specifies rate based on bed width. Uniformly wet entire bed width & root zone 6 to 8 inches deep at least 7 days before transplanting. Irrigate with 0.5 to 1.0 inch water 2 to 5 days after application.

Greenhouse Disease Control

R. A. Melanson, Extension Plant Pathologist, Mississippi State University; F. Louws, Plant Pathologist, NC State University; M. L. Lewis Ivey, Plant Pathologist, The Ohio State University; A. Keinath, Plant Pathologist, Clemson University

Note: Follow manufacturer's directions on label in all cases.

Caution: At the time this table was prepared, the entries were believed to be useful and accurate. However, labels change rapidly and errors are possible, so the user must follow all directions on the pesticide container. See product labels for application limits per crop/season.

Information in the following table must be used in the context of a total disease control program. For example, many diseases are controlled by the use of resistant varieties, crop rotation, sanitation, seed treatment, and cultural practices. Always use top-quality seed or plants obtained from reliable sources. Seeds are ordinarily treated by the seed producer for the control of seed decay and damping-off.

Most foliar diseases can be reduced or controlled by maintaining relative humidity under 90 percent, by keeping the air circulating in the house with a large overhead polytube, and by avoiding water on the leaves.

Caution: The risk of pesticide exposure in the greenhouse is high. Use protective clothing laundered daily or after each exposure. Ventilate during application and use appropriate personal protective equipment (PPE).

Table 10-57. Greenhouse Disease Control for Various Vegetable Crops

Disease/Location	Product[1]	FRAC Group	Rate of Formulation	Minimum Days Harvest	Minimum Days Reentry	Schedule and Remarks
Greenhouse						
Sanitation	Solarization	NA	140 degrees F, 4 to 8 hours for 7 days	—	—	Close greenhouse during hottest and sunniest part of summer for at least 1 week. Greenhouse must reach at least 140 degrees F each day. Remove debris and heat sensitive materials and keep greenhouse and contents moist. Will not control pests 0.5 inches or deeper in soil. Not effective against TMV.
	Added heat	NA	180 degrees F for 30 minutes	—	—	Remove all debris and heat-sensitive materials. Keep house and contents warm.
Soil						
Soilborne diseases and weeds		—	See soil fumigants table and check soil fumigant label if registered for greenhouse use.			Preplant soil treatment.
Basil						
Alternaria leaf spot (*Alternaria* spp.) Botrytis leaf blight (*Botrytis* spp.), Fusarium blight (*Fusarium* spp.)	cyprodinil + fludioxonil (Switch 62.5WG)	9 + 12	11 to 14 oz/acre	7	0.5	After 2 applications, alternate with another fungicide with a different mode of action for 2 applications. See label for application limits.
Downy mildew (*Peronospora belbahrii*)	cyazofamid (Ranman, Ranman SC400)	21	2.75 to 3.0 fl oz/acre	0	0.5	Do not exceed 27 fl oz per acre per growing season. See label for surfactant recommendations. Alternate applications with fungicides that have a different mode of action. Do not make more than 3 consecutive applications before switching to products that have a different mode of action for 3 applications before returning to Ranman/Ranman 400SC.
	mandipropamid (Micora, Revus)	40	8.0 fl oz/acre	1	4 hr	**Micora: For basil grown for transplants and retail sale to consumers only.** Do not make more than 2 applications per crop. See label for additional restrictions and recommendations. **Revus:** Do not exceed 32 fl oz of product per acre per season. See label for additional limits when producing multiple croppings per year.
Downy mildew, *Pythium* and *Rhizoctonia* root rots	phosphorous acid (various)	33	See label	See label	See label	See label for restrictions.
	potassium phosphite (various)	33	See label	See label	See label	See label for restrictions.
Cucurbits						
Angular leaf spot, downy mildew	copper, fixed (various)	M1	See label	See label	See label	**Some products are OMRI-listed.** See product label for complete application instructions, specific crop and disease labels, and greenhouse usage.
Downy mildew (*Pseudoperonospora cubensis*)	cymoxanil (Curzate 60DF)	27	3.2 to 5.0 oz/acre	3	12 hr	Always apply in a tank mix with the labeled rate of a protectant fungicide. Do not exceed 30 oz of product per 12-month period.
Alternaria leaf blight and spot (*A. cucumerina* and *A. alternata*), gummy stem blight (*Stagnosporopsis*[2]), powdery mildew (*S. fuliginea, E. cichoracearum*)	cyprodinil + fludioxonil (Switch 62.5WG)	9 + 12	11 to 14 oz/acre	1	0.5	After 2 applications, alternate with another fungicide with a different mode of action for 2 applications. See label for application limits.
Alternaria leaf blight and spot (*A. cucumerina* and *A. alternata*), anthracnose (*C. orbiculare*), Cercospora leaf spot (*C. citrullina*), gummy stem blight (*Stagnosporopsis*[2]), powdery mildew (*S. fuliginea, E. cichoracearum*), Septoria leaf blight (*S. cucurbitacearum*)	difenoconazole + cyprodinil (Inspire Super)	3 + 9	16 to 20 fl oz/acre	7	0.5	Do not apply more than 80 fl oz of product per acre per season. Make no more than 2 consecutive applications per season before alternating with fungicides that have a different mode of action.

Table 10-57. Greenhouse Disease Control for Various Vegetable Crops

Disease/Location	Product[1]	FRAC Group	Rate of Formulation	Minimum Days Harvest	Minimum Days Reentry	Schedule and Remarks
Cucurbits (continued)						
Alternaria leaf blight (*A. cucumerina*), anthracnose (*Colletotrichum* spp.), downy mildew (*P. cubensis*) **Suppression:** Phytophthora blight (*Phytophthora capsici*)	famoxadone + cymoxanil (Tanos)	11 + 27	8 oz/acre 8 to 10 oz/acre (for diseases listed under suppression)	3	0.5	Do not exceed 32 oz per acre of product per crop cycle or 72 oz per acre per 12-month period. Do not make more than 1 application of product before alternating with a fungicide that has a different mode of action. See label for tank mixing instructions. For suppression of foliar and fruit phases ONLY of Phytophthora blight.
Botrytis gray mold	fenhexamid (Decree 50 WDG)	17	1.5 lb/acre (stand-alone) 1.0 to 1.5 lb/acre (tank-mix)	0	0.5	**Labeled for cucumbers ONLY.** For use in transplant production and greenhouse production. Do not make more than 2 consecutive applications. See labels for additional tank-mixing instructions. Do not exceed 6.0 lb product per acre (transplants) or per acre per season (greenhouse production).
Alternaria leaf spot, anthracnose, Cercospora leaf spot, downy mildew, gummy stem blight, scab	mancozeb (various)	M3	See label	See label	See label	See product labels for complete application instructions, specific crop and disease labels, and greenhouse usage.
Alternaria leaf spot, Cercospora leaf spot, downy mildew, Phytopthora rot (*Phytopthora capsici*)	mancozeb + zoxamide (Gavel 75SF)	M3 + 22	1.4 to 2.0 lb/acre	5	2	Do not exceed 16 lb per acre per season.
Suppression: Downy mildew (*P. cubensis*), Phytophthora blight (*P. capsici*)	mandipropamid (Revus)	40	8.0 fl oz/acre (Revus)	0	4 hr	Do not exceed 32 fl oz of product per acre per season. See label for application instructions specific to each disease.
Alternaria leaf spot and blight, Botrytis gray mold, gummy stem blight (*Didymella*), powdery mildew, Sclerotinia stem rot	penthiopyrad (Fontelis)	7	0.375 to 0.5 fl oz/ gal to treat 1,360 sq ft	1	0.5	Do not exceed 67 fl oz of product per year. Make no more than 2 consecutive applications per season before alternating with fungicides that have a different mode of action. See label for cucurbit restrictions.
Botrytis gray mold, Corynespora leaf spot (*Corynespora cassicola*), early blight (*Alternaria* sp.), gummy stem blight (*Stagnosporopsis*[2]), powdery mildew (*Spthaerotheca* sp.), scab (*Cladosporium* sp.)	polyoxin D zinc salt (Affirm WDG, OSO 5%SC)	19	6.2 oz/acre(Affirm) 3.75 to 13.0 fl oz/acre (OSO)	0	4 hr	Check product labels for maximum limits of product per season. Alternate with fungicides that have a different mode of action. Check product labels for other tomato diseases also on the label.
Alternaria leaf spot, anthracnose, Cercospora leaf spot, downy mildew, gummy stem blight (GSB), scab, target spot	potassium phosphite + chlorothalonil (Catamaran)	33 + M5	4 pt/acre (anthracnose, downy mildew, target spot) 6 pt/acre (Alternaria and Cercospora leaf spot, GSB, scab)	1	0.5	Do not exceed 50 pt per acre per season. Phytotoxicity potential. Do not combine with other pesticides, surfactants or fertilizers. Do not apply on the day of harvest.
Damping off and root rots (*Phytophthora* spp, *Pythium* spp.)	promocarb hydrochloride (Previcur Flex)	28	See label	2	0.5	Product applied through a drip system or as a soil drench. Do not apply more than 4 applications of product after transplanting per crop cycle. Do not mix with other products. Phytotoxicity may occur if applied to dry growing media.
Powdery mildew	cyflufenamid (Torino)	U6	3.4 oz/acre	0	4	Do not make more than two applications per year. **Use with caution as resistance has been reported in Italy.**
	sulfur (various)	M2	See label	See label	See label	**Some products are OMRI-listed.** See product labels for complete application instructions, specific crop and disease labels, and greenhouse usage. Not all products are registered for use in all states.
	triflumizole (Procure 480SC, Terraguard SC)	3	4 to 8 fl oz/acre (Procure) 2 to 4 fl oz/100 gal (Terraguard)	1	0.5	**Procure:** Do not exceed 40 fl oz of product per acre per season. Label specifies the following powdery mildew species: *Erysiphe cichoracearum, Podosphaera xanthii*. **Not registered for use in all states.** **Terraguard SC:** For use only as a foliar spray. For use in commercial greenhouse production only. Can be used on greenhouse transplants. Do not exceed 40 fl oz of product per acre per cropping system. See label for additional application instructions. **Not registered for use in all states.**
	potassium bicarbonate (Milstop)	NC	1.25 to 5.0 lb/ 100 gal	0	1 hr	**OMRI-listed.** Do not exceed 0.5 lb of product per 4,350 sq ft or 1.15 lb product per 10,000 sq ft per application. Do not store unused spray solution. See label for additional diseases labeled.
Lettuce						
Downy mildew (*Bremia lactucae*)	cymoxanil (Curzate 60DF)	27	3.2 to 5.0 oz/acre (head lettuce) 5.0 oz/acre (leafy lettuce)	3 (head) 1 (leafy)	12 hr	For use on head and leaf lettuce. Use with the labeled rate of a protectant fungicide. Do not exceed 30 oz of product per 12-month period.

Table 10-57. Greenhouse Disease Control for Various Vegetable Crops

Disease/Location	Product[1]	FRAC Group	Rate of Formulation	Minimum Days Harvest	Minimum Days Reentry	Schedule and Remarks
Lettuce (continued)						
Alternaria leaf spot, *Botrytis* gray mold, Sclerotinia rot, basal rot (*Phoma*), Septoria leaf spot **Suppression:** Powdery mildew	cyprodinil + fludioxonil (Switch 62.5WG)	9 + 12	11 to 14 oz/acre	0	0.5	For use on head and leaf lettuce. After 2 applications, alternate with another fungicide with a different mode of action for 2 applications. See label for application limits.
Botrytis gray mold rot, drop rot, *Sclerotinia minor*, watery soft rot (*Sclerotinia sclerotiorum*)	dicloran (Botran 5F)	14	0.6 qt/acre (at planting) 0.6 to 1.8 qt/acre (prethinning) 1.8 to 3.2 qt/acre (post-thinning [direct seeded] and established transplants)	14	0.5	For use on head and leaf lettuce. See label for detailed instructions. Do not exceed 3.2 qt of product per acre per year or within 14 days of harvest. **Not registered for use in SC, TN, or VA.**
Downy mildew (*B. lactucae*), white rust (*Albugo occidentalis*)	famoxadone + cymoxanil (Tanos)	11 + 27	8 to 10 oz/acre	1	0.5	For use on head and leaf lettuce. Do not exceed 48 oz per acre of product per crop season. Do not make more than 1 application of product before alternating with a fungicide that has a different mode of action. See label for tank mixing instructions.
Botrytis gray mold	fenhexamid (Decree 50 WDG)	17	1.5 lb/acre (stand-alone) 1.0 to 1.5 lb/acre (tank-mix)	3	0.5	For use in transplant production and greenhouse production. Do not make more than 2 consecutive applications. See label for additional tank-mixing instructions. Do not exceed 3.0 lb product per acre (transplants) or per acre per crop (greenhouse production).
Anthracnose, downy mildew	mancozeb (various)	M3	See label	See label	See label	See product labels for complete application instructions, specific crop and disease labels, and greenhouse usage.
	mandipropamid (Micora, Revus)	40	5.5 to 8.0 fl oz/acre (Micora) 8.0 fl oz/acre (Revus)	— 1	4 hr	For use on head and leaf lettuce. **Micora:** For lettuce grown for transplants and retail sale to consumers only. Do not make more than 2 applications per crop. After making an application of Micora, alternate with a fungicide with a different mode of action. Apply in a tank-mix with another downy mildew fungicide with a different mode of action. **Revus:** Do not exceed 32 fl oz of product per acre per season. Do not make more than 2 applications before switching to a fungicide with a different mode of action.
Alternaria leaf spot (*Alternaria sonchi*), Botrytis gray mold, Cercospora leaf spot (*Cercospora* spp.), powdery mildew (*Golovinomyces cichoracearum*), Septoria leaf spot (*Septoria* spp.)	penthiopyrad (Fontelis)	7	14 to 24 fl oz/acre	3	0.5	For use on head and leaf lettuce. Do not exceed 72 fl oz of product per year. Make no more than 2 consecutive applications per season before alternating with fungicides that have a different mode of action. See label for additional diseases that may be controlled.
Alternaira leaf spot, anthracnose, *Botrytis*, Cercospora leaf spot, powdery mildew	potassium bicarbonate (Milstop)	NC	1.25 to 5.0 lbs/100 gal	0	1 hr	**OMRI-listed.** Do not exceed 0.5 lb of product per 4,350 sq ft or 1.15 lbs product per 10,000 sq ft per application. Do not store unused spray solution. See label for additional diseases labeled.
Damping off and root rots (*Phytophthora* spp, *Pythium* spp.)	promocarb hydrochloride (Previcur Flex)	28	See label	2	0.5	**For use on leaf lettuce only.** Product applied as a foliar treatment. Do not apply more than two applications of product per crop cycle. Do not mix with other products. Phytotoxicity may occur if applied to dry growing media.
Powdery mildew	sulfur (various)	M2	See label	See label	See label	Some products are OMRI-listed. See product labels for complete application instructions, specific crop and disease labels, and greenhouse usage. **Not all products are registered for use in all states.**
Alternaria leaf spot/black spot (*Alternaria* spp.), powdery mildew (*Erysiphe* spp.)	triflumizole (Procure 480SC)	3	6 to 8 fl oz/acre	0	0.5	For use on head and leaf lettuce. Do not exceed 18 fl oz of product per acre per season. **Product not registered for use in all states.**
Tomato (transplant production)						
Damping off (*Pythium* spp.)	cyazofamid (Ranman, Ranman 400SC)	21	3 fl oz/100 gal	—	0.5	**For transplant production only.** Apply as a soil drench. Do not use a surfactant. One fungicide application can be made to the seedling tray at planting or any time afterwards until one week before transplanting.
Late blight (*Phytophthora infestans*)	mandipropamid (Micora)	40	5.5 to 8.0 fl oz/acre	—	4 hr	**For tomatoes grown for transplants and retail sale to consumers only.** Do not make more than 2 applications per crop. Do not make more than 2 consecutive applications before switching to a fungicide from a different FRAC group.
Damping off and root rots (*Phytophthora* spp, *Pythium* spp.)	promocarb hydrochloride (Previcur Flex)	28	See label	—	0.5	**For preseeding and/or seedling treatment (before transplanting).** Do not mix with other products. See label for specific use directions. Phytotoxicity may occur if applied to dry growing media.
Bacterial canker, speck, and/or spot	*Streptomycin* sulfate (Agri-Mycin 17, Ag Streptomycin, Firewall 17 WP, Firewall 50 WP, Harbour)	25	See label	—	0.5	**For transplant production only.** Begin applications at the first true leaf stage. Repeat at 4 to 5 day intervals until transplanting in the field. Check product labels for specific diseases labeled and application rates. Firewall 50 WP has a maximum of 6 applications per year.

Table 10-57. Greenhouse Disease Control for Various Vegetable Crops

Disease/Location	Product[1]	FRAC Group	Rate of Formulation	Minimum Days Harvest	Minimum Days Reentry	Schedule and Remarks
Tomato (transplant production) (continued)						
Crown and basal rot (*Fusarium* spp., *Rhizoctonia solani*, *Sclerotinia* spp.), damping off (*Pythium* spp., *Rhizoctonia* spp.), downy mildew (*Peronospora* spp., *Plasmopara* spp.), spots and blights (*Alternaria* spp., *Cercospora* spp., *Phoma* spp., *Septoria* spp.), Phytophthora blight, powdery mildew (*Leveilula* spp. and *Oidiopsis* spp.), rots and blights (*Botrytis* spp.)	pyraclostrobin + boscalid (Pageant Intrinsic)	7 + 11	See label	—	0.5	**For transplant production only.** Begin applications at the first true leaf stage. Repeat at 4 to 5 day intervals until transplanting in the field. Check product labels for specific diseases labeled and application rates. Firewall 50 WP has a maximum of 6 applications per year.
Tomato (after transplanting in greenhouse)						
Anthracnose (*Colletotrichum* spp.), black mold (*Alternaria alternata*), early blight (*Alternaria solani*), gray leaf spot (*Stemphylium botryosum*), powdery mildew (*Leveilula taurica*), Septoria leaf spot (*Septoria lycopersici*), target spot (*Corynespora cassiicola*)	azoxystrobin + difenoconazole (Quadris Top)	11 + 9	8 fl oz/acre	0	0.5	*Not prohibited for greenhouse use.* **Do not use for transplant production.** Do not make more than 2 consecutive applications before switching to a fungicide with a different mode of action. Do not exceed 47 fl oz of product per acre per season. Do not apply until 21 days after transplanting or 35 days after seeding. Do not use with adjuvants or tank mix with any EC product on fresh market tomatoes. Plant injury may occur with the use of adjuvants. See label for specifics.
Botrytis gray mold, powdery mildew	Banda de Lupinus albus doce (BLAD) (Fracture)	BM01	24.4 to 36.6 fl oz/acre	1	4 hr	*Not prohibited for greenhouse use.* Do not make more than 2 sequential applications before alternating to a fungicide with a different mode of action.
Anthracnose, bacterial speck and spot, early blight, gray leaf mold, late blight, Septoria leaf spot	copper, fixed (various)	M1	See label	See label	See label	**Some products are OMRI-listed.** See product labels for complete application instructions, specific crop and disease labels, and greenhouse usage.
Late blight (*P. infestans*), Phytophthora blight (*P. capsici*)	cyazofamid (Ranman 400SC)	21	2.1 to 2.75 fl oz/acre (late blight) 2.75 fl oz/acre (Phytophthora blight)	0	0.5	Do not exceed 16.5 fl oz per acre per year. See label for surfactant recommendations. Alternate applications with fungicides that have a different mode of action. Do not make more than 3 consecutive applications before switching to products that have a different mode of action for 3 applications before returning to Ranman 400SC. See label for application instructions specific to target disease.
Late blight (*Phytophthora infestans*)	cymoxanil (Curzate 60DF)	27	3.2 to 5.0 oz/acre	3	12 hr	*Not prohibited for greenhouse use.* Use with the labeled rate of a protectant fungicide. Do not exceed 30 oz of product per 12-month period.
Early blight (*A. solani*), Botrytis gray mold, powdery mildew (*L. taurica*)	cyprodinil + fludioxonil (Switch 62.5WG)	9 + 12	11 to 14 oz/acre	0	0.5	*Not prohibited for greenhouse use.* After 2 applications, alternate with another fungicide with a different mode of action for 2 applications. See label for application limits. Do not apply to small tomatoes such as cherry or grape-type tomatoes in the greenhouse.
Anthracnose (*Colletotrichum* spp.), black mold (*A. alternata*), Botrytis gray mold, early blight (*A. solani*), gray leaf spot (*S. botryosum*), leaf mold (*Fulvia fulva*), powdery mildew (*L. taurica*), Septoria leaf spot, target spot (*C. cassiicola*)	difenoconazole + cyprodinil (Inspire Super)	3 + 9	16 to 20 fl oz/acre	0	0.5	*Not prohibited for greenhouse use.* Do not apply more than 47 fl oz of product per acre per season. Make no more than 2 consecutive applications per season before alternating with fungicides that have a different mode of action.
Phytophthora and Pythium root rots	etridiazole (Terramaster 4EC)	14	6 to 7 fl oz/acre	3	0.5	For application by drip irrigation. Apply as a 0.01% solution (6.5 fl oz/500 gal water) no sooner than 3 weeks after transplanting or a previous application. Do not exceed 27.4 fl oz of product per acre per crop season. Additional indoor restrictions regarding REI are provided on the label. **Product has a Section 24c registration for this use on greenhouse tomatoes in FL, KY, MS, OK, TN, TX, and VA. Product is not registered for this use in AL, GA, LA, NC, or SC.**
Anthracnose (*Colletotrichum* spp.), early blight (*A. solani*) late blight (*P. infestans*), leaf mold (*Cladosporium fulvum*), Septoria leaf spot (*S. lycopersici*), target spot (*C. cassiicola*) *Suppression:* Bacterial canker, bacterial speck, bacterial spot, buckeye rot (*Phytophthora* spp.)	famoxadone + cymoxanil (Tanos)	11 + 27	6 to 8 oz/acre (early blight) 8 oz/acre (other labeled diseases)	3	0.5	Do not exceed more than 72 oz/acre per crop cycle or 12-month period. Do not make more than 1 application of product before alternating with a fungicide that has a different mode of action. See label for tank mixing instructions.
Botrytis gray mold	fenhexamid (Decree 50 WDG)	17	1.5 lb/acre (stand-alone) 1.0 to 1.5 lb/acre (tank-mix)	0	0.5	Do not make more than 2 consecutive applications. See label for additional tank-mixing instructions. Do not exceed 6.0 lb product per acre per season for greenhouse production.

Table 10-57. Greenhouse Disease Control for Various Vegetable Crops

Disease/Location	Product[1]	FRAC Group	Rate of Formulation	Minimum Days Harvest	Minimum Days Reentry	Schedule and Remarks
Tomato (after transplanting in greenhouse) (continued)						
Damping off (*Pythium* spp.), root rots (*Phytophthora* spp.)	fosetyl-Al (various)	33	2.5 to 5.0 lb/acre	14	0.5	***Not prohibited for greenhouse use. For foliar application.*** Do not exceed 20 lb of product per acre per season. Phytotoxicity may occur if tank-mixed with copper products, if applied to plants with copper residues, or if mixed with adjuvants. Do not tank-mix with copper products. See label for additional restrictions and application instructions. **Products are not labeled for use on tomato in certain counties in AL, KY, LA, NC, and TN.**
Anthracnose, bacterial speck and spot, Botrytis, early blight, late blight, powdery mildew, and Rhizoctonia fruit rot	hydrogen dioxide (OxiDate)	NG	1/3 to 1 gal/100 gal water (foliar spray)	0	See label	See label for additional information regarding rate usage, including rates specific to non-foliar applications and other labeled diseases. **Toxic to bees and other beneficial insects exposed to direct contact on blooming crops.**
Anthracnose, bacterial speck and leaf spot, Botrytis gray mold, Cladosporium mold, early blight (*Alternaria*), *Fusarium*, late blight, *Pythium*, *Rhizoctonia*, powdery mildew	hydrogen peroxide + peroxyacetic acid (OxiDate 2.0)	NG	32 to 128 fl oz/100 gal water (foliar spray)	0	See label	See label for additional information regarding rate usage, including rates specific to non-foliar applications and other labeled diseases. Do not apply as a foliar spray sooner than at least 24 hrs following application of a metal-based product. Under some conditions, phytotoxicity may result when tank-mixed with metal-based chemicals. **Toxic to bees and other beneficial insects exposed to direct contact on blooming crops.**
Anthracnose, bacterial speck and spot, early blight, gray leaf spot, late blight, leaf mold, Septoria leaf spot	mancozeb (various)	M3	See label	See label	See label	See product labels for complete application instructions, specific crop and disease labels, and greenhouse usage.
Anthracnose, bacterial speck and spot, early blight, gray leaf spot, late blight, leaf mold, Septoria leaf spot	mancozeb + copper (ManKocide)	M3 + M1	1.7 lb/acre (processing) 1 to 3 lb/acre (fresh market)	5	2	***Not prohibited for greenhouse use.*** Do not exceed 58 lb product per acre per crop east of the Mississippi River or 42.66 lb product per acre per crop west of the Mississippi River. Phytotoxicity may occur when spray solution has a pH of less than 6.5 or when certain environmental conditions occur.
Anthracnose, bacterial speck and spot, buckeye rot, early blight, gray leaf spot, late blight, leaf mold, Septoria leaf spot	mancozeb + zoxamide (Gavel 75DF)	M3 + 22	1.5 to 2.0 lb/acre	5	2	***Not prohibited for greenhouse use.*** Do not exceed 8 lb per acre per season (west of the Mississippi River) or 16 lb per acre per season (east of the Mississippi River). For bacterial speck and spot, apply the full rate of product in a tank mix with a full rate of a fixed copper. See label for other application limits. **Product has a 2ee registration for anthracnose management in AL, AR, FL, GA, KY, LA, MS, NC, OK, SC, TN, and VA.**
Anthracnose (*Colletotrichum* spp.), black mold (*A. alternata*), early blight, (*A. solani*), gray leaf spot (*S. botryosum*), late blight (*P. infestans*), leaf mold (*F. fulva*), powdery mildew (*L. taurica*), Septoria leaf spot, target spot (*C. cassiicola*)	mandipropamid + difenoconazole (Revus Top)	3 + 40	5.5 to 7.0 fl oz/acre	1	0.5	***Not prohibited for greenhouse use.*** Do not make more than 2 consecutive applications per season before alternating with fungicides that have a different mode of action. Do not exceed 28 fl oz of product per acre per season.
Late blight (*P. infestans*)	mefenoxam + mancozeb (Ridomil Gold MZ WG)	4 + M3	2.5 lb/acre	5	2	***Not prohibited for greenhouse use.*** Do not exceed 7.5 lb product per acre per crop per season. Do not exceed 3 applications per season. Apply a protectant fungicide in between applications of product. See label for other restrictions.
Powdery mildew (*Leveilula* spp.), southern blight	myclobutanil (Rally 40WSP)	3	2.5 to 4 oz/acre	0	2	***Not prohibited for greenhouse use.*** Do not exceed 1.25 lb product per acre per crop. Do not exceed 21 days between applications.
Alternaria blights and leaf spots, black mold (*A. alternata*), early blight, Botrytis gray mold, powdery mildew (*L. taurica*), basal stem rot (*Sclerotium rolfsii*), Septoria leaf spot, target spot (*C. cassiicola*) *Suppression*: Anthracnose	penthiopyrad (Fontelis)	7	0.5 to 0.75 fl oz/gal per 1,360 sq ft	0	0.5	Do not exceed 72 fl oz of product per year. Do not make more than 2 consecutive applications per season before alternating with fungicides that have a different mode of action. See label for specific instructions for basal stem rot.
Root rot (*Phytophthora* spp.)	phosphorus acid (mono- and di-potassium salts) (various)	33	See label	See label	See label	See product labels for complete application instructions, specific crop and disease labels, and greenhouse usage.
Botrytis rot, early blight, powdery mildew (*L. taurica*, *Oidiopsis sipula*) *Suppression*: Anthracnose	polyoxin D zinc salt (Affirm WDG, OSO 5%SC)	19	6.2 oz/acre (Affirm) 3.75 to 13.0 fl oz/acre (OSO)	0	4 hr	Check products labels for maximum limits of product per season. Alternate with fungicides that have a different mode of action. Check product labels for other tomato diseases also on the label. **OSO 5%SC is not registered for use in AR, KY, MS, or OK.**

Table 10-57. Greenhouse Disease Control for Various Vegetable Crops

Disease/Location	Product[1]	FRAC Group	Rate of Formulation	Minimum Days Harvest	Minimum Days Reentry	Schedule and Remarks
Tomato (after transplanting in greenhouse) (continued)						
Alternaira leaf spot, anthracnose, *Botrytis*, Cercospora leaf spot, powdery mildew, Septoria leaf spot	potassium bicarbonate (Milstop, Carb-O-Nator)	NC	1.25 to 5.0 lbs/100 gal water (MilStop) 2.5 to 5.0 lb/100 gal water (Carb-O-Nator)	0	1 hr (MilStop) 4 hr (Carb-O-Nator)	**OMRI-listed.** Do not exceed 0.5 lbs of MilStop per 4,350 sq ft or 1.15 lbs MilStop per 10,000 sq ft per application. Do not exceed a mix rate of 5 lb of Carb-O-Nator per 100 gal water. Do not store unused spray solution. See label for additional diseases labeled.
On foliage: Early blight, gray leaf mold, gray leaf spot, late blight, Septoria leaf spot, target spot *On fruit:* Anthracnose, Alternaria fruit rot, *Botrytis*, late blight rot, Rhizoctonia rot	potassium phosphite + chlorothalonil (Catamaran)	33 + M5	4.5 to 5.5 pt/acre (foliage diseases) 7 pt/acre (fruit diseases)	0	0.5	*Not prohibited for greenhouse use.* Do not exceed 50 pt per acre per season. Phytotoxicity potential. Do not combine with other pesticides, surfactants or fertilizers. May be applied on the day of harvest.
Damping off and root rots (*Phytophthora* spp., *Pythium* spp.)	promocarb hydrochloride (Previcur Flex)	28	See label	5	0.5	Product applied through a drip system or as a soil drench. Do not apply more than 4 applications of product after transplanting per crop cycle. Do not mix with other products. Phytotoxicity may occur if applied to dry growing media.
Botrytis gray mold	pyraclostrobin + boscalid (Pageant Intrinsic)	7 + 11	23 oz/acre	0	0.5	Do not tank mix with adjuvants or other agricultural products. Do not exceed 69 oz per acre of product per crop cycle. Do not make more than 1 application of product before switching to a fungicide with a different mode of action.
Botrytis gray mold, early blight	pyrimethanil (Scala SC)	9	7 fl oz/acre	1	0.5	Plant injury may occur in non-ventilated houses; ventilation for at least 2 hours after application of product. Use only in a tank mix with another fungicide for early blight. Do not exceed 35 fl oz per acre per crop.
Powdery mildew	sulfur (various)	M2	See label	See label	See label	**Some products are OMRI-listed.** See product labels for complete application instructions, specific crop and disease labels, and greenhouse usage. Not all products are registered for use in all states.
	triflumizole (Terraguard SC)	3	2 to 4 fl oz/100 gal	1	0.5	For use only as a foliar spray. For use in commercial greenhouse production only. Can be used on greenhouse transplants. Do not exceed 40 fl oz of product per acre per cropping system. See label for additional application instructions. **Products not registered for use in all states.**
Anthracnose, early blight, Septoria leaf spot	zinc dimethyl-dithio-carbamate (Ziram 76DF)	M3	3 to 4 lb/acre	7	2	*Not prohibited for greenhouse use.* Do not use on cherry tomatoes. Do not exceed 23.7 lb per acre per crop cycle. May be mixed with copper fungicides.

[1] Products registered for field use may be used on greenhouse crops (but not transplants) unless excluded on the label. Always check the label before applying a product.

[2] Former names of pathogens listed in this table that may be still be listed on fungicide labels are as follows; *Golovinomyces* spp. (formerly *Erysiphe* spp.) or *Golovinomyces cichoracearum* (formerly *Erysiphe cichoracearum*); *Passalora fulva* (formerly *Cladosporium fulvum* and *Fulvia fulva*); *Stagnosporopsis* (formerly *Didymella*).

Seed Treatments

Seed sanitation to eradicate bacterial or viral plant pathogens: When treating vegetable seeds, it is critical to follow the directions exactly, because germination can be reduced by the treatment and/or the pathogen may not be completely eliminated. The effect of a treatment on germination should be determined on a small lot of seeds prior to treating large amounts of seed. Treatments should not be applied to: 1) pelleted seed, 2) previously treated seed, or 3) old or poor quality seed. A protective fungicide treatment (see below) can be applied to the seed following treatment for bacterial pathogens.

Seed treatments to prevent damping off diseases: Most commercially available vegetable seeds come treated with at least one fungicide and/or insecticide. Vegetable producers who would like to apply their own seed treatment should purchase non-treated seed. While many fungicides are labeled for use on vegetable seed, most fungicides are restricted to commercial treatment only and should not be applied by producers. Labeled fungicides can be applied to seed following treatment for bacterial pathogens (see above). **Do not use fungicide treated seed for food or feed.**

HOT WATER TREATMENT

By soaking seed in hot water, seedborne fungi and bacteria can be reduced, if not eradicated, from the seed coat. Hot water soaking will not kill pathogens associated with the embryo nor will it remove seedborne plant viruses from the seed surface.

1. Place seed loosely in a weighted cheesecloth or nylon bag.
2. Warm the seed by soaking it for 10 minutes in 100 degree Fahrenheit (37 Celsius) water.
3. Transfer the warmed seed into a water bath already heated to the temperature recommended for the vegetable seed being treated (see table below). The seeds should be completely emerged in the water for the recommended amount of time (see table below). Agitation of the water during the treatment process will help to maintain a uniform temperature in the water bath.
4. Transfer the hot water treated seed into a cold-water bath for five minutes to stop the heating action.
5. Remove seed from the cheesecloth or nylon bag and spread them evenly on clean paper towel or a sanitized drying screen to dry. Do not dry seed in areas where fungicides, pesticides or other chemicals are located.
6. Seed can be treated with a labeled fungicide to protect against damping off pathogens.
7.

8. CHLORINE BLEACH TREATMENT

9. Treating seeds with a solution of chlorine bleach can effectively remove bacterial pathogens and some viruses (i.e. Tobacco Mosaic Virus) that are borne on the surface of seeds.
10. Add 1 quart (946 ml) of Clorox® bleach to 5 quarts (4.7 L) of potable water.
11. Add a drop or two of liquid dish detergent or a commercial surfactant such as Activator 90 or Silwet to the disinfectant solution. Add seed to the disinfectant solution (1 pound of seed per 4 quarts of disinfectant solution) and agitate for 1 minute.
12. Prepare fresh disinfectant solution for each batch of seeds to be treated.
13. Rinse the seed in a cold water bath for 5 minutes to remove residual disinfectant.
14. Spread seeds evenly on clean paper towel or a sanitized drying screen to dry. Do not dry seed in area where fungicides, pesticides, or other chemicals are located.
15. Seed can be treated with a labeled fungicide to protect against damping off pathogens.

HYDROCHLORIC ACID TREATMENT

Tomato seed can be treated with a dilute solution of hydrochloric acid (HCl) solution to eliminate seedborne bacterial pathogens such as *Xanthomonas* spp. (Bacterial leaf spot), *Pseudomonas syringae* pv. *tomato* (Bacterial speck) and *Clavibacter michiganensis* subsp. *michiganensis* (Bacterial canker). Hydrochloric acid can also be used to remove TMV from the surface of tomato seed. **Do not use HCL-treated seed for food or animal feed.**

1. Prepare a 5% solution of HCl by adding one part acid to 19 parts potable water. Prepare the acid solution in a well-ventilated area and avoid direct skin contact with the acid.
2. Soak seeds for 6 hours with gentle agitation.
3. Carefully drain the acid off of the seed and rinse seed under running potable water for 30 minutes. Alternatively, rinse the seeds 10 to 12 times with potable water to remove residual acid.
4. Spread seeds evenly on clean paper towel or a sanitized drying screen to dry. Do not dry seed in area where fungicides, pesticides, or other chemicals are located.
5. Seed can be treated with a labeled fungicide to protect against damping off pathogens.

TRISODIUM PHOSPHATE TREATMENT

Tomato seed can be treated with trisodium phosphate (TSP) to eradicate seed-transmitted TMV. **Do not use TSP-treated seed for food or animal feed.**

1. Prepare a 10% solution of TSP (1 part TSP in 9 parts potable water). Trisodium phosphate is available at most home supply or paint stores. Avoid direct skin contact with the TSP solution.
2. Soak seed for 15 minutes in the disinfectant solution.
3. Rinse the seed in a cold water bath for 5 minutes to remove residual disinfectant.
4. Spread seeds evenly on clean paper towel or a sanitized drying screen to dry. Do not dry seed in area where fungicides, pesticides, or other chemicals are located.
5. Seed can be treated with a labeled fungicide to protect against damping off pathogens.

TESTING SEED GERMINATION AFTER SEED TREATMENTS
1. Randomly select 100 seeds from each seed lot.
2. Treat 50 seeds using one of the sanitizers described above.
3. After the treated seed has dried and before application of a protectant fungicide, plant the treated and non-treated seed separately in flats containing planting mix according to standard practice. Label each group as treated or non-treated.
4. Allow the seeds to germinate and grow until the first true leaf appears (to allow for differences in germination rates to be observed).
5. Count seedlings in each group separately.
6. Determine the percent germination for each group: # seedlings emerged ÷ # seeds planted x 100.
7. Compare percent germination between the treated and non-treated groups. Percent germination should be within 5% of each other.

Table 10-58. Recommended Temperatures and Treatment Times for Hot Water Disinfestation of Vegetable Seed

Vegetable Crop	Water Temperature (°F/°C)	Soaking Time (Minutes)
Broccoli	122/50	20 to 25
Brussels sprout	122/50	25
Cabbage	122/50	25
Carrot	122/50	15 to 20
Cauliflower	122/50	20
Celery	122/50	25
Chinese cabbage	122/50	20
Collard	122/50	20
Cucumber[1]	122/50	20
Eggplant	122/50	25
Garlic	120/49	20
Kale, Kohlrabi	122/50	20
Lettuce	118/48	30
Mint	112/44	10
Mustard, Cress, Radish	122/50	15
Onion	115/46	60
Pepper	125/51	30
Rape, Rutabaga	122/50	20
Shallot	115/46	60
Spinach	122/50	25
Tomato	122/50	25
Turnip	122/50	20

1 Cucurbits other than cucumbers can be severely damaged by hot water treatment and should be disinfested using chlorine bleach.

Table 10-59. Products for Seed Treatment

	42-S Thiram (thiram)	Allegiance (metalaxyl)	Acquire (metalaxyl)	Apron (mefenoxam)	Belmont 2.7FS (metalaxyl)	Botran 75W (dicloran)	Captan 400 (captan)	Cruiser Maxx (thiamethoxam + mefenoxam + fludioxonil)	Cruiser Maxx Potato (thiamethoxam + fludioxonil)	Dyna-Shield (fludioxonil)	Dividend Extreme (difenoconazole + mefenoxam)	Dynasty (azoxystrobin)	Emesto Silver (penflufen + prothioconazole)	Maxim 4FS (fludioxonil)	Maxim MZ (fludioxonil + mancozeb)	MetaStar 2E AG (metalaxyl)	MonCoat MZ (flutolanil + mancozeb)	Sebring 318S (metalaxyl)	Sebring 480S (metalaxyl)	Spirato 480 FS (fludioxonil)	Vitaflo-280 (carboxin + thiram)	Tops MZ-Gaucho (thiophanate-methyl + mancozeb)	Trilex (trifloxystrobin + metalaxyl)
Fungicide Group[F]	M3	4	4	4	4	14	M3	4 + 12	12	12	3 + 4	11	3	12	12 + M3	4	7 + M3	4	4	M3	7 + M3	1 + M3	11 + 4
Beans, Snap	X	X	X	X	X		X	X		X		X		X		X		X	X	X	X		X
Beans, Lima	X	X	X	X		X	X		X		X		X		X		X	X	X	X		X
Beets	X	X	X	X	X		X			X		X		X		X		X	X	X			
Broccoli	X		X	X	X		X			X		X		X				X	X	X			
Carrots	X	X	X	X	X					X		X		X		X		X		X			
Celery				X								X		X						X			
Chinese Cabbage	X		X	X	X					X		X						X		X			
Cole Crops	X		X	X	X		X			X		X		X						X			
Cucumbers	X	X	X	X	X		X			X		X		X	X			X	X	X			X
Eggplants	X		X	X	X					X		X		X				X		X			X
Garlic				X						X		X		X						X			
Greens, Mustard	X		X	X	X		X			X		X		X				X		X			
Greens, Turnip	X		X	X	X		X			X		X								X			

Table 10-59. Products for Seed Treatment

	42-S Thiram (thiram)	Allegiance (metalaxyl)	Acquire (metalaxyl)	Apron (mefenoxam)	Belmont 2.7FS (metalaxyl)	Botran 75W (dicloran)	Captan 400 (captan)	Cruiser Maxx (thiamethoxam + mefenoxam + fludioxonil)	Cruiser Maxx Potato (thiamethoxam + fludioxonil)	Dyna-Shield (fludioxonil)	Dividend Extreme (difenoconazole + mefenoxam)	Dynasty (azoxystrobin)	Emesto Silver (penflufen + prothioconazole)	Maxim 4FS (fludioxonil)	Maxim MZ (fludioxonil + mancozeb)	MetaStar 2E AG (metalaxyl)	MonCoat MZ (flutolanil + mancozeb)	Sebring 318S (metalaxyl)	Sebring 480S (metalaxyl)	Spirato 480 FS (fludioxonil)	Vitaflo-280 (carboxin + thiram)	Tops MZ-Gaucho (thiophanate-methyl + mancozeb)	Trilex (trifloxystrobin + metalaxyl)
Fungicide Group[F]	M3	4	4	4	4	14	M3	4 + 12	12	12	3 + 4	11	3	12	12 + M3	4	7 + M3	4	4	M3	7 + M3	1 + M3	11 + 4
Horseradish			X	X	X					X		X		X						X			
Leeks				X						X		X		X									
Lettuce	X		X	X	X					X		X		X				X		X			
Muskmelons	X		X	X	X		X			X		X		X				X		X			
Okra	X		X		X					X		X		X		X		X	X				
Onions, Dry	X		X	X	X					X		X		X						X			
Onions, Green	X		X	X	X					X		X		X						X			
Parsley			X	X	X					X		X		X				X		X			
Parsnips			X	X	X					X		X		X				X		X			
Peas	X	X	X	X	X		X	X		X				X				X	X	X			X
Peppers	X		X	X	X		X			X		X		X				X		X			X
Pumpkins	X		X	X	X		X			X		X		X				X		X			X
Radish	X		X	X	X		X			X		X		X				X		X			
Spinach	X		X	X	X		X			X		X		X		X		X	X	X			
Squash, Summer	X		X	X	X		X			X		X		X				X		X			X
Squash, Winter	X		X	X	X		X			X		X		X				X		X			X
Sweet Corn	X	X		X			X			X	X	X		X				X		X	X		
Sweet Potatoes				X		X				X		X		X				X		X			
Tomatoes	X	X	X	X	X					X		X		X				X		X			X
Watermelon	X	X	X	X	X		X			X		X		X				X		X			X
White Potatoes (Irish)							X		X[2]	X		X	X[3]	X	X		X[3]			X		X[4]	X

[F] To prevent resistance in pathogens, alternate fungicides within a group with fungicides in another group. Fungicides in the "M" group are generally considered "low risk" with no signs of resistance developing to the majority of fungicides.

[2] Registered for use in Florida and North Carolina only.

[3] Registered for use in North Carolina only.

[4] Registered for use in Alabama, Florida, Georgia, Mississippi, North Carolina, and South Carolina only.

Table 10-60. Biocontrol Agents and Disinfestants Registered for Seed Treatment

Crop	Actinovate STP (*Streptomyces lydicus* WYEC 108)	Kodiak HB (*Bacillus subtilis* GB03)	Mycostop (*Streptomyces griseoviridis* K61)	T-22 Planter Box (*Trichoderma harzianum*)	Yield Shield (*Bacillus pumilus* GB34)	Clorox	Hot water	Hydrochloric acid (HCl)	Oxidate 2.0 (hydrogen dioxide + peroxyacetic acid)	Trisodium phosphate (TSP)
		Biocontrol Agents					**Disinfestants**			
Beans, Snap		X	X	X	X	X				
Beans, Lima		X	X	X	X	X				
Beets		X	X	X	X					
Broccoli		X	X						X	
Carrots		X	X	X	X				X	
Celery		X	X						X	
Chinese Cabbage		X	X						X	
Cole Crops		X	X	X					X	
Cucumbers		X	X		X		X		X	
Eggplants		X	X	X			X		X	
Garlic		X	X		X				X	
Greens, Mustard		X	X				X		X	
Greens, Turnip		X	X				X		X	
Horseradish		X	X							
Leeks		X	X	X						
Lettuce		X	X	X			X		X	
Muskmelons		X	X		X		X			
Okra		X	X							
Onions, Dry		X	X	X	X				X	
Onions, Green		X	X	X	X				X	
Parsley		X	X	X			X			
Parsnips		X	X		X					
Peas		X	X	X	X	X				
Peppers		X	X	X			X		X	X
Pumpkins/Winter squash		X	X		X		X			X
Radish		X	X	X	X				X	
Spinach		X	X	X					X	
Squash, Summer		X	X		X		X			X
Sweet Corn		X	X		X					
Sweet Potatoes		X			X					
Tomatoes		X	X	X			X		X	X
Watermelon		X	X		X		X			
White Potatoes (Irish)					X					

Sanitation

Table 10-61. Sweetpotato Storage House Sanitation

Material	Rate per 1,000 Cubic Feet of Space	Methods and Remarks
Heat	140 degrees F 4 to 8 hr/day for 7 days or 180 degrees F for 30 minutes	See remarks under water, produce, and equipment sanitation. The storage house, ventilation system, and equipment must be very clean and moist during the procedure. *Caution:* rot-causing organisms inside a drain will probably not be exposed to lethal temperature.

Table 10-62. Water, Produce, and Equipment Sanitation

Medium	Sanitizer	Contact Time (minutes)	Target Rate (ppm)	Formulation	Method, Schedule, and Remarks
Wash water, dump tank water, or vegetable wash water*	calcium hypochlorite (Aquafit)	2	25	1 oz/200 gal	
	chlorine dioxide (Harvest Wash, ProOxine, Anthium Dioxide, Adox 750)	1 to 10	3 to 5	Varies between products; see product labels.	Maintain water pH between 6.0 and 10. Restricted to large operations. Requires automated and controlled injection systems. **NOTE: Chlorine dioxide is explosive.**
	chlorine gas (99.9%)	—	Contact supplier for rates.		Restricted to very large operations. Requires automated and controlled injection systems. Regulated by both the EPA (water) and FDA (food contact surfaces).
	hydrogen dioxide or peroxide (StorOx)	—	Varies based on method of application.		
	hydrogen peroxide + peroxyacetic acid	1	80 peroxyacetic acid	1 fl oz/16.4 gal	Contact times vary depending on the governing sanitary code. Post-sanitation rinse is not necessary.
	(BioSide HS)	1.5	80 peroxyacetic acid	1 fl oz/16.4 gal	
	(Keystone Fruit and Vegetable Wash)	0.75	88 to 130 (peroxyacetic acid, non-porous surfaces)	3 to 3.5 fl oz/16 gal	
	(PAA Sanitizer FP)	—	25 peroxyacetic acid	1 oz/20 gal	
	(Perasan A)	0.75	24 to 85 peroxyacetic acid	5.9 to 20.9 fl oz/100 gal	
	(SaniDate 5.0)	—	30 to 80	2.5 to 6.7 fl oz/100 gal	
	Tsunami 100	1.5	80	1 fl oz/16.4 gal	
	(Victory)	—	5 to 85 peroxyacetic acid (postharvest pathogens)	0.1 to 1 fl oz/16 gal	
	(VigorOx 15 F&V)	—	45 peroxyacetic acid (foodborne pathogens)	0.54 fl oz/16 gal	
	sodium hypochlorite (5.25%) (12.75%)	—	150 150	2.9 ml/L 1.18 ml/L	Maintain water pH between 6.0 and 7.5. Noxious chlorine gas can be released when the pH drops below 6.0. **NOTE: Household bleach is NOT registered for use with fresh produce.**
	sodium hypochlorite (Agclor 310) (Dibac) (Dynachlor) (Extract-2) (JP Optimum CRS) (Maxxum 700) (Zep FS Formula 4665)	— 2 2 2 — 2 2	65 to 400 25 25 25 25 25 25	0.5 to 3 gal/1000 gal 1 oz/20 gal 5 oz/200gal 5 oz/200gal 0.75 oz/10 gal 8 oz/200gal 5 oz/200gal	Monitor residual chlorine or change solution when it is visibly dirty. Rinse produce with potable water prior to packing.
Equipment** (conveyors, scrubbers, plastic harvest containers, peelers, field equipment, etc.)	calcium hypochlorite (Aquafit)	2	600 (porous surfaces)	3 oz/20 gal	Do not rinse or soak equipment overnight.
	chlorine dioxide (ProOxine, Sanogene, Anthium Dioxide, Adox 750)	1 to 10	10 to 20 (porous or non-porous surfaces) 500 (ceilings, floors and walls)	Varies between products; see product labels.	
	hydrogen dioxide or peroxide (StorOx)	—	1 to 3 ppm (1:300 to 1:100) (non-porous surfaces)	0.5 to 1.25 fl oz/gal	Apply until run-off. Requires a thorough post-sanitation rinse with potable water.
	hydrogen peroxide + peroxyacetic acid (BioSide HS) (Oxidate 2.0) (PAA Sanitizer FP) (Perasan A) (SaniDate 5.0) (VigorOx 15 F&V)	1 or more See label 1 or more 1 1 1	93 to 500 peroxyacetic acid 100 to 300 88 to 130 peroxyacetic acid (non-porous surfaces) 82 to 500 peroxyacetic acid ~128 85 peroxyacetic acid	0.7 to 3.8 fl oz/10 gal 1.25 to 1.5 fl oz/gal 1 to 1.5 fl oz/5 gal 1 to 6.1 oz/6 gal 1.6 fl oz/5 gal 3.1 fl oz/50 gal	Contact time varies depending on the governing sanitary code. Consult labels as some products require a post-disinfection rinse with potable water.

Table 10-62. Water, Produce, and Equipment Sanitation

Medium	Sanitizer	Contact Time (minutes)	Rate of Material to Use - Target Rate (ppm)	Rate of Material to Use - Formulation	Method, Schedule, and Remarks
Equipment** (conveyors, scrubbers, plastic harvest containers, peelers, field equipment, etc.) (continued)	sodium hypochlorite (5.25%)	2 2 —	100 to 200 (non-porous surfaces) 600 (porous surfaces) 1000 to 2000 (floors and walls)	1.9 to 3.8 ml/L 11.4 ml/L 1900 to 3800 ml/L	Noxious chlorine gas can be released when the pH drops below 6.0. Porous surfaces require a thorough post-disinfection rinse with potable water. Allow all surface types to air dry prior to re-use.
	sodium hypochlorite (12.75%)	2 2 —	100 to 200 (non-porous surfaces) 600 (porous surfaces) 1000 to 2000 (floors and walls)	0.78 to 1.56 ml/L 4.68 ml/L 780 to1560 ml/L	
	sodium hypochlorite (Agclor 310, Dibac, Dynachlor, Extract-2, JP Optimum CRS, Maxxum 700, Zep FS Formula 4665)	1	Varies based on method of application.		
	quaternary ammonia (DDAC) (KleenGrow)	10	—	1 fl oz/gal	Allow surfaces to air dry. If treated surfaces will contact food, thoroughly rinse surfaces with potable water.

* Recommendations are for potable water only. Recommended rates are not effective in reducing pathogen populations in non-potable water (i.e. surface or ground water).

** Recommendations are for potable water only. Always wash off organic debris and soil with water prior to sanitizing. Rates and contact time are dependent on surface type.

Various and Alternative Fungicides

Table 10-63. Various and Alternative Fungicides for Use on Vegetable Crops

Not all trade names are registered in all states. Check the label to confirm that the product is registered in your state and for your intended use.

Common Name	Trade Name(s)
copper octanoate (FRAC M1)	Camelot-O (SePRO)
copper (cuprous) oxide (FRAC M1)	Nordox (NORDOX Industrier AS)
	Nordox 75 WG (NORDOX Industrier AS)
copper sulfate (basic) (FRAC M1)	Basic Copper 53 (Albaugh)
	Cuprofix Ultra 40 Disperss (United Phosphorus)
	Cuproxat (NuFarm)
	Cuproxat FL (NuFarm)
copper sulfate pentahydrate (FRAC M1)	Magna-Bon CS 2005 (Magna-Bon II, LLC)
	Mastercop (Adama)
fludioxonil (FRAC 12)	Cannonball (Syngenta)
	Cannonball WG (Syngenta)
	Cannonball WP (Syngenta)
	Dyna-Shield Fludioxonil (Loveland Products)
	Fludioxonil 4L ST (Albaugh)
	Maxim 4FS (Syngenta)
	Maxim PSP (Syngenta)
	Scholar SC (Syngenta)
	Spirato 480 FS (Nufarm)
fosetyl-Al (Aluminum tris (O-ethyl phosphate)) (FRAC 33)	Aliette WDG Fungicide (Bayer)
	Linebacker WDG (NovaSource)
iprodione (FRAC 2)	Iprodione 4L AG (Arysta)
	Meteor (United Phosphorus)
	Nevado 4F (Adama)
	Rovral 4 Flowable Fungicide (FMC Corporation)
mancozeb (FRAC M3)	Dithane F-45 Rainshield (Dow)
	Dithane M45 (Dow)
	Fortuna 75 WDG (Agria Canada)
	Koverall (Cheminova, FMC Corporation)
	Manzate Max (United Phosphorus) Penncozeb 80WP (UPI)
	Manzate Pro-Stick (United Phosphorus)
	Penncozeb 75DF (United Phosphorus)
	Penncozeb 80WP (United Phosphorus)
	Potato Seed Treater PS (Loveland Products)
	Potato Seed Treater 6% (Loveland Products)
	Roper DF Rainshield (Loveland Products)
mefenoxam (FRAC 4)	Apron XL (Syngenta)

Table 10-63. Various and Alternative Fungicides for Use on Vegetable Crops

Not all trade names are registered in all states. Check the label to confirm that the product is registered in your state and for your intended use.

mefenoxam (FRAC 4) (continued)	Ridomil Gold GR *(Syngenta)*
	Ridomil Gold SL *(Syngenta)*
	Ultra Flourish *(Nufarm)*
myclobutanil (FRAC 3)	Rally 40WSP *(Dow)*
	Sonoma 25EW AG *(Albaugh)*
	Sonoma 40WSP *(Albaugh)*
phosphite, potassium (FRAC 33)	Helena Prophyt *(Helena)*
	Reveille *(Helena)*
phosphite (mono- and dibasic salts) (FRAC 33)	Helena Prophyt *(Helena)*
	Phostrol (Nufarm)
	Phostrol 500 (Nufarm)
phosphorous acid (mono- and di-potassium salts) (FRAC 33)	Alude (Nufarm)
	Confine Extra *(Winfield Solutions)*
	Fosphite Fungicide *(JH Biotech)*
	Fungi-Phite *(Verdesian Life Sciences)*
phosphorous acid (mono- and di-potassium salts) (FRAC 33) (continued)	K-Phite 7LP AG *(Plant Food Systems)*
	Rampart *(Loveland Products)*
	Reliant *(Quest Products)*
propamocarb hydrochloride (FRAC 28)	Previcur Flex *(Bayer)*
	Promess *(Agriphar)*
propiconazole (FRAC 3)	AmTide Propiconazole 41.8% EC *(AmTide)*
	Bumper 41.8 EC *(Adama)*
	Bumper ES *(Adama)*
	Fitness *(Loveland Products)*
	Mentor (Syngenta)
	Propi-star EC *(Albaugh)*
	Propicure 3.6 F *(United Supplies, Inc.)*
	Propimax EC *(Dow)*
	Shar-Shield PPZ *(Sharda USA)*
	Tide Propiconazole 41.8% EC *(Tide International)*
	Tilt *(Syngenta)*
	Topaz *(Winfield Solutions)*
sulfur (FRAC M2)	Cosavet-DF *(Sulphur Mills Limited)*
	CSC 80% Thiosperse *(Martin Resources)*
	CSC Dusting Sulfur *(Martin Resources)*
	CSC Thioben 90 *(Martin Resources)*
	CSC Wettable Sulfur *(Martin Resources)*
	Dusting Sulfur *(Loveland Products; Wilbur-Ellis)*
	First Choice Dusting Sulfur *(Loveland Products)*
	IAP Dusting Sulfur *(Independent Agribusiness Professionals)*
	InteGro Magic Sulfur Dust *(InteGro Inc.)*
	Kumulus DF *(Micro Flo and Wilbur-Ellis)*
	Liquid Sulfur Six *(Helena)*
	Micro Sulf *(Nufarm)*
	Microfine Sulfur *(Loveland Products)*
	Microthiol Disperss *(United Phosphorus)*
	Special Electric Sulfur *(Wilbur-Ellis)*
	Spray Sulfur *(Wilbur-Ellis)*
	Sulfur 6L *(Arysta)*
	Sulfur 90 W *(Drexel)*
	Sulfur DF *(Wilbur-Ellis)*
	THAT Flowable Sulfur *(Stoller Enterprises)*
	Thiolux *(Loveland Products)*
	Yellow Jacket Wettable Sulfur II *(Georgia Gulf Sulfur)*

Table 10-63. Various and Alternative Fungicides for Use on Vegetable Crops

Not all trade names are registered in all states. Check the label to confirm that the product is registered in your state and for your intended use.

tebuconazole (FRAC 3)	Monsoon *(Loveland Products)*
	Onset 3.6L *(Winfield Solutions)*
	Orius 3.6 F *(Adama)*
	Tebu-Crop 3.6 F *(Sharda USA)*
	Tebucon 3.6 F *(Repar Corp.)*
	Tebuconazole 3.6 F *(Solera Source Dynamics)*
	TebuStar 3.6L *(Albaugh)*
	Tebuzol 3.6F *(United Phosphorus)*
	Toledo 3.6F *(Rotam)*
thiophanate-methyl (FRAC 1)	Cercobin *(Cheminova)*
	Incognito 4.5 F *(makhteshim Agan of North)*
	Incognito 85 WDG *(MANA)*
	Thiophanate Methyl 85 WDG *(Makhteshim Agan of North))*
	T-Methyl 4.5 Ag *(Helena)*
	T-Methyl 4.5 F *(Nufarm)*
	T-Methyl 70W WSB *(Nufarm)*
	Topsin 4.5FL *(United Phosphorus)*
	Topsin M WSB *(United Phosphorus)*
	3336 EG (Cleary)
	3336 F (Cleary)
	3336 WP (Cleary)

Table 64. Biopesticides and Fungicide Alternatives for Vegetables

Active Ingredient	Product[1]	Target Diseases/Pests	PHI (days)	REI	Greenhouse Use	OMRI-Listed	Comments
allyl isothiocyanate	Dominus *(Isagro)*	Certain soil-borne fungi and nematodes	—	5 days	Yes	No	Preplant soil biofumigant. See label for other restrictions and application instructions. **Dominus is not registered for use in WV.**
acibenzolar-S-methyl	Actigard 50WG (Syngenta)	bacterial blights, downy mildew, powdery mildew (see label for crop-specific diseases)	See label	12 hr	No	No	Do not apply to plants stressed by heat, cold, or moisture extremes. **FRAC P01.**
Bacillus amyloliquefa- ciens strain D747	Double Nickel 55, Double Nickel LC *(Certis USA)*	Bacterial spots and speck, bacterial leaf spot, powdery mildew, white mold (timber rot), Botrytis gray mold, Alternaria leaf spot	0	4 hr	Yes	Yes	Multiple application methods. See label. **Double Nickel 55 is not registered for use in OK; Double Nickel LC is not registered for use in AR, KY, or OK. FRAC 44.**
Bacillus mycoides isolate J	LifeGard WG *(Certis USA)*	Bacterial spot and speck, downy mildew, early blight, late blight, powdery mildew, white mold and others (see label for crop specific diseases)	0	4 hr	Yes	Yes	**LifeGard is not registered for use in LA, NC, or WV. FRAC P06.**
Bacillus pumilus QST2808	Ballad Plus, Sonata *(Bayer)*	early blight, late blight, downy mildew, powdery mildew, leaf blights, rust	0	4 hr	Sonata	No	Products are not OMRI-listed, but labels state that they can be used for organic production. See labels for specifics of greenhouse use. Ballad Plus can be used on certain beans and sweet corn only.
Bacillus subtilis GB03	Companion (Liquid and WP formulations) *(Growth Products)*	Root and foliar diseases (see label for crop specific diseases)	0	See label	Yes	WP only	See product labels for instructions on various application uses. **Check state registration status prior to use. FRAC 44.**
Bacillus subtilis strain MBI 600	Subtilex NG *(BASF)*	*Rhizoctonia, Pythium,* and *Fusarium* diseases, powdery mildew, and gray mold	See label	4 hr	Yes	No	For post-plant applications to the soil/planting medium or as a foliar spray on cucurbits and fruiting vegetables. **FRAC 44.**
Bacillus subtilis strain QST 713	CEASE (BioWorks Inc.); Serenade ASO, Serenade MAX, Serenade Opti, Serenade Optimum, Serenade Soil *(Bayer)*	Various diseases - see label for crop-specific diseases	0	4 hr	CEASE, Serenade ASO	Yes	Works best when applied prior to disease development and used in an integrated program. See label for product-specific instructions regarding product application. **Serenade Optimum has a 2(ee) Recommendation for reduced rates on fruiting vegetables in FL and GA. FRAC 44.**

Table 64. Biopesticides and Fungicide Alternatives for Vegetables

Active Ingredient	Product[1]	Target Diseases/Pests	PHI (days)	REI	Greenhouse Use	OMRI-Listed	Comments
bacteriophage	AgriPhage (*Omni-Lytics*)	Bacterial spot and speck	0	4 hr	Yes (seedlings)	No	Product is strain specific (active against *Xanthomonas campestris* pv. *campestris* and *Pseudomonas syringae* pv. *tomato*) and is labeled for use on tomatoes and peppers. Do not tank-mix product with denaturing agents or copper salts. **Check state registration status prior to use.**
Banda de Lupinus albus doce (BLAD)	Fracture (*FMC Corporation*)	Botrytis gray mold and powdery mildew	1	4 hr	Yes	No	For use on tomato only. **Product has a 2(ee) Recommendation for use against southern blight in tomato in FL, GA, NC, and SC.** *FRAC BM01.*
Coniothyrium minitans strain CON/M/91-08	Contans WG (*Bayer, SipcamAdvan*)	*Sclerotinia sclerotiorum* and *S. minor*	0	4 hr	Yes	Yes	Apply to soil or potting medium. Do not tank-mix products with other fungicides or apply products 7 days before or after the application of other fungicides. **Tomato is not included in the list of fruiting vegetables on this label.**
copper	See disease control tables for individual crops.						
extract of *Reynoutria sachalinensis*	Regalia, Regalia Rx (*Marrone Bio Innovations*)	Certain bacterial and fungal diseases (see label for crop specific diseases)	0	4 hr	Regalia	Yes	**Regalia Rx is only labeled for use on corn.** *FRAC P5.*
Gliocladium cantenulatum strain J1446	PreStop (*AgBio, Inc.*)	Seed-borne, soilborne and wilt diseases and certain foliar diseases (*Botrytis* and *Didymella*)	See label	0	Yes	—	Product should not be tank-mixed with pesticides or fertilizers. See label for crop restrictions for foliar application. Product should not be applied as a foliar application after fruiting. **Check state registration status prior to use.**
Gliocladium virens GL-21	SoilGard (*Certis USA*)	Damping-off and root rots	0	0 hr	Yes	Yes	Do not apply in conjunction with chemical fungicides. **Product is not registered for use in AL, AR, KY, MS, OK, TN, or WV.** *FRAC BM02.*
harpin αβ protein	Employ, Messenger Gold, ProAct (*Plant Health Care, Inc*)	Nematode (suppression)	0	4 hr	See label	No	See label for instructions on various application uses. **Check state registration status prior to use.**
hydrogen dioxide	OxiDate (*BioSafe Systems, LLC*)	Various diseases – see label for crop specific diseases	0	See label	Yes	No	See label for instructions on various application uses.
hydrogen dioxide + peroxyacetic acid	OxiDate 2.0, TerraClean 5.0 (*BioSafe Systems, LLC*)	Various diseases – see label for crop specific diseases	0	See label	Oxidate 2.0	Yes	TerraClean is a soil treatment product. See label for instructions on various application uses.
Hydrogen peroxide + peroxyacetic acid	Rendition (*Certis USA*)	Various diseases - see label for crop-specific diseases	0	See label	See label	No	
milk	N/A	Viruses (*tomato mosaic virus* (ToMV) and *tobacco mosaic virus* (TMV))	Until spray dries	0	Yes	Yes	Spray plants until runoff. Dip hands every 5 min while handlings plants. Dip tools for 1 min; do not rinse. Use in combination with seed treatments and sanitation practices. **Sooty mold may develop on treated plants.**
Myrothecium verrucaria strain AARC-0255 fermentation solids and solubles	DiTera DF (*Valent*)	nematodes	—	4 hr	No	Yes	DiTera DF is a soil treatment product.
Neem oil (extract)	Trilogy (*Certis USA*), Triact 70 (*OHP*)	foliar fungal diseases - see label for specifics	0	4 hr	See label	Yes	May cause leaf burn; test a small number of plants before spraying entire crop. **Toxic to honey bees. Trilogy is not registered for use in OK, MS, or WV.**
Oils from cottonseed, corn, and garlic	Mildew Cure (*JH Biotech Inc*)	powdery mildew	See label	See label	Yes	Yes	May cause leaf burn; test a small number of plants before spraying entire crop. **Product is not registered for use in AL, LA, SC, or WV.**
Oil from soybean	Oleotrol-M (*NTS Research & Inc*)	Downy mildew, powdery mildew, Botrytis, rust, sour rot, gray mold	See label	See label	Yes	Yes	Tank-mix with a spreader-sticker. See product label for labeled crops.

Table 64. Biopesticides and Fungicide Alternatives for Vegetables

Active Ingredient	Product[1]	Target Diseases/Pests	PHI (days)	REI	Greenhouse Use	OMRI-Listed	Comments
Paecilomyces lilacinus	MeloCon WG (*Certis USA*)	nematodes (see label for specific species)	—	4 hr	No	Yes	Bionematicide. See product label for mixing restrictions and application instructions. **Product is not registered for use in KY, OK, TN, or WV.**
Phosphorous acid	See disease control tables for individual crops.						
Polyoxin D zinc salt	See disease control tables for individual crops.						
Potassium bicarbonate	Carb-O-Nator (*Certis USA*), Kaligreen,(*Otsuka AgriTechno Co, Ltd*), Milstop (*BioWorks, Inc*)	Various diseases - see label for crop-specific diseases	1 (Kaligreen) 0 (others)	1 hr (Milstop) 4 hr (others)	Carb-O-Nator, Milstop	Yes	See label for instructions on various application uses and for any instructions regarding spray solution pH. **FRAC NC (not classified).**
Potassium salts of fatty acids	M-Pede (*Gowan Company*)	powdery mildew	0	12 hr	Yes	Yes	See product label for notes regarding plant sensitivity.
Potassium silicate	Sil-MATRIX (*Certis USA*)	powdery mildew	0	4 hr	Yes	Yes	Avoid contact with glass. Tank-mix with a non-ionic surfactant for best results.
Streptomyces sp. Strain K61	Mycostop (*AgBio, Inc*)	Seed, root, and stem rots and wilt diseases caused by certain pathogens; suppression of certain diseases (see label for crop-specific diseases)		4 hr	Yes	Yes	Product can be incorporated into potting media, used as a seed treatment, or applied in-furrow to field soil, as a soil spray or drench or as a foliar application – see label for specific application instructions. Product should not be tank-mixed with pesticides or fertilizers.
Streptomyces lydicus WYEC 108	Actino-Iron, Actinovate AG (*Novozymes*)	Foliar diseases and/or damping-off and root rots, (see label for crop specific diseases)	See label	See label	Yes	Yes	See label for instructions on various application uses. Actino-Iron is a soil treatment product that includes iron, molybdenum, and humic acid.
Sulfur	See disease control tables for individual crops.						
Trichoderma harzianum Rifai strain KRL-AG2	RootShield WP (*Bioworks, Inc*),	root pathogens	See label	See label	See label	Yes	Product is for use in soil applications. Product should not be applied to chickpea. See label for instructions on various application uses.
Trichoderma harzianum Rifai strain T-22	RootShield AG, RootShield Granules (*BioWorks, Inc*)	Soilborne or root pathogens	See label	See label	See label	Yes	Products are for use in soil applications or seed treatments. Products should not be applied to chickpea. **RootShield AG should not be applied when aboveground harvestable food is present.** See product labels for instructions on various application uses and on compatibility with other products. **RootShield AG is not registered for use in AL, AR, KY, LA, MS, NC, OK, SC, TN, VA, or WV.**
Trichoderma harzianum Rifai strain T-22 + *T. virens* strain G-41	RootShield PLUS+ Granules, RootShield PLUS+ WP (*BioWorks, Inc*)	Soilborne or root pathogens	See label	See label	See label	Yes	Products are for use in soil applications or seed treatments. Products should not be applied to chickpea. **RootShield PLUS+ WP should not be applied when aboveground harvest- able food is present.** See product labels for instructions on various application uses and on compatibility with other products.
Trichoderma spp (*T. asperellum* strain ICC012 and *T. gamsii* strain ICC 080)	BIO-TAM 2.0 (*Marrone Bio Innovations*)	Certain fungal diseases (see label)	0	4 hr	Yes	Yes	See label for a list of incompatible fungicides.
Ulocladium oudemansii strain U3	BotryStop (*BioWorks, Inc*)	*Botrytis* spp. and *Sclerotinia* spp. (see label for crop-specific diseases)	0	4 hr	Yes	Yes	Product should be stored in a cool, dry place at or below 68°F.
Yeast extract hydrolysate from *Saccharomyces cerevisiae*	KeyPlex 350 OR (*KeyPlex*)	See label	See label	4 hr	See label	No	See product labels for instructions on various application uses. **Check state registration status prior to use.**

Fungicide Resistance Management

Fungicides are organized according to FRAC groups, chemical structure and Mode of Action (MoA). Fungicides within a given FRAC group control fungi in a similar manner and share the same risk for fungicide resistance development. Some fungicides are referred to as high- or at-risk fungicides because of their specific MoAs and therefore have a high risk for resistance development. Groups of fungicides, such as the QoI's (FRAC group 11) or Phenylamides (FRAC group 4) are prone to resistance development due to very specific MoAs. Fungicides in high- or at-risk groups should be rotated and/or tank-mixed with broad spectrum protectant fungicides (FRAC group M3 or M5) to delay the development of resistant strains of fungi. For more information on fungicide resistance management see: http://www.frac.info/

Table 10-65. Fungicide Modes of Action for Fungicide Resistance Management

FRAC Code	Fungicide Resistance Risk	Group Name	Example Active ingredients	Example Products
P1	Unknown	Benzo-thiadiazole (BTH)	Acibenzolar-S-methyl	Actigard
M1	Low	Inorganic copper	Fixed copper	Copper (various)
M2	Low	Inorganic sulfur	Sulfur	Sulfur (various)
M3	Low	Dithiocarbamates	Mancozeb	Mancozeb (various)
M5	Low	Chloronitriles	Chlorothalonil	Chlorothalonil (various)
1	High	Methyl benzimidazole carbamates (MBC)	Thiophanate-methyl	Topsin M
2	Medium to high	Dicarboximides	Iprodione	Rovral
3	Medium	Demethylation inhibitors (DMI)	Triflumizole Myclobutanil	Procure Rally
4	High	Phenylamide	Mefenoxam	Ridomil Gold
7	Medium to high	Succinate dehydrogenase inhibitors (SDHI)	Boscalid Penthiopyrad	Endura Fontelis
9	Medium	Anilino-pyrimidines (AP)	Pyrimethanil	Scala
11	High	Quinone outside inhibitors (QoI)	Pyraclostrobin Trifloxystrobin Azoxystrobin	Cabrio Flint Quadris
12	Low to medium	Phenylpyrroles (PP)	Fludioxinil	Maxim
13	Medium	Aza-naphthalenes	Quinoxyfen	Quintec
14	Low to medium	Aromatic hydrocarbons (AH)	Dichloran	Botran
19	Medium	Polyoxins	Polyoxin D	OSO
21	Medium to high	Quinone inside Inhibitors (QiI)	Cyazofamid	Ranman
22	Low to medium	Benzamides (toluamides)	Zoxamide	Gavel (contains zoxamide and mancozeb)
27	Low to medium	Cyanoacetamide-oximes	Cymoxanil	Curzate
28	Low to medium	Carbamates	Propamocarb	Presidio
29	Unknown	Dinitroanilines	Fluazinam	Omega
33	Low	Phosphonates	Fosetyl Al	Aliette
40	Low to medium	Carboxylic acid amides (CAA)	Dimethomorph Mandipropamid	Forum Revus
43	High	Benzamides	Fluopicolide	Presidio
45	Medium to high	Triazolo-pyrimidylamine	Ametoctradin	Zampro (contains ametoctradin and dimethomorph)
49	Medium to high	Piperidinyl-thiazole-isoxazoloines	Oxathiapiprolin	Orondis

INDEX

A

Abutilon, growth regulator, 411
Achillea, growth regulator, 411
Achmella Oleraea, growth regulator, 411
Agastache, growth regulator, 411
Ageratum, growth regulator, 411
Alcea Rosea, 412
Alfalfa
    lime and fertilizer, 44
    weed control, 334
Alternanthera, growth regulator, 412
Alyssum, growth regulator, 412
Amaryllis, growth regulator, 412
Anagallis, growth regulator, 412
Anemone, growth regulator, 412
Angelonia, growth regulator, 412
Animal damage control
    fish, 458
    general practices, 448
    rodenticides, 457
    suggestions, 453
Animal pests, description, 450
Ants
    control in commercial turf, 164
    home lawns, 187
    household, 144, 178
Apple
    disease control, 192
    fungicides, effectiveness, 192
    growth regulator, 408
    insect control, 193
    insecticides, effectiveness, 193
    lime and fertilizer, 45
    mite control, 193
    spray program, 190
    weed control, 326
Aquatic weeds, 391
Aquilegia, growth regulator, 412
Argyranthemum, growth regulator, 412
Artichoke, control, 399
Arthropod management
    Christmas trees, 159
    ornamentals, greenhouse, 147
    ornamentals, nurseries, landscape, 152
Asclepias, growth regulator, 412
Asparagus
    disease control, 534
    insect control, 97, 173
    lime and fertilizer, 53
    weed control, 355
Aster, growth regulator, 413
Asteriscus Maritimus, growth regulator, 413
Astilbe, growth regulator, 413
Azalea, growth regulator, 413, 438

B

Backpack sprayers, 24
Bacopa Sutera, growth regulator, 413
Bamboo, control, 399
Barley, see Small Grains
Bat, identification and control, 450, 453
Beans
    disease control, 535
    harvest aids, 407
    insect control, 97, 173
    weed control, 356

Beaver, identification, 450
Bed bug, 144, 179
Bedding plants, growth regulators, 414
Bees
    control, 144, 179
    home lawns, 187
    pesticide toxicity and bees, 72
    protecting honey bees, 7, 73
Beets
    insect control, 99, 173
    lime and fertilizer, 53
    weed control, 358
Begonia, growth regulator, 415
Bermudagrass
    weed control, 338
    lime and fertilizer, 48, 49, 50
Berries, control, 399
Bidens Ferulifolia, growth regulator, 415
Billbug, control in commercial turf, 164
Biosolids, use of municipal, 68
Blackberries (caneberry)
    fertilizer, 46
    management program, 200
    weed control, 319
Blueberries
    disease control, 195
    fertilizer, 46
    insect control, 195
    management program, 195
    weed control, 320
Bluegrass, lime and fertilizer, 44, 49, 50
Bobcat, identification, 450
Booklice, 144, 179
Bougainvillea, growth regulator, 415
Boxelder bug, 179
Brachyscome, growth regulator, 415
Bracteantha bracteata, growth regulator, 415
Brassicas, disease control, 539
Broadcast spreaders, 22
Broccoli
    disease control, 539
    insect control, 100, 173
    lime and fertilizer, 53
    weed control, 358
Bromeliad, growth regulator, 415
Browallia, growth regulator, 415
Brown dog tick, 179
Brussels sprouts
    disease control, 539
    insect control, 100, 173
    lime and fertilizer, 54
Bulb crops, growth regulator, 416
Bunch grapes
    insect control, 206
    lime and fertilizer, 46

C

Cabbage
    disease control, 539
    insect control, 100
    lime and fertilizer, 54
    weed control, 358
Caladium, growth regulator, 416
Calendula, growth regulator, 416
Calibrachoa, growth regulator, 416
Calla lily, growth regulator, 417

Campanula, growth regulator, 417
Calibration, 25
    broadcast spreaders, 28
    calibrating a sprayer, 25
    granular applicators, 27
    methods, 26
    variables, 29
Canna lily, growth regulator, 417
Caneberry
    Disease, insect control, 200
    management program, 200
Cantaloupe
    insect control, 105, 174
    lime and fertilizer, 55
    weed control, 358
Carpenter ant, 169
Carpenter bee, 170
Carpet beetle, 144, 180
Carrot
    disease control, 581
    insect control, 102, 131, 174
    lime and fertilizer, 54
    weed control, 360
Cattle
    insect control, 134
Cauliflower
    disease control, 539
    insect control, 100, 131, 173
    lime and fertilizer, 54
    weed control, 360
Celery
    insect control, 102, 131, 174
    lime and fertilizer, 54
    weed control, 361
Celosia, growth regulator, 417
Centaurea, growth regulator, 417
Centradenia, growth regulator, 417
Centipede
    household, 180
    licensed pest control, 144
Centipedegrass, fertilizer, 48
Chemical formulations, 32
Chemigation, 8
Chigger, 180
Chili peppers, harvest aids, 408
Chipmunk, identification, 450
Christmas cactus, growth regulator, 417
Christmas trees, arthropod control, 159
Chrysanthemum, growth regulator, 417
Chrysocephalum apiculatum, growth regulator, 418
Clarkia, growth regulator, 418
Clary Sage, weed control, 317
Clematis, growth regulator, 418
Cleome, growth regulator, 418
Clerodendrum, growth regulator, 418
Clothes moth
    household, 180
    licensed pest control, 144
Clover, lime and fertilizer, 44
Clover mite
    household, 181
    licensed pest control, 144
Cockroach
    household, 181
    licensed pest control, 144
Cole crops, weed control, 361
Coleus, growth regulator, 419

Collard
    disease control, 557
    insect control, 103, 174
    lime and fertilizer, 54
Columbine, growth regulator, 419
Compatibility test, herbicides with liquid fertilizers, 63
Coneflower, growth regulator, 419
Coreopsis, growth regulator, 419
Consolida, growth regulator, 419
Cornflower, growth regulator, 419
Corn
    disease control, 544
    foliar disease, 462
    insect control, field corn, 74
    insect control, sweet corn, 104, 131, 174
    lime and fertilizer, 42, 54
    nematode control, 462
    weed control, 252
    weed control, sweet corn, 362
Cosmos, growth regulator, 419
Cotton
    defoliation, growth regulators, 406
    insect control, 80
    insect resistance management, 83
    lime and fertilizer, 42
    nematode control, 464
    weed control, 264
Cotton rat, identification, 450
Cowbird, identification and control, 450, 453
Coyote, identification, 450
Cricket
    household, 182
    licensed pest control, 144
Crossandra, growth regulator, 419
Crow, identification and control, 450, 453
Cucumber
    disease control, 545
    insect control, 105, 131, 174
    insect control, greenhouse, 132
    lime and fertilizer, 54
    weed control, 364
Cucurbits, disease control, 545
Cuphea, growth regulator, 419

D

Daffodil, growth regulator, 420
Dahlia, growth regulator, 420
Deer, white-tailed, identification and control, 452, 457
Defoliants, cotton, 406
Delphinium, growth regulator, 420
Desiccants, preharvest, 407
Dianthus, growth regulator, 420
Diascia, growth regulator, 421
Dicentra spectabilis, growth regulator, 421
Dichondra argentea, growth regulator, 421
Digitalis, growth regulator, 421
Dilutions for liquids and dusts, 31
Disposal, of pesticides, 16
Dracaena, growth regulator, 421
Drift control, 17
Dusters, 22
Dusty miller, growth regulator, 421

E

Earwig
    household, 182
    licensed pest control, 144
Easter lily, growth regulator, 421
Eastern gammagrass, lime and fertilizer, 44
Eggplant
    disease control, 553
    growth regulator, 421
    insect control, 108, 131, 174
    lime and fertilizer, 54
    weed control, 366
Empty container, disposal, 17
Endive, lime and fertilizer, 54
English sparrow, identification and control, 450, 453
Environmental hazards, pesticide, 5
Equipment
    backpack sprayers, 24
    broadcast spreaders, 22
    calibrating, 25
    cleaning, 25
    dusters, 22
    granular applicators, 22
    nozzle types and materials, 30
    sprayers, 23
    wick applicators, 24
Erysimum, growth regulator, 421
Eupatorium, growth regulator, 421
Evolvulus, growth regulator, 421
Exacum, growth regulator, 422

F

Fatshedera, growth regulator, 422
Fertilizer
    compatibility with herbicides, 63
    field crops, 42
    lawns, 48
    livestock, poultry manure, 64
    mixed with herbicides, 62
    municipal and biosolids, 68
    nursery crops, 52
    organic alternatives, 69
    ornamental plants, 51
    pasture and hay crops, 42
    placement, 64
    rules and regulations, 57
    small fruit (berries), 46
    solubility of fertilizer materials, 61
    tree fruit, 45
    vegetable crops, 53
Fescue, fertilizer recommendations, 48
Field mouse, identification, 450
Fish, control, 458
Flaccidgrass, lime and fertilizer, 44
Flea
    household, 182
    licensed pest control, 144
Flies
    household, 183
    licensed pest control, 144
Floricultural crops, growth regulation in greenhouses, 411
Florida Betony, control, 399
Flowering and foliage plants, growth regulators, 422
Flying squirrel, identification, 450
Forest stands
    site preparation, 386
    weed control, 384
Fox, identification, 450, 451
Freesia, growth regulator, 423
Fruit crops
    fertilizer, 45, 46
    insect and disease control, 189
    weed control, 319, 326
Fuchsia, growth regulator, 423

G

Garden, vegetable, insect control, 173
Gardenia, growth regulator, 423
Garlic
    disease control, 555
    weed control, 367
Gaura, growth regulator, 423
Gazania, growth regulator, 423
Geranium, growth regulator, 424
Gerbera daisy, growth regulator, 424
Gladiolus, growth regulator, 424
Gloxinia, growth regulator, 424
Glyphosate formulations, 310
Goats, insect control, 136
Gomphrena, growth regulator, 424
GNSS, variable rate technology, 24
Grackle, identification and control, 450, 454
Grain sorghum, insect control, 77
Granular applicators, 22
Grape Ivy, growth regulator, 424
Grapes, weed control, 322
Grapes, bunch
    fertilizer, 46
    insect management, 206
Grapes, muscadine
    fertilizer, 46
    management program, 216
Gray fox, identification, 450
Gray squirrel, identification, 450
Greenhouse
    tobacco, fertilization, 43
    vegetables, insect control, 132
Greens, leafy
    lime and fertilizer, 54
    weed control, 368
Greens, mustard
    insect control, 113, 175
    weed control, 368
Ground pearl, commercial turf control, 165
Groundhog, identification, 450
Groundwater
    contamination potential, 11
    protecting, 8
Growth regulating chemicals
    apples, 408
    cotton, 406
    floricultural crops, 411
    fruiting vegetables, 441
    peanut, 441
    tobacco, 438, 440
    turfgrass, 442
    woody ornamentals, 438
Gulls, identification and control, 450, 454
Gypsophila, growth regulator, 424

H

Harvest aids and preharvest desiccants, 407

Hay crops and pastures
    fertilizer, 44
    weed control, 334
Hazards of pesticides, 5
Helenium amarum, growth regulator, 424
Helichrysum petiolare, growth regulator, 425
Herbicide resistance management, 311
Herbicides and nitrogen solution, mixing, 62
Herbicide, modes of action for hay crops, pastures, lawns and turf, 315
Hibiscus, growth regulator, 425
Holly, growth regulator, 425
Hollyhock, growth regulator, 425
Home vegetable garden, insects, 173
Honeysuckle, control, 399
Hornets, 144, 183
Horses, insect control, 138
Hosta, growth regulator, 425
House mouse, identification and control, 451
House rat, identification and control, 451
Household insect pests, 178
Hyacinth, growth regulator, 425
Hybrid lily, growth regulator, 425
Hydrangea, growth regulator, 425
Hypoestes, growth regulator, 426

I

Identification of insects, 39
Impatiens, growth regulator, 426
Industrial and household pests, 144
Insect control (see also specific crop)
    commercial turf, 164
    commercial vegetables, 97
    greenhouse vegetables, 132
    household pests, 144, 178
    identification of insects, 39
    industrial and household, 144
    lawns, turf (home), 187
    livestock and poultry, 134
    ornamentals, greenhouse, 147
    poultry, 134
    vegetable garden, 173
    wood and wood products, 169
Iochroma, growth regulator, 426
Ipomoea, growth regulator, 426
Iresine, growth regulator, 426
Irish potato
    disease control, 576
    insect control, 118, 175
    lime and fertilizer, 55
    harvest aids and desiccants, 408
    weed control, 371

J

Jacobinia pink, growth regulator, 426
Jerusalem cherry, growth regulator, 426

K

Kalanchoe, growth regulator, 427
Kale
    disease control, 557
    insect control, 103
    lime and fertilizer, 54
    weed control, 368

Kohlrabi
    insect control, 100
Kudzu, control, 399

L

Lachenalia, growth regulator, 427
Lamium, growth regulator, 427
Landscape flowers, fertilizer, 51
Lantana, growth regulator, 427
Larkspur, growth regulator, 429
Laurentia axillaris, growth regulator, 427
Lawns and turf
    disease control, 475
    growth regulators, 442
    insect control, 164, 187
    lime and fertilizer, 48
    weed control, 340
Leaching potential, 11
Lespedeza, weed control, 335
Lettuce
    disease control, 560
    insect control, 112, 175
    insect control, greenhouse, 132
    lime and fertilizer, 54
    weed control, 368
Liatris, growth regulator, 427
Lily
    Easter, growth regulator, 427
    hybrid, growth regulator, 428
    Oriental, growth regulator, 428
Lima bean, insect control, 97
Lime suggestions (see also specific crop)
    field crops, 42
    lawns, 48
    pasture and hay crops, 44
    vegetable crops, 53
Lipstick vine, growth regulator, 428
Lisianthus, growth regulator, 429
Livestock
    insect control, 134
    manure production rates, nutrient content, 64
Lobelia, growth regulator, 429
Lobularia, growth regulator, 429
Local need registrations, 2

M

Manure
    nutrient content, 64
    production rate, 64
Marigold, growth regulator, 429
Matthiola, growth regulator, 429
Meadow vole, identification, 451
Melons, lime and fertilizer, 55
Micronutrient fertilizers, composition, 57
Millipede, control
    commercial turf, 166
    household, 184
    licensed pest control, 144
Mixing, pesticide, 18
Mole, identification and control, 451, 454
Mole crickets, control
    commercial turf, 166
    home lawn, 188
Monarda, growth regulator, 429
Monstera, growth regulator, 430
Montbretia, growth regulator, 430
Mosquito control, 140, 144, 184

Mugwort, control, 400
Multiflora rose, control, 400
Muscadine grape
    fertilizer, 46
    fungicide effectiveness, 217
    insect management, 217
    management program, 216
Muskmelon
    insect control, 113
    weed control, 358
Muskrat, identification and control, 451, 454
Mustard greens
    insect control, 113, 175
    weed control, 368

N

Narsissus, growth regulator, 430
Nasturtium, growth regulator, 430
Nectarine, spray program, 221
Nematicides
    corn, 462
    cotton, 464
    peaches, 224
    turf, 498
Nemesia Fruticans, growth regulator, 430
Nepthytis, growth regulator, 430
New Guinea impatiens, growth regulator, 430
Nicotiana, growth regulator, 430
Nolana paradoxa, growth regulator, 430
Nozzle, discharge rates, 33
Nursery crops, fertilizer, 52
Nutria, identification, 451
Nutsedge, control, 400

O

Oats, see Small Grains
Okra
    disease control, 562
    insect control, 113, 131, 175
    lime and fertilizer, 55
    weed control, 369
Old house borer, 170
Onion
    disease control, 563
    insect control, 114, 175
    lime and fertilizer, 55
    weed control, 370
Opossum, identification, 451
Organic fertilizer, farm management alternatives, 69
Ornamental cabbage, kale, growth regulator, 430
Ornamental crops, growth regulators, 438
Ornamental peppers, vegetables, growth regulator, 430
Ornamental plants
    fertilizer, 51
    weed control, 352
Ornamentals, commercial greenhouse, Arthropod management, 147
Ornamentals, nurseries, landscape arthropod management, 152
Osteospermum, growth regulator, 431
Otacanthus, growth regulator, 431
Otomeria, growth regulator, 431
Oxallis, growth regulator, 431

P

Pantry pests, 184
Pansy, growth regulator, 431
Parsley,
 lime and fertilizer, 55
Parsnip
 lime and fertilizer, 55
Pastures, weed control, 334
Pea, cowpea
 insect control, 115
Pea, English,
 insect control, 115
 lime and fertilizer, 55
Pea, green
 weed control, 371
Pea, southern
 lime and fertilizer, 55
 weed control, 372
 harvest aids, 407
Peach
 disease control, 224
 insecticide effectiveness, 223
 lime and fertilizer, 45
 nematode control, 224
 spray program, 221
 weed control, 329
Peanut
 disease control, 464
 disease management calendar, 468
 growth regulators, 441
 insect control, 85
 lime and fertilizer, 42
 weed control, 277
Pecan
 disease control, 233
 insect control, 225
 weed control, 331
Pennisetum glaucum, growth regulator, 431
Pennisetum setaceum, growth regulator, 431
Pentas, growth regulator, 431
Pepino, growth regulator, 432
Pericallis, growth regulator, 432
Pepper
 disease control, 571
 growth regulator, 432
 insect control, 116, 131, 132, 175
 lime and fertilizer, 55
 weed control, 373
Perilla, growth regulator, 432
Pesticides
 disposal of containers, 16
 drift, controlling, 17
 hazard and toxicity, 5
 mixing and loading, 18
 recordkeeping, 18
 reducing risks to honey bees, 7, 73
 safe use, 4
 sprayers, 23
 toxicity to people, 5
Pests, household, 144, 178
Petunia, growth regulator, 432
Phalaenopsis, growth regulator, 432
Philodendron, growth regulator, 432
Phlox, growth regulator, 432
Pigeon, identification and control, 451, 454
Pilea, growth regulator, 432
Pillbugs, household, 185

Pine vole, identification, 451
Plant Disease and Insect Clinic, 38
Plant growth regulation
 apples, 408
 cotton, 406
 floricultural crops, 411
 peanut, 441
 tobacco, 438
 turfgrass, 442
 woody ornamentals, 438
Platycodon, growth regulator, 433
Plectranthus, growth regulator, 433
Plumbago auriculata, growth regulator, 433
Poinsettia, growth regulator, 433
Poison Ivy, Oak, control, 401
Porphyrocoma pohliana, growth regulator, 433
Pond dyes, 397
Portulaca, growth regulator, 433
Possum, identification, 451
Potato, Irish
 disease control, 576
 insect control, 118, 175
 lime and fertilizer, 55
 harvest aids and desiccants, 408
 weed control, 374
Potato, sweet
 disease control, 585
 insect control, 124, 131
 lime and fertilizer, 56
 storage house sanitation, 612
 weed control, 379
Pothos, growth regulator, 433
Poultry, insect control, 138
Preharvest intervals, 7
Preharvest intervals, pyrethroid insecticides in vegetable crops, 131
Primula acaulis, growth regulator, 433
Primula obconica, growth regulator, 433
Pumpkin
 insect control, 122, 175
 lime and fertilizer, 55
 weed control, 375
Purple coneflower, growth regulator, 434
Purple passion, growth regulator, 434

R

Rabbit identification and control, 451, 455
Raccoon identification and control, 451, 455
Radish
 disease control, 581
 insect control, 122, 131, 176
 lime and fertilizer, 56
 weed control, 377
Ranunculus, growth regulator, 434
Raspberries
 fertilizer, 47
 management program, 200
Rat, identification, 452
Rate tables
 dilutions, 31
 fertilizer, 35
 fumigant, 35
 granular application, 35
Recordkeeping, 18
Red bug (chigger), 180
Red fox identification and control, 451
Red-winged blackbird, control, 455

Relative soil leaching potential, 11
Restricted use pesticides, 2
Rodenticides, 457
Roof rat, identification and control, 452
Rose, growth regulator, 434
Rutabaga
 insect control, 173
 lime and fertilizer, 56
Rye, see Small Grains

S

Safety
 aerial application limitations, 7
 bee protection, 7, 72
 chemigation, 9
 disposal, 17
 drift, 17
 environmental hazards, 5
 Hazardous Chemicals Right-To-Know Act, 6
 groundwater contamination risk, 15
 local need registrations, 2
 mixing and loading, 18
 recordkeeping, 19
 restricted use pesticides, 2
 surface and groundwater, 8
 toxicity, 5
 worker protection standards, 6
Salvia, growth regulator, 434
Sanvitalia, growth regulation, 434
Scaevola Aemula, growth regulation, 434
Schefflera, growth regulator, 435
Schizanthus scoparis, growth regulator, 435
Scutellaria Javanica, growth regulator, 435
Shasta daisy, growth regulator, 435
Sheep, insect control, 136
Shrimp plant, growth regulator, 435
Shrubs, fertilizer, 51
Silverfish
 household, 185
 licensed pest control, 144
Skunk, identification and control, 452, 455
Slugs
 commercial turf, 167
 home lawns, 188
Small fruit, weed control, 319
Small grains (wheat, oats, grain sorghum, barley, rye)
 insect control, 78
 lime and fertilizer, 42, 43
 weed control, 309
Snails, control
 commercial turf, 167
 home lawns, 188
Snakes, identification and control, 452, 456
Snapdragon, growth regulator, 435
Snow pea, insect control, 115
Sod webworm, control,
 commercial turf, 167
 home lawns, 188
Soil leaching, pesticide, 11
Soil testing, 40
Solubility, fertilizer, 61
Sorghum
 insect control, 78
 lime and fertilizer, 43
 weed control, 285

Sowbugs
    commercial turf, 167
    household, 144
    licensed pest control, 144
Soybeans
    disease control, 468
    insect control, 87
    lime and fertilizer, 43
    weed control, 288
Spathiphyllum, growth regulator, 435
Spinach
    disease control, 583
    insect control, 123, 131, 176
    lime and fertilizer, 56
    weed control, 377
Spray program, see specific crop
Sprayer, calibration and components, 25
Springtails
    household, 186
    licensed pest control, 144
Squash
    insect control, 123, 131, 176
    lime and fertilizer, 55, 56
    weed control, 378
Starling, identification and control, 452, 456
Statice, growth regulation, 435
Stinging caterpillars, household, 186
Stokesia, growth regulation, 435
Storage, of pesticides, 16
Stored food pests (see pantry pests)
    household, 144
Strawberries
    disease control, 238
    fertilizer, 47
    insect management, 248
    weed control, 323
Streptocarpus, growth regulator, 435
Strobilanthes, growth regulator, 435
Sucker control, tobacco, 438
Sunflower
    growth regulators, 436
    harvest aids, 408
    weed control, 300
Surface water, protecting, 8
Sweet potato
    disease control, 585
    harvest aids, 408
    insect control, 124
    lime and fertilizer, 56
    storage house sanitation, 612
    weed control, 379
Swine, insect control, 137

T

Termites, 171
Ticks, 142, 144, 179
Tobacco, burley
    insect control, 90
    lime and fertilizer, 43
    sucker control, 440
    weed control, 301
Tobacco
    disease control, 473
    weed control, 301
Tobacco, flue-cured
    insect control, 90
    lime and fertilizer, 43
    sucker control, 438
    weed control, 301
    yellowing agents, 440
Tomatillo, growth regulator, 436
Tomato
    disease control, 590
    growth regulators, 436
    harvest aids, 408
    insect control, 125, 131, 132, 176
    lime and fertilizer, 56
    weed control, 380
Torenia, growth regulator, 436
Toxicity, pesticide, 5
Tree fruits
    apple, growth regulators, 408
    apple spray program, 190
    peach spray program, 221
    fertilizer, 45
    nectarine spray program, 221
    weed control, 326
Tree of Heaven, control, 401
Trees, fertilizer, 51
Triploid grass carp, control of weeds with, 393
Triticale, see Small Grains
Tropical plants, growth regulators, 436
Tulip, growth regulator, 436
Turfgrass
    disease control, 475
    growth regulators, 442
    insects, 164, 187
    lime and fertilizer, 48
    nematode control, 498
    weed control, 340
Turnip, insect control, 127, 131, 177
Turnip greens,
    insect control, 177
    weed control, 368

V

Vegetable crops
    disease control, 533
    harvest aids, 407
    insect control, 97, 132, 173
    lime and fertilizer, 53
    weed control, 355
Vegetables, fruiting, growth regulators, 441
Vegetable garden, insect control, 173
Verbena, growth regulator, 436
Veronica, growth regulator, 437
Vinca, growth regulator, 437
Viola, growth regulator, 437

W

Wandering Jew, growth regulator, 437
Wasps
    licensed pest control, 144, 186
    home lawn, 187
Water supplies, protection of, 8
Watermelon
    disease control, 545
    foliar disease spray program, 552
    insect control, 128, 131, 177
    lime and fertilizer, 56
    weed control, 382
Weed control (see also specific crop)
    aquatic weeds, 391
    field crops, 252
    forest site preparation, 386
    fruit crops, 319
    hay crops and pastures, 334
    lawns and turf, 340
    ornamentals, 352
    timber stand improvement, 386
    vegetable crops, 355
    woody plants, 402
Wheat, see Small Grains
Wheat leaf, disease control, 460
White-tailed deer, identification, control, 452, 457
Wick applicators, 24
Wine grape, see Bunch Grapes
Wood and wood products, insect control, 169
Woodpecker, identification, 452
Woody ornamental crops, growth regulators, 438
Woody plants, control, 402
Worker protection standards, 6

Y

Yams, see Sweet potato
Yellow jackets, household, 183
Yellowing agent, flue-cured tobacco, 440

Z

Zinnia, growth regulator, 437
Zoysiagrass, fertilizer, 48

ABBREVIATIONS

a.i.	active ingredient
A	aerosol, acre
AE	acid equivalent
AS	aqueous solution
B	bait
bu	bushel
cc	cubic centimeters
CEC	cation exchange capacity
conc	concentration
cu ft	cubic foot (feet)
D	dust
DF	dry flowable
EC or E	emulsifiable concentrate
ES	emulsifiable solution
F	Fahrenheit
F	flowable, liquid
FME	flowable microencapsulated
g	gram(s)
G	granule
gal	gallon(s)
gpa	gallons per acre
gpm	gallons per minute
hr	hour(s)
IE	invert emulsion
in	inch(es)
IU	international units
kg	kilogram(s)
L	liquid, flowable
LC	liquid concentrate
LD	lethal dose (e.g. LD 50)
lb	pound(s)
lb/gal	pounds per gallon
liq	liquid
LS	liquid solution
LV	low volume
ml	milliliter(s)
mm	millimeter(s)
mg	milligram(s)
min	minutes
mmd	mass median diameter
No	number
OC	oil concentration
OD	oil dispersible
OS	oil solution
oz	ounce(s)
P, PS	pellets
PO	pour on
ppm	parts per million
psi	pounds per square inch
pt	pint(s)
qt	quart(s)
RS	resin strip
S	sprayable, solution
SC	spray concentrate
SL	soluble liquid
SOL	solution
SP	soluble powder
sq ft	square foot (feet)
tbsp	tablespoon(s)
TM	trade mark
trt	treated
tsp	teaspoon(s)
ULV	ultra low volume
WDG	water dispersible granule
WDL	water dispersible liquid
WP or W	wettable powder
WS	water soluble
WSG	water soluble granule
WSL	water soluble liquid
WSP	water soluble powder
XLR	extra long residual
<	less than
>	more than

Help Make Extension Better—Contribute to Our Publications.

NC State Extension helps to strengthen North Carolina families and communities every day through our mission and outreach programs. Our publications and communications enhance Extension's statewide, regional, and county programmatic efforts. Your contribution will support the production of these publications, help empower people, and provide solutions.

Make a secure gift online: go.ncsu.edu/ExtPublications

☐ A check to support Extension publications for the total amount of $ _____ is enclosed.
   Please make checks payable to the North Carolina Agricultural Foundation, Inc. and note Account #011893.

☐ Please contact me about making my gift in bank drafts or appreciated stocks.

NAME _____

ADDRESS _____

CITY _____ STATE _____ ZIP _____

PHONE _____ EMAIL _____

Fundraising efforts for Extension Publications/Communications operate under the auspices of the North Carolina Agricultural Foundation, Inc., a 501(c)3 non-profit (Tax ID# 56-6049304). You will receive an official receipt for your tax-deductible donation.

Please contact cals_advancement_business@ncsu.edu or 919.515.2000 with questions regarding donations.

Please contact extension_publications@ncsu.edu with questions regarding publications.

Mail checks or contact information to

CALS Advancement
NC State University
Campus Box 7645
Raleigh, NC 27695-7645